CONVERSIONS BETWEEN U.S. CUSTOMARY UNITS AND SI UNITS

U.S. Customary unit		Times conversion factor		Equals SI unit	
		Accurate	Practical		
Moment of inertia (area)					
inch to fourth power	in.^{4}	416,231	416,000	millimeter to fourth power	mm^4
inch to fourth power	in.^{4}	0.416231×10^{-6}	0.416×10^{-6}	meter to fourth power	m^4
Moment of inertia (mass)					
slug foot squared	slug-ft^2	1.35582	1.36	kilogram meter squared	kg·m^2
Power					
foot-pound per second	ft-lb/s	1.35582	1.36	watt (J/s or N·m/s)	W
foot-pound per minute	ft-lb/min	0.0225970	0.0226	watt	W
horsepower (550 ft-lb/s)	hp	745.701	746	watt	W
Pressure; stress					
pound per square foot	psf	47.8803	47.9	pascal (N/m^2)	Pa
pound per square inch	psi	6894.76	6890	pascal	Pa
kip per square foot	ksf	47.8803	47.9	kilopascal	kPa
kip per square inch	ksi	6.89476	6.89	megapascal	MPa
Section modulus					
inch to third power	in.^{3}	16,387.1	16,400	millimeter to third power	mm^3
inch to third power	in.^{3}	16.3871×10^{-6}	16.4×10^{-6}	meter to third power	m^3
Velocity (linear)					
foot per second	ft/s	0.3048*	0.305	meter per second	m/s
inch per second	in./s	0.0254*	0.0254	meter per second	m/s
mile per hour	mph	0.44704*	0.447	meter per second	m/s
mile per hour	mph	1.609344*	1.61	kilometer per hour	km/h
Volume					
cubic foot	ft^3	0.0283168	0.0283	cubic meter	m^3
cubic inch	in.^{3}	16.3871×10^{-6}	16.4×10^{-6}	cubic meter	m^3
cubic inch	in.^{3}	16.3871	16.4	cubic centimeter (cc)	cm^3
gallon (231 in.^{3})	gal.	3.78541	3.79	liter	L
gallon (231 in.^{3})	gal.	0.00378541	0.00379	cubic meter	m^3

*An asterisk denotes an *exact* conversion factor

Note: To convert from SI units to USCS units, *divide* by the conversion factor

Temperature Conversion Formulas

$$T(°C) = \frac{5}{9}[T(°F) - 32] = T(K) - 273.15$$

$$T(K) = \frac{5}{9}[T(°F) - 32] + 273.15 = T(°C) + 273.15$$

$$T(°F) = \frac{9}{5}T(°C) + 32 = \frac{9}{5}T(K) - 459.67$$

The Science and Engineering of Materials

Sixth Edition, SI

Donald R. Askeland
University of Missouri—Rolla, Emeritus

Pradeep P. Fulay
University of Pittsburgh

Wendelin J. Wright
Bucknell University

SI Edition Prepared By:

D.K. Bhattacharya
Solid State Physics Laboratories, New Delhi

Australia • Brazil • Japan • Korea • Mexico • Singapore • Spain • United Kingdom • United States

CENGAGE
Learning™

**The Science and Engineering of Materials,
Sixth Edition, SI**
**Donald R. Askeland, Pradeep
P. Fulay, Wendelin J. Wright**
SI Edition Prepared by D.K. Bhattacharya

Publisher, Global Engineering:
Christopher M. Shortt

Acquisitions Editor, SI Edition: Swati Meherishi

Senior Developmental Editor: Hilda Gowans

Assistant Developmental Editor: Ojen Yumnam

Editorial Assistant: Tanya Altieri

Team Assistant: Carly Rizzo

Marketing Manager: Lauren Betsos

Media Editor: Chris Valentine

Director, Content and Media Production:
Tricia Boies

Content Project Manager: Darrell Frye

Production Service: RPK Editorial Services, Inc.

Copyeditor: Shelly Gerger-Knechtl

Proofreader: Martha McMaster/Erin Wagner

Indexer: Shelly Gerger-Knechtl

Compositor: Integra

Senior Art Director: Michelle Kunkler

Internal Design: Jennifer Lambert/jen2design

Cover Designer: Andrew Adams

Cover Image: © Sieu Ha/Antoine Kahn/
Princeton University

Text and Image Permissions Researcher:
Kristiina Paul

First Print Buyer: Arethea Thomas

100632339X

For product information and technology assistance,
contact us at **Cengage Learning Customer &
Sales Support, 1-800-354-9706.**

For permission to use material from this text or product,
submit all requests online at **www.cengage.com/
permissions.** Further permissions questions can be
emailed to **permissionrequest@cengage.com**

Library of Congress Control Number: 2010932702

ISBN-13: 978-0-495-66802-2

ISBN-10: 0-495-66802-8

Cengage Learning
200 First Stamford Place, Suite 400
Stamford, CT 06902
USA

Cengage Learning is a leading provider of customized
learning solutions with office locations around the globe,
including Singapore, the United Kingdom, Australia, Mexico,
Brazil, and Japan. Locate your local office at:
international.cengage.com/region.

Cengage Learning products are represented in Canada by
Nelson Education Ltd.

For your course and learning solutions, Please visit
login.cengage.com and log in to access instructor-specific
resources.

Purchase any of our products at your local college store or at
our preferred online store **www.cengagebrain.com.**

Printed in Canada
1 2 3 4 5 6 7 14 13 12 11 10

To Mary Sue and Tyler
–Donald R. Askeland

To Jyotsna, Aarohee, and Suyash
–Pradeep P. Fulay

To John, as we begin the next wonderful chapter in our life together
–Wendelin J. Wright

Contents

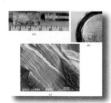

Preface to the SI Edition

This edition of ***The Science and Engineering of Materials*** has been adapted to incorporate the International System of Units (*Le Système International d'Unités* or SI) throughout the book.

Le Système International d'Unités

The United States Customary System (USCS) of units uses FPS (foot–pound–second) units (also called English or Imperial units). SI units are primarily the units of the MKS (meter–kilogram–second) system. However, CGS (centimeter–gram–second) units are often accepted as SI units, especially in textbooks.

Using SI Units in this Book

In this book, we have used both MKS and CGS units. USCS units or FPS units used in the US Edition of the book have been converted to SI units throughout the text and problems. However, in case of data sourced from hand-books, government standards, and product manuals, it is not only extremely difficult to convert all values to SI, it also encroaches upon the intellectual property of the source. Also, some quantities such as the ASTM grain size number and Jominy distances are generally computed in FPS units and would lose their relevance if converted to SI. Some data in figures, tables, examples, and references, therefore, remains in FPS units. For readers unfamiliar with the relationship between the FPS and the SI systems, conversion tables have been provided inside the front and back covers of the book.

To solve problems that require the use of sourced data, the sourced values can be converted from FPS units to SI units just before they are to be used in a calculation. To obtain standardized quantities and manufacturers' data in SI units, the readers may contact the appropriate government agencies or authorities in their countries/regions.

Instructor Resources

A Printed Instructor's Solution Manual in SI units is available on request. An electronic version of the Instructor's Solutions Manual, and PowerPoint slides of the figures from the SI text are available through www.login.cengage.com.

The readers' feedback on this SI Edition will be highly appreciated and will help us improve subsequent editions.

The Publishers

Preface

When the relationships between the structure, properties, and processing of materials are fully understood and exploited, materials become enabling—they are transformed from *stuff*, the raw materials that nature gives us, to *things*, the products and technologies that we develop as engineers. Any technologist can find materials properties in a book or search databases for a material that meets design specifications, but the ability to *innovate* and to incorporate materials *safely* in a design is rooted in an understanding of how to manipulate materials properties and functionality through the control of materials structure and processing techniques. The objective of this textbook, then, is to describe the foundations and applications of materials science for college-level engineering students as predicated upon the structure-processing-properties paradigm.

The challenge of any textbook is to provide the proper balance of breadth and depth for the subject at hand, to provide rigor at the appropriate level, to provide meaningful examples and up to date content, and to stimulate the intellectual excitement of the reader. Our goal here is to provide enough *science* so that the reader may understand basic materials phenomena, and enough *engineering* to prepare a wide range of students for competent professional practice.

Cover Art

2008 MRS Fall Meeting "Science As Art" Second Place Winner

Matthew J. Bierman, *University of Wisconsin-Madison*
Two Pine Trees
These pine tree lead sulfide nanowires obtain a complicated structure because only the trunk contains a screw dislocation that causes it to twist.

2008 MRS Spring Meeting "Science As Art", First Place Winner

S.K. Hark, *Chinese University of Hong Kong*
Field of Sunflowers
Amorphous SiOx nanowire bundles have an uncanny ability to self-assemble into various shapes, including one that strikingly resembles a sunflower. In these sunflowers, highly packed bundles form the disc florets and loosely packed ones around the rim of the disc form the ray florets. The SEM image shows a field of sunflowers. The grey scale image was mapped into pseudo-colors by graphic software. The nanowires grew out of the reaction of Si and oxygen, with molten Ga and Au acting as

catalysts. Each nanowire is about 10 nm in diameter and tens of micrometers in length.

2007 MRS Fall Meeting "Science As Art", First Place Winner

Fanny Beron, *École Polytechnique de Montréal*
Nano-Explosions
Color-enhanced scanning electron micrograph of an overflowed electrodeposited magnetic nanowire array (CoFeB), where the template has been subsequently completely etched. It's a reminder that nanoscale research can have unpredicted consequences at a high level. Fanny Beron, École Polytechnique de Montréal, Montréal, Canada.

2007 MRS Fall Meeting "Science As Art", Second Place Winner

Siddhartha Pathak, *Drexel University*
Layered steps in Lanthanum Cobaltite
The picture shows a colored image of the layered steps formed inside closed pores of La0.8Ca0.2CoO3, which were revealed due to fracture of the material. Credit: Siddhartha Pathak, Drexel University, Philadelphia, USA.

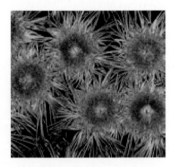

2007 MRS Spring Meeting "Science As Art", First Place Winner

Matthew Lloyd, *Cornell University*
Sunflowers
A Bouquet of Anthradithiophene.

2007 MRS Spring Meeting "Science As Art", Second Place Winner

Candace Lynch, *Air Force Research Laboratory*
GaAs Sea Creatures
This is an image of defects on a GaAs surface following hydride vapor phase epitaxy. The image was taken using a Nikon Optical Microscope with Nomarski contrast.

Audience and Prerequisites

This text is intended for an introductory science of materials class taught at the sophomore or junior level. A first course in college level chemistry is assumed, as is some coverage of first year college physics. A calculus course is helpful, but certainly not required. The text does not presume that students have taken other introductory engineering courses such as statics, dynamics, or mechanics of materials.

Changes to the Sixth Edition

Particular attention has been paid to revising the text for clarity and accuracy. New content has been added as described below.

New to this Edition
New content has been added to the text including enhanced crystallography descriptions and sections about the allotropes of carbon, nanoindentation, mechanical properties of bulk metallic glasses, mechanical behavior at small length scales, integrated circuit manufacturing, and thin film deposition. New problems have been added to the end of each chapter. New instructor supplements are also provided.

At the conclusion of the end-of-chapter problems, you will find a special section with problems that require the use of Knovel (www.knovel.com). Knovel is an online aggregator of engineering references including handbooks, encyclopedias, dictionaries, textbooks, and databases from leading technical publishers and engineering societies such as the American Society of Mechanical Engineers (ASME) and the American Institute of Chemical Engineers (AIChE.)

The Knovel problems build on material found in the textbook and require familiarity with online information retrieval. The problems are also available online for students at www.cengagebrain.com. In addition, the solutions are accessible by registered instructors, through login.cengage.com. If your institution does not have a subscription to Knovel or if you have any questions about Knovel, please contact

support@knovel.com
(866) 240-8174
(866) 324-5163

The Knovel problems were created by a team of engineers led by Sasha Gurke, senior vice president and co-founder of Knovel.

Supplements for the Instructor

Supplements to the text include the Instructor's Solutions Manual that provides complete solutions to selected problems, annotated PowerPoint™ slides, and an online Test Bank of potential exam questions. All instructor-specific resources can be accessed through login.cengage.com.

Acknowledgements

We thank all those who have contributed to the success of past editions and also the reviewers who provided detailed and constructive feedback on the fifth edition:

Deborah Chung, State University of New York, at Buffalo
Derrick R. Dean, University of Alabama at Birmingham
Angela L. Moran, U.S. Naval Academy
John R. Schlup, Kansas State University
Jeffrey Schott, University of Minnesota

We are grateful to the team at Cengage Learning who has carefully guided this sixth edition through all stages of the publishing process. In particular, we thank Christopher Carson, Executive Director of the Global Publishing Program at Cengage Learning, Christopher Shortt, Publisher for Global Engineering at Cengage Learning, Hilda Gowans, the Developmental Editor, Rose Kernan, the Production Editor, Kristiina Paul, the Permissions and Photo Researcher, and Lauren Betsos, the Marketing Manager. We also thank Jeffrey Florando of the Lawrence Livermore National Laboratory for input regarding portions of the manuscript and Venkat Balu for some of the new end-of-chapter problems in this edition.

Wendelin Wright thanks Particia Wright for assistance during the proofreading process and John Bravman for his feedback, contributed illustrations, patience, and constant support.

Donald R. Askeland
University of Missouri – Rolla, Emeritus

Pradeep P. Fulay
University of Pittsburgh

Wendelin J. Wright
Bucknell University

About the Authors

Donald R. Askeland is a Distinguished Teaching Professor Emeritus of Metallurgical Engineering at the University of Missouri–Rolla. He received his degrees from the Thayer School of Engineering at Dartmouth College and the University of Michigan prior to joining the faculty at the University of Missouri–Rolla in 1970. Dr. Askeland taught a number of courses in materials and manufacturing engineering to students in a variety of engineering and science curricula. He received a number of awards for excellence in teaching and advising at UMR. He served as a Key Professor for the Foundry Educational Foundation and received several awards for his service to that organization. His teaching and research were directed primarily to metals casting and joining, in particular lost foam casting, and resulted in over 50 publications and a number of awards for service and best papers from the American Foundry Society.

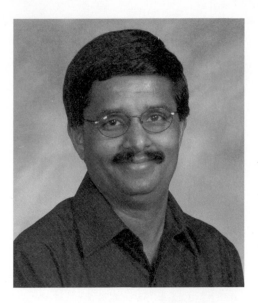

Pradeep P. Fulay is a Professor of Materials Science and Engineering at the University of Pittsburgh. He joined the University of Pittsburgh in 1989, was promoted to Associate Professor in 1994, and then to full professor in 1999. Dr. Fulay received a Ph.D. in Materials Science and Engineering from the University of Arizona (1989) and a B. Tech (1983) and M. Tech (1984) in Metallurgical Engineering from the Indian Institute of Technology Bombay (Mumbai) India.

He has authored close to 60 publications and has two U.S. patents issued. He has received the Alcoa Foundation and Ford Foundation research awards.

He has been an outstanding teacher and educator and was listed on the Faculty Honor Roll at the University of Pittsburgh (2001) for outstanding services and assistance. From 1992–1999, he was the William Kepler Whiteford Faculty Fellow at the University of Pittsburgh. From August to December 2002, Dr. Fulay was a visiting scientist at the Ford Scientific Research Laboratory in Dearborn, MI.

Dr. Fulay's primary research areas are chemical synthesis and processing of ceramics, electronic ceramics and magnetic materials, and development of smart materials and systems. Part of the MR fluids technology Dr. Fulay has developed is being transferred to industry. He was the Vice President (2001–2002) and President (2002–2003) of the Ceramic Educational Council and has been a Member of the Program Committee for the Electronics Division of the American Ceramic Society since 1996.

He has also served as an Associate Editor for the *Journal of the American Ceramic Society* (1994–2000). He has been the lead organizer for symposia on ceramics for sol-gel processing, wireless communications, and smart structures and sensors. In 2002, Dr. Fulay was elected as a Fellow of the American Ceramic Society. Dr. Fulay's research has been supported by National Science Foundation (NSF) and many other organizations.

Wendelin Wright will be appointed as an assistant professor of Mechanical Engineering at Bucknell University in the fall of 2010. At the time of publication, she is the Clare Boothe Luce Assistant Professor of Mechanical Engineering at Santa Clara University. She received her B.S., M.S., and Ph.D. (2003) in Materials Science and Engineering from Stanford University. Following graduation, she served a post–doctoral term at the Lawrence Livermore National Laboratory in the Manufacturing and Materials Engineering Division and returned to Stanford as an Acting Assistant Professor in 2005. She joined the Santa Clara University faculty in 2006.

Professor Wright's research interests focus on the mechanical behavior of materials, particularly of metallic glasses. She is the recipient of the 2003 Walter J. Gores Award for Excellence in Teaching, which is Stanford University's highest teaching honor, a 2005 Presidential Early Career Award for Scientists and Engineers, and a 2010 National Science Foundation CAREER Award.

In the fall of 2009, Professor Wright used *The Science and Engineering of Materials* as her primary reference text while taking and passing the Principles and Practices of Metallurgy exam to become a licensed Professional Engineer in California.

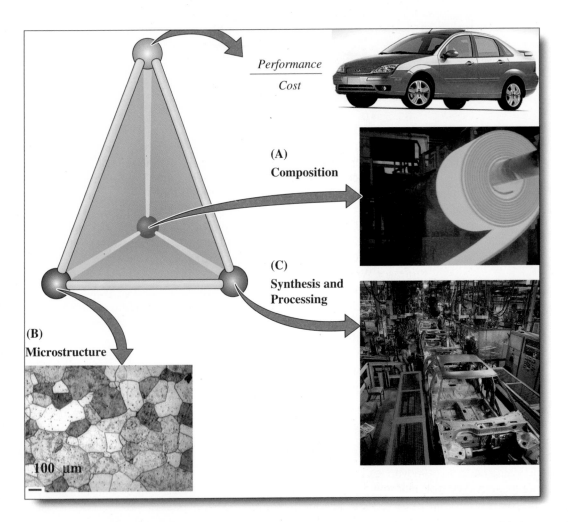

Performance / Cost

(A) Composition

(C) Synthesis and Processing

(B) Microstructure

100 μm

The principal goals of a materials scientist and engineer are to (1) make existing materials better and (2) invent or discover new phenomena, materials, devices, and applications. Breakthroughs in the materials science and engineering field are applied to many other fields of study such as biomedical engineering, physics, chemistry, environmental engineering, and information technology. The materials science and engineering tetrahedron shown here represents the heart and soul of this field. As shown in this diagram, a materials scientist and engineer's main objective is to develop materials or devices that have the best performance for a particular application. In most cases, the performance-to-cost ratio, as opposed to the performance alone, is of utmost importance. This concept is shown as the apex of the tetrahedron and the three corners are representative of A—the composition, B—the microstructure, and C—the synthesis and processing of materials. These are all interconnected and ultimately affect the performance-to-cost ratio of a material or a device. The accompanying micrograph shows the microstructure of stainless steel.

For materials scientists and engineers, materials are like a palette of colors to an artist. Just as an artist can create different paintings using different colors, materials scientists create and improve upon different materials using different elements of the periodic table, and different synthesis and processing routes. *(Car image courtesy of Ford Motor Company. Steel manufacturing image and car chassis image courtesy of Digital Vision/Getty Images. Micrograph courtesy of Dr. A.J. Deardo, Dr. M. Hua, and Dr. J. Garcia.)*

Introduction to Materials Science and Engineering

Have You Ever Wondered?

- *What do materials scientists and engineers study?*

- *How can steel sheet metal be processed to produce a high strength, lightweight, energy absorbing, malleable material used in the manufacture of car chassis?*

- *Can we make flexible and lightweight electronic circuits using plastics?*

- *What is a "smart material?"*

In this chapter, we will first introduce you to the field of materials science and engineering (MSE) using different real-world examples. We will then provide an introduction to the classification of materials. Although most engineering programs require students to take a materials science course, you should approach your study of materials science as more than a mere requirement. A thorough knowledge of materials science and engineering will make you a better engineer and designer. Materials science underlies all technological advances and an understanding of the basics of materials and their applications will not only make you a better engineer but also help you during the design process. In order to be a good designer, you must learn what materials will be appropriate to use in different applications. You need to be capable of choosing the right material for your application based on its properties, and you must recognize how and why these properties might change over time and due to processing. Any engineer can look up materials properties in a book or search databases for a material that meets design specifications, but the *ability to innovate* and to *incorporate materials safely* in a design is rooted in an understanding of how to manipulate materials properties and functionality through the control of the material's structure and processing techniques.

The most important aspect of materials is that they are enabling; materials make things happen. For example, in the history of civilization, materials such as stone, iron, and bronze played a key role in mankind's development. In today's fast-paced world, the discovery of silicon single crystals and an understanding of their properties have enabled the information age.

In this book, we provide compelling examples of real-world applications of engineered materials. The diversity of applications and the unique uses of materials illustrate why a good engineer needs to understand and know how to apply the principles of materials science and engineering.

1-1   What is Materials Science and Engineering?

Materials science and engineering (MSE) is an interdisciplinary field of science and engineering that studies and manipulates the composition and structure of materials across length scales to control materials properties through synthesis and processing. The term **composition** means the chemical make-up of a material. The term **structure** means a description of the arrangement of atoms, as seen at different levels of detail. Materials scientists and engineers deal not only with the development of materials but also with the **synthesis** and **processing** of materials and manufacturing processes related to the production of components. The term "synthesis" refers to how materials are made from naturally occurring or man-made chemicals. The term "processing" means how materials are shaped into useful components to cause changes in the properties of different materials. One of the most important functions of materials scientists and engineers is to establish the relationships between a material or a device's properties and performance and the microstructure of that material, its composition, and the way the material or the device was synthesized and processed. In **materials science**, the emphasis is on the underlying relationships between the synthesis and processing, structure, and properties of materials. In **materials engineering**, the focus is on how to translate or transform materials into useful devices or structures.

One of the most fascinating aspects of materials science involves the investigation of a material's structure. The structure of materials has a profound influence on many properties of materials, even if the overall composition does not change! For example, if you take a pure copper wire and bend it repeatedly, the wire not only becomes harder but also becomes increasingly brittle! Eventually, the pure copper wire becomes so hard and brittle that it will break! The electrical resistivity of the wire will also increase as we bend it repeatedly. In this simple example, take note that we did not change the material's composition (i.e., its chemical make-up). The changes in the material's properties are due to a change in its internal structure. If you look at the wire after bending, it will look the same as before; however, its structure has been changed at the microscopic scale. The structure at the microscopic scale is known as the **microstructure**. If we can understand what has changed microscopically, we can begin to discover ways to control the material's properties.

Let's examine one example using the **materials science and engineering tetrahedron** presented on the chapter opening page. Let's look at "sheet steels" used in the manufacture of car chassis (Figure 1-1). Steels, as you may know, have been used in manufacturing for more than a hundred years, but they probably existed in a crude form during the Iron Age, thousands of years ago. In the manufacture of automobile chassis, a material is needed that possesses extremely high strength but is formed easily into aerodynamic contours. Another consideration is fuel efficiency, so the sheet steel must also be thin and lightweight. The sheet steels also should be able to absorb significant amounts of energy in the event of a crash, thereby increasing vehicle safety. These are somewhat contradictory requirements.

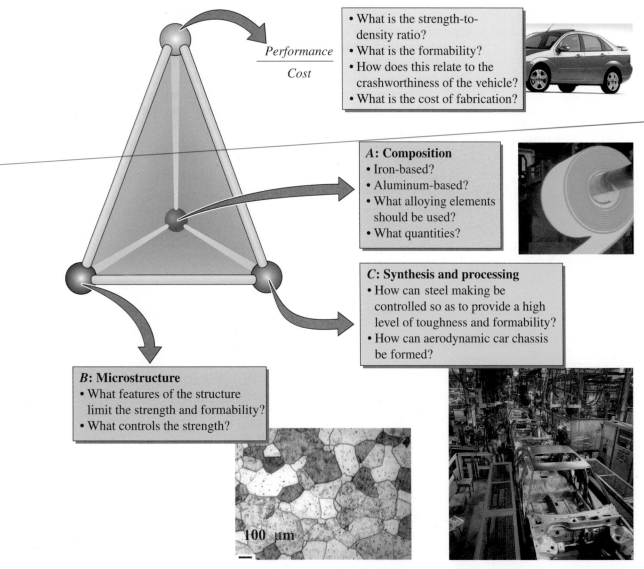

Figure 1-1 Application of the tetrahedron of materials science and engineering to sheet steels for automotive chassis. Note that the composition, microstructure, and synthesis-processing are all interconnected and affect the performance-to-cost ratio. *(Car image courtesy of Ford Motor Company. Steel manufacturing image and car chassis image courtesy of Digital Vision/Getty Images. Micrograph courtesy of Dr. A.J. Deardo, Dr. M. Hua, and Dr. J. Garcia.)*

Thus, in this case, materials scientists are concerned with the sheet steel's

- composition;
- strength;
- weight;
- energy absorption properties; and
- malleability (formability).

Materials scientists would examine steel at a microscopic level to determine if its properties can be altered to meet all of these requirements. They also would have to consider the cost of processing this steel along with other considerations. How can we shape such steel into a car chassis in a cost-effective way? Will the shaping process itself affect the mechanical properties of the steel? What kind of coatings can be developed to make the steel corrosion resistant? In some applications, we need to know if these steels could be welded easily. From this discussion, you can see that many issues need to be considered during the design and materials selection for any product.

Let's look at one more example of a class of materials known as semiconducting polymers (Figure 1-2). Many semiconducting polymers have been processed into light emitting diodes (LEDs). You have seen LEDs in alarm clocks, watches, and other displays. These displays often use inorganic compounds based on gallium arsenide (GaAs) and other materials. The advantage of using plastics for microelectronics is that they are lightweight and flexible. The questions materials scientists and engineers must answer with applications of semiconducting polymers are

- What are the relationships between the structure of polymers and their electrical properties?
- How can devices be made using these plastics?
- Will these devices be compatible with existing silicon chip technology?

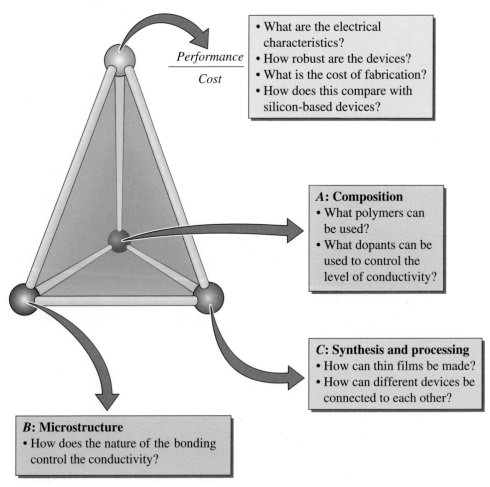

Figure 1-2   Application of the tetrahedron of materials science and engineering to semiconducting polymers for microelectronics.

- How robust are these devices?
- How will the performance and cost of these devices compare with traditional devices?

These are just a few of the factors that engineers and scientists must consider during the development, design, and manufacture of semiconducting polymer devices.

1-2   Classification of Materials

There are different ways of classifying materials. One way is to describe five groups (Table 1-1):

TABLE 1-1 ■ *Representative examples, applications, and properties for each category of materials*

	Examples of Applications	Properties
Metals and Alloys		
Copper	Electrical conductor wire	High electrical conductivity, good formability
Gray cast iron	Automobile engine blocks	Castable, machinable, vibration-damping
Alloy steels	Wrenches, automobile chassis	Significantly strengthened by heat treatment
Ceramics and Glasses		
SiO_2–Na_2O–CaO	Window glass	Optically transparent, thermally insulating
Al_2O_3, MgO, SiO_2	Refractories (i.e., heat-resistant lining of furnaces) for containing molten metal	Thermally insulating, withstand high temperatures, relatively inert to molten metal
Barium titanate	Capacitors for microelectronics	High ability to store charge
Silica	Optical fibers for information technology	Refractive index, low optical losses
Polymers		
Polyethylene	Food packaging	Easily formed into thin, flexible, airtight film
Epoxy	Encapsulation of integrated circuits	Electrically insulating and moisture-resistant
Phenolics	Adhesives for joining plies in plywood	Strong, moisture resistant
Semiconductors		
Silicon	Transistors and integrated circuits	Unique electrical behavior
GaAs	Optoelectronic systems	Converts electrical signals to light, lasers, laser diodes, etc.
Composites		
Graphite-epoxy	Aircraft components	High strength-to-weight ratio
Tungsten carbide-cobalt (WC-Co)	Carbide cutting tools for machining	High hardness, yet good shock resistance
Titanium-clad steel	Reactor vessels	Low cost and high strength of steel with the corrosion resistance of titanium

1. **metals** and **alloys**;
2. **ceramics**, **glasses**, and **glass-ceramics**;
3. **polymers** (plastics);
4. **semiconductors**; and
5. **composite** materials.

Materials in each of these groups possess different structures and properties. The differences in strength, which are compared in Figure 1-3, illustrate the wide range of properties from which engineers can select. Since metallic materials are extensively used for load-bearing applications, their mechanical properties are of great practical interest. We briefly introduce these here. The term "stress" refers to load or force per unit area. "Strain" refers to elongation or change in dimension divided by the original dimension. Application of "stress" causes "strain." If the strain goes away after the load or applied stress is removed, the strain is said to be "elastic." If the strain remains after the stress is removed, the strain is said to be "plastic." When the deformation is elastic, stress and strain are linearly related; the slope of the stress-strain diagram is known as the elastic or Young's modulus. The level of stress needed to initiate plastic deformation is known as the "yield strength." The maximum percent deformation that can be achieved is a measure of the ductility of a metallic material. These concepts are discussed further in Chapters 6 and 7.

Metals and Alloys   Metals and alloys include steels, aluminum, magnesium, zinc, cast iron, titanium, copper, and nickel. An alloy is a metal that contains additions of one or more metals or non-metals. In general, metals have good electrical and thermal conductivity. Metals and alloys have relatively high strength, high stiffness, ductility or formability, and shock resistance. They are particularly useful for structural or load-bearing applications. Although pure metals are occasionally used, alloys provide improvement in a particular desirable property or permit better combinations of properties.

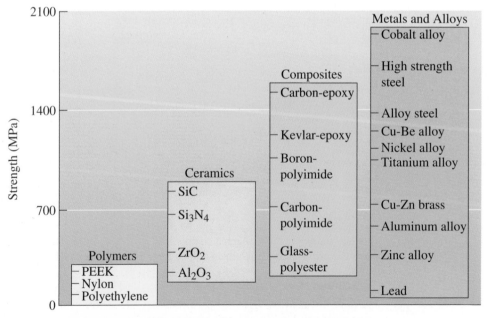

Figure 1-3   Representative strengths of various categories of materials.

Ceramics

Ceramics can be defined as inorganic crystalline materials. Beach sand and rocks are examples of naturally occurring ceramics. Advanced ceramics are materials made by refining naturally occurring ceramics and other special processes. Advanced ceramics are used in substrates that house computer chips, sensors and actuators, capacitors, wireless communications, spark plugs, inductors, and electrical insulation. Some ceramics are used as barrier coatings to protect metallic substrates in turbine engines. Ceramics are also used in such consumer products as paints, plastics, and tires, and for industrial applications such as the tiles for the space shuttle, a catalyst support, and the oxygen sensors used in cars. Traditional ceramics are used to make bricks, tableware, toilets, bathroom sinks, refractories (heat-resistant material), and abrasives. In general, due to the presence of porosity (small holes), ceramics do not conduct heat well; they must be heated to very high temperatures before melting. Ceramics are strong and hard, but also very brittle. We normally prepare fine powders of ceramics and convert these into different shapes. New processing techniques make ceramics sufficiently resistant to fracture that they can be used in load-bearing applications, such as impellers in turbine engines. Ceramics have exceptional strength under compression. Can you believe that an entire fire truck can be supported using four ceramic coffee cups?

Glasses and Glass-Ceramics

Glass is an amorphous material, often, but not always, derived from a molten liquid. The term "amorphous" refers to materials that do not have a regular, periodic arrangement of atoms. Amorphous materials will be discussed in Chapter 3. The fiber optics industry is founded on optical fibers based on high-purity silica glass. Glasses are also used in houses, cars, computer and television screens, and hundreds of other applications. Glasses can be thermally treated (tempered) to make them stronger. Forming glasses and nucleating (forming) small crystals within them by a special thermal process creates materials that are known as glass-ceramics. Zerodur™ is an example of a glass-ceramic material that is used to make the mirror substrates for large telescopes (e.g., the Chandra and Hubble telescopes). Glasses and glass-ceramics are usually processed by melting and casting.

Polymers

Polymers are typically organic materials. They are produced using a process known as **polymerization.** Polymeric materials include rubber (elastomers) and many types of adhesives. Polymers typically are good electrical and thermal insulators although there are exceptions such as the semiconducting polymers discussed earlier in this chapter. Although they have lower strength, polymers have a very good **strength-to-weight ratio.** They are typically not suitable for use at high temperatures. Many polymers have very good resistance to corrosive chemicals. Polymers have thousands of applications ranging from bullet-proof vests, compact disks (CDs), ropes, and liquid crystal displays (LCDs) to clothes and coffee cups. **Thermoplastic** polymers, in which the long molecular chains are not rigidly connected, have good ductility and formability; **thermosetting** polymers are stronger but more brittle because the molecular chains are tightly linked (Figure 1-4). Polymers are used in many applications, including electronic devices. Thermoplastics are made by shaping their molten form. Thermosets are typically cast into molds. **Plastics** contain additives that enhance the properties of polymers.

Semiconductors

Silicon, germanium, and gallium arsenide-based semiconductors such as those used in computers and electronics are part of a broader class of materials known as electronic materials. The electrical conductivity of semiconducting materials is between that of ceramic insulators and that of metallic conductors. Semiconductors have enabled the information age. In some semiconductors, the level of

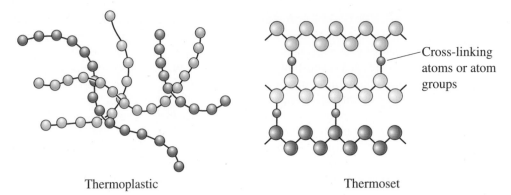

Thermoplastic   Thermoset

Figure 1-4   Polymerization occurs when small molecules, represented by the circles, combine to produce larger molecules, or polymers. The polymer molecules can have a structure that consists of many chains that are entangled but not connected (thermoplastics) or can form three-dimensional networks in which chains are cross-linked (thermosets).

conductivity can be controlled to enable electronic devices such as transistors and diodes that are used to build integrated circuits. In many applications, we need large single crystals of semiconductors. These are grown from molten materials. Often, thin films of semiconducting materials are also made using specialized processes.

Composite Materials   The main idea in developing composites is to blend the properties of different materials. These are formed from two or more materials, producing properties not found in any single material. Concrete, plywood, and fiberglass are examples of composite materials. Fiberglass is made by dispersing glass fibers in a polymer matrix. The glass fibers make the polymer stiffer, without significantly increasing its density. With composites, we can produce lightweight, strong, ductile, temperature-resistant materials or we can produce hard, yet shock-resistant, cutting tools that would otherwise shatter. Advanced aircraft and aerospace vehicles rely heavily on composites such as carbon fiber-reinforced polymers (Figure 1-5). Sports equipment such as bicycles, golf clubs, and tennis rackets also make use of different kinds of composite materials that are light and stiff.

Figure 1-5   The X-wing for advanced helicopters relies on a material composed of a carbon fiber-reinforced polymer. *(Courtesy of Sikorsky Aircraft Division – United Technologies Corporation.)*

1-3   Functional Classification of Materials

We can classify materials based on whether the most important function they perform is mechanical (structural), biological, electrical, magnetic, or optical. This classification of materials is shown in Figure 1-6. Some examples of each category are shown. These categories can be broken down further into subcategories.

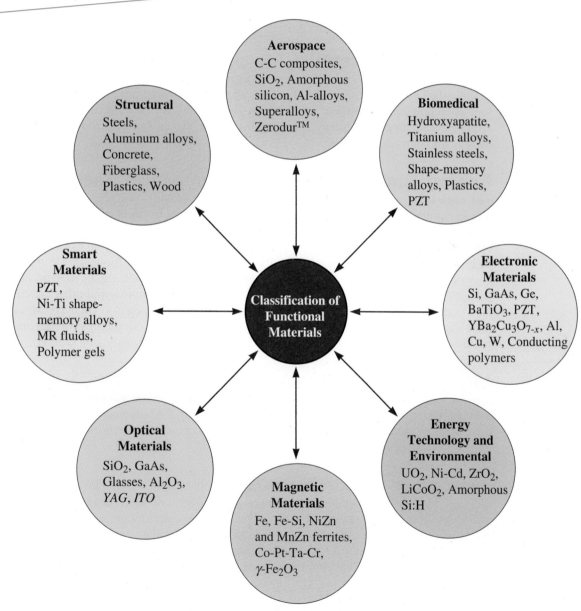

Figure 1-6   Functional classification of materials. Notice that metals, plastics, and ceramics occur in different categories. A limited number of examples in each category are provided.

Aerospace   Light materials such as wood and an aluminum alloy (that accidentally strengthened the engine even more by picking up copper from the mold used for casting) were used in the Wright brothers' historic flight. Today, NASA's space shuttle makes use of aluminum powder for booster rockets. Aluminum alloys, plastics, silica for space shuttle tiles, and many other materials belong to this category.

Biomedical   Our bones and teeth are made, in part, from a naturally formed ceramic known as hydroxyapatite. A number of artificial organs, bone replacement parts, cardiovascular stents, orthodontic braces, and other components are made using different plastics, titanium alloys, and nonmagnetic stainless steels. Ultrasonic imaging systems make use of ceramics known as PZT (lead zirconium titanate). Magnets used for magnetic resonance imaging make use of metallic niobium tin-based superconductors.

Electronic Materials   As mentioned before, semiconductors, such as those made from silicon, are used to make integrated circuits for computer chips. Barium titanate ($BaTiO_3$), tantalum oxide (Ta_2O_5), and many other dielectric materials are used to make ceramic capacitors and other devices. Superconductors are used in making powerful magnets. Copper, aluminum, and other metals are used as conductors in power transmission and in microelectronics.

Energy Technology and Environmental Technology

The nuclear industry uses materials such as uranium dioxide and plutonium as fuel. Numerous other materials, such as glasses and stainless steels, are used in handling nuclear materials and managing radioactive waste. New technologies related to batteries and fuel cells make use of many ceramic materials such as zirconia (ZrO_2) and polymers. Battery technology has gained significant importance owing to the need for many electronic devices that require longer lasting and portable power. Fuel cells will also be used in electric cars. The oil and petroleum industry widely uses zeolites, alumina, and other materials as catalyst substrates. They use Pt, Pt/Rh, and many other metals as catalysts. Many membrane technologies for purification of liquids and gases make use of ceramics and plastics. Solar power is generated using materials such as amorphous silicon (a:Si:H).

Magnetic Materials   Computer hard disks make use of many ceramic, metallic, and polymeric materials. Computer hard disks are made using alloys based on cobalt-platinum-tantalum-chromium (Co-Pt-Ta-Cr) alloys. Many magnetic ferrites are used to make inductors and components for wireless communications. Steels based on iron and silicon are used to make transformer cores.

Photonic or Optical Materials   Silica is used widely for making optical fibers. More than ten million kilometers of optical fiber have been installed around the world. Optical materials are used for making semiconductor detectors and lasers used in fiber optic communications systems and other applications. Similarly, alumina (Al_2O_3) and yttrium aluminum garnets (YAG) are used for making lasers. Amorphous silicon is used to make solar cells and photovoltaic modules. Polymers are used to make liquid crystal displays (LCDs).

Smart Materials   A **smart material** can sense and respond to an external stimulus such as a change in temperature, the application of a stress, or a change in humidity or chemical environment. Usually a smart material-based system consists of sensors and actuators that read changes and initiate an action. An example of a passively smart material is lead zirconium titanate (PZT) and shape-memory alloys. When properly processed, PZT can be subjected to a stress, and a voltage is generated. This effect is used to make such devices as spark generators for gas grills and sensors that can detect underwater objects such as fish and submarines. Other examples of smart materials include magnetorheological or MR fluids. These are magnetic paints that respond to magnetic fields. These materials are being used in suspension systems of automobiles, including models by General Motors, Ferrari, and Audi. Still other examples of smart materials and systems are photochromic glasses and automatic dimming mirrors.

Structural Materials   These materials are designed for carrying some type of stress. Steels, concrete, and composites are used to make buildings and bridges. Steels, glasses, plastics, and composites also are used widely to make automotives. Often in these applications, combinations of strength, stiffness, and toughness are needed under different conditions of temperature and loading.

1-4   Classification of Materials Based on Structure

As mentioned before, the term "structure" means the arrangement of a material's atoms; the structure at a microscopic scale is known as "microstructure." We can view these arrangements at different scales, ranging from a few angstrom units to a millimeter. We will learn in Chapter 3 that some materials may be **crystalline** (the material's atoms are arranged in a periodic fashion) or they may be amorphous (the arrangement of the material's atoms does not have long-range order). Some crystalline materials may be in the form of one crystal and are known as **single crystals.** Others consist of many crystals or **grains** and are known as **polycrystalline.** The characteristics of crystals or grains (size, shape, etc.) and that of the regions between them, known as the **grain boundaries**, also affect the properties of materials. We will further discuss these concepts in later chapters. A micrograph of stainless steel showing grains and grain boundaries is shown in Figure 1-1.

1-5   Environmental and Other Effects

The structure-property relationships in materials fabricated into components are often influenced by the surroundings to which the material is subjected during use. This can include exposure to high or low temperatures, cyclical stresses, sudden impact, corrosion, or oxidation. These effects must be accounted for in design to ensure that components do not fail unexpectedly.

Temperature   Changes in temperature dramatically alter the properties of materials (Figure 1-7). Metals and alloys that have been strengthened by certain heat treatments or forming techniques may suddenly lose their strength when heated. A tragic reminder of this is the collapse of the World Trade Center towers on

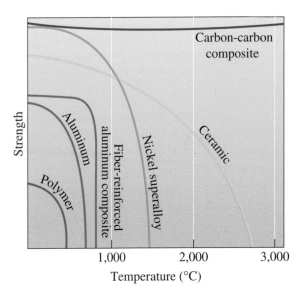

Figure 1-7
Increasing temperature normally reduces the strength of a material. Polymers are suitable only at low temperatures. Some composites, such as carbon-carbon composites, special alloys, and ceramics, have excellent properties at high temperatures.

September 11, 2001. Although the towers sustained the initial impact of the collisions, their steel structures were weakened by elevated temperatures caused by fire, ultimately leading to the collapse.

High temperatures change the structure of ceramics and cause polymers to melt or char. Very low temperatures, at the other extreme, may cause a metal or polymer to fail in a brittle manner, even though the applied loads are low. This low-temperature embrittlement was a factor that caused the *Titanic* to fracture and sink. Similarly, the 1986 *Challenger* accident, in part, was due to embrittlement of rubber O-rings. The reasons why some polymers and metallic materials become brittle are different. We will discuss these concepts in later chapters.

The design of materials with improved resistance to temperature extremes is essential in many technologies, as illustrated by the increase in operating temperatures of aircraft and aerospace vehicles (Figure 1-8). As higher speeds are attained, more heating of the vehicle skin occurs because of friction with the air. Also, engines operate more efficiently at higher temperatures. In order to achieve higher speed and better fuel economy,

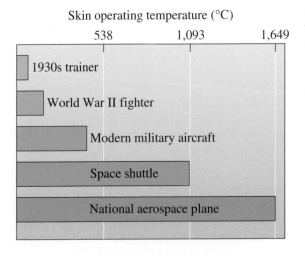

Figure 1-8
Skin operating temperatures for aircraft have increased with the development of improved materials. *(After M. Steinberg, Scientific American, October 1986.)*

Figure 1-9   NASA's X-43A unmanned aircraft is an example of an advanced hypersonic vehicle. *(Courtesy of NASA Dryden Flight Research Center (NASA-DFRC).)*

new materials have gradually increased allowable skin and engine temperatures. NASA's X-43A unmanned aircraft is an example of an advanced hypersonic vehicle (Figure 1-9). It sustained a speed of approximately Mach 10 (12,000 km/h) in 2004. Materials used include refractory tiles in a thermal protection system designed by Boeing and carbon-carbon composites.

Corrosion
Most metals and polymers react with oxygen or other gases, particularly at elevated temperatures. Metals and ceramics may disintegrate and polymers and non-oxide ceramics may oxidize. Materials also are attacked by corrosive liquids, leading to premature failure. The engineer faces the challenge of selecting materials or coatings that prevent these reactions and permit operation in extreme environments. In space applications, we may have to consider the effect of radiation.

Fatigue
In many applications, components must be designed such that the load on the material may not be enough to cause permanent deformation. When we load and unload the material thousands of times, even at low loads, small cracks may begin to develop, and materials fail as these cracks grow. This is known as **fatigue failure**. In designing load-bearing components, the possibility of fatigue must be accounted for.

Strain Rate
Many of you are aware of the fact that Silly Putty®, a silicone-(not silicon-) based plastic, can be stretched significantly if we pull it slowly (small rate of strain). If you pull it fast (higher rate of strain), it snaps. A similar behavior can occur with many metallic materials. Thus, in many applications, the level and rate of strain have to be considered.

In many cases, the effects of temperature, fatigue, stress, and corrosion may be interrelated, and other outside effects could affect the material's performance.

1-6   Materials Design and Selection

When a material is designed for a given application, a number of factors must be considered. The material must acquire the desired **physical** and **mechanical properties**, must be capable of being processed or manufactured into the desired shape, and must provide an economical solution to the design problem. Satisfying these requirements in a manner that protects the environment—perhaps by encouraging recycling of the materials—is also essential. In meeting these design requirements, the engineer may have to make a number of trade-offs in order to produce a serviceable, yet marketable, product.

As an example, material cost is normally calculated on a cost-per-kilogram basis. We must consider the **density** of the material, or its weight-per-unit volume, in our design and selection (Table 1-2). Aluminum may cost more than steel on a weight basis, but it is only one-third the density of steel. Although parts made from aluminum may have to be thicker, the aluminum part may be less expensive than the one made from steel because of the weight difference.

In some instances, particularly in aerospace applications, weight is critical, since additional vehicle weight increases fuel consumption. By using materials that are lightweight but very strong, aerospace or automobile vehicles can be designed to improve fuel utilization. Many advanced aerospace vehicles use composite materials instead of aluminum. These composites, such as carbon-epoxy, are more expensive than the traditional aluminum alloys; however, the fuel savings yielded by the higher strength-to-weight ratio of the composite (Table 1-2) may offset the higher initial cost of the aircraft. There are literally thousands of applications in which similar considerations apply. Usually the selection of materials involves trade-offs between many properties.

By this point of our discussion, we hope that you can appreciate that the properties of materials depend not only on composition but also on how the materials are made (synthesis and processing) and, most importantly, their internal structure. This is why it is not a good idea for an engineer to refer to a handbook and select a material for a given application. The handbooks may be a good starting point. A good engineer will consider: the effects of how the material was made, the exact composition of the candidate

TABLE 1-2 ■ *Strength-to-weight ratios of various materials*

Material	Strength (kg/m^2)	Density (g/cm^3)	Strength-to-weight ratio (cm)
Polyethylene	70×10^4	0.83	8.43×10^4
Pure aluminum	455×10^4	2.17	16.79×10^4
Al$_2$O$_3$	21×10^6	3.16	0.66×10^6
Epoxy	105×10^5	1.38	7.61×10^5
Heat-treated alloy steel	17×10^7	7.75	0.22×10^7
Heat-treated aluminum alloy	60×10^6	2.71	2.21×10^6
Carbon-carbon composite	42×10^6	1.80	2.33×10^6
Heat-treated titanium alloy	12×10^7	4.43	0.27×10^7
Kevlar-epoxy composite	46×10^6	1.47	3.13×10^6
Carbon-epoxy composite	56×10^6	1.38	4.06×10^6

material for the application being considered, any processing that may have to be done for shaping the material or fabricating a component, the structure of the material after processing into a component or device, the environment in which the material will be used, and the cost-to-performance ratio.

Earlier in this chapter, we had discussed the need for you to know the principles of materials science and engineering. If you are an engineer and you need to decide which materials you will choose to fabricate a component, the knowledge of principles of materials science and engineering will empower you with the fundamental concepts. These will allow you to make technically sound decisions in designing with engineered materials.

Summary

- Materials science and engineering (MSE) is an interdisciplinary field concerned with inventing new materials and devices and improving previously known materials by developing a deeper understanding of the microstructure-composition-synthesis-processing relationships.

- Engineered materials are materials designed and fabricated considering MSE principles.

- The properties of engineered materials depend upon their composition, structure, synthesis, and processing. An important performance index for materials or devices is their performance-to-cost ratio.

- The structure of a material refers to the arrangement of atoms or ions in the material.

- The structure at a microscopic level is known as the microstructure.

- Many properties of materials depend strongly on the structure, even if the composition of the material remains the same. This is why the structure-property or microstructure-property relationships in materials are extremely important.

- Materials are classified as metals and alloys; ceramics, glasses and glass-ceramics; composites; polymers; and semiconductors.

- Metals and alloys have good strength, good ductility, and good formability. Metals have good electrical and thermal conductivity. Metals and alloys play an indispensable role in many applications such as automotives, buildings, bridges, and aerospace.

- Ceramics are inorganic crystalline materials. They are strong, serve as good electrical and thermal insulators, are often resistant to damage by high temperatures and corrosive environments, but are mechanically brittle. Modern ceramics form the underpinnings of many microelectronic and photonic technologies.

- Glasses are amorphous, inorganic solids that are typically derived from a molten liquid. Glasses can be tempered to increase strength. Glass-ceramics are formed by annealing glasses to nucleate small crystals that improve resistance to fracture and thermal shock.

- Polymers have relatively low strength; however, the strength-to-weight ratio is very favorable. Polymers are not suitable for use at high temperatures. They have very good corrosion resistance, and—like ceramics—provide good electrical and thermal insulation. Polymers may be either ductile or brittle, depending on structure, temperature, and strain rate.

- Semiconductors possess unique electrical and optical properties that make them essential for manufacturing components in electronic and communication devices.

- Composites are made from different types of materials. They provide unique combinations of mechanical and physical properties that cannot be found in any single material.

- Functional classification of materials includes aerospace, biomedical, electronic, energy and environmental, magnetic, optical (photonic), and structural materials.

- Materials can also be classified as crystalline or amorphous. Crystalline materials may be single crystal or polycrystalline.

- Properties of materials can depend upon the temperature, level and type of stress applied, strain rate, oxidation and corrosion, and other environmental factors.

- Selection of a material having the needed properties and the potential to be manufactured economically and safely into a useful product is a complicated process requiring the knowledge of the structure-property-processing-composition relationships.

Glossary

Alloy A metallic material that is obtained by chemical combinations of different elements (e.g., steel is made from iron and carbon). Typically, alloys have better mechanical properties than pure metals.

Ceramics A group of crystalline inorganic materials characterized by good strength, especially in compression, and high melting temperatures. Many ceramics have very good electrical and thermal insulation behavior.

Composites A group of materials formed from mixtures of metals, ceramics, or polymers in such a manner that unusual combinations of properties are obtained (e.g., fiberglass).

Composition The chemical make-up of a material.

Crystalline material A material composed of one or many crystals. In each crystal, atoms or ions show a long-range periodic arrangement.

Density Mass per unit volume of a material, usually expressed in units of g/cm^3

Fatigue failure Failure of a material due to repeated loading and unloading.

Glass An amorphous material derived from the molten state, typically, but not always, based on silica.

Glass-ceramics A special class of materials obtained by forming a glass and then heat treating it to form small crystals.

Grain boundaries Regions between grains of a polycrystalline material.

Grains Crystals in a polycrystalline material.

Materials engineering An engineering oriented field that focuses on how to transform materials into a useful device or structure.

Materials science A field of science that emphasizes studies of relationships between the microstructure, synthesis and processing, and properties of materials.

Materials science and engineering (MSE) An interdisciplinary field concerned with inventing new materials and improving previously known materials by developing a deeper understanding of the microstructure-composition-synthesis-processing relationships between different materials.

Materials science and engineering tetrahedron A tetrahedron diagram showing how the performance-to-cost ratio of materials depends upon the composition, microstructure, synthesis, and processing.

Mechanical properties Properties of a material, such as strength, that describe how well a material withstands applied forces, including tensile or compressive forces, impact forces, cyclical or fatigue forces, or forces at high temperatures.

Metal An element that has metallic bonding and generally good ductility, strength, and electrical conductivity.

Microstructure The structure of a material at the microscopic length scale.

Physical properties Characteristics such as color, elasticity, electrical or thermal conductivity, magnetism, and optical behavior that generally are not significantly influenced by forces acting on a material.

Plastics Polymers containing other additives.

Polycrystalline material A material composed of many crystals (as opposed to a single-crystal material that has only one crystal).

Polymerization The process by which organic molecules are joined into giant molecules, or polymers.

Polymers A group of materials normally obtained by joining organic molecules into giant molecular chains or networks. Polymers are characterized by low strengths, low melting temperatures, and poor electrical conductivity.

Processing Different ways for shaping materials into useful components or changing their properties.

Semiconductors A group of materials having electrical conductivity between metals and typical ceramics (e.g., Si, GaAs).

Single crystal A crystalline material that is made of only one crystal (there are no grain boundaries).

Smart material A material that can sense and respond to an external stimulus such as change in temperature, application of a stress, or change in humidity or chemical environment.

Strength-to-weight ratio The strength of a material divided by its density; materials with a high strength-to-weight ratio are strong but lightweight.

Structure Description of the arrangements of atoms or ions in a material. The structure of materials has a profound influence on many properties of materials, even if the overall composition does not change.

Synthesis The process by which materials are made from naturally occurring or other chemicals.

Thermoplastics A special group of polymers in which molecular chains are entangled but not interconnected. They can be easily melted and formed into useful shapes. Normally, these polymers have a chainlike structure (e.g., polyethylene).

Thermosets A special group of polymers that decompose rather than melt upon heating. They are normally quite brittle due to a relatively rigid, three-dimensional network structure (e.g., polyurethane).

Problems

Section 1-1 What is Materials Science and Engineering?

1-1 Define materials science and engineering (MSE).

1-2 Define the following terms: (a) composition, (b) structure, (c) synthesis, (d) processing, and (e) microstructure.

1-3 Explain the difference between the terms materials science and materials engineering.

Section 1-2 Classification of Materials

Section 1-3 Functional Classification of Materials

Section 1-4 Classification of Materials Based on Structure

Section 1-5 Environmental and Other Effects

1-4 For each of the following classes of materials, give two *specific* examples that are a regular part of your life:
(a) metals;
(b) ceramics;
(c) polymers; and
(d) semiconductors.

Specify the object that each material is found in and explain why the material is used in each specific application. *Hint:* One example answer for part (a) would be that aluminum is a metal used in the base of some pots and pans for even heat distribution. It is also a lightweight metal that makes it useful in kitchen cookware. Note that in this partial answer to part (a), a specific metal is described for a specific application.

1-5 Describe the enabling materials property of each of the following and why it is so:
(a) silica tiles for the space shuttle;
(b) steel for I-beams in skyscrapers;
(c) a cobalt chrome molybdenum alloy for hip implants;
(d) polycarbonate for eyeglass lenses; and
(e) bronze for sculptures.

1-6 Describe the enabling materials property of each of the following and why it is so:
(a) aluminum for airplane bodies;
(b) polyurethane for teeth aligners (invisible braces);
(c) steel for the ball bearings in a bicycle's wheel hub;
(d) polyethylene terephthalate for water bottles; and
(e) glass for wine bottles.

1-7 Write one paragraph about why single-crystal silicon is currently the material of choice for microelecronics applications. Write a second paragraph about potential alternatives to single-crystal silicon for solar cell applications. Provide a list of the references or websites that you used. You must use at least three references.

1-8 Steel is often coated with a thin layer of zinc if it is to be used outside. What characteristics do you think the zinc provides to this coated, or galvanized, steel? What precautions should be considered in producing this product? How will the recyclability of the product be affected?

1-9 We would like to produce a transparent canopy for an aircraft. If we were to use a traditional window glass canopy, rocks or birds might cause it to shatter. Design a material that would minimize damage or at least keep the canopy from breaking into pieces.

1-10 Coiled springs ought to be very strong and stiff. Si_3N_4 is a strong, stiff material. Would you select this material for a spring? Explain.

1-11 Temperature indicators are sometimes produced from a coiled metal strip that uncoils a specific amount when the temperature increases. How does this work; from what kind of material would the indicator be made; and what are the important properties that the material in the indicator must possess?

1-12 You would like to design an aircraft that can be flown by human power nonstop for a distance of 30 km. What types of material properties would you recommend? What materials might be appropriate?

1-13 You would like to place a 1-m diameter microsatellite into orbit. The satellite will contain delicate electronic equipment that will send and receive radio signals from earth. Design the outer shell within which the electronic equipment is contained. What properties will be required, and what kind of materials might be considered?

1-14 What properties should the head of a carpenter's hammer possess? How would you manufacture a hammer head?

1-15 The hull of the space shuttle consists of ceramic tiles bonded to an aluminum skin. Discuss the design requirements of the shuttle hull that led to the use of this combination of materials. What problems in producing the hull might the designers and manufacturers have faced?

1-16 You would like to select a material for the electrical contacts in an electrical switching device that opens and closes frequently and forcefully. What properties should the contact material possess? What type of material might you recommend? Would Al_2O_3 be a good choice? Explain.

1-17 Aluminum has a density of 2.7 g/cm^3. Suppose you would like to produce a composite material based on aluminum having a density of 1.5 g/cm^3. Design a material that would have this density. Would introducing beads of polyethylene, with a density of 0.95 g/cm^3, into the aluminum be a likely possibility? Explain.

1-18 You would like to be able to identify different materials without resorting to chemical analysis or lengthy testing procedures. Describe some possible testing and sorting techniques you might be able to use based on the physical properties of materials.

1-19 You would like to be able to physically separate different materials in a scrap recycling plant. Describe some possible methods that might be used to separate materials such as polymers, aluminum alloys, and steels from one another.

1-20 Some pistons for automobile engines might be produced from a composite material containing small, hard silicon carbide particles in an aluminum alloy matrix. Explain what benefits each material in the composite may provide to the overall part. What problems might the different properties of the two materials cause in producing the part?

1-21 Investigate the origins and applications for a material that has been invented or discovered since you were born *or* investigate the development of a product or technology that has been invented since you were born that was made possible by the use of a novel material. Write one paragraph about this material or product. Provide a list of the references or websites that you used. You must use at least three references.

Ⓚ Knovel® **Problem**

All problems in the final section of each chapter require the use of the Knovel website (http://www.knovel.com/web/portal/browse).

These three problems are designed to provide an introduction to Knovel, its website, and the interactive tools available on it. For a detailed introduction describing the use of Knovel, please visit your textbook's website at: http://www.cengage.com/engineering/askeland and go to the Student Companion site.

K1-1 • Convert 7750 kg/m^3 to g/l using the Unit Converter.

• Using the Periodic Table, determine the atomic weight of magnesium.

• What is the name of Section 4 in *Perry's Chemical Engineers' Handbook (Seventh Edition)*?

• Find a book title that encompasses the fundamentals of chemistry as well as contains interactive tables of chemical data.

K1-2 • Using the basic search option in Knovel, find as much physical and thermodynamic data associated with ammonium nitrate as possible. What applications does this chemical have?

• Using the Basic Search, find the formula for the volume of both a sphere and a cylinder.

• Using the Data Search, produce a list of five chemicals with a boiling point between 300 and 400 K.

K1-3 • Using the Equation Plotter, determine the enthalpy of vaporization of pure acetic acid at 360 K.

• What is the pressure (in atm) of air with a temperature of 90°C and a water content of 10^{-2} kg water/kg air?

• Find three grades of polymers with a melting point greater than 325°C.

Diamond and graphite both consist of pure carbon, but their materials properties vary considerably. These differences arise from differences in the arrangements of the atoms in the solids and differences in the bonding between atoms. Covalent bonding in diamond leads to high strength and stiffness, excellent thermal conductivity, and poor electrical conductivity. (*Courtesy of Özer Öner/Shutterstock.*) The atoms in graphite are arranged in sheets. Within the sheets, the bonding between atoms is covalent, but between the sheets, the bonds are less strong. Thus graphite can easily be sheared off in sheets as occurs when writing with a pencil. (*Courtesy of Ronald van der Beek/Shutterstock.*) Graphite's thermal conductivity is much lower than that of diamond, and its electrical conductivity is much higher.

Atomic Structure

Have You Ever Wondered?

- *What is nanotechnology?*

- *Why is carbon, in the form of diamond, one of the hardest materials known, but as graphite is very soft and can be used as a solid lubricant?*

- *How is silica, which forms the main chemical in beach sand, used in an ultrapure form to make optical fibers?*

Materials scientists and engineers have developed a set of instruments in order to characterize the **structure** of materials at various length scales. We can examine and describe the structure of materials at five different levels:

1. atomic structure;
2. short- and long-range atomic arrangements;
3. nanostructure;
4. microstructure; and
5. macrostructure.

The features of the structure at each of these levels may have distinct and profound influences on a material's properties and behavior.

The goal of this chapter is to examine **atomic structure** (the nucleus consisting of protons and neutrons and the electrons surrounding the nucleus) in order to lay a foundation for understanding how atomic structure affects the properties, behavior, and resulting applications of engineering materials. We will see that the structure of atoms affects the types of bonds that hold materials together. These different types of bonds directly affect the suitability of materials for real-world engineering applications. The diameter of atoms typically is measured using the angstrom unit (Å or 10^{-10} m).

It also is important to understand how atomic structure and bonding lead to different atomic or ionic arrangements in materials. A close examination of atomic arrangements allows us to distinguish between materials that are **amorphous** (those that lack a long-range ordering of atoms or ions) or **crystalline** (those that exhibit periodic geometrical arrangements of atoms or ions.) Amorphous materials have only **short-range atomic arrangements**, while crystalline materials have short- and **long-range atomic arrangements**. In short-range atomic arrangements, the atoms or ions show a particular order only over relatively short distances (1 to 10 Å). For crystalline materials, the long-range atomic order is in the form of atoms or ions arranged in a three-dimensional pattern that repeats over much larger distances (from ~10 nm to cm.)

Materials science and engineering is at the forefront of **nanoscience** and **nanotechnology**. Nanoscience is the study of materials at the nanometer length scale, and nanotechnology is the manipulation and development of devices at the nanometer length scale. The **nanostructure** is the structure of a material at a **length scale** of 1 to 100 nm. Controlling nanostructure is becoming increasingly important for advanced materials engineering applications.

The **microstructure** is the structure of materials at a **length scale** of 100 to 100,000 nm or 0.1 to 100 micrometers (often written as μm and pronounced as "microns"). The microstructure typically refers to features such as the grain size of a crystalline material and others related to defects in materials. (A *grain* is a single crystal in a material composed of many crystals.)

Macrostructure is the structure of a material at a macroscopic level where the length scale is >100 μm. Features that constitute macrostructure include porosity, surface coatings, and internal and external microcracks.

We will conclude the chapter by considering some of the **allotropes** of carbon. We will see that, although both diamond and graphite are made from pure carbon, they have different materials properties. The key to understanding these differences is to understand how the atoms are arranged in each allotrope.

2-1   The Structure of Materials: Technological Relevance

In today's world, information technology (IT), biotechnology, energy technology, environmental technology, and many other areas require smaller, lighter, faster, portable, more efficient, reliable, durable, and inexpensive devices. We want batteries that are smaller, lighter, and last longer. We need cars that are relatively affordable, lightweight, safe, highly fuel efficient, and "loaded" with many advanced features, ranging from global positioning systems (GPS) to sophisticated sensors for airbag deployment.

Some of these needs have generated considerable interest in nanotechnology and **micro-electro-mechanical systems** (MEMS). As a real-world example of MEMS technology, consider a small accelerometer sensor obtained by the micro-machining of silicon (Si). This sensor is used to measure acceleration in automobiles. The information is processed to a central computer and then used for controlling airbag deployment. Properties and behavior of materials at these "micro" levels can vary greatly when compared to those in their "macro" or bulk state. As a result, understanding the nanostructure and microstructure are areas that have received considerable attention.

The applications shown in Table 2-1 and the accompanying figures (Figures 2-1 through 2-6) illustrate how important the different levels of structure are to materials behavior. The applications illustrated are broken out by their levels of structure and their length scales (the approximate characteristic length that is important for a given application). Examples of how such an application would be used within industry, as well as an illustration, are also provided.

TABLE 2-1 ■ *Levels of structure*

Level of Structure	Example of Technologies
Atomic Structure ($\sim 10^{-10}$ m or 1 Å)	*Diamond*: Diamond is based on carbon-carbon (C-C) covalent bonds. Materials with this type of bonding are expected to be relatively hard. Thin films of diamond are used for providing a wear-resistant edge in cutting tools.

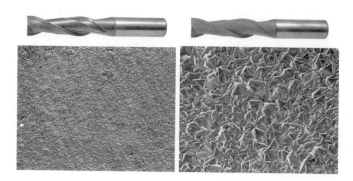

Figure 2-1   Diamond-coated cutting tools. *(Courtesy of OSG Tap & Die, Inc.)*

Atomic Arrangements: Long-Range Order (LRO) (~ 10 nm to cm)	*Lead-zirconium-titanate [Pb($Zr_x Ti_{1-x}$)O_3] or PZT*: When ions in this material are arranged such that they exhibit tetragonal and/or rhombohedral crystal structures, the material is piezoelectric (i.e., it develops a voltage when subjected to pressure or stress). PZT ceramics are used widely for many applications including gas igniters, ultrasound generation, and vibration control.

Figure 2-2   Piezoelectric PZT-based gas igniters. When the piezoelectric material is stressed (by applying a pressure), a voltage develops and a spark is created between the electrodes. *(Courtesy of Morgan Electro Ceramics, Ltd., UK.)*

TABLE 2-1 ■ *(Continued)*

Level of Structure	Example of Technologies
Atomic Arrangements: Short-Range Order (SRO) (1 to 10 Å)	*Ions in silica (SiO₂) glass* exhibit only a short-range order in which Si^{+4} and O^{-2} ions are arranged in a particular way (each Si^{+4} is bonded with 4 O^{-2} ions in a tetrahedral coordination, with each O^{-2} ion being shared by two tetrahedra). This order, however, is not maintained over long distances, thus making silica glass amorphous. Amorphous glasses based on silica and certain other oxides form the basis for the entire fiber-optic communications industry.

Figure 2-3
Optical fibers based on a form of silica that is amorphous. *(Nick Rowe/PhotoDisc Green/GettyImages.)*

Nanostructure ($\sim 10^{-9}$ to 10^{-7} m, 1 to 100 nm)	Nano-sized particles ($\sim$5–10 nm) of iron oxide are used in ferrofluids or liquid magnets. An application of these liquid magnets is as a cooling (heat transfer) medium for loudspeakers.

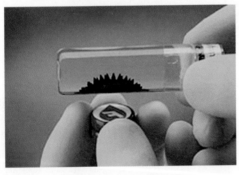

Figure 2-4
Ferrofluid. *(Courtesy of Ferro Tec, Inc.)*

Microstructure ($\sim > 10^{-7}$ to 10^{-4} m, 0.1 to 100 μm)	The mechanical strength of many metals and alloys depends very strongly on the grain size. The grains and grain boundaries in this accompanying micrograph of steel are part of the microstructural features of this crystalline material. In general, at room temperature, a finer grain size leads to higher strength. Many important properties of materials are sensitive to the microstructure.

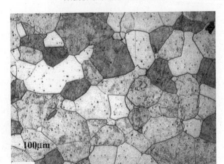

Figure 2-5
Micrograph of stainless steel showing grains and grain boundaries. *(Courtesy of Dr. A. J. Deardo, Dr. M. Hua and Dr. J. Garcia.)*

100μm

TABLE 2-1 ■ *(Continued)*

Level of Structure	Example of Technologies
Macrostructure ($\sim$>10^{-4} m, $\sim$>100,000 nm or 100 μm)	Relatively thick coatings, such as paints on automobiles and other applications, are used not only for aesthetics, but to provide corrosion resistance.

Figure 2-6
A number of organic and inorganic coatings protect the car from corrosion and provide a pleasing appearance. *(Courtesy of Lexus, a division of Toyota Motor Sales, U.S.A., Inc.)*

We now turn our attention to the details concerning the structure of atoms, the bonding between atoms, and how these form a foundation for the properties of materials. Atomic structure influences how atoms are bonded together. An understanding of this helps categorize materials as metals, semiconductors, ceramics, or polymers. It also permits us to draw some general conclusions concerning the mechanical properties and physical behaviors of these four classes of materials.

2-2   The Structure of the Atom

The concepts mentioned next are covered in typical introductory chemistry courses. We are providing a brief review. An atom is composed of a nucleus surrounded by electrons. The nucleus contains neutrons and positively charged protons and carries a net positive charge. The negatively charged electrons are held to the nucleus by an electrostatic attraction. The electrical charge q carried by each electron and proton is 1.60×10^{-19} coulomb (C).

The **atomic number** of an element is equal to the number of protons in each atom. Thus, an iron atom, which contains 26 protons, has an atomic number of 26. The atom as a whole is electrically neutral because the number of protons and electrons are equal.

Most of the mass of the atom is contained within the nucleus. The mass of each proton and neutron is 1.67×10^{-24} g, but the mass of each electron is only 9.11×10^{-28} g. The **atomic mass** M, which is equal to the total mass of the average number of protons and neutrons in the atom in atomic mass units, is also the mass in grams of the **Avogadro constant** N_A of atoms. The quantity $N_A = 6.022 \times 10^{23}$ atoms/mol is the number of atoms or molecules in a mole. Therefore, the atomic mass has units of g/mol. An alternative unit for atomic mass is the **atomic mass unit**, or amu, which is 1/12 the mass of carbon 12 (i.e., the carbon atom with twelve **nucleons**—six protons and six neutrons). As an example, one mole of iron contains 6.022×10^{23} atoms and has a mass of 55.847 g, or 55.847 amu. Calculations including a material's atomic mass and the Avogadro constant are helpful to understanding more about the structure of a material. Example 2-1 illustrates how to calculate the number of atoms for silver, a metal and a good electrical conductor. Example 2-2 illustrates an application to magnetic materials.

| **Example 2-1** | *Calculating the Number of Atoms in Silver* |

Calculate the number of atoms in 100 g of silver (Ag).

SOLUTION

The number of atoms can be calculated from the atomic mass and the Avogadro constant. From Appendix A, the atomic mass, or weight, of silver is 107.868 g/mol. The number of atoms is

$$\text{Number of Ag atoms} = \frac{(100\ \text{g})(6.022 \times 10^{23}\ \text{atoms/mol})}{107.868\ \text{g/mol}}$$
$$= 5.58 \times 10^{23}$$

| **Example 2-2** | *Iron-Platinum Nanoparticles for Information Storage* |

Scientists are considering using nanoparticles of such magnetic materials as iron-platinum (Fe-Pt) as a medium for ultra-high density data storage. Arrays of such particles potentially can lead to storage of trillions of bits of data per square centimeter—a capacity that will be 10 to 100 times higher than any other devices such as computer hard disks. If these scientists considered iron (Fe) particles that are 3 nm in diameter, what will be the number of atoms in one such particle?

SOLUTION

You will learn in a later chapter on magnetic materials that such particles used in recording media tend to be acicular (needle like). For now, let us assume the magnetic particles are spherical in shape.

The radius of a particle is 1.5 nm.

Volume of each iron magnetic nanoparticle $= (4/3)\pi(1.5 \times 10^{-7}\ \text{cm})^3$
$$= 1.4137 \times 10^{-20}\ \text{cm}^3$$

Density of iron $= 7.8\ \text{g/cm}^3$. Atomic mass of iron $= 55.847\ \text{g/mol}$.

Mass of each iron nanoparticle $= 7.8\ \text{g/cm}^3 \times 1.4137 \times 10^{-20}\ \text{cm}^3$
$$= 1.1027 \times 10^{-19}\ \text{g}$$

One mole or 55.847 g of Fe contains 6.022×10^{23} atoms; therefore, the number of atoms in one Fe nanoparticle will be 1189. This is a very small number of atoms. Compare this with the number of atoms in an iron particle that is 10 μm in diameter. Such larger iron particles often are used in breakfast cereals, vitamin tablets, and other applications.

2-3   The Electronic Structure of the Atom

Electrons occupy discrete energy levels within the atom. Each electron possesses a particular energy with no more than two electrons in each atom having the same energy. This also implies that there is a discrete energy difference between any two energy levels. Since each element possesses a different set of these energy levels, the differences between them also are unique. Both the energy levels and the differences between them are known with great precision for every element, forming the basis for many types of **spectroscopy**. Using a spectroscopic method, the identity of elements in a sample may be determined.

Quantum Numbers

The energy level to which each electron belongs is identified by four **quantum numbers**. The four quantum numbers are the principal quantum number n, the azimuthal or secondary quantum number l, the magnetic quantum number m_l, and the spin quantum number m_s.

The principal quantum number reflects the grouping of electrons into sets of energy levels known as shells. Azimuthal quantum numbers describe the energy levels within each shell and reflect a further grouping of similar energy levels, usually called orbitals. The magnetic quantum number specifies the orbitals associated with a particular azimuthal quantum number within each shell. Finally, the **spin quantum number** (m_s) is assigned values of $+1/2$ and $-1/2$, which reflect the two possible values of "spin" of an electron.

According to the **Pauli Exclusion Principle**, within each atom, no two electrons may have the same four quantum numbers, and thus, each electron is designated by a unique set of four quantum numbers. The number of possible energy levels is determined by the first three quantum numbers.

1. The principal quantum number n is assigned integer values 1, 2, 3, 4, 5, . . . that refer to the quantum shell to which the electron belongs. A **quantum shell** is a *set* of fixed energy levels to which electrons belong.

    Quantum shells are also assigned a letter; the shell for $n = 1$ is designated K, for $n = 2$ is L, for $n = 3$ is M, and so on. These designations were carried over from the nomenclature used in optical spectroscopy, a set of techniques that predates the understanding of quantized electronic levels.

2. The *number* of energy levels in *each* quantum shell is determined by the **azimuthal quantum number** l and the magnetic quantum number m_l. The azimuthal quantum numbers are assigned $l = 0, 1, 2, \ldots, n - 1$. For example, when $n = 2$, there are two azimuthal quantum numbers, $l = 0$ and $l = 1$. When $n = 3$, there are three azimuthal quantum numbers, $l = 0, l = 1$, and $l = 2$. The azimuthal quantum numbers are designated by lowercase letters; one speaks, for instance, of the *d* orbitals:

$$s \text{ for } l = 0 \quad d \text{ for } l = 2$$

$$p \text{ for } l = 1 \quad f \text{ for } l = 3$$

3. The number of values for the magnetic quantum number m_l gives the number of energy levels, or orbitals, for each azimuthal quantum number. The total number of magnetic quantum numbers for each l is $2l + 1$. The values for m_l are given by whole numbers between $-l$ and $+l$. For example, if $l = 2$, there are $2(2) + 1 = 5$ magnetic quantum numbers with values $-2, -1, 0, +1$, and $+2$. The combination of l and m_l specifies a particular orbital in a shell.

4. No more than two electrons with opposing electronic spins ($m_s = +1/2$ and $-1/2$) may be present in each orbital.

By carefully considering the possible numerical values for n, l, and m_l, the range of *possible* quantum numbers may be determined. For instance, in the K shell (that is, $n = 1$), there is just a single s orbital (as the only allowable value of l is 0 and m_l is 0). As a result, a K shell may contain no more than two electrons. As another example, consider an M shell. In this case, $n = 3$, so l takes values of 0, 1, and 2, (there are s, p, and d orbitals present). The values of m_l reflect that there is a single s orbital ($m_l = 0$, a single value), three p orbitals ($m_l = -1, 0, +1$, or three values), and five d orbitals ($m_l = -2, -1, 0, +1, +2$, or five discrete values).

The shorthand notation frequently used to denote the electronic structure of an atom combines the numerical value of the principal quantum number, the lowercase letter notation for the azimuthal quantum number, and a superscript showing the number of electrons in each type of orbital. The shorthand notation for neon, which has an atomic number of ten, is

$$1s^2 2s^2 2p^6$$

Deviations from Expected Electronic Structures

The energy levels of the quantum shells do not fill in strict numerical order. The **Aufbau Principle** is a graphical device that predicts deviations from the expected ordering of the energy levels. The Aufbau Principle is shown in Figure 2-7. To use the Aufbau Principle, write the possible combinations of the principal quantum number and azimuthal quantum number for each quantum shell. The combinations for each quantum shell should be written on a single line. As the principal quantum number increases by one, the number of combinations within each shell increases by one (i.e., each row is one entry longer than the prior row). Draw arrows through the rows on a diagonal from the upper right to the lower left as shown in Figure 2-7. By following the arrows, the order in which the energy levels of each quantum level are filled is predicted.

For example, according to the Aufbau Principle, the electronic structure of iron, atomic number 26, is

$$1s^2 2s^2 2p^6 3s^2 3p^6 4s^2 3d^6$$

Conventionally, the principal quantum numbers are arranged from lowest to highest when writing the electronic structure. Thus, the electronic structure of iron is written

$$1s^2 2s^2 2p^6 3s^2 3p^6 \qquad 3d^6 4s^2$$

The unfilled $3d$ level (there are five d orbitals, so in shorthand d^1, d^2, ..., d^{10} are possible) causes the magnetic behavior of iron.

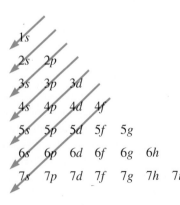

Figure 2-7
The Aufbau Principle. By following the arrows, the order in which the energy levels of each quantum level are filled is predicted: $1s$, $2s$, $2p$, $3s$, $3p$, etc. Note that the letter designations for $l = 4, 5, 6$ are g, h, and i.

Note that not all elements follow the Aufbau Principle. A few, such as copper, are exceptions. According to the Aufbau Principle, copper should have the electronic structure $1s^2 2s^2 2p^6 3s^2 3p^6 3d^9 4s^2$, but copper actually has the electronic structure

$$1s^2 2s^2 2p^6 3s^2 3p^6 3d^{10} \qquad \boxed{4s^1}$$

Generally, electrons will occupy each orbital of a given energy level singly before the orbitals are doubly occupied. For example, nitrogen has the electronic structure

$$1s^2 \qquad \boxed{2s^2 2p^3}$$

Each of the three p orbitals in the L shell contains one electron rather than one orbital containing two electrons, one containing one electron, and one containing zero electrons.

Valence

The **valence** of an atom is the number of electrons in an atom that participate in bonding or chemical reactions. Usually, the valence is the number of electrons in the outer s and p energy levels. The valence of an atom is related to the ability of the atom to enter into chemical combination with other elements. Examples of the valence are

$$Mg : 1s^2 \, 2s^2 \, 2p^6 \qquad \boxed{3s^2} \qquad valence = 2$$

$$Al : 1s^2 \, 2s^2 \, 2p^6 \qquad \boxed{3s^2 3p^1} \qquad valence = 3$$

$$Si : 1s^2 \, 2s^2 \, 2p^6 \qquad \boxed{3s^2 \, 3p^2} \qquad valence = 4$$

Valence also depends on the immediate environment surrounding the atom or the neighboring atoms available for bonding. Phosphorus has a valence of five when it combines with oxygen, but the valence of phosphorus is only three—the electrons in the $3p$ level—when it reacts with hydrogen. Manganese may have a valence of 2, 3, 4, 6, or 7!

Atomic Stability and Electronegativity

If an atom has a valence of zero, the element is inert (non-reactive). An example is argon, which has the electronic structure:

$$1s^2 \, 2s^2 \, 2p^6 \qquad \boxed{3s^2 \, 3p^6}$$

Other atoms prefer to behave as if their outer s and p levels are either completely full, with eight electrons, or completely empty. Aluminum has three electrons in its outer s and p levels. An aluminum atom readily gives up its outer three electrons to empty the $3s$ and $3p$ levels. The atomic bonding and the chemical behavior of aluminum are determined by how these three electrons interact with surrounding atoms.

On the other hand, chlorine contains seven electrons in the outer $3s$ and $3p$ levels. The reactivity of chlorine is caused by its desire to fill its outer energy level by accepting an electron.

Electronegativity describes the tendency of an atom to gain an electron. Atoms with almost completely filled outer energy levels—such as chlorine—are strongly electronegative and readily accept electrons. Atoms with nearly empty outer levels—such as sodium—readily give up electrons and have low electronegativity. High atomic number elements also have low electronegativity because the outer electrons are at a greater distance from the positive nucleus, so that they are not as strongly attracted to the atom. Electronegativities for some elements are shown in Figure 2-8. Elements with low electronegativity (i.e., <2.0) are sometimes described as **electropositive**.

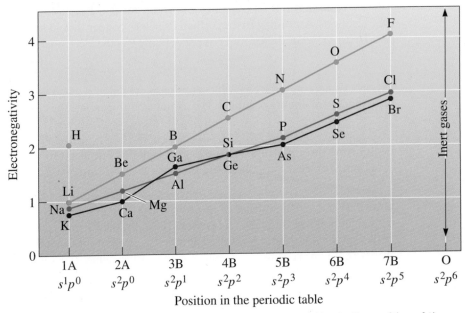

Figure 2-8 The electronegativities of selected elements relative to the position of the elements in the periodic table.

2-4   The Periodic Table

The periodic table contains valuable information about specific elements and can also help identify trends in atomic size, melting point, chemical reactivity, and other properties. The familiar periodic table (Figure 2-9) is constructed in accordance with the electronic structure of the elements. Not all elements in the periodic table are naturally occurring. Rows in the periodic table correspond to quantum shells, or principal quantum numbers. Columns typically refer to the number of electrons in the outermost s and p energy levels and correspond to the most common valence. In engineering, we are mostly concerned with

(a) polymers (plastics) (primarily based on carbon, which appears in Group 4B);

(b) ceramics (typically based on combinations of many elements appearing in Groups 1 through 5B, and such elements as oxygen, carbon, and nitrogen); and

(c) metallic materials (typically based on elements in Groups 1, 2 and transition metal elements).

Many technologically important semiconductors appear in Group 4B (e.g., silicon (Si), diamond (C), germanium (Ge)). Semiconductors also can be combinations of elements from Groups 2B and 6B (e.g., cadmium selenide (CdSe), based on cadmium (Cd) from Group 2 and selenium (Se) based on Group 6). These are known as **II–VI** (two-six) **semiconductors**. Similarly, gallium arsenide (GaAs) is a **III–V** (three-five) **semiconductor** based on gallium (Ga) from Group 3B and arsenic (As) from Group 5B. Many **transition elements** (e.g., titanium (Ti), vanadium (V), iron (Fe), nickel (Ni), cobalt (Co), etc.) are particularly useful for magnetic and optical materials due to their electronic configurations that allow multiple valences.

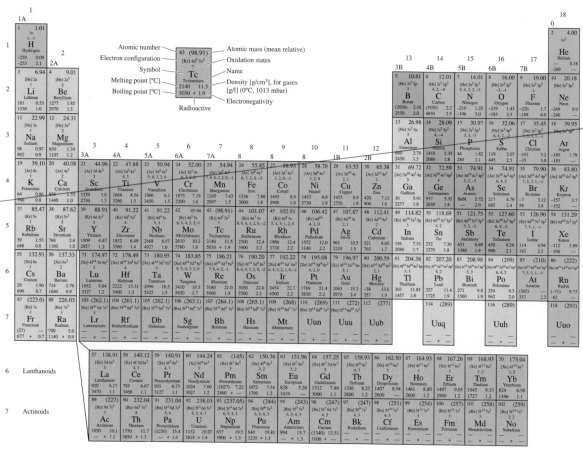

Figure 2-9 Periodic table of elements.

The ordering of the elements in the periodic table and the origin of the Aufbau Principle become even clearer when the rows for the Lathanoid and Actinoid series are inserted into their correct positions (see Figure 2-10) rather than being placed below the periodic table to conserve space. Figure 2-10 indicates the particular orbital being filled by each additional electron as the atomic number increases. Note that exceptions are indicated for those elements that do not follow the Aufbau Principle.

Trends in Properties

The periodic table contains a wealth of useful information (e.g., atomic mass, atomic number of different elements, etc.). It also points to trends in atomic size, melting points, and chemical reactivity. For example, carbon (in its diamond form) has the highest melting point (3550°C). Melting points of the elements below carbon decrease (i.e., silicon (Si) (1410°C), germanium (Ge) (937°C), tin (Sn) (232°C), and lead (Pb) (327°C)). Note that the melting temperature of Pb is higher than that of Sn. The periodic table indicates trends and not exact variations in properties.

We can discern trends in other properties from the periodic table. Diamond is a material with a very large bandgap (i.e., it is not a very effective conductor of electricity). This is consistent with the fact that carbon (in diamond form) has the highest melting point among Group 4B elements, which suggests the interatomic forces are strong (see Section 2-6). As we move down the column, the bandgap decreases (the

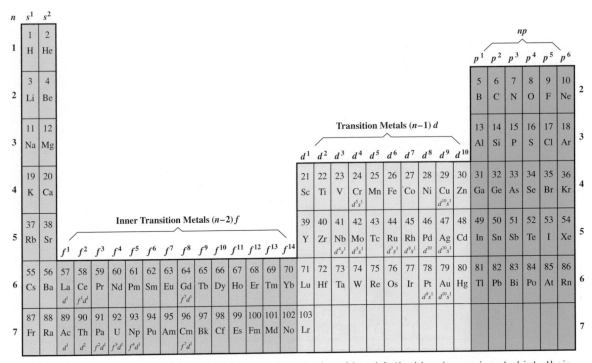

Figure 2-10 The periodic table for which the rows of the Lathanoid and Actinoid series are inserted into their correct positions. The column heading indicates the particular orbital being filled by each additional electron as the atomic number increases.

bandgaps of Si and Ge are 1.11 and 0.67 eV, respectively). Moving farther down, one form of tin is a semiconductor. Another form of tin is metallic. If we look at Group 1A, we see that lithium is highly electropositive (i.e., an element whose atoms want to participate in chemical interactions by donating electrons and are therefore highly reactive). Likewise, if we move down Column 1A, we can see that the chemical reactivity of elements decreases.

Thus, the periodic table gives us useful information about formulas, atomic numbers, and atomic masses of elements. It also helps us in predicting or rationalizing trends in properties of elements and compounds. This is why the periodic table is very useful to both scientists and engineers.

2-5 Atomic Bonding

There are four important mechanisms by which atoms are bonded in engineered materials. These are

1. **metallic bonds**;
2. **covalent bonds**;
3. **ionic bonds**; and
4. **van der Waals bonds**.

The first three types of bonds are relatively strong and are known as **primary bonds** (relatively strong bonds between adjacent atoms resulting from the transfer or sharing of outer orbital electrons). The van der Waals bonds are secondary bonds and originate from a different mechanism and are relatively weaker. Let's look at each of these types of bonds.

The Metallic Bond

The metallic elements have electropositive atoms that donate their valence electrons to form a "sea" of electrons surrounding the atoms (Figure 2-11). Aluminum, for example, gives up its three valence electrons, leaving behind a core consisting of the nucleus and inner electrons. Since three negatively charged electrons are missing from this core, it has a positive charge of three. The valence electrons move freely within the electron sea and become associated with several atom cores. The positively charged ion cores are held together by mutual attraction to the electrons, thus producing a strong metallic bond.

Because their valence electrons are not fixed in any one position, most pure metals are good electrical conductors of electricity at relatively low temperatures ($\sim T < 300$ K). Under the influence of an applied voltage, the valence electrons move, causing a current to flow if the circuit is complete.

Metals show good ductility since the metallic bonds are non-directional. There are other important reasons related to microstructure that can explain why metals actually exhibit *lower strengths* and *higher ductility* than what we may anticipate from their bonding. **Ductility** refers to the ability of materials to be stretched or bent permanently without breaking. We will discuss these concepts in greater detail in Chapter 6. In general, the melting points of metals are relatively high. From an optical properties viewpoint, metals make good reflectors of visible radiation. Owing to their electropositive character, many metals such as iron tend to undergo corrosion or oxidation. Many pure metals are good conductors of heat and are effectively used in many heat transfer applications. We emphasize that metallic bonding is *one of the factors* in our efforts to rationalize the trends

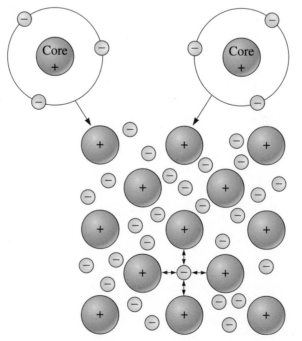

Figure 2-11
The metallic bond forms when atoms give up their valence electrons, which then form an electron sea. The positively charged atom cores are bonded by mutual attraction to the negatively charged electrons.

observed with respect to the properties of metallic materials. As we will see in some of the following chapters, there are other factors related to microstructure that also play a crucial role in determining the properties of metallic materials.

The Covalent Bond
Materials with **covalent bonding** are characterized by bonds that are formed by sharing of valence electrons among two or more atoms. For example, a silicon atom, which has a valence of four, obtains eight electrons in its outer energy shell by sharing its valence electrons with four surrounding silicon atoms, as in Figure 2-12(a) and (b). Each instance of sharing represents one covalent bond; thus, each silicon atom is bonded to four neighboring atoms by four covalent bonds. In order for the covalent bonds to be formed, the silicon atoms must be arranged so the bonds have a fixed **directional relationship** with one another. A directional relationship is formed when the bonds between atoms in a covalently bonded material form specific angles, depending on the material. In the case of silicon, this arrangement produces a tetrahedron, with angles of 109.5° between the covalent bonds [Figure 2-12(c)].

Covalent bonds are very strong. As a result, covalently bonded materials are very strong and hard. For example, diamond (C), silicon carbide (SiC), silicon nitride (Si_3N_4), and boron nitride (BN) all have covalent bonds. These materials also exhibit very high melting points, which means they could be useful for high-temperature applications. On the other hand, the high temperature needed for processing presents a challenge. The materials bonded in this manner typically have limited ductility because the bonds tend to be directional. The electrical conductivity of many covalently bonded materials (i.e., silicon, diamond, and many ceramics) is not high since the valence electrons are locked in bonds between atoms and are not readily available for conduction. With some of these materials such as Si, we can get useful and controlled levels of electrical conductivity by deliberately introducing small levels of other elements known as dopants. Conductive polymers are also a good example of

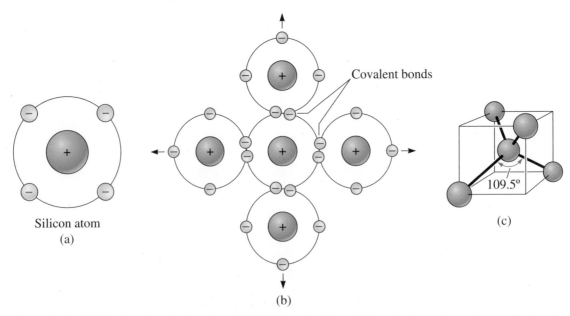

Figure 2-12   (a) Covalent bonding requires that electrons be shared between atoms in such a way that each atom has its outer *sp* orbitals filled. (b) In silicon, with a valence of four, four covalent bonds must be formed. (c) Covalent bonds are directional. In silicon, a tetrahedral structure is formed with angles of 109.5° required between each covalent bond.

covalently bonded materials that can be turned into semiconducting materials. The development of conducting polymers that are lightweight has captured the attention of many scientists and engineers for developing flexible electronic components.

We cannot simply predict whether or not a material will be high or low strength, ductile or brittle, simply based on the nature of bonding! We need additional information on the atomic, microstructure, and macrostructure of the material; however, the nature of bonding does point to a trend for materials with certain types of bonding and chemical compositions. Example 2-3 explores how one such bond of oxygen and silicon join to form silica.

Example 2-3   *How Do Oxygen and Silicon Atoms Join to Form Silica?*

Assuming that silica (SiO_2) has 100% covalent bonding, describe how oxygen and silicon atoms in silica (SiO_2) are joined.

SOLUTION

Silicon has a valence of four and shares electrons with four oxygen atoms, thus giving a total of eight electrons for each silicon atom. Oxygen has a valence of six and shares electrons with two silicon atoms, giving oxygen a total of eight electrons. Figure 2-13 illustrates one of the possible structures. Similar to silicon (Si), a tetrahedral structure is produced. We will discuss later in this chapter how to account for the ionic and covalent nature of bonding in silica.

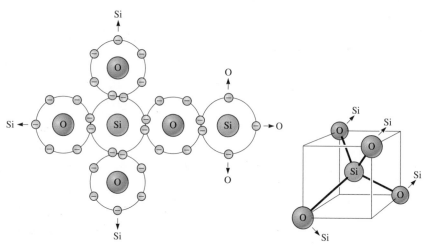

Figure 2-13   The tetrahedral structure of silica (SiO_2), which contains covalent bonds between silicon and oxygen atoms (for Example 2-3).

The Ionic Bond   When more than one type of atom is present in a material, one atom may donate its valence electrons to a different atom, filling the outer energy shell of the second atom. Both atoms now have filled (or emptied) outer energy levels, but both have acquired an electrical charge and behave as ions. The atom that contributes the electrons is left with a net positive charge and is called a **cation**, while the atom that accepts the electrons acquires a net negative charge and is called an **anion**. The oppositely charged ions are then attracted to one another and produce the **ionic bond**. For example, the attraction between sodium and chloride ions (Figure 2-14) produces sodium chloride (NaCl), or table salt.

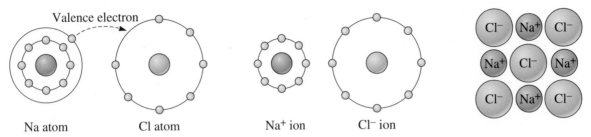

Figure 2-14   An ionic bond is created between two unlike atoms with different electronegativities. When sodium donates its valence electron to chlorine, each becomes an ion, attraction occurs, and the ionic bond is formed.

Van der Waals Bonding

The origin of van der Waals forces between atoms and molecules is quantum mechanical in nature and a meaningful discussion is beyond the scope of this book. We present here a simplified picture. If two electrical charges $+q$ and $-q$ are separated by a distance d, the dipole moment is defined as $q \times d$. Atoms are electrically neutral. Also, the centers of the positive charge (nucleus) and negative charge (electron cloud) coincide. Therefore, a neutral atom has no dipole moment. When a neutral atom is exposed to an internal or external electric field, the atom may become polarized (i.e., the centers of positive and negative charges separate). This creates or induces a dipole moment (Figure 2-15). In some molecules, the dipole moment does not have to be induced—it exists by virtue of the direction of bonds and the nature of atoms. These molecules are known as **polarized molecules**. An example of such a molecule that has a permanently built-in dipole moment is water (Figure 2-16).

Molecules or atoms in which there is either an induced or permanent dipole moment attract each other. The resulting force is known as the van der Waals force. Van der Waals forces between atoms and molecules have their origin in interactions between dipoles that are induced or in some cases interactions between permanent dipoles that are present in certain polar molecules. What is unique about these forces is they are present in every material.

There are three types of **van der Waals** interactions, namely London forces, Keesom forces, and Debye forces. If the interactions are between two dipoles that are induced in atoms or molecules, we refer to them as **London forces** (e.g., carbon tetrachloride) (Figure 2-15). When an induced dipole (that is, a dipole that is induced in what is otherwise a non-polar atom or molecule) interacts with a molecule that has a permanent dipole moment, we refer to this interaction as a **Debye interaction**. An example of Debye interaction would be forces between water molecules and those of carbon tetrachloride.

If the interactions are between molecules that are permanently polarized (e.g., water molecules attracting other water molecules or other polar molecules), we refer to these as **Keesom interactions**. The attraction between the positively charged regions of one

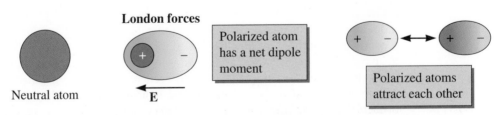

Figure 2-15   Illustration of London forces, a type of a van der Waals force, between atoms.

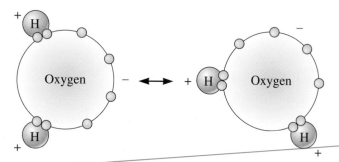

Figure 2-16
The Keesom interactions are formed as a result of polarization of molecules or groups of atoms. In water, electrons in the oxygen tend to concentrate away from the hydrogen. The resulting charge difference permits the molecule to be weakly bonded to other water molecules.

water molecule and the negatively charged regions of a second water molecule provides an attractive bond between the two water molecules (Figure 2-16).

The bonding between molecules that have a permanent dipole moment, known as the Keesom force, is often referred to as a **hydrogen bond**, where hydrogen atoms represent one of the polarized regions. Thus, hydrogen bonding is essentially a Keesom force and is a type of van der Waals force. The relatively strong Keesom force between water molecules is the reason why surface tension (72 mJ/m^2 or dyne/cm at room temperature) and the boiling point of water (100°C) are much higher than those of many organic liquids of comparable molecular weight (surface tension ~20 to 25 dyne/cm, boiling points up to 80°C).

Note that van der Waals bonds are **secondary bonds**, but the atoms within the molecule or group of atoms are joined by strong covalent or ionic bonds. Heating water to the boiling point breaks the van der Waals bonds and changes water to steam, but much higher temperatures are required to break the covalent bonds joining oxygen and hydrogen atoms.

Although termed "secondary," based on the bond energies, van der Waals forces play a very important role in many areas of engineering. Van der Waals forces between atoms and molecules play a vital role in determining the surface tension and boiling points of liquids. In materials science and engineering, the surface tension of liquids and the surface energy of solids come into play in different situations. For example, when we want to process ceramic or metal powders into dense solid parts, the powders often have to be dispersed in water or organic liquids. Whether we can achieve this dispersion effectively depends upon the surface tension of the liquid and the surface energy of the solid material. Surface tension of liquids also assumes importance when we are dealing with processing of molten metals and alloys (e.g., casting) and glasses.

Van der Waals bonds can dramatically change the properties of certain materials. For example, graphite and diamond have very different mechanical properties. In many plastic materials, molecules contain polar parts or side groups (e.g., cotton or cellulose, PVC, Teflon). Van der Waals forces provide an extra binding force between the chains of these polymers (Figure 2-17).

Polymers in which van der Waals forces are stronger tend to be relatively stiffer and exhibit relatively higher glass transition temperatures (T_g). The glass transition temperature is a temperature below which some polymers tend to behave as brittle materials (i.e., they show poor ductility). As a result, polymers with van der Waals bonding (in addition to the covalent bonds in the chains and side groups) are relatively brittle at room temperature (e.g., PVC). In processing such polymers, they need to be "plasticized" by adding other smaller polar molecules that interact with the polar parts of the long polymer chains, thereby lowering the T_g and enhancing flexibility.

Mixed Bonding
In most materials, bonding between atoms is a mixture of two or more types. Iron, for example, is bonded by a combination of metallic and covalent bonding that prevents atoms from packing as efficiently as we might expect.

(a)

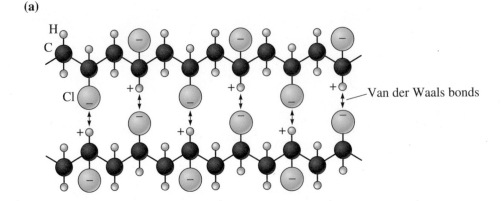

(b)

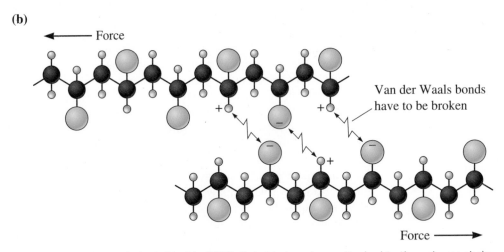

Figure 2-17   (a) In polyvinyl chloride (PVC), the chlorine atoms attached to the polymer chain have a negative charge and the hydrogen atoms are positively charged. The chains are weakly bonded by van der Waals bonds. This additional bonding makes PVC stiffer. (b) When a force is applied to the polymer, the van der Waals bonds are broken and the chains slide past one another.

Compounds formed from two or more metals (**intermetallic compounds**) may be bonded by a mixture of metallic and ionic bonds, particularly when there is a large difference in electronegativity between the elements. Because lithium has an electronegativity of 1.0 and aluminum has an electronegativity of 1.5, we would expect AlLi to have a combination of metallic and ionic bonding. On the other hand, because both aluminum and vanadium have electronegativities of 1.5, we would expect Al_3V to be bonded primarily by metallic bonds.

Many ceramic and semiconducting compounds, which are combinations of metallic and nonmetallic elements, have a mixture of covalent and ionic bonding. As the electronegativity difference between the atoms increases, the bonding becomes more ionic. The fraction of bonding that is covalent can be estimated from the following equation:

$$\text{Fraction covalent} = \exp(-0.25\Delta E^2) \qquad (2\text{-}1)$$

where ΔE is the difference in electronegativities. Example 2-4 explores the nature of the bonds found in silica.

| **Example 2-4** | *Determining if Silica is Ionically or Covalently Bonded* |

In a previous example, we used silica (SiO_2) as an example of a covalently bonded material. In reality, silica exhibits ionic and covalent bonding. What fraction of the bonding is covalent? Give examples of applications in which silica is used.

SOLUTION

From Figure 2-9, the electronegativity of silicon is 1.8 and that of oxygen is 3.5. The fraction of the bonding that is covalent is

$$\text{Fraction covalent} = \exp[-0.25(3.5 - 1.8)^2] = 0.486$$

Although the covalent bonding represents only about half of the bonding, the directional nature of these bonds still plays an important role in the eventual structure of SiO_2.

Silica has many applications. Silica is used for making glasses and optical fibers. We add nanoparticles of silica to tires to enhance the stiffness of the rubber. High-purity silicon (Si) crystals are made by reducing silica to silicon.

2-6   Binding Energy and Interatomic Spacing

Interatomic Spacing The equilibrium distance between atoms is caused by a balance between repulsive and attractive forces. In the metallic bond, for example, the attraction between the electrons and the ion cores is balanced by the repulsion between ion cores. Equilibrium separation occurs when the total interatomic energy (IAE) of the pair of atoms is at a minimum, or when no net force is acting to either attract or repel the atoms (Figure 2-18).

The **interatomic spacing** in a solid metal is *approximately* equal to the atomic diameter, or twice the atomic radius r. We cannot use this approach for ionically bonded materials, however, since the spacing is the sum of the two different ionic radii. Atomic and ionic radii for the elements are listed in Appendix B and will be used in the next chapter.

The minimum energy in Figure 2-18 is the **binding energy**, or the energy required to create or break the bond. Consequently, materials having a high binding energy also have a high strength and a high melting temperature. Ionically bonded materials have a particularly large binding energy (Table 2-2) because of the large difference in electronegativities between the ions. Metals have lower binding energies because the electronegativities of the atoms are similar.

Other properties can be related to the force-distance and energy-distance expressions in Figure 2-19. For example, the **modulus of elasticity** of a material (the slope (E) of the stress-strain curve in the elastic region, also known as Young's modulus) is related to the slope of the force-distance curve (Figure 2-19). A steep slope, which correlates with a higher binding energy and a higher melting point, means that a greater force is required to stretch the bond; thus, the material has a high modulus of elasticity.

An interesting point that needs to be made is that not all properties of engineered materials are microstructure sensitive. Modulus of elasticity is one such property. If we have two aluminum samples that have essentially the same chemical composition but different grain size, we expect that the modulus of elasticity of these samples will be about the same; however, **yield strengths**, the level of stress at which the material begins

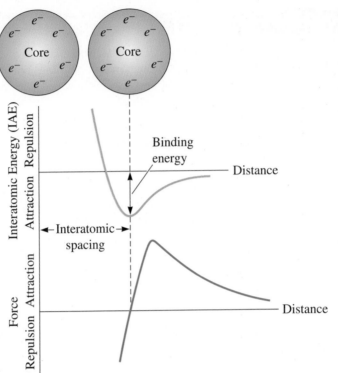

Figure 2-18
Atoms or ions are separated by an equilibrium spacing that corresponds to the minimum interatomic energy for a pair of atoms or ions (or when zero force is acting to repel or attract the atoms or ions).

TABLE 2-2 ■ Binding energies for the four bonding mechanisms

Bond	Binding Energy (kcal/mol)
Ionic	150–370
Covalent	125–300
Metallic	25–200
Van der Waals	<10

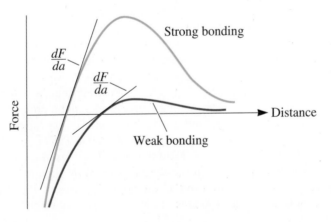

Figure 2-19
The force-distance (*F–a*) curve for two materials, showing the relationship between atomic bonding and the modulus of elasticity. A steep *dF/da* slope gives a high modulus.

to permanently deform, of these samples will be quite different. The yield strength, therefore, is a microstructure sensitive property. We will learn in subsequent chapters that, compared to other mechanical properties such as yield strength and tensile strength, the modulus of elasticity does not depend strongly on the microstructure. The modulus of elasticity can be linked directly to the stiffness of bonds between atoms. Thus, the modulus of elasticity depends primarily on the atoms that make up the material.

Another property that can be linked to the binding energy or interatomic force-distance curves is the **coefficient of thermal expansion** (CTE). The CTE, often denoted as α, is the fractional change in linear dimension of a material per degree of temperature. It can be written $\alpha = (1/L)(dL/dT)$, where L is length and T is temperature. The CTE is related to the strength of the atomic bonds. In order for the atoms to move from their equilibrium separation, energy must be supplied to the material. If a very deep interatomic energy (IAE) trough caused by strong atomic bonding is characteristic of the material (Figure 2-20), the atoms separate to a lesser degree and have a low, linear coefficient of thermal expansion. Materials with a low coefficient of thermal expansion maintain their dimensions more closely when the temperature changes. It is important to note that there are microstructural features (e.g., anistropy, or varying properties, in thermal expansion with different crystallographic directions) that also have a significant effect on the overall thermal expansion coefficient of an engineered material.

Materials that have very low expansion are useful in many applications where the components are expected to repeatedly undergo relatively rapid heating and cooling. For example, cordierite ceramics (used as catalyst support in catalytic converters in cars), ultra-low expansion (ULE) glasses, Visionware™, and other glass-ceramics developed by Corning, have very low thermal expansion coefficients. In the case of thin films or coatings on substrates, we are not only concerned about the actual values of thermal expansion coefficients but also the difference between thermal expansion coefficients between the substrate and the film or coating. Too much difference between these causes development of stresses that can lead to delamination or warping of the film or coating.

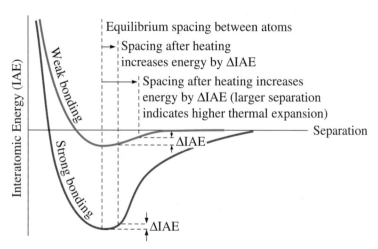

Figure 2-20 The interatomic energy (IAE)—separation curve for two atoms. Materials that display a steep curve with a deep trough have low linear coefficients of thermal expansion.

2-7   The Many Forms of Carbon: Relationships Between Arrangements of Atoms and Materials Properties

Carbon is one of the most abundant elements on earth. Carbon is an essential component of all living organisms, and it has enormous technological significance with a wide range of applications. For example, carbon dating is a process by which scientists measure the amount of a radioactive isotope of carbon that is present in fossils to determine their age, and now some of the most cutting-edge technologies exploit one of the world's strongest materials: carbon nanotubes. And of course, a small amount of carbon (e.g., 0.5 wt%) converts iron into steel.

Pure carbon exists as several **allotropes**, meaning that pure carbon exists in different forms (or has different arrangements of its atoms) depending on the temperature and pressure. We will learn more about allotropes in Chapter 3. Two allotropes of carbon are very familiar to us: diamond and graphite, while two other forms of carbon have been discovered much more recently: buckminsterfullerene also known as "buckyballs" and carbon nanotubes. In fact, there are other allotropes of carbon that will not be discussed here.

The allotropes of carbon all have the same composition—they are pure carbon— and yet they display dramatically different materials properties. The key to understanding these differences is to understand how the atoms are arranged in each allotrope.

In this chapter, we learned that carbon has an atomic number of six, meaning that it has six protons. Thus, a neutral carbon atom has six electrons. Two of these electrons occupy the innermost quantum shell (completely filling it), and four electrons occupy the quantum shell with the principal quantum number $n = 2$. Each carbon atom has four valence electrons and can share electrons with up to four different atoms. Thus, carbon atoms can combine with other elements as well as with other carbon atoms. This allows carbon to form many different compounds of varying structure as well as several allotropes of pure carbon.

Popular culture values one of the allotropes of carbon above all others—the diamond. Figure 2-21(a) is a diagram showing the repeat unit of the structure of diamond. Each sphere is an atom, and each line represents covalent bonds between carbon atoms. We will learn to make diagrams such as this in Chapter 3. Diamond is a crystal, meaning that its atoms are arranged in a regular, repeating array. Figure 2-21(a) shows that each carbon atom is bonded to four other carbon atoms. These bonds are covalent, meaning that each carbon atom shares each one of its outermost electrons with an adjacent carbon atom, so each atom has a full outermost quantum shell. Recall from Section 2-5 that covalent bonds are strong bonds. Figure 2-21(b) again shows the diamond structure, but the view has been rotated from Figure 2-21(a) by 45°.

Figure 2-21(c) is a micrograph of diamond that was acquired using an instrument known as a transmission electron microscope (TEM). A TEM does not simply take a picture of atoms; a TEM senses regions of electron intensity and, in so doing, maps the locations of the atoms.

The predominantly covalent bonding in diamond profoundly influences its macroscopic properties. Diamond is one of the highest melting-point materials known with a melting temperature of 3550°C (6420°F). This is due to the strong covalent bonding between atoms. Diamond also has one of the highest known thermal conductivities (2000 W/(m · K)). For comparison, aluminum (which is an excellent thermal conductor) has a thermal conductivity of only 238 W/(m · K). The high thermal conductivity of diamond is due to the rigidity of its covalently bonded structure. Diamond is the stiffest material with an elastic modulus of 1100 GPa. (In Chapter 6, we will learn more about the elastic modulus; for now, we will simply say that diamond is about ten times stiffer than

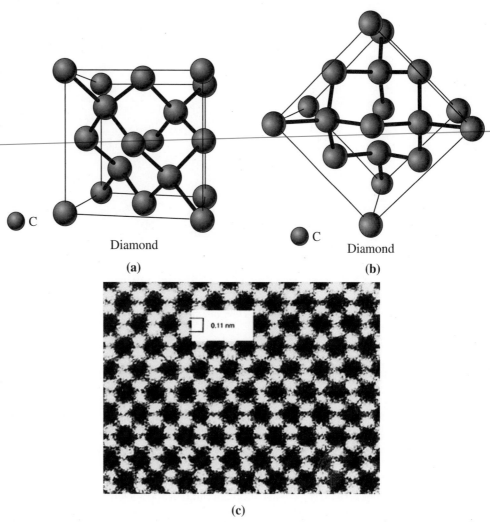

Figure 2-21   The structure of diamond. (a) The repeat unit of the diamond crystal, also known as the unit cell of diamond. (b) The same image as (a) rotated by 45°. (c) A transmission electron micrograph of diamond. *(Courtesy of Prof. Hiroshi Fujita.)*

titanium and more than fifteen times stiffer than aluminum.) As a material is heated, the atoms vibrate with more energy. When the bonds are stiff, the vibrations are transferred efficiently between atoms, thereby conducting heat. On the other hand, diamond is an electrical insulator. All of the valence electrons of each carbon atom are shared with the neighboring atoms, leaving no free electrons to conduct electricity. (Usually an electrical insulator is also a poor conductor of heat because it lacks free electrons, but diamond is an exception due to its extraordinary stiffness.) Diamond is one of the hardest substances known, which is why it is often used in cutting tools in industrial applications (the cutting surface needs to be harder than the material being cut).

Diamond's less illustrious (and lustrous) relative is graphite. Graphite, like pure diamond, contains only carbon atoms, but we know from the experience of writing with graphite pencils that the properties of graphite are significantly different from that of diamond. In graphite, the carbon atoms are arranged in layers. In each layer, the carbon atoms are arranged in a hexagonal pattern, as shown in Figure 2-22(a). Recall that in diamond, each carbon atom is covalently bonded to four others, but in each graphite layer, each

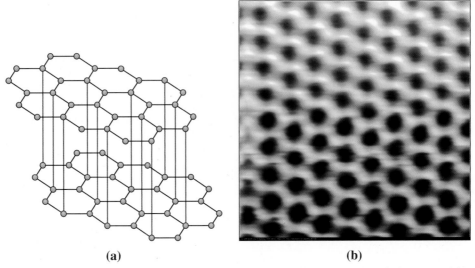

(a) **(b)**

Figure 2-22 The structure of graphite. (a) The carbon atoms are arranged in layers, and in each layer, the carbon atoms are arranged in a hexagonal pattern. (b) An atomic force micrograph of graphite. (*Courtesy of University of Augsburg.*)

carbon atom is bonded covalently to only three others. There is a fourth bond between the layers, but this is a much weaker van der Waals bond. Also, the spacing between the graphite layers is 2.5 times larger than the spacing between the carbon atoms in the plane.

Figure 2-22(b) is an image acquired using an instrument known as an atomic force microscope (AFM). An AFM scans a sharp tip over the surface of a sample. The deflection of the cantilever tip is tracked by a laser and position-sensitive photodetector. In contact mode AFM, an actuator moves the sample with respect to the tip in order to maintain a constant deflection. In this way, the surface of the sample is mapped as a function of height. Figure 2-22(b) shows the carbon atoms in a single graphite layer. Individual carbon atoms are visible. Again we see that although our ball and stick models of crystals may seem somewhat crude, they are, in fact, accurate representations of the atomic arrangements in materials.

Like diamond, graphite has a high melting point. To heat a solid to the point at which it becomes a liquid, the spacing between atoms must be increased. In graphite, it is not difficult to separate the individual layers, because the bonds between the layers are weak (in fact, this is what you do when you write with a graphite pencil—layers are separated and left behind on your paper), but each carbon atom has three strong bonds in the layer that cause the graphite to have a high melting point. Graphite has a lower density than diamond because of its layer structure—the atoms are not packed as closely together. Unlike diamond, graphite is electrically conductive. This is because the fourth electron of each carbon atom, which is not covalently bonded in the plane, is available to conduct electricity.

Buckminsterfullerene, an allotrope of carbon, was discovered in 1985. Each molecule of buckminsterfullerene or "buckyball" contains sixty carbon atoms and is known as C60. A buckyball can be envisioned by considering a two-dimensional pattern of twelve regular pentagons and twenty regular hexagons arranged as in Figure 2-23(a). If this pattern is folded into a three-dimensional structure by wrapping the center row into a circle and folding each end over to form end caps, then the polygons fit together perfectly— like a soccer ball! This is a highly symmetrical structure with sixty corners, and if a carbon atom is placed at each corner, then this is a model for the C60 molecule.

Buckminsterfullerene was named after the American mathematician and architect R. Buckminster Fuller who patented the design of the geodesic dome. Passing a large

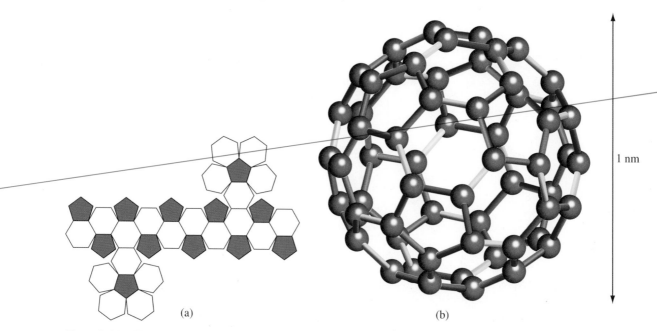

(a)   (b)

Figure 2-23   The structure of the buckyball. (a) The formation of a buckminsterfullerene from twelve regular pentagons and twenty regular hexagons. (b) A "ball and stick" model of the C60 molecule. (*Courtesy of Wendelin J. Wright.*)

current of about 150 A through a carbon rod creates buckyballs. Buckyballs are found in soot created in this fashion as well as in naturally occurring soot, like the carbon residue from a burning candle.

Figure 2-23(b) is a model of a single buckyball. Each of the 60 carbon atoms has two single bonds and one double bond. In fact, there are forms other than C60, e.g., C70, that form a class of carbon materials known generally as fullerenes. Buckyballs can enclose other atoms within them, appear to be quite strong, and have interesting magnetic and superconductive properties.

Carbon nanotubes, a fourth allotrope of carbon, can be envisioned as sheets of graphite rolled into tubes with hemispherical fullerene caps on the ends. A single sheet of graphite, known as graphene, can be rolled in different directions to produce nanotubes with different configurations, as shown in Figure 2-24(a). Carbon nanotubes may be single-walled or multi-walled. Multi-walled carbon nanotubes consist of multiple concentric nanotubes. Carbon nanotubes are typically 1 to 25 nm in diameter and are on the order of microns long. Carbon nanotubes with different configurations display different materials properties. For example, the electrical properties of nanotubes depend on the helicity and diameter of the nanotubes. Carbon nanotubes are currently being used as reinforcement to strengthen and stiffen polymers and as tips for atomic force microscopes. Carbon nanotubes also are being considered as possible conductors of electricity in advanced nanoelectronic devices. Figure 2-24(b) is an image of a single-walled carbon nanotube acquired using an instrument known as a scanning tunneling microscope (STM). An STM scans a sharp tip over the surface of a sample. A voltage is applied to the tip. Electrons from the tip tunnel or "leak" to the sample when the tip is in proximity to the atoms of the sample. The resulting current is a function of the tip to sample distance, and measurements of the current can be used to map the sample surface.

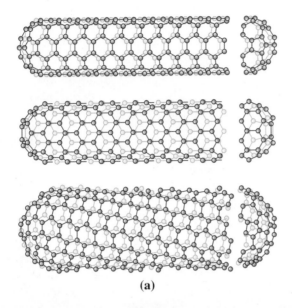

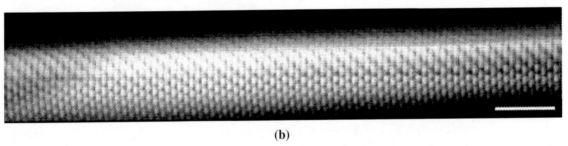

Figure 2-24 (a) Schematic diagrams of various configurations of single-walled carbon nanotubes. (*Courtesy of Figure 6 from Carbon Nanomaterials by Andrew R. Barron.*) (b) A scanning tunneling micrograph of a carbon nanotube. (Reprinted by permission from Macmillan Publishers Ltd: Nature. 2004 Nov 18; 432(7015), "Electrical generation and absorption of phonons in carbon nanotubes," LeRoy et al, copyright 2004.)

Summary

- Similar to composition, the structure of a material has a profound influence on the properties of a material.

- Structure of materials can be understood at various levels: atomic structure, long- and short-range atomic arrangements, nanostructure, microstructure, and macrostructure. Engineers concerned with practical applications need to understand the structure at both the micro and macro levels. Given that atoms and atomic arrangements constitute the building blocks of advanced materials, we need to understand the structure at an atomic level. There are many emerging novel devices centered on micro-electro-mechanical systems (MEMS) and nanotechnology. As a result, understanding the structure of materials at the nanoscale is also very important for some applications.

- The electronic structure of the atom, which is described by a set of four quantum numbers, helps determine the nature of atomic bonding, and, hence, the physical and mechanical properties of materials.

- Atomic bonding is determined partly by how the valence electrons associated with each atom interact. Types of bonds include metallic, covalent, ionic, and van der Waals. Most engineered materials exhibit mixed bonding.

- A metallic bond is formed as a result of atoms of low electronegativity elements donating their valence electrons and leading to the formation of a "sea" of electrons. Metallic bonds are non-directional and relatively strong. As a result, most pure metals show a high Young's modulus and ductility. They are good conductors of heat and electricity and reflect visible light.

- A covalent bond is formed when electrons are shared between two atoms. Covalent bonds are found in many polymeric and ceramic materials. These bonds are strong, and most inorganic materials with covalent bonds exhibit high levels of strength, hardness, and limited ductility. Most plastic materials based on carbon-carbon (C-C) and carbon-hydrogen (C-H) bonds show relatively lower strengths and good levels of ductility. Most covalently bonded materials tend to be relatively good electrical insulators. Some materials such as Si and Ge are semiconductors.

- The ionic bonding found in many ceramics is produced when an electron is "donated" from one electropositive atom to an electronegative atom, creating positively charged cations and negatively charged anions. As in covalently bonded materials, these materials tend to be mechanically strong and hard, but brittle. Melting points of ionically bonded materials are relatively high. These materials are typically electrical insulators. In some cases, though, the microstructure of these materials can be tailored so that significant ionic conductivity is obtained.

- The van der Waals bonds are formed when atoms or groups of atoms have a nonsymmetrical electrical charge, permitting bonding by an electrostatic attraction. The asymmetry in the charge is a result of dipoles that are induced or dipoles that are permanent.

- The binding energy is related to the strength of the bonds and is particularly high in ionically and covalently bonded materials. Materials with a high binding energy often have a high melting temperature, a high modulus of elasticity, and a low coefficient of thermal expansion.

- Not all properties of materials are microstructure sensitive, and modulus of elasticity is one such property.

- In designing components with materials, we need to pay attention to the base composition of the material. We also need to understand the bonding in the material and make efforts to tailor it so that certain performance requirements are met. Finally, the cost of raw materials, manufacturing costs, environmental impact, and factors affecting durability also must be considered.

- Carbon exists as several allotropes including diamond, graphite, buckminsterfullerene, and carbon nanotubes. All are composed of pure carbon, but their materials properties differ dramatically due to the different arrangements of atoms in their structures.

Glossary

Allotropy The characteristic of an element being able to exist in more than one crystal structure, depending on temperature and pressure.

Amorphous material A material that does not have long-range order for the arrangement of its atoms (e.g., silica glass).

Anion A negatively charged ion produced when an atom, usually of a non-metal, accepts one or more electrons.

Atomic mass The mass of the Avogadro constant of atoms, g/mol. Normally, this is the average number of protons and neutrons in the atom. Also called the atomic weight.

Atomic mass unit The mass of an atom expressed as 1/12 the mass of a carbon atom with twelve nucleons.

Atomic number The number of protons in an atom.

Aufbau Principle A graphical device used to determine the order in which the energy levels of quantum shells are filled by electrons.

Avogadro constant The number of atoms or molecules in a mole. The Avogadro constant is 6.022×10^{23} per mole.

Azimuthal quantum number A quantum number that designates different energy levels in principal shells. Also called the secondary quantum number.

Binding energy The energy required to separate two atoms from their equilibrium spacing to an infinite distance apart. The binding energy is a measure of the strength of the bond between two atoms.

Cation A positively charged ion produced when an atom, usually of a metal, gives up its valence electrons.

Coefficient of thermal expansion (CTE) The fractional change in linear dimension of a material per degree of temperature. A material with a low coefficient of thermal expansion tends to retain its dimensions when the temperature changes.

Composition The chemical make-up of a material.

Covalent bond The bond formed between two atoms when the atoms share their valence electrons.

Crystalline materials Materials in which atoms are arranged in a periodic fashion exhibiting a long-range order.

Debye interactions Van der Waals forces that occur between two molecules, with only one molecule having a permanent dipole moment.

Directional relationship The bonds between atoms in covalently bonded materials form specific angles, depending on the material.

Ductility The ability of materials to be permanently stretched or bent without breaking.

Electronegativity The relative tendency of an atom to accept an electron and become an anion. Strongly electronegative atoms readily accept electrons.

Electropositive The tendency for atoms to donate electrons, thus being highly reactive.

Glass transition temperature A temperature above which many polymers and inorganic glasses no longer behave as brittle materials. They gain a considerable amount of ductility above the glass transition temperature.

Hydrogen bond A Keesom interaction (a type of van der Waals bond) between molecules in which a hydrogen atom is involved (e.g., bonds *between* water molecules).

Interatomic spacing The equilibrium spacing between the centers of two atoms. In solid elements, the interatomic spacing equals the apparent diameter of the atom.

Intermetallic compound A compound such as Al_3V formed by two or more metallic atoms; bonding is typically a combination of metallic and ionic bonds.

Ionic bond The bond formed between two different atom species when one atom (the cation) donates its valence electrons to the second atom (the anion). An electrostatic attraction binds the ions together.

Keesom interactions Van der Waals forces that occur between molecules that have permanent dipole moments.

Length scale A relative distance or range of distances used to describe materials-related structure, properties or phenomena.

London forces Van der Waals forces that occur between molecules that do not have permanent dipole moments.

Long-range atomic arrangements Repetitive three-dimensional patterns with which atoms or ions are arranged in crystalline materials.

Macrostructure Structure of a material at a macroscopic level. The length scale is $\sim > 100,000$ nm. Typical features include porosity, surface coatings, and internal or external microcracks.

Magnetic quantum number A quantum number that describes the orbitals for each azimuthal quantum number.

Metallic bond The electrostatic attraction between the valence electrons and the positively charged ion cores.

Micro-electro-mechanical systems (MEMS) These consist of miniaturized devices typically prepared by micromachining.

Microstructure Structure of a material at a length scale of ~ 100 to $100,000$ nm.

Modulus of elasticity The slope of the stress-strain curve in the elastic region (E). Also known as Young's modulus.

Nanoscale A length scale of 1–100 nm.

Nanostructure Structure of a material at the nanoscale ($\sim$ length-scale 1–100 nm).

Nanotechnology An emerging set of technologies based on nanoscale devices, phenomena, and materials.

Nucleon A proton or neutron.

Pauli exclusion principle No more than two electrons in a material can have the same energy. The two electrons have opposite magnetic spins.

Polarized molecules Molecules that have developed a dipole moment by virtue of an internal or external electric field.

Primary bonds Strong bonds between adjacent atoms resulting from the transfer or sharing of outer orbital electrons.

Quantum numbers The numbers that assign electrons in an atom to discrete energy levels. The four quantum numbers are the principal quantum number n, the azimuthal quantum number l, the magnetic quantum number m_l, and the spin quantum number m_s.

Quantum shell A set of fixed energy levels to which electrons belong. Each electron in the shell is designated by four quantum numbers.

Secondary bond Weak bonds, such as van der Waals bonds, that typically join molecules to one another.

Short-range atomic arrangements Atomic arrangements up to a distance of a few nm.

Spectroscopy The science that analyzes the emission and absorption of electromagnetic radiation.

Spin quantum number A quantum number that indicates the spin of an electron.

Structure Description of spatial arrangements of atoms or ions in a material.

Transition elements A set of elements with partially filled d and f orbitals. These elements usually exhibit multiple valence and are useful for electronic, magnetic, and optical applications.

III–V semiconductor A semiconductor that is based on Group 3B and 5B elements (e.g., GaAs).

II–VI semiconductor A semiconductor that is based on Group 2B and 6B elements (e.g., CdSe).

Valence The number of electrons in an atom that participate in bonding or chemical reactions. Usually, the valence is the number of electrons in the outer s and p energy levels.

Van der Waals bond A secondary bond developed between atoms and molecules as a result of interactions between dipoles that are induced or permanent.

Yield strength The level of stress above which a material permanently deforms.

Problems

Section 2-1 The Structure of Materials— an Introduction

2-1 What is meant by the term *composition* of a material?

2-2 What is meant by the term *structure* of a material?

2-3 What are the different levels of structure of a material?

2-4 Why is it important to consider the structure of a material while designing and fabricating engineering components?

2-5 What is the difference between the microstructure and the macrostructure of a material?

Section 2-2 The Structure of the Atom

2-6 Using the densities and atomic weights given in Appendix A, calculate and compare the number of atoms per cubic centimeter in (i) lead and (ii) lithium.

2-7 (a) Using data in Appendix A, calculate the number of iron atoms in 1000 kg of iron.
(b) Using data in Appendix A, calculate the volume in cubic centimeters occupied by one mole of boron.

2-8 In order to plate a steel part having a surface area of 1250 cm^2 with a 0.005 cm thick layer of nickel: (a) How many atoms of nickel are required? (b) How many moles of nickel are required?

Section 2-3 The Electronic Structure of the Atom

2-9 Write the electron configuration for the element Tc.

2-10 Assuming that the Aufbau Principle is followed, what is the expected electronic configuration of the element with atomic number $Z = 116$?

2-11 Suppose an element has a valence of 2 and an atomic number of 27. Based only on the quantum numbers, how many electrons must be present in the $3d$ energy level?

Section 2-4 The Periodic Table

2-12 The periodic table of elements can help us better rationalize trends in properties of elements and compounds based on elements from different groups. Search the literature and obtain the coefficients of thermal expansion of elements from Group 4B. Establish a trend and see if it correlates with the melting temperatures and other properties (e.g., bandgap) of these elements.

2-13 Bonding in the intermetallic compound Ni_3Al is predominantly metallic. Explain why there will be little, if any, ionic bonding component. The electronegativity of nickel is about 1.8.

2-14 Plot the melting temperatures of elements in the 4A to 8–10 columns of the periodic table versus atomic number (i.e., plot melting temperatures of Ti through Ni, Zr through Pd, and Hf through Pt). Discuss these relationships, based on atomic bonding and binding energies: (a) as the atomic number increases in each row of the periodic table and (b) as the atomic number increases in each column of the periodic table.

2-15 Plot the melting temperature of the elements in the 1A column of the periodic table versus atomic number (i.e., plot melting temperatures of Li through Cs). Discuss this relationship, based on atomic bonding and binding energy.

Section 2-5 Atomic Bonding

2-16 Compare and contrast metallic and covalent primary bonds in terms of
(a) the nature of the bond,
(b) the valence of the atoms involved, and
(c) the ductility of the materials bonded in these ways.

2-17 What type of bonding does KCl have? Fully explain your reasoning by referring to the electronic structure and electronic properties of each element.

2-18 Calculate the fraction of bonding of MgO that is ionic.

2-19 What is the type of bonding in diamond? Are the properties of diamond commensurate with the nature of bonding?

2-20 What type of van der Waals forces acts between water molecules?

2-21 Explain the role of van der Waals forces in PVC plastic.

Section 2-6 Binding Energy and Interatomic Spacing

Section 2-7 The Many Forms of Carbon: Relationships Between Arrangements of Atoms and Materials Properties

2-22 Titanium is stiffer than aluminum, has a lower thermal expansion coefficient than aluminum, and has a higher melting temperature than aluminum. On the same graph, carefully and schematically draw the potential well curves for both metals. Be explicit in showing how the physical properties are manifested in these curves.

2-23 Beryllium and magnesium, both in the 2A column of the periodic table, are lightweight metals. Which would you expect to have the higher modulus of elasticity? Explain, considering binding energy and atomic radii and using appropriate sketches of force versus interatomic spacing.

2-24 Would you expect MgO or magnesium to have the higher modulus of elasticity? Explain.

2-25 Aluminum and silicon are side-by-side in the periodic table. Which would you expect to have the higher modulus of elasticity (E)? Explain.

2-26 Steel is coated with a thin layer of ceramic to help protect against corrosion. What do you expect to happen to the coating when the temperature of the steel is increased significantly? Explain.

Design Problems

2-27 You wish to introduce ceramic fibers into a metal matrix to produce a composite material, which is subjected to high forces and large temperature changes. What design parameters might you consider to ensure that the fibers will remain intact and provide strength to the matrix? What problems might occur?

2-28 Turbine blades used in jet engines can be made from such materials as nickel-based superalloys. We can, in principle, even use ceramic materials such as zirconia or other alloys based on steels. In some cases, the blades also may have to be coated with a thermal barrier coating (TBC) to minimize exposure of the blade material to high temperatures. What design parameters would you consider in selecting a material for the turbine blade and for the coating that would work successfully in a turbine engine? Note that different parts of the engine are exposed to different temperatures, and not all blades are exposed to relatively high operating temperatures. What problems might occur? Consider the factors such as temperature and humidity in the environment in which the turbine blades must function.

Ⓚ Knovel® Problems

K2-1 • A 5 cm-thick steel disk with an 200 cm diameter has been plated with a 0.00225 cm layer of zinc.
- What is the area of plating in cm^2?
- What is the weight of zinc required in kg? In g?
- How many moles of zinc are required?
- Name a few different methods used for zinc deposition on a steel substrate.
- Which method should be selected in this case?

Quartz also known as silica with the chemical formula SiO_2 is the mineral found in sand. Quartz is one of the most abundant materials on earth. On a weight basis, the cost of sand is very cheap; however, silica is the material that is refined to make electronics grade silicon, one of the most pure and nearly perfect materials in the world. Silicon wafers are the substrates for microchips, such as those found in the processor of your computer. While a ton of sand may cost only tens of dollars, a ton of silicon microchips is worth billions of dollars.

The image above shows a quartz crystal. A crystalline material is one in which the atoms are arranged in a regular, repeating array. The facets of the crystals reflect the long-range order of the atomic arrangements. (*Courtesy of Galyna Andrushko/Shutterstock.*)

Chapter 3

Atomic and Ionic Arrangements

Have You Ever Wondered?

- *What is amorphous silicon and how is it different from the silicon used to make computer chips?*

- *What are liquid crystals?*

- *If you were to pack a cubical box with uniform-sized spheres, what is the maximum packing possible?*

- *How can we calculate the density of different materials?*

Arrangements of atoms and ions play an important role in determining the microstructure and properties of a material. The main objectives of this chapter are to

(a) learn to classify materials based on atomic/ionic arrangements; and
(b) describe the arrangements in crystalline solids according to the concepts of the **lattice, basis**, and **crystal structure**.

For crystalline solids, we will illustrate the concepts of Bravais lattices, unit cells, and crystallographic directions and planes by examining the arrangements of atoms or ions in many technologically important materials. These include metals (e.g., Cu, Al, Fe, W, Mg, etc.), semiconductors (e.g., Si, Ge, GaAs, etc.), advanced ceramics (e.g., ZrO_2, Al_2O_3, $BaTiO_3$, etc.), ceramic superconductors, diamond, and other materials. We will develop the necessary nomenclature used to characterize atomic or ionic arrangements in crystalline materials. We will examine the use of **x-ray diffraction** (XRD), **transmission electron microscopy** (TEM), and **electron diffraction**. These techniques allow us to probe the arrangements of atoms/ions in different materials. We will present an overview of different types of **amorphous materials** such as amorphous silicon, metallic glasses, polymers, and inorganic glasses.

Chapter 2 highlighted how interatomic bonding influences certain properties of materials. This chapter will underscore the influence of atomic and ionic arrangements on the properties of engineered materials. In particular, we will concentrate on "perfect" arrangements of atoms or ions in crystalline solids.

The concepts discussed in this chapter will prepare us for understanding how *deviations* from these perfect arrangements in crystalline materials create what are described as **atomic level defects**. The term **defect** in this context refers to a lack of perfection in atomic or ionic order of crystals and not to any flaw or quality of an engineered material. In Chapter 4, we will describe how these atomic level defects actually enable the development of formable, strong steels used in cars and buildings, aluminum alloys for aircraft, solar cells and photovoltaic modules for satellites, and many other technologies.

3-1   Short-Range Order versus Long-Range Order

In different states of matter, we can find four types of atomic or ionic arrangements (Figure 3-1).

No Order   In monoatomic gases, such as argon (Ar) or plasma created in a fluorescent tubelight, atoms or ions have no orderly arrangement.

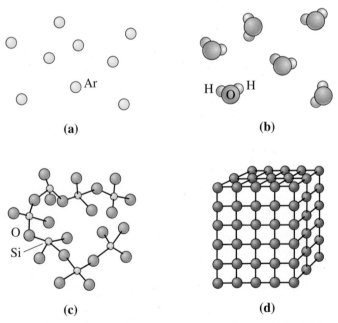

Figure 3-1   Levels of atomic arrangements in materials: (a) Inert monoatomic gases have no regular ordering of atoms. (b,c) Some materials, including water vapor, nitrogen gas, amorphous silicon, and silicate glass, have short-range order. (d) Metals, alloys, many ceramics and some polymers have regular ordering of atoms/ions that extends through the material.

Short-Range Order (SRO)

A material displays **short-range order (SRO)** if the special arrangement of the atoms extends only to the atom's nearest neighbors. Each water molecule in steam has short-range order due to the covalent bonds between the hydrogen and oxygen atoms; that is, each oxygen atom is joined to two hydrogen atoms, forming an angle of 104.5° between the bonds. There is no long-range order, however, because the water molecules in steam have no special arrangement with respect to each other's position.

A similar situation exists in materials known as inorganic glasses. In Chapter 2, we described the **tetrahedral structure** in silica that satisfies the requirement that four oxygen ions be bonded to each silicon ion [Figure 3-2(a)]. As will be discussed later, in a glass, individual tetrahedral units are joined together in a random manner. These tetrahedra may share corners, edges, or faces. Thus, beyond the basic unit of a $(SiO_4)^{4-}$ tetrahedron, there is no periodicity in their arrangement. In contrast, in quartz or other forms of crystalline silica, the $(SiO_4)^{4-}$ tetrahedra are indeed connected in different periodic arrangements.

Many polymers also display short-range atomic arrangements that closely resemble the silicate glass structure. Polyethylene is composed of chains of carbon atoms, with two hydrogen atoms attached to each carbon. Because carbon has a valence of four and the carbon and hydrogen atoms are bonded covalently, a tetrahedral structure is again produced [Figure 3-2(b)]. Tetrahedral units can be joined in a random manner to produce polymer chains.

Long-Range Order (LRO)

Most metals and alloys, semiconductors, ceramics, and some polymers have a crystalline structure in which the atoms or ions display **long-range order (LRO)**; the special atomic arrangement extends over much larger length scales $\sim$ >100 nm. The atoms or ions in these materials form a regular repetitive, grid-like pattern, in three dimensions. We refer to these materials as **crystalline materials**. If a crystalline material consists of only one large crystal, we refer to it as a *single crystal*. Single crystals are useful in many electronic and optical applications. For example, computer chips are made from silicon in the form of large (up to 30 cm diameter) single crystals [Figure 3-3(a)]. Similarly, many useful optoelectronic devices are made from crystals of lithium niobate ($LiNbO_3$). Single crystals can also be made as thin films and used for many electronic and other applications. Certain types of turbine blades may also be made from single crystals of nickel-based superalloys. **A polycrystalline material** is composed of many small crystals with varying orientations in space. These smaller crystals are known as **grains**. The borders between crystals, where the crystals are in misalignment, are known as **grain boundaries**. Figure 3-3(b) shows the microstructure of a polycrystalline stainless steel material.

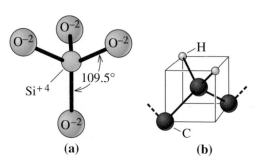

(a)   **(b)**

Figure 3-2
(a) Basic Si-O tetrahedron in silicate glass.
(b) Tetrahedral arrangement of C-H bonds in polyethylene.

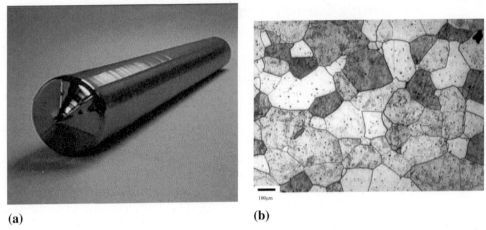

(a)   **(b)**

Figure 3-3   (a) Photograph of a silicon single crystal. (b) Micrograph of a polycrystalline stainless steel showing grains and grain boundaries (*Courtesy of Dr. A. J. Deardo, Dr. M. Hua and Dr. J. Garcia.*)

Many crystalline materials we deal with in engineering applications are polycrystalline (e.g., steels used in construction, aluminum alloys for aircrafts, etc.). We will learn in later chapters that many properties of polycrystalline materials depend upon the physical and chemical characteristics of both grains and grain boundaries. The properties of single crystal materials depend upon the chemical composition and specific directions within the crystal (known as the crystallographic directions). Long-range order in crystalline materials can be detected and measured using techniques such as **x-ray diffraction** or **electron diffraction** (see Section 3-9).

Liquid crystals (LCs) are polymeric materials that have a special type of order. Liquid crystal polymers behave as amorphous materials (liquid-like) in one state. When an external stimulus (such as an electric field or a temperature change) is provided, some polymer molecules undergo alignment and form small regions that are crystalline, hence the name "liquid crystals." These materials have many commercial applications in liquid crystal display (LCD) technology.

Figure 3-4 shows a summary of classification of materials based on the type of atomic order.

3-2   Amorphous Materials

Any material that exhibits only a short-range order of atoms or ions is an **amorphous material**; that is, a noncrystalline one. In general, most materials want to form periodic arrangements since this configuration maximizes the thermodynamic stability of the material. Amorphous materials tend to form when, for one reason or other, the kinetics of the process by which the material was made did not allow for the formation of periodic arrangements. **Glasses**, which typically form in ceramic and polymer systems, are good examples of amorphous materials. Similarly, certain types of polymeric or colloidal gels, or gel-like materials, are also considered amorphous. Amorphous materials often offer a unique blend of properties since the atoms or ions are not assembled into their "regular" and periodic arrangements. Note that often many engineered

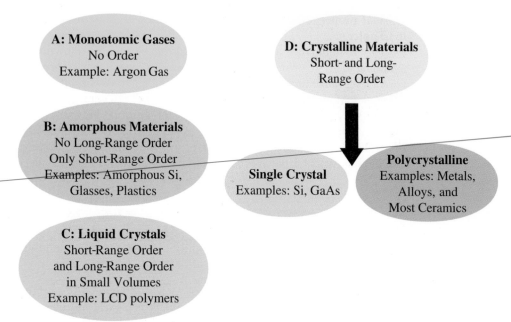

Figure 3-4   Classification of materials based on the type of atomic order.

materials labeled as "amorphous" may contain a fraction that is crystalline. Techniques such as electron diffraction and x-ray diffraction (see Section 3-9) cannot be used to characterize the short-range order in amorphous materials. Scientists use neutron scattering and other methods to investigate the short-range order in amorphous materials.

Crystallization of glasses can be controlled. Materials scientists and engineers, such as Donald Stookey, have developed ways of deliberately nucleating ultrafine crystals in amorphous glasses. The resultant materials, known as **glass-ceramics**, can be made up to ~99.9% crystalline and are quite strong. Some glass-ceramics can be made optically transparent by keeping the size of the crystals extremely small ($\sim$ <100 nm). The major advantage of glass-ceramics is that they are shaped using glass-forming techniques, yet they are ultimately transformed into crystalline materials that do not shatter like glass. We will consider this topic in greater detail in Chapter 9.

Similar to inorganic glasses, many plastics are amorphous. They do contain small portions of material that are crystalline. During processing, relatively large chains of polymer molecules get entangled with each other, like spaghetti. Entangled polymer molecules do not organize themselves into crystalline materials. During processing of polymeric beverage bottles, mechanical stress is applied to the preform of the bottle (e.g., the manufacturing of a standard 2-liter soft drink bottle using polyethylene terephthalate (PET plastic)). This process is known as **blow-stretch forming**. The radial (blowing) and longitudinal (stretching) stresses during bottle formation actually untangle some of the polymer chains, causing **stress-induced crystallization**. The formation of crystals adds to the strength of the PET bottles.

Compared to plastics and inorganic glasses, metals and alloys tend to form crystalline materials rather easily. As a result, special efforts must be made to quench the metals and alloys quickly in order to prevent crystallization; for some alloys, a cooling rate of $>10^6$°C/s is required to form **metallic glasses**. This technique of cooling metals and alloys very fast is known as **rapid solidification**. Many metallic glasses have both useful and unusual properties. The mechanical properties of metallic glasses will be discussed in Chapter 6.

To summarize, amorphous materials can be made by restricting the atoms/ions from assuming their "regular" periodic positions. This means that amorphous materials do not have a long-range order. This allows us to form materials with many different and unusual properties. Many materials labeled as "amorphous" can contain some level of crystallinity. Since atoms are assembled into nonequilibrium positions, the natural tendency of an amorphous material is to crystallize (i.e., since this leads to a thermodynamically more stable material). This can be done by providing a proper thermal (e.g., a silicate glass), thermal and mechanical (e.g., PET polymer), or electrical (e.g., liquid crystal polymer) driving force.

3-3   Lattice, Basis, Unit Cells, and Crystal Structures

A typical solid contains on the order of 10^{23} atoms/cm^3. In order to communicate the spatial arrangements of atoms in a crystal, it is clearly not necessary or practical to specify the position of each atom. We will discuss two complementary methodologies for simply describing the three-dimensional arrangements of atoms in a crystal. We will refer to these as the **lattice and basis concept** and the **unit cell** concept. These concepts rely on the principles of **crystallography**. In Chapter 2, we discussed the structure of the atom. An atom consists of a nucleus of protons and neutrons surrounded by electrons, but for the purpose of describing the arrangements of atoms in a solid, we will envision the atoms as hard spheres, much like ping-pong balls. We will begin with the lattice and basis concept.

A lattice is a collection of points, called **lattice points**, which are arranged in a periodic pattern so that the surroundings of each point in the lattice are identical. A lattice is a purely mathematical construct and is infinite in extent. A lattice may be one-, two-, or three-dimensional. In one dimension, there is only one possible lattice: It is a line of points with the points separated from each other by an equal distance, as shown in Figure 3-5(a). A group of one or more atoms located in a particular way with respect to each other and associated with each lattice point is known as the **basis** or **motif**. The basis must contain at least one atom, but it may contain many atoms of one or more types. A basis of one atom is shown in Figure 3-5(b). We obtain a **crystal structure** by placing the atoms of the basis on every lattice point (i.e., crystal structure = lattice + basis), as shown in Figure 3-5(c). A hypothetical one-dimensional crystal that has a basis of two different atoms is shown in Figure 3-5(d). The larger atom is located on every lattice point with the smaller atom located a fixed distance above each lattice point. Note that it is not necessary that one of the basis atoms be located on each lattice point, as shown in Figure 3-5(e). Figures 3-5(d) and (e) represent the same one-dimensional crystal; the atoms are simply shifted relative to one another. Such a shift does not change the atomic arrangements in the crystal.

There is only one way to arrange points in one dimension such that each point has identical surroundings—an array of points separated by an equal distance as discussed above. There are five distinct ways to arrange points in two dimensions such that each point has identical surroundings; thus, there are five two-dimensional lattices. There are only fourteen unique ways to arrange points in three dimensions. These unique three-dimensional arrangements of lattice points are known as the **Bravais lattices**, named after Auguste Bravais (1811–1863) who was an early French crystallographer.

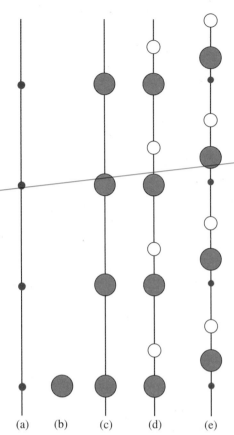

Figure 3-5
Lattice and basis. (a) A one-dimensional lattice. The lattice points are separated by an equal distance. (b) A basis of one atom. (c) A crystal structure formed by placing the basis of (b) on every lattice point in (a). (d) A crystal structure formed by placing a basis of two atoms of different types on the lattice in (a). (e) The same crystal as shown in (d); however, the basis has been shifted relative to each lattice point.

(a)   (b)   (c)   (d)   (e)

The fourteen Bravais lattices are shown in Figure 3-6. As stated previously, a lattice is infinite in extent, so a single unit cell is shown for each lattice. The unit cell is a subdivision of a lattice that still retains the overall characteristics of the entire lattice. Lattice points are located at the corners of the unit cells and, in some cases, at either the faces or the center of the unit cell.

The fourteen Bravais lattices are grouped into seven **crystal systems**. The seven crystal systems are known as cubic, tetragonal, orthorhombic, rhombohedral (also known as trigonal), hexagonal, monoclinic, and triclinic. Note that for the cubic crystal system, we have simple cubic (SC), face-centered cubic (FCC), and body-centered cubic (BCC) Bravais lattices. These names describe the arrangement of lattice points in the unit cell. Similarly, for the tetragonal crystal system, we have simple tetragonal and body-centered tetragonal lattices. Again remember that the concept of a lattice is mathematical and does not mention atoms, ions, or molecules. It is only when a basis is associated with a lattice that we can describe a crystal structure. For example, if we take the face-centered cubic lattice and position a basis of one atom on every lattice point, then the face-centered cubic crystal structure is reproduced.

Note that although we have only fourteen Bravais lattices, we can have an infinite number of bases. Hundreds of different crystal structures are observed in nature or can be synthesized. Many different materials can have the same crystal structure. For example, copper and nickel have the face-centered cubic crystal structure for which only one atom is associated with each lattice point. In more complicated structures, particularly polymer, ceramic, and biological materials, several atoms may be associated with each lattice point (i.e., the basis is greater than one), forming very complex unit cells.

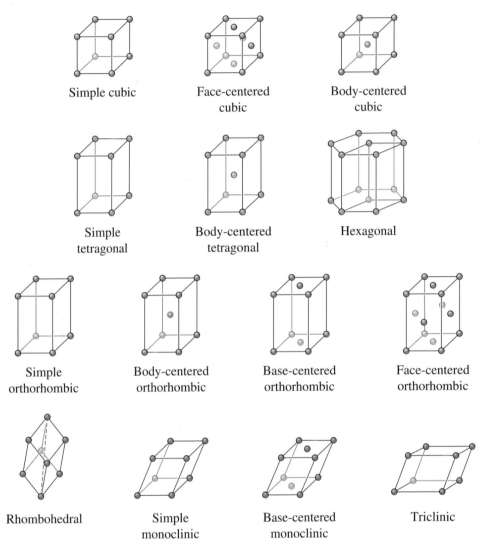

Simple cubic

Face-centered cubic

Body-centered cubic

Simple tetragonal

Body-centered tetragonal

Hexagonal

Simple orthorhombic

Body-centered orthorhombic

Base-centered orthorhombic

Face-centered orthorhombic

Rhombohedral

Simple monoclinic

Base-centered monoclinic

Triclinic

Figure 3-6   The fourteen types of Bravais lattices grouped in seven crystal systems. The actual unit cell for a hexagonal system is shown in Figures 3-8 and 3-13.

Unit Cell

Our goal is to develop a notation to model crystalline solids that simply and completely conveys how the atoms are arranged in space. The unit cell concept complements the lattice and basis model for representing a crystal structure. Although the methodologies of the lattice and basis and unit cell concepts are somewhat different, the end result—a description of a crystal—is the same.

Our goal in choosing a unit cell for a crystal structure is to find the single repeat unit that, when duplicated and translated, reproduces the entire crystal structure. For example, imagine the crystal as a three-dimensional puzzle for which each piece of the puzzle is exactly the same. If we know what one puzzle piece looks like, we know what the entire puzzle looks like, and we don't have to put the entire puzzle together to solve it. We just need one piece! To understand the unit cell concept, we start with the crystal. Figure 3-7(a) depicts a hypothetical two-dimensional crystal that consists of atoms all of the same type.

Next, we add a grid that mimics the symmetry of the arrangements of atoms. There is an infinite number of possibilities for the grid, but by convention, we usually choose the simplest. For the square array of atoms shown in Figure 3-7(a), we choose a

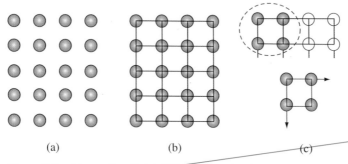

(a)   (b)   (c)

Figure 3-7   The unit cell. (a) A two-dimensional crystal. (b) The crystal with an overlay of a grid that reflects the symmetry of the crystal. (c) The repeat unit of the grid known as the unit cell. Each unit cell has its own origin.

square grid as is shown in Figure 3-7(b). Next, we select the repeat unit of the grid, which is also known as the unit cell. This is the unit that, when duplicated and translated by integer multiples of the axial lengths of the unit cell, recreates the entire crystal. The unit cell is shown in Figure 3-7(c); note that for each unit cell, there is only one quarter of an atom at each corner in two dimensions. We will always draw full circles to represent atoms, but it is understood that only the fraction of the atom that is contained inside the unit cell contributes to the total number of atoms per unit cell. Thus, there is 1/4 atom / corner ∗ 4 corners = 1 atom per unit cell, as shown in Figure 3-7(c). It is also important to note that, if there is an atom at one corner of a unit cell, there must be an atom at every corner of the unit cell in order to maintain the translational symmetry. Each unit cell has its own origin, as shown in Figure 3-7(c).

Lattice Parameters and Interaxial Angles   The **lattice parameters** are the axial lengths or dimensions of the unit cell and are denoted by convention as *a, b,* and *c*. The angles between the axial lengths, known as the interaxial angles, are denoted by the Greek letters α, β, and γ. By convention, α is the angle between the lengths *b* and *c*, β is the angle between *a* and *c*, and γ is the angle between *a* and *b*, as shown in Figure 3-8. (Notice that for each combination, there is a letter *a, b,* and *c* whether it be written in Greek or Roman letters.)

In a cubic crystal system, only the length of one of the sides of the cube need be specified (it is sometimes designated a_0). The length is often given in nanometers (nm) or angstrom (Å) units, where

$$1 \text{ nanometer (nm)} = 10^{-9} \text{ m} = 10^{-7} \text{ cm} = 10 \text{ Å}$$

$$1 \text{ angstrom (Å)} = 0.1 \text{ nm} = 10^{-10} \text{ m} = 10^{-8} \text{ cm}$$

The lattice parameters and interaxial angles for the unit cells of the seven crystal systems are presented in Table 3-1.

To fully define a unit cell, the lattice parameters or ratios between the axial lengths, interaxial angles, and atomic coordinates must be specified. In specifying atomic coordinates, whole atoms are placed in the unit cell. The coordinates are specified as fractions of the axial lengths. Thus, for the two-dimensional cell represented in Figure 3-7(c), the unit cell is fully specified by the following information:

Axial lengths: $a = b$

Interaxial angle: $\gamma = 90°$

Atomic coordinate: (0, 0)

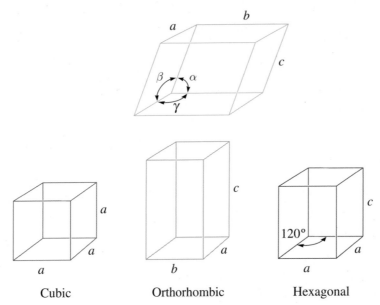

Figure 3-8 Definition of the lattice parameters and their use in cubic, orthorhombic, and hexagonal crystal systems.

Again, only 1/4 of the atom at each origin (0, 0) contributes to the number of atoms per unit cell; however, each corner acts as an origin and contributes 1/4 atom per corner for a total of one atom per unit cell. (Do you see why with an atom at (0, 0) of each unit cell it would be repetitive to also give the coordinates of (1, 0), (0, 1), and (1, 1)?)

Similarly, a cubic unit cell with an atom at each corner is fully specified by the following information:

$$\text{Axial lengths: } a = b = c$$

$$\text{Interaxial angles: } \alpha = \beta = \gamma = 90°$$

$$\text{Atomic coordinate: } (0, 0, 0)$$

TABLE 3-1 ■ *Characteristics of the seven crystal systems*

Structure	Axes	Angles between Axes	Volume of the Unit Cell
Cubic	$a = b = c$	All angles equal 90°.	a^3
Tetragonal	$a = b \neq c$	All angles equal 90°.	a^2c
Orthorhombic	$a \neq b \neq c$	All angles equal 90°.	abc
Hexagonal	$a = b \neq c$	Two angles equal 90°. The angle between a and b equals 120°.	$0.866a^2c$
Rhombohedral or trigonal	$a = b = c$	All angles are equal and none equals 90°.	$a^3\sqrt{1 - 3\cos^2\alpha + 2\cos^3\alpha}$
Monoclinic	$a \neq b \neq c$	Two angles equal 90°. One angle (β) is not equal to 90°.	$abc \sin \beta$
Triclinic	$a \neq b \neq c$	All angles are different and none equals 90°.	$abc\sqrt{1 - \cos^2\alpha - \cos^2\beta - \cos^2\gamma + 2\cos\alpha\cos\beta\cos\gamma}$

Now in three dimensions, each corner contributes 1/8 atom per each of the eight corners for a total of one atom per unit cell. Note that the number of atomic coordinates required is equal to the number of atoms per unit cell. For example, if there are two atoms per unit cell, with one atom at the corners and one atom at the body-centered position, two atomic coordinates are required: (0, 0, 0) and (1/2, 1/2, 1/2).

Number of Atoms per Unit Cell

Each unit cell contains a specific number of lattice points. When counting the number of lattice points belonging to each unit cell, we must recognize that, like atoms, lattice points may be shared by more than one unit cell. A lattice point at a corner of one unit cell is shared by seven adjacent unit cells (thus a total of eight cells); only one-eighth of each corner belongs to one particular cell. Thus, the number of lattice points from all corner positions in one unit cell is

$$\left(\frac{1/8 \text{ lattice point}}{\text{corner}} \right) \left(\frac{8 \text{ corners}}{\text{cell}} \right) = \frac{1 \text{ lattice point}}{\text{unit cell}}$$

Corners contribute 1/8 of a point, faces contribute 1/2, and body-centered positions contribute a whole point [Figure 3-9(a)].

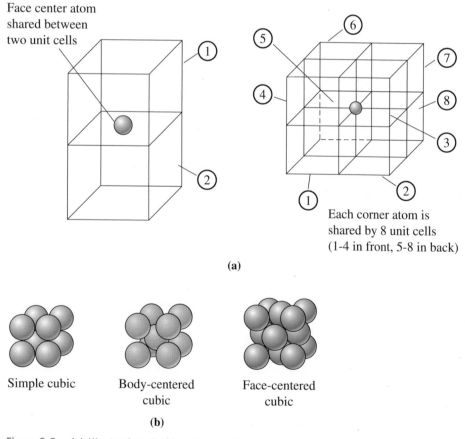

(a)

Simple cubic   Body-centered cubic   Face-centered cubic

(b)

Figure 3-9   (a) Illustration showing sharing of face and corner atoms. (b) The models for simple cubic (SC), body-centered cubic (BCC), and face-centered cubic (FCC) unit cells, assuming only one atom per lattice point.

The number of atoms per unit cell is the product of the number of atoms per lattice point and the number of lattice points per unit cell. The structures of simple cubic (SC), body-centered cubic (BCC), and face-centered cubic (FCC) unit cells (with one atom located at each lattice point) are shown in Figure 3-9(b). Example 3-1 illustrates how to determine the number of lattice points in cubic crystal systems.

Example 3-1   *Determining the Number of Lattice Points in Cubic Crystal Systems*

Determine the number of lattice points per cell in the cubic crystal systems. If there is only one atom located at each lattice point, calculate the number of atoms per unit cell.

SOLUTION

In the SC unit cell, lattice points are located only at the corners of the cube:

$$\frac{\text{lattice points}}{\text{unit cell}} = (8 \text{ corners})\left(\frac{1}{8}\right) = 1$$

In BCC unit cells, lattice points are located at the corners and the center of the cube:

$$\frac{\text{lattice points}}{\text{unit cell}} = (8 \text{ corners})\left(\frac{1}{8}\right) + (1 \text{ body-center})(1) = 2$$

In FCC unit cells, lattice points are located at the corners and faces of the cube:

$$\frac{\text{lattice points}}{\text{unit cell}} = (8 \text{ corners})\left(\frac{1}{8}\right) + (6 \text{ faces})\left(\frac{1}{2}\right) = 4$$

Since we are assuming there is only one atom located at each lattice point, the number of atoms per unit cell would be 1, 2, and 4, for the simple cubic, body-centered cubic, and face-centered cubic unit cells, respectively.

Example 3-2   *The Cesium Chloride Structure*

Crystal structures usually are assigned names of a representative element or compound that has that structure. Cesium chloride (CsCl) is an ionic, crystalline compound. A unit cell of the CsCl crystal structure is shown in Figure 3-10. Chlorine anions are located at the corners of the unit cell, and a cesium cation is located at the body-centered position of each unit cell. Describe this structure as a lattice and basis and also fully define the unit cell for cesium chloride.

SOLUTION

The unit cell is cubic; therefore, the lattice is either SC, FCC, or BCC. There are no atoms located at the face-centered positions; therefore, the lattice is either SC or BCC. Each Cl anion is surrounded by eight Cs cations at the body-centered positions

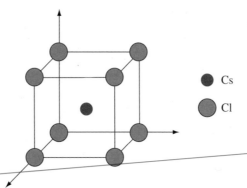

Figure 3-10
The CsCl crystal structure. *Note:* Ion sizes not to scale.

● Cs

◯ Cl

of the adjoining unit cells. Each Cs cation is surrounded by eight Cl anions at the corners of the unit cell. Thus, the corner and body-centered positions do not have identical surroundings; therefore, they both cannot be lattice points. The lattice must be simple cubic.

The simple cubic lattice has lattice points only at the corners of the unit cell. The cesium chloride crystal structure can be described as a simple cubic lattice with a basis of two atoms, Cl (0, 0, 0) and Cs (1/2, 1/2, 1/2). Note that the atomic coordinates are listed as fractions of the axial lengths, which for a cubic crystal structure are equal. The basis atom of Cl (0, 0, 0) placed on every lattice point (i.e., each corner of the unit cell) fully accounts for every Cl atom in the structure. The basis atom of Cs (1/2, 1/2, 1/2), located at the body-centered position with respect to each lattice point, fully accounts for every Cs atom in the structure.

Thus there are two atoms per unit cell in CsCl:

$$\frac{1 \text{ lattice point}}{\text{unit cell}} * \frac{2 \text{ atoms}}{\text{lattice point}} = \frac{2 \text{ atoms}}{\text{unit cell}}$$

To fully define a unit cell, the lattice parameters or ratios between the axial lengths, interaxial angles, and atomic coordinates must be specified. The CsCl unit cell is cubic; therefore,

Axial lengths: $a = b = c$

Interaxial angles: $\alpha = \beta = \gamma = 90°$

The Cl anions are located at the corners of the unit cell, and the Cs cations are located at the body-centered positions. Thus,

Atomic coordinates: Cl (0, 0, 0) and Cs (1/2, 1/2, 1/2)

Counting atoms for the unit cell,

$$\frac{8 \text{ corners}}{\text{unit cell}} * \frac{1/8 \text{ Cl atom}}{\text{corner}} + \frac{1 \text{ body-center}}{\text{unit cell}} * \frac{1 \text{ Cs atom}}{\text{body-center}} = \frac{2 \text{ atoms}}{\text{unit cell}}$$

As expected, the number of atoms per unit cell is the same regardless of the method used to count the atoms.

Atomic Radius versus Lattice Parameter

Directions in the unit cell along which atoms are in continuous contact are **close-packed directions**. In simple structures, particularly those with only one atom per lattice point, we use these directions to calculate the relationship between the apparent size of the atom and the size of the unit cell. By geometrically determining the length of the direction relative to the lattice parameters, and then adding the number of **atomic radii** along this direction, we can determine the desired relationship. Example 3-3 illustrates how the relationships between lattice parameters and atomic radius are determined.

Example 3-3   *Determining the Relationship between Atomic Radius and Lattice Parameters*

Determine the relationship between the atomic radius and the lattice parameter in SC, BCC, and FCC structures when one atom is located at each lattice point.

SOLUTION

If we refer to Figure 3-11, we find that atoms touch along the edge of the cube in SC structures. The corner atoms are centered on the corners of the cube, so

$$a_0 = 2r \tag{3-1}$$

In BCC structures, atoms touch along the body diagonal, which is $\sqrt{3}a_0$ in length. There are two atomic radii from the center atom and one atomic radius from each of the corner atoms on the body diagonal, so

$$a_0 = \frac{4r}{\sqrt{3}} \tag{3-2}$$

In FCC structures, atoms touch along the face diagonal of the cube, which is $\sqrt{2}a_0$ in length. There are four atomic radii along this length—two radii from the face-centered atom and one radius from each corner, so

$$a_0 = \frac{4r}{\sqrt{2}} \tag{3-3}$$

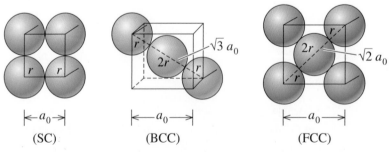

Figure 3-11   The relationships between the atomic radius and the lattice parameter in cubic systems (for Example 3-3).

The Hexagonal Lattice and Unit Cell

The image of the hexagonal lattice in Figure 3-6 reflects the underlying symmetry of the lattice, but unlike the other images in Figure 3-6, it does not represent the unit cell of the lattice. The hexagonal unit cell is shown in Figure 3-8. If you study the image of the hexagonal lattice in Figure 3-6, you can find the hexagonal unit cell. The lattice parameters for the hexagonal unit cell are

Axial lengths: $a = b \neq c$

Interaxial angles: $\alpha = \beta = 90°, \gamma = 120°$

When the atoms of the unit cell are located only at the corners, the atomic coordinate is (0, 0, 0).

Coordination Number

The **coordination number** is the number of atoms touching a particular atom, or the number of nearest neighbors for that particular atom. This is one indication of how tightly and efficiently atoms are packed together. For ionic solids, the coordination number of cations is defined as the number of nearest anions. The coordination number of anions is the number of nearest cations. We will discuss the crystal structures of different ionic solids and other materials in Section 3-7.

In cubic structures containing only one atom per lattice point, atoms have a coordination number related to the lattice structure. By inspecting the unit cells in Figure 3-12, we see that each atom in the SC structure has a coordination number of six, while each atom in the BCC structure has eight nearest neighbors. In Section 3-5, we will show that each atom in the FCC structure has a coordination number of twelve, which is the maximum.

Packing Factor

The **packing factor** or **atomic packing fraction** is the fraction of space occupied by atoms, assuming that the atoms are hard spheres. The general expression for the packing factor is

$$\text{Packing factor} = \frac{(\text{number of atoms/cell})(\text{volume of each atom})}{\text{volume of unit cell}} \tag{3-4}$$

Example 3-4 illustrates how to calculate the packing factor for the FCC unit cell.

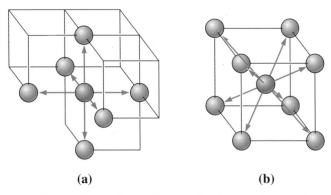

(a)   (b)

Figure 3-12   Illustration of the coordination number in (a) SC and (b) BCC unit cells. Six atoms touch each atom in SC, while eight atoms touch each atom in the BCC unit cell.

Example 3-4	*Calculating the Packing Factor*

Calculate the packing factor for the FCC unit cell.

SOLUTION

In the FCC unit cell, there are four lattice points per cell; if there is one atom per lattice point, there are also four atoms per cell. The volume of one atom is $4\pi r^3/3$ and the volume of the unit cell is a_0^3, where r is the radius of the atom and a_0 is the lattice parameter.

$$\text{Packing factor} = \frac{(4 \text{ atoms/cell})\left(\frac{4}{3}\pi r^3\right)}{a_0^3}$$

Since for FCC unit cells, $a_0 = 4r/\sqrt{2}$:

$$\text{Packing factor} = \frac{(4)\left(\frac{4}{3}\pi r^3\right)}{(4r/\sqrt{2})^3} = \frac{\pi}{\sqrt{18}} \cong 0.74$$

The packing factor of $\pi/\sqrt{18} \cong 0.74$ in the FCC unit cell is the most efficient packing possible. BCC cells have a packing factor of 0.68, and SC cells have a packing factor of 0.52. Notice that the packing factor is independent of the radius of atoms, as long as we assume that all atoms have a fixed radius. What this means is that it does not matter whether we are packing atoms in unit cells or packing basketballs or table tennis balls in a cubical box. The maximum achievable packing factor is $\pi/\sqrt{18}$! This discrete geometry concept is known as **Kepler's conjecture**. Johannes Kepler proposed this conjecture in the year 1611, and it remained an unproven conjecture until 1998 when Thomas C. Hales actually proved this to be true.

The FCC arrangement represents a **close-packed structure** (CP) (i.e., the packing fraction is the highest possible with atoms of one size). The SC and BCC structures are relatively open. We will see in the next section that it is possible to have a hexagonal structure that has the same packing efficiency as the FCC structure. This structure is known as the hexagonal close-packed structure (HCP). Metals with only metallic bonding are packed as efficiently as possible. Metals with mixed bonding, such as iron, may have unit cells with less than the maximum packing factor. No commonly encountered engineering metals or alloys have the SC structure, although this structure is found in ceramic materials.

Density   The theoretical **density** of a material can be calculated using the properties of the crystal structure. The general formula is

$$\text{Density } \rho = \frac{(\text{number of atoms/cell})(\text{atomic mass})}{(\text{volume of unit cell})(\text{Avogadro constant})} \tag{3-5}$$

If a material is ionic and consists of different types of atoms or ions, this formula will have to be modified to reflect these differences. Example 3-5 illustrates how to determine the density of BCC iron.

Example 3-5	*Determining the Density of BCC Iron*

Determine the density of BCC iron, which has a lattice parameter of 0.2866 nm.

SOLUTION

For a BCC cell,

$$\text{Atoms/cell} = 2$$
$$a_0 = 0.2866 \text{ nm} = 2.866 \times 10^{-8} \text{ cm}$$
$$\text{Atomic mass} = 55.847 \text{ g/mol}$$
$$\text{Volume of unit cell} = a_0 = (2.866 \times 10^{-8} \text{ cm})^3 = 23.54 \times 10^{-24} \text{ cm}^3/\text{cell}$$
$$\text{Avogadro constant } N_A = 6.022 \times 10^{23} \text{ atoms/mol}$$
$$\text{Density } \rho = \frac{(\text{number of atoms/cell})(\text{atomic mass of iron})}{(\text{volume of unit cell})(\text{Avogadro constant})}$$
$$\rho = \frac{(2)(55.847)}{(23.54 \times 10^{-24})(6.022 \times 10^{23})} = 7.879 \text{ g/cm}^3$$

The measured density is 7.870 g/cm^3. The slight discrepancy between the theoretical and measured densities is a consequence of defects in the material. As mentioned before, the term "defect" in this context means imperfections with regard to the atomic arrangement.

The Hexagonal Close-Packed Structure

The hexagonal close-packed structure (HCP) is shown in Figure 3-13. The lattice is hexagonal with a basis of two atoms of the same type: one located at (0, 0, 0) and one located at (2/3, 1/3, 1/2). (These coordinates are always fractions of the axial lengths *a, b,* and *c* even if the axial lengths are not equal.) The hexagonal lattice has one lattice point per unit cell located at the corners of the unit cell. In the HCP structure, two atoms are associated with every lattice point; thus, there are two atoms per unit cell.

An equally valid representation of the HCP crystal structure is a hexagonal lattice with a basis of two atoms of the same type: one located at (0, 0, 0) and one located at (1/3, 2/3, 1/2). The (2/3, 1/3, 1/2) and (1/3, 2/3, 1/2) coordinates are equivalent, meaning that they cannot be distinguished from one another.

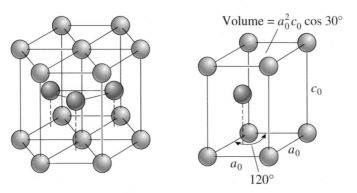

Figure 3-13 The hexagonal close-packed (HCP) structure (left) and its unit cell.

TABLE 3-2 ■ *Crystal structure characteristics of some metals at room temperature*

Structure	a_0 versus r	Atoms per Cell	Coordination Number	Packing Factor	Examples
Simple cubic (SC)	$a_0 = 2r$	1	6	0.52	Polonium (Po), α-Mo
Body-centered cubic (BCC)	$a_0 = 4r/\sqrt{3}$	2	8	0.68	Fe, W, Mo, Nb, Ta, K, Na, V, Cr
Face-centered cubic (FCC)	$a_0 = 4r/\sqrt{2}$	4	12	0.74	Cu, Au, Pt, Ag, Pb, Ni
Hexagonal close-packed (HCP)	$a_0 = 2r$ $c_0 \approx 1.633a_0$	2	12	0.74	Ti, Mg, Zn, Be, Co, Zr, Cd

In metals with an ideal HCP structure, the a_0 and c_0 axes are related by the ratio $c_0/a_0 = \sqrt{8/3} = 1.633$. Most HCP metals, however, have c_0/a_0 ratios that differ slightly from the ideal value because of mixed bonding. Because the HCP structure, like the FCC structure, has the most efficient packing factor of 0.74 and a coordination number of 12, a number of metals possess this structure. Table 3-2 summarizes the characteristics of crystal structures of some metals.

Structures of ionically bonded materials can be viewed as formed by the packing (cubic or hexagonal) of anions. Cations enter into the interstitial sites or holes that remain after the packing of anions. Section 3-7 discusses this in greater detail.

3-4   Allotropic or Polymorphic Transformations

Materials that can have more than one crystal structure are called allotropic or polymorphic. The term **allotropy** is normally reserved for this behavior in pure elements, while the term **polymorphism** is used for compounds. We discussed the allotropes of carbon in Chapter 2. Some metals, such as iron and titanium, have more than one crystal structure. At room temperature, iron has the BCC structure, but at higher temperatures, iron transforms to an FCC structure. These transformations result in changes in properties of materials and form the basis for the heat treatment of steels and many other alloys.

Many ceramic materials, such as silica (SiO_2) and zirconia (ZrO_2), also are polymorphic. A volume change may accompany the transformation during heating or cooling; if not properly controlled, this volume change causes the brittle ceramic material to crack and fail. For zirconia (ZrO_2), for instance, the stable form at room temperature ($\sim$25°C) is monoclinic. As we increase the temperature, more symmetric crystal structures become stable. At 1170°C, the monoclinic zirconia transforms into a tetragonal structure. The tetragonal form is stable up to 2370°C. At that temperature, zirconia transforms into a cubic form. The cubic form remains stable from 2370°C to a melting temperature of 2680°C. Zirconia also can have the orthorhombic form when high pressures are applied.

Ceramics components made from pure zirconia typically will fracture as the temperature is lowered and as zirconia transforms from the tetragonal to monoclinic form because of volume expansion (the cubic to tetragonal phase change does not cause much change in volume). As a result, pure monoclinic or tetragonal polymorphs of zirconia are not used. Instead, materials scientists and engineers have found that adding dopants such as yttria (Y_2O_3) make it possible to stabilize the cubic phase of zirconia, even at room temperature. This yttria stabilized zirconia (YSZ) contains up to 8 mol.% Y_2O_3. Stabilized zirconia formulations are used in many applications, including thermal barrier coatings (TBCs) for turbine blades and electrolytes for oxygen sensors and solid oxide fuel cells. Virtually every car made today uses an oxygen sensor that is made using stabilized zirconia compositions. Example 3-6 illustrates how to calculate volume changes in polymorphs of zirconia.

Example 3-6	*Calculating Volume Changes in Polymorphs of Zirconia*

Calculate the percent volume change as zirconia transforms from a tetragonal to monoclinic structure [9]. The lattice constants for the monoclinic unit cells are $a = 5.156$, $b = 5.191$, and $c = 5.304$ Å, respectively. The angle β for the monoclinic unit cell is 98.9°. The lattice constants for the tetragonal unit cell are $a = 5.094$ and $c = 5.304$ Å. [10] Does the zirconia expand or contract during this transformation? What is the implication of this transformation on the mechanical properties of zirconia ceramics?

SOLUTION

From Table 3-1, the volume of a tetragonal unit cell is given by

$$V = a^2c = (5.094)^2(5.304) = 137.63 \text{ Å}^3$$

and the volume of a monoclinic unit cell is given by

$$V = abc \sin \beta = (5.156)(5.191)(5.304) \sin(98.9) = 140.25 \text{ Å}^3$$

Thus, there is an expansion of the unit cell as ZrO_2 transforms from a tetragonal to monoclinic form.

The percent change in volume = (final volume − initial volume)/
(initial volume) * 100 = (140.25 − 137.63 Å³)/137.63 Å³ * 100 = 1.9%

Most ceramics are very brittle and cannot withstand more than a 0.1% change in volume. (We will discuss mechanical behavior of materials in Chapters 6, 7, and 8.) The conclusion here is that ZrO_2 ceramics cannot be used in their monoclinic form since, when zirconia does transform to the tetragonal form, it will most likely fracture. Therefore, ZrO_2 is often stabilized in a cubic form using different additives such as CaO, MgO, and Y_2O_3.

3-5 Points, Directions, and Planes in the Unit Cell

Coordinates of Points
We can locate certain points, such as atom positions, in the lattice or unit cell by constructing the right-handed coordinate system in Figure 3-14. Distance is measured in terms of the number of lattice parameters we must move in each of the x, y, and z coordinates to get from the origin to the point in question. The coordinates are written as the three distances, with commas separating the numbers.

Directions in the Unit Cell
Certain directions in the unit cell are of particular importance. **Miller indices** for directions are the shorthand notation used to describe these directions. The procedure for finding the Miller indices for directions is as follows:

1. Using a right-handed coordinate system, determine the coordinates of two points that lie on the direction.

2. Subtract the coordinates of the "tail" point from the coordinates of the "head" point to obtain the number of lattice parameters traveled in the direction of each axis of the coordinate system.

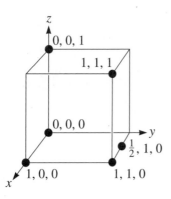

Figure 3-14
Coordinates of selected points in the unit cell. The number refers to the distance from the origin in terms of lattice parameters.

3. Clear fractions and/or reduce the results obtained from the subtraction to lowest integers.
4. Enclose the numbers in square brackets [ ]. If a negative sign is produced, represent the negative sign with a bar over the number.

Example 3-7 illustrates a way of determining the Miller indices of directions.

Example 3-7   *Determining Miller Indices of Directions*

Determine the Miller indices of directions A, B, and C in Figure 3-15.

SOLUTION

Direction A

1. Two points are 1, 0, 0, and 0, 0, 0
2. 1, 0, 0 − 0, 0, 0 = 1, 0, 0
3. No fractions to clear or integers to reduce
4. [100]

Direction B

1. Two points are 1, 1, 1 and 0, 0, 0
2. 1, 1, 1 − 0, 0, 0 = 1, 1, 1
3. No fractions to clear or integers to reduce
4. [111]

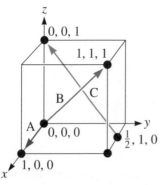

Figure 3-15
Crystallographic directions and coordinates (for Example 3-7).

Direction *C*

1. Two points are 0, 0, 1 and $\frac{1}{2}$, 1, 0

2. $0, 0, 1 - \frac{1}{2}, 1, 0 = -\frac{1}{2}, -1, 1$

3. $2 \left(-\frac{1}{2}, -1, 1 \right) = -1, -2, 2$

4. $[\bar{1}\bar{2}2]$

Several points should be noted about the use of Miller indices for directions:

1. Because directions are vectors, a direction and its negative are not identical; [100] is not equal to [$\bar{1}$00]. They represent the same line, but opposite directions.

2. A direction and its multiple are *identical*; [100] is the same direction as [200].

3. Certain groups of directions are *equivalent*; they have their particular indices because of the way we construct the coordinates. For example, in a cubic system, a [100] direction is a [010] direction if we redefine the coordinate system as shown in Figure 3-16. We may refer to groups of equivalent directions as **directions of a form** or **family**. The special brackets ⟨ ⟩ are used to indicate this collection of directions. All of the directions of the form ⟨110⟩ are listed in Table 3-3. We expect a material to have the same properties in each of these twelve directions of the form ⟨110⟩.

Significance of Crystallographic Directions   Crystallographic

directions are used to indicate a particular orientation of a single crystal or of an oriented polycrystalline material. Knowing how to describe these can be useful in many applications. Metals deform more easily, for example, in directions along which atoms are in closest contact. Another real-world example is the dependence of the magnetic properties of iron and other magnetic materials on the crystallographic directions. It is much easier to magnetize iron in the [100] direction compared to the [111] or [110] directions. This is why the grains in Fe-Si steels used in magnetic applications (e.g., transformer cores) are oriented in the [100] or equivalent directions.

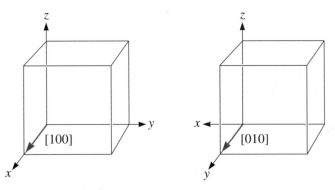

Figure 3-16   Equivalency of crystallographic directions of a form in cubic systems.

TABLE 3-3 ■ *Directions of the form* $\langle 110 \rangle$ *in cubic systems*

$$\langle 110 \rangle = \begin{cases} [110]\,[\bar{1}\bar{1}0] \\ [101]\,[\bar{1}0\bar{1}] \\ [011]\,[0\bar{1}\bar{1}] \\ [1\bar{1}0]\,[\bar{1}10] \\ [10\bar{1}]\,[\bar{1}01] \\ [01\bar{1}]\,[0\bar{1}1] \end{cases}$$

Repeat Distance, Linear Density, and Packing Fraction

Another way of characterizing directions is by the **repeat distance** or the distance between lattice points along the direction. For example, we could examine the [110] direction in an FCC unit cell (Figure 3-17); if we start at the 0, 0, 0 location, the next lattice point is at the center of a face, or a 1/2, 1/2, 0 site. The distance between lattice points is therefore one-half of the face diagonal, or $\frac{1}{2}\sqrt{2}a_0$. In copper, which has a lattice parameter of 0.3615 nm, the repeat distance is 0.2556 nm.

The **linear density** is the number of lattice points per unit length along the direction. In copper, there are two repeat distances along the [110] direction in each unit cell; since this distance is $\sqrt{2}a_0 = 0.5112$ nm, then

$$\text{Linear density} = \frac{2 \text{ repeat distances}}{0.5112 \text{ nm}} = 3.91 \text{ lattice points/nm}$$

Note that the linear density is also the reciprocal of the repeat distance.

Finally, we can compute the **packing fraction** of a particular direction, or the fraction actually covered by atoms. For copper, in which one atom is located at each lattice point, this fraction is equal to the product of the linear density and twice the atomic radius. For the [110] direction in FCC copper, the atomic radius $r = \sqrt{2}a_0/4 = 0.1278$ nm. Therefore, the packing fraction is

$$\begin{aligned} \text{Packing fraction} &= (\text{linear density})(2r) \\ &= (3.91)(2)(0.1278) \\ &= (1.0) \end{aligned}$$

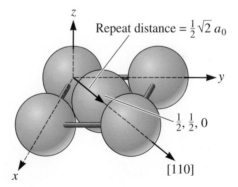

Figure 3-17
Determining the repeat distance, linear density, and packing fraction for a [110] direction in FCC copper.

Atoms touch along the [110] direction, since the [110] direction is close-packed in FCC metals.

Planes in the Unit Cell

Certain planes of atoms in a crystal also carry particular significance. For example, metals deform along planes of atoms that are most tightly packed together. The surface energy of different faces of a crystal depends upon the particular crystallographic planes. This becomes important in crystal growth. In thin film growth of certain electronic materials (e.g., Si or GaAs), we need to be sure the substrate is oriented in such a way that the thin film can grow on a particular crystallographic plane.

Miller indices are used as a shorthand notation to identify these important planes, as described in the following procedure.

1. Identify the points at which the plane intercepts the x, y, and z coordinates in terms of the number of lattice parameters. If the plane passes through the origin, the origin of the coordinate system must be moved to that of an adjacent unit cell.

2. Take reciprocals of these intercepts.

3. Clear fractions but do not reduce to lowest integers.

4. Enclose the resulting numbers in parentheses (). Again, negative numbers should be written with a bar over the number.

The following example shows how Miller indices of planes can be obtained.

Example 3-8   *Determining Miller Indices of Planes*

Determine the Miller indices of planes A, B, and C in Figure 3-18.

SOLUTION

Plane A

1. $x = 1, y = 1, z = 1$

2. $\dfrac{1}{x} = 1, \dfrac{1}{y} = 1, \dfrac{1}{z} = 1$

3. No fractions to clear

4. (111)

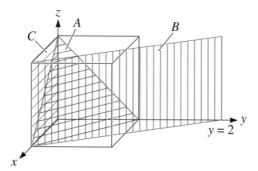

Figure 3-18
Crystallographic planes and intercepts (for Example 3-8).

Plane B

1. The plane never intercepts the z axis, so $x = 1$, $y = 2$, and $z = \infty$

2. $\dfrac{1}{x} = 1, \dfrac{1}{y} = \dfrac{1}{2}, \dfrac{1}{z} = 0$

3. Clear fractions: $\dfrac{1}{x} = 2, \dfrac{1}{y} = 1, \dfrac{1}{z} = 0$

4. (210)

Plane C

1. We must move the origin, since the plane passes through 0, 0, 0. Let's move the origin one lattice parameter in the y-direction. Then, $x = \infty$, $y = -1$, and $z = \infty$.

2. $\dfrac{1}{x} = 0, \dfrac{1}{y} = -1, \dfrac{1}{z} = 0$

3. No fractions to clear

4. $(0\bar{1}0)$

Several important aspects of the Miller indices for planes should be noted:

1. Planes and their negatives are identical (this was not the case for directions) because they are parallel. Therefore, $(020) = (0\bar{2}0)$.

2. Planes and their multiples are not identical (again, this is the opposite of what we found for directions). We can show this by defining planar densities and planar packing fractions. The **planar density** is the number of atoms per unit area with centers that lie on the plane; the packing fraction is the fraction of the area of that plane actually covered by these atoms. Example 3-9 shows how these can be calculated.

3. In each unit cell, **planes of a form** or **family** represent groups of equivalent planes that have their particular indices because of the orientation of the coordinates. We represent these groups of similar planes with the notation {}. The planes of the form {110} in cubic systems are shown in Table 3-4.

4. In cubic systems, a direction that has the same indices as a plane is perpendicular to that plane.

TABLE 3-4 ■ *Planes of the form {110} in cubic systems*

$$\{110\}\begin{cases} (110) \\ (101) \\ (011) \\ (1\bar{1}0) \\ (10\bar{1}) \\ (01\bar{1}) \end{cases}$$

Note: The negatives of the planes are not unique planes.

Example 3-9 *Calculating the Planar Density and Packing Fraction*

Calculate the planar density and planar packing fraction for the (010) and (020) planes in simple cubic polonium, which has a lattice parameter of 0.334 nm.

SOLUTION

The two planes are drawn in Figure 3-19. On the (010) plane, the atoms are centered at each corner of the cube face, with 1/4 of each atom actually in the face of the unit cell. Thus, the total atoms on each face is one. The planar density is

$$\text{Planar density (010)} = \frac{\text{atoms per face}}{\text{area of face}} = \frac{1 \text{ atom per face}}{(0.334)^2}$$
$$= 8.96 \text{ atoms/nm}^2 = 8.96 \times 10^{14} \text{ atoms/cm}^2$$

(020)

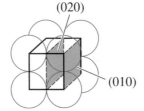

(010)

(010)   (020)

Figure 3-19
The planar densities of the (010) and (020) planes in SC unit cells are not identical (for Example 3-9).

The planar packing fraction is given by

$$\text{Packing fraction (010)} = \frac{\text{area of atoms per face}}{\text{area of face}} = \frac{(1 \text{ atom})(\pi r^2)}{(a_0)^2}$$
$$= \frac{\pi r^2}{(2r)^2} = 0.79$$

No atoms are centered on the (020) planes. Therefore, the planar density and the planar packing fraction are both zero. The (010) and (020) planes are not equivalent!

Construction of Directions and Planes To construct a direction or plane in the unit cell, we simply work backwards. Example 3-10 shows how we might do this.

Example 3-10 *Drawing a Direction and Plane*

Draw (a) the [1$\bar{2}$1] direction and (b) the ($\bar{2}$10) plane in a cubic unit cell.

SOLUTION

a. Because we know that we will need to move in the negative *y*-direction, let's locate the origin at 0, +1, 0. The "tail" of the direction will be located at this new origin. A second point on the direction can be determined by moving +1 in the *x*-direction, −2 in the *y*-direction, and +1 in the *z*-direction [Figure 3-20(a)].

b. To draw in the ($\bar{2}$10) plane, first take reciprocals of the indices to obtain the intercepts, that is

$$x = \frac{1}{-2} = -\frac{1}{2}; y = \frac{1}{1} = 1; z = \frac{1}{0} = \infty$$

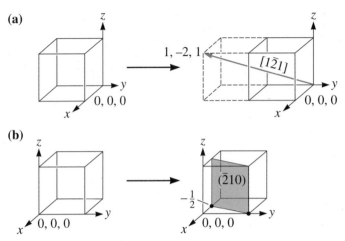

Figure 3-20 Construction of a (a) direction and (b) plane within a unit cell (for Example 3-10).

Since the *x*-intercept is in a negative direction, and we wish to draw the plane within the unit cell, let's move the origin +1 in the *x*-direction to 1, 0, 0.

Then we can locate the *x*-intercept at $-1/2$ and the *y*-intercept at +1. The plane will be parallel to the z-axis [Figure 3-20(b)].

Miller Indices for Hexagonal Unit Cells

A special set of **Miller-Bravais indices** has been devised for hexagonal unit cells because of the unique symmetry of the system (Figure 3-21). The coordinate system uses four axes instead of three, with the a_3 axis being redundant. The axes a_1, a_2, and a_3 lie in a plane that is perpendicular to the fourth axis. The procedure for finding the indices of planes is exactly the same as before, but four intercepts are required, giving indices of the form (*hkil*). Because of the redundancy of the a_3 axis and the special geometry of the system, the first three integers in the designation, corresponding to the a_1, a_2, and a_3 intercepts, are related by $h + k = -i$.

Directions in HCP cells are denoted with either the three-axis or four-axis system. With the three-axis system, the procedure is the same as for conventional Miller indices; examples of this procedure are shown in Example 3-11. A more complicated procedure,

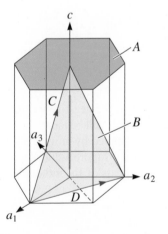

Figure 3-21
Miller-Bravais indices are obtained for crystallographic planes in HCP unit cells by using a four-axis coordinate system. The planes labeled *A* and *B* and the directions labeled *C* and *D* are those discussed in Example 3-11.

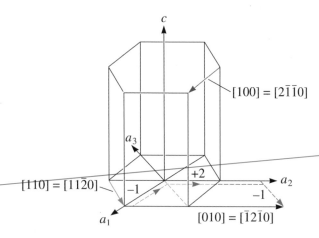

Figure 3-22
Typical directions in the HCP unit cell, using both three- and four-axis systems. The dashed lines show that the $[\bar{1}2\bar{1}0]$ direction is equivalent to a $[010]$ direction.

$[100] = [2\bar{1}\bar{1}0]$

$[110] = [11\bar{2}0]$

$[010] = [\bar{1}2\bar{1}0]$

by which the direction is broken up into four vectors, is needed for the four-axis system. We determine the number of lattice parameters we must move in each direction to get from the "tail" to the "head" of the direction, while for consistency still making sure that $h + k = -i$. This is illustrated in Figure 3-22, showing that the $[010]$ direction is the same as the $[\bar{1}2\bar{1}0]$ direction.

We can also convert the three-axis notation to the four-axis notation for directions by the following relationships, where h', k', and l' are the indices in the three-axis system:

$$\left.\begin{aligned} h &= \frac{1}{3}(2h' - k') \\[4pt] k &= \frac{1}{3}(2k' - h') \\[4pt] i &= -\frac{1}{3}(h' + k') \\[4pt] l &= l' \end{aligned}\right\} \tag{3-6}$$

After conversion, the values of h, k, i, and l may require clearing of fractions or reducing to lowest integers.

Example 3-11   *Determining the Miller-Bravais Indices for Planes and Directions*

Determine the Miller-Bravais indices for planes A and B and directions C and D in Figure 3-21.

SOLUTION

Plane A

1. $a_1 = a_2 = a_3 = \infty, c = 1$

2. $\dfrac{1}{a_1} = \dfrac{1}{a_2} = \dfrac{1}{a_3} = 0, \dfrac{1}{c} = 1$

3. No fractions to clear

4. (0001)

Plane B

1. $a_1 = 1, a_2 = 1, a_3 = -\dfrac{1}{2}, c = 1$

2. $\dfrac{1}{a_1} = 1, \dfrac{1}{a_2} = 1, \dfrac{1}{a_3} = -2, \dfrac{1}{c} = 1$

3. No fractions to clear.

4. $(11\bar{2}1)$

Direction C

1. Two points are 0, 0, 1 and 1, 0, 0.

2. 0, 0, 1 − 1, 0, 0 = −1, 0, 1

3. No fractions to clear or integers to reduce.

4. $[\bar{1}01]$ or $[\bar{2}113]$

Direction D

1. Two points are 0, 1, 0 and 1, 0, 0.

2. 0, 1, 0 − 1, 0, 0 = −1, 1, 0

3. No fractions to clear or integers to reduce.

4. $[\bar{1}10]$ or $[\bar{1}100]$

Close-Packed Planes and Directions

In examining the relationship between atomic radius and lattice parameter, we looked for close-packed directions, where atoms are in continuous contact. We can now assign Miller indices to these close-packed directions, as shown in Table 3-5.

We can also examine FCC and HCP unit cells more closely and discover that there is at least one set of close-packed planes in each. Close-packed planes are shown in Figure 3-23. Notice that a hexagonal arrangement of atoms is produced in two dimensions. The close-packed planes are easy to find in the HCP unit cell; they are the (0001) and (0002) planes of the HCP structure and are given the special name **basal planes**. In fact, we can build up an HCP unit cell by stacking together close-packed planes in an . . . *ABABAB* . . . **stacking sequence** (Figure 3-23). Atoms on plane *B*, the (0002) plane, fit into the valleys between atoms on plane *A*, the bottom (0001) plane. If another plane identical in orientation to plane *A* is placed in the valleys of plane *B* directly above plane *A*, the HCP structure is created. Notice that all of the possible close-packed planes are parallel to one another. Only the basal planes—(0001) and (0002)—are close-packed.

TABLE 3-5 ■ *Close-packed planes and directions*

Structure	Directions	Planes
SC	⟨100⟩	None
BCC	⟨111⟩	None
FCC	⟨110⟩	{111}
HCP	⟨100⟩, ⟨110⟩ or ⟨11$\bar{2}$0⟩	(0001), (0002)

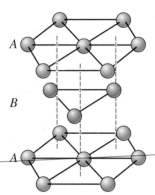

Figure 3-23
The *ABABAB* stacking sequence of close-packed planes produces the HCP structure.

From Figure 3-23, we find the coordination number of the atoms in the HCP structure. The center atom in a basal plane touches six other atoms in the same plane. Three atoms in a lower plane and three atoms in an upper plane also touch the same atom. The coordination number is twelve.

In the FCC structure, close-packed planes are of the form {111} (Figure 3-24). When parallel (111) planes are stacked, atoms in plane *B* fit over valleys in plane *A* and atoms in plane *C* fit over valleys in both planes *A* and *B*. The fourth plane fits directly over atoms in plane *A*. Consequently, a stacking sequence . . . *ABCABCABC* . . . is produced using the (111) plane. Again, we find that each atom has a coordination number of twelve.

Unlike the HCP unit cell, there are four sets of nonparallel close-packed planes—(111), $(11\bar{1})$, $(1\bar{1}1)$, and $(\bar{1}11)$—in the FCC cell. This difference between the FCC and HCP unit cells—the presence or absence of intersecting close-packed planes—affects the mechanical behavior of metals with these structures.

Isotropic and Anisotropic Behavior

Because of differences in atomic arrangement in the planes and directions within a crystal, some properties also vary with direction. A material is crystallographically **anisotropic** if its properties depend on the crystallographic direction along which the property is measured. For example, the modulus of elasticity of aluminum is 75.9 GPa in $\langle 111 \rangle$ directions, but only 63.4 GPa in $\langle 100 \rangle$ directions. If the properties are identical in all directions, the

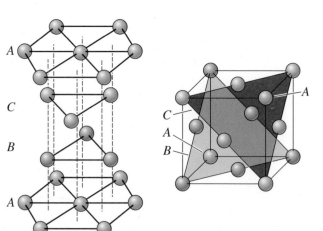

Figure 3-24
The *ABCABCABC* stacking sequence of close-packed planes produces the FCC structure.

material is crystallographically **isotropic**. Note that a material such as aluminum, which is crystallographically anisotropic, may behave as an isotropic material if it is in a polycrystalline form. This is because the random orientations of different crystals in a polycrystalline material will mostly cancel out any effect of the anisotropy as a result of crystal structure. In general, most polycrystalline materials will exhibit isotropic properties. Materials that are single crystals or in which many grains are oriented along certain directions (naturally or deliberately obtained by processing) will typically have anisotropic mechanical, optical, magnetic, and dielectric properties.

Interplanar Spacing   The distance between two adjacent parallel planes of atoms with the same Miller indices is called the **interplanar spacing** (d_{hkl}). The interplanar spacing in *cubic* materials is given by the general equation

$$d_{hkl} = \frac{a_0}{\sqrt{h^2 + k^2 + l^2}}, \tag{3-7}$$

where a_0 is the lattice parameter and h, k, and l represent the Miller indices of the adjacent planes being considered. The interplanar spacings for non-cubic materials are given by more complex expressions.

3-6   Interstitial Sites

In all crystal structures, there are small holes between the usual atoms into which smaller atoms may be placed. These locations are called **interstitial sites**.

      An atom, when placed into an interstitial site, touches two or more atoms in the lattice. This interstitial atom has a coordination number equal to the number of atoms it touches. Figure 3-25 shows interstitial locations in the SC, BCC, and FCC structures. The **cubic site**, with a coordination number of eight, occurs in the SC structure at the body-centered position. **Octahedral sites** give a coordination number of six (not eight). They are known as octahedral sites because the atoms contacting the interstitial atom form an octahedron. **Tetrahedral sites** give a coordination number of four. As an example, the octahedral sites in BCC unit cells are located at the faces of the cube; a small atom placed in the octahedral site touches the four atoms at the corners of the face, the atom at the center of the unit cell, plus another atom at the center of the adjacent unit cell, giving a coordination number of six. In FCC unit cells, octahedral sites occur at the center of each edge of the cube, as well as at the body center of the unit cell.

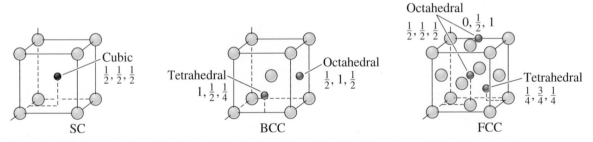

Figure 3-25   The location of the interstitial sites in cubic unit cells. Only representative sites are shown.

Example 3-12   *Calculating Octahedral Sites*

Calculate the number of octahedral sites that *uniquely* belong to one FCC unit cell.

SOLUTION

The octahedral sites include the twelve edges of the unit cell, with the coordinates

$$\frac{1}{2},0,0 \quad \frac{1}{2},1,0 \quad \frac{1}{2},0,1 \quad \frac{1}{2},1,1$$
$$0,\frac{1}{2},0 \quad 1,\frac{1}{2},0 \quad 1,\frac{1}{2},1 \quad 0,\frac{1}{2},1$$
$$0,0,\frac{1}{2} \quad 1,0,\frac{1}{2} \quad 1,1,\frac{1}{2} \quad 0,1,\frac{1}{2}$$

plus the center position, 1/2, 1/2, 1/2. Each of the sites on the edge of the unit cell is shared between four unit cells, so only 1/4 of each site belongs uniquely to each unit cell. Therefore, the number of sites belonging uniquely to each cell is

$$\frac{12 \text{ edges}}{\text{cell}} \cdot \frac{\frac{1}{4} \text{ site}}{\text{edge}} + \frac{1 \text{ body-center}}{\text{cell}} \cdot \frac{1 \text{ site}}{\text{body-center}} = 4 \text{ octahedral sites/cell}$$

Interstitial atoms or ions whose radii are slightly larger than the radius of the interstitial site may enter that site, pushing the surrounding atoms slightly apart. Atoms with radii smaller than the radius of the hole are not allowed to fit into the interstitial site because the ion would "rattle" around in the site. If the interstitial atom becomes too large, it prefers to enter a site having a larger coordination number (Table 3-6). Therefore,

TABLE 3-6 ■ *The coordination number and the radius ratio*

Coordination Number	Location of Interstitial	Radius Ratio	Representation
2	Linear	0–0.155	
3	Center of triangle	0.155–0.225	
4	Center of tetrahedron	0.225–0.414	
6	Center of octahedron	0.414–0.732	
8	Center of cube	0.732–1.000	

an atom with a radius ratio between 0.225 and 0.414 enters a tetrahedral site; if its radius is somewhat larger than 0.414, it enters an octahedral site instead.

Many ionic crystals (see Section 3-7) can be viewed as being generated by close packing of larger anions. Cations then can be viewed as smaller ions that fit into the interstitial sites of the close-packed anions. Thus, the radius ratios described in Table 3-6 also apply to the ratios of the radius of the cation to that of the anion. The packing in ionic crystals is not as tight as that in FCC or HCP metals.

3-7   Crystal Structures of Ionic Materials

Ionic materials must have crystal structures that ensure electrical neutrality, yet permit ions of different sizes to be packed efficiently. As mentioned before, ionic crystal structures can be viewed as close-packed structures of anions. Anions form tetrahedra or octahedra, allowing the cations to fit into their appropriate interstitial sites. In some cases, it may be easier to visualize coordination polyhedra of cations with anions going to the interstitial sites. Recall from Chapter 2 that very often in real materials with engineering applications, the bonding is never 100% ionic. We still use this description of the crystal structure, though, to discuss the crystal structure of most ceramic materials. The following factors need to be considered in order to understand crystal structures of ionically bonded solids.

Ionic Radii   The crystal structures of ionically bonded compounds often can be described by placing the anions at the normal lattice points of a unit cell, with the cations then located at one or more of the interstitial sites described in Section 3-6 (or vice versa). The ratio of the sizes of the ionic radii of anions and cations influences both the manner of packing and the coordination number (Table 3-6). Note that the radii of atoms and ions are different. For example, the radius of an oxygen atom is 0.6 Å; however, the radius of an oxygen anion (O^{2-}) is 1.32 Å. This is because an oxygen anion has acquired two additional electrons and has become larger. As a general rule, anions are larger than cations. Cations, having acquired a positive charge by losing electrons, are expected to be smaller. Strictly speaking, the radii of cations and anions also depend upon the coordination number. For example, the radius of an Al^{+3} ion is 0.39 Å when the coordination number is four (tetrahedral coordination); however, the radius of Al^{+3} is 0.53 Å when the coordination number is 6 (octahedral coordination). Also, note that the coordination number for cations is the number of nearest anions and vice versa. The radius of an atom also depends on the coordination number. For example, the radius of an iron atom in the FCC and BCC polymorphs is different! This tells us that atoms and ions are not "hard spheres" with fixed atomic radii. Appendix B in this book contains the atomic and ionic radii for different elements.

Electrical Neutrality   The overall material has to be electrically neutral. If the charges on the anion and the cation are identical and the coordination number for each ion is identical to ensure a proper balance of charge, then the compound will have a formula AX (A: cation, X: anion). As an example, each cation may be surrounded by six anions, while each anion is, in turn, surrounded by six cations. If the valence of the cation is +2 and that of the anion is −1, then twice as many anions must be present, and the formula is AX_2. The structure of the AX_2 compound must ensure that the coordination number of the cation is twice the coordination number of the anion. For example, each cation may have eight anion nearest neighbors, while only four cations touch each anion.

Connection between Anion Polyhedra

As a rule, the coordination polyhedra (formed by the close packing of anions) will share corners, as opposed to faces or edges. This is because in corner sharing polyhedra, electrostatic repulsion between cations is reduced considerably and this leads to the formation of a more stable crystal structure. A number of common structures in ceramic materials are described in the following discussions. Compared to metals, ceramic structures are more complex. The lattice constants of ceramic materials tend to be larger than those for metallic materials because electrostatic repulsion between ions prevents close packing of both anions and cations.

Example 3-13   *Radius Ratio for KCl*

For potassium chloride (KCl), (a) verify that the compound has the cesium chloride structure and (b) calculate the packing factor for the compound.

SOLUTION

a. From Appendix B, $r_{K^+} = 0.113$ nm and $r_{Cl^-} = 0.181$ nm, so

$$\frac{r_{K^+}}{r_{Cl^-}} = \frac{0.133}{0.181} = 0.735$$

Since $0.732 < 0.735 < 1.000$, the coordination number for each type of ion is eight, and the CsCl structure is likely.

b. The ions touch along the body diagonal of the unit cell, so

$$\sqrt{3}a_0 = 2r_{K^+} + 2r_{Cl^-} = 2(0.133) + 2(0.181) = 0.628 \text{ nm}$$

$$a_0 = 0.363 \text{ nm}$$

$$\text{Packing factor} = \frac{\frac{4}{3}\pi r_{K^+}^3 (1 \text{ K ion}) + \frac{4}{3}\pi r_{Cl^-}^3 (1 \text{ Cl ion})}{a_0^3}$$

$$= \frac{\frac{4}{3}\pi(0.133)^3 + \frac{4}{3}\pi(0.181)^3}{(0.363)^3} = 0.73$$

This structure is shown in Figure 3-10.

Sodium Chloride Structure

The radius ratio for sodium and chloride ions is $r_{Na^+}/r_{Cl^-} = 0.097$ nm$/0.181$ nm $= 0.536$; the sodium ion has a charge of $+1$; the chloride ion has a charge of -1. Therefore, based on the charge balance and radius ratio, each anion and cation must have a coordination number of six. The FCC structure, with Cl^{-1} ions at FCC positions and Na^+ at the four octahedral sites, satisfies these requirements (Figure 3-26). We can also consider this structure to be FCC with two ions—one Na^{+1} and one Cl^{-1}—associated with each lattice point. Many ceramics, including magnesium oxide (MgO), calcium oxide (CaO), and iron oxide (FeO) have this structure.

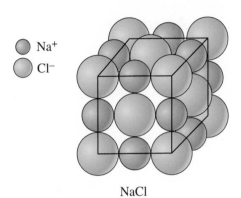

Na⁺

Cl⁻

NaCl

Figure 3-26
The sodium chloride structure, a FCC unit cell with two ions (Na⁺ and Cl⁻) per lattice point. *Note:* ion sizes not to scale.

Example 3-14   *Illustrating a Crystal Structure and Calculating Density*

Show that MgO has the sodium chloride crystal structure and calculate the density of MgO.

SOLUTION

From Appendix B, $r_{Mg^{+2}} = 0.066$ nm and $r_{O^{-2}} = 0.132$ nm, so

$$\frac{r_{Mg^{+2}}}{r_{O^{-2}}} = \frac{0.066}{0.132} = 0.50$$

Since $0.414 < 0.50 < 0.732$, the coordination number for each ion is six, and the sodium chloride structure is possible.

The atomic masses are 24.312 and 16.00 g/mol for magnesium and oxygen, respectively. The ions touch along the edge of the cube, so

$$a_0 = 2r_{Mg^{+2}} + 2r_{O^{-2}} = 2(0.066) + 2(0.132) = 0.396 \text{ nm} = 3.96 \times 10^{-8} \text{ cm}$$

$$\rho = \frac{(4 \text{ Mg}^{+2})(24.312) + (4 \text{ O}^{-2})(16.00)}{(3.96 \times 10^{-8} \text{ cm}^3)^3(6.022 \times 10^{23})} = 4.31 \text{ g/cm}^3$$

Zinc Blende Structure   Zinc blende is the name of the crystal structure adopted by ZnS. Although the Zn ions have a charge of +2 and S ions have a charge of −2, zinc blende (ZnS) cannot have the sodium chloride structure because

$$\frac{r_{Zn^{+2}}}{r_{S^{-2}}} = 0.074 \text{ nm}/0.184 \text{ nm} = 0.402$$

This radius ratio demands a coordination number of four, which in turn means that the zinc ions enter tetrahedral sites in a unit cell (Figure 3-27). The FCC structure, with S anions at the normal lattice points and Zn cations at half of the tetrahedral sites, can accommodate the restrictions of both charge balance and coordination number. A variety of materials, including the semiconductor GaAs and many other III–V semiconductors (Chapter 2), have this structure.

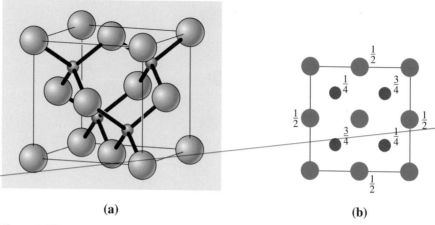

(a) (b)

Figure 3-27 (a) The zinc blende unit cell, (b) plan view. The fractions indicate the positions of the atoms out of the page relative to the height of one unit cell.

Example 3-15 | *Calculating the Theoretical Density of GaAs*

The lattice constant of gallium arsenide (GaAs) is 5.65 Å. Show that the theoretical density of GaAs is 5.33 g/cm^3.

SOLUTION

For the "zinc blende" GaAs unit cell, there are four Ga and four As atoms per unit cell.

From the periodic table (Chapter 2):
Each mole (6.022×10^{23} atoms) of Ga has a mass of 69.72 g. Therefore, the mass of four Ga atoms will be $4 \times 69.72 \ (6.022 \times 10^{23})$ g.
Each mole (6.022×10^{23} atoms) of As has a mass of 74.91 g. Therefore, the mass of four As atoms will be $4 \times 74.91 \ (6.022 \times 10^{23})$ g.
These atoms occupy a volume of $(5.65 \times 10^{-8})^3 \ cm^3$.

$$\text{density} = \frac{\text{mass}}{\text{volume}} = \frac{4(69.72 + 74.91)/(6.022 \times 10^{23})}{(5.65 \times 10^{-8})^3} = 5.33 \ g/cm^3$$

Therefore, the theoretical density of GaAs is 5.33 g/cm^3.

Fluorite Structure   The fluorite structure is FCC, with anions located at all eight of the tetrahedral positions (Figure 3-28). Thus, there are four cations and eight anions per cell, and the ceramic compound must have the formula AX_2, as in calcium fluoride, or CaF_2. In the designation AX_2, A is the cation and X is the anion. The coordination number of the calcium ions is eight, but that of the fluoride ions is four, therefore ensuring a balance of charge. One of the polymorphs of ZrO_2 known as cubic zirconia exhibits this crystal structure. Other compounds that exhibit this structure include UO_2, ThO_2, and CeO_2.

Corundum Structure   This is one of the crystal structures of alumina known as alpha alumina (α-Al_2O_3). In alumina, the oxygen anions pack in a hexagonal arrangement, and the aluminum cations occupy some of the available octahedral positions (Figure 3-29).

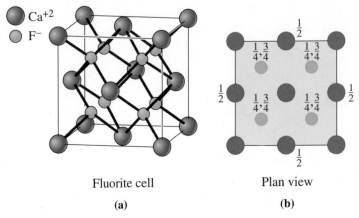

Fluorite cell

(a)

Plan view

(b)

Figure 3-28 (a) Fluorite unit cell, (b) plan view. The fractions indicate the positions of the atoms out of the page relative to the height of the unit cell.

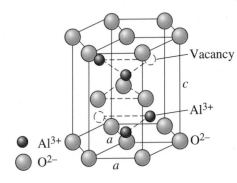

Figure 3-29
Corundum structure of alpha-alumina
(α-Al_2O_3).

Alumina is probably the most widely used ceramic material. Applications include, but are not limited to, spark plugs, refractories, electronic packaging substrates, and abrasives.

Example 3-16   *The Perovskite Crystal Structure*

Perovskite is a mineral containing calcium, titanium, and oxygen. The unit cell is cubic and has a calcium atom at each corner, an oxygen atom at each face center, and a titanium atom at the body-centered position. The atoms contribute to the unit cell in the usual way (1/8 atom contribution for each atom at the corners, etc.).

(a) Describe this structure as a lattice and a basis. (b) How many atoms of each type are there per unit cell? (c) An alternate way of drawing the unit cell of perovskite has calcium at the body-centered position of each cubic unit cell. What are the positions of the titanium and oxygen atoms in this representation of the unit cell? (d) By counting the number of atoms of each type per unit cell, show that the formula for perovskite is the same for both unit cell representations.

SOLUTION

(a) The lattice must belong to the cubic crystal system. Since different types of atoms are located at the corner, face-centered, and body-centered positions, the lattice must be simple cubic. The structure can be described as a simple cubic lattice

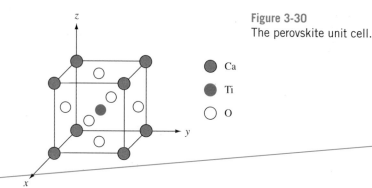

Figure 3-30
The perovskite unit cell.

with a basis of Ca $(0, 0, 0)$, Ti $(1/2, 1/2, 1/2)$, and O $(0, 1/2, 1/2)$, $(1/2, 0, 1/2)$, and $(1/2, 1/2, 0)$. The unit cell is shown in Figure 3-30.

(b) There are two methods for calculating the number of atoms per unit cell. Using the lattice and basis concept,

$$\frac{1 \text{ lattice point}}{\text{unit cell}} * \frac{5 \text{ atoms}}{\text{lattice point}} = \frac{5 \text{ atoms}}{\text{unit cell}}$$

Using the unit cell concept,

$$\frac{8 \text{ corners}}{\text{unit cell}} * \frac{1/8 \text{ Ca atom}}{\text{corner}} + \frac{1 \text{ body-center}}{\text{unit cell}} * \frac{1 \text{ Ti atom}}{\text{body-center}}$$
$$+ \frac{6 \text{ face–centers}}{\text{unit cell}} * \frac{1/2 \text{ O atom}}{\text{face-center}} = \frac{5 \text{ atoms}}{\text{unit cell}}$$

As expected, the number of atoms per unit cell is the same regardless of which method is used. The chemical formula for perovskite is $CaTiO_3$ (calcium titanate). Compounds with the general formula ABO_3 and this structure are said to have the perovskite crystal structure. One of the polymorphs of barium titanate, which is used to make capacitors for electronic applications, and one form of lead zirconate exhibit this structure.

(c) If calcium is located at the body-centered position rather than the corners of the unit cell, then titanium must be located at the corners of the unit cell, and the oxygen atoms must be located at the edge centers of the unit cell, as shown in Figure 3-31. Note that this is equivalent to shifting each atom in the basis given in part (a) by the

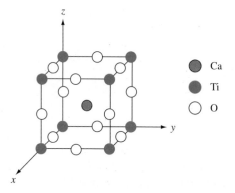

Figure 3-31
An alternate representation of the unit cell of perovskite.

vector [1/2 1/2 1/2]. The Ca atom is shifted from (0, 0, 0) to (1/2, 1/2, 1/2), and the Ti atom is shifted from (1/2, 1/2, 1/2) to (1, 1, 1), which is equivalent to the origin of an adjacent unit cell or (0, 0, 0). Note that the crystal has not been changed; only the coordinates of the atoms in the basis are different. Another lattice and basis description of perovskite is thus a simple cubic lattice with a basis of Ca (1/2, 1/2, 1/2), Ti (0, 0, 0), and O (1/2, 0, 0), (0, 1/2, 0), and (0, 0, 1/2).

Using the lattice and basis concept to count the number of atoms per unit cell,

$$\frac{1 \text{ lattice point}}{\text{unit cell}} * \frac{5 \text{ atoms}}{\text{lattice point}} = \frac{5 \text{ atoms}}{\text{unit cell}}$$

Using the unit cell concept,

$$\frac{1 \text{ body-center}}{\text{unit cell}} * \frac{1 \text{ Ca atom}}{\text{body-center}} + \frac{8 \text{ corners}}{\text{unit cell}} * \frac{1/8 \text{ Ti atom}}{\text{corner}}$$

$$+ \frac{12 \text{ edge centers}}{\text{unit cell}} * \frac{1/4 \text{ O atom}}{\text{edge-center}} = \frac{5 \text{ atoms}}{\text{unit cell}}$$

Again we find that the chemical formula is $CaTiO_3$.

3-8   Covalent Structures

Covalently bonded materials frequently have complex structures in order to satisfy the directional restraints imposed by the bonding.

Diamond Cubic Structure

Elements such as silicon, germanium (Ge), α-Sn, and carbon (in its diamond form) are bonded by four covalent bonds and produce a **tetrahedron** [Figure 3-32(a)]. The coordination number for each silicon atom is only four because of the nature of the covalent bonding.

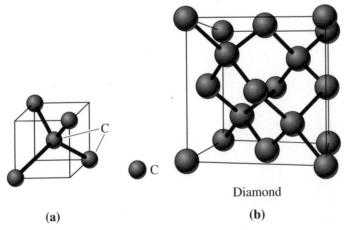

Diamond

(a)   (b)

Figure 3-32   (a) Tetrahedron and (b) the diamond cubic (DC) unit cell. This open structure is produced because of the requirements of covalent bonding.

As these tetrahedral groups are combined, a large cube can be constructed [Figure 3-32(b)]. This large cube contains eight smaller cubes that are the size of the tetrahedral cube; however, only four of the cubes contain tetrahedra. The large cube is the **diamond cubic** (DC) unit cell. The atoms at the corners of the tetrahedral cubes provide atoms at the regular FCC lattice points. Four additional atoms are present within the DC unit cell from the atoms at the center of the tetrahedral cubes. We can describe the DC crystal structure as an FCC lattice with two atoms associated with each lattice point (or a basis of 2). Therefore, there are eight atoms per unit cell.

Example 3-17	*Determining the Packing Factor for the Diamond Cubic Structure*

Describe the diamond cubic structure as a lattice and a basis and determine its packing factor.

SOLUTION

The diamond cubic structure is a face-centered cubic lattice with a basis of two atoms of the same type located at $(0, 0, 0)$ and $(1/4, 1/4, 1/4)$. The basis atom located at $(0, 0, 0)$ accounts for the atoms located at the FCC lattice points, which are $(0, 0, 0)$, $(0, 1/2, 1/2)$, $(1/2, 0, 1/2)$, and $(1/2, 1/2, 0)$ in terms of the coordinates of the unit cell. By adding the vector $[1/4\ 1/4\ 1/4]$ to each of these points, the four additional atomic coordinates in the interior of the unit cell are determined to be $(1/4, 1/4, 1/4)$, $(1/4, 3/4, 3/4)$, $(3/4, 1/4, 3/4)$, and $(3/4, 3/4, 1/4)$. There are eight atoms per unit cell in the diamond cubic structure:

$$\frac{4\ \text{lattice points}}{\text{unit cell}} * \frac{2\ \text{atoms}}{\text{lattice point}} = \frac{8\ \text{atoms}}{\text{unit cell}}$$

The atoms located at the $(1/4, 1/4, 1/4)$ type positions sit at the centers of tetrahedra formed by atoms located at the FCC lattice points. The atoms at the $(1/4, 1/4, 1/4)$ type positions are in direct contact with the four surrounding atoms. Consider the distance between the center of the atom located at $(0, 0, 0)$ and the center of the atom located at $(1/4, 1/4, 1/4)$. This distance is equal to one-quarter of the body diagonal or two atomic radii, as shown in Figure 3-33. Thus,

$$\frac{a_0\sqrt{3}}{4} = 2r$$

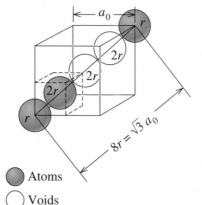

Figure 3-33
Determining the relationship between the lattice parameter and atomic radius in a diamond cubic cell (for Example 3-17).

or

$$a_0 = \frac{8r}{\sqrt{3}}$$

The packing factor is the ratio of the volume of space occupied by the atoms in the unit cell to the volume of the unit cell:

$$\text{Packing factor} = \frac{(8\text{ atoms/cell})\left(\frac{4}{3}\pi r^3\right)}{a_0^3}$$

$$\text{Packing factor} = \frac{(8\text{ atoms/cell})\left(\frac{4}{3}\pi r^3\right)}{(8r/\sqrt{3})^3}$$

$$\text{Packing factor} = 0.34$$

This is a relatively open structure compared to close-packed structures. In Chapter 5, we will learn that the openness of a structure is one of the factors that affects the rate at which different atoms can diffuse in a given material.

Example 3-18   *Calculating the Radius, Density, and Atomic Mass of Silicon*

The lattice constant of Si is 5.43 Å. Calculate the radius of a silicon atom and the theoretical density of silicon. The atomic mass of Si is 28.09 g/mol.

SOLUTION

Silicon has the diamond cubic structure. As shown in Example 3-17 for the diamond cubic structure,

$$r = \frac{a_0\sqrt{3}}{8}$$

Therefore, substituting $a_0 = 5.43$ Å, the radius of the silicon atom $= 1.176$ Å. This is the same radius listed in Appendix B. For the density, we use the same approach as in Example 3-15. Recognizing that there are eight Si atoms per unit cell, then

$$\text{density} = \frac{\text{mass}}{\text{volume}} = \frac{8(28.09)/(6.022 \times 10^{23})}{(5.43 \times 10^{-8}\text{ cm})^3} = 2.33\text{ g/cm}^3$$

This is the same density value listed in Appendix A.

Crystalline Silica

In a number of its forms, silica (or SiO_2) has a crystalline ceramic structure that is partly covalent and partly ionic. Figure 3-34 shows the crystal structure of one of the forms of silica, β-cristobalite, which is a complicated structure with an FCC lattice. The ionic radii of silicon and oxygen are 0.042 nm and 0.132 nm, respectively, so the radius ratio is $r_{Si^{+4}}/r_{O^{-2}} = 0.318$ and the coordination number is four.

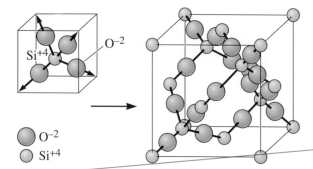

Figure 3-34
The silicon-oxygen tetrahedron and the resultant β-cristobalite form of silica.

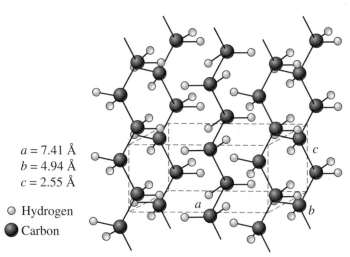

$a = 7.41$ Å
$b = 4.94$ Å
$c = 2.55$ Å

○ Hydrogen
● Carbon

Figure 3-35   The unit cell of crystalline polyethylene (not to scale).

Crystalline Polymers

A number of polymers may form a crystalline structure. The dashed lines in Figure 3-35 outline the unit cell for the lattice of polyethylene. Polyethylene is obtained by joining C_2H_4 molecules to produce long polymer chains that form an orthorhombic unit cell. Some polymers, including nylon, can have several polymorphic forms. Most engineered plastics are partly amorphous and may develop crystallinity during processing. It is also possible to grow single crystals of polymers.

Example 3-19	*Calculating the Number of Carbon and Hydrogen Atoms in Crystalline Polyethylene*

How many carbon and hydrogen atoms are in each unit cell of crystalline polyethylene? There are twice as many hydrogen atoms as carbon atoms in the chain. The density of polyethylene is about 0.9972 g/cm³.

SOLUTION

If we let x be the number of carbon atoms, then $2x$ is the number of hydrogen atoms. From the lattice parameters shown in Figure 3-35:

$$\rho = \frac{(x)(12 \text{ g/mol}) + (2x)(1 \text{ g/mol})}{(7.41 \times 10^{-8} \text{ cm})(4.94 \times 10^{-8} \text{ cm})(2.55 \times 10^{-8} \text{ cm})(6.022 \times 10^{23})}$$

$$0.9972 = \frac{14x}{56.2}$$

$$x = 4 \text{ carbon atoms per cell}$$

$$2x = 8 \text{ hydrogen atoms per cell}$$

3-9   Diffraction Techniques for Crystal Structure Analysis

A crystal structure of a crystalline material can be analyzed using **x-ray diffraction (XRD)** or electron diffraction. Max von Laue (1879–1960) won the Nobel Prize in 1914 for his discovery related to the diffraction of x-rays by a crystal. William Henry Bragg (1862–1942) and his son William Lawrence Bragg (1890–1971) won the 1915 Nobel Prize for their contributions to XRD.

When a beam of x-rays having a single wavelength on the same order of magnitude as the atomic spacing in the material strikes that material, x-rays are scattered in all directions. Most of the radiation scattered from one atom cancels out radiation scattered from other atoms; however, x-rays that strike certain crystallographic planes at specific angles are reinforced rather than annihilated. This phenomenon is called **diffraction**. The x-rays are diffracted, or the beam is reinforced, when conditions satisfy **Bragg's law**,

$$\sin \theta = \frac{\lambda}{2d_{hkl}} \tag{3-8}$$

where the angle θ is half the angle between the diffracted beam and the original beam direction, λ is the wavelength of the x-rays, and d_{hkl} is the interplanar spacing between the planes that cause constructive reinforcement of the beam (see Figure 3-36).

When the material is prepared in the form of a fine powder, there are always at least some powder particles (crystals or aggregates of crystals) with planes (hkl) oriented at the proper θ angle to satisfy Bragg's law. Therefore, a diffracted beam, making an angle of 2θ with the incident beam, is produced. In a *diffractometer*, a moving x-ray detector records the 2θ angles at which the beam is diffracted, giving a characteristic diffraction pattern (see Figure 3-37 on page 98). If we know the wavelength of the x-rays, we can determine the interplanar spacings and, eventually, the identity of the planes that cause the diffraction. In an XRD instrument, x-rays are produced by bombarding a metal target with a beam of high-energy electrons. Typically, x-rays emitted from copper have a wavelength $\lambda \cong 1.54060$ Å ($K\text{-}\alpha_1$ line) and are used.

In the Laue method, which was the first diffraction method ever used, the specimen is in the form of a single crystal. A beam of "white radiation" consisting of x-rays of different wavelengths is used. Each diffracted beam has a different wavelength. In the transmission Laue method, photographic film is placed behind the crystal. In the

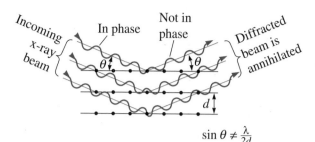

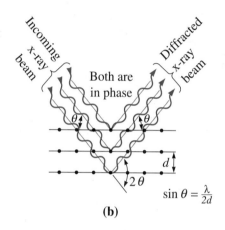

$$\sin\theta \neq \frac{\lambda}{2d}$$

(a)

$$\sin\theta = \frac{\lambda}{2d}$$

(b)

Figure 3-36
(a) Destructive and (b) reinforcing interactions between x-rays and the crystalline material. Reinforcement occurs at angles that satisfy Bragg's law.

back-reflection Laue method, the beams that are back diffracted are recorded on a film located between the source and sample. From the recorded diffraction patterns, the orientation and quality of the single crystal can be determined. It is also possible to determine the crystal structure using a rotating crystal and a fixed wavelength x-ray source.

Typically, XRD analysis can be conducted relatively rapidly ($\sim$30 minutes to 1 hour per sample), on bulk or powdered samples and without extensive sample preparation. This technique can also be used to determine whether the material consists of many grains oriented in a particular crystallographic direction (texture) in bulk materials and thin films. Typically, a well-trained technician can conduct the analysis as well as interpret the powder diffraction data rather easily. As a result, XRD is used in many industries as one tool for product quality control purposes. Analysis of single crystals and materials containing several phases can be more involved and time consuming.

To identify the crystal structure of a cubic material, we note the pattern of the diffracted lines—typically by creating a table of $\sin^2\theta$ values. By combining Equation 3-7 with Equation 3-8 for the interplanar spacing, we find that:

$$\sin^2\theta = \frac{\lambda^2}{4a_0^2}(h^2 + k^2 + l^2)$$

In simple cubic metals, all possible planes will diffract, giving an $h^2 + k^2 + l^2$ pattern of 1, 2, 3, 4, 5, 6, 8, ... In body-centered cubic metals, diffraction occurs only from planes having an even $h^2 + k^2 + l^2$ sum of 2, 4, 6, 8, 10, 12, 14, 16, ... For face-centered cubic metals, more destructive interference occurs, and planes having $h^2 + k^2 + l^2$ sums of 3, 4, 8, 11, 12, 16, . . . will diffract. By calculating the values of $\sin^2\theta$ and then finding the appropriate pattern, the crystal structure can be determined for metals having one of these simple structures, as illustrated in Example 3-20.

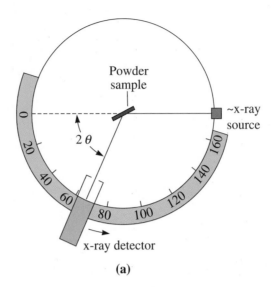

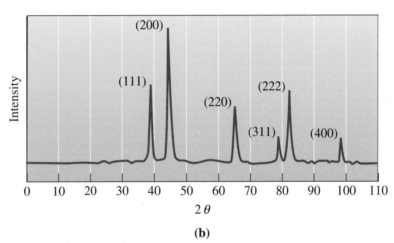

Figure 3-37   (a) Diagram of a diffractometer, showing powder sample, incident and diffracted beams. (b) The diffraction pattern obtained from a sample of gold powder.

Example 3-20   *Examining X-ray Diffraction Data*

The results of an x-ray diffraction experiment using x-rays with $\lambda = 0.7107$ Å (radiation obtained from a molybdenum (Mo) target) show that diffracted peaks occur at the following 2θ angles:

Peak	$2\theta\,(°)$	Peak	$2\theta\,(°)$
1	20.20	5	46.19
2	28.72	6	50.90
3	35.36	7	55.28
4	41.07	8	59.42

Determine the crystal structure, the indices of the plane producing each peak, and the lattice parameter of the material.

SOLUTION

We can first determine the $\sin^2\theta$ value for each peak, then divide through by the lowest denominator, 0.0308.

Peak	$2\theta(°)$	$\sin^2\theta$	$\sin^2\theta/0.0308$	$h^2 + k^2 + l^2$	(hkl)
1	20.20	0.0308	1	2	(110)
2	28.72	0.0615	2	4	(200)
3	35.36	0.0922	3	6	(211)
4	41.07	0.1230	4	8	(220)
5	46.19	0.1539	5	10	(310)
6	50.90	0.1847	6	12	(222)
7	55.28	0.2152	7	14	(321)
8	59.42	0.2456	8	16	(400)

When we do this, we find a pattern of $\sin^2\theta/0.0308$ values of 1, 2, 3, 4, 5, 6, 7, and 8. If the material were simple cubic, the 7 would not be present, because no planes have an $h^2 + k^2 + l^2$ value of 7. Therefore, the pattern must really be 2, 4, 6, 8, 10, 12, 14, 16, . . . and the material must be body-centered cubic. The (hkl) values listed give these required $h^2 + k^2 + l^2$ values.

We could then use 2θ values for any of the peaks to calculate the interplanar spacing and thus the lattice parameter. Picking peak 8:

$$2\theta = 59.42° \quad \text{or} \quad \theta = 29.71°$$

$$d_{400} = \frac{\lambda}{2\sin\theta} = \frac{0.7107}{2\sin(29.71)} = 0.71699 \text{ Å}$$

$$a_0 = d_{400}\sqrt{h^2 + k^2 + l^2} = (0.71699)(4) = 2.868 \text{ Å}$$

This is the lattice parameter for body-centered cubic iron.

Electron Diffraction and Microscopy

Louis de Broglie theorized that electrons behave like waves. In electron diffraction, we make use of high-energy ($\sim$100,000 to 400,000 eV) electrons. These electrons are diffracted from electron transparent samples of materials. The electron beam that exits from the sample is also used to form an image of the sample. Thus, transmission electron microscopy and electron diffraction are used for imaging microstructural features and determining crystal structures.

A 100,000 eV electron has a wavelength of about 0.004 nm! This ultra-small wavelength of high-energy electrons allows a **transmission electron microscope** (TEM) to simultaneously image the microstructure at a very fine scale. If the sample is too thick, electrons cannot be transmitted through the sample and an image or a diffraction pattern will not be observed. Therefore, in transmission electron microscopy and electron diffraction, the sample has to be made such that portions of it are electron transparent. A transmission electron microscope is the instrument used for this purpose. Figure 3-38 shows a TEM image and an electron diffraction pattern from an area of the sample. The large bright spots correspond to the grains of the matrix. The smaller spots originate from small crystals of another phase.

Another advantage to using a TEM is the high spatial resolution. Using TEM, it is possible to determine differences between different crystalline regions and between

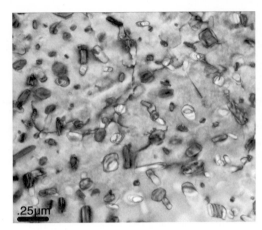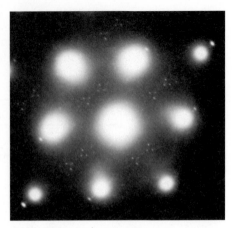

Figure 3-38 A TEM micrograph of an aluminum alloy (Al-7055) sample. The diffraction pattern at the right shows large bright spots that represent diffraction from the main aluminum matrix grains. The smaller spots originate from the nanoscale crystals of another compound that is present in the aluminum alloy. (*Courtesy of Dr. Jörg M.K. Wiezorek, University of Pittsburgh.*)

amorphous and crystalline regions at very small length scales (~1–10 nm). This analytical technique and its variations (e.g., high-resolution electron microscopy (HREM), scanning transmission electron microscopy (STEM), etc.) are also used to determine the orientation of different grains and other microstructural features discussed in later chapters. Advanced and specialized features associated with TEM also allow chemical mapping of elements in a given material. Some of the disadvantages associated with TEM include

(a) the time consuming preparation of samples that are almost transparent to the electron beam;

(b) considerable amount of time and skill are required for analysis of the data from a thin, three-dimensional sample, that is represented in a two-dimensional image and diffraction pattern;

(c) only a very small volume of the sample is examined; and

(d) the equipment is relatively expensive and requires great care in use.

In general, TEM has become a widely used and accepted research method for analysis of microstructural features at micro- and nano-length scales.

Summary

- Atoms or ions may be arranged in solid materials with either a short-range or long-range order.

- Amorphous materials, such as silicate glasses, metallic glasses, amorphous silicon, and many polymers, have only a short-range order. Amorphous materials form whenever the kinetics of a process involved in the fabrication of a material do not allow the atoms or ions to assume the equilibrium positions. These materials often offer novel properties. Many amorphous materials can be crystallized in a controlled fashion. This is the basis for the formation of glass-ceramics and strengthening of PET plastics used for manufacturing bottles.

- Crystalline materials, including metals and many ceramics, have both long- and short-range order. The long-range periodicity in these materials is described by the crystal structure.

- The atomic or ionic arrangements of crystalline materials are described by seven general crystal systems, which include fourteen specific Bravais lattices. Examples include simple cubic, body-centered cubic, face-centered cubic, and hexagonal lattices.

- A lattice is a collection of points organized in a unique manner. The basis or motif refers to one or more atoms associated with each lattice point. A crystal structure is defined by the combination of a lattice and a basis. Although there are only fourteen Bravais lattices, there are hundreds of crystal structures.

- A crystal structure is characterized by the lattice parameters of the unit cell, which is the smallest subdivision of the crystal structure that still describes the lattice. Other characteristics include the number of lattice points and atoms per unit cell, the coordination number (or number of nearest neighbors) of the atoms in the unit cell, and the packing factor of the atoms in the unit cell.

- Allotropic, or polymorphic, materials have more than one possible crystal structure. The properties of materials can depend strongly on the particular polymorph or allotrope. For example, the cubic and tetragonal polymorphs of barium titanate have very different properties.

- The atoms of metals having the face-centered cubic and hexagonal close-packed crystal structures are arranged in a manner that occupies the greatest fraction of space. The FCC and HCP structures achieve the closest packing by different stacking sequences of close-packed planes of atoms.

- The greatest achievable packing fraction with spheres of one size is 0.74 and is independent of the radius of the spheres (i.e., atoms and basketballs pack with the same efficiency as long as we deal with a constant radius of an atom and a fixed size basketball).

- Points, directions, and planes within the crystal structure can be identified in a formal manner by the assignment of coordinates and Miller indices.

- Mechanical, magnetic, optical, and dielectric properties may differ when measured along different directions or planes within a crystal; in this case, the crystal is said to be anisotropic. If the properties are identical in all directions, the crystal is isotropic. The effect of crystallographic anisotropy may be masked in a polycrystalline material because of the random orientation of grains.

- Interstitial sites, or holes between the normal atoms in a crystal structure, can be filled by other atoms or ions. The crystal structure of many ceramic materials can be understood by considering how these sites are occupied. Atoms or ions located in interstitial sites play an important role in strengthening materials, influencing the physical properties of materials, and controlling the processing of materials.

- Crystal structures of many ionic materials form by the packing of anions (e.g., oxygen ions (O^{-2})). Cations fit into coordination polyhedra formed by anions. These polyhedra typically share corners and lead to crystal structures. The conditions of charge neutrality and stoichiometry have to be balanced. Crystal structures of many ceramic materials (e.g., Al_2O_3, ZrO_2, $YBa_2Cu_3O_{7-x}$) can be rationalized from these considerations.

- Crystal structures of covalently bonded materials tend to be open. Examples include diamond cubic (e.g., Si, Ge).

- Although most engineered plastics tend to be amorphous, it is possible to have significant crystallinity in polymers, and it is also possible to grow single crystals of certain polymers.

- XRD and electron diffraction are used for the determination of the crystal structure of crystalline materials. Transmission electron microscopy can also be used for imaging of microstructural features in materials at smaller length scales.

Glossary

Allotropy The characteristic of an element being able to exist in more than one crystal structure, depending on temperature and pressure.

Amorphous materials Materials, including glasses, that have no long-range order or crystal structure.

Anisotropic Having different properties in different directions.

Atomic level defects Defects such as vacancies, dislocations, etc., occurring over a length scale comparable to a few interatomic distances.

Atomic radius The apparent radius of an atom, typically calculated from the dimensions of the unit cell, using close-packed directions (depends upon coordination number).

Basal plane The special name given to the close-packed plane in hexagonal close-packed unit cells.

Basis A group of atoms associated with a lattice point (same as motif).

Blow-stretch forming A process used to form plastic bottles.

Bragg's law The relationship describing the angle at which a beam of x-rays of a particular wavelength diffracts from crystallographic planes of a given interplanar spacing.

Bravais lattices The fourteen possible lattices that can be created in three dimensions using lattice points.

Close-packed directions Directions in a crystal along which atoms are in contact.

Close-packed (CP) structure Structures showing a packing fraction of 0.74 (FCC and HCP).

Coordination number The number of nearest neighbors to an atom in its atomic arrangement.

Crystal structure The arrangement of the atoms in a material into a regular repeatable lattice. A crystal structure is fully described by a lattice and a basis.

Crystal systems Cubic, tetragonal, orthorhombic, hexagonal, monoclinic, rhombohedral and triclinic arrangements of points in space that lead to fourteen Bravais lattices and hundreds of crystal structures.

Crystallography The formal study of the arrangements of atoms in solids.

Crystalline materials Materials comprising one or many small crystals or grains.

Crystallization The process responsible for the formation of crystals, typically in an amorphous material.

Cubic site An interstitial position that has a coordination number of eight. An atom or ion in the cubic site has eight nearest neighbor atoms or ions.

Defect A microstructural feature representing a disruption in the perfect periodic arrangement of atoms/ions in a crystalline material. This term is not used to convey the presence of a flaw in the material.

Density Mass per unit volume of a material, usually in units of g/cm^3.

Diamond cubic (DC) The crystal structure of carbon, silicon, and other covalently bonded materials.

Diffraction The constructive interference, or reinforcement, of a beam of x-rays or electrons interacting with a material. The diffracted beam provides useful information concerning the structure of the material.

Directions of a form or **directions of a family** Crystallographic directions that all have the same characteristics. Denoted by ⟨ ⟩ brackets.

Electron diffraction A method to determine the level of crystallinity at relatively small length scales. Usually conducted in a transmission electron microscope.

Glass-ceramics A family of materials typically derived from molten inorganic glasses and processed into crystalline materials with very fine grain size and improved mechanical properties.

Glasses Solid, non-crystalline materials (typically derived from the molten state) that have only short-range atomic order.

Grain A small crystal in a polycrystalline material.

Grain boundaries Regions between grains of a polycrystalline material.

Interplanar spacing Distance between two adjacent parallel planes with the same Miller indices.

Interstitial sites Locations between the "normal" atoms or ions in a crystal into which another—usually different—atom or ion is placed. Typically, the size of this interstitial location is smaller than the atom or ion that is to be introduced.

Isotropic Having the same properties in all directions.

Kepler's conjecture A conjecture made by Johannes Kepler in 1611 that stated that the maximum packing fraction with spheres of uniform size could not exceed $\pi/\sqrt{18}$. In 1998, Thomas Hales proved this to be true.

Lattice A collection of points that divide space into smaller equally sized segments.

Lattice parameters The lengths of the sides of the unit cell and the angles between those sides. The lattice parameters describe the size and shape of the unit cell.

Lattice points Points that make up the lattice. The surroundings of each lattice point are identical.

Linear density The number of lattice points per unit length along a direction.

Liquid crystals (LCs) Polymeric materials that are typically amorphous but can become partially crystalline when an external electric field is applied. The effect of the electric field is reversible. Such materials are used in liquid crystal displays.

Long-range order (LRO) A regular repetitive arrangement of atoms in a solid which extends over a very large distance.

Metallic glass Amorphous metals or alloys obtained using rapid solidification.

Miller-Bravais indices A special shorthand notation to describe the crystallographic planes in hexagonal close-packed unit cells.

Miller indices A shorthand notation to describe certain crystallographic directions and planes in a material. Denoted by [ ] brackets. A negative number is represented by a bar over the number.

Motif A group of atoms affiliated with a lattice point (same as basis).

Octahedral site An interstitial position that has a coordination number of six. An atom or ion in the octahedral site has six nearest neighbor atoms or ions.

Packing factor The fraction of space in a unit cell occupied by atoms.

Packing fraction The fraction of a direction (linear-packing fraction) or a plane (planar-packing factor) that is actually covered by atoms or ions. When one atom is located at each lattice point, the linear packing fraction along a direction is the product of the linear density and twice the atomic radius.

Planar density The number of atoms per unit area whose centers lie on the plane.

Planes of a form or **planes of a family** Crystallographic planes that all have the same characteristics, although their orientations are different. Denoted by { } braces.

Polycrystalline material A material comprising many grains.

Polymorphism Compounds exhibiting more than one type of crystal structure.

Rapid solidification A technique used to cool metals and alloys very quickly.

Repeat distance The distance from one lattice point to the adjacent lattice point along a direction.

Short-range order The regular and predictable arrangement of the atoms over a short distance—usually one or two atom spacings.

Stacking sequence The sequence in which close-packed planes are stacked. If the sequence is *ABABAB*, a hexagonal close-packed unit cell is produced; if the sequence is *ABCABCABC*, a face-centered cubic structure is produced.

Stress-induced crystallization The process of forming crystals by the application of an external stress. Typically, a significant fraction of many amorphous plastics can be crystallized in this fashion, making them stronger.

Tetrahedral site An interstitial position that has a coordination number of four. An atom or ion in the tetrahedral site has four nearest neighbor atoms or ions.

Tetrahedron The structure produced when atoms are packed together with a four-fold coordination.

Transmission electron microscopy (TEM) A technique for imaging and analysis of microstructures using a high energy electron beam.

Unit cell A subdivision of the lattice that still retains the overall characteristics of the entire lattice.

X-ray diffraction (XRD) A technique for analysis of crystalline materials using a beam of x-rays.

Problems

Section 3-1 Short-Range Order versus Long-Range Order

3-1 What is a "crystalline" material?

3-2 What is a single crystal?

3-3 State any two applications where single crystals are used.

3-4 What is a polycrystalline material?

3-5 What is a liquid crystal material?

3-6 What is an amorphous material?

3-7 Why do some materials assume an amorphous structure?

3-8 State any two applications of amorphous silicate glasses.

Section 3-2 Amorphous Materials: Principles and Technological Applications

3-9 What is meant by the term glass-ceramic?

3-10 Briefly compare the mechanical properties of glasses and glass-ceramics.

Section 3-3 Lattice, Unit Cells, Basis, and Crystal Structures

3-11 Define the terms lattice, unit cell, basis, and crystal structure.

3-12 Explain why there is no face-centered tetragonal Bravais lattice.

3-13 Calculate the atomic radius in cm for the following:
(a) BCC metal with $a_0 = 0.3294$ nm; and
(b) FCC metal with $a_0 = 4.0862$ Å.

3-14 Determine the crystal structure for the following:
(a) a metal with $a_0 = 4.9489$ Å, $r = 1.75$ Å, and one atom per lattice point; and
(b) a metal with $a_0 = 0.42906$ nm, $r = 0.1858$ nm, and one atom per lattice point.

3-15 The density of potassium, which has the BCC structure, is 0.855 g/cm^3. The atomic weight of potassium is 39.09 g/mol. Calculate
(a) the lattice parameter; and
(b) the atomic radius of potassium.

3-16 The density of thorium, which has the FCC structure, is 11.72 g/cm^3. The atomic weight of thorium is 232 g/mol. Calculate
(a) the lattice parameter; and
(b) the atomic radius of thorium.

3-17 A metal having a cubic structure has a density of 2.6 g/cm^3, an atomic weight of 87.62 g/mol, and a lattice parameter of 6.0849 Å. One atom is associated with each lattice point. Determine the crystal structure of the metal.

3-18 A metal having a cubic structure has a density of 1.892 g/cm^3, an atomic weight of

132.91 g/mol, and a lattice parameter of 6.13 Å. One atom is associated with each lattice point. Determine the crystal structure of the metal.

3-19 Indium has a tetragonal structure, with $a_0 = 0.32517$ nm and $c_0 = 0.49459$ nm. The density is 7.286 g/cm³, and the atomic weight is 114.82 g/mol. Does indium have the simple tetragonal or body-centered tetragonal structure?

3-20 Bismuth has a hexagonal structure, with $a_0 = 0.4546$ nm and $c_0 = 1.186$ nm. The density is 9.808 g/cm³, and the atomic weight is 208.98 g/mol. Determine

(a) the volume of the unit cell; and

(b) the number of atoms in each unit cell.

3-21 Gallium has an orthorhombic structure, with $a_0 = 0.45258$ nm, $b_0 = 0.45186$ nm, and $c_0 = 0.76570$ nm. The atomic radius is 0.1218 nm. The density is 5.904 g/cm³, and the atomic weight is 69.72 g/mol. Determine

(a) the number of atoms in each unit cell; and

(b) the packing factor in the unit cell.

3-22 Beryllium has a hexagonal crystal structure, with $a_0 = 0.22858$ nm and $c_0 = 0.35842$ nm. The atomic radius is 0.1143 nm, the density is 1.848 g/cm³, and the atomic weight is 9.01 g/mol. Determine

(a) the number of atoms in each unit cell; and

(b) the packing factor in the unit cell.

3-23 A typical paper clip weighs 0.59 g and consists of BCC iron. Calculate

(a) the number of unit cells; and

(b) the number of iron atoms in the paper clip. (See Appendix A for required data.)

3-24 Aluminum foil used to package food is approximately 0.0025 cm thick. Assume that all of the unit cells of the aluminum are arranged so that a_0 is perpendicular to the foil surface. For a 10 cm × 10 cm square of the foil, determine

(a) the total number of unit cells in the foil; and

(b) the thickness of the foil in number of unit cells. (See Appendix A.)

3-25 Rutile is the name given to a crystal structure commonly adopted by compounds of the form AB_2, where A represents a metal atom and B represents oxygen atoms. One form of

rutile has atoms of element A at the unit cell coordinates (0, 0, 0) and (1/2, 1/2, 1/2) and atoms of element B at (1/4, 1/4, 0), (3/4, 3/4, 0), (3/4, 1/4, 1/2), and (1/4, 3/4, 1/2). The unit cell parameters are $a = b \neq c$ and $\alpha = \beta = \gamma = 90°$. Note that the lattice parameter c is typically smaller than the lattice parameters a and b for the rutile structure.

(a) How many atoms of element A are there per unit cell?

(b) How many atoms of element B are there per unit cell?

(c) Is your answer to part (b) consistent with the stoichiometry of an AB_2 compound? Explain.

(d) Draw the unit cell for rutile. Use a different symbol for each type of atom. Provide a legend indicating which symbol represents which type of atom.

(e) For the simple tetragonal lattice, $a = b \neq c$ and $\alpha = \beta = \gamma = 90°$. There is one lattice point per unit cell located at the corners of the simple tetragonal lattice. Describe the rutile structure as a simple tetragonal lattice and a basis.

3-26 Consider the CuAu crystal structure. It can be described as a simple cubic lattice with a basis of Cu (0, 0, 0), Cu (1/2, 1/2, 0), Au (1/2, 0, 1/2), and Au (0, 1/2, 1/2).

(a) How many atoms of each type are there per unit cell?

(b) Draw the unit cell for CuAu. Use a different symbol for each type of atom. Provide a legend indicating which symbol represents which type of atom.

(c) Give an alternative lattice and basis representation for CuAu for which one atom of the basis is Au (0, 0, 0).

(d) A related crystal structure is that of Cu_3Au. This unit cell is similar to the face-centered cubic unit cell with Au at the corners of the unit cell and Cu at all of the face-centered positions. Describe this structure as a lattice and a basis.

(e) The Cu_3Au crystal structure is similar to the FCC crystal structure, but it does not have the face-centered cubic lattice. Explain briefly why this is the case.

3-27 Nanowires are high aspect-ratio metal or semiconducting wires with diameters on the order of 1 to 100 nm and typical lengths of 1 to 100 microns. Nanowires likely will be used in the future to create high-density electronic circuits.

Nanowires can be fabricated from ZnO. ZnO has the wurtzite structure. The wurtzite structure is a hexagonal lattice with four atoms per lattice point at Zn (0, 0, 0), Zn (2/3, 1/3, 1/2), O (0, 0, 3/8), and O (2/3, 1/3, 7/8).

(a) How many atoms are there in the conventional unit cell?

(b) If the atoms were located instead at Zn (0, 0, 0), Zn (1/3, 2/3, 1/2), O (0, 0, 3/8), and O (1/3, 2/3, 7/8), would the structure be different? Please explain.

(c) For ZnO, the unit cell parameters are $a = 3.24$ Å and $c = 5.19$ Å. (*Note*: This is not the ideal HCP c/a ratio.) A typical ZnO nanowire is 20 nm in diameter and 5 μm long. Assume that the nanowires are cylindrical. Approximately how many atoms are there in a single ZnO nanowire?

3-28 Calculate the atomic packing fraction for the hexagonal close-packed crystal structure for which $c = \sqrt{\dfrac{8}{3}}a$. Remember that the base of the unit cell is a parallelogram.

Section 3-4 Allotropic or Polymorphic Transformations

3-29 What is the difference between an allotrope and a polymorph?

3-30 What are the different polymorphs of zirconia?

3-31 Above 882°C, titanium has a BCC crystal structure, with $a = 0.332$ nm. Below this temperature, titanium has a HCP structure with $a = 0.2978$ nm and $c = 0.4735$ nm. Determine the percent volume change when BCC titanium transforms to HCP titanium. Is this a contraction or expansion?

3-32 α-Mn has a cubic structure with $a_0 = 0.8931$ nm and a density of 7.47 g/cm^3. β-Mn has a different cubic structure with $a_0 = 0.6326$ nm and a density of 7.26 g/cm^3. The atomic weight of manganese is

54.938 g/mol and the atomic radius is 0.112 nm. Determine the percent volume change that would occur if α-Mn transforms to β-Mn.

3-33 Calculate the theoretical density of the three polymorphs of zirconia. The lattice constants for the monoclinic form are $a = 5.156$, $b = 5.191$, and $c = 5.304$ Å, respectively. The angle β for the monoclinic unit cell is 98.9°. The lattice constants for the tetragonal unit cell are $a = 5.094$ and $c = 5.304$ Å, respectively. Cubic zirconia has a lattice constant of 5.124 Å.

3-34 From the information in this chapter, calculate the volume change that will occur when the cubic form of zirconia transforms into a tetragonal form.

3-35 Monoclinic zirconia cannot be used effectively for manufacturing oxygen sensors or other devices. Explain.

3-36 What is meant by the term stabilized zirconia?

3-37 State any two applications of stabilized zirconia ceramics.

Section 3-5 Points, Directions, and Planes in the Unit Cell

3-38 Explain the significance of crystallographic directions using an example of an application.

3-39 Why are Fe-Si alloys used in magnetic applications "grain oriented?"

3-40 How is the influence of crystallographic direction on magnetic properties used in magnetic materials for recording media applications?

3-41 Determine the Miller indices for the directions in the cubic unit cell shown in Figure 3-39.

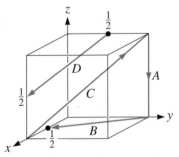

Figure 3-39   Directions in a cubic unit cell for Problem 3-41.

3-42 Determine the indices for the directions in the cubic unit cell shown in Figure 3-40.

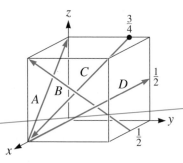

Figure 3-40 Directions in a cubic unit cell for Problem 3-42.

3-43 Determine the indices for the planes in the cubic unit cell shown in Figure 3-41.

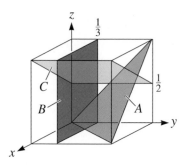

Figure 3-41 Planes in a cubic unit cell for Problem 3-43.

3-44 Determine the indices for the planes in the cubic unit cell shown in Figure 3-42.

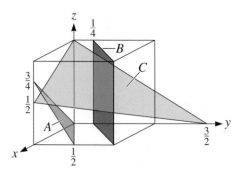

Figure 3-42 Planes in a cubic unit cell for Problem 3-44.

3-45 Determine the indices for the directions in the hexagonal lattice shown in Figure 3-43, using both the three-digit and four-digit systems.

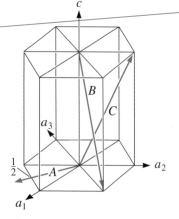

Figure 3-43 Directions in a hexagonal lattice for Problem 3-45.

3-46 Determine the indices for the directions in the hexagonal lattice shown in Figure 3-44, using both the three-digit and four-digit systems.

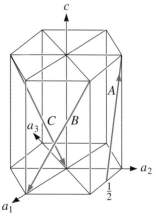

Figure 3-44 Directions in a hexagonal lattice for Problem 3-46.

3-47 Determine the indices for the planes in the hexagonal lattice shown in Figure 3-45.

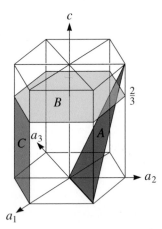

Figure 3-45 Planes in a hexagonal lattice for Problem 3-47.

3-48 Determine the indices for the planes in the hexagonal lattice shown in Figure 3-46.

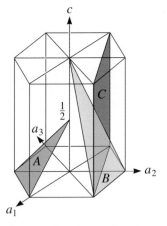

Figure 3-46 Planes in a hexagonal lattice for Problem 3-48.

3-49 Sketch the following planes and directions within a cubic unit cell:
(a) [101]   (b) [0$\bar{1}$0]   (c) [12$\bar{2}$]   (d) [301]
(e) [$\bar{2}$01]   (f) [2$\bar{1}$3]   (g) (0$\bar{1}\bar{1}$)   (h) (102)
(i) (002)   (j) (1$\bar{3}$0)   (k) ($\bar{2}$12)   (l) (3$\bar{1}\bar{2}$)

3-50 Sketch the following planes and directions within a cubic unit cell:
(a) [1$\bar{1}$0]   (b) [$\bar{2}\bar{2}$1]   (c) [410]   (d) [0$\bar{1}$2]
(e) [$\bar{3}\bar{2}$1]   (f) [1$\bar{1}$1]   (g) (11$\bar{1}$)   (h) (01$\bar{1}$)
(i) (030)   (j) ($\bar{1}$21)   (k) (11$\bar{3}$)   (l) (0$\bar{4}$1)

3-51 Sketch the following planes and directions within a hexagonal unit cell:
(a) [01$\bar{1}$0]   b) [11$\bar{2}$0]   (c) [$\bar{1}$011]
(d) (0003)   (e) ($\bar{1}$010)   (f) (01$\bar{1}$1)

3-52 Sketch the following planes and directions within a hexagonal unit cell:
(a) [$\bar{2}$110]   (b) [11$\bar{2}$1]   (c) [10$\bar{1}$0]
(d) (1$\bar{2}$10)   (e) ($\bar{1}\bar{1}$22)   (f) (12$\bar{3}$0)

3-53 What are the indices of the six directions of the form $\langle 110 \rangle$ that lie in the (11$\bar{1}$) plane of a cubic cell?

3-54 What are the indices of the four directions of the form $\langle 111 \rangle$ that lie in the ($\bar{1}$01) plane of a cubic cell?

3-55 Determine the number of directions of the form $\langle 110 \rangle$ in a tetragonal unit cell and compare to the number of directions of the form $\langle 110 \rangle$ in an orthorhombic unit cell.

3-56 Determine the angle between the [110] direction and the (110) plane in a tetragonal unit cell; then determine the angle between the [011] direction and the (011) plane in a tetragonal cell. The lattice parameters are $a_0 = 4.0$ Å and $c_0 = 5.0$ Å. What is responsible for the difference?

3-57 Determine the Miller indices of the plane that passes through three points having the following coordinates:
(a) 0, 0, 1; 1, 0, 0; and 1/2, 1/2, 0
(b) 1/2, 0, 1; 1/2, 0, 0; and 0, 1, 0
(c) 1, 0, 0; 0, 1, 1/2; and 1, 1/2, 1/4
(d) 1, 0, 0; 0, 0, 1/4; and 1/2, 1, 0

3-58 Determine the repeat distance, linear density, and packing fraction for FCC nickel, which has a lattice parameter of 0.35167 nm, in the [100], [110], and [111] directions. Which of these directions is close packed?

3-59 Determine the repeat distance, linear density, and packing fraction for BCC lithium, which has a lattice parameter of 0.35089 nm, in the [100], [110], and [111] directions. Which of these directions is close packed?

3-60 Determine the repeat distance, linear density, and packing fraction for HCP magnesium in the [2110] direction and the [11$\bar{2}$0] direction. The lattice parameters for HCP magnesium are given in Appendix A.

3-61 Determine the planar density and packing fraction for FCC nickel in the (100), (110), and (111) planes. Which, if any, of these planes are close packed?

3-62 Determine the planar density and packing fraction for BCC lithium in the (100), (110), and (111) planes. Which, if any, of these planes are close packed?

3-63 Suppose that FCC rhodium is produced as a 1 mm-thick sheet, with the (111) plane parallel to the surface of the sheet. How many (111) interplanar spacings d_{111} thick is the sheet? See Appendix A for necessary data.

3-64 In an FCC unit cell, how many d_{111} are present between the 0, 0, 0 point and the 1, 1, 1 point?

3-65 What are the stacking sequences in the FCC and HCP structures?

Section 3-6 Interstitial Sites

3-66 Determine the minimum radius of an atom that will just fit into
  (a) the tetrahedral interstitial site in FCC nickel; and
  (b) the octahedral interstitial site in BCC lithium.

3-67 What are the coordination numbers for octahedral and tetrahedral sites?

Section 3-7 Crystal Structures of Ionic Materials

3-68 What is the radius of an atom that will just fit into the octahedral site in FCC copper without disturbing the crystal structure?

3-69 Using the ionic radii given in Appendix B, determine the coordination number expected for the following compounds:
  (a) Y_2O_3 (b) UO_2 (c) BaO (d) Si_3N_4
  (e) GeO_2 (f) MnO (g) MgS (h) KBr

3-70 A particular unit cell is cubic with ions of type A located at the corners and face-centers of the unit cell and ions of type B located at the midpoint of each edge of the cube and at the body-centered position. The ions contribute to the unit cell in the usual way (1/8 ion contribution for each ion at the corners, etc.).
  (a) How many ions of each type are there per unit cell?
  (b) Describe this structure as a lattice and a basis. Check to be sure that the number of ions per unit cell given by your description of the structure as a lattice and a basis is consistent with your answer to part (a).
  (c) What is the coordination number of each ion?
  (d) What is the name commonly given to this crystal structure?

3-71 Would you expect NiO to have the cesium chloride, sodium chloride, or zinc blende structure? Based on your answer, determine
  (a) the lattice parameter;
  (b) the density; and
  (c) the packing factor.

3-72 Would you expect UO_2 to have the sodium chloride, zinc blende, or fluorite structure? Based on your answer, determine
  (a) the lattice parameter;
  (b) the density; and
  (c) the packing factor.

3-73 Would you expect BeO to have the sodium chloride, zinc blende, or fluorite structure? Based on your answer, determine
  (a) the lattice parameter;
  (b) the density; and
  (c) the packing factor.

3-74 Would you expect CsBr to have the sodium chloride, zinc blende, fluorite, or cesium chloride structure? Based on your answer, determine
  (a) the lattice parameter;
  (b) the density; and
  (c) the packing factor.

3-75 Sketch the ion arrangement of the (110) plane of ZnS (with the zinc blende structure) and compare this arrangement to that on the (110) plane of CaF_2 (with the fluorite structure). Compare the planar packing fraction on the (110) planes for these two materials.

3-76 MgO, which has the sodium chloride structure, has a lattice parameter of 0.396 nm. Determine the planar density and the planar packing fraction for the (111) and (222) planes of MgO. What ions are present on each plane?

3-77 Draw the crystal structure of the perovskite polymorph of PZT ($Pb(Zr_xTi_{1-x})O_3$, x: mole fraction of Zr^{+4}). Assume the two B-site cations occupy random B-site positions.

Section 3-8 Covalent Structures

3-78 Calculate the theoretical density of α-Sn. Assume α-Sn has the diamond cubic structure and obtain the atomic radius information from Appendix B.

3-79 Calculate the theoretical density of Ge. Assume Ge has the diamond cubic structure and obtain the radius information from Appendix B.

Section 3-9 Diffraction Techniques for Crystal Structure Analysis

3-80 A diffracted x-ray beam is observed from the (220) planes of iron at a 2θ angle of 99.1° when x-rays of 0.15418 nm wavelength are used. Calculate the lattice parameter of the iron.

3-81 A diffracted x-ray beam is observed from the (311) planes of aluminum at a 2θ angle of 78.3° when x-rays of 0.15418 nm wavelength are used. Calculate the lattice parameter of the aluminum.

3-82 Figure 3-47 shows the results of an x-ray diffraction experiment in the form of the intensity of the diffracted peak versus the 2θ diffraction angle. If x-rays with a wavelength of 0.15418 nm are used, determine

(a) the crystal structure of the metal;
(b) the indices of the planes that produce each of the peaks; and
(c) the lattice parameter of the metal.

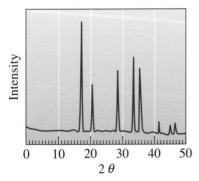

Figure 3-47   XRD pattern for Problem 3-82.

3-83 Figure 3-48 shows the results of an x-ray diffraction experiment in the form of the

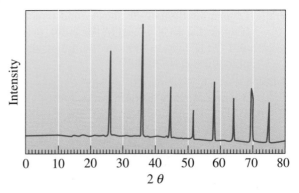

Figure 3-48   XRD pattern for Problem 3-83.

intensity of the diffracted peak versus the 2θ diffraction angle. If x-rays with a wavelength of 0.07107 nm are used, determine

(a) the crystal structure of the metal;
(b) the indices of the planes that produce each of the peaks; and
(c) the lattice parameter of the metal.

3-84 A sample of zirconia contains cubic and monoclinic polymorphs. What will be a good analytical technique to detect the presence of these two different polymorphs?

Design Problems

3-85 You would like to sort iron specimens, some of which are FCC and others BCC. Design an x-ray diffraction method by which this can be accomplished.

3-86 You want to design a material for making kitchen utensils for cooking. The material should be transparent and withstand repeated heating and cooling. What kind of materials could be used to design such transparent and durable kitchen-ware?

Computer Problems

Note: You should consult your instructor on the use of a computer language. In principle, it does not matter what computer language is used. Some suggestions are using C/C++ or Java. If these are unavailable, you can also solve most of these problems using spreadsheet software.

3-87 Table 3-1 contains formulas for the volume of different types of unit cells. Write a computer program to calculate the unit cell volume in the units of $Å^3$ and nm^3. Your program should prompt the user to input the (a) type of unit cell, (b) necessary lattice constants, and (c) angles. The program then should recognize the inputs made and use the appropriate formula for the calculation of unit cell volume.

3-88 Write a computer program that will ask the user to input the atomic mass, atomic radius, and cubic crystal structure for an element. The program output should be the packing fraction and the theoretical density.

K Knovel® **Problems**

K3-1 Determine the crystal lattice parameters and mass densities for GaN, GaP, GaAs, GaSb, InN, InP, InAs, and InSb semiconductors. Compare the data for lattice parameters from at least two different sources.

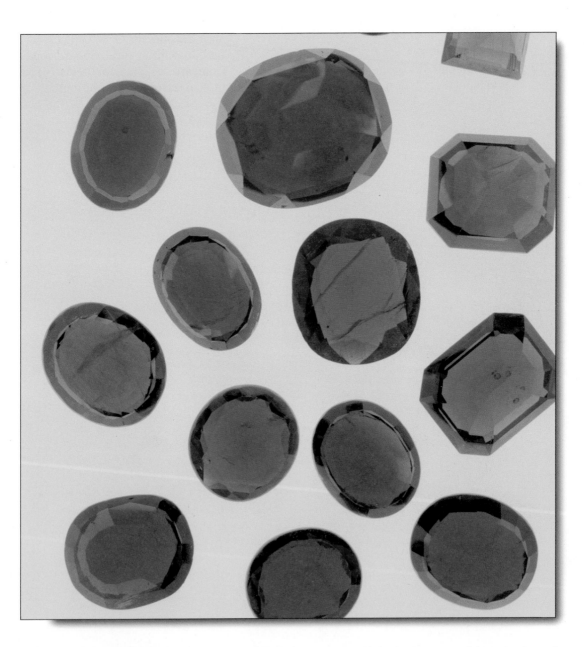

What makes a ruby red? The addition of about 1% chromium oxide in alumina creates defects. An electronic transition between defect levels produces the red ruby crystal. Similarly, incorporation of Fe^{+2} and Ti^{+4} makes the blue sapphire. *(Courtesy of Lawrence Lawry/ PhotoDisc/ Getty Images.)*

4

Imperfections in the Atomic and Ionic Arrangements

Have You Ever Wondered?

- *Why do silicon crystals used in the manufacture of semiconductor wafers contain trace amounts of dopants such as phosphorous or boron?*

- *What makes steel considerably harder and stronger than pure iron?*

- *What limits the current carrying capacity of a ceramic superconductor?*

- *Why do we use high-purity copper as a conductor in electrical applications?*

- *Why do FCC metals (such as copper and aluminum) tend to be more ductile than BCC and HCP metals?*

- *How can metals be strengthened?*

The arrangement of the atoms or ions in engineered materials contains imperfections or defects. These defects often have a profound effect on the properties of materials. In this chapter, we introduce the three basic types of imperfections: point defects, line defects (or dislocations), and surface defects. These imperfections only represent defects in or deviations from the perfect or ideal atomic or ionic arrangements expected in a given crystal structure. The material is not considered defective from a technological viewpoint. In many applications, the presence of such defects is useful. There are a few applications, though, where we strive to minimize a particular type of defect. For example, defects known as dislocations are useful for increasing the strength of metals and alloys; however, in single crystal silicon, used for manufacturing computer chips, the presence of dislocations is undesirable. Often the "defects" may be created intentionally to produce a desired set of electronic, magnetic, optical, or mechanical properties. For example, pure iron is relatively soft, yet, when we add a small amount of carbon, we create defects in the crystalline arrangement of iron and transform it into a plain carbon steel that exhibits considerably higher strength. Similarly, a crystal of pure alumina is transparent and colorless, but, when we add a small amount of chromium, it creates a special defect, resulting in a beautiful red ruby crystal. In the processing of Si crystals for microelectronics, we add very small concentrations of P or B atoms to Si. These additions create defects in the arrangement of atoms in silicon that impart special electrical properties to different parts of the silicon crystal. This, in turn, allows us to make useful devices such as transistors—the basic building blocks that enabled the development of modern computers and the information technology revolution. The effect of point defects, however, is not always desirable. When we want to use copper as a conductor for microelectronics, we use the highest purity available. This is because even small levels of impurities will cause an order of magnitude increase in the electrical resistivity of copper!

Grain boundaries, regions between different grains of a polycrystalline material, represent one type of defect. Ceramic superconductors, under certain conditions, can conduct electricity without any electrical resistance. Materials scientists and engineers have made long wires or tapes of such materials. They have also discovered that, although the current flows quite well within the grains of a polycrystalline superconductor, there is considerable resistance to the flow of current from one grain to another—across the grain boundary. On the other hand, the presence of grain boundaries actually helps strengthen metallic materials. In later chapters, we will show how we can control the concentrations of these defects through tailoring of composition or processing techniques. In this chapter, we explore the nature and effects of different types of defects.

4-1   Point Defects

Point defects are localized disruptions in otherwise perfect atomic or ionic arrangements in a crystal structure. Even though we call them point defects, the disruption affects a region involving several atoms or ions. These imperfections, shown in Figure 4-1, may be introduced by movement of the atoms or ions when they gain energy by heating, during processing of the material, or by the intentional or unintentional introduction of impurities.

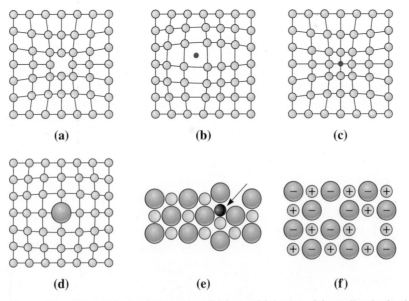

(a)   (b)   (c)

(d)   (e)   (f)

Figure 4-1   Point defects: (a) vacancy, (b) interstitial atom, (c) small substitutional atom, (d) large substitutional atom, (e) Frenkel defect, and (f) Schottky defect. All of these defects disrupt the perfect arrangement of the surrounding atoms.

Typically, **impurities** are elements or compounds that are present from raw materials or processing. For example, silicon crystals grown in quartz crucibles contain oxygen as an impurity. **Dopants**, on the other hand, are elements or compounds that are deliberately added, in known concentrations, at specific locations in the microstructure, with an intended beneficial effect on properties or processing. In general, the effect of impurities is deleterious, whereas the effect of dopants on the properties of materials is useful. Phosphorus (P) and boron (B) are examples of dopants that are added to silicon crystals to improve the electrical properties of pure silicon (Si).

A point defect typically involves one atom or ion, or a pair of atoms or ions, and thus is different from **extended defects**, such as dislocations or grain boundaries. An important "point" about point defects is that although the defect occurs at one or two sites, their presence is "felt" over much larger distances in the crystalline material.

Vacancies

A **vacancy** is produced when an atom or an ion is missing from its normal site in the crystal structure as in Figure 4-1(a). When atoms or ions are missing (i.e., when vacancies are present), the overall randomness or entropy of the material increases, which increases the thermodynamic stability of a crystalline material. All crystalline materials have vacancy defects. Vacancies are introduced into metals and alloys during solidification, at high temperatures, or as a consequence of radiation damage. Vacancies play an important role in determining the rate at which atoms or ions move around or diffuse in a solid material, especially in pure metals. We will see this in greater detail in Chapter 5.

At room temperature ($\sim$298 K), the concentration of vacancies is small, but the concentration of vacancies increases exponentially as the temperature increases, as shown by the following Arrhenius type behavior:

$$n_v = n \exp\left(\frac{-Q_v}{RT}\right) \tag{4-1}$$

where

n_v is the number of vacancies per cm^3;

n is the number of atoms per cm^3;

Q_v is the energy required to produce one mole of vacancies, in cal/mol or Joules/mol;

R is the gas constant, $1.987 \dfrac{cal}{mol \cdot K}$ or $8.314 \dfrac{Joules}{mol \cdot K}$; and

T is the temperature in Kelvin.

Due to the large thermal energy near the melting temperature, there may be as many as one vacancy per 1000 atoms. Note that this equation provides the equilibrium concentration of vacancies at a given temperature. It is also possible to retain the concentration of vacancies produced at a high temperature by quenching the material rapidly. Thus, in many situations, the concentration of vacancies observed at room temperature is not the equilibrium concentration predicted by Equation 4-1.

Example 4-1   *The Effect of Temperature on Vacancy Concentrations*

Calculate the concentration of vacancies in copper at room temperature (25°C). What temperature will be needed to heat treat copper such that the concentration of vacancies produced will be 1000 times more than the equilibrium concentration of vacancies at room temperature? Assume that 20,000 cal are required to produce a mole of vacancies in copper.

SOLUTION

The lattice parameter of FCC copper is 0.36151 nm. There are four atoms per unit cell; therefore, the number of copper atoms per cm^3 is

$$n = \frac{4 \text{ atoms/cell}}{(3.6151 \times 10^{-8} \text{ cm})^3} = 8.466 \times 10^{22} \text{ copper atoms/cm}^3$$

At room temperature, $T = 25 + 273 = 298$ K:

$$n_v = n \exp\left(\frac{-Q_v}{RT}\right)$$

$$= \left(8.466 \times 10^{22} \frac{\text{atoms}}{\text{cm}^3}\right) \exp\left[\frac{-20{,}000 \frac{\text{cal}}{\text{mol}}}{\left(1.987 \frac{\text{cal}}{\text{mol} \cdot \text{K}}\right)(298 \text{ K})}\right]$$

$$= 1.814 \times 10^8 \text{ vacancies/cm}^3$$

We wish to find a heat treatment temperature that will lead to a concentration of vacancies that is 1000 times higher than this number, or $n_v = 1.814 \times 10^{11}$ vacancies/cm³.

We could do this by heating the copper to a temperature at which this number of vacancies forms:

$$n_v = 1.814 \times 10^{11} = n \exp\left(\frac{-Q_v}{RT}\right)$$

$$= (8.466 \times 10^{22}) \exp(-20{,}000)/(1.987T)$$

$$\exp\left(\frac{-20{,}000}{1.987T}\right) = \frac{1.814 \times 10^{11}}{8.466 \times 10^{22}} = 0.214 \times 10^{-11}$$

$$\frac{-20{,}000}{1.987T} = \ln(0.214 \times 10^{-11}) = -26.87$$

$$T = \frac{20{,}000}{(1.987)(26.87)} = 375 \text{ K} = 102°\text{C}$$

By heating the copper slightly above 100°C, waiting until equilibrium is reached, and then rapidly cooling the copper back to room temperature, the number of vacancies trapped in the structure may be one thousand times greater than the equilibrium number of vacancies at room temperature. Thus, vacancy concentrations encountered in materials are often dictated by both thermodynamic and kinetic factors.

Example 4-2   *Vacancy Concentrations in Iron*

Calculate the theoretical density of iron, and then determine the number of vacancies needed for a BCC iron crystal to have a density of 7.874 g/cm³. The lattice parameter of iron is 2.866×10^{-8} cm.

SOLUTION

The theoretical density of iron can be calculated from the lattice parameter and the atomic mass. Since the iron is BCC, two iron atoms are present in each unit cell.

$$\rho = \frac{(2 \text{ atoms/cell})(55.847 \text{ g/mol})}{(2.866 \times 10^{-8} \text{ cm})^3(6.022 \times 10^{23} \text{ atoms/mol})} = 7.879 \text{ g/cm}^3$$

This calculation assumes that there are no imperfections in the crystal. Let's calculate the number of iron atoms and vacancies that would be present in each unit cell for a density of 7.874 g/cm^3:

$$\rho = \frac{(X \text{ atoms/cell})(55.847 \text{ g/mol})}{(2.866 \times 10^{-8} \text{ cm})^3(6.022 \times 10^{23} \text{ atoms/mol})} = 7.874 \text{ g/cm}^3$$

$$X \text{ atoms/cell} = \frac{(7.874 \text{ g/cm}^3)(2.866 \times 10^{-8} \text{ cm})^3(6.022 \times 10^{23} \text{ atoms/mol})}{(55.847 \text{ g/mol})} = 1.99878$$

There should be $2.00 - 1.99878 = 0.00122$ vacancies per unit cell. The number of vacancies per cm^3 is

$$\text{Vacancies/cm}^3 = \frac{0.00122 \text{ vacancies/cell}}{(2.866 \times 10^{-8} \text{ cm})^3} = 5.18 \times 10^{19}$$

Note that other defects such as grain boundaries in a polycrystalline material contribute to a density lower than the theoretical value.

Interstitial Defects

An **interstitial defect** is formed when an extra atom or ion is inserted into the crystal structure at a normally unoccupied position, as in Figure 4-1(b). The interstitial sites were illustrated in Table 3-6. Interstitial atoms or ions, although much smaller than the atoms or ions located at the lattice points, are still larger than the interstitial sites that they occupy; consequently, the surrounding crystal region is compressed and distorted. Interstitial atoms such as hydrogen are often present as impurities, whereas carbon atoms are intentionally added to iron to produce steel. For small concentrations, carbon atoms occupy interstitial sites in the iron crystal structure, introducing a stress in the localized region of the crystal in their vicinity. As we will see, the introduction of interstitial atoms is one important way of increasing the strength of metallic materials. Unlike vacancies, once introduced, the number of interstitial atoms or ions in the structure remains nearly constant, even when the temperature is changed.

Example 4-3   *Sites for Carbon in Iron*

In FCC iron, carbon atoms are located at *octahedral* sites, which occur at the center of each edge of the unit cell at sites such as (0, 0, 1/2) and at the center of the unit cell (1/2, 1/2, 1/2). In BCC iron, carbon atoms enter *tetrahedral* sites, such as (0, 1/2, 1/4). The lattice parameter is 0.3571 nm for FCC iron and 0.2866 nm for BCC iron. Assume that carbon atoms have a radius of 0.071 nm. (a) Would we expect a greater distortion of the crystal by an interstitial carbon atom in FCC or BCC iron? (b) What would be the atomic percentage of carbon in each type of iron if all the interstitial sites were filled?

SOLUTION

(a) We can calculate the size of the interstitial site in BCC iron at the $(0, 1/2, 1/4)$ location with the help of Figure 4-2(a). The radius R_{BCC} of the iron atom is

$$R_{BCC} = \frac{\sqrt{3}a_0}{4} = \frac{(\sqrt{3})(0.2866)}{4} = 0.1241 \text{ nm}$$

From Figure 4-2(a), we find that

$$(\tfrac{1}{2} a_0)^2 + (\tfrac{1}{4} a_0)^2 = (r_{interstitial} + R_{BCC})^2$$

$$(r_{interstitial} + R_{BCC})^2 = 0.3125a_0^2 = (0.3125)(0.2866 \text{ nm})^2 = 0.02567$$

$$r_{interstitial} = \sqrt{0.02567} - 0.1241 = 0.0361 \text{ nm}$$

For FCC iron, the interstitial site such as the $(0, 0, 1/2)$ lies along $\langle 001 \rangle$ directions. Thus, the radius of the iron atom and the radius of the interstitial site are [Figure 4-2(b)]:

$$R_{FCC} = \frac{\sqrt{2}a_0}{4} = \frac{(\sqrt{2})(0.3571)}{4} = 0.1263 \text{ nm}$$

$$2r_{interstitial} + 2R_{FCC} = a_0$$

$$r_{interstitial} = \frac{0.3571 - (2)(0.1263)}{2} = 0.0523 \text{ nm}$$

The interstitial site in BCC iron is smaller than the interstitial site in FCC iron. Although both are smaller than the carbon atom, carbon distorts the BCC crystal structure more than the FCC structure. As a result, fewer carbon atoms are expected to enter interstitial positions in BCC iron than in FCC iron.

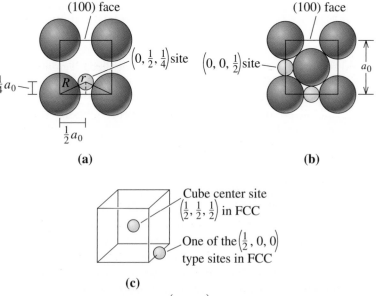

(a)   (b)

(c)

Figure 4-2   (a) The location of the $\left(0, \tfrac{1}{2}, \tfrac{1}{4}\right)$ interstitial site in BCC metals. (b) $\left(0, 0, \tfrac{1}{2}\right)$ site in FCC metals. (c) Edge centers and cube centers are some of the interstitial sites in the FCC structure (for Example 4-3).

(b) In BCC iron, two iron atoms are expected in each unit cell. We can find a total of 24 interstitial sites of the type $(1/4, 1/2, 0)$; however, since each site is located at a face of the unit cell, only half of each site belongs uniquely to a single cell. Thus, there are

$$(24 \text{ sites})(\tfrac{1}{2}) = 12 \text{ interstitial sites per unit cell}$$

If all of the interstitial sites were filled, the atomic percentage of carbon contained in the iron would be

$$\text{at } \% \text{ C} = \frac{12\,\text{C atoms}}{12\,\text{C atoms} + 2 \text{ Fe atoms}} \times 100 = 86\%$$

In FCC iron, four iron atoms are expected in each unit cell, and the number of octahedral interstitial sites is

$$(12 \text{ edges})(\tfrac{1}{4}) + 1 \text{ center} = 4 \text{ interstitial sites per unit cell [Figure 4 - 2(c)]}$$

Again, if all the octahedral interstitial sites were filled, the atomic percentage of carbon in the FCC iron would be

$$\text{at } \% \text{ C} = \frac{4\,\text{C atoms}}{4\,\text{C atoms} + 4\,\text{Fe atoms}} \times 100 = 50\%$$

As we will see in a later chapter, the maximum atomic percentage of carbon present in the two forms of iron under equilibrium conditions is

BCC: 1.0%

FCC: 8.9%

Because of the strain imposed on the iron crystal structure by the interstitial atoms—particularly in BCC iron—the fraction of the interstitial sites that can be occupied is quite small.

Substitutional Defects

A **substitutional defect** is introduced when one atom or ion is replaced by a different type of atom or ion as in Figure 4-1(c) and (d). The substitutional atoms or ions occupy the normal lattice site. Substitutional atoms or ions may either be larger than the normal atoms or ions in the crystal structure, in which case the surrounding interatomic spacings are reduced, or smaller causing the surrounding atoms to have larger interatomic spacings. In either case, the substitutional defects disturb the surrounding crystal. Again, the substitutional defect can be introduced either as an impurity or as a deliberate alloying addition, and, once introduced, the number of defects is relatively independent of temperature.

Examples of substitutional defects include incorporation of dopants such as phosphorus (P) or boron (B) into Si. Similarly, if we add copper to nickel, copper atoms will occupy crystallographic sites where nickel atoms would normally be present. The substitutional atoms will often increase the strength of the metallic material. Substitutional defects also appear in ceramic materials. For example, if we add MgO to NiO, Mg^{+2} ions occupy Ni^{+2} sites, and O^{-2} ions from MgO occupy O^{-2} sites of NiO. Whether atoms or ions go into interstitial or substitutional sites depends upon the size and valence of these guest atoms or ions compared to the size and valence of the host ions. The size of the available sites also plays a role in this as discussed in Chapter 3, Section 6.

4-2   Other Point Defects

An **interstitialcy** is created when an atom identical to those at the normal lattice points is located in an interstitial position. These defects are most likely to be found in crystal structures having a low packing factor.

A **Frenkel defect** is a vacancy-interstitial pair formed when an ion jumps from a normal lattice point to an interstitial site, as in Figure 4-1(e) leaving behind a vacancy. Although, this is usually associated with ionic materials, a Frenkel defect can occur in metals and covalently bonded materials. A **Schottky defect**, Figure 4-1(f), is unique to ionic materials and is commonly found in many ceramic materials. When vacancies occur in an ionically bonded material, a stoichiometric number of anions and cations must be missing from regular atomic positions if electrical neutrality is to be preserved. For example, one Mg^{+2} vacancy and one O^{-2} vacancy in MgO constitute a Schottky pair. In ZrO_2, for one Zr^{+4} vacancy, there will be two O^{-2} vacancies.

An important substitutional point defect occurs when an ion of one charge replaces an ion of a different charge. This might be the case when an ion with a valence of +2 replaces an ion with a valence of +1 (Figure 4-3). In this case, an extra positive charge is introduced into the structure. To maintain a charge balance, a vacancy might be created where a +1 cation normally would be located. Again, this imperfection is observed in materials that have pronounced ionic bonding.

Thus, in ionic solids, when point defects are introduced, the following rules have to be observed:

(a) a charge balance must be maintained so that the crystalline material as a whole is electrically neutral;

(b) a mass balance must be maintained; and

(c) the number of crystallographic sites must be conserved.

For example, in nickel oxide (NiO) if one oxygen ion is missing, it creates an oxygen ion vacancy (designated as $V_O^{\cdot\cdot}$). Each dot ($\cdot$) in the superscript position indicates an *effective* positive charge of one. To maintain stoichiometry, mass balance, and charge balance, we must also create a nickel ion vacancy (designated as V_{Ni}''). Each accent ($'$) in the superscript indicates an *effective* charge of -1.

We use the **Kröger-Vink notation** to write the defect chemistry equations. The main letter in this notation describes a vacancy or the name of the element. The superscript indicates the effective charge on the defect, and the subscript describes the location of the defect. A dot ($\cdot$) indicates an effective charge of +1 and an accent ($'$) represents an effective charge of -1. Sometimes x is used to indicate no net charge. Any free electrons or holes are indicated as e and h, respectively. (Holes will be discussed in Chapter 19.) Clusters of defects or defects that have association are shown in parentheses. Associated

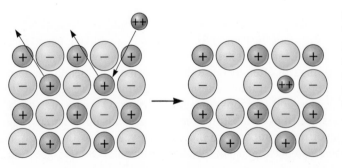

Figure 4-3
When a divalent cation replaces a monovalent cation, a second monovalent cation must also be removed, creating a vacancy.

defects, which can affect mass transport in materials, are sometimes neutral and hard to detect experimentally. Concentrations of defects are shown in square brackets.

The following examples illustrate the use of the Kröger-Vink notation for writing **defect chemical reactions**. Sometimes, it is possible to write multiple valid defect chemistry reactions to describe the possible defect chemistry. In such cases, it is necessary to take into account the energy that is needed to create different defects, and an experimental verification is necessary. This notation is useful in describing defect chemistry in semiconductors and many ceramic materials used as sensors, dielectrics, and in other applications.

Example 4-4   *Application of the Kröger-Vink Notation*

Write the appropriate defect reactions for (a) incorporation of magnesium oxide (MgO) in nickel oxide (NiO) and (b) formation of a Schottky defect in alumina (Al_2O_3).

SOLUTION

(a) MgO is the guest and NiO is the host material. We will assume that Mg^{+2} ions will occupy Ni^{+2} sites and oxygen anions from MgO will occupy O^{-2} sites of NiO.

$$MgO \xrightarrow{\text{NiO}} Mg_{Ni}^x + O_O^x$$

We need to ensure that the equation has charge, mass, and site balance. On the left-hand side, we have one Mg, one oxygen, and no net charge. The same is true on the right-hand side. The site balance can be a little tricky—one Mg^{+2} occupies one Ni^{+2} site. Since we are introducing MgO in NiO, we use one Ni^{+2} site, and, therefore, we must use one O^{-2} site. We can see that this is true by examining the right-hand side of this equation.

(b) A Schottky defect in alumina will involve two aluminum ions and three oxygen ions. When there is a vacancy at an aluminum site, a +3 charge is missing, and the site has an effective charge of −3. Thus V_{Al}''' describes one vacancy of Al^{+3}. Similarly, $V_O^{..}$ represents an oxygen ion vacancy. For site balance in alumina, we need to ensure that for every two aluminum ion sites used, we use three oxygen ion sites. Since we have vacancies, the mass on the right-hand side is zero, and so we write the left-hand side as null. Therefore, the defect reaction will be

$$\text{null} \xrightarrow{\text{Al}_2\text{O}_3} 2V_{Al}''' + 3V_O^{..}$$

Example 4-5   *Point Defects in Stabilized Zirconia for Solid Electrolytes*

Write the appropriate defect reactions for the incorporation of calcium oxide (CaO) in zirconia (ZrO_2) using the Kröger-Vink notation.

SOLUTION

We will assume that Ca^{+2} will occupy Zr^{+4} sites. If we send one Ca^{+2} to Zr^{+4}, the site will have an effective charge of −2 (instead of having a charge of +4, we have a charge of +2). We have used one Zr^{+4} site, and site balance requires *two oxygen sites*. We can send one O^{-2} from CaO to one of the O^{-2} sites in ZrO_2. The other oxygen site must be used and since mass balance must also be maintained, we will have to keep this site vacant (i.e., an oxygen ion vacancy will have to be created).

$$CaO \xrightarrow{ZrO_2} Ca_{Zr}'' + O_O^x + V_O^{\cdot\cdot}$$

The concentration of oxygen vacancies in ZrO_2 (i.e., $[V_O^{\cdot\cdot}]$) will increase with increasing CaO concentration. These oxygen ion vacancies make CaO stabilized ZrO_2 an ionic conductor. This allows the use of this type of ZrO_2 in oxygen sensors used in automotives and solid oxide fuel cells.

4-3   Dislocations

Dislocations are line imperfections in an otherwise perfect crystal. They typically are introduced into a crystal during solidification of the material or when the material is deformed permanently. Although dislocations are present in all materials, including ceramics and polymers, *they are particularly useful in explaining deformation and strengthening in metallic materials*. We can identify three types of dislocations: the screw dislocation, the edge dislocation, and the mixed dislocation.

Screw Dislocations   The **screw dislocation** (Figure 4-4) can be illustrated by cutting partway through a perfect crystal and then skewing the crystal by one atom spacing. If we follow a crystallographic plane one revolution around the axis on which the crystal was skewed, starting at point x and traveling equal atom spacings in each direction, we finish at point y one atom spacing below our starting point. If a screw dislocation were not present, the loop would close. The vector required to complete the loop is the **Burgers vector b**. If we continued our rotation, we would trace out a spiral path. The axis, or line around which we trace out this path, is the screw dislocation. The Burgers vector is parallel to the screw dislocation.

Edge Dislocations   An **edge dislocation** (Figure 4-5) can be illustrated by slicing partway through a perfect crystal, spreading the crystal apart, and partly filling

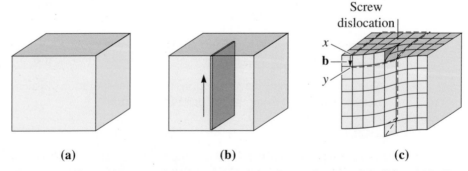

(a)   (b)   (c)

Figure 4-4   The perfect crystal (a) is cut and sheared one atom spacing, (b) and (c). The line along which shearing occurs is a screw dislocation. A Burgers vector **b** is required to close a loop of equal atom spacings around the screw dislocation.

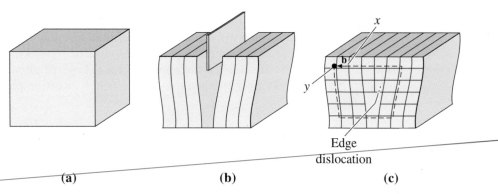

Edge
dislocation

(a)   **(b)**   **(c)**

Figure 4-5   The perfect crystal in (a) is cut and an extra half plane of atoms is inserted (b). The bottom edge of the extra half plane is an edge dislocation (c). A Burgers vector **b** is required to close a loop of equal atom spacings around the edge dislocation. (*Adapted from J.D. Verhoeven, Fundamentals of Physical Metallurgy, Wiley, 1975.*)

the cut with an extra half plane of atoms. The bottom edge of this inserted plane represents the edge dislocation. If we describe a clockwise loop around the edge dislocation, starting at point x and traveling an equal number of atom spacings in each direction, we finish at point y one atom spacing from the starting point. If an edge dislocation were not present, the loop would close. The vector required to complete the loop is, again, the Burgers vector. In this case, the Burgers vector is perpendicular to the dislocation. By introducing the dislocation, the atoms above the dislocation line are squeezed too closely together, while the atoms below the dislocation are stretched too far apart. The surrounding region of the crystal has been disturbed by the presence of the dislocation. [This is illustrated later in Figure 4-8(b).] A " ⊥ " symbol is often used to denote an edge dislocation. The long axis of the " ⊥ " points toward the extra half plane. Unlike an edge dislocation, a screw dislocation cannot be visualized as an extra half plane of atoms.

Mixed Dislocations

As shown in Figure 4-6, **mixed dislocations** have both edge and screw components, with a transition region between them. The Burgers vector, however, remains the same for all portions of the mixed dislocation.

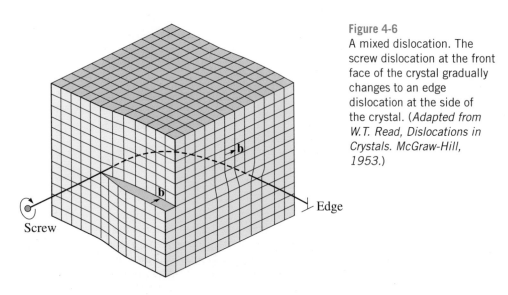

Figure 4-6
A mixed dislocation. The screw dislocation at the front face of the crystal gradually changes to an edge dislocation at the side of the crystal. (*Adapted from W.T. Read, Dislocations in Crystals. McGraw-Hill, 1953.*)

Stresses When discussing the motion of dislocations, we need to refer to the concept of stress, which will be covered in detail in Chapter 6. For now, it suffices to say that stress is force per unit area. Stress has units of N/m^2 known as the Pascal (Pa). A normal stress arises when the applied force acts perpendicular to the area of interest. A shear stress τ arises when the force acts in a direction parallel to the area of interest.

Dislocation Motion Consider the edge dislocation shown in Figure 4-7(a). A plane that contains both the dislocation line and the Burgers vector is known as a **slip plane**. When a sufficiently large shear stress acting parallel to the Burgers vector is applied to a crystal containing a dislocation, the dislocation can move through a process known as **slip**. The bonds across the slip plane between the atoms in the column to the right of the dislocation shown are broken. The atoms in the column to the right of the dislocation below the slip plane are shifted slightly so that they establish bonds with the atoms of the edge dislocation. In this way, the dislocation has shifted to the right [Figure 4-7(b)]. If this process continues, the dislocation moves through the crystal [Figure 4-7(c)] until it produces a step on the exterior of the crystal [Figure 4-7(d)] in the **slip direction** (which is parallel to the Burgers vector). (Note that the combination of a

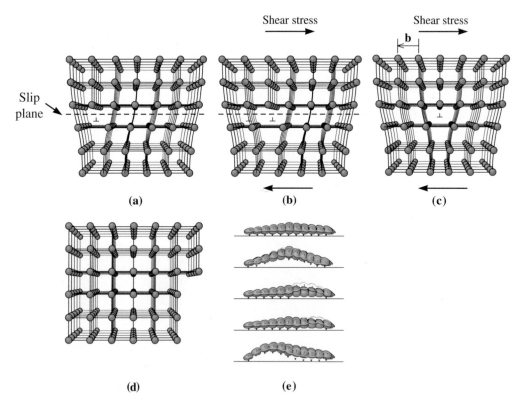

Figure 4-7 (a) When a shear stress is applied to the dislocation in (a), the atoms are displaced, (b) causing the dislocation to move one Burgers vector in the slip direction. (c) Continued movement of the dislocation eventually creates a step (d), and the crystal is deformed. (*Adapted from A.G. Guy, Essentials of Materials Science, McGraw-Hill, 1976.*) (e) The motion of a caterpillar is analogous to the motion of a dislocation.

slip plane and a slip direction comprises a **slip system**.) The top half of the crystal has been displaced by one Burgers vector relative to the bottom half; the crystal has been plastically (or permanently) deformed. This is the fundamental process that occurs many, many times as you bend a paper clip with your fingers. The plastic deformation of metals is primarily the result of the propagation of dislocations.

This process of progressively breaking and reforming bonds requires far less energy than the energy that would be required to instantaneously break all of the bonds across the slip plane. The crystal deforms via the propagation of dislocations because it is an energetically favorable process. Consider the motion by which a caterpillar moves [Figure 4-7(e)]. A caterpillar only lifts some of its legs at any given time rather than lifting all of its legs at one time in order to move forward. Why? Because lifting only some of its legs requires less energy; it is easier for the caterpillar to do. Another way to visualize this is to think about how you might move a large carpet that is positioned incorrectly in a room. If you want to reposition the carpet, instead of picking it up off the floor and moving it all at once, you might form a kink in the carpet and push the kink in the direction in which you want to move the carpet. The width of the kink is analogous to the Burgers vector. Again, you would move the carpet in this way because it requires less energy—it is easier to do.

Slip   Figure 4-8(a) is a schematic diagram of an edge dislocation that is subjected to a shear stress τ that acts parallel to the Burgers vector and perpendicular to the dislocation line. In this drawing, the edge dislocation is propagating in a direction opposite to the direction of propagation shown in Figure 4-7(a). A component of the shear stress must act parallel to the Burgers vector in order for the dislocation to move. The dislocation line moves in a direction parallel to the Burgers vector. Figure 4-8(b) shows a screw dislocation. For a screw dislocation, a component of the shear stress must act parallel to the Burgers vector (and thus the dislocation line) in order for the dislocation to move. The dislocation moves in a direction perpendicular to the Burgers vector, and the slip step that is produced is parallel to the Burgers vector. Since the Burgers vector of a screw dislocation is parallel to the dislocation line, specification of the Burgers vector and dislocation line does not define a slip plane for a screw dislocation.

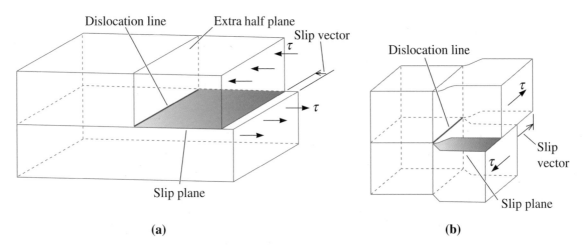

(a)   **(b)**

Figure 4-8   Schematic of the dislocation line, slip plane, and slip (Burgers) vector for (a) an edge dislocation and (b) a screw dislocation. (*Adapted from J.D. Verhoeven, Fundamentals of Physical Metallurgy, Wiley, 1975.*)

During slip, a dislocation moves from one set of surroundings to an identical set of surroundings. The **Peierls-Nabarro stress** (Equation 4-2) is required to move the dislocation from one equilibrium location to another,

$$\tau = c \exp\left(-kd/b\right) \tag{4-2}$$

where τ is the shear stress required to move the dislocation, d is the interplanar spacing between adjacent slip planes, b is the magnitude of the Burgers vector, and both c and k are constants for the material. The dislocation moves in a slip system that requires the least expenditure of energy. Several important factors determine the most likely slip systems that will be active:

1. The stress required to cause the dislocation to move increases exponentially with the length of the Burgers vector. Thus, the slip direction should have a small repeat distance or high linear density. The close-packed directions in metals and alloys satisfy this criterion and are the usual slip directions.

2. The stress required to cause the dislocation to move decreases exponentially with the interplanar spacing of the slip planes. Slip occurs most easily between planes of atoms that are smooth (so there are smaller "hills and valleys" on the surface) and between planes that are far apart (or have a relatively large interplanar spacing). Planes with a high planar density fulfill this requirement. Therefore, the slip planes are typically close-packed planes or those as closely packed as possible. Common slip systems in several materials are summarized in Table 4-1.

3. Dislocations do not move easily in materials such as silicon, which have covalent bonds. Because of the strength and directionality of the bonds, the materials typically fail in a brittle manner before the force becomes high enough to cause appreciable slip. Dislocations also play a relatively minor role in the deformation of polymers. Most polymers contain a substantial volume fraction of material that is amorphous and, therefore, does not contain dislocations. Permanent deformation in polymers primarily involves the stretching, rotation, and disentanglement of long chain molecules.

4. Materials with ionic bonding, including many ceramics such as MgO, also are resistant to slip. Movement of a dislocation disrupts the charge balance around the anions and cations, requiring that bonds between anions and cations be broken.

TABLE 4-1 ■ *Slip planes and directions in metallic structures*

Crystal Structure	Slip Plane	Slip Direction
BCC metals	{110}	⟨111⟩
	{112}	
	{123}	
FCC metals	{111}	⟨110⟩
HCP metals	{0001}	⟨100⟩
	{11$\bar{2}$0} ⎫	⟨110⟩
	{10$\bar{1}$0} ⎬ See Note	or ⟨10$\bar{2}$0⟩
	{10$\bar{1}$1} ⎭	
MgO, NaCl (ionic)	{110}	⟨110⟩
Silicon (covalent)	{111}	⟨110⟩

Note: These planes are active in some metals and alloys or at elevated temperatures.

During slip, ions with a like charge must also pass close together, causing repulsion. Finally, the repeat distance along the slip direction, or the Burgers vector, is larger than that in metals and alloys. Again, brittle failure of ceramic materials typically occurs due to the presence of flaws such as small pores before the applied level of stress is sufficient to cause dislocations to move. Ductility in ceramic materials can be obtained by

(a) phase transformations (known as transformation plasticity, an example is fully stabilized zirconia);

(b) mechanical twinning;

(c) dislocation motion; and

(d) grain boundary sliding.

We will discuss some of these concepts later in this chapter. Typically, higher temperatures and compressive stresses lead to higher ductility. Recently, it has been shown that certain ceramics such as strontium titanate ($SrTiO_3$) can exhibit considerable ductility. Under certain conditions, ceramics can exhibit very large deformations. This is known as superplastic behavior.

Example 4-6   *Dislocations in Ceramic Materials*

A sketch of a dislocation in magnesium oxide (MgO), which has the sodium chloride crystal structure and a lattice parameter of 0.396 nm, is shown in Figure 4-9. Determine the length of the Burgers vector.

SOLUTION

In Figure 4-9, we begin a clockwise loop around the dislocation at point x and then move equal atom spacings in the two horizontal directions and equal atom spacings in the two vertical directions to finish at point y. Note that it is necessary that the lengths of the two horizontal segments of the loop be equal and the lengths of the vertical segments be equal, but it is not necessary that the horizontal and vertical segments be equal in length to each other. The chosen loop must close in a perfect

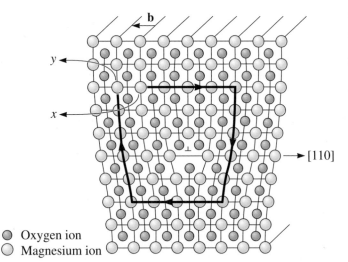

Figure 4-9
An edge dislocation in MgO showing the slip direction and Burgers vector (for Example 4-6). (*Adapted from W.D. Kingery, H.K. Bowen, and D.R. Uhlmann, Introduction to Ceramics, John Wiley, 1976.*)

● Oxygen ion
○ Magnesium ion

crystal. The vector **b** is the Burgers vector. Because **b** is parallel a [110] direction, it must be perpendicular to (110) planes. The length of **b** is the distance between two adjacent (110) planes. From Equation 3-7,

$$d_{110} = \frac{a_0}{\sqrt{h^2 + k^2 + l^2}} = \frac{0.396}{\sqrt{1^2 + 1^2 + 0^2}} = 0.280 \text{ nm}$$

The Burgers vector is a (110) direction that is 0.280 nm in length. Note, however, that two extra half planes of atoms make up the dislocation—one composed of oxygen ions and one of magnesium ions (Figure 4-9). This formula for calculating the magnitude of the Burgers vector will not work for non-cubic systems. It is better to consider the magnitude of the Burgers vector as equal to the repeat distance in the slip direction.

Example 4-7   *Burgers Vector Calculation*

Calculate the length of the Burgers vector in copper.

SOLUTION

Copper has an FCC crystal structure. The lattice parameter of copper (Cu) is 0.36151 nm. The close-packed directions, or the directions of the Burgers vector, are of the form ⟨110⟩. The repeat distance along the ⟨110⟩ directions is one-half the face diagonal, since lattice points are located at corners and centers of faces [Figure 4-10(a)].

$$\text{Face diagonal} = \sqrt{2}a_0 = (\sqrt{2})(0.36151) = 0.51125 \text{ nm}$$

The length of the Burgers vector, or the repeat distance, is

$$\mathbf{b} = \frac{1}{2}(0.51125)\text{nm} = 0.25563 \text{ nm}$$

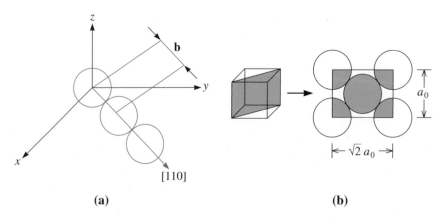

(a)   **(b)**

Figure 4-10   (a) Burgers vector for FCC copper. (b) The atom locations on a (110) plane in a BCC unit cell (for Examples 4-7 and 4-8, respectively).

Example 4-8   *Identification of Preferred Slip Planes*

The planar density of the (112) plane in BCC iron is 9.94×10^{14} atoms/cm^2. Calculate (a) the planar density of the (110) plane and (b) the interplanar spacings for both the (112) and (110) planes. On which plane would slip normally occur?

SOLUTION

The lattice parameter of BCC iron is 0.2866 nm or 2.866×10^{-8} cm. The (110) plane is shown in Figure 4-10(b), with the portion of the atoms lying within the unit cell being shaded. Note that one-fourth of the four corner atoms plus the center atom lie within an area of a_0 times $\sqrt{2}a_0$.

(a) The planar density is

$$\text{Planar density (110)} = \frac{\text{atoms}}{\text{area}} = \frac{2}{(\sqrt{2})(2.866 \times 10^{-8}\text{cm})^2}$$

$$= 1.72 \times 10^{15} \text{ atoms/cm}^2$$

$$\text{Planar density (112)} = 0.994 \times 10^{15} \text{ atoms/cm}^2 \text{ (from problem statement)}$$

(b) The interplanar spacings are

$$d_{110} = \frac{2.866 \times 10^{-8}}{\sqrt{1^2 + 1^2 + 0}} = 2.0266 \times 10^{-8} \text{ cm}$$

$$d_{112} = \frac{2.866 \times 10^{-8}}{\sqrt{1^2 + 1^2 + 2^2}} = 1.17 \times 10^{-8} \text{ cm}$$

The planar density is higher and the interplanar spacing is larger for the (110) plane than for the (112) plane; therefore, the (110) plane is the preferred slip plane.

When a metallic material is "etched" (a chemical reaction treatment that involves exposure to an acid or a base), the areas where dislocations intersect the surface of the crystal react more readily than the surrounding material. These regions appear in the microstructure as **etch pits**. Figure 4-11 shows the etch pit distribution on a surface of a silicon carbide (SiC) crystal. A **transmission electron microscope** (TEM) is used to observe dislocations. In a typical TEM image, dislocations appear as dark lines at very high magnifications as shown in Figure 4-12(a).

When thousands of dislocations move on the same plane, they produce a large step at the crystal surface. This is known as a **slip line** [Figure 4-12(b)]. A group of slip lines is known as a **slip band**.

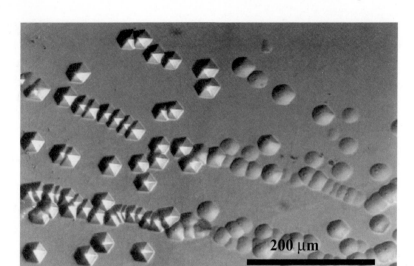

Figure 4-11   Optical image of etch pits in silicon carbide (SiC). The etch pits correspond to intersection points of pure edge dislocations with Burgers vector $\frac{a}{3} \langle 1\bar{1}20 \rangle$ and the dislocation line direction along [0001] (perpendicular to the etched surface). Lines of etch pits represent low angle grain boundaries. (*Courtesy of Dr. Marek Skowronski, Carnegie Mellon University.*)

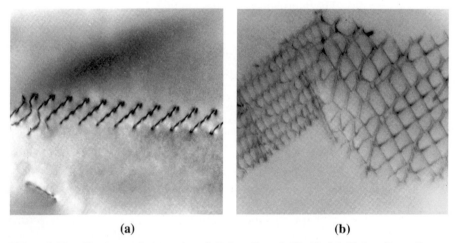

(a)   (b)

Figure 4-12   Electron micrographs of dislocations in Ti_3Al: (a) Dislocation pileups (×36,500). (b) Micrograph at ×100 showing slip lines in Al. (*Reprinted courtesy of Don Askeland.*)

4-4   Significance of Dislocations

Dislocations are most significant in metals and alloys since they provide a mechanism for plastic deformation, which is the cumulative effect of slip of numerous dislocations. **Plastic deformation** refers to irreversible deformation or change in shape that occurs when the force or stress that caused it is removed. This is because the applied stress causes dislocation motion that in turn causes permanent deformation. There are, however, other mechanisms that cause permanent deformation. We will examine these in later chapters. Plastic deformation is to be distinguished from **elastic deformation**, which is a temporary change in shape that occurs while a force or stress remains applied to a material. In elastic deformation, the shape change is a result of stretching of interatomic bonds, and no dislocation motion occurs. Slip can occur in some ceramics and polymers; however, other factors (e.g., porosity in ceramics, entanglement of chains in polymers, etc.) dominate the near room temperature mechanical behavior of polymers and ceramics. Amorphous materials such as silicate glasses do not have a periodic arrangement of ions and hence do not contain dislocations. *The slip process, therefore, is particularly important in understanding the mechanical behavior of metals.* First, slip explains why the strength of metals is much lower than the value predicted from the metallic bond. If slip occurs, only a tiny fraction of all of the metallic bonds across the interface need to be broken at any one time, and the force required to deform the metal is small. It can be shown that the actual strength of metals is 10^3 to 10^4 times *lower* than that expected from the strength of metallic bonds.

Second, slip provides ductility in metals. If no dislocations were present, an iron bar would be brittle and the metal could not be shaped by metalworking processes, such as forging, into useful shapes.

Third, we control the mechanical properties of a metal or alloy by interfering with the movement of dislocations. An obstacle introduced into the crystal prevents a dislocation from slipping unless we apply higher forces. Thus, the presence of dislocations helps strengthen metallic materials.

Enormous numbers of dislocations are found in materials. The **dislocation density**, or total length of dislocations per unit volume, is usually used to represent the amount of dislocations present. Dislocation densities of 10^6 cm/cm^3 are typical of the softest metals, while densities up to 10^{12} cm/cm^3 can be achieved by deforming the material.

Dislocations also influence electronic and optical properties of materials. For example, the resistance of pure copper increases with increasing dislocation density. We mentioned previously that the resistivity of pure copper also depends strongly on small levels of impurities. Similarly, we prefer to use silicon crystals that are essentially dislocation free since this allows the charge carriers such as electrons to move more freely through the solid. Normally, the presence of dislocations has a deleterious effect on the performance of photo detectors, light emitting diodes, lasers, and solar cells. These devices are often made from compound semiconductors such as gallium arsenide-aluminum arsenide (GaAs-AlAs), and dislocations in these materials can originate from concentration inequalities in the melt from which crystals are grown or stresses induced because of thermal gradients that the crystals are exposed to during cooling from the growth temperature.

4-5   Schmid's Law

We can understand the differences in behavior of metals that have different crystal structures by examining the force required to initiate the slip process. Suppose we apply a unidirectional force F to a cylinder of metal that is a single crystal (Figure 4-13). We can orient the slip plane and slip direction to the applied force by defining the angles λ and ϕ. The angle between the slip direction and the applied force is λ, and ϕ is the angle between the normal to the slip plane and the applied force. Note that the sum of angles ϕ and λ can be, but does not have to be, 90°.

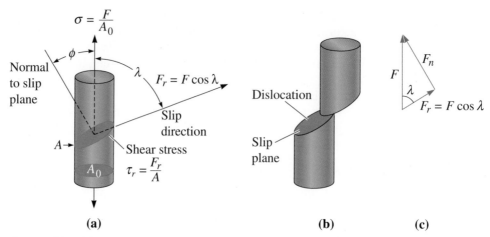

Figure 4-13 (a) A resolved shear stress τ is produced on a slip system. [Note: $(\phi + \lambda)$ does not have to equal 90°.] (b) Movement of dislocations on the slip system deforms the material. (c) Resolving the force.

In order for the dislocation to move in its slip system, a shear force acting in the slip direction must be produced by the applied force. This resolved shear force F_r is given by

$$F_r = F \cos \lambda$$

If we divide the equation by the area of the slip plane, $A = A_0/\cos\phi$, we obtain the following equation known as **Schmid's law**:

$$\tau_r = \sigma \cos \phi \cos \lambda \qquad (4\text{-}3)$$

where

$$\tau_r = \frac{F_r}{A} = \text{resolved shear } \textit{stress} \text{ in the slip direction}$$

and

$$\sigma = \frac{F}{A_0} = \text{normal } \textit{stress} \text{ applied to the cylinder}$$

Example 4-9   *Calculation of Resolved Shear Stress*

Apply Schmid's law for a situation in which the single crystal is oriented so that the slip plane is perpendicular to the applied tensile stress.

SOLUTION

Suppose the slip plane is perpendicular to the applied stress σ, as in Figure 4-14. Then, $\phi = 0$, $\lambda = 90°$, $\cos \lambda = 0$, and therefore $\tau_r = 0$. As noted before, the angles ϕ and λ can but do not always sum to 90°. Even if the applied stress σ is enormous, no resolved shear stress develops along the slip direction and the dislocation cannot move. (You could perform a simple experiment to demonstrate this with a deck of cards. If you push on the deck at an angle, the cards slide over one another, as in the slip process. If you push perpendicular to the deck, however, the cards do not slide.) Slip cannot occur if the slip system is oriented so that either λ or ϕ is 90°.

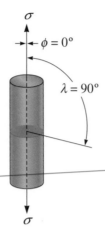

Figure 4-14
When the slip plane is perpendicular to the applied stress σ, the angle λ is 90°, and no shear stress is resolved.

The **critical resolved shear stress** τ_{crss} is the shear stress required for slip to occur. Thus slip occurs, causing the metal to plastically deform, when the *applied* stress (σ) produces a *resolved* shear stress (τ_r) that equals the critical resolved shear stress:

$$\tau_r = \tau_{crss} \tag{4-4}$$

Example 4-10 *Design of a Single Crystal Casting Process*

We wish to produce a rod composed of a single crystal of pure aluminum, which has a critical resolved shear stress of 1.02 MPa. We would like to orient the rod in such a manner that, when an axial stress of 3.45 MPa is applied, the rod deforms by slip in a 45° direction to the axis of the rod and actuates a sensor that detects the overload. Design the rod and a method by which it might be produced.

SOLUTION

Dislocations begin to move when the resolved shear stress τ_r equals the critical resolved shear stress, 1.02 MPa. From Schmid's law:

$$\tau_r = \sigma \cos \lambda \cos \phi \text{ or}$$

$$1.02 \text{ MPa} = (3.45 \text{ MPa}) \cos \lambda \cos \phi$$

Because we wish slip to occur at a 45° angle to the axis of the rod, $\lambda = 45°$, and

$$\cos \phi = \frac{1.02 \text{ MPa}}{3.45 \text{ MPa}} = 0.4186$$

$$\phi = 65.3°$$

Therefore, we must produce a rod that is oriented such that $\lambda = 45°$ and $\phi = 65.3°$. Note that ϕ and λ do not add to 90°.

We might do this by a solidification process. We could orient a seed crystal of solid aluminum at the bottom of a mold. Liquid aluminum could be introduced into the mold. The liquid solidifies at the seed crystal, and a single crystal rod of the proper orientation is produced.

4-6   Influence of Crystal Structure

We can use Schmid's law to compare the properties of metals having the BCC, FCC, and HCP crystal structures. Table 4-2 lists three important factors that we can examine. We must be careful to note, however, that this discussion describes the behavior of nearly perfect single crystals. Real engineering materials are seldom single crystals and always contain large numbers of defects. Also they tend to be polycrystalline. Since different crystals or grains are oriented in different random directions, we cannot apply Schmid's law to predict the mechanical behavior of polycrystalline materials.

Critical Resolved Shear Stress   If the critical resolved shear stress in a metal is very high, the applied stress σ must also be high in order for τ_r to equal τ_{crss}. A higher τ_{crss} implies a higher stress is necessary to plastically deform a metal, which in turn indicates the metal has a high strength! In FCC metals, which have close-packed {111} planes, the critical resolved shear stress is low—about 0.34 to 0.69 MPa in a perfect crystal. On the other hand, BCC crystal structures contain no close-packed planes, and we must exceed a higher critical resolved shear stress—on the order of 69 MPa in perfect crystals—before slip occurs. Thus, BCC metals tend to have high strengths and lower ductilities compared to FCC metals.

We would expect the HCP metals, because they contain close-packed basal planes, to have low critical resolved shear stresses. In fact, in HCP metals such as zinc that have a c/a ratio greater than or equal to the theoretical ratio of 1.633, the critical resolved shear stress is less than 0.69 MPa, just as in FCC metals. In HCP titanium, however, the c/a ratio is less than 1.633; the close-packed planes are spaced too closely together. Slip now occurs on planes such as $(10\bar{1}0)$, the "prism" planes or faces of the hexagon, and the critical resolved shear stress is then as great as or greater than in BCC metals.

Number of Slip Systems   If at least one slip system is oriented to give the angles λ and ϕ near 45°, then τ_r equals τ_{crss} at low applied stresses. Ideal HCP metals possess only one set of parallel close-packed planes, the (0001) planes, and three close-packed directions, giving three slip systems. Consequently, the probability of the close-packed planes and directions being oriented with λ and ϕ near 45° is very low. The HCP crystal may fail in a brittle manner without a significant amount of slip; however, in HCP metals with a low c/a ratio, or when HCP metals are properly alloyed, or when the temperature is increased, other slip systems become active, making these metals less brittle than expected.

TABLE 4-2 ■ *Summary of factors affecting slip in metallic structures*

Factor	FCC	BCC	HCP$\left(\dfrac{c}{a} \geq 1.633\right)$		
Critical resolved shear stress (MPa)	0.34–0.69	34–69	0.34–0.69[a]		
Number of slip systems	12	48	3[b]		
Cross-slip	Can occur	Can occur	Cannot occur[b]		
Summary of properties	Ductile	Strong	Relatively brittle		

[a]For slip on basal planes.
[b]By alloying or heating to elevated temperatures, additional slip systems are active in HCP metals, permitting cross-slip to occur and thereby improving ductility.

On the other hand, FCC metals contain four nonparallel close-packed planes of the form {111} and three close-packed directions of the form ⟨110⟩ within each plane, giving a total of twelve slip systems. At least one slip system is favorably oriented for slip to occur at low applied stresses, permitting FCC metals to have high ductilities.

Finally, BCC metals have as many as 48 slip systems that are nearly close-packed. Several slip systems are always properly oriented for slip to occur, allowing BCC metals to have ductility.

Cross-Slip

Consider a screw dislocation moving on one slip plane that encounters an obstacle and is blocked from further movement. This dislocation can shift to a second intersecting slip system, also properly oriented, and continue to move. This is called **cross-slip**. In many HCP metals, no cross-slip can occur because the slip planes are parallel (i.e., not intersecting). Therefore, polycrystalline HCP metals tend to be brittle. Fortunately, additional slip systems become active when HCP metals are alloyed or heated, thus improving ductility. Cross-slip is possible in both FCC and BCC metals because a number of intersecting slip systems are present. Consequently, cross-slip helps maintain ductility in these metals.

Example 4-11	*Ductility of HCP Metal Single Crystals and Polycrystalline Materials*

A single crystal of magnesium (Mg), which has the HCP crystal structure, can be stretched into a ribbon-like shape four to six times its original length; however, *polycrystalline* Mg and other metals with the HCP structure show limited ductilities. Use the values of critical resolved shear stress for metals with different crystal structures and the nature of deformation in polycrystalline materials to explain this observation.

SOLUTION

From Table 4-2, we note that for HCP metals such as Mg, the critical resolved shear stress is low (350–700 kPa). We also note that slip in HCP metals will occur readily on the basal plane—the primary slip plane. When a single crystal is deformed, assuming the basal plane is suitably oriented with respect to the applied stress, a large deformation can occur. This explains why single crystal Mg can be stretched into a ribbon four to six times the original length. When we have polycrystalline Mg, the deformation is not as simple. Each crystal must deform such that the strain developed in any one crystal is accommodated by its neighbors.

In HCP metals, there are no intersecting slip systems; thus, dislocations cannot glide from one slip plane in one crystal (grain) onto another slip plane in a neighboring crystal. As a result, polycrystalline HCP metals such as Mg show limited ductility.

4-7   Surface Defects

Surface defects are the boundaries, or planes, that separate a material into regions. For example, each region may have the same crystal structure but different orientations.

Material Surface

The exterior dimensions of the material represent surfaces at which the crystal abruptly ends. Each atom at the surface no longer has the proper coordination number, and atomic bonding is disrupted. The exterior surface may also be very rough, may contain tiny notches, and may be much more reactive than the bulk of the material.

In nanostructured materials, the ratio of the number of atoms or ions at the surface to that in the bulk is very high. As a result, these materials have a large surface area per unit mass. In petroleum refining and many other areas of technology, we make use of high surface area catalysts for enhancing the kinetics of chemical reactions. Similar to nanoscale materials, the surface area-to-volume ratio is high for porous materials, gels, and ultrafine powders. You will learn later that reduction in surface area is the thermodynamic driving force for **sintering** of ceramics and metal powders.

Grain Boundaries

The microstructure of many engineered ceramic and metallic materials consists of many grains. A **grain** is a portion of the material within which the arrangement of the atoms is nearly identical; however, the orientation of the atom arrangement, or crystal structure, is different for each adjoining grain. Three grains are shown schematically in Figure 4-15(a); the arrangement of atoms in each grain is identical but the grains are oriented differently. A **grain boundary**, the surface that separates the individual grains, is a narrow zone in which the atoms are not properly spaced. That is to say, the atoms are so close together at some locations in the grain boundary that they cause a region of compression, and in other areas they are so far apart that they cause a region of tension. Figure 4-15(b) shows a micrograph of grains in a stainless steel sample.

One method of controlling the properties of a material is by controlling the grain size. By reducing the grain size, we increase the number of grains and, hence, increase the amount of grain boundary area. Any dislocation moves only a short distance before encountering a grain boundary, and the strength of the metallic material is increased. The **Hall-Petch equation** relates the grain size to the **yield strength**,

$$\sigma_y = \sigma_0 + Kd^{-1/2} \tag{4-5}$$

where σ_y is the yield strength (the level of stress necessary to cause a certain amount of permanent deformation), d is the average diameter of the grains, and σ_0 and K are constants for the metal. Recall from Chapter 1 that the yield strength of a metallic material is the minimum level of stress that is needed to initiate plastic (permanent) deformation. Figure 4-16 shows

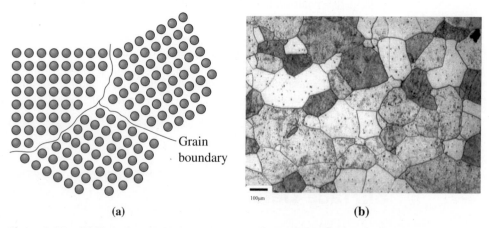

Grain boundary

(a)   (b)

Figure 4-15   (a) The atoms near the boundaries of the three grains do not have an equilibrium spacing or arrangement. (b) Grains and grain boundaries in a stainless steel sample. (*Micrograph courtesy of Dr. A. J. Deardo.*)

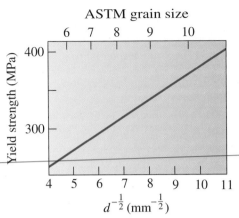

ASTM grain size

Figure 4-16
The effect of grain size on the yield strength of
steel at room temperature.

this relationship in steel. The Hall-Petch equation is not valid for materials with unusually large or ultrafine grains. In the chapters that follow, we will describe how the grain size of metals and alloys can be controlled through solidification, alloying, and heat treatment.

Example 4-12   *Design of a Mild Steel*

The yield strength of mild steel with an average grain size of 0.05 mm is 138 MPa. The yield stress of the same steel with a grain size of 0.007 mm is 276 MPa. What will be the average grain size of the same steel with a yield stress of 207 MPa? Assume the Hall-Petch equation is valid and that changes in the observed yield stress are due to changes in grain size.

SOLUTION

$$\sigma_y = \sigma_0 + Kd^{-1/2}$$

Thus, for a grain size of 0.05 mm, the yield stress is 138 MPa
    Using the Hall-Petch equation

$$138 = \sigma_0 + \frac{K}{\sqrt{0.05}}$$

For the grain size of 0.007 mm, the yield stress is 276 MPa. Therefore, again using the Hall-Petch equation:

$$276 = \sigma_0 + \frac{K}{\sqrt{0.007}}$$

Solving these two equations, $K = 18.44$ MPa $\cdot$ mm$^{1/2}$, and $\sigma_0 = 55.5$ MPa. Now we have the Hall-Petch equation as

$$\sigma_y = 55.5 + 18.44\, d^{-1/2}$$

If we want a yield stress of 207 MPa, the grain size should be 0.0148 mm.

Optical microscopy is one technique that is used to reveal microstructural features such as grain boundaries that require less than about 2000 magnification. The process of preparing a metallic sample and observing or recording its microstructure is called **metallography**. A sample of the material is sanded and polished to a mirror-like finish. The surface is then exposed to chemical attack, or *etching*, with grain boundaries being

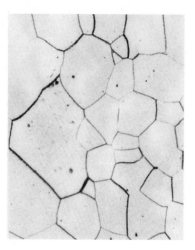

Figure 4-17
Microstructure of palladium ($\times$ 100). (*From ASM Handbook, Vol. 9, Metallography and Microstructure (1985), ASM International, Materials Park, OH 44073.*)

attacked more aggressively than the remainder of the grain. Light from an optical microscope is reflected or scattered from the sample surface, depending on how the surface is etched. When more light is scattered from deeply etched features such as the grain boundaries, these features appear dark (Figure 4-17). In ceramic samples, a technique known as **thermal grooving** is often used to observe grain boundaries. It involves polishing and heating a ceramic sample to temperatures below the sintering temperature (1300°C) for a short time. Sintering is a process for forming a dense mass by heating compacted powdered material.

One manner by which grain size is specified is the **ASTM grain size number** (**ASTM** is the American Society for Testing and Materials). The number of grains per square inch is determined from a photograph of the metal taken at a magnification of 100. The ASTM grain size number n is calculated as

$$N = 2^{n-1} \tag{4-6}$$

where N is the number of grains per square inch.

A large ASTM number indicates many grains, or a fine grain size, and correlates with high strengths for metals.

When describing a microstructure, whenever possible, it is preferable to use a micrometer marker or some other scale on the micrograph, instead of stating the magnification as $\times$, as in Figure 4-17. That way, if the micrograph is enlarged or reduced, the micrometer marker scales with it, and we do not have to worry about changes in the magnification of the original micrograph. A number of sophisticated **image analysis** programs are available. Using such programs, it is possible not only to obtain information on the ASTM grain size number but also quantitative information on average grain size, grain size distribution, porosity, second phases (Chapter 10), etc. A number of optical and scanning electron microscopes can be purchased with image analysis capabilities. The following example illustrates the calculation of the ASTM grain size number.

Example 4-13   *Calculation of ASTM Grain Size Number*

Suppose we count 16 grains per square inch in a photomicrograph taken at a magnification of 250. What is the ASTM grain size number?

SOLUTION

Consider one square inch from the photomicrograph taken at a magnification of 250. At a magnification of 100, the one square inch region from the 250 magnification image would appear as

$$1 \text{ in}^2 \left(\frac{100}{250} \right)^2 = 0.16 \text{ in}^2$$

and we would see

$$\frac{16 \text{ grains}}{0.16 \text{ in}^2} = 100 \text{ grains/in}^2$$

Substituting in Equation 4-6,

$$N = 100 \text{ grains/in}^2 = 2^{n-1}$$
$$\log 100 = (n - 1) \log 2$$
$$2 = (n - 1)(0.301)$$
$$n = 7.64$$

The approach discussed above to compute the ASTM grain size number is based on English units. In countries that use the metric system of units, alternate formulae have been developed to calculate the grain size number:

$$N_m = 8 \, (2^n) \tag{4-7}$$

where N_m is the number of grains per mm^2 at magnification $\times$ 1.

The value of n calculated from Equation 4-7 is slightly greater than that calculated using Equation 4-6, but the difference is negligible. The Swedish, Italian, Russian, French, and ISO standards use Equation 4-7. The German standard is also based on metric units but uses a different formulation:

$$n = 3.7 + 3.33 \log (N_{mG}) \tag{4-8}$$

where N_{mG} is the number of grains per cm^2 at magnification $\times$ 100.

Small Angle Grain Boundaries

A **small angle grain boundary** is an array of dislocations that produces a small misorientation between the adjoining crystals (Figure 4-18). Because the energy of the surface is less than that of a regular grain

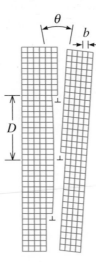

Figure 4-18
The small angle grain boundary is produced by an array of dislocations, causing an angular mismatch θ between the lattices on either side of the boundary.

boundary, the small angle grain boundaries are not as effective in blocking slip. Small angle boundaries formed by edge dislocations are called **tilt boundaries**, and those caused by screw dislocations are called **twist boundaries**.

Stacking Faults

Stacking faults, which occur in FCC metals, represent an error in the stacking sequence of close-packed planes. Normally, a stacking sequence of *ABC ABC ABC* is produced in a perfect FCC crystal. Suppose instead the following sequence is produced:

$$ABC \ \underset{\vee}{ABAB} \ CABC$$

In the portion of the sequence indicated, a type A plane replaces a type C plane. This small region, which has the HCP stacking sequence instead of the FCC stacking sequence, represents a stacking fault. Stacking faults interfere with the slip process.

Twin Boundaries

A **twin boundary** is a plane across which there is a special mirror image misorientation of the crystal structure (Figure 4-19). Twins can be produced

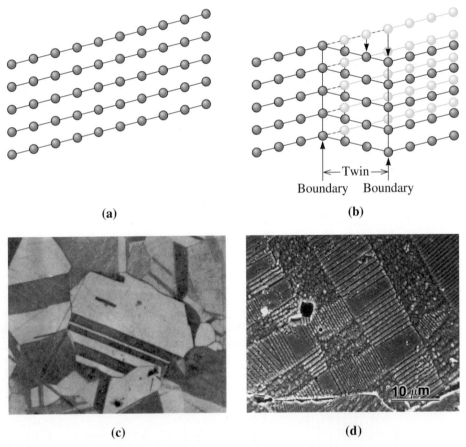

(a)

(b)

(c)

(d)

Figure 4-19   Application of stress to the (a) perfect crystal may cause a displacement of the atoms, (b) resulting in the formation of a twin. Note that the crystal has deformed as a result of twinning. (c) A micrograph of twins within a grain of brass (× 250). (d) Domains in ferroelectric barium titanate. *(Courtesy of Dr. Rodney Roseman, University of Cincinnati.)* Similar domain structures occur in magnetic materials.

TABLE 4-3 ■ *Energies of surface imperfections in selected metals*

Surface Imperfection (ergs/cm²)	Al	Cu	Pt	Fe
Stacking fault	200	75	95	—
Twin boundary	120	45	195	190
Grain boundary	625	645	1000	780

when a shear force, acting along the twin boundary, causes the atoms to shift out of position. Twinning occurs during deformation or heat treatment of certain metals. The twin boundaries interfere with the slip process and increase the strength of the metal. Twinning also occurs in some ceramic materials such as monoclinic zirconia and dicalcium silicate.

The effectiveness of the surface defects in interfering with the slip process can be judged from the surface energies (Table 4-3). The high-energy grain boundaries are much more effective in blocking dislocations than either stacking faults or twin boundaries.

Domain Boundaries   **Ferroelectrics** are materials that develop spontaneous and reversible dielectric polarization (e.g., PZT or BaTiO$_3$) (see Chapter 19). Magnetic materials also develop a magnetization in a similar fashion (e.g., Fe, Co, Ni, iron oxide, etc.) (see Chapter 20). These electronic and magnetic materials contain domains. A **domain** is a small region of the material in which the direction of magnetization or dielectric polarization remains the same. In these materials, many small domains form so as to minimize the total free energy of the material. Figure 4-19(d) shows an example of domains in tetragonal ferroelectric barium titanate. The presence of domains influences the dielectric and magnetic properties of many electronic and magnetic materials. We will discuss these materials in later chapters.

4-8   Importance of Defects

Extended and point defects play a major role in influencing mechanical, electrical, optical, and magnetic properties of engineered materials. In this section, we recapitulate the importance of defects on properties of materials. We emphasize that the effect of dislocations is most important in metallic materials.

Effect on Mechanical Properties via Control of the Slip Process   Any imperfection in the crystal raises the internal energy at the location of the imperfection. The local energy is increased because, near the imperfection, the atoms either are squeezed too closely together (compression) or are forced too far apart (tension).

A dislocation in an otherwise perfect metallic crystal can move easily through the crystal if the resolved shear stress equals the critical resolved shear stress. If the dislocation encounters a region where the atoms are displaced from their usual positions, however, a higher stress is required to force the dislocation past the region of high local energy; thus, the material is stronger. *Defects in materials, such as dislocations, point defects, and grain boundaries, serve as "stop signs" for dislocations.* They provide resistance to dislocation motion, and any mechanism that impedes dislocation motion makes a metal stronger. Thus, we can control the strength of a metallic material by controlling the number and type of imperfections. Three common strengthening mechanisms are based on the three categories of defects in crystals. Since dislocation motion is relatively easier in metals and alloys, these mechanisms typically work best for metallic materials. We need to keep in mind that very often the strength of ceramics in tension and at low temperatures is dictated by the level of porosity (presence of small holes). Polymers are often amorphous and hence dislocations play very little role in their mechanical behavior, as discussed in a later chapter. The strength of inorganic glasses (e.g., silicate float glass) depends on the distribution of flaws on the surface.

Strain Hardening

Dislocations disrupt the perfection of the crystal structure. In Figure 4-20, the atoms below the dislocation line at point *B* are compressed, while the atoms above dislocation *B* are too far apart. If dislocation *A* moves to the right and passes near dislocation *B*, dislocation *A* encounters a region where the atoms are not properly arranged. Higher stresses are required to keep the second dislocation moving; consequently, the metal must be stronger. Increasing the number of dislocations further increases the strength of the material since increasing the dislocation density causes more stop signs for dislocation motion. The dislocation density can be shown to increase markedly as we strain or deform a material. This mechanism of increasing the strength of a material by deformation is known as **strain hardening**, which is discussed in Chapter 8. We can also show that dislocation densities can be reduced substantially by heating a metallic material to a relatively high temperature (below the melting temperature) and holding it there for a long period of time. This heat treatment is known as **annealing** and is used to impart ductility to metallic materials. Thus, controlling the dislocation density is an important way of controlling the strength and ductility of metals and alloys.

Solid-Solution Strengthening

Any of the point defects also disrupt the perfection of the crystal structure. A solid solution is formed when atoms or ions of a guest element or compound are assimilated completely into the crystal structure of the host material. This is similar to the way salt or sugar in small concentrations dissolve in water. If dislocation *A* moves to the left (Figure 4-20), it encounters a disturbed crystal caused by the point defect; higher stresses are needed to continue slip of the dislocation. By intentionally introducing substitutional or interstitial atoms, we cause *solid-solution strengthening*, which is discussed in Chapter 10. This mechanism explains why plain carbon steel is stronger than pure Fe and why alloys of copper containing small concentrations of Be are much stronger than pure Cu. Pure gold or silver, both FCC metals with many active slip systems, are mechanically too soft.

Grain-Size Strengthening

Surface imperfections such as grain boundaries disturb the arrangement of atoms in crystalline materials. If dislocation *B* moves to the right (Figure 4-20), it encounters a grain boundary and is blocked. By increasing the number of grains or reducing the grain size, **grain-size strengthening** is achieved in metallic materials. Control of grain size will be discussed in a number of later chapters.

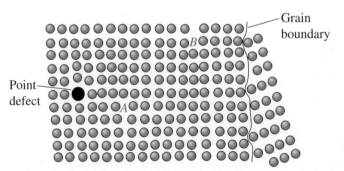

Figure 4-20
If the dislocation at point *A* moves to the left, it is blocked by the point defect. If the dislocation moves to the right, it interacts with the disturbed lattice near the second dislocation at point *B*. If the dislocation moves farther to the right, it is blocked by a grain boundary.

There are two more mechanisms for strengthening of metals and alloys. These are known as **second-phase strengthening** and **precipitation strengthening**. We will discuss these in later chapters.

Example 4-14   *Design/Materials Selection for a Stable Structure*

We would like to produce a bracket to hold ceramic bricks in place in a heat-treating furnace. The bracket should be strong, should possess some ductility so that it bends rather than fractures if overloaded, and should maintain most of its strength up to 600°C. Design the material for this bracket, considering the various crystal imperfections as the strengthening mechanism.

SOLUTION

In order to serve up to 600°C, the bracket should *not* be produced from a polymer. Instead, a metal or ceramic should be considered.

In order to have some ductility, dislocations must move and cause slip. Because slip in ceramics is difficult, the bracket should be produced from a metallic material. The metal should have a melting point well above 600°C; aluminum, with a melting point of 660°C, would not be suitable; iron, however, would be a reasonable choice.

We can introduce point, line, and surface imperfections into the iron to help produce strength, but we wish the imperfections to be stable as the service temperature increases. As shown in Chapter 5, grains can grow at elevated temperatures, reducing the number of grain boundaries and causing a decrease in strength. As indicated in Chapter 8, dislocations may be annihilated at elevated temperatures—again, reducing strength. The number of vacancies depends on temperature, so controlling these crystal defects may not produce stable properties.

The number of interstitial or substitutional atoms in the crystal does not, however, change with temperature. We might add carbon to the iron as interstitial atoms or substitute vanadium atoms for iron atoms at normal lattice points. These point defects continue to interfere with dislocation movement and help to keep the strength stable.

Of course, other design requirements may be important as well. For example, the steel bracket may deteriorate by oxidation or may react with the ceramic brick.

Effects on Electrical, Optical, and Magnetic Properties

In previous sections, we stated that the effect of point defects on the electrical properties of semiconductors is dramatic. The entire microelectronics industry critically depends upon the successful incorporation of substitutional dopants such as P, As, B, and Al in Si and other semiconductors. These dopant atoms allow us to have significant control of the electrical properties of semiconductors. Devices made from Si, GaAs, amorphous silicon (a:Si:H), etc., critically depend on the presence of dopant atoms. We can make *n*-type Si by introducing P atoms in Si. We can make *p*-type Si using B atoms. Similarly, a number of otherwise unsatisfied bonds in amorphous silicon are completed by incorporating H atoms.

The effect of defects such as dislocations on the properties of semiconductors is usually deleterious. Dislocations and other defects (including other point defects) can interfere with motion of charge carriers in semiconductors. This is why we make sure that the dislocation densities in single crystal silicon and other materials used in optical and electrical applications are very small. Point defects also cause increased resistivity in metals.

In some cases, defects can enhance certain properties. For example, incorporation of CaO in ZrO_2 causes an increase in the concentration of oxygen ion vacancies. This has a beneficial effect on the conductivity of zirconia and allows us to use such compositions for oxygen gas sensors and solid oxide fuel cells. Defects can convert many otherwise insulating dielectric materials into useful semiconductors! These are then used for many sensor applications (e.g., temperature, humidity, and gas sensors, etc.).

Addition of about 1% chromium oxide in alumina creates defects that make alumina ruby red. Similarly, incorporation of Fe^{+2} and Ti^{+4} makes the blue sapphire. Nanocrystals of materials such as cadmium sulfide (CdS) in inorganic glasses produce glasses that have a brilliant color. Nanocrystals of silver halide and other crystals also allow formation of photochromic and photosensitive glasses.

Many magnetic materials can be processed such that grain boundaries and other defects make it harder to reverse the magnetization in these materials. The magnetic properties of many commercial ferrites, used in magnets for loudspeakers and devices in wireless communication networks, depend critically on the distribution of different ions on different crystallographic sites in the crystal structure. As mentioned before, the presence of domains affects the properties of ferroelectric, ferromagnetic, and ferrimagnetic materials (Chapters 19 and 20).

Summary

- Imperfections, or defects, in a crystalline material are of three general types: point defects, line defects or dislocations, and surface defects.

- The number of vacancies depends on the temperature of the material; interstitial atoms (located at interstitial sites between the normal atoms) and substitutional atoms (which replace the host atoms at lattice points) are often deliberately introduced and are typically unaffected by changes in temperature.

- Dislocations are line defects which, when a force is applied to a metallic material, move and cause a metallic material to deform.

- The critical resolved shear stress is the stress required to move the dislocation.

- The dislocation moves in a slip system, composed of a slip plane and a slip direction. The slip direction is typically a close-packed direction. The slip plane is also normally close-packed or nearly close-packed.

- In metallic crystals, the number and type of slip directions and slip planes influence the properties of the metal. In FCC metals, the critical resolved shear stress is low and an optimum number of slip planes is available; consequently, FCC metals tend to be ductile. In BCC metals, no close-packed planes are available and the critical resolved shear stress is high; thus, BCC metals tend to be strong. The number of slip systems in HCP metals is limited, causing these metals typically to behave in a brittle manner.

- Point defects, which include vacancies, interstitial atoms, and substitutional atoms, introduce compressive or tensile strain fields that disturb the atomic arrangements in the surrounding crystal. As a result, dislocations cannot easily slip in the vicinity of point defects and the strength of the metallic material is increased.

- Surface defects include grain boundaries. Producing a very small grain size increases the amount of grain boundary area; because dislocations cannot easily pass through a grain boundary, the material is strengthened (Hall-Petch equation).

- The number and type of crystal defects control the ease of movement of dislocations and, therefore, directly influence the mechanical properties of the material.

- Defects in materials have a significant influence on their electrical, optical, and magnetic properties.

Glossary

Annealing A heat treatment that typically involves heating a metallic material to a high temperature for an extended period of time in order to lower the dislocation density and hence impart ductility.

ASTM American Society for Testing and Materials.

ASTM grain size number (*n*) A measure of the size of the grains in a crystalline material obtained by counting the number of grains per square inch at a magnification of 100.

Burgers vector The direction and distance that a dislocation moves in each step, also known as the slip vector.

Critical resolved shear stress The shear stress required to cause a dislocation to move and cause slip.

Cross-slip A change in the slip system of a dislocation.

Defect chemical reactions Reactions written using the Kröger-Vink notation to describe defect chemistry. The reactions must be written in such a way that mass and electrical charges are balanced and stoichiometry of sites is maintained. The existence of defects predicted by such reactions needs to be verified experimentally.

Dislocation A line imperfection in a crystalline material. Movement of dislocations helps explain how metallic materials deform. Interference with the movement of dislocations helps explain how metallic materials are strengthened.

Dislocation density The total length of dislocation line per cubic centimeter in a material.

Domain A small region of a ferroelectric, ferromagnetic, or ferrimagnetic material in which the direction of dielectric polarization (for ferroelectric) or magnetization (for ferromagnetic or ferrimagnetic) remains the same.

Dopants Elements or compounds typically added, in known concentrations and appearing at specific places within the microstructure, to enhance the properties or processing of a material.

Edge dislocation A dislocation introduced into the crystal by adding an "extra half plane" of atoms.

Elastic deformation Deformation that is fully recovered when the stress causing it is removed.

Etch pits Holes created at locations where dislocations meet the surface. These are used to examine the presence and density of dislocations.

Extended defects Defects that involve several atoms/ions and thus occur over a finite volume of the crystalline material (e.g., dislocations, stacking faults, etc.).

Ferroelectric A dielectric material that develops a spontaneous and reversible electric polarization (e.g., PZT, $BaTiO_3$).

Frenkel defect A pair of point defects produced when an ion moves to create an interstitial site, leaving behind a vacancy.

Grain One of the crystals present in a polycrystalline material.

Grain boundary A surface defect representing the boundary between two grains. The crystal has a different orientation on either side of the grain boundary.

Grain-size strengthening Strengthening of a material by decreasing the grain size and therefore increasing the grain boundary area. Grain boundaries resist dislocation motion, and thus, increasing the grain boundary area leads to increased strength.

Hall-Petch equation The relationship between yield strength and grain size in a metallic material—that is, $\sigma_y = \sigma_0 + Kd^{-1/2}$.

Image analysis A technique that is used to analyze images of microstructures to obtain quantitative information on grain size, shape, grain size distribution, etc.

Impurities Elements or compounds that find their way into a material, often originating from processing or raw materials and typically having a deleterious effect on the properties or processing of a material.

Interstitial defect A point defect produced when an atom is placed into the crystal at a site that is normally not a lattice point.

Interstitialcy A point defect caused when a "normal" atom occupies an interstitial site in the crystal.

Kröger-Vink notation A system used to indicate point defects in materials. The main body of the notation indicates the type of defect or the element involved. The subscript indicates the location of the point defect, and the superscript indicates the effective positive ($\cdot$) or negative ($'$) charge.

Metallography Preparation of a metallic sample of a material by polishing and etching so that the structure can be examined using a microscope.

Mixed dislocation A dislocation that contains partly edge components and partly screw components.

Peierls-Nabarro stress The shear stress, which depends on the Burgers vector and the interplanar spacing, required to cause a dislocation to move—that is, $\tau = c \exp(-kd/b)$.

Plastic deformation Permanent deformation of a material when a load is applied, then removed.

Point defects Imperfections, such as vacancies, that are located typically at one (in some cases a few) sites in the crystal.

Precipitation strengthening Strengthening of metals and alloys by formation of precipitates inside the grains. The small precipitates resist dislocation motion.

Schmid's law The relationship between shear stress, the applied stress, and the orientation of the slip system—that is, $\tau = \sigma \cos \lambda \cos \phi$.

Schottky defect A point defect in ionically bonded materials. In order to maintain a neutral charge, a stoichiometric number of cation and anion vacancies must form.

Screw dislocation A dislocation produced by skewing a crystal by one atomic spacing so that a spiral ramp is produced.

Second-phase strengthening A mechanism by which grains of an additional compound or phase are introduced in a polycrystalline material. These second phase crystals resist dislocation motion, thereby causing an increase in the strength of a metallic material.

Sintering A process for forming a dense mass by heating compacted powders.

Slip Deformation of a metallic material by the movement of dislocations through the crystal.

Slip band Collection of many slip lines, often easily visible.

Slip direction The direction in the crystal in which the dislocation moves. The slip direction is the same as the direction of the Burgers vector.

Slip line A visible line produced at the surface of a metallic material by the presence of several thousand dislocations.

Slip plane The plane swept out by the dislocation line during slip. Normally, the slip plane is a close-packed plane, if one exists in the crystal structure.

Slip system The combination of the slip plane and the slip direction.

Small angle grain boundary An array of dislocations causing a small misorientation of the crystal across the surface of the imperfection.

Stacking fault A surface defect in metals caused by the improper stacking sequence of close-packed planes.

Strain hardening Strengthening of a material by increasing the number of dislocations by deformation, or cold working. Also known as "work hardening."

Substitutional defect A point defect produced when an atom is removed from a regular lattice point and replaced with a different atom, usually of a different size.

Surface defects Imperfections, such as grain boundaries, that form a two-dimensional plane within the crystal.

Thermal grooving A technique used for observing microstructures in ceramic materials that involves heating a polished sample to a temperature slightly below the sintering temperature for a short time.

Tilt boundary A small angle grain boundary composed of an array of edge dislocations.

Transmission electron microscope (TEM) An instrument that, by passing an electron beam through a material, can detect microscopic structural features.

Twin boundary A surface defect across which there is a mirror image misorientation of the crystal structure. Twin boundaries can also move and cause deformation of the material.

Twist boundary A small angle grain boundary composed of an array of screw dislocations.

Vacancy An atom or an ion missing from its regular crystallographic site.

Yield strength The level of stress above which a material begins to show permanent deformation.

Problems

Section 4-1 Point Defects

4-1 Gold has 5.82×10^8 vacancies/cm^3 at equilibrium at 300 K. What fraction of the atomic sites is vacant at 600 K?

4-2 Calculate the number of vacancies per cm^3 expected in copper at 1080°C (just below the melting temperature). The energy for vacancy formation is 20,000 cal/mol.

4-3 The fraction of lattice points occupied by vacancies in solid aluminum at 660°C is 10^{-3}. What is the energy required to create vacancies in aluminum?

4-4 The density of a sample of FCC palladium is 11.98 g/cm^3, and its lattice parameter is 3.8902 Å. Calculate
(a) the fraction of the lattice points that contain vacancies; and
(b) the total number of vacancies in a cubic centimeter of Pd.

4-5 The density of a sample of HCP beryllium is 1.844 g/cm^3, and the lattice parameters are $a_0 = 0.22858$ nm and $c_0 = 0.35842$ nm. Calculate

(a) the fraction of the lattice points that contain vacancies; and
(b) the total number of vacancies in a cubic centimeter.

4-6 BCC lithium has a lattice parameter of 3.5089×10^{-8} cm and contains one vacancy per 200 unit cells. Calculate
(a) the number of vacancies per cubic centimeter; and
(b) the density of Li.

4-7 FCC lead has a lattice parameter of 0.4949 nm and contains one vacancy per 500 Pb atoms. Calculate
(a) the density; and
(b) the number of vacancies per gram of Pb.

4-8 Cu and Ni form a substitutional solid solution. This means that the crystal structure of a Cu-Ni alloy consists of Ni atoms substituting for Cu atoms in the regular atomic positions of the FCC structure. For a Cu-30% wt.% Ni alloy, what fraction of the atomic sites does Ni occupy?

4-9   A niobium alloy is produced by introducing tungsten substitutional atoms into the BCC structure; eventually an alloy is produced that has a lattice parameter of 0.32554 nm and a density of 11.95 g/cm³. Calculate the fraction of the atoms in the alloy that are tungsten.

4-10   Tin atoms are introduced into an FCC copper crystal, producing an alloy with a lattice parameter of 3.7589×10^{-8} cm and a density of 8.772 g/cm³. Calculate the atomic percentage of tin present in the alloy.

4-11   We replace 7.5 atomic percent of the chromium atoms in its BCC crystal with tantalum. X-ray diffraction shows that the lattice parameter is 0.29158 nm. Calculate the density of the alloy.

4-12   Suppose we introduce one carbon atom for every 100 iron atoms in an interstitial position in BCC iron, giving a lattice parameter of 0.2867 nm. For this steel, find the density and the packing factor.

4-13   The density of BCC iron is 7.882 g/cm³, and the lattice parameter is 0.2866 nm when hydrogen atoms are introduced at interstitial positions. Calculate
(a) the atomic fraction of hydrogen atoms; and
(b) the number of unit cells on average that contain hydrogen atoms.

Section 4-2 Other Point Defects

4-14   Suppose one Schottky defect is present in every tenth unit cell of MgO. MgO has the sodium chloride crystal structure and a lattice parameter of 0.396 nm. Calculate
(a) the number of anion vacancies per cm³; and
(b) the density of the ceramic.

4-15   ZnS has the zinc blende structure. If the density is 3.02 g/cm³ and the lattice parameter is 0.59583 nm, determine the number of Schottky defects
(a) per unit cell; and
(b) per cubic centimeter.

4-16   Suppose we introduce the following point defects.
(a) Mg^{2+} ions substitute for yttrium ions in Y_2O_3;

(b) Fe^{3+} ions substitute for magnesium ions in MgO;
(c) Li^{1+} ions substitute for magnesium ions in MgO; and
(d) Fe^{2+} ions replace sodium ions in NaCl.
What other changes in each structure might be necessary to maintain a charge balance? Explain.

4-17   Write down the defect chemistry equation for introduction of $SrTiO_3$ in $BaTiO_3$ using the Kröger-Vink notation.

Section 4-3 Dislocations

4-18   Draw a Burgers circuit around the dislocation shown in Figure 4-21. Clearly indicate the Burgers vector that you find. What type of dislocation is this? In what direction will the dislocation move due to the applied shear stress τ? Reference your answers to the coordinate axes shown.

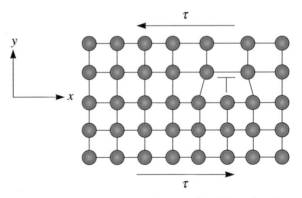

Figure 4-21   A schematic diagram of a dislocation for Problem 4-18.

4-19   What are the Miller indices of the slip directions:
(a) on the (111) plane in an FCC unit cell?
(b) on the (011) plane in a BCC unit cell?

4-20   What are the Miller indices of the slip planes in FCC unit cells that include the [101] slip direction?

4-21   What are the Miller indices of the {110} slip planes in BCC unit cells that include the [111] slip direction?

4-22 Calculate the length of the Burgers vector in the following materials:
(a) BCC niobium;
(b) FCC silver; and
(c) diamond cubic silicon.

4-23 Determine the interplanar spacing and the length of the Burgers vector for slip on the expected slip systems in FCC aluminum. Repeat, assuming that the slip system is a (110) plane and a [1$\bar{1}$1] direction. What is the ratio between the shear stresses required for slip for the two systems? Assume that $k = 2$ in Equation 4-2.

4-24 Determine the interplanar spacing and the length of the Burgers vector for slip on the (110)/[1$\bar{1}$1] slip system in BCC tantalum. Repeat, assuming that the slip system is a (111)/[1$\bar{1}$0] system. What is the ratio between the shear stresses required for slip for the two systems? Assume that $k = 2$ in Equation 4-2.

4-25 The crystal shown in Figure 4-22 contains two dislocations A and B. If a shear stress is applied to the crystal as shown, what will happen to dislocations A and B?

Section 4-4 Significance of Dislocations

4-28 What is meant by the terms plastic and elastic deformation?

4-29 Why is the theoretical strength of metals much higher than that observed experimentally?

4-30 How many grams of aluminum, with a dislocation density of 10^{10} cm/cm^3, are required to give a total dislocation length that would stretch from New York City to Los Angeles (4860 km)?

4-31 The distance from the Earth to the Moon is 384,000 km. If this were the total length of dislocation in a cubic centimeter of material, what would be the dislocation density? Compare your answer to typical dislocation densities for metals.

4-32 Why would metals behave as brittle materials without dislocations?

4-33 Why is it that dislocations play an important role in controlling the mechanical properties of metallic materials, however, they do not play a role in determining the mechanical properties of glasses?

4-34 Suppose you would like to introduce an interstitial or large substitutional atom into the crystal near a dislocation. Would the atom fit more easily above or below the dislocation line shown in Figure 4-7(c)? Explain.

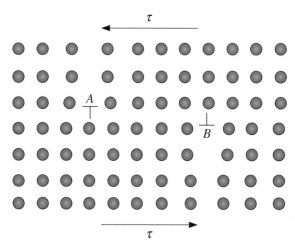

Figure 4-22 A schematic diagram of two dislocations for Problem 4-25.

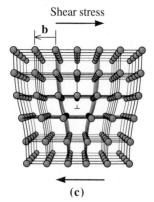

Figure 4-7(c) (Repeated for Problem 4-34).

4-26 Can ceramic and polymeric materials contain dislocations?

4-27 Why is it that ceramic materials are brittle?

4-35 Compare the c/a ratios for the following HCP metals, determine the likely slip processes in each, and estimate the approximate critical resolved shear stress. Explain. (See data in Appendix A.)

(a) Zinc
(b) Magnesium
(c) Titanium
(d) Zirconium
(e) Rhenium
(f) Beryllium

Section 4-5 Schmid's Law

4-36   A single crystal of an FCC metal is oriented so that the [001] direction is parallel to an applied stress of 35 MPa. Calculate the resolved shear stress acting on the (111) slip plane in the $[\bar{1}10]$, $[0\bar{1}1]$, and $[10\bar{1}]$ slip directions. Which slip system(s) will become active first?

4-37   A single crystal of a BCC metal is oriented so that the [001] direction is parallel to the applied stress. If the critical resolved shear stress required for slip is 83 MPa calculate the magnitude of the applied stress required to cause slip to begin in the $[1\bar{1}1]$ direction on the (110), (011), and $(10\bar{1})$ slip planes.

4-38   A single crystal of silver is oriented so that the (111) slip plane is perpendicular to an applied stress of 50 MPa. List the slip systems composed of close-packed planes and directions that may be activated due to this applied stress.

Section 4-6 Influence of Crystal Structure

4-39   Why is it that single crystal and polycrystalline copper are both ductile, however, only single crystal, but not polycrystalline, zinc can exhibit considerable ductility?

4-40   Why is it that cross slip in BCC and FCC metals is easier than in HCP metals? How does this influence the ductility of BCC, FCC, and HCP metals?

4-41   Arrange the following metals in the expected order of increasing ductility: Cu, Ti, and Fe.

Section 4-7 Surface Defects

4-42   The strength of titanium is found to be 948 MPa when the grain size is 17×10^{-6} m and 565 MPa when the grain size is 0.8×10^{-6} m. Determine

(a) the constants in the Hall-Petch equation; and

(b) the strength of the titanium when the grain size is reduced to 0.2×10^{-6} m.

4-43   A copper-zinc alloy has the following properties

Grain Diameter (mm)	Strength (MPa)
0.015	170 MPa
0.025	158 MPa
0.035	151 MPa
0.050	145 MPa

Determine

(a) the constants in the Hall-Petch equation and

(b) the grain size required to obtain a strength of 200 MPa.

4-44   For an ASTM grain size number of 8, calculate the number of grains per square inch

(a) at a magnification of 100 and

(b) with no magnification.

4-45   Determine the ASTM grain size number if 20 grains/square inch are observed at a magnification of 400.

4-46   Determine the ASTM grain size number if 25 grains/square inch are observed at a magnification of 50.

4-47   Determine the ASTM grain size number for the materials in Figure 4-17 and Figure 4-23.

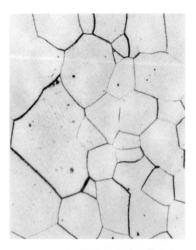

Figure 4-17   (Repeated for Problem 4-47) Microstructure of palladium (× 100). (*From ASM Handbook, Vol. 9, Metallography and Microstructure (1985), ASM International, Materials Park, OH 44073.*)

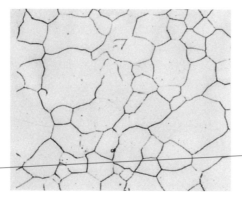

Figure 4-23 Microstructure of iron. *(From ASM Handbook, Vol. 9, Metallography and Microstructure (1985), ASM International, Materials Park, OH 44073.)*

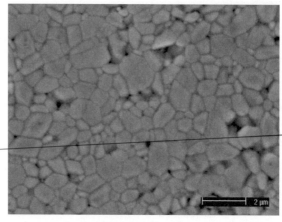

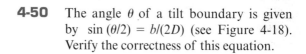

Figure 4-25 Microstructure of an alumina ceramic. *(Courtesy of Dr. Richard McAfee and Dr. Ian Nettleship.)*

4-48 Certain ceramics with special dielectric properties are used in wireless communication systems. Barium magnesium tantalate (BMT) and barium zinc tantalate (BZT) are examples of such materials. Determine the ASTM grain size number for a barium magnesium tantalate (BMT) ceramic microstructure shown in Figure 4-24.

4-50 The angle θ of a tilt boundary is given by $\sin(\theta/2) = b/(2D)$ (see Figure 4-18). Verify the correctness of this equation.

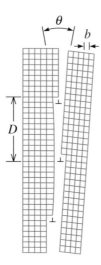

Figure 4-18 (Repeated for Problems 4-50, 4-51 and 4-52) The small angle grain boundary is produced by an array of dislocations, causing an angular mismatch θ between the lattices on either side of the boundary.

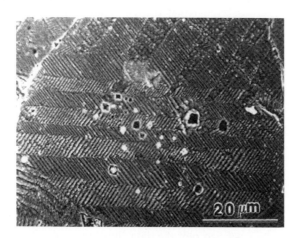

Figure 4-24 Microstructure of a barium magnesium tantalate (BMT) ceramic. *(Courtesy of H. Shivey.)*

4-51 Calculate the angle θ of a small-angle grain boundary in FCC aluminum when the dislocations are 5000 Å apart. (See Figure 4-18 and the equation in Problem 4-50.)

4-52 For BCC iron, calculate the average distance between dislocations in a small-angle grain boundary tilted 0.50°. (See Figure 4-18.)

4-49 Alumina is the most widely used ceramic material. Determine the ASTM grain size number for the polycrystalline alumina sample shown in Figure 4-25.

4-53 Why is it that a single crystal of a ceramic superconductor is capable of carrying much more current per unit area than a polycrystalline ceramic superconductor of the same composition?

Section 4-8 Importance of Defects

4-54 What makes plain carbon steel harder than pure iron?

4-55 Why is jewelry made from gold or silver alloyed with copper?

4-56 Why do we prefer to use semiconductor crystals that contain as small a number of dislocations as possible?

4-57 In structural applications (e.g., steel for bridges and buildings or aluminum alloys for aircraft), why do we use alloys rather than pure metals?

4-58 Do dislocations control the strength of a silicate glass? Explain.

4-59 What is meant by the term strain hardening?

4-60 To which mechanism of strengthening is the Hall-Petch equation related?

4-61 Pure copper is strengthened by the addition of a small concentration of Be. To which mechanism of strengthening is this related to?

Design Problems

4-62 The density of pure aluminum calculated from crystallographic data is expected to be 2.69955 g/cm^3.
  (a) Design an aluminum alloy that has a density of 2.6450 g/cm^3.
  (b) Design an aluminum alloy that has a density of 2.7450 g/cm^3.

4-63 You would like a metal plate with good weldability. During the welding process, the metal next to the weld is heated almost to the melting temperature and, depending on the welding parameters, may remain hot for some period of time. Design an alloy that will minimize the loss of strength in this "heat-affected zone" during the welding process.

4-64 We need a material that is optically transparent but electrically conductive. Such materials are used for touch screen displays. What kind of materials can be used

for this application? (*Hint*: Think about coatings of materials that can provide electronic or ionic conductivity; the substrate has to be transparent for this application.)

Computer Problems

4-65 *Temperature dependence of vacancy concentrations.* Write a computer program that will provide a user with the equilibrium concentration of vacancies in a metallic element as a function of temperature. The user should specify a meaningful and valid range of temperatures (e.g., 100 to 1200 K for copper). Assume that the crystal structure originally specified is valid for this range of temperature. Ask the user to input the activation energy for the formation of one mole of vacancies (Q_v). The program then should ask the user to input the density of the element and crystal structure (FCC, BCC, etc.). You can use character variables to detect the type of crystal structures (e.g., "F" or "f" for FCC, "B" or "b" for BCC, etc.). Be sure to pay attention to the correct units for temperature, density, etc. The program should ask the user if the temperature range that has been provided is in °C, °F, or K and convert the temperatures properly into K before any calculations are performed. The program should use this information to establish the number of atoms per unit volume and provide an output for this value. The program should calculate the equilibrium concentration of vacancies at different temperatures. The first temperature will be the minimum temperature specified and then temperatures should be increased by 100 K or another convenient increment. You can make use of any graphical software to plot the data showing the equilibrium concentration of vacancies as a function of temperature. Think about what scales will be used to best display the results.

4-66 *Hall-Petch equation.* Write a computer program that will ask the user to enter two sets of values of σ_y and grain size (*d*) for a metallic material. The program should then utilize the data to calculate and print the Hall-Petch equation. The program then should prompt the user to input another value of grain size and calculate the yield stress or vice versa.

4-67 *ASTM grain size number calculator.* Write a computer program that will ask the user to input the magnification of a micrograph of the sample for which the ASTM number is being calculated. The program should then ask the user for the number of grains counted and the area (in square inches) from which these grains were counted. The

program then should calculate the ASTM number, taking into consideration the fact that the micrograph magnification is not 100 and the area may not have been one square inch.

Ⓚ Knovel® **Problems**

K4-1 Describe the problems associated with metal impurities in silicon devices.

K4-2 What are the processes involved in the removal of metal impurities from silicon devices by gettering?

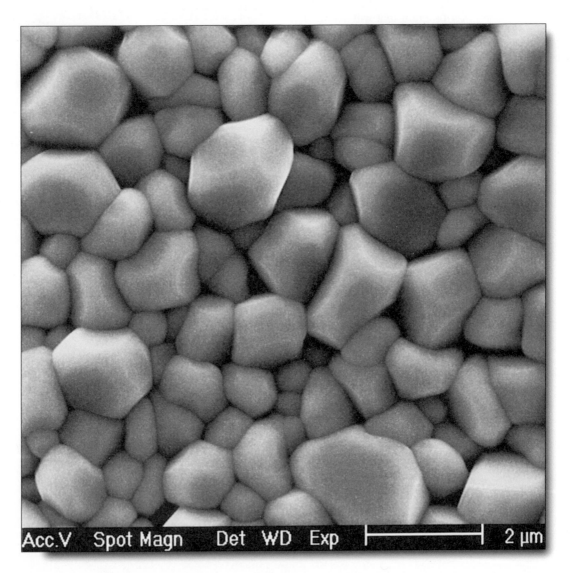

Sintered barium magnesium tantalate (BMT) ceramic microstructure. This ceramic material is useful in making electronic components used in wireless communications. The process of sintering is driven by the diffusion of atoms or ions. (*Courtesy of H. Shivey.*)

Atom and Ion Movements in Materials

Have You Ever Wondered?

- *Aluminum oxidizes more easily than iron, so why do we say aluminum normally does not "rust?"*

- *What kind of plastic is used to make carbonated beverage bottles?*

- *How are the surfaces of certain steels hardened?*

- *Why do we encase optical fibers using a polymeric coating?*

- *Who invented the first contact lens?*

- *How does a tungsten filament in a lightbulb fail?*

In Chapter 4, we learned that the atomic and ionic arrangements in materials are never perfect. We also saw that most materials are not pure elements; they are alloys or blends of different elements or compounds. Different types of atoms or ions typically "diffuse," or move within the material, so the differences in their concentration are minimized. **Diffusion** refers to an observable net flux of atoms or other species. It depends upon the concentration gradient and temperature. Just as water flows from a mountain toward the sea to minimize its gravitational potential energy, atoms and ions have a tendency to move in a predictable fashion to eliminate concentration differences and produce homogeneous compositions that make the material thermodynamically more stable.

In this chapter, we will learn that temperature influences the kinetics of diffusion and that a concentration difference contributes to the overall net flux of diffusing species. The goal of this chapter is to examine the principles and applications of diffusion in materials. We'll illustrate the concept of diffusion through examples of several real-world technologies dependent on the diffusion of atoms, ions, or molecules.

We will present an overview of Fick's laws that describe the diffusion process quantitatively. We will also see how the relative openness of different crystal structures and the size of atoms or ions, temperature, and concentration of diffusing species affect the rate at which diffusion occurs. We will discuss specific examples of how diffusion is used in the synthesis and processing of advanced materials as well as manufacturing of components using advanced materials.

5-1 Applications of Diffusion

Diffusion   Diffusion refers to the net flux of any species, such as ions, atoms, electrons, holes (Chapter 19), and molecules. The magnitude of this flux depends upon the concentration gradient and temperature. The process of diffusion is central to a large number of today's important technologies. In materials processing technologies, control over the diffusion of atoms, ions, molecules, or other species is key. There are hundreds of applications and technologies that depend on either enhancing or limiting diffusion. The following are just a few examples.

Carburization for Surface Hardening of Steels   Let's

say we want a surface, such as the teeth of a gear, to be hard; however, we do not want the entire gear to be hard. The carburization process can be used to increase surface hardness. In carburization, a source of carbon, such as a graphite powder or gaseous phase containing carbon, is diffused into steel components such as gears (Figure 5-1). In later chapters, you will learn how increased carbon concentration on the surface of the steel increases the steel's hardness. Similar to the introduction of carbon, we can also use a process known as **nitriding**, in which nitrogen is introduced into the surface of a metallic material. Diffusion also plays a central role in the control of the phase transformations

Figure 5-1   Furnace for heat treating steel using the carburization process. (*Courtesy of Cincinnati Steel Treating.*)

needed for the heat treatment of metals and alloys, the processing of ceramics, and the solidification and joining of materials (see Section 5-9).

Dopant Diffusion for Semiconductor Devices

The entire microelectronics industry, as we know it today, would not exist if we did not have a very good understanding of the diffusion of different atoms into silicon or other semiconductors. The creation of the *p-n junction* (Chapter 19) involves diffusing dopant atoms, such as phosphorous (P), arsenic (As), antimony (Sb), boron (B), aluminum (Al), etc., into precisely defined regions of silicon wafers. Some of these regions are so small that they are best measured in nanometers. A *p-n* junction is a region of the semiconductor, one side of which is doped with *n*-type dopants (e.g., As in Si) and the other side is doped with *p*-type dopants (e.g., B in Si).

Conductive Ceramics

In general, polycrystalline ceramics tend to be good insulators of electricity. Diffusion of ions, electrons, or holes also plays an important role in the electrical conductivity of many **conductive ceramics**, such as partially or fully stabilized zirconia (ZrO_2) or indium tin oxide (also commonly known as ITO). Lithium cobalt oxide ($LiCoO_2$) is an example of an ionically conductive material that is used in lithium ion batteries. These ionically conductive materials are used for such products as oxygen sensors in cars, touch-screen displays, fuel cells, and batteries. The ability of ions to diffuse and provide a pathway for electrical conduction plays an important role in enabling these applications.

Creation of Plastic Beverage Bottles

The occurrence of diffusion may not always be beneficial. In some applications, we may want to limit the occurrence of diffusion for certain species. For example, in the creation of certain plastic bottles, the diffusion of carbon dioxide (CO_2) must be minimized. This is one of the major reasons why we use polyethylene terephthalate (PET) to make bottles which ensure that the carbonated beverages they contain will not lose their fizz for a reasonable period of time!

Oxidation of Aluminum

You may have heard or know that aluminum does not "rust." In reality, aluminum oxidizes (rusts) more easily than iron; however, the aluminum oxide (Al_2O_3) forms a very protective but thin coating on the aluminum's surface preventing any further diffusion of oxygen and hindering further oxidation of the underlying aluminum. The oxide coating does not have a color and is thin and, hence, invisible. This is why we think aluminum does not rust.

Coatings and Thin Films

Coatings and thin films are often used to limit the diffusion of water vapor, oxygen, or other chemicals.

Thermal Barrier Coatings for Turbine Blades

In an aircraft engine, some of the nickel superalloy-based turbine blades are coated with ceramic oxides such as yttria stabilized zirconia (YSZ). These ceramic coatings protect the underlying alloy from high temperatures; hence, the name **thermal barrier coatings** (TBCs) (Figure 5-2). The diffusion of oxygen through these ceramic coatings and the subsequent oxidation of the underlying alloy play a major role in determining the lifetime and durability of the turbine blades. In Figure 5-2, EBPVD means electron beam physical vapor deposition. The bond coat is either a platinum or molybdenum-based alloy. It provides adhesion between the TBC and the substrate.

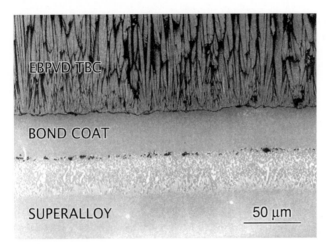

Figure 5-2
A thermal barrier coating on a nickel-based superalloy. (*Courtesy of Dr. F.S. Pettit and Dr. G.H. Meier, University of Pittsburgh.*)

Optical Fibers and Microelectronic Components   Optical fibers made from silica (SiO_2) are coated with polymeric materials to prevent diffusion of water molecules. This, in turn, improves the optical and mechanical properties of the fibers.

Example 5-1	*Diffusion of Ar/He and Cu/Ni*

Consider a box containing an impermeable partition that divides the box into equal volumes (Figure 5-3). On one side, we have pure argon (Ar) gas; on the other side, we have pure helium (He) gas. Explain what will happen when the partition is opened? What will happen if we replace the Ar side with a Cu single crystal and the He side with a Ni single crystal?

SOLUTION

Before the partition is opened, one compartment has no argon and the other has no helium (i.e., there is a concentration gradient of Ar and He). When the partition is opened, Ar atoms will diffuse toward the He side, and vice versa. This diffusion will continue until the entire box has a uniform concentration of both gases. There may be some density gradient driven convective flows as well. If we took random samples of different regions in this box after a few hours, we would find a statistically uniform concentration of Ar and He. Owing to their thermal energy, the Ar and He atoms will continue to move around in the box; however, there will be no concentration gradients.

If we open the hypothetical partition between the Ni and Cu single crystals at room temperature, we would find that, similar to the Ar/He situation, the

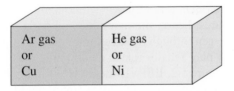

Figure 5-3
Illustration for diffusion of Ar/He and Cu/Ni (for Example 5-1).

concentration gradient exists but the temperature is too low to see any significant diffusion of Cu atoms into the Ni single crystal and vice versa. This is an example of a situation in which a concentration gradient exists; however, because of the lower temperature, the kinetics for diffusion are not favorable. Certainly, if we increase the temperature (say to 600°C) and wait for a longer period of time (e.g., ~24 hours), we would see diffusion of Cu atoms into the Ni single crystal and vice versa. After a very long time, the entire solid will have a uniform concentration of Ni and Cu atoms. The new solid that forms consists of Cu and Ni atoms completely dissolved in each other and the resultant material is termed a "solid solution," a concept we will study in greater detail in Chapter 10.

This example also illustrates something many of you may know by intuition. The diffusion of atoms and molecules occurs faster in gases and liquids than in solids. As we will see in Chapter 9 and other chapters, diffusion has a significant effect on the evolution of microstructure during the solidification of alloys, the heat treatment of metals and alloys, and the processing of ceramic materials.

5-2   Stability of Atoms and Ions

In Chapter 4, we showed that imperfections are present and also can be deliberately introduced into a material; however, these imperfections and, indeed, even atoms or ions in their normal positions in the crystal structures are not stable or at rest. Instead, the atoms or ions possess thermal energy, and they will move. For instance, an atom may move from a normal crystal structure location to occupy a nearby vacancy. An atom may also move from one interstitial site to another. Atoms or ions may jump across a grain boundary, causing the grain boundary to move.

The ability of atoms and ions to diffuse increases as the temperature, or thermal energy possessed by the atoms and ions, increases. The rate of atom or ion movement is related to temperature or thermal energy by the *Arrhenius equation*:

$$\text{Rate} = c_0 \exp\left(\frac{-Q}{RT}\right) \tag{5-1}$$

where c_0 is a constant, R is the gas constant $\left(1.987 \frac{\text{cal}}{\text{mol} \cdot \text{K}}\right)$, T is the absolute temperature (K), and Q is the **activation energy** (cal/mol) required to cause Avogadro's number of atoms or ions to move. This equation is derived from a statistical analysis of the probability that the atoms will have the extra energy Q needed to cause movement. The rate is related to the number of atoms that move.

We can rewrite the equation by taking natural logarithms of both sides:

$$\ln(\text{rate}) = \ln(c_0) - \frac{Q}{RT} \tag{5-2}$$

If we plot $\ln(\text{rate})$ of some reaction versus $1/T$ (Figure 5-4), the slope of the line will be $-Q/R$ and, consequently, Q can be calculated. The constant c_0 corresponds to the intercept at $\ln(c_0)$ when $1/T$ is zero.

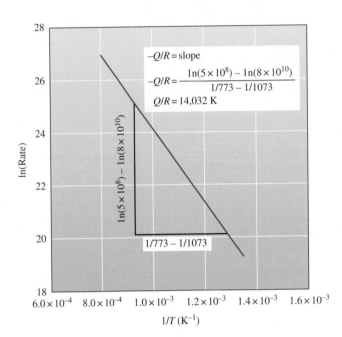

Figure 5-4
The Arrhenius plot of ln(rate) versus $1/T$ can be used to determine the activation energy for a reaction. The data from this figure is used in Example 5-2.

Svante August Arrhenius (1859–1927), a Swedish chemist who won the Nobel Prize in Chemistry in 1903 for his research on the electrolytic theory of dissociation applied this idea to the rates of chemical reactions in aqueous solutions. His basic idea of activation energy and rates of chemical reactions as functions of temperature has since been applied to diffusion and other rate processes.

Example 5-2   *Activation Energy for Interstitial Atoms*

Suppose that interstitial atoms are found to move from one site to another at the rates of 5×10^8 jumps/s at 500°C and 8×10^{10} jumps/s at 800°C. Calculate the activation energy Q for the process.

SOLUTION

Figure 5-4 represents the data on a ln(rate) versus $1/T$ plot; the slope of this line, as calculated in the figure, gives $Q/R = 14,032$ K, or $Q = 27,880$ cal/mol. Alternately, we could write two simultaneous equations:

$$\text{Rate}\left(\frac{\text{jumps}}{\text{s}}\right) = c_0 \exp\left(\frac{-Q}{RT}\right)$$

$$5 \times 10^8 \left(\frac{\text{jumps}}{\text{s}}\right) = c_0 \left(\frac{\text{jumps}}{\text{s}}\right) \exp\left[\frac{-Q\left(\frac{\text{cal}}{\text{mol}}\right)}{\left[1.987\left(\frac{\text{cal}}{\text{mol}\cdot\text{K}}\right)\right][(500 + 273)(\text{K})]}\right]$$

$$= c_0 \exp(-0.000651Q)$$

$$8 \times 10^{10}\left(\frac{\text{jumps}}{\text{s}}\right) = c_0\left(\frac{\text{jumps}}{\text{s}}\right)\exp\left[\frac{-Q\left(\frac{\text{cal}}{\text{mol}}\right)}{\left[1.987\left(\frac{\text{cal}}{\text{mol}\cdot\text{K}}\right)\right][(800 + 273)(\text{K})]}\right]$$

$$= c_0\exp(-0.000469Q)$$

Note the temperatures were converted into K. Since

$$c_0 = \frac{5 \times 10^8}{\exp(-0.000651Q)}\left(\frac{\text{jumps}}{\text{s}}\right)$$

then

$$8 \times 10^{10} = \frac{(5 \times 10^8)\exp(-0.000469Q)}{\exp(-0.000651Q)}$$

$$160 = \exp[(0.000651 - 0.000469)Q] = \exp(0.000182Q)$$

$$\ln(160) = 5.075 = 0.000182Q$$

$$Q = \frac{5.075}{0.000182} = 27{,}880 \text{ cal/mol}$$

5-3 Mechanisms for Diffusion

As we saw in Chapter 4, defects known as vacancies exist in materials. The disorder these vacancies create (i.e., increased entropy) helps minimize the free energy and, therefore, increases the thermodynamic stability of a crystalline material. In materials containing vacancies, atoms move or "jump" from one lattice position to another. This process, known as **self-diffusion**, can be detected by using radioactive tracers. As an example, suppose we were to introduce a radioactive isotope of gold (Au^{198}) onto the surface of standard gold (Au^{197}). After a period of time, the radioactive atoms would move into the standard gold. Eventually, the radioactive atoms would be uniformly distributed throughout the entire standard gold sample. Although self-diffusion occurs continually in all materials, its effect on the material's behavior is generally not significant.

Interdiffusion Diffusion of unlike atoms in materials also occurs (Figure 5-5). Consider a nickel sheet bonded to a copper sheet. At high temperatures, nickel atoms gradually diffuse into the copper, and copper atoms migrate into the nickel. Again, the nickel and copper atoms eventually are uniformly distributed. Diffusion of different atoms in different directions is known as **interdiffusion**. There are two important mechanisms by which atoms or ions can diffuse (Figure 5-6).

Vacancy Diffusion In self-diffusion and diffusion involving substitutional atoms, an atom leaves its lattice site to fill a nearby vacancy (thus creating a new vacancy at the original lattice site). As diffusion continues, we have counterflows of atoms

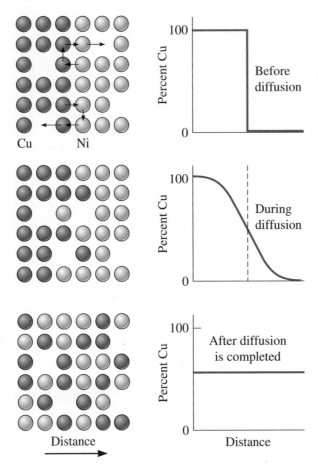

Figure 5-5
Diffusion of copper atoms into nickel. Eventually, the copper atoms are randomly distributed throughout the nickel.

and vacancies, called **vacancy diffusion**. The number of vacancies, which increases as the temperature increases, influences the extent of both self-diffusion and diffusion of substitutional atoms.

Interstitial Diffusion
When a small interstitial atom or ion is present in the crystal structure, the atom or ion moves from one interstitial site to another. No vacancies are required for this mechanism. Partly because there are many more interstitial sites than vacancies, **interstitial diffusion** occurs more easily than vacancy diffusion. Interstitial atoms that are relatively smaller can diffuse faster. In Chapter 3, we saw that

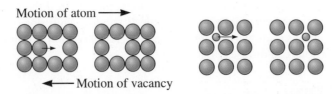

Figure 5-6   Diffusion mechanisms in materials: (a) vacancy or substitutional atom diffusion and (b) interstitial diffusion.

the structure of many ceramics with ionic bonding can be considered as a close packing of anions with cations in the interstitial sites. In these materials, smaller cations often diffuse faster than larger anions.

5-4 Activation Energy for Diffusion

A diffusing atom must squeeze past the surrounding atoms to reach its new site. In order for this to happen, energy must be supplied to allow the atom to move to its new position, as shown schematically for vacancy and interstitial diffusion in Figure 5-7. The atom is originally in a low-energy, relatively stable location. In order to move to a new location, the atom must overcome an energy barrier. The energy barrier is the activation energy Q. The thermal energy supplies atoms or ions with the energy needed to exceed this barrier. Note that the symbol Q is often used for activation energies for different processes (rate at which atoms jump, a chemical reaction, energy needed to produce vacancies, etc.), and we should be careful in understanding the specific process or phenomenon to which the general term for activation energy Q is being applied, as the value of Q depends on the particular phenomenon.

Normally, less energy is required to squeeze an interstitial atom past the surrounding atoms; consequently, activation energies are lower for interstitial diffusion than for vacancy diffusion. Typical values for activation energies for diffusion of different atoms in different host materials are shown in Table 5-1. We use the term **diffusion couple** to indicate a combination of an atom of a given element (e.g., carbon) diffusing in a host material (e.g., BCC Fe). A low-activation energy indicates easy diffusion. In self-diffusion, the activation energy is equal to the energy needed to create a vacancy and to cause the movement of the atom. Table 5-1 also shows values of D_0, which is the pre-exponential term and the constant c_0 from Equation 5-1, when the rate process is diffusion. We will see later that D_0 is the diffusion coefficient when $1/T = 0$.

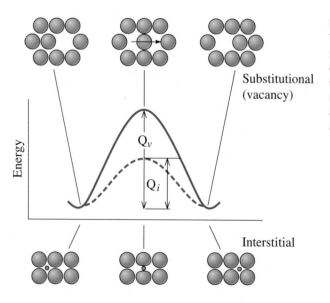

Figure 5-7
The activation energy Q is required to squeeze atoms past one another during diffusion. Generally, more energy is required for a substitutional atom than for an interstitial atom.

TABLE 5-1 ■ *Diffusion data for selected materials*

Diffusion Couple	Q (cal/mol)	D_0 (cm^2/s)
Interstitial diffusion:		
C in FCC iron	32,900	0.23
C in BCC iron	20,900	0.011
N in FCC iron	34,600	0.0034
N in BCC iron	18,300	0.0047
H in FCC iron	10,300	0.0063
H in BCC iron	3,600	0.0012
Self-diffusion (vacancy diffusion):		
Pb in FCC Pb	25,900	1.27
Al in FCC Al	32,200	0.10
Cu in FCC Cu	49,300	0.36
Fe in FCC Fe	66,700	0.65
Zn in HCP Zn	21,800	0.1
Mg in HCP Mg	32,200	1.0
Fe in BCC Fe	58,900	4.1
W in BCC W	143,300	1.88
Si in Si (covalent)	110,000	1800.0
C in C (covalent)	163,000	5.0
Heterogeneous diffusion (vacancy diffusion):		
Ni in Cu	57,900	2.3
Cu in Ni	61,500	0.65
Zn in Cu	43,900	0.78
Ni in FCC iron	64,000	4.1
Au in Ag	45,500	0.26
Ag in Au	40,200	0.072
Al in Cu	39,500	0.045
Al in Al$_2$O$_3$	114,000	28.0
O in Al$_2$O$_3$	152,000	1900.0
Mg in MgO	79,000	0.249
O in MgO	82,100	0.000043

From several sources, including Adda, Y. and Philibert, J., La Diffusion dans les Solides, Vol. 2, 1966.

5-5   Rate of Diffusion [Fick's First Law]

Adolf Eugen Fick (1829–1901) was the first scientist to provide a quantitative description of the diffusion process. Interestingly, Fick was also the first to experiment with contact lenses in animals and the first to implant a contact lens in human eyes in 1887–1888!

The rate at which atoms, ions, particles or other species diffuse in a material can be measured by the **flux** J. Here we are mainly concerned with diffusion of ions or atoms. The flux J is defined as the number of atoms passing through a plane of unit area per unit time (Figure 5-8). **Fick's first law** explains the net flux of atoms:

$$J = -D \frac{dc}{dx}$$

(5-3)

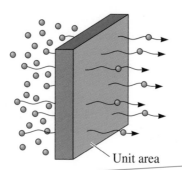

Figure 5-8
The flux during diffusion is defined as the number of atoms passing through a plane of unit area per unit time.

Unit area

where J is the flux, D is the **diffusivity** or **diffusion coefficient** $\left(\frac{cm^2}{s}\right)$, and dc/dx is the **concentration gradient** $\left(\frac{atoms}{cm^3 \cdot cm}\right)$. Depending upon the situation, concentration may be expressed as atom percent (at%), weight percent (wt%), mole percent (mol%), atom fraction, or mole fraction. The units of concentration gradient and flux will change accordingly.

Several factors affect the flux of atoms during diffusion. If we are dealing with diffusion of ions, electrons, holes, etc., the units of J, D, and $\frac{dc}{dx}$ will reflect the appropriate species that are being considered. The negative sign in Equation 5-3 tells us that the flux of diffusing species is from higher to lower concentrations, so that if the $\frac{dc}{dx}$ term is negative, J will be positive. Thermal energy associated with atoms, ions, etc., causes the random movement of atoms. At a microscopic scale, the thermodynamic driving force for diffusion is the concentration gradient. A net or an observable flux is created depending upon temperature and the concentration gradient.

Concentration Gradient
The concentration gradient shows how the composition of the material varies with distance: Δc is the difference in concentration over the distance Δx (Figure 5-9). A concentration gradient may be created when two materials of different composition are placed in contact, when a gas or liquid is in contact with a solid material, when nonequilibrium structures are produced in a material due to processing, and from a host of other sources.

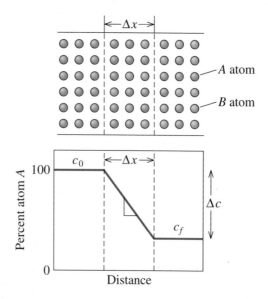

Figure 5-9
Illustration of the concentration gradient.

A atom

B atom

The flux at a particular temperature is constant only if the concentration gradient is also constant—that is, the compositions on each side of the plane in Figure 5-8 remain unchanged. In many practical cases, however, these compositions vary as atoms are redistributed, and thus the flux also changes. Often, we find that the flux is initially high and then gradually decreases as the concentration gradient is reduced by diffusion. The examples that follow illustrate calculations of flux and concentration gradients for diffusion of dopants in semiconductors and ceramics, but only for the case of constant concentration gradient. Later in this chapter, we will consider non-steady state diffusion with the aid of Fick's second law.

Example 5-3   *Semiconductor Doping*

One step in manufacturing transistors, which function as electronic switches in integrated circuits, involves diffusing impurity atoms into a semiconductor material such as silicon (Si). Suppose a silicon wafer 0.1 cm thick, which originally contains one phosphorus atom for every 10 million Si atoms, is treated so that there are 400 phosphorous (P) atoms for every 10 million Si atoms at the surface (Figure 5-10). Calculate the concentration gradient (a) in atomic percent/cm and (b) in $\frac{\text{atoms}}{\text{cm}^3 \cdot \text{cm}}$. The lattice parameter of silicon is 5.4307Å.

SOLUTION

a. First, let's calculate the initial and surface compositions in atomic percent:

$$c_i = \frac{1 \text{ P atom}}{10^7 \text{ atoms}} \times 100 = 0.00001 \text{ at\% P}$$

$$c_s = \frac{400 \text{ P atoms}}{10^7 \text{ atoms}} \times 100 = 0.004 \text{ at\% P}$$

$$\frac{\Delta c}{\Delta x} = \frac{0.00001 - 0.004 \text{ at\% P}}{0.1 \text{ cm}} = -0.0399 \frac{\text{at\% P}}{\text{cm}}$$

b. To find the gradient in terms of $\frac{\text{atoms}}{\text{cm}^3 \cdot \text{cm}}$, we must find the volume of the unit cell. The crystal structure of Si is diamond cubic (DC). The lattice parameter is 5.4307×10^{-8} cm. Thus,

$$V_{\text{cell}} = (5.4307 \times 10^{-8} \text{ cm})^3 = 1.6 \times 10^{-22} \frac{\text{cm}^3}{\text{cell}}$$

The volume occupied by 10^7 Si atoms, which are arranged in a DC structure with 8 atoms/cell, is

$$V = \left[\frac{10^7 \text{ atoms}}{8 \frac{\text{atoms}}{\text{cell}}}\right]\left[1.6 \times 10^{-22}\left(\frac{\text{cm}^3}{\text{cell}}\right)\right] = 2 \times 10^{-16} \text{ cm}^3$$

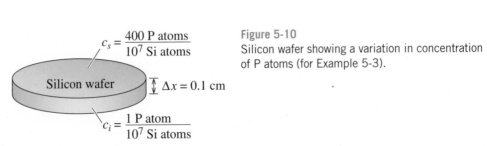

$$c_s = \frac{400 \text{ P atoms}}{10^7 \text{ Si atoms}}$$

Silicon wafer $\Delta x = 0.1$ cm

$$c_i = \frac{1 \text{ P atom}}{10^7 \text{ Si atoms}}$$

Figure 5-10
Silicon wafer showing a variation in concentration of P atoms (for Example 5-3).

The compositions in atoms/cm^3 are

$$c_i = \frac{1 \text{ P atom}}{2 \times 10^{-16} \text{ cm}^3} = 0.005 \times 10^{18} \text{ P}\left(\frac{\text{atoms}}{\text{cm}^3}\right)$$

$$c_s = \frac{400 \text{ P atoms}}{2 \times 10^{-16} \text{ cm}^3} = 2 \times 10^{18} \text{ P}\left(\frac{\text{atoms}}{\text{cm}^3}\right)$$

Thus, the composition gradient is

$$\frac{\Delta c}{\Delta x} = \frac{0.005 \times 10^{18} - 2 \times 10^{18} \text{ P}\left(\frac{\text{atoms}}{\text{cm}^3}\right)}{0.1 \text{ cm}}$$

$$= -1.995 \times 10^{19} \text{ P}\left(\frac{\text{atoms}}{\text{cm}^3 \cdot \text{cm}}\right)$$

Example 5-4   *Diffusion of Nickel in Magnesium Oxide (MgO)*

A 0.05 cm layer of magnesium oxide (MgO) is deposited between layers of nickel (Ni) and tantalum (Ta) to provide a diffusion barrier that prevents reactions between the two metals (Figure 5-11). At 1400°C, nickel ions diffuse through the MgO ceramic to the tantalum. Determine the number of nickel ions that pass through the MgO per second. At 1400°C, the diffusion coefficient of nickel ions in MgO is 9×10^{-12} cm^2/s, and the lattice parameter of nickel at 1400°C is 3.6×10^{-8} cm.

SOLUTION

The composition of nickel at the Ni/MgO interface is 100% Ni, or

$$c_{\text{Ni/MgO}} = \frac{4 \text{ Ni} \frac{\text{atoms}}{\text{unit cell}}}{(3.6 \times 10^{-8} \text{ cm})^3} = 8.573 \times 10^{22} \frac{\text{atoms}}{\text{cm}^3}$$

The composition of nickel at the Ta/MgO interface is 0% Ni. Thus, the concentration gradient is

$$\frac{\Delta c}{\Delta x} = \frac{0 - 8.573 \times 10^{22} \frac{\text{atoms}}{\text{cm}^3}}{0.05 \text{ cm}} = -1.715 \times 10^{24} \frac{\text{atoms}}{\text{cm}^3 \cdot \text{cm}}$$

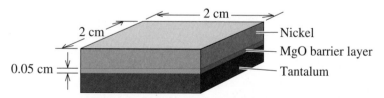

Figure 5-11   Diffusion couple (for Example 5-4).

The flux of nickel atoms through the MgO layer is

$$J = -D\frac{\Delta c}{\Delta x} = -(9 \times 10^{-12} \text{ cm}^2/\text{s})\left(-1.715 \times 10^{24}\tfrac{\text{atoms}}{\text{cm}^3 \cdot \text{cm}}\right)$$

$$J = 1.543 \times 10^{13}\tfrac{\text{Ni atoms}}{\text{cm}^2 \cdot \text{s}}$$

The total number of nickel atoms crossing the 2 cm × 2 cm interface per second is

$$\text{Total Ni atoms per second} = (J)(\text{Area}) = \left(1.543 \times 10^{13}\tfrac{\text{atoms}}{\text{cm}^2 \cdot \text{s}}\right)(2 \text{ cm})(2 \text{ cm})$$

$$= 6.17 \times 10^{13} \text{ Ni atoms/s}$$

Although this appears to be very rapid, in one second, the volume of nickel atoms removed from the Ni/MgO interface is

$$\frac{6.17 \times 10^{13}\tfrac{\text{Ni atoms}}{\text{s}}}{8.573 \times 10^{22}\tfrac{\text{Ni atoms}}{\text{cm}^3}} = 0.72 \times 10^{-9}\tfrac{\text{cm}^3}{\text{s}}$$

The thickness by which the nickel layer is reduced each second is

$$\frac{0.72 \times 10^{-9}\tfrac{\text{cm}^3}{\text{s}}}{4 \text{ cm}^2} = 1.8 \times 10^{-10}\tfrac{\text{cm}}{\text{s}}$$

For one micrometer (10^{-4} cm) of nickel to be removed, the treatment requires

$$\frac{10^{-4} \text{ cm}}{1.8 \times 10^{-10}\tfrac{\text{cm}}{\text{s}}} = 556{,}000 \text{ s} = 154 \text{ h}$$

5-6 Factors Affecting Diffusion

Temperature and the Diffusion Coefficient

The kinetics of diffusion are strongly dependent on temperature. The diffusion coefficient D is related to temperature by an Arrhenius-type equation,

$$D = D_0 \exp\left(\frac{-Q}{RT}\right) \tag{5-4}$$

where Q is the activation energy (in units of cal/mol) for diffusion of the species under consideration (e.g., Al in Si), R is the gas constant $\left(1.987\tfrac{\text{cal}}{\text{mol} \cdot \text{K}}\right)$, and T is the absolute temperature (K). D_0 is the pre-exponential term, similar to c_0 in Equation 5-1.

D_0 is a constant for a given diffusion system and is equal to the value of the diffusion coefficient at $1/T = 0$ or $T = \infty$. Typical values for D_0 are given in Table 5-1, while the temperature dependence of D is shown in Figure 5-12 for some metals and ceramics. Covalently bonded materials, such as carbon and silicon (Table 5-1), have unusually high

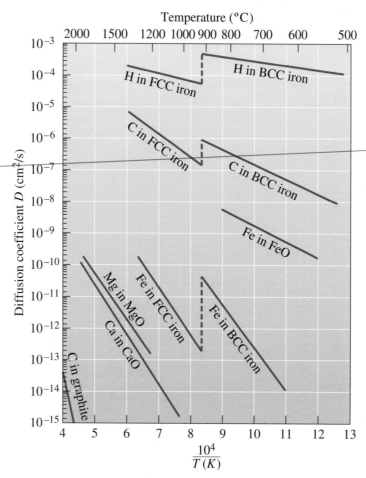

Figure 5-12 The diffusion coefficient *D* as a function of reciprocal temperature for some metals and ceramics. In this Arrhenius plot, *D* represents the rate of the diffusion process. A steep slope denotes a high activation energy.

activation energies, consistent with the high strength of their atomic bonds. Figure 5-13 shows the diffusion coefficients for different dopants in silicon.

In ionic materials, such as some of the oxide ceramics, a diffusing ion only enters a site having the same charge. In order to reach that site, the ion must physically squeeze past adjoining ions, pass by a region of opposite charge, and move a relatively long distance (Figure 5-14). Consequently, the activation energies are high and the rates of diffusion are lower for ionic materials than those for metals (Figure 5-15 on page 171). We take advantage of this in many situations. For example, in the processing of silicon (Si), we create a thin layer of silica (SiO_2) on top of a silicon wafer (Chapter 19). We then create a window by removing part of the silica layer. This window allows selective diffusion of dopant atoms such as phosphorus (P) and boron (B), because the silica layer is essentially impervious to the dopant atoms. Slower diffusion in most oxides and other ceramics is also an advantage in applications in which components are required to withstand high temperatures.

When the temperature of a material increases, the diffusion coefficient *D* increases (according to Equation 5-4) and, therefore, the flux of atoms increases as well. At higher

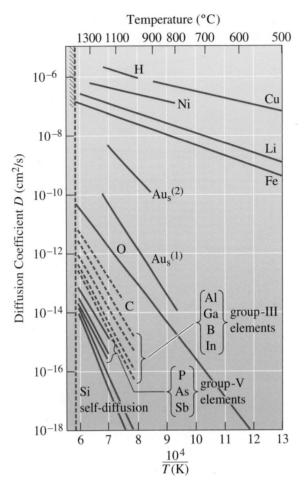

Temperature (°C)

Figure 5-13
Diffusion coefficients for different
dopants in silicon. (*From "Diffusion
and Diffusion Induced Defects
in Silicon," by U. Gösele.
In R. Bloor, M. Flemings, and
S. Mahajan (Eds.), Encyclopedia of
Advanced Materials, Vol. 1, 1994,
p. 631, Fig. 2. Copyright © 1994
Pergamon Press. Reprinted with
permission of the editor.*)

temperatures, the thermal energy supplied to the diffusing atoms permits the atoms to overcome the activation energy barrier and more easily move to new sites. At low temperatures—often below about 0.4 times the absolute melting temperature of the material—diffusion is very slow and may not be significant. For this reason, the heat treatment of metals and the processing of ceramics are done at high temperatures, where atoms move rapidly to complete reactions or to reach equilibrium conditions. Because less thermal energy is required to overcome the smaller activation energy barrier, a small activation energy Q increases the diffusion coefficient and flux. The following example illustrates how Fick's first law and concepts related to the temperature dependence of D can be applied to design an iron membrane.

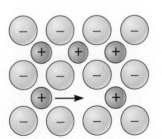

Figure 5-14
Diffusion in ionic compounds. Anions can only enter other anion sites. Smaller cations tend to diffuse faster.

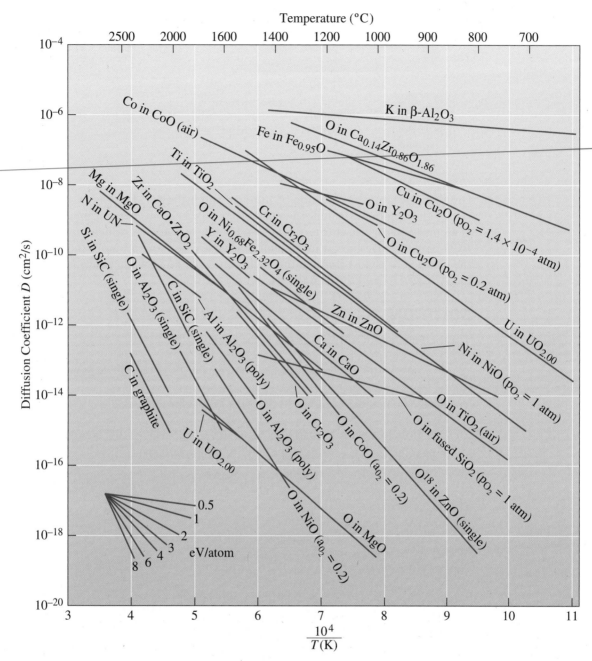

Figure 5-15   Diffusion coefficients of ions in different oxides. (*Adapted from* Physical Ceramics: Principles for Ceramic Science and Engineering, *by Y.M. Chiang, D. Birnie, and W.D. Kingery, Fig. 3-1. Copyright © 1997 John Wiley & Sons. This material is used by permission of John Wiley & Sons, Inc.*)

Example 5-5   *Design of an Iron Membrane*

An impermeable cylinder 3 cm in diameter and 10 cm long contains a gas that includes 0.5×10^{20} N atoms per cm^3 and 0.5×10^{20} H atoms per cm^3 on one side of an iron membrane (Figure 5-16). Gas is continuously introduced to the pipe to

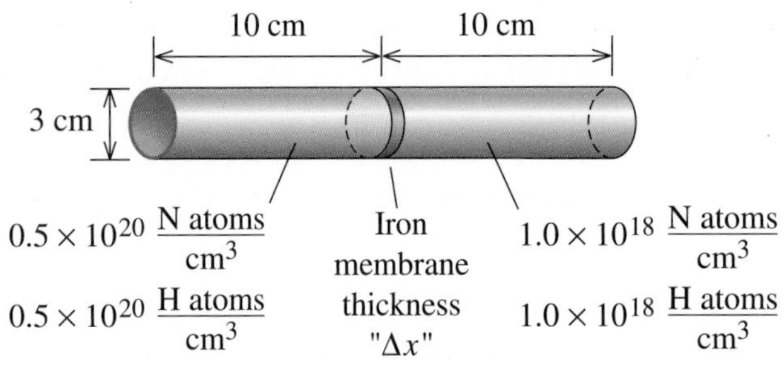

Figure 5-16
Design of an iron membrane (for Example 5-5).

10 cm   10 cm

3 cm

$0.5 \times 10^{20} \dfrac{\text{N atoms}}{\text{cm}^3}$

$0.5 \times 10^{20} \dfrac{\text{H atoms}}{\text{cm}^3}$

Iron membrane thickness "Δx"

$1.0 \times 10^{18} \dfrac{\text{N atoms}}{\text{cm}^3}$

$1.0 \times 10^{18} \dfrac{\text{H atoms}}{\text{cm}^3}$

ensure a constant concentration of nitrogen and hydrogen. The gas on the other side of the membrane includes a constant 1×10^{18} N atoms per cm^3 and 1×10^{18} H atoms per cm^3. The entire system is to operate at 700°C, at which iron has the BCC structure. Design an iron membrane that will allow no more than 1% of the nitrogen to be lost through the membrane each hour, while allowing 90% of the hydrogen to pass through the membrane per hour.

SOLUTION

The total number of nitrogen atoms in the container is

$$(0.5 \times 10^{20} \text{ N/cm}^3)(\pi/4)(3 \text{ cm})^2(10 \text{ cm}) = 35.343 \times 10^{20} \text{ N atoms}$$

The maximum number of atoms to be lost is 1% of this total, or

$$\text{N atom loss per h} = (0.01)\left(35.34 \times 10^{20}\right) = 35.343 \times 10^{18} \text{ N atoms/h}$$
$$\text{N atom loss per s} = (35.343 \times 10^{18} \text{ N atoms/h})/(3600 \text{ s/h})$$
$$= 0.0098 \times 10^{18} \text{ N atoms/s}$$

The flux is then

$$J = \frac{0.0098 \times 10^{18}(\text{ N atoms/s})}{\left(\dfrac{\pi}{4}\right)(3 \text{ cm})^2}$$
$$= 0.00139 \times 10^{18} \text{ N } \tfrac{\text{atoms}}{\text{cm}^2 \cdot \text{s}}$$

Using Equation 5-4 and values from Table 5-1, the diffusion coefficient of nitrogen in BCC iron at 700°C = 973 K is

$$D = D_0 \exp\left(\frac{-Q}{RT}\right)$$

$$D_N = 0.0047 \tfrac{\text{cm}^2}{\text{s}} \exp\left[\frac{-18{,}300 \tfrac{\text{cal}}{\text{mol}}}{1.987 \tfrac{\text{cal}}{\text{mol} \cdot \text{K}} (973 \text{ K})}\right]$$

$$= (0.0047)(7.748 \times 10^{-5}) = 3.64 \times 10^{-7} \tfrac{\text{cm}^2}{\text{s}}$$

From Equation 5-3:

$$J = -D\left(\frac{\Delta c}{\Delta x}\right) = 0.00139 \times 10^{18} \; \frac{\text{N atoms}}{\text{cm}^2 \cdot \text{s}}$$

$$\Delta x = -D\Delta c/J = \frac{\left[\left(-3.64 \times 10^{-7} \; \text{cm}^2/\text{s}\right)\left(1 \times 10^{18} - 50 \times 10^{18} \; \frac{\text{N atoms}}{\text{cm}^3}\right)\right]}{0.00139 \times 10^{18} \; \frac{\text{N atoms}}{\text{cm}^2 \cdot \text{s}}}$$

$\Delta x = 0.013$ cm = minimum thickness of the membrane

In a similar manner, the maximum thickness of the membrane that will permit 90% of the hydrogen to pass can be calculated as

$$\text{H atom loss per h} = (0.90)(35.343 \times 10^{20}) = 31.80 \times 10^{20}$$

$$\text{H atom loss per s} = 0.0088 \times 10^{20}$$

$$J = 0.125 \times 10^{18} \; \frac{\text{H atoms}}{\text{cm}^2 \cdot \text{s}}$$

From Equation 5-4 and Table 5-1,

$$D_{\text{H}} = 0.0012 \; \frac{\text{cm}^2}{\text{s}} \; \exp\left[\frac{-3{,}600 \; \frac{\text{cal}}{\text{mol}}}{1.987 \; \frac{\text{cal}}{\text{K} \cdot \text{mol}} \; (973 \; \text{K})}\right] = 1.86 \times 10^{-4} \; \text{cm}^2/\text{s}$$

Since

$$\Delta x = -D \, \Delta c/J$$

$$\Delta x = \frac{\left(-1.86 \times 10^{-4} \; \frac{\text{cm}^2}{\text{s}}\right)\left(-49 \times 10^{18} \; \frac{\text{H atoms}}{\text{cm}^3}\right)}{0.125 \times 10^{18} \; \frac{\text{H atoms}}{\text{cm}^2 \cdot \text{s}}}$$

$$= 0.073 \; \text{cm} = \text{maximum thickness}$$

An iron membrane with a thickness between 0.013 and 0.073 cm will be satisfactory.

Types of Diffusion

In **volume diffusion**, the atoms move through the crystal from one regular or interstitial site to another. Because of the surrounding atoms, the activation energy is large and the rate of diffusion is relatively slow.

Atoms can also diffuse along boundaries, interfaces, and surfaces in the material. Atoms diffuse easily by **grain boundary diffusion**, because the atom packing is disordered and less dense in the grain boundaries. Because atoms can more easily squeeze their way through the grain boundary, the activation energy is low (Table 5-2). **Surface diffusion** is easier still because there is even less constraint on the diffusing atoms at the surface.

Time

Diffusion requires time. The units for flux are $\frac{\text{atoms}}{\text{cm}^2 \cdot \text{s}}$. If a large number of atoms must diffuse to produce a uniform structure, long times may be required, even at high temperatures. Times for heat treatments may be reduced by using higher temperatures or by making the **diffusion distances** (related to Δx) as small as possible.

We find that some rather remarkable structures and properties are obtained if we prevent diffusion. Steels quenched rapidly from high temperatures to prevent

TABLE 5-2 ■ *The effect of the type of diffusion for thorium in tungsten and for self-diffusion in silver*

Diffusion Type	Diffusion Coefficient (*D*)			
	Thorium in Tungsten		Silver in Silver	
	D_0 (cm²/s)	Q (cal/mol)	D_0 (cm²/s)	Q (cal/mol)
Surface	0.47	66,400	0.068	8,900
Grain boundary	0.74	90,000	0.24	22,750
Volume	1.00	120,000	0.99	45,700

*Given by parameters for Equation 5-4.

diffusion form nonequilibrium structures and provide the basis for sophisticated heat treatments. Similarly, in forming metallic glasses, we have to quench liquid metals at a very high cooling rate. This is to avoid diffusion of atoms by decreasing their thermal energy and encouraging them to assemble into nonequilibrium amorphous arrangements. Melts of silicate glasses, on the other hand, are viscous and diffusion of ions through these is slow. As a result, we do not have to cool these melts very rapidly to attain an amorphous structure. There is a myth that many old buildings contain windowpanes that are thicker at the bottom than at the top because the glass has flowed down over the years. Based on kinetics of diffusion, it can be shown that even several hundred or thousand years will not be sufficient to cause such flow of glasses at near-room temperature. In certain thin film deposition processes such as sputtering, we sometimes obtain amorphous thin films if the atoms or ions are quenched rapidly after they land on the substrate. If these films are subsequently heated (after deposition) to sufficiently high temperatures, diffusion will occur and the amorphous thin films will eventually crystallize. In the following example, we examine different mechanisms for diffusion.

Example 5-6   *Tungsten Thorium Diffusion Couple*

Consider a diffusion couple between pure tungsten and a tungsten alloy containing 1 at% thorium. After several minutes of exposure at 2000°C, a transition zone of 0.01 cm thickness is established. What is the flux of thorium atoms at this time if diffusion is due to (a) volume diffusion, (b) grain boundary diffusion, and (c) surface diffusion? (See Table 5-2.)

SOLUTION

The lattice parameter of BCC tungsten is 3.165 Å. Thus, the number of tungsten atoms/cm³ is

$$\frac{\text{W atoms}}{\text{cm}^3} = \frac{2 \text{ atoms/cell}}{(3.165 \times 10^{-8})^3 \text{ cm}^3/\text{ cell}} = 6.3 \times 10^{22}$$

In the tungsten-1 at% thorium alloy, the number of thorium atoms is

$$c_{\text{Th}} = (0.01)(6.3 \times 10^{22}) = 6.3 \times 10^{20} \text{ Th} \frac{\text{atoms}}{\text{cm}^3}$$

In the pure tungsten, the number of thorium atoms is zero. Thus, the concentration gradient is

$$\frac{\Delta c}{\Delta x} = \frac{0 - 6.3 \times 10^{20} \frac{\text{atoms}}{\text{cm}^2}}{0.01 \text{ cm}} = -6.3 \times 10^{22} \text{ Th } \frac{\text{atoms}}{\text{cm}^3 \cdot \text{cm}}$$

a. Volume diffusion

$$D = 1.0 \frac{\text{cm}^2}{\text{s}} \exp\left(\frac{-120{,}000 \frac{\text{cal}}{\text{mol}}}{\left(1.987 \frac{\text{cal}}{\text{mol} \cdot \text{K}}\right)(2273 \text{ K})}\right) = 2.89 \times 10^{-12} \text{ cm}^2/\text{s}$$

$$J = -D \frac{\Delta c}{\Delta x} = -\left(2.89 \times 10^{-12} \frac{\text{cm}^2}{\text{s}}\right)\left(-6.3 \times 10^{22} \frac{\text{atoms}}{\text{cm}^3 \cdot \text{cm}}\right)$$

$$= 18.2 \times 10^{10} \frac{\text{Th atoms}}{\text{cm}^2 \cdot \text{s}}$$

b. Grain boundary diffusion

$$D = 0.74 \frac{\text{cm}^2}{\text{s}} \exp\left(\frac{-90{,}000 \frac{\text{cal}}{\text{mol}}}{\left(1.987 \frac{\text{cal}}{\text{mol} \cdot \text{K}}\right)(2273 \text{ K})}\right) = 1.64 \times 10^{-9} \text{ cm}^2/\text{s}$$

$$J = -\left(1.64 \times 10^{-9} \frac{\text{cm}^2}{\text{s}}\right)\left(-6.3 \times 10^{22} \frac{\text{atoms}}{\text{cm}^3 \cdot \text{cm}}\right) = 10.3 \times 10^{13} \frac{\text{Th atoms}}{\text{cm}^2 \cdot \text{s}}$$

c. Surface diffusion

$$D = 0.47 \frac{\text{cm}^2}{\text{s}} \exp\left(\frac{-66{,}400 \frac{\text{cal}}{\text{mol}}}{\left(1.987 \frac{\text{cal}}{\text{mol} \cdot \text{K}}\right)(2273 \text{ K})}\right) = 1.94 \times 10^{-7} \text{ cm}^2/\text{s}$$

$$J = -\left(1.94 \times 10^{-7} \frac{\text{cm}^2}{\text{s}}\right)\left(-6.3 \times 10^{22} \frac{\text{atoms}}{\text{cm}^3 \cdot \text{cm}}\right) = 12.2 \times 10^{15} \frac{\text{Th atoms}}{\text{cm}^2 \cdot \text{s}}$$

Dependence on Bonding and Crystal Structure

A number of factors influence the activation energy for diffusion and, hence, the rate of diffusion. Interstitial diffusion, with a low-activation energy, usually occurs much faster than vacancy, or substitutional diffusion. Activation energies are usually lower for atoms diffusing through open crystal structures than for close-packed crystal structures. Because the activation energy depends on the strength of atomic bonding, it is higher for diffusion of atoms in materials with a high melting temperature (Figure 5-17).

We also find that, due to their smaller size, cations (with a positive charge) often have higher diffusion coefficients than those for anions (with a negative charge). In sodium chloride, for instance, the activation energy for diffusion of chloride ions (Cl^-) is about twice that for diffusion of sodium ions (Na^+).

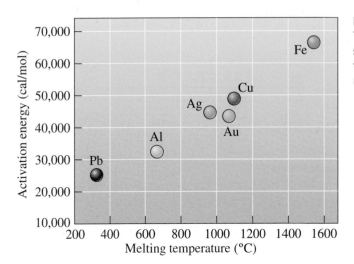

Figure 5-17
The activation energy for self-diffusion increases as the melting point of the metal increases.

Diffusion of ions also provides a transfer of electrical charge; in fact, the electrical conductivity of ionically bonded ceramic materials is related to temperature by an Arrhenius equation. As the temperature increases, the ions diffuse more rapidly, electrical charge is transferred more quickly, and the electrical conductivity is increased. As mentioned before, some ceramic materials are good conductors of electricity.

Dependence on Concentration of Diffusing Species and Composition of Matrix
The diffusion coefficient (D) depends not only on temperature, as given by Equation 5-4, but also on the concentration of diffusing species and composition of the matrix. The reader should consult higher-level textbooks for more information.

5-7   Permeability of Polymers

In polymers, we are most concerned with the diffusion of atoms or small molecules between the long polymer chains. As engineers, we often cite the permeability of polymers and other materials, instead of the diffusion coefficients. The **permeability** is expressed in terms of the volume of gas or vapor that can permeate per unit area, per unit time, or per unit thickness at a specified temperature and relative humidity. Polymers that have a polar group (e.g., ethylene vinyl alcohol) have higher permeability for water vapor than for oxygen gas. Polyethylene, on the other hand, has much higher permeability for oxygen than for water vapor. In general, the more compact the structure of polymers, the lesser the permeability. For example, low-density polyethylene has a higher permeability than high-density polyethylene. Polymers used for food and other applications need to have the appropriate barrier properties. For example, polymer films are typically used as packaging to store food. If air diffuses through the film, the food may spoil. Similarly, care has to be exercised in the storage of ceramic or metal powders that are sensitive to atmospheric water vapor, nitrogen, oxygen, or carbon dioxide. For example, zinc oxide powders used in rubbers, paints, and ceramics must be stored in polyethylene bags to avoid reactions with atmospheric water vapor.

Diffusion of some molecules into a polymer can cause swelling problems. For example, in automotive applications, polymers used to make o-rings can absorb considerable amounts of oil, causing them to swell. On the other hand, diffusion is required to enable dyes to uniformly enter many of the synthetic polymer fabrics. Selective diffusion through polymer membranes is used for desalinization of water. Water molecules pass through the polymer membrane, and the ions in the salt are trapped.

In each of these examples, the diffusing atoms, ions, or molecules penetrate between the polymer chains rather than moving from one location to another within the chain structure. Diffusion will be more rapid through this structure when the diffusing species is smaller or when larger voids are present between the chains. Diffusion through crystalline polymers, for instance, is slower than that through amorphous polymers, which have no long-range order and, consequently, have a lower density.

5-8   Composition Profile [Fick's Second Law]

Fick's second law, which describes the dynamic, or non-steady state, diffusion of atoms, is the differential equation

$$\frac{\partial c}{\partial t} = \frac{\partial}{\partial x}\left(D\frac{\partial c}{\partial x}\right) \tag{5-5}$$

If we assume that the diffusion coefficient D is not a function of location x and the concentration (c) of diffusing species, we can write a simplified version of Fick's second law as follows

$$\frac{\partial c}{\partial t} = D\left(\frac{\partial^2 c}{\partial x^2}\right) \tag{5-6}$$

The solution to this equation depends on the boundary conditions for a particular situation. One solution is

$$\frac{c_s - c_x}{c_s - c_0} = \mathrm{erf}\left(\frac{x}{2\sqrt{Dt}}\right) \tag{5-7}$$

where c_s is a constant concentration of the diffusing atoms at the surface of the material, c_0 is the initial uniform concentration of the diffusing atoms in the material, and c_x is the concentration of the diffusing atom at location x below the surface after time t. These concentrations are illustrated in Figure 5-18. In these equations we have assumed basically a one-dimensional model (i.e., we assume that atoms or other diffusing species are moving only in the direction x). The function "erf" is the error function and can be evaluated from Table 5-3 or Figure 5-19. Note that most standard spreadsheet

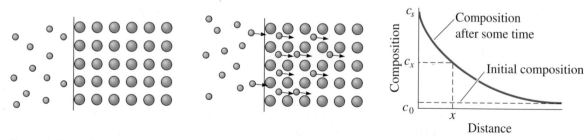

Figure 5-18   Diffusion of atoms into the surface of a material illustrating the use of Fick's second law.

TABLE 5-3 ■ *Error function values for Fick's second law*

Argument of the Error Function $\frac{x}{2\sqrt{Dt}}$	Value of the Error Function erf $\frac{x}{2\sqrt{Dt}}$
0	0
0.10	0.1125
0.20	0.2227
0.30	0.3286
0.40	0.4284
0.50	0.5205
0.60	0.6039
0.70	0.6778
0.80	0.7421
0.90	0.7969
1.00	0.8427
1.50	0.9661
2.00	0.9953

Note that error function values are available on many software packages found on personal computers.

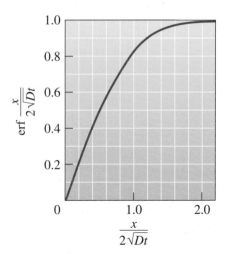

Figure 5-19 Graph showing the argument and value of the error function encountered in Fick's second law.

and other software programs available on a personal computer (e.g., Excel™) also provide error function values.

The mathematical definition of the error function is as follows:

$$\text{erf}(x) = \frac{2}{\sqrt{\pi}} \int_0^x \exp(-y^2)dy \qquad (5\text{-}8)$$

In Equation 5-8, y is known as the argument of the error function. We also define a complementary error function as follows:

$$\text{erfc}(x) = 1 - \text{erf}(x) \qquad (5\text{-}9)$$

This function is used in certain solution forms of Fick's second law.

As mentioned previously, depending upon the boundary conditions, different solutions (i.e., different equations) describe the solutions to Fick's second law. These solutions to Fick's second law permit us to calculate the concentration of one diffusing species as a function of time (t) and location (x). Equation 5-7 is *a possible solution* to Fick's law that describes the variation in concentration of different species near the surface of the material as a function of time and distance, provided that the diffusion coefficient D remains constant and the concentrations of the diffusing atoms at the surface (c_s) and at large distance (x) within the material (c_0) remain unchanged. Fick's second law can also assist us in designing a variety of materials processing techniques, including **carburization** and dopant diffusion in semiconductors as described in the following examples.

Example 5-7 *Design of a Carburizing Treatment*

The surface of a 0.1% C steel gear is to be hardened by carburizing. In gas carburizing, the steel gears are placed in an atmosphere that provides 1.2% C at the surface of the steel at a high temperature (Figure 5-1). Carbon then diffuses from the surface into the steel. For optimum properties, the steel must contain 0.45% C at a depth of 0.2 cm below the surface. Design a carburizing heat treatment that will produce these optimum properties. Assume that the temperature is high enough (at least 900°C) so that the iron has the FCC structure.

SOLUTION

Since the boundary conditions for which Equation 5-7 was derived are assumed to be valid, we can use this equation:

$$\frac{c_s - c_x}{c_s - c_0} = \text{erf}\left(\frac{x}{2\sqrt{Dt}}\right)$$

We know that $c_s = 1.2\%$ C, $c_0 = 0.1\%$ C, $c_x = 0.45\%$ C, and $x = 0.2$ cm. From Fick's second law:

$$\frac{c_s - c_x}{c_s - c_0} = \frac{1.2\% \text{ C} - 0.45\% \text{ C}}{1.2\% \text{ C} - 0.1\% \text{ C}} = 0.68 = \text{erf}\left(\frac{0.2 \text{ cm}}{2\sqrt{Dt}}\right) = \text{erf}\left(\frac{0.1 \text{ cm}}{\sqrt{Dt}}\right)$$

From Table 5-3, we find that

$$\frac{0.1 \text{ cm}}{\sqrt{Dt}} = 0.71 \text{ or } Dt = \left(\frac{0.1}{0.71}\right)^2 = 0.0198 \text{ cm}^2$$

Any combination of D and t with a product of 0.0198 cm² will work. For carbon diffusing in FCC iron, the diffusion coefficient is related to temperature by Equation 5-4:

$$D = D_0 \exp\left(\frac{-Q}{RT}\right)$$

From Table 5-1:

$$D = 0.23 \exp\left(\frac{-32,900 \text{ cal/mol}}{1.987 \frac{\text{cal}}{\text{mol} \cdot \text{K}} T \text{ (K)}}\right) = 0.23 \exp\left(\frac{-16,558}{T}\right)$$

Therefore, the temperature and time of the heat treatment are related by

$$t = \frac{0.0198 \text{ cm}^2}{D \frac{\text{cm}^2}{\text{s}}} = \frac{0.0198 \text{ cm}^2}{0.23 \exp(-16,558/T) \frac{\text{cm}^2}{\text{s}}} = \frac{0.0861}{\exp(-16,558/T)}$$

Some typical combinations of temperatures and times are

If $T = 900°C = 1173$ K, then $t = 116,273$ s $= 32.3$ h

If $T = 1000°C = 1273$ K, then $t = 38,362$ s $= 10.7$ h

If $T = 1100°C = 1373$ K, then $t = 14,876$ s $= 4.13$ h

If $T = 1200°C = 1473$ K, then $t = 6,560$ s $= 1.82$ h

The exact combination of temperature and time will depend on the maximum temperature that the heat treating furnace can reach, the rate at which parts must be produced, and the economics of the tradeoffs between higher temperatures versus longer times. Another factor to consider is changes in microstructure that occur in the rest of the material. For example, while carbon is diffusing into the surface, the rest of the microstructure can begin to experience grain growth or other changes.

Example 5-8 shows that one of the consequences of Fick's second law is that the same concentration profile can be obtained for different processing conditions, so long as the term Dt is constant. This permits us to determine the effect of temperature on the time required for a particular heat treatment to be accomplished.

Example 5-8   *Design of a More Economical Heat Treatment*

We find that 10 h are required to successfully carburize a batch of 500 steel gears at 900°C, where the iron has the FCC structure. We find that it costs $1000 per hour to operate the carburizing furnace at 900°C and $1500 per hour to operate the furnace at 1000°C. Is it economical to increase the carburizing temperature to 1000°C? What other factors must be considered?

SOLUTION

We again assume that we can use the solution to Fick's second law given by Equation 5-7:

$$\frac{c_s - c_x}{c_s - c_0} = \text{erf}\left(\frac{x}{2\sqrt{Dt}}\right)$$

Note that since we are dealing with only changes in heat treatment time and temperature, the term Dt must be constant.

The temperatures of interest are 900°C = 1173 K and 1000°C = 1273 K. To achieve the same carburizing treatment at 1000°C as at 900°C:

$$D_{1273}t_{1273} = D_{1173}t_{1173}$$

For carbon diffusing in FCC iron, the activation energy is 32,900 cal/mol. Since we are dealing with the ratios of times, it does not matter whether we substitute for the time in hours or seconds. It is, however, always a good idea to use units that balance out. Therefore, we will show time in seconds. Note that temperatures must be converted into Kelvin.

$$D_{1273}t_{1273} = D_{1173}t_{1173}$$

$$D = D_0 \exp(-Q/RT)$$

$$t_{1273} = \frac{D_{1173}t_{1173}}{D_{1273}}$$

$$= \frac{D_0 \exp\left(-\dfrac{32,900 \, \frac{\text{cal}}{\text{mol}}}{1.987 \, \frac{\text{cal}}{\text{mol}\cdot\text{K}} \, 1173\text{K}}\right)(10 \text{ h})(3600 \text{ s/h})}{D_0 \exp\left(-\dfrac{32,900 \, \frac{\text{cal}}{\text{mol}}}{1.987 \, \frac{\text{cal}}{\text{mol}\cdot\text{K}} \, 1273\text{K}}\right)}$$

$$t_{1273} = \frac{\exp(-14.1156)(10)(3600)}{\exp(-13.0068)}$$

$$= (10)(0.3299)(3600) \text{ s}$$

$$t_{1273} = 3.299 \text{ h} = 3 \text{ h and } 18 \text{ min}$$

Notice, we did not need the value of the pre-exponential term D_0, since it canceled out.

At 900°C, the cost per part is ($1000/h) (10 h)/500 parts = $20/part. At 1000°C, the cost per part is ($1500/h) (3.299 h)/500 parts = $9.90/part. Considering only the cost of operating the furnace, increasing the temperature reduces the heat-treating cost of the gears and increases the production rate. Another factor to consider is if the heat treatment at 1000°C could cause some other microstructural or other changes. For example, would increased temperature cause grains to grow significantly? If this is the case, we will be weakening the bulk of the material. How does the increased temperature affect the life of the other equipment such as the furnace itself and any accessories? How long would the cooling take? Will cooling from a higher temperature cause residual stresses? Would the product still meet all other specifications? These and other questions should be considered. The point is, as engineers, we need to ensure that the solution we propose is not only technically sound and economically sensible, it should recognize and make sense for the system as a whole. A good solution is often simple, solves problems for the system, and does not create new problems.

Example 5-9   *Silicon Device Fabrication*

Devices such as transistors are made by doping semiconductors. The diffusion coefficient of phosphorus (P) in Si is $D = 6.5 \times 10^{-13}$ cm^2/s at a temperature of 1100°C. Assume the source provides a surface concentration of 10^{20} atoms/cm^3 and the diffusion time is one hour. Assume that the silicon wafer initially contains no P.

Calculate the depth at which the concentration of P will be 10^{18} atoms/cm^3. State any assumptions you have made while solving this problem.

SOLUTION

We assume that we can use one of the solutions to Fick's second law (i.e., Equation 5-7):

$$\frac{c_s - c_x}{c_s - c_0} = \text{erf}\left(\frac{x}{2\sqrt{Dt}}\right)$$

We will use concentrations in atoms/cm^3, time in seconds, and D in $\frac{\text{cm}^2}{\text{s}}$. Notice that the left-hand side is dimensionless. Therefore, as long as we use concentrations in the same units for c_s, c_x, and c_0, it does not matter what those units are.

$$\frac{c_s - c_x}{c_s - c_0} = \frac{10^{20} \frac{\text{atoms}}{\text{cm}^3} - 10^{18} \frac{\text{atoms}}{\text{cm}^3}}{10^{20} \frac{\text{atoms}}{\text{cm}^3} - 0 \frac{\text{atoms}}{\text{cm}^3}} = 0.99$$

$$= \mathrm{erf}\left[\frac{x}{2\sqrt{\left(6.5 \times 10^{-13}\,\frac{cm^2}{s}\right)(3600\ s)}}\right]$$

$$= \mathrm{erf}\left(\frac{x}{9.67 \times 10^{-5}}\right)$$

From the error function values in Table 5-3 (or from your calculator/computer), If $\mathrm{erf}(z) = 0.99$, $z = 1.82$, therefore,

$$1.82 = \frac{x}{9.67 \times 10^{-5}}$$

or

$$x = 1.76 \times 10^{-4}\ cm$$

or

$$x = (1.76 \times 10^{-4}\ cm)\left(\frac{10^4\,\mu m}{cm}\right)$$

$$x = 1.76\ \mu m$$

Note that we have expressed the final answer in micrometers since this is the length scale that is appropriate for this application. The main assumptions we made are (1) the D value does not change while phosphorus (P) gets incorporated in the silicon wafer and (2) the diffusion of P is only in one dimension (i.e., we ignore any lateral diffusion).

Limitations to Applying the Error-Function Solution Given by Equation 5-7
Note that in the equation describing Fick's second law (Equation 5-7):

(a) It is assumed that D is independent of the concentration of the diffusing species;

(b) the surface concentration of the diffusing species (c_s) is always constant.

There are situations under which these conditions may not be met and hence the concentration profile evolution will not be predicted by the error-function solution shown in Equation 5-7. If the boundary conditions are different from the ones we assumed, different solutions to Fick's second law must be used.

5-9   Diffusion and Materials Processing

We briefly discussed applications of diffusion in processing materials in Section 5-1. Many important examples related to solidification, phase transformations, heat treatments, etc., will be discussed in later chapters. In this section, we provide more information to highlight the importance of diffusion in the processing of engineered materials. Diffusional processes become very important when materials are used or processed at elevated temperatures.

Melting and Casting

One of the most widely used methods to process metals, alloys, many plastics, and glasses involves melting and casting of materials into a desired shape. Diffusion plays a particularly important role in solidification of metals and alloys. During the growth of single crystals of semiconductors, for example, we must ensure that the differences in the diffusion of dopants in both the molten and solid forms are accounted for. This also applies for the diffusion of elements during the casting of alloys. Similarly, diffusion plays a critical role in the processing of glasses. In inorganic glasses, for instance, we rely on the fact that diffusion is slow and inorganic glasses do not crystallize easily. We will examine this topic further in Chapter 9.

Sintering

Although casting and melting methods are very popular for many manufactured materials, the melting points of many ceramic and some metallic materials are too high for processing by melting and casting. These relatively refractory materials are manufactured into useful shapes by a process that requires the consolidation of small particles of a powder into a solid mass (Chapter 15). **Sintering** is the high-temperature treatment that causes particles to join, gradually reducing the volume of pore space between them. Sintering is a frequent step in the manufacture of ceramic components (e.g., alumina, barium titanate, etc.) as well as in the production of metallic parts by **powder metallurgy**—a processing route by which metal powders are pressed and sintered into dense, monolithic components. A variety of composite materials such as tungsten carbide-cobalt based cutting tools, superalloys, etc., are produced using this technique. With finer particles, many atoms or ions are at the surface for which the atomic or ionic bonds are not satisfied. As a result, a collection of fine particles of a certain mass has higher energy than that for a solid cohesive material of the same mass. Therefore, the driving force for solid state sintering of powdered metals and ceramics is the *reduction in the total surface area* of powder particles. When a powdered material is compacted into a shape, the powder particles are in contact with one another at numerous sites, with a significant amount of pore space between them. In order to reduce the total energy of the material, atoms diffuse to the points of contact, bonding the particles together and eventually causing the pores to shrink.

Lattice diffusion from the bulk of the particles into the neck region causes densification. Surface diffusion, gas or vapor phase diffusion, and lattice diffusion from curved surfaces into the neck area between particles do not lead to densification (Chapter 15). If sintering is carried out over a long period of time, the pores may be eliminated and the material becomes dense (Figure 5-20). In Figure 5-21, particles of a

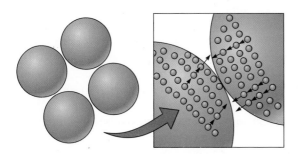
Compacted product

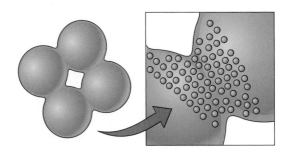
Partly sintered product

Figure 5-20 Diffusion processes during sintering and powder metallurgy. Atoms diffuse to points of contact, creating bridges and reducing the pore size.

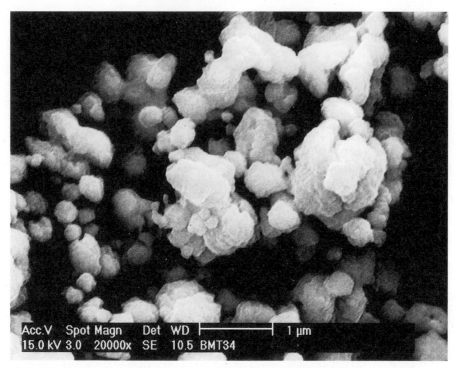

Figure 5-21   Particles of barium magnesium tantalate (BMT) (Ba(Mg$_{1/3}$ Ta$_{2/3}$)O$_3$) powder are shown. This ceramic material is useful in making electronic components known as dielectric resonators that are used for wireless communications. (*Courtesy of H. Shivey.*)

powder of a ceramic material known as barium magnesium tantalate (Ba(Mg$_{1/3}$Ta$_{2/3}$)O$_3$ or BMT) are shown. This ceramic material is useful in making electronic components known as *dielectric resonators* used in wireless communication systems. The microstructure of BMT ceramics is shown in Figure 5-22. These ceramics were produced by compacting the powders in a press and sintering the compact at a high temperature (~1500°C).

The extent and rate of sintering depends on (a) the initial density of the compacts, (b) temperature, (c) time, (d) the mechanism of sintering, (e) the average particle size, and (f) the size distribution of the powder particles. In some situations, a liquid phase forms in localized regions of the material while sintering is in process. Since diffusion of species, such as atoms and ions, is faster in liquids than in the solid state, the presence of a liquid phase can provide a convenient way for accelerating the sintering of many refractory metal and ceramic formulations. The process in which a small amount of liquid forms and assists densification is known as **liquid phase sintering**. For the liquid phase to be effective in enhancing sintering, it is important to have a liquid that can "wet" the grains, similar to how water wets a glass surface. If the liquid is non-wetting, similar to how mercury does not wet glass, then the liquid phase will not be helpful for enhancing sintering. In some cases, compounds are added to materials to cause the liquid phase to form at sintering temperatures. In other situations, impurities can react with the material and cause formation of a liquid phase. In most applications, it is desirable if the liquid phase is transient or converted into a crystalline material during cooling. This way a glassy and brittle amorphous phase does not remain at the grain boundaries.

When exceptionally high densities are needed, pressure (either uniaxial or isostatic) is applied while the material is being sintered. These techniques are known as

Figure 5-22   The microstructure of BMT ceramics obtained by compaction and sintering of BMT powders. (*Courtesy of H. Shivey.*)

hot pressing, when the pressure is unidirectional, or **hot isostatic pressing** (HIP), when the pressure is isostatic (i.e., applied in all directions). Many superalloys and ceramics such as lead lanthanum zirconium titanate (PLZT) are processed using these techniques. Hot isostatic pressing leads to high density materials with isotropic properties (Chapter 15).

Grain Growth

A polycrystalline material contains a large number of grain boundaries, which represent high-energy areas because of the inefficient packing of the atoms. A lower overall energy is obtained in the material if the amount of grain boundary area is reduced by grain growth. **Grain growth** involves the movement of grain boundaries, permitting larger grains to grow at the expense of smaller grains (Figure 5-23). If you have watched froth, you have probably seen the principle of grain growth! Grain growth is similar to the way smaller bubbles in the froth disappear at the expense of bigger bubbles. Another analogy is big fish getting bigger by eating small fish! For grain growth in materials, diffusion of atoms across the grain boundary is required, and, consequently, the growth of the grains is related to the activation energy needed for an atom to jump across the boundary. The increase in grain size can be seen from the sequence of micrographs for alumina ceramics shown in Figure 5-23. Another example for which grain growth plays a role is in the tungsten (W) filament in a lightbulb. As the tungsten filament gets hotter, the grains grow causing it to get weaker. This grain growth, vaporization of tungsten, and oxidation via reaction with remnant oxygen contribute to the failure of tungsten filaments in a lightbulb.

The **driving force** for grain growth is reduction in grain boundary area. Grain boundaries are defects and their presence causes the free energy of the material to increase. Thus, the thermodynamic tendency of polycrystalline materials is to transform

(a)   (b)

Figure 5-23   Grain growth in alumina ceramics can be seen from the scanning electron micrographs of alumina ceramics. (a) The left micrograph shows the microstructure of an alumina ceramic sintered at 1350°C for 15 hours. (b) The right micrograph shows a sample sintered at 1350°C for 30 hours. (*Courtesy of Dr. Ian Nettleship and Dr. Richard McAfee.*)

into materials that have a larger average grain size. High temperatures or low-activation energies increase the size of the grains. Many heat treatments of metals, which include holding the metal at high temperatures, must be carefully controlled to *avoid* excessive grain growth. This is because, as the average grain size grows, the grain-boundary area decreases, and there is consequently less resistance to motion of dislocations. As a result, the strength of a metallic material will decrease with increasing grain size. We have seen this concept before in the form of the Hall-Petch equation (Chapter 4). In **normal grain growth**, the average grain size increases steadily and the width of the grain size distribution is not affected severely. In **abnormal grain growth**, the grain size distribution tends to become bi-modal (i.e., we get a few grains that are very large and then we have a few grains that remain relatively small). Certain electrical, magnetic, and optical properties of materials also depend upon the grain size of materials. As a result, in the processing of these materials, attention has to be paid to factors that affect diffusion rates and grain growth.

Diffusion Bonding   A method used to join materials, called **diffusion bonding**, occurs in three steps (Figure 5-24). The first step forces the two surfaces together at a high temperature and pressure, flattening the surface, fragmenting impurities, and producing a high atom-to-atom contact area. As the surfaces remain pressed together at high temperatures, atoms diffuse along grain boundaries to the remaining voids; the atoms condense and reduce the size of any voids at the interface. Because grain boundary diffusion is rapid, this second step may occur very quickly. Eventually, however, grain growth isolates the remaining voids from the grain boundaries. For the third step—final elimination of the voids—volume diffusion, which is comparatively slow, must occur. The diffusion bonding process is often used for joining reactive metals such as titanium, for joining dissimilar metals and materials, and for joining ceramics.

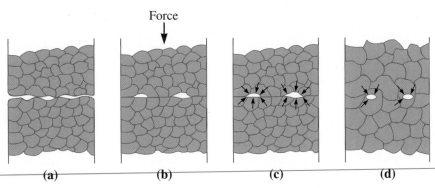

Figure 5-24   The steps in diffusion bonding: (a) Initially the contact area is small; (b) application of pressure deforms the surface, increasing the bonded area; (c) grain boundary diffusion permits voids to shrink; and (d) final elimination of the voids requires volume diffusion.

Summary

- The net flux of atoms, ions, etc., resulting from diffusion depends upon the initial concentration gradient.

- The kinetics of diffusion depend strongly on temperature. In general, diffusion is a thermally activated process and the dependence of the diffusion coefficient on temperature is given by the Arrhenius equation.

- The extent of diffusion depends on temperature, time, the nature and concentration of diffusing species, crystal structure, composition of the matrix, stoichiometry, and point defects.

- Encouraging or limiting the diffusion process forms the underpinning of many important technologies. Examples include the processing of semiconductors, heat treatments of metallic materials, sintering of ceramics and powdered metals, formation of amorphous materials, solidification of molten materials during a casting process, diffusion bonding, and barrier plastics, films, and coatings.

- Fick's laws describe the diffusion process quantitatively. Fick's first law defines the relationship between the chemical potential gradient and the flux of diffusing species. Fick's second law describes the variation of concentration of diffusing species under non-steady state diffusion conditions.

- For a particular system, the amount of diffusion is related to the term Dt. This term permits us to determine the effect of a change in temperature on the time required for a diffusion-controlled process.

- The two important mechanisms for atomic movement in crystalline materials are vacancy diffusion and interstitial diffusion. Substitutional atoms in the crystalline materials move by the vacancy mechanism.

- The rate of diffusion is governed by the Arrhenius relationship—that is, the rate increases exponentially with temperature. Diffusion is particularly significant at temperatures above about 0.4 times the melting temperature (in Kelvin) of the material.

- The activation energy Q describes the ease with which atoms diffuse, with rapid diffusion occurring for a low activation energy. A low-activation energy and rapid diffusion rate are obtained for (1) interstitial diffusion compared to vacancy diffusion, (2) crystal structures with a smaller packing factor, (3) materials with a low melting temperature or weak atomic bonding, and (4) diffusion along grain boundaries or surfaces.

- The total movement of atoms, or flux, increases when the concentration gradient and temperature increase.

- Diffusion of ions in ceramics is usually slower than that of atoms in metallic materials. Diffusion in ceramics is also affected significantly by non-stoichiometry, dopants, and the possible presence of liquid phases during sintering.

- Atom diffusion is of paramount importance because many of the materials processing techniques, such as sintering, powder metallurgy, and diffusion bonding, require diffusion. Furthermore, many of the heat treatments and strengthening mechanisms used to control structures and properties in materials are diffusion-controlled processes. The stability of the structure and the properties of materials during use at high temperatures depend on diffusion.

Glossary

Abnormal grain growth A type of grain growth observed in metals and ceramics. In this mode of grain growth, a bimodal grain size distribution usually emerges as some grains become very large at the expense of smaller grains. See "Grain growth" and "Normal grain growth."

Activation energy The energy required to cause a particular reaction to occur. In diffusion, the activation energy is related to the energy required to move an atom from one lattice site to another.

Carburization A heat treatment for steels to harden the surface using a gaseous or solid source of carbon. The carbon diffusing into the surface makes the surface harder and more abrasion resistant.

Concentration gradient The rate of change of composition with distance in a nonuniform material, typically expressed as $\frac{\text{atoms}}{\text{cm}^3 \cdot \text{cm}}$ or $\frac{\text{at\%}}{\text{cm}}$.

Conductive ceramics Ceramic materials that are good conductors of electricity as a result of their ionic and electronic charge carriers (electrons, holes, or ions). Examples of such materials are stabilized zirconia and indium tin oxide.

Diffusion The net flux of atoms, ions, or other species within a material caused by temperature and a concentration gradient.

Diffusion bonding A joining technique in which two surfaces are pressed together at high pressures and temperatures. Diffusion of atoms to the interface fills in voids and produces a strong bond between the surfaces.

Diffusion coefficient (D) A temperature-dependent coefficient related to the rate at which atoms, ions, or other species diffuse. The diffusion coefficient depends on temperature, the composition and microstructure of the host material and also the concentration of the diffusing species.

Diffusion couple A combination of elements involved in diffusion studies (e.g., if we are considering diffusion of Al in Si, then Al-Si is a diffusion couple).

Diffusion distance The maximum or desired distance that atoms must diffuse; often, the distance between the locations of the maximum and minimum concentrations of the diffusing atom.

Diffusivity Another term for the diffusion coefficient (D).

Driving force A cause that induces an effect. For example, an increased gradient in composition enhances diffusion; similarly reduction in surface area of powder particles is the driving force for sintering.

Fick's first law The equation relating the flux of atoms by diffusion to the diffusion coefficient and the concentration gradient.

Fick's second law The partial differential equation that describes the rate at which atoms are redistributed in a material by diffusion. Many solutions exist to Fick's second law; Equation 5-7 is one possible solution.

Flux The number of atoms or other diffusing species passing through a plane of unit area per unit time. This is related to the rate at which mass is transported by diffusion in a solid.

Grain boundary diffusion Diffusion of atoms along grain boundaries. This is faster than volume diffusion, because the atoms are less closely packed in grain boundaries.

Grain growth Movement of grain boundaries by diffusion in order to reduce the amount of grain boundary area. As a result, small grains shrink and disappear and other grains become larger, similar to how some bubbles in soap froth become larger at the expense of smaller bubbles. In many situations, grain growth is not desirable.

Hot isostatic pressing A sintering process in which a uniform pressure is applied in all directions during sintering. This process is used for obtaining very high densities and isotropic properties.

Hot pressing A sintering process conducted under uniaxial pressure, used for achieving higher densities.

Interdiffusion Diffusion of different atoms in opposite directions. Interdiffusion may eventually produce an equilibrium concentration of atoms within the material.

Interstitial diffusion Diffusion of small atoms from one interstitial position to another in the crystal structure.

Liquid phase sintering A sintering process in which a liquid phase forms. Since diffusion is faster in liquids, if the liquid can wet the grains, it can accelerate the sintering process.

Nitriding A process in which nitrogen is diffused into the surface of a material, such as a steel, leading to increased hardness and wear resistance.

Normal grain growth Grain growth that occurs in an effort to reduce grain boundary area. This type of grain growth is to be distinguished from abnormal grain growth in that the grain size distribution remains unimodal but the average grain size increases steadily.

Permeability A relative measure of the diffusion rate in materials, often applied to plastics and coatings. It is often used as an engineering design parameter that describes the effectiveness of a particular material to serve as a barrier against diffusion.

Powder metallurgy A method for producing monolithic metallic parts; metal powders are compacted into a desired shape, which is then heated to allow diffusion and sintering to join the powders into a solid mass.

Self-diffusion The random movement of atoms within an essentially pure material. No net change in composition results.

Sintering A high-temperature treatment used to join small particles. Diffusion of atoms to points of contact causes bridges to form between the particles. Further diffusion eventually fills in any remaining voids. The driving force for sintering is a reduction in total surface area of the powder particles.

Surface diffusion Diffusion of atoms along surfaces, such as cracks or particle surfaces.

Thermal barrier coatings (TBC) Coatings used to protect a component from heat. For example, some of the turbine blades in an aircraft engine are made from nickel-based superalloys and are coated with yttria stabilized zirconia (YSZ).

Vacancy diffusion Diffusion of atoms when an atom leaves a regular lattice position to fill a vacancy in the crystal. This process creates a new vacancy, and the process continues.

Volume diffusion Diffusion of atoms through the interior of grains.

Problems

Section 5-1 Applications of Diffusion

5-1 What is the driving force for diffusion?

5-2 In the carburization treatment of steels, what are the diffusing species?

5-3 Why do we use PET plastic to make carbonated beverage bottles?

5-4 Why is it that aluminum metal oxidizes more readily than iron but aluminum is considered to be a metal that usually does not "rust?"

5-5 What is a thermal barrier coating? Where are such coatings used?

Section 5-2 Stability of Atoms and Ions

5-6 What is a nitriding heat treatment?

5-7 A certain mechanical component is heat treated using carburization. A common engineering problem encountered is that we need to machine a certain part of the component and this part of the surface should not be hardened. Explain how we can achieve this objective.

5-8 Write down the Arrhenius equation and explain the different terms.

5-9 Atoms are found to move from one lattice position to another at the rate of $5 \times 10^5 \frac{\text{jumps}}{\text{s}}$ at 400°C when the activation energy for their movement is 30,000 cal/mol. Calculate the jump rate at 750°C.

5-10 The number of vacancies in a material is related to temperature by an Arrhenius equation. If the fraction of lattice points containing vacancies is 8×10^{-5} at 600°C, determine the fraction of lattice points containing vacancies at 1000°C.

5-11 The Arrhenius equation was originally developed for comparing rates of chemical reactions. Compare the rates of a chemical reaction at 20 and 100°C by calculating the ratio of the chemical reaction rates. Assume that the activation energy for liquids in which the chemical reaction is conducted is 10 kJ/mol and that the reaction is limited by diffusion.

Section 5-3 Mechanisms for Diffusion

5-12 What are the different mechanisms for diffusion?

5-13 Why is it that the activation energy for diffusion via the interstitial mechanism is smaller than those for other mechanisms?

5-14 How is self-diffusion of atoms in metals verified experimentally?

5-15 Compare the diffusion coefficients of carbon in BCC and FCC iron at the allotropic transformation temperature of 912°C and explain the difference.

5-16 Compare the diffusion coefficients for hydrogen and nitrogen in FCC iron at 1000°C and explain the difference.

Section 5-4 Activation Energy for Diffusion

5-17 Activation energy is sometimes expressed as (eV/atom). For example, see Figure 5-15 illustrating the diffusion coefficients of ions in different oxides. Convert eV/atom into J/mol.

5-18 The diffusion coefficient for Cr^{+3} in Cr_2O_3 is 6×10^{-15} cm²/s at 727°C and 1×10^{-9} cm²/s at 1400°C. Calculate
   (a) the activation energy and
   (b) the constant D_0.

5-19 The diffusion coefficient for O^{-2} in Cr_2O_3 is 4×10^{-15} cm²/s at 1150°C and 6×10^{-11} cm²/s at 1715°C. Calculate
   (a) the activation energy and
   (b) the constant D_0.

5-20 Without referring to the actual data, can you predict whether the activation energy for diffusion of carbon in FCC iron will be higher or lower than that in BCC iron? Explain.

Section 5-5 Rate of Diffusion (Fick's First Law)

5-21 Write down Fick's first law of diffusion. Clearly explain what each term means.

5-22 What is the difference between diffusivity and the diffusion coefficient?

5-23 A 1-mm-thick BCC iron foil is used to separate a region of high nitrogen gas concentration of 0.1 atomic percent from a region of low nitrogen gas concentration at 650°C.

If the flux of nitrogen through the foil is 10^{12} atoms/(cm$^2 \cdot$ s), what is the nitrogen concentration in the low concentration region?

5-24 A 0.2-mm-thick wafer of silicon is treated so that a uniform concentration gradient of antimony is produced. One surface contains 1 Sb atom per 10^8 Si atoms and the other surface contains 500 Sb atoms per 10^8 Si atoms. The lattice parameter for Si is 5.4307 Å (Appendix A). Calculate the concentration gradient in
(a) atomic percent Sb per cm and
(b) Sb $\frac{\text{atoms}}{\text{cm}^3 \cdot \text{cm}}$.

5-25 When a Cu-Zn alloy solidifies, one portion of the structure contains 25 atomic percent zinc and another portion 0.025 mm away contains 20 atomic percent zinc. The lattice parameter for the FCC alloy is about 3.63×10^{-8} cm. Determine the concentration gradient in
(a) atomic percent Zn per cm;
(b) weight percent Zn per cm; and
(c) Zn $\frac{\text{atoms}}{\text{cm}^3 \cdot \text{cm}}$.

5-26 A 0.0025-cm BCC iron foil is used to separate a high hydrogen content gas from a low hydrogen content gas at 650°C. 5×10^8 H atoms/cm^3 are in equilibrium on one side of the foil, and 2×10^3 H atoms/cm^3 are in equilibrium with the other side. Determine
(a) the concentration gradient of hydrogen and
(b) the flux of hydrogen through the foil.

5-27 A 1-mm-thick sheet of FCC iron is used to contain nitrogen in a heat exchanger at 1200°C. The concentration of N at one surface is 0.04 atomic percent, and the concentration at the second surface is 0.005 atomic percent. Determine the flux of nitrogen through the foil in N atoms/(cm$^2 \cdot$ s).

5-28 A 4-cm-diameter, 0.5-mm-thick spherical container made of BCC iron holds nitrogen at 700°C. The concentration at the inner surface is 0.05 atomic percent and at the outer surface is 0.002 atomic percent. Calculate the number of grams of nitrogen that are lost from the container per hour.

5-29 A BCC iron structure is to be manufactured that will allow no more than 50 g of hydrogen

to be lost per year through each square centimeter of the iron at 400°C. If the concentration of hydrogen at one surface is 0.05 H atom per unit cell and 0.001 H atom per unit cell at the second surface, determine the minimum thickness of the iron.

5-30 Determine the maximum allowable temperature that will produce a flux of less than 2000 H atoms/(cm$^2 \cdot$s) through a BCC iron foil when the concentration gradient is $-5 \times 10^{16} \frac{\text{atoms}}{\text{cm}^3 \cdot \text{cm}}$. (Note the negative sign for the flux.)

Section 5-6 Factors Affecting Diffusion

5-31 Write down the equation that describes the dependence of D on temperature.

5-32 In solids, the process of diffusion of atoms and ions takes time. Explain how this is used to our advantage while forming metallic glasses.

5-33 Why is it that inorganic glasses form upon relatively slow cooling of melts, while rapid solidification is necessary to form metallic glasses?

5-34 Use the diffusion data in the table below for atoms in iron to answer the questions that follow. Assume metastable equilibrium conditions and trace amounts of C in Fe. The gas constant in SI units is 8.314 J/(mol $\cdot$ K).

Diffusion Couple	Diffusion Mechanism	Q (J/mol)	D_0 (m^2/s)
C in FCC iron	Interstitial	1.38×10^5	2.3×10^{-5}
C in BCC iron	Interstitial	8.74×10^4	1.1×10^{-6}
Fe in FCC iron	Vacancy	2.79×10^5	6.5×10^{-5}
Fe in BCC iron	Vacancy	2.46×10^5	4.1×10^{-4}

(a) Plot the diffusion coefficient as a function of inverse temperature $(1/T)$ showing all four diffusion couples in the table.
(b) Recall the temperatures for phase transitions in iron, and for each case, indicate on the graph the temperature range over which the diffusion data is valid.
(c) Why is the activation energy for Fe diffusion higher than that for C diffusion in iron?

(d) Why is the activation energy for diffusion higher in FCC iron when compared to BCC iron?

(e) Does C diffuse faster in FCC Fe than in BCC Fe? Support your answer with a numerical calculation and state any assumptions made.

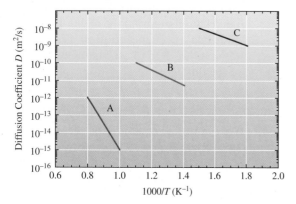

Figure 5-25   Plot for Problems 5-34 and 5-35.

5-35 The plot above has three lines representing grain boundary, surface, and volume self-diffusion in a metal. Match the lines labeled A, B, and C with the type of diffusion. Justify your answer by calculating the activation energy for diffusion for each case.

Section 5-7 Permeability of Polymers

5-36 What are barrier polymers?

5-37 What factors, other than permeability, are important in selecting a polymer for making plastic bottles?

5-38 Amorphous PET is more permeable to CO_2 than PET that contains microcrystallites. Explain why.

Section 5-8 Composition Profile (Fick's Second Law)

5-39 Transistors are made by doping single crystal silicon with different types of impurities to generate *n-* and *p-* type regions. Phosphorus (P) and boron (B) are typical *n-* and *p*-type dopant species, respectively. Assuming that a thermal treatment at 1100°C for 1 h is used to cause diffusion of the dopants, calculate the constant surface concentration of P and B needed to achieve a concentration of 10^{18} atoms/cm³ at a depth of 0.1 μm from the surface for both *n-* and *p*-type regions. The diffusion coefficients of P and B in single crystal silicon at 1100°C are 6.5×10^{-13} cm²/s and 6.1×10^{-13} cm²/s, respectively.

5-40 Consider a 2-mm-thick silicon (Si) wafer to be doped using antimony (Sb). Assume that the dopant source (gas mixture of antimony chloride and other gases) provides a constant concentration of 10^{22} atoms/m³. We need a dopant profile such that the concentration of Sb at a depth of 1 micrometer is 5×10^{21} atoms/m³. What is the required time for the diffusion heat treatment? Assume that the silicon wafer initially contains no impurities or dopants. Assume that the activation energy for diffusion of Sb in silicon is 380 kJ/mol and D_0 for Sb diffusion in Si is 1.3×10^{-3} m²/s. Assume $T = 1250°C$.

5-41 Consider doping of Si with gallium (Ga). Assume that the diffusion coefficient of gallium in Si at 1100°C is 7×10^{-13} cm²/s. Calculate the concentration of Ga at a depth of 2.0 micrometer if the surface concentration of Ga is 10^{23} atoms/cm³. The diffusion times are 1, 2, and 3 hours.

5-42 Compare the rate at which oxygen ions diffuse in alumina (Al_2O_3) with the rate at which aluminum ions diffuse in Al_2O_3 at 1500°C. Explain the difference.

5-43 A carburizing process is carried out on a 0.10% C steel by introducing 1.0% C at the surface at 980°C, where the iron is FCC. Calculate the carbon content at 0.01 cm, 0.05 cm, and 0.10 cm beneath the surface after 1 h.

5-44 Iron containing 0.05% C is heated to 912°C in an atmosphere that produces 1.20% C at the surface and is held for 24 h. Calculate the carbon content at 0.05 cm beneath the surface if
(a) the iron is BCC and
(b) the iron is FCC. Explain the difference.

5-45 What temperature is required to obtain 0.50% C at a distance of 0.5 mm beneath the surface of a 0.20% C steel in 2 h, when 1.10% C is present at the surface? Assume that the iron is FCC.

5-46 A 0.15% C steel is to be carburized at 1100°C, giving 0.35% C at a distance of 1 mm beneath the surface. If the surface composition is maintained at 0.90% C, what time is required?

5-47 A 0.02% C steel is to be carburized at 1200°C in 4 h, with the carbon content 0.6 mm beneath the surface reaching 0.45% C. Calculate the carbon content required at the surface of the steel.

5-48 A 1.2% C tool steel held at 1150°C is exposed to oxygen for 48 h. The carbon content at the steel surface is zero. To what depth will the steel be decarburized to less than 0.20% C?

5-49 A 0.80% C steel must operate at 950°C in an oxidizing environment for which the carbon content at the steel surface is zero. Only the outermost 0.02 cm of the steel part can fall below 0.75% C. What is the maximum time that the steel part can operate?

5-50 A steel with the BCC crystal structure containing 0.001% N is nitrided at 550°C for 5 h. If the nitrogen content at the steel surface is 0.08%, determine the nitrogen content at 0.25 mm from the surface.

5-51 What time is required to nitride a 0.002% N steel to obtain 0.12% N at a distance of 0.005 cm beneath the surface at 625°C? The nitrogen content at the surface is 0.15%.

5-52 We can successfully perform a carburizing heat treatment at 1200°C in 1 h. In an effort to reduce the cost of the brick lining in our furnace, we propose to reduce the carburizing temperature to 950°C. What time will be required to give us a similar carburizing treatment?

5-53 During freezing of a Cu-Zn alloy, we find that the composition is nonuniform. By heating the alloy to 600°C for 3 h, diffusion of zinc helps to make the composition more uniform. What temperature would be required if we wished to perform this homogenization treatment in 30 minutes?

5-54 To control junction depth in transistors, precise quantities of impurities are introduced at relatively shallow depths by ion implantation and diffused into the silicon substrate in a subsequent thermal treatment. This can be approximated as a finite source diffusion problem. Applying the appropriate boundary conditions, the solution to Fick's second law under these conditions is

$$c(x, t) = \frac{Q}{\sqrt{\pi Dt}} \exp\left(\frac{x^2}{4Dt}\right),$$

where Q is the initial surface concentration with units of atoms/cm^2. Assume that we implant 10^{14} atoms/cm^2 of phosphorus at the surface of a silicon wafer with a background boron concentration of 10^{16} atoms/cm^3 and this wafer is subsequently annealed at 1100°C. The diffusion coefficient (D) of phosphorus in silicon at 1100°C is 6.5×10^{-13} cm^2/s.

(a) Plot a graph of the concentration c (atoms/cm^3) versus x (cm) for anneal times of 5 minutes, 10 minutes, and 15 minutes.

(b) What is the anneal time required for the phosphorus concentration to equal the boron concentration at a depth of 1 μm?

Section 5-9 Diffusion and Materials Processing

5-55 Arrange the following materials in increasing order of self-diffusion coefficient: Ar gas, water, single crystal aluminum, and liquid aluminum at 700°C.

5-56 Most metals and alloys can be processed using the melting and casting route, but we typically do not choose to apply this method for the processing of specific metals (e.g., W) and most ceramics. Explain.

5-57 What is sintering? What is the driving force for sintering?

5-58 Why does grain growth occur? What is meant by the terms normal and abnormal grain growth?

5-59 Why is the strength of many metallic materials expected to decrease with increasing grain size?

5-60 A ceramic part made of MgO is sintered successfully at 1700°C in 90 minutes. To minimize thermal stresses during the process, we plan to reduce the temperature to 1500°C. Which will limit the rate at which sintering can be done: diffusion of magnesium ions or diffusion of oxygen ions? What time will be required at the lower temperature?

5-61 A Cu-Zn alloy has an initial grain diameter of 0.01 mm. The alloy is then heated to various temperatures, permitting grain growth to occur. The times required for the grains to grow to a diameter of 0.30 mm are shown below.

Temperature (°C)	Time (minutes)
500	80,000
600	3,000
700	120
800	10
850	3

Determine the activation energy for grain growth. Does this correlate with the diffusion of zinc in copper? (*Hint*: Note that rate is the reciprocal of time.)

5-62 What are the advantages of using hot pressing and hot isostatic pressing compared to using normal sintering?

5-63 A sheet of gold is diffusion-bonded to a sheet of silver in 1 h at 700°C. At 500°C, 440 h are required to obtain the same degree of bonding, and at 300°C, bonding requires 1530 years. What is the activation energy for the diffusion bonding process? Does it appear that diffusion of gold or diffusion of silver controls the bonding rate? (*Hint*: Note that rate is the reciprocal of time.)

⬡ Design Problems

5-64 Design a spherical tank, with a wall thickness of 2 cm that will ensure that no more than 50 kg of hydrogen will be lost per year. The tank, which will operate at 500°C, can be made of nickel, aluminum, copper, or iron. The diffusion coefficient of hydrogen and the cost per kilogram for each available material is listed below.

Diffusion Data Material	D_0 (cm²/s)	Q (cal/mol)	Cost ($/kg)
Nickel	0.0055	8,900	9.0
Aluminum	0.16	10,340	1.32
Copper	0.011	9,380	2.43
Iron (BCC)	0.0012	3,600	0.33

5-65 A steel gear initially containing 0.10% C is to be carburized so that the carbon content at a depth of 0.13 cm is 0.50% C. We can generate a carburizing gas at the surface that contains anywhere from 0.95% C to 1.15% C. Design an appropriate carburizing heat treatment.

5-66 When a valve casting containing copper and nickel solidifies under nonequilibrium conditions, we find that the composition of the alloy varies substantially over a distance of 0.005 cm. Usually we are able to eliminate this concentration difference by heating the alloy for 8 h at 1200°C; however, sometimes this treatment causes the alloy to begin to melt, destroying the part. Design a heat treatment that will eliminate the non-uniformity without melting. Assume that the cost of operating the furnace per hour doubles for each 100°C increase in temperature.

5-67 Assume that the surface concentration of phosphorus (P) being diffused in Si is 10^{21} atoms/cm³. We need to design a process, known as the pre-deposition step, such that the concentration of P (c_1 for step 1) at a depth of 0.25 μm is 10^{13} atoms/cm³. Assume that this is conducted at a temperature of 1000°C and the diffusion coefficient of P in Si at this temperature is 2.5×10^{-14} cm²/s. Assume that the process is carried out for a total time of 8 minutes. Calculate the concentration profile (i.e., c_1 as a function of depth, which in this case is given by the following equation). Notice the use of the complementary error function.

$$c_1(x, t_1) = c_s\left[1 - \text{erf}\left(\frac{x}{4Dt}\right)\right]$$

Use different values of x to generate and plot this profile of P during the pre-deposition step.

▲ Computer Problems

5-68 *Calculation of Diffusion Coefficients.* Write a computer program that will ask the user to provide the data for the activation energy Q and the value of D_0. The program should then ask the user to input a valid range of temperatures. The program, when executed, provides the values of D as a function of temperature in increments chosen by the user. The program should convert Q, D_0, and temperature to the proper units. For example, if the user provides temperatures in °F, the program should recognize that and convert the temperature into K. The program should also carry a cautionary statement about the standard assumptions made. For example, the program should caution the user that effects of any possible changes in crystal structure in the temperature range specified are not accounted for. Check the validity of your programs using examples in the book and also other problems that you may have solved using a calculator.

5-69 *Comparison of Reaction Rates.* Write a computer program that will ask the user to input the activation energy for a chemical reaction. The program should then ask the user to provide two temperatures for which the reaction rates need to be compared. Using the value of the gas constant and activation energy, the program should then provide a ratio of the reaction rates. The program should take into account different units for activation energy.

5-70 *Carburization Heat Treatment.* The program should ask the user to provide an input for the carbon concentration at the surface (c_s), and the concentration of carbon in the bulk (c_0). The program should also ask the user to provide the temperature and a value for D (or values for D_0 and Q, that will allow for D to be calculated). The program should then provide an output of the concentration profile in tabular form. The distances at which concentrations of carbon are to be determined can be provided by the user or defined by the person writing the program. The program should also be able to handle

calculation of heat treatment times if the user provides a level of concentration that is needed at a specified depth. This program will require calculation of the error function. The programming language or spreadsheet you use may have a built-in function that calculates the error function. If that is the case, use that function. You can also calculate the error function as the expansion of the following series:

$$\text{erf}(z) = 1 - \frac{e^{-z^2}}{\sqrt{\pi}z}\left(1 - \frac{1}{2z^2} - \frac{1*3}{(2z^2)^2} + \frac{1*3*5}{(2z^2)^3} + \cdots\right)$$

or use an approximation

$$\text{erf}(z) = 1 - \left(\left[\frac{1}{\sqrt{\pi}}\right]\frac{e^{-z^2}}{z}\right)$$

In these equations, z is the argument of the error function. Also, under certain situations, you will know the value of the error function (if all concentrations are provided) and you will have to figure out the argument of the error function. This can be handled by having part of the program compare different values for the argument of the error function and by minimizing the difference between the value of the error function you require and the value of the error function that was approximated.

Ⓚ Knovel® **Problems**

K5-1 Compare the carbon dioxide permeabilities of low-density polyethylene (LDPE), polypropylene, and polyethylene terephthalate (PET) films at room temperature.

Some materials can become brittle when temperatures are low and/or strain rates are high. The special chemistry of the steel used on the *Titanic* and the stresses associated with the fabrication and embrittlement of this steel when subjected to lower temperatures have been identified as factors contributing to the failure of the ship's hull. (*Hulton Archive/Getty Images.*)

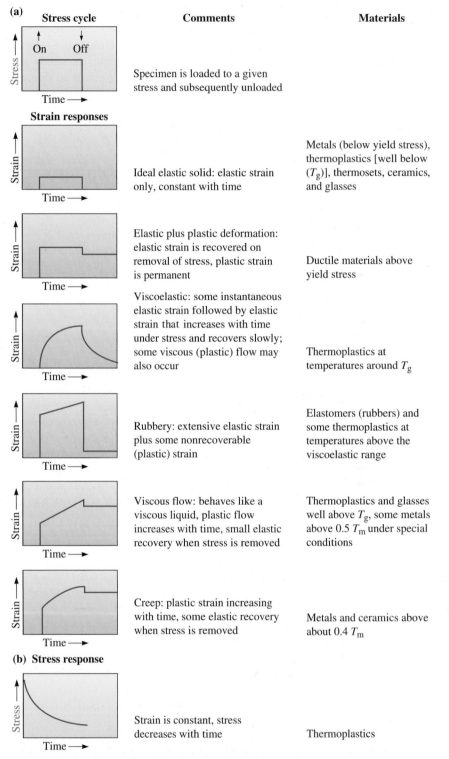

Figure 6-2 (a) Various types of strain response to an imposed stress where T_g = glass-transition temperature and T_m = melting point. *(Reprinted from* Materials Principles and Practice, *by C. Newey and G. Weaver (Eds.), 1991 p. 300, Fig. 6-9. Copyright © 1991 Butterworth-Heinemann. Reprinted with permission from Elsevier Science.)* (b) Stress relaxation in a viscoelastic material. *Note the vertical axis is stress.* Strain is constant.

While dealing with molten materials, liquids, and dispersions, such as paints or gels, a description of the resistance to flow under an applied stress is required. If the relationship between the applied shear stress τ and the **shear strain rate** ($\dot{\gamma}$) is linear, we refer to that material as **Newtonian**. The slope of the shear stress versus the steady-state shear strain rate curve is defined as the **viscosity** (η) of the material. Water is an example of a Newtonian material. The following relationship defines viscosity:

$$\tau = \eta\dot{\gamma} \tag{6-1}$$

The units of η are Pa·s (in the SI system) or Poise (P) or $\dfrac{g}{cm \cdot s}$ in the cgs system. Sometimes the term centipoise (cP) is used, 1 cP $= 10^{-2}$ P. Conversion between these units is given by 1 Pa·s $= 10$ P $= 1000$ cP.

The **kinematic viscosity** (v) is defined as

$$v = \eta/\rho \tag{6-2}$$

where viscosity (η) has units of Poise and density (ρ) has units of g/cm^3. The kinematic viscosity unit is Stokes (St) or equivalently cm^2/s. Sometimes the unit of centiStokes (cSt) is used; 1 cSt $= 10^{-2}$ St.

For many materials, the relationship between shear stress and shear strain rate is nonlinear. These materials are **non-Newtonian**. The stress versus steady state shear strain rate relationship in these materials can be described as

$$\tau = \eta\dot{\gamma}^m \tag{6-3}$$

where the exponent m is not equal to 1.

Non-Newtonian materials are classified as **shear thinning** (or pseudoplastic) or **shear thickening** (or dilatant). The relationships between the shear stress and shear strain rate for different types of materials are shown in Figure 6-3. If we take the slope of the line obtained by joining the origin to any point on the curve, we determine the **apparent viscosity** (η_{app}). Apparent viscosity as a function of steady-state shear strain rate is shown in Figure 6-4(a). The apparent viscosity of a Newtonian material will remain constant with changing shear strain rate. In shear thinning materials, the apparent viscosity decreases with increasing shear strain rate. In shear thickening materials, the apparent viscosity increases with increasing shear strain rate. If you have a can of paint sitting in

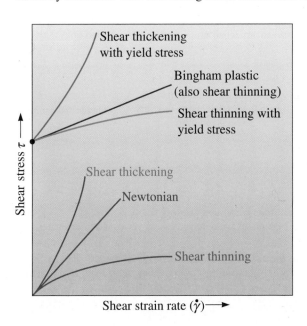

Figure 6-3
Shear stress-shear strain rate relationships for Newtonian and non-Newtonian materials.

(a)

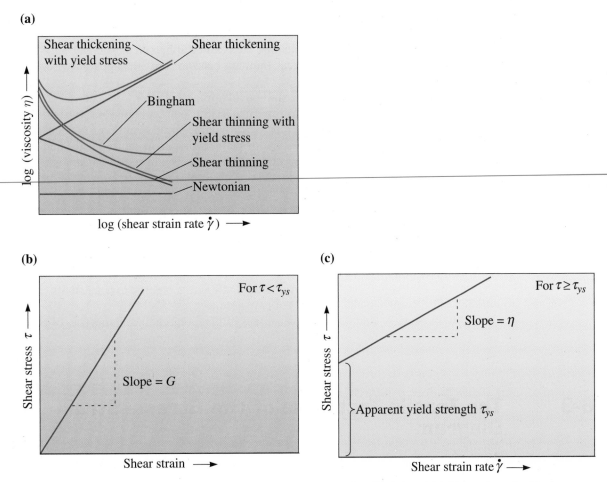

Figure 6-4   (a) Apparent viscosity as a function of shear strain rate ($\dot{\gamma}$). (b) and (c) Illustration of a Bingham plastic (Equations 6-4 and 6-5). Note the horizontal axis in (b) is shear strain.

storage, for example, the shear strain rate that the paint is subjected to is very small, and the paint behaves as if it is very viscous. When you take a brush and paint, the paint is subjected to a high shear strain rate. The paint now behaves as if it is quite thin or less viscous (i.e., it exhibits a small apparent viscosity). This is the shear thinning behavior. Some materials have "ideal plastic" behavior. For an ideal plastic material, the shear stress does not change with shear strain rate.

Many useful materials can be modeled as **Bingham plastics** and are defined by the following equations:

$$\tau = G \cdot \gamma \text{ (when } \tau \text{ is less than } \tau_{y \cdot s}) \tag{6-4}$$

$$\tau = \tau_{y \cdot s} + \eta \dot{\gamma} \text{ (when } \tau \geq \tau_{y \cdot s}) \tag{6-5}$$

This is illustrated in Figure 6-4(b) and 6-4(c). In these equations, $\tau_{y \cdot s}$ is the apparent **yield strength** obtained by interpolating the shear stress-shear strain rate data to zero shear strain rate. We define yield strength as the stress level that has to be exceeded so that the material deforms plastically. The existence of a true yield strength (sometimes also known as yield stress) has not been proven unambiguously for many plastics and dispersions such as paints. To prove the existence of a yield stress, separate measurements of stress versus strain are needed. For these materials, a critical yielding strain may be a better way to describe the mechanical behavior. Many ceramic slurries (dispersions such as those used in ceramic processing), polymer melts (used in polymer processing), paints and gels, and food products (yogurt, mayonnaise, ketchups, etc.) exhibit Bingham plastic-like behavior. Note that Bingham plastics exhibit shear thinning behavior (i.e., the apparent viscosity decreases with increasing shear rate).

Shear thinning materials also exhibit a **thixotropic behavior** (e.g., paints, ceramic slurries, polymer melts, gels, etc.). Thixotropic materials usually contain some type of network of particles or molecules. When a sufficiently large shear strain (i.e., greater than the critical yield strain) is applied, the thixotropic network or structure breaks and the material begins to flow. As the shearing stops, the network begins to form again, and the resistance to the flow increases. The particle or molecular arrangements in the newly formed network are different from those in the original network. Thus, the behavior of thixotropic materials is said to be time and deformation history dependent. Some materials show an increase in the apparent viscosity as a function of time and at a constant shearing rate. These materials are known as **rheopectic**.

The rheological properties of materials are determined using instruments known as a viscometer or a *rheometer*. In these instruments, a constant stress or constant strain rate is applied to the material being evaluated. Different geometric arrangements (e.g., cone and plate, parallel plate, Couette, etc.) are used.

In the sections that follow, we will discuss different mechanical properties of solid materials and some of the testing methods to evaluate these properties.

6-3   The Tensile Test: Use of the Stress–Strain Diagram

The tensile test is popular since the properties obtained can be applied to design different components. The tensile test measures the resistance of a material to a static or slowly applied force. The strain rates in a tensile test are typically small (10^{-4} to 10^{-2} s^{-1}). A test setup is shown in Figure 6-5; a typical specimen has a diameter of 1.263 cm and a gage length of 5 cm. The specimen is placed in the testing machine and a force F, called the **load**, is applied. A universal testing machine on which tensile and compressive tests can be

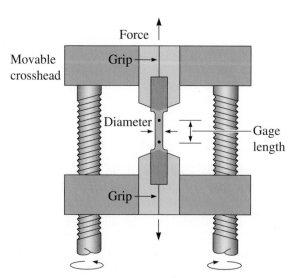

Figure 6-5
A unidirectional force is applied to a specimen in the tensile test by means of the moveable crosshead. The crosshead movement can be performed using screws or a hydraulic mechanism.

performed often is used. A **strain gage** or **extensometer** is used to measure the amount that the specimen stretches between the gage marks when the force is applied. Thus, the change in length of the specimen (Δl) is measured with respect to the original length (l_0). Information concerning the strength, Young's modulus, and ductility of a material can be obtained from such a tensile test. Typically, a tensile test is conducted on metals, alloys, and plastics. Tensile tests can be used for ceramics; however, these are not very popular because the sample may fracture while it is being aligned. The following discussion mainly applies to tensile testing of metals and alloys. We will briefly discuss the stress–strain behavior of polymers as well.

Figure 6-6 shows *qualitatively* the stress–strain curves for a typical (a) metal, (b) thermoplastic material, (c) elastomer, and (d) ceramic (or glass) under relatively small strain rates. The scales in this figure are qualitative and different for each material. In practice, the actual magnitude of stresses and strains will be very different. The temperature of the plastic material is assumed to be above its **glass-transition temperature** (T_g). The temperature of the metal is assumed to be room temperature. Metallic and thermoplastic materials show an initial elastic region followed by a non-linear plastic region. A separate curve for elastomers (e.g., rubber or silicones) is also included since the behavior of these materials is different from other polymeric materials. For elastomers, a large portion of the deformation is elastic and nonlinear. On the other hand, ceramics and glasses show only a linear elastic region and almost no plastic deformation at room temperature.

When a tensile test is conducted, the data recorded includes load or force as a function of change in length (Δl) as shown in Table 6-1 for an aluminum alloy test bar. These data are then subsequently converted into stress and strain. The stress-strain curve is analyzed further to extract properties of materials (e.g., Young's modulus, yield strength, etc.).

Engineering Stress and Strain   The results of a single test apply

to all sizes and cross-sections of specimens for a given material if we convert the force to

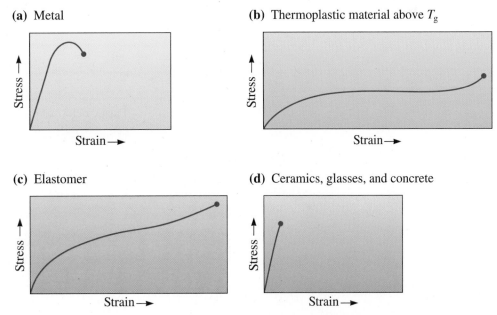

(a) Metal

(b) Thermoplastic material above T_g

(c) Elastomer

(d) Ceramics, glasses, and concrete

Figure 6-6   Tensile stress–strain curves for different materials. Note that these are *qualitative*. The magnitudes of the stresses and strains should not be compared.

TABLE 6-1 ■ *The results of a tensile test of a 1.263 cm diameter aluminum alloy test bar, initial length (l_0) = 5 cm*

Load (N)	Δl (cm)	Calculated	
		Stress (MPa)	Strain (cm/cm)
0	0.000	0	0
4450	0.0025	35.5	0.0005
13350	0.0075	106.5	0.0015
22240	0.0125	177.5	0.0025
31150	0.0175	248.6	0.0035
33360	0.075	266.2	0.0150
35140	0.2	280.4	0.0400
35580 (maximum load)	0.3	284	0.0600
35360	0.4	282.2	0.0800
33800 (fracture)	0.5125	269.8	0.1025

stress and the distance between gage marks to strain. **Engineering stress** and **engineering strain** are defined by the following equations:

$$\text{Engineering stress} = S = \frac{F}{A_0} \tag{6-6}$$

$$\text{Engineering stress} = e = \frac{\Delta l}{l_0} \tag{6-7}$$

where A_0 is the *original* cross-sectional area of the specimen before the test begins, l_0 is the *original* distance between the gage marks, and Δl is the change in length after force F is applied. The conversions from load and sample length to stress and strain are included in Table 6-1. The stress-strain curve (Figure 6-7) is used to record the results of a tensile test.

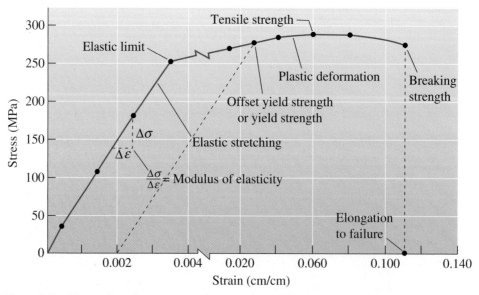

Figure 6-7   The engineering stress–strain curve for an aluminum alloy from Table 6-1.

| **Example 6-1** | *Tensile Testing of Aluminum Alloy* |

Convert the change in length data in Table 6-1 to engineering stress and strain and plot a stress-strain curve.

SOLUTION

For the 4450-N load:

$$S = \frac{F}{A_0} = \frac{4450 \text{ N}}{(\pi/4)(1.263 \text{ cm})^2} = 35.5 \text{ MPa}$$

$$e = \frac{\Delta l}{l_0} = \frac{0.0025 \text{ cm}}{5 \text{ cm}} = 0.0005 \text{ cm/cm}$$

The results of similar calculations for each of the remaining loads are given in Table 6-1 and are plotted in Figure 6-7.

Units   Many different units are used to report the results of the tensile test. The most common units for stress are pounds per square inch (psi) and MegaPascals (MPa). The units for strain include inch/inch, centimeter/centimeter, and meter/meter, and thus, strain is often written as unitless. The conversion factors for stress are summarized in Table 6-2. Because strain is dimensionless, no conversion factors are required to change the system of units.

TABLE 6-2 ▓ *Units and conversion factors*

1 pound (lb) = 4.448 Newtons (N)
1 psi = pounds per square inch
1 MPa = MegaPascal = MegaNewtons per square meter (MN/m^2)
                 = Newtons per square millimeter (N/mm^2) = 10^6 Pa
1 GPa = 1000 MPa = GigaPascal
1 ksi = 1000 psi = 6.895 MPa
1 psi = 0.006895 MPa
1 MPa = 0.145 ksi = 145 psi

| **Example 6-2** | *Design of a Suspension Rod* |

An aluminum rod is to withstand an applied force of 200,200 N. The engineering stress–strain curve for the aluminum alloy to be used is shown in Figure 6-7. To ensure safety, the maximum allowable stress on the rod is limited to 172 MPa, which is below the yield strength of the aluminum. The rod must be at least 375 cm long but must deform elastically no more than 0.625 cm when the force is applied. Design an appropriate rod.

SOLUTION

From the definition of engineering strain,

$$e = \frac{\Delta l}{l_0}$$

For a rod that is 375 cm long, the strain that corresponds to an extension of 0.625 cm is

$$e = \frac{0.625 \text{ cm}}{375 \text{ cm}} = 0.001667 \text{ cm/cm}$$

According to Figure 6-7, this strain is purely elastic, and the corresponding stress value is approximately 114.9 MPa, which is below the 172 Mpa limit. We use the definition of engineering stress to calculate the required cross-sectional area of the rod:

$$S = \frac{F}{A_0}$$

Note that the stress must not exceed 114.9 MPa, or consequently, the deflection will be greater than 0.625 cm. Rearranging,

$$A_0 = \frac{F}{S} = \frac{200{,}200 \text{ N}}{114.9 \text{ MPa}} = 17.42 \text{ cm}^2$$

The rod can be produced in various shapes, provided that the cross-sectional area is 17.42 cm². For a round cross section, the minimum diameter to ensure that the stress is not too high is

$$A_0 = \frac{\pi d^2}{4} = 17.42 \text{ cm}^2 \quad \text{or} \quad d = 4.71 \text{ cm}$$

Thus, one possible design that meets all of the specified criteria is a suspension rod that is 375 cm long with a diameter of 4.71 cm.

6-4   Properties Obtained from the Tensile Test

Yield Strength   As we apply stress to a material, the material initially exhibits elastic deformation. The strain that develops is completely recovered when the applied stress is removed. As we continue to increase the applied stress, the material eventually "yields" to the applied stress and exhibits both elastic and plastic deformation. The critical stress value needed to initiate plastic deformation is defined as the **elastic limit** of the material. In metallic materials, this is usually the stress required for dislocation motion, or slip, to be initiated. In polymeric materials, this stress will correspond to disentanglement of polymer molecule chains or sliding of chains past each other. The **proportional limit** is defined as the level of stress above which the relationship between stress and strain is not linear.

In most materials, the elastic limit and proportional limit are quite close; however, neither the elastic limit nor the proportional limit values can be determined precisely. Measured values depend on the sensitivity of the equipment used. We, therefore, define them at an **offset strain value** (typically, but not always, 0.002 or 0.2%). We then draw a line parallel to the linear portion of the engineering stress-strain curve starting at this offset value of strain. The stress value corresponding to the intersection of this line and the engineering stress-strain curve is defined as the **offset yield strength**, also often stated as the **yield strength**. The 0.2% offset yield strength for gray cast iron is 276 MPa as shown in Figure 6-8(a). Engineers normally prefer to use the offset yield strength for design purposes because it can be reliably determined.

For some materials, the transition from elastic deformation to plastic flow is rather abrupt. This transition is known as the **yield point phenomenon**. In these materials,

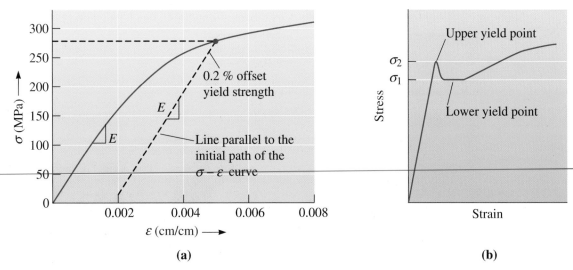

Figure 6-8   (a) Determining the 0.2% offset yield strength in gray cast iron, and (b) upper and lower yield point behavior in a low carbon steel.

as plastic deformation begins, the stress value drops first from the *upper yield point* (S_2) [Figure 6-8(b)]. The stress value then oscillates around an average value defined as the *lower yield point* (S_1). For these materials, the yield strength is usually defined from the 0.2% strain offset.

The stress-strain curve for certain low-carbon steels displays the yield point phenomenon [Figure 6-8(b)]. The material is expected to plastically deform at stress S_1; however, small interstitial atoms clustered around the dislocations interfere with slip and raise the yield point to S_2. Only after we apply the higher stress S_2 do the dislocations slip. After slip begins at S_2, the dislocations move away from the clusters of small atoms and continue to move very rapidly at the lower stress S_1.

When we design parts for load-bearing applications, we prefer little or no plastic deformation. As a result, we must select a material such that the design stress is considerably lower than the yield strength at the temperature at which the material will be used. We can also make the component cross-section larger so that the applied force produces a stress that is well below the yield strength. On the other hand, when we want to shape materials into components (e.g., take a sheet of steel and form a car chassis), we need to apply stresses that are well above the yield strength.

Tensile Strength   The stress obtained at the highest applied force is the **tensile strength** (S_{UTS}), which is the maximum stress on the engineering stress-strain curve. This value is also commonly known as the **ultimate tensile strength**. In many ductile materials, deformation does not remain uniform. At some point, one region deforms more than others and a large local decrease in the cross-sectional area occurs (Figure 6-9). This locally deformed region is called a "neck." This phenomenon is known as **necking**. Because the cross-sectional area becomes smaller at this point, a lower force is required to continue its deformation, and the engineering stress, calculated from the *original* area A_0, decreases. The tensile strength is the stress at which necking begins in ductile metals. In compression testing, the materials will bulge; thus necking is seen only in a tensile test.

Figure 6-10 shows typical yield strength values for various engineering materials. Ultra-pure metals have a yield strength of $\sim 1-10$ MPa. On the other hand, the yield strength of alloys is higher. Strengthening in alloys is achieved using different mechanisms

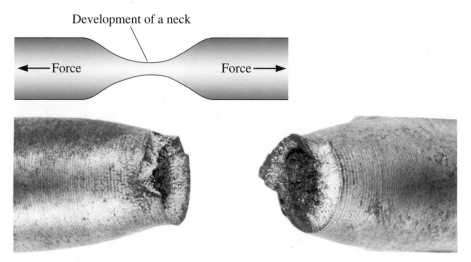

Figure 6-9 Localized deformation of a ductile material during a tensile test produces a necked region. The micrograph shows a necked region in a fractured sample. (*This article was published in Materials Principles and Practice, Charles Newey and Graham Weaver (Eds.), Figure 6.9, p. 300, Coyright Open University.*)

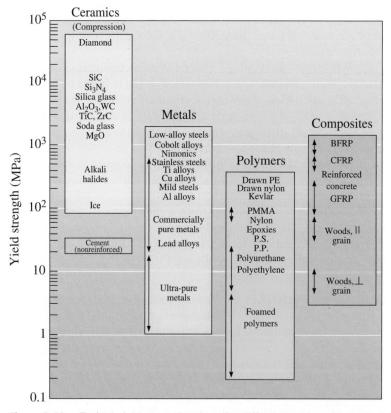

Figure 6-10 Typical yield strength values for different engineering materials. Note that values shown are in MPa. (*Reprinted from* Engineering Materials I, *Second Edition, M.F. Ashby and D.R.H. Jones, 1996, Fig. 8-12, p. 85. Copyright © 1996 Butterworth-Heinemann. Reprinted with permission from Elsevier Science.*)

described before (e.g., grain size, solid solution formation, strain hardening, etc.). The yield strength of a particular metal or alloy is usually the same for tension and compression. The yield strength of plastics and elastomers is generally lower than metals and alloys, ranging up to about 10–100 MPa. The values for ceramics (Figure 6-10) are for compressive strength (obtained using a hardness test). Tensile strength of most ceramics is much lower (~100–200 MPa). The tensile strength of glasses is about ~70 MPa and depends on surface flaws.

Elastic Properties

The modulus of elasticity, or *Young's modulus* (*E*), is the slope of the stress-strain curve in the elastic region. This relationship between stress and strain in the elastic region is known as **Hooke's law**:

$$E = \frac{S}{e}$$

(6-8)

The modulus is closely related to the binding energies of the atoms. (Figure 2-18). A steep slope in the force-distance graph at the equilibrium spacing (Figure 2-19) indicates that high forces are required to separate the atoms and cause the material to stretch elastically. Thus, the material has a high modulus of elasticity. Binding forces, and thus the modulus of elasticity, are typically higher for high melting point materials (Table 6-3). In metallic materials, the modulus of elasticity is considered a microstructure *insensitive* property since the value is dominated by the stiffness of atomic bonds. Grain size or other microstructural features do not have a very large effect on the Young's modulus. Note that Young's modulus does depend on such factors as orientation of a single crystal material (i.e., it depends upon crystallographic direction). For ceramics, the Young's modulus depends on the level of porosity. The Young's modulus of a composite depends upon the stiffness and amounts of the individual components.

The **stiffness** of a component is proportional to its Young's modulus. (The stiffness also depends on the component dimensions.) A component with a high modulus of elasticity will show much smaller changes in dimensions if the applied stress causes only elastic deformation when compared to a component with a lower elastic modulus. Figure 6-11 compares the elastic behavior of steel and aluminum. If a stress of 207 MPa is applied to each material, the steel deforms elastically 0.001 cm/cm; at the same stress, aluminum deforms 0.003 cm/cm. The elastic modulus of steel is about three times higher than that of aluminum.

Figure 6-12 shows the ranges of elastic moduli for various engineering materials. The modulus of elasticity of plastics is much smaller than that for metals or ceramics and glasses. For example, the modulus of elasticity of nylon is 2.7 GPa the modulus of glass fibers is 72 GPa. The Young's modulus of composites such as glass fiber-reinforced

TABLE 6-3 ■ *Elastic properties and melting temperature (T_m) of selected materials*

Material	T_m (°C)	E (MPa)	Poisson's ratio (ν)
Pb	327	13.8×10^3	0.45
Mg	650	44.8×10^3	0.29
Al	660	69.0×10^3	0.33
Cu	1085	124.8×10^3	0.36
Fe	1538	206.9×10^3	0.27
W	3410	408.2×10^3	0.28
Al_2O_3	2020	379.2×10^3	0.26
Si_3N_4		303.4×10^3	0.24

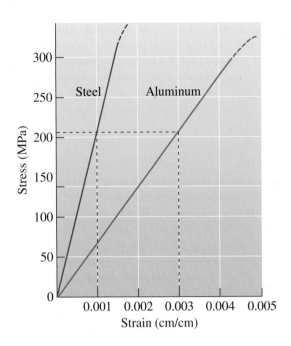

Figure 6-11
Comparison of the elastic behavior of steel and aluminum. For a given stress, aluminum deforms elastically three times as much as does steel (i.e., the elastic modulus of aluminum is about three times lower than that of steel).

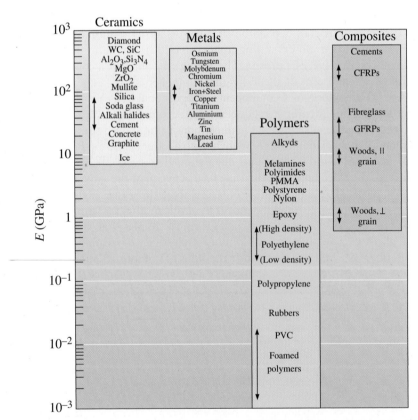

Figure 6-12   Range of elastic moduli for different engineering materials. Note: Values are shown in GPa. (*Reprinted from* Engineering Materials I, *Second Edition, M.F. Ashby and D.R.H. Jones, 1996, Fig. 3-5, p. 35, Copyright © 1996 Butterworth-Heinemann. Reprinted with permission from Elsevier Science.*)

composites (GFRC) or carbon fiber-reinforced composites (CFRC) lies between the values for the matrix polymer and the fiber phase (carbon or glass fibers) and depends upon their relative volume fractions. The Young's modulus of many alloys and ceramics is higher, generally ranging up to 410 GPa. Ceramics, because of the strength of ionic and covalent bonds, have the highest elastic moduli.

Poisson's ratio, ν, relates the longitudinal elastic deformation produced by a simple tensile or compressive stress to the lateral deformation that occurs simultaneously:

$$\nu = \frac{-e_{\text{lateral}}}{e_{\text{longitudinal}}} \tag{6-9}$$

For many metals in the elastic region, the Poisson's ratio is typically about 0.3 (Table 6-3). During a tensile test, the ratio increases beyond yielding to about 0.5, since during plastic deformation, volume remains constant. Some interesting structures, such as some honeycomb structures and foams, exhibit negative Poisson's ratios. *Note: Poisson's ratio should not be confused with the kinematic viscosity, both of which are denoted by the Greek letter v.*

The **modulus of resilience** (E_r), the area contained under the elastic portion of a stress-strain curve, is the elastic energy that a material absorbs during loading and subsequently releases when the load is removed. For linear elastic behavior:

$$E_r = \left(\frac{1}{2}\right) (\text{yield strength})(\text{strain at yielding}) \tag{6-10}$$

The ability of a spring or a golf ball to perform satisfactorily depends on a high modulus of resilience.

Tensile Toughness

The energy absorbed by a material prior to fracture is known as **tensile toughness** and is sometimes measured as the area under the true stress–strain curve (also known as the **work of fracture**). We will define true stress and true strain in Section 6-5. Since it is easier to measure engineering stress–strain, engineers often equate tensile toughness to the area under the engineering stress–strain curve.

Example 6-3 *Young's Modulus of an Aluminum Alloy*

Calculate the modulus of elasticity of the aluminum alloy for which the engineering stress–strain curve is shown in Figure 6-7. Calculate the length of a bar of initial length 125 cm when a tensile stress of 207 MPa is applied.

SOLUTION

When a stress of 248.6 MPa is applied, a strain of 0.0035 cm/cm is produced. Thus,

$$\text{Modulus of elasticity} = E = \frac{S}{e} = \frac{248.6 \text{ MPa}}{0.0035} = 71{,}028.6 \text{ MPa}$$

Note that any combination of stress and strain in the elastic region will produce this result. From Hooke's law,

$$e = \frac{S}{E} = \frac{207 \text{ MPa}}{71{,}028.6 \text{ MPa}} = 0.003 \text{ cm/cm}$$

From the definition of engineering strain,

$$e = \frac{\Delta l}{l_0}$$

Thus,

$$\Delta l = e(l_0) = 0.003 \text{ cm/cm } (125 \text{ cm}) = 0.375 \text{ cm}$$

When the bar is subjected to a stress of 207 MPa, the total length is given by

$$l = \Delta l + l_0 = 0.375 \text{ cm} + 125 \text{ cm} = 125.375 \text{ cm}$$

Ductility

Ductility is the ability of a material to be permanently deformed without breaking when a force is applied. There are two common measures of ductility. The **percent elongation** quantifies the permanent plastic deformation at failure (i.e., the elastic deformation recovered after fracture is not included) by measuring the distance between gage marks on the specimen before and after the test. Note that the strain after failure is smaller than the strain at the breaking point, because the elastic strain is recovered when the load is removed. The percent elongation can be written as

$$\% \text{ Elongation} = \frac{l_f - l_0}{l_0} \times 100 \qquad (6\text{-}11)$$

where l_f is the distance between gage marks after the specimen breaks.

A second approach is to measure the percent change in the cross-sectional area at the point of fracture before and after the test. The **percent reduction in area** describes the amount of thinning undergone by the specimen during the test:

$$\% \text{ Reduction in area} = \frac{A_0 - A_f}{A_0} \times 100 \qquad (6\text{-}12)$$

where A_f is the final cross-sectional area at the fracture surface.

Ductility is important to both designers of load-bearing components and manufacturers of components (bars, rods, wires, plates, I-beams, fibers, etc.) utilizing materials processing.

Example 6-4   *Ductility of an Aluminum Alloy*

The aluminum alloy in Example 6-1 has a final length after failure of 5.488 cm and a final diameter of 0.995 cm at the fractured surface. Calculate the ductility of this alloy.

SOLUTION

$$\% \text{ Elongation} = \frac{l_f - l_0}{l_0} \times 100 = \frac{5.488 - 5.000}{5.000} \times 100 = 9.76\%$$

$$\% \text{ Reduction in area} = \frac{A_0 - A_f}{A_0} \times 100$$

$$= \frac{(\pi/4)(1.263)^2 - (\pi/4)(0.995)^2}{(\pi/4)(1.263)^2} \times 100$$

$$= 37.9\%$$

The final length is less than 5.5125 cm (see Table 6-1) because, after fracture, the elastic strain is recovered.

Effect of Temperature
Mechanical properties of materials depend on temperature (Figure 6-13). Yield strength, tensile strength, and modulus of elasticity decrease at higher temperatures, whereas ductility commonly increases. A materials fabricator may wish to deform a material at a high temperature (known as *hot working*) to take advantage of the higher ductility and lower required stress.

We use the term "high temperature" here with a note of caution. A high temperature is defined relative to the melting temperature. Thus, 500°C is a high temperature for aluminum alloys; however, it is a relatively low temperature for the processing of steels. In metals, the yield strength decreases rapidly at higher temperatures due to a decreased dislocation density and an increase in grain size via grain growth (Chapter 5) or a related process known as recrystallization (as described later in Chapter 8). Similarly, any strengthening that may have occurred due to the formation of ultrafine precipitates may also decrease as the precipitates begin to either grow in size or dissolve into the matrix. We will discuss these effects in greater detail in later chapters. When temperatures are reduced, many, but not all, metals and alloys become brittle.

Increased temperatures also play an important role in forming polymeric materials and inorganic glasses. In many polymer-processing operations, such as extrusion or the stretch-blow process (Chapter 16), the increased ductility of polymers at higher temperatures is advantageous. Again, a word of caution concerning the use of the term "high temperature." For polymers, the term "high temperature" generally means a temperature

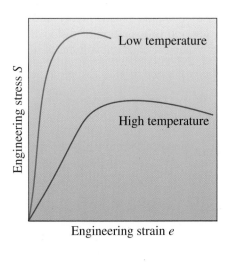

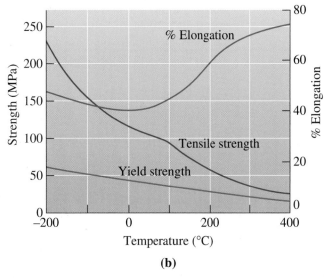

(a)   (b)

Figure 6-13   The effect of temperature (a) on the stress–strain curve and (b) on the tensile properties of an aluminum alloy.

higher than the glass-transition temperature (T_g). For our purpose, the glass-transition temperature is a temperature below which materials behave as brittle materials. Above the glass-transition temperature, plastics become ductile. The glass-transition temperature is not a fixed temperature, but depends on the rate of cooling as well as the polymer molecular weight distribution. Many plastics are ductile at room temperature because their glass-transition temperatures are *below* room temperature. To summarize, many polymeric materials will become harder and more brittle as they are exposed to temperatures that are below their glass-transition temperatures. The reasons for loss of ductility at lower temperatures in polymers and metallic materials are different; however, this is a factor that played a role in the failures of the *Titanic* in 1912 and the *Challenger* in 1986.

Ceramic and glassy materials are generally considered brittle at room temperature. As the temperature increases, glasses can become more ductile. As a result, glass processing (e.g., fiber drawing or bottle manufacturing) is performed at high temperatures.

6-5   True Stress and True Strain

The decrease in engineering stress beyond the tensile strength on an engineering stress–strain curve is related to the definition of engineering stress. We used the original area A_0 in our calculations, but this is not precise because the area continually changes. We define **true stress** and **true strain** by the following equations:

$$\text{True stress} = \sigma = \frac{F}{A} \tag{6-13}$$

$$\text{True strain} = \varepsilon = \int_{l_0}^{l} \frac{dl}{l} = \ln\left(\frac{l}{l_0}\right) \tag{6-14}$$

where A is the instantaneous area over which the force F is applied, l is the instantaneous sample length, and l_0 is the initial length. In the case of metals, plastic deformation is essentially a constant-volume process (i.e., the creation and propagation of dislocations results in a negligible volume change in the material). When the constant volume assumption holds, we can write

$$A_0 l_0 = Al \text{ or } A = \frac{A_0 l_0}{l} \tag{6-15}$$

and using the definitions of engineering stress S and engineering strain e, Equation 6-13 can be written as

$$\sigma = \frac{F}{A} = \frac{F}{A_0}\left(\frac{l}{l_0}\right) = S\left(\frac{l_0 + \Delta l}{l_0}\right) = S(1 + e) \tag{6-16}$$

It can also be shown that

$$\varepsilon = \ln(1 + e) \tag{6-17}$$

Thus, it is a simple matter to convert between the engineering stress–strain and true stress–strain systems. Note that the expressions in Equations 6-16 and 6-17 are not valid

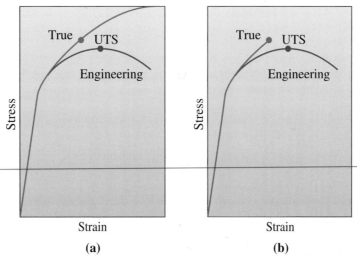

Figure 6-14   (a) The relation between the true stress–true strain diagram and engineering stress–engineering strain diagram. The curves are nominally identical to the yield point. The true stress corresponding to the ultimate tensile strength (UTS) is indicated. (b) Typically true stress–strain curves must be truncated at the true stress corresponding to the ultimate tensile strength, since the cross-sectional area at the neck is unknown.

after the onset of necking, because after necking begins, the distribution of strain along the gage length is not uniform. After necking begins, Equation 6-13 must be used to calculate the true stress and the expression

$$\varepsilon = \ln\left(\frac{A_0}{A}\right) \tag{6-18}$$

must be used to calculate the true strain. Equation 6-18 follows from Equations 6-14 and 6-15. After necking the instantaneous area A is the cross-sectional area of the neck A_{neck}.

The true stress–strain curve is compared to the engineering stress–strain curve in Figure 6-14(a). There is no maximum in the true stress–true strain curve.

Note that it is difficult to measure the instantaneous cross-sectional area of the neck. Thus, true stress–strain curves are typically truncated at the true stress that corresponds to the ultimate tensile strength, as shown in Figure 6-14(b).

Example 6-5   *True Stress and True Strain*

Compare engineering stress and strain with true stress and strain for the aluminum alloy in Example 6-1 at (a) the maximum load and (b) fracture. The diameter at maximum load is 1.243 cm and at fracture is 0.995 cm.

SOLUTION

(a) At the maximum load,

$$\text{Engineering stress } S = \frac{F}{A_0} = \frac{35{,}580 \text{ N}}{(\pi/4)(1.263 \text{ cm})^2} = 283.9 \text{ MPa}$$

$$\text{Engineering strain } e = \frac{\Delta l}{l_0} = \frac{5.3 - 5}{5} = 0.060 \text{ cm/cm}$$

$$\text{True stress} = \sigma = S(1 + e) = 283.9(1 + 0.060) = 300 \text{ MPa}$$

$$\text{True strain} = \ln(1 + e) = \ln(1 + 0.060) = 0.058 \text{ cm/cm}$$

The maximum load is the last point at which the expressions used here for true stress and true strain apply. Note that the same answers are obtained for true stress and strain if the instantaneous dimensions are used:

$$\sigma = \frac{F}{A} = \frac{35{,}580 \text{ N}}{(\pi/4)(1.243 \text{ cm})^2} = 3 \times 10^2 \text{ MPa}$$

$$\varepsilon = \ln\left(\frac{A_0}{A}\right) = \ln\left[\frac{(\pi/4)(1.263 \text{ cm}^2)}{(\pi/4)(1.243 \text{ cm}^2)}\right] = 0.058 \text{ cm/cm}$$

Up until the point of necking in a tensile test, the engineering stress is less than the corresponding true stress, and the engineering strain is greater than the corresponding true strain.

(b) At fracture,

$$S = \frac{F}{A_0} = \frac{33{,}800 \text{ N}}{(\pi/4)(1.263 \text{ cm})^2} = 269.7 \text{ MPa}$$

$$e = \frac{\Delta l}{l_0} = \frac{5.5125 - 5}{5} = 0.1025 \text{ cm/cm}$$

$$\sigma = \frac{F}{A_f} = \frac{35{,}580 \text{ N}}{(\pi/4)(0.995 \text{ cm})^2} = 434.5 \text{ MPa}$$

$$\varepsilon = \ln\left(\frac{A_0}{A_f}\right) = \ln\left[\frac{(\pi/4)(1.263 \text{ cm}^2)}{(\pi/4)(0.995 \text{ cm}^2)}\right] = \ln(1.601) = 0.476 \text{ cm/cm}$$

It was necessary to use the instantaneous dimensions to calculate the true stress and strain, since failure occurs past the point of necking. After necking, the true strain is greater than the corresponding engineering strain.

6-6   The Bend Test for Brittle Materials

In ductile metallic materials, the engineering stress–strain curve typically goes through a maximum; this maximum stress is the tensile strength of the material. Failure occurs at a lower engineering stress after necking has reduced the cross-sectional area supporting the load. In more brittle materials, failure occurs at the maximum load, where the tensile strength and breaking strength are the same (Figure 6-15).

In many brittle materials, the normal tensile test cannot easily be performed because of the presence of flaws at the surface. Often, just placing a brittle material in the grips of the tensile testing machine causes cracking. These materials may be tested using the **bend test** [Figure 6-16(a)]. By applying the load at three points and causing bending, a tensile force acts on the material opposite the midpoint. Fracture begins at this location. The **flexural strength**, or **modulus of rupture**, describes the material's strength:

$$\text{Flexural strength for three-point bend test } \sigma_{\text{bend}} = \frac{3FL}{2wh^2} \tag{6-19}$$

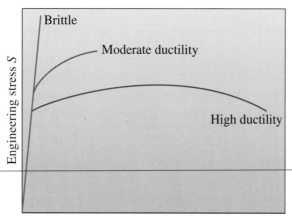

Figure 6-15
The engineering stress–strain behavior of brittle materials compared with that of more ductile materials.

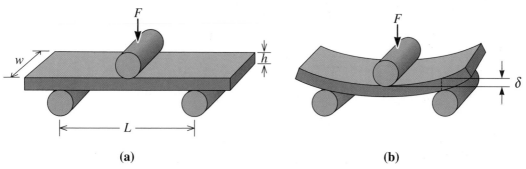

(a)   **(b)**

Figure 6-16   (a) The bend test often used for measuring the strength of brittle materials, and (b) the deflection δ obtained by bending.

where F is the fracture load, L is the distance between the two outer points, w is the width of the specimen, and h is the height of the specimen. The flexural strength has units of stress. The results of the bend test are similar to the stress-strain curves; however, the stress is plotted versus deflection rather than versus strain (Figure 6-17). The corresponding bending moment diagram is shown in Figure 6-18(a).

The modulus of elasticity in bending, or the **flexural modulus** (E_{bend}), is calculated as

$$\text{Flexural modulus } E_{bend} = \frac{L^3 F}{4wh^3\delta} \tag{6-20}$$

where δ is the deflection of the beam when a force F is applied.

This test can also be conducted using a setup known as the four-point bend test [Figure 6-18(b)]. The maximum stress or flexural stress for a four-point bend test is given by

$$\sigma_{bend} = \frac{3FL_1}{4wh^2} \tag{6-21}$$

for the specific case in which $L_1 = L/4$ in Figure 6-18(b).

Note that the derivations of Equations 6-19 through 6-21 assume a linear stress–strain response (and thus cannot be correctly applied to many polymers). The four-point bend

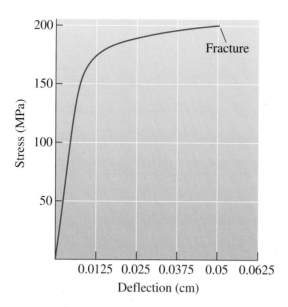

Figure 6-17
Stress-deflection curve for an MgO ceramic obtained from a bend test.

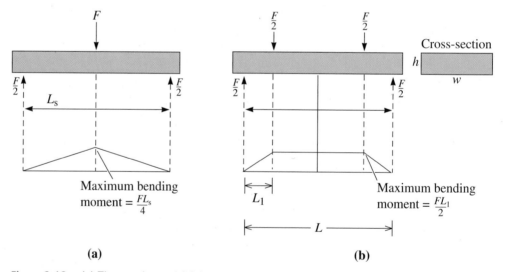

Figure 6-18   (a) Three-point and (b) four-point bend test setup.

test is better suited for testing materials containing flaws. This is because the bending moment between the inner platens is constant [Figure 6-18(b)]; thus samples tend to break randomly unless there is a flaw that locally raises the stress.

Since cracks and flaws tend to remain closed in compression, brittle materials such as concrete are often incorporated into designs so that only compressive stresses act on the part. Often, we find that brittle materials fail at much higher compressive stresses than tensile stresses (Table 6-4). This is why it is possible to support a fire truck on four coffee cups; however, ceramics have very limited mechanical toughness. Hence, when we drop a ceramic coffee cup, it can break easily.

TABLE 6-4 ■ *Comparison of the tensile, compressive, and flexural strengths of selected ceramic and composite materials*

Material	Tensile Strength (MPa)	Compressive Strength (MPa)	Flexural Strength (MPa)
Polyester—50% glass fibers	159	221	310
Polyester—50% glass fiber fabric	255	186[a]	317
Al_2O_3 (99% pure)	207	2586	345
SiC (pressureless-sintered)	172	3861	552

[a]*A number of composite materials are quite poor in compression.*

Example 6-6   *Flexural Strength of Composite Materials*

The flexural strength of a composite material reinforced with glass fibers is 310 MPa, and the flexural modulus is 124.1×10^3 MPa. A sample, which is 1.25 cm wide, 0.938 cm high, and 20 cm long, is supported between two rods 12.5 cm apart. Determine the force required to fracture the material and the deflection of the sample at fracture, assuming that no plastic deformation occurs.

SOLUTION

Based on the description of the sample, $w = 1.25$ cm, $h = 0.938$ cm, and $L = 10$ cm
From Equation 6-19:

$$310 \text{ MPa} = \frac{3FL}{2wh^2} = \frac{(3)(F)(12.5 \text{ cm})}{(2)(1.25 \text{ cm})(0.938 \text{ cm})^2} = 1.7 \times 10^5 F$$

$$F = \frac{310 \times 10^6}{1.7 \times 10^5} = 1823.5 \text{ N}$$

Therefore, the deflection, from Equation 6-20, is

$$124.1 \times 10^3 \text{ MPa} = \frac{L^3 F}{4wh^3\delta} = \frac{(12.5 \text{ cm})^3(1823.5 \text{ N})}{(4)(1.25 \text{ cm})(0.938 \text{ cm})^3\delta}$$

$$\delta = 0.0696 \text{ cm}$$

In this calculation, we assumed a linear relationship between stress and strain and also that there is no viscoelastic behavior.

6-7   Hardness of Materials

The **hardness test** measures the resistance to penetration of the surface of a material by a hard object. Hardness as a term is not defined precisely. Hardness, depending upon the context, can represent resistance to scratching or indentation and a qualitative measure of the strength of the material. In general, in **macrohardness** measurements, the load applied is ~2 N. A variety of hardness tests have been devised, but the most commonly used are

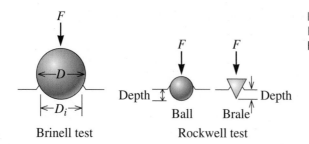

the Rockwell test and the Brinell test. Different indenters used in these tests are shown in Figure 6-19.

In the *Brinell hardness test*, a hard steel sphere (usually 10 mm in diameter) is forced into the surface of the material. The diameter of the impression, typically 2 to 6 mm, is measured and the Brinell hardness number (abbreviated as HB or BHN) is calculated from the following equation:

$$HB = \frac{2F}{\pi D\left[D - \sqrt{D^2 - D_i^2}\right]} \tag{6-22}$$

where F is the applied load in kilograms, D is the diameter of the indenter in millimeters, and D_i is the diameter of the impression in millimeters. The Brinell hardness has units of kg/mm^2.

The *Rockwell hardness test* uses a small-diameter steel ball for soft materials and a diamond cone, or Brale, for harder materials. The depth of penetration of the indenter is automatically measured by the testing machine and converted to a Rockwell hardness number (HR). Since an optical measurement of the indentation dimensions is not needed, the Rockwell test tends to be more popular than the Brinell test. Several variations of the Rockwell test are used, including those described in Table 6-5. A Rockwell C (HRC) test is used for hard steels, whereas a Rockwell F (HRF) test might be selected for aluminum. Rockwell tests provide a hardness number that has no units.

Hardness numbers are used primarily as a *qualitative* basis for comparison of materials, specifications for manufacturing and heat treatment, quality control, and

TABLE 6-5 ■ *Comparison of typical hardness tests*

Test	Indenter	Load	Application
Brinell	10-mm ball	3000 kg	Cast iron and steel
Brinell	10-mm ball	500 kg	Nonferrous alloys
Rockwell *A*	Brale	60 kg	Very hard materials
Rockwell *B*	1/16-in. ball	100 kg	Brass, low-strength steel
Rockwell *C*	Brale	150 kg	High-strength steel
Rockwell *D*	Brale	100 kg	High-strength steel
Rockwell *E*	1/8-in. ball	100 kg	Very soft materials
Rockwell *F*	1/16-in. ball	60 kg	Aluminum, soft materials
Vickers	Diamond square pyramid	10 kg	All materials
Knoop	Diamond elongated pyramid	500 g	All materials

correlation with other properties of materials. For example, Brinell hardness is related to the tensile strength of steel by the approximation:

$$\text{Tensile strength (psi)} = 500HB \qquad (6\text{-}23)$$

where HB has units of kg/mm^2 and 1 psi $= 6.895 \times 10^3$ Pa.

Hardness correlates well with wear resistance. A separate test is available for measuring the wear resistance. A material used in crushing or grinding of ores should be very hard to ensure that the material is not eroded or abraded by the hard feed materials. Similarly, gear teeth in the transmission or the drive system of a vehicle should be hard enough that the teeth do not wear out. Typically we find that polymer materials are exceptionally soft, metals and alloys have intermediate hardness, and ceramics are exceptionally hard. We use materials such as tungsten carbide-cobalt composite (WC-Co), known as "carbide," for cutting tool applications. We also use microcrystalline diamond or diamond-like carbon (DLC) materials for cutting tools and other applications.

The Knoop hardness (HK) test is a **microhardness test**, forming such small indentations that a microscope is required to obtain the measurement. In these tests, the load applied is less than 2 N. The Vickers test, which uses a diamond pyramid indenter, can be conducted either as a macro or microhardness test. Microhardness tests are suitable for materials that may have a surface that has a higher hardness than the bulk, materials in which different areas show different levels of hardness, or samples that are not macroscopically flat.

6-8   Nanoindentation

The hardness tests described in the previous section are known as macro or microhardness tests because the indentations have dimensions on the order of millimeters or microns. The advantages of such tests are that they are relatively quick, easy, and inexpensive. Some of the disadvantages are that they can only be used on macroscale samples and hardness is the only materials property that can be directly measured. Nanoindentation is hardness testing performed at the nanometer length scale. A small diamond tip is used to indent the material of interest. The imposed load and displacement are continuously measured with micro-Newton and sub-nanometer resolution, respectively. Both load and displacement are measured throughout the indentation process. Nanoindentation techniques are important for measuring the mechanical properties of thin films on substrates (such as for microelectronics applications) and nanophase materials and for deforming free-standing micro and nanoscale structures. Nanoindentation can be performed with high positioning accuracy, permitting indentations within selected grains of a material. Nanoindenters incorporate optical microscopes and sometimes a scanning probe microscope capability. Both hardness and elastic modulus are measured using nanoindentation.

Nanoindenter tips come in a variety of shapes. A common shape is known as the Berkovich indenter, which is a three-sided pyramid. An indentation made by a Berkovich indenter is shown in Figure 6-20. The indentation in the figure measures 12.5 μm on each side and about 1.6 μm deep.

The first step of a nanoindentation test involves performing indentations on a calibration standard. Fused silica is a common calibration standard, because it has homogeneous and well-characterized mechanical properties (elastic modulus $E = 72$ GPa and Poisson's ratio $\nu = 0.17$). The purpose of performing indentations on the calibration

Figure 6-20   An indentation in a $Zr_{41.2}Ti_{13.8}Cu_{12.5}Ni_{10.0}Be_{22.5}$ bulk metallic glass made using a Berkovich tip in a nanoindenter. (*Courtesy of Gang Feng, Villanova University.*)

standard is to determine the projected contact area of the indenter tip A_c as a function of indentation depth. For a perfect Berkovich tip,

$$A_c = 24.5\,h_c^2 \qquad (6\text{-}24)$$

This function relates the cross-sectional area of the indenter to the distance from the tip h_c that is in contact with the material being indented. No tip is perfectly sharp, and the tip wears and changes shape with each use. Thus, a calibration must be performed each time the tip is used as will be discussed below.

The total indentation depth h (as measured by the displacement of the tip) is the sum of the contact depth h_c and the depth h_s at the periphery of the indentation where the indenter does not make contact with the material surface, i.e.,

$$h = h_c + h_s \qquad (6\text{-}25)$$

as shown in Figure 6-21. The surface displacement term h_s is calculated according to

$$h_s = \varepsilon\,\frac{P_{\max}}{S} \qquad (6\text{-}26)$$

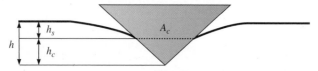

Figure 6-21   The relationship between the total indentation depth h, the contact depth h_c, and the displacement of the surface at the periphery of the indent h_s. The contact area A_c at a depth h_c is seen edge-on in this view. [(After W.C. Oliver and G.M. Pharr in the *Journal of Materials Research*, Volume 7, Number 6, p. 1573(1992). Reprinted by permission.)]

where P_{max} is the maximum load and ε is a geometric constant equal to 0.75 for a Berkovich indenter. S is the unloading stiffness.

In nanoindentation, the imposed load is measured as a function of indentation depth h, as shown in Figure 6-22. On loading, the deformation is both elastic and plastic. As the indenter is removed from the material, the recovery is elastic. The unloading stiffness is measured as the slope of a power law curve fit to the unloading curve at the maximum indentation depth. The reduced elastic modulus E_r is related to the unloading stiffness S according to

$$E_r = \frac{\sqrt{\pi}}{2\beta} \frac{S}{\sqrt{A_c}} \tag{6-27}$$

where β is a constant for the shape of the indenter being used ($\beta = 1.034$ for a Berkovich indenter). The reduced modulus E_r is given by

$$\frac{1}{E_r} = \frac{1 - \nu^2}{E} + \frac{1 - \nu_i^2}{E_i} \tag{6-28}$$

where E and ν are the elastic modulus and Poisson's ratio of the material being indented, respectively, and E_i and ν_i are the elastic modulus and Poisson's ratio of the indenter, respectively (for diamond, $E_i = 1.141$ TPa and $\nu_i = 0.07$). Since the elastic properties of the standard are known, the only unknown in Equation 6-27 for a calibration indent is A_c.

Using Equation 6-27, the projected contact area is calculated for a particular contact depth. When the experiment is subsequently performed on the material of interest, the same tip shape function is used to calculate the projected contact area for the same contact depth. Equation 6-27 is again used with the elastic modulus of the material being the unknown quantity of interest. (A Poisson's ratio must be assumed for the material being indented. As we saw in Table 6-3, typical values range from 0.2 to 0.4 with most metals having a Poisson's ratio of about 0.3. Errors in this assumption result in relatively little error in the elastic modulus measurement.)

The hardness of a material as determined by nanoindentation is calculated as

$$H = \frac{P_{max}}{A_c} \tag{6-29}$$

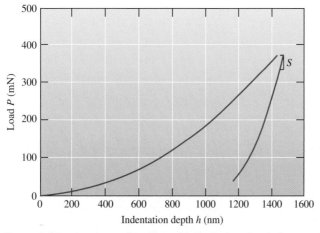

Figure 6-22   Load as a function of indentation depth for nanoindentation of MgO. The unloading stiffness S at maximum load is indicated.

Hardnesses (as determined by nanoindentation) are typically reported with units of GPa, and the results of multiple indentations are usually averaged to improve accuracy.

This analysis calculates the elastic modulus and hardness at the maximum load; however, an experimental technique known as dynamic nanoindentation is now usually employed. During dynamic nanoindentation, a small oscillating load is superimposed on the total load on the sample. In this way, the sample is continuously unloaded elastically as the total load is increased. This allows for continuous measurements of elastic modulus and hardness as a function of indentation depth.

This nanoindentation analysis was published in 1992 in the *Journal of Materials Research* and is known as the Oliver and Pharr method, named after Warren C. Oliver and George M. Pharr.

Example 6-7   *Nanoindentation of MgO*

Figure 6-22 shows the results of an indentation into single crystal (001) MgO using a diamond Berkovich indenter. The unloading stiffness at a maximum indentation depth of 1.45 μm is 1.8×10^6 N/m. A calibration on fused silica indicates that the projected contact area at the corresponding contact depth is 41 μm^2. The Poisson's ratio of MgO is 0.17. Calculate the elastic modulus and hardness of MgO.

SOLUTION

The projected contact area $A_c = 41 \ \mu\text{m}^2 \times \dfrac{(1 \text{ m})^2}{(10^6 \mu\text{m})^2} = 41 \times 10^{-12} \text{ m}^2.$

The reduced modulus is given by

$$E_r = \frac{\sqrt{\pi}}{2\beta} \frac{S}{\sqrt{A_c}} = \frac{\sqrt{\pi}(1.8 \times 10^6 \text{ N/m})}{2(1.034)\sqrt{(41 \times 10^{-12} \text{ m}^2)}} = 241 \times 10^9 \text{ N/m}^2 = 241 \text{ GPa}$$

Substituting for the Poisson's ratio of MgO and the elastic constants of diamond and solving for E,

$$\frac{1}{E_r} = \frac{1 - 0.17^2}{E} + \frac{1 - 0.07^2}{1.141 \times 10^{12} \text{ Pa}} = \frac{1}{241 \times 10^9 \text{ Pa}}$$

$$\frac{0.9711}{E} = \frac{1}{241 \times 10^9 \text{ Pa}} - \frac{1 - 0.07^2}{1.141 \times 10^{12} \text{ Pa}}$$

$$E = 296 \text{ GPa}$$

From Figure 6-22, the load at the indentation depth of 1.45 μm is 380 mN (380×10^{-3} N). Thus, the hardness is

$$H = \frac{P_{\text{max}}}{A_c} = \frac{380 \times 10^{-3} \text{ N}}{41 \times 10^{-12} \text{ m}^2} = 9.3 \times 10^9 \text{ Pa} = 9.3 \text{ GPa}$$

6-9   Strain Rate Effects and Impact Behavior

When a material is subjected to a sudden, intense blow, in which the strain rate ($\dot{\gamma}$ or $\dot{\varepsilon}$) is extremely rapid, it may behave in much more brittle a manner than is observed in the tensile test. This, for example, can be seen with many plastics and materials such as Silly Putty®. If you stretch a plastic such as polyethylene or Silly Putty® very slowly, the polymer molecules have time to disentangle or the chains to slide past each other and cause large plastic deformations. If, however, we apply an impact loading, there is insufficient time for these mechanisms to play a role and the materials break in a brittle manner. An **impact test** is often used to evaluate the brittleness of a material under these conditions. In contrast to the tensile test, in this test, the strain rates are much higher ($\dot{\varepsilon} \sim 10^3 \text{ s}^{-1}$).

Many test procedures have been devised, including the *Charpy* test and the *Izod* test (Figure 6-23). The Izod test is often used for plastic materials. The test specimen may be either notched or unnotched; V-notched specimens better measure the resistance of the material to crack propagation.

In the test, a heavy pendulum, starting at an elevation h_0, swings through its arc, strikes and breaks the specimen, and reaches a lower final elevation h_f. If we know the initial and final elevations of the pendulum, we can calculate the difference in potential energy. This difference is the **impact energy** absorbed by the specimen during failure. For the Charpy test, the energy is usually expressed in joules (J). The results of the Izod test are expressed in units of J/m. The ability of a material to withstand an impact blow is often referred to as the **impact toughness** of the material. As we mentioned before, in some situations, we consider the area under the true or engineering stress-strain curve

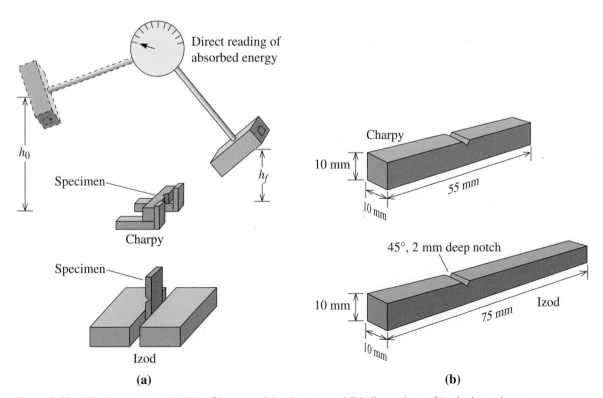

Figure 6-23   The impact test: (a) the Charpy and Izod tests, and (b) dimensions of typical specimens.

as a measure of **tensile toughness**. In both cases, we are measuring the energy needed to fracture a material. The difference is that, in tensile tests, the strain rates are much smaller compared to those used in an impact test. Another difference is that in an impact test we usually deal with materials that have a notch. **Fracture toughness** of a material is defined as the ability of a material containing flaws to withstand an applied load. We will discuss fracture toughness in Chapter 7.

6-10   Properties Obtained from the Impact Test

A curve showing the trends in the results of a series of impact tests performed on nylon at various temperatures is shown in Figure 6-24. In practice, the tests will be conducted at a limited number of temperatures.

Ductile to Brittle Transition Temperature (DBTT)   The **ductile to brittle transition temperature** is the temperature at which the failure mode of a material changes from ductile to brittle fracture. This temperature may be defined by the average energy between the ductile and brittle regions, at some specific absorbed energy, or by some characteristic fracture appearance. A material subjected to an impact blow during service should have a transition temperature *below* the temperature of the material's surroundings.

Not all materials have a distinct transition temperature (Figure 6-25). BCC metals have transition temperatures, but most FCC metals do not. FCC metals have high absorbed energies, with the energy decreasing gradually and, sometimes, even increasing as the temperature decreases. As mentioned before, the effect of this transition in steel may have contributed to the failure of the *Titanic*.

In polymeric materials, the ductile to brittle transition temperature is related closely to the glass-transition temperature and for practical purposes is treated as the same. As mentioned before, the transition temperature of the polymers used in booster rocket O-rings and other factors led to the *Challenger* disaster.

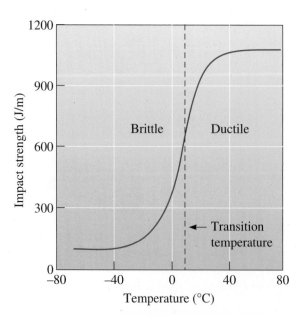

Figure 6-24
Results from a series of Izod impact tests for a tough nylon thermoplastic polymer.

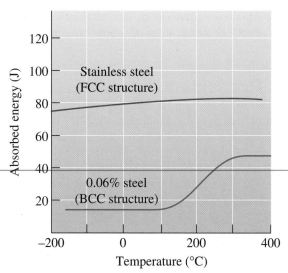

Figure 6-25
The Charpy V-notch properties for a BCC carbon steel and an FCC stainless steel. The FCC crystal structure typically leads to higher absorbed energies and no transition temperature.

Notch Sensitivity   Notches caused by poor machining, fabrication, or design concentrate stresses and reduce the toughness of materials. The **notch sensitivity** of a material can be evaluated by comparing the absorbed energies of notched versus unnotched specimens. The absorbed energies are much lower in notched specimens if the material is notch-sensitive. We will discuss in Section 7-7 how the presence of notches affect the behavior of materials subjected to cyclical stress.

Relationship to the Stress-Strain Diagram   The energy required to break a material during impact testing (i.e., the impact toughness) is not always related to the tensile toughness (i.e., the area contained under the true stress-true strain curve (Figure 6-26). As noted before, engineers often consider the area under the engineering stress–strain curve as tensile toughness. In general, metals with both high strength and high ductility have good tensile toughness; however, this is not always the case when the strain rates are high. For example, metals that show excellent tensile toughness may show brittle behavior under high strain rates (i.e., they may show poor impact toughness). Thus, the imposed strain rate can shift the ductile to brittle transition. Ceramics and many

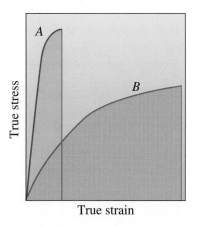

Figure 6-26
The area contained under the true stress-true strain curve is related to the tensile toughness. Although material *B* has a lower yield strength, it absorbs more energy than material *A*. The energies from these curves may not be the same as those obtained from impact test data.

composites normally have poor toughness, even though they have high strength, because they display virtually no ductility. These materials show both poor tensile toughness and poor impact toughness.

Use of Impact Properties
Absorbed energy and the DBTT are very sensitive to loading conditions. For example, a higher rate of application of energy to the specimen reduces the absorbed energy and increases the DBTT. The size of the specimen also affects the results; because it is more difficult for a thick material to deform, smaller energies are required to break thicker materials. Finally, the configuration of the notch affects the behavior; a sharp, pointed surface crack permits lower absorbed energies than does a V-notch. Because we often cannot predict or control all of these conditions, the impact test is a quick, convenient, and inexpensive way to compare different materials.

Example 6-8   *Design of a Sledgehammer*

Design a 3.6-kg sledgehammer for driving steel fence posts into the ground.

SOLUTION

First, we must consider the design requirements to be met by the sledgehammer. A partial list would include

1. The handle should be light in weight, yet tough enough that it will not catastrophically break.
2. The head must not break or chip during use, even in subzero temperatures.
3. The head must not deform during continued use.
4. The head must be large enough to ensure that the user does not miss the fence post, and it should not include sharp notches that might cause chipping.
5. The sledgehammer should be inexpensive.

Although the handle could be a lightweight, tough composite material (such as a polymer reinforced with Kevlar (a special polymer) fibers), a wood handle about 70 cm long would be much less expensive and would still provide sufficient toughness. As shown later in Chapter 17, wood can be categorized as a natural fiber-reinforced composite.

To produce the head, we prefer a material that has a low transition temperature, can absorb relatively high energy during impact, and yet also has enough hardness to avoid deformation. The toughness requirement would rule out most ceramics. A face-centered cubic metal, such as FCC stainless steel or copper, might provide superior toughness even at low temperatures; however, these metals are relatively soft and expensive. An appropriate choice might be a BCC steel. Ordinary steels are inexpensive, have good hardness and strength, and some have sufficient toughness at low temperatures.

In Appendix A, we find that the density of iron is 7.87 g/cm^3. We assume that the density of steel is about the same. The volume of steel required is $V = \dfrac{3.6 \times 10^3 \text{ g}}{7.87 \text{ g/cm}^3} =$ 0.46×10^3 cm^3. To ensure that we will hit our target, the head might have a cylindrical shape with a diameter of 6.25 cm. The length of the head would then be 15 cm.

6-11   Bulk Metallic Glasses and Their Mechanical Behavior

Metals, as they are found in nature, are crystalline; however, when particular multicomponent alloys are cooled rapidly, amorphous metals may form. Some alloys require cooling rates as high as 10^6 K/s in order to form an amorphous (or "glassy") structure, but recently, new compositions have been found that require cooling rates on the order of only a few degrees per second. This has enabled the production of so-called "bulk metallic glasses"—metallic glasses with thicknesses or diameters as large as 5 cm.

Before the development of bulk metallic glasses, amorphous metals were produced by a variety of rapid solidification techniques, including a process known as melt spinning. In melt spinning, liquid metal is poured onto chilled rolls that rotate and "spin off" thin ribbons on the order of 10 μm thick. It is difficult to perform mechanical testing on ribbons; thus, the development of bulk metallic glasses enabled mechanical tests that were not previously possible. Bulk metallic glasses can be produced by several methods. One method involves using an electric arc to melt elements of high purity and then suction casting or pour casting into cooled copper molds.

As shown in Figure 6-27, metallic glasses exhibit fundamentally different stress–strain behavior from other classes of materials. Most notably, metallic glasses are exceptionally high-strength materials. They typically have yield strengths on the order of 2 GPa, comparable to those of the highest strength steels, and yield strengths higher than 5 GPa have been reported for iron-based amorphous metal alloys. Since metallic glasses are not crystalline, they do not contain dislocations. As we learned in Chapter 4, dislocations lead to yield strengths that are lower than those that are theoretically predicted for perfect crystalline materials. The high strengths of metallic glasses are due to the lack of dislocations in the amorphous structure.

Despite their high strengths, typical bulk metallic glasses are brittle. Most metallic glasses exhibit nearly zero plastic strain in tension and only a few percent plastic strain

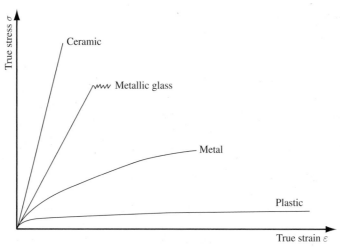

Figure 6-27   A schematic diagram of the compressive stress–strain behavior of various engineering materials including metallic glasses. Serrated flow is observed in the plastic region.

in compression (compared to tens of percent plastic strain for crystalline metals.) This lack of ductility is related to a lack of dislocations. Since metallic glasses do not contain dislocations, they do not work harden (i.e., the stress does not increase with strain after plastic deformation has commenced, as shown in Figure 6-27. Work hardening, also known as strain hardening, will be discussed in detail in Chapter 8). Thus, when deformation becomes localized, it intensifies and quickly leads to failure.

At room temperature, metallic glasses are permanently deformed through intense shearing in narrow bands of material about 10 to 100 nm thick. This creates shear offsets at the edges of the material, as shown in Figure 6-28. For the case of compression, this results in a decrease in length in the direction of the loading axis. As plasticity proceeds, more shear bands form and propagate across the sample. More shear bands form to accommodate the increasing plastic strain until finally one of these shear bands fails the

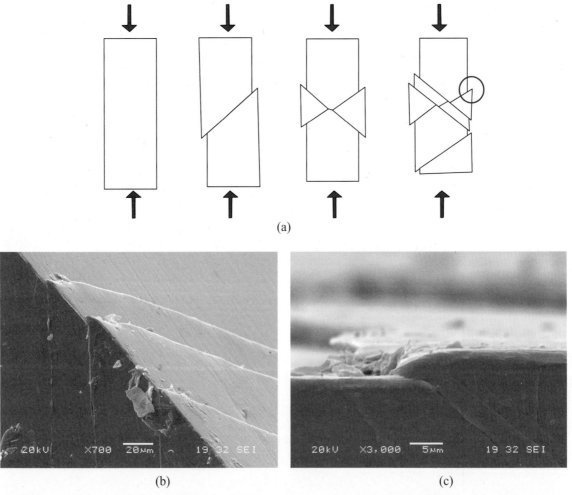

(a)

(b)   (c)

Figure 6-28   (a) A schematic diagram of shear band formation in a metallic glass showing successive stages of a compression test. (b) A scanning electron micrograph showing three shear bands in a $Zr_{41.2}Ti_{13.8}Cu_{12.5}Ni_{10.0}Be_{22.5}$ bulk metallic glass. The arrows indicate the loading direction. (c) A scanning electron micrograph of a shear band offset in a $Zr_{41.2}Ti_{13.8}Cu_{12.5}Ni_{10.0}Be_{22.5}$ bulk metallic glass. The location of such an offset is circled in (a). (*Photos courtesy of Wendelin Wright.*)

sample. The effects of shear banding can be observed in Figure 6-27. When a metallic-glass compression sample is deformed at a constant displacement rate, the load (or stress) drops as a shear band propagates. This leads to the "serrated" stress–strain behavior shown in the figure. Although some crystalline materials show serrated behavior, the origin of this phenomenon is completely different in amorphous metals. Current research efforts are aimed at understanding the shear banding process in metallic glasses and preventing shear bands from causing sample failure.

Metallic glasses have applications in sporting goods equipment, electronic casings, defense components, and as structural materials. Metallic glasses are also suitable as industrial coatings due to their high hardness and good corrosion resistance (both again due to a lack of dislocations in the structure). All potential applications must utilize metallic glasses well below their glass-transition temperatures since they will crystallize and lose their unique mechanical behavior at elevated temperatures.

6-12   Mechanical Behavior at Small Length Scales

With the development of thin films on substrates for microelectronics applications and the synthesis of structures with nanometer scale dimensions, materials scientists have observed that materials may display different mechanical properties depending on the length scale at which the materials are tested. Several experimental techniques have been developed in order to measure mechanical behavior at small length scales. Nanoindentation, which was discussed in Section 8, is used for this purpose. Another technique is known as wafer curvature analysis.

In the wafer curvature technique, the temperature of a thin film on a substrate is cycled, typically from room temperature to several hundred degrees Celsius. Since the thin film and the substrate have different coefficients of thermal expansion, they expand (or contract) at different rates. This induces stresses in the thin film. These stresses in the film are directly proportional to the change in curvature of the film–substrate system as the temperature is cycled. An example of a wafer curvature experiment for a 0.5 μm polycrystalline aluminum thin film on an oxidized silicon substrate is shown in Figure 6-29(a). Aluminum has a larger thermal expansion coefficient than silicon.

Figure 6-29(b) shows two cycles of the experiment. The experiment begins with the film–substrate system at room temperature. The aluminum has a residual stress of 30 MPa at room temperature due to cooling from the processing temperature. During the first cycle, the grains in the film grow as the temperature increases (Chapter 5). As the grains grow, the grain boundary area decreases, and the film densifies. At the same time, the aluminum expands more rapidly than the silicon, and it is constrained from expanding by the silicon substrate to which it is bonded. Thus, the aluminum is subjected to a state of compression. As the temperature increases, the aluminum plastically deforms. The stress does not increase markedly, because the temperature rise causes a decrease in strength.

As the system cools, the aluminum contracts more with a decrease in temperature than does the silicon. As the aluminum contracts, it first unloads elastically and then deforms plastically. As the temperature decreases, the stress continues to increase until the cycle is complete when room temperature is reached. A second thermal cycle is shown. Subsequent temperature cycles will be similar in shape, since the microstructure of the film does not change unless the highest temperature exceeds that of the previous cycle.

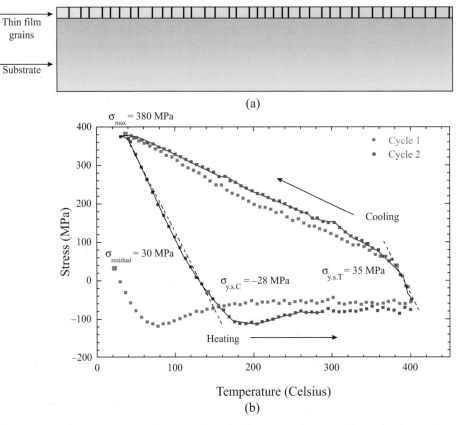

Figure 6-29 (a) A schematic diagram of a thin film on a substrate. The grains have diameters on the order of the film thickness. Note that thin films typically have thicknesses about 1/500th of the substrate thickness (some are far thinner), and so, this diagram is not drawn to scale. (b) Stress versus temperature for thermal cycling of a 0.5 μm polycrystalline aluminum thin film on an oxidized silicon substrate during a wafer curvature experiment.

Notice that the stress sustained by the aluminum at room temperature after cycling is 380 MPa! Pure bulk aluminum (aluminum with macroscopic dimensions) is able to sustain only a fraction of this stress without failing. Thus, we have observed a general trend in materials science—for crystalline metals, smaller is stronger!

In general, the key to the strength of a crystalline metal is dislocations. As the resistance to dislocation motion increases, the strength of the metal also increases. One mechanism for strengthening in micro and nanoscale crystalline metals is the grain size. The grain size of thin films tends to be on the order of the film thickness, as shown in Figure 6-29(a). As grain size decreases, yield strength increases (see Chapter 4, Section 7 for a discussion of the Hall-Petch equation).

Dislocations distort the surrounding crystal, increasing the strain energy in the atomic bonds. Thus, dislocations have energies associated with them. For all metals, as the dislocation density (or the amount of dislocation length per unit volume in the crystal) increases, this energy increases.

In thin films, dislocations may be pinned at the interface between the thin film and the substrate to which it is bonded. Thus, it is necessary that the dislocation increase in length along the interface in order for it to propagate and for the thin film to plastically deform. Increasing the dislocation length requires energy, and the stress required to cause the dislocation to propagate is higher than it would be for a dislocation that is not constrained by an interface. This stress is inversely proportional to the film thickness; thus, as film thickness decreases, the strength increases. When two surfaces constrain the dislocation, such as when a passivating layer (i.e., one that protects the thin film surface from oxidation or corrosion) is deposited on a thin film, the effect is even more pronounced. This inverse relationship between film strength and thickness is independent from the grain-size effect discussed previously. Remember: Any mechanism that interferes with the motion of dislocations makes a metal stronger.

In order to induce a nonuniform shape change in a material, such as bending a bar or indenting a material, dislocations must be introduced to the crystal structure. Such dislocations are called *geometrically necessary dislocations*. These dislocations exist in addition to the dislocations (known as *statistically stored dislocations*) that are produced by homogeneous strain; thus, the dislocation density is increased. At small length scales (such as for small indentations made using a nanoindenter), the density of the geometrically necessary dislocations is significant, but at larger length scales, the effect is diminished. Thus, the hardness of shallow indents is greater than the hardness of deep indents made in the same material. As you will learn in Chapter 8, as dislocation density increases, the strength of a metal increases. Dislocations act as obstacles to the propagation of other dislocations, and again, any mechanism that interferes with the motion of dislocations makes a metal stronger.

An increasingly common mechanical testing experiment involves fabricating compression specimens with diameters on the order of 1 micron using a tool known as the focused ion beam. Essentially, a beam of gallium ions is used to remove atoms from the surface of a material, thereby performing a machining process at the micron and sub-micron length scale. These specimens are then deformed under compression in a nanoindenter using a flat punch tip. The volume of such specimens is on the order of 2.5 μm^3. Extraordinary strengths have been observed in single-crystal pillars made from metals. This topic is an area of active research in the materials community.

Summary

- The mechanical behavior of materials is described by their mechanical properties, which are measured with idealized, simple tests. These tests are designed to represent different types of loading conditions. The properties of a material reported in various handbooks are the results of these tests. Consequently, we should always remember that handbook values are average results obtained from idealized tests and, therefore, must be used with some care.

- The tensile test describes the resistance of a material to a slowly applied tensile stress. Important properties include yield strength (the stress at which the material begins to permanently deform), tensile strength (the stress corresponding to the maximum applied load), modulus of elasticity (the slope of the elastic portion of the stress–strain curve), and % elongation and % reduction in area (both measures of the ductility of the material).

- The bend test is used to determine the tensile properties of brittle materials. A modulus of elasticity and a flexural strength (similar to a tensile strength) can be obtained.

- The hardness test measures the resistance of a material to penetration and provides a measure of the wear and abrasion resistance of the material. A number of hardness tests, including the Rockwell and Brinell tests, are commonly used. Often the hardness can be correlated to other mechanical properties, particularly tensile strength.

- Nanoindentation is a hardness testing technique that continuously measures the imposed load and displacement with micro-Newton and sub-nanometer resolution, respectively. Nanoindentation techniques are important for measuring the mechanical properties of thin films on substrates and nanophase materials and for deforming free-standing micro and nanoscale structures. Both hardness and elastic modulus are measured using nanoindentation.

- The impact test describes the response of a material to a rapidly applied load. The Charpy and Izod tests are typical. The energy required to fracture the specimen is measured and can be used as the basis for comparison of various materials tested under the same conditions. In addition, a transition temperature above which the material fails in a ductile, rather than a brittle, manner can be determined.

- Metallic glasses are amorphous metals. As such, they do not contain dislocations. A lack of dislocations leads to high strengths and low ductilities for these materials.

- Crystalline metals exhibit higher strengths when their dimensions are confined to the micro and nanoscale. Size-dependent mechanical behavior has critical implications for design and materials reliability in nanotechnology applications.

Glossary

Anelastic (viscoelastic) material A material in which the total strain developed has elastic and viscous components. Part of the total strain recovers similar to elastic strain. Some part, though, recovers over a period of time. Examples of viscoelastic materials include polymer melts and many polymers including Silly Putty®. Typically, the term anelastic is used for metallic materials.

Apparent viscosity Viscosity obtained by dividing shear stress by the corresponding value of the shear-strain rate for that stress.

Bend test Application of a force to a bar that is supported on each end to determine the resistance of the material to a static or slowly applied load. Typically used for brittle materials.

Bingham plastic A material with a mechanical response given by $\tau = G\gamma$ when $\tau < \tau_{y.s}$ and $\tau = \tau_{y.s} + \eta\dot{\gamma}$ when $\tau \geq \tau_{y.s}$.

Dilatant (shear thickening) Materials in which the apparent viscosity increases with increasing rate of shear.

Ductile to brittle transition temperature (DBTT) The temperature below which a material behaves in a brittle manner in an impact test; it also depends on the strain rate.

Ductility The ability of a material to be permanently deformed without breaking when a force is applied.

Elastic deformation Deformation of the material that is recovered instantaneously when the applied load is removed.

Elastic limit The magnitude of stress at which plastic deformation commences.

Elastic strain Fully and instantaneously recoverable strain in a material.

Elastomers Natural or synthetic plastics that are composed of molecules with spring-like coils that lead to large elastic deformations (e.g., natural rubber, silicones).

Engineering strain Elongation per unit length calculated using the original dimensions.

Engineering stress The applied load, or force, divided by the original area over which the load acts.

Extensometer An instrument to measure change in length of a tensile specimen, thus allowing calculation of strain. An extensometer is often a clip that attaches to a sample and elastically deforms to measure the length change.

Flexural modulus The modulus of elasticity calculated from the results of a bend test; it is proportional to the slope of the stress-deflection curve.

Flexural strength (modulus of rupture) The stress required to fracture a specimen in a bend test.

Fracture toughness The resistance of a material to failure in the presence of a flaw.

Glass-transition temperature (T_g) A temperature below which an otherwise ductile material behaves as if it is brittle. Usually, this temperature is not fixed and is affected by processing of the material.

Hardness test Measures the resistance of a material to penetration by a sharp object. Common hardness tests include the Brinell test, Rockwell test, Knoop test, and Vickers test.

Hooke's law The linear-relationship between stress and strain in the elastic portion of the stress-strain curve.

Impact energy The energy required to fracture a standard specimen when the load is applied suddenly.

Impact loading Application of stress at a very high strain rate ($\sim > 100\,s^{-1}$).

Impact test Measures the ability of a material to absorb the sudden application of a load without breaking. The Charpy and Izod tests are commonly used impact tests.

Impact toughness Energy absorbed by a material, usually notched, during fracture, under the conditions of the impact test.

Kinematic viscosity Ratio of viscosity and density, often expressed in centiStokes.

Load The force applied to a material during testing.

Macrohardness Bulk hardness of materials measured using loads >2 N.

Materials processing Manufacturing or fabrication methods used for shaping of materials (e.g., extrusion, forging).

Microhardness Hardness of materials typically measured using loads less than 2 N with a test such as the Knoop (HK).

Modulus of elasticity (E) Young's modulus, or the slope of the linear part of the stress–strain curve in the elastic region. It is a measure of the stiffness of the bonds of a material and is not strongly dependent upon microstructure.

Modulus of resilience (E_r) The maximum elastic energy absorbed by a material when a load is applied.

Nanoindentation Hardness testing performed at the nanometer length scale. The imposed load and displacement are measured with micro-Newton and sub-nanometer resolution, respectively.

Necking Local deformation causing a reduction in the cross-sectional area of a tensile specimen. Many ductile materials show this behavior. The engineering stress begins to decrease at the onset of necking.

Newtonian Materials in which the shear stress and shear strain rate are linearly related (e.g., light oil or water).

Non-Newtonian Materials in which the shear stress and shear strain rate are not linearly related; these materials are shear thinning or shear thickening (e.g., polymer melts, slurries, paints, etc.).

Notch sensitivity Measures the effect of a notch, scratch, or other imperfection on a material's properties such as toughness or fatigue life.

Offset strain value A value of strain (e.g., 0.002) used to obtain the offset yield stress.

Offset yield strength A stress value obtained graphically that describes the stress that gives no more than a specified amount of plastic deformation. Most useful for designing components. Also, simply stated as the yield strength.

Percent elongation The total percentage permanent increase in the length of a specimen due to a tensile test.

Percent reduction in area The total percentage permanent decrease in the cross-sectional area of a specimen due to a tensile test.

Plastic deformation or strain Permanent deformation of a material when a load is applied, then removed.

Poisson's ratio The negative of the ratio between the lateral and longitudinal strains in the elastic region.

Proportional limit A level of stress above which the relationship between stress and strain is not linear.

Pseudoplastics (shear thinning) Materials in which the apparent viscosity decreases with increasing rate of shear.

Rheopectic behavior Materials that show shear thickening and also an apparent viscosity that at a constant rate of shear increases with time.

Shear modulus (G) The slope of the linear part of the shear stress-shear strain curve.

Shear-strain rate Time derivative of shear strain. See "Strain rate."

Shear thickening (dilatant) Materials in which the apparent viscosity increases with increasing rate of shear.

Shear thinning (pseudoplastics) Materials in which the apparent viscosity decreases with increasing rate of shear.

Stiffness A measure of a material's resistance to elastic deformation. Stiffness is the slope of a load-displacement curve and is proportional to the elastic modulus. Stiffness depends on the geometry of the component under consideration, whereas the elastic or Young's modulus is a materials property. The inverse of stiffness is known as compliance.

Strain Elongation per unit length.

Strain gage A device used for measuring strain. A strain gage typically consists of a fine wire embedded in a polymer matrix. The strain gage is bonded to the test specimen and deforms as the specimen deforms. As the wire in the strain gage deforms, its resistance changes. The resistance change is directly proportional to the strain.

Strain rate The rate at which strain develops in or is applied to a material indicated; it is represented by $\dot{\varepsilon}$ or $\dot{\gamma}$ for tensile and shear-strain rates, respectively. Strain rate can have an effect on whether a material behaves in a ductile or brittle fashion.

Stress Force per unit area over which the force is acting.

Stress relaxation Decrease in stress for a material held under constant strain as a function of time, which is observed in viscoelastic materials. Stress relaxation is different from time dependent recovery of strain.

Tensile strength The stress that corresponds to the maximum load in a tensile test.

Tensile test Measures the response of a material to a slowly applied uniaxial force. The yield strength, tensile strength, modulus of elasticity, and ductility are obtained.

Tensile toughness The area under the true stress–true strain tensile test curve. It is a measure of the energy required to cause fracture under tensile test conditions.

Thixotropic behavior Materials that show shear thinning and also an apparent viscosity that at a constant rate of shear decreases with time.

True strain Elongation per unit length calculated using the instantaneous dimensions.

True stress The load divided by the instantaneous area over which the load acts.

Ultimate tensile strength (UTS) See Tensile strength.

Viscoelastic (or anelastic) material See Anelastic material.

Viscosity (η) Measure of the resistance to flow, defined as the ratio of shear stress to shear strain rate (units Poise or Pa-s).

Viscous material A viscous material is one in which the strain develops over a period of time and the material does not return to its original shape after the stress is removed.

Work of fracture Area under the stress–strain curve, considered as a measure of tensile toughness.

Yield point phenomenon An abrupt transition, seen in some materials, from elastic deformation to plastic flow.

Yield strength A stress value obtained graphically that describes no more than a specified amount of deformation (usually 0.002). Also known as the offset yield strength.

Young's modulus (E) The slope of the linear part of the stress–strain curve in the elastic region, same as modulus of elasticity.

Problems

Section 6-1 Technological Significance

6-1 Explain the role of mechanical properties in load-bearing applications using real-world examples.

6-2 Explain the importance of mechanical properties in functional applications (e.g., optical, magnetic, electronic, etc.) using real-world examples.

6-3 Explain the importance of understanding mechanical properties in the processing of materials.

Section 6-2 Terminology for Mechanical Properties

6-4 Define "engineering stress" and "engineering strain."

6-5 Define "modulus of elasticity."

6-6 Define "plastic deformation" and compare it to "elastic deformation."

6-7 What is strain rate? How does it affect the mechanical behavior of polymeric and metallic materials?

6-8 Why does Silly Putty® break when you stretch it very quickly?

6-9 What is a viscoelastic material? Give an example.

6-10 What is meant by the term "stress relaxation?"

6-11 Define the terms "viscosity," "apparent viscosity," and "kinematic viscosity."

6-12 What two equations are used to describe Bingham plastic-like behavior?

6-13 What is a Newtonian material? Give an example.

6-14 What is an elastomer? Give an example.

6-15 What is meant by the terms "shear thinning" and "shear thickening" materials?

6-16 Many paints and other dispersions are not only shear thinning, but also thixotropic. What does the term "thixotropy" mean?

6-17 Draw a schematic diagram showing the development of strain in an elastic and viscoelastic material. Assume that the load is applied at some time $t = 0$ and taken off at some time t.

Section 6-3 The Tensile Test: Use of the Stress-Strain Diagram

6-18 Draw qualitative engineering stress-engineering strain curves for a ductile polymer, a ductile metal, a ceramic, a glass, and natural rubber. Label the diagrams carefully. Rationalize your sketch for each material.

6-19 What is necking? How does it lead to reduction in engineering stress as true stress increases?

6-20 (a) Carbon nanotubes are one of the stiffest and strongest materials known to scientists and engineers. Carbon nanotubes have an elastic modulus of 1.1 TPa (1 TPa = 10^{12} Pa). If a carbon nanotube has a diameter of 15 nm, determine the engineering stress sustained by the nanotube when subjected to a tensile load of $4 \, \mu N$ ($1 \, \mu N = 10^{-6}$ N) along the length of the tube. Assume that the entire cross-sectional area of the nanotube is load bearing.

(b) Assume that the carbon nanotube is only deformed elastically (not plastically) under the load of 4 μN. The carbon nanotube has a length of 10 μm (1μm = 10^{-6} m). What is the tensile elongation (displacement) of the carbon nanotube in nanometers (1nm = 10^{-9} m)?

6-21 A 3780-N force is applied to a 0.375-cm-diameter nickel wire having a yield strength of 310 MPa and a tensile strength of 379 MPa. Determine
(a) whether the wire will plastically deform and
(b) whether the wire will experience necking.

6-22 (a) A force of 100,000 N is applied to an iron bar with a cross-sectional area of 10 mm $\times$ 20 mm and having a yield strength of 400 MPa and a tensile strength of 480 MPa. Determine whether the bar will plastically deform and whether the bar will experience necking.
(b) Calculate the maximum force that a 0.5-cm-diameter rod of Al_2O_3, having a yield strength of 241 MPa, can withstand with no plastic deformation. Express your answer in Newtons.

6-23 A force of 20,000 N will cause a 1 cm $\times$ 1 cm bar of magnesium to stretch from 10 cm to 10.045 cm. Calculate the modulus of elasticity.

6-24 A polymer bar's dimensions are 2.5 cm $\times$ 5 cm $\times$ 37.5 cm. The polymer has a modulus of elasticity of 4137 MPa. What force is required to stretch the bar elastically from 37.5 cm to 38.13 cm?

6-25 An aluminum plate 0.5 cm thick is to withstand a force of 50,000 N with no permanent deformation. If the aluminum has a yield strength of 125 MPa, what is the minimum width of the plate?

6-26 A steel cable 3.13 cm in diameter and 1500 cm long is to lift a 18,140-kg load. What is the length of the cable during lifting? The modulus of elasticity of the steel is 207×10^3 MPa.

Section 6-4 Properties Obtained from the Tensile Test and Section 6-5 True Stress and True Strain

6-27 Define "true stress" and "true strain." Compare with engineering stress and engineering strain.

6-28 Write down the formulas for calculating the stress and strain for a sample subjected to a tensile test. Assume the sample shows necking.

6-29 Derive the expression $\varepsilon = \ln (1 + e)$, where ε is the true strain and e is the engineering strain. Note that this expression is not valid after the onset of necking.

6-30 The following data were collected from a test specimen of cold-rolled and annealed brass. The specimen had an initial gage length l_0 of 35 mm and an initial cross-sectional area A_0 of 10.5 mm^2.

Load (N)	Δl (mm)
0	0.0000
66	0.0112
177	0.0157
327	0.0199
462	0.0240
797	1.72
1350	5.55
1720	8.15
2220	13.07
2690	22.77 (maximum load)
2410	25.25 (fracture)

(a) Plot the engineering stress–strain curve and the true stress–strain curve. Since the instantaneous cross-sectional area of the specimen is unknown past the point of necking, truncate the true stress–true strain data at the point that corresponds to the ultimate tensile strength. Use of a software graphing package is recommended.
(b) Comment on the relative values of true stress–strain and engineering stress–strain during the elastic loading and prior to necking.
(c) If the true stress–strain data were known past the point of necking, what might the curve look like?
(d) Calculate the 0.2% offset yield strength.

(e) Calculate the tensile strength.

(f) Calculate the elastic modulus using a linear fit to the appropriate data.

6-31 The following data were collected from a standard 1.263-cm-diameter test specimen of a copper alloy (initial length (l_0) = 5 cm):

Load (N)	Δl (cm)
0	0.00000
13,340	0.00418
26,690	0.00833
33,360	0.01043
40,030	0.0225
46,700	0.1
53,380	0.65
55,160	1.25 (maximum load)
50,710	2.55 (fracture)

After fracture, the total length was 7.535 cm and the diameter was 0.935 cm. Plot the engineering stress–strain curve and calculate

(a) the 0.2% offset yield strength;

(b) the tensile strength;

(c) the modulus of elasticity;

(d) the % elongation;

(e) the % reduction in area;

(f) the engineering stress at fracture; and

(g) the modulus of resilience.

6-32 The following data were collected from a 1-cm-diameter test specimen of polyvinyl chloride (l_0 = 5 cm):

Load (N)	Δl (cm)
0	0.00000
1334	0.01865
2669	0.0374
4003	0.05935
5338	0.08
6672	0.115
7384	0.175 (maximum load)
7117	0.235
6316	0.3 (fracture)

After fracture, the total length was 5.225 cm and the diameter was 0.983 cm. Plot the engineering stress–strain curve and calculate

(a) the 0.2% offset yield strength;

(b) the tensile strength;

(c) the modulus of elasticity;

(d) the % elongation;

(e) the % reduction in area;

(f) the engineering stress at fracture; and

(g) the modulus of resilience.

6-33 The following data were collected from a 12-mm-diameter test specimen of magnesium (l_0 = 30.00 mm):

Load (N)	Δl (mm)
0	0.0000
5,000	0.0296
10,000	0.0592
15,000	0.0888
20,000	0.15
25,000	0.51
26,500	0.90
27,000	1.50 (maximum load)
26,500	2.10
25,000	2.79 (fracture)

After fracture, the total length was 32.61 mm and the diameter was 11.74 mm. Plot the engineering stress–strain curve and calculate

(a) the 0.2% offset yield strength;

(b) the tensile strength;

(c) the modulus of elasticity;

(d) the % elongation;

(e) the % reduction in area;

(f) the engineering stress at fracture; and

(g) the modulus of resilience.

6-34 The following data were collected from a 20-mm-diameter test specimen of a ductile cast iron (l_0 = 40.00 mm):

Load (N)	Δl (mm)
0	0.0000
25,000	0.0185
50,000	0.0370
75,000	0.0555
90,000	0.20
105,000	0.60
120,000	1.56
131,000	4.00 (maximum load)
125,000	7.52 (fracture)

After fracture, the total length was 47.42 mm and the diameter was 18.35 mm. Plot the engineering stress–strain curve and the true stress–strain curve. Since the instantaneous cross-sectional area of the specimen is unknown past the point of necking, truncate the true stress–true strain data at the point that corresponds to the ultimate tensile strength. Use of a software graphing package is recommended. Calculate

(a) the 0.2% offset yield strength;
(b) the tensile strength;
(c) the modulus of elasticity, using a linear fit to the appropriate data;
(d) the % elongation;
(e) the % reduction in area;
(f) the engineering stress at fracture; and
(g) the modulus of resilience.

6-35   (a) A 1-cm-diameter, 30-cm-long titanium bar has a yield strength of 345 MPa, a modulus of elasticity of 110.3×10^3 MPa, and a Poisson's ratio of 0.30. Determine the length and diameter of the bar when a 2224 N load is applied.

(b) When a tensile load is applied to a 1.5-cm-diameter copper bar, the diameter is reduced to 1.498 cm. Determine the applied load, using the data in Table 6-3.

6-36   Consider the tensile stress–strain diagrams in Figure 6-30 labeled 1 and 2. These diagrams are typical of metals. Answer the following questions, and consider each part as a separate question that has no relationship to previous parts of the question.

(a) Samples 1 and 2 are identical except for the grain size. Which sample has the smaller grains? How do you know?

(b) Samples 1 and 2 are identical except that they were tested at different temperatures. Which was tested at the lower temperature? How do you know?

(c) Samples 1 and 2 are different materials. Which sample is tougher? Explain.

(d) Samples 1 and 2 are identical except that one of them is a pure metal and the other has a small percentage alloying addition. Which sample has been alloyed? How do you know?

(e) Given the stress–strain curves for materials 1 and 2, which material has the lower hardness value on the Brinell hardness scale? How do you know?

(f) Are the stress–strain curves shown true stress–strain or engineering stress–strain curves? How do you know?

(g) Which of the two materials represented by samples 1 and 2 would exhibit a higher shear yield strength? How do you know?

Section 6-6 The Bend Test for Brittle Materials

6-37   Define the term "flexural strength" and "flexural modulus."

6-38   Why is it that we often conduct a bend test on brittle materials?

6-39   A bar of Al_2O_3 that is 0.625 cm thick, 1.25 cm wide, and 22.5 cm long is tested in a three-point bending apparatus with the supports located 15 cm apart. The deflection of the center of the bar is measured as a function of the applied load. The data are shown below. Determine the flexural strength and the flexural modulus.

Force (N)	Deflection (cm)
64.5	0.00625
128.5	0.0125
193.0	0.01875
257.5	0.025
382.5	0.03725 (fracture)

6-40   A three-point bend test is performed on a block of ZrO_2 that is 20 cm long, 1.25 cm wide, and 0.625 cm thick and is resting on two supports 10 cm apart. When a force of 1780 N is applied, the specimen deflects 0.093 cm and breaks. Calculate

(a) the flexural strength and
(b) the flexural modulus, assuming that no plastic deformation occurs.

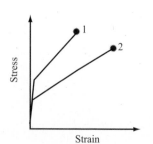

Figure 6-30
Stress–strain curves for Problem 6-36.

6-41 A three-point bend test is performed on a block of silicon carbide that is 10 cm long, 1.5 cm wide, and 0.6 cm thick and is resting on two supports 7.5 cm apart. The sample breaks when a deflection of 0.09 mm is recorded. The flexural modulus for silicon carbide is 480 GPa. Assume that no plastic deformation occurs. Calculate
(a) the force that caused the fracture and
(b) the flexural strength.

6-42 (a) A thermosetting polymer containing glass beads is required to deflect 0.5 mm when a force of 500 N is applied. The polymer part is 2 cm wide, 0.5 cm thick, and 10 cm long. If the flexural modulus is 6.9 GPa, determine the minimum distance between the supports. Will the polymer fracture if its flexural strength is 85 MPa? Assume that no plastic deformation occurs.
(b) The flexural modulus of alumina is 310.3×10^3 MPa, and its flexural strength is 317 MPa. A bar of alumina 0.75 cm thick, 2.5 cm wide, and 25 cm long is placed on supports 17.5 cm apart. Determine the amount of deflection at the moment the bar breaks, assuming that no plastic deformation occurs.

6-43 Ceramics are much stronger in compression than in tension. Explain why.

6-44 Dislocations have a major effect on the plastic deformation of metals, but do not play a major role in the mechanical behavior of ceramics. Why?

6-45 What controls the strength of ceramics and glasses?

Section 6-7 Hardness of Materials and Section 6-8 Nanoindentation

6-46 What does the term "hardness of a material" mean?

6-47 Why is hardest data difficult to correlate to mechanical properties of materials in a quantitative fashion?

6-48 What is the hardness material (natural or synthetic)? Is it diamond?

6-49 Explain the terms "macrohardness" and "microhardness."

6-50 A Brinell hardness measurement, using a 10-mm-diameter indenter and a 500-kg load, produces an indentation of 4.5 mm on an aluminum plate. Determine the Brinell hardness number (HB) of the metal.

6-51 When a 3000-kg load is applied to a 10-mm-diameter ball in a Brinell test of a steel, an indentation of 3.1 mm diameter is produced. Estimate the tensile strength of the steel.

6-52 Why is it necessary to perform calibrations on a standard prior to performing a nanoindentation experiment?

6-53 The elastic modulus of a metallic glass is determined to be 95 GPa using nanoindentation testing with a diamond Berkovich tip. The Poisson's ratio of the metallic glass is 0.36. The unloading stiffness as determined from the load-displacement data is 5.4×10^5 N/m. The maximum load is 120 mN. What is the hardness of the metallic glass at this indentation depth?

Section 6-9 Strain Rate Effects and Impact Behavior and Section 6-10 Properties from the Impact Test

6-54 The following data were obtained from a series of Charpy impact tests performed on four steels, each having a different manganese content. Plot the data and determine
(a) the transition temperature of each (defined by the mean of the absorbed energies in the ductile and brittle regions) and
(b) the transition temperature of each (defined as the temperature that provides 50 J of absorbed energy).

Test Temperature (°C)	Impact Energy (J)			
	0.30% Mn	0.39% Mn	1.01% Mn	1.55% Mn
−100	2	5	5	15
−75	2	5	7	25
−50	2	12	20	45
−25	10	25	40	70
0	30	55	75	110
25	60	100	110	135
50	105	125	130	140
75	130	135	135	140
100	130	135	135	140

6-55 Plot the transition temperature versus manganese content using the data in

Problem 6-54 and discuss the effect of manganese on the toughness of steel. What is the minimum manganese allowed in the steel if a part is to be used at 0°C?

6-56 The following data were obtained from a series of Charpy impact tests performed on four ductile cast irons, each having a different silicon content. Plot the data and determine

(a) the transition temperature of each (defined by the mean of the absorbed energies in the ductile and brittle regions) and

(b) the transition temperature of each (defined as the temperature that provides 10 J of absorbed energy).

Plot the transition temperature versus silicon content and discuss the effect of silicon on the toughness of the cast iron. What is the maximum silicon allowed in the cast iron if a part is to be used at 25°C?

Test Temperature (°C)	Impact Energy (J)			
	2.55% Si	2.85% Si	3.25% Si	3.63% Si
−50	2.5	2.5	2	2
−5	3	2.5	2	2
0	6	5	3	2.5
25	13	10	7	4
50	17	14	12	8
75	19	16	16	13
100	19	16	16	16
125	19	16	16	16

6-57 FCC metals are often recommended for use at low temperatures, particularly when any sudden loading of the part is expected. Explain.

6-58 A steel part can be made by powder metallurgy (compacting iron powder particles and sintering to produce a solid) or by machining from a solid steel block. Which part is expected to have the higher toughness? Explain.

6-59 What is meant by the term notch sensitivity?

6-60 What is the difference between a tensile test and an impact test? Using this, explain why the toughness values measured using impact

tests may not always correlate with tensile toughness measured using tensile tests.

6-61 A number of aluminum-silicon alloys have a structure that includes sharp-edged plates of brittle silicon in the softer, more ductile aluminum matrix. Would you expect these alloys to be notch-sensitive in an impact test? Would you expect these alloys to have good toughness? Explain your answers.

6-62 What is the ductile to brittle transition temperature (DBTT)?

6-63 How is tensile toughness defined in relation to the true stress–strain diagram? How is tensile toughness related to impact toughness?

6-11 Bulk Metallic Glasses and Their Mechanical Behavior

6-64 A load versus displacement diagram is shown in Figure 6-31 for a metallic glass. A metallic glass is a non-crystalline (amorphous) metal. The sample was tested in compression. *Therefore, even though the load and displacement values are plotted as positive, the sample length was shortened during the test.* The sample had a length in the direction of loading of 6 mm and a cross-sectional area of 4 mm². Numerical values for the load and displacement are given at the points marked with a circle and an X. The first data point is (0, 0). Sample failure is indicated with an X. Answer the following questions.

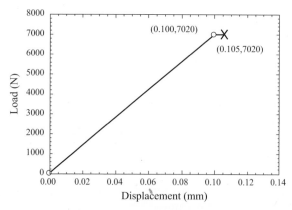

Figure 6-31 Load versus displacement for a metallic glass tested in compression for Problem 6-64.

(a) Calculate the elastic modulus.

(b) How does the elastic modulus compare to the modulus of steel?

(c) Calculate the engineering stress at the proportional limit.

(d) Consider your answer to part (c) to be the yield strength of the material. Is this a high yield stress or a low yield stress? Support your answer with an order of magnitude comparison for a typical polycrystalline metal.

(e) Calculate the true strain at the proportional limit. Remember that the length of the sample is decreasing in compression.

(f) Calculate the total true strain at failure.

(g) Calculate the work of fracture for this metallic glass based on engineering stress and strain.

6-12 Mechanical Behavior at Small Length Scales

6-65 Name a specific application for which understanding size-dependent mechanical

behavior may be important to the design process.

Ⓚ Knovel® **Problem**

K6-1 A 300 cm annealed rod with a cross-sectional area of 5.38 cm^2 was extruded from a 5083-O aluminum alloy and axially loaded. Under load, the length of the rod increased to 300.38 cm. No plastic deformation occurred.

(a) Find the modulus of elasticity of the material and calculate its allowable tensile stress, assuming it to be 50% of the tensile yield stress.

(b) Calculate the tensile stress and the axial load applied to the rod.

(c) Compare the calculated tensile stress with the allowable tensile stress, and find the absolute value of elongation of the rod for the allowable stress.

(a)

(b)

35KV X1000   0002   10 micrometer

(c)

The 316 stainless steel bolt failures shown here were due to mechanical fatigue. In this case, the bolts broke at the head-to-shank radius (see Figure (a)). An optical fractograph of one of the fracture surfaces (see Figure (b)) shows the fracture initiates at one location and propagates across the bolt until final failure occurred. Beach marks and striations (see Figure (c)), typical of fatigue fractures, are present on all of the fracture surfaces. Fatigue failure of threaded fasteners is most often associated with insufficient tightening of the fastener, resulting in flexing and subsequent fracture *(Images Courtesy of Corrosion Testing Laboratories, Bradley Kraritz, Richard Corbett, Albert Olszewski, and Robert R. Odle.)*

Mechanical Properties: Part Two

Have You Ever Wondered?

- *Why is it that glass fibers of different lengths have different strengths?*
- *Can a material or component ultimately fracture even if the overall stress does not exceed the yield strength?*
- *Why do aircraft have a finite service life?*
- *Why do materials ultimately fail?*

One goal of this chapter is to introduce the basic concepts associated with the fracture toughness of materials. In this regard, we will examine what factors affect the strength of glasses and ceramics and how the Weibull distribution quantitatively describes the variability in their strength. Another goal is to learn about time-dependent phenomena such as fatigue, creep, and stress corrosion. This chapter will review some of the basic testing procedures that engineers use to evaluate many of these properties and the failure of materials.

7-1   Fracture Mechanics

Fracture mechanics is the discipline concerned with the behavior of materials containing cracks or other small flaws. The term "flaw" refers to such features as small pores (holes), inclusions, or microcracks. The term "flaw" does *not* refer to atomic level defects such as vacancies or dislocations. What we wish to know is the maximum stress that a material can withstand if it contains flaws of a certain size and geometry. **Fracture toughness** measures the ability of a material containing a flaw to withstand an applied load. Note that this does *not* require a high strain rate (impact).

A typical fracture toughness test may be performed by applying a tensile stress to a specimen prepared with a flaw of known size and geometry (Figure 7-1). The stress applied to the material is intensified at the flaw, which acts as a *stress raiser*. For a simple case, the *stress intensity factor K* is

$$K = f\sigma\sqrt{\pi a} \tag{7-1}$$

where f is a geometry factor for the specimen and flaw, σ is the applied stress, and a is the flaw size [as defined in Figure 7-1]. If the specimen is assumed to have an "infinite" width, $f \cong 1.0$. For a small single-edge notch [Figure 7-1(a)], $f = 1.12$.

By performing a test on a specimen with a known flaw size, we can determine the value of K that causes the flaw to grow and cause failure. This critical stress intensity factor is defined as the *fracture toughness K_c*:

$$K_c = K \text{ required for a crack to propagate} \tag{7-2}$$

Fracture toughness depends on the thickness of the sample: as thickness increases, fracture toughness K_c decreases to a constant value (Figure 7-2). This constant is called the *plane strain fracture toughness K_{Ic}*. It is K_{Ic} that is normally reported as the property of a material. The value of K_{Ic} does not depend upon the thickness of the sample. Table 7-1 compares the value of K_{Ic} to the yield strength of several materials. Units for fracture toughness are MPa$\sqrt{\text{m}}$.

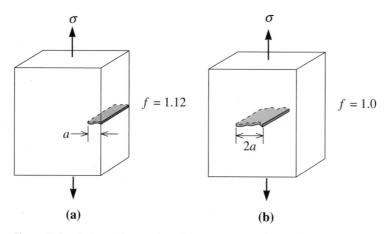

Figure 7-1   Schematic drawing of fracture toughness specimens with (a) edge and (b) internal flaws. The flaw size is defined differently for the two classes.

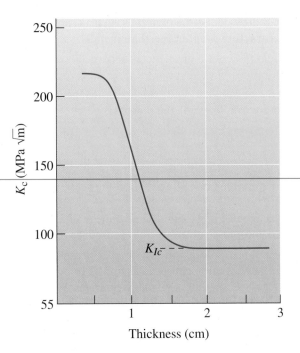

Figure 7-2
The fracture toughness K_c of a 2070 MPa yield strength steel decreases with increasing thickness, eventually leveling off at the plane strain fracture toughness K_{Ic}.

The ability of a material to resist the growth of a crack depends on a large number of factors:

1. Larger flaws reduce the permitted stress. Special manufacturing techniques, such as filtering impurities from liquid metals and hot pressing or hot isostatic pressing of powder particles to produce ceramic or superalloy components reduce flaw size and improve fracture toughness (Chapters 9 and 15).

2. The ability of a material to deform is critical. In ductile metals, the material near the tip of the flaw can deform, causing the tip of any crack to become blunt, reducing the stress intensity factor, and preventing growth of the crack. Increasing the strength of a given metal usually decreases ductility and gives a lower fracture

TABLE 7-1 ■ *The plane strain fracture toughness K_{Ic} of selected materials*

Material	Fracture Toughness K_{Ic} (MPa $\sqrt{m}$)	Yield Strength or Ultimate Strength (for Brittle Solids) (MPa)
Al-Cu alloy	24.2	455.1
	36.3	324.1
Ti-6% Al-4% V	54.9	896.4
	98.9	861.9
Ni-Cr steel	50.3	1641.0
	87.9	1420.4
Al_2O_3	1.8	206.9
Si_3N_4	4.9	551.6
Transformation toughened ZrO_2	11.0	413.7
Si_3N_4-SiC composite	56.0	827.4
Polymethyl methacrylate polymer	1.0	27.6
Polycarbonate polymer	3.3	57.9

toughness. (See Table 7-1.) Brittle materials such as ceramics and many polymers have much lower fracture toughnesses than metals.

3. Thicker, more rigid pieces of a given material have a lower fracture toughness than thin materials.

4. Increasing the rate of application of the load, such as in an impact test, typically reduces the fracture toughness of the material.

5. Increasing the temperature normally increases the fracture toughness, just as in the impact test.

6. A small grain size normally improves fracture toughness, whereas more point defects and dislocations reduce fracture toughness. Thus, a fine-grained ceramic material may provide improved resistance to crack growth.

7. In certain ceramic materials, we can take advantage of stress-induced transformations that lead to compressive stresses that cause increased fracture toughness.

Fracture testing of ceramics cannot be performed easily using a sharp notch, since formation of such a notch often causes the samples to break. We can use hardness testing to measure the fracture toughness of many ceramics. When a ceramic material is indented, tensile stresses generate secondary cracks that form at the indentation and the length of secondary cracks provides a measure of the toughness of the ceramic material. In some cases, an indentation created using a hardness tester is used as a starter crack for the bend test. In general, this direct-crack measurement method is better suited for comparison, rather than absolute measurements of fracture toughness values. The fracture toughness and fracture strength of many engineered materials are shown in Figure 7-3.

7-2   The Importance of Fracture Mechanics

The fracture mechanics approach allows us to design and select materials while taking into account the inevitable presence of flaws. There are three variables to consider: the property of the material (K_c or K_{Ic}), the stress σ that the material must withstand, and the size of the flaw a. If we know two of these variables, the third can be determined.

Selection of a Material   If we know the maximum size a of flaws in the material and the magnitude of the applied stress, we can select a material that has a fracture toughness K_c or K_{Ic} large enough to prevent the flaw from growing.

Design of a Component   If we know the maximum size of any flaw and the material (and therefore its K_c or K_{Ic} has already been selected), we can calculate the maximum stress that the component can withstand. Then we can size the part appropriately to ensure that the maximum stress is not exceeded.

Design of a Manufacturing or Testing Method   If the material has been selected, the applied stress is known, and the size of the component is fixed, we can calculate the maximum size of a flaw that can be tolerated. A nondestructive testing technique that detects any flaw greater than this critical size can help ensure that the part will function safely. In addition, we find that, by selecting the correct manufacturing process, we can produce flaws that are all smaller than this critical size.

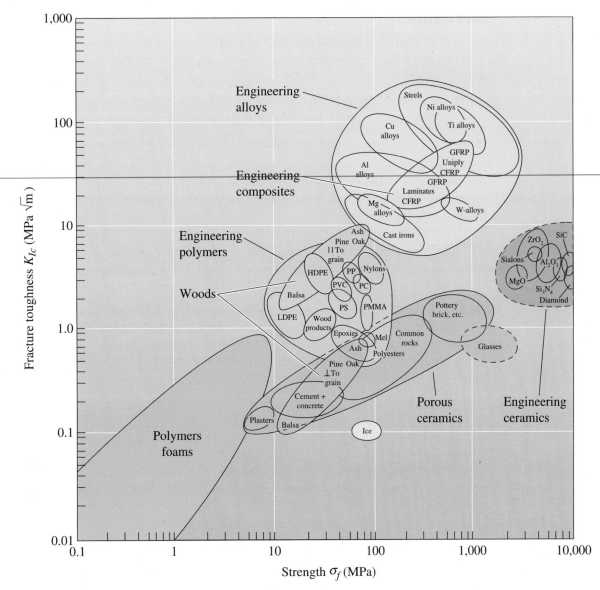

Figure 7-3 Fracture toughness versus strength of different engineered materials. *(Source: Adapted from Mechanical Behavior of Materials, by T. H. Courtney, 2000, p. 434, Fig. 9-18. Copyright © 2000 The McGraw-Hill Companies. Adapted with permission.)*

Example 7-1 *Design of a Nondestructive Test*

A large steel plate used in a nuclear reactor has a plane strain fracture toughness of 87.9 MPa and is exposed to a stress of 310 MPa during service. Design a testing or inspection procedure capable of detecting a crack at the edge of the plate before the crack is likely to grow at a catastrophic rate.

SOLUTION

We need to determine the minimum size of a crack that will propagate in the steel under these conditions. From Equation 7-1 assuming that $f = 1.12$ for a single-edge notch crack:

$$K_{Ic} = f\sigma\sqrt{\pi a}$$
$$87.9 = (1.12)(310)\sqrt{\pi a}$$
$$a = 2 \text{ cm}$$

A 2-cm-deep crack on the edge should be relatively easy to detect. Often, cracks of this size can be observed visually. A variety of other tests, such as dye penetrant inspection, magnetic particle inspection, and eddy current inspection, also detect cracks much smaller than this. If the growth rate of a crack is slow and inspection is performed on a regular basis, a crack should be discovered long before reaching this critical size.

Brittle Fracture   Any crack or imperfection limits the ability of a ceramic to withstand a tensile stress. This is because a crack (sometimes called a **Griffith flaw**) concentrates and magnifies the applied stress. Figure 7-4 shows a crack of length a at the surface of a brittle material. The radius r of the crack is also shown. When a tensile stress σ is applied, the actual stress at the crack tip is

$$\sigma_{actual} \cong 2\sigma\sqrt{a/r} \qquad (7\text{-}3)$$

For very thin cracks (r) or long cracks (a), the ratio σ_{actual}/σ becomes large, or the stress is intensified. If the stress (σ_{actual}) exceeds the yield strength, the crack grows and eventually causes failure, even though the nominal applied stress σ is small.

In a different approach, we recognize that an applied stress causes an elastic strain, related to the modulus of elasticity E of the material. When a crack propagates, this

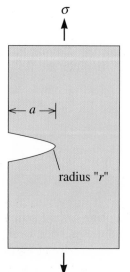

Figure 7-4
Schematic diagram of a Griffith flaw in a ceramic.

strain energy is released, reducing the overall energy. At the same time, however, two new surfaces are created by the extension of the crack; this increases the energy associated with the surface. By balancing the strain energy and the surface energy, we find that the critical stress required to propagate the crack is given by the Griffith equation,

$$\sigma_{\text{critical}} \cong \sqrt{\frac{2E\gamma}{\pi a}} \tag{7-4}$$

where a is the length of a surface crack (or one-half the length of an internal crack) and γ is the surface energy per unit area. Again, this equation shows that even small flaws severely limit the strength of the ceramic.

We also note that if we rearrange Equation 7-1, which described the stress intensity factor K, we obtain

$$\sigma = \frac{K}{f\sqrt{\pi a}} \tag{7-5}$$

This equation is similar in form to Equation 7-4. Each of these equations points out the dependence of the mechanical properties on the size of flaws present in the ceramic. Development of manufacturing processes (see Chapter 15) to minimize the flaw size becomes crucial in improving the strength of ceramics.

The flaws are most important when tensile stresses act on the material. Compressive stresses close rather than open a crack; consequently, ceramics often have very good compressive strengths.

Example 7-2   *Properties of SiAlON Ceramics*

Assume that an advanced ceramic sialon (acronym for SiAlON or silicon aluminum oxynitride), has a tensile strength of 414 MPa. Let us assume that this value is for a flaw-free ceramic. (In practice, it is almost impossible to produce flaw-free ceramics.) A thin crack 0.025 cm deep is observed before a sialon part is tested. The part unexpectedly fails at a stress of 3.5 MPa by propagation of the crack. Estimate the radius of the crack tip.

SOLUTION

The failure occurred because the 3.5 MPa applied stress, magnified by the stress concentration at the tip of the crack, produced an actual stress equal to the ultimate tensile strength. From Equation 7-3,

$$\sigma_{\text{actual}} = 2\sigma\sqrt{a/r}$$
$$414 \text{ MPa} = (2)(3.5 \text{ MPa})\sqrt{0.025 \text{ cm}/r}$$
$$\sqrt{0.025/r} = 60 \quad \text{or} \quad 0.025/r = 3492.8$$
$$r = 7.1 \times 10^{-6} \text{ cm} = 710\text{Å}$$

The likelihood of our being able to measure a radius of curvature of this size by any method of nondestructive testing is virtually zero. Therefore, although Equation 7-3 may help illustrate the factors that influence how a crack propagates in a brittle material, it does not help in predicting the strength of actual ceramic parts.

Example 7-3   *Design of a Ceramic Support*

Determine the minimum allowable thickness for a 7.5 cm wide plate made of sialon that has a fracture toughness of 9.9 MPa√m . The plate must withstand a tensile load of 177,920 N. The part will be nondestructively tested to ensure that no flaws are present that might cause failure. The minimum allowable thickness of the part will depend on the minimum flaw size that can be determined by the available testing technique. Assume that three nondestructive testing techniques are available. X-ray radiography can detect flaws larger than 0.05 cm; gamma-ray radiography can detect flaws larger than 0.02 cm; and ultrasonic inspection can detect flaws larger than 0.0125 cm. Assume that the geometry factor $f = 1.0$ for all flaws.

SOLUTION

For the given flaw sizes, we must calculate the minimum thickness of the plate that will ensure that these flaw sizes will not propagate. From Equation 7-5,

$$\sigma_{max} = \frac{K_{Ic}}{\sqrt{\pi a}} = \frac{F}{A_{min}}$$

$$A_{min} = \frac{F\sqrt{\pi a}}{K_{Ic}} = \frac{(177,920\ N)(\sqrt{\pi})(\sqrt{a})}{9.9\ MPa\sqrt{m}}$$

$$A_{min} = 0.03\sqrt{a}\ m^2 \text{ and thickness} = \left(\frac{0.03\ m^2}{0.075\ m}\right)\sqrt{a} = 0.4\sqrt{a}$$

Nondestructive Testing Method	Smallest Detectable Crack (cm)	Minimum Area (cm²)	Minimum Thickness (cm)	Maximum Stress (MPa)
X-ray radiography	0.05	6.7	0.89	270
γ-ray radiography	0.02	4.24	0.57	420
Ultrasonic inspection	0.0125	3.35	0.45	530

Our ability to detect flaws, coupled with our ability to produce a ceramic with flaws smaller than our detection limit, significantly affects the maximum stress than can be tolerated and, hence, the size of the part. In this example, the part can be smaller if ultrasonic inspection is available.

The fracture toughness is also important. Had we used Si_3N_4, with a fracture toughness of 3.3 MPa√m instead of the sialon, we could repeat the calculations and show that, for ultrasonic testing, the minimum thickness is 1.4 cm and the maximum stress is only 166 MPa.

7-3   Microstructural Features of Fracture in Metallic Materials

Ductile Fracture   Ductile fracture normally occurs in a **transgranular** manner (through the grains) in metals that have good ductility and toughness. Often, a considerable amount of deformation—including necking—is observed in the failed component.

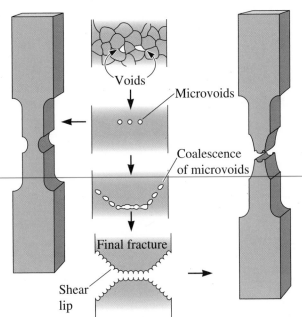

Figure 7-5
When a ductile material is pulled in a tensile test, necking begins and voids form—starting near the center of the bar—by nucleation at grain boundaries or inclusions. As deformation continues, a 45° shear lip may form, producing a final cup and cone fracture.

The deformation occurs before the final fracture. Ductile fractures are usually caused by simple overloads, or by applying too high a stress to the material.

In a simple tensile test, ductile fracture begins with the nucleation, growth, and coalescence of microvoids near the center of the test bar (Figure 7-5). **Microvoids** form when a high stress causes separation of the metal at grain boundaries or interfaces between the metal and small impurity particles (inclusions). As the local stress increases, the microvoids grow and coalesce into larger cavities. Eventually, the metal-to-metal contact area is too small to support the load and fracture occurs.

Deformation by slip also contributes to the ductile fracture of a metal. We know that slip occurs when the resolved shear stress reaches the critical resolved shear stress and that the resolved shear stresses are highest at a 45° angle to the applied tensile stress (Chapter 4, Schmid's Law).

These two aspects of ductile fracture give the failed surface characteristic features. In thick metal sections, we expect to find evidence of necking, with a significant portion of the fracture surface having a flat face where microvoids first nucleated and coalesced, and a small shear lip, where the fracture surface is at a 45° angle to the applied stress. The shear lip, indicating that slip occurred, gives the fracture a cup and cone appearance (Figure 6-9 and Figure 7-6). Simple macroscopic observation of this fracture may be sufficient to identify the ductile fracture mode.

Examination of the fracture surface at a high magnification—perhaps using a scanning electron microscope—reveals a dimpled surface (Figure 7-7). The dimples are traces of the microvoids produced during fracture. Normally, these microvoids are round, or equiaxed, when a normal tensile stress produces the failure [Figure 7-7(a)]; however, on the shear lip, the dimples are oval-shaped, or elongated, with the ovals pointing toward the origin of the fracture [Figure 7-7(b)].

In a thin plate, less necking is observed and the entire fracture surface may be a shear face. Microscopic examination of the fracture surface shows elongated dimples rather than equiaxed dimples, indicating a greater proportion of 45° slip than in thicker metals.

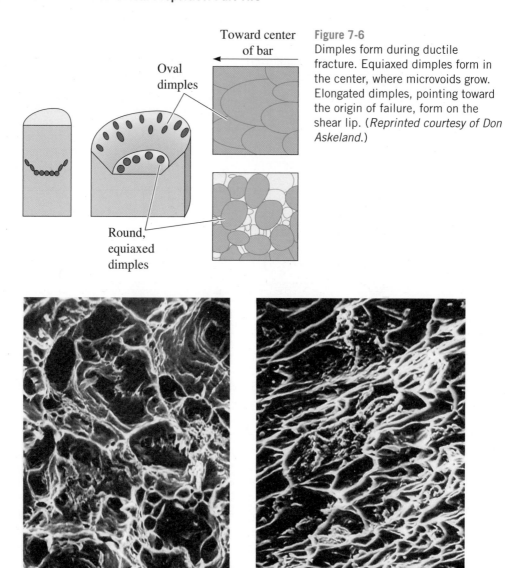

Figure 7-6
Dimples form during ductile fracture. Equiaxed dimples form in the center, where microvoids grow. Elongated dimples, pointing toward the origin of failure, form on the shear lip. (*Reprinted courtesy of Don Askeland.*)

Figure 7-7 Scanning electron micrographs of an annealed 1018 steel exhibiting ductile fracture in a tensile test. (a) Equiaxed dimples at the flat center of the cup and cone, and (b) elongated dimples at the shear lip (× 1250).

Example 7-4   *Hoist Chain Failure Analysis*

A chain used to hoist heavy loads fails. Examination of the failed link indicates considerable deformation and necking prior to failure. List some of the possible reasons for failure.

SOLUTION

This description suggests that the chain failed in a ductile manner by a simple tensile overload. Two factors could be responsible for this failure:

1. The load exceeded the hoisting capacity of the chain. Thus, the stress due to the load exceeded the ultimate tensile strength of the chain, permitting failure. Comparison of the load to the manufacturer's specifications will indicate that the chain was not intended for such a heavy load. This is the fault of the user!

2. The chain was of the wrong composition or was improperly heat treated. Consequently, the yield strength was lower than intended by the manufacturer and could not support the load. This may be the fault of the manufacturer!

Brittle Fracture

Brittle fracture occurs in high-strength metals and alloys or metals and alloys with poor ductility and toughness. Furthermore, even metals that are normally ductile may fail in a brittle manner at low temperatures, in thick sections, at high strain rates (such as impact), or when flaws play an important role. Brittle fractures are frequently observed when impact, rather than overload, causes failure.

In brittle fracture, little or no plastic deformation is required. Initiation of the crack normally occurs at small flaws, which cause a concentration of stress. The crack may move at a rate approaching the velocity of sound in the metal. Normally, the crack propagates most easily along specific crystallographic planes, often the {100} planes, by cleavage. In some cases, however, the crack may take an **intergranular** (along the grain boundaries) path, particularly when segregation (preferential separation of different elements) or inclusions weaken the grain boundaries.

Brittle fracture can be identified by observing the features on the failed surface. Normally, the fracture surface is flat and perpendicular to the applied stress in a tensile test. If failure occurs by cleavage, each fractured grain is flat and differently oriented, giving a crystalline or "rock candy" appearance to the fracture surface (Figure 7-8). Often, the layman claims that the metal failed because it crystallized. Of course, we know that the metal was crystalline to begin with and the surface appearance is due to the cleavage faces.

Another common fracture feature is the **Chevron pattern** (Figure 7-9), produced by separate crack fronts propagating at different levels in the material. A radiating pattern of surface markings, or ridges, fans away from the origin of the crack (Figure 7-10). The Chevron pattern is visible with the naked eye or a magnifying glass and helps us identify both the brittle nature of the failure process as well as the origin of the failure.

5 μm

Figure 7-8
Scanning electron micrograph of a brittle fracture surface of a quenched 1010 steel. *(Courtesy of C. W. Ramsay.)*

Figure 7-9
The Chevron pattern in a 1.25-cm-diameter quenched 4340 steel. The steel failed in a brittle manner by an impact blow. (*Reprinted courtesy of Don Askeland.*)

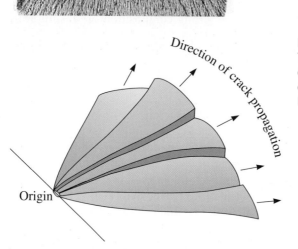

Figure 7-10
The Chevron pattern forms as the crack propagates from the origin at different levels. The pattern points back to the origin.

Direction of crack propagation

Origin

Example 7-5   *Automobile Axle Failure Analysis*

An engineer investigating the cause of an automobile accident finds that the right rear wheel has broken off at the axle. The axle is bent. The fracture surface reveals a Chevron pattern pointing toward the surface of the axle. Suggest a possible cause for the fracture.

SOLUTION

The evidence suggests that the axle did not break prior to the accident. The deformed axle means that the wheel was still attached when the load was applied. The Chevron pattern indicates that the wheel was subjected to an intense impact blow, which was transmitted to the axle. The preliminary evidence suggests that the driver lost control and crashed, and the force of the crash caused the axle to break. Further examination of the fracture surface, microstructure, composition, and properties may verify that the axle was manufactured properly.

7-4   Microstructural Features of Fracture in Ceramics, Glasses, and Composites

In ceramic materials, the ionic or covalent bonds permit little or no slip. Consequently, failure is a result of brittle fracture. Most crystalline ceramics fail by cleavage along widely spaced, closely packed planes. The fracture surface typically is smooth,

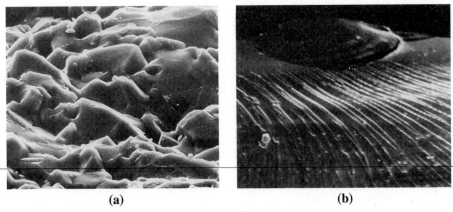

(a)   **(b)**

Figure 7-11   Scanning electron micrographs of fracture surfaces in ceramics. (a) The fracture surface of Al_2O_3, showing the cleavage faces ($\times 1250$) and (b) the fracture surface of glass, showing the mirror zone (top) and tear lines characteristic of conchoidal fracture ($\times 300$). *(Reprinted courtesy of Don Askeland.)*

and frequently no characteristic surface features point to the origin of the fracture [Figure 7-11(a)].

Glasses also fracture in a brittle manner. Frequently, a **conchoidal fracture** surface is observed. This surface contains a smooth mirror zone near the origin of the fracture, with tear lines comprising the remainder of the surface [Figure 7-11(b)]. The tear lines point back to the mirror zone and the origin of the crack, much like the chevron pattern in metals.

Polymers can fail by either a ductile or a brittle mechanism. Below the glass transition temperature (T_g), thermoplastic polymers fail in a brittle manner—much like a glass. Likewise, the hard thermoset polymers, which have a rigid, three-dimensional cross-linked structure (see Chapter 16), fail by a brittle mechanism. Some plastics with structures consisting of tangled but not chemically cross-linked chains fail in a ductile manner above the glass transition temperature, giving evidence of extensive deformation and even necking prior to failure. The ductile behavior is a result of sliding of the polymer chains, which is not possible in thermosetting polymers.

Fracture in fiber-reinforced composite materials is more complex. Typically, these composites contain strong, brittle fibers surrounded by a soft, ductile matrix, as in boron-reinforced aluminum. When a tensile stress is applied along the fibers, the soft aluminum deforms in a ductile manner, with void formation and coalescence eventually producing a dimpled fracture surface. As the aluminum deforms, the load is no longer transmitted effectively between the fibers and matrix; the fibers break in a brittle manner until there are too few of them left intact to support the final load.

Fracture is more common if the bonding between the fibers and matrix is poor. Voids can then form between the fibers and the matrix, causing pull-out. Voids can also form between layers of the matrix if composite tapes or sheets are not properly bonded, causing **delamination** (Figure 7-12). Delamination, in this context, means the layers of different materials in a composite begin to come apart.

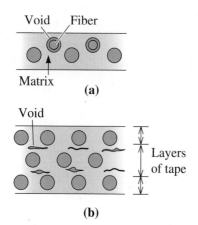

Figure 7-12
Fiber-reinforced composites can fail by several mechanisms. (a) Due to weak bonding between the matrix and fibers, voids can form, which then lead to fiber pull-out. (b) If the individual layers of the matrix are poorly bonded, the matrix may delaminate, creating voids.

Example 7-6 *Fracture in Composites*

Describe the difference in fracture mechanism between a boron-reinforced aluminum composite and a glass fiber-reinforced epoxy composite.

SOLUTION

In the boron-aluminum composite, the aluminum matrix is soft and ductile; thus, we expect the matrix to fail in a ductile manner. Boron fibers, in contrast, fail in a brittle manner. Both glass fibers and epoxy are brittle; thus the composite as a whole should display little evidence of ductile fracture.

7-5 Weibull Statistics for Failure Strength Analysis

We need a statistical approach when evaluating the strength of ceramic materials. The strength of ceramics and glasses depends upon the size and distribution of sizes of flaws. In these materials, flaws originate during processing as the ceramics are manufactured. The flaws can form during machining, grinding, etc. Glasses can also develop microcracks as a result of interaction with water vapor in air. If we test alumina or other ceramic components of different sizes and geometry, we often find a large scatter in the measured values—even if their nominal composition is the same. Similarly, if we are testing the strength of glass fibers of a given composition, we find that, on average, shorter fibers are stronger than longer fibers. The strength of ceramics and glasses depends upon the probability of finding a flaw that exceeds a certain critical size. This probability increases as the component size or fiber length increases. As a result, the strength of larger components or fibers is likely to be lower than that of smaller components or shorter fibers. In metallic or polymeric materials, which can exhibit relatively large plastic deformations, the effect of flaws and flaw size distribution is not felt to the extent it is in ceramics and glasses. In these materials, cracks initiating from flaws get blunted by plastic deformation. Thus, for ductile materials, the distribution of strength is narrow and close to a Gaussian distribution. The strength of ceramics and glasses, however, varies considerably (i.e., if we test a large number of identical samples of silica glass or alumina ceramic, the data will show a wide scatter owing to changes in distribution of flaw sizes). The strength of brittle materials,

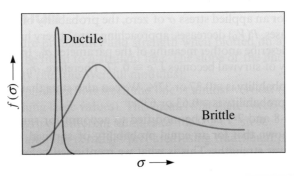

Figure 7-13
The Weibull distribution describes the fraction of the samples that fail at any given applied stress.

such as ceramics and glasses, is not Gaussian; it is given by the **Weibull distribution**. The Weibull distribution is an indicator of the variability of strength of materials resulting from a distribution of flaw sizes. This behavior results from critical sized flaws in materials with a distribution of flaw sizes (i.e., failure due to the weakest link of a chain).

The Weibull distribution shown in Figure 7-13 describes the fraction of samples that fail at different applied stresses. At low stresses, a small fraction of samples contain flaws large enough to cause fracture; most fail at an intermediate applied stress, and a few contain only small flaws and do not fail until large stresses are applied. To provide predictability, we prefer a very narrow distribution.

Consider a body of volume V with a distribution of flaws and subjected to a stress σ. If we assumed that the volume, V, was made up of n elements with volume V_0 and each element had the same flaw-size distribution, it can be shown that the survival probability, $P(V_0)$, (i.e., the probability that a brittle material will not fracture under the applied stress σ) is given by

$$P(V_0) = \exp\left[-\left(\frac{\sigma - \sigma_{\min}}{\sigma_0}\right)^m\right] \tag{7-6}$$

The probability of failure, $F(V_0)$, can be written as

$$F(V_0) = 1 - P(V_0) = 1 - \exp\left[-\left(\frac{\sigma - \sigma_{\min}}{\sigma_0}\right)^m\right] \tag{7-7}$$

In Equations 7-6 and 7-7, σ is the applied stress, σ_0 is a scaling parameter dependent on specimen size and shape, $\sigma_{\min}$ is the stress level below which the probability of failure is zero (i.e., the probability of survival is 1.0). In these equations, m is the **Weibull modulus**. In theory, Weibull modulus values can range from 0 to ∞. The Weibull modulus is a measure of the variability of the strength of the material.

The Weibull modulus m indicates the strength variability. For metals and alloys, the Weibull modulus is $\sim$100. For traditional ceramics (e.g., bricks, pottery, etc.), the Weibull modulus is less than 3. Engineered ceramics, in which the processing is better controlled and hence the number of flaws is expected to be less, have a Weibull modulus of 5 to 10.

Note that for ceramics and other brittle solids, we can assume $\sigma_{\min} = 0$. This is because there is no nonzero stress level for which we can claim a brittle material will not fail. For *brittle materials*, Equations 7-6 and 7-7 can be rewritten as follows:

$$P(V_0) = \exp\left[-\left(\frac{\sigma}{\sigma_0}\right)^m\right] \tag{7-8}$$

and

$$F(V_0) = 1 - P(V_0) = 1 - \exp\left[-\left(\frac{\sigma}{\sigma_0}\right)^m\right] \tag{7-9}$$

From Equation 7-8, for an applied stress σ of zero, the probability of survival is 1. As the applied stress σ increases, $P(V_0)$ decreases, approaching zero at very high values of applied stresses. We can also describe another meaning of the parameter σ_0. In Equation 7-8, when $\sigma = \sigma_0$, the probability of survival becomes $1/e \cong 0.37$. Therefore, σ_0 is the stress level for which the survival probability is $\cong 0.37$ or 37%. We can also state that σ_0 is the stress level for which the failure probability is $\cong 0.63$ or 63%.

Equations 7-8 and 7-9 can be modified to account for samples with different volumes. It can be shown that for an equal probability of survival, samples with larger volumes will have lower strengths. This is what we mentioned before (e.g., longer glass fibers will be weaker than shorter glass fibers).

The following examples illustrate how the Weibull plots can be used for analysis of mechanical properties of materials and design of components.

| **Example 7-7** | *Weibull Modulus for Steel and Alumina Ceramics* |

Figure 7-14 shows the log-log plots of the probability of failure and strength of a 0.2% plain carbon steel and an alumina ceramic prepared using conventional powder processing in which alumina powders are compacted in a press and sintered into a dense mass at high temperature. Also included is a plot for alumina ceramics prepared using special techniques that lead to much more uniform and controlled particle size. This in turn minimizes the flaws. These samples are labeled as controlled particle size (CPS). Comment on the nature of these graphs.

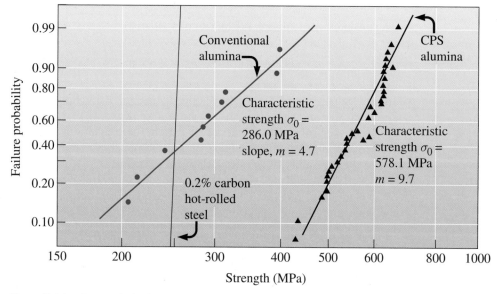

Figure 7-14   A cumulative plot of the probability that a sample will fail at any given stress yields the Weibull modulus or slope. Alumina produced by two different methods is compared with low carbon steel. Good reliability in design is obtained for a high Weibull modulus. *(Adapted from Mechanical Behavior of Materials, by M. A. Meyers and K. K. Chawla. Copyright © 1999 Prentice-Hall. Adapted with permission of Pearson Education, Inc., Upper Saddle River, NJ.)*

SOLUTION

The failure probability and strength when plotted on a log-log scale result in data that can be fitted to a straight line. The slope of the line provides us the measure of variability (i.e., the Weibull modulus).

For plain carbon steel, the line is almost vertical (i.e., the slope or m value is approaching large values). This means that there is very little variation (5 to 10%) in the strength of different samples of the 0.2% C steel.

For alumina ceramics prepared using traditional processing, the variability is high (i.e., m is low ~4.7).

For ceramics prepared using improved and controlled processing techniques, m is higher ~9.7 indicating a more uniform distribution of flaws. The characteristic strength (σ_0) is also higher (~578 MPa) suggesting fewer flaws that will lead to fracture.

Example 7-8   *Strength of Ceramics and Probability of Failure*

An advanced engineered ceramic has a Weibull modulus $m = 9$. The flexural strength is 250 MPa at a probability of failure $F = 0.4$. What is the level of flexural strength if the probability of failure has to be 0.1?

SOLUTION

We assume all samples tested had the same volume thus the size of the sample will not be a factor in this case. We can use the symbol V for sample volume instead of V_0. We are dealing with a brittle material so we begin with Equation 7-9:

$$F(V) = 1 - P(V) = 1 - \exp\left[-\left(\frac{\sigma}{\sigma_0}\right)^m\right]$$

or

$$1 - F(V) = \exp\left[-\left(\frac{\sigma}{\sigma_0}\right)^m\right]$$

Take the logarithm of both sides:

$$\ln[1 - F(V)] = -\left(\frac{\sigma}{\sigma_0}\right)^m$$

Take logarithms of both sides again,

$$\ln\{-\ln[1 - F(V)]\} = m(\ln \sigma - \ln \sigma_0) \tag{7-10}$$

We can eliminate the minus sign on the left-hand side of Equation 7-10 by rewriting it as

$$\ln\left\{\ln\left[\frac{1}{1 - F(V)}\right]\right\} = m(\ln \sigma - \ln \sigma_0) \tag{7-11}$$

For $F = 0.4$, $\sigma = 250$ MPa, and $m = 9$ in Equation 7-11, we have

$$\ln\left[\ln\left(\frac{1}{1-0.4}\right)\right] = 9(\ln 250 - \ln \sigma_0) \tag{7-12}$$

Therefore,

$$\ln[\ln(1/0.6)] = \ln[\ln(1.66667)] = \ln(0.510826) = -0.67173$$
$$= 9(5.52146 - \ln \sigma_0).$$

Therefore, $\ln \sigma_0 = 5.52146 + 0.07464 = 5.5961$. This gives us a value of $\sigma_0 = 269.4$ MPa. This is the characteristic strength of the ceramic. For a stress level of 269.4 MPa, the probability of survival is 0.37 (or the probability of failure is 0.63). As the required probability of failure (F) goes down, the stress level to which the ceramic can be subjected (σ) also goes down. Note that if Equation 7-12 is solved exactly for σ_0, a slightly different value is obtained.

Now, we want to determine the value of σ for $F = 0.1$. We know that $m = 9$ and $\sigma_0 = 269.4$ MPa, so we need to get the value of σ. We substitute these values into Equation 7-11:

$$\ln\left[\ln\left(\frac{1}{1-0.1}\right)\right] = 9(\ln \sigma - \ln 269.4)$$

$$\ln\left[\ln\left(\frac{1}{0.9}\right)\right] = 9(\ln \sigma - \ln 269.4)$$

$$\ln(\ln 1.11111) = \ln(0.105361) = -2.25037 = 9(\ln \sigma - 5.5962)$$

$$\therefore \quad -0.25004 = \ln \sigma - 5.5962, \text{ or}$$

$$\ln \sigma = 5.3462$$

or $\sigma = 209.8$ MPa. As expected, as we lowered the probability of failure to 0.1, we also decreased the level of stress that can be supported.

Example 7-9   *Weibull Modulus Parameter Determination*

Seven silicon carbide specimens were tested and the following fracture strengths were obtained: 23, 49, 34, 30, 55, 43, and 40 MPa. Estimate the Weibull modulus for the data by fitting the data to Equation 7-11. Discuss the reliability of the ceramic.

SOLUTION

First, we point out that for any type of statistical analysis, we need a large number of samples. Seven samples are not enough. The purpose of this example is to illustrate the calculation.

One simple though not completely accurate method for determining the behavior of the ceramic is to assign a numerical rank (1 to 7) to the specimens, with the specimen having the lowest fracture strength assigned the value 1. The total number of specimens is n (in our case, 7). The probability of failure F is then the numerical rank divided by $n + 1$ (in our case, 8). We can then plot $\ln\{\ln 1/[1 - F(V_0)]\}$ versus

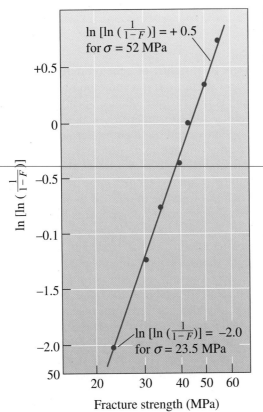

$$\ln \left[\ln \left(\tfrac{1}{1-F}\right)\right] = +0.5$$
for $\sigma = 52$ MPa

$$\ln \left[\ln \left(\tfrac{1}{1-F}\right)\right] = -2.0$$
for $\sigma = 23.5$ MPa

Fracture strength (MPa)

Figure 7-15
Plot of cumulative probability of failure versus fracture stress. Note the fracture strength is plotted on a log scale.

ln σ. The following table and Figure 7-15 show the results of these calculations. Note that σ is plotted on a log scale.

i^{th} Specimen	σ (MPa)	$F(V_0)$	$\ln\{\ln 1/[1 - F(V_0)]\}$
1	23	1/8 = 0.125	−2.013
2	30	2/8 = 0.250	−1.246
3	34	3/8 = 0.375	−0.755
4	40	4/8 = 0.500	−0.367
5	43	5/8 = 0.625	−0.019
6	49	6/8 = 0.750	+0.327
7	55	7/8 = 0.875	+0.732

The slope of the fitted line, or the Weibull modulus m, is (using the two points indicated on the curve):

$$m = \frac{0.5 - (-2.0)}{\ln(52) - \ln(23.5)} = \frac{2.5}{3.951 - 3.157} = 3.15$$

This low Weibull modulus of 3.15 suggests that the ceramic has a highly variable fracture strength, making it difficult to use reliably in load-bearing applications.

7-6   Fatigue

Fatigue is the lowering of strength or failure of a material due to repetitive stress which may be above or below the yield strength. It is a common phenomenon in load-bearing components in cars and airplanes, turbine blades, springs, crankshafts and other machinery, biomedical implants, and consumer products, such as shoes, that are subjected constantly to repetitive stresses in the form of tension, compression, bending, vibration, thermal expansion and contraction, or other stresses. These stresses are often *below* the yield strength of the material; however, when the stress occurs a sufficient number of times, it causes failure by fatigue! Quite a large fraction of components found in an automobile junkyard belongs to those that failed by fatigue. The possibility of a fatigue failure is the main reason why aircraft components have a finite life.

Fatigue failures typically occur in three stages. First, a tiny crack initiates or nucleates often at a time well after loading begins. Normally, nucleation sites are located at or near the surface, where the stress is at a maximum, and include surface defects such as scratches or pits, sharp corners due to poor design or manufacture, inclusions, grain boundaries, or dislocation concentrations. Next, the crack gradually propagates as the load continues to cycle. Finally, a sudden fracture of the material occurs when the remaining cross-section of the material is too small to support the applied load. Thus, components fail by fatigue because even though the overall applied stress may remain below the yield stress, at a local length scale, the stress intensity exceeds the tensile strength. For fatigue to occur, at least part of the stress in the material has to be tensile. We normally are concerned with fatigue of metallic and polymeric materials.

In ceramics, we normally do not consider fatigue since ceramics typically fail because of their low fracture toughness. Any fatigue cracks that may form will lower the useful life of the ceramic since it will cause lowering of the fracture toughness. In general, we design ceramics for static (and not cyclic) loading, and we factor in the Weibull modulus.

Polymeric materials also show fatigue failure. The mechanism of fatigue in polymers is different than that in metallic materials. In polymers, as the materials are subjected to repetitive stresses, considerable heating can occur near the crack tips and the interrelationships between fatigue and another mechanism, known as *creep* (discussed in Section 7-9), affect the overall behavior.

Fatigue is also important in dealing with composites. As fibers or other reinforcing phases begin to degrade as a result of fatigue, the overall elastic modulus of the composite decreases and this weakening will be seen before the fracture due to fatigue.

Fatigue failures are often easy to identify. The fracture surface—particularly near the origin—is typically smooth. The surface becomes rougher as the original crack increases in size and may be fibrous during final crack propagation. Microscopic and macroscopic examinations reveal a fracture surface including a beach mark pattern and striations (Figure 7-16). **Beach** or **clamshell marks** (Figure 7-17) are normally formed when the load is changed during service or when the loading is intermittent, perhaps permitting time for oxidation inside the crack. **Striations**, which are on a much finer scale, show the position of the crack tip after each cycle. Beach marks always suggest a fatigue failure, but—unfortunately—the absence of beach marks does not rule out fatigue failure.

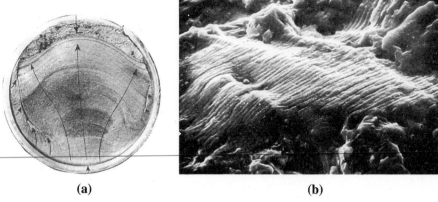

(a)   **(b)**

Figure 7-16   Fatigue fracture surface. (a) At low magnifications, the beach mark pattern indicates fatigue as the fracture mechanism. The arrows show the direction of growth of the crack front with the origin at the bottom of the photograph. *(Image* (a) is *from C. C. Cottell, "Fatigue Failures with Special Reference to Fracture Characteristics," Failure Analysis: The British Engine Technical Reports, American Society for Metals, 1981, p. 318.)* (b) At very high magnifications, closely spaced striations formed during fatigue are observed (× 1000). *(Reprinted courtesy of Don Askeland.)*

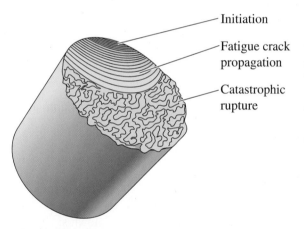

Initiation

Fatigue crack propagation

Catastrophic rupture

Figure 7-17
Schematic representation of a fatigue fracture surface in a steel shaft, showing the initiation region, the propagation of the fatigue crack (with beach markings), and catastrophic rupture when the crack length exceeds a critical value at the applied stress.

Example 7-10	*Fatigue Failure Analysis of a Crankshaft*

A crankshaft in a diesel engine fails. Examination of the crankshaft reveals no plastic deformation. The fracture surface is smooth. In addition, several other cracks appear at other locations in the crankshaft. What type of failure mechanism occurred?

SOLUTION

Since the crankshaft is a rotating part, the surface experiences cyclical loading. We should immediately suspect fatigue. The absence of plastic deformation supports our suspicion. Furthermore, the presence of other cracks is consistent with fatigue; the other cracks did not have time to grow to the size that produced catastrophic failure. Examination of the fracture surface will probably reveal beach marks or fatigue striations.

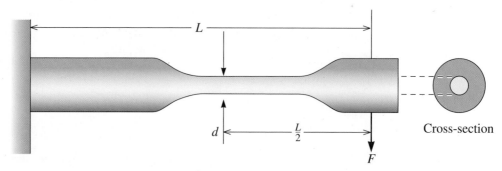

Figure 7-18   Geometry for the rotating cantilever beam specimen setup.

A conventional and older method used to measure a material's resistance to fatigue is the **rotating cantilever beam test** (Figure 7-18). One end of a machined, cylindrical specimen is mounted in a motor-driven chuck. A weight is suspended from the opposite end. The specimen initially has a tensile force acting on the top surface, while the bottom surface is compressed. After the specimen turns 90°, the locations that were originally in tension and compression have no stress acting on them. After a half revolution of 180°, the material that was originally in tension is now in compression. Thus, the stress at any one point goes through a complete sinusoidal cycle from maximum tensile stress to maximum compressive stress. The maximum stress acting on this type of specimen is given by

$$\pm \sigma = \frac{32\,M}{\pi d^3} \tag{7-13a}$$

In this equation, M is the bending moment at the cross-section, and d is the specimen diameter. The bending moment $M = F \cdot (L/2)$, and therefore,

$$\pm \sigma = \frac{16\,FL}{\pi d^3} = 5.09 \frac{FL}{d^3} \tag{7-13b}$$

where L is the distance between the bending force location and the support (Figure 7-18), F is the load, and d is the diameter.

Newer machines used for fatigue testing are known as direct-loading machines. In these machines, a servo-hydraulic system, an actuator, and a control system, driven by computers, applies a desired force, deflection, displacement, or strain. In some of these machines, temperature and atmosphere (e.g., humidity level) also can be controlled.

After a sufficient number of cycles in a fatigue test, the specimen may fail. Generally, a series of specimens are tested at different applied stresses. The results are presented as an **S-N curve** (also known as the **Wöhler curve**), with the stress (S) plotted versus the number of cycles (N) to failure (Figure 7-19).

7-7   Results of the Fatigue Test

The **fatigue test** can tell us how long a part may survive or the maximum allowable loads that can be applied without causing failure. The **endurance limit**, which is the stress below which there is a 50% probability that failure by fatigue will never occur, is our preferred design

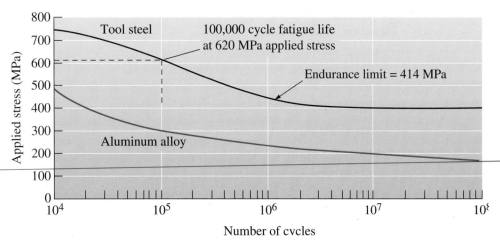

Figure 7-19 The stress-number of cycles to failure (S-N) curves for a tool steel and an aluminum alloy.

criterion. To prevent a tool steel part from failing (Figure 7-19), we must be sure that the applied stress is below 414 Mpa. The assumption of the existence of an endurance limit is a relatively older concept. Recent research on many metals has shown that probably an endurance limit does not exist. We also need to account for the presence of corrosion, occasional overloads, and other mechanisms that may cause the material to fail below the endurance limit. Thus, values for an endurance limit should be treated with caution.

Fatigue life tells us how long a component survives at a particular stress. For example, if the tool steel (Figure 7-19) is cyclically subjected to an applied stress of 620 Mpa, the fatigue life will be 100,000 cycles. Knowing the time associated with each cycle, we can calculate a fatigue life value in years. **Fatigue strength** is the maximum stress for which fatigue will not occur within a particular number of cycles, such as 500,000,000. The fatigue strength is necessary for designing with aluminum and polymers, which have no endurance limit.

In some materials, including steels, the endurance limit is approximately half the tensile strength. The ratio between the endurance limit and the tensile strength is known as the **endurance ratio**:

$$\text{Endurance ratio} = \frac{\text{endurance limit}}{\text{tensile strength}} \approx 0.5 \qquad (7\text{-}14)$$

The endurance ratio allows us to estimate fatigue properties from the tensile test. The endurance ratio values are ~0.3 to 0.4 for metallic materials other than low and medium strength steels. *Again, recall the cautionary note that research has shown that an endurance limit does not exist for many materials.*

Most materials are **notch sensitive**, with the fatigue properties particularly sensitive to flaws at the surface. Design or manufacturing defects concentrate stresses and reduce the endurance limit, fatigue strength, or fatigue life. Sometimes highly polished surfaces are prepared in order to minimize the likelihood of a fatigue failure. **Shot peening** is a process that is used very effectively to enhance fatigue life of materials. Small metal spheres are shot at the component. This leads to a residual compressive stress at the surface similar to **tempering** of inorganic glasses.

Example 7-11   *Design of a Rotating Shaft*

A solid shaft for a cement kiln produced from the tool steel in Figure 7-19 must be 240 cm long and must survive continuous operation for one year with an applied load of 55,600 N. The shaft makes one revolution per minute during operation. Design a shaft that will satisfy these requirements.

SOLUTION

The fatigue life required for our design is the total number of cycles N that the shaft will experience in one year:

$$N = (1 \text{ cycle/min})(60 \text{ min}/h)(24 \text{ } h/d)(365 \text{ } d/y).$$

$$N = 5.256 \times 10^5 \text{ cycles}/y$$

where y = year, d = day, and h = hour.

From Figure 7-19, the applied stress therefore, must be less than about 496 MPa. Using Equation 7-13, the diameter of the shaft is given by

$$\pm\sigma = \frac{16FL}{\pi d^3} = 5.09 \frac{FL}{d^3}$$

$$496 \text{ MPa} = \frac{(5.09)(55,600 \text{ N})(240 \text{ cm})}{d^3}$$

$$d = 11.1 \text{ cm}$$

A shaft with a diameter of 11.1 cm should operate for one year under these conditions; however, a significant margin of safety probably should be incorporated in the design. In addition, we might consider producing a shaft that would never fail.

Let us assume the factor of safety to be 2 (i.e., we will assume that the maximum allowed stress level will be $496/2 = 248$ MPa). The minimum diameter required to prevent failure would now be

$$248 \text{ MPa} = \frac{(5.09)(55,600 \text{ N})(240 \text{ cm})}{d^3}$$

$$d = 14 \text{ cm}$$

Selection of a larger shaft reduces the stress level and makes fatigue less likely to occur or delays the failure. Other considerations might, of course, be important. High temperatures and corrosive conditions are inherent in producing cement. If the shaft is heated or attacked by the corrosive environment, fatigue is accelerated. Thus, for applications involving fatigue of components, regular inspections of the components go a long way toward avoiding a catastrophic failure.

7-8   Application of Fatigue Testing

Components are often subjected to loading conditions that do not give equal stresses in tension and compression (Figure 7-20). For, example, the maximum stress during compression may be less than the maximum tensile stress. In other cases, the loading may be

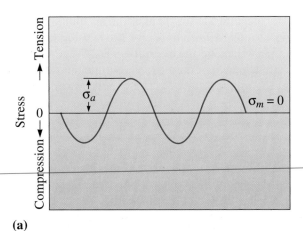

(a)

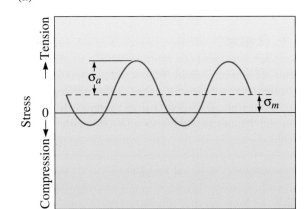

(b)

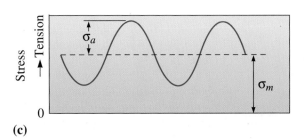

(c)

Figure 7-20
Examples of stress cycles. (a) Equal stress in tension and compression, (b) greater tensile stress than compressive stress, and (c) all of the stress is tensile.

between a maximum and a minimum tensile stress; here the S-N curve is presented as the stress amplitude versus the number of cycles to failure. *Stress amplitude* (σ_a) is defined as half of the difference between the maximum and minimum stresses, and *mean stress* (σ_m) is defined as the average between the maximum and minimum stresses:

$$\sigma_a = \frac{\sigma_{max} - \sigma_{min}}{2} \tag{7-15}$$

$$\sigma_m = \frac{\sigma_{max} + \sigma_{min}}{2} \tag{7-16}$$

A compressive stress is a "negative" stress. Thus, if the maximum tensile stress is 345 MPa and the minimum stress is a 69 MPa compressive stress, using Equations 7-15 and 7-16, the stress amplitude is 138 MPa, and the mean stress is 207 MPa.

As the mean stress increases, the stress amplitude must decrease in order for the material to withstand the applied stresses. The condition can be summarized by the Goodman relationship:

$$\sigma_a = \sigma_{fs}\left[1 - \left(\frac{\sigma_m}{\sigma_{UTS}}\right)\right] \tag{7-17}$$

where σ_{fs} is the desired fatigue strength for zero mean stress and σ_{UTS} is the tensile strength of the material. Therefore, in a typical rotating cantilever beam fatigue test, where the mean stress is zero, a relatively large stress amplitude can be tolerated without fatigue. If, however, an airplane wing is loaded near its yield strength, vibrations of even a small amplitude may cause a fatigue crack to initiate and grow.

Crack Growth Rate   In many cases, a component may not be in danger of failure even when a crack is present. To estimate when failure might occur, the rate of propagation of a crack becomes important. Figure 7-21 shows the crack growth rate versus the range of the stress intensity factor ΔK, which characterizes crack geometry and the stress amplitude. Below a threshold ΔK, a crack does not grow; for somewhat higher stress intensities, cracks grow slowly; and at still higher stress intensities, a crack grows at a rate given by

$$\frac{da}{dN} = C(\Delta K)^n \tag{7-18}$$

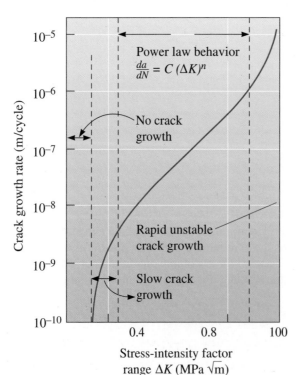

Figure 7-21
Crack growth rate versus stress intensity factor range for a high-strength steel. For this steel, $C = 1.62 \times 10^{-12}$ and $n = 3.2$ for the units shown.

In this equation, C and n are empirical constants that depend upon the material. Finally, when ΔK is still higher, cracks grow in a rapid and unstable manner until fracture occurs.

The rate of crack growth increases as a crack increases in size, as predicted from the stress intensity factor (Equation 7-1):

$$\Delta K = K_{max} - K_{min} = f\sigma_{max}\sqrt{\pi a} - f\sigma_{min}\sqrt{\pi a} = f\Delta\sigma\sqrt{\pi a} \qquad (7\text{-}19)$$

If the cyclical stress $\Delta\sigma \, (\sigma_{max} - \sigma_{min})$ is not changed, then as crack length a increases, ΔK and the crack growth rate da/dN increase. In using this expression, one should note that a crack will not propagate during compression. Therefore, if σ_{min} is compressive, or less than zero, σ_{min} should be set equal to zero.

Knowledge of crack growth rate is of assistance in designing components and in nondestructive evaluation to determine if a crack poses imminent danger to the structure. One approach to this problem is to estimate the number of cycles required before failure occurs. By rearranging Equation 7-18 and substituting for ΔK:

$$dN = \frac{da}{Cf^n\Delta\sigma^n\pi^{n/2}a^{n/2}}.$$

If we integrate this expression between the initial size of a crack and the crack size required for fracture to occur, we find that

$$N = \frac{2[(a_c)^{(2-n)/2} - (a_i)^{(2-n)/2}]}{(2 - n)Cf^n\Delta\sigma^n\pi^{n/2}} \qquad (7\text{-}20)$$

where a_i is the initial flaw size and a_c is the flaw size required for fracture. If we know the material constants n and C in Equation 7-18, we can estimate the number of cycles required for failure for a given cyclical stress (Example 7-12).

Example 7-12 *Design of a Fatigue Resistant Plate*

A high-strength steel plate (Figure 7-21), which has a plane strain fracture toughness of 80 MPa$\sqrt{\text{m}}$ is alternately loaded in tension to 500 MPa and in compression to 60 MPa. The plate is to survive for 10 years with the stress being applied at a frequency of once every 5 minutes. Design a manufacturing and testing procedure that ensures that the component will serve as intended. Assume a geometry factor $f = 1.0$ for all flaws.

SOLUTION

To design our manufacturing and testing capability, we must determine the maximum size of any flaws that might lead to failure within the 10 year period. The critical crack size using the fracture toughness and the maximum stress is

$$K_{Ic} = f\sigma\sqrt{\pi a_c}$$

$$80 \text{ MPa}\sqrt{\text{m}} = (1.0)(500 \text{ MPa})\sqrt{\pi a_c}$$

$$a_c = 0.0081 \text{ m} = 8.1 \text{ mm}$$

The maximum stress is 500 MPa; however, the minimum stress is zero, not 60 MPa in compression, because cracks do not propagate in compression. Thus, $\Delta\sigma$ is

$$\Delta\sigma = \sigma_{max} - \sigma_{min} = 500 - 0 = 500 \text{ MPa}$$

We need to determine the minimum number of cycles that the plate must withstand:

$$N = (1 \text{ cycle}/5 \text{ min})(60 \text{ min}/h)(24 \text{ } h/d)(365 \text{ } d/y)(10 \text{ } y)$$

$$N = 1{,}051{,}200 \text{ cycles}$$

If we assume that $f = 1.0$ for all crack lengths and note that $C = 1.62 \times 10^{-12}$ and $n = 3.2$ from Figure 7-21 in Equation 7-20, then

$$1{,}051{,}200 = \frac{2\left[(0.008)^{(2-3.2)/2} - (a_i)^{(2-3.2)/2}\right]}{(2-3.2)(1.62 \times 10^{-12})(1)^{3.2}(500)^{3.2}\pi^{3.2/2}}$$

$$1{,}051{,}200 = \frac{2\left[18 - a_i^{0.6}\right]}{(-1.2)(1.62 \times 10^{-12})(1)(4.332 \times 10^8)(6.244)}$$

$$a_i^{-0.6} = 18 + 2764 = 2782$$

$$a_i = 1.82 \times 10^{-6} \text{ m} = 0.00182 \text{ mm for surface flaws}$$

$$2a_i = 0.00364 \text{ mm for internal flaws}$$

The manufacturing process must produce surface flaws smaller than 0.00182 mm in length. In addition, nondestructive tests must be available to ensure that cracks approaching this length are not present.

Effects of Temperature   As the material's temperature increases, both fatigue life and endurance limit decrease. Furthermore, a cyclical temperature change encourages failure by thermal fatigue; when the material heats in a nonuniform manner, some parts of the structure expand more than others. This nonuniform expansion introduces a stress within the material, and when the structure later cools and contracts, stresses of the opposite sign are imposed. As a consequence of the thermally induced stresses and strains, fatigue may eventually occur. The frequency with which the stress is applied also influences fatigue behavior. In particular, high-frequency stresses may cause polymer materials to heat; at increased temperatures, polymers fail more quickly. Chemical effects of temperature (e.g., oxidation) must also be considered.

7-9   Creep, Stress Rupture, and Stress Corrosion

If we apply stress to a material at an elevated temperature, the material may stretch and eventually fail, even though the applied stress is *less* than the yield strength at that temperature. Time dependent permanent deformation under a constant load or constant stress and at high temperatures is known as **creep**. A large number of failures occurring in components used at high temperatures can be attributed to creep or a combination of creep and fatigue. Diffusion, dislocation glide or climb, or grain boundary sliding can contribute to the creep of metallic materials. Polymeric materials also show creep. In ductile metals and alloys subjected to creep, fracture is accompanied by necking, void nucleation and coalescence, or grain boundary sliding.

A material is considered failed by creep even if it has *not* actually fractured. When a material does creep and then ultimately breaks, the fracture is defined as *stress rupture*. Normally, ductile stress-rupture fractures include necking and the presence of many cracks that did not have an opportunity to produce final fracture. Furthermore, grains near the fracture surface tend to be elongated. Ductile stress-rupture failures generally occur at high creep rates and relatively low exposure temperatures and have short rupture times. Brittle stress-rupture failures usually show little necking and occur more often at smaller creep rates and high temperatures. Equiaxed grains are observed near the fracture surface. Brittle failure typically occurs by formation of voids at the intersection of three grain boundaries and precipitation of additional voids along grain boundaries by diffusion processes (Figure 7-22).

Stress Corrosion
Stress corrosion is a phenomenon in which materials react with corrosive chemicals in the environment. This leads to the formation of cracks and a lowering of strength. Stress corrosion can occur at stresses well below the yield strength of the metallic, ceramic, or glassy material due to attack by a corrosive medium. In metallic materials, deep, fine corrosion cracks are produced, even though the metal as a whole shows little uniform attack. The stresses can be either externally applied or stored residual stresses. Stress corrosion failures are often identified by microstructural examination of the surrounding metal. Ordinarily, extensive branching of the cracks along grain boundaries is observed (Figure 7-23). The location at which cracks initiated may be identified by the presence of a corrosion product.

Inorganic silicate glasses are especially prone to failure by reaction with water vapor. It is well known that the strength of silica fibers or silica glass products is very high when these materials are protected from water vapor. As the fibers or silica glass components get exposed to water vapor, corrosion reactions begin leading to formation of surface flaws, which ultimately cause the cracks to grow when stress is applied. As discussed in Chapter 5, polymeric coatings are applied to optical fibers to prevent them from reacting with water vapor. For bulk glasses, special heat treatments such as tempering are used. **Tempering** produces an overall compressive stress on the surface of glass. Thus, even if the glass surface reacts with water vapor, the cracks do not grow since the overall stress at the surface is compressive. If we create a flaw that will penetrate the compressive stress region on the surface, tempered glass will shatter. Tempered glass is used widely in building and automotive applications.

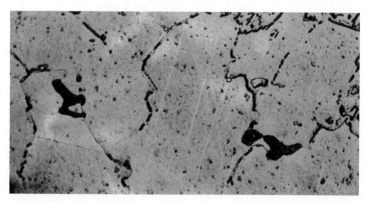

Figure 7-22   Creep cavities formed at grain boundaries in an austentic stainless steel ($\times 500$). *(From ASM Handbook, Vol. 7, Metallography and Microstructure (1972), ASM International, Materials Park, OH 44073.)*

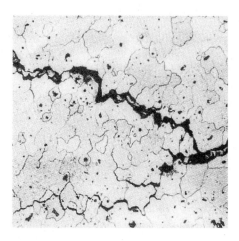

Figure 7-23
Micrograph of a metal near a stress corrosion fracture, showing the many intergranular cracks formed as a result of the corrosion process ($\times 200$). *(From* ASM Handbook, *Vol. 7, Metallography and Microstructure (1972), ASM International, Materials Park, OH 44073.)*

Example 7-13   *Failure Analysis of a Pipe*

A titanium pipe used to transport a corrosive material at 400°C is found to fail after several months. How would you determine the cause for the failure?

SOLUTION

Since a period of time at a high temperature was required before failure occurred, we might first suspect a creep or stress corrosion mechanism for failure. Microscopic examination of the material near the fracture surface would be advisable. If many tiny, branched cracks leading away from the surface are noted, stress corrosion is a strong possibility. If the grains near the fracture surface are elongated, with many voids between the grains, creep is a more likely culprit.

7-10   Evaluation of Creep Behavior

To determine the creep characteristics of a material, a constant stress is applied to a heated specimen in a **creep test**. As soon as the stress is applied, the specimen stretches elastically a small amount ε_0 (Figure 7-24), depending on the applied stress and the modulus of elasticity of the material at the high temperature. Creep testing can also be conducted under a constant load and is important from an engineering design viewpoint.

Dislocation Climb   High temperatures permit dislocations in a metal to **climb**. In climb, atoms move either to or from the dislocation line by diffusion, causing the dislocation to move in a direction that is perpendicular, not parallel, to the slip plane (Figure 7-25). The dislocation escapes from lattice imperfections, continues to slip, and causes additional deformation of the specimen even at low applied stresses.

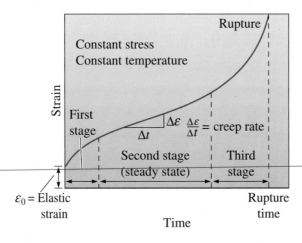

Figure 7-24
A typical creep curve showing the strain produced as a function of time for a constant stress and temperature.

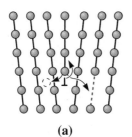 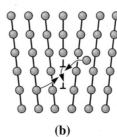

(a)   **(b)**

Figure 7-25
Dislocations can climb (a) when atoms leave the dislocation line to create interstitials or to fill vacancies or (b) when atoms are attached to the dislocation line by creating vacancies or eliminating interstitials.

Creep Rate and Rupture Times

During the creep test, strain or elongation is measured as a function of time and plotted to give the creep curve (Figure 7-24). In the first stage of creep of metals, many dislocations climb away from obstacles, slip, and contribute to deformation. Eventually, the rate at which dislocations climb away from obstacles equals the rate at which dislocations are blocked by other imperfections. This leads to the second stage, or steady-state, creep. The slope of the steady-state portion of the creep curve is the **creep rate**:

$$\text{Creep rate} = \frac{\Delta \text{ strain}}{\Delta \text{ time}} \qquad (7\text{-}21)$$

Eventually, during third stage creep, necking begins, the stress increases, and the specimen deforms at an accelerated rate until failure occurs. The time required for failure to occur is the **rupture time**. Either a higher stress or a higher temperature reduces the rupture time and increases the creep rate (Figure 7-26).

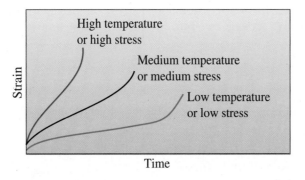

Figure 7-26
The effect of temperature or applied stress on the creep curve.

The creep rate and rupture time (t_r) follow an Arrhenius relationship that accounts for the combined influence of the applied stress and temperature:

$$\text{Creep rate} = C\sigma^n \exp\left(-\frac{Q_c}{RT}\right) \tag{7-22}$$

$$t_r = K\sigma^m \exp\left(\frac{Q_r}{RT}\right) \tag{7-23}$$

where R is the gas constant, T is the temperature in kelvin and C, K, n, and m are constants for the material. Q_c is the activation energy for creep, and Q_r is the activation energy for rupture. In particular, Q_c is related to the activation energy for self-diffusion when dislocation climb is important.

In crystalline ceramics, other factors—including grain boundary sliding and nucleation of microcracks—are particularly important. Often, a noncrystalline or glassy material is present at the grain boundaries; the activation energy required for the glass to deform is low, leading to high creep rates compared with completely crystalline ceramics. For the same reason, creep occurs at a rapid rate in ceramics glasses and amorphous polymers.

7-11   Use of Creep Data

The **stress-rupture curves**, shown in Figure 7-27(a), estimate the expected lifetime of a component for a particular combination of stress and temperature. The **Larson-Miller parameter**, illustrated in Figure 7-27(b), is used to consolidate the stress-temperature-rupture time relationship into a single curve. The Larson-Miller parameter $(L.M.)$ is

$$L.M. = \left(\frac{T}{1000}\right)(A + B \ln t) \tag{7-24}$$

where T is in kelvin, t is the time in hours, and A and B are constants for the material.

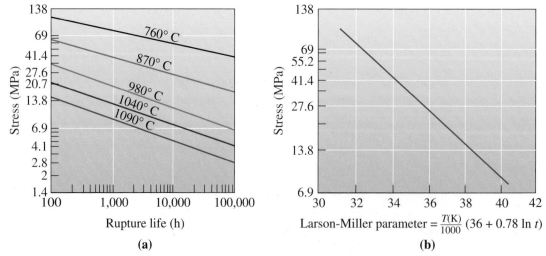

Figure 7-27   Results from a series of creep tests. (a) Stress-rupture curves for an iron-chromium-nickel alloy and (b) the Larson-Miller parameter for ductile cast iron.

Example 7-14   *Design of Links for a Chain*

Design a ductile cast iron chain (Figure 7-28) to operate in a furnace used to fire ceramic bricks. The chain will be used for five years at 600°C with an applied load of 22,240 N.

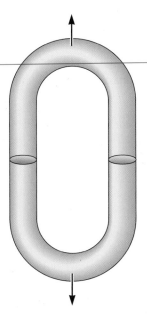

Figure 7-28
Sketch of chain link (for Example 7-14).

SOLUTION

From Figure 7-27(b), the Larson-Miller parameter for ductile cast iron is

$$L.M. = \frac{T(36 + 0.78 \ln t)}{1000}$$

The chain is to survive five years, or

$$t = (24 \, h/d)(365 \, d/y)(5 \, y) = 43,800 \, h$$

$$L.M. = \frac{(600 + 273)[36 + 0.78 \ln(43,800)]}{1000} = 38.7$$

From Figure 7-27(b), the applied stress must be no more than 12.4 MPa.

Let us assume a **factor of safety** of 2, this will mean the applied stress should not be more than $12.4/2 = 6.2$ MPa.

The total cross-sectional area of the chain required to support the 22,240 N load is

$$A = F/\sigma = \frac{22,240 \text{ N}}{6.2 \text{ MPa}} = 35.9 \text{ cm}^2$$

The cross-sectional area of each "half" of the iron link is then 17.95 cm² and, assuming a round cross section:

$$d^2 = (4/\pi)A = (4/\pi)(17.95) = 22.85$$
$$d = 4.78 \text{ cm}$$

Summary

- Toughness refers to the ability of materials to absorb energy before they fracture. Tensile toughness is equal to the area under the true stress-true strain curve. The impact toughness is measured using the impact test. This could be very different from the tensile toughness. Fracture toughness describes how easily a crack or flaw in a material propagates. The plane strain fracture toughness K_{Ic} is a common result of these tests.

- Weibull statistics are used to describe and characterize the variability in the strength of brittle materials. The Weibull modulus is a measure of the variability of the strength of a material.

- The fatigue test permits us to understand how a material performs when a cyclical stress is applied. Knowledge of the rate of crack growth can help determine fatigue life.

- Microstructural analysis of fractured surfaces can lead to better insights into the origin and cause of fracture. Different microstructural features are associated with ductile and brittle fracture as well as fatigue failure.

- The creep test provides information on the load-carrying ability of a material at high temperatures. Creep rate and rupture time are important properties obtained from these tests.

Glossary

Beach or clamshell marks Patterns often seen on a component subjected to fatigue. Normally formed when the load is changed during service or when the loading is intermittent, perhaps permitting time for oxidation inside the crack.

Chevron pattern A common fracture feature produced by separate crack fronts propagating at different levels in the material.

Climb Movement of a dislocation perpendicular to its slip plane by the diffusion of atoms to or from the dislocation line.

Conchoidal fracture Fracture surface containing a smooth mirror zone near the origin of the fracture with tear lines comprising the remainder of the surface. This is typical of amorphous materials.

Creep A time dependent, permanent deformation at high temperatures, occurring at constant load or constant stress.

Creep rate The rate at which a material deforms when a stress is applied at a high temperature.

Creep test Measures the resistance of a material to deformation and failure when subjected to a static load below the yield strength at an elevated temperature.

Delamination The process by which different layers in a composite will begin to debond.

Endurance limit An older concept that defined a stress below which a material will not fail in a fatigue test. Factors such as corrosion or occasional overloading can cause materials to fail at stresses below the assumed endurance limit.

Endurance ratio The endurance limit divided by the tensile strength of the material. The ratio is about 0.5 for many ferrous metals. See the cautionary note on endurance limit.

Factor of safety The ratio of the stress level for which a component is designed to the actual stress level experienced. A factor used to design load-bearing components. For example, the maximum load a component is subjected to 69 MPa. We design it (i.e., choose the material, geometry, etc.) such that it can withstand 138 MPa; in this case, the factor of safety is 2.0.

Fatigue life The number of cycles permitted at a particular stress before a material fails by fatigue.

Fatigue strength The stress required to cause failure by fatigue in a given number of cycles, such as 500 million cycles.

Fatigue test Measures the resistance of a material to failure when a stress below the yield strength is repeatedly applied.

Fracture mechanics The study of a material's ability to withstand stress in the presence of a flaw.

Fracture toughness The resistance of a material to failure in the presence of a flaw.

Griffith flaw A crack or flaw in a material that concentrates and magnifies the applied stress.

Intergranular In between grains or along the grain boundaries.

Larson-Miller parameter A parameter used to relate the stress, temperature, and rupture time in creep.

Microvoids Development of small holes in a material. These form when a high stress causes separation of the metal at grain boundaries or interfaces between the metal and inclusions.

Notch sensitivity Measures the effect of a notch, scratch, or other imperfection on a material's properties, such as toughness or fatigue life.

Rotating cantilever beam test A method for fatigue testing.

Rupture time The time required for a specimen to fail by creep at a particular temperature and stress.

S-N curve (also known as the Wöhler curve) A graph showing the relationship between the applied stress and the number of cycles to failure in fatigue.

Shot peening A process in which metal spheres are shot at a component. This leads to a residual compressive stress at the surface of a component and this enhances fatigue life.

Stress corrosion A phenomenon in which materials react with corrosive chemicals in the environment, leading to the formation of cracks and lowering of strength.

Stress-rupture curve A method of reporting the results of a series of creep tests by plotting the applied stress versus the rupture time.

Striations Patterns seen on a fractured surface of a fatigued sample. These are visible on a much finer scale than beach marks and show the position of the crack tip after each cycle.

Tempering A glass heat treatment that makes the glass safer; it does so by creating a compressive stress layer at the surface.

Toughness A qualitative measure of the energy required to cause fracture of a material. A material that resists failure by impact is said to be tough. One measure of toughness is the area under the true stress-strain curve (tensile toughness); another is the impact energy measured during an impact test (impact toughness). The ability of materials containing flaws to withstand load is known as fracture toughness.

Transgranular Meaning across the grains (e.g., a transgranular fracture would be fracture in which cracks go through the grains).

Weibull distribution A mathematical distribution showing the probability of failure or survival of a material as a function of the stress.

Weibull modulus (*m*) A parameter related to the Weibull distribution. It is an indicator of the variability of the strength of materials resulting from a distribution of flaw sizes.

Wöhler curve See S-N curve.

Problems

Section 7-1 Fracture Mechanics

Section 7-2 The Importance of Fracture Mechanics

7-1   Alumina (Al_2O_3) is a brittle ceramic with low toughness. Suppose that fibers of silicon carbide SiC, another brittle ceramic with low toughness, could be embedded within the alumina. Would doing this affect the toughness of the ceramic matrix composite? Explain.

7-2   A ceramic matrix composite contains internal flaws as large as 0.001 cm in length. The plane strain fracture toughness of the composite is $45\ MPa\sqrt{m}$, and the tensile strength is 550 MPa. Will the stress cause the composite to fail before the tensile strength is reached? Assume that $f = 1$.

7-3   An aluminum alloy that has a plane strain fracture toughness of $27.5\ MPa\sqrt{m}$ fails when a stress of 290 MPa is applied. Observation of the fracture surface indicates that fracture began at the surface of the part. Estimate the size of the flaw that initiated fracture. Assume that $f = 1.1$.

7-4   A polymer that contains internal flaws 1 mm in length fails at a stress of 25 MPa. Determine the plane strain fracture toughness of the polymer. Assume that $f = 1$.

7-5   A ceramic part for a jet engine has a yield strength of 517 MPa and a plane strain fracture toughness of $5.5\ MPa\sqrt{m}$. To be sure that the part does not fail, we plan to ensure that the maximum applied stress is only one-third of the yield strength. We use a nondestructive test that will detect any internal flaws greater than 0.125 cm long. Assuming that $f = 1.4$, does our nondestructive test have the required sensitivity? Explain.

7-6   A manufacturing process that unintentionally introduces cracks to the surface of a part was used to produce load-bearing components. The design requires that the component be able to withstand a stress of 450 MPa. The component failed catastrophically in service.

You are a failure analysis engineer who must determine whether the component failed due to an overload in service or flaws from the manufacturing process. The manufacturer claims that the components were polished to remove the cracks and inspected to ensure that no surface cracks were larger than 0.5 mm. The manufacturer believes the component failed due to operator error.

It has been independently verified that the 5-cm diameter part was subjected to a tensile load of 1 MN (10^6 N).

The material from which the component is made has a fracture toughness of $75\ MPa\sqrt{m}$ and an ultimate tensile strength of 600 MPa. Assume external cracks for which $f = 1.12$.

(a) Who is at fault for the component failure, the manufacturer or the operator? Show your work for both cases.

(b) In addition to the analysis that you presented in (a), what features might you look for on the fracture surfaces to support your conclusion?

7-7   Explain how the fracture toughness of ceramics can be obtained using hardness testing. Explain why such a method provides qualitative measurements.

Section 7-3 Microstructural Features of Fracture in Metallic Materials

Section 7-4 Microstructural Features of Fracture in Ceramics, Glasses, and Composites

7-8   Explain the terms intergranular and intragranular fractures. Use a schematic to show grains, grain boundaries, and a crack path that is typical of intergranular and intragranular fracture in materials.

7-9   What are the characteristic microstructural features associated with ductile fracture?

7-10 What are the characteristic microstructural features associated with a brittle fracture in a metallic material?

7-11 What materials typically show a conchoidal fracture?

7-12 Briefly describe how fiber-reinforced composite materials can fail.

7-13 Concrete has exceptional strength in compression, but it fails rather easily in tension. Explain why.

7-14 What controls the strength of glasses? What can be done to enhance the strength of silicate glasses?

Section 7-5 Weibull Statistics for Failure Strength Analysis

7-15 Sketch a schematic of the strength of ceramics and that of metals and alloys as a function of probability of failure. Explain the differences you anticipate.

7-16 Why does the strength of ceramics vary considerably with the size of ceramic components?

7-17 What parameter tells us about the variability of the strength of ceramics and glasses?

7-18 Why do glass fibers of different lengths have different strengths?

7-19 Explain the significance of the Weibull distribution.

7-20 *Turbochargers Are Us, a new start-up company, hires you to design their new turbocharger. They explain that they want to replace their metallic superalloy turbocharger with a high-tech ceramic that is much lighter for the same configuration. Silicon nitride may be a good choice, and you ask Ceramic Turbochargers, Inc. to supply you with bars made from a certain grade of Si_3N_4. They send you 25 bars that you break in three-point bending, using a 40-mm outer support span and a 20-mm inner loading span, obtaining the data in the following table.

(a) Calculate the bend strength using $\sigma_f = (1.5 \cdot F \cdot S)/(t^2 \cdot w)$, where F = load, S = support span ($a = 40$ mm loading span and $b = 20$ mm in this case), t = thickness, and w = width,

Bar	Width (mm)	Thickness (mm)	Load (N)
1	6.02	3.99	2510
2	6.00	4.00	2615
3	5.98	3.99	2375
4	5.99	4.04	2865
5	6.00	4.05	2575
6	6.01	4.00	2605
7	6.01	4.01	2810
8	5.95	4.02	2595
9	5.99	3.97	2490
10	5.98	3.96	2650
11	6.05	3.97	2705
12	6.00	4.05	2765
13	6.02	4.00	2680
14	5.98	4.01	2725
15	6.03	3.99	2830
16	5.95	3.98	2730
17	6.04	4.03	2565
18	5.96	4.01	2650
19	5.97	4.05	2650
20	6.02	4.00	2745
21	6.01	4.00	2895
22	6.00	3.99	2525
23	6.00	3.98	2660
24	6.04	3.95	2680
25	6.02	4.05	2640

and give the mean strength (50% probability of failure) and the standard deviation.

(b) Make a Weibull plot by ranking the strength data from lowest strength to highest strength. The lowest strength becomes $n = 1$, next $n = 2$, etc., until $n = N = 25$. Use $F = (n - 0.5)/N$ where n = rank of strength going from lowest strength $n = 1$ to highest strength $n = 25$ and $N = 25$. Note that F is the probability of failure. Plot $\ln [1/(1 - F)]$ as a function of $\ln \sigma_f$ and use a linear regression to find the slope of the line which is m: the Weibull modulus. Find the characteristic strength (σ_0) (63.2% probability of failure). (*Hint*: The characteristic strength is calculated easily by setting $\ln [1/(1 - F)] = 0$ once you know the equation of the line. A spreadsheet program (such as Excel) greatly facilitates the calculations).

*This problem was contributed by Dr. Raymond Cutler of Ceramatek Inc.

7.21 *Your boss asks you to calculate the design stress for a brittle nickel-aluminide rod she wants to use in a high-temperature design where the cylindrical component is stressed in tension. You decide to test rods of the same diameter and length as her design in tension to avoid the correction for the effective area (or effective volume in this case). You measure an average stress of 673 MPa and a Weibull modulus (m) of 14.7 for the nickel-aluminide rods. What is the design stress in MPa if 99.999% of the parts you build must be able to handle this stress without fracturing? (*Note*: The design stress is the stress you choose as the engineer so that the component functions as you want).

*This problem was contributed by Dr. Raymond Cutler of Ceramatek Inc.

Section 7-6 Fatigue

Section 7-7 Results of the Fatigue Test
Section 7-8 Application of Fatigue Testing

7-22 A cylindrical tool steel specimen that is 15 cm long and 0.625 cm in diameter rotates as a cantilever beam and is to be designed so that failure never occurs. Assuming that the maximum tensile and compressive stresses are equal, determine the maximum load that can be applied to the end of the beam. (See Figure 7-19.)

7-23 A 2-cm-diameter, 20-cm-long bar of an acetal polymer (Figure 7-29) is loaded on one end and is expected to survive one million cycles of loading, with equal maximum tensile and compressive stresses, during its lifetime. What is the maximum permissible load that can be applied?

7-24 A cyclical load of 6672 N is to be exerted at the end of a 25-cm-long aluminium beam (Figure 7-19). The bar must survive for at least 10^6 cycles. What is the minimum diameter of the bar?

7-25 A cylindrical acetal polymer bar 2 cm long and 1.5 cm in diameter is subjected to a vibrational load at one end of the bar at a frequency of 500 vibrations per minute, with a load of 50 N. How many hours will the part survive before breaking? (See Figure 7-29.)

7-26 Suppose that we would like a part produced from the acetal polymer shown in Figure 7-29 to survive for one million cycles under conditions that provide for equal compressive and tensile stresses. What is the fatigue strength, or maximum stress amplitude, required? What are the maximum stress, the minimum stress, and the mean stress on the part during its use? What effect would the frequency of the stress application have on your answers? Explain.

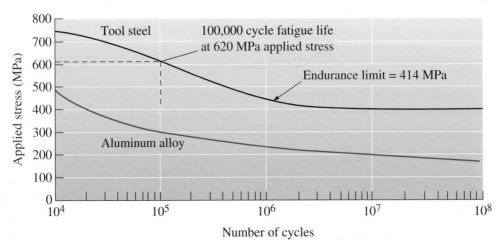

Figure 7-19   (Repeated for Problems 7-22 and 7-24) The stress-number of cycles to failure (S-N) curves for a tool steel and an aluminum alloy.

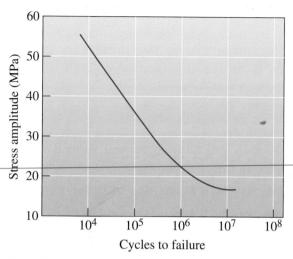

Figure 7-29   The S-N fatigue curve for an acetal polymer (for Problems 7-23, 7-25, and 7-26).

7-27   The high-strength steel in Figure 7-21 is subjected to a stress alternating at 200 revolutions per minute between 600 MPa and 200 MPa (both tension). Calculate the

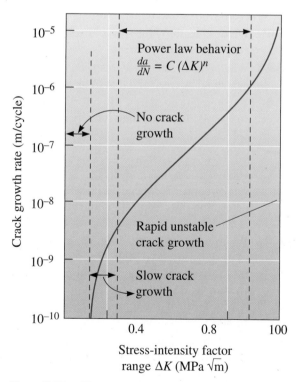

Figure 7-21   (Repeated for Problems 7-27 and 7-28) Crack growth rate versus stress intensity factor range for a high-strength steel. For this steel, $C = 1.62 \times 10^{-12}$ and $n = 3.2$ for the units shown.

growth rate of a surface crack when it reaches a length of 0.2 mm in both m/cycle and m/s. Assume that $f = 1.0$.

7-28   The high-strength steel in Figure 7-21, which has a critical fracture toughness of 80 MPa$\sqrt{m}$, is subjected to an alternating stress varying from −900 MPa (compression) to +900 MPa (tension). It is to survive for 10^5 cycles before failure occurs. Assume that $f = 1$. Calculate

   (a) the size of a surface crack required for failure to occur; and
   (b) the largest initial surface crack size that will permit this to happen.

7-29   The manufacturer of a product that is subjected to repetitive cycles has specified that the product should be removed from service when any crack reaches 15% of the critical crack length required to cause fracture.

   Consider a crack that is initially 0.02 mm long in a material with a fracture toughness of 55 MPa$\sqrt{m}$. The product is continuously cycled between compressive and tensile stresses of 300 MPa at a constant frequency. Assume external cracks for which $f = 1.12$. The materials constants for these units are $n = 3.4$ and $C = 2 \times 10^{-11}$.

   (a) What is the critical crack length required to cause fracture?
   (b) How many cycles will cause product failure?
   (c) If the product is removed from service as specified by the manufacturer, how much of the useful life of the product remains?

7-30   A material containing cracks of initial length 0.010 mm is subjected to alternating tensile stresses of 25 and 125 MPa for 350,000 cycles. The material is then subjected to alternating tensile and compressive stresses of 250 MPa. How many of the larger stress amplitude cycles can be sustained before failure? The material has a fracture toughness of 25 MPa$\sqrt{m}$ and materials constants of $n = 3.1$ and $C = 1.8 \times 10^{-10}$ for these units. Assume $f = 1.0$ for all cracks.

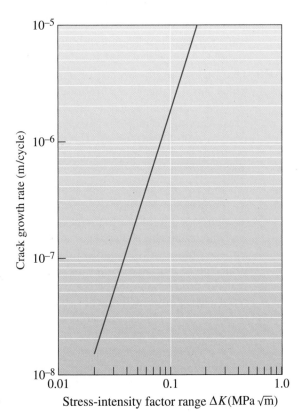

Figure 7-30   The crack growth rate for an acrylic polymer (for Problems 7-31, 7-32, and 7-33).

7-31 The acrylic polymer from which Figure 7-30 was obtained has a critical fracture toughness of 2 MPa$\sqrt{m}$. It is subjected to a stress alternating between -10 and $+10$ MPa. Calculate the growth rate of a surface crack when it reaches a length of 5×10^{-6} m if $f = 1.0$.

7-32 Calculate the constants C and n in Equation 7-18 for the crack growth rate of an acrylic polymer. (See Figure 7-30.)

7-33 The acrylic polymer from which Figure 7-30 was obtained is subjected to an alternating stress between 15 MPa and 0 MPa. The largest surface cracks initially detected by nondestructive testing are 0.001 mm in length. If the critical fracture toughness of the polymer is 2 MPa$\sqrt{m}$, calculate the number of cycles required before failure occurs. Let $f = 1.0$. (*Hint*: Use the results of Problem 7-32.)

7-34 Explain how fatigue failure occurs even if the material does not see overall stress levels higher than the yield strength.

7-35 Verify that integration of $da/dN = C(\Delta K)^n$ will give Equation 7-20.

7-36 What is shot peening? What is the purpose of using this process?

Section 7-9 Creep, Stress Rupture, and Stress Corrosion

Section 7-10 Evaluation of Creep Behavior

Section 7-11 Use of Creep Data

7-37 Why is creep accelerated by heat?

7-38 A child's toy was left at the bottom of a swimming pool for several weeks. When the toy was removed from the water, it failed after only a few hundred cycles of loading and unloading, even though it should have been able to withstand thousands of cycles. Speculate as to why the toy failed earlier than expected.

7-39 Define the term "creep" and differentiate creep from stress relaxation.

7-40 What is meant by the terms "stress rupture" and "stress corrosion?"

7-41 What is the difference between failure of a material by creep and that by stress rupture?

7-42 The activation energy for self-diffusion in copper is 49,300 cal/mol. A copper specimen creeps at $0.002\dfrac{cm/cm}{h}$ when a stress of 100 MPa is applied at 600°C. If the creep rate of copper is dependent on self-diffusion, determine the creep rate if the temperature is 800°C.

7-43 When a stress of 140 MPa is applied to a material heated to 900°C, rupture occurs in 25,000 h. If the activation energy for rupture is 35,000 cal/mol, determine the rupture time if the temperature is reduced to 800°C.

7-44 The following data were obtained from a creep test for a specimen having a gage length of 5 cm and an initial diameter of 1.5 cm. The initial stress applied to the material is 70 MPa. The diameter of the specimen at fracture is 1.3 cm.

Length Between Gage Marks (cm)	Time (h)
5.01	0
5.03	100
5.05	200
5.08	400
5.11	1000
5.19	2000
5.34	4000
5.48	6000
5.58	7000
5.75	8000 (fracture)

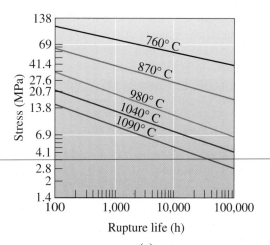

(a)

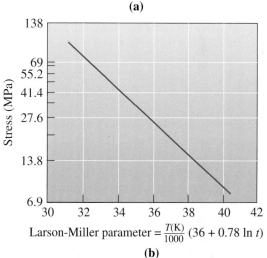

$$\text{Larson-Miller parameter} = \frac{T(K)}{1000} (36 + 0.78 \ln t)$$

(b)

Figure 7-27 (Repeated for Problems 7-46 through 7-52). Results from a series of creep tests. (a) Stress-rupture curves for an iron-chromium-nickel alloy and (b) the Larson-Miller parameter for ductile cast iron.

Determine

(a) the load applied to the specimen during the test;

(b) the approximate length of time during which linear creep occurs;

(c) the creep rate in $\dfrac{cm/cm}{h}$ and in %/h; and

(d) the true stress acting on the specimen at the time of rupture.

7-45 A stainless steel is held at 705°C under different loads. The following data are obtained:

Applied Stress (MPa)	Rupture Time (h)	Creep Rate (%/h)
106.9	1200	0.022
128.2	710	0.068
147.5	300	0.201
160.0	110	0.332

Determine the exponents n and m in Equations 7-22 and 7-23 that describe the dependence of creep rate and rupture time on applied stress.

7-46 Using the data in Figure 7-27 for an iron-chromium-nickel alloy, determine the activation energy Q_r and the constant m for rupture in the temperature range 980 to 1090°C.

7-47 A 2.5-cm-diameter bar of an iron-chromium-nickel alloy is subjected to a load of 11,000 N. How many days will the bar survive without rupturing at 980°C? (See Figure 7-27(a).)

7-48 A 5 mm × 20 mm bar of an iron-chrominum-nickel alloy is to operate at 1040°C for 10 years without rupturing. What is the maximum load that can be applied? (See Figure 7-27(a).)

7-49 An iron-chromium-nickel alloy is to withstand a load of 6600 N at 760°C for 6 years. Calculate the minimum diameter of the bar. (See Figure 7-27(a).)

7-50 A 3-cm-diameter bar of an iron-chromium-nickel alloy is to operate for 5 years under a load of 17,800 N. What is the maximum operating temperature? (See Figure 7-27(a).)

7-51 A 2.5-cm × 5 cm ductile cast-iron bar must operate for 9 years at 650°C. What is the maximum load that can be applied? (See Figure 7-27(b).)

7-52 A ductile cast-iron bar is to operate at a stress of 41 MPa for 1 year. What is the maximum allowable temperature? (See Figure 7-27(b).)

 Design Problems

7-53 A hook (Figure 7-31) for hoisting containers of ore in a mine is to be designed using a nonferrous (not based on iron) material. (A nonferrous material is used because iron and steel could cause a spark that would ignite explosive gases in the mine.) The hook must support a load of 111,200 N and a factor of safety of 2 should be used. We have determined that the cross-section labeled "?" is the most critical area; the rest of the device is already well overdesigned. Determine the design requirements for this device and, based on the mechanical property data given in Chapters 14 and 15 and the metal/alloy prices obtained from such sources as your local newspapers, the internet website of London Metal Exchange or *The Wall Street Journal*, design the hook and select an economical material for the hook.

7-54 A support rod for the landing gear of a private airplane is subjected to a tensile load during landing. The loads are predicted to be as high as 177,920 N. Because the rod is crucial and failure could lead to a loss of life, the rod is to be designed with a factor of safety of 4 (that is, designed so that the rod is capable of supporting loads four times as great as expected). Operation of the system also produces loads that may induce cracks in the rod. Our nondestructive testing equipment can detect any crack greater than 0.05 cm deep. Based on the materials given in Section 7-1, design the support rod and the material, and justify your answer.

7-55 A lightweight rotating shaft for a pump on the national aerospace plane is to be designed to support a cyclical load of 66,720 N during service. The maximum stress is the same in both tension and compression. The endurance limits or fatigue strengths for several candidate materials are shown below. Design the shaft, including an appropriate material, and justify your solution.

Material	Endurance Limit/ Fatigue Strength (MPa)
Al-Mn alloy	110
Al-Mg-Zn alloy	225
Cu-Be alloy	295
Mg-Mn alloy	80
Be alloy	180
Tungsten alloy	320

7-56 A ductile cast-iron bar is to support a load of 177,920 N in a heat-treating furnace used to make malleable cast iron. The bar is located in a spot that is continuously exposed to 500°C. Design the bar so that it can operate for at least 10 years without failing.

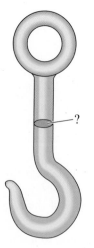

Figure 7-31
Schematic of a hook (for Problem 7-53).

ⓚ Knovel® **Problems**

K7-1 A hollow shaft made from AISI 4340 steel has an outer diameter D_o of 10 cm and an inner diameter D_i of 6.25 cm. The shaft rotates at 46 rpm for one hour during each day. It is supported by two bearings and loaded in the middle with a load W of 24,464 N. The distance between the bearings L is 195 cm. The maximum tensile stress due to bending for this type of cyclic loading is calculated using the following equation:

$$\sigma_m = \frac{8WLD_o}{\pi(D_o^4 - D_i^4)}$$

What is the stress ratio for this type of cyclic loading? Would this shaft last for one year assuming a safety factor of 2?

In applications such as the chassis formation of automobiles, metals and alloys are deformed. The mechanical properties of materials change during this process due to strain hardening. The strain hardening behavior of steels used in the fabrication of chassis influences the ability to form aerodynamic shapes. The strain hardening behavior is also important in improving the crashworthiness of vehicles. *(Courtesy of Digital Vision/ Getty Images.)*

Strain Hardening and Annealing

Have You Ever Wondered?

- *Why does bending a copper wire make it stronger?*

- *What type of steel improves the crashworthiness of cars?*

- *How are aluminum beverage cans made?*

- *Why do thermoplastics get stronger when strained?*

- *What is the difference between an annealed, tempered, and laminated safety glass?*

- *How is it that the strength of the metallic material around a weld can be lower than that of the surrounding material?*

In this chapter, we will learn how the strength of metals and alloys is influenced by mechanical processing and heat treatments. In Chapter 4, we learned about the different techniques that can strengthen metals and alloys (e.g., enhancing dislocation density, decreasing grain size, alloying, etc.). In this chapter, we will learn how to enhance the strength of metals and alloys using cold working, a process by which a metallic material is simultaneously deformed and strengthened. We will also see how hot working can be used to shape metals and alloys by deformation at high temperatures without strengthening. We will learn how the annealing heat treatment can be used to enhance ductility and counter the increase in hardness caused by cold working. The topics discussed in this chapter pertain particularly to metals and alloys.

What about polymers, glasses, and ceramics? Do they also exhibit strain hardening? We will show that the deformation of thermoplastic polymers often produces a strengthening effect, but the mechanism of deformation strengthening is completely different in polymers than in metallic materials. The strength of most brittle materials such as ceramics and glasses depends upon the flaws and flaw size distribution (Chapter 7). Therefore, inorganic glasses and ceramics do not respond well to strain hardening. We should consider different strategies to strengthen these materials. In this context, we will learn the principles of tempering and annealing of glasses.

We begin by discussing strain hardening in metallic materials in the context of stress-strain curves.

8-1 Relationship of Cold Working to the Stress-Strain Curve

A stress-strain curve for a ductile metallic material is shown in Figure 8-1(a). If we apply a stress S_1 that is greater than the yield strength S_y, it causes a permanent deformation or strain. When the stress is removed, a strain of e_1 remains. If we make a tensile test

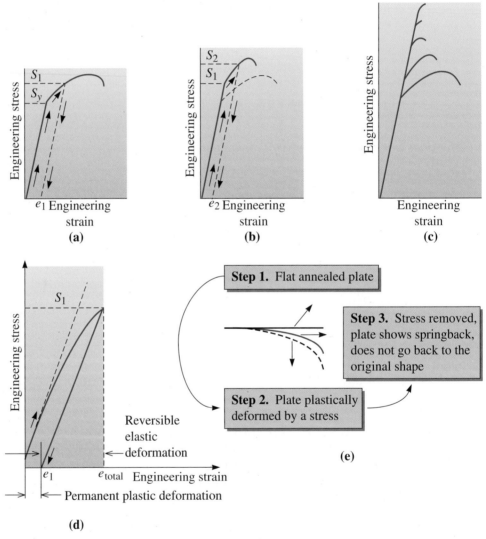

Figure 8-1 Development of strain hardening from the engineering stress-strain diagram. (a) A specimen is stressed beyond the yield strength S_y before the stress is removed. (b) Now the specimen has a higher yield strength and tensile strength, but lower ductility. (c) By repeating the procedure, the strength continues to increase and the ductility continues to decrease until the alloy becomes very brittle. (d) The total strain is the sum of the elastic and plastic components. When the stress is removed, the elastic strain is recovered, but the plastic strain is not. (e) Illustration of springback. *(Source: Reprinted from* Engineering Materials I, *Second Edition, M.F. Ashby, and D.R.H. Jones, 1996. Copyright © 1996 Butterworth-Heinemann. Reprinted with permission from Elsevier Science.)*

sample from the metallic material that had been previously stressed to S_1 and retest that material, we obtain the stress-strain curve shown in Figure 8-1(b). Our new test specimen would begin to deform plastically or flow at stress level S_1. We define the *flow stress* as the stress that is needed to initiate plastic flow in previously deformed material. Thus, S_1 is now the flow stress of the material. If we continue to apply a stress until we reach S_2 then release the stress and again retest the metallic material, the new flow stress is S_2. Each time we apply a higher stress, the flow stress and tensile strength increase, and the ductility decreases. We, eventually strengthen the metallic material until the flow stress, tensile, and breaking strengths are equal, and there is no ductility [Figure 8-1(c)]. At this point, the metallic material can be plastically deformed no further. Figures 8-1(d) and (e) are related to springback, a concept that is discussed later in this section.

By applying a stress that exceeds the original yield strength of the metallic material, we have **strain hardened** or **cold worked** the metallic material, while simultaneously deforming it. This is the basis for many manufacturing techniques, such as wire drawing. Figure 8-2 illustrates several manufacturing processes that make use of both cold-working and hot-working processes. We will discuss the difference between **hot working** and **cold working** later in this chapter. Many techniques for **deformation processing** are used to simultaneously shape and strengthen a material by cold working (Figure 8-2). For example, **rolling** is used to produce metal plate, sheet, or foil. **Forging** deforms the metal into a die cavity, producing relatively complex shapes such as automotive crankshafts or connecting rods. In **drawing**, a metallic rod is pulled through a die to produce a wire or fiber. In **extrusion**, a material is pushed through a die to form products of uniform cross-sections, including rods, tubes, or aluminum trims for doors or windows. *Deep drawing* is used to form the body of aluminum beverage cans. *Stretch forming* and *bending* are used to shape sheet material. Thus, cold working is an effective way of shaping metallic materials while simultaneously increasing their strength. The down side of this process is the loss of ductility. If you take a metal wire and bend it repeatedly, it will harden and eventually break because of strain hardening. Strain hardening is used in many products, especially those that are not going to be exposed to very high temperatures. For example, an aluminum beverage can derives almost 70% of its strength from strain hardening that occurs during its fabrication. Some of the strength of aluminum cans also comes from the alloying elements (e.g., Mg) added. Note that many of the processes, such as rolling, can be conducted using both cold and hot working. The pros and cons of using each will be discussed later in this chapter.

Strain-Hardening Exponent (*n*)   The response of a metallic material to cold working is given by the **strain-hardening exponent**, which is the slope of the plastic portion of the true stress-true strain curve. This relationship is governed by so-called power law behavior according to *true stress σ-true strain ε* curve in Figure 8-3 when a logarithmic scale is used

$$\sigma = K\varepsilon^n \tag{8-1}$$

or

$$\ln \sigma = \ln K + n \ln \varepsilon \tag{8-2}$$

The constant K (strength coefficient) is equal to the stress when $\varepsilon_t = 1$. Larger degrees of strengthening are obtained for a given strain as *n* increases as shown in Figure 8-3. For metals, strain hardening is the result of dislocation interaction and multiplication. The strain-hardening exponent is relatively low for HCP metals, but is higher for BCC and,

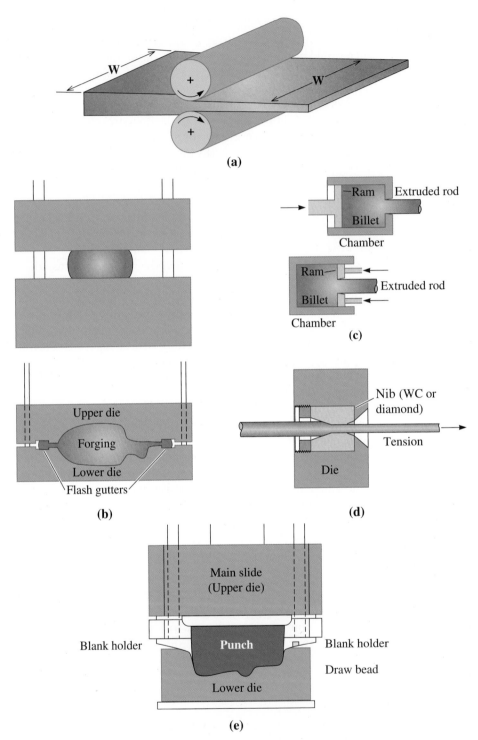

Figure 8-2   Manufacturing processes that make use of cold working as well as hot working. Common metalworking methods. (a) Rolling. (b) Forging (open and closed die). (c) Extrusion (direct and indirect). (d) Wire drawing. (e) Stamping. (*Adapted from Meyers, M. A., and Chawla, K. K., Mechanical behavior of materials, 2nd Edition. Cambridge University Press, Cambridge, England, 2009, Fig. 6.1. With permission of Cambridge University Press.*)

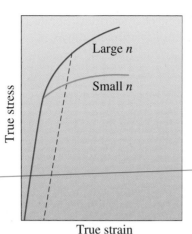

Figure 8-3
The true stress-true strain curves for metals with large and small strain-hardening exponents. Larger degrees of strengthening are obtained for a given strain for the metal with larger *n*.

particularly, for FCC metals (Table 8-1). Metals with a low strain-hardening exponent respond poorly to cold working. If we take a copper wire and bend it, the bent wire is stronger as a result of strain hardening.

Strain-Rate Sensitivity (*m*)

The **strain-rate sensitivity** (*m*) of stress is defined as

$$m = \left[\frac{\partial(\ln \sigma)}{\partial(\ln \dot{\varepsilon})} \right] \tag{8-3}$$

This describes how the flow stress changes with strain rate. The strain-rate sensitivity for crystalline metals is typically less than 0.1, but it increases with temperature. As mentioned before, the mechanical behavior of sheet steels under high strain rates ($\dot{\varepsilon}$) is important not only for shaping, but also for how well the steel will perform under high-impact loading. The crashworthiness of sheet steels is an important consideration for the automotive industry. Steels that harden rapidly under impact loading are useful in absorbing mechanical energy.

A positive value of *m* implies that a material will resist necking (Chapter 6). High values of *m* and *n* mean the material can exhibit better formability in stretching; however,

TABLE 8-1 ■ *Strain-hardening exponents and strength coefficients of typical metals and alloys*

Metal	Crystal Structure	*n*	K (MPa)
Titanium	HCP	0.05	1207
Annealed alloy steel	BCC	0.15	641
Quenched and tempered medium carbon steel	BCC	0.10	1572
Molybdenum	BCC	0.13	724
Copper	FCC	0.54	317
Cu-30% Zn	FCC	0.50	896
Austenitic stainless steel	FCC	0.52	1517

Adapted form G. Dieter, Mechanical Metallurgy, McGraw-Hill, 1961, and other sources.

these values do not affect the deep drawing characteristics. For deep drawing, the *plastic strain ratio* (*r*) is important. We define the plastic strain ratio as

$$r = \frac{\varepsilon_w}{\varepsilon_t} = \frac{\ln\left(\dfrac{w}{w_0}\right)}{\ln\left(\dfrac{h}{h_0}\right)} \tag{8-4}$$

In this equation, *w* and *h* correspond to the width and thickness of the material being processed, and the subscript zero indicates original dimensions. Forming limit diagrams are often used to better understand the **formability** of metallic materials. Overall, we define formability of a material as the ability of a material to maintain its integrity while being shaped. Formability of material is often described in terms of two strains—a major strain, always positive, and a minor strain that can be positive or negative. As illustrated in Figure 8-4, strain conditions on the left make circles stamped into a sample transform into ellipses; for conditions on the right, smaller circles stamped into samples become larger circles indicating stretching. The forming limit diagrams illustrate the specific regions over which the material can be processed without compromising mechanical integrity.

Springback

Another point to be noted is that when a metallic material is deformed using a stress above its yield strength to a higher level (S_1 in Figure 8-1(d)), the corresponding strain existing at stress S_1 is obtained by dropping a perpendicular line to the horizontal axis (point e_{total}). A strain equal to ($e_{total} - e_1$) is recovered since it is elastic in nature. The *elastic strain* that is recovered after a material has been *plastically* deformed is known as *springback* [Figure 8-1(e)]. The occurrence of springback is extremely important for the formation of automotive body panels from sheet steels along

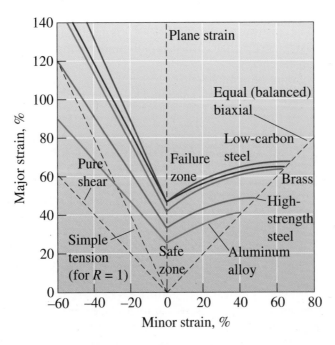

Figure 8-4
Forming limit diagram for different materials. (*Source: Reprinted from Metals Handbook—Desk Edition, Second Edition, ASM International, Materials Park, OH 44073–0002, p. 146, Fig. 5 © 1998 ASM International. Reprinted by permission.*)

with many other applications. This effect is also seen in polymeric materials processed, for example, by extrusion. This is because many polymers are viscoelastic, as discussed in Chapter 6.

It is possible to account for springback in designing components; however, variability in springback makes this very difficult. For example, an automotive supplier will receive coils of sheet steel from different steel manufacturers, and even though the specifications for the steel are identical, the springback variation in steels received from each manufacturer (or even for different lots from the same manufacturer) will make it harder to obtain cold worked components that have precisely the same shape and dimensions.

Bauschinger Effect   Consider a material that has been subjected to tensile plastic deformation. Then, consider two separate samples (A and B) of this material. Test sample A in tension, and sample B under compression. We notice that for the deformed material the flow stress in tension ($\sigma_{\text{flow, tension}}$) for sample A is greater than the compressive yield strength ($\sigma_{\text{flow, compression}}$) for sample B. This effect, in which a material subjected to tension shows a reduction in compressive strength, is known as the **Bauschinger effect**. Note that we are comparing the yield strength of a material under compression and tension after the material has been subjected to plastic deformation under a tensile stress. The Bauschinger effect is also seen on stress reversal. Consider a sample deformed under compression. We can then evaluate two separate samples C and D. The sample subjected to *compressive stress* (C) shows a higher flow stress than that for the sample D subjected to tensile stress. The Bauschinger effect plays an important role in mechanical processing of steels and other alloys.

8-2   Strain-Hardening Mechanisms

We obtain strengthening during deformation of a metallic material by increasing the number of dislocations. Before deformation, the dislocation density is about 10^6 cm of dislocation line per cubic centimeter of metal—a relatively small concentration of dislocations.

When we apply a stress greater than the yield strength, dislocations begin to slip (Schmid's Law, Chapter 4). Eventually, a dislocation moving on its slip plane encounters obstacles that pin the dislocation line. As we continue to apply the stress, the dislocation attempts to move by bowing in the center. The dislocation may move so far that a loop is produced (Figure 8-5). When the dislocation loop finally touches itself, a new dislocation is created. The original dislocation is still pinned and can create additional dislocation loops. This mechanism for generating dislocations is called a **Frank-Read source**; Figure 8-5(e) shows an electron micrograph of a Frank-Read source.

The dislocation density may increase to about 10^{12} cm of dislocation line per cubic centimeter of metal during strain hardening. As discussed in Chapter 4, dislocation motion is the mechanism for the plastic flow that occurs in metallic materials; however, when we have too many dislocations, they interfere with their own motion. An analogy for this is when we have too many people in a room, it is difficult for them to move around. The result is increased strength, but reduced ductility, for metallic materials that have undergone work hardening.

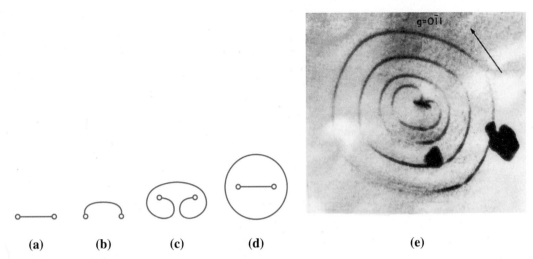

(a)   **(b)**   **(c)**   **(d)**   **(e)**

Figure 8-5   The Frank-Read source can generate dislocations. (a) A dislocation is pinned at its ends by lattice defects. (b) As the dislocation continues to move, the dislocation bows, (c) eventually bending back on itself. (d) Finally the dislocation loop forms, and a new dislocation is created. (e) Electron micrograph of a Frank-Read source (× 330,000). *(Adapted from Brittain, J., "Climb Sources in Beta Prime-NiAl,"* Metallurgical Transactions, *Vol. 6A, April 1975.)*

Ceramics contain dislocations and even can be strain hardened to a small degree; however, dislocations in ceramics are normally not very mobile. Polycrystalline ceramics also contain porosity. As a result, ceramics behave as brittle materials and significant deformation and strengthening by cold working are not possible. Likewise, covalently bonded materials such as silicon (Si) are too brittle to work harden appreciably. Glasses are amorphous and do not contain dislocations and therefore cannot be strain hardened.

Thermoplastics are polymers such as polyethylene, polystyrene, and nylon. These materials consist of molecules that are long spaghetti-like chains. Thermoplastics will strengthen when they are deformed. This is *not* strain hardening due to dislocation multiplication but, instead, strengthening of these materials involves alignment and possibly localized crystallization of the long, chainlike molecules. When a stress greater than the yield strength is applied to thermoplastic polymers such as polyethylene, the van der Waals bonds (Chapter 2) between the molecules in different chains are broken. The chains straighten and become aligned in the direction of the applied stress (Figure 8-6). The strength of the polymer, particularly in the direction of the applied stress, increases as a result of the alignment of polymeric chains in the direction of the applied stress. As discussed in previous chapters, the processing of polyethylene terephthalate (PET) bottles using the blow-stretch process involves such stress-induced crystallization. Thermoplastic polymers get stronger as a result of local alignment of polymer chains occurring as a result of applied stress. This strength increase is seen in the stress-strain curve of typical thermoplastics. Many techniques used for polymer processing are similar to those used for the fabrication of metallic materials. Extrusion, for example, is the most widely used polymer processing technique. Although many of these techniques share conceptual similarities, there are important differences between the mechanisms by which polymers become strengthened during their processing.

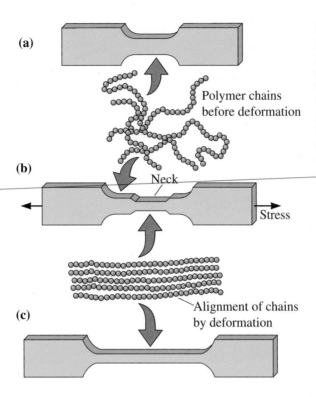

(a)

Polymer chains
before deformation

(b)

Neck

Stress

Alignment of chains
by deformation

(c)

Figure 8-6
In an undeformed thermoplastic
polymer tensile bar, (a) the polymer
chains are randomly oriented.
(b) When a stress is applied, a
neck develops as chains become
aligned locally. The neck continues
to grow until the chains in the
entire gage length have aligned.
(c) The strength of the polymer is
increased.

8-3   Properties versus Percent Cold Work

By controlling the amount of plastic deformation, we control strain hardening. We normally measure the amount of deformation by defining the percent cold work:

$$\text{Percent cold work} = \left[\frac{A_0 - A_f}{A_0} \right] \times 100 \qquad (8\text{-}5)$$

where A_0 is the original cross-sectional area of the metal and A_f is the final cross-sectional area after deformation. For the case of cold rolling, the percent reduction in thickness is used as the measure of cold work according to

$$\text{Percent reduction in thickness} = \left[\frac{t_0 - t_f}{t_0} \right] \times 100 \qquad (8\text{-}6)$$

where t_0 is the initial sheet thickness and t_f is the final thickness.

The effect of cold work on the mechanical properties of commercially pure copper is shown in Figure 8-7. As the cold work increases, both the yield and the tensile strength increase; however, the ductility decreases and approaches zero. The metal breaks if more cold work is attempted; therefore, there is a maximum amount of cold work or deformation that we can perform on a metallic material before it becomes too brittle and breaks.

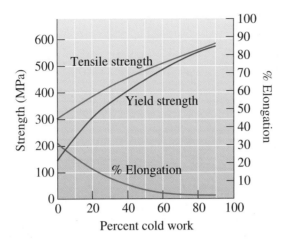

Figure 8-7
The effect of cold work on the mechanical properties of copper.

Example 8-1   *Cold Working a Copper Plate*

A 1-cm-thick copper plate is cold-reduced to 0.50 cm and later further reduced to 0.16 cm. Determine the total percent cold work and the tensile strength of the 0.16 cm plate. (See Figures 8-7 and 8-8.)

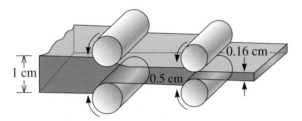

Figure 8-8
Diagram showing the rolling of a 1 cm plate to a 0.16 cm plate (for Example 8-1).

SOLUTION

Note that because the width of the plate does not change during rolling, the cold work can be expressed as the percentage reduction in the thickness t.

Our definition of cold work is the percentage change between the original and final cross-sectional areas; it makes no difference how many intermediate steps are involved. Thus, the total cold work is

$$\% \, CW = \left[\frac{t_0 - t_f}{t_0} \right] \times 100 = \left[\frac{1 \, cm - 0.16 \, cm}{1 \, cm} \right] \times 100 = 84\%$$

and, from Figure 8-7, the tensile strength is about 565 MPa.

We can predict the properties of a metal or an alloy if we know the amount of cold work during processing. We can then decide whether the component has adequate strength at critical locations.

When we wish to select a material for a component that requires certain minimum mechanical properties, we can design the deformation process. We first determine the necessary percent cold work and then, using the final dimensions we desire, calculate the original metal dimensions from the cold work equation.

| **Example 8-2** | *Design of a Cold Working Process* |

Design a manufacturing process to produce a 0.1-cm-thick copper plate having at least 448 MPa tensile strength, 414 MPa yield strength, and 5% elongation.

SOLUTION

From Figure 8-7, we need at least 35% cold work to produce a tensile strength of 448 MPa and 40% cold work to produce a yield strength of 414 MPa, but we need less than 45% cold work to meet the 5% elongation requirement. Therefore, any cold work between 40% and 45% gives the required mechanical properties.

To produce the plate, a cold-rolling process would be appropriate. The original thickness of the copper plate prior to rolling can be calculated from Equation 8-5, assuming that the width of the plate does not change. Because there is a range of allowable cold work—between 40% and 45%—there is a range of initial plate thicknesses:

$$\% \, CW_{min} = 40 = \left[\frac{t_{min} - 0.1 \, cm}{t_{min}} \right] \times 100, \quad \therefore \, t_{min} = 0.167 \, cm$$

$$\% \, CW_{max} = 45 = \left[\frac{t_{max} - 0.1 \, cm}{t_{max}} \right] \times 100, \quad \therefore \, t_{max} = 0.182 \, cm$$

To produce the 0.1-cm copper plate, we begin with a 0.167- to 0.182-cm copper plate in the softest possible condition, then cold roll the plate 40% to 45% to achieve the 0.1 cm thickness.

8-4   Microstructure, Texture Strengthening, and Residual Stresses

During plastic deformation using cold or hot working, a microstructure consisting of grains that are elongated in the direction of the applied stress is often produced (Figure 8-9).

Anisotropic Behavior   During deformation, grains rotate as well as elongate, causing certain crystallographic directions and planes to become aligned with the direction in which stress is applied. Consequently, preferred orientations, or textures, develop and cause anisotropic behavior.

In processes such as wire drawing and extrusion, a **fiber texture** is produced. The term "fibers" refers to the grains in the metallic material, which become elongated in a direction parallel to the axis of the wire or an extruded product. In BCC metals, <110> directions line up with the axis of the wire. In FCC metals, <111> or <100> directions are aligned. This gives the highest strength along the axis of the wire or the extrudate (product being extruded such as a tube), which is what we desire.

As mentioned previously, a somewhat similar effect is seen in thermoplastic materials when they are drawn into fibers or other shapes. The cause, as discussed before, is that polymer chains line up side-by-side along the length of fiber. The strength is greatest along the axis of the polymer fiber. This type of strengthening is also seen in PET plastic bottles made using the *blow-stretch process*. This process causes alignment

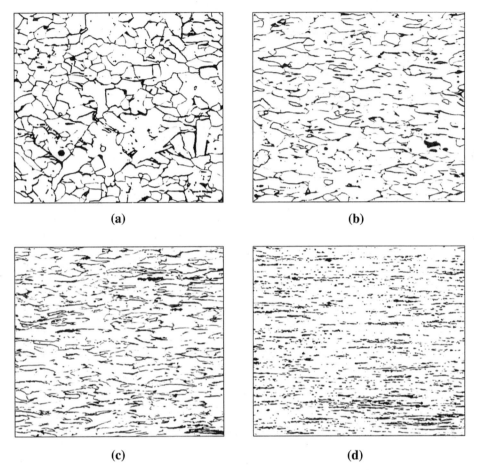

(a)   **(b)**

(c)   **(d)**

Figure 8-9   The fibrous grain structure of a low carbon steel produced by cold working:
(a) 10% cold work, (b) 30% cold work, (c) 60% cold work, and (d) 90% cold work ($\times$ 250).
(From ASM Handbook Vol. 9, Metallography and Microstructure, *(1985) ASM International,
Materials Park, OH 44073–0002. Used with permission.)*

of polymer chains along the radial and length directions, leading to increased strength
of PET bottles along those directions.

In processes such as rolling, grains become oriented in a preferred crystallographic direction and plane, giving a **sheet texture**. The properties of a rolled sheet or
plate depend on the direction in which the property is measured. Figure 8-10 summarizes the tensile properties of a cold-worked aluminim-lithium (Al-Li) alloy. For this alloy,
strength is highest parallel to the rolling direction, whereas ductility is highest at a 45°
angle to the rolling direction. The strengthening that occurs by the development of
anisotropy or of a texture is known as **texture strengthening**. As pointed out in Chapter 6,
the Young's modulus of materials also depends upon crystallographic directions in single crystals. For example, the Young's modulus of iron along [111] and [100] directions
is ~260 and 140 GPa, respectively. The dependence of yield strength on texture is even
stronger. Development of texture not only has an effect on mechanical properties but
also on magnetic and other properties of materials. For example, grain-oriented
magnetic steels made from about 3% Si and 97% Fe used in transformer cores are textured via thermo-mechanical processing so as to optimize their electrical and magnetic

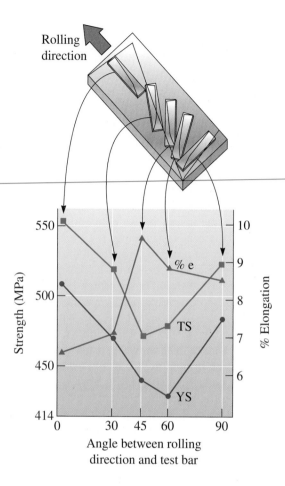

Figure 8-10
Anisotropic behavior in a rolled aluminum-lithium sheet material used in aerospace applications. The sketch relates the position of tensile bars to the mechanical properties that are obtained.

properties. Some common fiber (wire drawing) and sheet (rolling) textures with different crystal structures are shown in Table 8-2.

Texture Development in Thin Films
Orientation or crystallographic texture development also occurs in thin films. In the case of thin films, the texture is often a result of the mechanisms of the growth process and not a result of externally applied stresses. Sometimes, internally generated thermal stresses can play a role in the determination of thin-film crystallographic texture. Oriented thin films can offer better

TABLE 8-2 ■ *Common wire drawing and extrusion and sheet textures in materials*

Crystal Structure	Wire Drawing and Extrusion (Fiber Texture) (Direction Parallel to Wire Axis)	Sheet or Rolling Texture
FCC	<111> and <100>	{110} planes parallel to rolling plane <112> directions parallel to rolling direction
BCC	<110>	{001} planes parallel to rolling plane <110> directions parallel to rolling direction
HCP	<10$\bar{1}$0>	{0001} planes parallel to rolling plane <11$\bar{2}$0> directions parallel to rolling direction

electrical, optical, or magnetic properties. **Pole figure analysis**, a technique based on x-ray diffraction (XRD) (Chapter 3), or a specialized scanning electron microscopy technique known as **orientation microscopy** are used to identify textures in different engineered materials (films, sheets, single crystals, etc.).

Residual Stresses

A small portion of the applied stress is stored in the form of **residual stresses** within the structure as a tangled network of dislocations. The presence of dislocations increases the total internal energy of the structure. As the extent of cold working increases, the level of total internal energy of the material increases. Residual stresses generated by cold working may not always be desirable and can be relieved by a heat treatment known as a **stress-relief anneal** (Section 8-6). As will be discussed shortly, in some instances, we deliberately create residual compressive stresses at the surface of materials to enhance their mechanical properties.

The residual stresses are not uniform throughout the deformed metallic material. For example, high compressive residual stresses may be present at the surface of a rolled plate and high tensile stresses may be stored in the center. If we machine a small amount of metal from one surface of a cold-worked part, we remove metal that contains only compressive residual stresses. To restore the balance, the plate must distort. If there is a net compressive residual stress at the surface of a component, this may be beneficial to the mechanical properties since any crack or flaw on the surface will not likely grow. These are reasons why any residual stresses, originating from cold work or any other source, affect the ability of a part to carry a load (Figure 8-11). If a tensile stress is applied to a material that already contains tensile residual stresses, the total stress acting on the part is the sum of the applied and residual stresses. If, however, compressive stresses are stored at the surface of a metal part, an applied tensile stress must first balance the compressive residual stresses. Now the part may be capable of withstanding a larger than normal load. In Chapter 7, we learned that fatigue is a common mechanism of failure for load-bearing components. Sometimes, components that are subject to fatigue failure can be strengthened by **shot peening**. Bombarding the surface with steel shot propelled at a high velocity introduces compressive residual stresses at the surface that increase the resistance of the metal surface to fatigue failure (Chapter 7). The following example explains the use of shot peening.

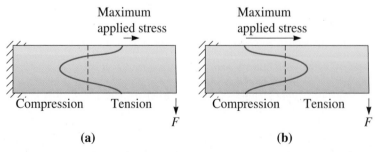

Figure 8-11   Residual stresses can be harmful or beneficial. (a) A bending force applies a tensile stress on the top of the beam. Since there are already tensile residual stresses at the top, the load-carrying characteristics are poor. (b) The top contains compressive residual stresses. Now the load-carrying characteristics are very good.

| **Example 8-3** | *Design of a Fatigue-Resistant Shaft* |

Your company has produced several thousand shafts that have a fatigue strength of 138 MPa. The shafts are subjected to high bending loads during rotation. Your sales engineers report that the first few shafts placed into service failed in a short period of time by fatigue. Design a process by which the remaining shafts can be salvaged by improving their fatigue properties.

SOLUTION

Fatigue failures typically begin at the surface of a rotating part; thus, increasing the strength at the surface improves the fatigue life of the shaft. A variety of methods might be used to accomplish this.

If the shaft is made from steel, we could carburize the surface of the part (Chapter 5). In carburizing, carbon is diffused into the surface of the shaft. After an appropriate heat treatment, the higher carbon content at the surface increases the strength of the surface and, perhaps more importantly, introduces *compressive* residual stresses at the surface.

We might consider cold working the shaft; cold working increases the yield strength of the metal and, if done properly, introduces compressive residual stresses. The cold work also reduces the diameter of the shaft and, because of the dimensional change, the shaft may not be able to perform its function.

Another alternative is to shot peen the shaft. Shot peening introduces local compressive residual stresses at the surface without changing the dimensions of the part. This process, which is also inexpensive, might be sufficient to salvage the remaining shafts.

Tempering and Annealing of Glasses   Residual stresses originating during the cooling of glasses are of considerable interest. We can deal with residual stresses in glasses in two ways. First, we can reheat the glass to a high temperature known as the *annealing point* ($\sim$450°C for silicate glasses with a viscosity of $\sim 10^{13}$ Poise) and let it cool slowly so that the outside and inside cool at about same rate. The resultant glass will have little or no residual stress. This process is known as **annealing**, and the resultant glass that is nearly stress-free is known as **annealed glass**. The purpose of annealing glasses and the process known as stress-relief annealing in metallic materials is the same (i.e., to remove or significantly lower the level of residual stress). The origin of residual stress, though, is different for these materials. Another option we have in glass processing is to conduct a heat treatment that leads to compressive stresses on the surface of a glass; this is known as **tempering**. The resultant glass is known as **tempered glass**. Tempered glass is obtained by heating glass to a temperature just below the annealing point, then, deliberately letting the surface cool more rapidly than the center. This leads to a uniform compressive stress at the surface of the glass. The center region remains under a tensile stress. It is also possible to exchange ions in the glass structure and to introduce a compressive stress. This is known as *chemical tempering*. In Chapter 7, we saw that the strength of glass depends on flaws on the surface. If we have a compressive stress at the surface of the glass and a tensile stress in the center (so as to have overall zero stress), the strength of glass is improved significantly. Any microcracks present will not grow readily, owing to the presence of a net compressive stress on the surface of the glass. If we, however, create a large impact, then the crack does penetrate through the region where the stresses are compressive, and the glass shatters.

Tempered glass has many uses. For example, side window panes and rear windshields of cars are made using tempered glass. Applications such as fireplace screens, ovens, shelving, furniture, and refrigerators also make use of tempered glass. For automobile windshields in the front, we make use of **laminated safety glass**. Front windshield glass is made from two annealed glass pieces laminated using a plastic known as polyvinyl butyral (PVB). If the windshield glass breaks, the laminated glass pieces are held together by PVB plastic. This helps minimize injuries to the driver and passengers. Also, the use of laminated safety glass reduces the chances of glass pieces cutting into the fabric of airbags that are probably being deployed simultaneously.

8-5   Characteristics of Cold Working

There are a number of advantages and limitations to strengthening a metallic material by cold working or strain hardening.

- We can simultaneously strengthen the metallic material and produce the desired final shape.
- We can obtain excellent dimensional tolerances and surface finishes by the cold working process.
- The cold-working process can be an inexpensive method for producing large numbers of small parts.
- Some metals, such as HCP magnesium, have a limited number of slip systems and are rather brittle at room temperature; thus, only a small degree of cold working can be accomplished.
- Ductility, electrical conductivity, and corrosion resistance are impaired by cold working. Since the extent to which electrical conductivity is reduced by cold working is less than that for other strengthening processes, such as introducing alloying elements (Figure 8-12), cold working is a satisfactory way to strengthen conductor materials, such as the copper wires used for transmission of electrical power.

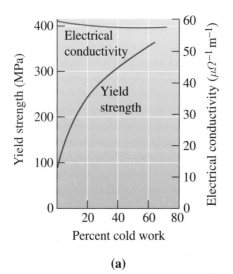

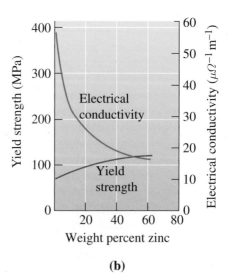

(a)   (b)

Figure 8-12   A comparison of strengthening copper by (a) cold working and (b) alloying with zinc. Note that cold working produces greater strengthening, yet has little effect on electrical conductivity.

- Properly controlled residual stresses and anisotropic behavior may be beneficial; however, if residual stresses are not properly controlled, the materials properties are greatly impaired.

- As will be seen in Section 8-6, since the effect of cold working is decreased or eliminated at higher temperatures, we cannot use cold working as a strengthening mechanism for components that will be subjected to high temperatures during service.

- Some deformation processing techniques can be accomplished only if cold working occurs. For example, wire drawing requires that a rod be pulled through a die to produce a smaller cross-sectional area (Figure 8-13). For a given draw force F_d, a different stress is produced in the original and final wire. The stress on the initial wire must exceed the yield strength of the metal to cause deformation. The stress on the final wire must be less than its yield strength to prevent failure. This is accomplished only if the wire strain hardens during drawing.

$$\text{Stress} = \frac{F_d}{\frac{\pi}{4}d_0^2} > \text{Original yield strength} \qquad \text{Stress} = \frac{F_d}{\frac{\pi}{4}d_f^2} < \text{Final yield strength}$$

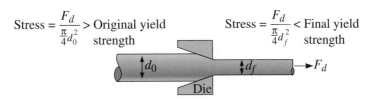

Figure 8-13   The wire-drawing process. The force F_d acts on both the original and final diameters. Thus, the stress produced in the final wire is greater than that in the original. If the wire did not strain harden during drawing, the final wire would break before the original wire was drawn through the die.

Example 8-4   *Design of a Wire-Drawing Process*

Design a process to produce 0.5-cm diameter copper wire. The mechanical properties of the copper are shown in Figure 8-7.

SOLUTION

Wire drawing is the obvious manufacturing technique for this application. To produce the copper wire as efficiently as possible, we make the largest reduction in the diameter possible. Our design must ensure that the wire strain hardens sufficiently during drawing to prevent the drawn wire from breaking.

As an example calculation, let's assume that the starting diameter of the copper wire is 1 cm and that the wire is in the softest possible condition. The cold work is

$$\% \text{ CW} = \left[\frac{A_0 - A_f}{A_0}\right] \times 100 = \left[\frac{(\pi/4)d_0^2 - (\pi/4)d_f^2}{(\pi/4)d_0^2}\right] \times 100$$

$$= \left[\frac{(1 \text{ cm})^2 - (0.5 \text{ cm})^2}{(1 \text{ cm})^2}\right] \times 100 = 75\%$$

From Figure 8-7, the initial yield strength with 0% cold work is 152 MPa. The final yield strength with 75% cold work is about 534 MPa (with very little ductility). The draw force required to deform the initial wire is

$$F = \sigma_y A_0 = (152 \text{ MPa})(\pi/4)(1 \text{ cm})^2 = 11,943 \text{ N}$$

TABLE 8-3 ■ *Mechanical properties of copper wire (see Example 8-4)*

d_0 (cm)	% CW	Yield Strength of Drawn Wire (MPa)	Force (N)	Draw stress on Drawn Wire (MPa)
0.625	36	400	4665	238
0.75	56	469	6718	342
0.875	67	510	9144	466
1	75	534	11,943	608

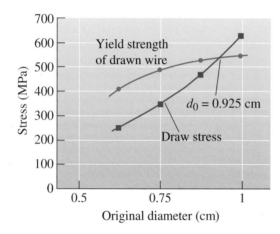

Figure 8-14
Yield strength and draw stress of wire (for Example 8-4).

The stress acting on the wire after passing through the die is

$$\sigma = \frac{F_d}{A_f} = \frac{11{,}943 \text{ N}}{(\pi/4)(0.5 \text{ cm})^2} = 608 \text{ MPa}$$

The applied stress of 608 MPa is greater than the 534 MPa yield strength of the drawn wire. Therefore, the wire breaks because the % elongation is almost zero.

We can perform the same set of calculations for other initial diameters, with the results shown in Table 8-3 and Figure 8-14.

The graph shows that the draw stress exceeds the yield strength of the drawn wire when the original diameter is about 0.925 cm. To produce the wire as efficiently as possible, the original diameter should be just under 0.925 cm.

8-6   The Three Stages of Annealing

Cold working is a useful strengthening mechanism, and it is an effective tool for shaping materials using wire drawing, rolling, extrusion, etc. Sometimes, cold working leads to effects that are undesirable. For example, the loss of ductility or development of residual stresses may not be desirable for certain applications. Since cold working or strain hardening results from increased dislocation density, we can assume that any treatment to rearrange or annihilate dislocations reverses the effects of cold working.

Annealing is a heat treatment used to eliminate some or all of the effects of cold working. Annealing at a low temperature may be used to eliminate the residual stresses produced during cold working without affecting the mechanical properties of the finished

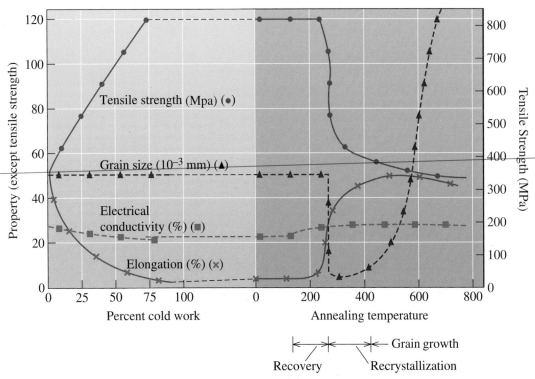

Figure 8-15 The effect of cold work on the properties of a Cu-35% Zn alloy and the effect of annealing temperature on the properties of a Cu-35% Zn alloy that is cold worked 75%.

part, or annealing may be used to completely eliminate the strain hardening achieved during cold working. In this case, the final part is soft and ductile but still has a good surface finish and dimensional accuracy. After annealing, additional cold work can be done since the ductility is restored; by combining repeated cycles of cold working and annealing, large total deformations may be achieved. There are three possible stages in the annealing process; their effects on the properties of brass are shown in Figure 8-15.

Note that the term "annealing" is also used to describe other thermal treatments. For example, glasses may be annealed, or heat treated, to eliminate residual stresses. Cast irons and steels may be annealed to produce the maximum ductility, even though no prior cold work was done to the material. These annealing heat treatments will be discussed in later chapters.

Recovery

The original cold-worked microstructure is composed of deformed grains containing a large number of tangled dislocations. When we first heat the metal, the additional thermal energy permits the dislocations to move and form the boundaries of a **polygonized subgrain structure** (Figure 8-16). The dislocation density, however, is virtually unchanged. This low temperature treatment removes the residual stresses due to cold working without causing a change in dislocation density and is called **recovery**.

The mechanical properties of the metal are relatively unchanged because the number of dislocations is not reduced during recovery. Since residual stresses are reduced or even eliminated when the dislocations are rearranged, recovery is often called a stress relief anneal. In addition, recovery restores high electrical conductivity to the metal, permitting us to manufacture copper or aluminum wire for transmission of electrical power that is strong yet still has high conductivity. Finally, recovery often improves the corrosion resistance of the material.

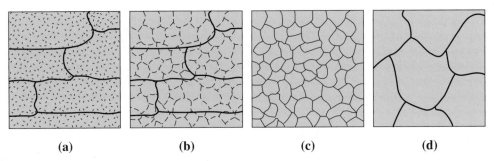

Figure 8-16 The effect of annealing temperature on the microstructure of cold-worked metals. (a) Cold worked, (b) after recovery, (c) after recrystallization, and (d) after grain growth.

Recrystallization When a cold-worked metal is heated above a certain temperature, rapid recovery eliminates residual stresses and produces the polygonized dislocation structure. New small grains then nucleate at the cell boundaries of the polygonized structure, eliminating most of the dislocations (Figure 8-16). Because the number of dislocations is greatly reduced, the recrystallized metal has low strength but high ductility. The temperature at which a microstructure of new grains that have very low dislocation density appears is known as the **recrystallization temperature**. The process of formation of new grains by heat treating a cold-worked material is known as **recrystallization**. As will be seen in Section 8-7, the recrystallization temperature depends on many variables and is not a fixed temperature.

Grain Growth At still higher annealing temperatures, both recovery and recrystallization occur rapidly, producing a fine recrystallized grain structure. If the temperature is high enough, the grains begin to grow, with favored grains consuming the smaller grains (Figure 8-17). This phenomenon, called *grain growth*, is driven by the reduction in grain boundary area and was described in Chapter 5. Illustrated for a

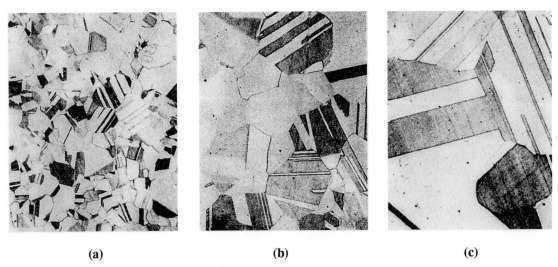

Figure 8-17 Photomicrographs showing the effect of annealing temperature on grain size in brass. Twin boundaries can also be observed in the structures. (a) Annealed at 400°C, (b) annealed at 650°C, and (c) annealed at 800°C (× 75) (*Adapted from Brick, R. and Phillips, A., The Structure and Properties of Alloys, 1949: McGraw-Hill.*)

copper-zinc alloy in Figure 8-15, grain growth is almost always undesirable. Remember that grain growth will occur in most materials if they are subjected to a high enough temperature and, as such, is not related to cold working. Thus, recrystallization or recovery are not needed for grain growth to occur.

You may be aware that incandescent lightbulbs contain filaments that are made from tungsten (W). The high operating temperature causes grain growth and is one of the factors leading to filament failure.

Ceramic materials, which normally do not show any significant strain hardening, show a considerable amount of grain growth (Chapter 5). Also, abnormal grain growth can occur in some materials as a result of formation of a liquid phase during sintering (see Chapter 15). Sometimes grain growth is desirable, as is the case for alumina ceramics for making optical materials used in lighting. In this application, we want very large grains since the scattering of light from grain boundaries has to be minimized. Some researchers have also developed methods for growing single crystals of ceramic materials using grain growth.

8-7   Control of Annealing

In many metallic materials applications, we need a combination of strength and toughness. Therefore, we need to design processes that involve shaping via cold working. We then need to control the annealing process to obtain the desired ductility. To design an appropriate annealing heat treatment, we need to know the recrystallization temperature and the size of the recrystallized grains.

Recrystallization Temperature   This is the temperature at which grains in the cold-worked microstructure begin to transform into new, equiaxed, and dislocation-free grains. The driving force for recrystallization is the difference between the internal energy a cold-worked material and that of a recrystallized material. It is important for us to emphasize that the recrystallization temperature is *not a* fixed temperature, like the melting temperature of a pure element, and is influenced by a variety of processing variables.

- The recrystallization temperature decreases when the amount of cold work increases. Greater amounts of cold work make the metal less stable and encourage nucleation of recrystallized grains. There is a minimum amount of cold work, about 30 to 40%, below which recrystallization will not occur.

- A smaller initial cold-worked grain size reduces the recrystallization temperature by providing more sites—the former grain boundaries—at which new grains can nucleate.

- Pure metals recrystallize at lower temperatures than alloys.

- Increasing the annealing time reduces the recrystallization temperature (Figure 8-18), since more time is available for nucleation and growth of the new recrystallized grains.

- Higher melting-point alloys have a higher recrystallization temperature. Since recrystallization is a diffusion-controlled process, the recrystallization temperature is roughly proportional to $0.4T_m$ (kelvin). Typical recrystallization temperatures for selected metals are shown in Table 8-4.

The concept of recrystallization temperature is very important since it also defines the boundary between cold working and hot working of a metallic material. If we conduct deformation (shaping) of a material above the recrystallization temperature, we refer to it as hot working. If we conduct the shaping or deformation at a temperature

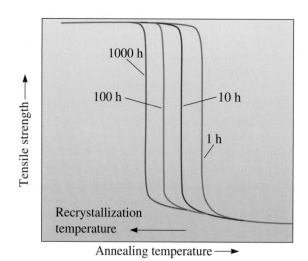

Figure 8-18
Longer annealing times reduce the recrystallization temperature. Note that the recrystallizatiion temperature is not a fixed temperature.

TABLE 8-4 ■ *Typical recrystallization temperatures for selected metals*

Metal	Melting Temperature (°C)	Recrystallization Temperature (°C)
Sn	232	−4
Pb	327	−4
Zn	420	10
Al	660	150
Mg	650	200
Ag	962	200
Cu	1085	200
Fe	1538	450
Ni	1453	600
Mo	2610	900
W	3410	1200

(STRUCTURE AND PROPERTIES OF ENGINEERING MATERIALS, 4TH EDITION by Brick, Pense, Gordon. Copyright 1977 by MCGRAW-HILL COMPANIES, INC. -BOOKS. Reproduced with permission of MCGRAW-HILL COMPANIES, INC. -BOOKS in the format Textbook via Copyright Clearance Center.)

below the recrystallization temperature, we refer to this as cold working. As can be seen from Table 8-4, for lead (Pb) or tin (Sn) deformed at 25°C, we are conducting hot working! This is why iron (Fe) can be cold worked at room temperature but not lead. For tungsten (W) being deformed at 1000°C, we are conducting cold working! In some cases, processes conducted above 0.6 times the melting temperature (T_m) of a metal (in K) are considered as hot working. Processes conducted below 0.3 times the melting temperature are considered cold working and processes conducted between 0.3 and 0.6 times T_m are considered **warm working**. These descriptions of ranges that define hot, cold, and warm working, however, are approximate and should be used with caution.

Recrystallized Grain Size

A number of factors influence the size of the recrystallized grains. Reducing the annealing temperature, the time required to heat to the annealing temperature, or the annealing time reduces grain size by minimizing the opportunity for grain growth. Increasing the initial cold work also reduces final grain size by providing a greater number of nucleation sites for new grains. Finally, the presence of a second phase in the microstructure may foster or hinder recrystallization and grain growth depending on its arrangement and size.

8-8   Annealing and Materials Processing

The effects of recovery, recrystallization, and grain growth are important in the processing and eventual use of a metal or an alloy.

Deformation Processing   By taking advantage of the annealing heat treatment, we can increase the total amount of deformation we can accomplish. If we are required to reduce a 12.5-cm-thick plate to a 0.125-cm-thick sheet, we can do the maximum permissible cold work, anneal to restore the metal to its soft, ductile condition, and then cold work again. We can repeat the cold work-anneal cycle until we approach the proper thickness. The final cold-working step can be designed to produce the final dimensions and properties required, as in the following example.

Example 8-5   *Design of a Process to Produce Copper Strip*

We wish to produce a 0.1-cm-thick, 6-cm-wide copper strip having at least 414 MPa yield strength and at least 5% elongation. We are able to purchase 6-cm-wide strip only in a thickness of 5 cm. Design a process to produce the product we need. Refer to Figure 8-7 as needed.

SOLUTION

In Example 8-2, we found that the required properties can be obtained with a cold work of 40 to 45%. Therefore, the starting thickness must be between 0.167 cm and 0.182 cm, and this starting material must be as soft as possible—that is, in the annealed condition. Since we are able to purchase only 5-cm-thick stock, we must reduce the thickness of the 5 cm strip to between 0.167 and 0.182 cm, then anneal the strip prior to final cold working. But can we successfully cold work from 5 cm to 0.182 cm?

$$\% \text{ CW} = \left[\frac{t_0 - t_f}{t_0}\right] \times 100 = \left[\frac{5 \text{ cm} - 0.182 \text{ cm}}{5 \text{ cm}}\right] \times 100 = 96.4\%$$

Based on Figure 8-7, a maximum of about 90% cold work is permitted. Therefore, we must do a series of cold work and anneal cycles. Although there are many possible combinations, one is as follows:

1. Cold work the 5 cm strip 80% to 1 cm:

$$80 = \left[\frac{t_0 - t_f}{t_0}\right] \times 100 = \left[\frac{5 \text{ cm} - t_f}{5 \text{ cm}}\right] \times 100 \text{ or } t_f = 1 \text{ cm}$$

2. Anneal the 1 cm strip to restore the ductility. If we don't know the recrystallization temperature, we can use the 0.4 T_m relationship to provide an estimate. The melting point of copper is 1085°C:

$$T_r \cong (0.4)(1085 + 273) = 543 \text{ K} = 270°C$$

3. Cold work the 1-cm-thick strip to 0.182 cm:

$$\% \, \text{CW} = \left[\frac{1 \text{ cm} - 0.182 \text{ cm}}{1 \text{ cm}} \right] \times 100 = 81.8\%$$

4. Again anneal the copper at 270°C to restore ductility.

5. Finally, cold work 45% from 0.182 cm to the final dimension of 0.1 cm. This process gives the correct final dimensions and properties.

High Temperature Service

As mentioned previously, neither strain hardening nor grain-size strengthening (Hall-Petch equation, Chapter 4) are appropriate for an alloy that is to be used at elevated temperatures, as in creep-resistant applications. When the cold-worked metal is placed into service at a high temperature, recrystallization immediately causes a catastrophic decrease in strength. In addition, if the temperature is high enough, the strength continues to decrease because of growth of the newly recrystallized grains.

Joining Processes

Metallic materials can be joined using processes such as welding. When we join a cold-worked metal using a welding process, the metal adjacent to the weld heats above the recrystallization and grain growth temperatures and subsequently cools slowly. This region is called the **heat-affected zone** (HAZ). The structure and properties in the heat-affected zone of a weld are shown in Figure 8-19. The mechanical properties are reduced catastrophically by the heat of the welding process.

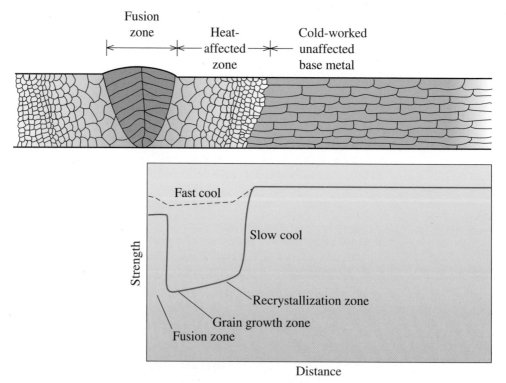

Figure 8-19   The structure and properties surrounding a fusion weld in a cold-worked metal. Only the right-hand side of the heat-affected zone is marked on the diagram. Note the loss in strength caused by recrystallization and grain growth in the heat-affected zone.

Welding processes, such as electron beam welding or laser welding, which provide high rates of heat input for brief times, and, thus, subsequent fast cooling, minimize the exposure of the metallic materials to temperatures above recrystallization and minimize this type of damage. Similarly a process known as friction stir welding provides almost no HAZ and is being commercially used for welding aluminum alloys. We will discuss metal-joining processes in greater detail in Chapter 9.

8-9   Hot Working

We can deform a metal into a useful shape by hot working rather than cold working. As described previously, hot working is defined as plastically deforming the metallic material at a temperature above the recrystallization temperature. During hot working, the metallic material is continually recrystallized (Figure 8-20). As mentioned before, at room temperature lead (Pb) is well above its recrystallization temperature of −4°C, and therefore, Pb does not strain harden and remains soft and ductile at room temperature.

Lack of Strengthening   No strengthening occurs during deformation by hot working; consequently, the amount of plastic deformation is almost unlimited. A very thick plate can be reduced to a thin sheet in a continuous series of operations. The first steps in the process are carried out well above the recrystallization temperature to take advantage of the lower strength of the metal. The last step is performed just above the recrystallization temperature, using a large percent deformation in order to produce the finest possible grain size.

Hot working is well suited for forming large parts, since the metal has a low yield strength and high ductility at elevated temperatures. In addition, HCP metals such as magnesium have more active slip systems at hot-working temperatures; the higher ductility permits larger deformations than are possible by cold working. The following example illustrates a design of a hot-working process.

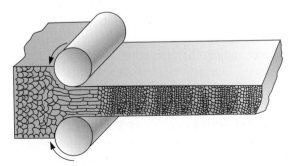

Figure 8-20   During hot working, the elongated, anisotropic grains immediately recrystallize. If the hot-working temperature is properly controlled, the final hot-worked grain size can be very fine.

Example 8-6	*Design of a Process to Produce a Copper Strip*

We want to produce a 0.1-cm-thick, 6-cm-wide copper strip having at least 414 MPa yield strength and at least 5% elongation. We are able to purchase 6-cm-wide strip only in thicknesses of 5 cm. Design a process to produce the product we need, but in fewer steps than were required in Example 8-5.

SOLUTION

In Example 8-5, we relied on a series of cold work-anneal cycles to obtain the required thickness. We can reduce the steps by hot rolling to the required intermediate thickness:

$$\% \text{ HW} = \left[\frac{t_0 - t_f}{t_0} \right] \times 100 = \left[\frac{5 \text{ cm} - 0.182 \text{ cm}}{5 \text{ cm}} \right] \times 100 = 96.4\%$$

$$\% \text{ HW} = \left[\frac{t_0 - t_f}{t_0} \right] \times 100 = \left[\frac{5 \text{ cm} - 0.167 \text{ cm}}{5 \text{ cm}} \right] \times 100 = 96.7\%$$

Note that the formulas for hot and cold work are the same.

Because recrystallization occurs simultaneously with hot working, we can obtain these large deformations and a separate annealing treatment is not required. Thus our design might be

1. Hot work the 5 cm strip 96.4% to the intermediate thickness of 0.182 cm.

2. Cold work 45% from 0.182 cm to the final dimension of 0.1 cm. This design gives the correct dimensions and properties.

Elimination of Imperfections

Some imperfections in the original metallic material may be eliminated or their effects minimized. Gas pores can be closed and welded shut during hot working—the internal lap formed when the pore is closed is eliminated by diffusion during the forming and cooling process. Composition differences in the metal can be reduced as hot working brings the surface and center of the plate closer together, thereby reducing diffusion distances.

Anisotropic Behavior

The final properties in hot-worked parts are *not* isotropic. The forming rolls or dies, which are normally at a lower temperature than the metal, cool the surface more rapidly than the center of the part. The surface then has a finer grain size than the center. In addition, a fibrous structure is produced because inclusions and second-phase particles are elongated in the working direction.

Surface Finish and Dimensional Accuracy

The surface finish formed during hot working is usually poorer than that obtained by cold working. Oxygen often reacts with the metal at the surface to form oxides, which are forced into the surface during forming. Hot worked steels and other metals are often subjected to a "pickling" treatment in which acids are used to dissolve the oxide scale. In some metals, such as tungsten (W) and beryllium (Be), hot working must be done in a protective atmosphere to prevent oxidation. Note that forming processes performed on Be-containing materials require protective measures, since the inhalation of Be-containing materials is hazardous.

The dimensional accuracy is also more difficult to control during hot working. A greater elastic strain must be considered, since the modulus of elasticity is lower at hot-working temperatures than at cold-working temperatures. In addition, the metal contracts as it cools from the hot-working temperature. The combination of elastic strain and thermal contraction requires that the part be made oversized during deformation; forming dies must be carefully designed, and precise temperature control is necessary if accurate dimensions are to be obtained.

Summary

- The properties of metallic materials can be controlled by combining plastic deformation and heat treatments.

- When a metallic material is deformed by cold working, strain hardening occurs as additional dislocations are introduced into the structure. Very large increases in strength may be obtained in this manner. The ductility of the strain hardened metallic material is reduced.

- Strain hardening, in addition to increasing strength and hardness, increases residual stresses, produces anisotropic behavior, and reduces ductility, electrical conductivity, and corrosion resistance.

- The amount of strain hardening is limited because of the simultaneous decrease in ductility; FCC metals typically have the best response to strengthening by cold working.

- Wire drawing, stamping, rolling, and extrusion are some examples of manufacturing methods for shaping metallic materials. Some of the underlying principles for these processes also can be used for the manufacturing of polymeric materials.

- Springback and the Bauschinger effect are very important in manufacturing processes for the shaping of steels and other metallic materials. Forming limit diagrams are useful in defining shaping processes for metallic materials.

- The strain-hardening mechanism is not effective at elevated temperatures, because the effects of the cold work are eliminated by recrystallization.

- Annealing of metallic materials is a heat treatment intended to eliminate all, or a portion of, the effects of strain hardening. The annealing process may involve as many as three steps.

- Recovery occurs at low temperatures, eliminating residual stresses and restoring electrical conductivity without reducing the strength. A "stress relief anneal" refers to recovery.

- Recrystallization occurs at higher temperatures and eliminates almost all of the effects of strain hardening. The dislocation density decreases dramatically during recrystallization as new grains nucleate and grow.

- Grain growth, which typically should be avoided, occurs at still higher temperatures. In cold-worked metallic materials, grain growth follows recovery and recrystallization. In ceramic materials, grain growth can occur due to high temperatures or the presence of a liquid phase during sintering.

- Hot working combines plastic deformation and annealing in a single step, permitting large amounts of plastic deformation without embrittling the material.

- Residual stresses in materials need to be controlled. In cold-worked metallic materials, residual stresses can be eliminated using a stress-relief anneal.

- Annealing of glasses leads to the removal of stresses developed during cooling. Thermal tempering of glasses is a heat treatment in which deliberate rapid cooling of the glass surface leads to a compressive stress at the surface. We use tempered or laminated glass in applications where safety is important.

- In metallic materials, compressive residual stresses can be introduced using shot peening. This treatment will lead to an increase in the fatigue life.

Glossary

Annealed glass Glass that has been treated by heating above the annealing point temperature (where the viscosity of glass becomes 10^{13} Poise) and then cooled slowly to minimize or eliminate residual stresses.

Annealing In the context of metals, annealing is a heat treatment used to eliminate part or all of the effects of cold working. For glasses, annealing is a heat treatment that removes thermally induced stresses.

Bauschinger effect A material previously plastically deformed under tension shows decreased flow stress under compression or vice versa.

Cold working Deformation of a metal below the recrystallization temperature. During cold working, the number of dislocations increases, causing the metal to be strengthened as its shape is changed.

Deformation processing Techniques for the manufacturing of metallic and other materials using such processes as rolling, extrusion, drawing, etc.

Drawing A deformation processing technique in which a material is pulled through an opening in a die (e.g., wire drawing).

Extrusion A deformation processing technique in which a material is pushed through an opening in a die. Used for metallic and polymeric materials.

Fiber texture A preferred orientation of grains obtained during the wire drawing process. Certain crystallographic directions in each elongated grain line up with the drawing direction, causing anisotropic behavior.

Formability The ability of a material to stretch and bend without breaking. Forming diagrams describe the ability to stretch and bend materials.

Frank-Read source A pinned dislocation that, under an applied stress, produces additional dislocations. This mechanism is at least partly responsible for strain hardening.

Heat-affected zone (HAZ) The volume of material adjacent to a weld that is heated during the welding process above some critical temperature at which a change in the structure, such as grain growth or recrystallization, occurs.

Hot working Deformation of a metal above the recrystallization temperature. During hot working, only the shape of the metal changes; the strength remains relatively unchanged because no strain hardening occurs.

Laminated safety glass Two pieces of annealed glass held together by a plastic such as polyvinyl butyral (PVB). This type of glass can be used in car windshields.

Orientation microscopy A specialized technique, often based on scanning electron microscopy, used to determined the crystallographic orientation of different grains in a polycrystalline sample.

Pole figure analysis A specialized technique based on x-ray diffraction, used for the determination of preferred orientation of thin films, sheets, or single crystals.

Polygonized subgrain structure A subgrain structure produced in the early stages of annealing. The subgrain boundaries are a network of dislocations rearranged during heating.

Recovery A low-temperature annealing heat treatment designed to eliminate residual stresses introduced during deformation without reducing the strength of the cold-worked material. This is the same as a stress-relief anneal.

Recrystallization A medium-temperature annealing heat treatment designed to eliminate all of the effects of the strain hardening produced during cold working.

Recrystallization temperature A temperature above which essentially dislocation-free and new grains emerge from a material that was previously cold worked. This depends upon the extent of cold work, time of heat treatment, etc., and is not a fixed temperature.

Residual stresses Stresses introduced in a material during processing. These can originate as a result of cold working or differential thermal expansion and contraction. A stress-relief anneal in metallic materials and the annealing of glasses minimize residual stresses. Compressive residual stresses deliberately introduced on the surface by the tempering of glasses or shot peening of metallic materials improve their mechanical properties.

Sheet texture A preferred orientation of grains obtained during the rolling process. Certain crystallographic directions line up with the rolling direction, and certain preferred crystallographic planes become parallel to the sheet surface.

Shot peening Introducing compressive residual stresses at the surface of a part by bombarding that surface with steel shot. The residual stresses may improve the overall performance of the material.

Strain hardening Strengthening of a material by increasing the number of dislocations by deformation. Also known as "work hardening."

Strain-hardening exponent (*n*) A parameter that describes the susceptibility of a material to cold working. It describes the effect that strain has on the resulting strength of the material. A material with a high strain-hardening coefficient obtains high strength with only small amounts of deformation or strain.

Strain rate The rate at which a material is deformed.

Strain-rate sensitivity (*m*) The rate at which stress changes as a function of strain rate. A material may behave much differently if it is slowly pressed into a shape rather than smashed rapidly into a shape by an impact blow.

Stress-relief anneal The recovery stage of the annealing heat treatment during which residual stresses are relieved without altering the strength and ductility of the material.

Tempered glass A glass, mainly for applications where safety is particularly important, obtained by either heat treatment and quenching or by the chemical exchange of ions. Tempering results in a net compressive stress at the surface of the glass.

Tempering In the context of glass making, tempering refers to a heat treatment that leads to a compressive stress on the surface of a glass. This compressive stress layer makes tempered glass safer. In the context of processing of metallic materials, tempering refers to a heat treatment used to soften the material and to increase its toughness.

Texture strengthening Increase in the yield strength of a material as a result of preferred crystallographic texture.

Thermomechanical processing Processes involved in the manufacturing of metallic components using mechanical deformation and various heat treatments.

Thermoplastics A class of polymers that consist of large, long spaghetti-like molecules that are intertwined (e.g., polyethylene, nylon, PET, etc.).

Warm working A term used to indicate the processing of metallic materials in a temperature range that is between those that define cold and hot working (usually a temperature between 0.3 to 0.6 of the melting temperature in K).

Problems

Section 8-1 Relationship of Cold Working to the Stress-Strain Curve

8-1 Using a stress-strain diagram, explain what the term "strain hardening" means.

8-2 What is meant by the term "springback?" What is the significance of this term from a manufacturing viewpoint?

8-3 What does the term "Bauschinger effect" mean?

8-4 What manufacturing techniques make use of the cold-working process?

8-5 Consider the tensile stress-strain curves in Figure 8-21 labeled 1 and 2 and answer the following questions. These curves are typical of metals. Consider each part as a separate question that has no relationship to previous parts of the question.

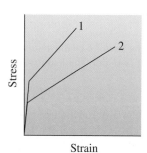

Figure 8-21   Stress-strain curves (for Problem 8-5).

(a) Which material has the larger work-hardening exponent? How do you know?

(b) Samples 1 and 2 are identical except that they were tested at different strain rates. Which sample was tested at the higher strain rate? How do you know?

(c) Assume that the two stress-strain curves represent successive tests of the same sample. The sample was loaded, then unloaded before necking began, and then the sample was reloaded. Which sample represents the first test: 1 or 2? How do you know?

8-6 Figure 8-22 is a plot of true stress versus true strain for a metal. For total imposed strains of $e = 0.1$, 0.2, 0.3 and 0.4, determine the elastic and plastic components of the strain. The modulus of elasticity of the metal is 100 GPa.

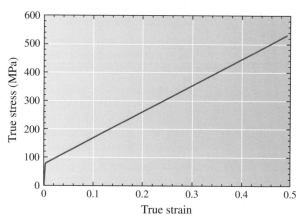

Figure 8-22   A true stress versus true strain curve for a metal (for Problem 8-6).

8-7 A 1.263-cm-diameter metal bar with a 5-cm gage length l_0 is subjected to a tensile test. The following measurements are made in the plastic region:

Force (N)	Change in Gage length (Δl) (cm)	Diameter (cm)
122,320	0.5258	1.2
120,100	1.107	1.1415
114,310	1.7493	1.0858

Determine the strain-hardening exponent for the metal. Is the metal most likely to be FCC, BCC, or HCP? Explain.

8-8 Define the following terms: strain-hardening exponent (n), strain-rate sensitivity (m), and plastic strain ratio (r). Use appropriate equations.

8-9 A 1.5-cm-diameter metal bar with a 3-cm gage length (l_0) is subjected to a tensile test. The following measurements are made:

Force (N)	Change in Gage length (cm)	Diameter (cm)
16,240	0.6642	1.2028
19,066	1.4754	1.0884
19,273	2.4663	0.9848

Determine the strain-hardening coefficient for the metal. Is the metal most likely to be FCC, BCC, or HCP? Explain.

8-10 What does the term "formability of a material" mean?

8-11 A true stress-true strain curve is shown in Figure 8-23. Determine the strain-hardening exponent for the metal.

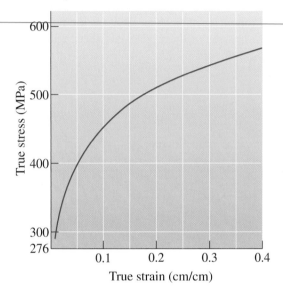

Figure 8-23 True stress-true strain curve (for Problem 8-11).

8-12 Figure 8-24 shows a plot of the natural log of the true stress versus the natural log of the true strain for a Cu-30% Zn sample tested in tension. Only the plastic portion of the stress-strain curve is shown.

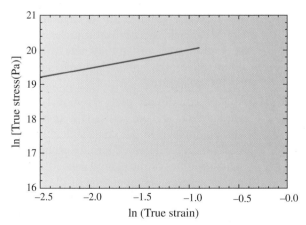

Figure 8-24 The natural log of the true stress versus the natural log of the true strain for a Cu-30% Zn sample tested in tension. Only the plastic portion of the stress-strain curve is shown.

Determine the strength coefficient K and the work-hardening exponent n.

8-13 A Cu-30% Zn alloy tensile bar has a strain hardening coefficient of 0.50. The bar, which has an initial diameter of 1 cm and an initial gage length of 3 cm, fails at an engineering stress of 120 MPa. After fracture, the gage length is 3.5 cm and the diameter is 0.926 cm. No necking occurred. Calculate the true stress when the true strain is 0.05 cm/cm.

Section 8-2 Strain-Hardening Mechanisms

8-14 Explain why many metallic materials exhibit strain hardening.

8-15 Does a strain-hardening mechanism depend upon grain size? Does it depend upon dislocation density?

8-16 Compare and contrast strain hardening with grain size strengthening. What causes resistance to dislocation motion in each of these mechanisms?

8-17 Strain hardening is normally not a consideration in ceramic materials. Explain why.

8-18 Thermoplastic polymers such as polyethylene show an increase in strength when subjected to stress. Explain how this strengthening occurs.

8-19 Bottles of carbonated beverages are made using PET plastic. Explain how stress-induced crystallization increases the strength of PET bottles made by the blow-stretch process (see Chapters 4 and 5).

Section 8-3 Properties versus Percent Cold Work

8-20 Write down the equation that defines percent cold work. Explain the meaning of each term.

8-21 A 0.625-cm-thick copper plate is to be cold worked 63%. Find the final thickness.

8-22 A 0.625-cm-diameter copper bar is to be cold worked 63% in tension. Find the final diameter.

8-23 A 5-cm-diameter copper rod is reduced to a 3.75-cm-diameter, then reduced again to a final diameter of 2.5 cm. In a second case, the 5-cm-diameter rod is reduced in one step from a 5 cm to a 2.5 cm diameter. Calculate the % CW for both cases.

8-24 A 3105 aluminum plate is reduced from 4.38 cm to 2.88 cm. Determine the final properties of the plate. Note 3105 designates a special composition of aluminum alloy. (See Figure 8-25.)

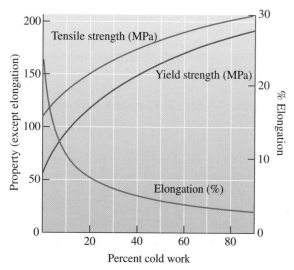

Figure 8-25   The effect of percent cold work on the properties of a 3105 aluminum alloy (for Problems 8-24, 8-26, and 8-30.)

8-25 A Cu-30% Zn brass bar is reduced from a 2.5-cm diameter to a 1.13-cm diameter. Determine the final properties of the bar. (See Figure 8-26.)

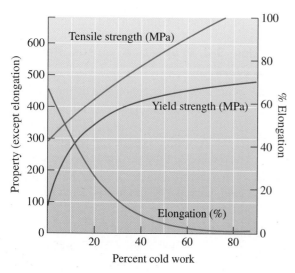

Figure 8-26   The effect of percent cold work on the properties of a Cu-30% Zn brass (for Problems 8-25 and 8-28).

8-26 A 3105 aluminum bar is reduced from a 2.5-cm diameter, to a 2-cm diameter, to a 1.5-cm diameter, to a final 1-cm diameter. Determine the % CW and the properties after each step of the process. Calculate the total percent cold work. Note 3105 designates a special composition of aluminum alloy. (See Figure 8-25.)

8-27 We want a copper bar to have a tensile strength of at least 483 MPa and a final diameter of 0.94 cm. What is the minimum diameter of the original bar? (See Figure 8-7.)

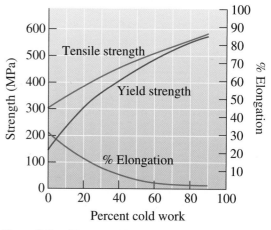

Figure 8-7   (Repeated for Problems 8-27 and 8-29) The effect of cold work on the mechanical properties of copper.

8-28 We want a Cu-30% Zn brass plate originally 3 cm thick to have a yield strength greater than 345 MPa and a % elongation of at least 10%. What range of final thicknesses must be obtained? (See Figure 8-26.)

8-29 We want a copper sheet to have at least 345 MPa yield strength and at least 10% elongation, with a final thickness of 0.3 cm What range of original thickness must be used? (See Figure 8-7.)

8-30 A 3105 aluminum plate previously cold worked 20% is 5 cm thick. It is then cold worked further to 3.25 cm. Calculate the total percent cold work and determine the final properties of the plate. (*Note*: 3105 designates a special composition of aluminum alloy.) (See Figure 8-25.)

8-31 An aluminum-lithium (Al-Li) strap 0.63 cm thick and 5 cm wide is to be cut from a rolled sheet, as described in Figure 8-10. The strap must be able to support a 156000 N load without plastic deformation. Determine the

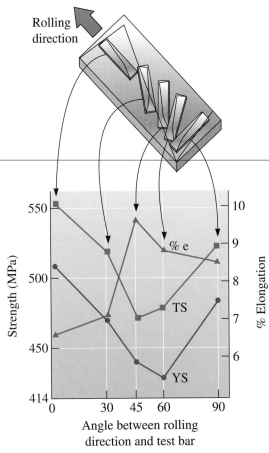

Figure 8-10 (Repeated for Problem 8-31) Anisotropic behavior in a rolled aluminum-lithium sheet material used in aerospace applications. The sketch relates the position of tensile bars to the mechanical properties that are obtained.

range of orientations from which the strap can be cut from the rolled sheet.

Section 8-4 Microstructure, Texture Strengthening, and Residual Stresses

8-32 Does the yield strength of metallic materials depend upon the crystallographic texture materials develop during cold working? Explain.

8-33 Does the Young's modulus of a material depend upon crystallographic directions in a crystalline material? Explain.

8-34 What do the terms "fiber texture" and "sheet texture" mean?

8-35 One of the disadvantages of the cold-rolling process is the generation of residual stresses. Explain how we can eliminate residual stresses in cold-worked metallic materials.

8-36 What is shot peening?

8-37 What is the difference between tempering and annealing of glasses?

8-38 How is laminated safety glass different from tempered glass? Where is laminated glass used?

8-39 What is thermal tempering? How is it different from chemical tempering? State applications of tempered glasses.

8-40 Explain factors that affect the strength of glass most. Explain how thermal tempering helps increase the strength of glass.

8-41 Residual stresses are not always undesirable. True or false? Justify your answer.

Section 8-5 Characteristics of Cold Working

8-42 Cold working cannot be used as a strengthening mechanism for materials that are going to be subjected to high temperatures during their use. Explain why.

8-43 Aluminum cans made by deep drawing derive considerable strength during their fabrication. Explain why.

8-44 Such metals as magnesium cannot be effectively strengthened using cold working. Explain why.

8-45 We want to draw a 0.75-cm-diameter copper wire having a yield strength of 138 MPa into a 0.625-cm diameter wire.
(a) Find the draw force, assuming no friction;
(b) Will the drawn wire break during the drawing process? Show why. (See Figure 8-7.)

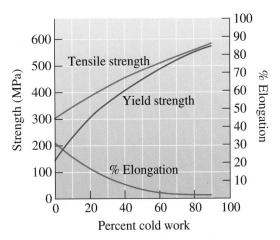

Figure 8-7 (Repeated for Problem 8-45) The effect of cold work on the mechanical properties of copper.

8-46 A 3105 aluminum wire is to be drawn to give a 1-mm-diameter wire having yield strength of 138 MPa. Note 3105 designates a special composition of aluminium alloy.
(a) Find the original diameter of the wire;
(b) calculate the draw force required; and
(c) determine whether the as-drawn wire will break during the process. (See Figure 8-25.)

Section 8-6 The Three Stages of Annealing

8-47 Explain the three stages of the annealing of metallic materials

8-48 What is the driving force for recrystallization?

8-49 In the recovery state, the residual stresses are reduced; however, the strength of the metallic material remains unchanged. Explain why.

8-50 What is the driving force for grain growth?

8-51 Treating grain growth as the third stage of annealing, explain its effect on the strength of metallic materials.

8-52 Why is it that grain growth is usually undesirable? Cite an example where grain growth is actually useful.

8-53 What are the different ways one can encounter grain growth in ceramics?

8-54 Are annealing and recovery always prerequisites to grain growth? Explain.

8-55 A titanium alloy contains a very fine dispersion of Er_2O_3 particles. What will be the effect of these particles on the grain growth temperature and the size of the grains at any particular annealing temperature? Explain.

8-56 Explain why a tungsten filament used in an incandescent lightbulb ultimately fails.

8-57 Samples of cartridge brass (Cu-30% Zn) were cold rolled and then annealed for one hour. The data shown in the table below were obtained.

Annealing Temperature (°C)	Grain Size (μm)	Yield Strength (MPa)
400	15	159
500	23	138
600	53	124
700	140	62
800	505	48

(a) Plot the yield strength and grain size as a function of annealing temperature on the same graph. Use two vertical axes, one for yield strength and one for grain size.
(b) For each temperature, state which stages of the annealing process occurred. Justify your answers by referring to features of the plot.

8-58 The following data were obtained when a cold-worked metal was annealed.
(a) Estimate the recovery, recrystallization, and grain growth temperatures;
(b) recommend a suitable temperature for a stress-relief heat treatment;
(c) recommend a suitable temperature for a hot-working process; and
(d) estimate the melting temperature of the alloy.

Annealing Temperature (°C)	Electrical Conductivity ($ohm^{-1} \cdot cm^{-1}$)	Yield Strength (MPa)	Grain Size (mm)
400	3.04×10^5	86	0.10
500	3.05×10^5	85	0.10
600	3.36×10^5	84	0.10
700	3.45×10^5	83	0.098
800	3.46×10^5	52	0.030
900	3.46×10^5	47	0.031
1000	3.47×10^5	44	0.070
1100	3.47×10^5	42	0.120

8-59 The following data were obtained when a cold-worked metallic material was annealed:
(a) Estimate the recovery, recrystallization, and grain growth temperatures;
(b) recommend a suitable temperature for obtaining a high-strength, high-electrical conductivity wire;
(c) recommend a suitable temperature for a hot-working process; and
(d) estimate the melting temperature of the alloy.

Annealing Temperature (°C)	Residual Stresses (MPa)	Tensile Strength (MPa)	Grain Size (cm)
250	145	359	0.0075
275	145	359	0.0075
300	35	359	0.0075
325	0	359	0.0075
350	0	234	0.0025
375	0	207	0.0025
400	0	186	0.0088
425	0	172	0.0018

8-60 What is meant by the term "recrystallization?" Explain why the yield strength of a metallic material goes down during this stage of annealing.

Section 8-7 Control of Annealing

8-61 How do we distinguish between the hot working and cold working of metallic materials?

8-62 Why is it that the recrystallization temperature is not a fixed temperature for a given material?

8-63 Why does increasing the time for a heat treatment mean recrystallization will occur at a lower temperature?

8-64 Two sheets of steel were cold worked 20% and 80%, respectively. Which one would likely have a lower recrystallization temperature? Why?

8-65 Give examples of two metallic materials for which deformation at room temperature will mean "hot working."

8-66 Give examples of two metallic materials for which mechanical deformation at 900°C will mean "cold working."

8-67 Consider the tensile stress-strain curves in Figure 8-21 labeled 1 and 2 and answer the following questions. These diagrams are typical of metals. Consider each part as a separate question that has no relationship to previous parts of the question.

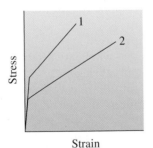

Figure 8-21
Stress-strain curves (for Problem 8-67).

(a) Which of the two materials represented by samples 1 and 2 can be cold rolled to a greater extent? How do you know?

(b) Samples 1 and 2 have the same composition and were processed identically, except that one of them was cold worked more than the other. The stress-strain curves were obtained after the samples were cold worked. Which sample has the lower recrystallization temperature: 1 or 2? How do you know?

(c) Samples 1 and 2 are identical except that they were annealed at different temperatures for the same period of time. Which sample was annealed at the higher temperature: 1 or 2? How do you know?

(d) Samples 1 and 2 are identical except that they were annealed at the same temperature for different periods of time. Which sample was annealed for the shorter period of time: 1 or 2? How do you know?

Section 8-8 Annealing and Materials Processing

8-68 Using the data in Table 8-4, plot the recrystallization temperature versus the melting temperature of each metal, using absolute temperatures (kelvin). Measure the slope and compare with the expected relationship between these two temperatures. Is our approximation a good one?

8-69 We wish to produce a 0.75-cm thick plate of 3105 aluminum having a tensile strength of at least 172 MPa and a % elongation of at

TABLE 8-4 ■ *Typical recrystallization temperatures for selected metals (Repeated for Problem 8-68)*

Metal	Melting Temperature (°C)	Recrystallization Temperature (°C)
SN	232	−4
Pb	327	−4
Zn	420	10
Al	660	150
Mg	650	200
Ag	962	200
Cu	1085	200
Fe	1538	450
Ni	1453	600
Mo	2610	900
W	3410	1200

least 5%. The original thickness of the plate is 7.5 cm. The maximum cold work in each step is 80%. Describe the cold working and annealing steps required to make this product. Compare this process with what you would recommend if you could do the initial deformation by hot working. (See Figure 8-25.)

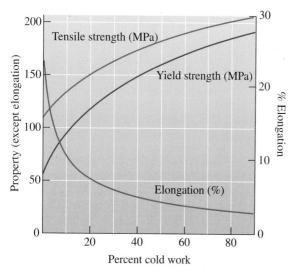

Figure 8-25   The effect of percent cold work on the properties of a 3105 aluminum alloy (for Problems 8-69.)

8-70 We wish to produce a 0.5-cm-diameter wire of copper having a minimum yield strength of 414 MPa and a minimum % elongation of 5%. The original diameter of the rod is 0.5 cm and the maximum cold work in each step is 80%. Describe a sequence of cold working and annealing steps to make this product. Compare this process with what you would recommend if you could do the initial deformation by hot working. (See Figure 8-7.)

8-71 What is a heat-affected zone? Why do some welding processes result in a joint where the material in the heat-affected zone is weaker than the base metal?

8-72 What welding techniques can be used to avoid loss of strength in the material in the heat-affected zone? Explain why these techniques are effective.

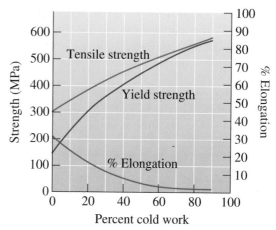

Figure 8-7   (Repeated for Problem 8-70) The effect of cold work on the mechanical properties of copper.

Section 8-9 Hot Working

8-73 The amount of plastic deformation that can be performed during hot working is almost unlimited. Justify this statement.

8-74 Compare and contrast hot working and cold working.

Design Problems

8-75 Design, using one of the processes discussed in this chapter, a method to produce each of the following products. Should the process include hot working, cold working, annealing, or some combination of these? Explain your decisions.
(a) paper clips;
(b) I-beams that will be welded to produce a portion of a bridge;
(c) copper tubing that will connect a water faucet to the main copper plumbing;
(d) the steel tape in a tape measure; and
(e) a head for a carpenter's hammer formed from a round rod.

8-76 We plan to join two sheets of cold-worked copper by soldering. Soldering involves heating the metal to a high enough temperature that a filler material melts and is drawn into the joint (Chapter 9). Design a soldering process that will not soften the copper. Explain. Could we use higher soldering temperatures if the sheet material were a Cu-30% Zn alloy? Explain.

8-77 We wish to produce a 1-mm-diameter copper wire having a minimum yield strength of 414 MPa and a minimum % elongation of 5%. We start with a 20-mm-diameter rod. Design the process by which the wire can be drawn. Include important details and explain.

 Computer Problems

8-78 *Plastic Strain Ratio.* Write a computer program that will ask the user to provide the initial and final dimensions (width and thickness) of a plate and provide the value of the plastic strain ratio.

8-79 *Design of a Wire Drawing Process.* Write a program that will effectively computerize the solution to solving Example 8-4. The program should ask the user to provide a value of the final diameter for the wire (e.g., 0.20 cm). The program should assume a reasonable value for the initial diameter (d_0) (e.g., 0.40 cm), and calculate the extent of cold work using the proper formula. Assume that the user has access to the yield strength versus % cold work curve and the user is then asked to enter the value of the yield strength for 0% cold work. Use this value to calculate the forces needed for drawing and the stress acting on the wire as it comes out of the die. The program should then ask the user to provide the value of the yield strength of the wire for the amount of cold work calculated for the assumed initial diameter and the final diameter needed. As in Example 8-4, the program should repeat these calculations until obtaining a value of d_0 that will be acceptable.

K Knovel® **Problem**

K8-1 An axially loaded compression member is cold formed from an AISI 1025 steel plate. A cross-section of the member is shown in Figure K8-1. The metal at the corners of the member was strengthened by cold work. Determine the ultimate and yield strengths of the material prior to forming; the yield strength of the material at the corners of the section after cold forming; and the average yield strength of the section, accounting for the flat portions and the corners strengthened by cold work. Assume a compact section and a reduction factor $\rho = 1.0$ (Figure K8-1).

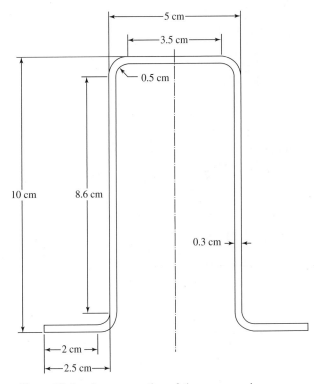

Figure K8-1   A cross-section of the compression member for Problem K8-1.

The photo on the left shows a casting process known as green sand molding (© *Peter Bowater/Alamy*). Clay-bonded sand is packed around a pattern in a two-part mold. The pattern is removed, leaving behind a cavity, and molten metal is poured into the cavity. Sand cores can produce internal cavities in the casting. After the metal solidifies, the part is shaken out from the sand. The photo on the right shows a process known as investment casting (© *Jim Powell/Alamy*). In investment casting, a pattern is made from a material that is easily melted such as wax. The pattern is then "invested" in a slurry, and a structure is built up around the pattern using particulate. The pattern is removed using heat, and molten metal is poured into the cavity. Unlike sand casting, the pattern is not reusable. Sand casting and investment casting are only two of a wide variety of casting processes.

Chapter

Principles of Solidification

Have You Ever Wondered?

- *Whether water really does "freeze" at 0°C and "boil" at 100°C?*

- *What is the process used to produce several million kilograms of steels and other alloys?*

- *Is there a specific melting temperature for an alloy or a thermoplastic material?*

- *What factors determine the strength of a cast product?*

- *Why are most glasses and glass-ceramics processed by melting and casting?*

Of all the processing techniques used in the manufacturing of materials, solidification is probably the most important. All metallic materials, as well as many ceramics, inorganic glasses, and thermoplastic polymers, are liquid or molten at some point during processing. Like water freezes to ice, molten materials solidify as they cool below their freezing temperature. In Chapter 3, we learned how materials are classified based on their atomic, ionic, or molecular order. During the solidification of materials that crystallize, the atomic arrangement changes from a short-range order (SRO) to a long-range order (LRO). The solidification of crystalline materials requires two steps. In the first step, ultra-fine crystallites, known as the nuclei of a solid phase, form from the liquid. In the second step, which can overlap with the first, the ultra-fine solid crystallites begin to grow as atoms from the liquid are attached to the nuclei until no liquid remains. Some materials, such as inorganic silicate glasses, will become solid without developing a long-range order (i.e., they remain amorphous). Many polymeric materials may develop partial crystallinity during solidification or processing.

The solidification of metallic, polymeric, and ceramic materials is an important process to study because of its effect on the properties of the materials involved. In this chapter, we will study the principles of solidification as they apply to pure metals. We will discuss solidification of alloys and more complex materials in subsequent chapters. We will first discuss the technological significance of solidification and then examine the mechanisms by which solidification occurs. This will be followed by an examination of the microstructure of cast metallic materials and its effect on the material's mechanical properties. We will also examine the role of casting as a materials shaping process. We will examine how techniques such as welding, brazing, and soldering are used for joining metals. Applications of the solidification process in single crystal growth and the solidification of glasses and polymers also will be discussed.

9-1   Technological Significance

The ability to use heat to produce, melt, and cast metals such as copper, bronze, and steel is regarded as an important hallmark in the development of mankind. The use of fire for reducing naturally occurring ores into metals and alloys led to the production of useful tools and other products. Today, thousands of years later, **solidification** is still considered one of the most important manufacturing processes. Several million kilograms of steel, aluminum alloys, copper, and zinc are being produced through the casting process. The solidification process is also used to manufacture specific components (e.g., aluminum alloys for automotive wheels). Industry also uses the solidification process as a **primary processing** step to produce metallic slabs or ingots (a simple, and often large casting that later is processed into useful shapes). The ingots or slabs are then hot and cold worked through **secondary processing** steps into more useful shapes (i.e., sheets, wires, rods, plates, etc.). Solidification also is applied when joining metallic materials using techniques such as welding, brazing, and soldering.

We also use solidification for processing inorganic glasses; silicate glass, for example, is processed using the float-glass process. High-quality optical fibers and other materials, such as fiberglass, also are produced from the solidification of molten glasses. During the solidification of inorganic glasses, amorphous rather than crystalline materials are produced. In the manufacture of glass-ceramics, we first shape the materials by casting amorphous glasses and then crystallize them using a heat treatment to enhance their strength. Many thermoplastic materials such as polyethylene, polyvinyl chloride (PVC), polypropylene, and the like are processed into useful shapes (i.e., fibers, tubes, bottles, toys, utensils, etc.) using a process that involves melting and solidification. Therefore, solidification is an extremely important technology used to control the properties of many melt-derived products as well as a tool for the manufacturing of modern engineered materials. In the sections that follow, we first discuss the nucleation and growth processes.

9-2   Nucleation

In the context of solidification, the term **nucleation** refers to the formation of the first nanocrystallites from molten material. For example, as water begins to freeze, nanocrystals, known as **nuclei**, form first. In a broader sense, the term nucleation refers to the initial stage of formation of one phase from another phase. When a vapor condenses into liquid, the nanoscale sized drops of liquid that appear when the condensation begins are referred to as nuclei. Later, we will also see that there are many systems in which the nuclei of a solid (β) will form from a second solid material (α) (i.e., α- to β-phase transformation). What is interesting about these transformations is that, in most engineered materials, many of them occur while the material is in the solid state (i.e., there is no melting involved). Therefore, although we discuss nucleation from a solidification perspective, it is important to note that the phenomenon of nucleation is general and is associated with phase transformations.

We expect a material to solidify when the liquid cools to just below its freezing (or melting) temperature, because the energy associated with the crystalline structure of the solid is then less than the energy of the liquid. This energy difference between the liquid and the solid is the free energy per unit volume ΔG_v and is the driving force for solidification.

When the solid forms, however, a solid-liquid interface is created (Figure 9-1(a)). A surface free energy σ_{sl} is associated with this interface. Thus, the total change in energy ΔG, shown in Figure 9-1(b), is

$$\Delta G = \tfrac{4}{3}\pi r^3 \Delta G_v + 4\pi r^2 \sigma_{sl} \tag{9-1}$$

where $\tfrac{4}{3}\pi r^3$ is the volume of a spherical solid of radius r, $4\pi r^2$ is the surface area of a spherical solid, σ_{sl} is the surface free energy of the solid-liquid interface (in this case), and ΔG_v is the free energy change per unit volume, which is negative since the phase transformation is assumed to be thermodynamically feasible. Note that σ_{sl} is not a strong function of r and is assumed constant. It has units of energy per unit area. ΔG_v also does not depend on r.

An **embryo** is a tiny particle of solid that forms from the liquid as atoms cluster together. The embryo is unstable and may either grow into a stable nucleus or redissolve.

In Figure 9-1(b), the top curve shows the parabolic variation of the total surface energy $(4\pi r^2 \cdot \sigma_{sl})$. The bottom most curve shows the total volume free energy change term $\left(\tfrac{4}{3}\pi r^3 \cdot \Delta G_v\right)$. The curve in the middle shows the variation of ΔG. It represents the

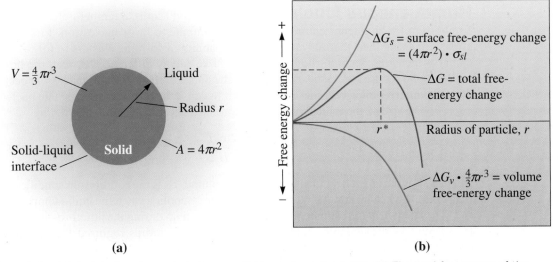

(a) **(b)**

Figure 9-1 (a) An interface is created when a solid forms from the liquid. (b) The total free energy of the solid-liquid system changes with the size of the solid. The solid is an embryo if its radius is less than the critical radius and is a nucleus if its radius is greater than the critical radius.

sum of the other two curves as given by Equation 9-1. At the temperature at which the solid and liquid phases are predicted to be in thermodynamic equilibrium (i.e., at the freezing temperature), the free energy of the solid phase and that of the liquid phase are equal ($\Delta G_v = 0$), so the total free energy change (ΔG) will be positive. When the solid is very small with a radius less than the **critical radius** for nucleation (r^*) (Figure 9-1(b)), further **growth** causes the total free energy to increase. The critical radius (r^*) is the minimum size of a crystal that must be formed by atoms clustering together in the liquid before the solid particle is stable and begins to grow.

The formation of embryos is a statistical process. Many embryos form and redissolve. If by chance, an embryo forms with a radius that is larger than r^*, further growth causes the total free energy to decrease. The new solid is then stable and sustainable since nucleation has occurred, and growth of the solid particle—which is now called a nucleus—begins. At the thermodynamic melting or freezing temperatures, the probability of forming stable, sustainable nuclei is extremely small. Therefore, solidification does not begin at the thermodynamic melting or freezing temperature. If the temperature continues to decrease below the equilibrium freezing temperature, the liquid phase that should have transformed into a solid becomes thermodynamically increasingly unstable. Because the temperature of the liquid is below the equilibrium freezing temperature, the liquid is considered undercooled. The **undercooling** (ΔT) is the difference between the equilibrium freezing temperature and the actual temperature of the liquid. As the extent of undercooling increases, the thermodynamic driving force for the formation of a solid phase from the liquid overtakes the resistance to create a solid-liquid interface.

This phenomenon can be seen in many other phase transformations. When one solid phase (α) transforms into another solid phase (β), the system has to be cooled to a temperature that is below the thermodynamic phase transformation temperature (at which the energies of the α and β phases are equal). When a liquid is transformed into a vapor (i.e., boiling water), a bubble of vapor is created in the liquid. In order to create the transformation though, we need to **superheat** the liquid above its boiling temperature! Therefore, we can see that liquids do not really freeze at their freezing temperature and do not really boil at their boiling point! We need to undercool the liquid for it to solidify and superheat it for it to boil!

Homogeneous Nucleation

As liquid cools to temperatures below the equilibrium freezing temperature, two factors combine to favor nucleation. First, since atoms are losing their thermal energy, the probability of forming clusters to form larger embryos increases. Second, the larger volume free energy difference between the liquid and the solid reduces the critical size (r^*) of the nucleus. **Homogeneous nucleation** occurs when the undercooling becomes large enough to cause the formation of a stable nucleus.

The size of the critical radius r^* for homogeneous nucleation is given by

$$r^* = \frac{2\sigma_{sl}T_m}{\Delta H_f \Delta T} \tag{9-2}$$

where ΔH_f is the latent heat of fusion per unit volume, T_m is the equilibrium solidification temperature in kelvin, and $\Delta T = (T_m - T)$ is the undercooling when the liquid temperature is T. The latent heat of fusion represents the heat given up during the liquid-to-solid transformation. As the undercooling increases, the critical radius required for nucleation decreases. Table 9-1 presents values for σ_{sl}, ΔH_f, and typical undercoolings observed experimentally for homogeneous nucleation.

The following example shows how we can calculate the critical radius of the nucleus for the solidification of copper.

TABLE 9-1 ■ *Values for freezing temperature, latent heat of fusion, surface energy, and maximum undercooling for selected materials*

Material	Freezing Temperature (T_m) (°C)	Heat of Fusion (ΔH_f) (J/cm³)	Solid-Liquid Interfacial Energy (σ_{sl}) (J/cm²)	Typical Undercooling for Homogeneous Nucleation (ΔT) (°C)
Ga	30	488	56×10^{-7}	76
Bi	271	543	54×10^{-7}	90
Pb	327	237	33×10^{-7}	80
Ag	962	965	126×10^{-7}	250
Cu	1085	1628	177×10^{-7}	236
Ni	1453	2756	255×10^{-7}	480
Fe	1538	1737	204×10^{-7}	420
NaCl	801			169
CsCl	645			152
H$_2$O	0			40

Example 9-1 *Calculation of Critical Radius for the Solidification of Copper*

Calculate the size of the critical radius and the number of atoms in the critical nucleus when solid copper forms by homogeneous nucleation. Comment on the size of the nucleus and assumptions we made while deriving the equation for the radius of the nucleus.

SOLUTION

From Table 9-1 for Cu:

$$\Delta T = 236°C \qquad T_m = 1085 + 273 = 1358 \text{ K}$$
$$\Delta H_f = 1628 \text{ J/cm}^3$$
$$\sigma_{sl} = 177 \times 10^{-7} \text{ J/cm}^2$$

Thus, r^* is given by

$$r^* = \frac{2\sigma_{sl}T_m}{\Delta H_f \Delta T} = \frac{(2)(177 \times 10^{-7})(1358)}{(1628)(236)} = 12.51 \times 10^{-8} \text{ cm}$$

Note that a temperature difference of 1°C is equal to a temperature change of 1 K, or $\Delta T = 236°C = 236$ K.

The lattice parameter for FCC copper is $a_0 = 0.3615$ nm $= 3.615 \times 10^{-8}$ cm. Thus, the unit cell volume is given by

$$V_{\text{unit cell}} = (a_0)^3 = (3.615 \times 10^{-8})^3 = 47.24 \times 10^{-24} \text{ cm}^3$$

The volume of the critical radius is given by

$$V_{r*} = \tfrac{4}{3}\pi r^3 = \left(\tfrac{4}{3}\pi\right)(12.51 \times 10^{-8})^3 = 8200 \times 10^{-24} \text{ cm}^3$$

The number of unit cells in the critical nucleus is

$$\frac{V_{\text{unit cell}}}{V_{r*}} = \frac{8200 \times 10^{-24}}{47.24 \times 10^{-24}} = 174 \text{ unit cells}$$

Since there are four atoms in each FCC unit cell, the number of atoms in the critical nucleus must be

$$(4 \text{ atoms/cell})(174 \text{ cells/nucleus}) = 696 \text{ atoms/nucleus}$$

In these types of calculations, we assume that a nucleus that is made from only a few hundred atoms still exhibits properties similar to those of bulk materials. This is not strictly correct and as such is a weakness of the classical theory of nucleation.

Heterogeneous Nucleation

From Table 9-1, we can see that water will not solidify into ice via homogeneous nucleation until we reach a temperature of −40°C (undercooling of 40°C)! Except in controlled laboratory experiments, homogeneous nucleation never occurs in liquids. Instead, impurities in contact with the liquid, either suspended in the liquid or on the walls of the container that holds the liquid, provide a surface on which the solid can form (Figure 9-2). Now, a radius of curvature greater than the critical radius is achieved with very little total surface between the solid and liquid. Relatively few atoms must cluster together to produce a solid particle that has the required radius of curvature. Much less undercooling is required to achieve the critical size, so nucleation occurs more readily. Nucleation on preexisting surfaces is known as **heterogeneous nucleation**. This process is dependent on the contact angle (θ) for the nucleating phase and the surface on which nucleation occurs. The same type of phenomenon occurs in solid-state transformations.

Rate of Nucleation

The *rate of nucleation* (the number of nuclei formed per unit time) is a function of temperature. Prior to solidification, of course, there is no nucleation and, at temperatures above the freezing point, the rate of nucleation is zero. As the temperature drops, the driving force for nucleation increases; however, as the temperature decreases, atomic diffusion becomes slower, hence slowing the nucleation process.

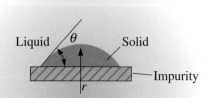

Figure 9-2
A solid forming on an impurity can assume the critical radius with a smaller increase in the surface energy. Thus, heterogeneous nucleation can occur with relatively low undercoolings.

Thus, a typical rate of nucleation reaches a maximum at some temperature below the transformation temperature. In heterogeneous nucleation, the rate of nucleation is dictated by the concentration of the nucleating agents. By considering the rates of nucleation and growth, we can predict the overall rate of a phase transformation.

9-3   Applications of Controlled Nucleation

Grain Size Strengthening   When a metal casting freezes, impurities in the melt and the walls of the mold in which solidification occurs serve as heterogeneous nucleation sites. Sometimes we intentionally introduce nucleating particles into the liquid. Such practices are called **grain refinement** or **inoculation**. Chemicals added to molten metals to promote nucleation and, hence, a finer grain size, are known as grain refiners or **inoculants**. For example, a combination of 0.03% titanium (Ti) and 0.01% boron (B) is added to many liquid-aluminum alloys. Tiny particles of an aluminum titanium compound (Al_3Ti) or titanium diboride (TiB_2) form and serve as sites for heterogeneous nucleation. Grain refinement or inoculation produces a large number of grains, each beginning to grow from one nucleus. The greater grain boundary area provides grain size strengthening in metallic materials. This was discussed using the Hall-Petch equation in Chapter 4.

Second-Phase Strengthening   In Chapters 4 and 5, we learned that in metallic materials, dislocation motion can be resisted by grain boundaries or the formation of ultra-fine precipitates of a second phase. Strengthening materials using ultra-fine precipitates is known as **dispersion strengthening** or **second-phase strengthening**; it is used extensively in enhancing the mechanical properties of many alloys. This process involves **solid-state phase transformations** (i.e., one solid transforming into another). The grain boundaries as well as atomic level defects within the grains of the parent phase (α) often serve as nucleation sites for heterogeneous nucleation of the new phase (β). This nucleation phenomenon plays a critical role in strengthening mechanisms. This will be discussed in Chapters 10 and 11.

Glasses   For rapid cooling rates and/or high viscosity melts, there may be insufficient time for nuclei to form and grow. When this happens, the liquid structure is locked into place and an amorphous—or glassy—solid forms. The complex crystal structure of many ceramic and polymeric materials prevents nucleation of a solid crystalline structure even at slow cooling rates. Some alloys with special compositions have sufficiently complex crystal structures, so they may form amorphous materials if cooled rapidly from the melt. These materials are known as metallic glasses. Typically, good metallic glass formers are multi-component alloys, often with large differences in the atomic sizes of the elemental constituents. This complexity limits the solid solubilities of the elements in the crystalline phases, thus requiring large chemical fluctuations to form the critical-sized crystalline nuclei. Metallic glasses were initially produced via **rapid solidification processing** in which cooling rates of $10^{6}°C^{-1}$ were attained by forming continuous, thin metallic ribbons about 0.0038 cm thick. (Heat can be extracted quickly from ribbons with a large surface area to volume ratio.)

Bulk metallic glasses with diameters greater than 2.5 cm are now produced using a variety of processing techniques for compositions that require cooling rates on the order of only tens of degrees per second. Many bulk metallic glass compositions have been discovered, including $Pd_{40}Ni_{40}P_{20}$ and $Zr_{41.2}Ti_{13.8}Cu_{12.5}Ni_{10.0}Be_{22.5}$. Many metallic glasses have strengths in excess of 1700 MPa while retaining fracture toughnesses of more than $10 \text{ MPa}\sqrt{m}$. Excellent corrosion resistance, magnetic properties, and other physical properties make these materials attractive for a wide variety of applications.

Other examples of materials that make use of controlled nucleation are colored glass and **photochromic glass** (glass that can change color or tint upon exposure to sunlight). In these otherwise amorphous materials, nanocrystallites of different materials are deliberately nucleated. The crystals are small and, hence, do not make the glass opaque. They do have special optical properties that make the glass brightly colored or photochromic.

Many materials formed from a vapor phase can be cooled quickly so that they do not crystallize and, therefore, are amorphous (i.e., amorphous silicon), illustrating that amorphous or non-crystalline materials do *not* always have to be formed from melts.

Glass-ceramics The term **glass-ceramics** refers to engineered materials that begin as amorphous glasses and end up as crystalline ceramics with an ultra-fine grain size. These materials are then nearly free from porosity, mechanically stronger, and often much more resistant to thermal shock. Nucleation does not occur easily in silicate glasses; however, we can help by introducing nucleating agents such as titania (TiO_2) and zirconia (ZrO_2). Engineered glass-ceramics take advantage of the ease with which glasses can be melted and formed. Once a glass is formed, we can heat it to deliberately form ultra-fine crystals, obtaining a material that has considerable mechanical toughness and thermal shock resistance. The crystallization of glass-ceramics continues until all of the material crystallizes (up to 99.9% crystallinity can be obtained). If the grain size is kept small ($\sim$ 50–100 nm), glass-ceramics can often be made transparent. All glasses eventually will crystallize as a result of exposure to high temperatures for long lengths of times. In order to produce a glass-ceramic, however, the crystallization must be carefully controlled.

9-4 Growth Mechanisms

Once the solid nuclei of a phase form (in a liquid or another solid phase), growth begins to occur as more atoms become attached to the solid surface. In this discussion, we will concentrate on the nucleation and growth of crystals from a liquid. The nature of the growth of the solid nuclei depends on how heat is removed from the molten material. Let's consider casting a molten metal in a mold, for example. We assume we have a nearly pure metal and not an alloy (as solidification of alloys is different in that in most cases, it occurs over a range of temperatures). In the solidification process, two types of heat must be removed: the specific heat of the liquid and the latent heat of fusion. The **specific heat** is the heat required to change the temperature of a unit weight of the material by one degree. The specific heat must be removed first, either by radiation into the surrounding atmosphere or by conduction into the surrounding mold, until the liquid cools to its freezing temperature. This is simply a cooling of the liquid from one temperature to a temperature at which nucleation begins.

We know that to melt a solid we need to supply heat. Therefore, when solid crystals form from a liquid, heat is generated! This type of heat is called the **latent heat of fusion** (ΔH_f). The latent heat of fusion must be removed from the solid-liquid interface before solidification is completed. The manner in which we remove the latent heat of fusion determines the material's growth mechanism and final structure of a casting.

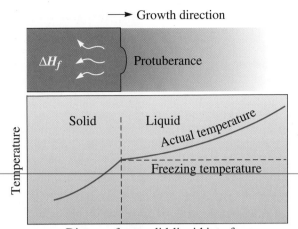

→ Growth direction

ΔH_f

Protuberance

Solid

Liquid

Actual temperature

Temperature

Freezing temperature

Distance from solid-liquid interface

Figure 9-3
When the temperature of the liquid is above the freezing temperature, a protuberance on the solid-liquid interface will not grow, leading to maintenance of a planar interface. Latent heat is removed from the interface through the solid.

Planar Growth

When a well-inoculated liquid (i.e., a liquid containing nucleating agents) cools under equilibrium conditions, there is no need for undercooling since heterogeneous nucleation can occur. Therefore, the temperature of the liquid ahead of the **solidification front** (i.e., solid-liquid interface) is greater than the freezing temperature. The temperature of the solid is at or below the freezing temperature. During solidification, the latent heat of fusion is removed by conduction from the solid-liquid interface. Any small protuberance that begins to grow on the interface is surrounded by liquid above the freezing temperature (Figure 9-3). The growth of the protuberance then stops until the remainder of the interface catches up. This growth mechanism, known as **planar growth**, occurs by the movement of a smooth solid-liquid interface into the liquid.

Dendritic Growth

When the liquid is not inoculated and the nucleation is poor, the liquid has to be undercooled before the solid forms (Figure 9-4). Under these conditions, a small solid protuberance called a **dendrite**, which forms at the interface, is encouraged to grow since the liquid ahead of the solidification front is undercooled. The word dendrite comes from the Greek word *dendron* that means tree. As the solid dendrite grows, the latent heat of fusion is conducted into the undercooled liquid, raising the temperature of the liquid toward the freezing temperature. Secondary and tertiary dendrite arms can also form on the primary stalks to speed the evolution of the latent heat. Dendritic growth continues until the undercooled liquid warms to the freezing temperature. Any remaining liquid then solidifies by planar growth. The difference between planar and dendritic growth arises because of the different sinks for the latent heat of fusion. The container or mold must absorb the heat in planar growth, but the undercooled liquid absorbs the heat in dendritic growth.

In pure metals, dendritic growth normally represents only a small fraction of the total growth and is given by

$$\text{Dendritic fraction} = f = \frac{c\Delta T}{\Delta H_f} \tag{9-3}$$

where c is the specific heat of the liquid. The numerator represents the heat that the undercooled liquid can absorb, and the latent heat in the denominator represents the total heat that must be given up during solidification. As the undercooling ΔT increases,

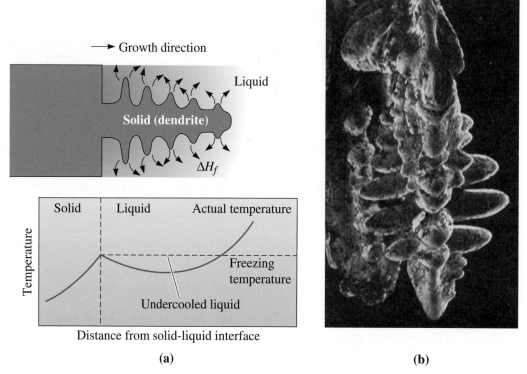

Figure 9-4   (a) If the liquid is undercooled, a protuberance on the solid-liquid interface can grow rapidly as a dendrite. The latent heat of fusion is removed by raising the temperature of the liquid back to the freezing temperature. (b) Scanning electron micrograph of dendrites in steel (× 15). (*Reprinted courtesy of Don Askeland.*)

more dendritic growth occurs. If the liquid is well-inoculated, undercooling is almost zero and growth would be mainly via the planar front solidification mechanism.

9-5   Solidification Time and Dendrite Size

The rate at which growth of the solid occurs depends on the cooling rate, or the rate of heat extraction. A higher cooling rate produces rapid solidification, or short solidification times. The time t_s required for a simple casting to solidify completely can be calculated using *Chvorinov's rule*:

$$t_s = B\left(\frac{V}{A}\right)^n \tag{9-4}$$

where V is the volume of the casting and represents the amount of heat that must be removed before freezing occurs, A is the surface area of the casting in contact with the mold and represents the surface from which heat can be transferred away from the casting, n is a constant (usually about 2), and B is the **mold constant**. The mold constant depends on the properties and initial temperatures of both the metal and the mold. This rule basically accounts for the geometry of a casting and the heat transfer conditions. The rule states that, for the same conditions, a casting with a small volume and relatively large surface area will cool more rapidly.

Example 9-2   *Redesign of a Casting for Improved Strength*

Your company currently is producing a disk-shaped brass casting 5 cm thick and 45 cm in diameter. You believe that by making the casting solidify 25% faster the improvement in the tensile properties of the casting will permit the casting to be made lighter in weight. Design the casting to permit this. Assume that the mold constant is 3.5 min/cm^2 for this process and $n = 2$.

SOLUTION

One approach would be to use the same casting process, but reduce the thickness of the casting. The thinner casting would solidify more quickly and, because of the faster cooling, should have improved mechanical properties. Chvorinov's rule helps us calculate the required thickness. If d is the diameter and x is the thickness of the casting, then the volume, surface area, and solidification time of the 5 cm thick casting are

$$V = (\pi/4)d^2x = (\pi/4)(45)^2(2) = 7955.4 \text{ cm}^3$$

$$A = 2(\pi/4)d^2 + \pi dx = 2(\pi/4)(45)^2 + \pi(45)(5) = 3889.2 \text{ cm}^2$$

$$t = \text{B}\left(\frac{V}{A}\right)^2 = (3.5)\left(\frac{7955.4}{3889.2}\right)^2 = 14.71 \text{ min}$$

The solidification time t_r of the redesigned casting should be 25% shorter than the current time:

$$t_r = 0.75t = (0.75)(14.71) = 11.03 \text{ min}$$

Since the casting conditions have not changed, the mold constant B is unchanged. The V/A ratio of the new casting is

$$t_r = \text{B}\left(\frac{V_r}{A_r}\right)^2 = (3.5)\left(\frac{V_r}{A_r}\right)^2 = 11.03 \text{ min}$$

$$\left(\frac{V_r}{A_r}\right)^2 = 3.1514 \text{ cm}^2 \quad \text{or} \quad \frac{V_r}{A_r} = 1.775 \text{ cm}$$

If x is the required thickness for our redesigned casting, then

$$\frac{V_r}{A_r} = \frac{(\pi/4)d^2x}{2(\pi/4)d^2 + \pi dx} = \frac{(\pi/4)(45)^2(x)}{2(\pi/4)(45)^2 + \pi(45)(x)} = 1.775 \text{ cm}$$

Therefore, $x = 4.22$ cm

This thickness provides the required solidification time, while reducing the overall weight of the casting by more than 15%.

Solidification begins at the surface, where heat is dissipated into the surrounding mold material. The rate of solidification of a casting can be described by how rapidly the thickness d of the solidified skin grows:

$$d = \text{k}_{\text{solidification}}\sqrt{t} - \text{c}_1 \tag{9-5}$$

where t is the time after pouring, $\text{k}_{\text{solidification}}$ is a constant for a given casting material and mold, and c_1 is a constant related to the pouring temperature.

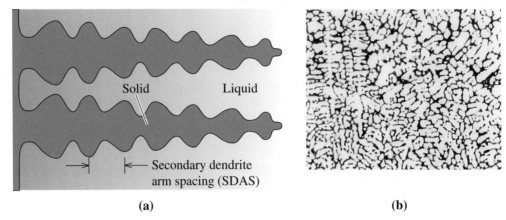

(a)   **(b)**

Figure 9-5   (a) The secondary dendrite arm spacing (SDAS). (b) Dendrites in an aluminum alloy (× 50). *(From* ASM Handbook, *Vol. 9, Metallography and Microstructure (1985), ASM International, Materials Park, OH 44073-0002.)*

Effect on Structure and Properties
The solidification time affects the size of the dendrites. Normally, dendrite size is characterized by measuring the distance between the secondary dendrite arms (Figure 9-5). The **secondary dendrite arm spacing** (SDAS) is reduced when the casting freezes more rapidly. The finer, more extensive dendritic network serves as a more efficient conductor of the latent heat to the undercooled liquid. The SDAS is related to the solidification time by

$$SDAS = kt_s^m \tag{9-6}$$

where m and k are constants depending on the composition of the metal. This relationship is shown in Figure 9-6 for several alloys. Small secondary dendrite arm spacings are associated with higher strengths and improved ductility (Figure 9-7).

Rapid solidification processing is used to produce exceptionally fine secondary dendrite arm spacings; a common method is to produce very fine liquid droplets that freeze into solid particles. This process is known as spray atomization. The tiny droplets freeze at a rate of about 10^4°C/s, producing powder particles that range from ~5–100 μm. This cooling rate is not rapid enough to form a metallic glass, but does produce a fine dendritic structure. By carefully consolidating the solid droplets by powder metallurgy processes, improved properties in the material can be obtained. Since the particles are

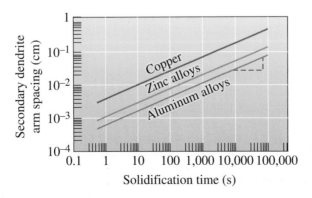

Figure 9-6
The effect of solidification time on the secondary dendrite arm spacings of copper, zinc, and aluminum.

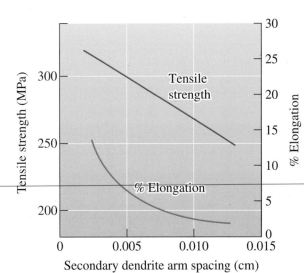

Figure 9-7
The effect of the secondary dendrite arm spacing on the mechanical properties of an aluminum casting alloy.

derived from a melt, many complex alloy compositions can be produced in the form of chemically homogenous powders.

The following three examples discuss how Chvorinov's rule, the relationship between SDAS and the time of solidification, and the SDAS and mechanical properties can be used to design casting processes.

Example 9-3	*Secondary Dendrite Arm Spacing for Aluminum Alloys*

Determine the constants in the equation that describe the relationship between secondary dendrite arm spacing and solidification time for aluminum alloys (Figure 9-6).

SOLUTION

We could obtain the value of SDAS at two times from the graph and calculate k and m using simultaneous equations; however, if the scales on the ordinate and abscissa are equal for powers of ten (as in Figure 9-6), we can obtain the slope m from the log-log plot by directly measuring the slope of the graph. In Figure 9-6, we can mark five equal units on the vertical scale and 12 equal units on the horizontal scale. The slope is

$$m = \frac{5}{12} = 0.42$$

The constant k is the value of SDAS when $t_s = 1$ s, since

$$\log \text{SDAS} = \log k + m \log t_s$$

If $t_s = 1$ s, m log $t_s = 0$, and SDAS = k, from Figure 9-6:

$$k = 7 \times 10^{-4} \frac{\text{cm}}{\text{s}}$$

Example 9-4   *Time of Solidification*

A 10-cm-diameter aluminum bar solidifies to a depth of 1.25 cm beneath the surface in 5 minutes. After 20 minutes, the bar has solidified to a depth of 3.75 cm. How much time is required for the bar to solidify completely?

SOLUTION

From our measurements, we can determine the constants $k_{solidification}$ and c_1 in Equation 9-5:

$$1.25 \text{ cm} = k_{solidification}\sqrt{(5 \text{ min})} - c_1 \quad \text{or} \quad c_1 = k\sqrt{5} - 1.25$$

$$3.75 \text{ cm} = k_{solidification}\sqrt{(20 \text{ min})} - c_1 = k\sqrt{20} - (k\sqrt{5} - 1.25)$$

$$3.75 = k_{solidification}(\sqrt{20} - \sqrt{5}) + 1.25$$

$$k_{solidification} = \frac{3.75 - 1.25}{4.472 - 2.236} = 1.118\frac{\text{cm}}{\sqrt{\text{min}}}$$

$$c_1 = (1.118)\sqrt{5} - 1.25 = 1.25 \text{ cm}$$

Solidification is complete when $d = 5$ cm (half the diameter, since freezing is occurring from all surfaces):

$$5 = 1.118\sqrt{t} - 1.25$$

$$\sqrt{t} = \frac{5 + 1.25}{1.118} = 5.59$$

$$t = 31.25 \text{ min}$$

In actual practice, we would find that the total solidification time is somewhat longer than 31.25 min. As solidification continues, the mold becomes hotter and is less effective in removing heat from the casting.

Example 9-5   *Design of an Aluminum Alloy Casting*

Design the thickness of an aluminum alloy casting with a length of 30 cm a width of 20 cm and a tensile strength of 280 MPa. The mold constant in Chvorinov's rule for aluminum alloys cast in a sand mold is 7.2 min/cm². Assume that data shown in Figures 9-6 and 9-7 can be used.

SOLUTION

In order to obtain a tensile strength of 280 MPa, a secondary dendrite arm spacing of about 0.007 cm is required (see Figure 9-7). From Figure 9-6 we can determine that the solidification time required to obtain this spacing is about 300 s or 5 minutes.

From Chvorinov's rule

$$t_s = B\left(\frac{V}{A}\right)^2$$

where $B = 7.2 \text{ min/cm}^2$ and x is the thickness of the casting. Since the length is 30 cm and the width is 20 cm.

$$V = (20)(30)(x) = 600x$$

$$A = (2)(20)(30) + (2)(x)(20) + (2)(x)(30) = 100x + 1200$$

$$5 \text{ min} = (7.2 \text{ min/cm}^2)\left(\frac{600x}{100x + 1200}\right)^2$$

$$\frac{600x}{100x + 1200} = \sqrt{(5/7.2)} = 0.83$$

$$600x = 13.83x + 996$$

$$x = 1.93 \text{ cm}$$

9-6 Cooling Curves

We can summarize our discussion at this point by examining cooling curves. A cooling curve shows how the temperature of a material (in this case, a pure metal) decreases with time [Figure 9-8 (a) and (b)]. The liquid is poured into a mold at the pouring temperature, point A. The difference between the pouring temperature and the freezing temperature is the superheat. The specific heat is extracted by the mold until the liquid reaches the freezing temperature (point B). If the liquid is not well-inoculated, it must be undercooled

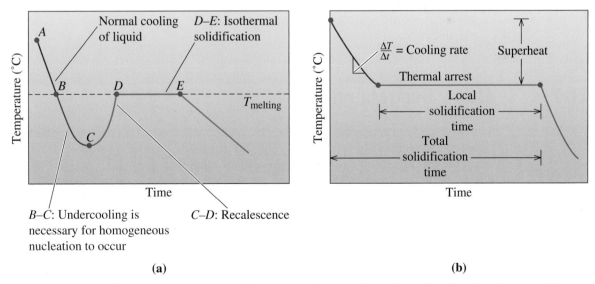

(a)   **(b)**

Figure 9-8   (a) Cooling curve for a pure metal that has not been well-inoculated. The liquid cools as specific heat is removed (between points A and B). Undercooling is thus necessary (between points B and C). As the nucleation begins (point C), latent heat of fusion is released causing an increase in the temperature of the liquid. This process is known as recalescence (point C to point D). The metal continues to solidify at a constant temperature (T_{melting}). At point E, solidification is complete. The solid casting continues to cool from this point. (b) Cooling curve for a well-inoculated, but otherwise pure, metal. No undercooling is needed. Recalescence is not observed. Solidification begins at the melting temperature.

(point B to C). The slope of the cooling curve before solidification begins is the cooling rate $\frac{\Delta T}{\Delta t}$. As nucleation begins (point C), latent heat of fusion is given off, and the temperature rises. This increase in temperature of the undercooled liquid as a result of nucleation is known as **recalescence** (point C to D). Solidification proceeds isothermally at the melting temperature (point D to E) as the latent heat given off from continued solidification is balanced by the heat lost by cooling. This region between points D and E, where the temperature is constant, is known as the **thermal arrest**. A thermal arrest, or plateau, is produced because the evolution of the latent heat of fusion balances the heat being lost because of cooling. At point E, solidification is complete, and the solid casting cools from point E to room temperature.

If the liquid is well-inoculated, the extent of undercooling and recalescence is usually very small and can be observed in cooling curves only by very careful measurements. If effective heterogeneous nuclei are present in the liquid, solidification begins at the freezing temperature [Figure 9-8 (b)]. The latent heat keeps the remaining liquid at the freezing temperature until all of the liquid has solidified and no more heat can be evolved. Growth under these conditions is planar. The **total solidification time** of the casting is the time required to remove both the specific heat of the liquid and the latent heat of fusion. Measured from the time of pouring until solidification is complete, this time is given by Chvorinov's rule. The **local solidification time** is the time required to remove only the latent heat of fusion at a particular location in the casting; it is measured from when solidification begins until solidification is completed. The local solidification times (and the total solidification times) for liquids solidified via undercooled and inoculated liquids will be slightly different.

We often use the terms "melting temperature" and "freezing temperature" while discussing solidification. It would be more accurate to use the term "melting temperature" to describe when a solid turns completely into a liquid. For pure metals and compounds, this happens at a fixed temperature (assuming fixed pressure) and without superheating. "Freezing temperature" or "freezing point" can be defined as the temperature at which solidification of a material is complete.

9-7 Cast Structure

In manufacturing components by casting, molten metals are often poured into molds and permitted to solidify. The mold produces a finished shape, known as a *casting*. In other cases, the mold produces a simple shape called an **ingot**. An ingot usually requires extensive plastic deformation before a finished product is created. A *macrostructure* sometimes referred to as the **ingot structure**, consists of as many as three regions (Figure 9-9). (Recall that in Chapter 2 we used the term "macrostructure" to describe the structure of a material at a macroscopic scale. Hence, the term "ingot structure" may be more appropriate.)

Chill Zone   The **chill zone** is a narrow band of randomly oriented grains at the surface of the casting. The metal at the mold wall is the first to cool to the freezing temperature. The mold wall also provides many surfaces at which heterogeneous nucleation takes place.

Columnar Zone   The **columnar zone** contains elongated grains oriented in a particular crystallographic direction. As heat is removed from the casting by the mold material, the grains in the chill zone grow in the direction opposite to that of the heat

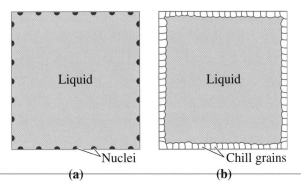

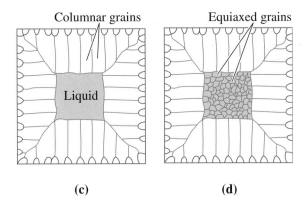

Figure 9-9
Development of the ingot structure of a casting during solidification: (a) nucleation begins, (b) the chill zone forms, (c) preferred growth produces the columnar zone, and (d) additional nucleation creates the equiaxed zone.

flow, or from the coldest toward the hottest areas of the casting. This tendency usually means that the grains grow perpendicular to the mold wall.

Grains grow fastest in certain crystallographic directions. In metals with a cubic crystal structure, grains in the chill zone that have a $\langle 100 \rangle$ direction perpendicular to the mold wall grow faster than other less favorably oriented grains (Figure 9-10). Eventually, the grains in the columnar zone have $\langle 100 \rangle$ directions that are parallel to one another, giving the columnar zone anisotropic properties. This formation of the columnar zone is

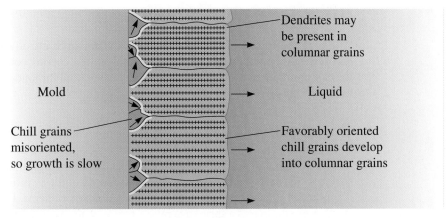

Figure 9-10   Competitive growth of the grains in the chill zone results in only those grains with favorable orientations developing into columnar grains.

influenced primarily by growth—rather than nucleation—phenomena. The grains may be composed of many dendrites if the liquid is originally undercooled. The solidification may proceed by planar growth of the columnar grains if no undercooling occurs.

Equiaxed Zone   Although the solid may continue to grow in a columnar manner until all of the liquid has solidified, an equiaxed zone frequently forms in the center of the casting or ingot. The **equiaxed zone** contains new, randomly oriented grains, often caused by a low pouring temperature, alloying elements, or grain refining or inoculating agents. Small grains or dendrites in the chill zone may also be torn off by strong convection currents that are set up as the casting begins to freeze. These also provide heterogeneous nucleation sites for what ultimately become equiaxed grains. These grains grow as relatively round, or equiaxed, grains with a random orientation, and they stop the growth of the columnar grains. The formation of the equiaxed zone is a nucleation-controlled process and causes that portion of the casting to display isotropic behavior.

By understanding the factors that influence solidification in different regions, it is possible to produce castings that first form a "skin" of a chill zone and then dendrites. Metals and alloys that show this macrostructure are known as **skin-forming alloys**. We also can control the solidification such that no skin or advancing dendritic arrays of grains are seen; columnar to equiaxed switchover is almost at the mold walls. The result is a casting with a macrostructure consisting predominantly of equiaxed grains. Metals and alloys that solidify in this fashion are known as **mushy-forming alloys** since the cast material seems like a mush of solid grains floating in a liquid melt. Many aluminum and magnesium alloys show this type of solidification. Often, we encourage an all-equiaxed structure and thus create a casting with isotropic properties by effective grain refinement or inoculation. In a later section, we will examine one case (turbine blades) where we control solidification to encourage all columnar grains and hence anisotropic behavior.

Cast ingot structure and microstructure are important particularly for components that are directly cast into a final shape. In many situations though, as discussed in Section 9-1, metals and alloys are first cast into ingots, and the ingots are subsequently subjected to thermomechanical processing (e.g., rolling, forging etc.). During these steps, the cast macrostructure is broken down and a new microstructure will emerge, depending upon the thermomechanical process used (Chapter 8).

9-8   Solidification Defects

Although there are many defects that potentially can be introduced during solidification, shrinkage and porosity deserve special mention. If a casting contains pores (small holes), the cast component can fail catastrophically when used for load-bearing applications (e.g., turbine blades).

Shrinkage   Almost all materials are more dense in the solid state than in the liquid state. During solidification, the material contracts, or shrinks, as much as 7% (Table 9-2).

Often, the bulk of the **shrinkage** occurs as **cavities**, if solidification begins at all surfaces of the casting, or *pipes*, if one surface solidifies more slowly than the others (Figure 9-11). The presence of such pipes can pose problems. For example, if in the production of zinc ingots a shrinkage pipe remains, water vapor can condense in it. This water can lead to an explosion if the ingot gets introduced in a furnace in which zinc is being remelted for such applications as hot-dip galvanizing.

TABLE 9-2 ■ *Shrinkage during solidification for selected materials*

Material	Shrinkage (%)
Al	7.0
Cu	5.1
Mg	4.0
Zn	3.7
Fe	3.4
Pb	2.7
Ga	+3.2 (expansion)
H_2O	+8.3 (expansion)
Low-carbon steel	2.5–3.0
High-carbon steel	4.0
White Cast Iron	4.0–5.5
Gray Cast Iron	+1.9 (expansion)

Note: Some data from DeGarmo, E. P., Black, J. T., and Koshe, R. A. Materials and Processes in Manufacturing, Prentice Hall, 1997.

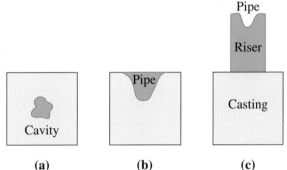

Figure 9-11
Several types of macroshrinkage can occur, including cavities and pipes. Risers can be used to help compensate for shrinkage.

A common technique for controlling **cavity** and **pipe shrinkage** is to place a **riser**, or an extra reservoir of metal, adjacent and connected to the casting. As the casting solidifies and shrinks, liquid metal flows from the riser into the casting to fill the shrinkage void. We need only to ensure that the riser solidifies after the casting and that there is an internal liquid channel that connects the liquid in the riser to the last liquid to solidify in the casting. Chvorinov's rule can be used to help design the size of the riser. The following example illustrates how risers can be designed to compensate for shrinkage.

Example 9-6   *Design of a Riser for a Casting*

Design a cylindrical riser, with a height equal to twice its diameter, that will compensate for shrinkage in a 2 cm × 8 cm × 16 cm, casting (Figure 9-12).

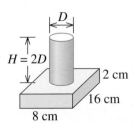

Figure 9-12
The geometry of the casting and riser (for Example 9-6).

SOLUTION

We know that the riser must freeze after the casting. To be conservative, we typically require that the riser take 25% longer to solidify than the casting. Therefore,

$$t_r = 1.25 t_c \text{ or } B\left(\frac{V}{A}\right)_r^2 = 1.25 \, B\left(\frac{V}{A}\right)_c^2$$

The subscripts r and c stand for riser and casting, respectively. The mold constant B is the same for both casting and riser, so

$$\left(\frac{V}{A}\right)_r = \sqrt{1.25}\left(\frac{V}{A}\right)_c$$

The volume of the casting is

$$V_c = (2 \text{ cm})(8 \text{ cm})(16 \text{ cm}) = 256 \text{ cm}^3$$

The area of the riser adjoined to the casting must be subtracted from the total surface area of the casting in order to calculate the surface area of the casting in contact with the mold:

$$A_c = (2)(2 \text{ cm})(8 \text{ cm}) + (2)(2 \text{ cm})(16 \text{ cm}) + (2)(8 \text{ cm})(16 \text{ cm}) - \frac{\pi D^2}{4} = 352 \text{ cm}^2 - \frac{\pi D^2}{4}$$

where D is the diameter of the cylindrical riser. We can write equations for the volume and area of the cylindrical riser, noting that the cylinder height $H = 2D$:

$$V_r = \frac{\pi D^2}{4} H = \frac{\pi D^2}{4}(2D) = \frac{\pi D^3}{2}$$

$$A_r = \frac{\pi D^2}{4} + \pi D H = \frac{\pi D^2}{4} + \pi D(2D) = \frac{9}{4}\pi D^2$$

where again we have not included the area of the riser adjoined to the casting in the area calculation. The volume to area ratio of the riser is given by

$$\left(\frac{V}{A}\right)_r = \frac{(\pi D^3/2)}{(9\pi D^2/4)} = \frac{2}{9}D$$

and must be greater than that of the casting according to

$$\left(\frac{V}{A}\right)_r = \frac{2}{9}D > \sqrt{1.25}\left(\frac{V}{A}\right)_c$$

Substituting,

$$\frac{2}{9}D > \sqrt{1.25}\left(\frac{256 \text{ cm}^3}{352 \text{ cm}^2 - \pi D^2/4}\right)$$

Solving for the smallest diameter for the riser:

$$D = 3.78 \text{ cm}$$

Although the volume of the riser is less than that of the casting, the riser solidifies more slowly because of its compact shape.

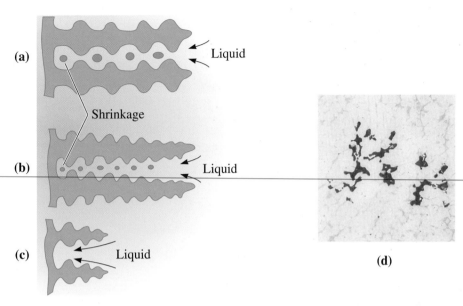

Figure 9-13   (a) Shrinkage can occur between the dendrite arms. (b) Small secondary dendrite arm spacings result in smaller, more evenly distributed shrinkage porosity. (c) Short primary arms can help avoid shrinkage. (d) Interdendritic shrinkage in an aluminum alloy is shown ($\times 80$). (*Reprinted courtesy of Don Askeland.*)

Interdendritic Shrinkage

This consists of small shrinkage pores between dendrites (Figure 9-13). This defect, also called **microshrinkage** or **shrinkage porosity**, is difficult to prevent by the use of risers. Fast cooling rates may reduce problems with **interdendritic shrinkage**; the dendrites may be shorter, permitting liquid to flow through the dendritic network to the solidifying solid interface. In addition, any shrinkage that remains may be finer and more uniformly distributed.

Gas Porosity

Many metals dissolve a large quantity of gas when they are molten. Aluminum, for example, dissolves hydrogen. When the aluminum solidifies, however, the solid metal retains in its crystal structure only a small fraction of the hydrogen since the solubility of the solid is remarkably lower than that of the liquid (Figure 9-14). The excess hydrogen that cannot be incorporated in the solid metal or alloy crystal structure forms bubbles that may be trapped in the solid metal, producing **gas porosity**. The amount of gas that can be dissolved in molten metal is given by **Sievert's law**:

$$\text{Percent of gas} = K\sqrt{p_{\text{gas}}} \tag{9-7}$$

where p_{gas} is the partial pressure of the gas in contact with the metal and K is a constant which, for a particular metal-gas system, increases with increasing temperature. We can minimize gas porosity in castings by keeping the liquid temperature low, by adding materials to the liquid to combine with the gas and form a solid, or by ensuring that the partial pressure of the gas remains low. The latter may be achieved by placing the molten metal in a vacuum chamber or bubbling an inert gas through the metal. Because p_{gas} is low in the vacuum, the gas leaves the metal, enters the vacuum, and is

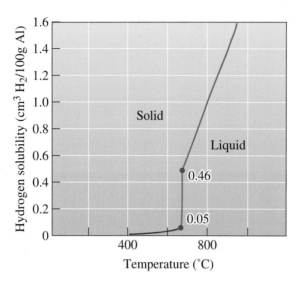

Figure 9-14
The solubility of hydrogen gas in aluminum when the partial pressure of $H_2 = 1$ atm.

carried away. **Gas flushing** is a process in which bubbles of a gas, inert or reactive, are injected into a molten metal to remove undesirable elements from molten metals and alloys. For example, hydrogen in aluminum can be removed using nitrogen or chlorine. The following example illustrates how a degassing process can be designed.

Example 9-7   *Design of a Degassing Process for Copper*

After melting at atmospheric pressure, molten copper contains 0.01 weight percent oxygen. To ensure that your castings will not be subject to gas porosity, you want to reduce the weight percent to less than 0.00001% prior to pouring. Design a degassing process for the copper.

SOLUTION

We can solve this problem in several ways. In one approach, the liquid copper is placed in a vacuum chamber; the oxygen is then drawn from the liquid and carried away into the vacuum. The vacuum required can be estimated from Sievert's law:

$$\frac{\% \, O_{initial}}{\% \, O_{vacuum}} = \frac{K\sqrt{p_{\,initial}}}{K\sqrt{p_{\,vacuum}}} = \sqrt{\left(\frac{1 \text{ atm}}{p_{vacuum}}\right)}$$

$$\frac{0.01\%}{0.00001\%} = \sqrt{\left(\frac{1}{p_{vacuum}}\right)}$$

$$\frac{1 \text{ atm}}{p_{vacuum}} = (1000)^2 \quad \text{or} \quad p_{vacuum} = 10^{-6} \text{ atm}$$

Another approach would be to introduce a copper-15% phosphorous alloy. The phosphorous reacts with oxygen to produce P_2O_5, which floats out of the liquid, by the reaction:

$$5O + 2P \rightarrow P_2O_5$$

Typically, about 0.01 to 0.02% P must be added remove the oxygen.

In the manufacturing of **stainless steel**, a process known as **argon oxygen decarburization** (AOD) is used to lower the carbon content of the melt without oxidizing chromium or nickel. In this process, a mixture of argon (or nitrogen) and oxygen gases is forced into molten stainless steel. The carbon dissolved in the molten steel is oxidized by the oxygen gas via the formation of carbon monoxide (CO) gas; the CO is carried away by the inert argon (or nitrogen) gas bubbles. These processes need very careful control since some reactions (e.g., oxidation of carbon to CO) are exothermic (generate heat).

9-9   Casting Processes for Manufacturing Components

Figure 9-15 summarizes four of the dozens of commercial casting processes. In some processes, the molds can be reused; in others, the mold is expendable. **Sand casting** processes include green sand molding, for which silica (SiO_2) sand grains bonded with wet clay are packed around a removable pattern. Ceramic casting processes use a fine-grained ceramic material as the mold, as slurry containing the ceramic may be poured around a reusable pattern, which is removed after the ceramic hardens. In **investment casting**, the ceramic slurry of a material such as colloidal silica (consisting of ceramic nanoparticles) coats a wax pattern. After the ceramic hardens (i.e., the colloidal silica dispersion gels), the wax is melted and drained from the ceramic shell, leaving behind a cavity that is then filled with molten metal. After the metal solidifies, the mold is broken to remove the part. The investment casting process, also known as the **lost wax process**, is suited for generating complex shapes. Dentists and jewelers originally used the precision investment casting process. Currently, this process is used to produce such components as turbine blades, titanium heads of golf clubs, and parts for knee and hip prostheses. In another process known as the **lost foam process**, polystyrene beads, similar to those used to make coffee cups or packaging materials, are used to produce a foam pattern. Loose sand is compacted around the pattern to produce a mold. When molten metal is poured into the mold, the polymer foam pattern melts and decomposes, with the metal taking the place of the pattern.

In the permanent mold and pressure die casting processes, a cavity is machined from metallic material. After the liquid poured into the cavity solidifies, the mold is opened, the casting is removed, and the mold is reused. The processes using metallic molds tend to give the highest strength castings because of the rapid solidification. Ceramic

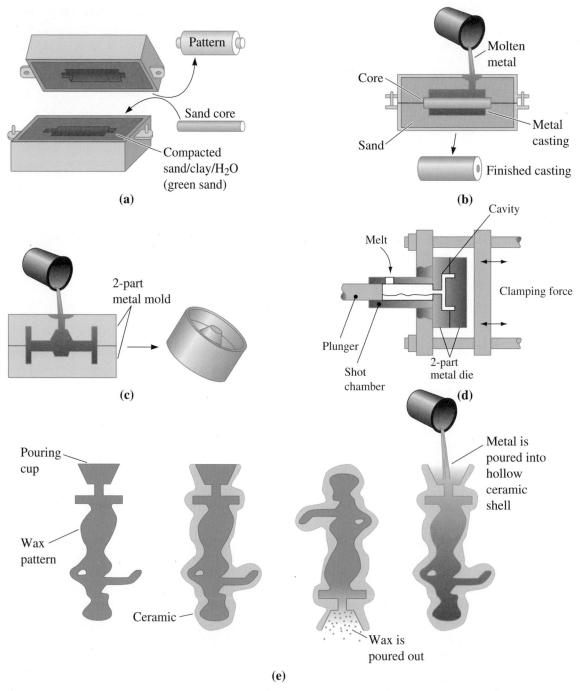

Figure 9-15 Four typical casting processes: (a) and (b) Green sand molding, in which clay-bonded sand is packed around a pattern. Sand cores can produce internal cavities in the casting. (c) The permanent mold process, in which metal is poured into an iron or steel mold. (d) Die casting, in which metal is injected at high pressure into a steel die. (e) Investment casting, in which a wax pattern is surrounded by a ceramic; after the wax is melted and drained, metal is poured into the mold.

molds, including those used in investment casting, are good insulators and give the slow-est-cooling and lowest-strength castings. Millions of truck and car pistons are made in foundries using permanent mold casting. Good surface finish and dimensional accuracy are the advantages of **permanent mold casting** techniques. High mold costs and limited complexity in shape are the disadvantages.

In **pressure die casting**, molten metallic material is forced into the mold under high pressures and is held under pressure during solidification. Many zinc, aluminum, and magnesium-based alloys are processed using pressure die casting. Extremely smooth surface finishes, very good dimensional accuracy, the ability to cast intricate shapes, and high production rates are the advantages of the pressure die casting process. Since the mold is metallic and must withstand high pressures, the dies used are expensive and the technique is limited to smaller sized components.

9-10   Continuous Casting and Ingot Casting

As discussed in the prior section, casting is a tool used for the manufacturing of components. It is also a process for producing ingots or slabs that can be further processed into different shapes (e.g., rods, bars, wires, etc.). In the steel industry, millions of kilograms of steels are produced using blast furnaces, electric arc furnaces and other processes. Figure 9-16 shows

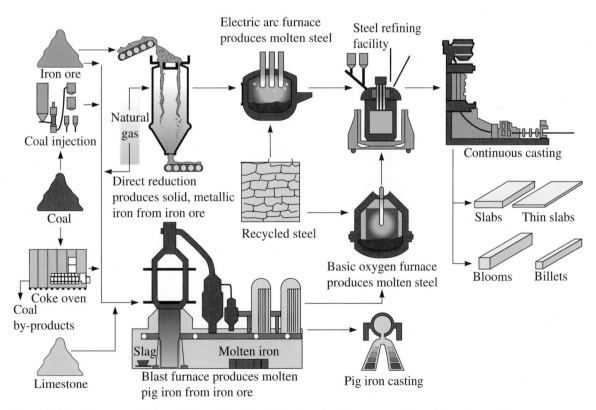

Figure 9-16   Summary of steps in the extraction of steels using iron ores, coke and limestone. *(Source: www.steel.org. Used with permission of the American Iron and Steel Institute.)*

a summary of steps to extract steels using iron ores, coke, and limestone. Although the details change, most metals and alloys (e.g., copper and zinc) are extracted from their ores using similar processes. Certain metals, such as aluminum, are produced using an electrolytic process since aluminum oxide is too stable and cannot be readily reduced to aluminum metal using coke or other reducing agents.

In many cases, we begin with scrap metals and recyclable alloys. In this case, the scrap metal is melted and processed, removing the impurities and adjusting the composition. Considerable amounts of steel, aluminum, zinc, stainless steel, titanium, and many other materials are recycled every year.

In **ingot casting**, molten steels or alloys obtained from a furnace are cast into large molds. The resultant castings, called ingots, are then processed for conversion into useful shapes via thermomechanical processing, often at another location. In the **continuous casting** process, the idea is to go from molten metallic material to some more useful "semi-finished" shape such as a plate, slab, etc. Figure 9-17 illustrates a common method for producing steel plate and bars. The liquid metal is fed from a holding vessel (a tundish) into a water-cooled oscillating copper mold, which rapidly cools the surface of the steel. The partially solidified steel is withdrawn from the mold at the same rate that additional liquid steel is introduced. The center of the steel casting finally solidifies well after the casting exits the mold. The continuously cast material is then cut into appropriate lengths by special cutting machines.

Continuous casting is cost effective for processing many steels, stainless steels, and aluminum alloys. Ingot casting is also cost effective and used for many steels where a continuous caster is not available or capacity is limited and for alloys of non-ferrous metals (e.g., zinc, copper) where the volumes are relatively small and the capital expenditure needed for a continuous caster may not be justified. Also, not all alloys can be cast using the continuous casting process.

The secondary processing steps in the processing of steels and other alloys are shown in Figure 9-18.

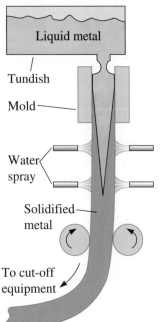

Figure 9-17
Vertical continuous casting, used in producing many steel products. Liquid metal contained in the tundish partially solidifies in a mold.

Liquid metal

Tundish

Mold

Water spray

Solidified metal

To cut-off equipment

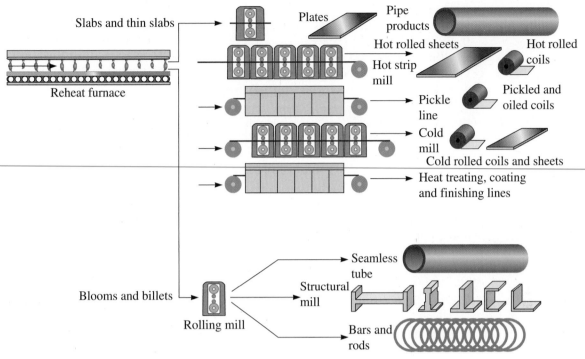

Figure 9-18 Secondary processing steps in processing of steel and alloys. *(Source www.steel.org. Used with permission of the American Iron and Steel Institute.)*

Example 9-8 *Design of a Continuous Casting Machine*

Figure 9-19 shows a method for continuous casting of 0.625-cm-thick, 120-cm-wide aluminum plate that is subsequently rolled into aluminum foil. The liquid aluminum is introduced between two large steel rolls that slowly turn. We want the aluminum to be completely solidified by the rolls just as the plate emerges from the machine. The rolls act as a permanent mold with a mold constant B of about 0.8 min/cm^2 when the aluminum is poured at the proper superheat. Design the rolls required for this process.

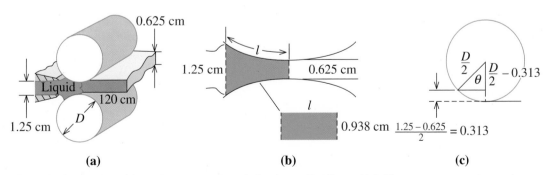

Figure 9-19 Horizontal continuous casting of aluminum (for Example 9-8).

SOLUTION

It would be helpful to simplify the geometry so that we can determine a solidification time for the casting. Let's assume that the shaded area shown in Figure 9-19(b) represents the casting and can be approximated by the average thickness times a length and width. The average thickness is $(1.25\text{ cm} + 0.625\text{ cm})/2 = 0.938$ cm. Then

$$V = (\text{thickness})(\text{length})(\text{width}) = 0.938lw$$

$$A = 2\,(\text{length})(\text{width}) = 2lw$$

$$\frac{V}{A} = \frac{0.938lw}{2lw} = 0.469\text{ cm}$$

Only the area directly in contact with the rolls is used in Chyorinov's rule, since little or no heat is transferred from other surfaces. The solidification time should be

$$t_s = B\left(\frac{V}{A}\right)^2 = (0.8)(0.469)^2 = 0.1758\text{ min}$$

For the plate to remain in contact with the rolls for this period of time, the diameter of the rolls and the rate of rotation of the rolls must be properly specified. Figure 9-19(c) shows that the angle θ between the points where the liquid enters and exits the rolls is

$$\cos\theta = \frac{(D/2) - 0.313}{(D/2)} = \frac{D - 0.626}{D}$$

The surface velocity of the rolls is the product of the circumference and the rate of rotation of the rolls, $v = \pi DR$, where R has units of revolutions/minute. The velocity v is also the rate at which we can produce the aluminum plate. The time required for the rolls to travel the distance l must equal the required solidification time:

$$t = \frac{l}{v} = 0.1758\text{ min}$$

The length l is the fraction of the roll diameter that is in contact with the aluminum during freezing and can be given by

$$l = \frac{\pi D\theta}{360}$$

Note that θ has units of degrees. Then, by substituting for l and v in the equation for the time:

$$t = \frac{l}{v} = \frac{\pi D\theta}{360\pi DR} = \frac{\theta}{360\,R} = 0.1758\text{ min}$$

$$R = \frac{\theta}{(360)(0.1758)} = .0158\,\theta\text{ rev/min}$$

A number of combinations of D and R provide the required solidification rate. Let's calculate θ for several diameters and then find the required R.

D (cm)	θ (°)	l (cm)	$R = 0.0159\theta$ (rev/min)	$v = \pi DR$ (cm/min)
60	8.28	4.34	0.131	24.82
90	6.76	5.31	0.107	30.4
120	5.85	6.13	0.092	35.07
150	5.23	6.85	0.083	39.22

As the diameter of the rolls increases, the contact area (l) between the rolls and the metal increases. This, in turn, permits a more rapid surface velocity (v) of the rolls and increases the rate of production of the plate. Note that the larger diameter rolls do not need to rotate as rapidly to achieve these higher velocities.

In selecting our final design, we prefer to use the largest practical roll diameter to ensure high production rates. As the rolls become more massive, however, they and their supporting equipment become more expensive.

In actual operation of such a continuous caster, faster speeds could be used, since the plate does not have to be completely solidified at the point where it emerges from the rolls.

9-11 Directional Solidification [DS], Single Crystal Growth, and Epitaxial Growth

There are some applications for which a small equiaxed grain structure in the casting is not desired. Castings used for blades and vanes in turbine engines are an example (Figure 9-20). These castings are often made of titanium, cobalt, or nickel-based super alloys using precision investment casting.

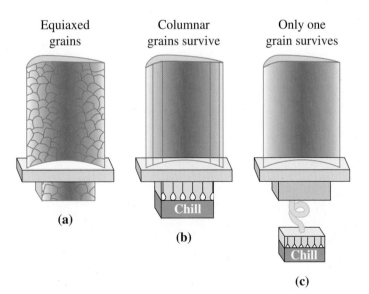

Figure 9-20 Controlling grain structure in turbine blades: (a) conventional equiaxed grains, (b) directionally solidified columnar grains, and (c) a single crystal.

In conventionally cast parts, an equiaxed grain structure is often produced; however, blades and vanes for turbine and jet engines fail along transverse grain boundaries. Better creep and fracture resistance are obtained using the **directional solidification** (DS) growth technique. In the DS process, the mold is heated from one end and cooled from the other, producing a columnar microstructure with all of the grain boundaries running in the longitudinal direction of the part. No grain boundaries are present in the transverse direction [Figure 9-20(b)].

Still better properties are obtained by using a *single crystal* (SC) technique, Solidification of columnar grains again begins at a cold surface; however, due to a helical cavity in the mold between the heat sink and the main mold cavity, only one columnar grain is able to grow to the main body of the casting [Figure 9-20(c)]. The single-crystal casting has no grain boundaries, so its crystallographic planes and directions can be directed in an optimum orientation.

Single Crystal Growth

One of the most important applications of solidification is the growth of single crystals. Polycrystalline materials cannot be used effectively in many electronic and optical applications. Grain boundaries and other defects interfere with the mechanisms that provide useful electrical or optical functions. For example, in order to utilize the semiconducting behavior of doped silicon, high-purity single crystals must be used. The current technology for silicon makes use of large (up to 300 mm. diameter) single crystals. Typically, a large crystal of the material is grown [Figure 9-21(a)]. The large crystal is then cut into silicon wafers that are only a few millimeters thick [Figure 9-21(b)]. The *Bridgman* and *Czochralski processes* are some of the popular methods used for growing single crystals of silicon, GaAs, lithium niobate ($LiNbO_3$), and many other materials.

Crystal growth furnaces containing molten materials must be maintained at a precise and stable temperature. Often, a small crystal of a predetermined crystallographic orientation is used as a "seed." Heat transfer is controlled so that the entire melt crystallizes into a single crystal. Typically, single crystals offer considerably improved, controllable,

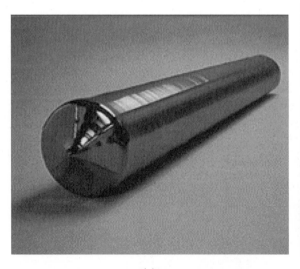
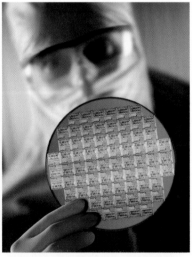

(a)   (b)

Figure 9-21   (a) Silicon single crystal (*Courtesy of Dr. A. J. Deardo, Dr. M. Hua and Dr. J. Garcia*) and (b) silicon wafer. (*Steve McAlister/Stockbyte/Getty Images.*)

and predictable properties at a higher cost than polycrystalline materials. With a large demand, however, the cost of single crystals may not be a significant factor compared to the rest of the processing costs involved in making novel and useful devices.

Epitaxial Growth   There are probably over a hundred processes for the deposition of thin films materials. In some of these processes, there is a need to control the texture or crystallographic orientation of the polycrystalline material being deposited; others require a single crystal film oriented in a particular direction. If this is the case, we can make use of a substrate of a known orientation. Epitaxy is the process by which one material is made to grow in an oriented fashion using a substrate that is crystallographically matched with the material being grown. If the lattice matching between the substrate and the film is good (within a few %), it is possible to grow highly oriented or single crystal thin films. This is known as **epitaxial growth**.

9-12   Solidification of Polymers and Inorganic Glasses

Many polymers do not crystallize, but solidify, when cooled. In these materials, the thermodynamic driving force for crystallization may exist; however, the rate of nucleation of the solid may be too slow or the complexity of the polymer chains may be so great that a crystalline solid does not form. Crystallization in polymers is almost never complete and is significantly different from that of metallic materials, requiring long polymer chains to become closely aligned over relatively large distances. By doing so, the polymer grows as **lamellar**, or plate-like, crystals (Figure 9-22). The region between each lamella contains polymer chains arranged in an amorphous manner. In addition, bundles of lamellae grow from a common nucleus, but the crystallographic orientation of the lamellae within any one bundle is different from that in another. As the bundles grow, they may produce a

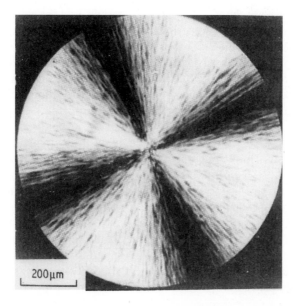

Figure 9-22
A spherulite in polystyrene. *(From R. Young and P. Lovell,* Introduction to Polymers, *2nd Ed., Chapman & Hall 1991).*

200μm

spheroidal shape called a **spherulite**. The spherulite is composed of many individual bundles of differently oriented lamellae. Amorphous regions are present between the individual lamellae, bundles of lamellae, and individual spherulites.

Many polymers of commercial interest develop crystallinity during their processing. Crystallinity can originate from cooling as discussed previously, or from the application of stress. For example, we have learned how PET plastic bottles are prepared using the blow-stretch process (Chapter 3) and how they can develop considerable crystallinity during formation. This crystallization is a result of the application of stress, and thus, is different from that encountered in the solidification of metals and alloys. In general, polymers such as nylon and polyethylene crystallize more easily compared to many other thermoplastics.

Inorganic glasses, such as silicate glasses, also do not crystallize easily for kinetic reasons. While the thermodynamic driving force exists, similar to the solidification of metals and alloys, the melts are often too viscous and the diffusion is too slow for crystallization to proceed during solidification. The float-glass process is used to melt and cast large flat pieces of glasses. In this process, molten glass is made to float on molten tin. As discussed in Chapter 7, since the strength of inorganic glasses depends critically on surface flaws produced by the manufacturing process or the reaction with atmospheric moisture, most glasses are strengthened using tempering. When safety is not a primary concern, annealing is used to reduce stresses. Long lengths of glass fibers, such as those used with fiber optics, are produced by melting a high-purity glass rod known as a **preform**. As mentioned earlier, careful control of nucleation in glasses can lead to glass-ceramics, colored glasses, and photochromic glasses (glasses that can change their color or tint upon exposure to sunlight).

9-13   Joining of Metallic Materials

In **brazing**, an alloy, known as a filler, is used to join one metal to itself or to another metal. The brazing filler metal has a melting temperature above about 450°C. **Soldering** is a brazing process in which the filler has a melting temperature below 450°C. Lead-tin and antimony-tin alloys are the most common materials used for soldering. Currently, there is a need to develop lead-free soldering materials due to the toxicity of lead. Alloys being developed include those that are based on Sn-Cu-Ag. In brazing and soldering, the metallic materials being joined do not melt; only the filler material melts. For both brazing and soldering, the composition of the filler material is different from that of the base material being joined. Various aluminum-silicon, copper, magnesium, and precious metals are used for brazing.

Solidification is also important in the joining of metals through **fusion welding**. In the fusion-welding processes, a portion of the metals to be joined is melted and, in many instances, additional molten filler metal is added. The pool of liquid metal is called the **fusion zone** (Figures 9-23 and 9-24). When the fusion zone subsequently solidifies, the original pieces of metal are joined together. During solidification of the fusion zone, nucleation is not required. The solid simply begins to grow from existing grains, frequently in a columnar manner.

The structure and properties of the fusion zone depend on many of the same variables as in a metal casting. Addition of inoculating agents to the fusion zone reduces the grain size. Fast cooling rates or short solidification times promote a finer microstructure and improved properties. Factors that increase the cooling rate include

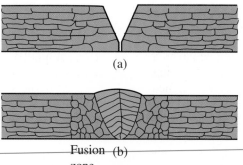

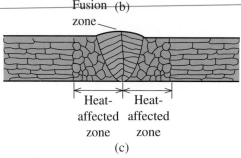

Figure 9-23
A schematic diagram of the fusion zone and solidification of the weld during fusion welding: (a) initial prepared joint, (b) weld at the maximum temperature, with joint filled with filler metal, and (c) weld after solidification.

increased thickness of the metal, smaller fusion zones, low original metal temperatures, and certain types of welding processes. Oxyacetylene welding, for example, uses a relatively low-intensity heat source; consequently, welding times are long and the surrounding solid metal, which becomes very hot, is not an effective heat sink. Arc-welding processes provide a more intense heat source, thus reducing heating of the surrounding metal and providing faster cooling. Laser welding and electron-beam welding are exceptionally intense heat sources and produce very rapid cooling rates and potentially strong welds. The friction stir welding process has been developed for Al and Al-Li alloys for aerospace applications.

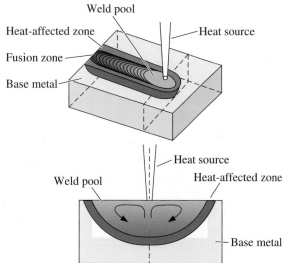

Figure 9-24
Schematic diagram showing interaction between the heat source and the base metal. Three distinct regions in the weldment are the fusion zone, the heat-affected zone, and the base metal. *(Reprinted with permission from "Current Issues and Problems in Welding Science," by S.A. David and T. DebRoy, 1992, Science, 257, pp. 497–502, Fig. 2. Copyright © 1992 American Association for the Advancement of Science.)*

Summary

- Transformation of a liquid to a solid is probably the most important phase transformation in applications of materials science and engineering.

- Solidification plays a critical role in the processing of metals, alloys, thermoplastics, and inorganic glasses. Solidification is also important in techniques used for the joining of metallic materials.

- Nucleation produces a critical-size solid particle from the liquid melt. Formation of nuclei is determined by the thermodynamic driving force for solidification and is opposed by the need to create the solid-liquid interface. As a result, solidification may not occur at the freezing temperature.

- Homogeneous nucleation requires large undercoolings of the liquid and is not observed in normal solidification processing. By introducing foreign particles into the liquid, nuclei are provided for heterogeneous nucleation. This is done in practice by inoculation or grain refining. This process permits the grain size of the casting to be controlled.

- Rapid cooling of the liquid can prevent nucleation and growth, producing amorphous solids, or glasses, with unusual mechanical and physical properties. Polymeric, metallic, and inorganic materials can be made in the form of glasses.

- In solidification from melts, the nuclei grow into the liquid melt. Either planar or dendritic modes of growth may be observed. In planar growth, a smooth solid-liquid interface grows with little or no undercooling of the liquid. Special directional solidification processes take advantage of planar growth. Dendritic growth occurs when the liquid is undercooled. Rapid cooling, or a short solidification time, produces a finer dendritic structure and often leads to improved mechanical properties of a metallic casting.

- Chvorinov's rule, $t_s = B(V/A)^n$, can be used to estimate the solidification time of a casting. Metallic castings that have a smaller interdendritic spacing and finer grain size have higher strengths.

- Cooling curves indicate the pouring temperature, any undercooling and recalescence, and time for solidification.

- By controlling nucleation and growth, a casting may be given a columnar grain structure, an equiaxed grain structure, or a mixture of the two. Isotropic behavior is typical of the equiaxed grains, whereas anisotropic behavior is found in columnar grains.

- Porosity and cavity shrinkage are major defects that can be present in cast products. If present, they can cause cast products to fail catastrophically.

- In commercial solidification processing methods, defects in a casting (such as solidification shrinkage or gas porosity) can be controlled by proper design of the casting and riser system or by appropriate treatment of the liquid metal prior to casting.

- Sand casting, investment casting, and pressure die casting are some of the processes for casting components. Ingot casting and continuous casting are employed in the production and recycling of metals and alloys.

- The solidification process can be carefully controlled to produce directionally solidified materials as well as single crystals. Epitaxial processes make use of crystal structure match between the substrate and the material being grown and are useful for making electronic and other devices.

Glossary

Argon oxygen decarburization (AOD) A process to refine stainless steel. The carbon dissolved in molten stainless steel is reduced by blowing argon gas mixed with oxygen.

Brazing An alloy, known as a filler, is used to join two materials to one another. The composition of the filler, which has a melting temperature above 450°C, is quite different from the metal being joined.

Cavities Small holes present in a casting.

Cavity shrinkage A large void within a casting caused by the volume contraction that occurs during solidification.

Chill zone A region of small, randomly oriented grains that forms at the surface of a casting as a result of heterogeneous nucleation.

Chvorinov's rule The solidification time of a casting is directly proportional to the square of the volume-to-surface area ratio of the casting.

Columnar zone A region of elongated grains having a preferred orientation that forms as a result of competitive growth during the solidification of a casting.

Continuous casting A process to convert molten metal or an alloy into a semi-finished product such as a slab.

Critical radius (r^*) The minimum size that must be formed by atoms clustering together in the liquid before the solid particle is stable and begins to grow.

Dendrite The treelike structure of the solid that grows when an undercooled liquid solidifies.

Directional solidification (DS) A solidification technique in which cooling in a given direction leads to preferential growth of grains in the opposite direction, leading to an anisotropic and an oriented microstructure.

Dispersion strengthening Increase in strength of a metallic material by generating resistance to dislocation motion by the introduction of small clusters of a second material. (Also called second-phase strengthening.)

Embryo A particle of solid that forms from the liquid as atoms cluster together. The embryo may grow into a stable nucleus or redissolve.

Epitaxial growth Growth of a single-crystal thin film on a crystallographically matched single-crystal substrate.

Equiaxed zone A region of randomly oriented grains in the center of a casting produced as a result of widespread nucleation.

Fusion welding Joining process in which a portion of the materials must melt in order to achieve good bonding.

Fusion zone The portion of a weld heated to produce all liquid during the welding process. Solidification of the fusion zone provides joining.

Gas flushing A process in which a stream of gas is injected into a molten metal in order to eliminate a dissolved gas that might produce porosity.

Gas porosity Bubbles of gas trapped within a casting during solidification, caused by the lower solubility of the gas in the solid compared with that in the liquid.

Glass-ceramics Polycrystalline, ultra-fine grained ceramic materials obtained by controlled crystallization of amorphous glasses.

Grain refinement The addition of heterogeneous nuclei in a controlled manner to increase the number of grains in a casting.

Growth The physical process by which a new phase increases in size. In the case of solidification, this refers to the formation of a stable solid as the liquid freezes.

Heterogeneous nucleation Formation of critically-sized solid from the liquid on an impurity surface.

Homogeneous nucleation Formation of critically sized solid from the liquid by the clustering together of a large number of atoms at a high undercooling (without an external interface).

Ingot A simple casting that is usually remelted or reprocessed by another user to produce a more useful shape.

Ingot casting Solidification of molten metal in a mold of simple shape. The metal then requires extensive plastic deformation to create a finished product.

Ingot structure The macrostructure of a casting, including the chill zone, columnar zone, and equiaxed zone.

Inoculants Materials that provide heterogeneous nucleation sites during the solidification of a material.

Inoculation The addition of heterogeneous nuclei in a controlled manner to increase the number of grains in a casting.

Interdendritic shrinkage Small pores between the dendrite arms formed by the shrinkage that accompanies solidification. Also known as microshrinkage or shrinkage porosity.

Investment casting A casting process that is used for making complex shapes such as turbine blades, also known as the lost wax process.

Lamellar A plate-like arrangement of crystals within a material.

Latent heat of fusion (ΔH_f) The heat evolved when a liquid solidifies. The latent heat of fusion is related to the energy difference between the solid and the liquid.

Local solidification time The time required for a particular location in a casting to solidify once nucleation has begun.

Lost foam process A process in which a polymer foam is used as a pattern to produce a casting.

Lost wax process A process in which a wax pattern is used to cast a metal.

Microshrinkage Small, frequently isolated pores between the dendrite arms formed by the shrinkage that accompanies solidification. Also known as microshrinkage or shrinkage porosity.

Mold constant (B) A characteristic constant in Chvorinov's rule.

Mushy-forming alloys Alloys with a cast macrostructure consisting predominantly of equiaxed grains. They are known as such since the cast material seems like a mush of solid grains floating in a liquid melt.

Nucleation The physical process by which a new phase is produced in a material. In the case of solidification, this refers to the formation of small, stable solid particles in the liquid.

Nuclei Small particles of solid that form from the liquid as atoms cluster together. Because these particles are large enough to be stable, growth of the solid can begin.

Permanent mold casting A casting process in which a mold can be used many times.

Photochromic glass Glass that changes color or tint upon exposure to sunlight.

Pipe shrinkage A large conical-shaped void at the surface of a casting caused by the volume contraction that occurs during solidification.

Planar growth The growth of a smooth solid-liquid interface during solidification when no undercooling of the liquid is present.

Preform A component from which a fiber is drawn or a bottle is made.

Pressure die casting A casting process in which molten metal is forced into a die under pressure.

Primary processing Process involving casting of molten metals into ingots or semi-finished useful shapes such as slabs.

Rapid solidification processing Producing unique material structures by promoting unusually high cooling rates during solidification.

Recalescence The increase in temperature of an undercooled liquid metal as a result of the liberation of heat during nucleation.

Riser An extra reservoir of liquid metal connected to a casting. If the riser freezes after the casting, the riser can provide liquid metal to compensate for shrinkage.

Sand casting A casting process using sand molds.

Secondary dendrite arm spacing (SDAS) The distance between the centers of two adjacent secondary dendrite arms.

Secondary processing Processes such as rolling, extrusion, etc. used to process ingots or slabs and other semi-finished shapes.

Shrinkage Contraction of a casting during solidification.

Shrinkage porosity Small pores between the dendrite arms formed by the shrinkage that accompanies solidification. Also known as microshrinkage or interdendritic porosity.

Sievert's law The amount of a gas that dissolves in a metal is proportional to the square root of the partial pressure of the gas in the surroundings.

Skin-forming alloys Alloys whose microstructure shows an outer skin of small grains in the chill zone followed by dendrites.

Soldering Soldering is a joining process in which the filler has a melting temperature below 450°C; no melting of the base materials occurs.

Solidification front Interface between a solid and liquid.

Solidification process Processing of materials involving solidification (e.g., single crystal growth, continuous casting, etc.).

Solid-state phase transformation A change in phase that occurs in the solid state.

Specific heat The heat required to change the temperature of a unit weight of the material one degree.

Spherulites Spherical-shaped crystals produced when certain polymers solidify.

Stainless steel A corrosion resistant alloy made from Fe-Cr-Ni-C.

Superheat The difference between the pouring temperature and the freezing temperature.

Thermal arrest A plateau on the cooling curve during the solidification of a material caused by the evolution of the latent heat of fusion during solidification. This heat generation balances the heat being lost as a result of cooling.

Total solidification time The time required for the casting to solidify completely after the casting has been poured.

Undercooling The temperature to which the liquid metal must cool below the equilibrium freezing temperature before nucleation occurs.

Problems

Section 9-1 Technological Significance

9-1 Give examples of materials based on inorganic glasses that are made by solidification.

9-2 What do the terms "primary" and "secondary processing" mean?

9-3 Why are ceramic materials not prepared by melting and casting?

Section 9-2 Nucleation

9-4 Define the following terms: nucleation, embryo, heterogeneous nucleation, and homogeneous nucleation.

9-5 Does water freeze at 0°C and boil at 100°C? Explain.

9-6 Does ice melt at 0°C? Explain.

9-7 Assume that instead of a spherical nucleus, we had a nucleus in the form of a cube of length (x). Calculate the critical dimension x^* of the cube necessary for nucleation. Write down an equation similar to Equation 9-1 for a cubical nucleus, and derive an expression for x^* similar to Equation 9-2.

9-8 Why is undercooling required for solidification? Derive an equation showing the total free energy change as a function of undercooling when the nucleating solid has the critical nucleus radius r^*.

9-9 Why is it that nuclei seen experimentally are often sphere-like but faceted? Why are they sphere-like and not like cubes or other shapes?

9-10 Explain the meaning of each term in Equation 9-2.

9-11 Suppose that liquid nickel is undercooled until homogeneous nucleation occurs. Calculate
(a) the critical radius of the nucleus required and
(b) the number of nickel atoms in the nucleus.
Assume that the lattice parameter of the solid FCC nickel is 0.356 nm.

9-12 Suppose that liquid iron is undercooled until homogeneous nucleation occurs. Calculate
(a) the critical radius of the nucleus required and
(b) the number of iron atoms in the nucleus.
Assume that the lattice parameter of the solid BCC iron is 2.92 Å.

9-13 Suppose that solid nickel was able to nucleate homogeneously with an undercooling of only 22°C. How many atoms would have to group together spontaneously for this occur? Assume that the lattice parameter of the solid FCC nickel is 0.356 nm.

9-14 Suppose that solid iron was able to nucleate homogeneously with an undercooling of only 15°C. How many atoms would have to group together spontaneously for this to occur? Assume that the lattice parameter of the solid BCC iron is 2.92 Å.

Section 9-3 Applications of Controlled Nucleation

9-15 Explain the term inoculation.

9-16 Explain how aluminum alloys can be strengthened using small levels of titanium and boron additions.

9-17 Compare and contrast grain size strengthening and strain hardening mechanisms.

9-18 What is second-phase strengthening?

9-19 Why is it that many inorganic melts solidify into amorphous materials more easily compared to those of metallic materials?

9-20 What is a glass-ceramic? How are glass-ceramics made?

9-21 What is photochromic glass?

9-22 What is a metallic glass?

9-23 How do machines in ski resorts make snow?

Section 9-4 Growth Mechanisms

9-24 What are the two steps encountered in the solidification of molten metals? As a function of time, can they overlap with one another?

9-25 During solidification, the specific heat of the material and the latent heat of fusion need to be removed. Define each of these terms.

9-26 Describe under what conditions we expect molten metals to undergo dendritic solidification.

9-27 Describe under what conditions we expect molten metals to undergo planar front solidification.

9-28 Use the data in Table 9-1 and the specific heat data given below to calculate the undercooling required to keep the dendritic fraction at 0.5 for each metal.

Metal	Specific Heat (J/(cm³ · K))
Bi	1.27
Pb	1.47
Cu	3.48
Ni	4.75

9-29 Calculate the fraction of solidification that occurs dendritically when silver nucleates
(a) at 10°C undercooling;
(b) at 100°C undercooling; and
(c) homogeneously.
The specific heat of silver is 3.25 J/(cm³·°C).

9-30 Calculate the fraction of solidification that occurs dendritically when iron nucleates
(a) at 10°C undercooling;
(b) at 100°C undercooling; and
(c) homogeneously.
The specific heat of iron is 5.78 J/(cm³·°C)

9-31 Analysis of a nickel casting suggests that 28% of the solidification process occurred in a dendritic manner. Calculate the temperature at which nucleation occurred. The specific heat of nickel is 4.1 J/(cm³·°C).

Section 9-5 Solidification Time and Dendrite Size

9-32 Write down Chvorinov's rule and explain the meaning of each term.

9-33 Find the mold constant B and exponent n in Chvorinov's rule using the following data and a log–log plot:

Shape	Dimensions (cm)	Solidification Time (s)
Cylinder	Radius = 10, Length = 30	5000
Sphere	Radius = 9	1800
Cube	Length = 6	200
Plate	Length = 30, Width = 20, Height = 1	40

9-34 A 5 cm cube solidifies in 4.6 min. Assume $n = 2$. Calculate
(a) the mold constant in Chvorinov's rule and
(b) the solidification time for a 1.25 cm × 12.5 cm × 15 cm bar cast under the same conditions.

9-35 A 5-cm-diameter sphere solidifies in 1050 s. Calculate the solidification time for a 0.3 cm × 10 cm × 20 cm plate cast under the same conditions. Assume that $n = 2$.

9-36 Find the constants B and n in Chvorinov's rule by plotting the following data on a log-log plot:

Casting Dimensions (cm)	Solidification Time (min)
1.25 × 20 × 30	3.48
5 × 7.5 × 25	15.78
6.25 cube	10.17
2.5 × 10 × 22.5	8.13

9-37 Find the constants B and n in Chvorinov's rule by plotting the following data on a log-log plot:

Casting Dimensions (cm)	Solidification Time (s)
1 × 1 × 6	28.58
2 × 4 × 4	98.30
4 × 4 × 4	155.89
8 × 6 × 5	306.15

9-38 A 7.5 cm-diameter casting was produced. The times required for the solid-liquid interface to reach different distances beneath the casting surface were measured and are shown in the following table:

Distance from Surface (cm)	Time (s)
0.25	32.6
0.75	73.5
1.25	130.6
1.875	225.0
2.5	334.9

Determine
(a) the time at which solidification begins at the surface and
(b) the time at which the entire casting is expected to be solid.
(c) Suppose the center of the casting actually solidified in 720 s. Explain why this time might differ from the time calculated in part (b).

9-39 An aluminum alloy plate with dimensions 20 cm × 10 cm × 2 cm needs to be cast with a secondary dendrite arm spacing of 10^{-2} cm (refer to Figure 9-6). What mold constant B is required (assume $n = 2$)?

9-40 Figure 9-5(b) shows a micrograph of an aluminum alloy. Estimate
(a) the secondary dendrite arm spacing and
(b) the local solidification time for that area of the casting.

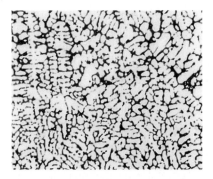

Figure 9-5 (Repeated for Problem 9-40) (b) Dendrites in an aluminum alloy (× 50). *(From ASM Handbook, Vol. 9, Metallography and Microstructure (1985), ASM International, Materials Park, OH 44073-0002.)*

9-41 Figure 9-25 shows a photograph of FeO dendrites that have precipitated from an oxide glass (an undercooled liquid). Estimate the secondary dendrite arm spacing.

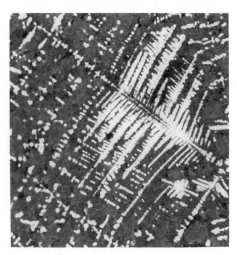

Figure 9-25 Micrograph of FeO dendrites in an oxide glass (×450) (for Problem 9-41) *(Courtesy of C.W. Ramsay, University of Missouri—Rolla.)*

9-42 Find the constants k and m relating the secondary dendrite arm spacing to the local solidification time by plotting the following data on a log-log plot:

Solidification Time (s)	SDAS (cm)
156	0.0176
282	0.0216
606	0.0282
1356	0.0374

9-43 Figure 9-26 shows dendrites in a titanium powder particle that has been rapidly solidified. Assuming that the size of the titanium dendrites is related to solidification time by the same relationship as in aluminum, estimate the solidification time of the powder particle.

Figure 9-26 Dendrites in a titanium powder particle produced by rapid solidification processing (×2200) (for Problem 9-43). *(From J.D. Ayers and K. Moore, "Formation of Metal Carbide Powder by Spark Machining of Reactive Metals," in Metallurgical Transactions, Vol. 15A, June 1984, p. 1120.)*

9-44 The secondary dendrite arm spacing in an electron-beam weld of copper is 9.5×10^{-4} cm. Estimate the solidification time of the weld.

Section 9-6 Cooling Curves

9-45 Sketch a cooling curve for a pure metal and label the different regions carefully.

9-46 What is meant by the term recalescence?

9-47 What is thermal arrest?

9-48 What is meant by the terms "local" and "total solidification" times?

9-49 A cooling curve is shown Figure 9-27.

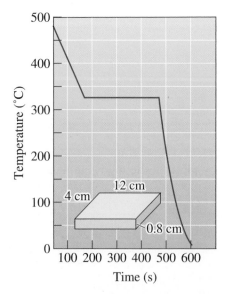

Figure 9-27 Cooling curve (for Problem 9-49).

Determine

(a) the pouring temperature;
(b) the solidification temperature;
(c) the superheat;
(d) the cooling rate, just before solidification begins;
(e) the total solidification time;
(f) the local solidification time; and
(g) the probable identity of the metal.
(h) If the cooling curve was obtained at the center of the casting sketched in the figure, determine the mold constant, assuming that $n = 2$.

9-50 A cooling curve is shown in Figure 9-28.

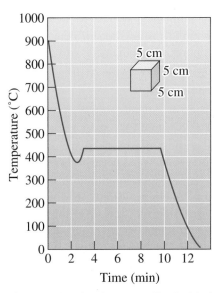

Figure 9-28 Cooling curve (for Problem 9-50).

Determine

(a) the pouring temperature;
(b) the solidification temperature;
(c) the superheat;
(d) the cooling rate, just before solidification begins;
(e) the total solidification time;
(f) the local solidification time;
(g) the undercooling; and
(h) the probable identity of the metal.
(i) If the cooling curve was obtained at the center of the casting sketched in the figure, determine the mold constant, assuming that $n = 2$.

9-51 Figure 9-29 shows the cooling curves obtained from several locations within a cylindrical aluminum casting. Determine the local solidification times and the SDAS at each location, then plot the tensile strength versus distance from the casting surface. Would you recommend that the casting be designed so that a large or small amount of material must be machined from the surface during finishing? Explain.

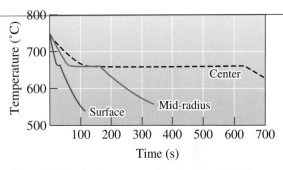

Figure 9-29 Cooling curves (for Problem 9-51).

Section 9-7 Cast Structure

9-52 What are the features expected in the macrostructure of a cast component? Explain using a sketch.

9-53 In cast materials, why does solidification almost always begin at the mold walls?

9-54 Why is it that forged components do not show a cast ingot structure?

Section 9-8 Solidification Defects

9-55 What type of defect in a casting can cause catastrophic failure of cast components such as turbine blades? What precautions are taken to prevent porosity in castings?

9-56 In general, compared to components prepared using forging, rolling, extrusion, etc., cast products tend to have lower fracture toughness. Explain why this may be the case.

9-57 What is a riser? Why should it freeze after the casting?

9-58 Calculate the volume, diameter, and height of the cylindrical riser required to prevent shrinkage in a $2.5 \text{ cm} \times 15 \text{ cm} \times 15 \text{ cm}$ casting if the H/D of the riser is 1.0.

9-59 Calculate the volume, diameter, and height of the cylindrical riser required to prevent shrinkage in a $10 \text{ cm} \times 25 \text{ cm} \times 50 \text{ cm}$ casting if the H/D of the riser is 1.5.

9-60 Figure 9-30 shows a cylindrical riser attached to a casting. Compare the solidification times for each casting section and the riser and determine whether the riser will be effective.

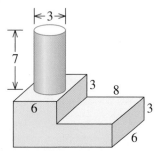

Figure 9-30   Step-block casting (for Problem 9-60).

9-61 Figure 9-31 shows a cylindrical riser attached to a casting. Compare the solidification times for each casting section and the riser and determine whether the riser will be effective.

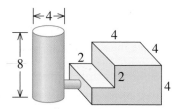

Figure 9-31   Step-block casting (for Problem 9-61).

9-62 A 10 cm-diameter sphere of liquid copper is allowed to solidify, producing a spherical shrinkage cavity in the center of the casting. Compare the volume and diameter of the shrinkage cavity in the copper casting to that obtained when a 10 cm sphere of liquid iron is allowed to solidify.

9-63 A 10 cm cube of a liquid metal is allowed to solidify. A spherical shrinkage cavity with a diameter of 3.725 cm is observed in the solid casting. Determine the percent volume change that occurs during solidification.

9-64 A 2 cm × 4 cm × 6 cm magnesium casting is produced. After cooling to room temperature, the casting is found to weigh 80 g. Determine
(a) the volume of the shrinkage cavity at the center of the casting and
(b) the percent shrinkage that must have occurred during solidification.

9-65 A 5 cm × 20 cm × 25 cm iron casting is produced and, after cooling to room temperature, is found to weigh 19 kg Determine
(a) the percent of shrinkage that must have occurred during solidification and
(b) the number of shrinkage pores in the casting if all of the shrinkage occurs as pores with a diameter of 0.125 cm.

9-66 Give examples of materials that expand upon solidification.

9-67 How can gas porosity in molten alloys be removed or minimized?

9-68 In the context of stainless steel making, what is argon oxygen decarburization?

9-69 Liquid magnesium is poured into a 2 cm × 2 cm × 24 cm mold and, as a result of directional solidification, all of the solidification shrinkage occurs along the 24 cm length of the casting. Determine the length of the casting immediately after solidification is completed.

9-70 A liquid cast iron has a density of 7.65 g/cm^3. Immediately after solidification, the density of the solid cast iron is found to be 7.71 g/cm^3. Determine the percent volume change that occurs during solidification. Does the cast iron expand or contract during solidification?

9-71 Molten copper at atmospheric pressure contains 0.01 wt% oxygen. The molten copper is placed in a chamber that is pumped down to 1 Pa to remove gas from the melt prior to pouring into the mold. Calculate the oxygen content of the copper melt after it is subjected to this degassing treatment.

9-72 From Figure 9-14, find the solubility of hydrogen in liquid aluminum just before solidification begins when the partial pressure of hydrogen is 1 atm. Determine the solubility of hydrogen (in $cm^3/100$ g Al) at the same temperature if the partial pressure were reduced to 0.01 atm.

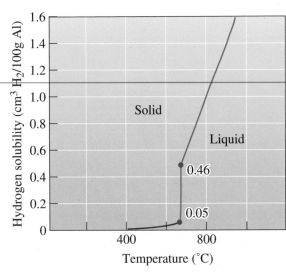

Figure 9-14 (Repeated for Problem 9-72). The solubility of hydrogen gas in aluminum when the partial pressure of $H_2 = 1$ atm.

9-73 The solubility of hydrogen in liquid aluminum at 715°C is found to be 1 $cm^3/100$ g Al. If all of this hydrogen precipitated as gas bubbles during solidification and remained trapped in the casting, calculate the volume percent gas in the solid aluminium.

Section 9-9 Casting Processes for Manufacturing Components

9-74 Explain the green sand molding process.

9-75 Why is it that castings made from pressure die casting are likely to be stronger than those made using the sand casting process?

9-76 An alloy is cast into a shape using a sand mold and a metallic mold. Which casting is expected to be stronger and why?

9-77 What is investment casting? What are the advantages of investment casting? Explain why this process is often used to cast turbine blades.

9-78 Why is pressure a key ingredient in the pressure die casting process?

Section 9-10 Continuous Casting and Ingot Casting

9-79 What is an ore?

9-80 Explain briefly how steel is made, starting with iron ore, coke, and limestone.

9-81 Explain how scrap is used for making alloys.

9-82 What is an ingot?

9-83 Why has continuous casting of steels and other alloys assumed increased importance?

9-84 What are some of the steps that follow the continuous casting process?

Section 9-11 Directional Solidification (DS), Single-Crystal Growth, and Epitaxial Growth

9-85 Define the term directional solidification.

9-86 Explain the role of nucleation and growth in growing single crystals.

Section 9-12 Solidification of Polymers and Inorganic Glasses

9-87 Why do most plastics contain amorphous and crystalline regions?

9-88 What is a spherulite?

9-89 How can processing influence crystallinity of polymers?

9-90 Explain why silicate glasses tend to form amorphous glasses, however, metallic melts typically crystallize easily.

Section 9-13 Joining of Metallic Materials

9-91 Define the terms brazing and soldering.

9-92 What is the difference between fusion welding and brazing and soldering?

9-93 What is a heat affected zone?

9-94 Explain why, while using low intensity heat sources, the strength of the material in a weld region can be reduced.

9-95 Why do laser and electron-beam welding processes lead to stronger welds?

⬡ Design Problems

9-96 Aluminum is melted under conditions that give 0.06 cm^3 H_2 per 100 g of aluminium. We have found that we must have no more than 0.002 cm³ H₂ per 100 g of aluminum in order to prevent the formation of hydrogen gas bubbles during solidification. Design a treatment process for the liquid aluminum that will ensure that hydrogen porosity does not form.

9-97 When two 1.25-cm-thick copper plates are joined using an arc-welding process, the fusion zone contains dendrites having a SDAS of 0.006 cm; however, this process produces large residual stresses in the weld. We have found that residual stresses are low when the welding conditions produce a SDAS of more than 0.02 cm. Design a process by which we can accomplish low residual stresses. Justify your design.

9-98 Design an efficient riser system for the casting shown in Figure 9-32. Be sure to include a sketch of the system, along with appropriate dimensions.

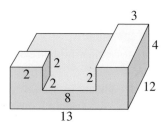

Figure 9-32 Casting to be risered (for Problem 9-98).

9-99 Design a process that will produce a steel casting having uniform properties and high strength. Be sure to include the microstructure features you wish to control and explain how you would do so.

9-100 Molten aluminum is to be injected into a steel mold under pressure (die casting). The casting is essentially a 30-cm-long, 5-cm-diameter cylinder with a uniform wall thickness, and it must have a minimum tensile strength of 276 MPa Based on the properties given in Figure 9-7, design the casting and process.

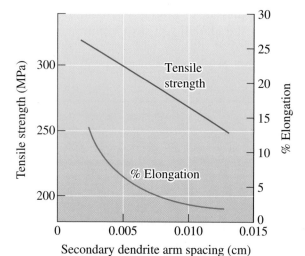

Figure 9-7 (Repeated for Problem 9-100). The effect of the secondary dendrite arm spacing on the properties of an aluminum casting alloy.

▲ Computer Problems

9-101 *Critical Radius for Homogeneous Nucleation.* Write a computer program that will allow calculation of the critical radius for nucleation (r^*). The program should ask the user to provide inputs for values for σ_{sl}, T_m, undercooling (ΔT), and enthalpy of fusion ΔH_f. Please be sure to have the correct prompts in the program to have the values entered in correct units.

9-102 *Free Energy for Formation of Nucleus of Critical Size via Heterogeneous Nucleation.* When nucleation occurs heterogeneously, the free energy for a nucleus of a critical size ($\Delta G^{*\text{hetero}}$) is given by

$$\Delta G^{*\text{hetero}} = \Delta G^{*\text{homo}} f(\theta), \text{ where}$$

$$f(\theta) = \frac{(2 + \cos\theta)(1 - \cos\theta)^2}{4}$$

and $\Delta G^{*\text{homo}}$ is given by $\dfrac{16\pi\sigma_{sl}^3}{3\Delta G_v^2}$,

which is the free energy for homogeneous nucleation of a nucleus of a critical size. If the contact angle (θ) of the phase that is nucleating on the pre-existing surface is 180°, there is no wetting, and the value of the function $f(\theta)$ is 1. The free energy of forming a nucleus of a critical radius is the same as that for homogeneous nucleation. If the nucleating phase wets the solid completely (i.e., $\theta = 0$), then $f(\theta) = 0$, and there is no barrier for nucleation. Write a computer program that will ask the user to provide the values of parameters needed to calculate the free energy for formation of a nucleus via homogeneous nucleation. The program should then calculate the value of $\Delta G^{*\text{hetero}}$ as a function of the contact angle (θ) ranging from 0 to 180°. Examine the variation of the free energy values as a function of contact angle.

9-103 *Chvorinov's Rule.* Write a computer program that will calculate the time of solidification for a casting. The program should ask the user to enter the volume of the casting and surface area from which heat transfer will occur and the mold constant. The program should then use Chvorinov's rule to calculate the time of solidification.

Ⓚ Knovel® **Problems**

K9-1   What is chilled white iron and what is it used for?

K9-2   What kinds of defects may exist in chilled white iron?

Josiah Willard Gibbs (1839–1903) was a brilliant American physicist and mathematician who conducted some of the most important pioneering work related to thermodynamic equilibrium. (*Courtesy of the University of Pennsylvania Library.*)

10

Solid Solutions and Phase Equilibrium

Have You Ever Wondered?

- *Is it possible for the solid, liquid, and gaseous forms of a material to coexist?*

- *What material is used to make red light-emitting diodes used in many modern product displays?*

- *When an alloy such as brass solidifies, which element solidifies first—copper or zinc?*

We have seen that the strength of metallic materials can be enhanced using

(a) grain size strengthening (Hall-Petch equation);

(b) cold working or strain hardening;

(c) formation of small particles of second phases; and

(d) additions of small amounts of elements.

When small amounts of elements are added, a solid material known as a solid solution may form. A **solid solution** contains two or more types of atoms or ions that are dispersed uniformly throughout the material. The impurity or **solute** atoms may occupy regular lattice sites in the crystal or interstitial sites. By controlling the amount of these point defects via the composition, the mechanical and other properties of solid solutions can be manipulated. For example, in metallic materials, the point defects created by the impurity or solute atoms disturb the atomic arrangement in the crystalline material and interfere with the movement of dislocations. The point defects cause the material to be solid-solution strengthened.

The introduction of alloying elements or impurities during processing changes the composition of the material and influences its solidification behavior. In this chapter, we will examine this effect by introducing the concept of an equilibrium phase diagram. For now, we consider a "phase" as a unique form in which a material exists.

We will define the term "phase" more precisely later in this chapter. A phase diagram depicts the stability of different phases for a set of elements (e.g., Al and Si). From the phase diagram, we can predict how a material will solidify under equilibrium conditions. We can also predict what phases will be expected to be thermodynamically stable and in what concentrations such phases should be present.

Therefore, the major objectives of this chapter are to explore

1. the formation of solid solutions;
2. the effects of solid-solution formation on the mechanical properties of metallic materials;
3. the conditions under which solid solutions can form;
4. the development of some basic ideas concerning phase diagrams; and
5. the solidification process in simple alloys.

10-1   Phases and the Phase Diagram

Pure metallic elements have engineering applications; for example, ultra-high purity copper (Cu) or aluminum (Al) is used to make microelectronic circuitry. In most applications, however, we use **alloys**. We define an "alloy" as a material that exhibits properties of a metallic material and is made from multiple elements. A *plain carbon steel* is an alloy of iron (Fe) and carbon (C). Corrosion-resistant **stainless steels** are alloys that usually contain iron (Fe), carbon (C), chromium (Cr), nickel (Ni), and some other elements. Similarly, there are alloys based on aluminum (Al), copper (Cu), cobalt (Co), nickel (Ni), titanium (Ti), zinc (Zn), and zirconium (Zr). There are two types of alloys: **single-phase alloys** and **multiple phase alloys**. In this chapter, we will examine the behavior of single-phase alloys. As a first step, let's define a "phase" and determine how the **phase rule** helps us to determine the state—solid, liquid, or gas—in which a pure material exists.

A **phase** can be defined as any portion, including the whole, of a system which is physically homogeneous within itself and bounded by a surface that separates it from any other portions. For example, water has three phases—liquid water, solid ice, and steam. A phase has the following characteristics:

1. the same structure or atomic arrangement throughout;
2. roughly the same composition and properties throughout; and
3. a definite interface between the phase and any surrounding or adjoining phases.

For example, if we enclose a block of ice in a vacuum chamber [Figure 10-1(a)], the ice begins to melt, and some of the water vaporizes. Under these conditions, we have three phases coexisting: solid H_2O, liquid H_2O, and gaseous H_2O. Each of these forms of H_2O is a distinct phase; each has a unique atomic arrangement, unique properties, and a definite boundary between each form. In this case, the phases have identical compositions.

Phase Rule   Josiah Willard Gibbs (1839–1903) was a brilliant American physicist and mathematician who conducted some of the most important pioneering work related to thermodynamic equilibrium.

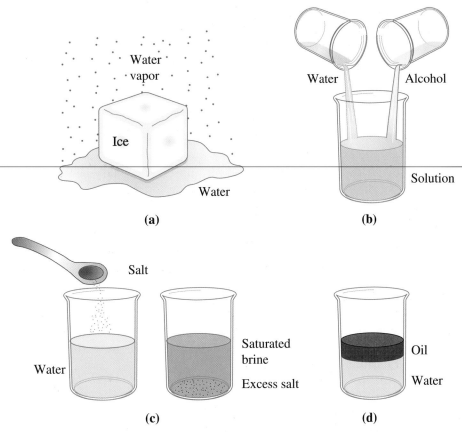

Figure 10-1   Illustration of phases and solubility: (a) The three forms of water—gas, liquid, and solid—are each a phase. (b) Water and alcohol have unlimited solubility. (c) Salt and water have limited solubility. (d) Oil and water have virtually no solubility.

Gibbs developed the **phase rule** in 1875–1876. It describes the relationship between the number of components and the number of phases for a given system and the conditions that may be allowed to change (e.g., temperature, pressure, etc.). It has the general form:

$$2 + C = F + P \text{ (when temperature and pressure both can vary)} \qquad (10\text{-}1)$$

A useful mnemonic (something that will help you remember) for the Gibbs phase rule is to start with a numeric and follow with the rest of the terms alphabetically (i.e., C, F, and P) using all positive signs. In the phase rule, C is the number of chemically independent components, usually elements or compounds, in the system; F is the number of degrees of freedom, or the number of variables (such as temperature, pressure, or composition), that are allowed to change independently without changing the number of phases in equilibrium; and P is the number of phases present (please do not confuse P with "pressure"). The constant "2" in Equation 10-1 implies that both the temperature and pressure are allowed to change. The term "chemically independent" refers to the number of different elements or compounds needed to specify a system. For example, water (H_2O) is considered as a one component system, since the concentrations of H and O in H_2O cannot be independently varied.

It is important to note that the Gibbs phase rule assumes thermodynamic equilibrium and, more often than not in materials processing, we encounter conditions in which equilibrium is *not* maintained. Therefore, you should not be surprised to see that the number and compositions of phases seen in practice are dramatically different from those predicted by the Gibbs phase rule.

Another point to note is that phases do not always have to be solid, liquid, and gaseous forms of a material. An element, such as iron (Fe), can exist in FCC and BCC crystal structures. These two solid forms of iron are two different phases of iron that will be stable at different temperatures and pressure conditions. Similarly, ice, itself, can exist in several crystal structures. Carbon can exist in many forms (e.g., graphite or diamond). These are only two of the many possible phases of carbon as we saw in Chapter 2.

As an example of the use of the phase rule, let's consider the case of pure magnesium (Mg). Figure 10-2 shows a **unary** ($C = 1$) **phase diagram** in which the lines divide the liquid, solid, and vapor phases. This unary phase diagram is also called a pressure-temperature or **P-T diagram**. In the unary phase diagram, there is only one component; in this case, magnesium (Mg). Depending on the temperature and pressure, however, there may be one, two, or even three *phases* present at any one time: solid magnesium, liquid magnesium, and magnesium vapor. Note that at atmospheric pressure (one atmosphere, given by the dashed line), the intersection of the lines in the phase diagram give the usual melting and boiling temperatures for magnesium. At very low pressures, a solid such as magnesium (Mg) can *sublime*, or go directly to a vapor form without melting, when it is heated.

Suppose we have a pressure and temperature that put us at point A in the phase diagram (Figure 10-2). At this point, magnesium is all liquid. The number of phases is one (liquid). The phase rule tells us that there are two degrees of freedom. From Equation 10-1:

$$2 + C = F + P, \quad \text{therefore, } 2 + 1 = F + 1 \text{ (i.e., } F = 2)$$

What does this mean? Within limits, as seen in Figure 10-2, we can change the pressure, the temperature, or both, and still be in an all-liquid portion of the diagram. Put another way, we must fix both the temperature and the pressure to know precisely where we are in the liquid portion of the diagram.

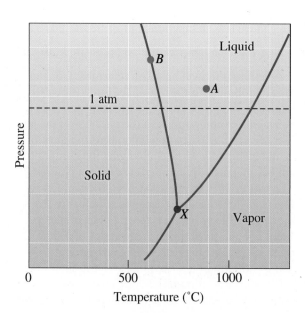

Figure 10-2
Schematic unary phase diagram for magnesium, showing the melting and boiling temperatures at one atmosphere pressure. On this diagram, point X is the triple point.

Consider point B, the boundary between the solid and liquid portions of the diagram. The number of components, C, is still one, but at point B, the solid and liquid coexist, or the number of phases P is two. From the phase rule Equation 10-1,

$$2 + C = F + P, \quad \text{therefore, } 2 + 1 = F + 2 \text{ (i.e, } F = 1)$$

or there is only one degree of freedom. For example, if we change the temperature, the pressure must also be adjusted if we are to stay on the boundary where the liquid and solid coexist. On the other hand, if we fix the pressure, the phase diagram tells us the temperature that we must have if solid and liquid are to coexist.

Finally, at point X, solid, liquid, and vapor coexist. While the number of components is still one, there are three phases. The number of degrees of freedom is zero:

$$2 + C = F + P, \quad \text{therefore, } 2 + 1 = F + 3 \text{ (i.e., } F = 0)$$

Now we have no degrees of freedom; all three phases coexist only if both the temperature and the pressure are fixed. A point on the phase diagram at which the solid, liquid, and gaseous phases coexist under equilibrium conditions is the **triple point** (Figure 10-2). In the following two examples, we see how some of these ideas underlying the Gibbs phase rule can be applied.

Example 10-1   *Design of an Aerospace Component*

Because magnesium (Mg) is a low-density material ($\rho_{Mg} = 1.738 \text{ g/cm}^3$), it has been suggested for use in an aerospace vehicle intended to enter outer space. Is this a good design?

SOLUTION

The pressure is very low in space. Even at relatively low temperatures, solid magnesium can begin to transform to a vapor, causing metal loss that could damage a space vehicle. In addition, solar radiation could cause the vehicle to heat, increasing the rate of magnesium loss.

A low-density material with a higher boiling point (and, therefore, lower vapor pressure at any given temperature) might be a better choice. At atmospheric pressure, aluminum boils at 2494°C and beryllium (Be) boils at 2770°C, compared with the boiling temperature of 1107°C for magnesium. Although aluminum and beryllium are somewhat denser than magnesium, either might be a better choice. Given the toxic effects of Be and many of its compounds when in powder form, we may want to consider aluminum first.

There are other factors to consider. In load-bearing applications, we should not only look for density but also for relative strength. Therefore, the ratio of Young's modulus to density or yield strength to density could be a better parameter to compare different materials. In this comparison, we will have to be aware that yield strength, for example, depends strongly on microstructure and that the strength of aluminum can be enhanced using aluminum alloys, while keeping the density about the same. Other factors such as oxidation during reentry into Earth's atmosphere may be applicable and will also have to be considered.

10-2   Solubility and Solid Solutions

Often, it is beneficial to know how much of each material or component we can combine without producing an additional phase. When we begin to combine different components or materials, as when we add alloying elements to a metal, solid or liquid solutions can form. For example, when we add sugar to water, we form a sugar solution. When we diffuse a small number of phosphorus (P) atoms into single crystal silicon (Si), we produce a solid solution of P in Si (Chapter 5). In other words, we are interested in the **solubility** of one material in another (e.g., sugar in water, copper in nickel, phosphorus in silicon, etc.).

Unlimited Solubility   Suppose we begin with a glass of water and a glass of alcohol. The water is one phase, and the alcohol is a second phase. If we pour the water into the alcohol and stir, only one phase is produced [Figure 10-1(b)]. The glass contains a solution of water and alcohol that has unique properties and composition. Water and alcohol are soluble in each other. Furthermore, they display **unlimited solubility**. Regardless of the ratio of water and alcohol, only one phase is produced when they are mixed together.

Similarly, if we were to mix any amounts of liquid copper and liquid nickel, only one liquid phase would be produced. This liquid alloy has the same composition and properties everywhere [Figure 10-3(a)] because nickel and copper have unlimited liquid solubility.

If the liquid copper-nickel alloy solidifies and cools to room temperature while maintaining thermal equilibrium, only one solid phase is produced. After solidification, the copper and nickel atoms do not separate but, instead, are randomly located within the FCC crystal structure. Within the solid phase, the structure, properties, and composition are uniform and no interface exists between the copper and nickel atoms. Therefore, copper and nickel also have unlimited solid solubility. The solid phase is a solid solution of copper and nickel [Figure 10-3(b)].

A solid solution is *not* a mixture. A mixture contains more than one type of phase, and the characteristics of each phase are retained when the mixture is formed. In contrast to this, the components of a solid solution completely dissolve in one another and do not retain their individual characteristics.

Another example of a system forming a solid solution is that of barium titanate ($BaTiO_3$) and strontium titanate ($SrTiO_3$), which are compounds found in the BaO–TiO_2–SrO ternary system. We use solid solutions of $BaTiO_3$ with $SrTiO_3$ and other oxides to make electronic components such as capacitors. Millions of multilayer capacitors are made each year using such materials (Chapter 19).

Many compound semiconductors that share the same crystal structure readily form solid solutions with 100% solubility. For example, we can form solid solutions of gallium arsenide (GaAs) and aluminum arsenide (AlAs). The most commonly used red LEDs for in displays are made using solid solutions based on the GaAs-GaP system. Solid solutions can be formed using more than two compounds or elements.

Limited Solubility   When we add a small quantity of salt (one phase) to a glass of water (a second phase) and stir, the salt dissolves completely in the water. Only one phase—salty water or brine—is found. If we add too much salt to the water, the excess salt sinks to the bottom of the glass [Figure 10-1(c)]. Now we have two phases—water that is saturated with salt plus excess solid salt. We find that salt has a **limited solubility** in water.

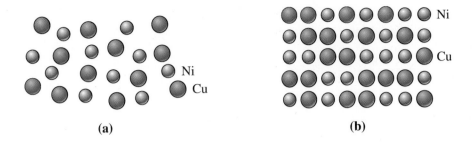

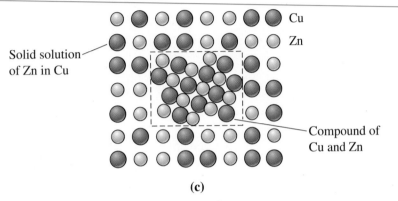

Figure 10-3   (a) Liquid copper and liquid nickel are completely soluble in each other. (b) Solid copper-nickel alloys display complete solid solubility with copper and nickel atoms occupying random lattice sites. (c) In copper-zinc alloys containing more than 30% Zn, a second phase forms because of the limited solubility of zinc in copper.

If we add a small amount of liquid zinc to liquid copper, a single liquid solution is produced. When that copper-zinc solution cools and solidifies, a single solid solution having an FCC structure results, with copper and zinc atoms randomly located at the normal lattice points. If the liquid solution contains more than about 30% Zn, some of the excess zinc atoms combine with some of the copper atoms to form a CuZn compound [Figure 10-3(c)]. Two solid phases now coexist: a solid solution of copper saturated with about 30% Zn plus a CuZn compound. The solubility of zinc in copper is limited. Figure 10-4 shows a portion of the Cu-Zn phase diagram illustrating the solubility of zinc in copper at low temperatures. The solubility increases with increasing temperature. This is similar to how we can dissolve more sugar or salt in water by increasing the temperature.

In Chapter 5, we examined how silicon (Si) can be doped with phosphorous (P), boron (B), or arsenic (As). All of these dopant elements exhibit limited solubility in Si (i.e., at small concentrations they form a solid solution with Si). Thus, solid solutions are produced even if there is limited solubility. We do not need 100% solid solubility to form solid solutions. Note that solid solutions may form either by substitutional or interstitial mechanisms. The guest atoms or ions may enter the host crystal structure at regular crystallographic positions or the interstices.

In the extreme case, there may be almost no solubility of one material in another. This is true for oil and water [Figure 10-1(d)] or for copper-lead (Cu-Pb) alloys. Note that even though materials do not dissolve into one another, they can be dispersed into one another. For example, oil-like phases and aqueous liquids can be mixed, often using surfactants (soap-like molecules), to form emulsions. Immiscibility, or lack of solubility, is seen in many molten and solid ceramic and metallic materials.

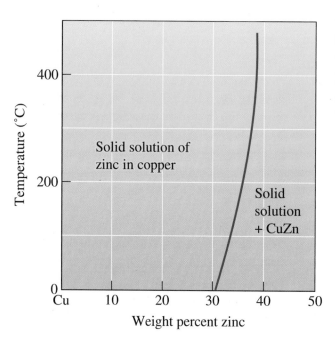

Figure 10-4
The solubility of zinc in copper. The solid line represents the solubility limit; when excess zinc is added, the solubility limit is exceeded and two phases coexist.

Polymeric Systems

We can process polymeric materials to enhance their usefulness by employing a concept similar to the formation of solid solutions in metallic and ceramic systems. We can form materials that are known as **copolymers** that consist of different monomers. For example, acrylonitrile (A), butadiene (B), and styrene (S) monomers can be made to react to form a copolymer known as ABS. This resultant copolymer is similar to a solid solution in that it has the functionalities of the three monomers from which it is derived, blending their properties. Similar to the Cu-Ni or $BaTiO_3$-$SrTiO_3$ solid solutions, we will not be able to separate out the acrylonitrile, butadiene, or styrene from an ABS plastic. Injection molding is used to convert ABS into telephones, helmets, steering wheels, and small appliance cases. Figure 10-5 illustrates the properties of different copolymers in the ABS system. Note that this is *not* a phase diagram. Dylark™ is another example of a copolymer. It is formed using maleic anhydride and a styrene monomer. The Dylark™ copolymer, with carbon black for UV protection, reinforced with fiberglass, and toughened with rubber, has been used for instrument panels in many automobiles (Chapter 16).

10-3 Conditions for Unlimited Solid Solubility

In order for an alloy system, such as copper-nickel to have unlimited solid solubility, certain conditions must be satisfied. These conditions, the **Hume-Rothery** rules, are as follows:

1. *Size factor*: The atoms or ions must be of similar size, with no more than a 15% difference in atomic radius, in order to minimize the lattice strain (i.e., to minimize, at an atomic level, the deviations caused in interatomic spacing).

2. *Crystal structure*: The materials must have the same crystal structure; otherwise, there is some point at which a transition occurs from one phase to a second phase with a different structure.

3. *Valence*: The ions must have the same valence; otherwise, the valence electron difference encourages the formation of compounds rather than solutions.

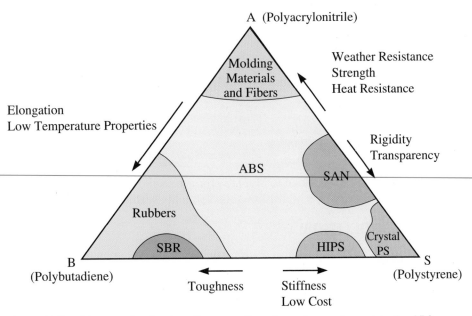

A (Polyacrylonitrile)

Molding
Materials
and Fibers

Weather Resistance
Strength
Heat Resistance

Elongation
Low Temperature Properties

Rigidity
Transparency

ABS

SAN

Rubbers

SBR

HIPS

Crystal
PS

B
(Polybutadiene)

S
(Polystyrene)

Toughness   Stiffness
Low Cost

Figure 10-5   Diagram showing how the properties of copolymers formed in the ABS system vary. This is not a phase diagram. *(From* STRONG, A. BRENT, PLASTICS: MATERIALS AND PROCESSING, 2nd, ©2000. Electronically reproduced by permission of Pearson Education, Inc., Upper Saddle River, New Jersey.*)*

4. *Electronegativity*: The atoms must have approximately the same electronegativity. Electronegativity is the affinity for electrons (Chapter 2). If the electronegativities differ significantly, compounds form—as when sodium and chloride ions combine to form sodium chloride.

Hume-Rothery's conditions must be met, but they are not necessarily sufficient, for two metals (e.g., Cu and Ni) or compounds (e.g., $BaTiO_3$-$SrTiO_3$) to have unlimited solid solubility.

Figure 10-6 shows schematically the two-dimensional structures of MgO and NiO. The Mg^{+2} and Ni^{+2} ions are similar in size and valence and, consequently, can

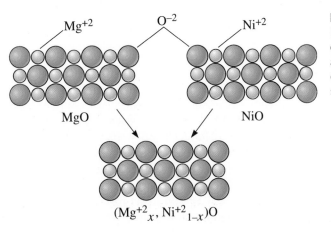

Mg^{+2} O^{-2} Ni^{+2}

MgO   NiO

$(Mg^{+2}_{x}, Ni^{+2}_{1-x})O$

Figure 10-6
MgO and NiO have similar crystal structures, ionic radii, and valences; thus the two ceramic materials can form solid solutions.

replace one another in a sodium chloride (NaCl) crystal structure (Chapter 3), forming a complete series of solid solutions of the form $(Mg_x^{+2}Ni_{1-x}^{+2})O$, where x = the mole fraction of Mg^{+2} or MgO.

The solubility of interstitial atoms is always limited. Interstitial atoms are much smaller than the atoms of the host element, thereby violating the first of Hume-Rothery's conditions.

Example 10-2   *Ceramic Solid Solutions of MgO*

NiO can be added to MgO to produce a solid solution. What other ceramic systems are likely to exhibit 100% solid solubility with MgO?

SOLUTION

In this case, we must consider oxide additives that have metal cations with the same valence and ionic radius as the magnesium cations. The valence of the magnesium ion is +2, and its ionic radius is 0.66 Å. From Appendix B, some other possibilities in which the cation has a valence of +2 include the following:

	$r(Å)$	$\left[\dfrac{r_{ion} - r_{Mg^{+2}}}{r_{Mg^{+2}}}\right] \times 100\%$	Crystal Structure
Cd^{+2} in CdO	$r_{Cd^{+2}} = 0.97$	47	NaCl
Ca^{+2} in CaO	$r_{Ca^{+2}} = 0.99$	50	NaCl
Co^{+2} in CoO	$r_{Co^{+2}} = 0.72$	9	NaCl
Fe^{+2} in FeO	$r_{Fe^{+2}} = 0.74$	12	NaCl
Sr^{+2} in SrO	$r_{Sr^{+2}} = 1.12$	70	NaCl
Zn^{+2} in ZnO	$r_{Zn^{+2}} = 0.74$	12	NaCl

The percent difference in ionic radii and the crystal structures are also shown and suggest that the FeO-MgO system will probably display unlimited solid solubility. The CoO and ZnO systems also have appropriate radius ratios and crystal structures.

10-4   Solid-Solution Strengthening

In metallic materials, one of the important effects of solid-solution formation is the resultant **solid-solution strengthening** (Figure 10-7). This strengthening, via solid-solution formation, is caused by increased resistance to dislocation motion. This is one of the important reasons why brass (Cu-Zn alloy) is stronger than pure copper. We will learn later that carbon also plays another role in the strengthening of steels by forming iron carbide (Fe_3C) and other phases (Chapter 12). Jewelry could be made out from pure gold or silver; however, pure gold and pure silver are extremely soft and malleable. Jewelers add copper to gold and silver so that the jewelry will retain its shape.

In the copper-nickel (Cu-Ni) system, we intentionally introduce a solid substitutional atom (nickel) into the original crystal structure (copper). The copper-nickel alloy is stronger than pure copper. Similarly, if less than 30% Zn is added to copper, the zinc

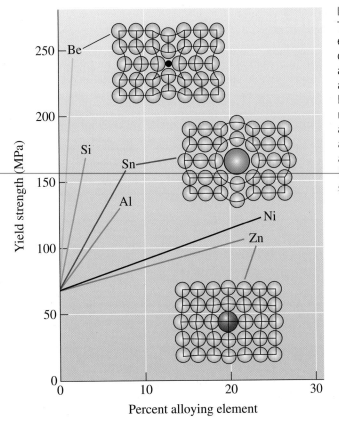

Figure 10-7
The effects of several alloying elements on the yield strength of copper. Nickel and zinc atoms are about the same size as copper atoms, but beryllium and tin atoms are much different from copper atoms. Increasing both the atomic size difference and the amount of alloying element increases solid-solution strengthening.

behaves as a substitutional atom that strengthens the copper-zinc alloy, as compared with pure copper.

Recall from Chapter 7 that the strength of ceramics is mainly dictated by the distribution of flaws; solid-solution formation does not have a strong effect on their mechanical properties. This is similar to why strain hardening was not much of a factor in enhancing the strength of ceramics or semiconductors such as silicon (Chapter 8). As discussed before, solid-solution formation in ceramics and semiconductors (such as Si, GaAs, etc.) has considerable influence on their magnetic, optical, and dielectric properties. The following discussion related to mechanical properties, therefore, applies mainly to metals.

Degree of Solid-Solution Strengthening
The degree of solid-solution strengthening depends on two factors. First, a large difference in atomic size between the original (host or solvent) atom and the added (guest or solute) atom increases the strengthening effect. A larger size difference produces a greater disruption of the initial crystal structure, making slip more difficult (Figure 10-7).

Second, the greater the amount of alloying element added, the greater the strengthening effect (Figure 10-7). A Cu-20% Ni alloy is stronger than a Cu-10% Ni alloy. Of course, if too much of a large or small atom is added, the solubility limit may be exceeded and a different strengthening mechanism, **dispersion strengthening**, is produced. In dispersion strengthening, the interface between the host phase and guest phase resists dislocation motion and contributes to strengthening. This mechanism is discussed further in Chapter 11.

Example 10-3 *Solid-Solution Strengthening*

From the atomic radii, show whether the size difference between copper atoms and alloying atoms accurately predicts the amount of strengthening found in Figure 10-7.

SOLUTION

The atomic radii and percent size difference are shown below.

Metal	Atomic Radius (Å)	$\left[\dfrac{r_{atom} - r_{Cu}}{r_{Cu}}\right] \times 100\%$
Cu	1.278	0
Zn	1.332	+4.2
Sn	1.405	+9.9
Al	1.432	+12.1
Ni	1.243	−2.7
Si	1.176	−8.0
Be	1.143	−10.6

For atoms larger than copper—namely, zinc, tin, and aluminum—increasing the size difference generally increases the strengthening effect. Likewise for smaller atoms, increasing the size difference increases strengthening.

Effect of Solid-Solution Strengthening on Properties

The effects of solid-solution strengthening on the properties of a metal include the following (Figure 10-8):

1. The yield strength, tensile strength, and hardness of the alloy are greater than those of the pure metals. This is one reason why we most often use alloys rather than pure

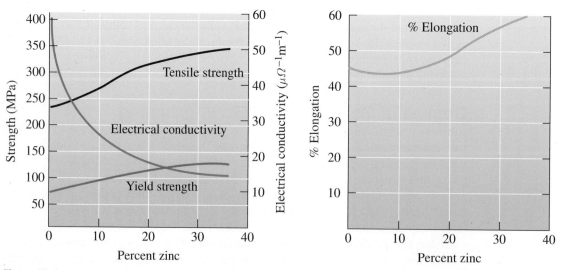

Figure 10-8
The effect of additions of zinc to copper on the properties of the solid-solution-strengthened alloy. The increase in % elongation with increasing zinc content is *not* typical of solid-solution strengthening.

metals. For example, small concentrations of Mg are added to aluminum to provide higher strength to the aluminum alloys used in making aluminum beverage cans.

2. Almost always, the ductility of the alloy is less than that of the pure metal. Only rarely, as in copper-zinc alloys, does solid-solution strengthening increase both strength and ductility.

3. Electrical conductivity of the alloy is much lower than that of the pure metal (Chapter 19). This is because electrons are scattered by the atoms of the alloying elements more so than the host atoms. Solid-solution strengthening of copper or aluminum wires used for transmission of electrical power is not recommended because of this pronounced effect. Electrical conductivity of many alloys, although lower than pure metals, is often more stable as a function of temperature.

4. The resistance to creep and strength at elevated temperatures is improved by solid-solution strengthening. Many high-temperature alloys, such as those used for jet engines, rely partly on extensive solid-solution strengthening.

10-5 Isomorphous Phase Diagrams

A **phase diagram** shows the phases and their compositions at any combination of temperature and alloy composition. When only two elements or two compounds are present in a material, a **binary phase diagram** can be constructed. **Isomorphous phase diagrams** are found in a number of metallic and ceramic systems. In the isomorphous systems, which include the copper-nickel and NiO-MgO systems [Figure 10-9(a) and (b)], only one solid phase forms; the two components in the system display complete solid solubility. As shown in the phase diagrams for the $CaO \cdot SiO_2 \cdot SrO$ and thallium-lead (Tl-Pb) systems, it is possible to have phase diagrams show a minimum or maximum point, respectively [Figure 10-9(c) and (d)]. Notice the horizontal scale can represent either mole% or weight% of one of the components. We can also plot atomic% or mole fraction of one of the components. Also, notice that the $CaO \cdot SiO_2$ and $SrO \cdot SiO_2$ diagram could be plotted as a *ternary phase diagram*. A ternary phase diagram is a phase diagram for systems consisting of three components. Here, we represent it as a *pseudo-binary diagram* (i.e., we assume that this is a diagram that represents phase equilibria between $CaO \cdot SiO_2$ and $SrO \cdot SiO_2$). In a pseudo-binary diagram, we represent equilibria between three or more components using two compounds. Ternary phase diagrams are often encountered in ceramic and metallic systems.

More recently, considerable developments have been made in phase diagrams using computer databases containing thermodynamic properties of different elements and compounds. There are several valuable pieces of information to be obtained from phase diagrams, as follows.

Liquidus and Solidus Temperatures   We define the **liquidus temperature** as the temperature above which a material is completely liquid. The upper curve in Figure 10-9(a), known as the **liquidus**, represents the liquidus temperatures for copper-nickel alloys of different compositions. We must heat a copper-nickel alloy above the liquidus temperature to produce a completely liquid alloy that can then be cast into a useful shape. The liquid alloy begins to solidify when the temperature cools to the liquidus temperature. For the Cu-40% Ni alloy in Figure 10-9(a), the liquidus temperature is 1280°C.

The **solidus temperature** is the temperature below which the alloy is 100% solid. The lower curve in Figure 10-9(a), known as the **solidus**, represents the solidus temperatures for Cu-Ni alloys of different compositions. A copper-nickel alloy is not completely solid

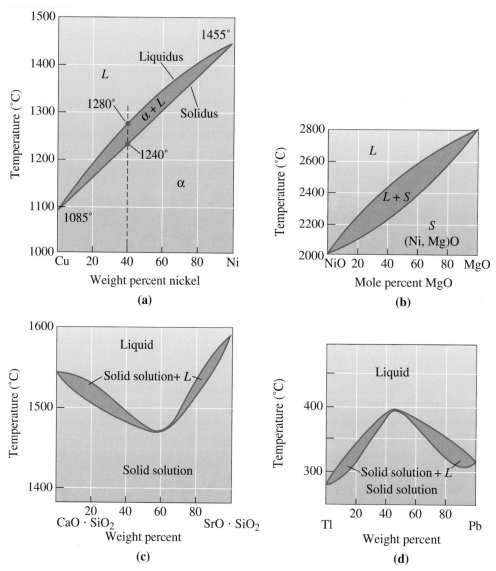

Figure 10-9   (a) and (b) The equilibrium phase diagrams for the Cu-Ni and NiO-MgO systems. The liquidus and solidus temperatures are shown for a Cu-40% Ni alloy. (c) and (d) Systems with solid-solution maxima and minima. *(Adapted from* Introduction to Phase Equilibria, *by C.G. Bergeron, and S.H. Risbud. Copyright © 1984 American Ceramic Society. Adapted by permission.)*

until the material cools below the solidus temperature. If we use a copper-nickel alloy at high temperatures, we must be sure that the service temperature is below the solidus so that no melting occurs. For the Cu-40% Ni alloy in Figure 10-9(a), the solidus temperature is 1240°C.

Copper-nickel alloys melt and freeze over a range of temperatures between the liquidus and the solidus. The temperature difference between the liquidus and the solidus is the **freezing range** of the alloy. Within the freezing range, two phases coexist: a liquid and a solid. The solid is a solution of copper and nickel atoms and is designated as the α phase. For the Cu-40% Ni alloy (α phase) in Figure 10-9(a), the freezing range is $1280 - 1240 = 40$°C. Note that pure metals solidify at a fixed temperature (i.e., the freezing range is zero degrees).

Phases Present Often we are interested in which phases are present in an alloy at a particular temperature. If we plan to make a casting, we must be sure that the metal is initially all liquid; if we plan to heat treat an alloy component, we must be sure that no liquid forms during the process. Different solid phases have different properties. For example, BCC Fe (indicated as the α phase on the iron-carbon phase diagram) is ferromagnetic; however, FCC iron (indicated as the γ phase on the Fe-C diagram) is not.

The phase diagram can be treated as a road map; if we know the coordinates—temperature and alloy composition—we can determine the phases present, assuming we know that thermodynamic equilibrium exists. There are many examples of technologically important situations in which we do not want equilibrium phases to form. For example, in the formation of silicate glass, we want an amorphous glass and not crystalline SiO_2 to form. When we harden steels by quenching them from a high temperature, the hardening occurs because of the formation of nonequilibrium phases. In such cases, phase diagrams will **not** provide all of the information we need. In these cases, we need to use special diagrams that take into account the effect of time (i.e., kinetics) on phase transformations. We will examine the use of such diagrams in later chapters.

The following two examples illustrate the applications of some of these concepts.

Example 10-4 *NiO-MgO Isomorphous System*

From the phase diagram for the NiO-MgO binary system [Figure 10-9(b)], describe a composition that can melt at 2600°C but will not melt when placed into service at 2300°C.

SOLUTION

The material must have a liquidus temperature below 2600°C, but a solidus temperature above 2300°C. The NiO-MgO phase diagram [Figure 10-9(b)] permits us to choose an appropriate composition.

To identify a composition with a liquidus temperature below 2600°C, there must be less than 60 mol% MgO in the refractory. To identify a composition with a solidus temperature above 2300°C, there must be at least 50 mol% MgO present. Consequently, we can use any composition between 50 mol% MgO and 60 mol% MgO.

Example 10-5 *Design of a Composite Material*

One method to improve the fracture toughness of a ceramic material (Chapter 7) is to reinforce the ceramic matrix with ceramic fibers. A materials designer has suggested that Al_2O_3 could be reinforced with 25% Cr_2O_3 fibers, which would interfere with the propagation of any cracks in the alumina. The resulting composite is expected to operate under load at 2000°C for several months.

Criticize the appropriateness of this design.

SOLUTION

Since the composite will operate at high temperatures for a substantial period of time, the two phases—the Cr_2O_3 fibers and the Al_2O_3 matrix—must not react with

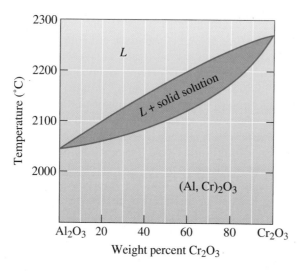

Figure 10-10
The Al_2O_3-Cr_2O_3 phase diagram (for Example 10-5).

one another. In addition, the composite must remain solid to at least 2000°C. The phase diagram in Figure 10-10 permits us to consider this choice for a composite.

Pure Cr_2O_3, pure Al_2O_3, and Al_2O_3-25% Cr_2O_3 have solidus temperatures above 2000°C; consequently, there is no danger of melting any of the constituents; however, Cr_2O_3 and Al_2O_3 display unlimited solid solubility. At the high service temperature, 2000°C, Al^{3+} ions will diffuse from the matrix into the fibers, replacing Cr^{3+} ions in the fibers. Simultaneously, Cr^{3+} ions will replace Al^{3+} ions in the matrix. Long before several months have elapsed, these diffusion processes will cause the fibers to completely dissolve into the matrix. With no fibers remaining, the fracture toughness will again be poor.

Composition of Each Phase

For each phase, we can specify a composition, expressed as the percentage of each element in the phase. Usually the composition is expressed in weight percent (wt%). When only one phase is present in the alloy or a ceramic solid solution, the composition of the phase equals the overall composition of the material. If the original composition of a single phase alloy or ceramic material changes, then the composition of the phase must also change.

When two phases, such as liquid and solid, coexist, their compositions differ from one another and also differ from the original overall composition. In this case, if the original composition changes slightly, the composition of the two phases is unaffected, provided that the temperature remains constant.

This difference is explained by the Gibbs phase rule. In this case, unlike the example of pure magnesium (Mg) described earlier, we keep the pressure fixed at one atmosphere, which is normal for binary phase diagrams. The phase rule given by Equation 10-1 can be rewritten as

$$1 + C = F + P \quad \text{(for constant pressure)} \tag{10-2}$$

where, again, C is the number of independent chemical components, P is the number of phases (*not pressure*), and F is the number of degrees of freedom. We now use the number 1 instead of the number 2 because we are holding the pressure constant. This reduces the number of degrees of freedom by one. The pressure is typically, although not necessarily, one atmosphere. In a binary system, the number of components C is two; the degrees of freedom that we have include changing the temperature and changing the composition of the phases present. We can apply this form of the phase rule to the Cu-Ni system, as shown in Example 10-6.

Example 10-6	*Gibbs Rule for an Isomorphous Phase Diagram*

Determine the degrees of freedom in a Cu-40% Ni alloy at (a) 1300°C, (b) 1250°C, and (c) 1200°C. Use Figure 10-9(a).

SOLUTION

This is a binary system ($C = 2$). The two components are Cu and Ni. We will assume constant pressure. Therefore, Equation 10-2 ($1 + C = F + P$) can be used as follows.

(a) At 1300°C, $P = 1$, since only one phase (liquid) is present; $C = 2$, since both copper and nickel atoms are present. Thus,

$$1 + C = F + P \quad \therefore 1 + 2 = F + 1 \text{ or } F = 2$$

We must fix both the temperature and the composition of the liquid phase to completely describe the state of the copper-nickel alloy in the liquid region.

(b) At 1250°C, $P = 2$, since both liquid and solid are present; $C = 2$, since copper and nickel atoms are present. Now,

$$1 + C = F + P \quad \therefore 1 + 2 = F + 2 \text{ or } F = 1$$

If we fix the temperature in the two-phase region, the compositions of the two phases are also fixed. Alternately, if the composition of one phase is fixed, the temperature and composition of the second phase are automatically fixed.

(c) At 1200°C, $P = 1$, since only one phase (solid) is present; $C = 2$, since both copper and nickel atoms are present. Again,

$$1 + C = F + P \quad \therefore 1 + 2 = F + 1 \text{ or } F = 2$$

and we must fix both temperature and composition to completely describe the state of the solid.

Because there is only one degree of freedom in a two-phase region of a binary phase diagram, the compositions of the two phases are always fixed when we specify the temperature. This is true even if the overall composition of the alloy changes. Therefore, we can use a tie line to determine the composition of the two phases. **A tie line** is a horizontal line within a two-phase region drawn at the temperature of interest (Figure 10-11). In an isomorphous system, the tie line connects the liquidus and solidus points at the specified temperature. The ends of the tie line represent the compositions of the two phases in equilibrium. Tie lines are not used in single-phase regions because we do not have two phases to "tie" in.

For any alloy with an overall or bulk composition lying between c_L and c_S, the composition of the liquid is c_L and the composition of the solid α is c_S.

The following example illustrates how the concept of a tie line is used to determine the composition of different phases in equilibrium.

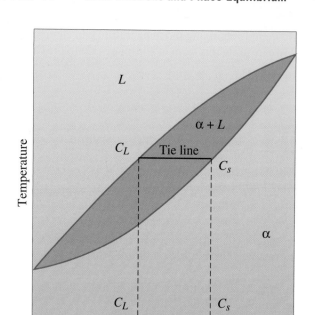

Figure 10-11
A hypothetical binary phase diagram between two elements A and B. When an alloy is present in a two-phase region, a tie line at the temperature of interest fixes the composition of the two phases. This is a consequence of the Gibbs phase rule, which provides only one degree of freedom in the two-phase region.

Example 10-7   *Compositions of Phases in the Cu-Ni Phase Diagram*

Determine the composition of each phase in a Cu-40% Ni alloy at 1300°C, 1270°C, 1250°C, and 1200°C. (See Figure 10-12.)

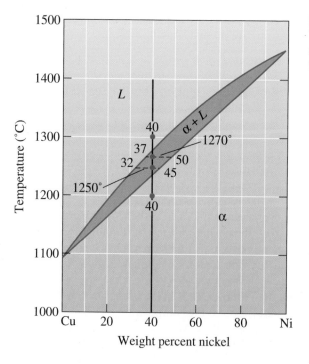

Figure 10-12
Tie lines and phase compositions for a Cu-40% Ni alloy at several temperatures (for Example 10-7).

SOLUTION

The vertical line at 40% Ni represents the overall composition of the alloy.

- **1300°C**: Only liquid is present. The liquid must contain 40% Ni, the overall composition of the alloy.
- **1270°C**: Two phases are present. A horizontal line within the $\alpha + L$ field is drawn. The endpoint at the liquidus, which is in contact with the liquid region, is at 37% Ni. The endpoint at the solidus, which is in contact with the α region, is at 50% Ni. Therefore, the liquid contains 37% Ni, and the solid contains 50% Ni.
- **1250°C**: Again two phases are present. The tie line drawn at this temperature shows that the liquid contains 32% Ni, and the solid contains 45% Ni.
- **1200°C**: Only solid α is present, so the solid must contain 40% Ni.

In Example 10-7, we find that, in the two-phase region, solid α contains more nickel and the liquid L contains more copper than the overall composition of the alloy. Generally, the higher melting point element (in this case, nickel) is concentrated in the first solid that forms.

Amount of Each Phase (the Lever Rule)   Lastly, we are
interested in the relative amounts of each phase present in the alloy. These amounts are normally expressed as weight percent (wt%). We express absolute amounts of different phases in units of mass or weight (grams, kilograms, etc.). The following example illustrates the rationale for the **lever rule**.

Example 10-8 *Application of the Lever Rule*

Calculate the amounts of α and L at 1250°C in the Cu-40% Ni alloy shown in Figure 10-13.

SOLUTION

Let's say that x = mass fraction of the alloy that is solid α. Since we have only two phases, the balance of the alloy must be in the liquid phase (L). Thus, the mass fraction of liquid will be $1 - x$. Consider 100 grams of the alloy. This alloy will

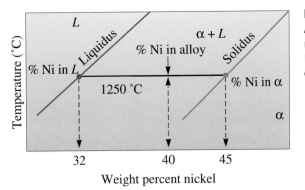

Figure 10-13
A tie line at 1250°C in the copper-nickel system that is used in Example 10-8 to find the amount of each phase.

consist of 40 grams of nickel at all temperatures. At 1250°C, let us write an equation that will represent the mass balance for nickel. At 1250°C, we have $100x$ grams of the α phase. We have $100(1 - x)$ grams of liquid.

$$\text{Total mass of nickel in 100 grams of the alloy} = \text{mass of nickel in liquid} + \text{mass of nickel in } \alpha$$

$$\therefore 100 \times (\% \text{ Ni in alloy}) = [(100)(1 - x)](\% \text{ Ni in } L) + (100)(x)(\% \text{ Ni in } \alpha)$$
$$\therefore (\% \text{ Ni in alloy}) = (\% \text{ Ni in } L)(1 - x) + (\% \text{ Ni in } \alpha)(x)$$

By multiplying and rearranging,

$$x = \frac{(\% \text{ Ni in alloy}) - (\% \text{ Ni in } L)}{(\% \text{ Ni in } \alpha) - (\% \text{ Ni in } L)}$$

From the phase diagram at 1250°C:

$$x = \frac{40 - 32}{45 - 32} = \frac{8}{13} = 0.62$$

If we convert from mass fraction to mass percent, the alloy at 1250°C contains 62% α and 38% L. Note that the concentration of nickel in the α phase (at 1250°C) is 45%, and the concentration of nickel in the liquid phase (at 1250°C) is 32%.

To calculate the amounts of liquid and solid, we construct a lever on our tie line, with the fulcrum of our lever being the original composition of the alloy. The leg of the lever *opposite* to the composition of the phase, the amount of which we are calculating, is divided by the total length of the lever to give the amount of that phase. In Example 10-8, note that the denominator represents the total length of the tie line and the numerator is the portion of the lever that is *opposite* the composition of the solid we are trying to calculate.

The lever rule in general can be written as

$$\text{Phase percent} = \frac{\text{opposite arm of lever}}{\text{total length of tie line}} \times 100 \tag{10-3}$$

We can work the lever rule in any two-phase region of a binary phase diagram. The lever rule calculation is not used in single-phase regions because the answer is trivial (there is 100% of that phase present). The lever rule is used to calculate the relative fraction or % of a phase in a two-phase mixture. The end points of the tie line we use give us the composition (i.e., the chemical concentration of different components) of each phase.

The following example reinforces the application of the lever rule for calculating the amounts of phases for an alloy at different temperatures. This is one way to track the solidification behavior of alloys, something we did not see in Chapter 9.

Example 10-9 *Solidification of a Cu-40% Ni Alloy*

Determine the amount of each phase in the Cu-40% Ni alloy shown in Figure 10-12 at 1300°C, 1270°C, 1250°C, and 1200°C.

SOLUTION

- **1300°C**: There is only one phase, so 100% L.

- **1270°C**: $\% \, L = \dfrac{50 - 40}{50 - 37} \times 100 = 77\%$

  $\% \, \alpha = \dfrac{40 - 37}{50 - 37} \times 100 = 23\%$

- **1250°C**: $\% \, L = \dfrac{45 - 40}{45 - 32} \times 100 = 38\%$

  $\% \, \alpha = \dfrac{40 - 32}{45 - 32} \times 100 = 62\%$

- **1200°C**: There is only one phase, so 100% α.

Note that at each temperature, we can determine the composition of the phases in equilibrium from the ends of the tie line drawn at that temperature.

This may seem a little odd at first. How does the α phase change its composition? The liquid phase also changes its composition, and the amounts of each phase change with temperature as the alloy cools from the liquidus to the solidus.

Sometimes we wish to express composition as atomic percent (at%) rather than weight percent (wt%). For a Cu-Ni alloy, where M_{Cu} and M_{Ni} are the molecular weights, the following equations provide examples for making these conversions:

$$\text{at\% Ni} = \left(\frac{\dfrac{\text{wt\% Ni}}{M_{Ni}}}{\dfrac{\text{wt\% Ni}}{M_{Ni}} + \dfrac{\text{wt\% Cu}}{M_{Cu}}} \right) \times 100 \qquad (10\text{-}4)$$

$$\text{wt\% Ni} = \left(\frac{(\text{at\% Ni}) \times (M_{Ni})}{\text{at\% Ni} \times M_{Ni} + \text{at\% Cu} \times M_{Cu}} \right) \times 100 \qquad (10\text{-}5)$$

10-6 Relationship Between Properties and the Phase Diagram

We have previously mentioned that a copper-nickel alloy will be stronger than either pure copper or pure nickel because of solid solution strengthening. The mechanical properties of a series of copper-nickel alloys can be related to the phase diagram as shown in Figure 10-14.

The strength of copper increases by solid-solution strengthening until about 67% Ni is added. Pure nickel is solid-solution strengthened by the addition of copper until

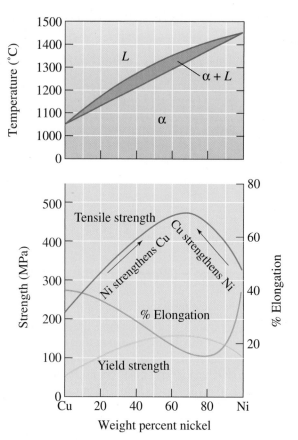

Figure 10-14
The mechanical properties of
copper-nickel alloys. Copper is
strengthened by up to 67% Ni,
and nickel is strengthened by up to
33% Cu.

33% Cu is added. The maximum strength is obtained for a Cu 67% Ni alloy, known as *Monel*. The maximum is closer to the pure nickel side of the phase diagram because pure nickel is stronger than pure copper.

Example 10-10   *Design of a Melting Procedure for a Casting*

You need to produce a Cu-Ni alloy having a minimum yield strength of 138 MPa, a minimum tensile strength of 414 MPa, and a minimum % elongation of 20%. You have in your inventory a Cu-20% Ni alloy and pure nickel. Design a method for producing castings having the required properties.

SOLUTION

From Figure 10-14, we determine the required composition of the alloy. To meet the required yield strength, the alloy must contain between 40 and 90% Ni; for the tensile strength, 40 to 88% Ni is required. The required % elongation can be obtained for alloys containing less than 60% Ni or more than 90% Ni. To satisfy all of these conditions, we could use Cu-40% to 60% Ni.

We prefer to select a low nickel content, since nickel is more expensive than copper. In addition, the lower nickel alloys have a lower liquidus, permitting castings to be made with less energy. Therefore, a reasonable alloy is Cu-40% Ni.

To produce this composition from the available melting stock, we must blend some of the pure nickel with the Cu-20% Ni ingot. Assume we wish to produce 10 kg of the alloy. Let x be the mass of Cu-20% Ni alloy we will need. The mass of pure nickel needed will be $10 - x$. Since the final alloy consists of 40% Ni, the total mass of nickel needed will be

$$(10\,kg)\left(\frac{40\%\ Ni}{100\%}\right) = 4\,kg\ Ni$$

Now let's write a mass balance for nickel. The sum of the nickel from the Cu-20% Ni alloy and the pure nickel must be equal to the total nickel in the Cu-40% Ni alloy being produced:

$$(x\,kg)\left(\frac{20\%\ Ni}{100\%}\right) + (10 - x\,kg)\left(\frac{100\%\ Ni}{100\%}\right) = 4\,kg\ Ni$$

$$0.2x + 10 - x = 4$$

$$6 = 0.8x$$

$$x = 7.5\,kg$$

Therefore, we need to melt 7.5 kg of Cu-20% Ni with 2.5 kg of pure nickel to produce the required alloy. We would then heat the alloy above the liquidus temperature, which is 1280°C for the Cu-40% Ni alloy, before pouring the liquid metal into the appropriate mold.

We need to conduct such calculations for many practical situations dealing with the processing of alloys, because when we make them, we typically use new and recycled materials.

10-7   Solidification of a Solid-Solution Alloy

When an alloy such as Cu-40% Ni is melted and cooled, solidification requires both nucleation and growth. Heterogeneous nucleation permits little or no undercooling, so solidification begins when the liquid reaches the liquidus temperature (Chapter 9). The phase diagram (Figure 10-15), with a tie line drawn at the liquidus temperature, indicates that the *first solid to form* has a composition of Cu-52% Ni.

Two conditions are required for growth of the solid α. First, growth requires that the latent heat of fusion (ΔH_f), which evolves as the liquid solidifies, be removed from the solid–liquid interface. Second, unlike the case of pure metals, diffusion must occur so that the compositions of the solid and liquid phases follow the solidus and liquidus curves during cooling. The latent heat of fusion (ΔH_f) is removed over a range of temperatures so that the cooling curve shows a change in slope, rather than a flat plateau (Figure 10-16). Thus, as we mentioned before in Chapter 9, the solidification of alloys is different from that of pure metals.

At the start of freezing, the liquid contains Cu-40% Ni, and the first solid contains Cu-52% Ni. Nickel atoms must have diffused to and concentrated at the first solid to form. After cooling to 1250°C, solidification has advanced, and the phase diagram tells us that now all of the liquid must contain 32% Ni and all of the solid must contain 45% Ni. On cooling from the liquidus to 1250°C, some nickel atoms must diffuse from the first solid to the new solid, reducing the nickel in the first solid. Additional nickel atoms diffuse from the solidifying liquid to the new solid. Meanwhile, copper atoms have concentrated—by diffusion—into the remaining liquid. This process must continue until we reach the solidus temperature, where the last liquid to freeze, which contains Cu-28% Ni, solidifies and forms a solid containing Cu-40% Ni. Just below the solidus, all of the solid must contain a uniform concentration of 40% Ni throughout.

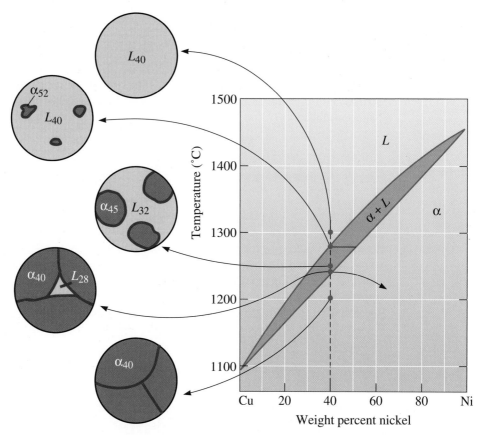

Figure 10-15 The change in structure of a Cu-40% Ni alloy during equilibrium solidification. The nickel and copper atoms must diffuse during cooling in order to satisfy the phase diagram and produce a uniform equilibrium structure.

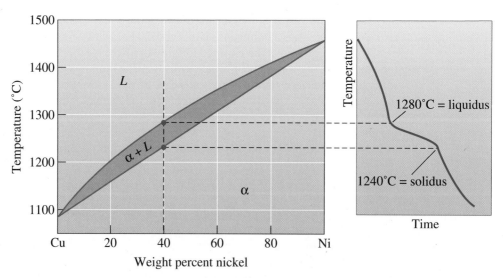

Figure 10-16 The cooling curve for an isomorphous alloy during solidification. We assume that cooling rates are low so that thermal equilibrium is maintained at each temperature. The changes in slope of the cooling curve indicate the liquidus and solidus temperatures, in this case, for a Cu-40% Ni alloy.

In order to achieve this equilibrium final structure, the cooling rate must be extremely slow. Sufficient time must be permitted for the copper and nickel atoms to diffuse and produce the compositions given by the phase diagram. In many practical casting situations, the cooling rate is too rapid to permit equilibrium. Therefore, in most castings made from alloys, we expect chemical segregation. We saw in Chapter 9 that porosity is a defect that can be present in many cast products. Another such defect often present in cast products is chemical **segregation**. This is discussed in detail in the next section.

10-8 Nonequilibrium Solidification and Segregation

In Chapter 5, we examined the thermodynamic and kinetic driving forces for diffusion. We know that diffusion occurs fastest in gases, followed by liquids, and then solids. We also saw that increasing the temperature enhances diffusion rates. When cooling is too rapid for atoms to diffuse and produce equilibrium conditions, nonequilibrium structures are produced in the casting. Let's see what happens to our Cu-40% Ni alloy on rapid cooling.

Again, the first solid, containing 52% Ni, forms on reaching the liquidus temperature (Figure 10-17). On cooling to 1260°C, the tie line tells us that the liquid contains 34% Ni and the solid that forms at that temperature contains 46% Ni. Since diffusion occurs

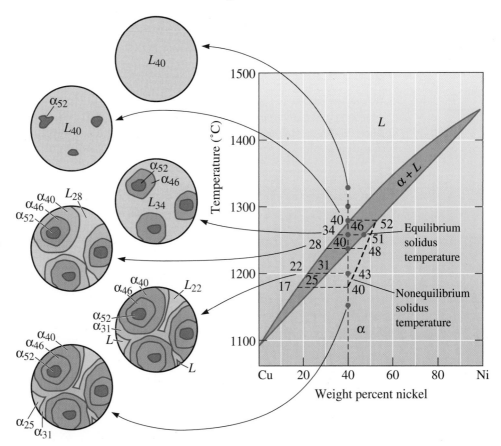

Figure 10-17 The change in structure of a Cu-40% Ni alloy during nonequilibrium solidification. Insufficient time for diffusion in the solid produces a segregated structure. Notice the nonequilibrium solidus curve.

rapidly in liquids, we expect the tie line to predict the liquid composition accurately; however, diffusion in solids is comparatively slow. The first solid that forms still has about 52% Ni, but the new solid contains only 46% Ni. We might find that the average composition of the solid is 51% Ni. This gives a different nonequilibrium solidus than that given by the phase diagram. As solidification continues, the nonequilibrium solidus line continues to separate from the equilibrium solidus.

When the temperature reaches 1240°C (the equilibrium solidus), a significant amount of liquid remains. The liquid will not completely solidify until we cool to 1180°C, where the nonequilibrium solidus intersects the original composition of 40% Ni. At that temperature, liquid containing 17% Ni solidifies, giving solid containing 25% Ni. The last liquid to freeze therefore contains 17% Ni, and the last solid to form contains 25% Ni. The average composition of the solid is 40% Ni, but the composition is not uniform.

The actual location of the nonequilibrium solidus line and the final nonequilibrium solidus temperature depend on the cooling rate. Faster cooling rates cause greater departures from equilibrium. The following example illustrates how we can account for the changes in composition under nonequilibrium conditions.

Example 10-11 *Nonequilibrium Solidification of Cu-Ni Alloys*

Calculate the composition and amount of each phase in a Cu-40% Ni alloy that is present under the nonequilibrium conditions shown in Figure 10-17 at 1300°C, 1280°C, 1260°C, 1240°C, 1200°C, and 1150°C. Compare with the equilibrium compositions and amounts of each phase.

SOLUTION

We use the tie line to the equilibrium solidus temperature to curve to calculate compositions and percentages of phases as per the lever rule. Similarly, the nonequilibrium solidus temperature curve is used to calculate percentages and concentrations of different phases formed under nonequilibrium conditions.

Temperature	Equilibrium		Nonequilibrium	
1300°C	*L*: 40% Ni 100% *L*		*L*: 40% Ni 100% *L*	
1280°C	*L*: 40% Ni 100% *L*		*L*: 40% Ni 100% *L*	
1260°C	*L*: 34% Ni $\dfrac{46-40}{46-34} = 50\%\ L$		*L*: 34% Ni $\dfrac{51-40}{51-34} = 65\%\ L$	
	α: 46% Ni $\dfrac{40-34}{46-34} = 50\%\ \alpha$		α: 51% Ni $\dfrac{40-34}{51-34} = 35\%\ \alpha$	
1240°C	*L*: 28% Ni ~ 0% *L*		*L*: 28% Ni $\dfrac{48-40}{48-28} = 40\%\ L$	
	α: 40% Ni 100% α		α: 48% Ni $\dfrac{40-28}{48-28} = 60\%\ \alpha$	
1200°C	α: 40% Ni 100% α		*L*: 22% Ni $\dfrac{43-40}{43-22} = 14\%\ L$	
			α: 43% Ni $\dfrac{40-22}{43-22} = 86\%\ \alpha$	
1150°C	α: 40% Ni 100% α		α: 40% Ni 100% α	

Microsegregation

The nonuniform composition produced by nonequilibrium solidification is known as segregation. **Microsegregation**, also known as **interdendritic segregation** and **coring**, occurs over short distances, often between small dendrite arms. The centers of the dendrites, which represent the first solid to freeze, are rich in the higher melting point element in the alloy. The regions between the dendrites are rich in the lower melting point element, since these regions represent the last liquid to freeze. The composition and properties of the α phase (in the case of Cu-Ni alloys) differ from one region to the next, and we expect the casting to have poorer properties as a result.

Microsegregation can cause **hot shortness**, or melting of the lower melting point interdendritic material at temperatures below the equilibrium solidus. When we heat the Cu-40% Ni alloy to 1225°C, below the equilibrium solidus but above the nonequilibrium solidus, the low nickel regions between the dendrites melt.

Homogenization

We can reduce the interdendritic segregation and problems with hot shortness by means of a **homogenization heat treatment**. If we heat the casting to a temperature below the nonequilibrium solidus, the nickel atoms in the centers of the dendrites diffuse to the interdendritic regions; copper atoms diffuse in the opposite direction [Figure 10-18(a)]. Since the diffusion distances are relatively short, only a few hours are required to eliminate most of the composition differences. The homogenization time is related to

$$t = c \frac{(\text{SDAS})^2}{D_s} \tag{10-6}$$

where SDAS is the secondary dendrite arm spacing, D_s is the rate of diffusion of the solute in the matrix, and c is a constant. A small SDAS reduces the diffusion distance and permits short homogenization times.

Macrosegregation

There exists another type of segregation, known as **macrosegregation**, which occurs over a large distance, between the surface and the center of the casting, with the surface (which freezes first) containing slightly more than the average amount of the higher melting point metal. We cannot eliminate macrosegregation by a homogenization treatment, because the diffusion distances are too great. Macrosegregation can be reduced by hot working, which was discussed in Chapter 8. This is because in hot working, we are basically breaking down the cast macrostructure.

Rapidly Solidified Powders

In applications in which porosity, microsegregation, and macrosegregation must be minimized, powders of complex alloys are prepared using **spray atomization** [Figure 10-18(b)]. In spray atomization, homogeneous melts of complex compositions are prepared and sprayed through a ceramic nozzle. The melt stream is broken into finer droplets and quenched using argon (Ar) or nitrogen (N$_2$) gases (gas atomization) or water (water atomization). The molten droplets solidify rapidly, generating powder particles ranging from $\sim$10–100 μm in size. Since the solidification of droplets occurs very quickly, there is very little time for diffusion, and therefore, chemical segregation does not occur. Many complex nickel- and cobalt-based superalloys and stainless steel powders are examples of materials prepared using this technique. The spray atomized powders are blended and formed into desired shapes. The techniques used in processing such powders include sintering (Chapter 5), **hot pressing** (HP) and **hot isostatic pressing** (HIP). In HIP, sintering is conducted under an isostatic pressure ($\sim$ 170 MPa) using, for example, argon gas. Very large ($\sim$ up to 6 cm diameter, several meters long) and smaller components can be

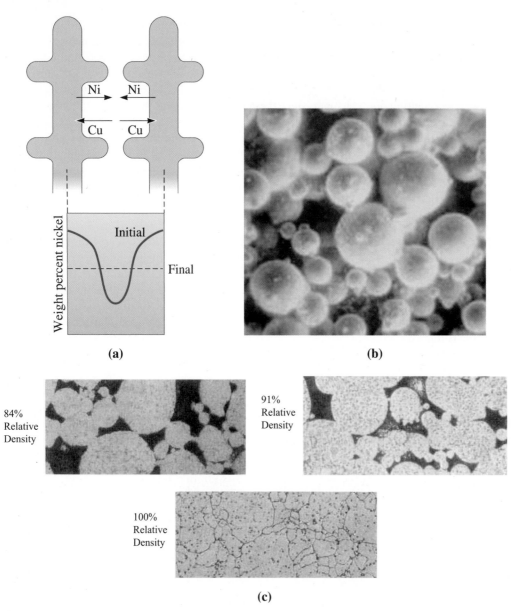

(a)

(b)

(c)

Figure 10-18 (a) Microsegregation between dendrites can be reduced by a homogenization heat treatment. Counterdiffusion of nickel and copper atoms may eventually eliminate the composition gradients and produce a homogeneous composition. (b) Spray atomized powders of superalloys. (c) Progression of densification in low carbon Astroalloy sample processed using HIP. *(Micrographs courtesy of J. Staite, Hann, B. and Rizzo, F., Crucible Compaction Metals.)*

processed using HIP. Smaller components such as disks that hold turbine blades can be machined from these. The progression of densification in a low carbon Astroalloy sample processed using spray atomized powders is shown in Figure 10-18(c).

Hot pressing is sintering under a uniaxial pressure and is used in the production of smaller components of materials that are difficult to sinter otherwise (Chapter 15). The HIP and hot pressing techniques are used for both metallic and ceramic powder materials.

Summary

- A phase is any portion, including the whole, of a system that is physically homogeneous within it and bounded by a surface that separates it from any other portions.

- A phase diagram typically shows phases that are expected to be present in a system under thermodynamic equilibrium conditions. Sometimes metastable phases may also be shown.

- Solid solutions in metallic or ceramic materials exist when elements or compounds with similar crystal structures form a single phase that is chemically homogeneous.

- Solid-solution strengthening is accomplished in metallic materials by the formation of solid solutions. The point defects created restrict dislocation motion and cause strengthening.

- The degree of solid-solution strengthening increases when (1) the amount of the alloying element increases and (2) the atomic size difference between the host material and the alloying element increases.

- The amount of alloying element (or compound) that we can add to produce solid-solution strengthening is limited by the solubility of the alloying element or compound in the host material. The solubility is limited when (1) the atomic size difference is more than about 15%, (2) the alloying element (or compound) has a different crystal structure than the host element (or compound), and (3) the valence and electronegativity of the alloying element or constituent ions are different from those of the host element (or compound).

- In addition to increasing strength and hardness, solid-solution strengthening typically decreases ductility and electrical conductivity of metallic materials. An important function of solid-solution strengthening is to provide good high-temperature properties to the alloy.

- A phase diagram in which constituents exhibit complete solid solubility is known as an isomorphous phase diagram.

- As a result of solid-solution formation, solidification begins at the liquidus temperature and is completed at the solidus temperature; the temperature difference over which solidification occurs is the freezing range.

- In two-phase regions of the phase diagram, the ends of a tie line fix the composition of each phase, and the lever rule permits the amount of each phase to be calculated.

- Microsegregation and macrosegregation occur during solidification. Microsegregation, or coring, occurs over small distances, often between dendrites. The centers of the dendrites are rich in the higher melting point element, whereas interdendritic regions, which solidify last, are rich in the lower melting point element.

- Homogenization can reduce microsegregation.

- Macrosegregation describes differences in composition over long distances, such as between the surface and center of a casting. Hot working may reduce macrosegregation.

Glossary

Alloy A material made from multiple elements that exhibits properties of a metallic material.

Binary phase diagram A phase diagram for a system with two components.

Copolymer A polymer that is formed by combining two or more different types of monomers, usually with the idea of blending the properties affiliated with individual polymers.

Coring Chemical segregation in cast products, also known as microsegregation or interdendritic segregation. The centers of the dendrites are rich in the higher melting point element, whereas interdendritic regions, which solidify last, are rich in the lower melting point element.

Dispersion strengthening Strengthening, typically used in metallic materials, by the formation of ultra-fine dispersions of a second phase. The interface between the newly formed phase and the parent phase provides additional resistance to dislocation motion, thereby causing strengthening of metallic materials (Chapter 11).

Freezing range The temperature difference between the liquidus and solidus temperatures.

Gibbs phase rule Describes the number of degrees of freedom, or the number of variables that must be fixed to specify the temperature and composition of a phase ($2 + C = F + P$, where pressure and temperature can change, $1 + C = F + P$, where pressure or temperature is constant).

Homogenization heat treatment The heat treatment used to reduce the microsegregation caused by nonequilibrium solidification. This heat treatment cannot eliminate macrosegregation.

Hot isostatic pressing (HIP) Sintering of metallic or ceramic powders, conducted under an isostatic pressure.

Hot pressing (HP) Sintering of metal or ceramic powders under a uniaxial pressure; used for production of smaller components of materials that are difficult to sinter otherwise.

Hot shortness Melting of the lower melting point nonequilibrium material that forms due to segregation, even though the temperature is below the equilibrium solidus temperature.

Hume-Rothery rules The conditions that an alloy or ceramic system must meet if the system is to display unlimited solid solubility. The Hume-Rothery rules are necessary but are not sufficient for materials to show unlimited solid solubility.

Interdendritic segregation See "Coring."

Isomorphous phase diagram A phase diagram in which the components display unlimited solid solubility.

Lever rule A technique for determining the amount of each phase in a two-phase system.

Limited solubility When only a certain amount of a solute material can be dissolved in a solvent material.

Liquidus Curves on phase diagrams that describe the liquidus temperatures of all possible alloys.

Liquidus temperature The temperature at which the first solid begins to form during solidification.

Macrosegregation The presence of composition differences in a material over large distances caused by nonequilibrium solidification. The only way to remove this type of segregation is to break down the cast structure by hot working.

Microsegregation See "Coring."

Multiple-phase alloy An alloy that consists of two or more phases.

Phase Any portion, including the whole of a system, which is physically homogeneous within it and bounded by a surface so that it is separate from any other portions.

Phase diagrams Diagrams showing phases present under equilibrium conditions and the phase compositions at each combination of temperature and overall composition. Sometimes phase diagrams also indicate metastable phases.

Phase rule See Gibbs phase rule.

P-T diagram A diagram describing thermodynamic stability of phases under different temperature and pressure conditions (same as a unary phase diagram).

Segregation The presence of composition differences in a material, often caused by insufficient time for diffusion during solidification.

Single-phase alloy An alloy consisting of one phase.

Solid solution A solid phase formed by combining multiple elements or compounds such that the overall phase has a uniform composition and properties that are different from those of the elements or compounds forming it.

Solid-solution strengthening Increasing the strength of a metallic material via the formation of a solid solution.

Solidus Curves on phase diagrams that describe the solidus temperature of all possible alloys.

Solidus temperature The temperature below which all liquid has completely solidified.

Solubility The amount of one material that will completely dissolve in a second material without creating a second phase.

Spray atomization A process in which molten alloys or metals are sprayed using a ceramic nozzle. The molten material stream is broken using a gas (e.g., Ar, N_2) or water. This leads to fine droplets that solidify rapidly, forming metal or alloy powders with ~ 10–100 μm particle size range.

Tie line A horizontal line drawn in a two-phase region of a phase diagram to assist in determining the compositions of the two phases.

Triple point A pressure and temperature at which three phases of a single material are in equilibrium.

Unary phase diagram A phase diagram in which there is only one component.

Unlimited solubility When the amount of one material that will dissolve in a second material without creating a second phase is unlimited.

Problems

10-1 Explain the principle of grain-size strengthening. Does this mechanism work at high temperatures? Explain.

10-2 Explain the principle of strain hardening. Does this mechanism work at high temperatures? Explain.

10-3 What is the principle of solid-solution strengthening? Does this mechanism work at high temperatures? Explain.

10-4 What is the principle of dispersion strengthening?

Section 10-1 Phases and Phase Diagrams

10-5 What does the term "phase" mean?

10-6 What are the different phases of water?

10-7 Ice has been known to exist in different polymorphs. Are these different phases of water?

10-8 Write down the Gibbs phase rule, assuming temperature and pressure are allowed to change. Explain clearly the meaning of each term.

10-9 What is a phase diagram?

10-10 The unary phase diagram for SiO_2 is shown in Figure 10-19. Locate the triple point where solid, liquid, and vapor coexist and give the temperature and the type of solid present. What do the other "triple" points indicate?

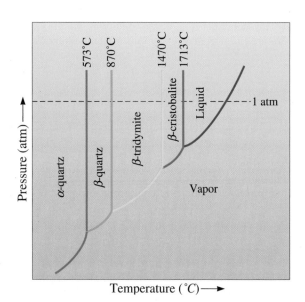

Figure 10-19 Pressure-temperature diagram for SiO_2. The dotted line shows one atmosphere pressure. (For Problem 10-10.)

10-11 Figure 10-20 shows the unary phase diagram for carbon. Based on this diagram,

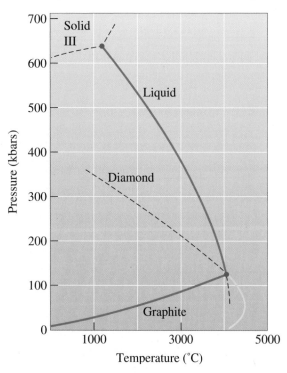

Figure 10-20 Unary phase diagram for carbon. Region for diamond formation is shown with a dotted line (for Problem 10-11). *(Adapted from* Introduction to Phase Equilibria, *by C.G. Bergeron and S.H. Risbud, Fig. 2-11. Copyright © 1984 American Ceramic Society. Adapted by permission.)*

under what conditions can carbon in the form of graphite be converted into diamond?

10-12 Natural diamond is formed approximately 120 to 200 km below the earth's surface under high pressure and high temperature conditions. Assuming that the average density of the earth is 5500 kg/m³, use this information and the unary phase diagram for C (Figure 10–20) to calculate the range of the earth's geothermal gradient (rate of increase of temperature with depth). Estimate the pressure below the earth's surface as ρgh where ρ is density, g is gravity, and h is depth. Note that 10 kbar $= 10^9$ Pa.

Section 10-2 Solubility and Solid Solutions

10-13 What is a solid solution?

10-14 How can solid solutions form in ceramic systems?

10-15 Do we need 100% solid solubility to form a solid solution of one material in another?

10-16 Small concentrations of lead zirconate ($PbZrO_3$) are added to lead titanate ($PbTiO_3$). Draw a schematic of the resultant solid-solution crystal structure that is expected to form. This material, known as lead zirconium titanate (better known as PZT), has many applications ranging from spark igniters to ultrasound imaging. See Section 3-7 for information on the perovskite crystal structure.

10-17 Can solid solutions be formed between three elements or three compounds?

10-18 What is a copolymer? What is the advantage to forming copolymers?

10-19 Is copolymer formation similar to solid-solution formation?

10-20 What is the ABS copolymer? State some of the applications of this material.

Section 10-3 Conditions for Unlimited Solid Solubility

10-21 Briefly state the Hume-Rothery rules and explain the rationale.

10-22 Can the Hume-Rothery rules apply to ceramic systems? Explain.

10-23 Based on Hume-Rothery's conditions, which of the following systems would be expected to display unlimited solid solubility? Explain.

(a) Au-Ag; (b) Al-Cu; (c) Al-Au; (d) U-W; (e) Mo-Ta; (f) Nb-W; (g) Mg-Zn; and (h) Mg-Cd.

10-24 Identify which of the following oxides when added to $BaTiO_3$ are likely to exhibit 100% solid solubility: (a) $SrTiO_3$; (b) $CaTiO_3$; (c) $ZnTiO_3$; and (d) $BaZrO_3$. All of these oxides have a perovskite crystal structure.

Section 10-4 Solid-Solution Strengthening

10-25 Suppose 1 at% of the following elements is added to copper (forming a separate alloy with each element) without exceeding the solubility limit. Which one would be expected to give the higher strength alloy? Are any of the alloying elements expected to have unlimited solid solubility in copper? (a) Au; (b) Mn; (c) Sr; (d) Si; and (e) Co.

10-26 Suppose 1 at% of the following elements is added to aluminum (forming a separate alloy with each element) without exceeding the solubility limit. Which one would be expected to give the smallest reduction in electrical conductivity? Are any of the alloy elements expected to have unlimited solid solubility in aluminum? (a) Li; (b) Ba; (c) Be; (d) Cd; and (e) Ga.

10-27 Which of the following oxides is expected to have the largest solid solubility in Al_2O_3? (a) Y_2O_3; (b) Cr_2O_3; and (c) Fe_2O_3.

10-28 What is the role of small concentrations of Mg in aluminum alloys used to make beverage cans?

10-29 Why do jewelers add small amounts of copper to gold and silver?

10-30 Why is it not a good idea to use solid-solution strengthening as a mechanism to increase the strength of copper for electrical applications?

10-31 Determine the degrees of freedom under the following conditions:

(a) Tl-20 wt% Pb at 325°C and 400°C;

(b) Tl-40 wt% Pb at 325°C and 400°C;

(c) Tl-90 wt% Pb at 325°C and 400°C.

Refer to the phase diagram in Figure 10-9(d).

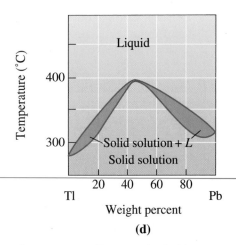

Figure 10-9(d) (Repeated for Problems 10-31 and 10-32.) The Tl-Pb phase diagram.

Section 10-5 Isomorphous Phase Diagrams

10-32 Determine the composition range in which the Tl-Pb alloy at 350°C is (a) fully liquid; (b) fully solid; and (c) partly liquid and partly solid.

Refer to Figure 10-9(d) for the Tl-Pb phase diagram. Further, determine the amount of liquid and solid solution for Tl-25 wt% Pb and Tl-75 wt% Pb at 350°C and also the wt% Pb in the liquid and solid solution for both of the alloy compositions.

10-33 Determine the liquidus temperature, solidus temperature, and freezing range for the following NiO-MgO ceramic compositions: (a) NiO-30 mol% MgO; (b) NiO-45 mol% MgO; (c) NiO-60 mol% MgO; and (d) NiO-85 mol% MgO. [See Figure 10-9(b).]

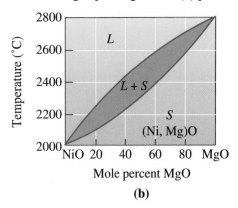

Figure 10-9(b) (Repeated for Problems 10-33, 10-35, 10-38, 10-42, 10-43 and 10-45.) The equilibrium phase diagram for the NiO-MgO system.

10-34 Determine the liquidus temperature, solidus temperature, and freezing range for the following MgO-FeO ceramic compositions:

(a) MgO-25 wt% FeO;

(b) MgO-45 wt% FeO;

(c) MgO-65 wt% FeO; and

(d) MgO-80 wt% FeO.

(See Figure 10-21.)

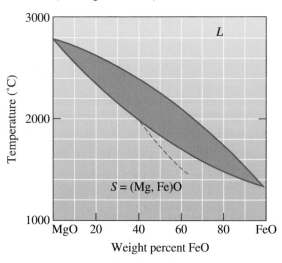

Figure 10-21 The equilibrium phase diagram for the MgO-FeO system. The dotted curve shows the nonequilibrium solidus. (For Problems 10-34, 10-36, 10-44, 10-45, 10-53, 10-57, and 10-63.)

10-35 Determine the phases present, the compositions of each phase, and the amount of each phase in mol% for the following NiO-MgO ceramics at 2400°C: (a) NiO-30 mol% MgO; (b) NiO-45 mol% MgO; (c) NiO-60 mol% MgO; and (d) NiO-85 mol% MgO. [See Figure 10-9(b).]

10-36 Determine the phases present, the compositions of each phase, and the amount of each phase in wt% for the following MgO-FeO ceramics at 2000°C: (i) MgO-25 wt% FeO; (ii) MgO-45 wt% FeO; (iii) MgO-60 wt% FeO; and (iv) MgO-80 wt% FeO. (See Figure 10-21.)

10-37 Consider a ceramic composed of 30 mol% MgO and 70 mol% FeO. Calculate the composition of the ceramic in wt%.

10-38 A NiO-20 mol% MgO ceramic is heated to 2200°C. Determine (a) the composition of the solid and liquid phases in both mol% and wt%; (b) the amount of each phase in mol% and wt%; and (c) assuming that the density

of the solid is 6.32 g/cm³ and that of the liquid is 7.14 g/cm³, determine the amount of each phase in vol%. [See Figure 10-9(b).]

10-39 A Nb-60 wt% W alloy is heated to 2800°C. Determine (a) the composition of the solid and liquid phases in both wt% and at%; (b) the amount of each phase in both wt% and at%; and (c) assuming that the density of the solid is 16.05 g/cm³ and that of the liquid is 13.91 g/cm³, determine the amount of each phase in vol%. (See Figure 10-22.)

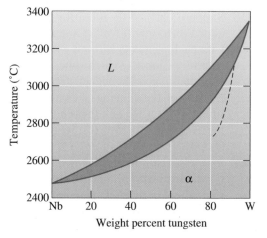

Figure 10-22 The equilibrium phase diagram for the Nb-W system. The dotted curve shows nonequilibrium solidus. (Repeated for Problems 10-39, 10-46, 10-47, 10-48, 10-49, 10-54, 10-56, 10-59, and 10-64.)

10-40 How many grams of nickel must be added to 500 grams of copper to produce an alloy that has a liquidus temperature of 1350°C? What is the ratio of the number of nickel atoms to copper atoms in this alloy? [See Figure 10-9(a).]

10-41 How many grams of nickel must be added to 500 grams of copper to produce an alloy that contains 50 wt% α at 1300°C? [See Figure 10-9(a).]

10-42 How many grams of MgO must be added to 1 kg of NiO to produce a ceramic that has a solidus temperature of 2200°C? [See Figure 10-9(b).]

10-43 How many grams of MgO must be added to 1 kg of NiO to produce a ceramic that contains 25 mol% solid at 2400°C? [See Figure 10-9(b).]

10-44 We would like to produce a solid MgO-FeO ceramic that contains equal mol percentages

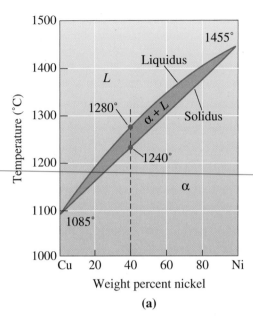

1455°

Liquidus

L

1280°

α + L

Solidus

1240°

α

1085°

Figure 10-9(a) (Repeated for Problems 10-40 and 10-41

of MgO and FeO at 1200°C. Determine the wt% FeO in the ceramic. (See Figure 10-21.)

10-45 We would like to produce a MgO-FeO ceramic that is 30 wt% solid at 2000°C. Determine the composition of the ceramic in wt%. (See Figure 10-21.)

10-46 A Nb-W alloy held at 2800°C is partly liquid and partly solid. (a) If possible, determine the composition of each phase in the alloy, and (b) if possible, determine the amount of each phase in the alloy. (See Figure 10-22.)

10-47 A Nb-W alloy contains 55% α at 2600°C. Determine (a) the composition of each phase, and (b) the composition of the alloy. (See Figure 10-22.)

10-48 Suppose a 544-kg bath of a Nb-40 wt% W alloy is held at 2800°C. How many kilograms of tungsten can be added to the bath before any solid forms? How many kilograms of tungsten must be added to cause the entire bath to be solid? (See Figure 10-22.)

10-49 A fiber-reinforced composite material is produced, in which tungsten fibers are embedded in a Nb matrix. The composite is composed of 70 vol% tungsten. (a) Calculate the wt% of tungsten fibers in the composite, and (b) suppose the composite is heated to 2600°C and held for several years. What happens to the fibers? Explain. (See Figure 10-22.)

10-50 Suppose a crucible made of pure nickel is used to contain 500 g of liquid copper at 1150°C. Describe what happens to the system as it is held at this temperature for several hours. Explain [See Figure 10-9(a).]

Section 10-6 Relationship between Properties and the Phase Diagram

10-51 What is brass? Explain which element strengthens the matrix for this alloy.

10-52 What is the composition of the Monel alloy?

Section 10-7 Solidification of a Solid-Solution Alloy

10-53 Equal moles of MgO and FeO are combined and melted. Determine (a) the liquidus temperature, the solidus temperature, and the freezing range of the ceramic, and (b) determine the phase(s) present, their composition(s), and their amount(s) at 1800°C. (See Figure 10-21.)

10-54 Suppose 75 cm^3 of Nb and 45 cm^3 of W are combined and melted. Determine (a) the liquidus temperature, the solidus temperature, and the freezing range of the alloy, and (b) the phase(s) present, their composition(s), and their amount(s) at 2800°C. [See Figure 10-22.]

10-55 A NiO-60 mol% MgO ceramic is allowed to solidify. Determine (a) the composition of the first solid to form, and (b) the composition of the last liquid to solidify under equilibrium conditions. (See Figure 10-9(b).)

10-56 A Nb-35% W alloy is allowed to solidify. Determine (a) the composition of the first solid to form, and (b) the composition of the last liquid to solidify under equilibrium conditions. (See Figure 10-22.)

10-57 For equilibrium conditions and a MgO-65 wt% FeO ceramic, determine (a) the liquidus temperature; (b) the solidus temperature; (c) the freezing range; (d) the composition of the first solid to form during solidification; (e) the composition of the last liquid to solidify; (f) the phase(s) present, the composition of the phase(s), and the amount of the phase(s) at 1800°C; and (g) the phase(s) present, the composition of the phase(s), and the amount of the phase(s) at 1600°C. (See Figure 10-21.)

10-58   Figure 10-23 shows the cooling curve for a NiO-MgO ceramic. Determine (a) the liquidus temperature; (b) the solidus temperature; (c) the freezing range; (d) the pouring temperature; (e) the superheat; (f) the local solidification time; (g) the total solidification time; and (h) the composition of the ceramic.

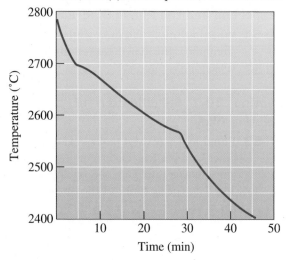

Figure 10-23   Cooling curve for a NiO-MgO ceramic (for Problem 10-58).

10-59   For equilibrium conditions and a Nb-80 wt% W alloy, determine (a) the liquidus temperature; (b) the solidus temperature; (c) the freezing range; (d) the composition of the first solid to form during solidification; (e) the composition of the last liquid to solidify; (f) the phase(s) present, the composition of the phase(s), and the amount of the phase(s) at 3000°C; and (g) the phase(s) present, the composition of the phase(s), and the amount of the phase(s) at 2800°C. (See Figure 10-22.)

10-60   Figure 10-24 shows the cooling curve for a Nb-W alloy. Determine (a) the liquidus temperature; (b) the solidus temperature; (c) the freezing range; (d) the pouring temperature; (e) the superheat; (f) the local solidification time; (g) the total solidification time; and (h) the composition of the alloy.

10-61   Cooling curves are shown in Figure 10-25 for several Mo-V alloys. Based on these curves, construct the Mo-V phase diagram.

Section 10-8 Nonequilibrium Solidification and Segregation

10-62   What are the origins of chemical segregation in cast products?

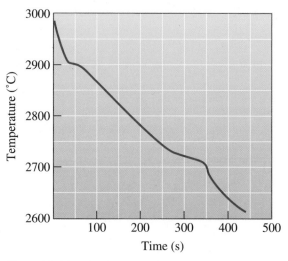

Figure 10-24   Cooling curve for a Nb-W alloy (for Problem 10-60).

10-63   For the nonequilibrium conditions shown for the MgO-65 wt% FeO ceramic, determine (a) the liquidus temperature; (b) the nonequilibrium solidus temperature; (c) the freezing range; (d) the composition of the first solid to form during solidification; (e) the composition of the last liquid to solidify; (f) the phase(s) present, the composition of the phase(s), and the amount of the phase(s) at 1800°C; and (g) the phase(s) present, the composition of the phase(s), and the amount of the phase(s) at 1600°C. (See Figure 10-21.)

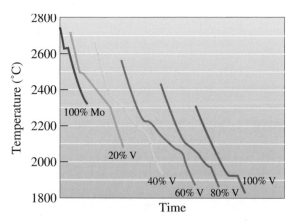

Figure 10-25   Cooling curves for a series of Mo-V alloys (for Problem 10-61).

10-64   For the nonequilibrium conditions shown for the Nb-80 wt% W alloy, determine (a) the liquidus temperature; (b) the nonequilibrium solidus temperature; (c) the freezing range; (d) the composition of the first solid to form during solidification; (e) the composition of the last liquid to solidify; (f) the phase(s) present, the composition of the phase(s), and the amount of the phase(s) at 3000°C; and (g) the phase(s) present, the composition of the phase(s), and the amount of the phase(s) at 2800°C. (See Figure 10-22.)

10-65 How can microsegregation be removed?

10-66 What is macrosegregation? Is there a way to remove it without breaking up the cast structure?

10-67 What is homogenization? What type of segregation can it remove?

10-68 A copper-nickel alloy that solidifies with a secondary dendrite arm spacing (SDAS) of 0.001 cm requires 15 hours of homogenization heat treatment at 1100°C. What is the homogenization time required for the same alloy with a SDAS of 0.01 cm and 0.0001 cm? If the diffusion coefficient of Ni in Cu at 1100°C is 3×10^{-10} cm^2/s, calculate the constant c in the homogenization time equation. What assumption is made in this calculation?

10-69 What is spray atomization? Can it be used for making ceramic powders?

10-70 Suppose you are asked to manufacture a critical component based on a nickel-based superalloy. The component must not contain any porosity and it must be chemically homogeneous. What manufacturing process would you use for this application? Why?

10-71 What is hot pressing? How is it different from hot isostatic pressing?

Design Problems

10-72 Homogenization of a slowly cooled Cu-Ni alloy having a secondary dendrite arm spacing of 0.025 cm requires 8 hours at 1000°C. Design a process to produce a homogeneous structure in a more rapidly cooled Cu-Ni alloy having a SDAS of 0.005 cm.

10-73 Design a process to produce a NiO-60% MgO refractory with a structure that is 40% glassy phase at room temperature. Include all relevant temperatures.

10-74 Design a method by which glass beads (having a density of 2.3 g/cm^3) can be

uniformly mixed and distributed in a Cu-20% Ni alloy (density of 8.91 g/cm^3).

10-75 Suppose that MgO contains 5 mol% NiO. Design a solidification purification method that will reduce the NiO to less than 1 mol% in the MgO.

Computer Problems

10-76 *Gibbs Phase Rule.* Write a computer program that will automate the Gibbs phase rule calculation. The program should ask the user for information on whether the pressure and temperature or only the pressure is to be held constant. The program then should use the correct equation to calculate the appropriate variable the user wants to know. The user will provide inputs for the number of components. Then, if the user wishes to provide the number of phases present, the program should calculate the degrees of freedom and vice-versa.

10-77 *Conversion of Wt% to At% for a Binary System.* Write a computer program that will allow conversion of wt% into at%. The program should ask the user to provide appropriate formula weights of the elements/compounds. (See Equations 10-4 and 10-5.)

10-78 *Hume-Rothery Rules.* Write a computer program that will predict whether or not there will likely be 100% solid solubility between two elements. The program should ask the user to provide the user with information on crystal structures of the elements or compounds, radii of different/atoms or ions involved, and valence and electronegativity values. You will have to make assumptions as to how much difference in values of electronegativity might be acceptable. The program should then use the Hume-Rothery rules and provide the user with guidance on the possibility of forming a system that shows 100% solid solubility.

 Knovel® Problems

K10-1 What is the solidus temperature for a silicon-germanium system containing 30 wt% Si?

Soldering plays a key role in processing of printed circuit boards and other devices for microelectronics. This process often makes use of alloys based on lead and tin. Specific compositions of these alloys known as the eutectic compositions, melt at constant temperature (like pure elements). These compositions are also mechanically strong since their microstructure comprises an intimate mixture of two distinct phases. The enhancement of mechanical properties by dispersing one phase in another phase via the formation of eutectics is the central theme for this chapter. *(Courtesy of Photolink/Photodisc Green/Getty Images.)*

Dispersion Strengthening and Eutectic Phase Diagrams

Have You Ever Wondered?

- *Why did some of the earliest glassmakers use plant ash to make glass?*
- *What alloys are most commonly used for soldering?*
- *What is fiberglass?*
- *Is there an alloy that freezes at a constant temperature?*
- *What is Pyrex® glass?*

When the solubility of a material is exceeded by adding too much of an alloying element or compound, a second phase forms and a two-phase material is produced. The boundary between the two phases, known as the **interphase interface**, is a surface where the atomic arrangement is not perfect. In metallic materials, this boundary interferes with the slip or movement of dislocations, causing strengthening. The general term for such strengthening by the introduction of a second phase is known as dispersion strengthening. In this chapter, we first discuss the fundamentals of dispersion strengthening to determine the microstructure we should aim to produce. Next, we examine the types of reactions that produce multiple phase alloys. Finally, we examine methods to achieve dispersion strengthening through control of the solidification process. In this context, we will examine phase diagrams that involve the formation of multiple phases. We will concentrate on eutectic phase diagrams.

We will conclude the chapter by examining the vapor-liquid-solid method of nanowire growth, which can be understood by considering eutectic phase diagrams.

11-1   Principles and Examples of Dispersion Strengthening

Most engineered materials are composed of more than one phase, and many of these materials are designed to provide improved strength. In simple **dispersion-strengthened** alloys, small particles of one phase, usually very strong and hard, are introduced into a second phase, which is weaker but more ductile. The soft phase, usually continuous and present in larger amounts, is called the **matrix**. The hard-strengthening phase may be called the **dispersed phase** or the **precipitate**, depending on how the alloy is formed. In some cases, a phase or a mixture of phases may have a very characteristic appearance—in these cases, this phase or phase mixture may be called a **microconstituent.** For dispersion strengthening to occur, the dispersed phase or precipitate must be small enough to provide effective obstacles to dislocation movement, thus providing the strengthening mechanism.

In most alloys, dispersion strengthening is produced by phase transformations. In this chapter, we will concentrate on a solidification transformation by which a liquid freezes to simultaneously form two solid phases. This will be called a **eutectic** reaction and is of particular importance in cast irons and many aluminum alloys. In the next chapter, we will discuss the **eutectoid** reaction, by which one solid phase reacts to simultaneously form two different solid phases; this reaction is key in the control of properties of steels. In Chapter 12, we will also discuss **precipitation** (or **age**) **hardening**, which produces precipitates by a sophisticated heat treatment.

When increased strength and toughness are the goals of incorporating a dispersed phase, the guidelines below should be followed (Figure 11-1).

1. The matrix should be soft and ductile, while the dispersed phase should be hard and strong. The dispersed phase particles interfere with slip, while the matrix provides at least some ductility to the overall alloy.

2. The hard dispersed phase should be discontinuous, while the soft, ductile matrix should be continuous. If the hard and brittle dispersed phase were continuous, cracks could propagate through the entire structure.

3. The dispersed phase particles should be small and numerous, increasing the likelihood that they interfere with the slip process since the area of the interphase interface is increased significantly.

4. The dispersed phase particles should be round, rather than needle-like or sharp edged, because the rounded shape is less likely to initiate a crack or to act as a notch.

5. Higher concentrations of the dispersed phase increase the strength of the alloy.

11-2   Intermetallic Compounds

An **intermetallic compound** contains two or more metallic elements, producing a new phase with its own composition, crystal structure, and properties. Intermetallic compounds are almost always very hard and brittle. Intermetallics or intermetallic compounds are similar to ceramic materials in terms of their mechanical properties.

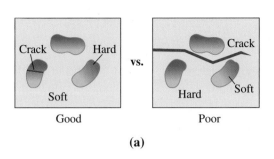

Figure 11-1
Considerations for effective dispersion strengthening: (a) The precipitate phase should be hard and discontinuous, while the matrix should be continuous and soft, (b) the dispersed phase particles should be small and numerous, (c) the dispersed phase particles should be round rather than needle-like, and (d) larger amounts of the dispersed phase increase strengthening.

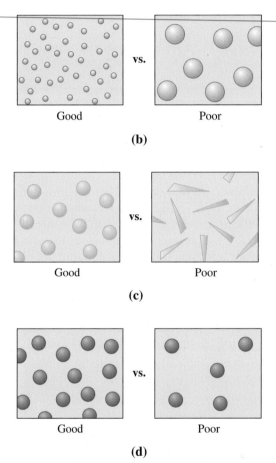

Our interest in intermetallics is two-fold. First, often dispersion-strengthened alloys contain an intermetallic compound as the dispersed phase. Secondly, many intermetallic compounds, on their own (and not as a second phase), are being investigated and developed for high temperature applications. In this section, we will discuss properties of intermetallics as stand-alone materials. In the sections that follow, we will discuss how intermetallic phases help strengthen metallic materials. Table 11-1 summarizes the properties of some intermetallic compounds.

Stoichiometric intermetallic compounds have a fixed composition. Steels are often strengthened by a stoichiometric compound, iron carbide (Fe_3C), which has a fixed ratio of three iron atoms to one carbon atom. Stoichiometric intermetallic compounds are

TABLE 11-1 ■ *Properties of some intermetallic compounds*

Intermetallic Compound	Crystal Structure	Melting Temperature (°C)	Density $\left(\frac{g}{cm^3}\right)$	Young's Modulus (GPa)
FeAl	Ordered BCC	1250–1400	5.6	263
NiAl	Ordered FCC (*B2)	1640	5.9	206
Ni₃Al	Ordered FCC (*L1₂)	1390	7.5	337
TiAl	Ordered tetragonal (*L1₀)	1460	3.8	94
Ti₃Al	Ordered HCP	1600	4.2	210
MoSi₂	Tetragonal	2020	6.31	430

*Also known as. (Adapted from Meyers, M. A., and Chawla, K. K., Mechanical behavior of materials, 2nd Edition. Cambridge University Press, Cambridge, England, 2009, Table 12.2. With permission of Cambridge University Press.)

represented in the phase diagram by a vertical line [Figure 11-2(a)]. An example of a useful intermetallic compound is molybdenum disilicide ($MoSi_2$). This material is used for making heating elements for high temperature furnaces. At high temperatures ($\sim$1000 to 1600°C), $MoSi_2$ shows outstanding oxidation resistance. At low temperatures ($\sim$500°C and below), $MoSi_2$ is brittle and shows catastrophic oxidation known as pesting.

Nonstoichiometric intermetallic compounds have a range of compositions and are sometimes called **intermediate solid solutions**. In the molybdenum–rhodium system, the γ phase is a nonstoichiometric intermetallic compound [Figure 11-2(b)]. Because the molybdenum–rhodium atom ratio is not fixed, the γ phase can contain from 45 wt% to 83 wt% Rh at 1600°C. Precipitation of the nonstoichiometric intermetallic copper aluminide $CuAl_2$ causes strengthening in a number of important aluminum alloys.

Properties and Applications of Intermetallics
Intermetallics such as Ti_3Al and Ni_3Al maintain their strength and even develop usable ductility at elevated temperatures (Figure 11-3). Lower ductility, though, has impeded further development of

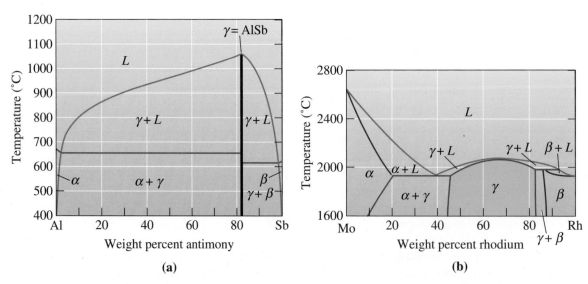

Figure 11-2 (a) The aluminum-antimony phase diagram includes a stoichiometric intermetallic compound γ. (b) The molybdenum-rhodium phase diagram includes a nonstoichiometric intermetallic compound γ.

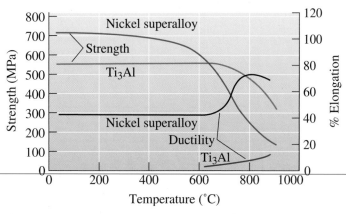

Figure 11-3   The strength and ductility of the intermetallic compound Ti_3Al compared with that of a conventional nickel superalloy. The Ti_3Al maintains its strength to higher temperatures than does the nickel superalloy.

these materials. It has been shown that the addition of small levels of boron (B) (up to 0.2%) can enhance the ductility of polycrystalline Ni_3Al. Enhanced ductility levels could make it possible for intermetallics to be used in many high temperature and load-bearing applications. Ordered compounds of NiAl and Ni_3Al are also candidates for supersonic aircraft, jet engines, and high-speed commercial aircraft. Not all applications of intermetallics are structural. Intermetallics based on silicon (e.g., platinum silicide) play a useful role in microelectronics and certain intermetallics such as Nb_3Sn are useful as superconductors (Chapter 19).

11-3   Phase Diagrams Containing Three-Phase Reactions

Many binary systems produce phase diagrams more complicated than the isomorphous phase diagrams discussed in Chapter 10. The systems we will discuss here contain reactions that involve three separate phases. Five such reactions are defined in Figure 11-4. Each of these reactions can be identified in a phase diagram by the following procedure.

1. Locate a horizontal line on the phase diagram. The horizontal line, which indicates the presence of a three-phase reaction, represents the temperature at which the reaction occurs under equilibrium conditions.

2. Locate three distinct points on the horizontal line: the two endpoints plus a third point, in between the two endpoints of the horizontal line. This third point represents the composition at which the three-phase reaction occurs. In Figure 11-4, the point in between has been shown at the center; however, on a real phase diagram, this point is not necessarily at the center.

3. Look immediately above the in-between point and identify the phase or phases present; look immediately below the point in between the end points and identify the phase or phases present. Then write the reaction from the phase(s) above the point that are transforming to the phase(s) below the point. Compare this reaction with those in Figure 11-4 to identify the reaction.

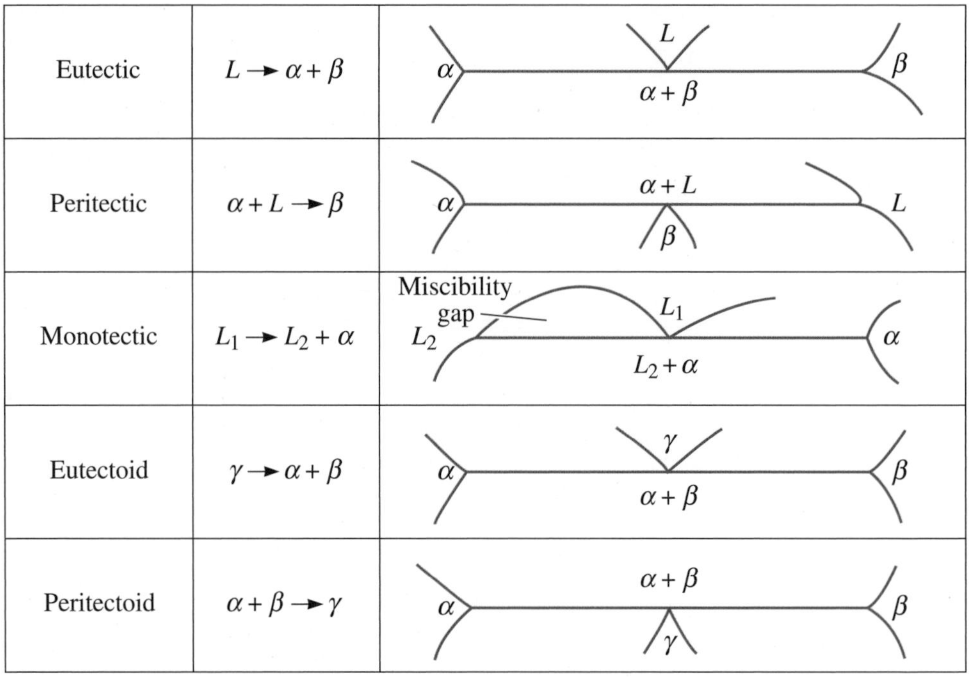

Eutectic	$L \rightarrow \alpha + \beta$	
Peritectic	$\alpha + L \rightarrow \beta$	
Monotectic	$L_1 \rightarrow L_2 + \alpha$	
Eutectoid	$\gamma \rightarrow \alpha + \beta$	
Peritectoid	$\alpha + \beta \rightarrow \gamma$	

Figure 11-4   The five most important three-phase reactions in binary phase diagrams.

Example 11-1   *Identifying Three-phase Reactions*

Consider the binary phase diagram in Figure 11-5. Identify the three-phase reactions that occur.

SOLUTION

We find horizontal lines at 1150°C, 920°C, 750°C, 450°C, and 300°C. For 1150°C: This reaction occurs at 15% B (i.e., the in-between point is at 15% B). $\delta + L$ are present above the point, and γ is present below. The reaction is

$$\delta + L \rightarrow \gamma \quad \text{a } peritectic$$

920°C: This reaction occurs at 40% B:

$$L_1 \rightarrow \gamma + L_2, \quad \text{a } monotectic$$

750°C: This reaction occurs at 70% B:

$$L \rightarrow \gamma + \beta, \quad \text{a } eutectic$$

450°C: This reaction occurs at 20% B:

$$\gamma \rightarrow \alpha + \beta, \quad \text{a } eutectoid$$

300°C: This reaction occurs at 50% B:

$$\alpha + \beta \rightarrow \mu, \quad \text{a } peritectoid$$

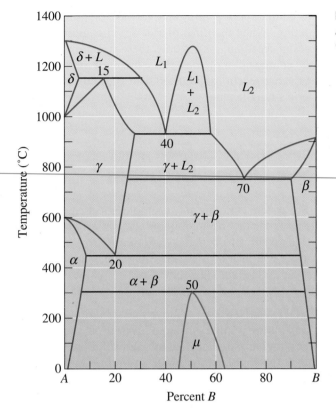

Figure 11-5
A hypothetical phase diagram
(for Example 11-1).

The eutectic, **peritectic,** and **monotectic** reactions are part of the solidification process. Alloys used for casting or soldering often take advantage of the low melting point of the eutectic reaction. The phase diagram of monotectic alloys contains a dome, or a **miscibility gap**, in which two liquid phases coexist. In the copper-lead system, the monotectic reaction produces small globules of dispersed lead, which improve the machinability of the copper alloy. Peritectic reactions lead to nonequilibrium solidification and segregation.

In many systems, there is a **metastable miscibility gap**. In this case, the immiscibility dome extends into the sub-liquidus region. In some cases, the entire miscibility gap is metastable (i.e., the immiscibility dome is completely under the liquidus). These systems form such materials as Vycor™ and Pyrex® glasses, also known as phase separated glasses. R. Roy was the first scientist to describe the underlying science for the formation of these glasses using the concept of a metastable miscibility gap existing below the liquidus.

The eutectoid and **peritectoid** reactions are completely solid-state reactions. The eutectoid reaction forms the basis for the heat treatment of several alloy systems, including steel (Chapter 12). The peritectoid reaction is extremely slow, often producing undesirable, nonequilibrium structures in alloys. As noted in Chapter 5, the rate of diffusion of atoms in solids is much smaller than in liquids.

Each of these three-phase reactions occurs at a fixed temperature and composition. The Gibbs phase rule for a three-phase reaction is (at a constant pressure),

$$1 + C = F + P$$
$$F = 1 + C - P = 1 + 2 - 3 = 0 \tag{11-1}$$

since there are two components C in a binary phase diagram and three phases P are involved in the reaction. When the three phases are in equilibrium during the reaction, there are no degrees of freedom. As a result, these reactions are called invariant. The temperature and the composition of each phase involved in the three-phase reaction are fixed. Note that of the five reactions discussed here, only eutectic and eutectoid reactions can lead to dispersion strengthening.

11-4   The Eutectic Phase Diagram

The lead–tin (Pb–Sn) system contains only a simple eutectic reaction (Figure 11-6). This alloy system is the basis for the most common alloys used for soldering. As mentioned before, because of the toxicity of Pb, there is an intense effort underway to replace lead in Pb–Sn solders. We will continue to use a Pb–Sn system, though, to discuss the eutectic phase diagram. Let's examine four classes of alloys in this system.

Solid-Solution Alloys
Alloys that contain 0 to 2% Sn behave exactly like the copper-nickel alloys; a single-phase solid solution α forms during solidification (Figure 11-7). These alloys are strengthened by solid-solution strengthening, strain hardening, and controlling the solidification process to refine the grain structure.

Alloys That Exceed the Solubility Limit
Alloys containing between 2% and 19% Sn also solidify to produce a single solid solution α; however, as the

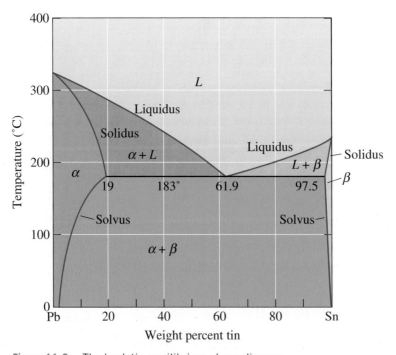

Figure 11-6   The lead–tin equilibrium phase diagram.

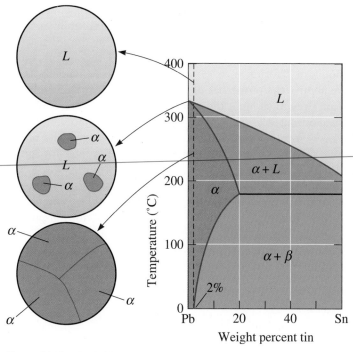

Figure 11-7 Solidification and microstructure of a Pb-2% Sn alloy. The alloy is a single-phase solid solution.

alloy continues to cool, a solid-state reaction occurs, permitting a second solid phase (β) to precipitate from the original α phase (Figure 11-8).

On this phase diagram, the α is a solid solution of tin in lead; however, the solubility of tin in the α solid solution is limited. At 0°C, only 2% Sn can dissolve in α. As the

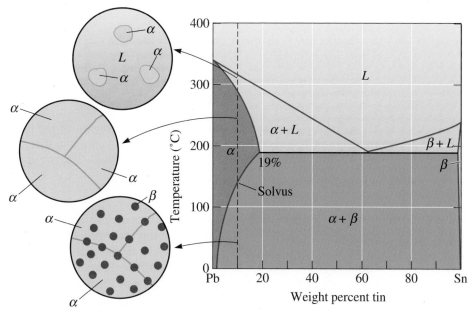

Figure 11-8 Solidification, precipitation, and microstructure of a Pb-10% Sn alloy. Some dispersion strengthening occurs as the β solid precipitates.

temperature increases, more tin dissolves into the lead until, at 183°C, the solubility of tin in lead has increased to 19% Sn. This is the maximum solubility of tin in lead. The solubility of tin in solid lead at any temperature is given by the solvus curve. Any alloy containing between 2% and 19% Sn cools past the solvus, the solubility limit is exceeded, and a small amount of β forms.

We control the properties of this type of alloy by several techniques, including solid-solution strengthening of the α portion of the structure, controlling the microstructure produced during solidification, and controlling the amount and characteristics of the β phase. These types of compositions, which form a single solid phase at high temperatures and two solid phases at lower temperatures, are suitable for age or precipitate hardening. In Chapter 12, we will learn how nonequilibrium processes are needed to make precipitation hardened alloys. A vertical line on a phase diagram (e.g., Figure 11-8) that shows a specific composition is known as an **isopleth**. Determination of reactions that occur upon the cooling of a particular composition is known as an **isoplethal study**. The following example illustrates how certain calculations related to the composition of phases and their relative concentrations can be performed.

Example 11-2   *Phases in the Lead–Tin (Pb–Sn) Phase Diagram*

Determine (a) the solubility of tin in solid lead at 100°C, (b) the maximum solubility of lead in solid tin, (c) the amount of β that forms if a Pb-10% Sn alloy is cooled to 0°C, (d) the masses of tin contained in the α and β phases, and (e) the mass of lead contained in the α and β phases. Assume that the total mass of the Pb-10% Sn alloy is 100 grams.

SOLUTION

The phase diagram we need is shown in Figure 11-8. All percentages shown are weight %.

(a) The 100°C temperature intersects the solvus curve at 6% Sn. The solubility of tin (Sn) in lead (Pb) at 100°C, therefore, is 6%.

(b) The maximum solubility of lead (Pb) in tin (Sn), which is found from the tin-rich side of the phase diagram, occurs at the eutectic temperature of 183°C and is 97.5% Sn or 2.5% Pb.

(c) At 0°C, the 10% Sn alloy is in the $\alpha + \beta$ region of the phase diagram. By drawing a tie line at 0°C and applying the lever rule, we find that

$$\% \ \beta = \frac{10 - 2}{100 - 2} \times 100 = 8.2\%$$

Note that the tie line intersects the solvus curve for solubility of Pb in Sn at a non-zero concentration of Sn. We cannot read this accurately from the diagram; however, we assume that the right-hand point for the tie line is 100% Sn. The percent of α would be $(100 - \% \ \beta) = 91.8\%$. This means if we have 100 g of the 10% Sn alloy, it will consist of 8.2 g of the β phase and 91.8 g of the α phase.

(d) Note that 100 g of the alloy will consist of 10 g of Sn and 90 g of Pb. The Pb and Sn are distributed in two phases (i.e., α and β). The mass of Sn in the α phase = 2% Sn × 91.8 g of α phase = 0.02 × 91.8 g = 1.836 g. Since tin (Sn) appears in both the α and

β phases, the mass of Sn in the β phase will be $= (10 - 1.836)$ g $= 8.164$ g. Note that in this case, the β phase at 0°C is nearly pure Sn.

(e) Let's now calculate the mass of lead in the two phases. The mass of Pb in the α phase will be equal to the mass of the α phase minus the mass of Sn in the α phase $= 91.8$ g $- 1.836$ g $= 89.964$ g. We could have also calculated this as

$$\text{Mass of Pb in the } \alpha \text{ phase} = 98\% \text{ Pb} \times 91.8 \text{ g of } \alpha \text{ phase} = 0.98 \times 91.8 \text{ g}$$
$$= 89.964 \text{ g}$$

We know the total mass of the lead (90 g), and we also know the mass of lead in the α phase. Thus, the mass of Pb in the β phase $= 90 - 89.964 = 0.036$ g. This is consistent with what we said earlier (i.e., the β phase, in this case, is almost pure tin).

Eutectic Alloys

The alloy containing 61.9% Sn has the eutectic composition (Figure 11-9). The word eutectic comes from the Greek word *eutectos* that means easily fused. Indeed, in a binary system showing one eutectic reaction, an alloy with a eutectic composition has the lowest melting temperature. This is the composition for which there is no freezing range (i.e., solidification of this alloy occurs at one temperature, 183°C in the Pb–Sn system). Above 183°C, the alloy is all liquid and, therefore, must contain 61.9% Sn. After the liquid cools to 183°C, the eutectic reaction begins:

$$L_{61.9\% \text{ Sn}} \rightarrow \alpha_{19\% \text{ Sn}} + \beta_{97.5\% \text{ Sn}}$$

Two solid solutions—α and β—are formed during the eutectic reaction. The compositions of the two solid solutions are given by the ends of the eutectic line.

During solidification, growth of the eutectic requires both removal of the latent heat of fusion and redistribution of the two different atom species by diffusion. Since

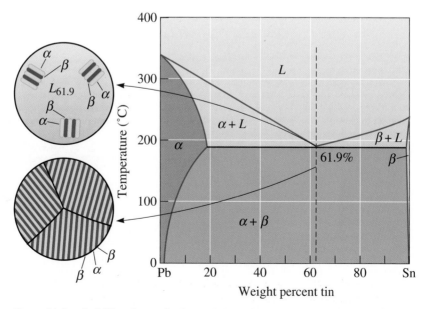

Figure 11-9   Solidification and microstructure of the eutectic alloy Pb-61.9% Sn.

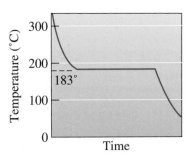

Figure 11-10
The cooling curve for an eutectic alloy is a simple thermal arrest, since eutectics freeze or melt at a single temperature.

solidification occurs completely at 183°C, the cooling curve (Figure 11-10) is similar to that of a pure metal; that is, a thermal arrest or plateau occurs at the eutectic temperature. In Chapter 9, we stated that alloys solidify over a range of temperatures (between the liquidus and solidus) known as the freezing range. Eutectic compositions are an exception to this rule since they transform from a liquid to a solid at a constant temperature (i.e., the eutectic temperature).

As atoms are redistributed during eutectic solidification, a characteristic microstructure develops. In the lead-tin system, the solid α and β phases grow from the liquid in a **lamellar**, or plate-like, arrangement (Figure 11-11). The lamellar structure permits the lead and tin atoms to move through the liquid, in which diffusion is rapid, without having to move an appreciable distance. This lamellar structure is characteristic of numerous other eutectic systems.

The product of the eutectic reaction has a characteristic arrangement of the two solid phases called the **eutectic microconstituent**. In the Pb-61.9% Sn alloy, 100% of the eutectic microconstituent is formed, since all of the liquid goes through the reaction. The following example shows how the amounts and compositions of the phases present in a eutectic alloy can be calculated.

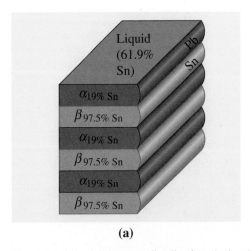

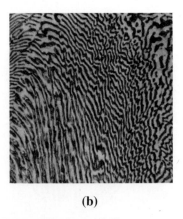

(a)   (b)

Figure 11-11   (a) Atom redistribution during lamellar growth of a lead–tin eutectic. Tin atoms from the liquid preferentially diffuse to the β plates, and lead atoms diffuse to the α plates. (b) Photomicrograph of the lead-tin eutectic microconstituent ($\times$ 400). *(Reprinted Courtesy of Don Askeland.)*

Example 11-3 *Amount of Phases in the Eutectic Alloy*

(a) Determine the amount and composition of each phase in 200 g of a lead-tin alloy of eutectic composition immediately after the eutectic reaction has been completed. (b) Calculate the mass of phases present. (c) Calculate the masses of lead and tin in each phase.

SOLUTION

(a) The eutectic alloy contains 61.9% Sn. We work the lever law at a temperature just below the eutectic—say, at 182°C, since that is the temperature at which the eutectic reaction is just completed. The fulcrum of our lever is 61.9% Sn. The composition of the α is Pb-19% Sn, and the composition of the β is Pb-97.5% Sn. The ends of the tie line coincide approximately with the ends of the eutectic line.

$$\alpha: (\text{Pb} - 19\% \text{ Sn}) \% \ \alpha = \frac{97.5 - 61.9}{97.5 - 19.0} \times 100 = 45.35\%$$

$$\beta: (\text{Pb} - 97.5\% \text{ Sn}) \% \ \beta = \frac{61.9 - 19.0}{97.5 - 19.0} \times 100 = 54.65\%$$

Alternately, we could state that the weight fraction of the α phase is 0.4535 and that of the β phase is 0.5465.

A 200 g sample of the alloy would contain a total of $200 \times 0.6190 = 123.8$ g Sn and a balance of 76.2 g lead. The total mass of lead and tin cannot change as a result of conservation of mass. What changes is the mass of lead and tin in the different phases.

(b) At a temperature just below the eutectic:

The mass of the α phase in 200 g of the alloy

$$= \text{mass of the alloy} \times \text{fraction of the } \alpha \text{ phase}$$
$$= 200 \text{ g} \times 0.4535 = 90.7 \text{ g}$$

The amount of the β phase in 200 g of the alloy

$$= (\text{mass of the alloy} - \text{mass of the } \alpha \text{ phase})$$
$$= 200.0 \text{ g} - 90.7 = 109.3 \text{ g}$$

We could have also written this as

Amount of β phase in 200 g of the alloy

$$= \text{mass of the alloy} \times \text{fraction of the } \beta \text{ phase}$$
$$= 200 \text{ g} \times 0.5465 = 109.3 \text{ g}$$

Thus, at a temperature just below the eutectic (i.e., at 182°C), the alloy contains 109.3 g of the β phase and 90.7 g of the α phase.

(c) Now let's calculate the masses of lead and tin in the α and β phases:

Mass of Pb in the α phase $= $ mass of the α phase in 200 g
$$\times \text{ (wt. fraction Pb in } \alpha)$$

Mass Pb in the α phase $= 90.7 \text{ g} \times (1 - 0.19) = 73.5 \text{ g}$

Mass of Sn in the α phase = mass of the α phase − mass of Pb in the α phase

Mass of Sn in the α phase = (90.7 − 73.5 g) = 17.2 g

Mass of Pb in the β phase = mass of the β phase in 200 g × (wt. fraction Pb in β)

Mass of Pb in the β phase = (109.3 g) × (1 − 0.975) = 2.7 g

Mass of Sn in the β phase = total mass of Sn − mass of Sn in the α phase

Mass of Sn in the β phase = 123.8 g − 17.2 g = 106.6 g

Notice that we could have obtained the same result by considering the total lead mass balance as follows:

Total mass of lead in the alloy = mass of lead in the α phase
+ mass of lead in the β phase

76.2 g = 73.5 g + mass of lead in the β phase

Mass of lead in the β phase = 76.2 − 73.5 g = 2.7 g

Figure 11-12 summarizes the various concentrations and masses. This analysis confirms that most of the lead in the eutectic alloy gets concentrated in the α phase. Most of the tin gets concentrated in the β phase.

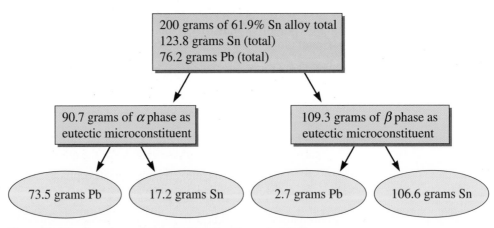

Figure 11-12   Summary of calculations (for Example 11-3).

Hypoeutectic and Hypereutectic Alloys   A **hypoeutectic alloy**
is an alloy with a composition between that of the left-hand end of the tie line defining the eutectic reaction and the eutectic composition. As a hypoeutectic alloy containing between 19% and 61.9% Sn cools, the liquid begins to solidify at the liquidus temperature, producing solid α; however, solidification is completed by going through the eutectic

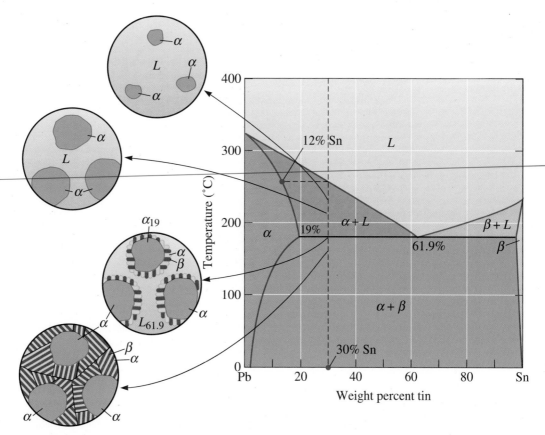

Figure 11-13   The solidification and microstructure of a hypoeutectic alloy (Pb-30% Sn).

reaction (Figure 11-13). This solidification sequence occurs for compositions in which the vertical line corresponding to the original composition of the alloy crosses both the liquidus and the eutectic.

An alloy composition between that of the right-hand end of the tie line defining the eutectic reaction and the eutectic composition is known as a **hypereutectic alloy**. In the Pb–Sn system, any composition between 61.9% and 97.5% Sn is hypereutectic.

Let's consider a hypoeutectic alloy containing Pb-30% Sn and follow the changes in structure during solidification (Figure 11-13). On reaching the liquidus temperature of 260°C, solid α containing about 12% Sn nucleates. The solid α grows until the alloy cools to just above the eutectic temperature. At 184°C, we draw a tie line and find that the solid α contains 19% Sn and the remaining liquid contains 61.9% Sn. We note that at 184°C, the liquid contains the eutectic composition! When the alloy is cooled below 183°C, all of the remaining liquid goes through the eutectic reaction and transforms to a lamellar mixture of α and β. The microstructure shown in Figure 11-14(a) results. Notice that the eutectic microconstituent surrounds the solid α that formed between the liquidus and eutectic temperatures. The eutectic microconstituent is continuous and the primary phase is dispersed between the colonies of the eutectic microconstituent.

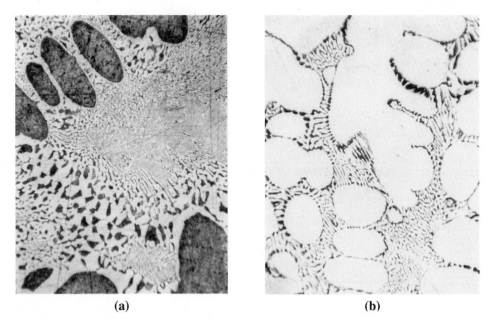

(a)   **(b)**

Figure 11-14   (a) A hypoeutectic lead-tin alloy. (b) A hypereutectic lead-tin alloy. The dark constituent is the lead-rich solid α, the light constituent is the tin-rich solid β, and the fine plate structure is the eutectic ($\times$ 400). *(Micrographs reprinted courtesy of Don Askeland.)*

Example 11-4	*Determination of Phases and Amounts in a Pb-30% Sn Hypoeutectic Alloy*

For a Pb-30% Sn alloy, determine the phases present, their amounts, and their compositions at 300°C, 200°C, 184°C, 182°C, and 0°C.

SOLUTION

Temperature (°C)	Phases	Compositions	Amounts
300	L	L: 30% Sn	$L = 100\%$
200	$\alpha + L$	L: 55% Sn	$L = \dfrac{30 - 18}{55 - 18} \times 100 = 32\%$
		α: 18% Sn	$\alpha = \dfrac{55 - 30}{55 - 18} \times 100 = 68\%$
184	$\alpha + L$	L: 61.9% Sn	$L = \dfrac{30 - 19}{61.9 - 19} \times 100 = 26\%$
		α: 19% Sn	$\alpha = \dfrac{61.9 - 30}{61.9 - 19} \times 100 = 74\%$
182	$\alpha + \beta$	α: 19% Sn	$\alpha = \dfrac{97.5 - 30}{97.5 - 19} \times 100 = 86\%$
		β: 97.5% Sn	$\beta = \dfrac{30 - 19}{97.5 - 19} \times 100 = 14\%$
0	$\alpha + \beta$	α: 2% Sn	$\alpha = \dfrac{100 - 30}{100 - 2} \times 100 = 71\%$
		β: 100% Sn	$\beta = \dfrac{30 - 2}{100 - 2} \times 100 = 29\%$

Note that in these calculations, the fractions have been rounded off to the nearest %. This can pose problems if we were to calculate masses of different phases, in that you may not be able to preserve mass balance. It is usually a good idea not to round these percentages if you are going to perform calculations concerning amounts of different phases or masses of elements in different phases.

We call the solid α phase that forms when the liquid cools from the liquidus to the eutectic the **primary** or **proeutectic microconstituent**. This solid α does not take part in the eutectic reaction. Thus, the morphology and appearance of this α phase is distinct from that of the α phase that appears in the eutectic microconstituent. Often we find that the amounts and compositions of the microconstituents are of more use to us than the amounts and compositions of the phases.

Example 11-5	*Microconstituent Amount and Composition for a Hypoeutectic Alloy*

Determine the amounts and compositions of each microconstituent in a Pb-30% Sn alloy immediately after the eutectic reaction has been completed.

SOLUTION

This is a hypoeutectic composition. Therefore, the microconstituents expected are primary α and eutectic. Note that we still have only two phases (α and β).

We can determine the amounts and compositions of the microconstituents if we look at how they form. The *primary* α microconstituent is all of the solid α that forms before the alloy cools to the eutectic temperature; the eutectic microconstituent is all of the liquid that goes through the eutectic reaction. At a temperature just above the eutectic—say, 184°C—the amounts and compositions of the two phases are

$$\alpha: 19\% \ \text{Sn}, \ \%\,\alpha = \frac{61.9 - 30}{61.9 - 19} \times 100 = 74\% = \% \text{ primary } \alpha$$

$$L: 61.9\% \ \text{Sn}, \ \%\,L = \frac{30 - 19}{61.9 - 19} \times 100 = 26\% = \% \text{ eutectic at } 182°C$$

Thus, the primary alpha microconstituent is obtained by determining the amount of α present at the temperature just above the eutectic. The amount of eutectic microconstituent at a temperature *just below* the eutectic (e.g., 182°C) is determined by calculating the amount of liquid *just above* the eutectic temperature (e.g., at 184°C), since all of this liquid of eutectic composition is transformed into the eutectic microconstituent. Note that at the eutectic temperature (183°C), the eutectic reaction is in progress (formation of the proeutectic α is complete); hence, the amount of the eutectic microconstituent at 183°C will change with time (starting at 0% and ending at 26% eutectic, in this case). Please be certain that you understand this example since many students tend to miss how the calculation is performed.

When the alloy cools below the eutectic to 182°C, all of the liquid at 184°C transforms to eutectic and the composition of the eutectic microconstituent is 61.9% Sn. The solid α present at 184°C remains unchanged after cooling to 182°C and is the primary microconstituent.

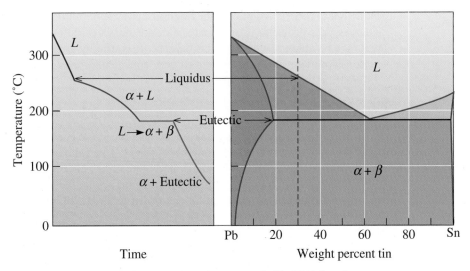

Figure 11-15   The cooling curve for a hypoeutectic Pb-30% Sn alloy.

The cooling curve for a hypoeutectic alloy is a composite of those for solid-solution alloys and "straight" eutectic alloys (Figure 11-15). A change in slope occurs at the liquidus as primary α begins to form. Evolution of the latent heat of fusion slows the cooling rate as the solid α grows. When the alloy cools to the eutectic temperature, a thermal arrest is produced as the eutectic reaction proceeds at 183°C. The solidification sequence is similar in a hypereutectic alloy, giving the microstructure shown in Figure 11-14(b).

11-5   Strength of Eutectic Alloys

Each phase in the eutectic alloy is, to some degree, solid-solution strengthened. In the lead–tin system, α, which is a solid solution of tin in lead, is stronger than pure lead (Chapter 10). Some eutectic alloys can be strengthened by cold working. We also control grain size by adding appropriate inoculants or grain refiners during solidification. Finally, we can influence the properties by controlling the amount and microstructure of the eutectic.

Eutectic Colony Size   Eutectic colonies each nucleate and grow independently. Within each colony, the orientation of the lamellae in the eutectic microconstituent is identical. The orientation changes on crossing a colony boundary [Figure 11-16(a)]. We can refine the eutectic colonies and improve the strength of the eutectic alloy by inoculation (Chapter 9).

Interlamellar Spacing   The **interlamellar spacing** of a eutectic is the distance from the center of one α lamella to the center of the next α lamella [Figure 11-16(b)]. A small interlamellar spacing indicates that the amount of α to β interface area is large. A small interlamellar spacing therefore increases the strength of the eutectic.

The interlamellar spacing is determined primarily by the growth rate of the eutectic,

$$\gamma = cR^{-1/2} \tag{11-2}$$

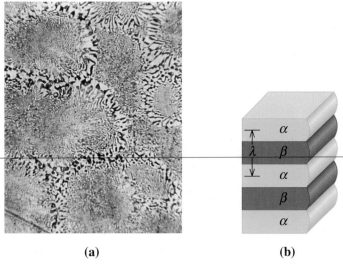

(a)   (b)

Figure 11-16   (a) Colonies in the lead–tin eutectic ($\times$ 300). (b) The interlamellar spacing in a eutectic microstructure.

where R is the growth rate (cm/s) and c is a constant. The interlamellar spacing for the lead–tin eutectic is shown in Figure 11-17. We can increase the growth rate R, and consequently reduce the interlamellar spacing by increasing the cooling rate or reducing the solidification time. The following example demonstrates how the solidification of a Pb–Sn alloy can be controlled.

Amount of Eutectic

We also control the properties by the relative amounts of the primary microconstituent and the eutectic. In the lead–tin system, the amount of the eutectic microconstituent changes from 0% to 100% when the tin content increases from 19% to 61.9%. With increasing amounts of the stronger eutectic microconstituent, the strength of the alloy increases (Figure 11-18). Similarly, when we increase the lead added to tin from 2.5% to 38.1% Pb, the amount of primary β in the hypereutectic alloy decreases, the amount of the strong eutectic increases, and the strength increases. When both individual phases have about the same strength, the eutectic alloy is expected to have the highest strength due to effective dispersion strengthening.

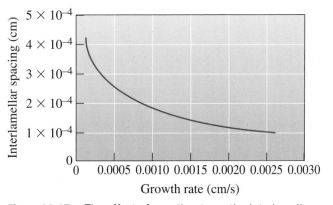

Figure 11-17   The effect of growth rate on the interlamellar spacing in the lead–tin eutectic.

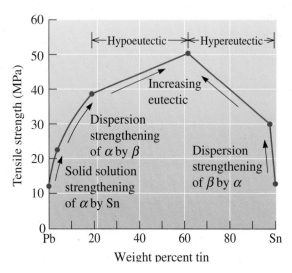

Figure 11-18
The effect of the composition and strengthening mechanism on the tensile strength of lead-tin alloys.

Microstructure of the Eutectic

Not all eutectics give a lamellar structure. The shapes of the two phases in the microconstituent are influenced by the cooling rate, the presence of impurity elements, and the nature of the alloy.

The aluminum-silicon eutectic phase diagram (Figure 11-19) forms the basis for a number of important commercial alloys. The silicon portion of the eutectic grows as thin, flat plates that appear needle-like in a photomicrograph [Figure 11-20(a)]. The brittle silicon platelets concentrate stresses and reduce ductility and toughness.

The eutectic microstructure in aluminum-silicon alloys is altered by modification. **Modification** causes the silicon phase to grow as thin, interconnected rods between aluminum dendrites [Figure 11-20(b)], improving both tensile strength and percent elongation. In two dimensions, the modified silicon appears to be composed of small, round particles. Rapidly cooled alloys, such as those used for die casting, are modified naturally during solidification. At slower cooling rates, however, about 0.02% Na or 0.01% Sr must be added to cause modification.

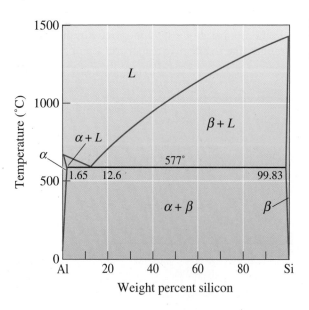

Figure 11-19
The aluminum-silicon phase diagram.

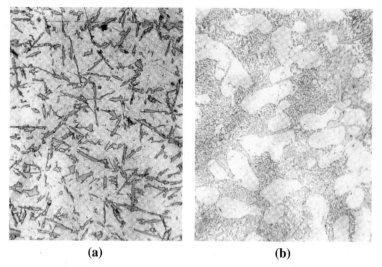

(a)   (b)

Figure 11-20   Typical eutectic microstructures: (a) needle-like silicon plates in the aluminum–silicon eutectic (× 100) and (b) rounded silicon rods in the modified aluminum–silicon eutectic (× 100). *(Reprinted courtesy of Don Askeland.)*

The shape of the primary phase is also important. Often the primary phase grows in a dendritic manner; decreasing the secondary dendrite arm spacing of the primary phase may improve the properties of the alloy. In hypereutectic aluminum-silicon alloys, coarse β is the primary phase [Figure 11-21(a)]. Because β is hard, the hypereutectic alloys are wear-resistant and are used to produce automotive engine parts, but the coarse β causes poor machinability and gravity segregation (where the primary β floats to the surface of the casting during freezing). Addition of 0.05% phosphorus (P) encourages nucleation of primary silicon, refines its size, and minimizes its deleterious qualities [Figure 11-21(b)]. The two examples that follow show how eutectic compositions can be designed to achieve certain levels of mechanical properties.

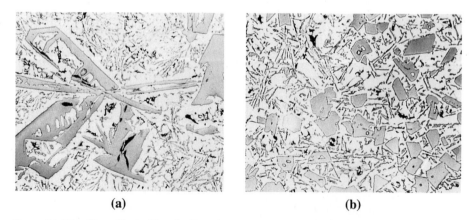

(a)   (b)

Figure 11-21   The effect of hardening with phosphorus on the microstructure of hypereutectic aluminum-silicon alloys: (a) coarse primary silicon and (b) fine primary silicon, as refined by phosphorus addition (× 75). *(From ASM Handbook, Vol. 7, (1972), ASM International, Materials Park, OH 44073.)*

Example 11-6 *Design of Materials for a Wiping Solder*

One way to repair dents in a metal is to wipe a partly liquid-partly solid material into the dent, then allow this filler material to solidify. For our application, the wiping material should have the following specifications: (1) a melting temperature below 230°C, (2) a tensile strength in excess of 41 MPa, (3) be 60% to 70% liquid during application, and (4) the lowest possible cost. Design an alloy and repair procedure that will meet these specifications. You may wish to consider a Pb–Sn alloy.

SOLUTION

Let's see if one of the Pb–Sn alloys will satisfy these conditions. First, the alloy must contain more than 40% Sn in order to have a melting temperature below 230°C (Figure 11-6). This low temperature will make it easier for the person doing the repairs to apply the filler.

Second, Figure 11-18 indicates that the tin content must be between 23% and 80% to achieve the required 41 MPa tensile strength. In combination with the first requirement, any alloy containing between 40 and 80% Sn will be satisfactory.

Third, the cost of tin is about $5500/ton whereas that of lead is $550/ton. Thus, an alloy of Pb-40% Sn might be the most economical choice. There are considerations, as well, such as: What is the geometry? Can the alloy flow well under that geometry (i.e., the viscosity of the molten metal)?

Finally, the filler material must be at the correct temperature in order to be 60% to 70% liquid. As the calculations below show, the temperature must be between 200°C and 210°C:

$$\% \, L_{200} = \frac{40 - 18}{55 - 18} \times 100 = 60\%$$

$$\% \, L_{210} = \frac{40 - 17}{50 - 17} \times 100 = 70\%$$

Our recommendation, therefore, is to use a Pb-40% Sn alloy applied at 205°C, a temperature at which there will be 65% liquid and 35% primary α. As mentioned before, we should also pay attention to the toxicity of lead and any legal liabilities the use of such materials may cause. A number of new lead-free solders have been developed.

Example 11-7 *Design of a Wear Resistant Part*

Design a lightweight, cylindrical component that will provide excellent wear resistance at the inner wall, yet still have reasonable ductility and toughness overall. Such a product might be used as a cylinder liner in an automotive engine.

SOLUTION

Many wear resistant parts are produced from steels, which have a relatively high density, but the hypereutectic Al-Si alloys containing primary β may provide the wear resistance that we wish at one-third the weight of the steel.

Since the part to be produced is cylindrical in shape, centrifugal casting (Figure 11-22) might be a unique method for producing it. In centrifugal casting,

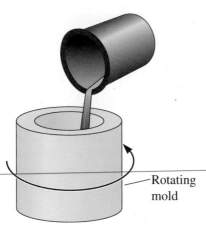

Figure 11-22
Centrifugal casting of a hypereutectic Al-Si alloy: (a) Liquid alloy is poured into a rotating mold, and (b) the solidified casting is hypereutectic at the inner diameter and eutectic at the outer diameter (for Example 11-7).

Rotating mold

(a)

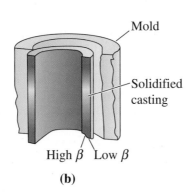

Mold

Solidified casting

High β Low β

(b)

liquid metal is poured into a rotating mold and the centrifugal force produces a hollow shape. In addition, material that has a high density is spun to the outside wall of the casting, while material that has a lower density than the liquid migrates to the inner wall.

When we centrifugally cast a hypereutectic Al-Si alloy, primary β nucleates and grows. The density of the β phase (if we assume it to be same as that of pure Si) is, according to Appendix A, 2.33 g/cm^3, compared with a density near 2.7 g/cm^3 for aluminum. As the primary β particles precipitate from the liquid, they are spun to the inner surface. The result is a casting that is composed of eutectic microconstituent (with reasonable ductility) at the outer wall and a hypereutectic composition, containing large amounts of primary β, at the inner wall.

A typical alloy used to produce aluminum engine components is Al-17% Si. From Figure 11-19, the total amount of primary β that can form is calculated at 578°C, just above the eutectic temperature:

$$\% \text{ Primary } \beta = \frac{17 - 12.6}{99.83 - 12.6} \times 100 = 5.0\%$$

Although only 5.0% primary β is expected to form, the centrifugal action can double or triple the amount of β at the inner wall of the casting.

11-6   Eutectics and Materials Processing

Manufacturing processes take advantage of the low melting temperature associated with the eutectic reaction. The Pb–Sn alloys are the basis for a series of alloys used to produce filler materials for soldering (Chapter 9). If, for example, we wish to join copper pipe, individual segments can be joined by introducing the low melting point eutectic Pb–Sn alloy into the joint [Figure 11-23(a)]. The copper is heated to just above the eutectic temperature. The heated copper melts the Pb–Sn alloy, which is then drawn into the thin gap by capillary action. When the Pb–Sn alloy cools and solidifies, the copper is joined. The possibility of corrosion of such pipes and the introduction of lead (Pb) into water must also be considered.

Many casting alloys are also based on eutectic alloys. Liquid can be melted and poured into a mold at low temperatures, reducing energy costs involved in melting, minimizing casting defects such as gas porosity, and preventing liquid metal-mold reactions. Cast iron (Chapter 13) and many aluminum alloys (Chapter 14) are eutectic alloys.

Although most of this discussion has been centered around metallic materials, it is important to recognize that eutectics are very important in many ceramic systems (Chapter 15). Formation of eutectics played a role in the successful formation of glass-like materials known as the Egyptian faience. The sands of the Nile River Valley contained appreciable amounts of limestone ($CaCO_3$). Plant ash contains considerable amounts of potassium and sodium oxide and is used to cause the sand to melt at lower temperatures by the formation of eutectics.

Silica and alumina are the most widely used ceramic materials. Figure 11-23(b) shows a phase diagram for the Al_2O_3-SiO_2 system. Notice the eutectic at ~1587°C. The dashed lines on this diagram show metastable extensions of the liquidus and metastable miscibility gaps. As mentioned before, the existence of these gaps makes it possible to make technologically useful products such as Vycor™ and Pyrex® glasses. A Vycor™ glass is made by first melting (approximately at 1500°C) silica (63%), boron oxide (27%), sodium oxide (7%), and alumina (3%). The glass is then formed into the desired shapes. During glass formation, the glass phase separates (because of the metastable miscibility gap) into boron oxide rich and silica rich regions. The boron oxide rich regions are dissolved using an acid. The porous object is sintered to form Vycor™ glass that contains 95 wt% silica, 4% boron oxide, and 1% sodium oxide. It would be very difficult to achieve a high silica glass such as this without resorting to the technique described above. Pyrex® glasses contain about 80 wt% silica, 13% boron oxide, 4% sodium oxide, and 2% alumina. These are used widely in making laboratory ware (i.e., beakers, etc.) and household products.

Figure 1-23(c) shows a binary phase diagram for the CaO-SiO_2 system. Compositions known as E-glass or S-glass are used to make the fibers that go into fiber-reinforced plastics. These glasses are made by melting silica sand, limestone, and boric acid at about 1260°C. The glass is then drawn into fibers. The E-glass (the letter "E" stands for "electrical," as the glass was originally made for electrical insulation) contains approximately 52–56 wt% silica, 12–16% Al_2O_3, 5–10% B_2O_3, 0–5% MgO, 0–2% Na_2O, and 0–2% K_2O. The S-glass (the letter "S" represents "strength") contains approximately 65 wt% silica, 12–25% Al_2O_3, 10% MgO, 0–2% Na_2O, and 0–2% K_2O.

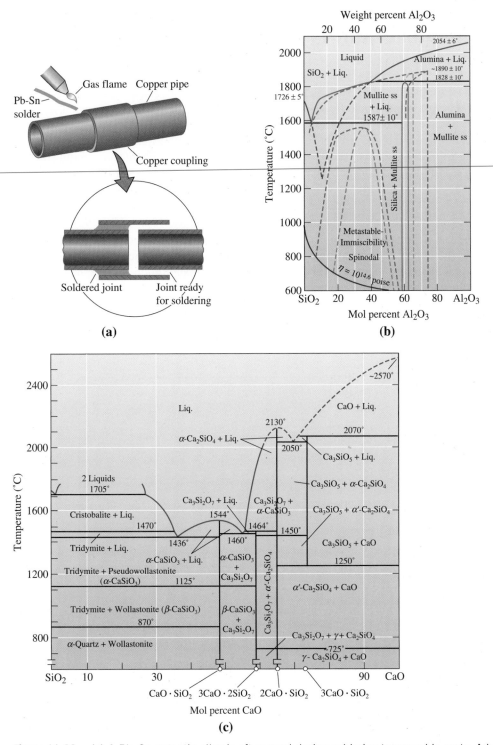

Figure 11-23 (a) A Pb–Sn eutectic alloy is often used during soldering to assemble parts. A heat source, such as a gas flame, heats both the parts and the filler material. The filler is drawn into the joint and solidifies. (b) A phase diagram for Al_2O_3-SiO_2. (Adapted from *Introduction to Phase Equilibria in Ceramics,* by Bergeron, C.G. and Risbud, S.H., The American Ceramic Society, Inc., 1984, page 44.) (c) A phase diagram for the CaO-SiO_2 system. (*Adapted from* Introduction to Phase Equilibria, *by C.G. Bergeron and S.H. Risbud, pp. 44 and 45, Figs. 3-36 and 3-37. Copyright © 1984 American Ceramic Society Adapted by permission.*).

11-7   Nonequilibrium Freezing in the Eutectic System

Suppose we have an alloy, such as Pb-15% Sn, that ordinarily solidifies as a solid solution alloy. The last liquid should freeze near 230°C, well above the eutectic; however, if the alloy cools too quickly, a nonequilibrium solidus curve is produced (Figure 11-24). The primary α continues to grow until, just above 183°C, the remaining nonequilibrium liquid contains 61.9% Sn. This liquid then transforms to the eutectic microconstituent, surrounding the primary α. For the conditions shown in Figure 11-24, the amount of nonequilibrium eutectic is

$$\% \text{ eutectic} = \frac{15 - 10}{61.9 - 10} \times 100 = 9.6\%$$

When heat treating an alloy such as Pb-15% Sn, we must keep the maximum temperature below the eutectic temperature of 183°C to prevent hot shortness or partial melting (Chapter 10). This concept is very important in the precipitation, or age, hardening of metallic alloys such as those in the Al-Cu system.

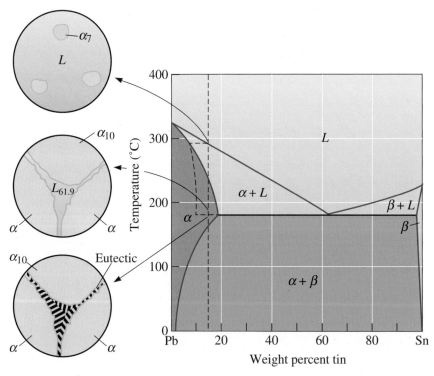

Figure 11-24   Nonequilibrium solidification and microstructure of a Pb-15% Sn alloy. A nonequilibrium eutectic microconstituent can form if the solidification is too rapid.

11-8   Nanowires and the Eutectic Phase Diagram

Nanowires are cylinders or "wires" of material with diameters on the order of 10 to 100 nm. Nanowires have generated great technological interest due to their electrical, mechanical, chemical, and optical properties. Potential applications include biological and chemical sensors and electrical conductors in nanoelectronic devices. A common method of fabricating nanowire materials is through the technique known as Vapor–Liquid–Solid (VLS) growth. The growth of silicon nanowires can be understood by considering the gold–silicon binary phase diagram in Figure 11-25(a). Notice that there is no mutual solubility between gold and silicon.

In one example of VLS silicon nanowire growth, the first step is to deposit a thin layer of pure gold on a substrate. When the substrate is heated, the gold dewets from the substrate or "balls up," forming a series of gold nanoparticles on the substrate surface, as shown in Figure 11-25(b). These gold nanoparticles, also known as the nanowire catalysts, act as the template for silicon nanowire growth. Silane gas SiH_4, the "vapor" in the VLS process for silicon nanowire growth, then flows through a chamber holding the substrate with the

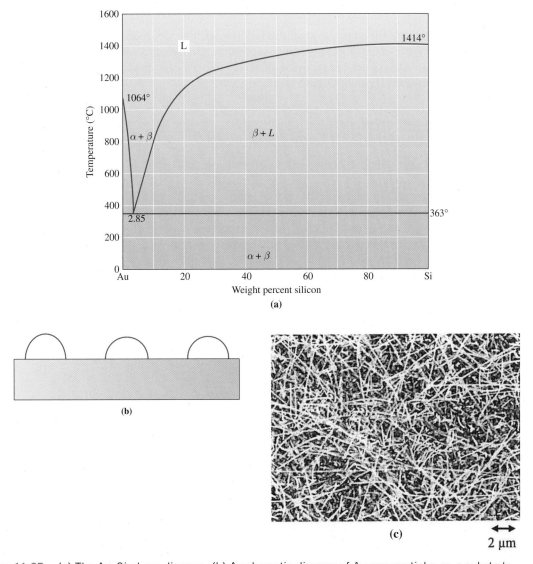

Figure 11-25   (a) The Au–Si phase diagram. (b) A schematic diagram of Au nanoparticles on a substrate. The drawing is not to scale. (c) A scanning electron micrograph of silicon nanowires. *(Reprinted figure with permission from "Diameter-Independent Kinetics in the Vapor-Liquid-Solid Growth of Si Nanowires" by S. Kodambaka, J. Tersolf, M. C. Reuter, and F. M. Ross, in Physical Review Letters 96, p. 096105-2 (2006). Copyright 2006 by the American Physical Society.) (Courtesy of Wendelyn J. Wright.)*

gold nanoparticles. The substrate is heated to a temperature above the gold–silicon eutectic temperature of 363°C but below the melting temperature of the gold (1064°C). The silane gas decomposes according to $SiH_4 \rightarrow Si + 2H_2$. The silicon that is produced from the decomposition of the silane adsorbs to the gold nanoparticles and diffuses into them, causing the gold nanoparticles to begin to melt (thus forming the "liquid" of VLS growth). The gold nanoparticles are held at a constant temperature, the silicon content increases, and the gold–silicon alloy enters a two-phase region in the gold–silicon phase diagram, as shown in Figure 11-25(a). Silicon continues to diffuse into the gold nanoparticles until the nanoparticles are completely molten. The nanoparticles retain their shape as roughly hemispherical balls due to surface-energy considerations between the nanoparticles and the substrate (think of oil droplets on a Teflon® pan). As the silicon content increases due to the diffusion of silicon into the nanoparticles, a solid phase begins to form. The solid phase that forms is the silicon nanowire. It is pure silicon. The nanowire has the same diameter as the gold nanoparticle. The liquid phase that is present rides atop the silicon nanowire as it grows upward from the substrate. The nanowire continues to grow in length while silane gas is supplied to the substrate. When the nanowire growth is complete, the temperature is reduced, and the liquid phase atop the silicon nanowire solidifies on cooling through the eutectic temperature. Silicon nanowires are shown in Figure 11-25(c).

Example 11-8   *Growth of Silicon Nanowires*

Consider gold (Au) nanoparticles on a silicon (Si) substrate held at 600°C over which silane gas flows. (a) When a nanoparticle consists of 1 wt% Si, describe the phases that are present in the nanoparticle, the relative amount of each phase, and the composition of each phase. Make the necessary composition approximations from the phase diagram in Figure 11-25 (a). (b) At what percentage of silicon will the nanoparticle become completely molten? (c) At what percentage of silicon will the molten nanoparticle begin to solidify again? (d) What are the final compositions of the nanowire and the solidified catalyst?

SOLUTION

(a) The nanoparticle consists of a solid and a liquid phase at 1 wt% Si. The solid is 100% Au. According to Figure 11-25(a), the liquid contains approximately 2.4 wt% Si and 97.6 wt% Au. (The wt% silicon must be less than 2.85% according to the eutectic point on the phase diagram.)

Using the lever rule to determine the relative amounts of each phase:

$$\% \text{ Solid} = \frac{2.4 - 1.0}{2.4 - 0} = 58\% \qquad \% \text{ Liquid} = \frac{1.0 - 0}{2.4 - 0} = 42\%$$

Thus, the nanoparticle consists of 58% solid (100% Au) and 42% liquid (2.4 wt% Si and 97.6 wt% Au).

(b) The island will completely melt when the silicon content reaches approximately 2.4 wt%.

(c) The island will begin to solidify again when the silicon content reaches approximately 7 wt%.

(d) The solid phase is 100% Si. The liquid phase is approximately 7 wt% Si and 93 wt% Au.

Summary

- By producing a material containing two or more phases, dispersion strengthening is obtained. In metallic materials, the boundary between the phases impedes the movement of dislocations and improves strength. Introduction of multiple phases may provide other benefits, including improvement of the fracture toughness of ceramics and polymers.

- For optimum dispersion strengthening, particularly in metallic materials, a large number of small, hard, discontinuous dispersed particles should form in a soft, ductile matrix to provide the most effective obstacles to dislocations. Round dispersed phase particles minimize stress concentrations, and the final properties of the alloy can be controlled by the relative amounts of these and the matrix.

- Intermetallic compounds, which normally are strong but brittle, are frequently introduced as dispersed phases.

- Phase diagrams for materials containing multiple phases normally contain one or more three-phase reactions.

- The eutectic reaction permits liquid to solidify as an intimate mixture of two phases. By controlling the solidification process, we can achieve a wide range of properties. Some of the factors that can be controlled include the grain size or secondary dendrite arm spacings of primary microconstituents, the colony size of the eutectic microconstituent, the interlamellar spacing within the eutectic microconstituent, the microstructure, or shape, of the phases within the eutectic microconstituent, and the amount of the eutectic microconstituent that forms.

- The eutectoid reaction causes a solid to transform to a mixture of two other solids. As shown in the next chapter, heat treatments to control the eutectoid reaction provide an excellent basis for dispersion strengthening.

- Nanowires can be grown from eutectic systems through a process known as Vapor–Liquid–Solid (VLS) growth. In one example of VLS silicon nanowire growth, silane gas is passed over gold catalysts. Silicon is deposited on the gold, forming a binary system. As the silicon concentration increases, the catalysts become molten, and as the silicon concentration increases further, a solid phase, which is the silicon nanowire, forms. The diameter of the nanowires is controlled by the catalyst diameter, and the nanowire length is controlled by the growth time.

Glossary

Age hardening A strengthening mechanism that relies on a sequence of solid-state phase transformations in generating a dispersion of ultra-fine particles of a second phase. Age hardening is a form of dispersion strengthening. Also called precipitation hardening (Chapter 12).

Dispersed phase A solid phase that forms from the original matrix phase when the solubility limit is exceeded.

Dispersion strengthening Increasing the strength of a material by forming more than one phase. By proper control of the size, shape, amount, and individual properties of the phases, excellent combinations of properties can be obtained.

Eutectic A three-phase invariant reaction in which one liquid phase solidifies to produce two solid phases.

Eutectic microconstituent A characteristic mixture of two phases formed as a result of the eutectic reaction.

Eutectoid A three-phase invariant reaction in which one solid phase transforms to two different solid phases.

Hyper- A prefix indicating that the composition of an alloy is more than the composition at which a three-phase reaction occurs.

Hypereutectic alloy An alloy composition between that of the right-hand end of the tie line defining the eutectic reaction and the eutectic composition.

Hypo- A prefix indicating that the composition of an alloy is less than the composition at which a three-phase reaction occurs.

Hypoeutectic alloy An alloy composition between that of the left-hand end of the tie line defining the eutectic reaction and the eutectic composition.

Interlamellar spacing The distance between the center of a lamella or plate of one phase and the center of the adjacent lamella or plate of the same phase.

Intermediate solid solution A nonstoichiometric intermetallic compound displaying a range of compositions.

Intermetallic compound A compound formed of two or more metals that has its own unique composition, structure, and properties.

Interphase interface The boundary between two phases in a microstructure. In metallic materials, this boundary resists dislocation motion and provides dispersion strengthening and precipitation hardening.

Isopleth A line on a phase diagram that shows constant chemical composition.

Isoplethal study Determination of reactions and microstructural changes that are expected while studying a particular chemical composition in a system.

Lamella A thin plate of a phase that forms during certain three-phase reactions, such as the eutectic or eutectoid.

Matrix The continuous solid phase in a complex microstructure. Solid dispersed phase particles may form within the matrix.

Metastable miscibility gap A miscibility gap that extends below the liquidus or exists completely below the liquidus. Two liquids that are immiscible continue to exist as liquids and remain unmixed. These systems form the basis for Vycor™ and Pyrex® glasses.

Microconstituent A phase or mixture of phases in an alloy that has a distinct appearance. Frequently, we describe a microstructure in terms of the microconstituents rather than the actual phases.

Miscibility gap A region in a phase diagram in which two phases, with essentially the same structure, do not mix, or have no solubility in one another.

Modification Addition of alloying elements, such as sodium or strontium, which change the microstructure of the eutectic microconstituent in aluminum-silicon alloys.

Monotectic A three-phase reaction in which one liquid transforms to a solid and a second liquid on cooling.

Nanowires are cylinders or "wires" of material with diameters on the order of 10 to 100 nm.

Nonstoichiometric intermetallic compound A phase formed by the combination of two components into a compound having a structure and properties different from either component. The nonstoichiometric compound has a variable ratio of the components present in the compound (see also intermediate solid solution).

Peritectic A three-phase reaction in which a solid and a liquid combine to produce a second solid on cooling.

Peritectoid A three-phase reaction in which two solids combine to form a third solid on cooling.

Precipitate A solid phase that forms from the original matrix phase when the solubility limit is exceeded. We often use the term precipitate, as opposed to dispersed phase particles, for alloys formed by precipitation or age hardening. In most cases, we try to control the formation of the precipitate second phase particles to produce the optimum dispersion strengthening or age hardening. (Also called the dispersed phase.)

Precipitation hardening A strengthening mechanism that relies on a sequence of solid-state phase transformations in generating a dispersion of ultra-fine precipitates of a second phase (Chapter 12). It is a form of dispersion strengthening. Also called age hardening.

Primary microconstituent The microconstituent that forms before the start of a three-phase reaction. Also called the proeutectic microconstituent.

Solvus A solubility curve that separates a single solid-phase region from a two solid-phase region in the phase diagram.

Stoichiometric intermetallic compound A phase formed by the combination of two components into a compound having a structure and properties different from either component. The stoichiometric intermetallic compound has a fixed ratio of the components present in the compound.

Problems

Section 11-1 Principles and Examples of Dispersion Strengthening

11-1 What are the requirements of a matrix and precipitate for dispersion strengthening to be effective?

Section 11-2 Intermetallic Compounds

11-2 What is an intermetallic compound? How is it different from other compounds? For example, other than the obvious difference in composition, how is TiAl different from, for example, Al_2O_3?

11-3 Explain clearly the two different ways in which intermetallic compounds can be used.

11-4 What are some of the major problems in the utilization of intermetallics for high temperature applications?

Section 11-3 Phase Diagrams Containing Three-Phase Reactions

11-5 Define the terms eutectic, eutectoid, peritectic, peritectoid, and monotectic reactions.

11-6 What is an invariant reaction? Show that for a two-component system the number of degrees of freedom for an invariant reaction is zero.

11-7 A hypothetical phase diagram is shown in Figure 11-26.

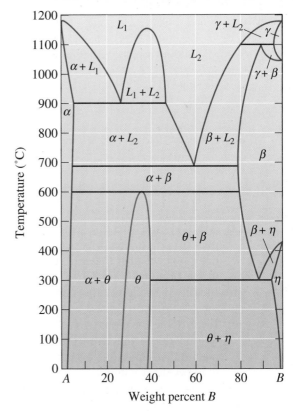

Figure 11-26 Hypothetical phase diagram (for Problem 11-7).

(a) Are any intermetallic compounds present? If so, identify them and determine whether they are stoichiometric or nonstoichiometric.

(b) Identify the solid solutions present in the system. Is either material A or B allotropic? Explain.

(c) Identify the three-phase reactions by writing down the temperature, the reaction in equation form, the composition of each phase in the reaction, and the name of the reaction.

11-8   The Cu-Zn phase diagram is shown in Figure 11-27.

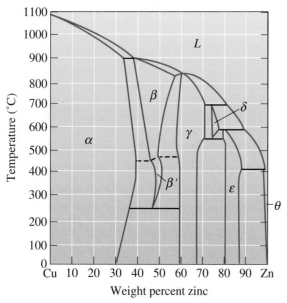

Figure 11-27   The copper-zinc phase diagram (for Problem 11-8).

(a) Are any intermetallic compounds present? If so, identify them and determine whether they are stoichiometric or nonstoichiometric.

(b) Identify the solid solutions present in the system.

(c) Identify the three-phase reactions by writing down the temperature, the reaction in equation form, and the name of the reaction.

11-9   The Al-Li phase diagram is shown in Figure 11-28.

(a) Are any intermetallic compounds present? If so, identify them and determine whether they are stoichiometric or nonstoichiometric. Determine the formula for each compound.

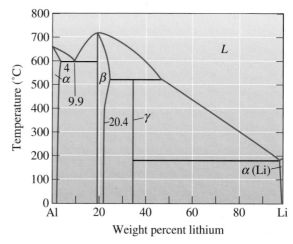

Figure 11-28   The aluminum-lithium phase diagram (for Problems 11-9 and 11-25).

(b) Identify the three-phase reactions by writing down the temperature, the reaction in equation form, the composition of each phase in the reaction, and the name of the reaction.

11-10   An intermetallic compound is found for 38 wt% Sn in the Cu-Sn phase diagram. Determine the formula for the compound.

11-11   An intermetallic compound is found for 10 wt% Si in the Cu-Si phase diagram. Determine the formula for the compound.

Section 11-4 The Eutectic Phase Diagram

11-12   Consider a Pb-15% Sn alloy. During solidification, determine

(a) the composition of the first solid to form;

(b) the liquidus temperature, solidus temperature, solvus temperature, and freezing range of the alloy;

(c) the amounts and compositions of each phase at 260°C;

(d) the amounts and compositions of each phase at 183°C; and

(e) the amounts and compositions of each phase at 25°C.

11-13   Consider an Al-12% Mg alloy (Figure 11-29). During solidification, determine

(a) the composition of the first solid to form;

(b) the liquidus temperature, solidus temperature, solvus temperature, and freezing range of the alloy;

(c) the amounts and compositions of each phase at 525°C;

(d) the amounts and compositions of each phase at 450°C; and

(e) the amounts and compositions of each phase at 25°C.

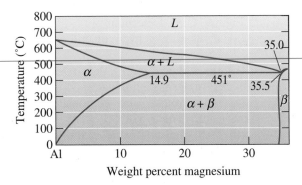

Weight percent magnesium

Figure 11-29   Portion of the aluminum-magnesium phase diagram (for Problems 11-13 and 11-26).

11-14 Consider a Pb-35% Sn alloy. Determine

(a) if the alloy is hypoeutectic or hypereutectic;

(b) the composition of the first solid to form during solidification;

(c) the amounts and compositions of each phase at 184°C;

(d) the amounts and compositions of each phase at 182°C;

(e) the amounts and compositions of each microconstituent at 182°C; and

(f) the amounts and compositions of each phase at 25°C.

11-15 Consider a Pb-70% Sn alloy. Determine

(a) if the alloy is hypoeutectic or hypereutectic;

(b) the composition of the first solid to form during solidification;

(c) the amounts and compositions of each phase at 184°C;

(d) the amounts and compositions of each phase at 182°C;

(e) the amounts and compositions of each microconstituent at 182°C; and

(f) the amounts and compositions of each phase at 25°.

11-16 (a) Sketch a typical eutectic phase diagram with components A and B having similar melting points. B is much more soluble in A (maximum = 15%) than A is in B (maximum = 5%), and the eutectic composition occurs near 40% B. The eutectic temperature is 2/3 of the melting point. Label the axes of the diagram. Label all the phases. Use α and β to denote the solid phases. (b) For an overall composition of 60% B, list the sequence of phases found as the liquid is slowly cooled to room temperature.

11-17 The copper-silver phase diagram is shown in Figure 11-30. Copper has a higher melting point than silver. Refer to the silver-rich solid phase as gamma (γ) and the copper-rich solid phase as delta (δ). Denote the liquid as L.

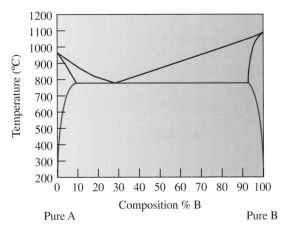

Composition % B

Pure A   Pure B

Figure 11-30   A phase diagram for elements A and B (for Problem 11-17).

(a) For an overall composition of 60% B (40% A) at a temperature of 800°C, what are the compositions and amounts of the phases present?

(b) For an overall composition of 30% B (70% A) at a temperature of 1000°C, what are the compositions and amounts of the phases present?

(c) Draw a schematic diagram illustrating the final microstructure of a material with a composition of 50% B (50% A) cooled to 200°C from the liquid state. Label each phase present.

11-18 Calculate the total % β and the % eutectic microconstituent at room temperature for the following lead-tin alloys: 10% Sn, 20% Sn, 50% Sn, 60% Sn, 80% Sn, and 95% Sn. Using Figure 11-18, plot the strength of the alloys versus the % β and the % eutectic and explain your graphs.

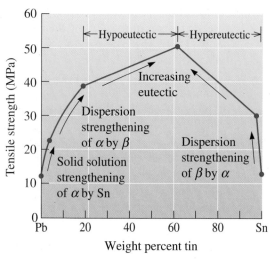

Figure 11-18   (Repeated for Problem 11-18.) The effect of the composition and strengthening mechanism on the tensile strength of lead-tin alloys.

11-19   Consider an Al-4% Si alloy (Figure 11-19). Determine
    (a) if the alloy is hypoeutectic or hypereutectic;
    (b) the composition of the first solid to form during solidification;
    (c) the amounts and compositions of each phase at 578°C;

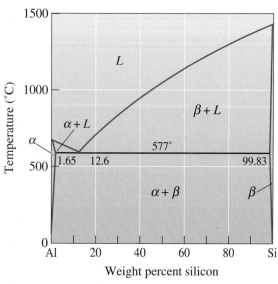

Figure 11-19   (Repeated for Problems 11-19, 11-20 and 11-29.) The aluminum-silicon phase diagram.

    (d) the amounts and compositions of each phase at 576°C, the amounts and compositions of each microconstituent at 576°C; and
    (e) the amounts and compositions of each phase at 25°C.

11-20   Consider an Al-25% Si alloy (see Figure 11-19). Determine
    (a) if the alloy is hypoeutectic or hypereutectic;
    (b) the composition of the first solid to form during solidification;
    (c) the amounts and compositions of each phase at 578°C;
    (d) the amounts and compositions of each phase at 576°C;
    (e) the amounts and compositions of each microconstituent at 576°C; and
    (f) the amounts and compositions of each phase at 25°C.

11-21   A Pb–Sn alloy contains 45% α and 55% β at 100°C. Determine the composition of the alloy. Is the alloy hypoeutectic or hypereutectic?

11-22   An Al-Si alloy contains 85% α and 15% β at 500°C. Determine the composition of the alloy. Is the alloy hypoeutectic or hypereutectic?

11-23   A Pb–Sn alloy contains 23% primary α and 77% eutectic microconstituent immediately after the eutectic reaction has been completed. Determine the composition of the alloy.

11-24   An Al-Si alloy contains 15% primary β and 85% eutectic microconstituent immediately after the eutectic reaction has been completed. Determine the composition of the alloy.

11-25   Observation of a microstructure shows that there is 28% eutectic and 72% primary β in an Al-Li alloy (Figure 11-28). Determine the composition of the alloy and whether it is hypoeutectic or hypereutectic.

11-26   Write the eutectic reaction that occurs, including the compositions of the three phases in equilibrium, and calculate the amount of α and β in the eutectic microconstituent in the Mg-Al system (Figure 11-29).

11-27 Calculate the total amount of α and β and the amount of each microconstituent in a Pb-50% Sn alloy at 182°C. What fraction of the total α in the alloy is contained in the eutectic microconstituent?

11-28 Figure 11-31 shows a cooling curve for a Pb–Sn alloy. Determine
  (a) the pouring temperature;
  (b) the superheat;
  (c) the liquidus temperature;
  (d) the eutectic temperature;
  (e) the freezing range;
  (f) the local solidification time;
  (g) the total solidification time; and
  (h) the composition of the alloy.

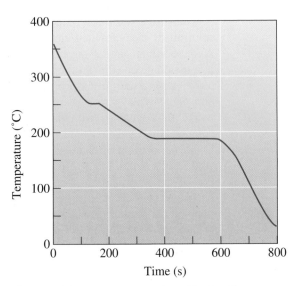

Figure 11-31 Cooling curve for a Pb–Sn alloy (for Problem 11-28).

11-29 Figure 11-32 shows a cooling curve for an Al-Si alloy and Figure 11-19 shows the binary phase diagram for this system. Determine
  (a) the pouring temperature;
  (b) the superheat;
  (c) the liquidus temperature;
  (d) the eutectic temperature;
  (e) the freezing range;
  (f) the local solidification time;
  (g) the total solidification time; and
  (h) the composition of the alloy.

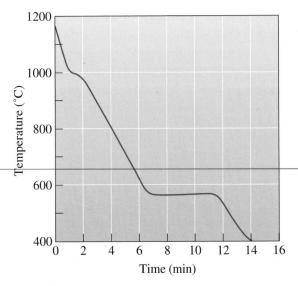

Figure 11-32 Cooling curve for an Al-Si alloy (for Problem 11-29).

11-30 Draw the cooling curves, including appropriate temperatures, expected for the following Al-Si alloys:
  (a) Al-4% Si;
  (b) Al-12.6% Si;
  (c) Al-25% Si; and
  (d) Al-65% Si.

11-31 Cooling curves are obtained for a series of Cu-Ag alloys (Figure 11-33). Use this data to produce the Cu-Ag phase diagram. The maximum solubility of Ag in Cu is 7.9%, and the maximum solubility of Cu in Ag is 8.8%. The solubilities at room temperature are near zero.

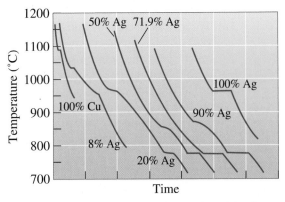

Figure 11-33 Cooling curves for a series of Cu-Ag alloys (for Problem 11-31).

Section 11-5 Strength of Eutectic Alloys

11-32 In regards to eutectic alloys, what does the term "modification" mean? How does it help properties of the alloy?

11-33 For the Pb–Sn system, explain why the tensile strength is a maximum at the eutectic composition.

11-34 Does the shape of the proeutectic phase have an effect on the strength of eutectic alloys? Explain.

11-35 The binary phase diagram for the silver (Ag) and germanium (Ge) system is shown in Figure 11-34.

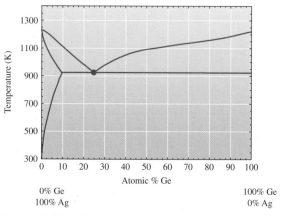

Figure 11-34 The silver–germanium phase diagram (for Problem 11-35).

(a) Schematically draw the phase diagram and label the phases present in each region of the diagram. Denote α as the Ag-rich solid phase and β as the Ge-rich solid phase. Use L to denote the liquid phase.

(b) For an overall composition of 80% Ge (20% Ag) at a temperature of 700 K, what are the compositions and amounts of the phases present?

(c) What is the transformation in phases that occurs on solidification from the melt at the point marked with a circle? What is the special name given to this transformation?

(d) Draw a schematic diagram illustrating the final microstructure of 15% Ge (85% Ag) cooled slowly to 300 K from the liquid state.

(e) Consider two tensile samples at room temperature. One is pure Ag and one is Ag with 2% Ge. Which sample would you expect to be stronger?

11-36 The copper-silver phase diagram is shown in Figure 11-35. Copper has a higher melting point than silver.

(a) Is copper element A or element B as labeled in the phase diagram?

(b) Schematically draw the phase diagram and label all phases present in each region (single phase and two phase) of the phase diagram by writing directly on your sketch. Denote the silver-rich solid phase as gamma (γ) and the copper-rich solid phase as delta (δ). Denote the liquid as L.

(c) At 600°C, the solid solution of element A in element B is stronger than the solid solution of element B in element A. Assume similar processing conditions. Is a material cooled from the liquid to 600°C with a composition of 90% A and 10% B likely to be stronger or weaker than a material with the eutectic composition? Explain your answer fully.

(d) Upon performing mechanical testing, your results indicate that your assumption of similar processing conditions in part (c) was wrong and that the material that you had assumed to be stronger is in fact weaker. Give an example of a processing condition and a description of the associated microstructure that could have led to this discrepancy.

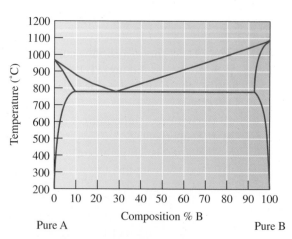

Figure 11-35 A phase diagram for elements A and B (for Problem 11-36).

Section 11-6 Eutectics and Materials Processing

11-37 Explain why Pb–Sn alloys are used for soldering.

11-38 Refractories used in steel making include silica brick that contain very small levels of alumina (Al_2O_3). The eutectic temperature in this system is about 1587°C. Silica melts at about 1725°C. Explain what will happen to the load bearing capacity of the bricks if a small amount of alumina gets incorporated into the silica bricks.

11-39 The Fe-Fe$_3$C phase diagram exhibits a eutectic near the composition used for cast irons. Explain why it is beneficial for ferrous alloys used for casting to have a eutectic point.

Section 11-7 Nonequilibrium Freezing in the Eutectic System

11-40 What is hot shortness? How does it affect the temperature at which eutectic alloys can be used?

Section 11-8 Nanowires and the Eutectic Phase Diagram

11-41 Explain the vapor-liquid-solid mechanism of nanowire growth.

Design Problems

11-42 Design a processing method that permits a Pb-15% Sn alloy solidified under non-equilibrium conditions to be hot worked.

11-43 Design a eutectic diffusion bonding process to join aluminum to silicon. Describe the changes in microstructure at the interface during the bonding process.

11-44 Design an Al-Si brazing alloy and process that will be successful in joining an Al-Mn alloy that has a liquidus temperature of 659°C and a solidus temperature of 656°C. Brazing, like soldering, involves introducing a liquid filler metal into a joint without melting the metals that are to be joined.

Computer Problems

11-45 Write a computer program to assist you in solving problems such as those illustrated in Examples 11-2 or 11-3. The input will be, for example, the bulk composition of the alloy, the mass of the alloy, and the atomic masses of the elements (or compounds) forming the binary system. The program should then prompt the user to provide temperature, number of phases, and the composition of the phases. The program should provide the user with the outputs for the fractions of phases present and the total masses of the phases, as well as the mass of each element in different phases. Start with a program that will provide a solution for Example 11-3 and then extend it to Example 11-4.

Ⓚ Knovel® Problems

K11-1 Find an iron-titanium phase diagram and identify the temperatures and binary alloy compositions for the three-phase points for all eutectic reactions.

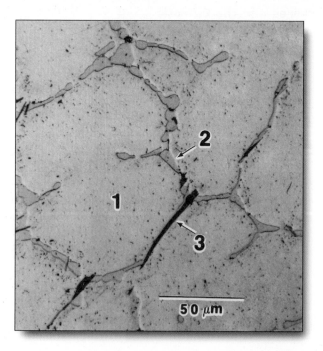

On December 17, 1903, the Wright brothers flew the first controllable airplane. This historic first flight lasted only 12 seconds and covered 36.6 m, but changed the world forever. *(Topical Press Agency/Hulton Archive/Getty Images.)* The micrograph of the alloy used in the Wright brothers airplane is also shown. *(Courtesy of Dr. Frank Gayle, NTST.)* What was not known at the time was that the aluminum alloy engine that they used was inadvertently strengthened by precipitation hardening!

12

Dispersion Strengthening by Phase Transformations and Heat Treatment

Have You Ever Wondered?

- *Who invented and flew the first controllable airplane?*
- *How do engineers strengthen aluminum alloys used in aircrafts?*
- *Can alloys remember their shape?*
- *Why do some steels become very hard upon quenching from high temperatures?*
- *What alloys are used to make orthodontic braces?*
- *Would it be possible to further enhance the strength and, hence, the dent resistance of sheet steels after the car chassis is made?*
- *What are smart materials?*

In Chapter 11, we examined in detail how second-phase particles can increase the strength of metallic materials. We also saw how dispersion-strengthened materials are prepared and what pathways are possible for the formation of second phases during the solidification of alloys, especially eutectic alloys. In this chapter, we will further discuss dispersion strengthening as we describe a variety of solid-state transformation processes, including precipitation or age hardening and the eutectoid reaction. We also examine how nonequilibrium phase transformations—in particular, the martensitic reaction—can provide mechanisms for strengthening.

As we discuss these strengthening mechanisms, keep in mind the characteristics that produce the most desirable dispersion strengthening, as discussed in Chapter 11:

- The matrix should be relatively soft and ductile and the precipitate, or second phase, should be strong;
- the precipitate should be round and discontinuous;
- the second-phase particles should be small and numerous; and
- in general, the more precipitate we have, the stronger the alloy will be.

As in Chapter 11, we will concentrate on how these reactions influence the strength of the materials and how heat treatments can influence other properties. We will begin with the nucleation and growth of second-phase particles in solid-state phase transformations.

12-1   Nucleation and Growth in Solid-State Reactions

In Chapter 9, we discussed nucleation of a solid from a melt. We also discussed the concepts of supersaturation, undercooling, and homogeneous and heterogeneous nucleation. Let's now see how these concepts apply to solid-state phase transformations such as the eutectoid reaction. In order for a precipitate of phase β to form from a solid matrix of phase α, both nucleation and growth must occur. The total change in free energy required for nucleation of a spherical solid precipitate of radius r from the matrix is

$$\Delta G = \tfrac{4}{3}\pi r^3 \Delta G_{v(\alpha \to \beta)} + 4\pi r^2 \sigma_{\alpha\beta} + \tfrac{4}{3}\pi r^3 \varepsilon \qquad (12\text{-}1)$$

The first two terms include the free energy change per unit volume (ΔG_v) and the energy change needed to create the unit area of the $\alpha - \beta$ interface ($\sigma_{\alpha\beta}$), just as in solidification (Equation 9-1). The third term takes into account the **strain energy** per unit volume (ε), the energy required to permit a precipitate to fit into the surrounding matrix during the nucleation and growth of the precipitate. The precipitate does not occupy the same volume that is displaced, so additional energy is required to accommodate the precipitate in the matrix.

Nucleation   As in solidification, nucleation occurs most easily on surfaces already present in the structure, thereby minimizing the surface energy term. Thus, the precipitates heterogeneously nucleate most easily at grain boundaries and other defects.

Growth   Growth of the precipitates normally occurs by long-range diffusion and redistribution of atoms. Diffusing atoms must be detached from their original locations (perhaps at lattice points in a solid solution), move through the surrounding material to the nucleus, and be incorporated into the crystal structure of the precipitate. In some cases, the diffusing atoms might be so tightly bonded within an existing phase that the detachment process limits the rate of growth. In other cases, attaching the diffusing atoms to the precipitate—perhaps because of the lattice strain—limits growth. This result sometimes leads to the formation of precipitates that have a special relationship to the matrix structure that minimizes the strain. In most cases, however, the controlling factor is the diffusion step.

Kinetics   The overall rate, or *kinetics,* of the transformation process depends on both nucleation and growth. If more nuclei are present at a particular temperature, growth occurs from a larger number of sites and the phase transformation is completed in a shorter period of time. At higher temperatures, the diffusion coefficient is higher, growth rates are more rapid, and again we expect the transformation to be completed in a shorter time, assuming an equal number of nuclei.

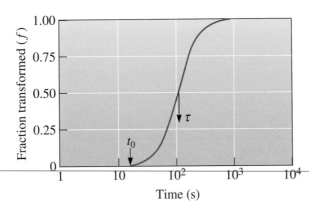

Figure 12-1
Sigmoidal curve showing the rate of
transformation of FCC iron at a
constant temperature. The
incubation time t_0 and the time τ
for 50% transformation are also
shown.

The rate of transformation is given by the Avrami equation (Equation 12-2), with the fraction transformed, f, related to time, t, by

$$f = 1 - \exp(-ct^n) \qquad (12\text{-}2)$$

where c and n are constants for a particular temperature. This **Avrami relationship**, shown in Figure 12-1, produces a sigmoidal, or S-shaped, curve. This equation can describe most solid-state phase transformations. An incubation time, t_0, during which no observable transformation occurs, is the time required for nucleation. Initially, the transformation occurs slowly as nuclei form.

Incubation is followed by rapid growth as atoms diffuse to the growing precipitate. Near the end of the transformation, the rate again slows as the source of atoms available to diffuse to the growing precipitate is depleted. The transformation is 50% complete in time τ. The rate of transformation is often given by the reciprocal of τ:

$$\text{Rate} = 1/\tau \qquad (12\text{-}3)$$

Effect of Temperature

In many phase transformations, the material undercools below the temperature at which the phase transformation occurs under equilibrium conditions. Because both nucleation and growth are temperature-dependent, the rate of phase transformation depends on the undercooling (ΔT). The rate of nucleation is low for small undercoolings (since the thermodynamic driving force is low) and increases for larger undercoolings as the thermodynamic driving force increases at least up to a certain point (since diffusion becomes slower as temperature decreases). At the same time, the growth rate of the new phase decreases continuously (because of slower diffusion), as the undercooling increases. The growth rate follows an *Arrhenius relationship* (recall, Equation 5-1):

$$\text{Growth rate} = A \exp\left(\frac{-Q}{RT}\right), \qquad (12\text{-}4)$$

where Q is the activation energy (in this case, for the phase transformation), R is the gas constant, T is the temperature, and A is a constant.

Figure 12-2 shows sigmoidal curves at different temperatures for the recrystallization of copper; as the temperature increases, the rate of recrystallization of copper *increases*.

At any particular temperature, the overall rate of transformation is the product of the nucleation and growth rates. In Figure 12-3(a), the combined effect of the nucleation and growth rates is shown. A maximum transformation rate may be observed at a critical

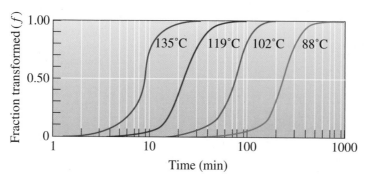

Figure 12-2   The effect of temperature on the recrystallization of cold-worked copper.

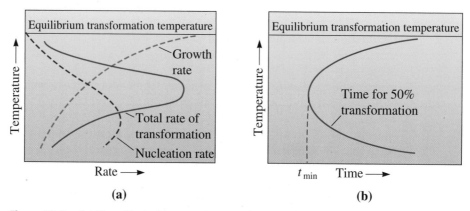

Figure 12-3   (a) The effect of temperature on the rate of a phase transformation is the product of the growth rate and nucleation rate contributions, giving a maximum transformation rate at a critical temperature. (b) Consequently, there is a minimum time (t_{min}) required for the transformation, given by the "C-curve."

undercooling. The time required for transformation is inversely related to the rate of transformation; Figure 12-3(b) describes the time (on a log scale) required for the transformation. This C-shaped curve is common for many transformations in metals, ceramics, glasses, and polymers. Notice that the time required at a temperature corresponding to the equilibrium phase transformation would be ∞ (i.e., the phase transformation will not occur). This is because there is no undercooling and, hence, the rate of homogenous nucleation is zero.

In some processes, such as the recrystallization of a cold-worked metal, we find that the transformation rate continually decreases with decreasing temperature. In this case, nucleation occurs easily, and diffusion—or growth—predominates (i.e., the growth is the rate limiting step for the transformation). The following example illustrates how the activation energy for a solid-state phase transformation such as recrystallization can be obtained from data related to the kinetics of the process.

Example 12-1   *Activation Energy for the Recrystallization of Copper*

Determine the activation energy for the recrystallization of copper from the sigmoidal curves in Figure 12-2.

SOLUTION

The rate of transformation is the reciprocal of the time τ required for half of the transformation to occur. From Figure 12-2, the times required for 50% transformation at several different temperatures can be calculated:

$T(°C)$	$T(K)$	τ (min)	Rate (min^{-1})
135	408	9	0.111
119	392	22	0.045
102	375	80	0.0125
88	361	250	0.0040

The rate of transformation is an Arrhenius equation, so a plot of ln (rate) versus $1/T$ (Figure 12-4 and Equation 12-4) allows us to calculate the constants in the equation. Taking the natural log of both sides of Equation 12-4:

$$\ln(\text{Growth rate}) = \ln A - \frac{Q}{RT}$$

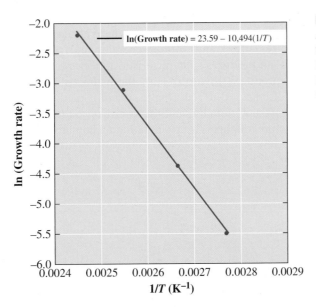

Figure 12-4
Arrhenius plot of transformation rate versus reciprocal temperature for recrystallization of copper (for Example 12-1).

Thus, if we plot ln (Growth rate) as a function of $1/T$, we expect a straight line with a slope of $-Q/R$. A linear regression to the data is shown in Figure 12-4, which indicates that the slope $-Q/R = -10,494$ K. Therefore,

$$Q = 10,494 \text{ K} \times 1.987 \text{ cal}/(\text{mol} \cdot \text{K}) = 20,852 \text{ cal/mol}$$

and the constant A is calculated as

$$\ln A = 23.59$$
$$A = \exp(23.59) = 1.76 \times 10^{10} \text{ min}^{-1}$$

In this particular example, the rate at which the reaction occurs *increases* as the temperature *increases,* indicating that the reaction may be dominated by diffusion.

12-2   Alloys Strengthened by Exceeding the Solubility Limit

In Chapter 11, we learned that lead-tin (Pb-Sn) alloys containing about 2 to 19% Sn can be dispersion strengthened because the solubility of tin in lead is exceeded.

A similar situation occurs in aluminum-copper alloys. For example, the Al-4% Cu alloy (shown in Figure 12-5) is 100% α above 500°C. The α phase is a solid solution of aluminum containing copper up to 5.65 wt%. On cooling below the solvus temperature, a second phase, θ, precipitates. The θ phase, which is the hard, brittle intermetallic compound $CuAl_2$, provides dispersion strengthening. Applying the lever rule to the phase diagram shown in Figure 12-5, we can show that at 200°C and below, in a 4% Cu alloy, only about 7.5% of the final structure is θ. We must control the precipitation of the second phase to satisfy the requirements of good dispersion strengthening.

Widmanstätten Structure

The second phase may grow so that certain planes and directions in the precipitate are parallel to preferred planes and directions in the matrix, creating a basket-weave pattern known as the **Widmanstätten structure**. This growth mechanism minimizes strain and surface energies and permits faster growth rates. Widmanstätten growth produces a characteristic appearance for the precipitate. When a needle-like shape is produced [Figure 12-6(a)], the Widmanstätten precipitate may encourage the nucleation of cracks, thus reducing the ductility of the material. Conversely, some of these structures make it more difficult for cracks, once formed, to propagate, therefore providing good fracture toughness. Certain titanium alloys and ceramics obtain toughness in this way.

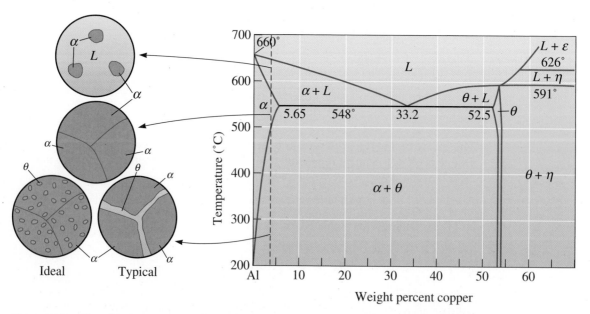

Figure 12-5   The aluminum-copper phase diagram and the microstructures that may develop during cooling of an Al-4% Cu alloy.

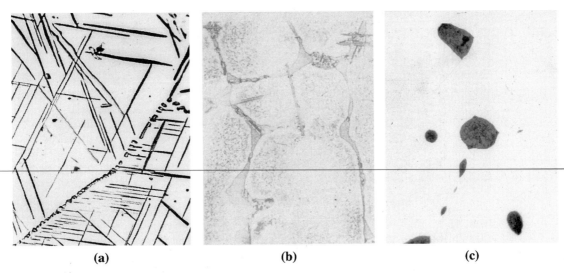

Figure 12-6   (a) Widmanstätten needles in a Cu-Ti alloy (× 420). (*From* ASM Handbook, *Vol. 9, Metallography and Microstructure (1985), ASM International, Materials Park, OH 44073-0002.*) (b) Continuous θ precipitate in an Al-4% Cu alloy, caused by slow cooling (× 500). (c) Precipitates of lead at grain boundaries in copper (× 500). (*Micrographs (b) and (c) reprinted courtesy of Don Askeland.*)

Interfacial Energy Relationships

We expect the precipitate to have a spherical shape in order to minimize surface energy; however, when the precipitate forms at an interface, the precipitate shape is also influenced by the **interfacial energy** of the boundary between the matrix grains and the precipitate. Assuming that the second phase is nucleating at the grain boundaries, the interfacial surface energies of the matrix-precipitate boundary (γ_{mp}) and the grain boundary energy of the matrix ($\gamma_{m,gb}$) fix a **dihedral angle** θ between the matrix-precipitate interface that, in turn, determines the shape of the precipitate (Figure 12-7). The relationship is

$$\gamma_{m,gb} = 2\gamma_{mp} \cos \frac{\theta}{2}$$

Note that this equation cannot be used when the dihedral angle is 0° or 180°.

If the precipitate phase completely wets the grain (similar to how water wets glass), then the dihedral angle is zero, and the second phase grows as a continuous layer along the grain boundaries of the matrix phase. If the dihedral angle is small, the precipitate may be continuous. If the precipitate is also hard and brittle, the thin film that surrounds the matrix grains causes the alloy to be very brittle [Figure 12-6(b)].

Matrix
grain

Precipitate
phase

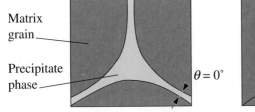 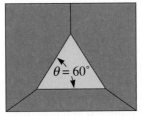 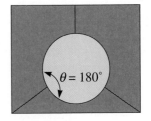

Figure 12-7   The effect of surface energy and the dihedral angle on the shape of a precipitate.

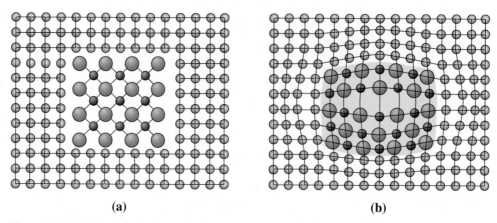

(a)   **(b)**

Figure 12-8   (a) A noncoherent precipitate has no relationship with the crystal structure of the surrounding matrix. (b) A coherent precipitate forms so that there is a definite relationship between the precipitate's and the matrix's crystal structures.

On the other hand, discontinuous and even spherical precipitates form when the dihedral angle is large [Figure 12-6(c)]. This occurs if the precipitate phase does not wet the matrix.

Coherent Precipitate   Even if we produce a uniform distribution of discontinuous precipitate, the precipitate may not significantly disrupt the surrounding matrix structure [Figure 12-8(a)]. Consequently, the precipitate blocks slip only if it lies directly in the path of the dislocation.

   When a **coherent precipitate** forms, the planes of atoms in the crystal structure of the precipitate are related to—or even continuous with—the planes in the crystal structure of the matrix [Figure 12-8(b)]. Now a widespread disruption of the matrix crystal structure is created, and the movement of a dislocation is impeded even if the dislocation merely passes near the coherent precipitate. A special heat treatment, such as age hardening, may produce the coherent precipitate.

12-3   Age or Precipitation Hardening

Age hardening, or **precipitation hardening**, is produced by a sequence of phase transformations that leads to a uniform dispersion of nanoscale, coherent precipitates in a softer, more ductile matrix. The inadvertent occurrence of this process may have helped the Wright brothers, who, on December 17, 1903, made the first controllable flight that changed the world forever (See chapter opener image.) Gayle and co-workers showed that the aluminum alloy used by the Wright brothers for making the engine of the first airplane ever flown picked up copper from the casting mold. The age hardening occurred inadvertently as the mold remained hot during the casting process. The application of age hardening started with the Wright brothers' historic flight and, even today, aluminum alloys used for aircrafts are strengthened using this technique. Age or precipitation hardening is probably one of the earliest examples of nanostructured materials that have found widespread applications.

12-4   Applications of Age-Hardened Alloys

Before we examine the details of the mechanisms of phase transformations that are needed for age hardening to occur, let's examine some of the applications of this technique. A major advantage of precipitation hardening is that it can be used to increase the yield strength of many metallic materials via relatively simple heat treatments and without creating significant changes in density. Thus, the strength-to-density ratio of an alloy can be improved substantially using age hardening. For example, the yield strength of an aluminum alloy can be increased from about 138 MPa to 414 MPa as a result of age hardening.

Nickel-based super alloys (alloys based on Ni, Cr, Al, Ti, Mo, and C) are precipitation hardened by precipitation of a Ni₃Al-like γ' phase that is rich in Al and Ti. Similarly, titanium alloys (e.g., Ti – 6% Al – 4% V), stainless steels, Be-Cu and many steels are precipitation hardened and used for a variety of applications.

New sheet-steel formulations are designed such that precipitation hardening occurs in the material while the paint on the chassis is being "baked" or cured (~100°C). These **bake-hardenable steels** are just one example of steels that take advantage of the strengthening effect provided by age-hardening mechanisms.

A weakness associated with this mechanism is that age-hardened alloys can be used over a limited range of temperatures. At higher temperatures, the precipitates formed initially begin to grow and eventually dissolve if the temperatures are high enough (Section 12-8). This is where alloys in which dispersion strengthening is achieved by using a second phase that is insoluble are more effective than age-hardened alloys.

12-5   Microstructural Evolution in Age or Precipitation Hardening

How do precipitates form in precipitation hardening? How do they grow or age? Can the precipitates grow too much, or overage, so that they cannot provide maximum dispersion strengthening? Answers to these questions can be found by following the microstructural evolution in the sequence of phase transformations that are necessary for age hardening.

Let's use Al-Cu as an archetypal system to illustrate these ideas. The Al-4% Cu alloy is a classic example of an age-hardenable alloy. There are three steps in the age-hardening heat treatment (Figure 12-9).

Step 1: Solution Treatment   In the **solution treatment**, the alloy is first heated above the solvus temperature and held until a homogeneous solid solution α is produced. This step dissolves the θ phase precipitate and reduces any microchemical segregation present in the original alloy.

We could heat the alloy to just below the solidus temperature and increase the rate of homogenization; however, the presence of a nonequilibrium eutectic microconstituent may cause melting (hot shortness, Chapter 10). Thus, the Al-4% Cu alloy is solution treated between 500°C and 548°C, that is, between the solvus and the eutectic temperatures.

Step 2: Quench   After solution treatment, the alloy, which contains only α in its structure, is rapidly cooled, or quenched. The atoms do not have time to diffuse to potential nucleation sites, so the θ does not form. After the quench, the structure is a

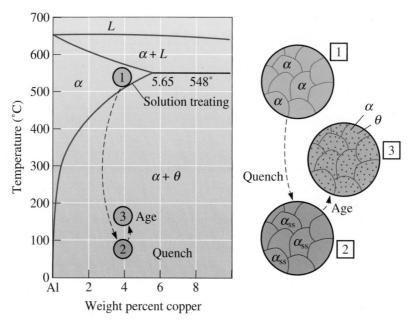

Figure 12-9   The aluminum-rich end of the aluminum-copper phase diagram showing the three steps in the age-hardening heat treatment and the microstructures that are produced.

supersaturated solid solution α_{ss} containing excess copper, and it is not an equilibrium structure. It is a metastable structure. This situation is effectively the same as undercooling of water, molten metals, and silicate glasses (Chapter 9). The only difference is we are dealing with materials in their solid state.

Step 3: Age

Finally, the supersaturated α is heated at a temperature below the solvus temperature. At this *aging* temperature, atoms diffuse only short distances. Because the supersaturated α is metastable, the extra copper atoms diffuse to numerous nucleation sites and precipitates grow. Eventually, if we hold the alloy for a sufficient time at the aging temperature, the equilibrium $\alpha + \theta$ structure is produced. Note that even though the structure that is formed has two equilibrium phases (i.e., $\alpha + \theta$), the morphology of the phases is different from the structure that would have been obtained by the slow cooling of this alloy (Figure 12-5). When we go through the three steps described previously, we produce the θ phase in the form of ultra-fine uniformly dispersed second-phase precipitate particles. This is what we need for effective precipitation strengthening.

The following two examples illustrate the effect of quenching on the composition of phases and a design for an age-hardening treatment.

Example 12-2   *Composition of Al-4% Cu Alloy Phases*

Compare the composition of the α solid solution in the Al-4% Cu alloy at room temperature when the alloy cools under equilibrium conditions with that when the alloy is quenched.

SOLUTION

From Figure 12-9, a tie line can be drawn at room temperature. The composition of the α determined from the tie line is about 0.02% Cu; however, the composition of the α after quenching is still 4% Cu. Since α contains more than the equilibrium copper content, the α is supersaturated with copper.

Example 12-3   *Design of an Age-Hardening Treatment*

A portion of the magnesium-aluminum phase diagram is shown in Figure 12-10. Suppose a Mg-8% Al alloy is responsive to an age-hardening heat treatment. Design a heat treatment for the alloy.

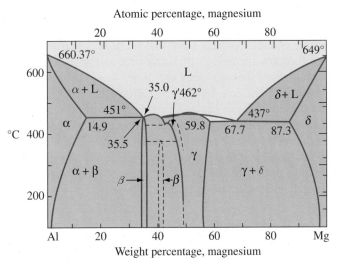

Figure 12-10   The aluminum-magnesium phase diagram.

SOLUTION

Step 1: Solution-treat at a temperature between the solvus and the eutectic to avoid hot shortness. Thus, heat between 340°C and 451°C.

Step 2: Quench to room temperature fast enough to prevent the precipitate phase β from forming.

Step 3: Age at a temperature below the solvus, that is, below 340°C, to form a fine dispersion of the β phase.

Nonequilibrium Precipitates during Aging
During aging of aluminum-copper alloys, a continuous series of other precursor precipitate phases forms prior to the formation of the equilibrium θ phase. This is fairly common in precipitation-hardened alloys. The simplified diagram in Figure 12-9 does not show these intermediate phases. At the start of aging, the copper atoms concentrate on {100} planes in the α matrix and produce very thin precipitates called **Guinier-Preston** (GP) **zones**. As

Figure 12-11
An electron micrograph of aged Al-15% Ag showing coherent γ' plates and round GP zones (×40,000). *(Courtesy of J. B. Clark.)*

aging continues, more copper atoms diffuse to the precipitate and the GP-I zones thicken into thin disks, or GP-II zones. With continued diffusion, the precipitates develop a greater degree of order and are called θ'. Finally, the stable θ precipitate is produced.

The nonequilibrium precipitates—GP-I, GP-II, and θ'—are coherent precipitates. The strength of the alloy increases with aging time as these coherent phases grow in size during the initial stages of the heat treatment. When these coherent precipitates are present, the alloy is in the aged condition. Figure 12-11 shows the structure of an aged Al-Ag alloy. This important development in the microstructure evolution of precipitation-hardened alloys is the reason the time for heat treatment during aging is very important.

When the stable noncoherent θ phase precipitates, the strength of the alloy begins to decrease. Now the alloy is in the overaged condition. The θ still provides some dispersion strengthening, but with increasing time, the θ grows larger and even the simple dispersion-strengthening effect diminishes.

12-6   Effects of Aging Temperature and Time

The properties of an age-hardenable alloy depend on both aging temperature and aging time (Figure 12-12). At 260°C, diffusion in the Al-4% Cu alloy is rapid, and precipitates quickly form. The strength reaches a maximum after less than 0.1 h exposure. Over-aging occurs if the alloy is held for longer than 0.1 h (6 minutes).

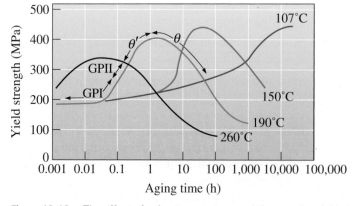

Figure 12-12   The effect of aging temperature and time on the yield strength of an Al-4% Cu alloy.

At 190°C, which is a typical aging temperature for many aluminum alloys, a longer time is required to produce the optimum strength; however, there are benefits to using the lower temperature. First, the maximum strength increases as the aging temperature decreases. Second, the alloy maintains its maximum strength over a longer period of time. Third, the properties are more uniform. If the alloy is aged for only 10 min at 260°C, the surface of the part reaches the proper temperature and strengthens, but the center remains cool and ages only slightly. The example that follows illustrates the effect of aging heat treatment time on the strength of aluminum alloys.

Example 12-4	*Effect of Aging Heat-Treatment Time on the Strength of Aluminum Alloys*

The operator of a furnace left for his hour lunch break without removing the Al-4% Cu alloy from the furnace used for the aging treatment. Compare the effect on the yield strength of the extra hour of aging for the aging temperatures of 190°C and 260°C.

SOLUTION

At 190°C, the peak strength of 400 MPa occurs at 2 h (Figure 12-12). After 3 h, the strength is essentially the same.

At 260°C, the peak strength of 340 MPa occurs at 0.06 h; however, after 1 h, the strength decreases to 250 MPa.

Thus, the higher aging temperature gives a lower peak strength and makes the strength more sensitive to aging time.

Aging at either 190°C or 260°C is called **artificial aging** because the alloy is heated to produce precipitation. Some solution-treated and quenched alloys age at room temperature; this is called **natural aging**. Natural aging requires long times—often several days—to reach maximum strength; however, the peak strength is higher than that obtained in artificial aging, and no overaging occurs.

An interesting observation made by Dr. Gayle and his coworkers at NIST is a striking example of the difference between natural aging and artificial aging. Dr. Gayle and coworkers analyzed the aluminum alloy of the engine used in the Wright brothers' airplane. They found two interesting things. First, they found that the original alloy had undergone precipitation hardening as a result of being held in the mold for a period of time and at a temperature that was sufficient to cause precipitation hardening. Second, since the alloy was cast in 1903 until about 1993 when the research was done (almost ninety years), the alloy had continued to age naturally! This could be seen from two different size distributions for the precipitate particles using transmission electron microscopy. In some aluminum alloys (designated as T4) used to make tapered poles or fasteners, it may be necessary to refrigerate the alloy prior to forming to avoid natural aging at room temperature. If not, the alloy would age at room temperature, become harder, and not be workable!

12-7   Requirements for Age Hardening

Not all alloys are age hardenable. Four conditions must be satisfied for an alloy to have an age-hardening response during heat treatment:

1. The alloy system must display decreasing solid solubility with decreasing temperature. In other words, the alloy must form a single phase on heating above the solvus line, then enter a two-phase region on cooling.

2. The matrix should be relatively soft and ductile, and the precipitate should be hard and brittle. In most age hardenable alloys, the precipitate is a hard, brittle intermetallic compound.

3. The alloy must be quenchable. Some alloys cannot be cooled rapidly enough to suppress the formation of the precipitate. Quenching may, however, introduce residual stresses that cause distortion of the part (Chapter 8). To minimize residual stresses, aluminum alloys are quenched in hot water at about 80°C.

4. A coherent precipitate must form.

As mentioned before in Section 12-4, a number of important alloys, including certain stainless steels and alloys based on aluminum, magnesium, titanium, nickel, chromium, iron, and copper, meet these conditions and are age hardenable.

12-8   Use of Age-Hardenable Alloys at High Temperatures

Based on our previous discussion, we would not select an age-hardened Al-4% Cu alloy for use at high temperatures. At service temperatures ranging from 100°C to 500°C, the alloy overages and loses its strength. Above 500°C, the second phase redissolves in the matrix, and we do not even obtain dispersion strengthening. In general, the aluminum age-hardenable alloys are best suited for service near room temperature; however, some magnesium alloys may maintain their strength to about 250°C and certain nickel superalloys resist overaging at 1000°C.

We may also have problems when welding age-hardenable alloys (Figure 12-13). During welding, the metal adjacent to the weld is heated. The *heat-affected zone* (HAZ) contains two principal zones. The lower temperature zone near the unaffected base metal is exposed to temperatures just below the solvus and may overage. The higher temperature zone is solution treated, eliminating the effects of age hardening. If the solution-treated zone cools slowly, stable θ may form at the grain boundaries, embrittling the weld area. Very fast welding processes such as electron-beam welding, complete reheat treatment of the area after welding, or welding the alloy in the solution-treated condition improve the quality of the weld (Chapter 9). Welding of nickel-based superalloys strengthened by precipitation hardening does not pose such problems since the precipitation process is sluggish and the welding process simply acts as a solution and quenching treatment. The process of friction stir welding has also been recently applied to welding of Al and Al-Li alloys for aerospace and aircraft applications.

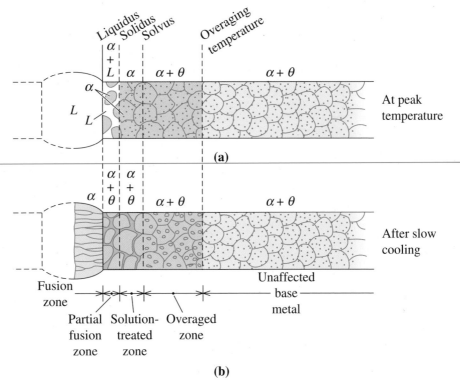

Figure 12-13   Microstructural changes that occur in age-hardened alloys during fusion welding: (a) microstructure in the weld at the peak temperature, and (b) microstructure in the weld after slowly cooling to room temperature.

12-9   The Eutectoid Reaction

In Chapter 11, we defined the eutectoid as a solid-state reaction in which one solid phase transforms to two other solid phases:

$$S_1 \rightarrow S_2 + S_3 \qquad (12\text{-}6)$$

As an example of how we can use the eutectoid reaction to control the microstructure and properties of an alloy, let's examine the technologically important portion of the iron-iron carbide (Fe-Fe$_3$C) phase diagram (Figure 12-14), which is the basis for steels and cast irons. The formation of the two solid phases (α and Fe$_3$C) permits us to obtain dispersion strengthening. The ability to control the occurrence of the eutectoid reaction (this includes either making it happen, slowing it down, or avoiding it all together) is probably the most important step in the thermomechanical processing of steels. On the Fe-Fe$_3$C diagram, the eutectoid temperature is known as the A_1 temperature. The boundary between austenite (γ) and the two-phase field consisting of ferrite (α) and austenite is known as the A_3. The boundary between austenite (γ) and the two-phase field consisting of cementite (Fe$_3$C) and austenite is known as the A_{cm}.

We normally are not interested in the carbon-rich end of the Fe-C phase diagram and this is why we examine the Fe-Fe$_3$C diagram as part of the Fe-C binary phase diagram.

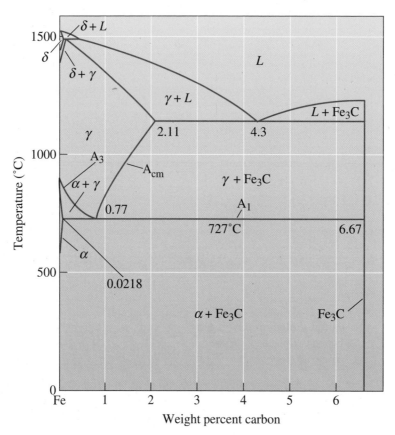

Figure 12-14   The Fe-Fe₃C phase diagram (a portion of the Fe-C diagram). The vertical line at 6.67% C is the stoichiometric compound Fe₃C.

Solid Solutions

Iron goes through two allotropic transformations (Chapter 3) during heating or cooling. Immediately after solidification, iron forms a BCC structure called δ-ferrite. On further cooling, the iron transforms to a FCC structure called γ, or **austenite**. Finally, iron transforms back to the BCC structure at lower temperatures; this structure is called α, or **ferrite**. Both of the ferrites (α and δ) and the austenite are solid solutions of interstitial carbon atoms in iron. Normally, when no specific reference is made, the term ferrite refers to the α ferrite, since this is the phase we encounter more often during the heat treatment of steels. Certain ceramic materials used in magnetic applications are also known as ferrites (Chapter 20) but are not related to the ferrite phase in the Fe-Fe₃C system.

Because interstitial holes in the FCC crystal structure are somewhat larger than the holes in the BCC crystal structure, a greater number of carbon atoms can be accommodated in FCC iron. Thus, the maximum solubility of carbon in austenite is 2.11% C, whereas the maximum solubility of carbon in BCC iron is much lower (i.e., ~0.0218% C in α and 0.09% C in δ). The solid solutions of carbon in iron are relatively soft and ductile, but are stronger than pure iron due to solid-solution strengthening by the carbon.

Compounds

A stoichiometric compound Fe₃C, or **cementite**, forms when the solubility of carbon in solid iron is exceeded. The Fe₃C contains 6.67% C, is extremely hard and brittle (like a ceramic material), and is present in all commercial steels. By properly controlling the amount, size, and shape of Fe₃C, we control the degree of dispersion strengthening and the properties of the steel.

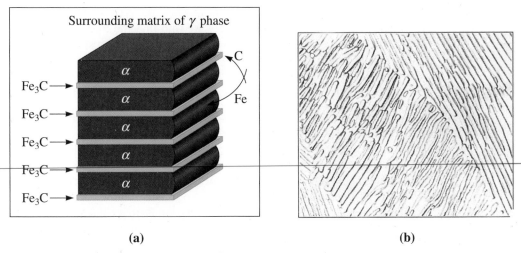

(a)   **(b)**

Figure 12-15   Growth and structure of pearlite: (a) redistribution of carbon and iron, and (b) micrograph of the pearlite lamellae (× 2000). *(From ASM Handbook, Vol. 7, Metallography and Microstructure (1972), ASM International, Materials Park, OH 44073-0002.)*

The Eutectoid Reaction

If we heat an alloy containing the eutectoid composition of 0.77% C above 727°C, we produce a structure containing only austenite grains. When austenite cools to 727°C, the eutectoid reaction begins:

$$\gamma(0.77\% \text{ C}) \rightarrow \alpha(0.0218\% \text{ C}) + \text{Fe}_3\text{C}(6.67\% \text{ C}) \tag{12-7}$$

As in the eutectic reaction, the two phases that form have different compositions, so atoms must diffuse during the reaction (Figure 12-15). Most of the carbon in the austenite diffuses to the Fe_3C, and most of the iron atoms diffuse to the α. This redistribution of atoms is easiest if the diffusion distances are short, which is the case when the α and Fe_3C grow as thin lamellae, or plates.

Pearlite

The lamellar structure of α and Fe_3C that develops in the iron-carbon system is called **pearlite**, which is a microconstituent in steel. This was so named because a polished and etched pearlite shows the colorfulness of mother-of-pearl. The lamellae in pearlite are much finer than the lamellae in the lead-tin eutectic because the iron and carbon atoms must diffuse through solid austenite rather than through liquid. One way to think about pearlite is to consider it as a metal-ceramic nanocomposite. The following example shows the calculation of the amounts of the phases in the pearlite microconstituent.

Example 12-5   *Phases and Composition of Pearlite*

Calculate the amounts of ferrite and cementite present in pearlite.

SOLUTION

Since pearlite must contain 0.77% C, using the lever rule:

$$\% \ \alpha = \frac{6.67 - 0.77}{6.67 - 0.0218} \times 100 = 88.7\%$$

$$\% \ \text{Fe}_3\text{C} = \frac{0.77 - 0.0218}{6.67 - 0.0218} \times 100 = 11.3\%$$

In Example 12-5, we saw that most of the pearlite is composed of ferrite. In fact, if we examine the pearlite closely, we find that the Fe_3C lamellae are surrounded by α. The pearlite structure, therefore, provides dispersion strengthening—the continuous ferrite phase is relatively soft and ductile and the hard, brittle cementite is dispersed.

Primary Microconstituents

Hypoeutectoid steels contain less than 0.77% C, and hypereutectoid steels contain more than 0.77% C. Ferrite is the primary or proeutectoid microconstituent in hypoeutectoid alloys, and cementite is the primary or proeutectoid microconstituent in hypereutectoid alloys. If we heat a hypoeutectoid alloy containing 0.60% C above 750°C, only austenite remains in the microstructure. Figure 12-16 shows what happens when the austenite cools. Just below 750°C, ferrite nucleates and grows, usually at the austenite grain boundaries. Primary ferrite continues to grow until the temperature falls to 727°C. The remaining austenite at that temperature is now surrounded by ferrite and has changed in composition from 0.60% C to 0.77% C. Subsequent cooling to below 727°C causes all of the remaining austenite to transform to pearlite by the eutectoid reaction. The structure contains two phases—ferrite and cementite—arranged as two microconstituents—primary ferrite and pearlite. The final microstructure contains islands of pearlite surrounded by the primary ferrite [Figure 12-17(a)]. This structure permits the alloy to be strong, due to the dispersion-strengthened pearlite, yet ductile, due to the continuous primary ferrite.

In hypereutectoid alloys, the primary phase is Fe_3C, which forms at the austenite grain boundaries. After the austenite cools through the eutectoid reaction, the steel contains hard, brittle cementite surrounding islands of pearlite [Figure 12-17(b)]. Now, because the hard, brittle microconstituent is continuous, the steel is also brittle. Fortunately, we can improve the microstructure and properties of the hypereutectoid steels by heat treatment. The following example shows the calculation for the amounts and compositions of phases and microconstituents in a plain carbon steel.

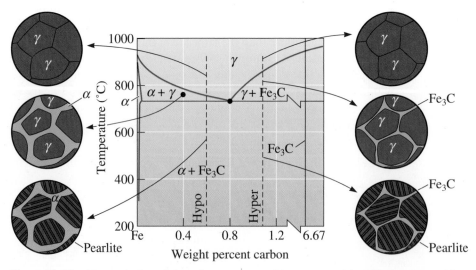

Figure 12-16   The evolution of the microstructure of hypoeutectoid and hypereutectoid steels during cooling, in relationship to the Fe-Fe_3C phase diagram.

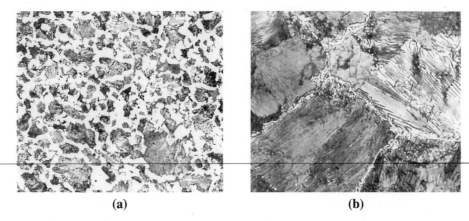

(a)   **(b)**

Figure 12-17   (a) A hypoeutectoid steel showing primary α (white) and pearlite ($\times$ 400). (b) A hypereutectoid steel showing primary Fe$_3$C surrounding pearlite ($\times$ 800). *(From* ASM Handbook, *Vol. 7, (1972), ASM International, Materials Park, OH 44073-0002.)*

Example 12-6   *Phases In Hypoeutectoid Plain Carbon Steel*

Calculate the amounts and compositions of phases and microconstituents in a Fe-0.60% C alloy at 726°C.

SOLUTION

The phases are ferrite and cementite. Using a tie line and working the lever law at 726°C, we find

$$\alpha(0.0218\% \text{ C}) \quad \% \, \alpha = \left[\frac{6.67 - 0.60}{6.67 - 0.0218} \right] \times 100 = 91.3\%$$

$$\text{FeC}(6.67\% \text{ C}) \quad \% \, \text{Fe}_3\text{C} = \left[\frac{0.60 - 0.0218}{6.67 - 0.0218} \right] \times 100 = 8.7\%$$

The microconstituents are primary ferrite and pearlite. If we construct a tie line just above 727°C, we can calculate the amounts and compositions of ferrite and austenite just before the eutectoid reaction starts. All of the austenite at that temperature will have the eutectoid composition (i.e., it will contain 0.77% C) and will transform to pearlite; all of the proeutectoid ferrite will remain as primary ferrite.

$$\text{Primary } \alpha(0.0218\% \text{ C}) \, \% \, \text{Primary } \alpha = \left[\frac{0.77 - 0.60}{0.77 - 0.0218} \right] \times 100 = 22.7\%$$

$$\text{Austentite just above 727°C} = \text{Pearlite}: 0.77\% \text{ C}$$

$$\% \, \text{Pearlite} = \left[\frac{0.60 - 0.0218}{0.77 - 0.0218} \right] \times 100 = 77.3\%$$

12-10   Controlling the Eutectoid Reaction

We control dispersion strengthening in the eutectoid alloys in much the same way that we did in eutectic alloys (Chapter 11).

Controlling the Amount of the Eutectoid   By changing the composition of the alloy, we change the amount of the hard second phase. As the carbon content of steel increases towards the eutectoid composition of 0.77% C, the amounts of Fe_3C and pearlite increase, thus increasing the strength. This strengthening effect eventually peaks, and the properties level out or even decrease when the carbon content is too high (Table 12-1).

Controlling the Austenite Grain Size   We can increase the number of pearlite colonies by reducing the prior austenite grain size, usually by using low temperatures to produce the austenite. Typically, we can increase the strength of the alloy by reducing the initial austenite grain size, thus increasing the number of colonies. Pearlite grows as grains or *colonies*. Within each colony, the orientation of the lamellae is identical. The colonies nucleate most easily at the grain boundaries of the original austenite grains.

Controlling the Cooling Rate   By increasing the cooling rate during the eutectoid reaction, we reduce the distance that the atoms are able to diffuse. Consequently, the lamellae produced during the reaction are finer or more closely spaced. By producing fine pearlite, we increase the strength of the alloy (Table 12-1 and Figure 12-18).

Controlling the Transformation Temperature   The solid-state eutectoid reaction is rather slow, and the steel may cool below the equilibrium eutectoid temperature before the transformation begins (i.e., the austenite phase can be undercooled). Lower transformation temperatures give a finer, stronger structure (Figure 12-19), influence the time required for transformation, and even alter the arrangement of the two phases. This information is contained in the **time-temperature-transformation** (TTT)

TABLE 12-1 ■ *The effect of carbon on the strength of steels*

	Slow Cooling (Coarse Pearlite)			Fast Cooling (Fine Pearlite)		
Carbon %	Yield Strength (MPa)	Tensile Strength (MPa)	% Elongation	Yield Strength (MPa)	Tensile Strength (MPa)	% Elongation
0.20	295	394	36.5	347	441	36.0
0.40	353	519	30.0	374	590	28.0
0.60	372	626	23.0	420	776	18.0
0.80	376	615	25.0	524	1010	11.0
0.95	379	657	13.0	500	1014	9.5

After Metals Progress Materials and Processing Databook, *1981.*

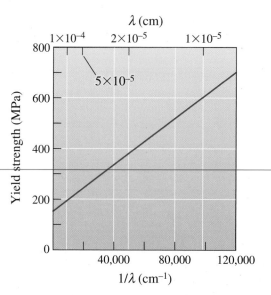

λ (cm)

1/λ (cm⁻¹)

Figure 12-18
The effect of interlamellar spacing (λ) on the yield strength of pearlite.

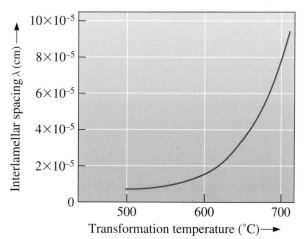

Transformation temperature (°C) →

Figure 12-19
The effect of the austenite transformation temperature on the interlamellar spacing in pearlite.

diagram (Figure 12-20). This diagram, also called the **isothermal transformation (IT)** diagram or the **C-curve**, permits us to predict the structure, properties, and heat treatment required in steels.

The shape of the TTT diagram is a consequence of the kinetics of the eutectoid reaction and is similar to the diagram shown by the Avrami relationship (Figure 12-3). At any particular temperature, a sigmoidal curve describes the rate at which the austenite transforms to a mixture of ferrite and cementite (Figure 12-21). An incubation time is required for nucleation. The P_s (pearlite start) curve represents the time at which austenite starts to transform to ferrite and cementite via the eutectoid transformation. The sigmoidal curve also gives the time at which the transformation is complete; this time is given by the P_f (pearlite finish) curve. When the temperature decreases from 727°C, the rate of nucleation increases, while the rate of growth of the microconstituent decreases. As in Figure 12-3, a maximum transformation rate, or minimum transformation time, is found; the maximum rate of transformation occurs near 550°C for a eutectoid steel (Figure 12-20).

Two types of microconstituents are produced as a result of the transformation. Pearlite (*P*) forms above 550°C, and bainite (*B*) forms at lower temperatures.

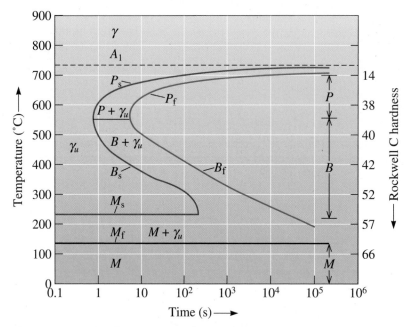

Figure 12-20 The time-temperature-transformation (TTT) diagram for a eutectoid steel, where P = Pearlite, B = Bainite, and M = Martensite. The subscripts "s" and "f" indicate the start and finish of a transformation. γ_u is unstable austenite.

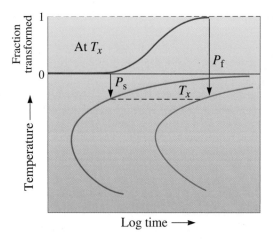

Figure 12-21
The sigmoidal curve is related to the start and finish times on the TTT diagram for steel. In this case, austenite is transforming to pearlite.

Nucleation and growth of phases in pearlite: If we quench to just below the eutectoid temperature, the austenite is only slightly undercooled. Long times are required before stable nuclei for ferrite and cementite form. After the phases that form pearlite nucleate, atoms diffuse rapidly and *coarse* pearlite is produced; the transformation is complete at the pearlite finish (P_f) time. Austenite quenched to a lower temperature is more highly undercooled. Consequently, nucleation occurs more rapidly and the P_s is shorter. Diffusion is also slower, however, so atoms diffuse only short distances and *fine* pearlite is produced. Even though growth rates are slower, the overall time required for the transformation is reduced because of the shorter incubation time. Finer pearlite forms in shorter times as we reduce the isothermal transformation temperature to about 550°C, which is the *nose*, or *knee,* of the TTT curve (Figure 12-20).

Nucleation and growth of phases in bainite: At a temperature just below the nose of the TTT diagram, diffusion is very slow and total transformation times increase. In addition, we find a

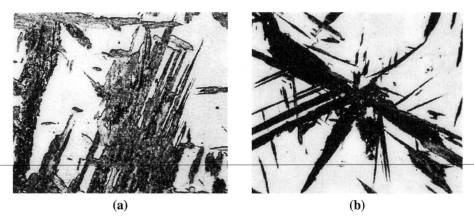

(a) **(b)**

Figure 12-22 (a) Upper bainite (gray, feathery plates) (×600). (b) Lower bainite (dark needles) (×400). *(From ASM Handbook, Vol. 8, (1973), ASM International, Materials Park, OH 44073-0002.)*

different microstructure! At low transformation temperatures, the lamellae in pearlite would have to be extremely thin and, consequently, the boundary area between the ferrite and Fe_3C lamellae would be very large. Because of the energy associated with the ferrite-cementite interface, the total energy of the steel would have to be very high. The steel can reduce its internal energy by permitting the cementite to precipitate as discrete, rounded particles in a ferrite matrix. This new microconstituent, or arrangement of ferrite and cementite, is called **bainite**. Transformation begins at a bainite start (B_s) time and ends at a bainite finish (B_f) time.

The times required for austenite to begin and finish its transformation to bainite increase and the bainite becomes finer as the transformation temperature continues to decrease. The bainite that forms just below the nose of the curve is called coarse bainite, upper bainite, or feathery bainite. The bainite that forms at lower temperatures is called fine bainite, lower bainite, or acicular bainite. Figure 12-22 shows typical microstructures of bainite. Note that the morphology of bainite depends on the heat treatment used.

Figure 12-23 shows the effect of transformation temperature on the properties of eutectoid (0.77% C) steel. As the temperature decreases, there is a general trend toward

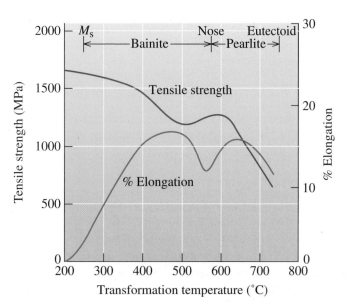

Figure 12-23
The effect of transformation temperature on the properties of a eutectoid steel.

higher strength and lower ductility due to the finer microstructure that is produced. The following two examples illustrate how we can design heat treatments of steels to produce desired microstructures and properties.

Example 12-7   *Design of a Heat Treatment to Generate the Pearlite Microstructure*

Design a heat treatment to produce the pearlite structure shown in Figure 12-15(b).

SOLUTION

First, we need to determine the interlamellar spacing of the pearlite. If we count the number of lamellar spacings in the upper right of Figure 12-15(b), remembering that the interlamellar spacing is measured from one α plate to the next α plate, we find 14 spacings over a 2 cm distance. Due to the factor of 2000 magnification, this 2 cm distance is actually 0.001 cm. Thus,

$$\lambda = \left[\frac{0.001 \text{ cm}}{14 \text{ spacings}} \right] = 7.14 \times 10^{-5} \text{ cm}$$

If we assume that the pearlite is formed by an isothermal transformation, we find from Figure 12-19 that the transformation temperature must have been approximately 700°C. From the TTT diagram (Figure 12-20), our heat treatment should be

1. Heat the steel to about 750°C and hold—perhaps for 1 h—to produce all austenite. A higher temperature may cause excessive growth of austenite grains.
2. Quench to 700°C and hold for at least 10^5 s (the P_f time). We assume here that the steel cools instantly to 700°C. In practice, this does not happen, and thus, the transformation does not occur at one temperature. We may need to use the continuous cooling transformation diagrams to be more precise (See Chapter 13).
3. Cool to room temperature.

The steel should have a hardness of HRC 14 (Figure 12-20) and a yield strength of about 200 MPa.

Example 12-8   *Heat Treatment To Generate the Bainite Microstructure*

Excellent combinations of hardness, strength, and toughness are obtained from bainite. One heat treatment facility austenitized a eutectoid steel at 750°C, quenched and held the steel at 250°C for 15 min, and finally permitted the steel to cool to room temperature. Was the required bainitic structure produced?

SOLUTION

Let's examine the heat treatment using Figure 12-20. After heating at 750°C, the microstructure is 100% γ. After quenching to 250°C, unstable austenite remains for slightly more than 100 s, when fine bainite begins to grow. After 15 min, or 900 s, about 50% fine bainite has formed, and the remainder of the steel still contains unstable austenite. As we will see later, the unstable austenite transforms to martensite when the steel is cooled to room temperature, and the final structure is a mixture of bainite and hard, brittle martensite. The heat treatment was not successful! The heat treatment facility should have held the steel at 250°C for at least 10^4 s, or about 3 h.

12-11   The Martensitic Reaction and Tempering

Martensite is a phase that forms as the result of a diffusionless solid-state transformation. In this transformation, there is no diffusion and, hence, it does not follow the Avrami transformation kinetics. The growth rate in **martensitic transformations** (also known as **displacive** or **athermal transformations**) is so high that nucleation becomes the controlling step.

Cobalt, for example, transforms from a FCC to a HCP crystal structure by a slight shift in the atom locations that alters the stacking sequence of close-packed planes. Because the reaction does not depend on diffusion, the martensite reaction is an athermal transformation—that is, the reaction depends only on the temperature, not on the time. The martensite reaction often proceeds rapidly, at speeds approaching the velocity of sound in the material.

Many other alloys (such as Cu-Zn-Al, Cu-Al-Ni, and Ni-Ti) and ceramic materials show martensitic phase transformations. These transformations can also be driven by the application of mechanical stress. Other than the martensite that forms in certain types of steels, the Ni-Ti alloy, known as **nitinol**, is perhaps the best-known example of alloys that make use of martensitic phase transformations. These materials can remember their shape and are known as shape-memory alloys (SMAs). (See Section 12-12.)

Martensite in Steels   In steels with less than about 0.2% C, upon quenching, the FCC austenite can transform to a nonequilibrium supersaturated BCC martensite structure. In higher carbon steels, the martensite reaction occurs as FCC austenite transforms to BCT (body-centered tetragonal) martensite. The relationship between the FCC austenite and the BCT martensite [Figure 12-24(a)] shows that carbon atoms in the $(1/2, 0, 0)$ type of interstitial sites in the FCC cell can be trapped during the transformation to the body-centered structure, causing the tetragonal structure to be produced. As the carbon content of the steel increases, a greater number of carbon atoms are trapped in these sites, thereby increasing the difference in length between the a- and c-axes of the martensite [Figure 12-24(b)].

The steel must be quenched, or rapidly cooled, from the stable austenite region to prevent the formation of pearlite, bainite, or primary microconstituents. The martensite reaction begins in an eutectoid steel when austenite cools below 220°C, the martensite start (M_s) temperature (Figure 12-20). The amount of martensite increases as the temperature decreases. When the temperature passes below the martensite finish temperature (M_f), the steel should contain 100% martensite. At any intermediate temperature, the amount of martensite does not change as the time at that temperature increases.

Owing to the conservation of mass, the composition of martensite must be the same as that of the austenite from which it forms. There is no long-range diffusion during the transformation that can change the composition. Thus, in iron-carbon alloys, the initial austenite composition and the final martensite composition are the same. The following example illustrates how heat treatment is used to produce a dual-phase steel.

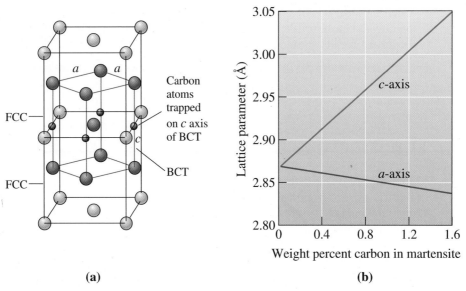

(a)   **(b)**

Figure 12-24   (a) The unit cell of BCT martensite is related to the FCC austenite unit cell. (b) As the percentage of carbon increases, more interstitial sites are filled by the carbon atoms, and the tetragonal structure of the martensite becomes more pronounced.

Example 12-9   *Design of a Heat Treatment for a Dual-Phase Steel*

Unusual combinations of properties can be obtained by producing a steel with a microstructure containing 50% ferrite and 50% martensite. The martensite provides strength, and the ferrite provides ductility and toughness. Design a heat treatment to produce a dual phase steel in which the composition of the martensite is 0.60% C.

SOLUTION

To obtain a mixture of ferrite and martensite, we need to heat treat a hypoeutectoid steel into the $\alpha + \gamma$ region of the phase diagram. The steel is then quenched, permitting the γ portion of the structure to transform to martensite.

The heat treatment temperature is fixed by the requirement that the martensite contain 0.60% C. From the solubility line between the γ and the $\alpha + \gamma$ regions, we find that 0.60% C is obtained in austenite when the temperature is about 750°C. To produce 50% martensite, we need to select a steel that gives 50% austenite when the steel is held at 750°C. If the carbon content of the steel is x, then

$$\% \, \gamma = \left[\frac{(x - 0.02)}{(0.60 - 0.02)} \right] \times 100 = 50 \text{ or } x = 0.31\% \text{ C}$$

Our final design is

1. Select a hypoeutectoid steel containing 0.31% C.

2. Heat the steel to 750°C and hold (perhaps for 1 h, depending on the thickness of the part) to produce a structure containing 50% ferrite and 50% austenite, with 0.60% C in the austenite.

3. Quench the steel to room temperature. The austenite transforms to martensite, also containing 0.60% C.

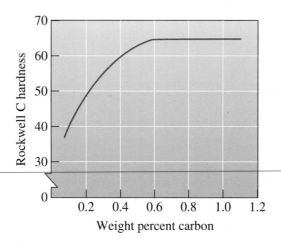

Figure 12-25
The effect of carbon content on the hardness of martensite in steels.

Properties of Steel Martensite

Martensite in steels is very hard and brittle, just like ceramics. The BCT crystal structure has no close-packed slip planes in which dislocations can easily move. The martensite is highly supersaturated with carbon, since iron normally contains less than 0.0218% C at room temperature, and martensite contains the amount of carbon present in the steel. Finally, martensite has a fine grain size and an even finer substructure within the grains.

The structure and properties of steel martensites depend on the carbon content of the alloy (Figure 12-25). When the carbon content is low, the martensite grows in a "lath" shape, composed of bundles of flat, narrow plates that grow side by side [Figure 12-26(a)]. This martensite is not very hard. At a higher carbon content, plate martensite grows, in which flat, narrow plates grow individually rather than as bundles [Figure 12-26(b)]. The hardness is much greater in the higher carbon, plate martensite structure, partly due to the greater distortion, or large c/a ratio, of the crystal structure.

Tempering of Steel Martensite

Martensite is not an equilibrium phase. This is why it does not appear on the Fe-Fe$_3$C phase diagram (Figure 12-14). When martensite in a steel is heated below the eutectoid temperature, the thermodynamically

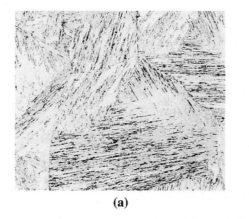

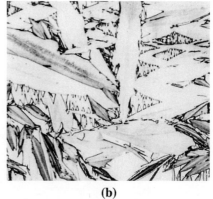

(a)   (b)

Figure 12-26 (a) Lath martensite in low-carbon steel ($\times$ 80). (b) Plate martensite in high-carbon steel ($\times$ 400). *(From* ASM Handbook, *Vol. 8, (1973), ASM International, Materials Park, OH 44073-0002.)*

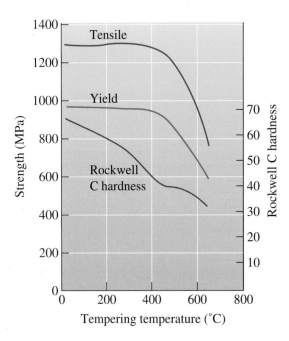

Figure 12-27
Effect of tempering temperature on the properties of a eutectoid steel.

stable α and Fe_3C phases precipitate. This process is called **tempering**. The decomposition of martensite in steels causes the strength and hardness of the steel to decrease while the ductility and impact properties are improved (Figure 12-27). Note that the term tempering here is different from the term we used for tempering of silicate glasses. In both tempering of glasses and tempering of steels, however, the key result is an increase in the toughness of the material.

At low tempering temperatures, the martensite may form two transition phases—a lower carbon martensite and a very fine nonequilibrium ε-carbide, or $Fe_{2.4}C$. The steel is still strong, brittle, and perhaps even harder than before tempering. At higher temperatures, the stable α and Fe_3C form, and the steel becomes softer and more ductile. If the steel is tempered just below the eutectoid temperature, the Fe_3C becomes very coarse, and the dispersion-strengthening effect is greatly reduced. By selecting the appropriate tempering temperature, a wide range of properties can be obtained. The product of the tempering process is a microconstituent called tempered martensite (Figure 12-28).

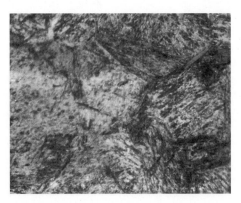

Figure 12-28
Tempered martensite in steel ($\times$ 500).
(From ASM Handbook, Vol. 9, Metallography and Microstructure (1985), ASM International Materials Park, OH 44073-0002.)

Martensite in Other Systems   The characteristics of the martensitic reaction are different in other alloy systems. For example, martensite can form in iron-based alloys that contain little or no carbon by a transformation of the FCC crystal structure to a BCC crystal structure. In certain high-manganese steels and stainless steels, the FCC structure changes to a HCP crystal structure during the martensitic transformation. In addition, the martensitic reaction occurs during the transformation of many polymorphic ceramic materials, including ZrO_2, and even in some crystalline polymers. Thus, the terms martensitic reaction and martensite are rather generic. In the context of steel properties, microstructure, and heat treatment, the term "martensite" refers to the hard and brittle BCT phase obtained upon the quenching of steels.

The properties of martensite in other alloys are also different from the properties of steel martensite. In titanium alloys, BCC titanium transforms to a HCP martensite structure during quenching; however, the titanium martensite is softer and weaker than the original structure. The martensite that forms in other alloys can also be tempered. The martensite produced in titanium alloys can be reheated to permit the precipitation of a second phase. Unlike the case of steel, however, the tempering process *increases,* rather than decreases, the strength of the titanium alloy.

12-12   The Shape-Memory Alloys [SMAs]

The **shape-memory effect** is a unique property possessed by some alloys that undergo the martensitic reaction. These alloys can be processed using a sophisticated thermomechanical treatment to produce a martensitic structure. At the end of the treatment process, the material is deformed to a predetermined shape. The metal can then be deformed into a second shape, but when the temperature is increased, the metal changes back to its original shape! Orthodontic braces, blood-clot filters, engines, antennas for cellular phones, frames for eyeglasses and actuators for smart systems have been developed using these materials. Flaps that change direction of airflow depending upon temperature have been developed and used for air conditioners.

More recently, a special class of materials known as ferromagnetic shape-memory alloys also has been developed. Examples of ferromagnetic shape-memory alloys include Ni_2MnGa, Fe-Pd, and Fe_3Pt. Unlike Ni-Ti, these materials show a shape-memory effect in response to a magnetic field. Most commercial shape-memory alloys including Ni-Ti are not ferromagnetic. We discussed in Chapter 6 that many polymers are viscoelastic, and the viscous component is recovered over time. Thus, many polymers do have a memory of their shape! Recently, researchers have developed new shape-memory plastics.

Shape-memory alloys exhibit a memory that can be triggered by stress or temperature change. **Smart materials** are materials that can sense an external stimulus (such as stress, temperature change, magnetic field, etc.) and undergo some type of change. Actively smart materials can even initiate a response (i.e., they function as a sensor and an actuator). Shape-memory alloys are a family of passively smart materials in that they merely sense a change in stress or temperature.

Shape-memory alloys also show a **superelastic** behavior. Recoverable strains up to 10% are possible. This is why shape-memory alloys have been used so successfully in such applications as orthodontic wires, eyeglass frames, and antennas for cellular phones. In these applications, we make use of the superelastic (and not the shape memory) effect.

Example 12-10 *Design of a Coupling for Tubing*

At times, you need to join titanium tubing in the field. Design a method for doing this quickly.

SOLUTION

Titanium is quite reactive and, unless special welding processes are used, may be contaminated. In the field, we may not have access to these processes. Therefore, we wish to make the joint without resorting to high-temperature processes.

We can take advantage of the shape-memory effect for this application (Figure 12-29). Ahead of time, we can set a Ni-Ti coupling into a small diameter, then deform it into a larger diameter in the martensitic state. In the field, the coupling, which is in the martensitic state, is slipped over the tubing and heated (at a low enough temperature so that the titanium tubing is not contaminated). The coupling contracts back to its predetermined shape as a result of the shape-memory effect, producing a strong mechanical bond to join the tubes.

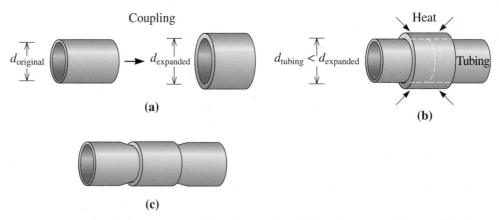

Figure 12-29 Use of shape-memory alloys for coupling tubing: A memory alloy coupling is expanded (a) so it fits over the tubing, (b) when the coupling is reheated, it shrinks back to its original diameter, (c) squeezing the tubing for a tight fit (for Example 12-10).

Summary

- Solid-state phase transformations, which have a profound effect on the structure and properties of a material, can be controlled by proper heat treatments. These heat treatments are designed to provide an optimum distribution of two or more phases in the microstructure. Dispersion strengthening permits a wide variety of structures and properties to be obtained.

- These transformations typically require both nucleation and growth of new phases from the original structure. The kinetics of the phase transformation help us understand the mechanisms that control the reaction and the rate at which the reaction occurs, enabling us to design the heat treatment to produce the desired microstructure. Reference to appropriate phase diagrams also helps us select the necessary compositions and temperatures.

- Age hardening, or precipitation hardening, is one powerful method for controlling the optimum dispersion strengthening in many metallic alloys. In age hardening, a very fine widely dispersed coherent precipitate is allowed to precipitate by a heat treatment that includes (a) solution treating to produce a single-phase solid solution, (b) quenching to retain that single phase, and (c) aging to permit a precipitate to form. In order for age hardening to occur, the phase diagram must show decreasing solubility of the solute in the solvent as the temperature decreases.

- The eutectoid reaction can be controlled to permit one type of solid to transform to two different types of solid. The kinetics of the reaction depends on the nucleation of the new solid phases and the diffusion of the different atoms in the material to permit the growth of the new phases.

- The most widely used eutectoid reaction occurs in producing steels from iron-carbon alloys. Either pearlite or bainite can be produced as a result of the eutectoid reaction in steel. In addition, primary ferrite or primary cementite may be present, depending on the carbon content of the alloy. The trick is to formulate a microstructure that consists of the right mix of metal-like phases that are tough and ceramic-like phases that are hard and brittle.

- Factors that influence the mechanical properties of the microconstituent produced by the eutectoid reaction include (a) the composition of the alloy (amount of eutectoid microconstituent), (b) the grain size of the original solid, the eutectoid microconstituent, and any primary microconstituents, (c) the fineness of the structure within the eutectoid microconstituent (interlamellar spacing), (d) the cooling rate during the phase transformation, and (e) the temperature at which the transformation occurs (the amount of undercooling).

- A martensitic reaction occurs with no long-range diffusion. Again, the best known transformation occurs in steels:

  - The amount of martensite that forms depends on the temperature of the transformation (an athermal reaction).

  - Martensite is very hard and brittle, with the hardness determined primarily by the carbon content.

  - The amount and composition of the martensite are the same as the austenite from which it forms.

- Martensite can be tempered. During tempering, a dispersion-strengthened structure is produced. In steels, tempering reduces the strength and hardness but improves the ductility and toughness.

- Since optimum properties are obtained through heat treatment, we must remember that the structure and properties may change when the material is used at or exposed to elevated temperatures. Overaging or overtempering occur as a natural extension of the phenomena governing these transformations when the material is placed into service.

- Shape-memory alloys (e.g., Ni-Ti) are a class of smart materials that can remember their shape. They also exhibit superelastic behavior.

Glossary

Age hardening A special dispersion-strengthening heat treatment. By solution treatment, quenching, and aging, a coherent precipitate forms that provides a substantial strengthening effect. (Also known as precipitation hardening.)

Artificial aging Reheating a solution-treated and quenched alloy to a temperature below the solvus in order to provide the thermal energy required for a precipitate to form.

Athermal transformation When the amount of the transformation depends only on the temperature, not on the time (same as martensitic transformation or displacive transformation).

Austenite The name given to the FCC crystal structure of iron and iron-carbon alloys.

Avrami relationship Describes the fraction of a transformation that occurs as a function of time. This describes most solid-state transformations that involve diffusion; thus martensitic transformations are not described.

Bainite A two-phase microconstituent, containing ferrite and cementite, that forms in steels that are isothermally transformed at relatively low temperatures.

Bake-hardenable steels These are steels that can show an increase in their yield stress as a result of precipitation hardening that can occur at fairly low temperatures ($\sim$100°C), conditions that simulate baking of paints on cars. This additional increase leads to better dent resistance.

Cementite The hard, brittle ceramic-like compound Fe_3C that, when properly dispersed, provides the strengthening in steels.

Coherent precipitate A precipitate with a crystal structure and atomic arrangement that have a continuous relationship with the matrix from which the precipitate is formed. The coherent precipitate provides excellent disruption of the atomic arrangement in the matrix and provides excellent strengthening.

Dihedral angle The angle that defines the shape of a precipitate particle in the matrix. The dihedral angle is determined by the relative surface energies of the grain boundary energy of the matrix and the matrix-precipitate interfacial energy.

Displacive transformation A phase transformation that occurs via small displacements of atoms or ions and without diffusion. Same as athermal or martensitic transformation.

Ferrite The name given to the BCC crystal structure of iron that can occur as α or δ. This is not to be confused with ceramic ferrites, which are magnetic materials.

Guinier-Preston (GP) zones Clusters of atoms that precipitate from the matrix in the early stages of the age-hardening process. Although the GP zones are coherent with the matrix, they are too small to provide optimum strengthening.

Interfacial energy The energy associated with the boundary between two phases.

Isothermal transformation When the amount of a transformation at a particular temperature depends on the time permitted for the transformation.

Martensite A metastable phase formed in steel and other materials by a diffusionless, athermal transformation.

Martensitic transformation A phase transformation that occurs without diffusion. Same as athermal or displacive transformation. These occur in steels, Ni-Ti, and many ceramic materials.

Natural aging When a coherent precipitate forms from a solution treated and quenched age-hardenable alloy at room temperature, providing optimum strengthening.

Nitinol A nickel-titanium shape memory alloy.

Pearlite A two-phase lamellar microconstituent, containing ferrite and cementite, that forms in steels cooled in a normal fashion or isothermally transformed at relatively high temperatures.

Precipitation hardening See age hardening.

Shape-memory effect The ability of certain materials to develop microstructures that, after being deformed, can return the material to its initial shape when heated (e.g. Ni-Ti alloys).

Smart materials Materials that can sense an external stimulus (e.g., stress, pressure, temperature change, magnetic field, etc.) and initiate a response. Passively smart materials can sense external stimuli; actively smart materials have sensing and actuation capabilities.

Solution treatment The first step in the age-hardening heat treatment. The alloy is heated above the solvus temperature to dissolve any second phase and to produce a homogeneous single-phase structure.

Strain energy The energy required to permit a precipitate to fit into the surrounding matrix during nucleation and growth of the precipitate.

Superelastic behavior Shape-memory alloys deformed above a critical temperature show a large reversible elastic deformation as a result of a stress-induced martensitic transformation.

Supersaturated solid solution The solid solution formed when a material is rapidly cooled from a high-temperature single-phase region to a low-temperature two-phase region without the second phase precipitating. Because the quenched phase contains more alloying element than the solubility limit, it is supersaturated in that element.

Tempering A heat treatment used to reduce the hardness of martensite by permitting the martensite to begin to decompose to the equilibrium phases. This leads to increased toughness.

Time-temperature-transformation (TTT) diagram The TTT diagram describes the time required at any temperature for a phase transformation to begin and end. The TTT diagram assumes that the temperature is constant during the transformation.

Widmanstätten structure The precipitation of a second phase from the matrix when there is a fixed crystallographic relationship between the precipitate and matrix crystal structures. Often needle-like or plate-like structures form in the Widmanstätten structure.

Problems

Section 12-1 Nucleation and Growth in Solid-State Reactions

12-1 (a) Determine the critical nucleus size $r*$ for homogeneous nucleation for precipitation of phase β in a matrix of phase α. *Hint:* The critical nucleus size occurs at the maximum in the expression for $\Delta G(r)$ in Equation 12-1. (b) Plot the total free energy change ΔG as a function of the radius of the precipitate. (c) Comment on the value of $r*$ for homogeneous nucleation for solid-state precipitation when compared to the liquid to solid transformation.

12-2 How is the equation for nucleation of a phase in the solid state different from that for a liquid to solid transformation?

12-3 Determine the constants c and n in Equation 12-2 that describe the rate of crystallization of polypropylene at 140°C. (See Figure 12-30.)

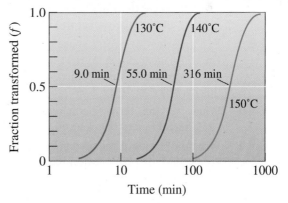

Figure 12-30 The effect of temperature on the crystallization of polypropylene (for Problems 12-3 and 12-5).

12-4 Determine the constants c and n in Equation 12-2 that describe the rate of recrystallization of copper at 135°C. (See Figure 12-2.)

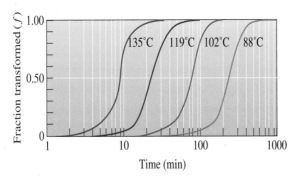

Figure 12-2   (Repeated for Problem 12-4.) The effect of temperature on the recrystallization of cold-worked copper.

12-5   Determine the constants c and n in Equation 12-2 that describe the rate of crystallization of polypropylene at 150°C. (See Figure 12-30.)

12-6   Most solid-state phase transformations follow the Avrami equation. True or false? Discuss briefly.

12-7   What step controls the rate of recrystallization of a cold-worked metal?

Section 12-2 Alloys Strengthened By Exceeding the Solubility Limit

12-8   What are the different ways by which a second phase can be made to precipitate in a two-phase microstructure?

12-9   Explain why the second phase in Al-4% Cu alloys nucleates and grows along the grain boundaries when cooled slowly. Is this usually desirable?

12-10   What do the terms "coherent" and "incoherent" precipitates mean?

12-11   What properties of the precipitate phase are needed for precipitation hardening? Why?

12-12   Electromigration (diffusion of atoms/ions due to momentum transfer from high energy electrons) leads to voids in aluminum interconnects used in many semiconductor metallization processes and thus is a leading cause of device reliability issues in the industry. Propose an additive to Al that can help mitigate this issue. Please refer to the appropriate phase diagram to justify your answers.

Section 12-3 Age or Precipitation Hardening

12-13   What is the principle of precipitation hardening?

12-14   What is the difference between precipitation hardening and dispersion strengthening?

12-15   What is a supersaturated solution? How do we obtain supersaturated solutions during precipitation hardening? Why is the formation of a supersaturated solution necessary?

12-16   Why do the precipitates formed during precipitation hardening form throughout the microstructure and not just at grain boundaries?

12-17   On aging for longer times, why do the second-phase precipitates grow? What is the driving force? Compare this with driving forces for grain growth and solid-state sintering.

Section 12-4 Applications of Age-Hardened Alloys

12-18   Why is precipitation hardening an attractive mechanism of strengthening for aircraft materials?

12-19   Why are most precipitation-hardened alloys suitable only for low-temperature applications?

Section 12-5 Microstructural Evolution in Age or Precipitation Hardening

12-20   Explain the three basic steps encountered during precipitation hardening.

12-21   Explain how hot shortness can occur in precipitation-hardened alloys.

12-22   In precipitation hardening, does the phase that provides strengthening form directly from the supersaturated matrix phase? Explain.

12-23   (a)   Recommend an artificial age-hardening heat treatment for a Cu-1.2% Be alloy. (See Figure 12-33 on page 489.) Include appropriate temperatures.

(b)   Compare the amount of the γ_2 precipitate that forms by artificial aging at 400°C with the amount of the precipitate that forms by natural aging.

12-24   Suppose that age hardening is possible in the Al-Mg system. (See Figure 12-10.)

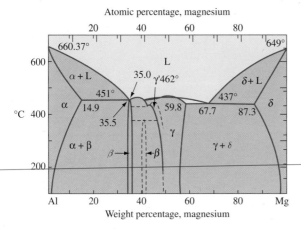

Atomic percentage, magnesium

Figure 12-10 (Repeated for Problem 12-24.) Portion of the aluminum-magnesium phase diagram.

(a) Recommend an artificial age-hardening heat treatment for each of the following alloys, and

(b) compare the amount of the β precipitate that forms from your treatment of each alloy: (i) Al-4% Mg (ii) Al-6% Mg (iii) Al-12% Mg.

(c) Testing of the alloys after the heat treatment reveals that little strengthening occurs as a result of the heat treatment. Which of the requirements for age hardening is likely not satisfied?

12-25 An Al-2.5% Cu alloy is solution-treated, quenched, and overaged at 230°C to produce a stable microstructure. If the θ precipitates as spheres with a diameter of 9000 Å and a density of 4.26 g/cm^3, determine the number of precipitate particles per cm^3. (See Figure 12-5.)

Section 12-6 Effects of Aging Temperature and Time

12-26 What is aging? Why is this step needed in precipitation hardening?

12-27 What is overaging?

12-28 What do the terms "natural aging" and "artificial aging" mean?

12-29 In the plane flown by the Wright brothers, how was the alloy precipitation strengthened?

12-30 Why did the work of Dr. Gayle and coworkers reveal two sets of precipitates in the alloy that was used to make the Wright brothers' plane?

12-31 Based on the principles of age hardening of Al-Cu alloys, rank the following Al-Cu alloys from highest to lowest for maximum yield strength achievable by age hardening and longest to shortest time required at 190°C to achieve the maximum yield strength: Al-2 wt% Cu, Al-3 wt% Cu, and Al-4 wt% Cu. Refer to the Al-Cu phase diagram (See Figure 12-9 repeated on the next page).

12-32 What analytical techniques would you use to characterize the nanoscale precipitates in an alloy?

12-33 Why do we have to keep some aluminum alloys at low temperatures until they are ready for forming steps?

Section 12-7 Requirements for Age Hardening

12-34 Can all alloy compositions be strengthened using precipitation hardening? Can we use this mechanism for the strengthening of ceramics, glasses, or polymers?

12-35 A conductive copper wire is to be made. Would you choose precipitation hardening as a way of strengthening this wire? Explain.

12-36 Figure 12-31 shows a hypothetical phase diagram. Determine whether each of the following alloys might be good candidates

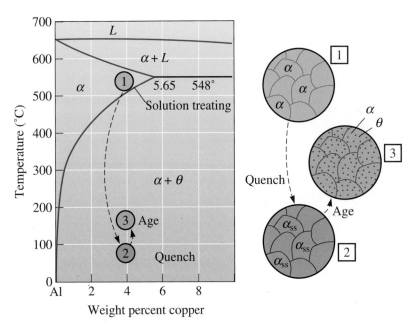

Figure 12-9 (Repeated for Problem 12-31) The aluminum-rich end of the aluminum-copper phase diagram showing the three steps in the age-hardening heat treatment and the microstructures that are produced.

for age hardening, and explain your answer. For those alloys that might be good candidates, describe the heat treatment required, including recommended temperatures.

(a) A-10% B
(b) A-20% B
(c) A-55% B
(d) A-87% B
(e) A-95% B.

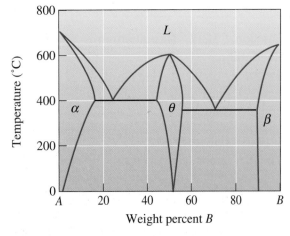

Figure 12-31 Hypothetical phase diagram (for Problem 12-36).

Section 12-8 Use of Age-Hardenable Alloys at High Temperatures

12-37 What is the major limitation of the use for precipitation-hardened alloys?

12-38 Why is it that certain aluminum (not nickel-based) alloys strengthened using age hardening can lose their strength on welding?

12-39 Would you choose a precipitation-hardened alloy to make an aluminum alloy baseball bat?

12-40 What type of dispersion strengthened alloys can retain their strength up to ~1000°C?

Section 12-9 The Eutectoid Reaction

12-41 Write down the eutectoid reaction in the Fe-Fe$_3$C system.

12-42 Sketch the microstructure of pearlite formed by the slow cooling of a steel with the eutectoid composition.

12-43 Compare and contrast eutectic and eutectoid reactions.

12-44 What are the solubilities of carbon in the α, δ, and γ forms of iron?

12-45 Define the following terms: ferrite, austenite, pearlite, and cementite.

12-46 The pearlite microstructure is similar to a ceramic-metal nanocomposite. True or false. Comment.

12-47 What do the terms "hypoeutectoid" and "hypereutectoid" steels mean?

12-48 What is the difference between a microconstituent and a phase?

12-49 For an Fe-0.35% C alloy, determine
(a) the temperature at which austenite first begins to transform on cooling;
(b) the primary microconstituent that forms;
(c) the composition and amount of each phase present at 728°C;
(d) the composition and amount of each phase present at 726°C; and
(e) the composition and amount of each microconstituent present at 726°C.

12-50 For an Fe-1.15% C alloy, determine
(a) the temperature at which austenite first begins to transform on cooling;
(b) the primary microconstitutent that forms;
(c) the composition and amount of each phase present at 728°C;
(d) the composition and amount of each phase present at 726°C; and
(e) the composition and amount of each microconstituent present at 726°C.

12-51 A steel contains 8% cementite and 92% ferrite at room temperature. Estimate the carbon content of the steel. Is the steel hypoeutectoid or hypereutectoid?

12-52 A steel contains 18% cementite and 82% ferrite at room temperature. Estimate the carbon content of the steel. Is the steel hypoeutectoid or hypereutectoid?

12-53 A steel contains 18% pearlite and 82% primary ferrite at room temperature. Estimate the carbon content of the steel. Is the steel hypoeutectoid or hypereutectoid?

12-54 A steel contains 94% pearlite and 6% primary cementite at room temperature. Estimate the carbon content of the steel. Is the steel hypoeutectoid or hypereutectoid?

12-55 A steel contains 55% α and 45% γ at 750°C. Estimate the carbon content of the steel.

12-56 A steel contains 96% γ and 4% Fe_3C at 800°C. Estimate the carbon content of the steel.

12-57 A steel is heated until 40% austenite, with a carbon content of 0.5%, forms. Estimate the temperature and the overall carbon content of the steel.

12-58 A steel is heated until 85% austenite, with a carbon content of 1.05%, forms. Estimate the temperature and the overall carbon content of the steel.

12-59 The carbon steels listed in the table below were soaked at 1000°C for 1 hour to form austenite and were cooled slowly, under equilibrium conditions to room temperature.

Refer to the Fe-Fe$_3$C phase diagram to answer the following questions for each of the carbon steel compositions listed in the table below.
(a) Determine the amounts of the phases present;
(b) determine the C content of each phase; and
(c) plot the C content of pearlite, α, and Fe$_3$C versus yield strength. Based on the graph, discuss the factors influencing yield strength in steels with <1% C.

Carbon %	Yield Strength (MPa)
0.2	295
0.4	353
0.6	372
0.8	376
0.95	379

12-60 A faulty thermocouple in a carburizing heat-treatment furnace leads to unreliable temperature measurement during the process. Microstructure analysis from the carburized steel that initially contained 0.2% carbon revealed that the surface had 93% pearlite and 7% primary Fe$_3$C, while at a depth of 0.5 cm, the microconstituents were 99% pearlite and 1% primary Fe$_3$C.

Estimate the temperature at which the heat treatment was carried out if the carburizing heat treatment was carried out for four hours. For interstitial carbon diffusion in FCC iron, the diffusion coefficient $D_0 = 2.3 \times 10^{-5}$ m^2/s and the activation energy $Q = 1.38 \times 10^5$ J/mol. It may be helpful to refer to Sections 5-6 and 5-8.

12-61 Determine the eutectoid temperature, the composition of each phase in the eutectoid reaction, and the amount of each phase present in the eutectoid microconstituent for the following systems. For the metallic systems, comment on whether you expect the eutectoid microconstituent to be ductile or brittle.
(a) ZrO$_2$-CaO (See Figure 12-32.)
(b) Cu-Al at 11.8% Al [See Figure 12-33(c).]
(c) Cu-Zn at 47% Zn [See Figure 12-33(a).]
(d) Cu-Be [See Figure 12-33(d).]

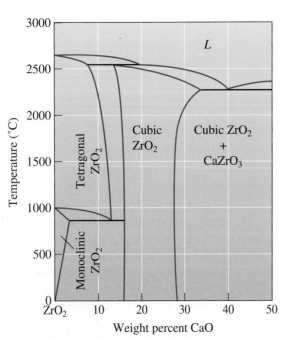

Figure 12-32 The ZrO$_2$-CaO phase diagram. A polymorphic phase transformation occurs for pure ZrO$_2$. Adding 16 to 26% CaO produces a single cubic zirconia phase at all temperatures (for Problem 12-61).

Section 12-10 Controlling the Eutectoid Reaction

12-62 Why are the distances between lamellae formed in a eutectoid reaction typically separated by distances smaller than those formed in eutectic reactions?

12-63 Compare the interlamellar spacing and the yield strength when a eutectoid steel is isothermally transformed to pearlite at
(a) 700°C and
(b) 600°C.

12-64 Why is it that a eutectoid steel exhibits different yield strengths and % elongations, depending upon if it was cooled slowly or relatively fast?

12-65 What is a TTT diagram?

12-66 Sketch and label clearly different parts of a TTT diagram for a plain carbon steel with 0.77% carbon.

12-67 On the TTT diagram, what is the difference between the γ and γ_u phases?

12-68 How is it that bainite and pearlite do not appear in the Fe-Fe$_3$C diagram? Are these phases or microconstituents?

12-69 Why is it that we cannot make use of TTT diagrams for describing heat treatment profiles in which samples are cooled over a period of time (i.e., why are TTT diagrams suitable only for isothermal transformations)?

12-70 What is bainite? Why do steels containing bainite exhibit higher levels of toughness?

12-71 An isothermally transformed eutectoid steel is found to have a yield strength of 410 MPa. Estimate
(a) the transformation temperature and
(b) the interlamellar spacing in the pearlite.

12-72 Determine the required transformation temperature and microconstituent if a eutectoid steel is to have the following hardness values:
(a) HRC 38   (b) HRC 42
(c) HRC 48   (d) HRC 52.

12-73 Describe the hardness and microstructure in a eutectoid steel that has been heated to 800°C for 1 h, quenched to 350°C and held for 750 s, and finally quenched to room temperature.

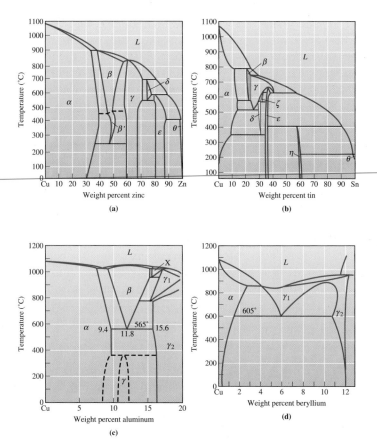

Figure 12-33 Binary phase diagrams for the (a) copper-zinc, (b) copper-tin, (c) copper-aluminum, and (d) copper-berrylium systems (for Problems 12-23 and 12-61).

12-74 Describe the hardness and microstructure in a eutectoid steel that has been heated to 800°C, quenched to 650°C, held for 500 s, and finally quenched to room temperature.

12-75 Describe the hardness and microstructure in a eutectoid steel that has been heated to 800°C, quenched to 300°C and held for 10 s, and finally quenched to room temperature.

12-76 Describe the hardness and microstructure in a eutectoid steel that has been heated to 800°C, quenched to 300°C and held for 10 s, quenched to room temperature, and then reheated to 400°C before finally cooling to room temperature again.

Section 12-11 The Martensitic Reaction and Tempering

12-77 What is the difference between solid-state phase transformations such as the eutectoid reaction and the martensitic phase transformation?

12-78 What is the difference between isothermal and athermal transformations?

12-79 What step controls the rate of martensitic phase transformations?

12-80 Why does martensite not appear on the Fe-Fe$_3$C phase diagram (Figure 12-34)?

12-81 Does martensite in steels have a fixed composition? What do the properties of martensite depend upon?

12-82 Can martensitic phase transformations occur in other alloys and ceramics?

12-83 Compare the mechanical properties of martensite, pearlite, and bainite formed from the eutectoid steel composition.

12-84 A steel containing 0.3% C is heated to various temperatures above the eutectoid temperature, held for 1 h, and then quenched to room temperature. Using Figure 12-34, determine the amount, composition, and hardness of any martensite that forms when the heating temperature is

(a) 728°C   (b) 750°C

(c) 790°C   (d) 850°C.

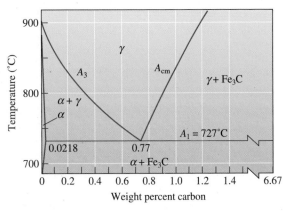

Figure 12-34   The eutectoid portion of the Fe-Fe₃C phase diagram (for Problems 12-80, 12-84, 12-85, 12-86 and 12-87).

12-85 A steel containing 0.95% C is heated to various temperatures above the eutectoid temperature, held for 1 h, and then quenched to room temperature. Using Figure 12-34, determine the amount and the composition of any martensite that forms when the heating temperature is

(a) 728°C   (b) 750°C

(c) 780°C   (d) 850°C.

12-86 A steel microstructure contains 75% martensite and 25% ferrite; the composition of the martensite is 0.6% C. Using Figure 12-34, determine

(a) the temperature from which the steel was quenched and

(b) the carbon content of the steel.

12-87 A steel microstructure contains 92% martensite and 8% Fe₃C; the composition of the martensite is 1.10% C. Using Figure 12-34, determine

(a) the temperature from which the steel was quenched and

(b) the carbon content of the steel.

12-88 A steel containing 0.8% C is quenched to produce all martensite. Estimate the

volume change that occurs, assuming that the lattice parameter of the austenite is 3.6 Å. Does the steel expand or contract during quenching?

12-89 Describe the complete heat treatment required to produce a quenched and tempered eutectoid steel having a tensile strength of at least 862 MPa. Include appropriate temperatures.

12-90 Describe the complete heat treatment required to produce a quenched and tempered eutectoid steel having an HRC hardness of less than 50. Include appropriate temperatures.

12-91 In eutectic alloys, the eutectic microconstituent is generally the continuous one, but in the eutectoid structures, the primary microconstituent is normally continuous. By describing the changes that occur with decreasing temperature in each reaction, explain why this difference is expected.

12-92 What is the tempering of steels? Why is tempering necessary?

12-93 What phases are formed by the decomposition of martensite?

12-94 What is tempered martensite?

12-95 If tempering results in the decomposition of martensite, why should we form martensite in the first place?

12-96 Describe the changes in properties that occur upon the tempering of a eutectoid steel.

Section 12-12 Shape-Memory Alloys (SMAs)

12-97 What is the principle by which shape-memory alloys display a memory effect?

12-98 Give examples of materials that display a shape-memory effect.

12-99 What are some of the applications of shape-memory alloys?

Design Problems

12-100 You wish to attach aluminum sheets to the frame of the twenty-fourth floor of a skyscraper. You plan to use rivets made of an age-hardenable aluminum, but the rivets must be soft and ductile in order to close. After the sheets are attached, the rivets must be very strong. Design a method for producing, using, and strengthening the rivets.

12-101 Design a process to produce a polypropylene polymer with a structure that is 75% crystalline. Figure 12-30 will provide appropriate data.

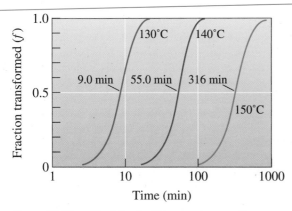

Figure 12-30 The effect of temperature on the crystallization of polypropylene. (Repeated for Problem 12-101.)

12-102 An age-hardened, Al-Cu bracket is used to hold a heavy electrical-sensing device on the outside of a steel-making furnace. Temperatures may exceed 200°C. Is this a good design? Explain. If it is not, design an appropriate bracket and explain why your choice is acceptable.

12-103 You use an arc-welding process to join a eutectoid steel. Cooling rates may be very high following the joining process. Describe what happens in the heat-affected area of the weld and discuss the problems that might occur. Design a joining process that may minimize these problems.

▲ Computer Problems

12-104 *Calculation of Phases and Micro-constituents in Plain Carbon Steels* Write a computer program that will calculate the amounts of phases in plain carbon steels in compositions that range from 0 to 1.5% carbon. Assume room temperature for your calculations. The program should ask the user to provide a value for the carbon concentration. The program should make use of this input and provide the amount of phases (ferrite, Fe_3C) in the steel using the lever rule (similar to Example 12-6). The program should also tell the user (as part of the output) whether the steel is eutectoid, hypoeutectoid, or hypereutectoid. Depending upon the composition, the program should also make available the amounts of microconstituents present (e.g., how much pearlite). The program should also print an appropriate descriptive message that describes the expected microstructure (assume slow cooling). For example, if the composition chosen is that of a hypoeutectoid steel, the program should output a message stating that the microstructure will primarily consist of α grains and pearlite.

Ⓚ Knovel® Problems

K12-1 What are the effects of thermal cycling on NiTi shape-memory alloys (SMA)?

K12-2 Some precipitation-hardened iron super-alloys are used for high-temperature service. Describe heat treatment procedures used for these steels.

Steels constitute the most widely used family of materials for structural, load-bearing applications. Most buildings, bridges, tools, automobiles, and numerous other applications make use of ferrous alloys. With a range of heat treatments that can provide a wide assortment of microstructures and properties, steels are probably the most versatile family of engineering materials.

The Golden Gate Bridge in California is made from steel. The steel is painted to increase its corrosion resistance. (The color is famously known as International Orange.) Each of the main cables that suspend the road over the San Francisco Bay contains 27,572 wires for a total length of approximately 129,000 km (*Courtesy of Chee-Onn Leong/Shutterstock.*)

13

Heat Treatment of Steels and Cast Irons

Have You Ever Wondered?

- *What is the most widely used engineered material?*
- *What makes stainless steels "stainless?"*
- *What is the difference between cast iron and steel?*
- *Are stainless steels ferromagnetic?*

Ferrous alloys, which are based on iron-carbon alloys, include plain carbon steels, alloy and tool steels, stainless steels, and cast irons. These are the most widely used materials in the world. In the history of civilization, these materials made their mark by defining the *Iron Age*. Steels typically are produced in two ways: by refining iron ore or by recycling scrap steel.

In producing primary steel, iron ore (processed to contain 50 to 70% iron oxide, Fe_2O_3 or Fe_3O_4) is heated in a blast furnace in the presence of coke (a form of carbon) and oxygen. The coke reduces the iron oxide into a crude molten iron known as **hot metal or pig iron**. At about ~1600°C, this material contains about 95% iron; 4% carbon; 0.3 to 0.9% silicon; 0.5% manganese; and 0.025 to 0.05% of sulfur, phosphorus, and titanium. Slag is a byproduct of the blast furnace process. It contains silica, CaO, and other impurities in the form of a silicate melt.

Because the liquid pig iron contains a large amount of carbon, oxygen is blown into it in the *basic oxygen furnace* (BOF) to eliminate the excess carbon and produce liquid steel. Steel has a carbon content up to a maximum of ~2% on a weight basis. In the second method, scrap is often melted in an **electric arc furnace** in which the heat of the are melts the scrap. Many alloy and specialty steels, such as stainless steels, are produced using electric melting. Molten steels (including stainless steels) often undergo further refining. The goal here is to reduce the levels of impurities such as phosphorus, sulfur, etc. and to bring the carbon to a desired level.

Molten steel is poured into molds to produce finished steel castings or cast into shapes that are later processed through metal-forming techniques such as rolling or forging. In the latter case, the steel is either poured into large ingot molds or is continuously cast into regular shapes.

All of the strengthening mechanisms discussed in the previous chapter apply to at least some of the ferrous alloys. In this chapter, we will discuss how to use the eutectoid reaction to control the structure and properties of steels through heat treatment and alloying. We will also examine two special classes of ferrous alloys: stainless steels and cast irons.

13-1   Designations and Classification of Steels

The dividing point between "steels" and "cast irons" is 2.11% C, where the eutectic reaction becomes possible. For steels, we concentrate on the eutectoid portion of the diagram (Figure 13-1) in which the solubility lines and the eutectoid isotherm are specially identified. The A_3 shows the temperature at which ferrite starts to form on cooling; the A_{cm} shows the temperature at which cementite starts to form; and the A_1 is the eutectoid temperature.

Almost all of the heat treatments of steel are directed toward producing the mixture of ferrite and cementite that gives the proper combination of properties. Figure 13-2 shows the three important microconstituents, or arrangements of ferrite and cementite, that are usually sought. Pearlite is a microconstituent consisting of a lamellar mixture of ferrite and cementite. In bainite, which is obtained by transformation of austenite at a large undercooling, the cementite is more rounded than in pearlite. Tempered martensite, a mixture of very fine and nearly round cementite in ferrite, forms when martensite is reheated following its formation.

Designations   The AISI (American Iron and Steel Institute) and SAE (Society of Automotive Engineers) provide designation systems (Table 13-1) that use a four-or five-digit number. The first two numbers refer to the major alloying elements present, and the last two or three numbers refer to the percentage of carbon. An AISI 1040 steel is a plain carbon steel with 0.40% C. An SAE 10120 steel is a plain carbon steel containing 1.20% C. An AISI 4340 steel is an alloy steel containing 0.40% C. Note that the American Society for Testing of Materials (ASTM) has a different way of classifying steels. The ASTM has a list of specifications that describe steels suitable for different applications. The example below illustrates the use of AISI numbers.

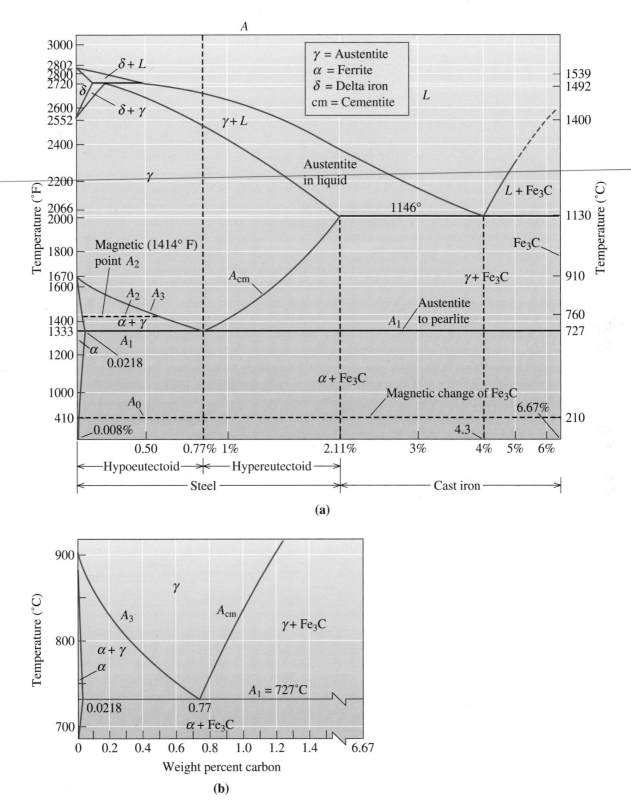

Figure 13-1   (a) The Fe-Fe₃C phase diagram. (b) An expanded version of the eutectoid portion of the Fe-Fe₃C diagram, adapted from several sources.

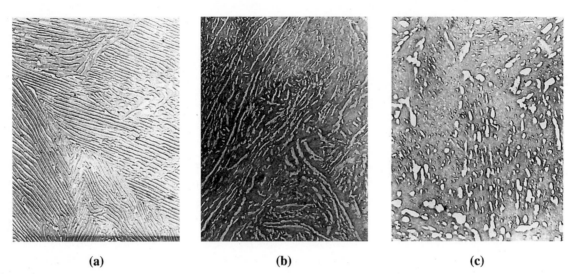

(a) **(b)** **(c)**

Figure 13-2   Micrographs of (a) pearlite, (b) bainite, and (c) tempered martensite, illustrating the differences in cementite size and shape among the three microconstituents (× 7500). (*From* The Making, Shaping and Treating of Steel, 10th Ed. *Courtesy of the Association of Iron and Steel Engineers.*)

TABLE 13-1 ■ *Compositions of selected AISI-SAE steels*

AISI-SAE Number	% C	% Mn	% Si	% Ni	% Cr	Others
1020	0.18–0.23	0.30–0.60				
1040	0.37–0.44	0.60–0.90				
1060	0.55–0.65	0.60–0.90				
1080	0.75–0.88	0.60–0.90				
1095	0.90–1.03	0.30–0.50				
1140	0.37–0.44	0.70–1.00				0.08–0.13% S
4140	0.38–0.43	0.75–1.00	0.15–0.30		0.80–1.10	0.15–0.25% Mo
4340	0.38–0.43	0.60–0.80	0.15–0.30	1.65–2.00	0.70–0.90	0.20–0.300% Mo
4620	0.17–0.22	0.45–0.65	0.15–0.30	1.65–2.00		0.20–0.30% Mo
52100	0.98–1.10	0.25–0.45	0.15–0.30		1.30–1.60	
8620	0.18–0.23	0.70–0.90	0.15–0.30	0.40–0.70	0.40–0.60	0.15–0.25% Y
9260	0.56–0.64	0.75–1.00	1.80–2.20			

Example 13-1 *Design of a Method to Determine AISI Number*

An unalloyed steel tool used for machining aluminum automobile wheels has been found to work well, but the purchase records have been lost and you do not know the steel's composition. The microstructure of the steel is tempered martensite. Assume that you cannot estimate the composition of the steel from the structure. Design a treatment that may help determine the steel's carbon content.

SOLUTION

Assume that there is no access to equipment that would permit you to analyze the chemical composition directly. Since the entire structure of the steel is a very fine tempered martensite, we can do a simple heat treatment to produce a structure that can be analyzed more easily. This can be done in two different ways.

The first way is to heat the steel to a temperature just below the A_1 temperature and hold for a long time. The steel overtempers and large Fe_3C spheres form in a ferrite matrix. We then estimate the amount of ferrite and cementite and calculate the carbon content using the lever law. If we measure 16% Fe_3C using this method, the carbon content is

$$\% \ Fe_3C = \left[\frac{(x - 0.0218)}{(6.67 - 0.0218)} \right] \times 100 = 16 \ \text{ or } \ x = 1.086\% \ C$$

A better approach, however, is to heat the steel above the A_{cm} to produce all austenite. If the steel then cools slowly, it transforms to pearlite and a primary microconstituent. If, when we do this, we estimate that the structure contains 95% pearlite and 5% primary Fe_3C, then

$$\% \ Pearlite = \left[\frac{(6.67 - x)}{(6.67 - 0.77)} \right] \times 100 = 95 \ \text{ or } \ x = 1.065\% \ C$$

The carbon content is on the order of 1.065 to 1.086%, consistent with a 10110 steel. In this procedure, we assumed that the weight and volume percentages of the microconstituents are the same, which is nearly the case in steels.

Classifications
Steels can be classified based on their composition or the way they have been processed. Carbon steels contain up to ~2% carbon. These steels may also contain other elements, such as Si (maximum 0.6%), Cu (up to 0.6%), and Mn (up to 1.65%). Decarburized steels contain less than 0.005% C. Ultra-low carbon steels contain a maximum of 0.03% carbon. They also contain very low levels of other elements such as Si and Mn. Low-carbon steels contain 0.04 to 0.15% carbon. These low-carbon steels are used for making car bodies and hundreds of other applications. Mild steel contains 0.15 to 0.3% carbon. This steel is used in buildings, bridges, piping, etc. Medium-carbon steels contain 0.3 to 0.6% carbon. These are used in making machinery, tractors, mining equipment, etc. High-carbon steels contain above 0.6% carbon. These are used in making springs, railroad car wheels, and the like. Note that cast irons are Fe-C alloys containing 2 to 4% carbon.

Alloy steels are compositions that contain more significant levels of alloying elements. We will discuss the effect of alloying elements later in this chapter. They improve the hardenability of steels. The AISI defines alloy steels as steels that exceed one or more of the following composition limits: ≥1.65% Mn, 0.6% Si, or 0.6% Cu. The total carbon content is up to 1%, and the total alloying element content is below 5%. A material is also an alloy steel if a definite concentration of alloying elements, such as Ni, Cr, Mo, Ti, etc., is specified. These steels are used for making tools (hammers, chisels, etc.) and also in making parts such as axles, shafts, and gears.

Certain specialty steels may contain higher levels of sulfur (>0.1%) or lead (~0.15 to 0.35%) to provide machinability. These, however, cannot be welded easily. Recently, researchers have developed "green steel" in which lead, an environmental toxin, is replaced with tin (Sn) and/or antimony (Sb). Steels can also be classified based on their processing. For example, the term "concast steels" refers to continuously cast steels.

Galvanized steels have a zinc coating for corrosion resistance (Chapter 23). Similarly, tin-plated steel is used to make corrosion-resistant cans and other products. Tin is deposited using electroplating—a process known as "continuous web electrodeposition." "E-steels" are steels that are melted using an electric furnace, while "B-steels" contain a small (0.0005 to 0.003%), yet significant, concentration of boron. Recently, a "germ-resistant" coated stainless steel has been developed.

13-2   Simple Heat Treatments

Four simple heat treatments—process annealing, annealing, normalizing, and spheroidizing—are commonly used for steels (Figure 13-3). These heat treatments are used to accomplish one of three purposes: (1) eliminating the effects of cold work, (2) controlling dispersion strengthening, or (3) improving machinability.

Process Annealing—Eliminating Cold Work   The recrystallization heat treatment used to eliminate the effect of cold working in steels with less than about 0.25% C is called a **process anneal**. The process anneal is done 80°C to 170°C below the A_1 temperature. The intent of the process anneal treatment for steels is similar to the annealing of inorganic glasses in that the main idea is to significantly reduce or eliminate residual stresses.

Annealing and Normalizing—Dispersion Strengthening
Steels can be dispersion-strengthened by controlling the fineness of pearlite. The steel is initially heated to produce homogeneous austenite (FCC γ phase), a step called **austenitizing**. **Annealing**, or a full anneal, allows the steel to cool slowly in a furnace, producing coarse pearlite. **Normalizing** allows the steel to cool more rapidly, in air, producing fine pearlite. Figure 13-4 shows the typical properties obtained by annealing and normalizing plain carbon steels.

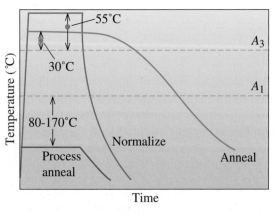

(a) Hypoeutectoid

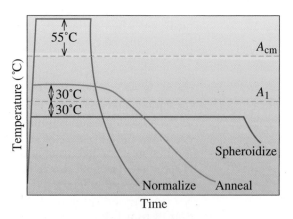

(b) Hypereutectoid

Figure 13-3   Schematic summary of the simple heat treatments for (a) hypoeutectoid steels and (b) hypereutectoid steels.

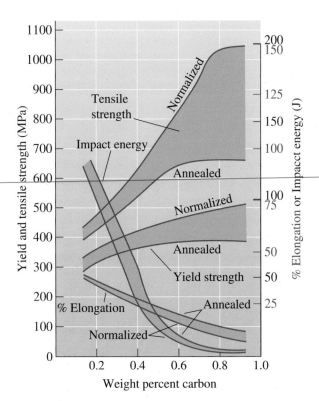

Figure 13-4
The effect of carbon and heat treatment on the properties of plain carbon steels.

For annealing, austenitizing of hypoeutectoid steels is conducted about 30°C above the A_3, producing 100% γ; however, austenitizing of a hypereutectoid steel is done at about 30°C above the A_1, producing austenite and Fe_3C. This process prevents the formation of a brittle, continuous film of Fe_3C at the grain boundaries that occurs on slow cooling from the 100% γ region. In both cases, the slow furnace cool and coarse pearlite provide relatively low strength and good ductility.

For normalizing, austenitizing is done at about 55°C above the A_3 or A_{cm}; the steel is then removed from the furnace and cooled in air. The faster cooling gives fine pearlite and provides higher strength.

Spheroidizing—Improving Machinability

Steels that contain a large concentration of Fe_3C have poor machining characteristics. It is possible to transform the morphology of Fe_3C using *spheroidizing*. During the spheroidizing treatment, which requires several hours at about 30°C below the A_1, the Fe_3C phase morphology changes into large, spherical particles in order to reduce boundary area. The microstructure, known as **spheroidite**, has a continuous matrix of soft, machinable ferrite (Figure 13-5). After machining, the steel is given a more sophisticated heat treatment to produce the required properties. A similar microstructure occurs when martensite is tempered just below the A_1 for long periods of time. As noted before, alloying elements such as Pb and S are also added to improve machinability of steels and, more recently, lead-free "green steels" that have very good machinability have been developed.

The following example shows how different heat treatment conditions can be developed for a given composition of steel.

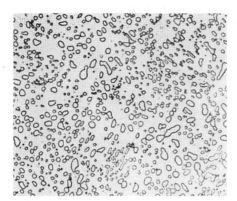

Figure 13-5
The microstructure of spheroidite with Fe_3C particles dispersed in a ferrite matrix ($\times$ 850). (*From* ASM Handbook, *Vol. 7, (1972), ASM International, Materials Park, OH 44073-0002.*)

Example 13-2   *Determination of Heat Treating Temperatures*

Recommend temperatures for the process annealing, annealing, normalizing, and spheroidizing of 1020, 1077, and 10120 steels.

SOLUTION

From Figure 13-1, we find the critical A_1, A_3, or A_{cm}, temperatures for each steel. We can then specify the heat treatment based on these temperatures.

Steel Type	1020	1077	10120
Critical temperatures	$A_1 = 727°C$ $A_3 = 830°C$	$A_1 = 727°C$	$A_1 = 727°C$ $A_{cm} = 895°C$
Process annealing	727 – (80 to 170) = 557°C to 647°C	Not done	Not done
Annealing	830 + 30 = 860°C	727 + 30 = 757°C	727 + 30 = 757°C
Normalizing	830 + 55 = 885°C	727 + 55 = 782°C	895 + 55 = 950°C
Spheroidizing	Not done	727 – 30 = 697°C	727 – 30 = 697°C

13-3   Isothermal Heat Treatments

The effect of transformation temperature on the properties of a 1080 (eutectoid) steel was discussed in Chapter 12. As the isothermal transformation temperature decreases, pearlite becomes progressively finer before bainite begins to form. At very low temperatures, martensite is obtained.

Austempering and Isothermal Annealing   The isothermal transformation heat treatment used to produce bainite, called **austempering**, simply involves austenitizing the steel, quenching to some temperature below the nose of the TTT curve, and holding, at that temperature until all of the austenite transforms to bainite (Figure 13-6).

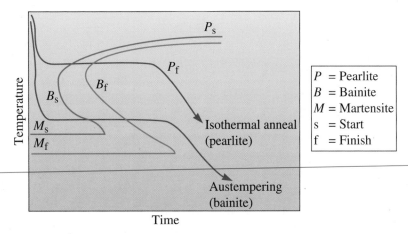

Figure 13-6 The austempering and isothermal anneal heat treatments in a 1080 steel.

Annealing and normalizing are usually used to control the fineness of pearlite; however, pearlite formed by an **isothermal anneal** (Figure 13-6) may give more uniform properties, since the cooling rates and microstructure obtained during annealing and normalizing vary across the cross-section of the steel. *Note that the TTT diagrams only describe isothermal heat treatments (i.e., we assume that the sample begins and completes heat treatment at a given temperature).* Thus, we cannot exactly describe heat treatments by superimposing cooling curves on a TTT diagram such as those shown in Figure 13-6.

Effect of Changes in Carbon Concentration on the TTT Diagram

In either a hypoeutectoid or a hypereutectoid steel, the TTT diagram must reflect the possible formation of a primary phase. The isothermal transformation diagrams for a 1050 and a 10110 steel are shown in Figure 13-7. The most remarkable change is the presence of a "wing" that begins at the nose of the curve and becomes asymptotic to the A_3 or A_{cm} temperature. The wing represents the ferrite start (F_s) time in hypoeutectoid steels or the cementite start (C_s) time in hypereutectoid steels.

When a 1050 steel is austenitized, quenched, and held between the A_1 and the A_3, primary ferrite nucleates and grows. Eventually, equilibrium amounts of ferrite and austenite result. Similarly, primary cementite nucleates and grows to its equilibrium amount in a 10110 steel held between the A_{cm} and A_1 temperatures.

If an austenitized 1050 steel is quenched to a temperature between the nose and the A_1 temperatures, primary ferrite again nucleates and grows until reaching the equilibrium amount. The remainder of the austenite then transforms to pearlite. A similar situation, producing primary cementite and pearlite, is found for the hypereutectoid steel.

If we quench the steel below the nose of the curve, only bainite forms, regardless of the carbon content of the steel. If the steels are quenched to temperatures below the M_s, martensite will form. The following example shows how the phase diagram and TTT diagram can guide development of the heat treatment of steels.

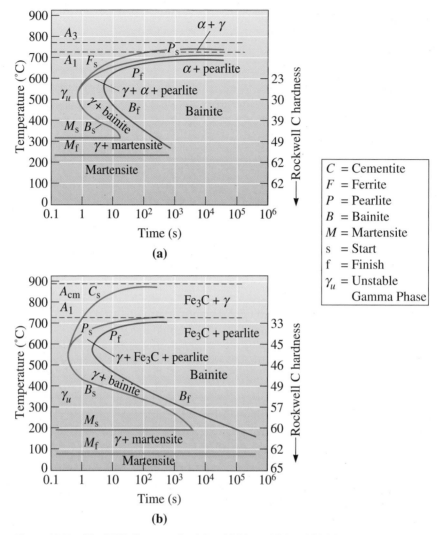

Figure 13-7   The TTT diagrams for (a) a 1050 and (b) a 10110 steel. Note $\gamma_u =$ unstable austenite.

Example 13-3   *Design of a Heat Treatment for an Axle*

A heat treatment is needed to produce a uniform microstructure and hardness of HRC 23 in a 1050 steel axle.

SOLUTION

We might attempt this task in several ways. We could austenitize the steel, then cool at an appropriate rate by annealing or normalizing to obtain the correct hardness. By doing this, however, we find that the structure and hardness vary from the surface to the center of the axle.

A better approach is to use an isothermal heat treatment. From Figure 13-7, we find that a hardness of HRC 23 is obtained by transforming austenite to a mixture of ferrite and pearlite at 600°C. From Figure 13-1, we find that the A_3 temperature is 770°C. Therefore, our heat treatment is

1. Austenitize the steel at 770 + (30 to 55) = 800°C to 825°C, holding for 1 h and obtaining 100% γ.

2. Quench the steel to 600°C and hold for a minimum of 10 s. Primary ferrite begins to precipitate from the unstable austenite after about 1.0 s. After 1.5 s, pearlite begins to grow, and the austenite is completely transformed to ferrite and pearlite after about 10 s. After this treatment, the microconstituents present are

$$\text{Primary } \alpha = \left[\frac{(0.77 - 0.5)}{(0.77 - 0.218)} \right] \times 100 = 36\%$$

$$\text{Pearlite} = \left[\frac{(0.5 - 0.0218)}{(0.77 - 0.0218)} \right] \times 100 = 64\%$$

3. Cool in air to room temperature, preserving the equilibrium amounts of primary ferrite and pearlite. The microstructure and hardness are uniform because of the isothermal anneal.

Interrupting the Isothermal Transformation

Complicated microstructures are produced by interrupting the isothermal heat treatment. For example, we could austenitize the 1050 steel (Figure 13-8) at 800°C, quench to 650°C and hold for 10 s (permitting some ferrite and pearlite to form), then quench to 350°C and hold for 1 h (3600 s). Whatever unstable austenite remained before quenching to 350°C transforms to bainite. The final structure is ferrite, pearlite, and bainite. We could complicate the treatment further by interrupting the treatment at 350°C after 1 min (60 s) and quenching. Any austenite remaining after 1 min at 350°C forms martensite. The final structure now contains ferrite, pearlite, bainite, and martensite. Note that each time we change the temperature, we start at zero time! In practice, temperatures cannot be changed instantaneously (i.e., we cannot go instantly from 800 to 650 or 650 to 350°C). This is why it is better to use the continuous cooling transformation (CCT) diagrams.

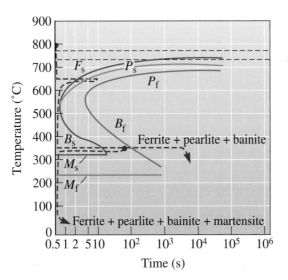

Figure 13-8
Producing complicated structures by interrupting the isothermal heat treatment of a 1050 steel.

13-4   Quench and Temper Heat Treatments

Quenching hardens most steels and tempering increases the toughness. This has been known for perhaps thousands of years. For example, a series of such heat treatments has been used for making Damascus steel and Japanese Samurai swords. We can obtain an exceptionally fine dispersion of Fe_3C and ferrite (known as tempered martensite) if we first quench the austenite to produce martensite and then temper. The tempering treatment controls the final properties of the steel (Figure 13-9). Note that this is different from a spheroidizing heat treatment (Figure 13-5). The following example shows how a combination of heat treatments is used to obtain steels with desired properties.

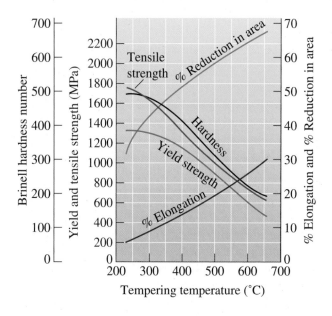

Figure 13-9
The effect of tempering temperature on the mechanical properties of a 1050 steel.

Example 13-4   *Design of a Quench and Temper Treatment*

A rotating shaft that delivers power from an electric motor is made from a 1050 steel. Its yield strength should be at least 1000 MPa, yet it should also have at least 15% elongation in order to provide toughness. Design a heat treatment to produce this part.

SOLUTION

We are not able to obtain this combination of properties by annealing or normalizing (Figure 13-4); however, a quench and temper heat treatment produces a microstructure that can provide both strength and toughness. Figure 13-9 shows that the yield strength exceeds 1000 MPa if the steel is tempered below 460°C, whereas the elongation exceeds 15% if tempering is done above 425°C. The A_3 temperature for the steel is 770°C. A possible heat treatment is

1. Austenitize above the A_3 temperature of 770°C for 1 h. An appropriate temperature may be 770 + 55 = 825°C.

2. Quench rapidly to room temperature. Since the M_f is about 250°C, martensite will form.

3. Temper by heating the steel to 440°C. Normally, 1 h will be sufficient if the steel is not too thick.

4. Cool to room temperature.

Retained Austenite

There is a large volume expansion when martensite forms from austenite. As the martensite plates form during quenching, they surround and isolate small pools of austenite (Figure 13-10), which deform to accommodate the lower density martensite. As the transformation progresses, however, for the remaining pools of austenite to transform, the surrounding martensite must deform. Because the strong martensite resists the transformation, either the existing martensite cracks or the austenite remains trapped in the structure as **retained austenite**. Retained austenite can be a serious problem. Martensite softens and becomes more ductile during tempering. After tempering, the retained austenite cools below the M_s and M_f temperatures and transforms to martensite, since the surrounding **tempered martensite** can deform. But now the steel contains more of the hard, brittle martensite! A second tempering step may be needed to eliminate the martensite formed from the retained austenite. Retained austenite is also more of a problem for high-carbon steels. The martensite start and finish temperatures are reduced when the carbon content increases (Figure 13-11). High-carbon steels must be refrigerated to produce all martensite.

Residual Stresses and Cracking

Residual stresses are also produced because of the volume change or because of cold working. A stress-relief anneal can be used to remove or minimize residual stresses due to cold working. Stresses are also induced because of thermal expansion and contraction. In steels, there is one more mechanism that causes stress. When steels are quenched, the surface of the quenched steel cools rapidly and transforms to martensite. When the austenite in the center later transforms, the hard surface is placed in tension, while the center is compressed. If the residual stresses exceed the yield strength, **quench cracks** form at the surface (Figure 13-12). To avoid this, we can first cool to just above the M_s and hold until the temperature equalizes in the steel; subsequent quenching permits all of the steel to transform to martensite at about the same time. This heat treatment is called

Figure 13-10
Retained austenite (white) trapped between martensite needles (black) ($\times$ 1000). (*From* ASM Handbook, *Vol. 8, (1973), ASM International, Materials Park, OH 44073-0002.*)

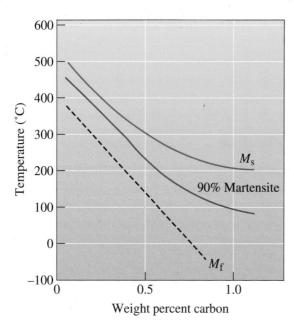

Figure 13-11
Increasing carbon reduces the M_s and M_f temperatures in plain carbon steels.

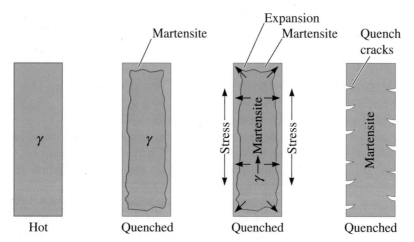

Figure 13-12   Formation of quench cracks caused by residual stresses produced during quenching. The figure illustrates the development of stresses as the austenite transforms to martensite during cooling.

marquenching or **martempering** (Figure 13-13). Note that as discussed presently, strictly speaking, the CCT diagrams should be used to examine non-isothermal heat treatments.

Quench Rate

In using the TTT diagram, we assumed that we could cool from the austenitizing temperature to the transformation temperature instantly. Because this does not occur in practice, undesired microconstituents may form during the quenching process. For example, pearlite may form as the steel cools past the nose of the curve, particularly because the time of the nose is less than one second in plain carbon steels.

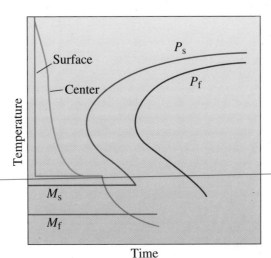

Figure 13-13
The marquenching heat treatment, designed to reduce residual stresses and quench cracking.

TABLE 13-2 ■ *The H coefficient, or severity of the quench, for several quenching media*

Medium	H Coefficient	Cooling Rate at the Center of a 2.5-cm Bar (°C/s)
Oil (no agitation)	0.25	18
Oil (agitation)	1.0	45
H_2O (no agitation)	1.0	45
H_2O (agitation)	4.0	190
Brine (no agitation)	2.0	90
Brine (agitation)	5.0	230

The rate at which the steel cools during quenching depends on several factors. First, the surface always cools faster than the center of the part. In addition, as the size of the part increases, the cooling rate at any location is slower. Finally, the cooling rate depends on the temperature and heat transfer characteristics of the quenching medium (Table 13-2). Quenching in oil, for example, produces a lower H coefficient, or slower cooling rate, than quenching in water or brine. The H coefficient is equivalent to the heat transfer coefficient. Agitation helps break the vapor blanket (e.g., when water is the quenching medium) and improves the overall heat transfer rate by bringing cooler liquid into contact with the parts being quenched.

Continuous Cooling Transformation Diagrams

We can develop a *continuous cooling transformation* (CCT) diagram by determining the microstructures produced in the steel at various rates of cooling. The CCT curve for a 1080 steel is shown in Figure 13-14. The CCT diagram differs from the TTT diagram in that longer times are required for transformations to begin and no bainite region is observed.

If we cool a 1080 steel at 5°C/s, the CCT diagram tells us that we obtain coarse pearlite; we have annealed the steel. Cooling at 35°C/s gives fine pearlite and is a

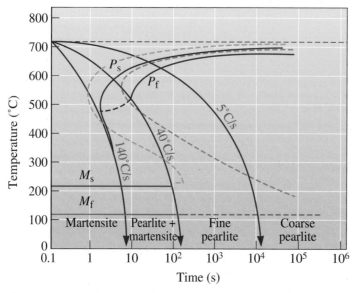

Figure 13-14   The CCT diagram (solid lines) for a 1080 steel compared with the TTT diagram (dashed lines).

normalizing heat treatment. Cooling at 100°C/s permits pearlite to start forming, but the reaction is incomplete and the remaining austenite changes to martensite. We obtain 100% martensite and thus are able to perform a quench and temper heat treatment, only if we cool faster than 140°C/s. Other steels, such as the low-carbon steel in Figure 13-15 have more complicated CCT diagrams. In various handbooks, you can find a compilation of TTT and CCT diagrams for different grades of steels.

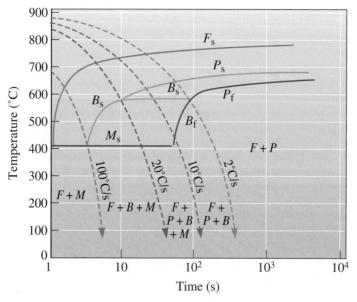

Figure 13-15   The CCT diagram for a low-alloy, 0.2% C steel.

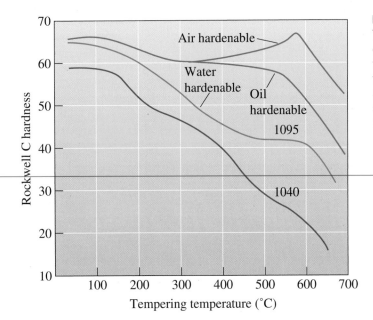

Figure 13-19
The effect of alloying elements on the phases formed during the tempering of steels. The air-hardenable steel shows a secondary hardening peak.

13-6   Application of Hardenability

A **Jominy test** (Figure 13-20) is used to compare hardenabilities of steels. A steel bar 10 cm long and 2.5 cm in diameter is austenitized, placed into a fixture, and sprayed at one end with water. This procedure produces a range of cooling rates—very fast at the quenched end, almost air cooling at the opposite end. After the test, hardness measurements are made along the test specimen and plotted to produce a **hardenability curve** (Figure 13-21). The distance from the quenched end is the **Jominy distance** and is related to the cooling rate (Table 13-3).

Virtually any steel transforms to martensite at the quenched end. Thus, the hardness at zero Jominy distance is determined solely by the carbon content of the steel. At larger Jominy distances, there is a greater likelihood that bainite or pearlite will form instead of martensite. An alloy steel with a high hardenability (such as 4340) maintains a

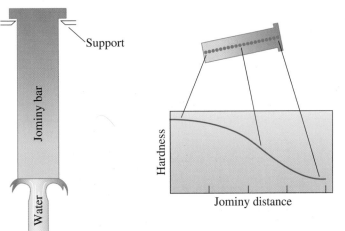

Figure 13-20
The set-up for the Jominy test used for determining the hardenability of a steel.

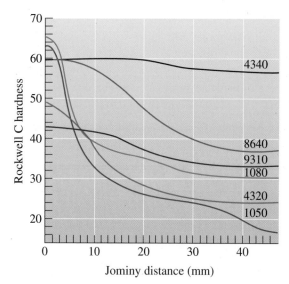

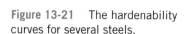

Figure 13-21   The hardenability curves for several steels.

rather flat hardenability curve; a plain carbon steel (such as 1050) has a curve that drops off quickly. The hardenability is determined primarily by the alloy content of the steel.

We can use hardenability curves in selecting or replacing steels in practical applications. The fact that two different steels cool at the same rate if quenched under identical conditions helps in this selection process. The Jominy test data are used as shown in the following example.

TABLE 13-3 ■ *The relationship between cooling rate and Jominy distance*

Jominy Distance (mm)	Cooling Rate (°C/s)
1.6	315
3.1	110
4.7	50
6.3	36
7.8	28
9.4	22
10.9	17
12.5	15
15.6	10
18.8	8
25.0	5
31.3	3
37.5	2.8
43.8	2.5
56.3	2.2

| **Example 13-5** | *Design of a Wear-Resistant Gear* |

A gear made from 9310 steel, which has an as-quenched hardness at a critical location of HRC 40, wears at an excessive rate. Tests have shown that an as-quenched hardness of at least HRC 50 is required at that critical location. Design a steel that would be appropriate.

SOLUTION

We know that if different steels of the same size are quenched under identical conditions, their cooling rates or Jominy distances are the same. From Figure 13-21, a hardness of HRC 40 in a 9310 steel corresponds to a Jominy distance of 15.6 mm (10°C/s). If we assume the same Jominy distance, the other steels shown in Figure 13-21 have the following hardnesses at the critical location:

> 1050 HRC 28
> 1080 HRC 36
> 4320 HRC 31
> 8640 HRC 52
> 4340 HRC 60

Both the 8640 and 4340 steels are appropriate. The 4320 steel has too low a carbon content to ever reach HRC 50; the 1050 and 1080 have enough carbon, but the hardenability is too low. In Table 13-1, we find that the 86xx steels contain less alloying elements than the 43xx steels; thus the 8640 steel is probably less expensive than the 4340 steel and might be our best choice. We must also consider other factors such as durability.

In another simple technique, we utilize the severity of the quench and the Grossman chart (Figure 13-22) to determine the hardness at the *center* of a round bar. The bar diameter and H coefficient, or severity of the quench in Table 13-2, give the Jominy distance at the center of the bar. We can then determine the hardness from the hardenability curve of the steel. (See Example 13-6.)

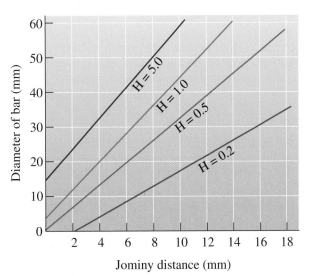

Figure 13-22
The Grossman chart used to determine the hardenability at the center of a steel bar for different quenchants.

Example 13-6 *Design of a Quenching Process*

Design a quenching process to produce a minimum hardness of HRC 40 at the center of a 3.75 cm diameter 4320 steel bar.

SOLUTION

Several quenching media are listed in Table 13-2. We can find an approximate H coefficient for each of the quenching media, then use Figure 13-22 to estimate the Jominy distance in a 3.75 cm diameter bar for each media. Finally, we can use the hardenability curve (Figure 13-21) to find the hardness in the 4320 steel. The results are listed below.

	H Coefficient	Jominy Distance (mm)	HRC
Oil (no agitation)	0.25	17.2	28
Oil (agitation)	1.00	9.4	39
H_2O (no agitation)	1.00	9.4	39
H_2O (agitation)	4.00	6.3	44
Brine (no agitation)	2.00	7.8	42
Brine (agitation)	5.00	4.7	46

The last three methods, based on brine or agitated water, are satisfactory. Using an unagitated brine quenchant might be least expensive, since no extra equipment is needed to agitate the quenching bath; however, H_2O is less corrosive than the brine quenchant.

13-7   Specialty Steels

There are many special categories of steels, including tool steels, interstitial-free steels, high-strength-low-alloy (HSLA) steels, dual-phase steels, and maraging steels.

Tool steels are usually high-carbon steels that obtain high hardnesses by a quench and temper heat treatment. Their applications include cutting tools in machining operations, dies for die casting, forming dies, and other uses in which a combination of high strength, hardness, toughness, and temperature resistance is needed.

Alloying elements improve the hardenability and high-temperature stability of the tool steels. The water-hardenable steels such as 1095 must be quenched rapidly to produce martensite and also soften rapidly even at relatively low temperatures. Oil-hardenable steels form martensite more easily, temper more slowly, but still soften at high temperatures. The air-hardenable and special tool steels may harden to martensite while cooling in air. In addition, these steels may not soften until near the A_1 temperature. In fact, the highly alloyed tool steels may pass through a **secondary hardening peak** near 500°C as the normal cementite dissolves and hard alloy carbides precipitate (Figure 13-19). The alloy carbides are particularly stable, resist growth or spheroidization, and are important in establishing the high-temperature resistance of these steels.

High-strength-low-alloy (HSLA) steels are low-carbon steels containing small amounts of alloying elements. The HSLA steels are specified on the basis of yield strength with grades up to 552 MPa; the steels contain the least amount of alloying element that still provides the proper yield strength without heat treatment. In these steels, careful processing permits precipitation of carbides and nitrides of Nb, V, Ti, or Zr, which provide dispersion strengthening and a fine grain size.

Dual-phase steels contain a uniform distribution of ferrite and martensite, with the dispersed martensite providing yield strengths of 414 MPa to 1000 MPa. These low-carbon steels do not contain enough alloying elements to have good hardenability using the normal quenching processes. But when the steel is heated into the ferrite-plus-austenite portion of the phase diagram, the austenite phase becomes enriched in carbon, which provides the needed hardenability. During quenching, only the austenite portion transforms to martensite [Figure 13-23(a)].

The microstructure of **TRIP steels** [Figure 13-23(b)] consists of a continuous ferrite matrix and a dispersion of a harder second phase (martensite and/or bainite). In addition, the microstructure consists of retained austenite. TRIP steels exhibit better ductility and formability at a given strength level because of the transformation of retained austenite to martensite during plastic deformation. Transformation induced plasticity (TRIP) steels are useful for more complex shapes.

Maraging steels are low-carbon, highly alloyed steels. The steels are austenitized and quenched to produce a soft martensite that contains less than 0.3% C. When the martensite is aged at about 500°C, intermetallic compounds such as Ni_3Ti, Fe_2Mo, and Ni_3Mo precipitate.

Interstitial-free steels are steels containing Nb and Ti. They react with C and S to form precipitates of carbides and sulfides. Thus, virtually no carbon remains in the ferrite. These steels are very formable and therefore attractive for the automobile industry.

Grain-oriented steels containing silicon are used as soft magnetic materials and are used in transformer cores. Nearly pure iron powder (known as carbonyl iron), obtained by the decomposition of iron pentacarbonyl [$Fe(CO)_5$] and sometimes a reducing heat treatment, is used to make magnetic materials. Pure iron powder is also used as an additive for food supplements in breakfast cereals and other iron-fortified food products under the name reduced iron.

As mentioned before, many steels are also coated, usually to provide good corrosion protection. *Galvanized* steel is coated with a thin layer of zinc (Chapter 23), *terne* steel is coated with lead, and other steels are coated with aluminum or tin.

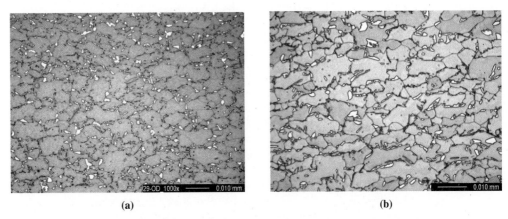

(a)   **(b)**

Figure 13-23   (a): Microstructure of a dual-phase steel, showing islands of white martensite in a light gray ferrite matrix. (*From G. Speich, "Physical Metallurgy of Dual-Phase Steels," Fundamentals of Dual-Phase Steels, The Metallurgical Society of AIME, 1981.*) (b) Microstructure of a TRIP steel, showing ferrite (light gray) + bainite (black along grain boundaries) + retained austenite (white). (*Courtesy of D. P. Hoydick, D. M. Haezebrouck, and E. A. Silva, United States Steel Corporation Research and Technology Center, 2005.*)

13-8   Surface Treatments

We can, by proper heat treatment, produce a structure that is hard and strong at the surface, so that excellent wear and fatigue resistance are obtained, but at the same time gives a soft, ductile, tough core that provides good resistance to impact failure. We have seen principles of carburizing in Chapter 5, when we discussed diffusion. In this section, we see this and other similar processes.

Selectively Heating the Surface

We could begin by rapidly heating the surface of a medium-carbon steel above the A_3 temperature (the center remains below the A_1). After the steel is quenched, the center is still a soft mixture of ferrite and pearlite, while the surface is martensite (Figure 13-24). The depth of the martensite layer is the **case depth**. Tempering produces the desired hardness at the surface. We can provide local heating of the surface by using a gas flame, an induction coil, a laser beam, or an electron beam. We can, if we wish, harden only selected areas of the surface that are most subject to failure by fatigue or wear.

Carburizing and Nitriding

These techniques involve controlled diffusion of carbon and nitrogen, respectively (Chapter 5). For best toughness, we start with a low-carbon steel. In **carburizing**, carbon is diffused into the surface of the steel at a temperature above the A_3 (Figure 13-25). A high carbon content is produced at the surface due to rapid diffusion and the high solubility of carbon in austenite. When the steel is then quenched and tempered, the surface becomes a high-carbon tempered martensite, while the ferritic center remains soft and ductile. The thickness of the hardened surface, again called the case depth, is much smaller in carburized steels than in flame- or induction-hardened steels.

Nitrogen provides a hardening effect similar to that of carbon. In **cyaniding**, the steel is immersed in a liquid cyanide bath that permits both carbon and nitrogen to diffuse into the steel. In **carbonitriding**, a gas containing carbon monoxide and ammonia is generated, and both carbon and nitrogen diffuse into the steel. Finally, only nitrogen diffuses into the surface from a gas in **nitriding**. Nitriding is carried out below the A_1 temperature.

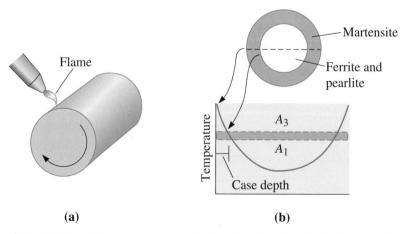

Figure 13-24   (a) Surface hardening by localized heating. (b) Only the surface heats above the A_1 temperature and is quenched to martensite.

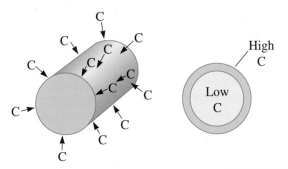

Figure 13-25
Carburizing of a low-carbon steel to produce a high-carbon, wear-resistant surface.

In each of these processes, compressive residual stresses are introduced at the surface, providing excellent fatigue resistance (Chapter 7) in addition to the good combination of hardness, strength, and toughness. The following example explains considerations for heat treatments such as quenching and tempering and surface hardening.

Example 13-7	*Design of Surface-Hardening Treatments for a Drive Train*

Design the materials and heat treatments for an automobile axle and drive gear (Figure 13-26).

SOLUTION

Both parts require good fatigue resistance. The gear also should have a good hardness to avoid wear, and the axle should have good overall strength to withstand bending and torsional loads. Both parts should have good toughness. Finally, since millions of these parts will be made, they should be inexpensive.

Quenched and tempered alloy steels might provide the required combination of strength and toughness; however, the alloy steels are expensive. An alternative approach for each part is described below.

The axle might be made from a forged 1050 steel containing a matrix of ferrite and pearlite. The axle could be surface-hardened, perhaps by moving the axle through an induction coil to selectively heat the surface of the steel above the A_3 temperature (about 770°C). After the coil passes any particular location of the axle,

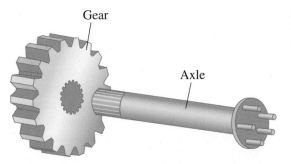

Figure 13-26
Sketch of axle and gear assembly (for Example 13-7).

the cold interior quenches the surface to martensite. Tempering then softens the martensite to improve ductility. This combination of carbon content and heat treatment meets our requirements. The plain carbon steel is inexpensive; the core of ferrite and pearlite produces good toughness and strength; and the hardened surface provides good fatigue and wear resistance.

The gear is subject to more severe loading conditions, for which the 1050 steel does not provide sufficient toughness, hardness, and wear resistance. Instead, we might carburize a 1010 steel for the gear. The original steel contains mostly ferrite, providing good ductility and toughness. By performing a gas carburizing process above the A_3 temperature (about 860°C), we introduce about 1.0% C in a very thin case at the surface of the gear teeth. This high-carbon case, which transforms to martensite during quenching, is tempered to control the hardness. Now we obtain toughness due to the low-carbon ferrite core, wear resistance due to the high-carbon surface, and fatigue resistance due to the high-strength surface containing compressive residual stresses introduced during carburizing. In addition, the plain carbon 1010 steel is an inexpensive starting material that is easily forged into a near-net shape prior to heat treatment.

13-9   Weldability of Steel

In Chapter 9, we discussed welding and other joining processes. We noted that steels are the most widely used structural materials. In bridges, buildings, and many other applications, steels must be welded. The structural integrity of steel structures not only depends upon the strength of the steel but also the strength of the welded joints. This is why the weldability of steel is always an important consideration.

Many low-carbon steels weld easily. Welding of medium- and high-carbon steels is comparatively more difficult since martensite can form in the heat-affected zone rather easily, thereby causing a weldment with poor toughness. Several strategies such as preheating the material or minimizing incorporation of hydrogen have been developed to counter these problems. The incorporation of hydrogen causes the steel to become brittle. In low-carbon steels, the strength of the welded regions in these materials is higher than the base material. This is due to the finer pearlite microstructure that forms during cooling of the heat-affected zone. Retained austenite along ferrite grain boundaries also limits recrystallization and thus helps retain a fine grain size, which contributes to the strength of the welded region. During welding, the metal nearest the weld heats above the A_1 temperature and austenite forms (Figure 13-27). During cooling, the austenite in this heat-affected zone transforms to a new structure, depending on the cooling rate and the CCT diagram for the steel. Plain low-carbon steels have such a low hardenability that normal cooling rates seldom produce martensite; however, an alloy steel may have to be preheated to slow down the cooling rate or post-heated to temper any martensite that forms.

A steel that is originally quenched and tempered has two problems during welding. First, the portion of the heat-affected zone that heats above the A_1 may form martensite after cooling. Second, a portion of the heat-affected zone below the A_1 may overtemper. Normally, we should not weld a steel in the quenched and tempered condition. The following example shows how the heat-affected zone microstructure can be accounted for using CCT diagrams.

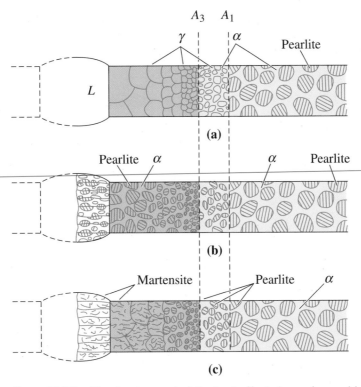

Figure 13-27 The development of the heat-affected zone in a weld: (a) the structure at the maximum temperature, (b) the structure after cooling in a steel of low hardenability, and (c) the structure after cooling in a steel of high hardenability.

Example 13-8 *Structures of Heat-Affected Zones*

Compare the structures in the heat-affected zones of welds in 1080 and 4340 steels if the cooling rate in the heat-affected zone is 5°C/s.

SOLUTION

From the CCT diagrams, Figures 13-14 and 13-16, the cooling rate in the weld produces the following structures:

> 1080: 100% pearlite
> 4340: Bainite and martensite

The high hardenability of the alloy steel reduces the weldability, permitting martensite to form and embrittle the weld.

13-10 Stainless Steels

Stainless steels are selected for their excellent resistance to corrosion. All true stainless steels contain a minimum of about 11% Cr, which permits a thin, protective surface layer of chromium oxide to form when the steel is exposed to oxygen. The chromium is

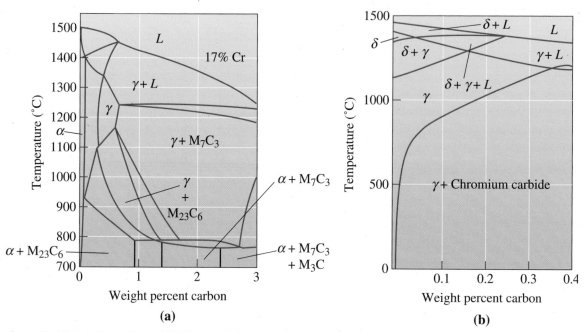

Figure 13-28   (a) The effect of 17% chromium on the iron-carbon phase diagram. At low carbon contents, ferrite is stable at all temperatures. Note that "M" stands for "metal" such as Cr and Fe or other alloying additions. (b) A section of the iron-chromium-nickel-carbon phase diagram at a constant 18% Cr-8% Ni. At low carbon contents, austenite is stable at room temperature.

what makes stainless steels stainless. Chromium is also a *ferrite stabilizing element*. Figure 13-28(a) illustrates the effect of chromium on the iron-carbon phase diagram. Chromium causes the austenite region to shrink, while the ferrite region increases in size. For high-chromium, low-carbon compositions, ferrite is present as a single phase up to the solidus temperature.

There are several categories of stainless steels based on crystal structure and strengthening mechanism. Typical properties are included in Table 13-4.

Ferritic Stainless Steels
Ferritic stainless steels contain up to 30% Cr and less than 0.12% C. Because of the BCC structure, the ferritic stainless steels have good strengths and moderate ductilities derived from solid-solution strengthening and strain hardening. Ferritic stainless steels are ferromagnetic. They are not heat treatable. They have excellent corrosion resistance, moderate formability, and are relatively inexpensive.

Martensitic Stainless Steels
From Figure 13-28(a), we find that a 17% Cr-0.5% C alloy heated to 1200°C forms 100% austenite, which transforms to martensite on quenching in oil. The martensite is then tempered to produce high strengths and hardnesses [Figure 13-29(a)]. The chromium content is usually less than 17% Cr; otherwise, the austenite field becomes so small that very stringent control over both the austenitizing temperature and carbon content is required. Lower chromium contents also permit the carbon content to vary from about 0.1% to 1.0%, allowing martensites of different hardnesses to be produced. The combination of hardness, strength, and corrosion resistance makes the alloys attractive for applications such as high-quality knives, ball bearings, and valves.

TABLE 13-4 ■ *Typical compositions and properties of stainless steels*

Steel	% C	% Cr	% Ni	Others	Tensile Strength (MPa)	Yield Strength (MPa)	% Elongation	Condition
Austenitic								
201	0.15	17	5	6.5% Mn	655	310	40	Annealed
304	0.08	19	10		517	207	30	Annealed
					1276	965	9	Cold-worked
304L	0.03	19	10		517	207	30	Annealed
316	0.08	17	12	2.5% Mo	517	207	30	Annealed
321	0.08	18	10	0.4% Ti	586	241	55	Annealed
347	0.08	18	11	0.8% Nb	621	241	50	Annealed
Ferritic								
430	0.12	17			448	207	22	Annealed
442	0.12	20			517	276	20	Annealed
Martensitic								
416	0.15	13		0.6% Mo	1241	965	18	Quenched and tempered
431	0.20	16	2		1379	1034	16	Quenched and tempered
440C	1.10	17		0.7% Mo	1965	1896	2	Quenched and tempered
Precipitation hardening								
17-4	0.07	17	4	0.4% Nb	1310	1172	10	Age-hardened
17-7	0.09	17	7	1.0% Al	1655	1586	6	Age-hardened

Austenitic Stainless Steels

Nickel, which is an *austenite stabilizing element*, increases the size of the austenite field, while nearly eliminating ferrite from the iron-chromium-carbon alloys [Figure 13-28(b)]. If the carbon content is below about 0.03%, the carbides do not form and the steel is virtually all austenite at room temperature [Figure 13-29(b)].

The FCC austenitic stainless steels have excellent ductility, formability, and corrosion resistance. Strength is obtained by extensive solid-solution strengthening, and

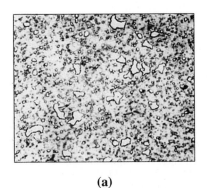
(a)

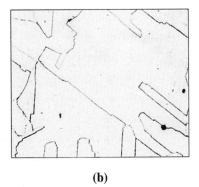
(b)

Figure 13-29 (a) Martensitic stainless steel containing large primary carbides and small carbides formed during tempering (× 350). (b) Austenitic stainless steel (× 500). (*From* ASM Handbook, *Vols. 7 and 8, (1972, 1973), ASM International, Materials Park, OH 44073-0002.*)

the austenitic stainless steels may be cold worked to higher strengths than the ferritic stainless steels. These are not ferromagnetic, which is an advantage for many applications. For example, cardivascular stents are often made from 316 stainless steels. The steels have excellent low-temperature impact properties, since they have no transition temperature. Unfortunately, the high-nickel and chromium contents make the alloys expensive. The 304 alloy containing 18% Cr and 8% nickel (also known as 18-8 stainless) is the most widely used grade of stainless steel. Although stainless, this alloy can undergo **sensitization**. When heated to a temperature of ~480–860°C, chromium carbides precipitate along grain boundaries rather than within grains. This causes chromium depletion in the interior of the grains and this will cause the stainless steel to corrode very easily.

Precipitation-Hardening (PH) Stainless Steels

The precipitation-hardening (or PH) stainless steels contain Al, Nb, or Ta and derive their properties from solid-solution strengthening, strain hardening, age hardening, and the martensitic reaction. The steel is first heated and quenched to permit the austenite to transform to martensite. Reheating permits precipitates such as Ni_3Al to form from the martensite. High strengths are obtained even with low carbon contents.

Duplex Stainless Steels

In some cases, mixtures of phases are deliberately introduced into the stainless steel structure. By appropriate control of the composition and heat treatment, a **duplex stainless steel** containing approximately 50% ferrite and 50% austenite can be produced. This combination provides a set of mechanical properties, corrosion resistance, formability, and weldability not obtained in any one of the usual stainless steels.

Most stainless steels are recyclable and the following example shows how differences in properties can be used to separate different types of stainless steels.

Example 13-9   *Design of a Test to Separate Stainless Steels*

In order to efficiently recycle stainless steel scrap, we wish to separate the high-nickel stainless steel from the low-nickel stainless steel. Design a method for doing this.

SOLUTION

Performing a chemical analysis on each piece of scrap is tedious and expensive. Sorting based on hardness might be less expensive; however, because of the different types of treatments—such as annealing, cold working, or quench and tempering—the hardness may not be related to the steel composition.

The high-nickel stainless steels are ordinarily austenitic, whereas the low-nickel alloys are ferritic or martensitic. An ordinary magnet will be attracted to the low-nickel ferritic and martensitic steels, but will not be attracted to the high-nickel austenitic steel. We might specify this simple and inexpensive magnetic test for our separation process.

13-11   Cast Irons

Cast irons are iron-carbon-silicon alloys, typically containing 2–4% C and 0.5–3% Si, that pass through the eutectic reaction during solidification. The microstructures of the five important types of cast irons are shown schematically in Figure 13-30.

Eutectic Reaction in Cast Irons
Based on the Fe-Fe$_3$C phase diagram (dashed lines in Figure 13-31), the eutectic reaction that occurs in Fe-C alloys at 1140°C is

$$L \rightarrow \gamma + Fe_3C \qquad (13\text{-}1)$$

This reaction produces **white cast iron**, with a microstructure composed of Fe$_3$C and pearlite. The Fe-Fe$_3$C system, however, is really a metastable phase diagram. Under truly equilibrium conditions, the eutectic reaction is

$$L \rightarrow \gamma + graphite \qquad (13\text{-}2)$$

The Fe-C phase diagram is shown as solid lines in Figure 13-31. When the stable $L \rightarrow \gamma +$ graphite eutectic reaction occurs at 1146°C, gray, ductile, or compacted graphite cast iron forms.

In Fe-C alloys, the liquid easily undercools 6°C (the temperature difference between the stable and metastable eutectic temperatures), and white iron forms. Adding about 2% silicon to the iron increases the temperature difference between the eutectics, permitting larger undercoolings to be tolerated and more time for the stable graphite eutectic to nucleate and grow. Silicon, therefore, is a *graphite stabilizing* element. Elements such as chromium and

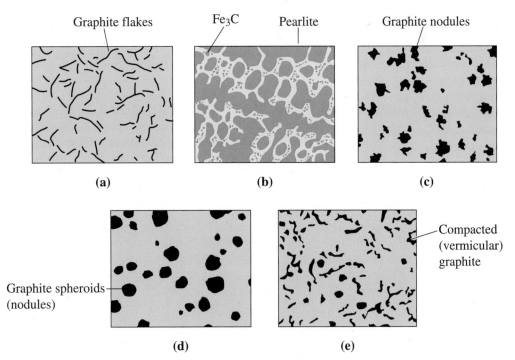

Figure 13-30   Schematic drawings of the five types of cast iron: (a) gray iron, (b) white iron, (c) malleable iron, (d) ductile iron, and (e) compacted graphite iron.

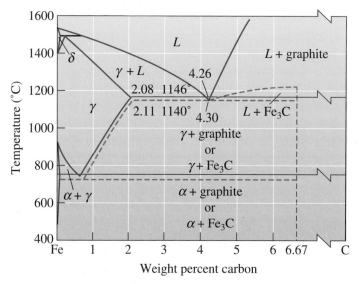

Figure 13-31 The iron-carbon phase diagram showing the relationship between the stable iron-graphite equilibria (solid lines) and the metastable iron-cementite reactions (dashed lines).

bismuth have the opposite effect and encourage white cast iron formation. We can also introduce inoculants, such as silicon (as Fe-Si ferrosilicon), to encourage the nucleation of graphite, or we can reduce the cooling rate of the casting to provide more time for the growth of graphite.

Silicon also reduces the amount of carbon contained in the eutectic. We can take this effect into account by defining the **carbon equivalent** (CE):

$$CE = \% \, C + \tfrac{1}{3} \% \, Si \tag{13-3}$$

The eutectic composition is always near 4.3% CE. A high carbon equivalent encourages the growth of the graphite eutectic.

Eutectoid Reaction in Cast Irons

The matrix structure and properties of each type of cast iron are determined by how the austenite transforms during the eutectoid reaction. In the Fe-Fe$_3$C phase diagram used for steels, the austenite transformed to ferrite and cementite, often in the form of pearlite; however, silicon also encourages the *stable* eutectoid reaction:

$$\gamma \longrightarrow \alpha + graphite \tag{13-4}$$

Under equilibrium conditions, carbon atoms diffuse from the austenite to existing graphite particles, leaving behind the low-carbon ferrite. The transformation diagram (Figure 13-32) describes how the austenite might transform during heat treatment. **Annealing** (or furnace cooling) of cast iron gives a soft ferritic matrix (not coarse pearlite as in steels!). Normalizing, or air cooling, gives a pearlitic matrix. The cast irons can also be austempered to produce bainite or can be quenched to martensite and tempered. Austempered ductile iron, with strengths of up to 1379 MPa, is used for high-performance gears.

Gray cast iron contains small, interconnected graphite flakes that cause low strength and ductility. This is the most widely used cast iron and is named for the dull gray color of the fractured surface. Gray cast iron contains many clusters, or **eutectic cells**, of interconnected graphite flakes (Figure 13-33). The point at which the flakes are

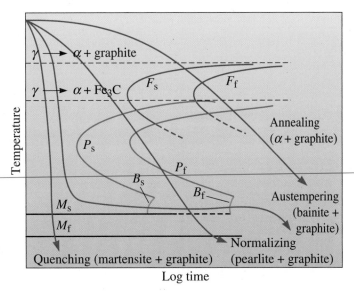

Figure 13-32   The transformation diagram for austenite in a cast iron.

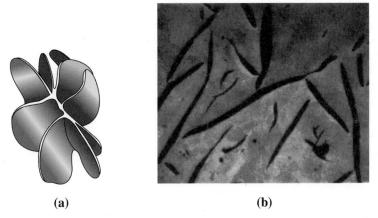

   (a)   **(b)**

Figure 13-33   (a) Sketch and (b) micrograph of the flake graphite in gray cast iron ($\times$ 100) *(Reprinted courtesy of Don Askeland.)*

connected is the original graphite nucleus. Inoculation helps produce smaller eutectic cells, thus improving strength. The gray irons are specified by a class number of 20 to 80. A class 20 gray iron has a nominal tensile strength of 138 MPa. In thick castings, coarse graphite flakes and a ferrite matrix produce tensile strengths as low as 83 MPa (Figure 13-34), whereas in thin castings, fine graphite and pearlite form and give tensile strengths near 276 MPa. Higher strengths are obtained by reducing the carbon equivalent, by alloying, or by heat treatment. Although the graphite flakes concentrate stresses and cause low strength and ductility, gray iron has a number of attractive properties, including high compressive strength, good machinability, good resistance to sliding wear, good resistance to thermal fatigue, good thermal conductivity, and good vibration damping.

  White cast iron is a hard, brittle alloy containing massive amounts of Fe_3C. A fractured surface of this material appears white, hence the name. A group of highly alloyed white irons are used for their hardness and resistance to abrasive wear. Elements such as

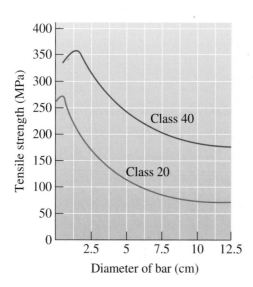

Figure 13-34
The effect of the cooling rate or casting size on the tensile properties of two gray cast irons.

chromium, nickel, and molybdenum are added so that, in addition to the alloy carbides formed during solidification, martensite is formed during subsequent heat treatment.

Malleable cast iron, formed by the heat treatment of white cast iron, produces rounded clumps of graphite. It exhibits better ductility than gray or white cast irons. It is also very machinable. Malleable iron is produced by heat treating unalloyed 3% carbon equivalent (2.5% C, 1.5% Si) white iron. During the heat treatment, the cementite formed during solidification decomposes and graphite clumps, or nodules, are produced. The nodules, or temper carbon, often resemble popcorn. The rounded graphite shape permits a good combination of strength and ductility. The production of malleable iron requires several steps (Figure 13-35). Graphite nodules nucleate as the white iron is slowly heated. During **first stage graphitization** (FSG), cementite decomposes to the stable austenite and graphite phases as the carbon in Fe_3C diffuses to the graphite nuclei. Following FSG, the austenite transforms during cooling. Figure 13-36 shows the microstructures of the original white iron (a) and the two types of malleable iron that can be produced (b and c). To make *ferritic malleable iron*, the casting is cooled slowly

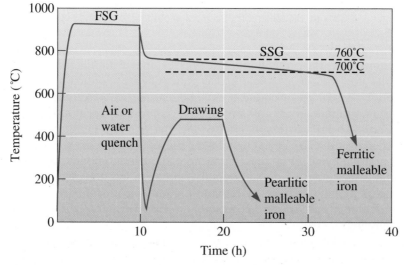

Figure 13-35   The heat treatments for ferritic and pearlitic malleable irons.

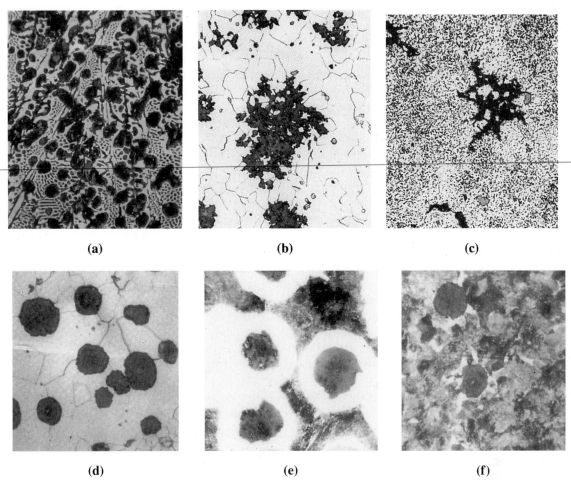

(a) (b) (c)

(d) (e) (f)

Figure 13-36 (a) White cast iron prior to heat treatment (× 100). (b) Ferritic malleable iron with graphite nodules and small MnS inclusions in a ferrite matrix (× 200). (c) Pearlitic malleable iron drawn to produce a tempered martensite matrix (× 500). (*Images (b) and (c) are from Metals Handbook, Vols. 7 and 8, (1972, 1973), ASM International, Materials Park, OH 44073-0002.*) (d) Annealed ductile iron with a ferrite matrix (× 250). (e) As-cast ductile iron with a matrix of ferrite (white) and pearlite (× 250). (f) Normalized ductile iron with a pearlite matrix (× 250). (*Images (a), (d), (e) and (f) are reprinted courtesy of Don Askeland.*)

through the eutectoid temperature range to cause **second stage graphitization** (SSG). The ferritic malleable iron has good toughness compared with that of other irons because its low-carbon equivalent reduces the transition temperature below room temperature. *Pearlitic malleable iron* is obtained when austenite is cooled in air or oil to form pearlite or martensite. In either case, the matrix is hard and brittle. The iron is then drawn at a temperature below the eutectoid. **Drawing** is a heat treatment that tempers the martensite or spheroidizes the pearlite. A higher drawing temperature decreases strength and increases ductility and toughness.

Ductile or **nodular cast iron** contains spheroidal graphite particles. Ductile iron is produced by treating liquid iron with a carbon equivalent of near 4.3% with magnesium, which causes spheroidal graphite (called nodules) to grow during solidification, rather than during a lengthy heat treatment. Several steps are required to produce this iron. These include desulfurization, **nodulizing**, and inoculation. In desulfurization, any sulfur and oxygen in the liquid metal is removed by adding desulfurizing agents such as calcium oxide (CaO). In nodulizing, Mg is added, usually in a dilute form such as a MgFeSi alloy. If pure Mg is added, the nodulizing reaction is very violent, since the boiling point of Mg is much lower than the temperature of the liquid iron, and most of the Mg will be lost. A residual of about 0.03% Mg must be in the liquid iron after treatment in order for spheroidal graphite to grow. Finally, **inoculation** with FeSi compounds to cause heterogeneous nucleation of the graphite is essential; if inoculation is not effective, white iron will form instead of ductile iron. The nodulized and inoculated iron must then be poured into molds within a few minutes to avoid fading. **Fading** occurs by the gradual, nonviolent loss of Mg due to vaporization and/or reaction with oxygen, resulting in flake or compacted graphite instead of spheroidal graphite. In addition, the inoculant effect will also fade, resulting in white iron.

Compared with gray iron, ductile cast iron has excellent strength and ductility. Due to the higher silicon content (typically around 2.4%) in ductile irons compared with 1.5% Si in malleable irons, the ductile irons are stronger but not as tough as malleable irons.

Compacted graphite cast iron contains rounded but interconnected graphite also produced during solidification. The graphite shape in compacted graphite cast iron is intermediate between flakes and spheres with numerous rounded rods of graphite that are interconnected to the nucleus of the eutectic cell. This compacted graphite, sometimes called **vermicular graphite**, also forms when ductile iron fades. The compacted graphite permits strengths and ductilities that exceed those of gray cast iron, but allows the iron to retain good thermal conductivity and vibration damping properties. The treatment for the compacted graphite iron is similar to that for ductile iron; however, only about 0.015% Mg is introduced during nodulizing. A small amount of titanium (Ti) is added to ensure the formation of the compacted graphite.

Typical properties of cast irons are given in Table 13-5.

TABLE 13-5 ■ *Typical properties of cast irons*

	Tensile Strength (MPa)	Yield Strength (MPa)	% E	Notes
Gray irons:				
Class 20	83–276	—	—	
Class 40	193–372	—	—	
Class 60	303–455	—	—	
Malleable irons:				
32510	345	224	10	Ferritic
35018	365	241	18	Ferritic
50005	483	345	5	Pearlitic
70003	586	483	3	Pearlitic
90001	724	621	1	Pearlitic
Ductile irons:				
60–40–18	414	276	18	Annealed
65–45–12	448	310	12	As-cast ferritic
80–55–06	552	379	6	As-cast pearlitic
100–70–03	689	483	3	Normalized
120–90–02	827	621	2	Quenched and tempered
Compacted graphite irons:				
Low strength	276	193	5	90% Ferritic
High strength	448	379	1	80% Pearlitic

Summary

- The properties of steels, determined by dispersion strengthening, depend on the amount, size, shape, and distribution of cementite. These factors are controlled by alloying and heat treatment.
- A process anneal recrystallizes cold-worked steels.
- Spheroidizing produces large, spherical Fe_3C and good machinability in high-carbon steels.
- Annealing, involving a slow furnace cool after austenitizing, produces a coarse pearlitic structure containing lamellar Fe_3C.
- Normalizing, involving an air cool after austenitizing, produces a fine pearlitic structure and higher strength compared with annealing.
- In isothermal annealing, pearlite with a uniform interlamellar spacing is obtained by transforming the austenite at a constant temperature.
- Austempering is used to produce bainite, containing rounded Fe_3C, by an isothermal transformation.
- Quench and temper heat treatments require the formation and decomposition of martensite, providing exceptionally fine dispersions of round Fe_3C.
- We can better understand heat treatments by use of TTT diagrams, CCT diagrams, and hardenability curves.
- The TTT diagrams describe how austenite transforms to pearlite and bainite at a constant temperature.
- The CCT diagrams describe how austenite transforms during continuous cooling. These diagrams give the cooling rates needed to obtain martensite in quench and temper treatments.
- The hardenability curves compare the ease with which different steels transform to martensite.
- Alloying elements increase the times required for transformations in the TTT diagrams, reduce the cooling rates necessary to produce martensite in the CCT diagrams, and improve the hardenability of the steel.
- Specialty steels and heat treatments provide unique properties or combinations of properties. Of particular importance are surface-hardening treatments, such as carburizing, that produce an excellent combination of fatigue and impact resistance. Stainless steels, which contain a minimum of 11% Cr, have excellent corrosion resistance.
- Cast irons, by definition, undergo the eutectic reaction during solidification. Depending on the composition and treatment, either γ and Fe_3C or γ and graphite form during freezing.

Glossary

Annealing (cast iron) A heat treatment used to produce a ferrite matrix in a cast iron by the transformation of austenite via furnace cooling.

Annealing (steel) A heat treatment used to produce a soft, coarse pearlite in steel by austenitizing, then furnace cooling.

Ausforming A thermomechanical heat treatment in which austenite is plastically deformed below the A_1 temperature, then permitted to transform to bainite or martensite.

Austempering The isothermal heat treatment by which austenite transforms to bainite.

Austenitizing Heating a steel or cast iron to a temperature at which homogeneous austenite can form. Austenitizing is the first step in most of the heat treatments for steels and cast irons.

Carbon equivalent (CE) Carbon plus one-third of the silicon in a cast iron.

Carbonitriding Hardening the surface of steel with carbon and nitrogen obtained from a special gas atmosphere.

Carburizing A group of surface-hardening techniques by which carbon diffuses into steel.

Case depth The depth below the surface of a steel to which hardening occurs by surface hardening and carburizing processes.

Cast iron Ferrous alloys containing sufficient carbon so that the eutectic reaction occurs during solidification.

Compacted graphite cast iron A cast iron treated with small amounts of magnesium and titanium to cause graphite to grow during solidification as an interconnected, coral-shaped precipitate, giving properties midway between gray and ductile iron.

Cyaniding Hardening the surface of steel with carbon and nitrogen obtained from a bath of liquid cyanide solution.

Drawing Reheating a malleable iron in order to reduce the amount of carbon combined as cementite by spheroidizing pearlite, tempering martensite, or graphitizing both.

Dual-phase steels Special steels treated to produce martensite dispersed in a ferrite matrix.

Ductile cast iron Cast iron treated with magnesium to cause graphite to precipitate during solidification as spheres, permitting excellent strength and ductility. (Also known as *nodular* cast iron.)

Duplex stainless steel A special class of stainless steels containing a microstructure of ferrite and austenite.

Electric arc furnace A furnace used to melt steel scrap using electricity. Often, specialty steels are made using electric arc furnaces.

Eutectic cell A cluster of graphite flakes produced during solidification of gray iron that are all interconnected to a common nucleus.

Fading The loss of the nodulizing or inoculating effect in cast irons as a function of time, permitting undesirable changes in microstructure and properties.

First stage graphitization (FSG) The first step in the heat treatment of a malleable iron, during which the carbides formed during solidification are decomposed to graphite and austenite.

Gray cast iron Cast iron which, during solidification, contains graphite flakes, causing low strength and poor ductility. This is the most widely used type of cast iron.

Hardenability The ease with which a steel can be quenched to form martensite. Steels with high hardenability form martensite even on slow cooling.

Hardenability curves Graphs showing the effect of the cooling rate on the hardness of as-quenched steel.

Hot metal The molten iron produced in a blast furnace, also known as pig iron. It contains about 95% iron, 4% carbon, 0.3–0.9% silicon, 0.5% manganese, and 0.025–0.05% each of sulfur, phosphorus, and titanium.

Inoculation The addition of an agent to molten cast iron that provides nucleation sites at which graphite precipitates during solidification.

Interstitial-free steels These are steels containing Nb and Ti. They react with C and S to form precipitates of carbides and sulfides, leaving the ferrite nearly free of interstitial elements.

Isothermal annealing Heat treatment of a steel by austenitizing, cooling rapidly to a temperature between the A_1 and the nose of the TTT curve, and holding until the austenite transforms to pearlite.

Jominy distance The distance from the quenched end of a Jominy bar. The Jominy distance is related to the cooling rate.

Jominy test The test used to evaluate hardenability. An austenitized steel bar is quenched at one end only, thus producing a range of cooling rates along the bar.

Malleable cast iron Cast iron obtained by a lengthy heat treatment, during which cementite decomposes to produce rounded clumps of graphite. Good strength, ductility, and toughness are obtained as a result of this structure.

Maraging steels A special class of alloy steels that obtain high strengths by a combination of the martensitic and age-hardening reactions.

Marquenching Quenching austenite to a temperature just above the M_S and holding until the temperature is equalized throughout the steel before further cooling to produce martensite. This process reduces residual stresses and quench cracking. (Also known as *martempering*.)

Nitriding Hardening the surface of steel with nitrogen obtained from a special gas atmosphere.

Nodulizing The addition of magnesium to molten cast iron to cause the graphite to precipitate as spheres rather than as flakes during solidification.

Normalizing A simple heat treatment obtained by austenitizing and air cooling to produce a fine pearlitic structure. This can be done for steels and cast irons.

Pig iron The molten iron produced in a blast furnace also known as hot metal. It contains about 95% iron, 4% carbon, 0.3–0.9% silicon, 0.5% manganese, and 0.025–0.05% each of sulfur, phosphorus, and titanium.

Process anneal A low-temperature heat treatment used to eliminate all or part of the effect of cold working in steels.

Quench cracks Cracks that form at the surface of a steel during quenching due to tensile residual stresses that are produced because of the volume change that accompanies the austenite-to-martensite transformation.

Retained austenite Austenite that is unable to transform into martensite during quenching because of the volume expansion associated with the reaction.

Second stage graphitization (SSG) The second step in the heat treatment of malleable irons that are to have a ferritic matrix. The iron is cooled slowly from the first stage graphitization temperature so that austenite transforms to ferrite and graphite rather than to pearlite.

Secondary hardening peak Unusually high hardness in a steel tempered at a high temperature caused by the precipitation of alloy carbides.

Sensitization When heated to a temperature of ~480–860°C, chromium carbides precipitate along grain boundaries rather than within grains, causing chromium depletion in the interior. This causes stainless steel to corrode very easily.

Spheroidite A microconstituent containing coarse spherical cementite particles in a matrix of ferrite, permitting excellent machining characteristics in high-carbon steels.

Stainless steels A group of ferrous alloys that contain at least 11% Cr, providing extraordinary corrosion resistance.

Tempered martensite The microconstituent of ferrite and cementite formed when martensite is tempered.

Tool steels A group of high-carbon steels that provide combinations of high hardness, toughness, and resistance to elevated temperatures.

TRIP steels A group of steels with a microstructure that consists of a continuous ferrite matrix, a harder second phase (martensite and/or bainite), and retained austenite. TRIP stands for transformation induced plasticity.

Vermicular graphite The rounded, interconnected graphite that forms during the solidification of cast iron. This is the intended shape in compacted graphite iron, but it is a defective shape in ductile iron.

White cast iron Cast iron that produces cementite rather than graphite during solidification. The white irons are hard and brittle.

Problems

Section 13-1 Designations and Classification of Steels

13-1 What is the difference between cast iron and steels?

13-2 What do A_1, A_3, and A_{cm} temperatures refer to? Are these temperatures constant?

13-3 Calculate the amounts of ferrite, cementite, primary microconstituent, and pearlite in the following steels:
(a) 1015
(b) 1035
(c) 1095
(d) 10130

13-4 Estimate the AISI-SAE number for steels having the following microstructures:
(a) 38% pearlite-62% primary ferrite
(b) 93% pearlite-7% primary cementite
(c) 97% ferrite-3% cementite
(d) 86% ferrite-14% cementite

13-5 What do the terms low-, medium-, and high-carbon steels mean?

13-6 Two samples of steel contain 93% pearlite. Estimate the carbon content of each sample if one is known to be hypoeutectoid and the other hypereutectoid.

Section 13-2 Simple Heat Treatments

Section 13-3 Isothermal Heat Treatments

13-7 Explain the following heat treatments: (a) process anneal, (b) austenitizing, (c) annealing, (d) normalizing, and (e) quenching.

13-8 Complete the following table:

	1035 Steel	10115 Steel
A_1 temperature		
A_3 or A_{cm} temperature		
Full annealing temperature		
Normalizing temperature		
Process annealing temperature		
Spheroidizing temperature		

13-9 Explain why, strictly speaking, TTT diagrams can be used for isothermal treatments only.

13-10 Determine the constants c and n in the Avrami relationship (Equation 12-2) for the transformation of austenite to pearlite for a 1050 steel. Assume that the material has been subjected to an isothermal heat treatment at 600°C and make a log–log plot of f versus t given the following information:

$$f = 0.2 \text{ at } t = 2 \text{ s};$$
$$f = 0.5 \text{ at } t = 4 \text{ s; and}$$
$$f = 0.8 \text{ at } t = 7 \text{ s.}$$

13-11 In a pearlitic 1080 steel, the cementite platelets are 4×10^{-5} cm thick, and the ferrite platelets are 14×10^{-5} cm thick. In a spheroidized 1080 steel, the cementite spheres are 4×10^{-3} cm in diameter. Estimate the total interface area between the ferrite and cementite in a cubic centimeter of each steel. Determine the percent reduction in surface area when the pearlitic steel is spheroidized. The density of ferrite is 7.87 g/cm^3 and that of cementite is 7.66 g/cm^3.

13-12 Describe the microstructure present in a 1050 steel after each step in the following heat treatments:
(a) heat at 820°C, quench to 650°C and hold for 90 s, and quench to 25°C;
(b) heat at 820°C, quench to 450°C and hold for 90 s, and quench to 25°C;
(c) heat at 820°C, and quench to 25°C;
(d) heat at 820°C, quench to 720°C and hold for 100 s, and quench to 25°C;
(e) heat at 820°C, quench to 720°C and hold for 100 s, quench to 400°C and hold for 500 s, and quench to 25°C;
(f) heat at 820°C, quench to 720°C and hold for 100 s, quench to 400°C and hold for 10 s, and quench to 25°C; and
(g) heat at 820°C, quench to 25°C, heat to 500°C and hold for 10^3 s, and air cool to 25°C.

13-13 Describe the microstructure present in a 10110 steel after each step in the following heat treatments:

(a) heat to 900°C, quench to 400°C and hold for 10^3 s, and quench to 25°C;

(b) heat to 900°C, quench to 600°C and hold for 50 s, and quench to 25°C;

(c) heat to 900°C and quench to 25°C;

(d) heat to 900°C, quench to 300°C and hold for 200 s, and quench to 25°C;

(e) heat to 900°C, quench to 675°C and hold for 1 s, and quench to 25°C;

(f) heat to 900°C, quench to 675°C and hold for 1 s, quench to 400°C and hold for 900 s, and slowly cool to 25°C;

(g) heat to 900°C, quench to 675°C and hold for 1 s, quench to 300°C and hold for 10^3 s, and air cool to 25°C; and

(h) heat to 900°C, quench to 300°C and hold for 100 s, quench to 25°C, heat to 450°C for 3600 s, and cool to 25°C.

13-14 Recommend appropriate isothermal heat treatments to obtain the following, including appropriate temperatures and times:

(a) an isothermally annealed 1050 steel with HRC 23;

(b) an isothermally annealed 10110 steel with HRC 40;

(c) an isothermally annealed 1080 steel with HRC 38;

(d) an austempered 1050 steel with HRC 40;

(e) an austempered 10110 steel with HRC 55; and

(f) an austempered 1080 steel with HRC 50.

13-15 Compare the minimum times required to isothermally anneal the following steels at 600°C. Discuss the effect of the carbon content of the steel on the kinetics of nucleation and growth during the heat treatment.

(a) 1050

(b) 1080

(c) 10110.

Section 13-4 Quench and Temper Heat Treatments

13-16 Explain the following terms: (a) quenching, (b) tempering, (c) retained austenite, and (d) marquenching/martempering.

13-17 Typical media used for quenching include air, brine (10% salt in water), water, and various oils.

(a) Rank the four media in order of the cooling rate from fastest to slowest.

(b) Describe a situation when quenching in air would be undesirable.

(c) During quenching in liquid media, typically either the part being cooled or the bath is agitated. Explain why.

13-18 We wish to produce a 1050 steel that has a Brinell hardness of at least 330 and an elongation of at least 15%.

(a) Recommend a heat treatment, including appropriate temperatures, that permits this to be achieved. Determine the yield strength and tensile strength that are obtained by this heat treatment.

(b) What yield and tensile strengths would be obtained in a 1080 steel using the same heat treatment? See Figure 12-27.

(c) What yield strength, tensile strength and % elongation would be obtained in the 1050 steel if it were normalized? See Figure 13-4.

13-19 We wish to produce a 1050 steel that has a tensile strength of at least 1207 MPa and a reduction in area of at least 50%.

(a) Recommend a heat treatment, including appropriate temperatures, that permits this to be achieved. Determine the Brinell hardness number, % elongation, and yield strength that are obtained by this heat treatment.

(b) What yield strength and tensile strength would be obtained in a 1080 steel using the same heat treatment?

(c) What yield strength, tensile strength, and % elongation would be obtained in the 1050 steel if it were annealed?

13-20 A 1030 steel is given an improper quench and temper heat treatment, producing a final structure composed of 60% martensite and 40% ferrite. Estimate the carbon content of the martensite and the austenitizing temperature that was used. What austenitizing temperature would you recommend?

13-21 A 1050 steel should be austenitized at 820°C, quenched in oil to 25°C, and tempered at 400°C for an appropriate time.

(a) What yield strength, hardness, and % elongation would you expect to obtain from this heat treatment?

(b) Suppose the actual yield strength of the steel is found to be 862 MPa. What might have gone wrong in the heat treatment to cause this low strength?

(c) Suppose the Brinell hardness is found to be HB 525. What might have gone wrong in the heat treatment to cause this high hardness?

13-22 A part produced from a low-alloy, 0.2% C steel (Figure 13-15) has a microstructure containing ferrite, pearlite, bainite, and martensite after quenching. What microstructure would be obtained if we used a 1080 steel? What microstructure would be obtained if we used a 4340 steel?

13-23 Fine pearlite and a small amount of martensite are found in a quenched 1080 steel. What microstructure would be expected if we used a low-alloy, 0.2% C steel? What microstructure would be expected if we used a 4340 steel? See Figures 13-14, 13-15, and 13-16.

13-24 Predict the phases formed when a bar of 1080 steel is quenched from slightly above the eutectoid temperature under the following conditions:

(a) oil (without agitation);

(b) oil (with agitation);

(c) water (with agitation); and

(d) brine (no agitation).

Suggest a quenching medium if we wish to obtain coarse pearlite.

Refer to Figure 13-14 and Table 13-2 for this problem.

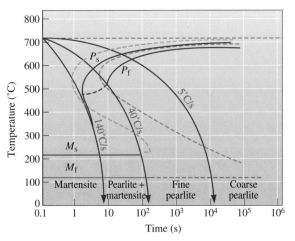

Figure 13-14 (Repeated for Problem 13-24.) The CCT diagram (solid lines) for a 1080 steel compared with the TTT diagram (dashed lines).

Section 13-5 Effect of Alloying Elements

Section 13-6 Application of Hardenability

13-25 Explain the difference between hardenability and hardness. Explain using a sketch how hardenability of steels is measured.

13-26 We have found that a 1070 steel, when austenitized at 750°C, forms a structure containing pearlite and a small amount of grain-boundary ferrite that gives acceptable

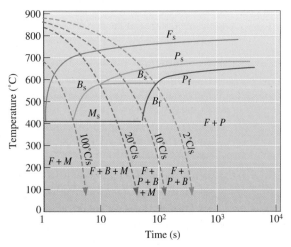

Figure 13-15 (Repeated for Problem 13-22.) The CCT diagram for a low alloy, 0.2% C steel.

strength and ductility. What changes in the microstructure, if any, would be expected if the 1070 steel contained an alloying element such as Mo or Cr? Explain.

13-27 Using the TTT diagrams, compare the hardenabilities of 4340 and 1050 steels by determining the times required for the isothermal transformation of ferrite and pearlite (F_s, P_s, and P_f) to occur at 650°C.

13-28 We would like to obtain a hardness of HRC 38 to 40 in a quenched steel. What range of cooling rates would we have to obtain for the following steels? Are some steels inappropriate?

(a) 4340
(b) 8640
(c) 9310
(d) 4320
(e) 1050
(f) 1080.

13-29 A steel part must have an as-quenched hardness of HRC 35 in order to avoid excessive-wear rates during use. When the part is made from 4320 steel, the hardness is only HRC 32. Determine the hardness if the part were made under identical conditions, but with the following steels. Which, if any, of these steels would be better choices than 4320?

(a) 4340
(b) 8640
(c) 9310
(d) 1050
(e) 1080.

13-30 A part produced from a 4320 steel has a hardness of HRC 35 at a critical location after quenching. Determine

(a) the cooling rate at that location, and
(b) the microstructure and hardness that would be obtained if the part were made of a 1080 steel.

13-31 A 1080 steel is cooled at the fastest possible rate that still permits all pearlite to form. What is this cooling rate? What Jominy distance and hardness are expected for this cooling rate?

13-32 Determine the hardness and microstructure at the center of a 3.75 cm-diameter 1080 steel bar produced by quenching in

(a) unagitated oil;
(b) unagitated water; and
(c) agitated brine.

13-33 A 5 cm-diameter bar of 4320 steel is to have a hardness of at least HRC 35. What is the minimum severity of the quench (H coefficient)? What type of quenching medium would you recommend to produce the desired hardness with the least chance of quench cracking?

13-34 A steel bar is to be quenched in agitated water. Determine the maximum diameter of the bar that will produce a minimum hardness of HRC 40 if the bar is

(a) 1050
(b) 1080
(c) 4320
(d) 8640
(e) 4340.

13-35 The center of a 2.5 cm-diameter bar of 4320 steel has a hardness of HRC 40. Determine the hardness and microstructure at the center of a 5 cm bar of 1050 steel quenched in the same medium.

Section 13-7 Specialty Steels
Section 13-8 Surface Treatments
Section 13-9 Weldability of Steel

13-36 What is the principle of the surface hardening of steels using carburizing and nitriding?

13-37 A 1010 steel is to be carburized using a gas atmosphere that produces 1.0% C at the surface of the steel. The case depth is defined as the distance below the surface that contains at least 0.5% C. If carburizing is done at 1000°C, determine the time required to produce a case depth of 0.025 cm (See Chapter 5 for a review.)

13-38 A 1015 steel is to be carburized at 1050°C for 2 h using a gas atmosphere that produces 1.2% C at the surface of the steel. Plot the percent carbon versus the distance from the surface of the steel. If the steel is slowly cooled after carburizing, determine the amount of each phase and microconstituent at 0.005 cm intervals from the surface. (See Chapter 5.)

13-39 Why is it that the strength of the heat-affected zone is higher for low-carbon steels? What is the role of retained austenite in this case?

13-40 Why is it easy to weld low-carbon steels, but difficult to weld high-carbon steels?

13-41 A 1050 steel is welded. After cooling, hardnesses in the heat-affected zone are obtained at various locations from the edge of the fusion zone. Determine the hardnesses expected at each point if a 1080 steel were welded under the same conditions. Predict the microstructure at each location in the as-welded 1080 steel.

Distance from Edge of Fusion Zone	Hardness in 1050 Weld
0.05 mm	HRC 50
0.10 mm	HRC 40
0.15 mm	HRC 32
0.20 mm	HRC 28

Section 13-10 Stainless Steels

13-42 What is a stainless steel? Why are stainless steels stainless?

13-43 We wish to produce a martensitic stainless steel containing 17% Cr. Recommend a carbon content and austenitizing temperature that would permit us to obtain 100% martensite during the quench. What microstructure would be produced if the martensite were then tempered until the equilibrium phases formed?

13-44 Occasionally, when an austenitic stainless steel is welded, the weld deposit may be slightly magnetic. Based on the Fe-Cr-Ni-C phase diagram [Figure 13-28(b)], what phase would you expect is causing the magnetic behavior? Why might this phase have formed? What could you do to restore the nonmagnetic behavior?

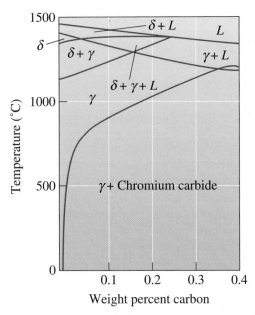

Figure 13-28 (Repeated for Problem 13-44.) (b) A section of the iron-chromium-nickel-carbon phase diagram at a constant 18% Cr-8% Ni. At low carbon contents, austenite is stable at room temperature.

Section 13-11 Cast Irons

13-45 Define cast iron using the Fe-Fe$_3$C phase diagram.

13-46 Compare the eutectic temperatures of a Fe-4.3% C cast iron with a Fe-3.6% C-2.1% Si alloy. Which alloy is expected to be more machinable and why?

13-47 What are the different types of cast irons? Explain using a sketch.

13-48 A bar of a class 40 gray iron casting is found to have a tensile strength of 345 MPa. Why is the tensile strength greater than that given by the class number? What do you think is the diameter of the test bar?

13-49 You would like to produce a gray iron casting that freezes with no primary austenite or graphite. If the carbon content in the iron is 3.5%, what percentage of silicon must you add?

 Design Problems

13-50 We would like to produce a 5 cm-thick steel wear plate for a rock-crushing unit.

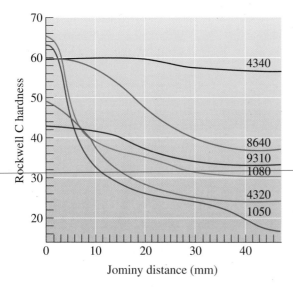

Figure 13-21   (Repeated for Problem 13-50.) The hardenability curves for several steels.

To avoid frequent replacement of the wear plate, the hardness should exceed HRC 38 within 0.625 cm of the steel surface. The center of the plate should have a hardness of no more than HRC 32 to ensure some toughness. We have only a water quench available to us. Design the plate, assuming that we only have the steels given in Figure 13-21 available to us.

13-51 A quenched and tempered 10110 steel is found to have surface cracks that cause the heat-treated part to be rejected by the customer. Why did the cracks form? Design a heat treatment, including appropriate temperatures and times, that will minimize these problems.

13-52 Design a corrosion-resistant steel to use for a pump that transports liquid helium at 4 K in a superconducting magnet.

13-53 Design a heat treatment for a hook made from a 2.5 cm-diameter steel rod having a microstructure containing a mixture of ferrite, bainite, and martensite after quenching. Estimate the mechanical properties of your hook.

13-54 Design an annealing treatment for a 1050 steel. Be sure to include details of temperatures, cooling rates, microstructures, and properties.

13-55 Design a process to produce a 0.5-cm-diameter steel shaft having excellent toughness, yet excellent wear and fatigue resistance. The surface hardness should be at least HRC 60, and the hardness 0.01 cm beneath the surface should be approximately HRC 50. Describe the process, including details of the heat-treating atmosphere, the composition of the steel, temperatures, and times.

▲ Computer Problems

13-56 *Empirical Relationships for Transformations in Steels.* Heat treatment of steels depends upon knowing the A_1, A_3, A_{cm}, M_s, and M_f temperatures. Steel producers often provide their customers with empirical formulas that define these temperatures as a function of alloying element concentrations. For example, one such formula described in the literature is

$$M_s(\text{in °C}) = 561 - 474(\% \text{ C}) - 33$$
$$(\% \text{ Mn}) - 17(\% \text{ Ni})$$
$$- 17(\% \text{ Cr}) - 21(\% \text{ Mo})$$

Write a computer program that will ask the user to provide information for the concentrations of different alloying elements and provide the M_s and M_f temperatures. Assume that the M_f temperature is 215°C below M_s.

Ⓚ Knovel® Problems

K13-1 Find a time-temperature-transformation (TTT) diagram for 1340 steel. Assume that the steel is held at 720°C (1330°F) for 1 h, quenched to 425°C (800°F) and held for 1000 s at this temperature, and then cooled in air to room temperature.
  - (a) What is the microstructure and hardness of the steel after the heat treatment?
  - (b) What are the designation, hardenability, and applications of this steel?
  - (c) What are the mechanical properties of this steel delivered as hot rolled, normalized, and annealed when tempered at 425°C (800°F) and oil-quenched?

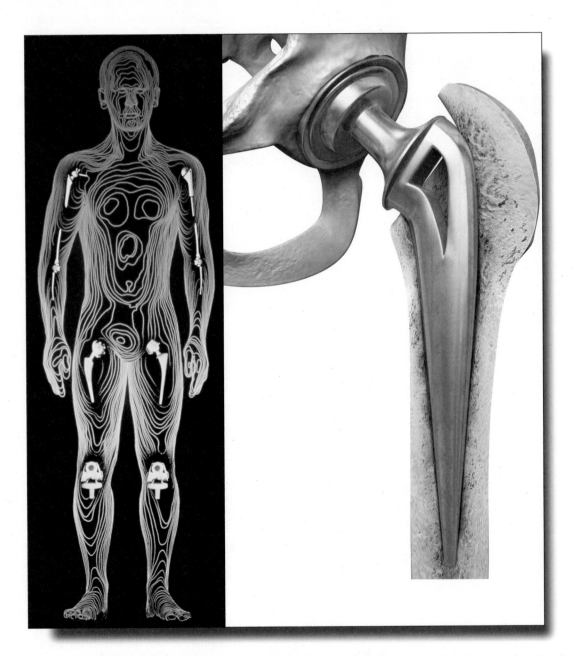

Increasingly, artificial implants are being incorporated into the human body. Some examples of commonly replaced joints are shown in the diagram on the left (*Courtesy of PhotoLink/PhotoDisc Green/Getty Images*). Every year, approximately 500,000 people receive a hip implant such as that shown in the photo on the right (©*MedicalRF.com/Alamy*). The implant must be made from a biocompatible material that has strong fracture toughness and an excellent fatigue life, and it must be corrosion resistant and possess a stiffness similar to that of bone. Nonferrous alloys based on titanium are often used for this application, as are cobalt-chrome alloys.

14

Nonferrous Alloys

Have You Ever Wondered?

- *In the history of mankind, which came first: copper or steel?*

- *Why does the Statue of Liberty have a green patina?*

- *What materials are used to manufacture biomedical implants for hip prostheses?*

- *Why is tungsten used for lightbulb filaments?*

- *Are some metals toxic?*

- *What materials are used as "catalysts" in the automobile catalytic converter? What do they "convert?"*

Nonferrous alloys (i.e., alloys of elements other than iron) include, but are not limited to, alloys based on aluminum, copper, nickel, cobalt, zinc, precious metals (such as Pt, Au, Ag, Pd), and other metals (e.g., Nb, Ta, W). In this chapter, we will briefly explore the properties and applications of Cu, Al, and Ti alloys in load-bearing applications. We will not discuss the electronic, magnetic, and other applications of nonferrous alloys.

In many applications, weight is a critical factor. To relate the strength of the material to its weight, a **specific strength**, or strength-to-weight ratio, is defined:

$$\text{Specific strength} = \frac{\text{strength}}{\text{density}} \qquad (14\text{-}1)$$

Table 14-1 compares the specific strength of steel, some high-strength nonferrous alloys, and polymer-matrix composites. Another factor to consider in designing with nonferrous metals is their cost, which also varies considerably. Table 14-1 gives the approximate price of different materials. One should note, however, that materials prices fluctuate with global economic trends and that the price of the material is only a small portion of the cost of a part. Fabrication and finishing, not to mention marketing and distribution, often contribute much more to the overall cost of a part. Composites based on carbon and other fibers also have significant advantages

TABLE 14-1 ■ *Specific strength and cost of nonferrous alloys, steels, and polymer composites*

Metal	Density (g/cm³)	Tensile Strength (MPa)	Specific Strength (cm)	Cost per kg ($)[c]
Aluminum	2.70	572	21.5×10^5	1.32
Beryllium	1.85	379	20.5×10^5	771.60
Copper	8.93	207–483	11.8×10^5	1.57
Lead	11.36	69	0.5×10^5	0.99
Magnesium	1.74	379	21.8×10^5	3.31
Nickel	8.90	1241	14.0×10^5	9.04
Titanium	4.51	1103	24.5×10^5	8.82
Tungsten	19.25	1034	5.5×10^5	8.82
Zinc	7.13	517	7.3×10^5	0.88
Steels	~7.87	1379	17.5×10^5	0.22
Aramid/epoxy (Kevlar, vol. fraction of fibers 0.6, longitudinal tension)	1.4	1379	10.0×10^6	—
Aramid/epoxy (Kevlar, vol. fraction of fibers 0.6, transverse tension)[a]	1.4	30	2.15×10^4	—
Glass/epoxy (Vol. fraction of E-glass fibers 0.6, longitudinal tension)[b]	2.1	1034	5.0×10^6	—
Glass/epoxy (Vol. fraction of E-glass fibers 0.6, transverse tension)	2.1	48	23.3×10^4	—

[a] *Data for composites from Harper, C.A.,* Handbook of Materials Product Design, *3rd ed. 2001: McGraw-Hill. Commodity composites are relatively inexpensive; high-performance composites are expensive.*
[b] *Properties of composites are highly anisotropic. This is taken care of during fabrication though.*
[c] *Costs based on average prices for the years 1998 to 2002.*

with respect to their specific strength. Their properties are typically anisotropic; however, the temperature at which they can be used is limited. In practice, to overcome the anisotropy, composites are often made in many layers. The directions of fibers are changed in different layers so as to minimize the anisotropy in properties.

14-1   Aluminum Alloys

Aluminum is the third most plentiful element on earth (next to oxygen and silicon), but, until the late 1800s, was expensive and difficult to produce. The 2.7-kg cap installed on the top of the Washington Monument in 1884 was one of the largest aluminum parts made at that time.

General Properties and Uses of Aluminum   Aluminum has
a density of 2.70 g/cm³, or one-third the density of steel, and a modulus of elasticity of 69×10^3 MPa. Although aluminum alloys have lower tensile properties than steel, their specific strength (or strength-to-weight ratio) is excellent. The Wright brothers used an Al-Cu alloy for their engine for this very reason (Chapter 12). Aluminum can be formed easily, it has high thermal and electrical conductivity, and does not show a ductile-to-brittle transition at low temperatures. It is nontoxic and can be recycled with only about 5%

TABLE 14-2 ■ *The effect of strengthening mechanisms in aluminum and aluminum alloys*

Material	Tensile Strength (MPa)	Yield Strength (MPa)	% Elongation	Ratio of Alloy-to-Metal Yield Strengths
Pure Al	45	17	60	1.0
Commercially pure Al (at least 99% pure)	90	35	45	2.0
Solid-solution-strengthened Al alloy	110	41	35	2.4
Cold-worked Al	166	152	15	8.8
Dispersion-strengthened Al alloy	290	152	35	8.8
Age-hardened Al alloy	572	503	11	29.2

of the energy that was needed to make it from alumina. This is why the recycling of aluminum is so successful. Aluminum's beneficial physical properties include its non-ferromagnetic behavior and resistance to oxidation and corrosion. Aluminum does not display a true endurance limit, however, so failure by fatigue may eventually occur, even at low stresses. Because of its low melting temperature, aluminum does not perform well at elevated temperatures. Finally, aluminum alloys have low hardness, leading to poor wear resistance. Aluminum responds readily to strengthening mechanisms. Table 14-2 compares the strength of pure, annealed aluminum with that of alloys strengthened by various techniques. The alloys may be 30 times stronger than pure aluminum.

About 25% of the aluminum produced today is used in the transportation industry, another 25% in the manufacture of beverage cans and other packaging, about 15% in construction, 15% in electrical applications, and 20% in other applications. About 91 kg of aluminum was used in an average car made in the United States in 2010. Aluminum reacts with oxygen, even at room temperature, to produce an extremely thin aluminum-oxide layer that protects the underlying metal from many corrosive environments (Chapter 23). We should be careful, though, not to generalize this behavior. For example, aluminum powder (because it has a high surface area), when present in the form of an oxidizer, such as ammonium perchlorate and iron oxide as catalysts, serves as the fuel for solid rocket boosters (SRBs). These boosters use ~91,000 kg of atomized aluminum powder every time the space shuttle takes off and can generate enough force for the shuttle to reach a speed of ~4800 km/h. New developments related to aluminum include the development of aluminum alloys containing higher Mg concentrations for use in making automobiles. There is also interest in developing processes that will directly transform molten Al into sheet or other solid products.

Designation
Aluminum alloys can be divided into two major groups: wrought and casting alloys, depending on their method of fabrication. **Wrought alloys**, which are shaped by plastic deformation, have compositions and microstructures significantly different from casting alloys, reflecting the different requirements of the manufacturing process. Within each major group, we can divide the alloys into two subgroups: heat-treatable and non heat-treatable alloys.

Aluminum alloys are designated by the numbering system shown in Table 14-3. The first number specifies the principle alloying elements, and the remaining numbers refer to the specific composition of the alloy.

The degree of strengthening is given by the **temper designation** T or H, depending on whether the alloy is heat treated or strain hardened (Table 14-4). Other designations indicate whether the alloy is annealed (O), solution treated (W), or used in the as-fabricated

TABLE 14-3 ■ Designation system for aluminum alloys

Wrought alloys:

1xxx[a]	Commercially pure Al (>99% Al)	Not age hardenable
2xxx	Al-Cu and Al-Cu-Li	Age hardenable
3xxx	Al-Mn	Not age hardenable
4xxx	Al-Si and Al-Mg-Si	Age hardenable if magnesium is present
5xxx	Al-Mg	Not age hardenable
6xxx	Al-Mg-Si	Age hardenable
7xxx	Al-Mg-Zn	Age hardenable
8xxx	Al-Li, Sn, Zr, or B	Age hardenable
9xxx	Not currently used	

Casting alloys:

1xx.x.[b]	Commercially pure Al	Not age hardenable
2xx.x.	Al-Cu	Age hardenable
3xx.x.	Al-Si-Cu or Al-Mg-Si	Some are age hardenable
4xx.x.	Al-Si	Not age hardenable
5xx.x.	Al-Mg	Not age hardenable
7xx.x.	Al-Mg-Zn	Age hardenable
8xx.x.	Al-Sn	Age hardenable
9xx.x.	Not currently used	

[a]The first digit shows the main alloying element, the second digit shows modification, and the last two digits shows the decimal % of the Al concentration (e.g., 1060: will be 99.6% Al alloy).
[b]Last digit indicates product form, 1 or 2 is ingot (depends upon purity) and 0 is for casting.

TABLE 14-4 ■ Temper designations for aluminum alloys

F   As-fabricated (hot worked, forged, cast, etc.)
O   Annealed (in the softest possible condition)
H   Cold-worked
      H1x—cold-worked only. (x refers to the amount of cold work and strengthening.)
      H12—cold work that gives a tensile strength midway between the O and H14 tempers.
      H14—cold work that gives a tensile strength midway between the O and H18 tempers.
      H16—cold work that gives a tensile strength midway between the H14 and H18 tempers.
      H18—cold work that gives about 75% reduction.
      H19—cold work that gives a tensile strength greater than 14 MPa of that obtained by the
        H18 temper.
      H2x—cold worked and partly annealed.
      H3x—cold worked and stabilized at a low temperature to prevent age hardening of the structure.
W   Solution treated
T   Age hardened
      T1—cooled from the fabrication temperature and naturally aged.
      T2—cooled from the fabrication temperature, cold worked, and naturally aged.
      T3—solution treated, cold worked, and naturally aged.
      T4—solution treated and naturally aged.
      T5—cooled from the fabrication temperature and artificially aged.
      T6—solution treated and artificially aged.
      T7—solution treated and stabilized by overaging.
      T8—solution treated, cold worked, and artificially aged.
      T9—solution treated, artificially aged, and cold worked.
      T10—cooled from the fabrication temperature, cold worked, and artificially aged.

TABLE 14-5 ■ *Properties of typical aluminum alloys*

Alloy		Tensile Strength (MPa)	Yield Strength (MPa)	% Elongation	Applications
Non heat-treatable wrought alloys:					
1100-O	>99% Al	90	35	40	Electrical components, foil,
1100-H18		166	152	10	food processing,
3004-O	1.2% Mn-1.0% Mg	179	69	25	beverage can bodies,
3004-H18		283	248	9	architectural uses,
4043-O	5.2% Si	145	69	22	filler metal for welding,
4043-H18		283	269	1	beverage can tops, and
5182-O	4.5% Mg	290	131	25	marine components
5182-H19		421	393	4	
Heat-treatable wrought alloys:					
2024-T4	4.4% Cu	469	324	20	Truck wheels, aircraft skins,
2090-T6	2.4% Li-2.7% Cu	552	517	6	pistons, canoes, railroad
4032-T6	12% Si-1% Mg	379	317	9	cars, and aircraft frames
6061-T6	1% Mg-0.6% Si	310	276	15	
7075-T6	5.6% Zn-2.5% Mg	572	503	11	
Casting alloys:					
201-T6	4.5% Cu	483	434	7	Transmission housings,
319-F	6% Si-3.5% Cu	186	124	2	general purpose castings,
356-T6	7% Si-0.3% Mg	228	166	3	aircraft fittings, motor
380-F	8.5% Si-3.5% Cu	317	159	3	housings, automotive
390-F	17% Si-4.5% Cu	283	241	1	engines, food-handling
443-F	5.2% Si (sand cast)	131	55	8	equipment, and marine
	(permanent mold)	159	62	10	fittings
	(die cast)	228	110	9	

condition (F). The numbers following the T or H indicate the amount of strain hardening, the exact type of heat treatment, or other special aspects of the processing of the alloy. Typical alloys and their properties are included in Table 14-5.

Wrought Alloys

The 1xxx, 3xxx, 5xxx, and most of the 4xxx wrought alloys are not age hardenable. The 1xxx and 3xxx alloys are single-phase alloys except for the presence of small amounts of inclusions or intermetallic compounds (Figure 14-1). Their properties are controlled by strain hardening, solid-solution strengthening, and grain-size control. Because the solubilities of the alloying elements in aluminum are small at room temperature, the degree of solid-solution strengthening is limited. The 5xxx alloys contain two phases at room temperature—α, a solid solution of magnesium in aluminum, and Mg_2Al_3, a hard, brittle intermetallic compound (Figure 14-2). The aluminum-magnesium alloys are strengthened by a fine dispersion of Mg_2Al_3, as well as by strain hardening, solid-solution strengthening, and grain-size control. Because Mg_2Al_3 is not coherent, age-hardening treatments are not possible. The 4xxx series alloys also contain two phases, α and nearly pure silicon, β (Chapter 11). Alloys that contain both silicon and magnesium can be age hardened by permitting Mg_2Si to precipitate. The 2xxx, 6xxx, and 7xxx alloys are age-hardenable alloys. Although excellent specific strengths are obtained for these alloys, the amount of precipitate that can form is limited. In addition, they cannot be used

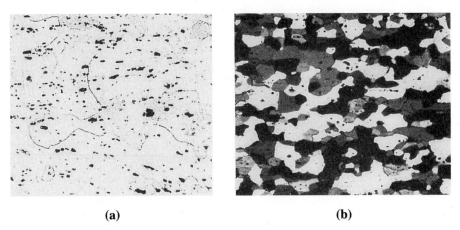

(a)   **(b)**

Figure 14-1   (a) FeAl₃ inclusions in annealed 1100 aluminum (×350). (b) Mg₂Si precipitates in annealed 5457 aluminum alloy (×75). *(From* ASM Handbook, *Vol. 7, (1972), ASM International, Materials Park, OH 44073–0002.)*

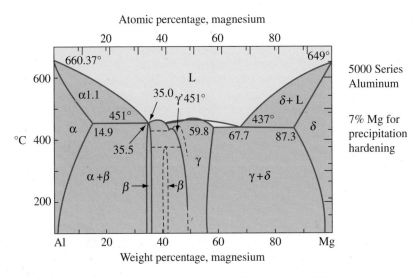

Figure 14.2   Aluminum–magnesium phase diagram. Dotted lines indicate uncertain solubility limits. (*American Society for Metals Handbook, 8th ed., Vol. 8, Metallography, Structures and Phase Diagrams. Metals Park, Ohio, 1973, p. 261. Reprinted with permission of ASM International.® All rights reserved. www.asminternational.org.*)

at temperatures above approximately 175°C in the aged condition. Alloy 2024 is the most widely used aircraft alloys. There is also a renewed interest in the development of precipitation-hardened Al-Li alloys due to their high Young's modulus and low density; however, high processing costs, anisotropic properties, and lower fracture toughness have proved to be limiting factors. Al-Li alloys are used to make space shuttle fuel tanks.

Casting Alloys

Many of the common aluminum casting alloys shown in Table 14-5 contain enough silicon to cause the eutectic reaction, giving the alloys low melting points, good fluidity, and good castability. **Fluidity** is the ability of the liquid metal to flow through a mold without prematurely solidifying, and **castability** refers to the ease with which a good casting can be made from the alloy.

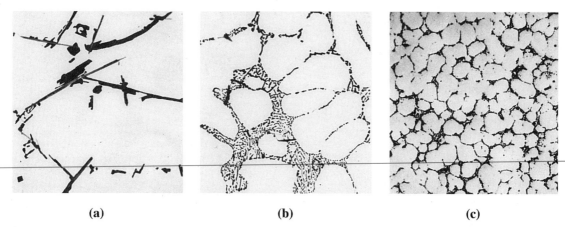

(a)   **(b)**   **(c)**

Figure 14-3   (a) Sand cast 443 aluminum alloy containing coarse silicon and inclusions. (b) Permanent mold 443 alloy containing fine dendrite cells and fine silicon due to faster cooling, (c) Die cast 443 alloy with a still finer microstructure ($\times$ 350). (*From* ASM Handbook, *Vol. 7, (1972), ASM International, Materials Park, OH 44073–0002.*)

The properties of the aluminum-silicon alloys are controlled by solid-solution strengthening of the α aluminum matrix, dispersion strengthening by the β phase, and solidification, which controls the primary grain size and shape as well as the nature of the eutectic microconstituent. Fast cooling obtained in die casting or permanent mold casting (Chapter 9) increases strength by refining grain size and the eutectic microconstituent (Figure 14-3). Grain refinement using boron and titanium additions, modification using sodium or strontium to change the eutectic structure, and hardening with phosphorus to refine the primary silicon (Chapter 9) are all done in certain alloys to improve the microstructure and, thus, the degree of dispersion strengthening. Many alloys also contain copper, magnesium, or zinc, thus permitting age hardening. The following examples illustrate applications of aluminum alloys.

Example 14-1   *Strength-to-Weight Ratio in Design*

A steel cable 1.25 cm in diameter has a yield strength of 483 MPa. The density of steel is about 7.87 g/cm^3. Based on the data in Table 14-5, determine (a) the maximum load that the steel cable can support without yielding, (b) the diameter of a cold-worked aluminum-manganese alloy (3004-H18) required to support the same load as the steel, and (c) the weight per meter of the steel cable versus the aluminum alloy cable.

SOLUTION

a. Load $= F = (S_y \times A) = 483$ MPa $\left(\dfrac{\pi}{4}\right)(1.25 \text{ cm})^2 = 59{,}297$ N

b. The yield strength of the aluminum alloy is 248 MPa. Thus,

$$A = \frac{\pi}{4} d^2 = \frac{F}{S_y} = \frac{59{,}297 \text{ N}}{248 \text{ MPa}} = 2.39 \text{ cm}^2$$

$$d = 1.744 \text{ cm}$$

c. Density of steel $= \rho = 7.87$ g/cm^3

Density of aluminum $= \rho = 2.70$ g/cm^3

Weight of steel $= Al\rho = \dfrac{\pi}{4}(1.25\,\text{cm})^2(100)(7.87) = 966.183$ g/m

Weight of aluminum $= Al\rho = \dfrac{\pi}{4}(1.744)^2(12)(2.7) = 645.24$ g/m

Although the yield strength of the aluminum is lower than that of the steel and the cable must be larger in diameter, the aluminum cable weighs only about two-thirds as much as the steel cable. When comparing materials, a proper safety factor should also be included during design.

Example 14-2   *Design of an Aluminum Recycling Process*

Design a method for recycling aluminum alloys used for beverage cans.

SOLUTION

Recycling aluminum is advantageous because only a fraction (about 5%) of the energy required to produce aluminum from Al_2O_3 is required. Recycling beverage cans presents several difficulties, however, because the beverage cans are made from two aluminum alloys (3004 for the main body, and 5182 for the lids) having different compositions (Table 14-5). The 3004 alloy has the exceptional formability needed to perform the deep drawing process; the 5182 alloy is harder and permits the pull-tops to function properly. When the cans are remelted, the resulting alloy contains both Mg and Mn and is not suitable for either application.

One approach to recycling the cans is to separate the two alloys from the cans. The cans are shredded, then heated to remove the lacquer that helps protect the cans during use. We could then further shred the material at a temperature at which the 5182 alloy begins to melt. The 5182 alloy has a wider freezing range than the 3004 alloy and breaks into very small pieces; the more ductile 3004 alloy remains in larger pieces. The small pieces of 5182 can therefore be separated by passing the material through a screen. The two separated alloys can then be melted, cast, and rolled into new can stock.

An alternative method would be to simply remelt the cans. Once the cans have been remelted, we could bubble chlorine gas through the liquid alloy. The chlorine reacts selectively with the magnesium, removing it as a chloride. The remaining liquid can then be adjusted to the proper composition and be recycled as 3004 alloy.

Example 14-3   *Design/Materials Selection for a Cryogenic Tank*

Design the material to be used to contain liquid hydrogen fuel for the space shuttle.

SOLUTION

Liquid hydrogen is stored below $-253°C$; therefore, our tank must have good cryogenic properties. The tank is subjected to high stresses, particularly when the shuttle is inserted into orbit, and it should have good fracture toughness to minimize the chances of catastrophic failure. Finally, it should be light in weight to permit higher payloads or less fuel consumption.

Lightweight aluminum would appear to be a good choice. Aluminum does not show a ductile to brittle transition. Because of its good ductility, we expect aluminum to also have good fracture toughness, particularly when the alloy is in the annealed condition.

One of the most common cryogenic aluminum alloys is 5083-O. Aluminum-lithium alloys are also being considered for low temperature applications to take advantage of their even lower density.

14-2   Magnesium and Beryllium Alloys

Magnesium, which is often extracted electrolytically from concentrated magnesium chloride in seawater, is lighter than aluminum with a density of 1.74 g/cm^3, and it melts at a slightly lower temperature than aluminum. In many environments, the corrosion resistance of magnesium approaches that of aluminum; however, exposure to salts, such as that near a marine environment, causes rapid deterioration. Although magnesium alloys are not as strong as aluminum alloys, their specific strengths are comparable. Consequently, magnesium alloys are used in aerospace applications, high-speed machinery, and transportation and materials handling equipment.

Magnesium, however, has a low modulus of elasticity (44.8×10^6 MPa or 45 GPa) and poor resistance to fatigue, creep, and wear. Magnesium also poses a hazard during casting and machining, since it combines easily with oxygen and burns. Finally, the response of magnesium to strengthening mechanisms is relatively poor.

Structure and Properties   Magnesium, which has the HCP structure, is less ductile than aluminum. The alloys do have some ductility, however, because alloying increases the number of active slip planes. Some deformation and strain hardening can be accomplished at room temperature, and the alloys can be readily deformed at elevated temperatures. Strain hardening produces a relatively small effect in pure magnesium because of the low strain-hardening coefficient (Chapter 8).

As in aluminum alloys, the solubility of alloying elements in magnesium at room temperature is limited, causing only a small degree of solid-solution strengthening. The solubility of many alloying elements increases with temperature, however, as shown in the Mg-Al phase diagram (Figure 14-4). Therefore, alloys may be strengthened by either dispersion strengthening or age hardening. Some age-hardened magnesium alloys, such as those containing Zr, Th, Ag, or Ce, have good resistance to overaging at temperatures as high as 300°C. Alloys containing up to 9% Li have exceptionally light weight. Properties of typical magnesium alloys are listed in Table 14-6.

Advanced magnesium alloys include those with very low levels of impurities and those containing large amounts (>5%) of cerium and other rare earth elements. These alloys form a protective MgO film that improves corrosion resistance. Rapid solidification processing permits larger amounts of alloying elements to be dissolved in the magnesium, further improving corrosion resistance. Improvements in strength, particularly at high temperatures, can be obtained by introducing ceramic particles or fibers such as silicon carbide into the metal.

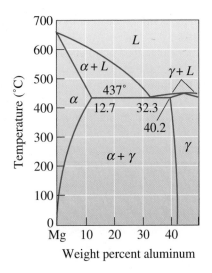

Figure 14-4
The magnesium-aluminum phase diagram.

TABLE 14-6 ■ *Properties of typical magnesium alloys*

Alloy	Composition	Tensile Strength (MPa)	Yield Strength (MPa)	% Elongation
Pure Mg:				
Annealed		159	90	3–15
Cold-worked		179	117	2–10
Casting alloys:				
AM 100-T6	10% Al-0.1% Mn	276	152	1
AZ81A-T4	7.6% Al-0.7% Zn	276	83	15
ZK61A-T6	6% Zn-0.7% Zr	310	193	10
Wrought alloys:				
AZ80A-T5	8.5% Al-0.5% Zn	379	276	7
ZK40A-T5	4% Zn-0.45% Zr	276	255	4
HK31A-H24	3% Th-0.6% Zr	262	207	8

Beryllium is lighter than aluminum, with a density of 1.848 g/cm^3, yet it is stiffer than steel, with a modulus of elasticity of 290×10^3 MPa. Beryllium alloys, which have yield strengths of 207 to 345 MPa, have high specific strengths and maintain both strength and stiffness to high temperatures. Instrument grade beryllium is used in inertial guidance systems where the elastic deformation must be minimal; structural grades are used in aerospace applications; and nuclear applications take advantage of the transparency of beryllium to electromagnetic radiation. Unfortunately, beryllium is expensive (Table 14-1), brittle, reactive, and toxic. Its production is quite complicated, and hence, the applications of Be alloys are very limited. Beryllium oxide (BeO), which is also toxic *in a powder form*, is used to make high-thermal conductivity ceramics.

14-3 Copper Alloys

Copper occurs in nature as elemental copper and was extracted successfully from minerals long before iron, since the relatively lower temperatures required for the extraction could be achieved more easily. Copper is typically produced by a pyrometallurgical (high temperature) process. The copper ore containing high-sulfur contents is concentrated, then converted into a molten immiscible liquid containing copper sulfide-iron sulfide and is known as a copper matte. This is done in a flash smelter. In a separate reactor, known as a copper converter, oxygen introduced to the matte converts the iron sulfide to iron oxide and the copper sulfide to an impure copper called **blister copper**, which is then purified electrolytically. Other methods for copper extraction include leaching copper from low-sulfur ores with a weak acid, then electrolytically extracting the copper from the solution.

Copper-based alloys have higher densities than steels. Although the yield strength of some alloys is high, their specific strength is typically less than that of aluminum or magnesium alloys. These alloys have better resistance to fatigue, creep, and wear than the lightweight aluminum and magnesium alloys. Many of these alloys have excellent ductility, corrosion resistance, electrical and thermal conductivity, and most can easily be joined or fabricated into useful shapes. Applications for copper-based alloys include electrical components (such as wire), pumps, valves, and plumbing parts, where these properties are used to advantage.

Copper alloys also are unusual in that they may be selected to produce an appropriate decorative color. Pure copper is red; zinc additions produce a yellow color; and nickel produces a silver color. Copper can corrode easily, forming a basic copper sulfate $[CuSO_4 \cdot 3Cu(OH)_2]$. This is a green compound that is insoluble in water (but soluble in acids). This green patina provides an attractive finish for many applications. The Statue of Liberty is green because of the green patina of the oxidized copper skin that covers the steel structure.

The wide variety of copper-based alloys takes advantage of all of the strengthening mechanisms that we have discussed. The effects of these strengthening mechanisms on the mechanical properties are summarized in Table 14-7.

Copper containing less than 0.1% impurities is used for electrical and microelectronics applications. Small amounts of cadmium, silver, and Al_2O_3 improve hardness without significantly impairing conductivity. The single-phase copper alloys are strengthened by cold working. Examples of this effect are shown in Table 14-7. FCC copper has excellent ductility and a high strain-hardening coefficient.

TABLE 14-7 ■ *Properties of typical copper alloys obtained by different strengthening mechanisms*

Material	Tensile Strength (MPa)	Yield Strength (MPa)	% Elongation	Strengthening Mechanism
Pure Cu, annealed	209	33	60	None
Commercially pure Cu, annealed to coarse grain size	221	69	55	Solid Solution
Commercially pure Cu, annealed to fine grain size	234	76	55	Grain size
Commercially pure Cu, cold-worked 70%	393	365	4	Strain hardening
Annealed Cu-35% Zn	324	103	62	Solid solution
Annealed Cu-10% Sn	455	193	68	Solid solution
Cold-worked Cu-35% Zn	676	434	3	Solid solution + strain hardening
Age-hardened Cu-2% Be	1310	1207	4	Age hardening
Quenched and tempered Cu-Al	759	414	5	Martensitic reaction
Cast manganese bronze	490	193	30	Eutectoid reaction

Solid-Solution Strengthened Alloys
A number of copper-based alloys contain large quantities of alloying elements, yet remain single phase. Important binary phase diagrams are shown in Figure 14-5. The copper-zinc, or **brass**, alloys with less than 40% Zn form single-phase solid solutions of zinc in copper. The

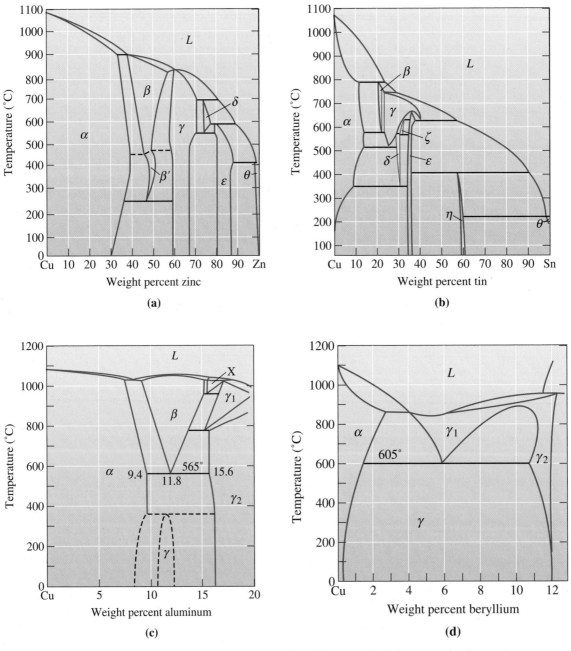

Figure 14-5   Binary phase diagrams for the (a) copper-zinc, (b) copper-tin, (c) copper-aluminum, and (d) copper-beryllium systems.

mechanical properties—even elongation—increase as the zinc content increases. These alloys can be cold formed into rather complicated yet corrosion-resistant components. **Bronzes** are generally considered alloys of copper containing tin and can certainly contain other elements. Manganese bronze is a particularly high-strength alloy containing manganese as well as zinc for solid-solution strengthening.

Tin bronzes, often called phosphor bronzes, may contain up to 10% Sn and remain single phase. The phase diagram predicts that the alloy will contain the Cu_3Sn (ε) compound; however, the kinetics of the reaction are so slow that the precipitate may not form.

Alloys containing less than about 9% Al or less than 3% Si are also single phase. These aluminum bronzes and silicon bronzes have good forming characteristics and are often selected for their good strength and excellent toughness.

Age-Hardenable Alloys

A number of copper-based alloys display an age-hardening response, including zirconium-copper, chromium-copper, and beryllium-copper. The copper-beryllium alloys are used for their high strength, high stiffness (making them useful as springs and fine wires), and nonsparking qualities (making them useful for tools to be used near flammable gases and liquids).

Phase Transformations

Aluminum bronzes that contain over 9% Al can form the β phase on heating above 565°C, the eutectoid temperature [Figure 14-5(c)]. On subsequent cooling, the eutectoid reaction produces a lamellar structure (like pearlite) that contains a brittle γ_2 compound. The low-temperature peritectoid reaction, $\alpha + \gamma_2 \rightarrow \gamma$, normally does not occur. The eutectoid product is relatively weak and brittle, but we can rapidly quench the β phase to produce martensite, or β', which has high strength and low ductility. When β' is subsequently tempered, a combination of high strength, good ductility, and excellent toughness is obtained as fine platelets of α precipitate from the β'.

Lead-Copper Alloys

Virtually any of the wrought alloys may contain up to 4.5% Pb. The lead forms a monotectic reaction with copper and produces tiny lead spheres as the last liquid to solidify. The lead improves machining characteristics. Use of lead-copper alloys, however, has a major environmental impact and, consequently, new alloys that are lead free have been developed. The following two examples illustrate the use of copper-based alloys.

Example 14-4	*Design/Materials Selection for an Electrical Switch*

Design the contacts for a switch or relay that opens and closes a high-current electrical circuit.

SOLUTION

When the switch or relay opens and closes, contact between the conductive surfaces can cause wear and result in poor contact and arcing. A high hardness would minimize wear, but the contact materials must allow the high current to pass through the connection without overheating or arcing.

Therefore, our design must provide for both good electrical conductivity and good wear resistance. A relatively pure copper alloy dispersion strengthened with a hard phase that does not disturb the copper lattice would, perhaps, be ideal. In a Cu-Al$_2$O$_3$ alloy, hard ceramic-oxide particles provide wear resistance but do not interfere with the electrical conductivity of the copper matrix.

Example 14-5   *Design of a Heat Treatment for a Cu-Al Alloy Gear*

Design the heat treatment required to produce a high-strength aluminum-bronze gear containing 10% Al.

SOLUTION

The aluminum bronze can be strengthened by a quench and temper heat treatment. We must heat above 900°C to obtain 100% β for a Cu-10% Al alloy [Figure 14-5(c)]. The eutectoid temperature for the alloy is 565°C. Therefore, our recommended heat treatment is

1. Heat the alloy to 950°C and hold to produce 100% β.

2. Quench the alloy to room temperature to cause β to transform to martensite, β', which is supersaturated in copper.

3. Temper below 565°C; a temperature of 400°C might be suitable. During tempering, the martensite transforms to α and γ_2. The amount of the γ_2 that forms at 400°C is

$$\% \, \gamma_2 = \frac{10 - 9.4}{15.6 - 9.4} \times 100 = 9.7\%$$

4. Cool rapidly to room temperature so that the equilibrium γ does not form.

   Note that if tempering were carried out below about 370°C, γ would form rather than γ_2.

14-4   Nickel and Cobalt Alloys

Nickel and cobalt alloys are used for corrosion protection and for high-temperature resistance, taking advantage of their high melting points and high strengths. Nickel is FCC and has good formability; cobalt is an allotropic metal, with an FCC structure above 417°C and an HCP structure at lower temperatures. Special cobalt alloys are used for exceptional wear resistance and, because of resistance to human body fluids, for prosthetic devices. Typical alloys and their applications are listed in Table 14-8.

In Chapter 10, we saw how rapidly solidified powders of nickel- and cobalt-based superalloys can be formed using spray atomization followed by hot isostatic pressing. These materials are used to make the rings that retain turbine blades, as well as for turbine blades for aircraft engines. In Chapter 12, we discussed applications of shape-memory alloys based on Ni-Ti. Iron, nickel, and cobalt are ferromagnetic. Certain Fe-Ni- and Fe-Co-based alloys form very good magnetic materials (Chapter 20). A Ni-36% Fe alloy (Invar) displays practically no expansion during heating; this effect is exploited in producing bimetallic composite materials. Cobalt is used in WC-Co cutting tools.

TABLE 14-8 ■ *Compositions, properties, and applications for selected nickel and cobalt alloys*

Material	Tensile Strength (MPa)	Yield Strength (MPa)	% Elongation	Strengthening Mechanism	Applications
Pure Ni (99.9% Ni)	345	110	45	Annealed	Corrosion resistance
	655	621	4	Cold worked	Corrosion resistance
Ni-Cu alloys:					
Monel 400 (Ni-31.5% Cu)	538	269	37	Annealed	Valves, pumps, heat exchangers
Monel K-500 (Ni-29.5% Cu-2.7% Al-0.6% Ti)	1034	759	30	Aged	Shafts, springs, impellers
Ni superalloys:					
Inconel 600 (Ni-15.5% Cr-8% Fe)	621	200	49	Carbides	Heat-treatment equipment
Hastelloy B-2 (Ni-28% Mo)	896	414	61	Carbides	Corrosion resistance
DS-Ni (Ni-2% ThO$_2$)	490	331	14	Dispersion	Gas turbines
Fe-Ni superalloys:					
Incoloy 800 (Ni-46% Fe-21% Cr)	614	283	37	Carbides	Heat exchangers
Co superalloys:					
Stellite 6B (60% Co-30% Cr-4.5% W)	1220	710	4	Carbides	Abrasive wear resistance

Nickel and Monel

Nickel and its alloys have excellent corrosion resistance and forming characteristics. When copper is added to nickel, the maximum strength is obtained near 60% Ni. A number of alloys, called **Monels**, with approximately this composition are used for their strength and corrosion resistance in salt water and at elevated temperatures. Some of the Monels contain small amounts of aluminum and titanium. These alloys show an age-hardening response by the precipitation of γ', a coherent Ni$_3$Al or Ni$_3$Ti precipitate that nearly doubles the tensile properties. The precipitates resist overaging at temperatures up to 425°C (Figure 14-6).

Superalloys

Superalloys are nickel, iron-nickel, and cobalt alloys that contain large amounts of alloying elements intended to produce a combination of high strength at elevated temperatures, resistance to creep at temperatures up to 1000°C, and resistance to corrosion. These excellent high-temperature properties are obtained even though the melting temperatures of the alloys are about the same as that for steels. Typical applications include vanes and blades for turbine and jet engines, heat exchangers, chemical reaction vessel components, and heat-treating equipment.

To obtain high strength and creep resistance, the alloying elements must produce a strong, stable microstructure at high temperatures. Solid-solution strengthening, dispersion strengthening, and precipitation hardening are generally employed.

Solid-Solution Strengthening

Large additions of chromium, molybdenum, and tungsten and smaller additions of tantalum, zirconium, niobium, and boron provide solid-solution strengthening. The effects of solid-solution strengthening

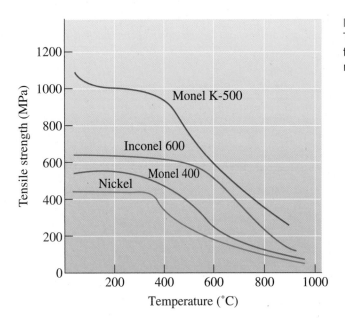

Figure 14-6
The effect of temperature on the tensile strength of several nickel-based alloys.

are stable and, consequently, make the alloy resistant to creep, particularly when large atoms such as molybdenum and tungsten (which diffuse slowly) are used.

Carbide Dispersion Strengthening

All alloys contain a small amount of carbon which, by combining with other alloying elements, produces a network of fine, stable carbide particles. The carbide network interferes with dislocation movement and prevents grain boundary sliding. The carbides include TiC, BC, ZrC, TaC, Cr_7C_3, $Cr_{23}C_6$, Mo_6C, and W_6C, although often they are more complex and contain several alloying elements. Stellite 6B, a cobalt-based superalloy, has unusually good wear resistance at high temperatures due to these carbides.

Precipitation Hardening

Many of the nickel and nickel-iron superalloys that contain aluminum and titanium form the coherent precipitate γ' (Ni_3Al or Ni_3Ti) during aging. The γ' particles (Figure 14-7) have a crystal structure and lattice parameter similar to that of the nickel matrix; this similarity leads to a low surface energy and minimizes overaging of the alloys, providing good strength and creep resistance even at high temperatures.

By varying the aging temperature, precipitates of various sizes can be produced. Small precipitates, formed at low aging temperatures, can grow between the larger precipitates produced at higher temperatures, therefore increasing the volume percentage of the γ' and further increasing the strength [Figure 14-7(b)].

The high temperature use of superalloys can be improved when a ceramic coating is used. One method for doing this is to first coat the superalloy with a metallic bond coat composed of a complex NiCoCrAlY, alloy and then apply an outer thermal barrier coating of a stabilized ZrO_2-ceramic (Chapter 5). The coating helps reduce oxidation of the superalloy and permits jet engines to operate at higher temperatures and with greater efficiency. The next example shows the application of a nickel-based superalloy.

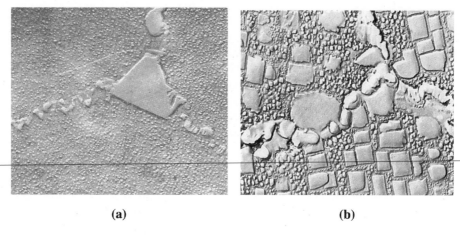

(a)   (b)

Figure 14-7   (a) Microstructure of a superalloy, with carbides at the grain boundaries and γ' precipitates in the matrix ($\times$ 15,000). (b) Microstructure of a superalloy aged at two temperatures, producing both large and small cubical γ' precipitates ($\times$ 10,000). (*From ASM* Handbook, *Vol. 9, Metallography and Microstructure (1985), ASM International, Materials Park, OH 44073–0002.*)

Example 14-6   *Design/Materials Selection for a High-Performance Jet Engine Turbine Blade*

Design a nickel-based superalloy for producing turbine blades for a gas turbine aircraft engine that will have a particularly long creep-rupture time at temperatures approaching 1100°C.

SOLUTION

First, we need a very stable microstructure. Addition of aluminum or titanium permits the precipitation of up to 60 vol% of the γ' phase during heat treatment and may permit the alloy to operate at temperatures approaching 0.85 times the absolute melting temperature. Addition of carbon and alloying elements such as tantalum and hafnium permits the precipitation of alloy carbides that prevent grain boundaries from sliding at high temperatures. Other alloying elements, including molybdenum and tungsten, provide solid-solution strengthening.

Second, we might produce a directionally solidified or even single-crystal turbine blade (Chapter 9). In directional solidification, only columnar grains form during freezing, eliminating transverse grain boundaries that might nucleate cracks. In a single crystal, no grain boundaries are present. We might use the investment casting process, being sure to pass the liquid superalloy through a filter to trap any tiny inclusions before the metal enters the ceramic investment mold.

We would then heat treat the casting to ensure that the carbides and γ' precipitate have the correct size and distribution. Multiple aging temperatures might be used to ensure that the largest possible volume percent γ' is formed.

Finally, the blade might contain small cooling channels along its length. Air for combustion in the engine can pass through these channels, providing active cooling to the blade, before reacting with fuel in the combustion chamber.

14-5   Titanium Alloys

Titanium is produced from TiO_2 by the Kroll process. The TiO_2 is converted to $TiCl_4$ (titanium tetrachloride, also informally known as *tickle!*), which is subsequently reduced to titanium metal by sodium or magnesium. The resultant titanium sponge is then consolidated, alloyed as necessary, and processed using vacuum arc melting. A more cost-effective process involves producing titanium sponge directly from TiO_2. Titanium provides excellent corrosion resistance, high specific strength, and good high-temperature properties. Strengths up to1380 MPa, coupled with a density of 4.505 g/cm^3, provide excellent mechanical properties. An adherent, protective TiO_2 film provides excellent resistance to corrosion and contamination below 535°C. Above 535°C, the oxide film breaks down, and small atoms such as carbon, oxygen, nitrogen, and hydrogen embrittle the titanium.

Titanium's excellent corrosion resistance allows applications in chemical processing equipment, marine components, and biomedical implants such as hip prostheses. Titanium is an important aerospace material, finding applications as airframe and jet engine components. When it is combined with niobium, a superconductive intermetallic compound is formed; when it is combined with nickel, the resulting alloy displays the shape-memory effect; when it is combined with aluminum, a new class of intermetallic alloys is produced, as discussed in Chapter 11. Titanium alloys are used for sports equipment such as the heads of golf clubs.

Titanium is allotropic, with the HCP crystal structure (α) at low temperatures and a BCC structure (β) above 882°C. Alloying elements provide solid-solution strengthening and change the allotropic transformation temperature. The alloying elements can be divided into four groups (Figure 14-8). Additions such as tin and zirconium provide solid-solution strengthening without affecting the transformation temperature. Aluminum, oxygen, hydrogen, and other α-stabilizing elements increase the temperature at which α transforms to β. Beta stabilizers such as vanadium, tantalum, molybdenum, and niobium lower the transformation temperature, even causing β to be stable at room temperature. Finally, manganese, chromium, and iron produce a eutectoid reaction, reducing the temperature at which the $\alpha - \beta$ transformation occurs and producing a two-phase structure at room temperature. Several categories of titanium and its alloys are listed in Table 14-9.

Commercially Pure Titanium
Unalloyed titanium is used for its superior corrosion resistance. Impurities, such as oxygen, increase the strength of the titanium (Figure 14-9) but reduce corrosion resistance. Applications include heat exchangers, piping, reactors, pumps, and valves for the chemical and petrochemical industries.

Alpha Titanium Alloys
The most common of the all-α alloys contains 5% Al and 2.5% Sn, which provide solid-solution strengthening to the HCP α. The α alloys are annealed at high temperatures in the β region. Rapid cooling gives an acicular, or Widmanstätten, α-grain structure (Figure 14-10) that provides good resistance to fatigue. Furnace cooling gives a more platelike α structure that provides better creep resistance.

Beta Titanium Alloys
Although large additions of vanadium or molybdenum produce an entirely β structure at room temperature, none of the β alloys actually are alloyed to that extent. Instead, they are rich in β stabilizers, so that rapid cooling

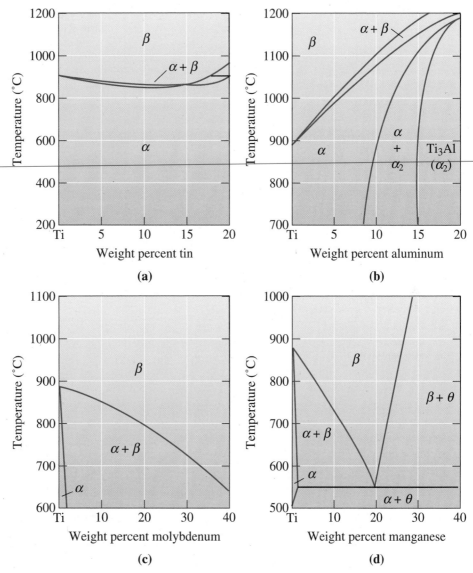

Figure 14-8 Portions of the phase diagrams for (a) titanium-tin, (b) titanium-aluminum, (c) titanium-molybdenum, and (d) titanium-manganese.

TABLE 14-9 ■ *Properties of selected titanium alloys*

Material	Tensile Strength (MPa)	Yield Strength (MPa)	% Elongation
Commercially pure Ti:			
99.5% Ti	241	172	24
99.0% Ti	552	483	15
Alpha Ti alloys:			
5% Al-2.5% Sn	862	779	15
Beta Ti alloys:			
13% V-11% Cr-3% Al	1289	1214	5
Alpha-beta Ti alloys:			
6% Al-4% V	1034	965	8

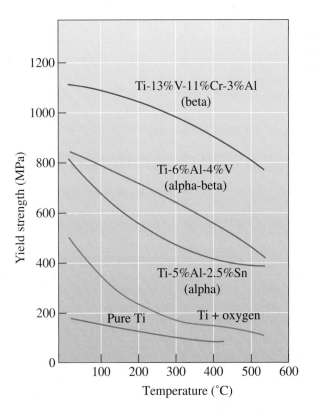

Figure 14-9
The effect of temperature on the
yield strength of selected titanium
alloys.

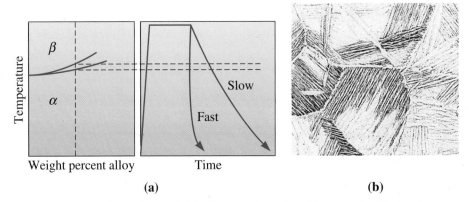

Figure 14-10 (a) Annealing and (b) microstructure of rapidly cooled α titanium
(× 100). Both the grain boundary precipitate and the Widmanstätten plates are α.
(From ASM Handbook, Vol. 7, (1972), ASM International, Materials Park, OH 44073.)

produces a metastable structure composed of all β. Strengthening is obtained both from the large amount of solid-solution strengthening alloying elements and by aging the metastable β structure to permit α to precipitate. Applications include high-strength fasteners, beams, and other fittings for aerospace applications.

Alpha-Beta Titanium Alloys

With proper balancing of the α and β stabilizers, a mixture of α and β is produced at room temperature. Ti-6% Al-4% V, an example of this approach, is by far the most common of all the titanium alloys. Because

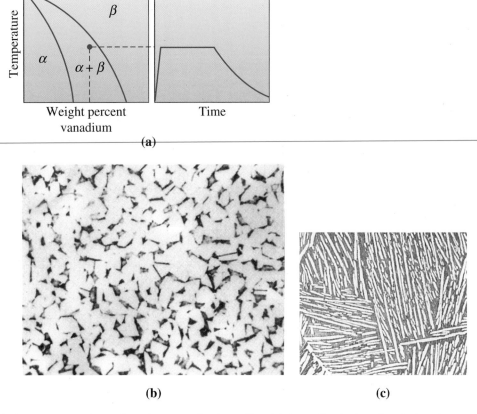

Figure 14-11   Annealing of an α–β titanium alloy. (a) Annealing is done just below the α–β transformation temperature, (b) slow cooling gives equiaxed α grains ($\times$ 250), and (c) rapid cooling yields acicular α grains ($\times$ 2500). *(From ASM Handbook, Vol. 7, (1972), ASM International, Materials Park, OH 44073–0002.)*

the alloys contain two phases, heat treatments can be used to control the microstructure and properties.

Annealing provides a combination of high ductility, uniform properties, and good strength. The alloy is heated just below the β-transus temperature, permitting a small amount of α to remain and prevent grain growth (Figure 14-11). Slow cooling causes equiaxed α grains to form; the equiaxed structure provides good ductility and formability while making it difficult for fatigue cracks to nucleate. Faster cooling, particularly from above the α-β transus temperature, produces an acicular—or "basketweave"—α phase [Figure 14-11(c)]. Although fatigue cracks may nucleate more easily in this structure, cracks must follow a tortuous path along the boundaries between α and β. This condition results in a low-fatigue crack growth rate, good fracture toughness, and good resistance to creep.

Two possible microstructures can be produced when the β phase is quenched from a high temperature. The phase diagram in Figure 14-12 includes a dashed martensite start line, which provides the basis for a quench and temper treatment. The β transforms to titanium martensite (α') in an alloy that crosses the M_s line on cooling. The titanium martensite is a relatively soft supersaturated phase. When α' is reheated, tempering occurs by the precipitation of β from the supersaturated α':

$$\alpha' \rightarrow \alpha + \beta \text{ precipitates}$$

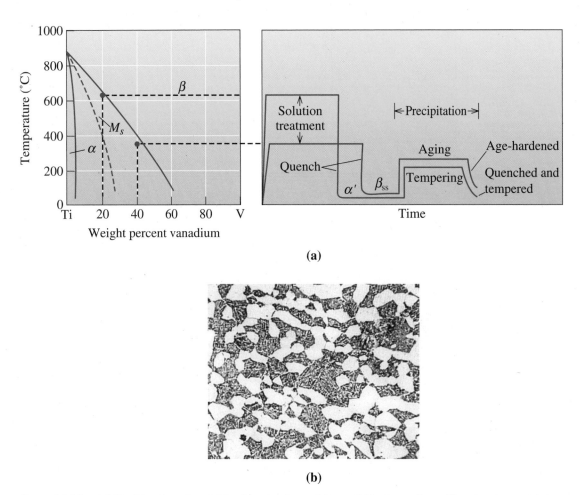

(a)

(b)

Figure 14-12 (a) Heat treatment and (b) microstructure of the α-β titanium alloys. The structure contains primary α (large white grains) and a dark β matrix with needles of α formed during aging ($\times$ 250). (*From ASM Handbook, Vol. 7, (1972), ASM International, Materials Park, OH 44073.*)

Fine β precipitates initially increase the strength compared with the α', opposite to what is found when a steel martensite is tempered; however, softening occurs when tempering is done at too high a temperature.

More highly alloyed α-β compositions are age hardened. When the β in these alloys is quenched, β_{ss}, which is supersaturated in titanium, remains. When β_{ss} is aged, α precipitates in a Widmanstätten structure, (Figure 14-12):

$$\beta_{ss} \rightarrow \beta + \alpha \text{ precipitates}$$

The formation of this structure leads to improved strength and fracture toughness. Components for airframes, rockets, jet engines, and landing gear are typical applications for the heat-treated α-β alloys. Some alloys, including the Ti-6% Al-4% V alloy, are superplastic and can be deformed as much as 1000%. This alloy is also used for making implants for human bodies. Titanium alloys are considered **biocompatible** (i.e., they are not rejected by the body). By developing porous coatings of bone-like ceramic compositions known as hydroxyapatite, it is possible to make titanium implants **bioactive** (i.e., the natural bone can grow into the hydroxyapatite coating). The following three examples illustrate applications of titanium alloys.

Example 14-7   *Design of a Heat Exchanger*

Design a 150-cm-diameter, 900-cm-long heat exchanger for the petrochemical industry (Figure 14-13).

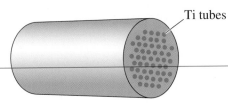

Ti tubes

Figure 14-13
Sketch of a heat exchanger using titanium tubes (for Example 14-7).

SOLUTION

The heat exchanger must meet several design criteria. It must have good corrosion resistance to handle aggressive products of the chemical refinery; it must operate at relatively high temperatures; it must be easily formed into the sheet and tubes from which the heat exchanger will be fabricated; and it must have good weldability for joining the tubes to the body of the heat exchanger.

Provided that the maximum operating temperature is below 535°C so that the oxide film is stable, titanium might be a good choice to provide corrosion resistance at elevated temperatures. A commercially pure titanium provides the best corrosion resistance.

Pure titanium also provides superior forming and welding characteristics and would, therefore, be our most logical selection. If pure titanium does not provide sufficient strength, an alternative is an α titanium alloy, still providing good corrosion resistance, forming characteristics, and weldability but also somewhat improved strength.

Example 14-8   *Design of a Connecting Rod*

Design a high-performance connecting rod for the engine of a racing automobile (Figure 14-14).

Figure 14-14
Sketch of connecting rod (for Example 14-8).

SOLUTION

A high-performance racing engine requires materials that can operate at high temperatures and stresses while minimizing the weight of the engine. In normal automobiles, the connecting rods are often a forged steel or a malleable cast iron. We might be able to save considerable weight by replacing these parts with titanium.

To achieve high strengths, we might consider an α-β titanium alloy. Because of its availability, the Ti-6% A- 4% V alloy is a good choice. The alloy is heated to about 1065°C, which is in the all-β portion of the phase diagram. On quenching, a titanium martensite forms; subsequent tempering produces a microstructure containing β precipitates in an α matrix.

When the heat treatment is performed in the all-β region, the tempered martensite has an acicular structure, which reduces the rate of growth of any fatigue cracks that might develop.

Example 14-9   *Materials for Hip Prosthesis*

What type of a material would you choose for an implant to be used for a total hip replacement implant?

SOLUTION

A hip prosthesis is intended to replace part of the worn out or damaged femur bone. (See the chapter opener image.) The implant has a metal head and fits down the cavity of the femur. We need to consider the following factors: biocompatibility, corrosion resistance, high fracture toughness, excellent fatigue life (so that implants last for many years since it is difficult to do the surgery as patients get older), and wear resistance. We also need to consider the stiffness. If the alloy chosen is too stiff compared to the bone, most of the load will be carried by the implant. This leads to weakening of the remaining bone and, in turn, can make the implant loose. Thus, we need a material that has a high tensile strength, corrosion resistance, biocompatibility, and fracture toughness. These requirements suggest 316 stainless steel or Ti-6% Al-4% V. Neither of these materials are ferromagnetic, and both are opaque to x-rays. This is good for magnetic resonance and x-ray imaging. Titanium alloys are not very hard and can wear out. Stainless steels are harder, but they are much stiffer than bone. Titanium is biocompatible and would be a better choice. Perhaps a composite material in which the stem is made from a Ti-6% Al-4% V alloy and the head is made from a wear resistant, corrosion resistant, and relatively tough ceramic, such as alumina, may be an answer. The inside of the socket could be made from an ultra-high density (ultra-high molecular weight) polyethylene that has a very low friction coefficient. The surface of the implant could be made porous so as to encourage the bone to grow. Another option is to coat the implant with a material like porous hydroxyapatite to encourage bone growth.

14-6   Refractory and Precious Metals

The **refractory metals**, which include tungsten, molybdenum, tantalum, and niobium (or columbium), have exceptionally high melting temperatures (above 1925°C) and, consequently, have the potential for high temperature service. Applications include filaments

TABLE 14-10 ■ *Properties of some refractory metals*

| Metal | Melting Temperature (°C) | Density (g/cm³) | T = 1000°C | | Transition Temperature (°C) |
			Tensile Strength (MPa)	Yield Strength (MPa)	
Nb	2468	8.57	117	55	−140
Mo	2610	10.22	345	207	30
Ta	2996	16.6	186	166	−270
W	3410	19.25	455	103	300

for lightbulbs, rocket nozzles, nuclear power generators, tantalum- and niobium-based electronic capacitors, and chemical processing equipment. The metals, however, have a high density, limiting their specific strengths (Table 14-10).

Oxidation

The refractory metals begin to oxidize between 200°C and 425°C and are rapidly contaminated or embrittled. Consequently, special precautions are required during casting, hot working, welding, or powder metallurgy. The metals must also be protected during service at elevated temperatures. For example, the tungsten filament in a lightbulb is protected by a vacuum.

For some applications, the metals may be coated with a silicide or aluminide coating. The coating must (a) have a high melting temperature, (b) be compatible with the refractory metal, (c) provide a diffusion barrier to prevent contaminants from reaching the underlying metal, and (d) have a coefficient of thermal expansion similar to that of the refractory metal. Coatings are available that protect the metal to about 1650°C. In some applications, such as capacitors for cellular phones, the formation of oxides is useful since we want to make use of the oxide as a nonconducting material.

Forming Characteristics

The refractory metals, which have the BCC crystal structure, display a ductile-to-brittle transition temperature. Because the transition temperatures for niobium and tantalum are below room temperature, these two metals can be readily formed. Annealed molybdenum and tungsten, however, normally have a transition temperature above room temperature, causing them to be brittle. Fortunately, if these metals are hot worked to produce a fibrous microstructure, the transition temperature is lowered and the forming characteristics are improved.

Alloys

Large increases in both room-temperature and high-temperature mechanical properties are obtained by alloying. Tungsten alloyed with hafnium, rhenium, and carbon can operate up to 2100°C. These alloys typically are solid-solution strengthened; in fact, tungsten and molybdenum form a complete series of solid solutions, much like copper and nickel. Some alloys, such as W-2% ThO_2, are dispersion strengthened by oxide particles during their manufacture by powder metallurgy processes. Composite materials, such as niobium reinforced with tungsten fibers, may also improve high-temperature properties.

Precious Metals

These include gold, silver, palladium, platinum, and rhodium. As their name suggests, these are precious and expensive. From an engineering viewpoint, these materials resist corrosion and make very good conductors of electricity.

As a result, alloys of these materials are often used as electrodes for devices. These electrodes are formed using thin-film deposition (e.g., sputtering or electroplating) or screen printing of metal powder dispersions/pastes. Nanoparticles of Pt/Rh/Pd (loaded onto a ceramic support) are also used as catalysts in automobiles. These metals facilitate the oxidation of CO to CO_2 and NO_x to N_2 and O_2. They are also used as catalysts in petroleum refining.

Summary

- The "light metals" include low-density alloys based on aluminum, magnesium, and beryllium. Aluminum alloys have a high specific strength due to their low density and, as a result, find many aerospace applications. Excellent corrosion resistance and electrical conductivity of aluminum also provide for a vast number of applications. Aluminum and magnesium are limited to use at low temperatures because of the loss of their mechanical properties as a result of overaging or recrystallization. Copper alloys (brasses and bronzes) are also used in many structural and other applications. Titanium alloys have intermediate densities and temperature resistance, along with excellent corrosion resistance, leading to applications in aerospace, chemical processing, and biomedical devices.

- Nickel and cobalt alloys, including superalloys, provide good properties at even higher temperatures. Combined with their good corrosion resistance, these alloys find many applications in aircraft engines and chemical processing equipment.

Glossary

Bioactive A material that is not rejected by the human body and eventually becomes part of the body (e.g., hydroxyapatite).

Biocompatible A material that is not rejected by the human body.

Blister copper An impure form of copper obtained during the copper refining process.

Brass A group of copper-based alloys, normally containing zinc as the major alloying element.

Bronze Generally, copper alloys containing tin, but can contain other elements.

Castability The ease with which a metal can be poured into a mold to make a casting without producing defects or requiring unusual or expensive techniques to prevent casting problems.

Fluidity The ability of liquid metal to fill a mold cavity without prematurely freezing.

Monel The copper-nickel alloy, containing approximately 60% Ni, that gives the maximum strength in the binary alloy system.

Nonferrous alloy An alloy based on some metal other than iron.

Refractory metals Metals having a melting temperature above 1925°C.

Specific strength The ratio of strength to density. Also called the strength-to-weight ratio.

Superalloys A group of nickel, iron-nickel, and cobalt-based alloys that have exceptional heat resistance, creep resistance, and corrosion resistance.

Temper designation A shorthand notation using letters and numbers to describe the processing of an alloy. H tempers refer to cold-worked alloys; T tempers refer to age-hardening treatments.

Wrought alloys Alloys that are shaped by a deformation process.

Problems

14-1 In some cases, we may be more interested in cost per unit volume than in cost per unit weight. Rework Table 14-1 to show the cost in terms of $/cm^3. Does this change/alter the relationship between the different materials?

14-2 Determine the specific strength of the following metals and alloys (use the densities of the major metal component as an approximation of the alloy density where required):

Alloy / Metal	Tensile Strength (MPa)
1100-H18	165
5182-O	290
2024-T4	469
2090-T6	552
201-T6	483
ZK40A-T5	276
Age hardened Cu-2% Be	1310
Alpha Ti alloy	862
W	455
Ta	186

Section 14-1 Aluminum Alloys

14-3 Structural steels have traditionally been used in shipbuilding; however, with increasing fuel costs, it is desirable to find alternative lower weight materials. Some of the key properties required for ship-building materials are high yield strength, high corrosion resistance, and low cost. Discuss the benefits and disadvantages of aluminum alloys as a replacement for structural steels in ships.

14-4 Assuming that the density remains unchanged, compare the specific strength of the 2090-T6 aluminum alloy to that of a die cast 443-F aluminum alloy. If you considered the actual density, do you think the difference between the specific strengths would increase or become smaller? Explain.

14-5 Explain why aluminum alloys containing more than about 15% Mg are not used.

14-6 Would you expect a 2024-T9 aluminum alloy to be stronger or weaker than a 2024-T6 alloy? Explain.

14-7 Estimate the tensile strength expected for the following aluminum alloys:
(a) 1100-H14
(b) 5182-H12
(c) 3004-H16.

14-8 A 1-cm-diameter steel cable with a yield strength of 480 MPa needs to be replaced to reduce the overall weight of the cable. Which of the following aluminum alloys could be a potential replacement?
(a) 3004-H18 ($S_y = 248$ MPa)
(b) 1100-H18 ($S_y = 151$ MPa)
(c) 4043-H18 ($S_y = 269$ MPa)
(d) 5182-O ($S_y = 131$ MPa).

The density of the steel used in the cable was 7.87 g/cm^3, and assume that the density of all the aluminum alloys listed above is 2.7 g/cm^3.

14-9 Suppose, by rapid solidification from the liquid state, that a supersaturated Al-7% Li alloy can be produced and subsequently aged. Compare the amount of β that will form in this alloy with that formed in a 2090 alloy.

14-10 Determine the amount of $Mg_2Al_3(\beta)$ expected to form in a 5182-O aluminum alloy (See Figure 14-2).

14-11 A 5182-O aluminum alloy part that had been exposed to salt water showed severe corrosion along the grain boundaries. Explain this observation based on the expected phases at room temperature in this alloy. (See Figure 14-2.)

Section 14-2 Magnesium and Beryllium Alloys

14-12 From the data in Table 14-6, estimate the ratio by which the yield strength of magnesium can be increased by alloying and

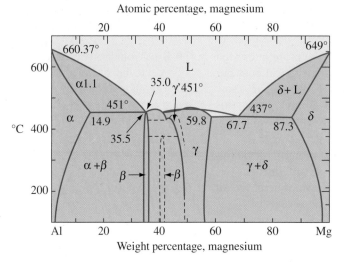

Figure 14-2 (Repeated for Problems 14-10 and 14-11.) Portion of the aluminum-magnesium phase diagram.

heat treatment and compare with that of aluminum alloys.

14-13 Suppose a 60-cm-long round bar is to support a load of 181 kg without any permanent deformation. Calculate the minimum diameter of the bar if it is made from

(a) AZ80A-T5 magnesium alloy and
(b) 6061-T6 aluminum alloy.
Calculate the weight of the bar and the approximate cost (based on pure Mg and Al) in each case.

14-14 A 10-m rod 0.5 cm in diameter must elongate no more than 2 mm under load. Determine the maximum force that can be applied if the rod is made from

(a) aluminum,
(b) magnesium, and
(c) beryllium.

Section 14-3 Copper Alloys

14-15 (a) Explain how pure copper is made. (b) What are some of the important properties of copper? (c) What is brass? (d) What is bronze? (e) Why is the Statue of Liberty green?

14-16 We say that copper can contain up to 40% Zn or 9% Al and still be single phase. How do we explain this statement in view of the phase diagram in Figure 14-5(a)?

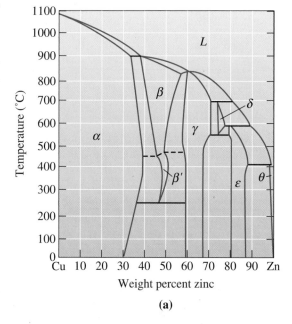

Figure 14-5 (a) (Repeated for Problem 14-16.) Binary phase diagram for the copper-zinc system.

14-17 Compare the percent increase in the yield strength of commercially pure annealed aluminum, magnesium, and copper by strain hardening. Explain the differences observed.

14-18 We would like to produce a quenched and tempered aluminum bronze containing

13% Al. Recommend a heat treatment, including appropriate temperatures. Calculate the amount of each phase after each step of the treatment.

14-19 A number of casting alloys have very high lead contents; however, the Pb content in wrought alloys is comparatively low. Why isn't more lead added to the wrought alloys? What precautions must be taken when a leaded wrought alloy is hot worked or heat treated?

14-20 Would you expect the fracture toughness of quenched and tempered aluminum bronze to be high or low? Would there be a difference in the resistance of the alloy to crack nucleation compared with crack growth? Explain.

Section 14-4 Nickel and Cobalt Alloys

14-21 Based on the photomicrograph in Figure 14-7(a), would you expect the γ' precipitate or the carbides to provide a greater strengthening effect in superalloys at low temperatures? Explain.

14-22 The density of Ni_3Al is 7.5 g/cm^3. Suppose a Ni-5 wt% Al alloy is heat treated so that all of the aluminum reacts with nickel to produce Ni_3Al. Determine the volume percentage of the Ni_3Al precipitate in the nickel matrix.

Section 14-5 Titanium Alloys

14-23 When steel is joined using arc welding, only the liquid-fusion zone must be

Figure 14-7 (a) (Repeated for Problem 14-21.) Microstructure of a superalloy with carbides at the grain boundaries and γ' precipitates in the matrix ($\times$ 15,000). (*From* ASM Handbook, *Vol. 9,* Metallography and Microstructure (1985), ASM International, Materials Park, OH 44073–0002.)

protected by a gas or flux. When titanium is welded, both the front and back sides of the welded metal must be protected. Why must these extra precautions be taken when joining titanium?

14-24 Both a Ti-15% V alloy and a Ti-35% V alloy are heated to the lowest temperature at which all β just forms. They are then quenched and reheated to 300°C. Describe the changes in microstructure during the heat treatment for each alloy, including the amount of each phase. What is the matrix and what is the precipitate in each case? Which is an age-hardening process? Which is a quench and temper process? [See Figure 14-12(a).]

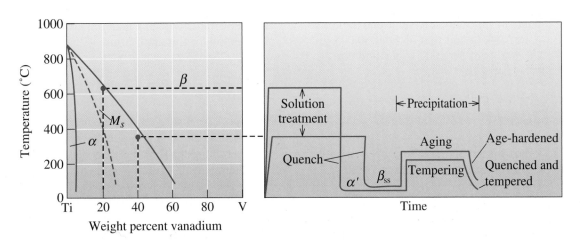

Figure 14-12 (Repeated for Problem 14-24) Heat treatment of the α-β titanium alloys.

14-25 Determine the specific strength of the strongest Al, Mg, Cu, Ti, and Ni alloys. Use the densities of the pure metals in your calculations. Try to explain their order.

14-26 Based on the phase diagrams, estimate the solubilities of Ni, Zn, Al, Sn, and Be in copper at room temperature. Are these solubilities expected in view of Hume-Rothery's conditions for solid solubility? Explain.

Section 14-6 Refractory and Precious Metals

14-27 What is a refractory metal or an alloy? What is a precious metal?

14-28 The temperature of a coated tungsten part is increased. What happens when the protective coating on a tungsten part expands more than the tungsten? What happens when the protective coating on a tungsten part expands less than the tungsten?

14-29 For what applications are Pt, Rh, Pd, and Ag used?

14-30 Nanoparticles (3 nm in diameter) of platinum (Pt) with a total weight of 1 milligram are used in automobile catalytic converters to facilitate oxidation reactions. As Pt is an expensive metal, a method to coat iron (Fe) nanoparticles with Pt is being explored to reduce the cost of the catalyst metal. If 3 nm diameter Fe particles are being coated with 1 nm of Pt to achieve the same effective surface area with fewer particles and less Pt, calculate the difference in cost and weight of the catalyst. Assume that the cost of Pt is \$40/g and that of Fe is \$0.005/g. The density of Fe is 7.87 g/cm^3 and that of Pt is 21.45 g/cm^3.

◆ Design Problems

14-31 A part for an engine mount for a private aircraft must occupy a volume of 60 cm^3 with a minimum thickness of 0.5 cm and a minimum width of 4 cm. The load on the part during service may be as much as 75,000 N. The part is expected to remain below 100°C during service. Design a material and its treatment that will perform satisfactorily in this application.

14-32 You wish to design the rung on a ladder. The ladder should be light in weight so that it can be easily transported and used. The rungs on the ladder should be 0.625 cm × 2.5 cm and are 30 cm long. Design a material and its processing for the rungs.

14-33 We have determined that we need an alloy having a density of 2.3 ± 0.05 g/cm^3 that must be strong, yet still have some ductility. Design a material and its processing that might meet these requirements.

14-34 We wish to design a mounting device that will position and aim a laser for precision cutting of a composite material. What design requirements might be important? Design a material and its processing that might meet these requirements.

14-35 Design a nickel-titanium alloy that will produce 60 volume percent Ni_3Ti precipitate in a pure nickel matrix.

14-36 An actuating lever in an electrical device must open and close almost instantly and carry a high current when closed. What design requirements would be important for this application? Design a material and its processing to meet these requirements.

14-37 A fan blade in a chemical plant must operate at temperatures as high as 400°C under rather corrosive conditions. Occasionally, solid material is ingested and impacts the fan. What design requirements would be important? Design a material and its processing for this application.

▲ Computer Problems

14-38 *Database Identification System for Alloys.* Write a computer program that will ask the user to input a three- or four-digit code for aluminum alloys (Table 14-3). You will have to ask the user to provide one digit at a time or figure out a way of comparing different digits in a string. This will be followed by a letter and a number (e.g., T and 4).

The program should then provide the user with some more detailed information about the type of alloy. For example, if the user enters 2024 and then T4, the program should provide an output that will specify that (**a**) the alloy is wrought type, (**b**) Cu is the major alloying element (since the first digit is 2), and (**c**) it is naturally aged. Do not make the program too complex. The main idea here is for you to see how databases for alloys are designed.

Ⓚ Knovel® **Problems**

K14-1 Not all aluminum alloys are easily weldable. Provide a few designations of readily weldable aluminum alloys. List some welding techniques recommended for aluminum alloys.

K14-2 List several wrought aluminum alloys with tensile strengths between 70 and 100 MPa.

Ceramics play a critical role in a wide array of electronic, magnetic, optical, and energy related technologies. Many advanced ceramics play a very important role in providing thermal insulation and high-temperature properties. Applications of advanced ceramics range from credit cards, houses for silicon chips, tiles for the space shuttle, medical imaging, optical fibers that enable communication, and safe and energy efficient glasses. Traditional ceramics serve as refractories for metals processing and consumer applications. (*Image courtesy of NASA.*)

Ceramic Materials

Have You Ever Wondered?

- *What is the magnetic strip on a credit card made from?*
- *What material is used to protect the space shuttle from high temperatures during re-entry?*
- *What ceramic material is commonly added to paints?*
- *What ceramic material is found in bone and teeth?*
- *What are spark plugs made from?*
- *Is there a ceramic in processed milk?*
- *What ceramic gemstone is the hardest naturally occurring material?*

The goal of this chapter is to closely examine the synthesis, processing, and applications of **ceramic** materials. Ceramics have been used for many thousands of years. Most ceramics exhibit good strength under compression; however, they exhibit virtually no ductility under tension. The family of ceramic materials includes polycrystalline and single-crystal inorganic materials, inorganic glasses, and glass-ceramics. We have discussed many of these materials in previous chapters.

In Chapters 2 and 3, we learned about the bonding in ceramic materials, the crystal structures of technologically useful ceramics, and the arrangements of ions in glasses. We begin with a discussion that summarizes the classification and applications of ceramics.

15-1   Applications of Ceramics

There are many different ways to classify ceramics. One way is to define ceramics based on their class of chemical compounds (e.g., oxides, carbides, nitrides, sulfides, fluorides, etc.). Another way, which we will use here, is to classify ceramics by their major function (Table 15-1).

Ceramics are used in a wide range of technologies such as refractories, spark plugs, dielectrics in capacitors, sensors, abrasives, and magnetic recording media. The space shuttle makes use of ~25,000 reusable, lightweight, highly porous ceramic tiles that protect the aluminum frame from the heat generated during re-entry into the earth's atmosphere. These tiles are made from high-purity silica fibers and colloidal silica coated with a borosilicate glass. Ceramics also appear in nature as oxides and in natural materials; the human body has the amazing ability to make hydroxyapatite, a ceramic found in bones and teeth. Ceramics are also used as coatings. **Glazes** are ceramic coatings applied to glass objects; **enamels** are ceramic coatings applied to metallic objects. Let's follow the classification shown in Table 15-1 and take note of different applications. Alumina and silica are the most widely used ceramic materials and, as you will notice, there are numerous applications listed in Table 15-1 that depend upon the use of these two ceramics.

The following is a brief summary of applications of some of the more widely used ceramic materials:

- *Alumina* (Al_2O_3) is used to contain molten metal or in applications where a material must operate at high temperatures with high strength. Alumina is also used as a low dielectric constant substrate for electronic packaging that houses silicon chips. One classic application is insulators in spark plugs. Some unique applications are being found in dental and medical use. Chromium-doped alumina is used for making lasers. Fine particles of alumina are also used as catalyst supports.

- *Diamond* (C) is the hardest naturally occurring material. Industrial diamonds are used as abrasives for grinding and polishing. Diamond and diamond-like coatings prepared using chemical vapor deposition processes are used to make abrasion-resistant coatings for many different applications (e.g., cutting tools). It is, of course, also used in jewelry.

- *Silica* (SiO_2) is probably the most widely used ceramic material. Silica is an essential ingredient in glasses and many glass-ceramics. Silica-based materials are used in thermal insulation, refractories, abrasives, as fiber-reinforced composites, and laboratory glassware. In the form of long continuous fibers, silica is used to make optical fibers for communications. Powders made using fine particles of silica are used in tires, paints, and many other applications.

- *Silicon carbide* (SiC) provides outstanding oxidation resistance at temperatures even above the melting point of steel. SiC often is used as a coating for metals, carbon-carbon composites, and other ceramics to provide protection at these extreme temperatures. SiC is also used as an abrasive in grinding wheels and as particulate and fibrous reinforcement in both metal matrix and ceramic matrix composites. It is also used to make heating elements for furnaces. SiC is a semiconductor and is a very good candidate for high-temperature electronics.

TABLE 15-1 ■ *Functional classification of ceramics**

Function	Application	Examples of Ceramics
Electrical	Capacitor dielectrics	$BaTiO_3$, $SrTiO_3$, Ta_2O_5
	Microwave dielectrics	$Ba(Mg_{1/3}Ta_{2/3})O_3$, $Ba(Zn_{1/3}Ta_{2/3})O_3$
		$BaTi_4O_9$, $Ba_2Ti_9O_{20}$, $Zr_xSn_{1-x}TiO_4$, Al_2O_3
	Conductive oxides	In-doped SnO_2 (*ITO*)
	Superconductors	$YBa_2Cu_3O_{7-x}$ (*YBCO*)
	Electronic packaging	Al_2O_3
	Insulators	Porcelain
	Solid-oxide fuel cells	ZrO_2, $LaCrO_3$
	Piezoelectric	$Pb(Zr_xTi_{1-x})O_3$ (*PZT*), $Pb(Mg_{1/3}Nb_{2/3})O_3$
	Electro-optical	*PLZT*, $LiNbO_3$
Magnetic	Recording media	γ-Fe_2O_3, CrO_2 ("chrome" cassettes)
	Ferrofluids, credit cards	Fe_3O_4
	Circulators, isolators	Nickel zinc ferrite
	Inductors, magnets	Manganese zinc ferrite
Optical	Fiber optics	Doped SiO_2
	Glasses	SiO_2 based
	Lasers	Al_2O_3, yttrium aluminum garnate (*YAG*)
	Lighting	Al_2O_3, glasses
Automotive	Oxygen sensors, fuel cells	ZrO_2
	Catalyst support	Cordierite
	Spark plugs	Al_2O_3
	Tires	SiO_2
	Windshields/windows	SiO_2 based glasses
Mechanical/Structural	Cutting tools	WC-Co cermets
		Silicon-aluminum-oxynitride (*Sialon*)
		Al_2O_3
	Composites	SiC, Al_2O_3, silica glass fibers
	Abrasives	SiC, Al_2O_3, diamond, BN, $ZrSiO_4$
Biomedical	Implants	Hydroxyapatite
	Dentistry	Porcelain, Al_2O_3
	Ultrasound imaging	*PZT*
Construction	Buildings	Concrete
		Glass
		Sanitaryware
Others	Defense applications	*PZT*, B_4C
	Armor materials	
	Sensors	SnO_2
	Nuclear	UO_2
		Glasses for waste disposal
	Metals processing	Alumina and silica-based refractories, oxygen sensors, casting molds, etc.
Chemical	Catalysis	Various oxides (Al_2O_3, ZrO_2, ZnO, TiO_2)
	Air, liquid filtration	
	Sensors	
	Paints, rubber	
Domestic	Tiles, sanitaryware, whiteware, kitchenware, pottery, art, jewelry	Clay, alumina, and silica-based ceramics, glass-ceramics, diamond, ruby, cubic zirconia, and other crystals

*Acronyms are indicated in italics.

- *Silicon nitride* (Si_3N_4) has properties similar to those of SiC, although its oxidation resistance and high temperature strength are somewhat lower. Both silicon nitride and silicon carbide are likely candidates for components for automotive and gas turbine engines, permitting higher operating temperatures and better fuel efficiencies with less weight than traditional metals and alloys.

- *Titanium dioxide* (TiO_2) is used to make electronic ceramics such as $BaTiO_3$. The largest uses are as a white pigment to make paints and to whiten milk. Titania is used in certain glass-ceramics as a nucleating agent. Fine particles of TiO_2 are used to make suntan lotions that provide protection against ultraviolet rays.

- *Zirconia* (ZrO_2) is used to make many other ceramics such as zircon. Zirconia is also used to make oxygen gas sensors that are used in automotives and to measure dissolved oxygen in molten steels. Zirconia is used as an additive in many electronic ceramics as well as a refractory material. The cubic form of zirconia single crystals is used to make jewelry items.

15-2   Properties of Ceramics

The properties of some ceramics are summarized in Table 15-2. Mechanical properties of some structural ceramics are summarized in Table 15-3.

Take note of the high melting temperatures and high compressive strengths of ceramics. As mentioned in Chapter 6, the weight of an entire firetruck can be supported on four ceramic coffee cups. We should also remember that the tensile and flexural strength values show considerable variation since the strength of ceramics is dependent on the distribution of flaw sizes and is not affected by dislocation motion. We discussed the Weibull distribution and the strength of ceramics and glasses in Chapter 7. Also note that, contrary to common belief, ceramics are not always brittle. At low strain rates and at high temperatures, many ceramics with a very fine grain size indeed show superplastic behavior.

TABLE 15-2 ■ *Properties of commonly encountered polycrystalline ceramics*

Material	Melting Point (°C)	Thermal Expansion Coefficient ($\times 10^{-6}$ cm/cm)/°C	Knoop Hardness (HK) (100 g)
Al_2O_3	2000	~6.8	2100
BN	2732	0.57[a], −0.46[b]	5000
SiC	2700	~3.7	2500
Diamond		1.02	7000
Mullite	1810	4.5	—
TiO_2	1840	8.8	—
Cubic ZrO_2	2700	10.5	—

[a]Perpendicular to pressing direction.
[b]Parallel to pressing direction.

TABLE 15-3 ■ *Mechanical properties of selected advanced ceramics*

Material	Density (g/cm^3)	Tensile Strength (MPa)	Flexural Strength (MPa)	Compressive Strength (MPa)	Young's Modulus (MPa)	Fracture Toughness (MPa $\sqrt{m}$)
Al$_2$O$_3$	3.98	207	552	3025	386×10^3	5.5
SiC (sintered)	3.1	172	552	3860	414×10^3	4.4
Si$_3$N$_4$ (reaction bonded)	2.5	138	241	1030	207×10^3	3.3
Si$_3$N$_4$ (hot pressed)	3.2	552	896	3450	310×10^3	5.5
Sialon	3.24	414	965	3450	310×10^3	9.9
ZrO$_2$ (partially stabilized)	5.8	448	690	1860	207×10^3	11.0
ZrO$_2$ (transformation toughened)	5.8	345	793	1725	200×10^3	12.1

15-3 Synthesis and Processing of Ceramic Powders

Ceramic materials melt at high temperatures, and they exhibit brittle behavior under tension. As a result, the casting and thermomechanical processing used widely for metals, alloys, and thermoplastics cannot be applied when processing ceramics. Inorganic glasses, though, make use of lower melting temperatures due to the formation of eutectics in the float-glass process (to be discussed in Section 15-5). Since melting, casting, and thermomechanical processing are not viable options for polycrystalline ceramics, we typically process ceramics into useful shapes starting with ceramic powders. A **powder** is a collection of fine particles. The step of making a ceramic powder is defined here as the **synthesis** of ceramics. We begin with a ceramic powder and get it ready for shaping by crushing, grinding, separating impurities, blending different powders, drying, and **spray drying** to form soft agglomerates. Different techniques such as compaction, **tape casting,** extrusion, and **slip casting** are then used to convert properly processed powders into a desired shape to form what is known as a **green ceramic**. A green ceramic has not yet been sintered. The steps of converting a ceramic powder (or mixture of powders) into a useful shape are known as **powder processing**. The green ceramic is then consolidated further using a high-temperature treatment known as **sintering** or firing. In this process, the green ceramic is heated to a high temperature using a controlled heat treatment and atmosphere, so that a dense material is obtained. The ceramic may be then subjected to additional operations such as grinding, polishing, or machining as needed for the final application. In some cases, leads will be attached or coatings or electrodes will be deposited. These general steps encountered in the synthesis and processing of ceramics are summarized in Figure 15-1.

Ceramic powders prepared using conventional or chemical processes are shaped using the techniques in Figure 15-2. We emphasize that very similar processes are used for metal and alloy powders, a route known as **powder metallurgy**. Powders consist of particles that are loosely bonded, and powder processing involves the consolidation of these powders into a desired shape. Often, the ceramic powders need to be converted into soft agglomerates by spraying a slurry of the powder through a nozzle into a chamber (spray dryer) in the presence of hot air. This process leads to the formation of soft agglomerates that flow into the dies used for powder compaction; this is known as **spray drying**.

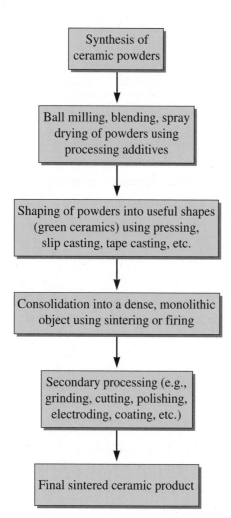

Figure 15-1
Typical steps encountered in the processing of ceramics.

Compaction and Sintering

One of the most cost-effective ways to produce thousands of relatively small, simple shapes (<15 cm) is compaction and sintering. Many electronic and magnetic ceramics, **cermet** (e.g., WC-Co) cutting tool bits, and other materials are processed using this technique. Fine powders can be spray dried, forming soft agglomerates that flow and compact well. The different steps of uniaxial compaction, in which the compacting force is applied in one direction, are shown in Figure 15-3(a). As an example, the microstructure of a barium magnesium tantalate (BMT) ceramic prepared using compaction and sintering is shown in Figure 15-3(b). Sintering involves different mass-transport mechanisms [Figure 15-3(c)]. The driving force for sintering is the reduction in the surface area of a powder (Chapter 5). With sintering, the grain boundary and bulk (volume) diffusion contribute to densification (increase in density). Surface diffusion and evaporation condensation can cause grain growth, but they do not cause densification.

The compaction process can be completed within one minute for smaller parts; thus, uniaxial compaction is well suited for making a large number of smaller and simple shapes. Compaction is used to create what we call "green ceramics"; these have respectable strengths and can be handled and machined. In some cases, very large pieces (up to 100 cm in diameter and 200 to 240 cm long) can be produced using a process called

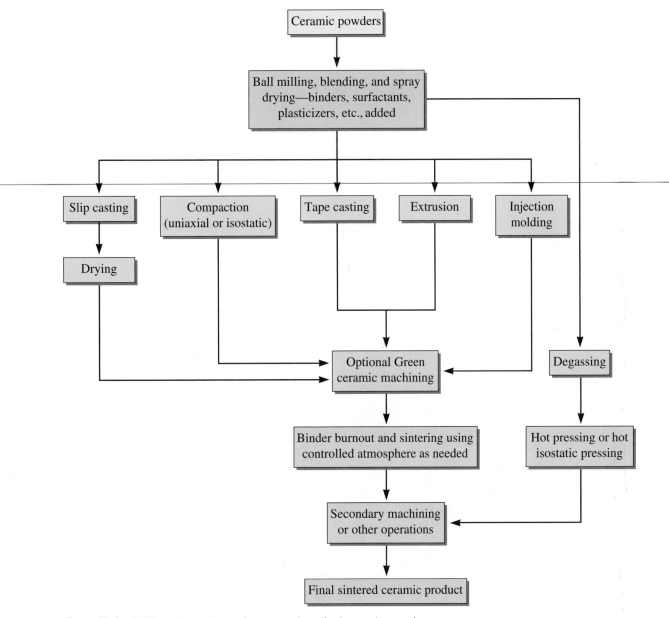

Figure 15-2   Different techniques for processing of advanced ceramics.

cold isostatic pressing (CIP), where pressure is applied using oil. Such large pieces then are sintered with or without pressure. Cold isostatic pressing is used for achieving a higher green ceramic density or where the compaction of more complex shapes is required.

In some cases, parts may be produced under conditions in which sintering is conducted using applied pressure. This technique, known as **hot pressing**, is used for refractory and covalently bonded ceramics that do not show good pressureless sintering behavior. Similarly, large pieces of metals and alloys compacted using CIP can be sintered under pressure in a process known as **hot isostatic pressing** (HIP) (Chapter 5). In hot pressing or HIP, the applied pressure acts against the internal pore pressure and enhances densification without causing grain growth. Hot pressing or hot isostatic pressing also are used for

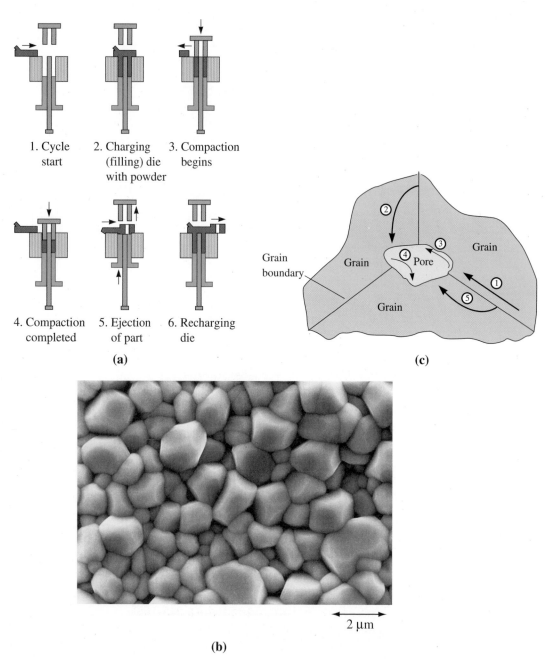

1. Cycle start
2. Charging (filling) die with powder
3. Compaction begins
4. Compaction completed
5. Ejection of part
6. Recharging die

(a)

(c)

2 μm

(b)

Figure 15-3 (a) Uniaxial powder compaction showing the die-punch assembly during different stages. Typically, for small parts these stages are completed in less than a minute. (*From* Materials and Processes in Manufacturing, *Eighth Edition, by E.P. DeGarmo, J.T. Black, and R.A. Koshe, Fig. 16-4. © 1997 John Wiley & Sons, Inc. Reproduced with permission of John Wiley & Sons, Inc.*) (b) Microstructure of a barium magnesium tantalate (BMT) ceramic prepared using compaction and sintering. (Courtesy of Heather Shivey.) (c) Different diffusion mechanisms involved in sintering. The grain boundary and bulk diffusion (1, 2 and 5) to the neck contribute to densification. Evaporation condensation (4) and surface diffusion (3) do not contribute to densification. (*From* Physical Ceramics: Principles for Ceramic Science and Engineering, *by Y.M. Chiang, D. Birnie, and W.D. Kingery, Fig. 5-40. © 1997 John Wiley & Sons, Inc. Reproduced with permission of John Wiley & Sons, Inc.*)

making ceramics or metallic parts with almost no porosity. Some recent innovative processes that make use of microwaves (similar to the way food gets heated in microwave ovens) have also been developed for the drying and sintering of ceramic materials.

Some ceramics, such as silicon nitride (Si_3N_4), are produced by **reaction bonding**. Silicon is formed into a desired shape and then reacted with nitrogen to form the nitride. Reaction bonding, which can be done at lower temperatures, provides better dimensional control compared with hot pressing; however, lower densities and degraded mechanical properties are obtained. As a comparison, the effect of processing on silicon nitride ceramics is shown in Table 15-4.

Tape Casting

A technique known as **tape casting** is used for the production of thin ceramic tapes (~3 to 100 μm). A slurry is cast with the help of a blade onto a plastic substrate. The green tape is then subjected to sintering. Many commercially important electronic packages based on alumina substrates and millions of barium titanate capacitors are made using this type of tape casting process.

Slip Casting

Slip casting typically uses an aqueous slurry of ceramic powder. The slurry, known as the **slip**, is poured into a plaster of Paris ($CaSO_4 : 2H_2O$) mold (Figure 15-4). As the water from the slurry begins to move out by capillary action, a thick mass builds along the mold wall. When sufficient product thickness is built, the rest of the slurry is poured out (this is called *drain casting*). It is also possible to continue to pour more slurry in to form a solid piece (this is called *solid casting*). Pressure may also be used to inject the slurry into polymer molds. The green ceramic is then dried and "fired" or sintered at a high temperature. Slip casting is widely used to make ceramic art (figurines and statues), sinks, and other ceramic sanitaryware such as toilets.

Extrusion and Injection Molding

Extrusion and injection molding are popular techniques used for making furnace tubes, bricks, tiles, and insulators. The extrusion process uses a viscous, dough-like mixture of ceramic particles containing a binder and other additives. This mixture has a clay-like consistency, which is then fed to an extruder where it is mixed well in a machine known as a pug mill, sheared, deaerated, and then injected into a die where a continuous shape of green ceramic is produced by the extruder. This material is cut at appropriate lengths and then dried and sintered. Cordierite ceramics used for making catalyst honeycomb structures are made using the extrusion process.

Injection molding of ceramics is similar to injection molding of polymers (Chapter 16). Ceramic powder is mixed with a thermoplastic plasticizer and other additives. The mixture is then extruded and injected into a die. Ceramic injection molding is better suited for complex shapes. The polymer contained in the injection-molded ceramic is burnt off, and the rest of the ceramic body is sintered at a high temperature.

TABLE 15-4 ■ *Properties of Si_3N_4 processed using different techniques*

Process	Compressive Strength (MPa)	Flexural Strength (MPa)
Slip casting	138	69
Reaction bonding	772	207
Hot pressing	345	862

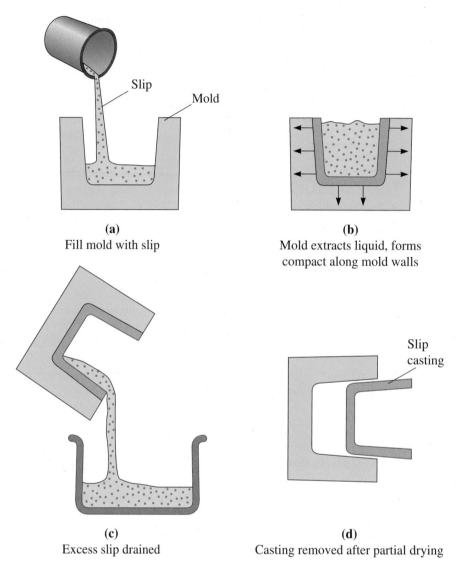

(a)
Fill mold with slip

(b)
Mold extracts liquid, forms
compact along mold walls

(c)
Excess slip drained

(d)
Casting removed after partial drying

Figure 15-4 Steps in slip casting of ceramics. (*From* Modern Ceramic Engineering, *by D.W. Richerson, p. 462, Fig. 10-34. Copyright © 1992 Marcel Dekker. Reprinted by permission.*)

15-4 Characteristics of Sintered Ceramics

For sintered ceramics, the average grain size, grain size distribution, and the level and type of porosity are important. Similarly, depending upon the application, second phases at grain boundaries and orientation effects (due to extrusion) also should be considered.

Grains and Grain Boundaries
The average grain size is often closely related to the primary particle size. An exception to this is if there is grain growth due to long sintering times or exaggerated or abnormal grain growth (Chapter 5). Typically, ceramics with a small grain size are stronger than coarse-grained ceramics. Finer grain

sizes help minimize stresses that develop at grain boundaries due to anisotropic expansion and contraction. Normally, starting with finer ceramic raw materials produces a fine grain size. Magnetic, dielectric, and optical properties of ceramic materials depend upon the average grain size and, in these applications, grain size must be controlled properly. Although we have not discussed this here in detail, in certain applications, it is important to use single crystals of ceramic materials so as to avoid the deleterious grain boundaries that are always present in polycrystalline ceramics.

Porosity   Pores represent the most important defect in polycrystalline ceramics. The presence of pores is usually detrimental to the mechanical properties of bulk ceramics, since pores provide a pre-existing location from which a crack can grow. The presence of pores is one of the reasons why ceramics show such brittle behavior under tensile loading. Since there is a distribution of pore sizes and the overall level of porosity changes, the mechanical properties of ceramics vary. This variability is measured using Weibull statistics (Chapter 7). The presence of pores, on the other hand, may be useful for increasing resistance to thermal shock. In certain applications, such as filters for hot metals or for liquids or gases, the presence of interconnected pores is desirable.

Pores in a ceramic may be either interconnected or closed. The **apparent porosity** measures the interconnected pores and determines the permeability, or the ease with which gases and fluids seep through the ceramic component. The apparent porosity is determined by weighing the dry ceramic (W_d), then reweighing the ceramic both when it is suspended in water (W_s) and after it is removed from the water (W_w). Using units of grams and cm^3:

$$\text{Apparent porosity} = \frac{W_w - W_d}{W_w - W_s} \times 100 \qquad (15\text{-}1)$$

The **true porosity** includes both interconnected and closed pores. The true porosity, which better correlates with the properties of the ceramic, is

$$\text{True porosity} = \frac{\rho - B}{\rho} \times 100 \qquad (15\text{-}2)$$

where

$$B = \frac{W_d}{W_w - W_s} \qquad (15\text{-}3)$$

B is the **bulk density**, and ρ is the true density or specific gravity of the ceramic. The bulk density is the mass of the ceramic divided by its volume. The following example illustrates how porosity levels in ceramics are determined.

Example 15-1 *Silicon Carbide Ceramics*

Silicon carbide particles are compacted and fired at a high temperature to produce a strong ceramic shape. The specific gravity of SiC is 3.2 g/cm^3. The ceramic shape subsequently is weighed when dry (360 g), after soaking in water (385 g), and while suspended in water (224 g). Calculate the apparent porosity, the true porosity, and the fraction of the pore volume that is closed.

SOLUTION

$$\text{Apparent porosity} = \frac{W_w - W_d}{W_w - W_s} \times 100 = \frac{385 - 360}{385 - 224} \times 100 = 15.5\%$$

$$\text{Bulk density} = B = \frac{W_d}{W_w - W_s} = \frac{360}{385 - 224} = 2.24\,\text{g/cm}^3$$

Note that the denominator is equal numerically to the volume of the ceramic, since the density of water is 1 g/cm^3.

$$\text{True porosity} = \frac{\rho - B}{\rho} \times 100 = \frac{3.2 - 2.24}{3.2} \times 100 = 30\%$$

The closed-pore percentage is the true porosity minus the apparent porosity, or 30 − 15.5 = 14.5%. Thus,

$$\text{Fractional closed pores} = \frac{14.5}{30} = 0.483$$

15-5   Inorganic Glasses

In Chapter 3, we discussed amorphous materials and the concept of short- versus long-range order in terms of atomic or ionic arrangements in noncrystalline materials. The most important of the noncrystalline materials are glasses, especially those based on silica. Of course, there are glasses based on other compounds (e.g., sulfides, fluorides, and various alloys). A **glass** is a metastable material that has hardened and become rigid without crystallizing. A glass in some ways resembles an undercooled liquid. Below the **glass transition temperature** T_g (Figure 15-5), the rate of volume contraction on cooling is reduced, and the material can be considered a "glass" rather than an "undercooled

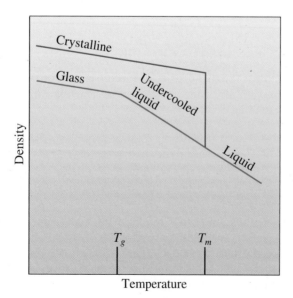

Figure 15-5
When silica crystallizes on cooling, an abrupt change in the density is observed. For glassy silica, however, the change in slope at the glass transition temperature indicates the formation of a glass from the undercooled liquid. Glass does not have a fixed T_m or T_g. Crystalline materials have a fixed T_m, and they do not have a T_g.

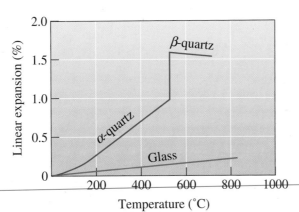

Figure 15-6
The expansion of quartz. In addition to the regular—almost linear—expansion, a large, abrupt expansion accompanies the α- to β-quartz transformation. Glasses expand uniformly.

TABLE 15-5 ■ *Division of the oxides into glass formers, intermediates, and modifiers*

Glass Formers	Intermediates	Modifiers
B_2O_3	TiO_2	Y_2O_3
SiO_2	ZnO	MgO
GeO_2	PbO_2	CaO
P_2O_5	Al_2O_3	PbO
V_2O_3	BeO	Na_2O

liquid." Joining silica tetrahedra or other ionic groups produces a solid, but noncrystalline, network structure (Chapter 3).

Silicate Glasses
The silicate glasses are the most widely used. *Fused silica*, formed from pure SiO_2, has a high melting point, and the dimensional changes during heating and cooling are small (Figure 15-6). Generally, however, the silicate glasses contain additional oxides (Table 15-5). While oxides such as silica behave as **glass formers,** an **intermediate** oxide (such as lead oxide or aluminum oxide) does not form a glass by itself but is incorporated into the network structure of the glass formers. A third group of oxides, the *modifiers,* break up the network structure and eventually cause the glass to devitrify, or crystallize.

Modified Silicate Glasses
Modifiers break up the silica network if the oxygen-to-silicon ratio (O:Si) increases significantly. When Na_2O is added, for example, the sodium ions enter holes within the network rather than becoming part of the network; however, the oxygen ion that enters with the Na_2O does become part of the network (Figure 15-7). When this happens, there are not enough silicon ions to combine with the extra oxygen ions and keep the network intact. Eventually, a high O:Si ratio causes the remaining silica tetrahedra to form chains, rings, or compounds, and the silica no longer transforms to a glass. When the O:Si ratio is above about 2.5, silica glasses are difficult to form; above a ratio of three, a glass forms only when special precautions are taken, such as the use of rapid cooling rates.

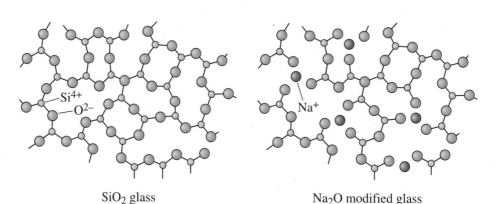

SiO$_2$ glass   Na$_2$O modified glass

Figure 15-7   The effect of Na$_2$O on the silica glass network. Sodium oxide is a modifier, disrupting the glassy network and reducing the ability to form a glass.

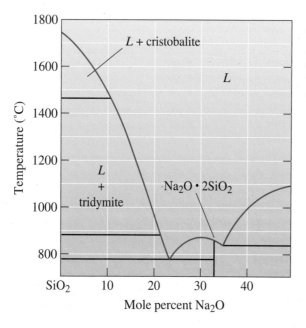

Figure 15-8
The SiO$_2$-Na$_2$O phase diagram. Additions of soda (Na$_2$O) to silica dramatically reduce the melting temperature of silica by forming eutectics.

Modification also lowers the melting point and viscosity of silica, making it possible to produce glass at lower temperatures. As you can see in Figure 15-8, the addition of Na$_2$O produces eutectics with very low melting temperatures. Adding CaO, which reduces the solubility of the glass in water, further modifies these glasses. The example that follows shows how to design a glass.

Example 15-2   *Design of a Glass*

We produce good chemical resistance in a glass when we introduce B$_2$O$_3$ into silica. To ensure that we have good glass-forming tendencies, we wish the O:Si ratio to be no more than 2.5, but we also want the glassware to have a low melting temperature to make the glass-forming process easier and more economical. Design such a glass.

SOLUTION

Because B_2O_3 reduces the melting temperature of silica, we would like to add as much as possible. We also, however, want to ensure that the O:Si ratio is no more than 2.5, so the amount of B_2O_3 is limited. As an example, let us determine the amount of B_2O_3 we must add to obtain an O:Si ratio of exactly 2.5. Let f_B be the mole fraction of B_2O_3 added to the glass, and $(1 - f_B)$ be the mole fraction of SiO_2:

$$\frac{O}{Si} = \frac{\left(3\,\dfrac{O\ ions}{B_2O_3}\right)(f_B) + \left(2\,\dfrac{O\ ions}{SiO_2}\right)(1 - f_B)}{\left(1\,\dfrac{Si\ ion}{SiO_2}\right)(1 - f_B)} = 2.5$$

$$3f_B + 2 - 2f_B = 2.5 - 2.5f_B \text{ or } f_B = 0.143$$

Therefore, we must produce a glass containing no more than 14.3 mol% B_2O_3. In weight percent:

$$\text{wt\% } B_2O_3 = \frac{(f_B)(69.62\,\text{g/mol})}{(f_B)(69.62\,\text{g/mol}) + (1 - f_B)(60.08\,\text{g/mol})} \times 100$$

$$\text{wt\% } B_2O_3 = \frac{(0.143)(69.62)}{(0.143)(69.62) + (0.857)(60.08)} \times 100 = 16.2$$

Glasses are manufactured into useful articles at a high temperature by controlling the viscosity so that the glass can be shaped without breaking. Figure 15-9 helps us understand the processing in terms of the viscosity ranges.

1. *Liquid range*. Sheet and plate glass are produced when the glass is in the molten state. Techniques include rolling the molten glass through water-cooled rolls or floating the molten glass over a pool of liquid tin (Figure 15-10). The liquid-tin process produces an exceptionally smooth surface on the glass. The development of the float-glass process was a genuine breakthrough in the area of glass processing. The basic float-glass composition has been essentially unchanged for many years (Table 15-6 on page 587).

Some glass shapes, including large optical mirrors, are produced by casting the molten glass into a mold, then ensuring that cooling is as slow as possible to minimize residual stresses and avoid cracking of the glass part. Glass fibers may be produced by drawing the liquid glass through small openings in a platinum die [Figure 15-11(c)]. Typically, many fibers are produced simultaneously for a single die.

2. *Working range*. Shapes such as those of containers or lightbulbs can be formed by pressing, drawing, or blowing glass into molds (Figure 15-11). A hot *gob* of liquid glass may be pre-formed into a crude shape (**parison**), then pressed or blown into a heated die to produce the final shape. The glass is heated to the working range so that the glass is formable, but not "runny."

3. *Annealing range*. Some ceramic parts may be annealed to reduce residual stresses introduced during forming. Large glass castings, for example, are often annealed and slowly cooled to prevent cracking. Some glasses may be heat treated to cause **devitrification**, or the precipitation of a crystalline phase from the glass.

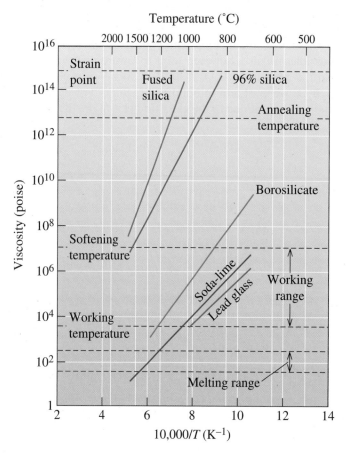

Figure 15-9
The effect of temperature and composition on the viscosity of glass.

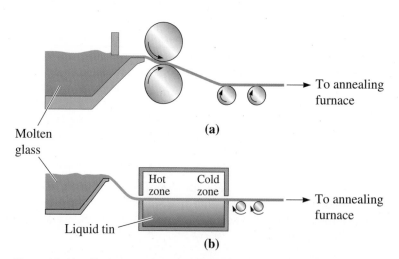

Figure 15-10   Techniques for manufacturing sheet and plate glass: (a) rolling and (b) floating the glass on molten tin.

TABLE 15-6 ■ *Compositions of typical glasses (in weight percent)*

Glass	SiO$_2$	Al$_2$O$_3$	CaO	Na$_2$O	B$_2$O$_3$	MgO	PbO	Others
Fused silica	99							
Vycor™	96				4			
Pyrex™	81	2		4	12			
Glass jars	74	1	5	15		4		
Window glass	72	1	10	14		2		
Plate glass/Float glass	73	1	13	13				
Lightbulbs	74	1	5	16		4		
Fibers	54	14	16		10	4		
Thermometer	73	6		10	10			
Lead glass	67			6			17	10% K$_2$O
Optical flint	50			1			19	13% BaO, 8% K$_2$O, ZnO
Optical crown	70			8		10		2% BaO, 8% K$_2$O
E-glass fibers	55	15	20		10			
S-glass fibers	65	25				10		

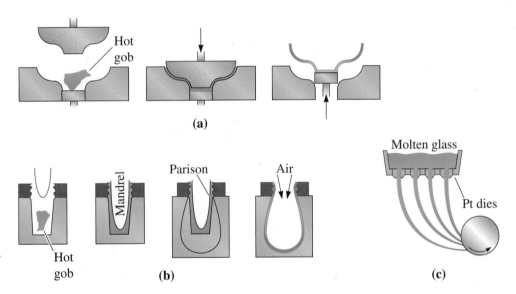

Figure 15-11 Techniques for forming glass products: (a) pressing, (b) press and blow process, and (c) drawing of fibers.

Tempered glass is produced by quenching the surface of plate glass with air, causing the surface layers to cool and contract. When the center cools, its contraction is restrained by the already rigid surface, which is placed in compression (Figure 15-12). Tempered glass is capable of withstanding much higher tensile stresses and impact blows than untempered glass. Tempered glass is used in car and home windows, shelving for refrigerators, ovens, furniture, and many other applications where safety is important.

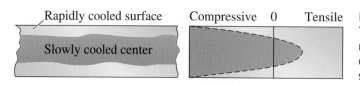

Figure 15-12
Tempered glass is cooled rapidly to produce compressive residual stresses at the surface.

Laminated glass, consisting of two annealed glass pieces with a polymer (such as polyvinyl butyral or PVB) in between, is used to make car windshields.

Glass Compositions

Pure SiO_2 must be heated to very high temperatures to obtain viscosities that permit economical forming. Most commercial glasses are based on silica (Table 15-5) modifiers such as soda (Na_2O) to break down the network structure and form eutectics with low melting temperatures, whereas lime (CaO) is added to reduce the solubility of the glasses in water. The most common commercial glass contains approximately 75% SiO_2, 15% Na_2O, and 10% CaO and is known as soda lime glass.

Borosilicate glasses, which contain about 15% B_2O_3, have excellent chemical and dimensional stability. Their uses include laboratory glassware (Pyrex™), glass-ceramics, and containers for the disposal of radioactive nuclear waste. Calcium aluminoborosilicate glass—or E-glass—is used as a general-purpose fiber for composite materials, such as fiberglass. Aluminosilicate glass, with 20% Al_2O_3 and 12% MgO, and high-silica glasses, with 3% B_2O_3, are excellent for high-temperature resistance and for protection against heat or thermal shock. The S-glass, a magnesium aluminosilicate, is used to produce high-strength fibers for composite materials. Fused silica, or virtually pure SiO_2, has the best resistance to high temperature, thermal shock, and chemical attack, although it is also expensive.

Special optical qualities can also be obtained, including sensitivity to light. Photochromic glass, which is darkened by the ultraviolet portion of sunlight, is used for sunglasses. Photosensitive glass darkens permanently when exposed to ultraviolet light; if only selected portions of the glass are exposed and then immersed in hydrofluoric acid, etchings can be produced. Polychromatic glasses are sensitive to all light, not just ultraviolet radiation. Similarly, nanosized crystals of semiconductors such as cadmium sulfide (CdS) are nucleated in silicate glasses in a process known as *striking*. These glasses exhibit lively colors and have useful optical properties.

15-6   Glass-Ceramics

Glass-ceramics are crystalline materials that are derived from amorphous glasses. Usually, glass-ceramics have a substantial level of crystallinity ($\sim > 70$–99%). The formation of glass-ceramics was discovered serendipitously by Don Stookey. With glass-ceramics, we can take advantage of the formability and density of glass. Also, a product that contains very low porosity can be obtained by producing a shape with conventional glass-forming techniques, such as pressing or blowing.

The first step in producing a glass-ceramic is to ensure that crystallization does not occur during cooling from the forming temperature. A continuous and isothermal cooling transformation diagram, much like the CCT and TTT diagrams for steels, can be constructed for silicate-based glasses. Figure 15-13(a) shows a TTT diagram for a glass. If glass cools too slowly, a transformation line is crossed; nucleation and growth of the crystals

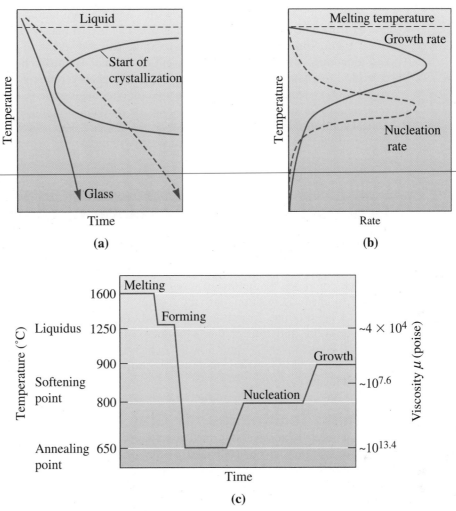

Figure 15-13 Producing a glass-ceramic: (a) Cooling must be rapid to avoid the start of crystallization. (b) The rate of nucleation of precipitates is high at low temperatures, whereas the rate of growth of the precipitates is high at higher temperatures. (c) A typical heat-treatment profile for glass-ceramics fabrication, illustrated here for Li_2O-Al_2O_3-SiO_2 glasses.

begin, but in an uncontrolled manner. Addition of modifying oxides to glass, much like addition of alloying elements to steel, shifts the transformation curve to longer times and prevents devitrification even at slow cooling rates. As noted in previous chapters, strictly speaking, we should make use of CCT (and not TTT) diagrams for this discussion.

Nucleation of the crystalline phase is controlled in two ways. First, the glass contains agents, such as TiO_2, that react with other oxides and form phases that provide the nucleation sites. Second, a heat treatment is designed to provide the appropriate number of nuclei; the temperature should be relatively low in order to maximize the rate of nucleation [Figure 15-13(b)]. The overall rate of crystallization depends on the growth rate of the crystals once nucleation occurs; thus, higher temperatures are then required to maximize the growth rate. Consequently, a heat-treatment schedule similar to that shown in Figure 15-13(c) for the Li_2O-Al_2O_3-SiO_2 glass-ceramics (Pyroceram™) can be used. The low-temperature step provides nucleation sites, and the high-temperature step speeds the rate of growth of the crystals. As much as 99% of the part may crystallize.

This special structure of glass-ceramics can provide good mechanical strength and toughness, often with a low coefficient of thermal expansion and high-temperature corrosion resistance. Perhaps the most important glass-ceramic is based on the Li_2O-Al_2O_3-SiO_2 system. These materials are used for cooking utensils (Corning Ware™) and ceramic tops for stoves. Other glass-ceramics are used in communications, computer, and optical applications.

15-7   Processing and Applications of Clay Products

Crystalline ceramics are often manufactured into useful articles by preparing a shape, or compact, composed of the raw materials in a fine powder form. The powders are then bonded by chemical reaction, partial or complete **vitrification** (melting), or sintering.

Clay products form a group of traditional ceramics used for producing pipe, brick, cookware, and other common products. Clay, such as kaolinite, and water serve as the initial binder for the ceramic powders, which are typically silica. Other materials, such as feldspar $[(K, Na)_2O \cdot Al_2O_3 \cdot 6SiO_2]$, serve as fluxing (glass-forming) agents during later heat treatment.

Forming Techniques for Clay Products   The powders, clay, flux, and water are mixed and formed into a shape. Dry or semi-dry mixtures are mechanically pressed into "green" (unbaked) shapes of sufficient strength to be handled. For more uniform compaction of complex shapes, isostatic pressing may be done; the powders are placed into a rubber mold and subjected to high pressures through a gas or liquid medium. Higher moisture contents permit the powders to be more plastic or formable. **Hydroplastic forming** processes, including extrusion, jiggering (forming of clay in a rotating mold using a profile tool to form the inner surface), and hand working, can be applied to these plastic mixes. Ceramic slurries containing large amounts of organic plasticizers (substances that impart properties such as softness), rather than water, can be injected into molds.

Still higher moisture contents permit the formation of a slip, or pourable slurry, containing fine ceramic powder. The slip is poured into a porous mold. The water in the slip nearest to the mold wall is drawn into the mold, leaving behind a soft solid that has a low moisture content. When enough water has been drawn from the slip to produce a desired thickness of solid, the remaining liquid slip is poured from the mold, leaving behind a hollow shell (Figure 15-4). Slip casting is used in manufacturing washbasins and other commercial products. After forming, the ceramic bodies—or greenware—are still weak, contain water or other lubricants, and are porous, and subsequent drying and firing are required.

Drying and Firing of Clay Products   During drying, excess moisture is removed and large dimensional changes occur. Initially, the water between the clay platelets—the interparticle water—evaporates and provides most of the shrinkage. Relatively little dimensional change occurs as the remaining water between the pores evaporates. The

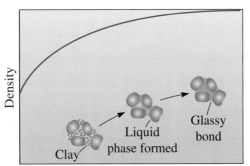

temperature and humidity are controlled to provide uniform drying throughout the part, thus minimizing stresses, distortion, and cracking.

The rigidity and strength of a ceramic part are obtained during **firing**. During heating, the clay dehydrates, eliminating the hydrated water that is part of the kaolinite crystal structure, and vitrification, or melting, begins (Figure 15-14). Impurities and the fluxing agent react with the ceramic particles (SiO_2) and clay, producing a low melting-point liquid phase at the grain surfaces. The liquid helps eliminate porosity and, after cooling, changes to a rigid glass that binds the ceramic particles. This glassy phase provides a **ceramic bond,** but it also causes additional shrinkage of the entire ceramic body.

The grain size of the final part is determined primarily by the size of the original powder particles. Furthermore, as the amount of **flux** increases, the melting temperature decreases; more glass forms, and the pores become rounder and smaller. A smaller initial grain size accelerates this process by providing more surface area at which vitrification can occur.

Applications of Clay Products   Many structural clay products and whitewares are produced using these processes. Brick and tile used for construction are pressed or extruded into shape, dried, and fired to produce the ceramic bond. Higher firing temperatures or finer original particle sizes produce more vitrification, less porosity, and higher density. The higher density improves mechanical properties but reduces the insulating qualities of the brick or tile.

Earthenware are porous clay bodies fired at relatively low temperatures. Little vitrification occurs; the porosity is very high and interconnected, and earthenware ceramics may leak. Consequently, these products must be covered with an impermeable glaze.

Higher firing temperatures, which provide more vitrification and less porosity, produce stoneware. The stoneware, which is used for drainage and sewer pipe, contains only 2% to 4% porosity. Ceramics known as china and porcelain require even higher firing temperatures to cause complete vitrification and virtually no porosity.

15-8   Refractories

Refractory materials are important components of the equipment used in the production, refining, and handling of metals and glasses, for constructing heat-treating furnaces, and for other high-temperature processing equipment. The **refractories** must survive at high temperatures without being corroded or weakened by the surrounding environment. Typical

TABLE 15-7 ■ *Compositions of typical refractories (weight percents)*

Refractory	SiO_2	Al_2O_3	MgO	Fe_2O_3	Cr_2O_3
Acidic					
Silica	95–97				
Superduty firebrick	51–53	43–44			
High-alumina firebrick	10–45	50–80			
Basic					
Magnesite			83–93	2–7	
Olivine	43		57		
Neutral					
Chromite	3–13	12–30	10–20	12–25	30–50
Chromite-magnesite	2–8	20–24	30–39	9–12	30–50

From Ceramic Data Book, *Cahners Publishing Co., 1982.*

refractories are composed of coarse oxide particles bonded by a finer refractory material. The finer material melts during firing, providing bonding. In some cases, refractory bricks contain about 20% to 25% apparent porosity to provide improved thermal insulation.

Refractories are often divided into three groups—acid, basic, and neutral—based on their chemical behavior (Table 15-7).

Acid Refractories

Common acidic refractories include silica, alumina, and fireclay (an impure kaolinite). Pure silica is sometimes used to contain molten metal. In some applications, the silica may be bonded with small amounts of boron oxide, which melts and produces the ceramic bond. When a small amount of alumina is added to silica, the refractory contains a very low melting-point eutectic microconstituent (Figure 15-15) and is not suited for refractory applications at temperatures above about 1600°C, a temperature often required for steel making. When larger amounts of alumina are added, the microstructure contains increasing amounts of mullite, $3Al_2O_3 \cdot 2SiO_2$, which has a high melting temperature. These fireclay refractories are generally relatively weak, but they are inexpensive. Alumina concentrations above about 50% constitute the high-alumina refractories.

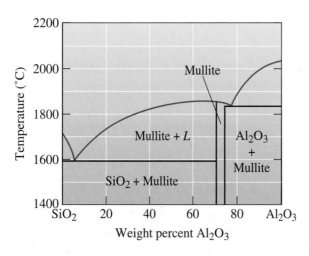

Figure 15-15
A simplified SiO_2-Al_2O_3 phase diagram, the basis for alumina silicate refractories.

Basic Refractories A number of refractories are based on MgO (magnesia, or periclase). Pure MgO has a high melting point, good refractory properties, and good resistance to attack by the basic environments often found in steel making processes. Olivine refractories contain forsterite, or Mg_2SiO_4, and also have high melting points. Other magnesia refractories may include CaO or carbon. Typically, the basic refractories are more expensive than the acid refractories.

Neutral Refractories These refractories, which include chromite and chromite-magnesite, might be used to separate acid and basic refractories, preventing them from attacking one another.

Special Refractories Carbon, or graphite, is used in many refractory applications, particularly when oxygen is not present. Other refractory materials include zirconia (ZrO_2), zircon ($ZrO_2 \cdot SiO_2$), and a variety of nitrides, carbides, and borides. Most of the carbides, such as TiC and ZrC, do not resist oxidation well, and their high temperature applications are best suited to reducing conditions. Silicon carbide is an exception, however; when SiC is oxidized at high temperatures, a thin layer of SiO_2 forms at the surface, protecting the SiC from further oxidation up to about 1500°C. Nitrides and borides also have high melting temperatures and are less susceptible to oxidation. Some of the oxides and nitrides are candidates for use in jet engines.

15-9   Other Ceramic Materials

In addition to their use in producing construction materials, appliances, structural materials, and refractories, ceramics find a host of other applications, including the following.

Cements Ceramic raw materials are joined using a binder that does not require firing or sintering in a process called **cementation**. A chemical reaction converts a liquid resin to a solid that joins the particles. In the case of sodium silicate, the introduction of CO_2 gas acts as a catalyst to dehydrate the sodium silicate solution into a glassy material:

$$x Na_2O \cdot y SiO_2 \cdot H_2O + CO_2 \rightarrow \text{glass (not balanced)}$$

Figure 15-16 on the next page shows silica sand grains used to produce molds for metal casting. The liquid sodium silicate coats the sand grains and provides bridges between the sand grains. Introduction of the CO_2 converts the bridges to a solid, joining the sand grains.

Fine alumina powder dispersions catalyzed with phosphoric acid produce an aluminum phosphate cement:

$$Al_2O_3 + 2H_3PO_4 \rightarrow 2AlPO_4 + 3H_2O$$

When alumina particles are bonded with the aluminum phosphate cement, refractories capable of operating at temperatures as high as 1650°C are produced.

Plaster of paris, or gypsum, is another material that is hardened by a cementation reaction:

$$CaSO_4 \cdot \tfrac{1}{2}H_2O + \tfrac{3}{2}H_2O \rightarrow CaSO_4 \cdot 2H_2O$$

Figure 15-16
A micrograph of silica sand grains bonded with sodium silicate through the cementation mechanism ($\times$ 60).
(*Reprinted courtesy of Don Askeland.*)

When the liquid slurry reacts, interlocking solid crystals of gypsum ($CaSO_4 \cdot 2H_2O$) grow with very small pores between the crystals. Larger amounts of water in the original slurry provide more porosity and decrease the strength of the final plaster. One of the important uses of this material is for construction of walls in buildings.

The most common and important of the cementation reactions occurs in Portland cement, which is used to produce concrete. (See Chapter 18.)

Coatings
Ceramics are often used to provide protective coatings to other materials. Common commercial coatings include glazes and enamels. Glazes are applied to the surface of a ceramic material to seal a permeable clay body, to provide protection and decoration, or for special purposes. Enamels are applied to metal surfaces. The enamels and glazes are clay products that vitrify easily during firing. A common composition is $CaO \cdot Al_2O_3 \cdot 2SiO_2$.

Special colors can be produced in glazes and enamels by the addition of other minerals. Zirconium silicate gives a white glaze, cobalt oxide makes the glaze blue, chromium oxide produces green, lead oxide gives a yellow color, and a red glaze may be produced by adding a mixture of selenium and cadmium sulfides.

One of the problems encountered with a glaze or enamel is surface cracking, or crazing, which occurs when the glaze has a coefficient of thermal expansion different than that of the underlying material. This is frequently the most important factor in determining the composition of the coating.

Special coatings are used for advanced ceramics and high-service temperature metals. SiC coatings are applied to carbon-carbon composite materials to improve their oxidation resistance. Zirconia coatings are applied to nickel-based superalloys to provide thermal barriers that protect the metal from melting or adverse reactions.

Thin Films and Single Crystals
Thin films of many complex and multi-component ceramics are produced using different techniques such as sputtering, sol-gel, and chemical-vapor deposition (CVD). Usually, the thickness of such films is 0.05 to 10 μm and more likely greater than 2 μm. Many functional electronic ceramic thin films are prepared and integrated onto silicon wafers, glasses, and other substrates. For example, the magnetic strips on credit cards use iron oxide (γ-Fe_2O_3 or Fe_3O_4) thin films for storing data. Indium tin oxide (ITO), a conductive and transparent material, is coated on glass and used in applications such as touch-screen displays (Table 15-1). Many other

coatings are used on glass to make the glass energy efficient. Recently, a self-cleaning glass using a TiO_2 coating has been developed. Similarly, thin films of ceramics, such as lead zirconium titanate (PZT), lanthanum-doped PZT (PLZT), and $BaTiO_3$-$SrTiO_3$ solid solutions, can be prepared and used. Often, the films develop an orientation, or texture, which may be advantageous for a given application. Single crystals of ceramics [e.g., SiO_2 or quartz, lithium niobate ($LiNbO_3$), sapphire, or yttrium aluminum garnate] are used in many electrical and electro-optical applications. These crystals are grown from melts using techniques similar to those described in Chapter 9.

Fibers
Fibers are produced from ceramic materials for several uses: as a reinforcement in composite materials, for weaving into fabrics, or for use in fiber-optic systems. Borosilicate glass fibers, the most commonly produced fibers, provide strength and stiffness in fiberglass. Fibers can be produced from a variety of other ceramics, including alumina, silicon carbide, silica, and boron carbide. The sol-gel process is also used to produce commercial fibers for many applications.

A special type of fibrous material is the silica tile used to provide the thermal protection system for NASA's space shuttle. Silica fibers are bonded with colloidal silica to produce an exceptionally lightweight tile with densities as low as 0.144 g/cm^3; the tile is coated with special high-emissivity glazes to permit protection up to 1300°C.

Joining and Assembly of Ceramic Components
Ceramics are often made as monolithic components rather than assemblies of numerous components. When two ceramic parts are placed in contact under a load, stress concentrations at the brittle surface are created, leading to an increased probability of failure. In addition, methods for joining ceramic parts into a larger assembly are limited. The brittle ceramics cannot be joined by fusion welding or deformation bonding processes. At low temperatures, adhesive bonding using polymer materials may be accomplished; ceramic cements may be used at higher temperatures. Diffusion bonding and brazing can be used to join ceramics and to join ceramics to metals.

Summary

- Ceramics are inorganic materials that have high hardnesses and high melting points. These include single crystal and polycrystalline ceramics, glasses, and glass-ceramics. Typical ceramics are electrical and thermal insulators with good chemical stability and good strength in compression.

- Polycrystalline ceramics exhibit brittle behavior, partly because of porosity. Because most polycrystalline ceramics cannot plastically deform (unless special conditions with respect to temperature and strain rates are met), the porosity limits the ability of the material to withstand a tensile load.

- Ceramics play a critical role in a wide array of electronic, magnetic, optical, and energy related technologies. Many advanced ceramics play a very important role in providing thermal insulation and high-temperature properties. Applications of advanced ceramics range from credit cards, houses for silicon chips, tiles for the space shuttle, medical imaging, optical fibers that enable communication, and safe and energy efficient glasses. Traditional ceramics serve as refractories for metals processing and consumer applications.

- Ceramics processing is commonly conducted using compaction and sintering. For specialized applications, isostatic compaction, hot pressing, and hot isostatic pressing (HIP) are used, especially to achieve higher densification levels.

- Tape casting, slip casting, extrusion, and injection molding are some of the other techniques used to form green ceramics into different shapes. These processes are then followed by a burnout step in which binders and plasticizers are burnt off, and the resultant ceramic is sintered.

- Many silicates and other ceramics form glasses rather easily, since the kinetics of crystallization are sluggish. Glasses can be formed as sheets using float-glass or as fibers and other shapes. Silicate glasses are used in a significant number of applications that include window glass, windshields, fiber optics, and fiberglass.

- Glass-ceramics are formed using controlled crystallization of inorganic glasses. These materials are used widely for kitchenware and many other applications.

- Ceramics, in the form of fibers, thin films, coatings, and single crystals, have many different applications.

Glossary

Apparent porosity The percentage of a ceramic body that is composed of interconnected porosity.

Bulk density The mass of a ceramic body per unit volume, including closed and interconnected porosity.

Cementation Bonding ceramic raw materials using binders that form a glass or gel without firing at high temperatures.

Ceramic An inorganic material with a high melting temperature. Usually hard and brittle.

Ceramic bond Bonding ceramic materials by permitting a glassy product to form at high firing temperatures.

Cermet A ceramic-metal composite (e.g., WC-Co) providing a good combination of hardness with other properties such as toughness.

Cold isostatic pressing (CIP) A powder-shaping technique in which hydrostatic pressure is applied during compaction. This is used for achieving a higher green ceramic density or compaction of more complex shapes.

Devitrification The crystallization of glass.

Enamel A ceramic coating on metal.

Firing Heating a ceramic body at a high temperature to cause a ceramic bond to form.

Flux Additions to ceramic raw materials that promote vitrification.

Glass An amorphous material formed by cooling of a melt.

Glass-ceramics Ceramic shapes formed in the glassy state and later allowed to crystallize during heat treatment to achieve improved strength and toughness.

Glass formers Oxides with a high bond strength that easily produce a glass during processing.

Glass transition temperature The temperature below which an undercooled liquid becomes a glass. This is not a fixed temperature.

Glaze A ceramic coating applied to glass. The glaze contains glassy and crystalline ceramic phases.

Green ceramic A ceramic that has been shaped into a desired form but has not yet been sintered.

Hot isostatic pressing (HIP) A powder-processing technique in which large pieces of metals, alloys, and ceramics can be produced using sintering under a hydrostatic pressure generated by a gas.

Hot pressing A processing technique in which sintering is conducted under uniaxial pressure.

Hydroplastic forming A number of processes by which a moist ceramic clay body is formed into a useful shape.

Injection molding A processing technique in which a thermoplastic mass (loaded with ceramic powder) is mixed in an extruder-like setup and then injected into a die to form complex parts. In the case of ceramics, the thermoplastic is burnt off. Also, widely used for thermoplastics (Chapter 16).

Intermediates Oxides that, when added to a glass, help to extend the glassy network, although the oxides normally do not form a glass themselves.

Laminated glass Annealed glass with a polymer (e.g., polyvinyl butyral, PVB) sandwiched in between, used for car windshields.

Parison A crude glassy shape that serves as an intermediate step in the production process. The parison is later formed into a finished product.

Powder A collection of fine particles.

Powder metallurgy Powder processing routes used for converting metal and alloy powders into useful shapes.

Powder processing Unit operations conducted to convert powders into useful shapes (e.g., pressing, tape casting, etc.).

Reaction bonding A ceramic processing technique by which a shape is made using one material that is later converted into a ceramic material by reaction with a gas.

Refractories A group of ceramic materials capable of withstanding high temperatures for prolonged periods of time.

Sintering A process in which a material is heated to a high temperature so as to densify it.

Slip A liquid slurry that is poured into a mold. When the slurry begins to harden at the mold surface, the remaining liquid slurry is decanted, leaving behind a hollow ceramic casting.

Slip casting Forming a hollow ceramic part by introducing a pourable slurry into a mold. The water in the slurry is extracted into the porous mold, leaving behind a drier surface. Excess slurry can then be poured out.

Spray drying A slurry of a ceramic powder is sprayed into a large chamber in the presence of hot air. This leads to the formation of soft agglomerates that can flow well into the dies used during powder compaction.

Synthesis Steps conducted to make a ceramic powder.

Tape casting A process for making thin sheets of ceramics using a ceramic slurry consisting of binders, plasticizers, etc. The slurry is cast with the help of a blade onto a plastic substrate. The resultant green tape is then dried, cut, and machined and used to make electronic ceramic and other devices.

Tempered glass A high-strength glass that has a surface layer where the stress is compressive, induced thermally during cooling or by the chemical diffusion of ions.

True porosity The percentage of a ceramic body that is composed of both closed and interconnected porosity.

Vitrification Formation of a glass. To devitrify means to crystallize a glass.

Problems

Section 15-1 Applications of Ceramics

15-1 What are the primary types of atomic bonds in ceramics?

15-2 Explain the meaning of the following terms: ceramics, inorganic glasses, and glass-ceramics.

15-3 Explain why ceramics typically are processed as powders. How is this similar to or different from the processing of metals?

15-4 What do the terms "glaze" and "enamel" mean?

15-5 What material is used to make the tiles that provide thermal protection in NASA's space shuttle?

15-6 Which ceramic materials are most widely used?

15-7 Explain how ceramic materials can be classified in different ways.

15-8 State any one application of the following ceramics: (a) alumina, (b) silica, (c) barium titanate, (d) zirconia, (e) boron carbide, and (f) diamond.

Section 15-2 Properties of Ceramics

15-9 What are some of the typical characteristics of ceramic materials?

15-10 Why is the tensile strength of ceramics much lower than the compressive strength?

15-11 Plastic deformation due to dislocation motion is important in metals; however, this is not a very important consideration for the properties of ceramics and glasses. Explain.

15-12 Can ceramic materials show superplastic behavior or are they always brittle? Explain.

15-13 Explain why ceramics tend to show wide scatter in their mechanical properties.

Section 15-3 Synthesis and Processing of Ceramic Powders

15-14 What is the driving force for sintering?

15-15 What is the driving force for grain growth?

15-16 What mechanisms of diffusion play the most important role in the solid-state sintering of ceramics?

15-17 Explain the use of the following processes (use a sketch as needed): (a) uniaxial compaction and sintering, (b) hot pressing, (c) HIP, and (d) tape casting.

Section 15-4 Characteristics of Sintered Ceramics

15-18 What are some of the important characteristics of sintered ceramics?

15-19 What typical density levels are obtained in sintered ceramics?

15-20 What do the terms "apparent porosity" and "true porosity" of ceramics mean?

15-21 The specific gravity of Al_2O_3 is 3.96 g/cm^3. A ceramic part is produced by sintering alumina powder. It weighs 80 g when dry, 92 g after it has soaked in water, and 58 g when suspended in water. Calculate the apparent porosity, the true porosity, and the closed porosity.

15-22 Silicon carbide (SiC) has a specific gravity of 3.1 g/cm^3. A sintered SiC part is produced, occupying a volume of 500 cm^3 and weighing 1200 g. After soaking in water, the part weighs 1250 g. Calculate the bulk density, the true porosity, and the volume fraction of the total porosity that consists of closed pores.

15-23 A sintered zirconium oxide (ZrO_2) ceramic has a true porosity of 28%, and the closed pore fraction is 0.5. If the weight after soaking in water is 760 g, what is the dry weight of the ceramic? The specific gravity of ZrO_2 is 5.68 g/cm^3.

Section 15-5 Inorganic Glasses

15-24 What is the main reason why glass formation is easy in silicate systems?

15-25 Can glasses be formed using metallic materials?

15-26 Define the terms "glass formers," "intermediates," and "modifiers."

15-27 What does the term "glass transition temperature" mean? Is this a fixed temperature for a given composition of glass?

15-28 How many grams of BaO can be added to 1 kg of SiO_2 before the O:Si ratio exceeds 2.5 and glass-forming tendencies are poor? Compare this with the case when Li_2O is added to SiO_2.

15-29 Calculate the O:Si ratio when 30 wt% Y_2O_3 is added to SiO_2. Will this material have good glass-forming tendencies?

15-30 A silicate glass has 10 mol% Na_2O and 5 mol% CaO. Calculate the O:Si ratio and the composition in wt% for this glass.

15-31 A borosilicate glass (82% SiO_2, 2% Al_2O_3, 4% Na_2O, 12% B_2O_3) has a density of 2.23 g/cm^3, while a fused silica glass (assume

100% SiO_2) has a density of 2.2 g/cm^3. Explain why the density of the borosilicate glass is different from the weighted average of the densities of its components. The densities of Al_2O_3, Na_2O and B_2O_3 are 3.98 g/cm^3, 2.27 g/cm^3 and 2.5 g/cm^3, respectively.

15-32 Rank the following glasses by softening temperature (highest to lowest):
 (a) Pyrex^{TM} (81% SiO_2, 2% Al_2O_3, 4% Na_2O, 12% B_2O_3);
 (b) Lime glass (72% SiO_2, 1% Al_2O_3, 10% CaO, 14% Na_2O); and
 (c) Vycor^{TM} (96% SiO_2, 4% B_2O_3).

Section 15-6 Glass-Ceramics

15-33 How is a glass-ceramic different from a glass and a ceramic?

15-34 What are the advantages of using glass-ceramics as compared to either glasses or ceramics?

15-35 Draw a typical heat-treatment profile encountered in the processing of glass-ceramics.

15-36 What are some of the important applications of glass-ceramics?

Design Problems

15-37 Design a silica soda-lime glass that can be cast at 1300°C and that will have an O:Si ratio of less than 2.3 to ensure good glass-forming tendencies. To ensure adequate viscosity, the casting temperature should be at least 100°C above the liquidus temperature.

Computer Problems

15-38 *Sintering Profile*. Write a computer program that will ask the user to provide temperatures during different stages of sintering. The program should ask the user to provide starting temperatures, temperatures for binder burnout, sintering-hold temperatures, and final temperatures. A time for sintering must also be provided by the user. The program should ask the user to provide the heating rates for the binder burnout and the stage between binder burnout and sintering, as well as the cooling rate. The program should then provide temperatures for any given time and declare the sintering cycle stage in the furnace.

15-39 *Apparent and Bulk Density*. Write a program that will ask the user to provide the information needed to calculate the apparent and bulk density of ceramic parts. The program should also calculate the interconnected and closed porosity.

Knovel® Problems

K15-1 What are typical grain sizes and sintering parameters for nanoceramics?

K15-2 What are the advantages of sol-gel coatings over other ceramic coatings?

K15-3 What are typical values for the percent linear thermal expansion of refractories? Using interactive graphs, determine which refractory experiences a greater linear thermal expansion at 1200°C: zirconia or alumina?

Polymers can be processed into a variety of shapes through numerous manufacturing techniques. This photo shows the extrusion of a molten thermoplastic tube. Before the polymer fully solidifies, hot air is blown through the tube, as it moves vertically upward, causing it to expand up to several times in diameter. The flattened tube is then sealed on one side with heat and cut on the other to form plastic bags or simply cut along the length to form plastic sheet (©Bob Masini/Phototake).

Chapter 16

Polymers

Have You Ever Wondered?

- *What are compact discs (CDs) made from?*
- *What is Silly Putty® made from?*
- *What polymer is used in chewing gum?*
- *Which was the first synthetic fiber ever made?*
- *Why are some plastics "dishwasher safe" and some not?*
- *What are bulletproof vests made from?*
- *What polymer is used for non-stick cookware?*

In this chapter, we will examine the structure, properties, and processing of polymers. The suffix *mer* means "unit." Thus, the term polymer means "many units," and in this context, the term mer refers to a unit group of atoms or molecules that defines a characteristic arrangement for a polymer. **Polymers** consist of chains of molecules. The chains have average molecular weights that range from 10,000 to more than one million g/mol built by joining many mers through chemical bonding to form giant molecules known as macromolecules. Molecular weight is defined as the sum of atomic masses in each molecule. *Polymerization* is the process by which small molecules consisting of one unit (known as a **monomer**) or a few units (known as **oligomers**) are chemically joined to create these giant molecules. Polymerization normally begins with the production of long chains in which the atoms are strongly joined by covalent bonding. Most polymers are organic, meaning that they are carbon-based; however, polymers can be inorganic (e.g., silicones based on a Si-O network).

      Plastics are materials that are composed principally of naturally occurring and modified or artificially made polymers often containing additives such as fibers, fillers, pigments, and the like that further enhance their properties. Plastics include thermoplastics, thermosets, and elastomers (natural or synthetic). In this book, we use the terms plastic and polymers interchangeably.

      Plastics are used in an amazing number of applications including clothing, toys, home appliances, structural and decorative items, coatings, paints, adhesives, automobile tires, biomedical materials, car bumpers and interiors, foams, and packaging.

Polymers are often used in composites, both as fibers and as a matrix. Liquid crystal displays (LCDs) are based on polymers. We also use polymers in photochromic lenses. Plastics are often used to make electronic components because of their insulating ability and low dielectric constant. More recently, significant developments have occurred in the area of flexible electronic devices based on the useful piezoelectric, semiconducting, optical and electro-optical properties seen in some polymers. Polymers such as polyvinyl acetate (PVA) are water-soluble. Many such polymers can be dissolved in water or organic solvents to be used as binders, surfactants, or plasticizers in processing ceramics, semiconductors, and as additives to many consumer products. Polyvinyl butyral (PVB), a polymer, makes up part of the laminated glass used for car windshields (Chapter 15). Polymers are probably used in more technologies than any other class of materials.

Commercial—or standard commodity—polymers are lightweight, corrosion-resistant materials with low strength and stiffness, and they are not suitable for use at high temperatures. These polymers are, however, relatively inexpensive and are readily formed into a variety of shapes, ranging from plastic bags to mechanical gears to bathtubs. *Engineering polymers* are designed to give improved strength or better performance at elevated temperatures. These materials are produced in relatively small quantities and often are expensive. Some of the engineering polymers can perform at temperatures as high as 350°C; others—usually in a fiber form—have strengths that are greater than that of steel.

Polymers also have many useful physical properties. Some polymers, such as acrylics like Plexiglas™ and Lucite™, are transparent and can be substituted for glasses. Although most polymers are electrical insulators, special polymers (such as the acetals) and polymer-based composites possess useful electrical conductivity. Teflon™ has a low coefficient of friction and is the coating for nonstick cookware. Polymers also resist corrosion and chemical attack.

16-1   Classification of Polymers

Polymers are classified in several ways: by how the molecules are synthesized, by their molecular structure, or by their chemical family. One way to classify polymers is to state if the polymer is a **linear polymer** or a **branched polymer** (Figure 16-1). A linear polymer consists of spaghetti-like molecular chains. In a branched polymer, there are primary polymer chains and secondary offshoots of smaller chains that stem from these main chains. Note that even though we say "linear," the chains are actually not in the form of straight lines. A better method to describe polymers is in terms of their mechanical and thermal behavior. Table 16-1 compares the three major polymer categories.

Thermoplastics are composed of long chains produced by joining together monomers; they typically behave in a plastic, ductile manner. The chains may or may not have branches. Individual chains are intertwined. There are relatively weak van der Waals bonds between atoms of different chains. This is somewhat similar to a few trees that are tangled up together. The trees may or may not have branches, and they are not physically connected to each other. The chains in thermoplastics can be untangled by application of a tensile stress. Thermoplastics can be amorphous or crystalline. Upon heating, thermoplastics soften and melt. They are processed into shapes by heating to elevated temperatures. Thermoplastics are easily recycled.

Thermosetting polymers are composed of long chains (linear or branched) of molecules that are strongly cross-linked to one another to form three-dimensional network

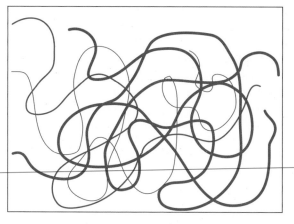

(a) Linear unbranched

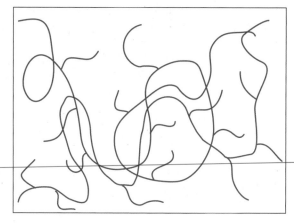

(b) Linear branched

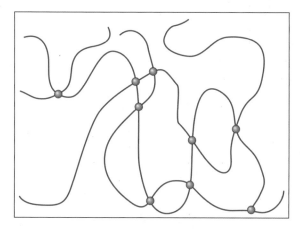

(c) Thermoset-unbranched

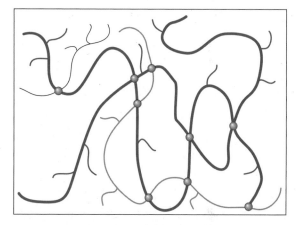

(d) Thermoset-branched

Figure 16-1   Schematic showing linear and branched polymers. Note that branching can occur in any type of polymer (e.g., thermoplastics, thermosets, and elastomers). (a) Linear unbranched polymer: notice chains are not straight lines and not connected. Different polymer chains are shown using different shades designed to show clearly that each chain is not connected to another. (b) Linear branched polymer: chains are not connected; however, they have branches. (c) Thermoset polymer without branching: chains are connected to one another by covalent bonds, but they do not have branches. Joining points are highlighted with solid circles. (d) Thermoset polymer that has branches and chains that are interconnected via covalent bonds: different chains and branches are shown in different shades for better contrast. Places where chains are actually chemically bonded are shown with solid circles.

TABLE 16-1 ■ *Comparison of the three polymer categories*

Behavior	General Structure	Example
Thermoplastic	Flexible linear chains (straight or branched)	Polyethylene
Thermosetting	Rigid three-dimensional network (chains may be linear or branched)	Polyurethanes
Elastomers	Thermoplastics or lightly cross-linked thermosets, consist of spring-like molecules	Natural rubber

structures. Network or thermosetting polymers are like a bunch of strings that are knotted to one another in several places and not just tangled up. Each string may have other side strings attached to it. Thermosets are generally stronger, but more brittle, than thermoplastics. Thermosets do not melt upon heating but begin to decompose. They cannot easily be reprocessed after the cross-linking reaction has occurred, and hence, recycling is difficult.

Elastomers These are known as rubbers. They sustain elastic deformations greater than 200%. These may be thermoplastics or lightly cross-linked thermosets. The polymer chains consist of coil-like molecules that can reversibly stretch by applying a force.

Thermoplastic elastomers are a special group of polymers. They have the processing ease of thermoplastics and the elastic behavior of elastomers.

Representative Structures

Figure 16-2 shows three ways we can represent a segment of polyethylene, the simplest of the thermoplastics. The polymer chain consists of a backbone of carbon atoms; two hydrogen atoms are bonded to each carbon atom in the chain. The chain twists and turns throughout space. As shown in the figure, polyethylene has no branches and hence is a linear thermoplastic. The simple two-dimensional model in Figure 16-2(c) includes the essential elements of the polymer structure and will be used to describe the various polymers. The single lines (—) between the carbon atoms and between the carbon and hydrogen atoms represent a single covalent bond. Two parallel lines (=) represent a double covalent bond between atoms. A number of polymers include ring structures such as the benzene ring found in polystyrene and other polymers (Figure 16-3). Molecules that contain the six-membered benzene ring are known as aromatics.

In the structure shown in Figure 16-2(c) if we replace one of the hydrogen atoms in C_2H_4 with CH_3, a benzene ring, or chlorine, we get the structure of polypropylene, polystyrene, and polyvinyl chloride (PVC), respectively. If we replaced all H atoms in the

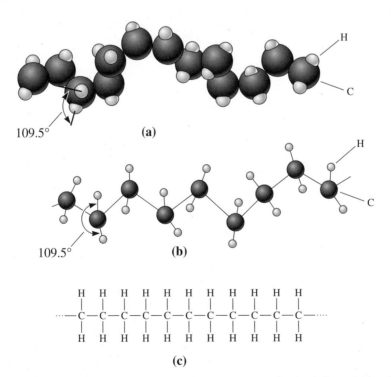

Figure 16-2   Three ways to represent the structure of polyethylene: (a) a solid three-dimensional model, (b) a three-dimensional "space" model, and (c) a simple two-dimensional model.

C_2H_4 groups with fluorine (F), we get the structure of polytetrafluoroethylene or Teflon™. Like many other discoveries, Teflon™ was discovered by accident. Many polymer structures can thus be derived from the structure of polyethylene. The following example shows how different types of polymers are used.

Example 16-1 | *Design/Materials Selection for Polymer Components*

Design the type of polymer material you might select for the following applications: a surgeon's glove, a beverage container, and a pulley.

SOLUTION

The glove must be capable of stretching a great deal in order to slip onto the surgeon's hand, yet it must conform tightly to the hand to permit the maximum sensation of touch during surgery. A material that undergoes a large amount of elastic strain—particularly with relatively little applied stress—might be appropriate; this requirement describes an elastomer.

The beverage container should be easily and economically produced. It should have some ductility and toughness so that it does not accidentally shatter and leak the contents. If the beverage is carbonated, diffusion of CO_2 is a major concern (Chapter 5). A thermoplastic such as polyethylene terephthalate (PET) will have the necessary formability and strength needed for this application. It is also easily recycled.

The pulley will be subjected to some stress and wear as a belt passes over it. A relatively strong, rigid, hard material is required to prevent wear, so a thermosetting polymer might be most appropriate.

16-2 Addition and Condensation Polymerization

Polymerization by **addition** and **condensation** are the two main ways to create a polymer. The polymers derived from these processes are known as addition and condensation polymers, respectively. The formation of the most common polymer, polyethylene (PE) from ethylene molecules, is an example of addition or chain-growth polymerization. Ethylene, a gas, is the monomer (single unit) and has the formula C_2H_4. The two carbon atoms are joined by a double covalent bond. Each carbon atom shares two of its electrons with the second carbon atom, and two hydrogen atoms are bonded covalently to each of the carbon atoms (Figure 16-4).

H   H
| |
C=C   Ethylene monomer
| |
H   H

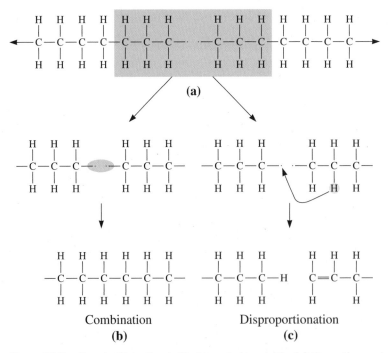

H   H   H   H   H   H
| | | | | |
·—C—C—⟨ ⟩—C—C—⟨ ⟩—C—C—· Ethylene
| | | | | | repeat
H   H   H   H   H   H   units

Figure 16-4
The addition reaction for producing polyethylene from ethylene molecules. The unsaturated double bond in the monomer is broken to produce active sites, which then attract additional repeat units to either end to produce a chain.

In the presence of an appropriate combination of heat, pressure, and catalysts, the double bond between the carbon atoms is broken and replaced with a single covalent bond. The ends of the monomer are now *free radicals*; each carbon atom has an unpaired electron that it may share with other free radicals. Addition polymerization occurs because the original monomer contains a double covalent bond between the carbon atoms. The double bond is an **unsaturated bond**. After changing to a single bond, the carbon atoms are still joined, but they become active; other **repeat units** or mers can be added to produce the polymer chain.

Termination of Addition Polymerization

We need polymers that have a controlled average molecular weight and molecular weight distribution. Thus, the polymerization reactions must have an "off" switch as well! The chains may be terminated by two mechanisms (Figure 16-5). First, the ends of two growing chains may be

Figure 16-5 Termination of polyethylene chain growth: (a) the active ends of two chains come into close proximity, (b) the two chains undergo combination and become one large chain, and (c) rearrangement of a hydrogen atom and creation of a double covalent bond by disproportionation cause termination of two chains.

joined. This process, called *combination*, creates a single large chain from two smaller chains. Second, the active end of one chain may remove a hydrogen atom from a second chain by a process known as *disproportionation*. This reaction terminates two chains, rather than combining two chains into one larger chain. Sometimes, compounds known as terminators are added to end polymerization reactions. In general, for thermoplastics, a higher average molecular weight leads to a higher melting temperature and higher Young's modulus for the polymer (Section 16-3). The following example illustrates an addition polymerization reaction.

Example 16-2 *Calculation of Initiator Required*

Calculate the amount of benzoyl peroxide [$(C_6H_5CO)_2O_2$] initiator required to produce 1 kg of polyethylene with an average molecular weight of 200,000 g/mol. Each benzoyl peroxide molecule produces two free radicals that are each capable of initiating a polyethylene chain. Assume that 20% of the initiator actually is effective and that all termination occurs by the combination mechanism.

SOLUTION

One benzoyl peroxide molecule produces two free radicals that initiate two chains that then combine to form one polyethylene chain. Thus, there is a 1:1 ratio between benzoyl peroxide molecules and polyethylene chains. If the initiator were 100% effective, one molecule of benzoyl peroxide would be required per polyethylene chain.

To determine the amount of benzoyl peroxide required, the number of chains with an average molecular weight of 200,000 g/mol in 1 kg of polyethylene must be calculated.

The molecular weight of ethylene = (2 C atoms)(12 g/mol) + (4 H atoms) (1 g/mol) = 28 g/mol. The number of molecules per chain (also known as the degree of polymerization) is given by

$$\frac{200,000 \text{ g/mol}}{28 \text{ g/mol}} = 7143 \text{ ethylene molecules/chain}$$

The total number of monomers required to form 1 kg of polyethylene is

$$\frac{(1000 \text{ g})(6.022 \times 10^{23} \text{ monomers/mol})}{28 \text{ g/mol}} = 215 \times 10^{23} \text{ monomers}$$

Thus, the number of polyethylene chains in 1 kg is

$$\frac{215 \times 10^{23} \text{ ethylene molecules}}{7143 \text{ ethylene molecules/chain}} = 3.0 \times 10^{21} \text{ chains}$$

Since the benzoyl peroxide initiator is only 20% effective, $5(3.0 \times 10^{21}) = 1.5 \times 10^{22}$ benzoyl peroxide molecules are required to initiate 3.0×10^{21} chains.

The molecular weight of benzoyl peroxide is (14 C atoms)(12 g/mol) + (10 H atoms) (1 g/mol) + (4 O atoms)(16 g/mol) = 242 g/mol. Therefore, the amount of initiator required is

$$\frac{1.5 \times 10^{22} \text{ molecules } (242 \text{ g/mol})}{6.022 \times 10^{23} \text{ molecules/mol}} = 6.0 \text{ g}$$

Condensation Polymerization

Polymer chains can also form by condensation reactions, or *step-growth* polymerization, producing structures and properties that resemble those of addition polymers. In condensation polymerization, a relatively small molecule (such as water, ethanol, methanol, etc.) is formed as a result of the polymerization reaction. This mechanism may often involve different monomers as starting or precursor molecules. The polymerization of dimethyl terephthalate and ethylene glycol (also used as radiator coolant) to produce *polyester* is an important example (Figure 16-6). During polymerization, a hydrogen atom on the end of the ethylene glycol monomer combines with an OCH$_3$ (methoxy) group from the dimethyl terephthalate. A byproduct, methyl alcohol (CH$_3$OH), is "condensed" off, and the two monomers combine to produce a larger molecule. Each of the monomers in this example is bifunctional, meaning that both ends of the monomer may react, and the condensation polymerization can continue by the same reaction. Eventually, a long polymer chain—a polyester—is produced. The length of the polymer chain depends on the ease with which the monomers can diffuse to the ends and undergo the condensation reaction. Chain growth ceases when no more monomers reach the end of the chain to continue the reaction. Condensation polymerization reactions also occur in sol-gel processing of ceramic materials.

The following example describes the discovery of nylon and calculations related to condensation polymerization.

Dimethyl terephthalate   Ethylene glycol

Repeat unit for polyethylene terephthalate   Methyl alcohol
          (PET polymer)   (byproduct)

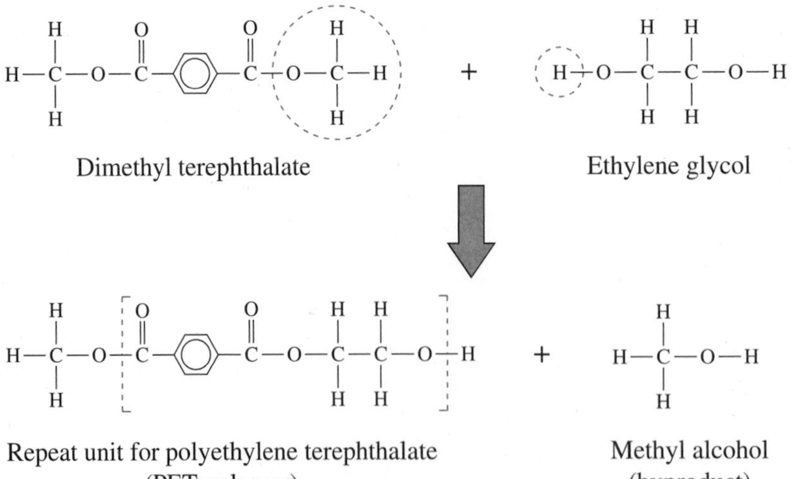

Figure 16-6   The condensation reaction for polyethylene terephthalate (PET), which is a common polyester. The OCH$_3$ group and a hydrogen atom are removed from the monomers, permitting the two monomers to join and producing methyl alcohol as a byproduct.

Example 16-3 *Condensation Polymerization of 6,6-Nylon*

Nylon was first reported by Wallace Hume Carothers, of du Pont in about 1934. In 1939, Charles Stine, also from du Pont, reported the discovery of this first synthetic fiber to a group of 3000 women gathered for the New York World's Fair. The first application was nylon stockings. Today nylon is used in hundreds of applications. Prior to nylon, Carothers had discovered neoprene (an elastomer).

The linear polymer 6,6-nylon is to be produced by combining 1000 g of hexamethylene diamine with adipic acid. A condensation reaction then produces the polymer. The molecular structures of the monomers are shown below. The linear nylon chain is produced when a hydrogen atom from the hexamethylene diamine combines with an OH group from adipic acid to form a water molecule.

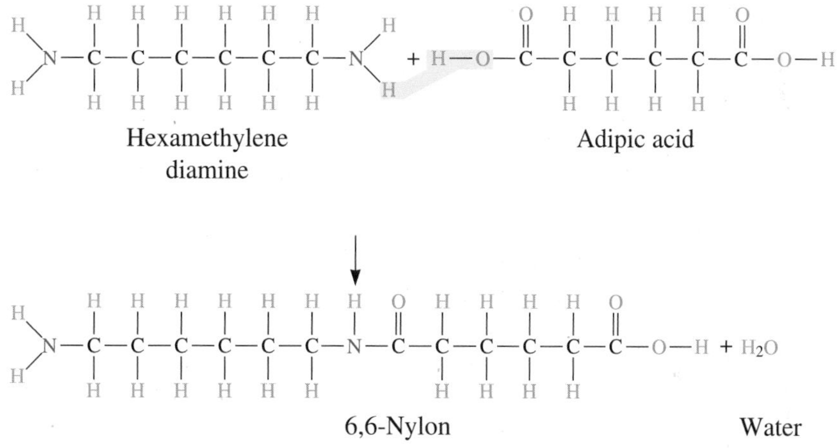

Hexamethylene
diamine

Adipic acid

6,6-Nylon

Water

Note that the reaction can continue at both ends of the new molecule; consequently, long chains may form. This polymer is called 6,6-nylon because both monomers contain six carbon atoms.

How many grams of adipic acid are needed, and how much 6,6-nylon is produced, assuming 100% efficiency?

SOLUTION

The molecular weights of hexamethylene diamine, adipic acid, and water are 116, 146, and 18 g/mol, respectively. The number of moles of hexamethylene diamine is equal to the number of moles of adipic acid:

$$\frac{1000 \text{ g}}{116 \text{ g/mol}} = 8.62 \text{ moles} = \frac{x \text{ g}}{146 \text{ g/mol}}$$

$$x = 1259 \text{ g of adipic acid required}$$

One water molecule is lost when hexamethylene diamine reacts with adipic acid. Each time a monomer is added to the chain, one molecule of water is lost. Thus, when a long chain forms, there are (on average) two water molecules released for each repeat unit of the chain (each repeat unit being formed from two monomers). Thus, the number of moles of water lost is 2 (8.62 moles) = 17.24 moles or

$$17.24 \text{ moles } H_2O \ (18 \text{ g/mol}) = 310 \text{ g } H_2O$$

The total amount of nylon produced is

$$1000 \text{ g} + 1259 \text{ g} - 310 \text{ g} = 1949 \text{ g}$$

16-3   Degree of Polymerization

Polymers, unlike organic or inorganic compounds, do not have a fixed molecular weight. For example, polyethylene may have a molecular weight that ranges from ~25,000 to 6 million! The average length of a linear polymer is represented by the **degree of polymerization**, or the number of repeat units in the chain. The degree of polymerization can also be defined as

$$\text{Degree of polymerization} = \frac{\text{average molecular weight of polymer}}{\text{molecular weight of repeat unit}} \qquad (16\text{-}1)$$

If the polymer contains only one type of monomer, the molecular weight of the repeat unit is that of the monomer. If the polymer contains more than one type of monomer, the molecular weight of the repeat unit is the sum of the molecular weights of the monomers, less the molecular weight of the byproduct.

The lengths of the chains in a linear polymer vary considerably. Some may be quite short due to early termination; others may be exceptionally long. We can define an average molecular weight in two ways.

The *weight average molecular weight* is obtained by dividing the chains into size ranges and determining the fraction of chains having molecular weights within that range. The weight average molecular weight $\overline{M}_w$ is

$$\overline{M}_w = \sum f_i M_i \qquad (16\text{-}2)$$

where M_i is the mean molecular weight of each range and f_i is the weight fraction of the polymer having chains within that range.

The *number average molecular weight* $\overline{M}_n$ is based on the number fraction, rather than the weight fraction, of the chains within each size range. It is always smaller than the weight average molecular weight and is given by

$$\overline{M}_n = \sum x_i M_i \qquad (16\text{-}3)$$

where M_i is again the mean molecular weight of each size range, but x_i is the fraction of the total number of chains within each range. Either $\overline{M}_w$ or $\overline{M}_n$ can be used to calculate the degree of polymerization.

The two following examples illustrate these concepts.

Example 16-4 *Degree of Polymerization for 6,6-Nylon*

Calculate the degree of polymerization if 6,6-nylon has a molecular weight of 120,000 g/mol.

SOLUTION

The reaction by which 6,6-nylon is produced was described in Example 16-3. Hexamethylene diamine and adipic acid combine and release a molecule of water. When a long chain forms, there is, on average, one water molecule released for each reacting molecule. The molecular weights are 116 g/mol for hexamethylene diamine, 146 g/mol for adipic acid, and 18 g/mol for water. The repeat unit for 6,6-nylon is

The molecular weight of the repeat unit is the sum of the molecular weights of the two monomers, minus that of the two water molecules that are evolved:

$$M_{\text{repeat unit}} = 116 + 146 - 2(18) = 226 \text{ g/mol}$$

$$\text{Degree of polymerization} = \frac{120,000}{226} = 531$$

The degree of polymerization refers to the total number of repeat units in the chain.

Example 16-5 *Number and Weight Average Molecular Weights*

We have a polyethylene sample containing 4000 chains with molecular weights between 0 and 5000 g/mol, 8000 chains with molecular weights between 5000 and 10,000 g/mol, 7000 chains with molecular weights between 10,000 and 15,000 g/mol, and 2000 chains with molecular weights between 15,000 and 20,000 g/mol. Determine both the number and weight average molecular weights.

SOLUTION

First we need to determine the number fraction x_i and weight fraction f_i for each of the four ranges. For x_i, we simply divide the number in each range by 21,000, which is the total number of chains. To find f_i, we first multiply the number of chains by the mean molecular weight of the chains in each range, giving the "weight" of each group, then find f_i by dividing by the total weight of 192.5×10^6. We can then use Equations 16-2 and 16-3 to find the molecular weights.

Number of Chains	Mean M per Chain	x_i	$x_i M_i$	Weight	f_i	$f_i M_i$
4000	2500	0.191	477.5	10×10^6	0.0519	129.75
8000	7500	0.381	2857.5	60×10^6	0.3118	2338.50
7000	12,500	0.333	4162.5	87.5×10^6	0.4545	5681.25
2000	17,500	0.095	1662.5	35×10^6	0.1818	3181.50
$\Sigma = 21{,}000$		$\Sigma = 1.00$	$\Sigma = 9160$	$\Sigma = 192.5 \times 10^6$	$\Sigma = 1$	$\Sigma = 11{,}331$

$$\overline{M}_n = \sum x_i M_i = 9160 \ \text{g/mol}$$

$$\overline{M}_w = \sum f_i M_i = 11{,}331 \ \text{g/mol}$$

The weight average molecular weight is larger than the number average molecular weight.

16-4 Typical Thermoplastics

Some of the mechanical properties of typical thermoplastics are shown in Table 16-2. Table 16-3 shows the repeat units and applications for several thermoplastics formed by addition polymerization.

TABLE 16-2 ■ *Properties of selected thermoplastics*

	Tensile Strength (MPa)	% Elongation	Elastic Modulus (MPa)	Density (g/cm³)	Izod Impact (J/cm)
Polyethylene (PE):					
Low-density	21	800	276	0.92	4.9
High-density	38	130	1241	0.96	2.2
Ultrahigh molecular weight	48	350	690	0.934	16.2
Polyvinyl chloride (PVC)	62	100	4140	1.40	
Polypropylene (PP)	41	700	1517	0.90	0.5
Polystyrene (PS)	55	60	3103	1.06	0.2
Polyacrylonitrile (PAN)	62	4	4000	1.15	2.6
Polymethyl methacrylate (PMMA) (acrylic, Plexiglas)	83	5	3100	1.22	0.3
Polychlorotrifluoroethylene	41	250	2070	2.15	1.4
Polytetrafluoroethylene (PTFE, Teflon)	48	400	550	2.17	1.6
Polyoxymethylene (POM) (acetal)	83	75	3590	1.42	1.2
Polyamide (PA) (nylon)	83	300	3450	1.14	1.1
Polyester (PET)	72	300	4140	1.36	0.3
Polycarbonate (PC)	76	130	2760	1.20	8.6
Polyimide (PI)	117	10	2070	1.39	0.8
Polyetheretherketone (PEEK)	70	150	3790	1.31	0.9
Polyphenylene sulfide (PPS)	66	2	3310	1.30	0.3
Polyether sulfone (PES)	84	80	2410	1.37	0.9
Polyamide-imide (PAI)	186	15	5030	1.39	2.2

TABLE 16-3 ■ Repeat units and applications for selected addition thermoplastics

Polymer	Repeat Unit	Application	Polymer	Repeat Unit	Application
Polyethylene (PE)		Packing films, wire insulation, squeeze bottles, tubing, household items	Polyacrylonitrile (PAN)		Textile fibers, precursor for carbon fibers, food container
Polyvinyl chloride (PVC)		Pipe, valves, fittings, floor tile, wire insulation, vinyl automobile roofs	Polymethyl methacrylate (PMMA) (acrylic-Plexiglas)		Windows, windshields, coatings, hard contact lenses, lighted signs
Polypropylene (PP)		Tanks, carpet fibers, rope, packaging			
Polystyrene (PS)		Packaging and insulation foams, lighting panels, appliance components, egg cartons	Polychlorotri-fluoroethylene		Valve components, gaskets, tubing, electrical insulation
			Polytetrafluoroethylene (Teflon) (PTFE)		Seals, valves, nonstick coatings

Thermoplastics with Complex Structures

A large number of polymers, which typically are used for special applications and in relatively small quantities, are formed from complex monomers, often by the condensation mechanism. Oxygen, nitrogen, sulfur, and benzene rings (or aromatic groups) may be incorporated into the chain. Table 16-4 shows the repeat units and typical applications for a number of these complex polymers. Polyoxymethylene, or acetal, is a simple example in which the backbone of the polymer chain contains alternating carbon and oxygen atoms. A number of these polymers, including polyimides and polyetheretherketone (PEEK), are important aerospace materials. Because bonding within the chains is stronger compared to the simpler addition polymers, rotation and sliding of the chains is more difficult, leading to higher strengths, higher stiffnesses, and higher melting points. In some cases, good impact properties can be gained from these complex chains, with polycarbonates being particularly remarkable. Polycarbonates (Lexan™, Merlon™, and Sparlux™) are used to make bulletproof windows, compact discs for data storage, and in many other applications.

TABLE 16-4 ■ *Repeat units and applications for complex thermoplastics*

Polymer	Repeat Unit	Applications
Polyoxymethylene (acetal)(POM)		Plumbing fixtures, pens, bearings, gears, fan blades
Polyamide (nylon) (PA)		Bearings, gears, fibers, rope, automotive components, electrical components
Polyester (PET)		Fibers, photographic film, recording tape, boil-in bag containers, beverage containers
Polycarbonate (PC)		Electrical and appliance housings, automotive components, football helmets, returnable bottles, compact discs (CDs)
Polyimide (PI)		Adhesives, circuit boards, fibers for space shuttle
Polyetheretherketone (PEEK)		High-temperature electrical insulation and coatings
Polyphenylene sulfide (PPS)		Coatings, fluid-handling components, electronic components, hair dryer components
Polyether sulfone (PES)		Electrical components, coffeemakers, hair dryers, microwave oven components
Polyamide-imide (PAI)		Electronic components, aerospace and automotive applications

16-5   Structure—Property Relationships in Thermoplastics

Degree of Polymerization

In general, for a given type of thermoplastic (e.g., polyethylene), the tensile strength, creep resistance, impact toughness, wear resistance, and melting temperature all increase with increasing average molecular weight or degree of polymerization. The increases in these properties are not linear. As the average molecular weight increases, the melting temperature increases, and this makes the processing more difficult. In fact, we can make use of a bimodal molecular weight distribution in polymer processing. One component has lower molecular weight and helps melting, thereby making processing easier.

Effect of Side Groups

In polyethylene, the linear chains easily rotate and slide when stress is applied, and no strong polar bonds are formed between the chains; thus, polyethylene has a low strength.

Vinyl compounds have one of the hydrogen atoms replaced with a different atom or atom group. When R in the side group is chlorine, we produce polyvinyl chloride (PVC); when the side group is CH_3, we produce polypropylene (PP); addition of a benzene ring as a side group gives polystyrene (PS); and a CN group produces polyacrylonitrile (PAN).

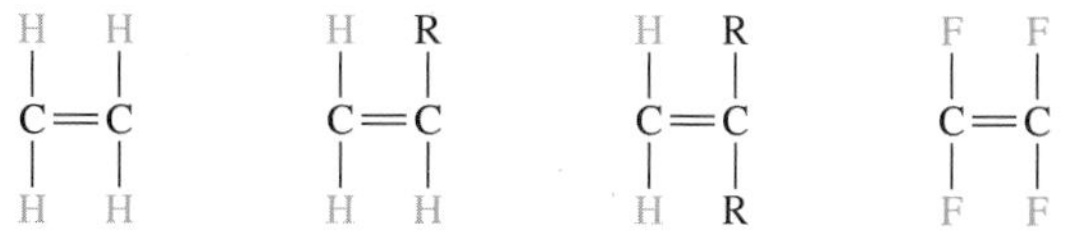

| Ethylene | Vinyl compound | Vinylidene compound | Tetrafluoro-ethylene |

When two of the hydrogen atoms are replaced, the monomer is a *vinylidene compound*, important examples of which include polyvinylidene chloride (the basis for Saran Wrap™) and polymethyl methacrylate (acrylics such as Lucite™ and Plexiglas™). Generally, a head-to-tail arrangement of the repeat units in the polymers is obtained (Figure 16-7). The head-to-tail arrangement is most typical.

The effects of adding other atoms or atom groups to the carbon backbone in place of hydrogen atoms are illustrated by the typical properties given in Table 16-2. Larger atoms such as chlorine or groups of atoms such as methyl (CH_3) and benzene make it more difficult for the chains to rotate, uncoil, disentangle, and deform by viscous flow when a stress is applied or when the temperature is increased. This condition leads to higher strengths, stiffnesses, and melting temperatures than those for polyethylene. The chlorine atom in PVC and the carbon-nitrogen group in PAN are strongly attracted by

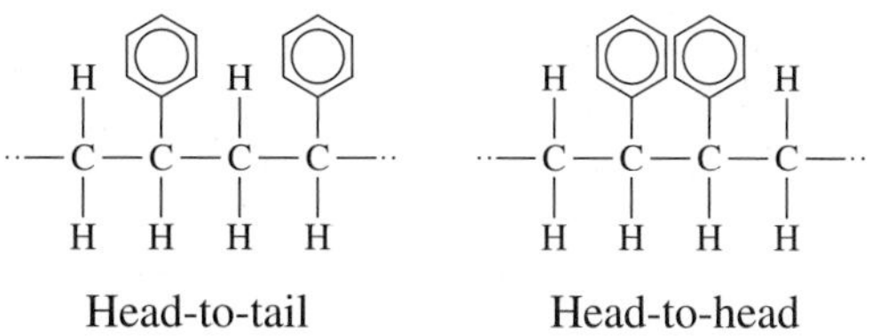

Head-to-tail   Head-to-head

Figure 16-7
Head-to-tail versus head-to-head arrangement of repeat units. The head-to-tail arrangement is most typical.

hydrogen bonding to hydrogen atoms on adjacent chains. This, for example, is the reason why PVC is more rigid than many other polymers. The way to get around the rigidity of PVC is to add low molecular weight compounds such as phthalate esters, known as plasticizers. When PVC contains these compounds, the glass transition temperature is lowered. This makes PVC more ductile and workable; such PVC is known as vinyl (not to be confused with the vinyl group mentioned here and in other places). PVC is used to make three-ring binders, pipes, tiles, and clear Tygon™ tubing.

In polytetrafluoroethylene (PTFE or Teflon™), all four hydrogen atoms in the polyethylene structure are replaced by fluorine. The monomer again is symmetrical, and the strength of the polymer is not much greater than that of polyethylene. The C-F bond permits PTFE to have a high melting point with the added benefit of low friction and nonstick characteristics that make the polymer useful for bearings and cookware. Teflon™ was invented by accident by Roy Plunkett, who was working with tetrafluoroethylene gas. He found a tetrafluoroethylene gas cylinder that had no pressure (and, thus, seemed empty) but was heavier than usual. The gas inside had polymerized into solid Teflon™!

Branching prevents dense packing of the chains, thereby reducing the density, stiffness, and strength of the polymer. Low-density polyethylene (LPDE), which has many branches, is weaker than high-density polyethylene (HDPE), which has virtually no branching (Table 16-2).

Crystallization and Deformation   *Crystallinity* is important in polymers since it affects mechanical and optical properties. Crystallinity evolves in the processing of polymers as a result of temperature changes and applied stress (e.g., formation of PET bottles discussed in previous chapters). If crystalline regions become too large, they begin to scatter light and make the plastic translucent. In certain special polymers, localized regions crystallize in response to an applied electric field, and this is the principle by which liquid-crystal displays work. We will discuss this in detail in the following section. Crystallization of the polymer also helps to increase density, resistance to chemical attack, and mechanical properties—even at higher temperatures—because of the stronger bonding between the chains. In addition, the deformation that straightens and aligns the chains, leading to crystallization, also produces a preferred orientation. Deformation of a polymer is often used in producing fibers having mechanical properties in the direction of the fiber that exceed those of many metals and ceramics. In fact, this texture strengthening played a key role in the discovery of nylon fibers. In previous chapters, we have seen how PET bottles develop a biaxial texture and strength along the radial and length directions.

Tacticity   When a polymer is formed from nonsymmetrical repeat units, the structure and properties are determined by the location of the nonsymmetrical atoms or atom groups. This condition is called **tacticity**, or stereoisomerism. In the syndiotactic arrangement, the atoms or atom groups alternately occupy positions on opposite sides of the linear chain. The atoms are all on the same side of the chain in *isotactic* polymers, whereas the arrangement of the atoms is random in *atactic* polymers (Figure 16-8).

The atactic structure, which is the least regular and least predictable, tends to give poor packing, low density, low strength and stiffness, and poor resistance to heat or chemical attack. Atactic polymers are more likely to have an amorphous structure. An example of the importance of tacticity occurs in polypropylene. Atactic polypropylene is an amorphous wax-like polymer with poor mechanical properties, whereas isotactic polypropylene may crystallize and is one of the most widely used commercial polymers.

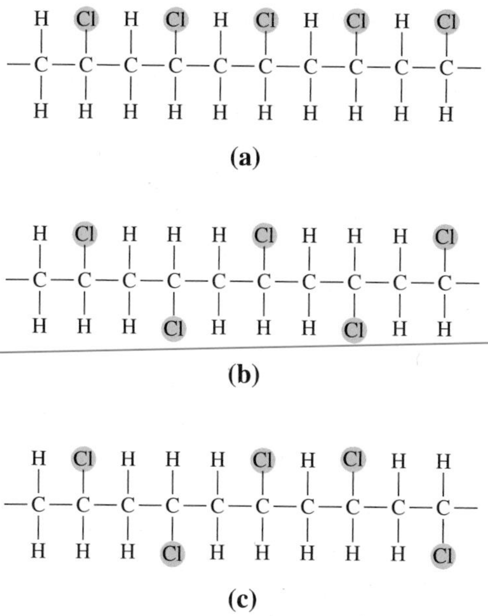

H Cl H Cl H Cl H Cl H Cl
| | | | | | | | | |
—C—C—C—C—C—C—C—C—C—C—
| | | | | | | | | |
H H H H H H H H H H

(a)

H Cl H H H Cl H H H Cl
| | | | | | | | | |
—C—C—C—C—C—C—C—C—C—C—
| | | | | | | | | |
H H H Cl H H H Cl H H

(b)

H Cl H H H Cl H Cl H H
| | | | | | | | | |
—C—C—C—C—C—C—C—C—C—C—
| | | | | | | | | |
H H H Cl H H H H H Cl

(c)

Figure 16-8
Three possible arrangements of nonsymmetrical monomers: (a) isotactic, (b) syndiotactic, and (c) atactic.

Copolymers

Similar to the concept of solid solutions or the idea of composites, linear-addition chains composed of two or more types of molecules can be arranged to form **copolymers**. This is a very powerful way to blend properties of different polymers. The arrangement of the monomers in a copolymer may take several forms (Figure 16-9). These include alternating, random, block, and grafted copolymers. ABS, composed of acrylonitrile, butadiene (a synthetic elastomer), and styrene, is one of the most common polymer materials (Figure 16-10). Styrene and acrylonitrile form a linear copolymer (SAN) that serves as a matrix. Styrene and butadiene also form a linear copolymer, BS rubber, which acts as the filler material. The combination of the two copolymers gives ABS an excellent combination of strength, rigidity, and toughness. Another common copolymer contains repeat units of ethylene and propylene. Whereas polyethylene and polypropylene are both easily crystallized, the copolymer remains amorphous. When this copolymer is cross-linked, it behaves as an elastomer. Dylark™ is based on a copolymer of maleic anhydride and styrene. Styrene provides toughness, while maleic anhydride provides high temperature properties. Carbon black (for protection from ultraviolet rays and enhancing stiffness), rubber (for toughness), and glass fibers (for stiffness) are added to the Dylark™ copolymer. It has been used to make instrument panels for car dashboards. The Dylark™ plastic is then coated with vinyl, which provides a smooth and soft finish.

Blending and Alloying

We can improve the mechanical properties of many of the thermoplastics by blending or alloying. By mixing an immiscible elastomer with the thermoplastic, we produce a two-phase polymer, as found in ABS. The elastomer does not enter the structure as a copolymer but, instead, helps to absorb energy and improve toughness. Polycarbonates used to produce transparent aircraft canopies are also toughened by elastomers in this manner.

Liquid-Crystalline Polymers

Some of the complex thermoplastic chains become so stiff that they act as rigid rods, even at high temperatures. These materials are **liquid-crystalline polymers** (LCPs). Some aromatic polyesters and aromatic

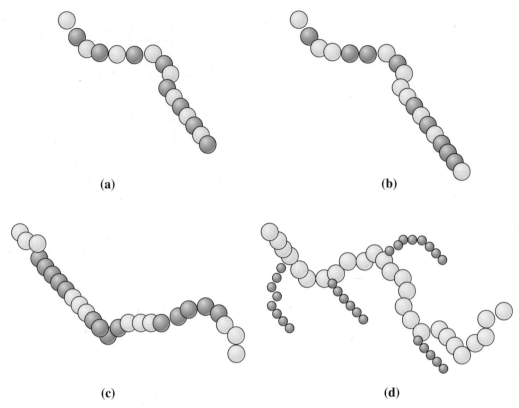

Figure 16-9 Four types of copolymers: (a) alternating monomers, (b) random monomers, (c) block copolymers, and (d) grafted copolymers. Circles of different colors or sizes represent different monomers.

polyamides (or **aramids**) are examples of liquid-crystalline polymers and are used as high-strength fibers (as will be discussed in Chapter 17). Kevlar^{TM}, an aromatic polyamide, is the most familiar of the LCPs and is used as a reinforcing fiber for aerospace applications and for bulletproof vests. Liquid-crystal polymers are also used to make electronic displays.

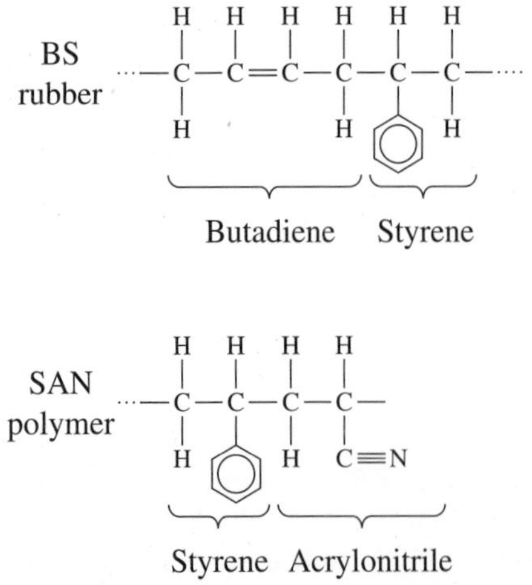

Figure 16-10
Copolymerization produces the polymer ABS, which consists of two copolymers, SAN and BS, grafted together.

16-6   Effect of Temperature on Thermoplastics

Properties of thermoplastics change depending upon temperature. We need to know how these changes occur because this can help us (a) better design components and (b) guide the type of processing techniques that need to be used. Several critical temperatures and structures, summarized in Figures 16-11 and 16-12, may be observed.

Thermoplastics can be amorphous or crystalline once they cool below the melting temperature (Figure 16-11). Most often, engineered thermoplastics have both amorphous and crystalline regions. The crystallinity in thermoplastics can be introduced by temperature (slow cooling) or by the application of stress that can untangle chains (**stress-induced crystallization**). Similar to dispersion strengthening of metallic materials, the formation of crystalline regions in an otherwise amorphous matrix helps increase the strength of thermoplastics. In typical thermoplastics, bonding within the chains is covalent, but the long coiled chains are held to one another by weak van der Waals bonds and by entanglement. When a tensile stress is applied to the thermoplastic, the weak bonding between the chains can be overcome, and the chains can rotate and slide relative to one another. The ease with which the chains slide depends on both temperature and the polymer structure.

Degradation Temperature   At very high temperatures, the covalent bonds between the atoms in the linear chain may be destroyed, and the polymer may burn or char. In thermoplastics, decomposition occurs in the liquid state; in thermosets, the decomposition occurs in the solid state. This temperature T_d (not shown in Figure 16-12), is the **degradation** (or decomposition) **temperature**. When plastics burn, they create smoke,

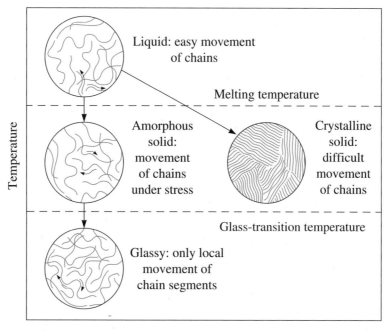

Figure 16-11   The effect of temperature on the structure and behavior of thermoplastics.

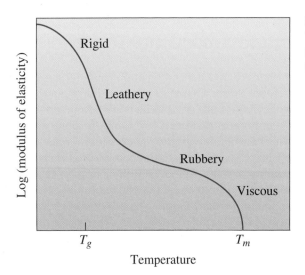

Figure 16-12
The effect of temperature on the modulus of elasticity for an amorphous thermoplastic. Note that T_g and T_m are not fixed.

which is dangerous. Some materials (such as limestone, talc, alumina, etc.) added to thermoplastics are thermal or heat stabilizers. They absorb heat and protect the polymer matrix. Fire retardant additives include hydrated alumina, antimony compounds, and halogen compounds (e.g., $MgBr$, PCl_5). Some additives retard fire by excluding oxygen but generate dangerous gases and are not appropriate for certain applications.

Exposure to other forms of chemicals or energy (e.g., oxygen, ultraviolet radiation, and attack by bacteria) also cause a polymer to degrade or **age** slowly, even at low temperatures. Carbon black (up to ~3%) is one of the commonly used additives that helps improve the resistance of plastics to ultraviolet degradation.

Liquid Polymers

Thermoplastics usually do not melt at a precise temperature. Instead there is a range of temperatures over which melting occurs. The approximate melting ranges of typical polymers are included in Table 16-5. At or above the melting temperature T_m, bonding between the twisted and intertwined chains is weak. If a force is applied, the chains slide past one another, and the polymer flows with virtually no elastic strain. The strength and modulus of elasticity are nearly zero, and the polymer is suitable for casting and many forming processes. Most thermoplastic melts are shear thinning (i.e., their apparent viscosity decreases within an increase in the steady-state shear rate).

Rubbery and Leathery States

Below the melting temperature, the polymer chains are still twisted and intertwined. These polymers have an amorphous structure. Just below the melting temperature, the polymer behaves in a *rubbery* manner. When stress is applied, both elastic and plastic deformation of the polymer occurs. When the stress is removed, the elastic deformation is quickly recovered, but the polymer is permanently deformed due to the movement of the chains. Large permanent elongations can be achieved, permitting the polymer to be formed into useful shapes by molding and extrusion.

At lower temperatures, bonding between the chains is stronger, the polymer becomes stiffer and stronger, and a *leathery* behavior is observed. Many of the commercial polymers, including polyethylene, have a useable strength in this condition.

TABLE 16-5 ■ *Melting, glass-transition, and processing temperature ranges (°C) for selected thermoplastics and elastomers*

Polymer	Melting Temperature Range	Glass-Transition Temperature Range (T_g)	Processing Temperature Range
Addition polymers			
Low-density (LD) polyethylene	98–115	−90 to −25	149–232
High-density (HD) polyethylene	130–137	−110	177–260
Polyvinyl chloride	175–212	87	
Polypropylene	160–180	−25 to −20	190–288
Polystyrene	240	85–125	
Polyacrylonitrile	320	107	
Polytetrafluoroethylene (Teflon)	327		
Polychlorotrifluoroethylene	220		
Polymethyl methacrylate (acrylic)		90–105	
Acrylonitrile butadiene styrene (ABS)	110–125	100	177–260
Condensation polymers			
Acetal	181	−85	
6,6-nylon	243–260	49	260–327
Cellulose acetate	230		
Polycarbonate	230	149	271–300
Polyester	255	75	
Polyethylene terephthalate (PET)	212–265	66–80	227–349
Elastomers			
Silicone		−123	
Polybutadiene	120	−90	
Polychloroprene	80	−50	
Polyisoprene	30	−73	

Glassy State Below the **glass-transition temperature** T_g, the linear amorphous polymer becomes hard, brittle, and glass-like. This is again not a fixed temperature but a range of temperatures. When the polymer cools below the glass-transition temperature, certain properties—such as density or modulus of elasticity—change at a different rate (Figure 16-13).

Although glassy polymers have poor ductility and formability, they do have good strength, stiffness, and creep resistance. A number of important polymers, including polystyrene and polyvinyl chloride, have glass-transition temperatures above room temperature (Table 16-5).

The glass-transition temperature is typically about 0.5 to 0.75 times the absolute melting temperature T_m. Polymers such as polyethylene, which have no complicated side groups attached to the carbon backbone, have low glass-transition temperatures (even below room temperature) compared with polymers such as polystyrene, which have more complicated side groups.

As pointed out in Chapter 6, many thermoplastics become brittle at lower temperatures. The brittleness of the polymer used for some of the O-rings ultimately caused the 1986 *Challenger* disaster. The lower temperatures that existed during the launch time caused the embrittlement of the rubber O-rings used for the booster rockets.

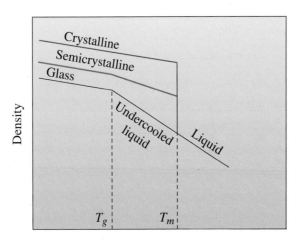

Figure 16-13
The relationship between the density and temperature of a polymer shows the melting and glass-transition temperatures. Note that T_g and T_m are not fixed; rather, they are ranges of temperatures.

Observing and Measuring Crystallinity in Polymers

Many thermoplastics partially crystallize when cooled below the melting temperature, with the chains becoming closely aligned over appreciable distances. A sharp increase in the density occurs as the coiled and intertwined chains in the liquid are rearranged into a more orderly, close-packed structure (Figure 16-13).

One model describing the arrangement of the chains in a crystalline polymer is shown in Figure 16-14. In this *folded chain* model, the chains loop back on themselves, with each loop being approximately 100 carbon atoms long. The folded chain extends in three dimensions, producing thin plates or lamellae. The crystals can take various forms, with the spherulitic shape shown in Figure 16-15(a) being particularly common. The crystals have a unit cell that describes the regular packing of the chains. The crystal structure for polyethylene, shown in Figure 16-15(b), describes one such unit cell. Crystal structures for several polymers are listed in Table 16-6. Some polymers are polymorphic, having more than one crystal structure.

Even in crystalline polymers, there are always thin regions between the lamellae, as well as between spherulites, that are amorphous transition zones. The weight percentage of the structure that is crystalline can be calculated from the density of the polymer:

$$\% \text{ Crystalline} = \frac{\rho_c(\rho - \rho_a)}{\rho(\rho_c - \rho_a)} \times 100 \qquad (16\text{-}4)$$

where ρ is the measured density of the polymer, ρ_a is the density of amorphous polymer, and ρ_c is the density of completely crystalline polymer. Similarly, x-ray diffraction (XRD) can be used to measure the level of crystallinity and determine lattice constants for single crystal polymers.

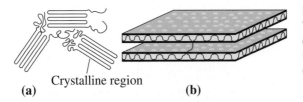

Crystalline region
(a)   **(b)**

Figure 16-14
The folded chain model for crystallinity in polymers, shown in (a) two dimensions and (b) three dimensions.

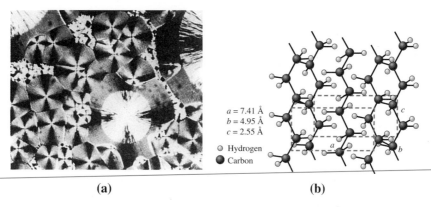

(a)   **(b)**

Figure 16-15 (a) Photograph of spherulitic crystals in an amorphous matrix of nylon ($\times$ 200). *(From R. Brick, A. Pense and R. Gordon,* Structure and Properties of Engineering Materials, *4th Ed., McGraw-Hill, 1997.)* (b) The unit cell of crystalline polyethylene.

TABLE 16-6 ■ *Crystal structures of several polymers*

Polymer	Crystal Structure	Lattice Parameters (nm)
Polyethylene	Orthorhombic	$a_0 = 0.741$ $b_0 = 0.495$ $c_0 = 0.255$
Polypropylene	Orthorhombic	$a_0 = 1.450$ $b_0 = 0.569$ $c_0 = 0.740$
Polyvinyl chloride	Orthorhombic	$a_0 = 1.040$ $b_0 = 0.530$ $c_0 = 0.510$
Polyisoprene (cis)	Orthorhombic	$a_0 = 1.246$ $b_0 = 0.886$ $c_0 = 0.810$

As the side groups get more complex, it becomes harder to crystallize thermoplastics. For example, polyethylene (H as side group) can be crystallized more easily than polystyrene (benzene ring as side group). High-density polyethylene (HDPE) has a higher level of crystallinity and, therefore, a higher density (0.97 g/cm^3) than low-density polyethylene (LDPE), which has a density of 0.92 g/cm^3. The crystallinity and, hence, the density in LDPE is lower, since the polymer is branched. Thus, branched polymers show lower levels of crystallinity. A completely crystalline polymer would not display a glass-transition temperature; however, the amorphous regions in semicrystalline polymers do transform to a glassy material below the glass-transition temperature (Figure 16-13). Such polymers as acetal, nylon, HDPE, and polypropylene are referred to as crystalline even though the level of crystallinity may be moderate. The following examples show how properties of plastics can be accounted for in different applications.

Example 16-6 *Design of a Polymer Insulation Material*

A storage tank for liquid hydrogen will be made from metal, but we wish to coat the metal with a 3-mm thick polymer as an intermediate layer between the metal and additional insulation layers. The temperature of the intermediate layer may drop to −80°C. Choose a material for this layer.

SOLUTION

We want the material to have reasonable ductility. As the temperature of the tank changes, stresses develop in the coating due to differences in thermal expansion, and we do not want the polymer to fail due to these stresses. A material that has good ductility and/or can undergo large elastic strains is needed. We therefore would prefer either a thermoplastic that has a glass-transition temperature below −80°C or an elastomer, also with a glass-transition temperature below −80°C. Of the polymers listed in Table 16-2, thermoplastics such as polyethylene and acetal are satisfactory. Suitable elastomers include silicone and polybutadiene.

We might prefer one of the elastomers, for they can accommodate thermal stress by elastic, rather than plastic, deformation.

Example 16-7   *Impact Resistant Polyethylene*

A new grade of flexible, impact resistant polyethylene for use as a thin film requires a density of 0.88 to 0.915 g/cm^3. Design the polyethylene required to produce these properties. The density of amorphous polyethylene is about 0.87 g/cm^3.

SOLUTION

To produce the required properties and density, we must control the percent crystallinity of the polyethylene. We can use Equation 16-4 to determine the crystallinity that corresponds to the required density range. To do so, however, we must know the density of completely crystalline polyethylene. We can use the data in Figure 16-15 and Table 16-6 to calculate this density if we recognize that there are two polyethylene repeat units in each unit cell.

$$\rho_c = \frac{(4\ C)(12\ g/mol) + (8\ H)(1\ g/mol)}{(7.41 \times 10^{-8}\ cm)(4.95 \times 10^{-8}\ cm)(2.55 \times 10^{-8}\ cm)(6.022 \times 10^{23}\ atoms/mol)}$$

$$= 0.9942\ g/cm^3$$

We know that $\rho_a = 0.87$ g/cm^3 and that ρ varies from 0.88 to 0.915 g/cm^3. The required crystallinity then varies from

$$\%\ crystalline = \frac{(0.9942)(0.88 - 0.87)}{(0.88)(0.9942 - 0.87)} \times 100 = 9.1$$

$$\%\ crystalline = \frac{(0.9942)(0.915 - 0.87)}{(0.915)(0.9942 - 0.87)} \times 100 = 39.4$$

Therefore, we must be able to process the polyethylene to produce a range of crystallinity between 9.2 and 39.4%.

16-7   **Mechanical Properties of Thermoplastics**

Most thermoplastics (molten and solid) exhibit a non-Newtonian and **viscoelastic behavior**. The behavior is non-Newtonian (i.e., the stress and strain are not linearly related for most parts of the stress-strain curve). The viscoelastic behavior means when an

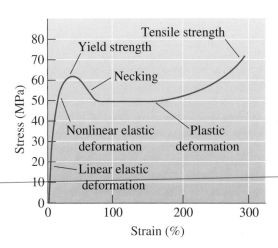

Figure 16-16
The engineering stress-strain curve for 6,6-nylon, a typical thermoplastic polymer.

external force is applied to a thermoplastic polymer, both elastic and plastic (or viscous) deformation occurs. The mechanical behavior is closely tied to the manner in which the polymer chains move relative to one another under load. The deformation process depends on both time and the rate at which the load is applied. Figure 16-16 shows a stress-strain curve for 6,6-nylon.

Elastic Behavior

Elastic deformation in thermoplastics is the result of two mechanisms. An applied stress causes the covalent bonds within the chain to stretch and distort, allowing the chains to elongate elastically. When the stress is removed, recovery from this distortion is almost instantaneous. This behavior is similar to that in metals and ceramics, which also deform elastically by the stretching of metallic, ionic, or covalent bonds. In addition, entire segments of the polymer chains may be distorted; when the stress is removed, the segments move back to their original positions over a period of time—often hours or even months. This time-dependent, or viscoelastic, behavior may contribute to some nonlinear elastic behavior.

Plastic Behavior of Amorphous Thermoplastics

These polymers deform plastically when the stress exceeds the yield strength. Unlike deformation in the case of metals, however, plastic deformation is not a consequence of dislocation movement. Instead, chains stretch, rotate, slide, and disentangle under load to cause permanent deformation. The drop in the stress beyond the yield point can be explained by this phenomenon. Initially, the chains may be highly tangled and intertwined. When the stress is sufficiently high, the chains begin to untangle and straighten. Necking also occurs, permitting continued sliding of the chains at a lesser stress. Eventually, however, the chains become almost parallel and close together; stronger van der Waals bonding between the more closely aligned chains requires higher stresses to complete the deformation and fracture process (Figure 16-17). This type of crystallization due to orientation played an important role in the discovery of nylon as a material to make strong fibers.

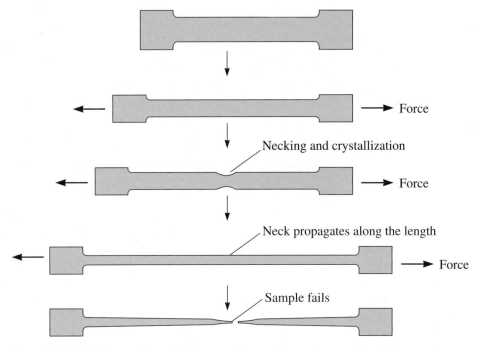

Figure 16-17   Necks can be stable in amorphous polymers because local alignment strengthens the necked region.

Example 16-8   *Comparing Mechanical Properties of Thermoplastics*

Compare the mechanical properties of low-density (LD) polyethylene, high-density (HD) polyethylene, polyvinyl chloride, polypropylene, and polystyrene, and explain their differences in terms of their structures.

SOLUTION

Let us look at the maximum tensile strength and modulus of elasticity for each polymer.

Polymer	Tensile Strength (MPa)	Modulus of Elasticity (MPa)	Structure
LD polyethylene	21	276	Highly branched, amorphous structure with symmetrical monomers
HD polyethylene	38	1241	Amorphous structure with symmetrical monomers but little branching
Polypropylene	41	1517	Amorphous structure with small methyl side groups
Polystyrene	55	3103	Amorphous structure with benzene side groups
Polyvinyl chloride	62	4137	Amorphous structure with large chlorine atoms as side groups

We can conclude that

1. Branching, which reduces the density and close packing of chains, reduces the mechanical properties of polyethylene.

2. Adding atoms or atom groups other than hydrogen to the chain increases strength and stiffness. The methyl group in polypropylene provides some improvement; the benzene ring of styrene provides higher strength and stiffness; and the chlorine atom in polyvinyl chloride provides a large increase in these properties.

Creep and Stress Relaxation

Thermoplastics exhibit creep, a time-dependent permanent deformation with constant stress or load (Figures 16-18 and 16-19).

They also show **stress relaxation** (i.e., under a constant strain, the stress level decreases with time) (Chapter 6). Stress relaxation, like creep, is a consequence of the viscoelastic behavior of the polymer. Perhaps the most familiar example of this behavior is a rubber band (an elastomer) stretched around a pile of books. Initially, the tension in the rubber band is high when the rubber band is taut. After several weeks, the strain in the rubber band is unchanged (it still completely encircles the books), but the stress will have decreased. Similarly, the nylon strings in tennis rackets are pulled at a higher tension initially since this tension (i.e., stress) decreases with time.

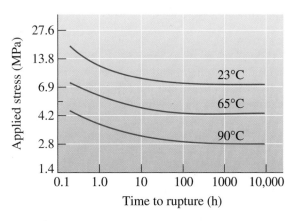

Figure 16-18
The effect of temperature on the stress-rupture behavior of high-density polyethylene.

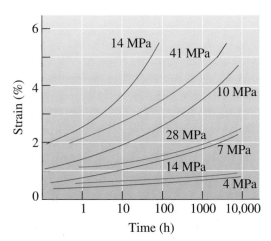

Figure 16-19
Creep curves for acrylic (PMMA) (colored lines) and polypropylene (black lines) at 20°C and several applied stresses.

In a simple model, the rate at which stress relaxation occurs is related to the **relaxation time** λ, which is considered a property of the polymer (more complex models consider a distribution of relaxation times). The stress after time t is given by

$$\sigma = \sigma_0 \exp(-t/\lambda) \tag{16-5}$$

where λ_0 is the original stress. The *relaxation time*, in turn, depends on the viscosity and, thus, the temperature:

$$\lambda = \lambda_0 \exp(Q/RT) \tag{16-6}$$

where σ_0 is a constant and Q is the activation energy related to the ease with which polymer chains slide past each other. Relaxation of the stress occurs more *rapidly* at *higher temperatures* and for polymers with a low viscosity.

The following example shows how stress relaxation can be accounted for while designing with polymers.

Example 16-9   *Determination of Initial Stress in a Polymer*

A band of polyisoprene is to hold together a bundle of steel rods for up to one year. If the stress on the band is less than 10 MPa, the band will not hold the rods tightly. Determine the initial stress that must be applied to a polyisoprene band when it is slipped over the steel. A series of tests showed that an initial stress of 7 MPa decreased to 6.8 MPa after six weeks.

SOLUTION

Although the strain on the elastomer band may be constant, the stress will decrease over time due to stress relaxation. We can use Equation 16-5 and our initial tests to determine the relaxation time for the polymer:

$$\sigma = \sigma_0 \exp\left(-\frac{t}{\lambda}\right)$$

$$6.8 = 7 \exp\left(-\frac{6}{\lambda}\right)$$

$$-\frac{6}{\lambda} = \ln\left(\frac{6.8}{7}\right) = \ln(0.98) = -0.0305$$

$$\lambda = \frac{6}{0.0305} = 197 \text{ weeks}$$

Now that we know the relaxation time, we can determine the stress that must be initially placed onto the band so that it will still be stressed to 10 MPa after 1 year (52 weeks).

$$10 = \sigma_0 \exp(-52/197) = \sigma_0 \exp(-0.264) = 0.768\sigma_0$$

$$\sigma_0 = \frac{10}{0.768} = 13 \text{ MPa}$$

The polyisoprene band must be made significantly undersized so it can slip over the materials it is holding together with a tension of 13 MPa. After one year, the stress will still be 10 MPa.

TABLE 16-7 ■ *Deflection temperatures for selected polymers for a 1.8 MPa load*

Polymer	Deflection Temperature (°C)
Polyester	40
Polyethylene (ultra-high density)	40
Polypropylene	60
Phenolic	80
Polyamide (6,6-nylon)	90
Polystyrene	100
Polyoxymethylene (acetal)	130
Polyamide-imide	280
Epoxy	290

One more practical measure for high temperature and creep properties of a polymer is the heat **deflection temperature** or heat **distortion temperature** under load, which is the temperature at which a given deformation of a beam occurs for a standard load. A high deflection temperature indicates good resistance to creep and permits us to compare various polymers. The deflection temperatures for several polymers are shown in Table 16-7, which gives the temperature required to cause a 0.025 cm deflection for a 1.8 MPa load at the center of a bar resting on supports 10 cm apart. A polymer is "dishwasher safe" if it has a heat distortion temperature greater than ~50°C.

Impact Behavior
Viscoelastic behavior also helps us understand the impact properties of polymers. At very high rates of strain, as in an impact test, there is insufficient time for the chains to slide and cause plastic deformation. For these conditions, the thermoplastics behave in a brittle manner and have poor impact values. Polymers may have a transition temperature. At low temperatures, brittle behavior is observed in an impact test, whereas more ductile behavior is observed at high temperatures, where the chains move more easily. These effects of temperature and strain rate are similar to those seen in metals that exhibit a ductile-to-brittle transition temperature; however, the mechanisms are different.

Deformation of Crystalline Polymers
A number of polymers are used in the crystalline state. As we discussed earlier, however, the polymers are never completely crystalline. Instead, small regions—between crystalline lamellae and between crystalline spherulites—are amorphous transition regions. Polymer chains in the crystalline region extend into these amorphous regions as tie chains. When a tensile load is applied to the polymer, the crystalline lamellae within the spherulites slide past one another and begin to separate as the tie chains are stretched. The folds in the lamellae tilt and become aligned with the direction of the tensile load. The crystalline lamellae break into smaller units and slide past one another, until eventually the polymer is composed of small aligned crystals joined by tie chains and oriented parallel to the tensile load. The spherulites also change shape and become elongated in the direction of the applied stress. With continued stress, the tie chains disentangle or break, causing the polymer to fail.

Crazing
Crazing occurs in thermoplastics when localized regions of plastic deformation occur in a direction perpendicular to that of the applied stress. In transparent thermoplastics, such as some of the glassy polymers, the craze produces a translucent or

opaque region that looks like a crack. The craze can grow until it extends across the entire cross section of the polymer part. The craze is not a crack, and, in fact, it can continue to support an applied stress. The process is similar to that for the plastic deformation of the polymer, but the process can proceed even at a low stress over an extended length of time. Crazing can lead to brittle fracture of the polymer and is often assisted by the presence of a solvent (known as solvent crazing).

Blushing   **Blushing** or whitening refers to failure of a plastic because of localized crystallization (due to repeated bending, for example) that ultimately causes voids to form.

16-8   Elastomers [Rubbers]

A number of natural and synthetic polymers called elastomers display a large amount (>200%) of elastic deformation when a force is applied. Rubber bands, automobile tires, O-rings, hoses, and insulation for electrical wires are common uses for these materials. Crude natural rubber, which is an elastomer, can erase pencil marks; hence, elastomers got the name rubber.

Geometric Isomers   Some monomers that have different structures, even though they have the same composition, are called **geometric isomers**. Isoprene, or natural rubber, is an important example (Figure 16-20). The monomer includes two double bonds between carbon atoms; this type of monomer is called a **diene**. Polymerization occurs by breaking both double bonds, creating a new double bond at the center of the molecule and active sites at both ends.

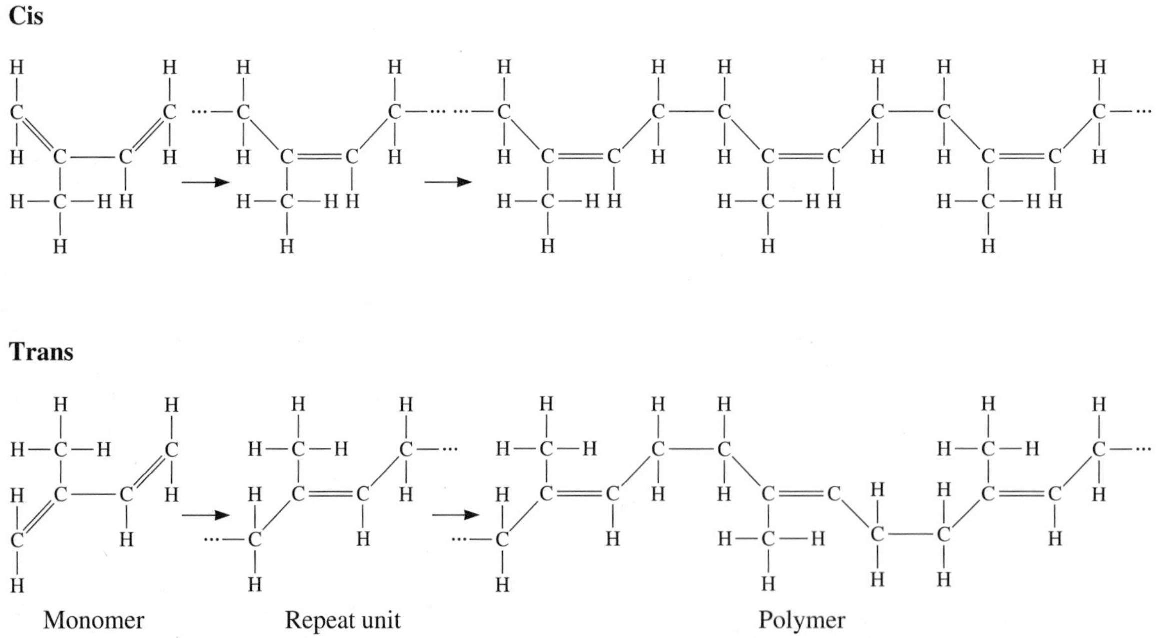

Figure 16-20   The cis and trans structures of isoprene. The cis form is useful for producing the isoprene elastomer.

In the *trans* form of isoprene, the hydrogen atom and the methyl group at the center of the repeat unit are located on opposite sides of the newly formed double bond. This arrangement leads to relatively straight chains; the polymer crystallizes and forms a hard rigid polymer called *gutta percha*. This is used to make golf balls and shoe soles.

In the *cis* form, however, the hydrogen atom and the methyl group are located on the same side of the double bond. This different geometry causes the polymer chains to develop a highly coiled structure, preventing close packing and leading to an amorphous, rubbery polymer. If a stress is applied to cis-isoprene, the polymer behaves in a viscoelastic manner. The chains uncoil and bonds stretch, producing elastic deformation, but the chains also slide past one another, producing nonrecoverable plastic deformation. The polymer behaves as a thermoplastic rather than an elastomer.

Cross-Linking

We prevent viscous plastic deformation while retaining large elastic deformation by **cross-linking** the chains (Figure 16-21). **Vulcanization**, which uses sulfur atoms, is a common method for cross-linking. Strands of sulfur atoms link the polymer chains as the polymer is processed and shaped at temperatures of about 120 to 180°C (Figure 16-22). The cross-linking steps may include rearranging a hydrogen atom and replacing one or more of the double bonds with single bonds. The cross-linking process is not reversible; consequently, the elastomer cannot be easily recycled.

The stress-strain curve for an elastomer is shown in Figure 16-23. Virtually all of the curve represents elastic deformation; elastomers display a nonlinear elastic behavior. Initially, the modulus of elasticity decreases because of the uncoiling of the chains; however, after the chains have been extended, further elastic deformation occurs by the stretching of the bonds, leading to a higher modulus of elasticity.

The number of cross-links (or the amount of sulfur added to the material) determines the elasticity of the rubber. Low sulfur additions leave the rubber soft and flexible, as in elastic bands or rubber gloves. Increasing the sulfur content restricts the uncoiling of the chains, and the rubber becomes harder, more rigid, and brittle, as in rubber used for

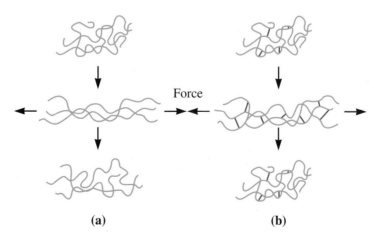

Force

(a)   (b)

Figure 16-21   (a) When the elastomer contains no cross-links, the application of a force causes both elastic and plastic deformation; after the load is removed, the elastomer is permanently deformed. (b) When cross-linking occurs, the elastomer still may undergo large elastic deformation; however, when the load is removed, the elastomer returns to its original shape.

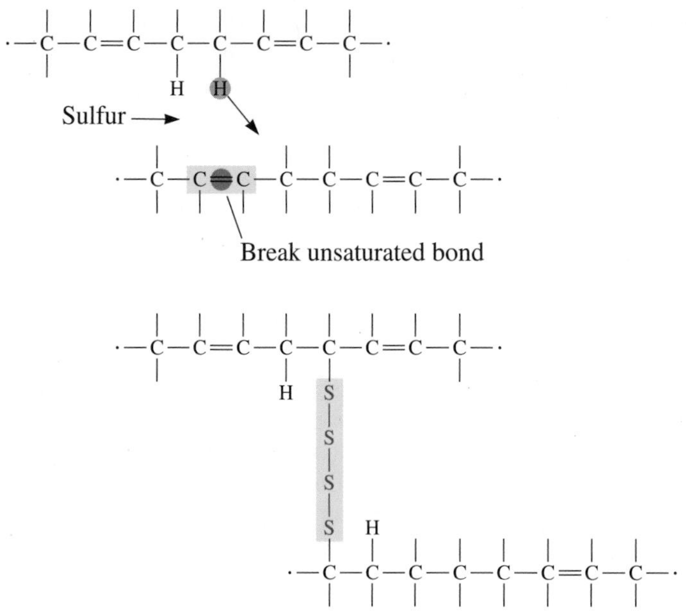

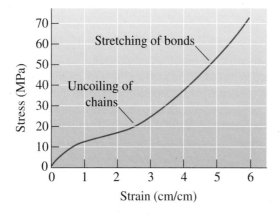

Figure 16-22
Cross-linking of polyisoprene chains may occur by introducing strands of sulfur atoms. Sites for attachment of the sulfur strands occur by rearrangement or loss of a hydrogen atom and the breaking of an unsaturated bond.

Figure 16-23
The stress-strain curve for an elastomer. Virtually all of the deformation is elastic; therefore, the modulus of elasticity varies as the strain changes.

motor mounts. Typically, 0.5 to 5% sulfur is added to provide cross-linking in elastomers. Many more *efficient vulcanizing* (EV) systems, which are sulfur free, also have been developed and used in recent years.

Typical Elastomers

Elastomers, which are amorphous polymers, do not easily crystallize during processing. They have a low glass-transition temperature, and chains can easily be deformed elastically when a force is applied. The typical elastomers (Tables 16-8 and 16-9) meet these requirements. Polyisoprene is a natural rubber. Polychloroprene, or Neoprene, is a common material for hoses and electrical insulation. Many of the important synthetic elastomers are copolymers. Polybutadiene (butadiene rubber or Buna-S) is similar to polyisoprene, but the repeat unit has four carbon atoms consisting of one double bond. This is a relatively low-cost rubber, but

TABLE 16-8 ■ *Repeat units and applications for selected elastomers*

Polymer	Repeat Unit	Applications
Polyisoprene		Tires, golf balls, shoe soles
Polybutadiene (or butadiene rubber or Buna-S)		Industrial tires, toughening other elastomers, inner tubes of tires, weatherstripping, steam hoses
Polyisobutylene (or butyl rubber)		
Polychloroprene (Neoprene)		Hoses, cable sheathing
Butadiene-styrene (BS or SBR rubber)		Tires
Butadiene-acrylonitrile (Buna-N)		Gaskets, fuel hoses
Silicones		Gaskets, seals

the resistance to solvents is poor. As a result, it is used as a toughening material to make other elastomers. Butadiene-styrene rubber (BSR or BS), which is also one of the components of ABS (Figure 16-10), is used for automobile tires. Butyl rubber is different from polybutadiene. Butyl rubber, or polyisobutadiene, is used to make inner tubes for tires, vibration mounts, and weather-stripping material. Silicones are another

TABLE 16-9 ■ *Properties of selected elastomers*

	Tensile Strength (MPa)	% Elongation	Density (g/cm^3)
Polyisoprene	21	800	0.93
Polybutadiene	24		0.94
Polyisobutylene	28	350	0.92
Polychloroprene	24	800	1.24
Butadiene-styrene	21	2000	1.0
Butadiene-acrylonitrile	5	400	1.0
Silicones	7	700	1.5
Thermoplastic elastomers	35	1300	1.06

important elastomer based on chains composed of silicon and oxygen atoms. Silly Putty® was invented by James Wright of General Electric. It is made using hydroxyl terminated polydimethyl siloxane, boric oxide, and some other additives. At slow strain rates, you can stretch it significantly, while if you pull it fast, it snaps. The silicone rubbers (also known as polysiloxanes) provide high temperature resistance, permitting use of the elastomer at temperatures as high as 315°C. Low molecular weight silicones form liquids and are known as silicon oils. Silicones can also be purchased as a two-part system that can be molded and cured. Chewing gum contains a base that is made from natural rubber, styrene butadiene, or polyvinyl acetate (PVA).

Thermoplastic Elastomers (TPEs)   This is a special group of polymers that do not rely on cross-linking to produce a large amount of elastic deformation. Figure 16-24 shows the structure of a styrene-butadiene block copolymer engineered so that the styrene repeat units are located only at the ends of the chains. Approximately 25% of the chain is composed of styrene. The styrene ends of several chains form spherical-shaped domains. The styrene has a high glass-transition temperature; consequently, the domains are strong and rigid and tightly hold the chains together. Rubbery areas containing butadiene repeat units are located between the styrene domains; these portions of the polymer have a glass-transition temperature below room temperature and therefore behave in a soft, rubbery manner. Elastic deformation occurs by recoverable movement of the chains; sliding of the chains at normal temperatures is prevented by the styrene domains.

    The styrene-butadiene block copolymers differ from the BS rubber discussed earlier in that cross-linking of the butadiene monomers is not necessary and, in fact, is undesirable. When the thermoplastic elastomer is heated, the styrene heats above the glass-transition temperature, the domains are destroyed, and the polymer deforms in a viscous manner—that is, it behaves as any other thermoplastic, making fabrication very easy. When the polymer cools, the domains reform, and the polymer reverts to its elastomeric characteristics. The thermoplastic elastomers consequently behave as ordinary thermoplastics at elevated temperatures and as elastomers at low temperatures. This behavior also permits thermoplastic elastomers to be more easily recycled than conventional elastomers. A useful fluoroelastomer for high temperature and corrosive environments is Viton™. It is used for seals, O-rings, and other applications.

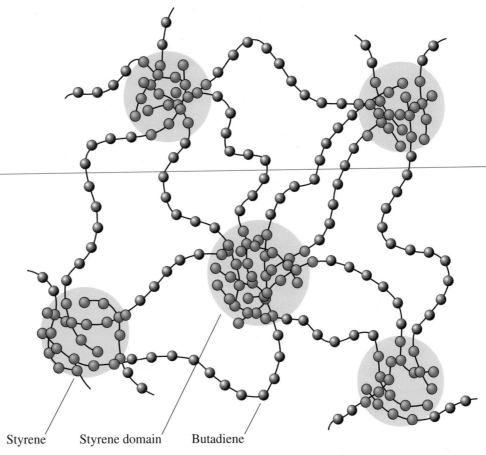

Styrene   Styrene domain   Butadiene

Figure 16-24 The structure of the styrene-butadiene copolymer in a thermoplastic elastomer. The glassy nature of the styrene domains provides elastic behavior without cross-linking of the butadiene.

16-9 Thermosetting Polymers

Thermosets are highly cross-linked polymer chains that form a three-dimensional network structure. Because the chains cannot rotate or slide, these polymers possess good strength, stiffness, and hardness. Thermosets also have poor ductility and impact properties and a high glass-transition temperature. In a tensile test, thermosetting polymers display the same behavior as a brittle metal or ceramic.

Thermosetting polymers often begin as linear chains. Depending on the type of repeat units and the degree of polymerization, the initial polymer may be either a solid or a liquid resin; in some cases, a two- or three-part liquid resin is used (as in the case of the two tubes of epoxy glue that we often use). Heat, pressure, mixing of the various resins, or other methods initiate the cross-linking process. Cross-linking is not reversible; once formed, the thermosets cannot be reused or conveniently recycled.

The functional groups for a number of common thermosetting polymers are summarized in Table 16-10, and representative properties are given in Table 16-11. A functional group is a defined arrangement of atoms with a particular set of properties.

TABLE 16-10 ■ *Functional units and applications for selected thermosets*

Polymer	Functional Units	Typical Applications
Phenolics		Adhesives, coatings, laminates
Amines		Adhesives, cookware, electrical moldings
Polyesters		Electrical moldings, decorative laminates, polymer matrix in fiberglass
Epoxies		Adhesives, electrical moldings, matrix for composites
Urethanes		Fibers, coatings, foams, insulation
Silicone		Adhesives, gaskets, sealants

Phenolics

Phenolics, the most commonly used thermosets, are often used as adhesives, coatings, laminates, and molded components for electrical or motor applications.

Bakelite™ is one of the common phenolic thermosets. A condensation reaction joining phenol and formaldehyde molecules produces the initial linear phenolic resin. This

TABLE 16-11 ■ *Properties of typical thermosetting polymers*

	Tensile Strength (MPa)	% Elongation	Elastic Modulus (MPa)	Density (g/cm^3)
Phenolics	62	2	9	1.27
Amines	69	1	11	1.50
Polyesters	90	3	5	1.28
Epoxies	103	6	4	1.25
Urethanes	69	6		1.30
Silicone	28	0	8	1.55

process continues until a linear phenol-formaldehyde chain is formed; however, the phenol is trifunctional. After the chain has formed, there is a third location on each phenol ring that provides a site for cross-linking with the adjacent chains.

Amines   Amino resins, produced by combining urea or melamine monomers with formaldehyde, are similar to the phenolics. The monomers are joined by a formaldehyde link to produce linear chains. Excess formaldehyde provides the cross-linking needed to give strong, rigid polymers suitable for adhesives, laminates, molding materials for cookware, and electrical hardware such as circuit breakers, switches, outlets, and wall plates.

Urethanes   Depending on the degree of cross-linking, the urethanes behave as thermosetting polymers, thermoplastics, or elastomers. These polymers find applications as fibers, coatings, and foams for furniture, mattresses, and insulation.

Polyesters   Polyesters form chains from acid and alcohol molecules by a condensation reaction, giving water as a byproduct. When these chains contain unsaturated bonds, a styrene molecule may provide cross-linking. Polyesters are used as molding materials for a variety of electrical applications, decorative laminates, boats and other marine equipment, and as a matrix for composites such as fiberglass.

Epoxies   Epoxies are thermosetting polymers formed from molecules containing a tight C—O—C ring. During polymerization, the C—O—C rings are opened and the bonds are rearranged to join the molecules. The most common of the commercial epoxies is based on bisphenol A, to which two epoxide units have been added. These molecules are polymerized to produce chains and then co-reacted with curing agents that provide cross-linking. Epoxies are used as adhesives, rigid molded parts for electrical applications, automotive components, circuit boards, sporting goods, and a matrix for high-performance fiber-reinforced composite materials for aerospace applications.

Polyimides   Polyimides display a ring structure that contains a nitrogen atom. One special group, the bismaleimides (BMI), is important in the aircraft and aerospace industry. They can operate continuously at temperatures of 175°C and do not decompose until reaching 460°C.

Interpenetrating Polymer Networks   Some special polymer materials can be produced when linear thermoplastic chains are intertwined through a thermosetting framework, forming **interpenetrating polymer networks**. For example, nylon, acetal, and polypropylene chains can penetrate into a cross-linked silicone thermoset. In more advanced systems, two interpenetrating thermosetting framework structures can be produced.

16-10   Adhesives

Adhesives are polymers used to join other polymers, metals, ceramics, composites, or combinations of these materials. Adhesives are used for a variety of applications. The most critical of these are the "structural adhesives," which find use in the automotive, aerospace, appliance, electronics, construction, and sporting equipment areas.

Chemically Reactive Adhesives
These adhesives include polyurethane, epoxy, silicone, phenolics, anaerobics, and polyimides. One-component systems consist of a single polymer resin cured by exposure to moisture, heat, or—in the case of anaerobics—the absence of oxygen. Two-component systems (such as epoxies) cure when two resins are combined.

Evaporation or Diffusion Adhesives
The adhesive is dissolved in either an organic solvent or water and is applied to the surfaces to be joined. When the carrier evaporates, the remaining polymer provides the bond. Water-base adhesives are preferred from the standpoint of environmental and safety considerations. The polymer may be completely dissolved in water or may consist of latex, a stable dispersion of polymer in water. A number of elastomers, vinyls, and acrylics are used.

Hot-Melt Adhesives
These thermoplastics and thermoplastic elastomers melt when heated. On cooling, the polymer solidifies and joins the materials. Typical melting temperatures of commercial hot-melts are about 80°C to 110°C, which limits the elevated-temperature use of these adhesives. High-performance hot-melts, such as polyamides and polyesters, can be used up to 200°C.

Pressure-Sensitive Adhesives
These adhesives are primarily elastomers or elastomer copolymers produced as films or coatings. Pressure is required to cause the polymer to stick to the substrate. They are used to produce electrical and packaging tapes, labels, floor tiles, wall coverings, and wood-grained textured films. Removable pressure-sensitive adhesives are used for medical applications such as bandages and transdermal drug delivery.

Conductive Adhesives
A polymer adhesive may contain a filler material such as silver, copper, or aluminum flakes or powders to provide electrical and thermal conductivity. In some cases, thermal conductivity is desired but electrical conductivity is not; alumina, boron nitride, and silica may be used as fillers to provide this combination of properties.

16-11   Polymer Processing and Recycling

There are a number of methods for producing polymer shapes, including molding, extrusion, and manufacture of films and fibers. The techniques used to form the polymers depend to a large extent on the nature of the polymer—in particular, whether it is thermoplastic or thermosetting. The greatest variety of techniques are used to form the thermoplastics. The polymer is heated to near or above the melting temperature so that it becomes rubbery or liquid. The polymer is then formed in a mold or die to produce the required shape. Thermoplastic elastomers can be formed in the same manner. In these processes, scrap can be easily recycled, and waste is minimized. Fewer forming techniques are used for the thermosetting polymers because, once cross-linking has occurred, the thermosetting polymers are no longer capable of being formed. Elastomers are processed in high-shear equipment such as a Banbury mixer. Carbon black and other additives are added. The heating from viscoelastic deformation can begin to cross-link the material prematurely. After the mixing step, a curing agent (e.g., zinc oxide) is added. The material discharged from the mixer is pliable and is processed using a short extruder, molded using a two-roll mill, or applied on parts by dip coating. This processing of elastomers is known as **compounding** of rubber.

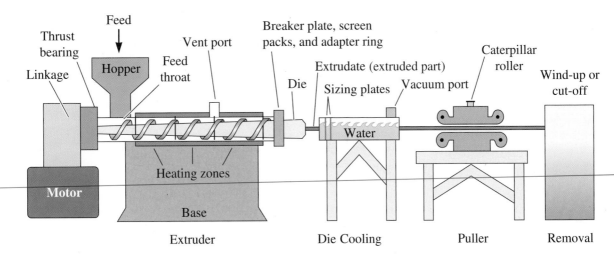

Figure 16-25   Schematic of an extruder used for polymer processing. *(Adapted from* Plastics: Materials and Processing, Second Edition, *by A. Brent Strong, p. 382, Fig. 11-1. Copyright © 2000 Prentice Hall. Adapted with permission of Pearson Education, Inc., Upper Saddle River, NJ.)*

The following are some of the techniques mainly used for processing of polymers; most of these, you will note, apply only to thermoplastics.

Extrusion

This is the most widely used technique for processing thermoplastics. Extrusion can serve two purposes. First, it provides a way to form certain simple shapes continuously (Figure 16-25). Second, extrusion provides an excellent mixer for additives (e.g., carbon black, fillers, etc.) when processing polymers that ultimately may be formed using some other method. A screw mechanism consisting of one or a pair of screws (twin screw) forces heated thermoplastic and additives through a die opening to produce solid shapes, films, sheets, tubes, and pipes (Figure 16-25). An industrial extruder can be up to 18 to 21 m long, 60 cm in diameter, and consist of different heating or cooling zones. Since thermoplastics show shear thinning behavior and are viscoelastic, the control of both temperature and viscosity is critical in polymer extrusion. One special extrusion process for producing films is illustrated in Figure 16-26. Extrusion also can be used to coat wires and cables with either thermoplastics or elastomers.

Blow Molding

A hollow preform of a thermoplastic called a **parison** is introduced into a die by gas pressure and expanded against the walls of the die (Figure 16-27). This process is used to produce plastic bottles, containers, automotive fuel tanks, and other hollow shapes.

Injection Molding

Thermoplastics heated above the melting temperature using an extruder are forced into a closed die to produce a molding. This process is similar to die casting of molten metals. A plunger or a special screw mechanism applies pressure to force the hot polymer into the die. A wide variety of products, ranging from cups, combs, and gears to garbage cans, can be produced in this manner.

Thermoforming

Thermoplastic polymer sheets heated to the plastic region can be formed over a die to produce such diverse products as egg cartons and decorative panels. The forming can be done using matching dies, a vacuum, or air pressure.

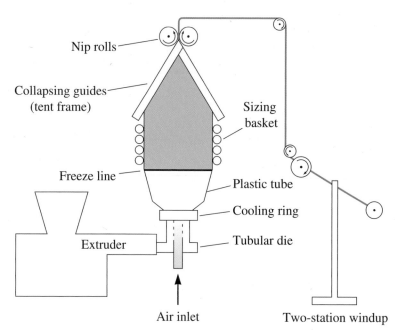

Nip rolls

Collapsing guides
(tent frame)

Sizing
basket

Freeze line

Plastic tube

Cooling ring

Extruder

Tubular die

Air inlet

Two-station windup

Figure 16-26   One technique by which polymer films (used in the manufacture of garbage bags, for example) can be produced. The film is extruded in the form of a bag, which is separated by air pressure until the polymer cools. *(Adapted from* Plastics: Materials and Processing, Second Edition, *by A. Brent Strong, p. 397, Fig. 11-8. Copyright © 2000 Prentice Hall. Adapted with permission of Pearson Education, Inc., Upper Saddle River, NJ.)*

Calendaring

In a calendar, molten plastic is poured into a set of rolls with a small opening. The rolls, which may be embossed with a pattern, squeeze out a thin sheet of the polymer—often, polyvinyl chloride. Typical products include vinyl floor tile and shower curtains.

Spinning

Filaments, fibers, and yarns may be produced by spinning. The molten thermoplastic polymer is forced through a die containing many tiny holes. The die, called a **spinnerette**, can rotate and produce a yarn. For some materials, including nylon, the fiber may subsequently be stretched to align the chains parallel to the axis of the fiber; this process increases the strength of the fibers.

Casting

Many polymers can be cast into molds to solidify. The molds may be plate glass for producing individual thick plastic sheets or moving stainless steel belts for continuous casting of thinner sheets. *Rotational molding* is a special casting process in which molten polymer is poured into a mold rotating about two axes. Centrifugal action forces the polymer against the walls of the mold, producing a thin shape such as a camper top.

Compression Molding

Thermoset moldings are most often formed by placing the solid material before cross-linking into a heated die. Application of high pressure and temperature causes the polymer to melt, fill the die, and immediately begin to harden. Small electrical housings as well as fenders, hoods, and side panels for automobiles can be produced by this process (Figure 16-28).

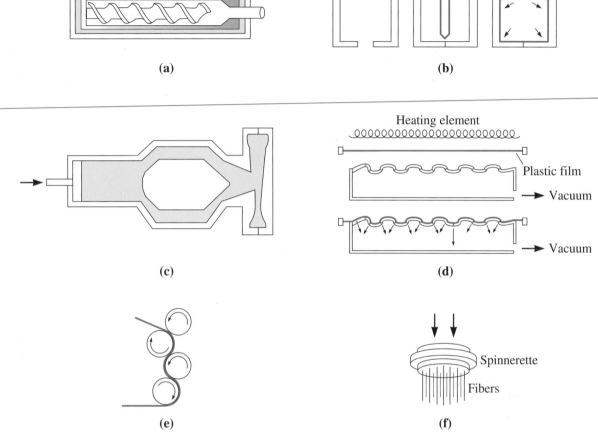

Figure 16-27 Typical forming processes for thermoplastics: (a) extrusion, (b) blow molding, (c) injection molding (the extrusion barrel is not shown), (d) thermoforming, (e) calendaring, and (f) spinning.

Transfer Molding

A double chamber is used in the transfer molding of thermosetting polymers. The polymer is heated under pressure in one chamber. After melting, the polymer is injected into the adjoining die cavity. This process permits some of the advantages of injection molding to be used for thermosetting polymers (Figure 16-28).

Reaction Injection Molding (RIM)

Thermosetting polymers in the form of liquid resins are first injected into a mixer and then directly into a heated mold to produce a shape. Forming and curing occur simultaneously in the mold. In reinforced-reaction injection molding (RRIM), a reinforcing material consisting of particles or short fibers is introduced into the mold cavity and is impregnated by the liquid resins to produce a composite material. Automotive bumpers, fenders, and furniture parts are made using this process.

Foams

Foamed products can be produced using polystyrene, urethanes, polymethyl methacrylate, and a number of other polymers. The polymer is produced in the form of tiny

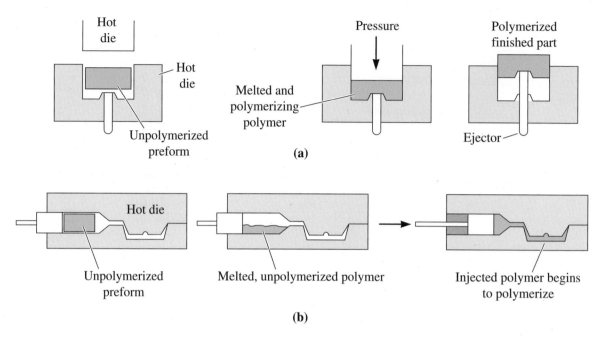

Figure 16-28   Typical forming processes for thermosetting polymers: (a) compression molding and (b) transfer molding.

beads, often containing a blowing agent such as pentane. During the pre-expansion process, the bead increases in diameter by as many as 50 times. The pre-expanded beads are then injected into a die, with the individual beads fusing together, using steam, to form exceptionally lightweight products with densities of perhaps only 0.02 g/cm^3. Expandable polystyrene (EPS) cups, packaging, and insulation are some of the applications for foams. Engine blocks for many automobiles are made using a pattern of expanded polystyrene beads.

Example 16-10   *Insulation Boards for Houses*

You want to design a material that can be used for making insulation boards that are approximately 120 cm wide and 240 cm tall. The material must provide good thermal insulation. What material would you choose?

SOLUTION

Glasses tend to be good insulators of heat; however, they will be heavy, more expensive, and prone to fracture. Polymers are lightweight and can be produced inexpensively, and they can be good thermal insulators. We can use foamed polystyrene since the air contained in the beads adds significantly to their effectiveness as thermal insulators. For better mechanical properties, we may want to produce foams that have relatively high density (compared to foams that are used to make coffee cups). Finally, from a safety viewpoint, we want to be sure that some fire and flame retardants are added to the foams. Such panels are made using expanded polystyrene beads containing pentane. A molding process is used to make the foams. The sheets can be cut into the required sizes using a heated metal wire.

Recycling of Plastics Recycling is a very important issue, and a full discussion of the entire process is outside the scope of this book. It is critical to remember, however, that recycling plays an important role in our everyday lives. Material is recycled in many ways. For example, part of the polymer that is scrap from a manufacturing process (known as regrind) is used by recycling plants. The recycling of thermoplastics is relatively easy and practiced widely. Note that many of the everyday plastic products you encounter (bags, soda bottles, yogurt containers, etc.) have numbers stamped on them. For PET products (recycling symbol "PETE" because of trademark issues), the number is 1. For HDPE, vinyl (recycling symbol V), LDPE, PP, and PS the numbers are 2, 3, 4, 5, and 6, respectively. Other plastics are marked number 7.

Thermosets and elastomers are more difficult to recycle, although they can still be used. For example, tires can be shredded and used to make safer playground surfaces or roads.

Despite enormous recycling efforts, a large portion of the materials in landfills today is plastics (the largest portion is paper). Given the limited amount of petroleum, the threat of global warming, and a need for a cleaner and safer environment, careful use and recycling makes sense for all materials.

Summary

- Polymers are made from large macromolecules produced by the joining of smaller molecules, called monomers, using addition or condensation polymerization reactions. *Plastics* are materials that are based on polymeric compounds, and they contain other additives that improve their properties. Compared with most metals and ceramics, plastics have low strength, stiffness, and melting temperatures; however, they also have low density and good chemical resistance. Plastics are used in many diverse technologies.

- Thermoplastics have chains that are not chemically bonded to each other, permitting the material to be easily formed into useful shapes, to have good ductility, and to be economically recycled. Thermoplastics can have an amorphous structure, which provides low strength and good ductility when the ambient temperature is above the glass-transition temperature. The polymers are more rigid and brittle when the temperature falls below the glass-transition temperature. Many thermoplastics can also partially crystallize during cooling or by application of a stress. This increases their strength.

- The thermoplastic chains can be made more rigid and stronger by using nonsymmetrical monomers that increase the bonding strength between the chains and make it more difficult for the chains to disentangle when stress is applied. In addition, many monomers containing atoms or groups of atoms other than carbon produce more rigid chains; this structure also produces high-strength thermoplastics.

- Elastomers are thermoplastics or lightly cross-linked thermosets that exhibit greater than 200% elastic deformation. Chains are cross-linked using vulcanization. The cross-linking makes it possible to obtain very large elastic deformations without permanent plastic deformation. Increasing the number of cross-links increases the stiffness and reduces the amount of elastic deformation of the elastomers.

- Thermoplastic elastomers combine features of both thermoplastics and elastomers. At high temperatures, these polymers behave as thermoplastics and are plastically formed into shapes; at low temperatures, they behave as elastomers.

- Thermosetting polymers are highly cross-linked into a three-dimensional network structure. Typically, high glass-transition temperatures, good strength, and brittle behavior result. Once cross-linking occurs, these polymers cannot be easily recycled.

- Manufacturing processes depend on the behavior of the polymers. Processes such as extrusion, injection molding, thermoforming, casting, and spinning are made possible by the viscoelastic behavior of thermoplastics. The non-reversible bonding in thermosetting polymers limits their processing to fewer techniques, such as compression molding, transfer molding, and reaction-injection molding.

Glossary

Addition polymerization Process by which polymer chains are built up by adding monomers together without creating a byproduct.

Aging Slow degradation of polymers as a result of exposure to low levels of heat, oxygen, bacteria, or ultraviolet rays.

Aramids Polyamide polymers containing aromatic groups of atoms in the linear chain.

Blushing A thermoplastic bent repeatedly leads to crystallization of small volumes of material; this creates voids that ultimately cause the material to fail.

Branched polymer Any polymer comprising chains that consist of a main chain and secondary chains that branch off from the main chain.

Compounding Processing of elastomers in a device known as a Banbury mixer followed by forming using extrusion, molding, or dip coating.

Condensation polymerization A polymerization mechanism in which a small molecule (e.g., water, methanol, etc.) forms as a byproduct.

Copolymer An addition polymer produced by joining more than one type of monomer.

Crazing Localized plastic deformation in a polymer. A craze may lead to the formation of cracks in the material.

Cross-linking Attaching chains of polymers together via permanent chemical bonds to produce a three-dimensional network polymer.

Deflection temperature The temperature at which a polymer will deform a given amount under a standard load. (Also called the *distortion* temperature.)

Degradation temperature The temperature above which a polymer burns, chars, or decomposes.

Degree of polymerization The average molecular weight of the polymer divided by the molecular weight of the monomer.

Diene A group of monomers that contain two double-covalent bonds. These monomers are often used in producing elastomers.

Elastomers Polymers (thermoplastics or lightly cross-linked thermosets) that have an elastic deformation $> 200\%$.

Geometric isomer A molecule that has the same composition as, but a structure different from, a second molecule.

Glass-transition temperature (T_g) The temperature range below which the amorphous polymer assumes a rigid glassy structure.

Interpenetrating polymer networks Polymer structures produced by intertwining two separate polymer structures or networks.

Linear polymer Any polymer in which molecules are in the form of spaghetti-like chains.

Liquid-crystalline polymers Exceptionally stiff polymer chains that act as rigid rods, even at high temperatures.

Mer A unit group of atoms and molecules that defines a characteristic arrangement for a polymer. A polymer can be thought of as a material made by combining several mers or units.

Monomer The molecule from which a polymer is produced.

Oligomer Low molecular weight molecules. These may contain two (dimers) or three (trimers) mers.

Parison A hot glob of soft or molten polymer that is blown or formed into a useful shape.

Plastic A predominantly polymeric material containing other additives.

Polymer Polymers are materials made from giant (or macromolecular), chain-like molecules having average molecular weights from 10,000 to more than 1,000,000 g/mol built by the joining of many mers or units by chemical bonds. Polymers are usually, but not always, carbon based.

Relaxation time A property of a polymer that is related to the rate at which stress relaxation occurs.

Repeat unit The structural unit from which a polymer is built. Also called a *mer*.

Spinnerette An extrusion die containing many small openings through which hot or molten polymer is forced to produce filaments. Rotation of the spinnerette twists the filaments into a yarn.

Stress-induced crystallization The process of forming crystals by the application of an external stress. Typically, a significant fraction of many amorphous plastics can be crystallized in this fashion, making them stronger.

Stress relaxation A reduction of the stress acting on a material over a period of time at a constant strain due to viscoelastic deformation.

Tacticity Describes the location in the polymer chain of atoms or atom groups in nonsymmetrical monomers.

Thermoplastic elastomers Polymers that behave as thermoplastics at high temperatures, but as elastomers at lower temperatures.

Thermoplastics Linear or branched polymers in which chains of molecules are not interconnected.

Thermosetting polymers Polymers that are heavily cross-linked to produce a strong three-dimensional network structure.

Unsaturated bond The double- or even triple-covalent bond joining two atoms together in an organic molecule. When a single covalent bond replaces the unsaturated bond, polymerization can occur.

Viscoelastic behavior The deformation of a material by elastic deformation and viscous flow when stress is applied.

Vulcanization Cross-linking elastomer chains by introducing sulfur or other chemicals.

Problems

Section 16-1 Classification of Polymers

16-1 What are linear and branched polymers? Can thermoplastics be branched?

16-2 Define (a) a thermoplastic, (b) thermosetting plastics, (c) elastomers, and (d) thermoplastic elastomers.

16-3 For what electrical and optical applications are polymers used? Explain using examples.

16-4 What are the major advantages of plastics compared to ceramics, glasses, and metallic materials?

Sections 16-2 Addition Polymerization

16-5 What do the terms condensation polymerization, addition polymerization, initiator, and terminator mean?

16-6 Kevlar ($C_{14}H_{10}N_2O_2$) is used in various applications from tires to body armor due to its high strength-to-weight ratio. The polymer is produced from the monomers paraphenylene diamine ($C_6H_8N_2$) and terephthaloyl chloride ($C_8H_4Cl_2O_2$) by the following reaction:

H_2N—⟨⟩—NH_2 (paraphenylene diamine)

+ Cl—CO—⟨⟩—CO—Cl (terephthaloyl chloride) → [—HN—⟨⟩—NH—CO—⟨⟩—CO—]

+ 2HCl

(a) What type of polymerization does the above reaction represent?

(b) Assuming 100% efficiency, calculate the weight of terephthaloyl chloride required to completely combine with 1 kg of paraphenylene diamine.

(c) How much Kevlar is produced?

Section 16-3 Degree of Polymerization

Section 16-4 Typical Thermoplastics

16-7 Explain why low-density polyethylene is good for making grocery bags, but super high molecular weight polyethylene must be used in applications for which strength and high wear resistance are needed.

16-8 Calculate the number of chains in a 5-m-long PVC pipe with an inner diameter of 5 cm and a thickness of 0.5 cm if the degree of polymerization is 1000. Assume that the chains are equal in length. The density of PVC is 1.4 g/cm³. The repeat unit of PVC is shown in Table 16-3.

16-9 The molecular weight of polymethyl methacrylate (see Table 16-3) is 250,000 g/mol. If all of the polymer chains are the same length,

(a) calculate the degree of polymerization, and

(b) the number of chains in 1 g of the polymer.

16-10 The degree of polymerization of polytetrafluoroethylene (see Table 16-3) is 7500. If all of the polymer chains are the same length, calculate

(a) the molecular weight of the chains, and

(b) the total number of chains in 1000 g of the polymer.

16-11 A polyethylene rope weighs 15.12 g/cm. If each chain contains 7000 repeat units,

(a) calculate the number of polyethylene chains in a 3-m length of rope, and

(b) the total length of chains in the rope, assuming that carbon atoms in each chain are approximately 0.15 nm apart and the length of one repeat unit is 0.24495 nm.

16-12 Analysis of a sample of polyacrylonitrile (see Table 16-3) shows that there are six lengths of chains, with the following number of chains of each length. Determine

(a) the weight average molecular weight and degree of polymerization, and

(b) the number average molecular weight and degree of polymerization.

Number of Chains	Mean Molecular Weight of Chains (g/mol)
10,000	3,000
18,000	6,000
17,000	9,000
15,000	12,000
9,000	15,000
4,000	18,000

Section 16-5 Structure—Property Relationships in Thermoplastics

Section 16-6 Effect of Temperature on Thermoplastics

Section 16-7 Mechanical Properties of Thermoplastics

16-13 Explain what the following terms mean: decomposition temperature, heat distortion temperature, glass-transition temperature, and melting temperature. Why is it that thermoplastics do not have a fixed melting or glass-transition temperature?

16-14 Using Table 16-5, plot the relationship between the glass-transition temperatures and the melting temperatures of the addition thermoplastics. What is the approximate relationship between these two critical temperatures? Do the condensation thermoplastics and the elastomers also follow the same relationship?

16-15 List the addition polymers in Table 16-5 that might be good candidates for making a bracket that holds a sideview mirror onto the outside of an automobile, assuming that temperatures frequently fall below zero degrees Celsius. Explain your choices.

16-16 Based on Table 16-5, which of the elastomers might be suited for use as a gasket in a pump for liquid CO_2 at $-78°C$? Explain.

16-17 How do the glass-transition temperatures of polyethylene, polypropylene, and polymethyl methacrylate compare? Explain their differences, based on the structure of the monomer.

16-18 Which of the addition polymers in Table 16-5 are used in their leathery condition at room temperature? How is this condition expected to affect their mechanical properties compared with those of glassy polymers?

16-19 What factors influence the crystallinity of polymers? Explain the development and role of crystallinity in PET and nylon.

16-20 Describe the relative tendencies of the following polymers to crystallize. Explain your answer.
(a) Branched polyethylene versus linear polyethylene;
(b) polyethylene versus polyethylene-polypropylene copolymer;
(c) isotactic polypropylene versus atactic polypropylene; and
(d) polymethyl methacrylate versus acetal (polyoxymethylene).

16-21 The crystalline density of polypropylene is 0.946 g/cm^3, and its amorphous density is 0.855 g/cm^3. What is the weight percent of the structure that is crystalline in a polypropylene that has a density of 0.9 g/cm^3?

16-22 If the strain rate in a polymer can be represented by

$$d\varepsilon/dt = \sigma/\eta + (1/E)\,d\sigma/dt$$

where ε = strain, σ = stress, η = viscosity, E = modulus of elasticity, and t = time, derive Equation 16-5, assuming constant strain (i.e., $d\varepsilon = 0$). What is the relaxation time (λ) in Equation 16-5 a function of?

16-23 A polymer component that needs to maintain a stress level above 10 MPa for proper functioning of an assembly is scheduled to be replaced every two years as preventative maintenance. If the initial stress was 18 MPa and dropped to 15 MPa after one year of operation (assuming constant strain), will the part replacement under the specified preventative schedule successfully prevent a failure?

16-24 Explain the meaning of these terms: creep, stress relaxation, crazing, blushing, environmental stress cracking, and aging of polymers.

16-25 A stress of 17 MPa is applied to a polymer serving as a fastener in a complex assembly. At a constant strain, the stress drops to 16.6 MPa after 100 h. If the stress on the part must remain above 14.5 MPa in order for the part to function properly, determine the life of the assembly.

16-26 A stress of 7 MPa is applied to a polymer that operates at a constant strain; after six months, the stress drops to 5.9 MPa. For a particular application, a part made of the same polymer must maintain a stress of 6.2 MPa after 12 months. What should be the original stress applied to the polymer for this application?

16-27 Data for the rupture time of polyethylene are shown in Figure 16-18. At an applied stress of 4.8 MPa, the figure indicates that the polymer ruptures in 0.2 h at 90°C but survives 10,000 h at 65°C. Assuming that

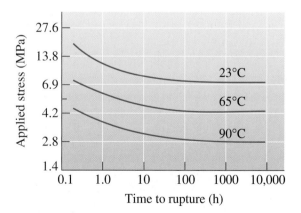

Figure 16-18 (Repeated for Problem 16-27) The effect of temperature on the stress-rupture behavior of high-density polyethylene.

the rupture time is related to the viscosity, calculate the activation energy for the viscosity of the polyethylene and estimate the rupture time at 23°C.

Section 16-8 Elastomers (Rubbers)

16-28 The polymer ABS can be produced with varying amounts of styrene, butadiene, and acrylonitrile monomers, which are present in the form of two copolymers: BS rubber and SAN.
(a) How would you adjust the composition of ABS if you wanted to obtain good impact properties?
(b) How would you adjust the composition if you wanted to obtain good ductility at room temperature?
(c) How would you adjust the composition if you wanted to obtain good strength at room temperature?

16-29 Figure 16-23 shows the stress-strain curve for an elastomer. From the curve, calculate and plot the modulus of elasticity versus strain and explain the results.

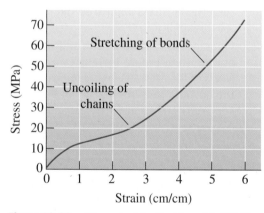

Figure 16-23   (Repeated for Problem 16-29) The stress-strain curve for an elastomer. Virtually all of the deformation is elastic; the modulus of elasticity varies as the strain changes.

Section 16-9 Thermosetting Polymers

16-30 Explain the term thermosetting polymer. A thermosetting polymer cannot be produced using only adipic acid and ethylene glycol. Explain why.

16-31 Explain why the degree of polymerization is not usually used to characterize thermosetting polymers.

16-32 Defend or contradict the choice to use the following materials as hot-melt adhesives for an application in which the assembled part is subjected to impact loading:
(a) polyethylene;
(b) polystyrene;
(c) styrene-butadiene thermoplastic elastomer;
(d) polyacrylonitrile; and
(e) polybutadiene.

16-33 Compare and contrast properties of thermoplastics, thermosetting materials, and elastomers.

⬡ Design Problems

16-34 Figure 16-29 shows the behavior of polypropylene, polyethylene, and acetal at two temperatures. We would like to produce a 30-cm-long rod of a polymer that will operate at 40°C for 6 months under a constant load of 230 kg. Design the material and size of the rod such that no more than 5% elongation will occur by creep.

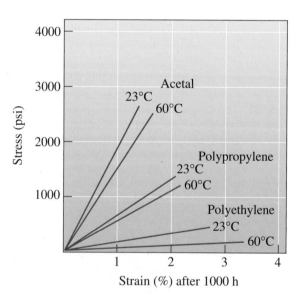

Figure 16-29   The effect of applied stress on the percent creep strain for three polymers (for Problem 16-34).

16-35 Design a polymer material that might be used to produce a 7.5-cm-diameter gear to be used to transfer energy from a low-power electric motor. What are the design requirements? What class of polymers (thermoplastics, thermosets, elastomers) might be most appropriate? What particular polymer might you first consider? What additional information concerning the application and polymer properties do you need to know to complete your design?

16-36 Design a polymer material and a forming process to produce the case for a personal computer. What are the design and forming requirements? What class of polymers might be most appropriate? What particular polymer might you first consider? What additional information do you need to know?

16-37 What kind of polymer can be used to line the inside of the head of a hip prosthesis implant? Discuss what requirements would be needed for this type of polymer.

 Computer Problems

16-38 *Polymer Molecular Weight Distribution.* The following data were obtained for polyethylene. Determine the average molecular weight and degree of polymerization.

Molecular Weight Range (g/mol)	f_i	x_i
0–3,000	0.01	0.03
3,000–6,000	0.08	0.10
6,000–9,000	0.19	0.22
9,000–12,000	0.27	0.36
12,000–15,000	0.23	0.19
15,000–18,000	0.11	0.07
18,000–21,000	0.06	0.02
21,000–24,000	0.05	0.01

Write a computer program or use a spreadsheet program to solve this problem.

Ⓚ Knovel® Problems

K16-1 Provide an example of a thermoplastic elastomer (TPE) based on an interpenetrating polymer network (IPN). How is it made?

K16-2 How does chain length affect the glass-transition temperature of polymers? Find an equation describing this relationship.

K16-3 Hindered phenolics are common organic antioxidants. For what classes of plastics are they used?

In a drive to increase efficiency and thereby decrease fuel costs, aerospace companies look to incorporate lighter and stronger materials into airplanes. In 2009, Boeing completed the first successful test flight of the Boeing 787 Dreamliner. One of the primary innovations of the Boeing 787 is the extensive use of composite materials; composites are formed by incorporating multiple phase components in a material in such a way that the properties of the resultant material are unique and not otherwise attainable. Composite materials comprise half of the Dreamliner's total weight. For example, the fuselage of the Boeing 787 is made from carbon fiber-reinforced plastic. Carbon fiber-reinforced plastic is a composite of carbon fiber in an epoxy matrix. The carbon fibers impart strength and stiffness to the material, and the epoxy matrix binds the fibers together. (*AFP/Getty Images.*)

Chapter 17

Composites: Teamwork and Synergy in Materials

Have You Ever Wondered?

- *What are some of the naturally occurring composites?*

- *Why is abalone shell, made primarily of calcium carbonate, so much stronger than chalk, which is also made of calcium carbonate?*

- *What sporting gear applications make use of composites?*

- *Why are composites finding increased usage in aircraft and automobiles?*

Composites are produced when two or more materials or phases are used together to give a combination of properties that cannot be attained otherwise. Composite materials may be selected to give unusual combinations of stiffness, strength, weight, high-temperature performance, corrosion resistance, hardness, or conductivity. Composites highlight how different materials can work in synergy. Abalone shell, wood, bone, and teeth are examples of naturally occurring composites. Microstructures of selected composites are shown in Figure 17-1. An example of a material that is a composite at the macroscale is steel-reinforced concrete. Microscale composites include such materials as carbon or glass fiber-reinforced plastics (CFRP or GFRP). These composites offer significant gains in specific strengths and are finding increasing usage in airplanes, electronic components, automotives, and sporting equipment.

As mentioned in Chapters 12 and 13, dispersion-strengthened (like steels) and precipitation-hardened alloys are examples of traditional materials that are **nanocomposites**. In a nanocomposite, the **dispersed phase** consists of nanoscale particles and is distributed in a **matrix phase**. Essentially, the same concept has been applied in developing what are described as **hybrid organic-inorganic nanocomposites**. These are materials in which the microstructure of the composites consists of an inorganic part or block and an organic block. The idea is similar to the formation of block copolymers (Chapter 16). These and other

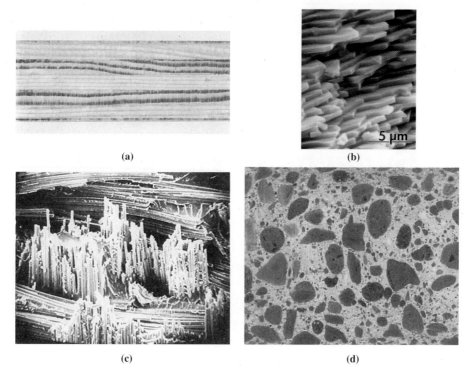

(a)   (b)

(c)   (d)

Figure 17-1   Some examples of composite materials: (a) plywood is a laminar composite of layers of wood veneer, (b) abalone shell is a composite of aragonitic ($CaCO_3$-orthorhombic) platelets surrounded by 10 nm thick layers of proteinaceous organic matrix. (*Courtesy of Professor Mehmet Sarikaya, University of Washington, Seattle.*) (c) Fiberglass is a fiber-reinforced composite containing stiff, strong glass fibers in a softer polymer matrix ($\times 175$), and (d) concrete is a particulate composite containing coarse sand or gravel in a cement matrix (reduced 50%). (*Images (a), (c), and (d) reprinted courtesy of Don Askeland.*)

functional composites can provide unusual combinations of electronic, magnetic, or optical properties. For example, a porous dielectric material prepared using phase-separated inorganic glasses exhibits a dielectric constant that is lower than that for the same material with no porosity. Space shuttle tiles made from silica fibers are lightweight because they consist of air and silica fibers and exhibit a low thermal conductivity. The two phases in these examples are ceramic and air. Many glass-ceramics are nanoscale composites of different ceramic phases. Many plastics can be considered composites as well. For example, Dylark™ is a composite of maleic anhydride-styrene copolymer. It contains carbon black for stiffness and protection against ultraviolet rays. It also contains glass fibers for increased Young's modulus and rubber for toughness. Epoxies may be filled with silver to increase thermal conductivity. Some dielectric materials are made using multiple phases such that the overall dielectric properties of interest (e.g., the dielectric constant) do not change appreciably with temperature (within a certain range). Some composite structures may consist of different materials arranged in different layers. This leads to what are known as functionally graded materials and structures. For example, a yttria stabilized zirconia (YSZ) coating on a turbine blade will have other layers in between that provide bonding with the turbine blade material. The YSZ coating itself contains levels of porosity that are essential for providing protection against high temperatures (Chapter 5). Similarly, coatings on glass are examples of composite structures. Thus, the *concept* of using composites is a generic one and can be applied at the macro, micro, and nano length scales.

In composites, the properties and volume fractions of individual phases are important. The connectivity of phases is also very important. Usually the matrix phase is the continuous phase, and the other phase is said to be the dispersed phase. Thus, terms such as "metal-matrix" indicate a metallic material used to form the continuous phase.

Connectivity describes how the two or more phases are connected in the composite. Composites are often classified based on the shape or nature of the dispersed phase (e.g., particle-reinforced, whisker-reinforced, or fiber-reinforced composites). **Whiskers** are like fibers, but their length is much smaller. The bonding between the particles, whiskers, or fibers and the matrix is also very important. In structural composites, polymeric molecules known as "coupling agents" are used. These molecules form bonds with the dispersed phase and become integrated into the continuous matrix phase as well.

In this chapter, we will primarily focus on composites used in structural or mechanical applications. Composites can be placed into three categories—particulate, fiber, and laminar—based on the shapes of the materials (Figure 17-1). Concrete, a mixture of cement and gravel, is a particulate composite; fiberglass, containing glass fibers embedded in a polymer, is a fiber-reinforced composite; and plywood, having alternating layers of wood veneer, is a laminar composite. If the reinforcing particles are uniformly distributed, particulate composites have isotropic properties; fiber composites may be either isotropic or anisotropic; laminar composites always display anisotropic behavior.

17-1   Dispersion-Strengthened Composites

A special group of dispersion-strengthened nanocomposite materials containing particles 10 to 250 nm in diameter is classified as particulate composites. The **dispersoids,** usually a metallic oxide, are introduced into the matrix by means other than traditional phase transformations (Chapters 12 and 13). Even though the small particles are not coherent with the matrix, they block the movement of dislocations and produce a pronounced strengthening effect.

At room temperature, the dispersion-strengthened composites may be weaker than traditional age-hardened alloys, which contain a coherent precipitate. Because the composites do not catastrophically soften by overaging, overtempering, grain growth, or coarsening of the dispersed phase, the strength of the composite decreases only gradually with increasing temperature (Figure 17-2 on the next page). Furthermore, their creep resistance is superior to that of metals and alloys.

The dispersoid must have a low solubility in the matrix and must not chemically react with the matrix, but a small amount of solubility may help improve the bonding between the dispersant and the matrix. Copper oxide (Cu_2O) dissolves in copper at high temperatures; thus, the Cu_2O-Cu system would not be effective. Al_2O_3 does not dissolve in aluminum; the Al_2O_3-Al system does give an effective dispersion-strengthened material.

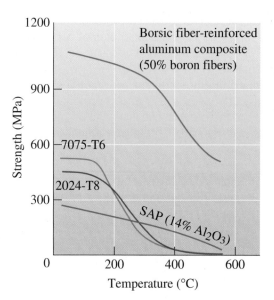

Figure 17-2
Comparison of the yield strength of dispersion-strengthened sintered aluminum powder (SAP) composite with that of two conventional two-phase high-strength aluminum alloys. The composite has benefits above about 300°C. A fiber-reinforced aluminum composite is shown for comparison.

Illustrations of Dispersion-Strengthened Composites

Table 17-1 lists some materials of interest. Perhaps the classic example is the sintered aluminum powder (SAP) composite. SAP has an aluminum matrix strengthened by up to 14% Al_2O_3. The composite is formed by powder metallurgy. In one method, aluminum and alumina powders are blended, compacted at high pressures, and sintered. In a second technique, the aluminum powder is treated to add a continuous oxide film on each particle. When the powder is compacted, the oxide film fractures into tiny flakes that are surrounded by the aluminum metal during sintering.

Another important group of dispersion-strengthened composites includes thoria dispersed (TD) metals such as TD-nickel (Figure 17-3). TD-nickel can be produced by internal oxidation. Thorium is present in nickel as an alloying element. After a powder compact is made, oxygen is allowed to diffuse into the metal, react with the thorium, and produce thoria (ThO_2). The following example illustrates calculations related to a dispersion-strengthened composite.

TABLE 17-1 ■ *Applications of selected dispersion-strengthened composites*

System	Applications
Ag-CdO	Electrical contact materials
Al-Al_2O_3	Possible use in nuclear reactors
Be-BeO	Aerospace and nuclear reactors
Co-ThO_2, Y_2O_3	Possible creep-resistant magnetic materials
Ni-20% Cr-ThO_2	Turbine engine components
Pb-PbO	Battery grids
Pt-ThO_2	Filaments, electrical components
W-ThO_2, ZrO_2	Filaments, heaters

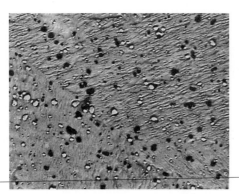

Figure 17-3
Electron micrograph of TD-nickel. The
dispersed ThO_2 particles have a diameter
of 300 nm or less ($\times 2000$). (*From* Oxide
Dispersion Strengthening, *p. 714, Gordon
and Breach, 1968. © AIME.*)

Example 17-1 *TD-Nickel Composite*

Suppose 2 wt% ThO_2 is added to nickel. Each ThO_2 particle has a diameter of 1000 Å.
How many particles are present in each cubic centimeter?

SOLUTION

The densities of ThO_2 and nickel are 9.69 and 8.9 g/cm³, respectively. The volume
fraction is

$$f_{ThO_2} = \frac{\dfrac{2}{9.69}}{\dfrac{2}{9.69} + \dfrac{98}{8.9}} = 0.0184$$

Therefore, there is 0.0184 cm³ of ThO_2 per cm³ of composite. The volume of each
ThO_2 sphere is

$$V_{ThO_2} = \tfrac{4}{3}\pi r^3 = \tfrac{4}{3}\pi(0.5 \times 10^{-5}\text{ cm})^3 = 0.524 \times 10^{-15}\text{ cm}^3$$

$$\text{Concentration of ThO}_2\text{ particles} = \frac{0.0184}{0.524 \times 10^{-15}} = 35.1 \times 10^{12}\text{ particles/cm}^3$$

17-2   Particulate Composites

The particulate composites are designed to produce unusual combinations of properties
rather than to improve strength. The particulate composites contain large amounts of
coarse particles that do not block slip effectively.

Rule of Mixtures   Certain properties of a particulate composite depend only
on the relative amounts and properties of the individual constituents. The **rule of mixtures**

Figure 17-4
Microstructure of tungsten carbide—20% cobalt-
cemented carbide ($\times$ 1300). (*From ASM
Handbook, Vol. 7, (1972), ASM International,
Materials Park, OH 44073-0002.*)

can accurately predict these properties. The density of a particulate composite, for example, is

$$\rho_c = \sum (f_i \rho_i) = f_1 \rho_1 + f_2 \rho_2 + \cdots + f_n \rho_n \tag{17-1}$$

where ρ_c is the density of the composite, $\rho_1, \rho_2, \ldots, \rho_n$ are the densities of each constituent in the composite, and $f_1, f_2, \ldots, f_n$ are the volume fractions of each constituent. Note that the connectivity of different phases (i.e., how the dispersed phase is arranged with respect to the continuous phase) is also very important for many properties.

Cemented Carbides

Cemented carbides, or cermets, contain hard ceramic particles dispersed in a metallic matrix (Chapter 15). Tungsten carbide inserts used for cutting tools in machining operations are typical of this group. Tungsten carbide (WC) is a hard, stiff, high-melting temperature ceramic. To improve toughness, tungsten carbide particles are combined with cobalt powder and pressed into powder compacts. The compacts are heated above the melting temperature of the cobalt. The liquid cobalt surrounds each of the solid tungsten carbide particles (Figure 17-4). After solidification, the cobalt serves as the binder for tungsten carbide and provides good impact resistance. Other carbides, such as TaC and TiC, may also be included in the cermet. The following example illustrates the calculation of density for cemented carbide.

Example 17-2   *Cemented Carbides*

A cemented carbide cutting tool used for machining contains 75 wt% WC, 15 wt% TiC, 5 wt% TaC, and 5 wt% Co. Estimate the density of the composite.

SOLUTION

First, we must convert the weight percentages to volume fractions. The densities of the components of the composite are

$$\rho_{WC} = 15.77 \, \text{g/cm}^3 \qquad \rho_{TiC} = 4.94 \, \text{g/cm}^3$$
$$\rho_{TaC} = 14.5 \, \text{g/cm}^3 \qquad \rho_{Co} = 8.83 \, \text{g/cm}^3$$

$$f_{WC} = \frac{75/15.77}{75/15.77 + 15/4.94 + 5/14.5 + 5/8.83} = \frac{4.76}{8.70} = 0.546$$

$$f_{TiC} = \frac{15/4.94}{8.70} = 0.349$$

$$f_{TaC} = \frac{5/14.5}{8.70} = 0.040$$

$$f_{Co} = \frac{5/8.90}{8.70} = 0.065$$

From the rule of mixtures, the density of the composite is

$$\rho_c = \sum (f_i \rho_i) = (0.546)(15.77) + (0.349)(4.94) + (0.040)(14.5)$$
$$+ (0.065)(8.83)$$
$$= 11.5 \, g/cm^3$$

Abrasives

Grinding and cutting wheels are formed from alumina (Al_2O_3), silicon carbide (SiC), and cubic boron nitride (CBN). To provide toughness, the abrasive particles are bonded by a glass or polymer matrix. Diamond abrasives are typically bonded with a metal matrix. As the hard particles wear, they fracture or pull out of the matrix, exposing new cutting surfaces.

Electrical Contacts

Materials used for electrical contacts in switches and relays must have a good combination of wear resistance and electrical conductivity. Otherwise, the contacts erode, causing poor contact and arcing. Tungsten-reinforced silver provides this combination of characteristics. A tungsten powder compact is made using conventional powder metallurgy processes (Figure 17-5) to produce high interconnected

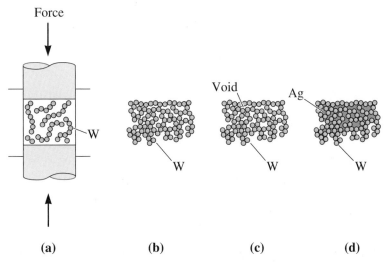

Figure 17-5 The steps in producing a silver-tungsten electrical composite: (a) Tungsten powders are pressed, (b) a low-density compact is produced, (c) sintering joins the tungsten powders, and (d) liquid silver is infiltrated into the pores between the particles.

porosity. Liquid silver is then vacuum infiltrated to fill the interconnected voids. Both the silver and the tungsten are continuous. Thus, the pure silver efficiently conducts current while the hard tungsten provides wear resistance.

Example 17-3   *Silver-Tungsten Composite*

A silver-tungsten composite for an electrical contact is produced by first making a porous tungsten powder metallurgy compact, then infiltrating pure silver into the pores. The density of the tungsten compact before infiltration is 14.5 g/cm^3. Calculate the volume fraction of porosity and the final weight percent of silver in the compact after infiltration.

SOLUTION

The densities of pure tungsten f_W and pure silver f_{Ag} are 19.25 g/cm^3 and 10.49 g/cm^3, respectively. We can assume that the density of a pore is zero, so from the rule of mixtures:

$$\rho_c = f_W\rho_W + f_{pore}\rho_{pore}$$
$$14.5 = f_W(19.25) + f_{pore}(0)$$
$$f_W = 0.75$$
$$f_{pore} = 1 - 0.75 = 0.25$$

After infiltration, the volume fraction of silver equals the volume fraction of pores:

$$f_{Ag} = f_{pore} = 0.25$$

$$\text{wt\% Ag} = \frac{(0.25)(10.49)}{(0.25)(10.49) + (0.75)(19.25)} \times 100 = 15\%$$

This solution assumes that all of the pores are open, or interconnected.

Polymers   Many engineering polymers that contain fillers and extenders are particulate composites. A classic example is carbon black in vulcanized rubber. Carbon black consists of tiny carbon spheroids only 5 to 500 nm in diameter. The carbon black improves the strength, stiffness, hardness, wear resistance, resistance to degradation due to ultraviolet rays, and heat resistance of the rubber. Nanoparticles of silica are added to rubber tires to enhance their stiffness.

Extenders, such as calcium carbonate ($CaCO_3$), solid glass spheres, and various clays, are added so that a smaller amount of the more expensive polymer is required. The extenders may stiffen the polymer, increase the hardness and wear resistance, increase thermal conductivity, or improve resistance to creep; however, strength and ductility normally decrease (Figure 17-6). Introducing hollow glass spheres may impart the same changes in properties while significantly reducing the weight of the composite. Other special properties can be obtained. Elastomer particles are introduced into polymers to

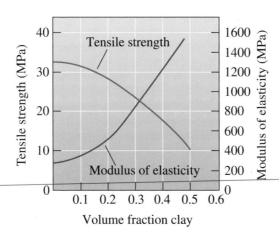

Figure 17-6
The effect of clay on the properties of polyethylene.

improve toughness. Polyethylene may contain metallic powders, such as lead, to improve the absorption of fission products in nuclear applications. The design of a polymer composite is illustrated in the example that follows.

Example 17-4   *Design of a Particulate Polymer Composite*

Design a clay-filled polyethylene composite suitable for injection molding of inexpensive components. The final part must have a tensile strength of at least 21 MPa and a modulus of elasticity of at least 552 MPa. Polyethylene costs approximately $1.1 per kg, and clay costs approximately $0.1 per kg. The density of polyethylene is 0.95 g/cm^3 and that of clay is 2.4 g/cm^3.

SOLUTION

From Figure 17-6, a volume fraction of clay below 0.35 is required to maintain a tensile strength greater than 21 MPa, whereas a volume fraction of at least 0.2 is needed for the minimum modulus of elasticity. For lowest cost, we use the maximum allowable clay, or a volume fraction of 0.35.

In 1000 cm^3 of composite parts, there are 350 cm^3 of clay and 650 cm^3 of polyethylene in the composite, or

$$(350 \text{ cm}^3)(2.4 \text{ g/cm}^3) = 840 \text{ g clay}$$

$$(650 \text{ cm}^3)(0.95 \text{ g/cm}^3) = 617.50 \text{ g polyethylene}$$

The cost of materials is

$$(840 \text{ g clay})(\$0.1/\text{kg}) = \$0.084$$

$$(617.50 \text{ g PE})(\$1.1/\text{kg}) = \$0.68$$

$$\text{total} = \$0.764 \text{ per } 1000 \text{ cm}^3$$

Suppose that weight is critical. The composite's density is

$$\rho_c = (0.35)(2.4) + (0.65)(0.95) = 1.46 \text{ g/cm}^3$$

We may wish to sacrifice some of the economic savings in order to obtain lighter weight. If we use only 0.2 volume fraction clay, then (using the same method as above) we find that we need 480 g clay and 760 g polyethylene.
The cost of materials is now

$$(480\,\text{g clay})(\$0.1/\text{kg}) = \$0.048$$
$$(760\,\text{kg PE})(\$1.1/\text{kg}) = \$0.836$$
$$\text{Total} = \$0.88\,\text{per}\,1000\,\text{cm}^3$$

The density of the composite is

$$\rho_c = (0.2)(2.4) + (0.8)(0.95) = 1.24\,\text{g/cm}^3$$

The material costs about 15% more, but there is a weight savings of 15%.

Cast Metal Particulate Composites   Aluminum castings containing dispersed SiC particles for automotive applications, including pistons and connecting rods, represent an important commercial application for particulate composites (Figure 17-7). With special processing, the SiC particles can be wet by the liquid, helping to keep the ceramic particles from sinking during freezing.

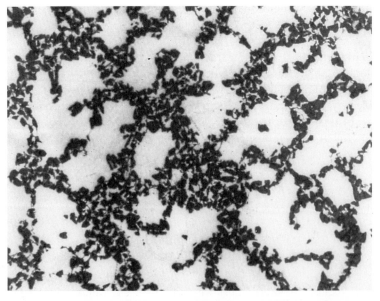

Figure 17-7   Microstructure of an aluminum casting alloy reinforced with silicon carbide particles. In this case, the reinforcing particles have segregated to interdendritic regions of the casting (× 125). (*Courtesy of David Kennedy and Lester B. Knight, Cast Metals, Inc.*)

17-3   Fiber-Reinforced Composites

Most fiber-reinforced composites provide improved strength, fatigue resistance, Young's modulus, and strength-to-weight ratio by incorporating strong, stiff, but brittle fibers into a softer, more ductile matrix. The matrix material transmits the force to the fibers, which carry most of the applied force. The matrix also provides protection for the fiber surface and minimizes diffusion of species such as oxygen or moisture that can degrade the mechanical properties of fibers. The strength of the composite may be high at both room temperature and elevated temperatures (Figure 17-2).

Many types of reinforcing materials are employed. Straw has been used to strengthen mud bricks for centuries. Steel-reinforcing bars are introduced into concrete structures. Glass fibers in a polymer matrix produce fiberglass for transportation and aerospace applications. Fibers made of boron, carbon, polymers (e.g., aramids, Chapter 16), and ceramics provide exceptional reinforcement in advanced composites based on matrices of polymers, metals, ceramics, and even intermetallic compounds.

The Rule of Mixtures in Fiber-Reinforced Composites

As for particulate composites, the rule of mixtures always predicts the density of fiber-reinforced composites:

$$\rho_c = f_m \rho_m + f_f \rho_f \qquad (17\text{-}2)$$

where the subscripts m and f refer to the matrix and the fiber. Note that $f_m = 1 - f_f$.

In addition, the rule of mixtures accurately predicts the electrical and thermal conductivity of fiber-reinforced composites along the fiber direction if the fibers are *continuous* and *unidirectional*:

$$k_c = f_m k_m + f_f k_f \qquad (17\text{-}3)$$

$$\sigma_c = f_m \sigma_m + f_f \sigma_f \qquad (17\text{-}4)$$

where k is the thermal conductivity and σ is the electrical conductivity. Thermal or electrical energy can be transferred through the composite at a rate that is proportional to the volume fraction of the conductive material. In a composite with a metal matrix and ceramic fibers, the bulk of the energy would be transferred through the matrix; in a composite consisting of a polymer matrix containing metallic fibers, energy would be transferred through the fibers.

When the fibers are not continuous or unidirectional, the simple rule of mixtures may not apply. For example, in a metal fiber-polymer matrix composite, electrical conductivity would be low and would depend on the length of the fibers, the volume fraction of the fibers, and how often the fibers touch one another. This is expressed using the concept of connectivity of phases.

Modulus of Elasticity   The rule of mixtures is used to predict the modulus of elasticity when the fibers are continuous and unidirectional. Parallel to the fibers, the modulus of elasticity may be as high as

$$E_{c,\|} = f_m E_m + f_f E_f \qquad (17\text{-}5)$$

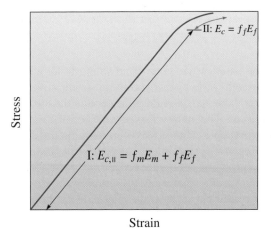

Figure 17-8
The stress-strain curve for a fiber-reinforced composite. At low stresses (region I), the modulus of elasticity is given by the rule of mixtures. At higher stresses (region II), the matrix deforms and the rule of mixtures is no longer obeyed.

In the figure: $II: E_c = f_f E_f$ and $I: E_{c,\parallel} = f_m E_m + f_f E_f$

When the applied stress is very large, the matrix begins to deform and the stress-strain curve is no longer linear (Figure 17-8). Since the matrix now contributes little to the stiffness of the composite, the modulus can be approximated by

$$E_{c,\parallel} = f_f E_f \qquad (17\text{-}6)$$

When the load is applied perpendicular to the fibers, each component of the composite acts independently of the other. The modulus of the composite is now

$$\frac{1}{E_{c,\perp}} = \frac{f_m}{E_m} + \frac{f_f}{E_f} \qquad (17\text{-}7)$$

Again, if the fibers are not continuous and unidirectional, the rule of mixtures does not apply.

The following examples further illustrate these concepts.

Example 17-5 *Rule of Mixtures for Composites: Stress Parallel to Fibers*

Derive the rule of mixtures (Equation 17-5) for the modulus of elasticity of a fiber-reinforced composite when a stress (σ) is applied along the axis of the fibers.

SOLUTION

The total force acting on the composite is the sum of the forces carried by each constituent:

$$F_c = F_m + F_f$$

Since $F = \sigma A$

$$\sigma_c A_c = \sigma_m A_m + \sigma_f A_f$$

$$\sigma_c = \sigma_m \left(\frac{A_m}{A_c}\right) + \sigma_f \left(\frac{A_f}{A_c}\right)$$

If the fibers have a uniform cross-section, the area fraction equals the volume fraction f:

$$\sigma_c = \sigma_m f_m + \sigma_f f_f$$

From Hooke's law, $\sigma = \varepsilon E$. Therefore,

$$E_{c,\parallel} \varepsilon_c = E_m \varepsilon_m f_m + E_f \varepsilon_f f_f$$

If the fibers are rigidly bonded to the matrix, both the fibers and the matrix must stretch equal amounts (iso-strain conditions):

$$\varepsilon_c = \varepsilon_m = \varepsilon_f$$

$$E_{c,\parallel} = f_m E_m + f_f E_f$$

Example 17-6 | *Modulus of Elasticity for Composites: Stress Perpendicular to Fibers*

Derive the equation for the modulus of elasticity of a fiber-reinforced composite when a stress is applied perpendicular to the axis of the fiber (Equation 17-7).

SOLUTION

In this example, the strains are no longer equal; instead, the weighted sum of the strains in each component equals the total strain in the composite, whereas the stresses in each component are equal (iso-stress conditions):

$$\varepsilon_c = f_m \varepsilon_m + f_f \varepsilon_f$$

$$\frac{\sigma_c}{E_c} = f_m\left(\frac{\sigma_m}{E_m}\right) + f_f\left(\frac{\sigma_f}{E_f}\right)$$

Since $\sigma_c = \sigma_m = \sigma_f$,

$$\frac{1}{E_{c,\perp}} = \frac{f_m}{E_m} + \frac{f_f}{E_f}$$

Strength of Composites   The tensile strength of a fiber-reinforced composite (TS_c) depends on the bonding between the fibers and the matrix. The rule of mixtures is sometimes used to approximate the tensile strength of a composite containing continuous, parallel fibers:

$$TS_c = f_f TS_f + f_m S_m, \tag{17-8}$$

where TS_f is the tensile strength of the fiber and σ_m is the stress acting on the matrix when the composite is strained to the point where the fiber fractures. Thus, S_m is *not* the actual tensile strength of the matrix. Other properties, such as ductility, impact properties, fatigue properties, and creep properties, are difficult to predict even for unidirectionally aligned fibers.

Example 17-7   *Boron Aluminum Composites*

Boron coated with SiC (or Borsic) reinforced aluminum containing 40 vol% fibers is an important high-temperature, lightweight composite material. Estimate the density, modulus of elasticity, and tensile strength parallel to the fiber axis. Also estimate the modulus of elasticity perpendicular to the fibers.

SOLUTION

The properties of the individual components are shown here.

Material	Density (ρ) (g/cm^3)	Modulus of Elasticity (E) (MPa)	Tensile Strength (TS) (MPa)
Fibers	2.36	379×10^3	2760
Aluminum	2.70	69×10^3	35

From the rule of mixtures:

$$\rho_c = (0.6)(2.7) + (0.4)(2.36) = 2.56 \text{ g/cm}^3$$

$$E_{c,\parallel} = (0.6)(69 \times 10^3) + (0.4)(379 \times 10^3) = 193 \times 10^3 \text{ MPa}$$

$$TS_c = (0.6)(35) + (0.4)(2760) = 1125 \text{ MPa}$$

Note that the tensile strength calculation is only an approximation. Perpendicular to the fibers:

$$\frac{1}{E_{c,\perp}} = \frac{0.6}{69 \times 10^3} + \frac{0.4}{379 \times 10^3} = 0.00976 \times 10^{-3}$$

$$E_{c,\perp} = 102.5 \times 10^3 \text{ MPa}$$

The actual modulus and strength parallel to the fibers are shown in Figure 17-9. The calculated modulus of elasticity (193×10^3 MPa) is exactly the same as the measured modulus. The estimated strength (1125 MPa) is substantially higher than the actual strength (about 896 MPa). We also note that the modulus of elasticity is very anisotropic, with the modulus perpendicular to the fibers being only half the modulus parallel to the fibers.

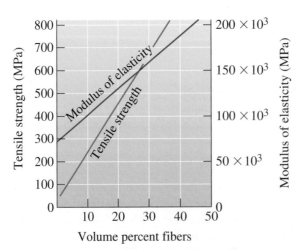

Figure 17-9
The influence of volume percent boron-coated SiC (Borsic) fibers on the properties of Borsic-reinforced aluminum parallel to the fibers (for Example 17-7).

| **Example 17-8** | *Nylon-Glass Fiber Composites* |

Glass fibers in nylon provide reinforcement. If the nylon contains 30 vol% E-glass, what fraction of the force applied parallel to the fiber axis is carried by the glass fibers?

SOLUTION

The modulus of elasticity for each component of the composite is

$$E_{glass} = 72.4 \times 10^3 \text{ MPa} \qquad E_{nylon} = 2.8 \times 10^3 \text{ MPa}$$

Both the nylon and the glass fibers have equal strain if the bonding is good, so

$$\varepsilon_c = \varepsilon_m = \varepsilon_f$$

$$\varepsilon_m = \frac{\sigma_m}{E_m} = \varepsilon_f = \frac{\sigma_f}{E_f}$$

$$\frac{\sigma_f}{\sigma_m} = \frac{E_f}{E_m} = \frac{72.4 \times 10^3}{2.8 \times 10^3} = 25.86$$

Thus the fraction of the force borne by the fibers is given by

$$\text{Fraction} = \frac{F_f}{F_f + F_m} = \frac{\sigma_f A_f}{\sigma_f A_f + \sigma_m A_m} = \frac{\sigma_f (0.3)}{\sigma_f (0.3) + \sigma_m (0.7)}$$

$$= \frac{0.3}{0.3 + 0.7(\sigma_m/\sigma_f)} = \frac{0.3}{0.3 + 0.7(1/26.25)} = 0.92$$

where F_f is the force carried by the fibers and F_m is the force carried by the matrix. Almost all of the load is carried by the glass fibers.

17-4 Characteristics of Fiber-Reinforced Composites

Many factors must be considered when designing a fiber-reinforced composite, including the length, diameter, orientation, amount, and properties of the fibers; the properties of the matrix; and the bonding between the fibers and the matrix.

Fiber Length and Diameter

Fibers can be short, long, or even continuous. Their dimensions are often characterized by the **aspect ratio** l/d, where l is the fiber length and d is the diameter. Typical fibers have diameters varying from 10 μm (10×10^{-4} cm) to 150 μm (150×10^{-4} cm).

The strength of a composite improves when the aspect ratio is large. Fibers often fracture because of surface imperfections. Making the diameter as small as possible gives the fiber less surface area and, consequently, fewer flaws that might propagate during processing or under a load. We also prefer long fibers. The ends of a fiber carry less of the load than the remainder of the fiber; consequently, the fewer the ends, the higher the load-carrying ability of the fibers (Figure 17-10).

In many fiber-reinforced systems, discontinuous fibers with an aspect ratio greater than some critical value are used to provide an acceptable compromise between processing

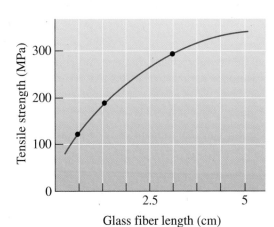

Figure 17-10
Increasing the length of chopped E-glass fibers in an epoxy matrix increases the strength of the composite. In this example, the volume fraction of glass fibers is about 0.5.

ease and properties. A critical fiber length l_c for any given fiber diameter d can be determined according to

$$l_c = \frac{TS_f d}{2\tau_i} \qquad (17\text{-}9)$$

where TS_f is the tensile strength of the fiber and τ_i is related to the strength of the bond between the fiber and the matrix, or the stress at which the matrix begins to deform. If the fiber length l is smaller than l_c, little reinforcing effect is observed; if l is greater than about $15l_c$, the fiber behaves almost as if it were continuous. The strength of the composite can be estimated from

$$\sigma_c = f_f TS_f \left(1 - \frac{l_c}{2l} \right) + f_m S_m \qquad (17\text{-}10)$$

where S_m is the stress on the matrix when the fibers break.

Amount of Fiber
A greater volume fraction of fibers increases the strength and stiffness of the composite, as we would expect from the rule of mixtures. The maximum volume fraction is about 80%, beyond which fibers can no longer be completely surrounded by the matrix.

Orientation of Fibers
The reinforcing fibers may be introduced into the matrix in a number of orientations. Short, randomly oriented fibers having a small aspect ratio—typical of fiberglass—are easily introduced into the matrix and give relatively isotropic behavior in the composite.

Long, or even continuous, unidirectional arrangements of fibers produce anisotropic properties, with particularly good strength and stiffness parallel to the fibers. These fibers are often designated as 0° plies, indicating that all of the fibers are aligned with the direction of the applied stress. Unidirectional orientations provide poor properties if the load is perpendicular to the fibers (Figure 17-11).

One of the unique characteristics of fiber-reinforced composites is that their properties can be tailored to meet different types of loading conditions. Long, continuous fibers can be introduced in several directions within the matrix (Figure 17-12); in orthogonal arrangements (0°/90° plies), good strength is obtained in two perpendicular directions. More complicated arrangements (such as 0°/±45°/90° plies) provide reinforcement in multiple directions.

Fibers can also be arranged in three-dimensional patterns. In even the simplest of fabric weaves, the fibers in each individual layer of fabric have some small degree of

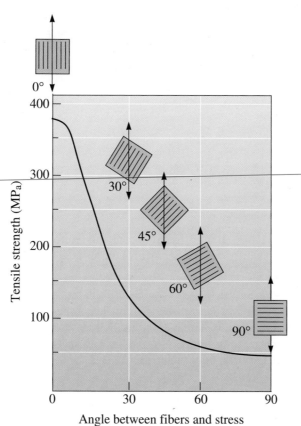

Figure 17-11
Effect of fiber orientation on the tensile strength of E-glass fiber-reinforced epoxy composites.

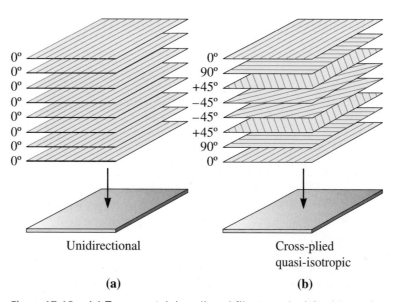

Figure 17-12   (a) Tapes containing aligned fibers can be joined to produce a multi-layered unidirectional composite structure. (b) Tapes containing aligned fibers can be joined with different orientations to produce a quasi-isotropic composite. In this case, a 0°/±45°/90° composite is formed.

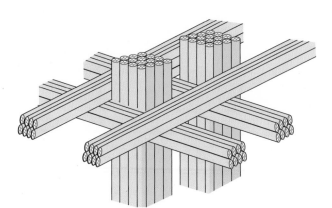

Figure 17-13
A three-dimensional weave for fiber-reinforced composites.

orientation in a third direction. Better three-dimensional reinforcement occurs when the fabric layers are knitted or stitched together. More complicated three-dimensional weaves can also be used (Figure 17-13).

Fiber Properties

In most fiber-reinforced composites, the fibers are strong, stiff, and lightweight. If the composite is to be used at elevated temperatures, the fiber should also have a high melting temperature. Thus the **specific strength** and **specific modulus** of the fiber are important characteristics:

$$\text{Specific strength} = \frac{TS}{\rho} \tag{17-11}$$

$$\text{Specific modulus} = \frac{E}{\rho} \tag{17-12}$$

where TS is the tensile strength, ρ is the density, and E is the modulus of elasticity.

Properties of typical fibers are shown in Table 17-2 and Figure 17-14. Note in Table 17-2, the units of density are g/cm^3. The highest specific modulus is usually found in

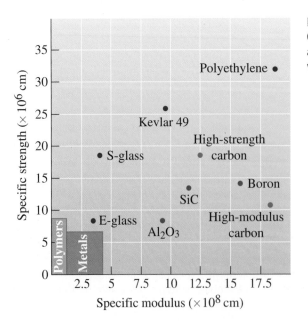

Figure 17-14
Comparison of the specific strength and specific modulus of fibers versus metals and polymers.

TABLE 17-2 ■ *Properties of selected reinforcing materials**

Material	Density (ρ) (g/cm^3)	Tensile Strength (TS) (MPa)	Modulus of Elasticity (E) ($\times 10^3$ MPa)	Melting Temperature (°C)	Specific Modulus ($\times 10^7$ cm)	Specific Strength ($\times 10^6$ cm)
Polymers:						
Kevlar^{TM}	1.44	4480	124	500	86.8	31.3
Nylon	1.14	83	3.5	249	2.5	7.3
Polyethylene	0.97	21–48	0.3–0.7	147	17.8	34.3
Metals:						
Be composites	1.83	276–345	303	1277	193.8	7.0
Boron	2.36	3450	379	2030	161.8	11.8
W	19.40	4000	407	3410	21.3	2.0
Glass:						
E-glass	2.55	3450	72.4	<1725	28.5	14.0
S-glass	2.50	4480	86.9	<1725	35.0	18.0
Carbon:						
HS (high strength)	1.75	5650	276	3700	158.8	32.5
HM (high modulus)	1.90	1860	531	3700	280.0	9.8
Ceramics:						
Al_2O_3	3.95	2070	379	2015	97.0	5.3
B_4C	2.36	2280	483	2450	206.0	9.8
SiC	3.00	3930	483	2700	118.3	13.3
ZrO_2	4.84	2070	345	2677	71.5	4.3
Whiskers:						
Al_2O_3	3.96	20700	428	1982	108.5	52.5
Cr	7.20	8900	241	1890	33.5	12.3
Graphite	1.66	20700	703	3700	425.0	125.5
SiC	3.18	20700	483	2700	152.0	65.5
Si_3N_4	3.18	13800	379		119.5	43.8

materials having a low atomic number and covalent bonding, such as carbon and boron. These two elements also have a high strength and melting temperature.

Aramid fibers, of which Kevlar^{TM} is the best known example, are aromatic polyamide polymers strengthened by a backbone containing benzene rings (Figure 17-15) and are examples of liquid-crystalline polymers in that the polymer chains are rod-like and very stiff. Specially prepared polyethylene fibers are also available. Both the aramid and polyethylene fibers have excellent strength and stiffness but are limited to low temperature use. Because of their lower density, polyethylene fibers have superior specific strength and specific modulus.

Ceramic fibers and whiskers, including alumina and silicon carbide, are strong and stiff. Glass fibers, which are the most commonly used, include pure silica, S-glass (25% Al_2O_3, 10% MgO, balance SiO_2), and E-glass (18% CaO, 15% Al_2O_3, balance SiO_2). Although they are considerably denser than the polymer fibers, the ceramics can be used at much higher temperatures. Beryllium and tungsten, although metallically bonded, have a high modulus that makes them attractive fiber materials for certain applications. The following example discusses issues related to designing with composites.

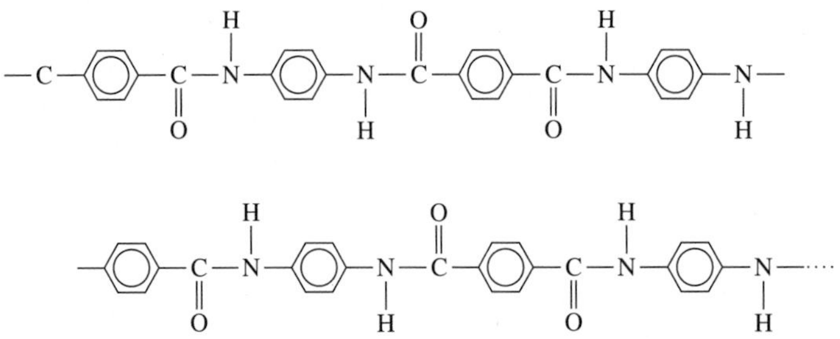

Figure 17-15   The structure of Kevlar™. The fibers are joined by secondary bonds between oxygen and hydrogen atoms on adjoining chains.

Example 17-9   *Design of an Aerospace Composite*

We are now using a 7075-T6 aluminum alloy (modulus of elasticity of 69×10^3 MPa) to make a 227-kg panel on a commercial aircraft. Experience has shown that each kilogram reduction in weight on the aircraft reduces the fuel consumption by 4200 liters each year. Design a material for the panel that will reduce weight, yet maintain the same specific modulus, and will be economical over a 10-year lifetime of the aircraft.

SOLUTION

There are many possible materials that might be used to provide a weight savings. As an example, let's consider using a boron fiber-reinforced Al-Li alloy in the T6 condition. Both the boron fiber and the lithium alloying addition increase the modulus of elasticity; the boron and the Al-Li alloy also have densities less than that of typical aluminum alloys.

The specific modulus of the current 7075-T6 alloy is

$$\text{Specific modulus} = \frac{(69 \times 10^3 \text{ MPa})}{\left(2.7 \frac{\text{g}}{\text{cm}^3}\right)}$$

$$= 2.61 \times 10^8 \text{ cm}$$

The density of the boron fibers is approximately 2.36 g/cm^3 and that of a typical Al-Li alloy is approximately 2.5 g/cm^3. If we use 0.6 volume fraction boron fibers in the composite, then the density, modulus of elasticity, and specific modulus of the composite are

$$\rho_c = (0.6)(2.36) + (0.4)(2.5) = 2.416 \text{ g/cm}^3$$

$$E_{c,\parallel} = (0.6)(379 \times 10^3 \text{ MPa}) + (0.4)(76 \times 10^3 \text{ MPa}) = 258 \times 10^3 \text{ MPa}$$

$$\text{Specific modulus} = \frac{258 \times 10^3 \text{ MPa}}{2.416 \text{ g/cm}^3} = 10.9 \times 10^8 \text{ cm}$$

If the specific modulus is the only factor influencing the design of the component, the thickness of the part might be reduced by 75%, giving a component weight of 56.8 kg rather than 227 kg. The weight savings would then be 170.2 kg, or (4200 liters/kg)(170.2 kg) = 714,840 liters per year. At $0.50 per liter, about $357,000 in fuel savings could be realized each year, or $3.57 million over the 10-year aircraft lifetime.

This is certainly an optimistic comparison, since strength or fabrication factors may not permit the part to be made as thin as suggested. In addition, the high cost of boron fibers (over \$660/kg) and higher manufacturing costs of the composite compared with those of 7075 aluminum would reduce cost savings.

Matrix Properties

The matrix supports the fibers and keeps them in the proper position, transfers the load to the strong fibers, protects the fibers from damage during manufacture and use of the composite, and prevents cracks in the fiber from propagating throughout the entire composite. The matrix usually provides the major control over electrical properties, chemical behavior, and elevated temperature use of the composite.

Polymer matrices are particularly common. Most polymer materials—both thermoplastics and thermosets—are available in short glass fiber-reinforced grades. These composites are formed into useful shapes by the processes described in Chapter 16. Sheet-molding compounds (SMCs) and bulk-molding compounds (BMCs) are typical of this type of composite. Thermosetting aromatic polyimides are used for somewhat higher temperature applications.

Metal-matrix composites (MMCs) include aluminum, magnesium, copper, nickel, and intermetallic compound alloys reinforced with ceramic and metal fibers. A variety of aerospace and automotive applications utilize MMCs. The metal matrix permits the composite to operate at high temperatures, but producing the composite is often more difficult and expensive than producing the polymer-matrix materials.

The ceramic-matrix composites (CMCs) have good properties at elevated temperatures and are lighter in weight than the high-temperature metal-matrix composites. In a later section, we discuss how to develop toughness in CMCs.

Bonding and Failure

Particularly in polymer and metal-matrix composites, good bonding must be obtained between the various constituents. The fibers must be firmly bonded to the matrix material if the load is to be properly transmitted from the matrix to the fibers. In addition, the fibers may pull out of the matrix during loading, reducing the strength and fracture resistance of the composite if bonding is poor. Figure 17-16 illustrates poor bonding of carbon fibers in a copper matrix. In some cases, special coatings may be used to improve bonding. Glass fibers are coated with a silane coupling or "keying" agent (called **sizing**) to improve bonding and moisture resistance in fiberglass composites. Carbon fibers similarly are coated with an organic material to improve bonding. Boron fibers can be coated with silicon carbide or boron nitride to improve bonding with an aluminum matrix; in fact, these fibers are called Borsic fibers to reflect the presence of the silicon carbide (SiC) coating.

Another property that must be considered when combining fibers into a matrix is the similarity between the coefficients of thermal expansion for the two materials. If the fiber expands and contracts at a rate much different from that of the matrix, fibers may break or bonding can be disrupted, causing premature failure.

In many composites, individual plies or layers of fabric are joined. Bonding between these layers must also be good or another problem—**delamination**—may occur. Delamination has been suspected as a cause in some accidents involving airplanes using composite-based structures. The layers may tear apart under load and cause failure. Using composites with a three-dimensional weave will help prevent delamination.

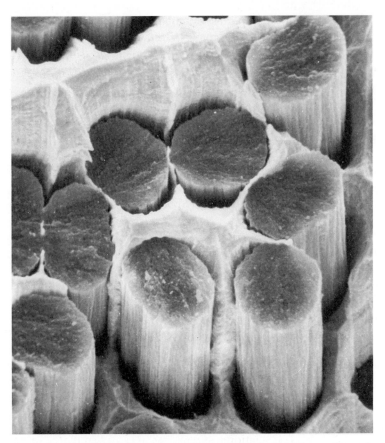

Figure 17-16 Scanning electron micrograph of the fracture surface of a silver-copper alloy reinforced with carbon fibers. Poor bonding causes much of the fracture surface to follow the interface between the metal matrix and the carbon tows (× 3000). (*From ASM Handbook, Vol. 9, Metallography and Microstructure (1985), ASM International, Materials Park, OH 44073-0002.*)

17-5 Manufacturing Fibers and Composites

Producing a fiber-reinforced composite involves several steps, including producing the fibers, arranging the fibers into bundles or fabrics, and introducing the fibers into the matrix.

Making the Fiber
Metallic fibers, glass fibers, and many polymer fibers (including nylon, aramid, and polyacrylonitrile) can be formed by drawing processes, as described in Chapter 8 (wire drawing of metal) and Chapter 16 (using the spinnerette for polymer fibers).

Boron, carbon, and ceramics are too brittle and reactive to be worked by conventional drawing processes. Boron fiber is produced by **chemical vapor deposition** (CVD) [Figure 17-17(a)]. A very fine, heated tungsten filament is used as a substrate, passing through a seal into a heated chamber. Vaporized boron compounds such as BCl_3 are introduced into the chamber, decompose, and permit boron to precipitate onto the tungsten wire (Figure 17-18). SiC fibers are made in a similar manner, with carbon fibers as the substrate for the vapor deposition of silicon carbide.

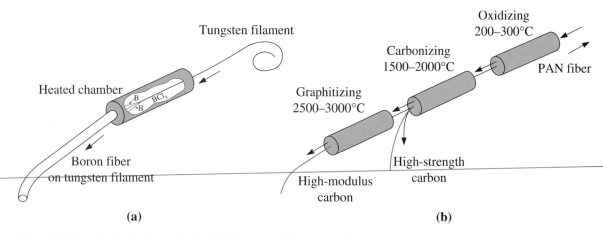

(a)   **(b)**

Figure 17-17   Methods for producing (a) boron and (b) carbon fibers.

Carbon fibers are made by **carbonizing**, or pyrolizing, an organic filament, which is more easily drawn or spun into thin, continuous lengths [Figure 17-17(b)]. The organic filament, known as a **precursor**, is often rayon (a cellulosic polymer), polyacrylonitrile (PAN), or pitch (various aromatic organic compounds). High temperatures decompose the organic polymer, driving off all of the elements but carbon. As the carbonizing temperature increases from 1000°C to 3000°C, the tensile strength decreases while the modulus of elasticity increases (Figure 17-19). Drawing the carbon filaments at critical times during carbonizing may produce desirable preferred orientations in the final carbon filament.

Whiskers are single crystals with aspect ratios of 20 to 1000. Because the whiskers contain no mobile dislocations, slip cannot occur, and they have exceptionally high strengths. Because of the complex processing required to produce whiskers, their cost may be quite high.

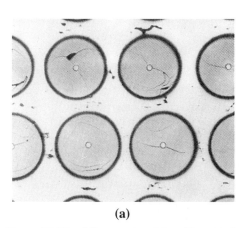

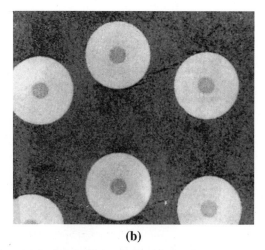

(a)   **(b)**

Figure 17-18   Micrographs of two fiber-reinforced composites: (a) In Borsic fiber-reinforced aluminum, the fibers are composed of a thick layer of boron deposited on a small-diameter tungsten filament ($\times$ 1000). (*From ASM Handbook, Vol. 9, Metallography and Microstructure (1985), ASM International, Materials Park, OH 44073-0002.*) (b) In this microstructure of a ceramic-fiber–ceramic-matrix composite, silicon carbide fibers are used to reinforce a silicon nitride matrix. The SiC fiber is vapor-deposited on a small carbon precursor filament ($\times$ 125). (*Courtesy of Dr. R.T. Bhatt, NASA Lewis Research Center.*)

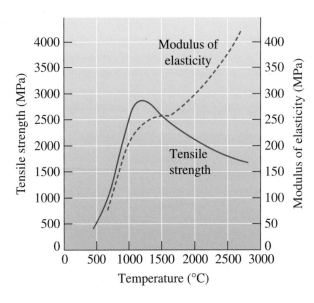

Figure 17-19
The effect of heat-treatment temperature on the strength and modulus of elasticity of carbon fibers.

Arranging the Fibers

Exceptionally fine filaments are bundled together as rovings, yarns, or tows. In **yarns**, as many as 10,000 filaments are twisted together to produce the fiber. A **tow** contains a few hundred to more than 100,000 untwisted filaments (Figure 17-20). **Rovings** are untwisted bundles of filaments, yarns, or tows.

Often, fibers are chopped into short lengths of 1 cm or less. These fibers, also called **staples**, are easily incorporated into the matrix and are typical of the sheet-molding and bulk-molding compounds for polymer-matrix composites. The fibers often are present in the composite in a random orientation.

Long or continuous fibers for polymer-matrix composites can be processed into mats or fabrics. *Mats* contain non-woven, randomly oriented fibers loosely held together by a polymer resin. The fibers can also be woven, braided, or knitted into two-dimensional or three-dimensional fabrics. The fabrics are then impregnated with a polymer resin. The resins at this point in the processing have not yet been completely polymerized; these mats or fabrics are called **prepregs**.

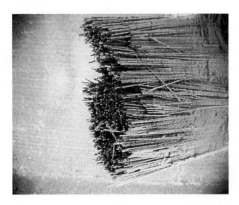

Figure 17-20
A scanning electron micrograph of a carbon tow containing many individual carbon filaments (× 200). (*Reprinted courtesy of Don Askeland.*)

When unidirectionally aligned fibers are to be introduced into a polymer matrix, **tapes** may be produced. Individual fibers can be unwound from spools onto a mandrel, which determines the spacing of the individual fibers, and prepregged with a polymer resin. These tapes, only one fiber diameter thick, may be up to 120 cm wide. Figure 17-21 illustrates that tapes can also be produced by covering the fibers with upper and lower layers of metal foil that are then joined by diffusion bonding.

Producing the Composite

A variety of methods for producing composite parts are used, depending on the application and materials. Short fiber-reinforced composites are normally formed by mixing the fibers with a liquid or plastic matrix, then using relatively conventional techniques such as injection molding for polymer-base composites or casting for metal-matrix composites. Polymer matrix composites can also be produced by a spray-up method, in which short fibers mixed with a resin are sprayed against a form and cured.

Special techniques, however, have been devised for producing composites using continuous fibers, either in unidirectionally aligned, mat, or fabric form (Figure 17-22). In hand lay-up techniques, the tapes, mats, or fabrics are placed against a form, saturated with a polymer resin, rolled to ensure good contact and freedom from porosity, and finally cured. Fiberglass car and truck bodies might be made in this manner, which is generally slow and labor intensive.

Tapes and fabrics can also be placed in a die and formed by bag molding. High-pressure gases or a vacuum are introduced to force the individual plies together so that good bonding is achieved during curing. Large polymer matrix components for the skins of military aircraft have been produced by these techniques. In matched die molding, short fibers or mats are placed into a two-part die; when the die is closed, the composite shape is formed.

Filament winding is used to produce products such as pressure tanks and rocket motor castings (Figure 17-23). Fibers are wrapped around a form or mandrel to gradually build up a hollow shape that may be even several meters in thickness. The filament can be dipped in the polymer-matrix resin prior to winding, or the resin can be impregnated around the fiber during or after winding. Curing completes the production of the composite part.

Pultrusion is used to form a simple, shaped product with a constant cross section, such as round, rectangular, pipe, plate, or sheet shapes (Figure 17-24). Fibers or mats are drawn from spools, passed through a polymer resin bath for impregnation, and gathered together to produce a particular shape before entering a heated die for curing. Curing of the resin is accomplished almost immediately, so a continuous product is produced.

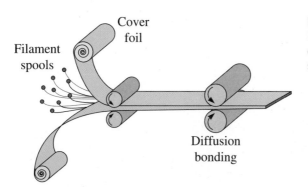

Figure 17-21
Production of fiber tapes by encasing fibers between metal cover sheets by diffusion bonding.

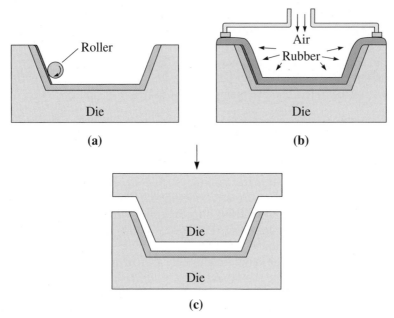

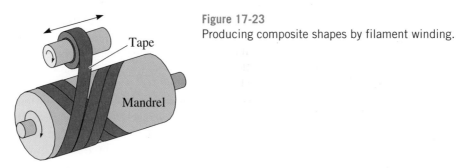

Figure 17-22 Producing composite shapes in dies by (a) hand lay-up, (b) pressure bag molding, and (c) matched die molding.

Figure 17-23
Producing composite shapes by filament winding.

The pultruded stock can subsequently be formed into somewhat more complicated shapes, such as fishing poles, golf club shafts, and ski poles.

Metal-matrix composites with continuous fibers are more difficult to produce than are the polymer-matrix composites. Casting processes that force liquid around the fibers using capillary rise, pressure casting, vacuum infiltration, or continuous casting are used. Various solid-state compaction processes can also be used.

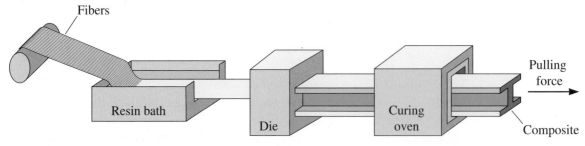

Figure 17-24 Producing composite shapes by pultrusion.

17-6   Fiber-Reinforced Systems and Applications

Before completing our discussion of fiber-reinforced composites, let's look at the behavior and applications of some of the most common of these materials. Figure 17-25 compares the specific modulus and specific strength of several composites with those of metals and polymers. Note that the values in this figure are lower than those in Figure 17-14, since we are now looking at the composite, not just the fiber.

Advanced Composites   The term advanced composites is often used when the composite is intended to provide service in very critical applications, as in the aerospace industry (Table 17-3). The advanced composites normally are polymer–matrix composites reinforced with high-strength polymer, metal, or ceramic fibers. Carbon fibers are used extensively where particularly good stiffness is required; aramid—and, to an even greater extent, polyethylene—fibers are better suited to high-strength applications in which toughness and damage resistance are more important. Unfortunately, the polymer fibers lose their strength at relatively low temperatures, as do all of the polymer matrices (Figure 17-26).

The advanced composites are also frequently used for sporting goods. Tennis rackets, golf clubs, skis, ski poles, and fishing poles often contain carbon or aramid fibers because the higher stiffness provides better performance. In the case of golf clubs, carbon fibers allow less weight in the shaft and therefore more weight in the head. Fabric reinforced with polyethylene fibers is used for lightweight sails for racing yachts.

A unique application for aramid fiber composites is armor. Tough Kevlar[TM] composites provide better ballistic protection than do other materials, making them suitable for lightweight, flexible bulletproof clothing.

Hybrid composites are composed of two or more types of fibers. For instance, Kevlar[TM] fibers may be mixed with carbon fibers to improve the toughness of a stiff

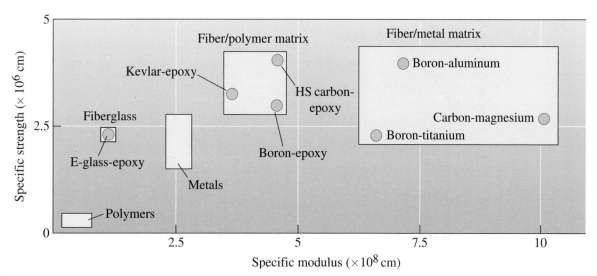

Figure 17-25   A comparison of the specific modulus and specific strength of several composite materials with those of metals and polymers.

TABLE 17-3 ■ *Examples of fiber-reinforced materials and applications*

Material	Applications
Borsic aluminum	Fan blades in engines, other aircraft and aerospace applications
Kevlar™-epoxy and Kevlar™-polyester	Aircraft, aerospace applications (including space shuttle), boat hulls, sporting goods (including tennis rackets, golf club shafts, fishing rods), flak jackets
Graphite-polymer	Aerospace and automotive applications, sporting goods
Glass-polymer	Lightweight automotive applications, water and marine applications, corrosion-resistant applications, sporting goods equipment, aircraft and aerospace components

composite, or Kevlar™ may be mixed with glass fibers to improve stiffness. Particularly good tailoring of the composite to meet specific applications can be achieved by controlling the amounts and orientations of each fiber.

Tough composites can also be produced if careful attention is paid to the choice of materials and processing techniques. Better fracture toughness in the rather brittle composites can be obtained by using long fibers, amorphous (such as polyetheretherketone known as PEEK or polyphenylene sulfide known as PPS) rather than crystalline or cross-linked matrices, thermoplastic-elastomer matrices, or interpenetrating network polymers.

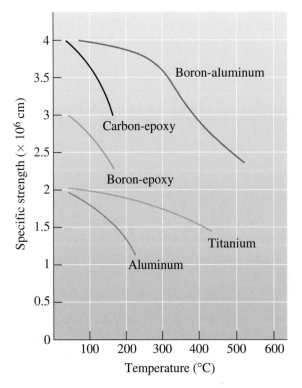

Figure 17-26
The specific strength versus temperature for several composites and metals.

Metal-Matrix Composites

Metal-matrix composites, strengthened by metal or ceramic fibers, provide high temperature resistance. Aluminum reinforced with borsic fibers has been used extensively in aerospace applications, including struts for the space shuttle. Copper-based alloys have been reinforced with SiC fibers, producing high-strength propellers for ships.

Aluminum is commonly used as the matrix in metal-matrix composites. Al_2O_3 fibers reinforce the pistons for some diesel engines; SiC fibers and whiskers are used in aerospace applications, including stiffeners and missile fins; and carbon fibers provide reinforcement for the aluminum antenna mast of the Hubble telescope. Polymer fibers, because of their low melting or degradation temperatures, normally are not used in a metallic matrix. *Polymets,* however, are produced by hot extruding aluminum powder and high melting-temperature liquid-crystalline polymers. A reduction of 1000 to 1 during the extrusion process elongates the polymer into aligned filaments and bonds the aluminum powder particles into a solid matrix.

Metal-matrix composites may find important applications in components for rocket or aircraft engines. Superalloys reinforced with metal fibers (such as tungsten) or ceramic fibers (such as SiC or B_4N) maintain their strength at higher temperatures, permitting jet engines to operate more efficiently. Similarly, titanium and titanium aluminides reinforced with SiC fibers are candidates for turbine blades and disks.

A unique application for metal-matrix composites is in the superconducting wire required for fusion reactors. The intermetallic compound Nb_3Sn has good superconducting properties but is very brittle. To produce Nb_3Sn wire, pure niobium wire is surrounded by copper as the two metals are formed into a wire composite (Figure 17-27). The niobium-copper composite wire is then coated with tin. The tin diffuses through the copper and reacts with the niobium to produce the intermetallic compound. Niobium titanium systems are also used.

Ceramic-Matrix Composites

Composites containing ceramic fibers in a ceramic matrix are also finding applications. Two important uses will be discussed to illustrate the unique properties that can be obtained with these materials.

Carbon-carbon (C-C) composites are used for extraordinary temperature resistance in aerospace applications. Carbon-carbon composites can operate at temperatures of up to 3000°C and, in fact, are stronger at high temperatures than at low temperatures (Figure 17-28). Carbon-carbon composites are made by forming a polyacrylonitrile or carbon fiber fabric into a mold, then impregnating the fabric with an organic resin, such as a phenolic. The part is pyrolyzed to convert the phenolic resin to carbon. The composite, which is still soft and porous, is impregnated and pyrolyzed several more times, continually increasing the density, strength, and stiffness. Finally, the part is coated with silicon carbide to protect the carbon-carbon composite from oxidation. Strengths of 2070 MPa and stiffnesses of 345×10^3 MPa can be obtained. Carbon-carbon composites have been used as nose cones and leading edges of high-performance aerospace vehicles such as the space shuttle and as brake discs on racing cars and commercial jet aircraft.

Ceramic-fiber–ceramic-matrix composites provide improved strength and fracture toughness compared with conventional ceramics (Table 17-4). Fiber reinforcements improve the toughness of the ceramic matrix in several ways. First, a crack moving through the matrix encounters a fiber; if the bonding between the matrix and the fiber is poor, the crack is forced to propagate around the fiber in order to continue the fracture process. In addition, poor bonding allows the fiber to begin to pull out of the matrix

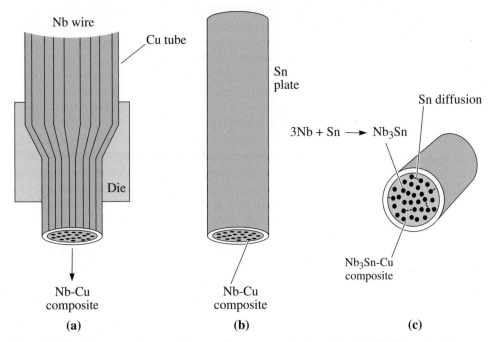

Figure 17-27   The manufacture of composite superconductor wires: (a) Niobium wire is surrounded with copper during forming. (b) Tin is plated onto Nb-Cu composite wire. (c) Tin diffuses to niobium to produce the Nb_3Sn-Cu composite.

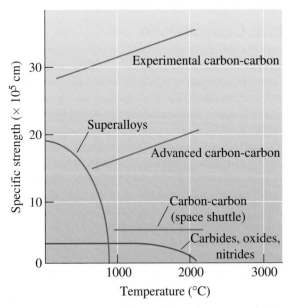

Figure 17-28   A comparison of the specific strengths of various carbon-carbon composites with that of other high-temperature materials relative to temperature.

[Figure 17-29(a)]. Both processes consume energy, thereby increasing fracture toughness. Finally, as a crack in the matrix begins, unbroken fibers may bridge the crack and make it more difficult for the crack to open [Figure 17-29(b)].

Unlike polymer and metal matrix composites, poor bonding—rather than good bonding—is required! Consequently, control of the interface structure is crucial. In a glass-ceramic (based on $Al_2O_3 \cdot SiO_2 \cdot Li_2O$) reinforced with SiC fibers, an interface layer containing carbon and NbC is produced that makes debonding of the fiber from the matrix easy. If, however, the composite is heated to a high temperature, the interface is oxidized; the oxide occupies a large volume, exerts a clamping force on the fiber, and prevents easy pull-out. Fracture toughness is then decreased. The following example illustrates some of the cost and property issues that come up while working with composites.

Example 17-10 *Design of a Composite Strut*

Design a unidirectional fiber-reinforced epoxy-matrix strut having a round cross-section. The strut is 3 m long and, when a force of 2224 N is applied, it should stretch no more than 0.25 cm. We want to ensure that the stress acting on the strut is less than the yield strength of the epoxy matrix, 83 MPa. If the fibers should happen to break, the strut will stretch an extra amount but may not catastrophically fracture. Epoxy costs about \$1.8/kg and has a modulus of elasticity of 3450 MPa.

SOLUTION

Suppose that the strut were made entirely of epoxy (that is, no fibers):

$$\varepsilon_{max} = \frac{0.25\,cm}{300\,cm} = 0.00083\,cm/cm$$

$$\sigma_{max} = E\varepsilon = (3450)(0.00083) = 2.9\,MPa$$

TABLE 17-4 ■ *Effect of SiC-reinforcement fibers on the properties of selected ceramic materials*

Material	Flexural Strength (MPa)	Fracture Toughness (MPa $\sqrt{m}$)
Al_2O_3	550	5.5
Al_2O_3/SiC	790	8.8
SiC	500	4.4
SiC/SiC	760	25.3
ZrO_2	210	5.5
ZrO_2/SiC	450	22.2
Si_3N_4	470	4.4
Si_3N_4/SiC	790	56.0
Glass	60	1.1
Glass/SiC	830	18.7
Glass ceramic	210	2.2
Glass ceramic/SiC	830	17.6

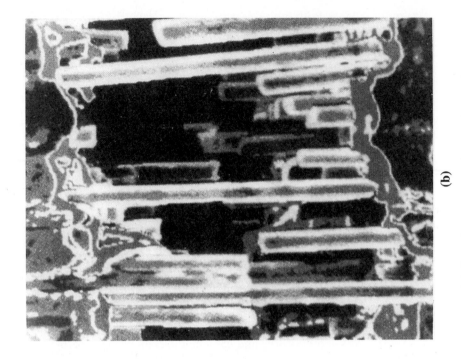

(b)

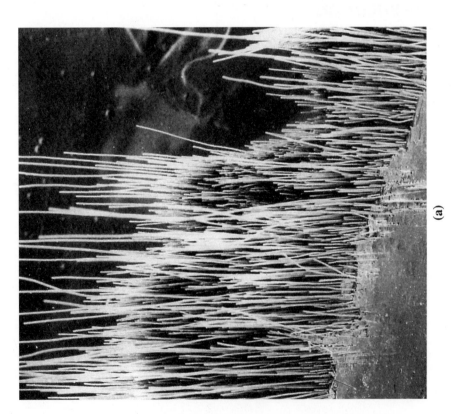

(a)

Figure 17-29   Two failure modes in ceramic-ceramic composites: (a) Extensive pull-out of SiC fibers in a glass matrix provides good composite toughness (×20). (*From ASM Handbook, Vol. 9, Metallography and Microstructure (1985), ASM International, Materials Park, OH 44073-0002.*) (b) Bridging of some fibers across a crack enhances the toughness of a ceramic-matrix composite (unknown magnification). (*From Journal of Metals, May 1991.*)

$$A_{strut} = \frac{F}{\sigma} = \frac{2224 \text{ N}}{2.9 \text{ MPa}} = 7.67 \text{ cm}^2 \text{ or } d = 3.12 \text{ cm}$$

Since $\rho_{epoxy} = 1.25 \text{ g/cm}^3$

$$\text{Weight}_{strut} = (1.25)(\pi)(3.12/2)^2(300) = 2.87 \text{ kg}$$
$$\text{Cost}_{strut} = (2.87 \text{ kg})(\$1.8/\text{kg}) = \$5.17$$

With no reinforcement, the strut is large and heavy; the materials cost is high due to the large amount of epoxy needed.

In a composite, the maximum strain is still 0.00083 cm/cm. If we make the strut as small as possible—that is, it operates at 83 MPa—then the minimum modulus of elasticity E_c of the composite is

$$E_c > \frac{\sigma}{\varepsilon_{max}} = \frac{83}{0.00083} = 100 \times 10^3 \text{ MPa}$$

Let's look at several possible composite systems. The modulus of glass fibers is less than 100×10^3 MPa; therefore, glass reinforcement is not a possible choice.

For high modulus carbon fibers, $E = 531 \times 10^3$ MPa; the density is 1.9 g/cm^3 and the cost is about \$66/kg. The minimum volume fraction of carbon fibers needed to give a composite modulus of 100×10^3 MPa is

$$E_c = f_c(531 \times 10^3) + (1 - f_c)(3.45 \times 10^3) > 100 \times 10^3$$
$$f_c = 0.183$$

The volume fraction of epoxy remaining is 0.817. An area of 0.817 times the total cross-sectional area of the strut must support a 2224-N load with no more than 83 MPa if all of the fibers should fail:

$$A_{epoxy} = 0.817 A_{total} = \frac{F}{\sigma} = \frac{2224 \text{ N}}{83 \text{ MPa}} = 0.268 \text{ cm}^2$$

$$A_{total} = \frac{0.268}{0.817} = 0.328 \text{ cm}^2 \text{ or } d = 0.646 \text{ cm}$$

$$\text{Volume}_{strut} = (0.328 \text{ cm}^2)(300 \text{ cm}) = 98.4 \text{ cm}^3$$
$$\text{Weight}_{strut} = \rho \text{Volume}_{strut} = [(1.9)(0.183) + (1.25)(0.817)](98.4)$$
$$= 0.135 \text{ kg}$$

To calculate the weight fraction of carbon, consider 1 cm^3 of composite. Then the weight fraction of the carbon is given by

$$\text{Weight fraction}_{carbon} = \frac{(0.183)(1.9)}{(0.183)(1.9) + (0.817)(1.25)} = 0.254$$

and the total weight of the carbon in the composite is

$$\text{Weight}_{carbon} = (0.254)(0.135 \text{ kg}) = 0.034 \text{ kg}$$

Similarly, the weight fraction of epoxy is calculated as

$$\text{Weight fraction}_{epoxy} = \frac{(0.183)(1.25)}{(0.183)(1.9) + (0.817)(1.25)} = 1 - 0.254 = 0.746$$

and the total weight of the epoxy in the composite is

$$\text{Weight}_{\text{epoxy}} = (0.746)(0.135 \text{ kg}) = 0.101 \text{ kg}$$

Therefore,

$$\text{Cost}_{\text{strut}} = (0.034 \text{ kg})(\$66/\text{kg}) + (0.101 \text{ kg})(\$1.80/\text{kg}) = \$2.42$$

The carbon fiber-reinforced strut is less than one-quarter the diameter of an all-epoxy structure, with only 5% of the weight and half of the cost.

We might also repeat these calculations using Kevlar^{TM} fibers, with a modulus of 124×10^3 MPa, a density of 1.44 g/cm^3, and a cost of about $44/kg. By doing so, we would find that a volume fraction of 0.8 fibers is required. Note that 0.8 volume fraction is at the maximum of fiber volume that can be incorporated into a matrix. We would also find that the required diameter of the strut is 1.288 cm and that the strut weighs 0.573 kg and costs $20.94. The modulus of the Kevlar^{TM} is not high enough to offset its high cost.

Although the carbon fibers are the most expensive, they permit the lightest weight and the lowest material cost strut. (This calculation does not, however, take into consideration the costs of manufacturing the strut.) Our design, therefore, is to use a 0.646-cm-diameter strut containing 0.183 volume fraction high modulus carbon fiber.

17-7   Laminar Composite Materials

Laminar composites include very thin coatings, thicker protective surfaces, claddings, bimetallics, laminates, and a host of other applications. In addition, the fiber-reinforced composites produced from tapes or fabrics can be considered partly laminar. Many laminar composites are designed to improve corrosion resistance while retaining low cost, high strength, or light weight. Other important characteristics include superior wear or abrasion resistance, improved appearance, and unusual thermal expansion characteristics.

Rule of Mixtures   Some properties of the laminar composite materials parallel to the lamellae are estimated from the rule of mixtures. The density, electrical and thermal conductivity, and modulus of elasticity parallel to the lamellae can be calculated with little error using the following formulas:

$$\text{Density} = \rho_{c,\parallel} = \sum (f_i \rho_i) \tag{17-13}$$

$$\text{Electrical conductivity} = \sigma_{c,\parallel} = \sum (f_i \sigma_i)$$

$$\text{Thermal conductivity} = k_{c,\parallel} = \sum (f_i k_i) \tag{17-14}$$

$$\text{Modulus of elasticity} = E_{c,\parallel} = \sum (f_i E_i)$$

The laminar composites are very anisotropic. The properties perpendicular to the lamellae are

$$\text{Electrical conductivity} = \frac{1}{\sigma_{c,\perp}} = \sum\left(\frac{f_i}{\sigma_i}\right)$$

$$\text{Thermal conductivity} = \frac{1}{k_{c,\perp}} = \sum\left(\frac{f_i}{k_i}\right) \qquad (17.15)$$

$$\text{Modulus of elasticity} = \frac{1}{E_{c,\perp}} = \sum\left(\frac{f_i}{E_i}\right)$$

Many other properties, such as corrosion and wear resistance, depend primarily on only one of the components of the composite, so the rule of mixtures is not applicable.

Producing Laminar Composites

Several methods are used to produce laminar composites, including a variety of deformation and joining techniques used primarily for metals (Figure 17-30).

Individual plies are often joined by *adhesive bonding*, as is the case in producing plywood. Polymer-matrix composites built up from several layers of fabric or tape prepregs are also joined by adhesive bonding; a film of unpolymerized material is placed between each layer of prepreg. When the layers are pressed at an elevated temperature, polymerization is completed and the prepregged fibers are joined to produce composites that may be dozens of layers thick.

Most of the metallic laminar composites, such as claddings and bimetallics, are produced by *deformation bonding*, such as hot- or cold-roll bonding. The pressure exerted by the rolls breaks up the oxide film at the surface, brings the surfaces into atom-to-atom contact, and permits the two surfaces to be joined. Explosive bonding also can be used. An explosive charge provides the pressure required to join metals. This process is particularly well suited for joining very large plates that will not fit into a rolling mill. Very simple laminar composites, such as coaxial cable, are produced by coextruding two materials through a die in such a way that the soft material surrounds

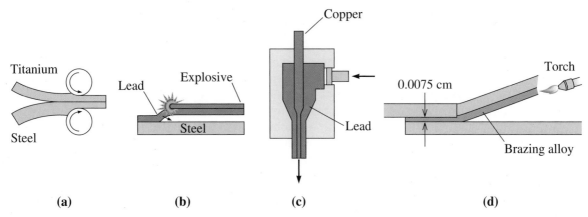

Figure 17-30   Techniques for producing laminar composites: (a) roll bonding, (b) explosive bonding, (c) coextrusion, and (d) brazing.

the harder material. Metal conductor wire can be coated with an insulating thermoplastic polymer in this manner.

Brazing can join composite plates (Chapter 9). The metallic sheets are separated by a very small clearance—preferably, about 0.0075 cm—and heated above the melting temperature of the brazing alloy. The molten brazing alloy is drawn into the thin joint by capillary action.

17-8 Examples and Applications of Laminar Composites

The number of laminar composites is so varied and their applications are so numerous that we cannot make generalizations concerning their behavior. Instead we will examine the characteristics of a few commonly used examples.

Laminates

Laminates are layers of materials joined by an organic adhesive. In laminated safety glass, a plastic adhesive, such as polyvinyl butyral (PVB), joins two pieces of glass; the adhesive prevents fragments of glass from flying about when the glass is broken (Chapter 15). Laminates are used for insulation in motors, for gears, for printed circuit boards, and for decorative items such as Formica® countertops and furniture.

Microlaminates include composites composed of alternating layers of aluminum sheet and fiber-reinforced polymer. *Arall* (aramid-aluminum laminate) and *Glare* (glass-aluminum laminate) are two examples. In Arall, an aramid fiber such as Kevlar™ is prepared as a fabric or unidirectional tape, impregnated with an adhesive, and laminated between layers of aluminum alloy (Figure 17-31). The composite laminate has an unusual combination of strength, stiffness, corrosion resistance, and light weight. Fatigue resistance is improved, since the interface between the layers may block cracks. Glare has similar properties and is used in the fuselage of the Airbus 380. Compared with polymer-matrix composites, the microlaminates have good resistance to lightning-strike damage (which is important in aerospace applications), are formable and machinable, and are easily repaired.

Clad Metals

Clad materials are metal-metal composites. A common example of **cladding** is United States silver coinage. A Cu-80% Ni alloy is bonded to both sides of a Cu-20% Ni alloy. The ratio of thicknesses is about 1/6:2/3:1/6. The high-nickel alloy is a silver color, while the predominantly copper core provides low cost.

Clad materials provide a combination of good corrosion resistance with high strength. *Alclad* is a clad composite in which commercially pure aluminum is bonded to higher strength aluminum alloys. The pure aluminum protects the higher strength alloy from corrosion. The thickness of the pure aluminum layer is about 1% to 15% of the total thickness. Alclad is used in aircraft construction, heat exchangers, building construction, and storage tanks, where combinations of corrosion resistance, strength, and light weight are desired.

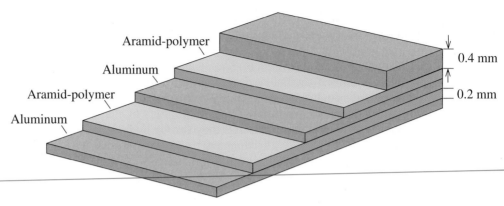

Figure 17-31 Schematic diagram of an aramid-aluminum laminate, Arall, which has potential for aerospace applications.

Bimetallics

Laminar composites made from two metals with different coefficients of thermal expansion are used as temperature indicators and controllers. If two pieces of metal are heated, the metal with the higher coefficient of thermal expansion becomes longer. If the two pieces of metal are rigidly bonded together, the difference in their coefficients causes the strip to bend and produce a curved surface. The amount of movement depends on the temperature. By measuring the curvature or deflection of the strip, we can determine the temperature. Likewise, if the free end of the strip activates a relay, the strip can turn on or off a furnace or air conditioner to regulate temperature. Metals selected for **bimetallics** must have (a) very different coefficients of thermal expansion, (b) expansion characteristics that are reversible and repeatable, and (c) a high modulus of elasticity, so that the bimetallic device can do work. Often the low-expansion strip is made from Invar, an iron-nickel alloy, whereas the high-expansion strip may be brass, Monel, or pure nickel.

Bimetallics can act as circuit breakers as well as thermostats; if a current passing through the strip becomes too high, heating causes the bimetallic to deflect and break the circuit.

Multilayer Capacitors

A laminar geometry is used to make enormous numbers of multilayer capacitors. Their structure comprises thin sheets of $BaTiO_3$-based ceramics separated by Ag/Pd or Ni electrodes (Chapter 19).

17-9   Sandwich Structures

Sandwich materials have thin layers of a facing material joined to a lightweight filler material, such as a polymer foam. Neither the filler nor the facing material is strong or rigid, but the composite possesses both properties. A familiar example is corrugated cardboard. A corrugated core of paper is bonded on either side to flat, thick paper. Neither the corrugated core nor the facing paper is rigid, but the combination is.

Another important example is the honeycomb structure used in aircraft applications. A **honeycomb** is produced by gluing thin aluminum strips at selected locations. The honeycomb material is then expanded to produce a very low-density cellular panel that, by itself, is unstable (Figure 17-32). When an aluminum facing sheet is adhesively bonded to either side of the honeycomb, however, a very stiff, rigid, strong, and exceptionally lightweight sandwich with a density as low as 0.04 g/cm^3 is obtained.

The honeycomb cells can have a variety of shapes, including hexagonal, square, rectangular, and sinusoidal, and they can be made from aluminum, fiberglass, paper, aramid polymers, and other materials. The honeycomb cells can be filled with foam or fiberglass to provide excellent sound and vibration absorption. Figure 17-33 describes one method by which honeycomb can be fabricated.

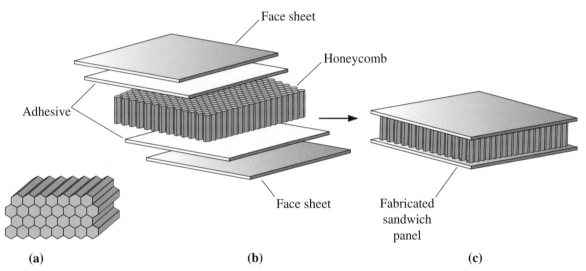

Figure 17-32   (a) A hexagonal cell honeycomb core, (b) can be joined to two face sheets by means of adhesive sheets, (c) producing an exceptionally lightweight yet stiff, strong honeycomb sandwich structure.

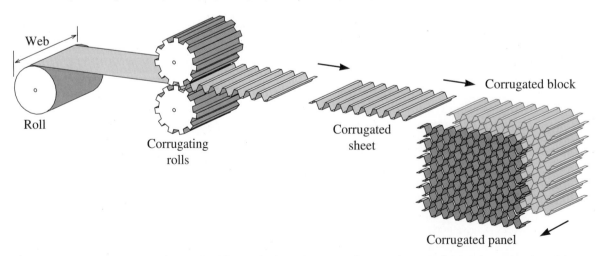

Figure 17-33   In the corrugation method for producing a honeycomb core, the material (such as aluminum) is corrugated between two rolls. The corrugated sheets are joined together with adhesive and then cut to the desired thickness.

Summary

- Composites are composed of two or more materials or phases joined or connected in such a way so as to give a combination of properties that cannot be attained otherwise. The volume fraction and the connectivity of the phases or materials in a composite and the nature of the interface between the dispersed phase and matrix are very important in determining the properties.

- Virtually any combination of metals, polymers, and ceramics is possible. In many cases, the rule of mixtures can be used to estimate the properties of the composite.

- Composites have many applications in construction, aerospace, automotive, sports, microelectronics and other industries.

- Dispersion-strengthened materials contain oxide particles in a metal matrix. The small stable dispersoids interfere with slip, providing good mechanical properties at elevated temperatures.

- Particulate composites contain particles that impart combinations of properties to the composite. Metal-matrix composites contain ceramic or metallic particles that provide improved strength and wear resistance and ensure good electrical conductivity, toughness, or corrosion resistance. Polymer-matrix composites contain particles that enhance stiffness, heat resistance, or electrical conductivity while maintaining light weight, ease of fabrication, or low cost.

- Fiber-reinforced composites provide improvements in strength, stiffness, or high-temperature performance in metals and polymers and impart toughness to ceramics. Fibers typically have low densities, giving high specific strength and specific modulus, but they often are very brittle. Fibers can be continuous or discontinuous. Discontinuous fibers with a high aspect ratio l/d produce better reinforcement.

- Fibers are introduced into a matrix in a variety of orientations. Random orientations and isotropic behavior are obtained using discontinuous fibers; unidirectionally aligned fibers produce composites with anisotropic behavior with large improvements in strength and stiffness parallel to the fiber direction. Properties can be tailored to meet the imposed loads by orienting the fibers in multiple directions.

- Laminar composites are built of layers of different materials. These layers may be sheets of different metals, with one metal providing strength and the other providing hardness or corrosion resistance. Layers may also include sheets of fiber-reinforced material bonded to metal or polymer sheets or even to fiber-reinforced sheets having different fiber orientations. The laminar composites are always anisotropic.

- Sandwich materials, including honeycombs, are exceptionally lightweight laminar composites, with solid facings joined to an almost hollow core.

Glossary

Aramid fibers Polymer fibers, such as Kevlar^{TM}, formed from polyamides, which contain the benzene ring in the backbone of the polymer.

Aspect ratio The length of a fiber divided by its diameter.

Bimetallic A laminar composite material produced by joining two strips of metal with different thermal expansion coefficients, making the material sensitive to temperature changes.

Brazing A process in which a liquid filler metal is introduced by capillary action between two solid materials that are to be joined. Solidification of the brazing alloy provides the bond.

Carbonizing Driving off the non-carbon atoms from a polymer fiber, leaving behind a carbon fiber of high strength. Also known as pyrolizing.

Cemented carbides Particulate composites containing hard ceramic particles bonded with a soft metallic matrix. The composite combines high hardness and cutting ability, yet still has good shock resistance.

Chemical vapor deposition (CVD) Method for manufacturing materials by condensing the material from a vapor onto a solid substrate.

Cladding A laminar composite produced when a corrosion-resistant or high-hardness layer is formed onto a less expensive or higher strength backing.

Delamination Separation of individual plies of a fiber-reinforced composite.

Dispersed phase The phase or phases that are distributed throughout a continuous matrix of the composite.

Dispersoids Oxide particles formed in a metal matrix that interfere with dislocation movement and provide strengthening, even at elevated temperatures.

Filament winding Process for producing fiber-reinforced composites in which continuous fibers are wrapped around a form or mandrel. The fibers may be prepregged or the filament-wound structure may be impregnated to complete the production of the composite.

Honeycomb A lightweight but stiff assembly of aluminum strip joined and expanded to form the core of a sandwich structure.

Hybrid organic-inorganic composites Nanocomposites consisting of structures that are partly organic and partly inorganic, often made using sol-gel processing.

Matrix phase The continuous phase in a composite. The composites are named after the continuous phase (e.g., polymer-matrix composites).

Nanocomposite A material in which the dispersed phase has dimensions on the order of nanometers and is distributed in the continuous matrix.

Precursor A starting chemical (e.g., a polymer fiber) that is carbonized to produce carbon fibers.

Prepregs Layers of fibers in unpolymerized resins. After the prepregs are stacked to form a desired structure, polymerization joins the layers together.

Pultrusion A method for producing composites containing mats or continuous fibers.

Rovings Untwisted bundles of filaments, yarns, or tows.

Rule of mixtures The statement that the properties of a composite material are a function of the volume fraction of each material in the composite.

Sandwich A composite material constructed of a lightweight, low-density material surrounded by dense, solid layers. The sandwich combines overall light weight with excellent stiffness.

Sizing Coating glass fibers with an organic material to improve bonding and moisture resistance in fiberglass.

Specific modulus The modulus of elasticity of a material divided by the density.

Specific strength The tensile or yield strength of a material divided by the density.

Staples Fibers chopped into short lengths.

Tapes A strip of prepreg that is only one filament thick. The filaments may be unidirectional or woven. Several layers of tapes can be joined to produce a composite structure.

Tow A bundle of untwisted filaments.

Whiskers Very fine fibers grown in a manner that produces single crystals with no mobile dislocations, thus giving nearly theoretical strengths.

Yarns Continuous fibers produced from a group of twisted filaments.

Problems

Section 17-1 Dispersion-Strengthened Composites

17-1 What is a composite?

17-2 What do the properties of composite materials depend upon?

17-3 Give examples for which composites are used for load bearing applications.

17-4 Give two examples for which composites are used for non-structural applications.

17-5 What is a dispersion-strengthened composite? How is it different from a particle-reinforced composite?

17-6 What is a nanocomposite? How can certain steels containing ferrite and martensite be described as composites? Explain.

17-7 A tungsten matrix with 20% porosity is infiltrated with silver. Assuming that the pores are interconnected, what is the density of the composite before and after infiltration with silver? The density of pure tungsten is 19.25 g/cm^3 and that of pure silver is 10.49 g/cm^3.

17-8 Nickel containing 2 wt% thorium is produced in powder form, consolidated into a part, and sintered in the presence of oxygen, causing all of the thorium to produce ThO_2 spheres 80 nm in diameter. Calculate the number of spheres per cm^3. The density of ThO_2 is 9.69 g/cm^3.

17-9 Spherical aluminum powder (SAP) 0.002 mm in diameter is treated to create a thin oxide layer and is then used to produce a SAP dispersion-strengthened material containing 10 vol% Al_2O_3. Calculate the average thickness of the oxide film prior to compaction and sintering of the powders into the part.

17-10 Yttria (Y_2O_3) particles 750 Å in diameter are introduced into tungsten by internal oxidation. Measurements using an electron microscope show that there are 5×10^{14} oxide particles per cm^3. Calculate the wt% Y originally in the alloy. The density of Y_2O_3 is 5.01 g/cm^3.

17-11 With no special treatment, aluminum is typically found to have an Al_2O_3 layer that is 3 nm thick. If spherical aluminum powder prepared with a total diameter of 0.01 mm is used to produce SAP dispersion-strengthened aluminum, calculate the volume percent Al_2O_3 in the material and the number of oxide particles per cm^3. Assume that the oxide breaks into disk-shaped flakes 3 nm thick and 3×10^{-4} mm in diameter. Compare the number of oxide particles per cm^3 with the number of solid solution atoms per cm^3 when 3 at% of an alloying element is added to aluminum.

Section 17-2 Particulate Composites

17-12 What is a particulate composite?

17-13 What is a cermet? What is the role of WC and Co in a cermet?

17-14 Calculate the density of a cemented carbide, or cermet, based on a titanium matrix if the composite contains 50 wt% WC, 22 wt% TaC, and 14 wt% TiC. (See Example 17-2 for densities of the carbides.)

17-15 Spherical silica particles (100 nm in diameter) are added to vulcanized rubber in tires to improve stiffness. If the density of the vulcanized rubber matrix is 1.1 g/cm^3, the density of silica is 2.5 g/cm^3, and the tire has a porosity of 4.5%, calculate the number of silica particles lost when a tire wears down 0.4 cm in thickness. The density of the tire is 1.2 g/cm^3; the overall tire diameter is 63 cm; and it is 10 cm wide.

17-16 A typical grinding wheel is 22.5 cm in diameter, 2.5 cm thick, and weighs 2.7 kg. The wheel contains SiC (density of 3.2 g/cm^3) bonded by silica glass (density of 2.5 g/cm^3); 5 vol% of the wheel is porous. The SiC is in the form of 0.04 cm cubes. Calculate

(a) the volume fraction of SiC particles in the wheel and

(b) the number of SiC particles lost from the wheel after it is worn to a diameter of 20 cm.

17-17 An electrical contact material is produced by infiltrating copper into a porous tungsten carbide (WC) compact. The density of the final composite is 12.3 g/cm³. Assuming that all of the pores are filled with copper, calculate

(a) the volume fraction of copper in the composite,

(b) the volume fraction of pores in the WC compact prior to infiltration, and

(c) the original density of the WC compact before infiltration.

17-18 An electrical contact material is produced by first making a porous tungsten compact that weighs 125 g. Liquid silver is introduced into the compact; careful measurement indicates that 105 g of silver is infiltrated. The final density of the composite is 13.8 g/cm³. Calculate the volume fraction of the original compact that is interconnected porosity and the volume fraction that is closed porosity (no silver infiltration).

17-19 How much clay must be added to 10 kg of polyethylene to produce a low-cost composite having a modulus of elasticity greater than 827 MPa and a tensile strength greater than 14 MPa? The density of the clay is 2.4 g/cm³ and that of the polyethylene is 0.92 g/cm³. (See Figure 17-6.)

17-20 We would like to produce a lightweight epoxy part to provide thermal insulation. We have available hollow glass beads for which the outside diameter is 0.16 cm and the wall thickness is 0.0025 cm. Determine the weight and number of beads that must be added to the epoxy to produce a one-pound composite with a density of 0.65g/cm³. The density of the glass is 2.5 g/cm³ and that of the epoxy is 1.25 g/cm³.

Section 17-3 Fiber Reinforced Composites

Section 17-4 Characteristics of Fiber-Reinforced Composites

17-21 What is a fiber-reinforced composite?

17-22 What fiber-reinforcing materials are commonly used?

17-23 In a fiber-reinforced composite, what is the role of the matrix?

17-24 What do the terms CFRP and GFRP mean?

17-25 Explain briefly how the volume of fiber, fiber orientation, and fiber strength and modulus affect the properties of fiber-reinforced composites.

17-26 Five kilograms of continuous boron fibers are introduced in a unidirectional orientation into 8 kg of an aluminum matrix. Calculate

(a) the density of the composite,

(b) the modulus of elasticity parallel to the fibers, and

(c) the modulus of elasticity perpendicular to the fibers.

17-27 We want to produce 4.5 kg of a continuous unidirectional fiber-reinforced composite of HS carbon in a polyimide matrix that has a modulus of elasticity of at least 172×10^3 MPa parallel to the fibers. How many kilograms of fibers are required? See Chapter 16 for properties of polyimide.

17-28 Carbon nanotubes (CNTs) with low weight (density = 1.3 g/cm³), high tensile strength (50 GPa), and modulus of elasticity (1 TPa) in the axial direction have been touted as the strongest material yet discovered. Calculate the specific strength for CNTs. How does this value compare to that of graphite whiskers given in Table 17-2? If a composite was made using an alumina (Al_2O_3) matrix with 1% volume CNT fibers, what fraction of the load would the CNT fibers carry? In practice, ceramic matrix CNT composites do not exhibit the expected improvement in mechanical properties. What could be some possible reasons for this?

17-29 An epoxy matrix is reinforced with 40 vol% E-glass fibers to produce a 2-cm-diameter composite that is to withstand a load of 25,000 N that is applied parallel to the fiber length. Calculate the stress acting on each fiber. The elastic modulus of the epoxy is 2.76×10^3 MPa.

17-30 A titanium alloy with a modulus of elasticity of 110×10^3 MPa is used to make a 454 kg part for a manned space vehicle. Determine the weight of a part having the same modulus of elasticity parallel to the fibers, if the part is made from

(a) aluminum reinforced with boron fibers and

(b) polyester (with a modulus of 4482 MPa) reinforced with high modulus carbon fibers.

(c) Compare the specific modulus for all three materials.

17-31 Short, but aligned, Al_2O_3 fibers with a diameter of 20 μm are introduced into a 6,6-nylon matrix. The strength of the bond between the fibers and the matrix is estimated to be 7 MPa. Calculate the critical fiber length and compare with the case when 1 μm alumina whiskers are used instead of the coarser fibers. What is the minimum aspect ratio in each case?

17-32 We prepare several epoxy-matrix composites using different lengths of 3-μm-diameter ZrO_2 fibers and find that the strength of the composite increases with increasing fiber length up to 5 mm. For longer fibers, the strength is virtually unchanged. Estimate the strength of the bond between the fibers and the matrix.

17-33 Glass fibers with a diameter 50 μm are introduced into a 6,6 nylon matrix. Determine the critical fiber length at which the fiber behaves as if it is continuous. The strength of the bond between the fibers and matrix is 10 MPa. The relevant materials properties are as follows:

$$E_{glass} = 72 \, GPa, \ E_{nylon} = 2.75 \, GPa,$$
$$TS_{glass} = 3.4 \, GPa, \text{ and } TS_{nylon} = 827 \, MPa.$$

17-34 A copper–silver bimetallic wire, 1 cm in diameter, is prepared by co–extrusion with copper as the core and silver as the outer layer. The desired properties along the axis parallel to the length of the bimetallic wire are as follows:

(a) Thermal conductivity > 410 W/(m·K);

(b) Electrical conductivity $> 60 \times 10^6 \ \Omega^{-1} \cdot m^{-1}$; and

(c) Weight < 750 g/m.

Determine the allowed range of the diameter of the copper core.

	Copper	Silver
Density (g/cm^3)	8.96	10.49
Electrical conductivity ($\Omega^{-1} \cdot m^{-1}$)	59×10^6	63×10^6
Thermal conductivity [W/m · K]	401	429

17-35 What is a coupling agent? What is "sizing" as it relates to the production of glass fibers?

Section 17-5 Manufacturing Fibers and Composites

17-36 Explain briefly how boron and carbon fibers are made.

17-37 Explain briefly how continuous-glass fibers are made.

17-38 What is the difference between a fiber and a whisker?

17-39 In one polymer-matrix composite, as produced, discontinuous glass fibers are introduced directly into the matrix; in a second case, the fibers are first "sized." Discuss the effect this difference might have on the critical fiber length and the strength of the composite.

17-40 Explain why bonding between carbon fibers and an epoxy matrix should be excellent, whereas bonding between silicon nitride fibers and a silicon carbide matrix should be poor.

17-41 A polyimide matrix with an elastic modulus of 2.07×10^3 MPa is to be reinforced with 70 vol% carbon fibers to give a minimum modulus of elasticity of 276×10^3 MPa. Recommend a process for producing the carbon fibers required. Estimate the tensile strength of the fibers that are produced.

Section 17-6 Fiber-Reinforced Systems and Applications

17-42 Explain briefly in what sporting equipment composite materials are used. What is the main reason why composites are used in these applications?

17-43 What are the advantages of using ceramic-matrix composites?

Section 17-7 Laminar Composite Materials

Section 17-8 Examples and Applications of Laminar Composites

Section 17-9 Sandwich Structures

17-44 What is a laminar composite?

17-45 A microlaminate, Arall, is produced using five sheets of 0.4-mm-thick aluminum and four sheets of 0.2-mm-thick epoxy reinforced with unidirectionally aligned Kevlar^{TM} fibers. The volume fraction of Kevlar^{TM} fibers in these intermediate sheets is 55%. The elastic modulus of the epoxy is 0.5×10^6 psi. Calculate the modulus of elasticity of the microlaminate parallel and perpendicular to the unidirectionally aligned Kevlar^{TM} fibers. What are the principle advantages of the Arall material compared with those of unreinforced aluminum?

17-46 A laminate composed of 0.1-mm-thick aluminum sandwiched around a 2-cm-thick layer of polystyrene foam is produced as an insulation material. Calculate the thermal conductivity of the laminate parallel and perpendicular to the layers. The thermal conductivity of aluminum is $0.57 \dfrac{\text{cal}}{\text{cm} \cdot \text{s} \cdot \text{K}}$ and that of the foam is $0.000077 \dfrac{\text{cal}}{\text{cm} \cdot \text{s} \cdot \text{K}}$.

17-47 A 0.01-cm-thick sheet of a polymer with a modulus of elasticity of 4.8×10^3 MPa is sandwiched between two 4-mm-thick sheets of glass with a modulus of elasticity of 83×10^3 MPa. Calculate the modulus of elasticity of the composite parallel and perpendicular to the sheets.

17-48 A U.S. quarter is 2.3 cm in diameter and is about 0.16 cm in thick. Assuming copper costs about $2.43 per kg and nickel costs about $9 per kg, compare the material cost in a composite quarter versus a quarter made entirely of nickel.

17-49 Calculate the density of a honeycomb structure composed of the following elements: The two 2-mm-thick cover sheets are produced using an epoxy matrix prepreg containing 55 vol% E-glass fibers. The aluminum honeycomb is 2 cm thick; the cells are in the shape of 0.5-cm squares and the walls of the cells are 0.1 mm thick. The density of the epoxy is 1.25 g/cm^3. Compare the weight of a 1 m $\times$ 2 m panel of the honeycomb with a solid aluminum panel of the same dimensions.

Design Problems

17-50 Design the materials and processing required to produce a discontinuous, but aligned, fiber reinforced fiberglass composite that will form the hood of a sports car. The composite should provide a density of less than 1.6 g/cm^3 and a strength of 138 MPa. Be sure to list all of the assumptions you make in creating your design.

17-51 Design an electrical-contact material and a method for producing the material that will result in a density of no more than 6 g/cm^3, yet at least 50 vol% of the material will be conductive.

17-52 What factors will have to be considered in designing a bicycle frame using an aluminum frame and a frame made using C-C composite?

Computer Problems

17-53 *Properties of a Laminar Composite.* Write a computer program (or use spreadsheet software) to calculate the properties of a laminar composite. For example, if the user provides the value of the thermal conductivity of each phase and the corresponding volume fraction, the program should provide the value of the effective thermal conductivity. Properties parallel and perpendicular to the lamellae should be calculated.

K Knovel® **Problems**

K17-1 Using the rule of mixtures, calculate the density of polypropylene containing 30 vol% talc filler. Compare the calculated density with the typical density for this composite and explain the discrepancy, if any. Assume polypropylene has 40–50% crystallinity.

Construction materials, such as steels, concrete, wood, and glasses, along with composites, such as fiberglass, play a major role in the development and maintenance of a nation's infrastructure. Although many of our current construction materials are well developed and the technologies for producing them are mature, several new directions can be seen in construction materials. The use of smart sensors and actuators that can help control structures or monitor the health of structures is one such area. The increased use of composites and polymeric adhesives is one more dimension that has developed considerably. Processes such as welding and galvanizing as well as phenomena such as corrosion play important roles in the reliability and durability of structures. *(Martin Puddy/Stone/Getty Images)*

Construction Materials

Have You Ever Wondered?

* *What are the most widely used manufactured construction materials?*

* *What is the difference between cement and concrete?*

* *What construction material is a composite made by nature?*

* *What is reinforced concrete?*

A number of important materials are used in the construction of buildings, highways, bridges, and much of our country's infrastructure. In this chapter, we examine three of the most important of these materials: wood, concrete, and asphalt. The field of construction materials is indeed very important for engineers, especially civil engineers and highway engineers. The properties and processing of steels used for making reinforced concrete, ceramics (e.g., sand, lime, concrete), plastics (e.g., epoxies and polystyrene foams), glasses, and composites (e.g., fiberglass) play a critical role in the development and use of construction materials. Another area in civil engineering that is becoming increasingly important is related to the use of sensors and actuators in buildings and bridges. Advanced materials developed for microelectronic and optical applications play an important role in this area.

For example, smart bridges and buildings that make use of optical fiber sensors currently are being developed. These sensors can monitor the health of the structures on a continuous basis and thus can provide early warnings of any potential problems. Similarly, researchers in areas of smart structures are also working on many other ideas using sensors that can detect things such as the formation of ice. If ice is detected, the system can start to spray salt water to prevent or delay freezing. Sensors such as this are also installed on steep driveways in some commercial parking garages where activation of the snow/ice sensors initiates heating of that part of the driveway. Similarly, we now have many smart coatings on glasses that can deflect heat and make buildings energy efficient.

There are new coatings that have resulted in self-cleaning glasses. New technologies are also being implemented to develop "green buildings."

There are many other areas of materials that relate to structures. For example, the corrosion of bridges and the limitations it poses on the bridge's life expectancy is a major cost for any nation. Strategies using galvanized steels and the proper paints to protect against corrosion are crucial aspects of bridges design. Similarly, many material joining techniques, such as welding, play a very important role in the construction of bridges and buildings. Long-term environmental impacts of the materials used must be considered (e.g., what are the best materials to use for water pipes, insulation, fire retardancy, etc.?). The goal of this chapter is to present a summary of the properties of wood, concrete, and asphalt.

18-1   The Structure of Wood

Wood, a naturally occuring composite, is one of our most familiar materials. Although it is not a "high-tech" material, we are literally surrounded by it in our homes and value it for its beauty and durability. In addition, wood is a strong, lightweight material that still dominates much of the construction industry.

We can consider wood to be a complex fiber-reinforced composite composed of long, unidirectionally aligned, tubular polymer cells in a polymer matrix. Furthermore, the polymer tubes are composed of bundles of partially crystalline, cellulose fibers aligned at various angles to the axes of the tubes. This arrangement provides excellent tensile properties in the longitudinal direction.

Wood consists of four main constituents. **Cellulose** fibers make up about 40% to 50% of wood. Cellulose is a naturally occurring thermoplastic polymer with a degree of polymerization of about 10,000. The structure of cellulose is shown in Figure 18-1. About 25% to 35% of a tree is **hemicellulose**, a polymer having a degree of polymerization of about 200. Another 20% to 30% of a tree is **lignin**, a low molecular weight, organic cement that bonds the various constituents of the wood. Finally, **extractives** are organic impurities such as oils, which provide color to the wood or act as preservatives against the environment and insects, and inorganic minerals such as silica, which dull saw blades during the cutting of wood. As much as 10% of the wood may be extractives.

There are three important levels in the structure of wood: the fiber structure, the cell structure, and the macrostructure (Figure 18-2).

Fiber Structure
The basic component of wood is cellulose, $C_6H_{10}O_5$, arranged in polymer chains that form long fibers. Much of the fiber length is crystalline,

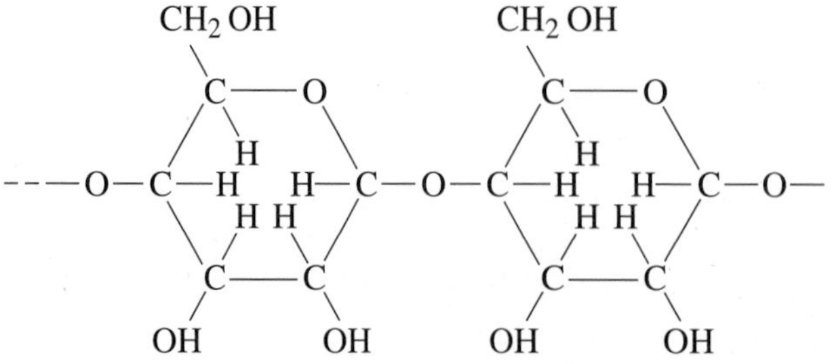

Figure 18-1
The structure of the cellulose filaments in wood.

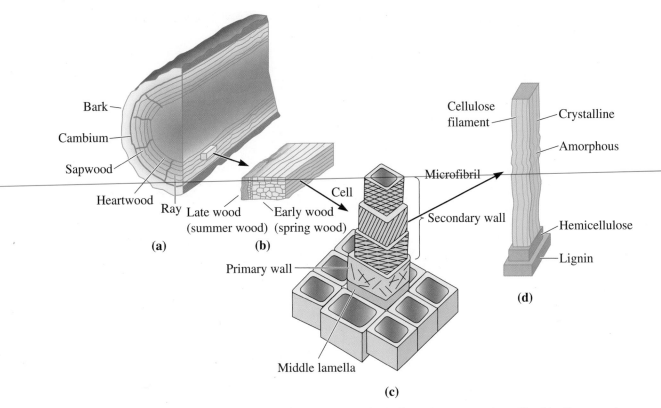

Figure 18-2 The structure of wood: (a) the macrostructure, including a layer structure outlined by the annual growth rings, (b) detail of the cell structure within one annual growth ring, (c) the structure of a cell, including several layers composed of microfibrils of cellulose fibers, hemicellulose fibers, and lignin, and (d) the microfibril's aligned, partly crystalline cellulose chains.

with the crystalline regions separated by small lengths of amorphous cellulose. A bundle of cellulose chains are encased in a layer of randomly oriented, amorphous hemicellulose chains. Finally, the hemicellulose is covered with lignin [Figure 18-2(d)]. The entire bundle, consisting of cellulose chains, hemicellulose chains, and lignin, is called a **microfibril**; it can have a virtually infinite length.

Cell Structure

The tree is composed of elongated cells, often having an aspect ratio of 100 or more, that constitute about 95% of the solid material in wood. The hollow cells are composed of several layers built up from the microfibrils [Figure 18-2(c)]. The first, or primary, wall of the cell contains randomly oriented microfibrils. As the cell walls thicken, three more distinct layers are formed. The outer and inner walls contain microfibrils oriented in two directions that are not parallel to the cell. The middle wall, which is the thickest, contains microfibrils that are unidirectionally aligned, usually at an angle not parallel to the axis of the cell.

Macrostructure

A tree is composed of several layers [Figure 18-2(a)]. The outer layer, or *bark*, protects the tree. The **cambium**, just beneath the bark, contains new growing cells. The **sapwood** contains a few hollow living cells that store nutrients and serve as the conduit for water. Finally, the **heartwood**, which contains only dead cells, provides most of the mechanical support for the tree.

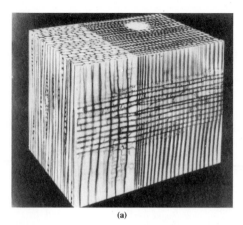 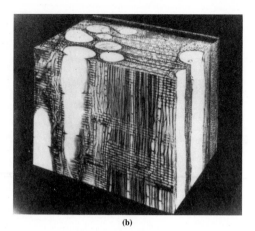

Figure 18-3   The cellular structure in (a) softwood and (b) hardwood. Softwoods contain larger, longer cells than hardwoods. The hardwoods, however, contain large-diameter vessels. Water is transported through softwoods by the cells and through hardwoods by the vessels. *(From J.M. Dinwoodie,* Wood: Nature's Cellular Polymeric Fiber-Composite, *The Institute of Metals, 1989.)*

The tree grows when new elongated cells develop in the cambium. Early in the growing season, the cells are large; later they have a smaller diameter, thicker walls, and a higher density. This difference between the early (or *spring*) wood and the late (or *summer*) wood permits us to observe annual growth rings [Figure 18-2(b)]. In addition, some cells grow in a radial direction; these cells, called *rays*, provide the storage and transport of food.

Hardwood Versus Softwood   The hardwoods are deciduous trees such as oak, ash, hickory, elm, beech, birch, walnut, and maple. In these trees, the elongated cells are relatively short, with a diameter of less than 0.1 mm and a length of less than 1 mm. Contained within the wood are longitudinal pores, or vessels, which carry water through the tree (Figure 18-3).

The softwoods are the conifers, evergreens such as pine, spruce, hemlock, fir, spruce, and cedar, and have similar structures. In softwoods, the cells tend to be somewhat longer than in the hardwoods. The hollow center of the cells is responsible for transporting water. In general, the density of softwoods tends to be lower than that of hardwoods because of a greater percentage of void space.

18-2   Moisture Content and Density of Wood

The material making up the individual cells in virtually all woods has essentially the same density—about 1.45 g/cm³; however, wood contains void space that causes the actual density to be much lower. The density of wood depends primarily on the species of the tree (or the amount of void space peculiar to that species) and the percentage of water in the wood (which depends on the amount of drying and on the relative humidity to which the wood is exposed during use). Completely dry wood varies in density from about 0.3 to 0.8 g/cm³, with hardwoods having higher densities than softwoods. The measured

TABLE 18-1 ■ *Properties of typical woods*

Wood	Density (for 12% Water) (g/cm³)	Modulus of Elasticity (MPa)
Cedar	0.32	7,600
Pine	0.35	8,300
Fir	0.48	13,800
Maple	0.48	10,350
Birch	0.62	13,800
Oak	0.68	12,400

density is normally higher due to the water contained in the wood. The percentage water is given by

$$\% \text{ Water} = \frac{\text{weight of water}}{\text{weight of dry wood}} \times 100 \qquad (18\text{-}1)$$

On the basis of this definition, it is possible to describe a wood as containing more than 100% water. The water is contained both in the hollow cells or vessels, where it is not tightly held, and in the cellulose structure in the cell walls, where it is more tightly bonded to the cellulose fibers. While a large amount of water is stored in a live tree, the amount of water in the wood after the tree is harvested depends, eventually, on the humidity to which the wood is exposed; higher humidity increases the amount of water held in the cell walls. The density of a wood is usually given at a moisture content of 12%, which corresponds to 65% humidity. The density and modulus of elasticity parallel to the grain of several common woods are included in Table 18-1 for this typical water content.

The following example illustrates the calculation for the density of the wood.

Example 18-1 *Density of Dry and Wet Wood*

A green wood has a density of 0.86 g/cm³ and contains 175% water. Calculate the density of the wood after it has completely dried.

SOLUTION

A 100-cm³ sample of the wood would have a mass of 86 g. From Equation 18-1, we can calculate the weight (or in this case, mass) of the dry wood to be

$$\% \text{ Water} = \frac{\text{weight of water}}{\text{weight of dry wood}} \times 100 = 175$$

$$= \frac{\text{green weight} - \text{dry weight}}{\text{dry weight}} \times 100 = 175$$

Solving for the dry weight of the wood:

$$\text{Dry weight of wood} = \frac{(100)(\text{green weight})}{275}$$

$$= \frac{(100)(86)}{275} = 31.3 \text{ g}$$

$$\text{Density of dry wood} = \frac{31.3 \text{ g}}{100 \text{ cm}^3} = 0.313 \text{ g/cm}^3$$

18-3   Mechanical Properties of Wood

The strength of a wood depends on its density, which in turn depends on both the water content and the type of wood. As a wood dries, water is eliminated first from the vessels and later from the cell walls. As water is removed from the vessels, practically no change in the strength or stiffness of the wood is observed (Figure 18-4). On continued drying to less than about 30% water, there is water loss from the actual cellulose fibers. This loss permits the individual fibers to come closer together, increasing the bonding between the fibers and the density of the wood and, thereby, increasing the strength and stiffness of the wood.

The type of wood also affects the density. Because they contain less of the higher-density late wood, softwoods typically are less dense and therefore have lower strengths than hardwoods. In addition, the cells in softwoods are larger, longer, and more open than those in hardwoods, leading to lower density.

The mechanical properties of wood are highly anisotropic. In the longitudinal direction (Figure 18-5), an applied tensile load acts parallel to the microfibrils and cellulose chains in the middle section of the secondary wall. These chains are strong—because they are mostly crystalline—and are able to carry a relatively high load. In the radial and

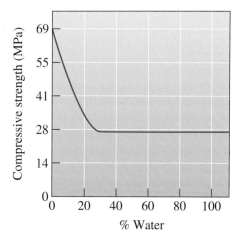

Figure 18-4
The effect of the percentage of water in a typical wood on the compressive strength parallel to the grain.

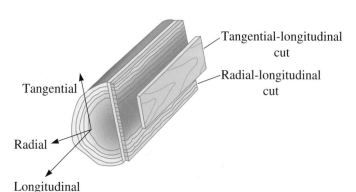

Figure 18-5
The different directions in a log. Because of differences in cell orientation and the grain, wood displays anisotropic behavior.

TABLE 18-2 ■ *Anisotropic behavior of several woods (at 12% moisture)*

	Tensile Strength Longitudinal (MPa)	Tensile Strength Radial (MPa)	Compressive Strength Longitudinal (MPa)	Compressive Strength Radial (MPa)
Beech	86	7	50	7
Elm	121	4.6	38	4.8
Maple	108	8	54	10
Oak	78	6.5	43	6
Cedar	46	2.2	42	6.3
Fir	78	2.7	37.6	4.2
Pine	73	2.1	33	3
Spruce	59	2.6	39	4

tangential directions, however, the weaker bonds between the microfibrils and cellulose fibers may break, resulting in very low tensile properties. Similar behavior is observed in compression and bending loads. Because of the anisotropic behavior, most lumber is cut in a tangential-longitudinal or radial-longitudinal manner. These cuts maximize the longitudinal behavior of the wood.

Wood has poor properties in compression and bending (which produces a combination of compressive and tensile forces). In compression, the fibers in the cells tend to buckle, causing the wood to deform and break at low stresses. Unfortunately, most applications for wood place the component in compression or bending and therefore do not take full advantage of the engineering properties of the material. Similarly, the modulus of elasticity is highly anisotropic; the modulus perpendicular to the grain is about 1/20th that given in Table 18-1 parallel to the grain. Table 18-2 compares the tensile and compressive strengths parallel and perpendicular to the cells for several woods.

Clear wood, free of imperfections such as knots, has a specific strength and specific modulus that compare well with those of other common construction materials (Table 18-3). Wood also has good toughness, largely due to the slight misorientation of the cellulose fibers in the middle layer of the secondary wall. Under load, the fibers straighten, permitting some ductility and energy absorption. The mechanical properties of wood also depend on imperfections in the wood. Clear wood may have a longitudinal tensile strength of 69 to 138 MPa. Less expensive construction lumber, which usually contains many imperfections, may have a tensile strength below 34 MPa. The knots also disrupt the grain of the wood in the vicinity of the knot, causing the cells to be aligned perpendicular to the tensile load.

TABLE 18-3 ■ *Comparison of the specific strength and specific modulus of wood with those of other common construction materials*

Material	Specific Strength ($\times 10^5$ cm)	Specific Modulus ($\times 10^7$ cm)
Clear wood	17.5	23.8
Aluminum	12.5	26.3
1020 steel	5	26.3
Copper	3.8	13.8
Concrete	1.5	8.8

After F.F. Wangaard, "Wood: Its Structure and Properties," J. Educ. Models for Mat. Sci. and Engr., Vol. 3, No. 3, 1979.

18-4 Expansion and Contraction of Wood

Like other materials, wood changes dimensions when heated or cooled. Dimensional changes in the longitudinal direction are very small in comparison with those in metals, polymers, and ceramics; however, the dimensional changes in the radial and tangential directions are greater than those for most other materials.

In addition to dimensional changes caused by temperature fluctuations, the moisture content of the wood causes significant changes in dimension. Again, the greatest changes occur in the radial and tangential directions, where the moisture content affects the spacing between the cellulose chains in the microfibrils. The change in dimensions Δx in wood in the radial and tangential directions is approximated by

$$\Delta x = x_0[c(M_f - M_i)] \tag{18-2}$$

where x_0 is the initial dimension, M_i is the initial water content, M_f is the final water content, and c is a coefficient that describes the dimensional change and can be measured in either the radial or the tangential direction. Table 18-4 includes the dimensional coefficients for several woods. In the longitudinal direction, no more than 0.1% to 0.2% change is observed.

During the initial drying of wood, the large dimensional changes perpendicular to the cells may cause warping and even cracking. In addition, when the wood is used, its water content may change, depending on the relative humidity in the environment. As the wood gains or loses water during use, shrinkage or swelling continues to occur. If a wood construction does not allow movement caused by moisture fluctuations, warping and cracking can occur—a particularly severe condition in large expanses of wood, such as the floor of a large room. Excessive expansion may cause large bulges in the floor; excessive shrinkage may cause large gaps between individual planks of the flooring.

TABLE 18-4 ■ Dimensional coefficient c (cm/cm·% H₂O) for several woods

Wood	Radial	Tangential
Beech	0.00190	0.00431
Elm	0.00144	0.00338
Maple	0.00165	0.00353
Oak	0.00183	0.00462
Cedar	0.00111	0.00234
Fir	0.00155	0.00278
Pine	0.00141	0.00259
Spruce	0.00148	0.00263

18-5   Plywood

The anisotropic behavior of wood can be reduced and wood products can be made in larger sizes by producing plywood. Thin layers of wood, called **plies,** are cut from logs—normally, softwoods. The plies are stacked together with the grains between adjacent plies oriented at 90° angles; usually an odd number of plies is used. Ensuring that these angles are as precise as possible is important to ensure that the plywood does not warp or twist when the moisture content in the material changes. The individual plies are generally bonded to one another using a thermosetting phenolic resin. The resin is introduced between the plies, which are then pressed together while hot to cause the resin to polymerize.

Similar wood products are also produced as "laminar" composite materials. The facing (visible) plies may be of a more expensive hardwood with the center plies of a less expensive softwood. Wood particles can be compacted into sheets and laminated between two wood plies, producing a particle board. Wood plies can be used as the facings for honeycomb materials.

18-6   Concrete Materials

An **aggregate** is a combination of gravel, sand, crushed stones, or slag. A **mortar** is made by mixing cement, water, air, and fine aggregate. Concrete contains all of the ingredients of the mortar and coarse aggregates. **Cements** are inorganic materials that set and harden after being mixed into a paste using water. **Concrete** is a particulate composite in which both the particulate and the matrix are ceramic materials. In concrete, sand and a coarse aggregate are bonded in a matrix of **Portland cement**. A cementation reaction between water and the minerals in the cement provides a strong matrix that holds the aggregate in place and provides good compressive strength to the concrete.

Cements   Cements are classified as hydraulic and nonhydraulic. **Hydraulic cements** set and harden under water. **Nonhydraulic cements** (e.g., lime, CaO) cannot harden under water and require air for hardening. Portland cement is the most widely used and manufactured construction material. It was patented by Joseph Aspdin in 1824 and is named as such after the limestone cliffs on the Isle of Portland in England.

Hydraulic cement is made from calcium silicates with an approximate composition of CaO (~60 to 65%), SiO_2 (~20 to 25%), and iron oxide and alumina (~7 to 12%). The cement binder, which is very fine in size, is composed of various ratios of $3CaO \cdot Al_2O_3$, $2CaO \cdot SiO_2$, $3CaO \cdot SiO_2$, $4CaO \cdot Al_2O_3 \cdot Fe_2O_3$, and other minerals. In the cement terminology, CaO, SiO_2, Al_2O_3, and Fe_2O_3 are often indicated as C, S, A, and F, respectively. Thus, C_3S means $3CaO\text{-}SiO_2$. When water is added to the cement, a hydration reaction occurs, producing a solid gel that bonds the aggregate particles. Possible reactions include

$$3CaO \cdot Al_2O_3 + 6H_2O \rightarrow Ca_3Al_2(OH)_{12} + \text{heat}$$
$$3CaO + SiO_2 + (x + 1)H_2O \rightarrow Ca_2SiO_4 \cdot xH_2O + Ca(OH)_2 + \text{heat}$$

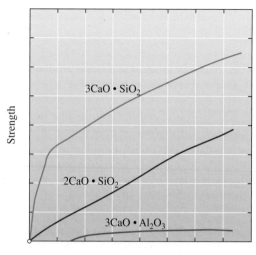

Figure 18-6
The rate of hydration of the minerals in Portland cement. (Based on Lea, Chemistry of Cement and Concrete, p. 286.)

After hydration, the cement provides the bond for the aggregate particles. Consequently, enough cement must be added to coat all of the aggregate particles. The cement typically constitutes on the order of 15 vol% of the solids in the concrete.

The composition of the cement helps determine the rate of curing and the final properties of the concrete, as shown in Figure 18-6. Nearly complete curing of the concrete is normally expected within 28 days (Figure 18-7), although some additional curing may continue for years.

There are about ten general types of cements used. Some are shown in Table 18-5. In large structures such as dams, curing should be slow in order to avoid excessive heating

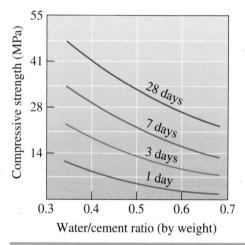

Figure 18-7
The compressive strength of concrete increases with time. After 28 days, the concrete approaches its maximum strength.

TABLE 18-5 ■ *Types of Portland cements*

	Approximate Composition				
	3C · S	2C · S	3C · A	4C · A · F	Characteristics
Type I	55	20	12	9	General purpose
Type II	45	30	7	12	Low rate of heat generation, moderate resistance to sulfates
Type III	65	10	12	8	Rapid setting
Type IV	25	50	5	13	Very low rate of heat generation
Type V	40	35	3	14	Good sulfate resistance

TABLE 18-6 ■ *Characteristics of concrete materials*

Material	True Density	
Cement	3 g/cm^3	1 sack = 42.6 kg
Sand	2.6 g/cm^3	
Aggregate	2.7 g/cm^3	Normal
	1.3 g/cm^3	Lightweight slag
	0.5 g/cm^3	Lightweight vermiculite
	4.5 g/cm^3	Heavy·Fe_3O_4
	6.2 g/cm^3	Heavy ferrophosphorus
Water	1 g/cm^3	1.05×10^{-3} liters/cm^3

caused by the hydration reaction. These cements typically contain low percentages of $3CaO \cdot SiO_2$, such as in Types II and IV. Some construction jobs, however, require that concrete forms be removed and reused as quickly as possible; cements for these purposes may contain large amounts of $3CaO \cdot SiO_2$, as in Type III.

The composition of the cement also affects the resistance of the concrete to the environment. For example, sulfates in the soil may attack the concrete; using higher proportions of $4CaO \cdot Al_2O_3 \cdot Fe_2O_3$ and $2CaO \cdot SiO_2$ helps produce concretes more resistant to sulfates, as in Type V.

Sand   Chemically, sand is predominantly silica (SiO_2). Sands are composed of fine mineral particles, typically of the order of 0.1 to 1.0 mm in diameter. They often contain at least some adsorbed water, which should be taken into account when preparing a concrete mix. The sand helps fill voids between the coarser aggregate, giving a high packing factor, reducing the amount of open (or interconnected) porosity in the finished concrete, and reducing problems with disintegration of the concrete due to repeated freezing and thawing during service.

Aggregate   Coarse aggregate is composed of gravel and rock. Aggregate must be clean, strong, and durable. Aggregate particles that have an angular rather than a round shape provide strength due to mechanical interlocking between particles, but angular particles also provide more surface on which voids or cracks may form. It is normally preferred that the aggregate size be large; this condition also minimizes the surface area at which cracks or voids form. The size of the aggregate must, of course, be matched to the size of the structure being produced; aggregate particles should not be any larger than about 20% of the thickness of the structure.

In some cases, special aggregates may be used. Lightweight concretes can be produced by using mineral slags, which are produced during steel making operations; these concretes have improved thermal insulation. Particularly heavy concretes can be produced using dense minerals or even metal shot; these heavy concretes can be used in building nuclear reactors to better absorb radiation. The densities of several aggregates are included in Table 18-6.

18-7   Properties of Concrete

Many factors influence the properties of concrete. Some of the most important are the water-cement ratio, the amount of air entrainment, and the type of aggregate.

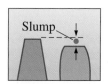

Water-Cement Ratio

The ratio of water to cement affects the behavior of concrete in several ways:

1. A minimum amount of water must be added to the cement to ensure that all of it undergoes the hydration reaction. Too little water therefore causes low strength. Normally, however, other factors such as workability place the lower limit on the water–cement ratio.

2. A high water–cement ratio improves the **workability** of concrete—that is, how easily the concrete slurry can fill all of the space in the form. Air pockets or interconnected porosity caused by poor workability reduce the strength and durability of the concrete structure. Workability can be measured by the *slump test*. For example, a wet concrete shape 30 cm tall is produced (Figure 18-8) and is permitted to stand under its own weight. After some period of time, the shape deforms. The reduction in height of the form is the **slump**. A minimum water–cement ratio of about 0.4 (by weight) is usually required for workability. A larger slump, caused by a higher water–cement ratio, indicates greater workability. Slumps of 2.5 to 15 cm are typical; high slumps are needed for pouring narrow or complex forms, while low slumps may be satisfactory for large structures such as dams.

3. Increasing the water–cement ratio beyond the minimum required for workability decreases the compressive strength of the concrete. This strength is usually measured by determining the stress required to crush a concrete cylinder 15 cm in diameter and 30 cm tall. Figure 18-9 shows the effect of the water–cement ratio on concrete's strength.

4. High water–cement ratios increase the shrinkage of concrete during curing, creating a danger of cracking.

Because of the different effects of the water–cement ratio, a compromise between strength, workability, and shrinkage may be necessary. A weight ratio of 0.45 to 0.55 is typical. To maintain good workability, organic plasticizers may be added to the mix with little effect on strength.

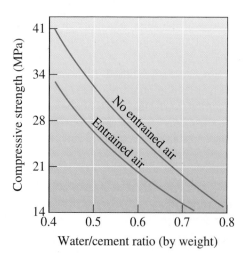

Figure 18-9
The effect of the water–cement ratio and entrained air on the 28-day compressive strength of concrete.

Air-Entrained Concrete

Almost always, a small amount of air is entrained into the concrete during pouring. For coarse aggregate, such as 4 cm rock, 1% by volume of the concrete may be air. For finer aggregate, such as 1.3 cm gravel, 2.5% air may be trapped.

We sometimes intentionally entrain air into concrete—sometimes as much as 8% for fine gravel. The entrained air improves workability of the concrete and helps minimize problems with shrinkage and freeze–thaw conditions. Air-entrained concrete has a lower strength, however. (See Figure 18-9.)

Type and Amount of Aggregate

The size of the aggregate affects the concrete mix. Figure 18-10 shows the amount of water per cubic yard of concrete required to produce the desired slump, or workability; more water is required for smaller aggregates. Figure 18-11 shows the amount of aggregate that should be present in the concrete mix. The volume ratio of aggregate in the concrete is based on the bulk density of the aggregate, which is about 60% of the true density shown in Table 18-6.

The examples that follow show how to calculate the contents for a concrete mixture.

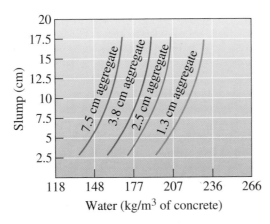

Figure 18-10
The amount of water per cubic yard of concrete required to give the desired workability (or slump) depends on the size of the coarse aggregate.

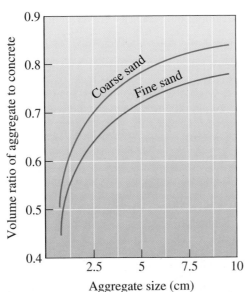

Figure 18-11
The volume ratio of aggregate to concrete depends on the sand and aggregate sizes. Note that the volume ratio uses the bulk density of the aggregate—about 60% of the true density.

Example 18-2   *Composition of Concrete*

Determine the amounts of water, cement, sand, and aggregate in 4 cubic meters of concrete, assuming that we want to obtain a water/cement ratio of 0.4 (by weight) and that the cement/sand/aggregate ratio is 1:2.5:4 (by weight). A "normal" aggregate will be used, containing 1% water, and the sand contains 4% water. Assume that no air is entrained into the concrete.

SOLUTION

One method by which we can calculate the concrete mix is to first determine the volume of each constituent based on one sack (42.6 kg) of cement. We should remember that after the concrete is poured, there are no void spaces between the various constituents; therefore, we need to consider the true density—not the bulk density—of the constituents in our calculations.

For each sack of cement we use, the volume of materials required is

$$\text{Cement} = \frac{42.6 \text{ kg/sack}}{3 \text{ g/cm}^3} = 0.014 \text{ m}^3$$

$$\text{Sand} = \frac{2.5 \times 42.6 \text{ kg cement}}{2.6 \text{ g/cm}^3} = 0.041 \text{ m}^3$$

$$\text{Gravel} = \frac{4 \times 42.6 \text{ kg cement}}{2.7 \text{ g/cm}^3} = 0.063 \text{ m}^3$$

$$\text{Water} = \frac{0.4 \times 42.6 \text{ kg cement}}{1 \text{ g/cm}^3} = 0.017 \text{ m}^3$$

Total volume of concrete $= 0.135 \text{ m}^3$/sack of cement
Therefore, in 4 m^3 (or 135 ft^3), we need

$$\text{Cement} = \frac{4 \text{ m}^3}{0.135 \text{ m}^3/\text{sack}} = 30 \text{ sacks}$$

$$\text{Sand} = (30 \text{ sacks})(42.6 \text{ kg/sack})(2.5 \text{ sand/cement}) = 3195 \text{ kg}$$

$$\text{Gravel} = (30 \text{ sacks})(42.6 \text{ kg/sack})(4 \text{ gravel/cement}) = 5112 \text{ kg}$$

$$\text{Water} = (30 \text{ sacks})(42.6 \text{ kg/sack})(0.4 \text{ water/cement}) = 511 \text{ kg}$$

The sand contains 4% water and the gravel contains 1% water. To obtain the weight of the *wet* sand and gravel, we must adjust for the water content of each:

$$\text{Sand} = (3195 \text{ kg dry})/0.96 = 3323 \text{ kg and water} = 128 \text{ kg}$$

$$\text{Gravel} = (5112 \text{ kg dry})/0.99 = 5163 \text{ kg and water} = 51 \text{ kg}$$

Therefore, we actually only need to add:

$$\text{Water} = 511 \text{ kg} - 128 \text{ kg} - 51 \text{ kg} = 332 \text{ kg}$$

$$= \frac{(332 \text{ kg})}{1 \text{ g/cm}^3} = 332,000 \text{ cm}^3 = 332 \text{ liters}$$

Accordingly, we recommend that 30 sacks of cement, 3323 kg of sand, and 5163 kg of gravel be combined with 332 liters of water.

Example 18-3 *Design of a Concrete Mix for a Retaining Wall*

Design a concrete mix that will provide a 28-day compressive strength of 28 MPa in a concrete intended for producing a 13 cm-thick retaining wall 180 cm high. We expect to have about 2% air entrained in the concrete, although we will not intentionally entrain air. The aggregate contains 1% moisture, and we have only coarse sand containing 5% moisture available.

SOLUTION

Some workability of the concrete is needed to ensure that the form will fill properly with the concrete. A slump of 8 cm might be appropriate for such an application. The wall thickness is 13 cm. To help minimize cost, we would use a large aggregate. A 2.5 cm diameter aggregate size would be appropriate (about 1/5 of the wall thickness).

To obtain the desired workability of the concrete using 2.5 cm aggregate, we should use about 189 kg of water per cubic meter (Figure 18-10).

To obtain the 28 MPa compressive strength after 28 days (assuming no intentional entrained air), we need a water–cement weight ratio of 0.57 (Figure 18-9).

Consequently, the weight of cement required per cubic meter of concrete is (189 kg water/0.57 water–cement) = 332 kg cement.

Because our aggregate size is 2.5 cm and we have only coarse sand available, the volume ratio of the aggregate to the concrete is 0.7 (Figure 18-11). Thus, the amount of aggregate required per meter of concrete is 0.7 m^3; however, this amount is in terms of the bulk density of the aggregate. Because the bulk density is about 60% of the true density, the actual volume occupied by the aggregate in the concrete is 0.7 m$^3 \times 0.6 = 0.42$ m^3.

Let's determine the volume of each constituent per cubic meter (27 ft^3) of concrete in order to calculate the amount of sand required:

$$\text{Water} = 189\,\text{kg}/(1\,\text{g/cm}^3) = 0.19\,\text{m}^3$$
$$\text{Cement} = 332\,\text{kg}/(3\,\text{g/cm}^3) = 0.11\,\text{m}^3$$
$$\text{Aggregate} = 0.42\,\text{m}^3$$
$$\text{Air} = 0.02 \times 1\,\text{m}^3 = 0.02\,\text{m}^3$$
$$\text{Sand} = 1 - 0.19 - 0.11 - 0.42 - 0.02 = 0.26\,\text{m}^3$$

Or converting to other units, assuming that the aggregate and sand are dry:

$$\text{Water} = 0.19\,\text{m}^3 \times \frac{1\,\text{liter}}{1000\,\text{cc}} = 190\,\text{liter}s$$
$$\text{Cement} = 332\,\text{kg}/(42.6\,\text{kg/sack}) = 8\,\text{sacks}$$
$$\text{Aggregate} = 0.42\,\text{m}^3 \times 2.7\,\text{g/cm}^3 = 1134\,\text{kg}$$
$$\text{Sand} = 0.26\,\text{m}^3 \times 2.6\,\text{g/m}^3 = 676\,\text{kg}$$

The aggregate and the sand are wet. Thus, the actual amounts of aggregate and sand needed are

$$\text{Aggregate} = 1134/0.99 = 1145\,\text{kg}\,(11\,\text{kg water})$$
$$\text{Sand} = 676\,\text{kg}/0.95 = 710\,\text{kg}\,(34\,\text{kg water})$$

The actual amount of water needed is

$$\text{Water} = 190\,\text{liters} - \frac{(11\,\text{kg} + 34\,\text{kg})(1\,\text{liter}/1000\,\text{cc})}{1\,\text{g/cm}^3} = 145\,\text{liters}$$

Thus, for each cubic yard of concrete, we will combine eight sacks of cement, 1145 kg of aggregate, 710 kg of sand, and 145 liters of water. This should give us a slump of 8 cm (the desired workability) and a compressive strength of 28 MPa after 28 days.

18-8   Reinforced and Prestressed Concrete

Concrete, like other ceramic-based materials, develops good compressive strength. Due to the porosity and interfaces present in the brittle structure, however, it has very poor tensile properties. Several methods are used to improve the load-bearing capability of concrete in tension.

Reinforced Concrete   Steel rods (known as rebar), wires, or mesh are frequently introduced into concrete to provide improvement in resisting tensile and bending forces. The tensile stresses are transferred from the concrete to the steel, which has good tensile properties. Polymer fibers, which are less likely to corrode, also can be used as reinforcement. Under flexural stresses, the steel supports the part that is in tension; the part that is under compression is supported by the concrete.

Prestressed Concrete   Instead of simply being laid as reinforcing rods in a form, the steel initially can be pulled in tension between an anchor and a jack, thus remaining under tension during the pouring and curing of the concrete. After the concrete sets, the tension on the steel is released. The steel then tries to relax from its stretched condition, but the restraint caused by the surrounding concrete places the concrete in compression. Now higher tensile and bending stresses can be applied to the concrete because of the compressive residual stresses introduced by the pretensioned steel. In order to permit the external tension to be removed in a timely manner, the early-setting Type III cements are often used for these applications.

Poststressed Concrete   An alternate method of placing concrete under compression is to place hollow tubes in the concrete before pouring. After the concrete cures, steel rods or cables running through the tubes then can be pulled in tension, acting against the concrete. As the rods are placed in tension, the concrete is placed in compression. The rods or cables then are secured permanently in their stretched condition.

18-9   Asphalt

Asphalt is a composite of aggregate and **bitumen**, which is a thermoplastic polymer most frequently obtained from petroleum. Asphalt is an important material for paving roads. The properties of the asphalt are determined by the characteristics of the aggregate and binder, their relative amounts, and additives.

The aggregate, as in concrete, should be clean and angular and should have a distribution of grain sizes to provide a high packing factor and good mechanical interlocking between the aggregate grains (Figure 18-12). The binder, composed of thermoplastic chains, bonds the aggregate particles. The binder has a relatively narrow useful temperature range, being brittle at sub-zero temperatures and beginning to melt at relatively low temperatures. Additives such as gasoline or kerosene can be used to modify the binder, permitting it to liquefy more easily during mixing and causing the asphalt to cure more rapidly after application.

The ratio of binder to aggregate is important. Just enough binder should be added so that the aggregate particles touch, but voids are minimized. Excess binder permits viscous deformation of the asphalt under load. Approximately 5 to 10% bitumen is present in a typical asphalt. Some void space is also required—usually, about 2 to 5%. When the asphalt is compressed, the binder can squeeze into voids, rather than be squeezed from the surface of the asphalt and lost. Too much void space, however, permits water to enter the structure; this increases the rate of deterioration of the asphalt and may embrittle the binder. Emulsions of asphalt are used for sealing driveways. The aggregate for asphalt is typically sand and fine gravel; however, there is some interest in using recycled glass products as the aggregate. **Glasphalt** provides a useful application for crushed glass. Similarly, there are also applications for materials developed using asphalt and shredded rubber tires.

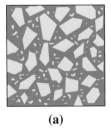

(a)

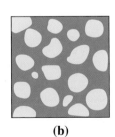

(b)

Figure 18-12
The ideal structure of asphalt (a) compared with the undesirable structure (b) in which round grains, a narrow distribution of grains, and excess binder all reduce the strength of the final material.

Summary

- Construction materials play a vital role in the infrastructure of any nation. Concrete, wood, asphalt, glasses, composites, and steels are some of the most commonly encountered materials.

- Advanced materials used to make sensors and actuators also play an increasingly important role in monitoring the structural health of buildings and bridges. There are also new coatings and thin films on glasses that have contributed to more energy-efficient buildings.

- Wood is a natural fiber-reinforced polymer composite material. Cellulose fibers constitute aligned cells that provide excellent reinforcement in longitudinal directions in wood, but give poor strength and stiffness in directions perpendicular to the cells and fibers. The properties of wood therefore are highly anisotropic and depend on the species of the tree and the amount of moisture present in the wood. Wood has good tensile strength but poor compressive strength.

- Concrete is a particulate composite. In concrete, ceramic particles such as sand and gravel are used as filler in a ceramic-cement matrix. The water–cement ratio is a particularly important factor governing the behavior of the concrete. This behavior can be modified by entraining air and by varying the composition of the cement and aggregate materials. Concrete has good compressive strength but poor tensile strength.

- Asphalt also is a particulate composite, using the same type of aggregates as in concrete, but with an organic, polymer binder.

Glossary

Aggregate A combination of gravel, sand, crushed stones, or slag.

Bitumen The organic binder, composed of low melting point polymers and oils, for asphalt.

Cambium The layer of growing cells in wood.

Cellulose A naturally occurring polymer fiber that is the major constituent of wood. Cellulose has a high degree of polymerization.

Cements Inorganic materials that set and harden after being mixed into a paste using water.

Concrete A composite material that consists of a binding medium in which particles of aggregate are dispersed.

Extractives Impurities in wood.

Glasphalt Asphalt in which the aggregate includes recycled glass.

Heartwood The center of a tree comprised of dead cells, which provides mechanical support to a tree.

Hemicellulose A naturally occurring polymer fiber that is an important constituent of wood. It has a low degree of polymerization.

Hydraulic cement A cement that sets and hardens under water.

Lignin The polymer cement in wood that bonds the cellulose fibers in the wood cells.

Microfibril Bundles of cellulose and other polymer chains that serve as the fiber reinforcement in wood.

Mortar A mortar is made by mixing cement, water, air, and fine aggregates. Concrete contains all of the ingredients of the mortar and coarse aggregates.

Nonhydraulic Cement Cements that cannot harden under water and require air for hardening.

Plies The individual sheet of wood veneer from which plywood is constructed.

Portland cement A hydraulic cement made from calcium silicates; the approximate composition is CaO (~60–65%), SiO_2 (~20–25%), and iron oxide and alumina (~7 to 12%).

Sapwood Hollow, living cells in wood that store nutrients and conduct water.

Slump The decrease in height of a standard concrete form when the concrete settles under its own weight.

Workability The ease with which a concrete slurry fills all of the space in a form.

Problems

Section 18-1 The Structure of Wood

Section 18-2 Moisture Content and Density of Wood

Section 18-3 Mechanical Properties of Wood

Section 18-4 Expansion and Contraction of Wood

Section 18-5 Plywood

18-1 Table 18-1 lists the densities for typical woods. Calculate the densities of the woods after they are completely dried and at 100% water content.

18-2 A sample of wood with the dimensions 8 cm × 10 cm × 30 cm has a dry density of 0.35 g/cm^3.
(a) Calculate the number of gallons of water that must be absorbed by the sample to contain 120% water.
(b) Calculate the density after the wood absorbs this amount of water.

18-3 The density of a sample of oak is 0.90 g/cm^3. Calculate
(a) the density of completely dry oak and
(b) the percent water in the original sample.

18-4 A green wood with a density of 0.82 g/cm^3 contains 150% water. The compressive strength of this wood is 27 MPa. After several days of drying, the compressive strength increases to 41 MPa. What is the water content and density of the dried wood? Refer to Figure 18-4.

18-5 Boards of oak 0.5 cm thick, 1 m long, and 0.25 m wide are used as flooring for a 10 m × 10 m area. If the floor was laid at a moisture content of 25% and the expected moisture could be as high as 45%, determine the dimensional change in the floor parallel to and perpendicular to the length of the boards. The boards were cut from logs with a tangential-longitudinal cut.

18-6 Boards of maple 3 cm thick, 15 cm wide, and 40 cm long are used as the flooring for a 18 m × 18 m hall. The boards were cut from logs with a tangential-longitudinal cut. The floor is laid when the boards have a moisture content of 12%. After some particularly humid days, the moisture content in the boards increases to 45%. Determine the dimensional change in the flooring parallel to the boards and perpendicular to the boards. What will happen to the floor? How can this problem be corrected?

18-7 A wall 9 m long is built using radial-longitudinal cuts of 13 cm wide pine with the boards arranged in a vertical fashion. The wood contains a moisture content of 55% when the wall is built;

however, the humidity level in the room is maintained to give 45% moisture in the wood. Determine the dimensional changes in the wood boards, and estimate the size of the gaps that will be produced as a consequence of these changes.

Section 18-6 Concrete Materials

Section 18-7 Properties of Concrete

Section 18-8 Reinforced and Prestressed Concrete

Section 18-9 Asphalt

18-8   Determine the amounts of water, cement, and sand in 10 m³ of concrete if the cement–sand–aggregate ratio is 1:2.5:4.5 and the water–cement ratio is 0.4. Assume that no air is entrained into the concrete. The sand used for this mixture contains 4 wt% water, and the aggregate contains 2 wt% water.

18-9   Calculate the amount of cement, sand, aggregate, and water needed to create a concrete mix with a 28-day compressive strength of 34 MPa for a 10 m × 10 m × 0.25 m structure given the following conditions: allowed slump = 10 cm and only 3.8 cm aggregate with 2% moisture and coarse sand with 4% moisture are available for this project. Assume no air entrainment in your calculations.

18-10   We have been asked to prepare 80 m³ of normal concrete using a volume ratio of cement–sand–coarse aggregate of 1:2:4. The water–cement ratio (by weight) is to be 0.5. The sand contains 6 wt% water, and the coarse aggregate contains 3 wt% water. No entrained air is expected.

(a)   Determine the number of sacks of cement that must be ordered, the kilograms of sand and aggregate required, and the amount of water needed.

(b)   Calculate the total weight of the concrete per cubic meter.

(c)   What is the weight ratio of cement–sand–coarse aggregate?

18-11   We plan to prepare 10 m³ of concrete using a 1:2.5:4.5 weight ratio of cement-sand-coarse aggregate. The water–cement ratio (by weight) is 0.45. The sand contains 3 wt% water; the coarse aggregate contains 2 wt% water; and 5% entrained air is expected. Determine the number of sacks of cement, kilograms of sand and coarse aggregate, and liters of water required.

⬡ Design Problems

18-12   A wooden structure is functioning in an environment controlled at 65% humidity. Design a wood support column that is to hold a compressive load of 90,000 N. The distance from the top to the bottom of the column should be 240 ± 0.63 cm when the load is applied.

18-13   Design a wood floor that will be 15 m by 50 ft and will be in an environment in which humidity changes will cause a fluctuation of plus or minus 5% water in the wood. We want to minimize any buckling or gap formation in the floor.

18-14   We would like to produce a concrete that is suitable for use in building a large structure in a sulfate environment. For these situations, the maximum water–cement ratio should be 0.45 (by weight). The compressive strength of the concrete after 28 days should be at least 28 MPa. We have an available coarse aggregate containing 2% moisture in a variety of sizes, and both fine and coarse sand containing 4% moisture. Design a concrete that will be suitable for this application.

18-15   We would like to produce a concrete sculpture. The sculpture will be as thin as 8 cm in some areas and should be light in weight, but it must have a 28-day compressive strength of at least 14 MPa. Our available aggregate contains 1% moisture, and our sands contain 5% moisture. Design a concrete that will be suitable for this application.

18-16 The binder used in producing asphalt has a density of about 1.3 g/cm^3. Design an asphalt, including the weight and volumes of each constituent, that might be suitable for use as pavement. Assume that the sands and aggregates are the same as those for a normal concrete.

18-17 *Concrete Canoe Design.* Describe what novel materials can be used to make a concrete canoe.

Knovel® Problems

K18-1 What is the difference in lignin content between hardwood and softwood species? What is the chemical structure of lignin?

K18-2 The equilibrium moisture content of wood is the moisture content at which the wood is neither gaining nor losing moisture. What is the moisture content of wood at 21°C with a relative humidity of 50%?

K18-3 What measures are used to control cracking in reinforced concrete?

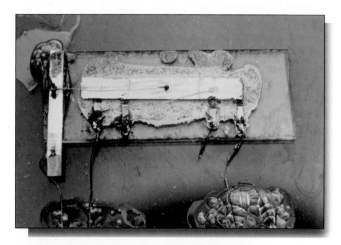

The first integrated circuit was fabricated in 1958 by Jack Kilby, who was then an engineer at Texas Instruments. The first integrated circuit functioned as a phase-shift oscillator and consisted of transistor, resistor, and capacitor devices as shown on the left (*Courtesy of Texas Instruments.*). For the first time, multiple electronic devices were incorporated on a single substrate. For this innovation, Jack Kilby was awarded the Nobel Prize in Physics in 2000.

Today's microchips may contain hundreds of millions of transistors. Dynamic random access memory (DRAM) chips have passed the billion-transistor milestone. This integration of devices is made possible by sophisticated manufacturing equipment and techniques. The image on the right shows the multiple layers of copper interconnects that are needed to make connections between transistors and other circuit elements. The insulating oxide between the wires has been removed. The transistors are not visible in this image. (*Courtesy of IBM.*)

Electronic Materials

Have You Ever Wondered?

- *Is diamond a good conductor of electricity?*
- *How many devices are there in a single microchip?*
- *How are thin films less than 1 micron thick deposited on a substrate?*
- *How does a microwave oven heat food?*

Silicon-based microelectronics are a ubiquitous part of modern life. With microchips in items from laundry machines and microwaves to cell phones, in MP3 players and from personal computers to the world's fastest super computers, silicon was the defining material of the later 20th century and will dominate computer-based and information-related technologies for the foreseeable future.

While silicon is the substrate or base material of choice for most devices, microelectronics include materials of nearly every class, including metals such as copper and gold, other semiconductors such as gallium arsenide, and insulators such as silicon dioxide. Even semiconducting polymers are finding applications in such devices as light-emitting diodes.

In this chapter, we will discuss the principles of electrical conductivity in metals, semiconductors, insulators, and ionic materials. We will see that a defining difference between metals and semiconductors is that as temperature increases, the resistivity of a metal increases, while the resistivity of a semiconductor decreases. This critical difference arises from the **band structure** of these materials. The band structure consists of the array of energy levels that are available to or forbidden for electrons to occupy and determines the electronic behavior of a solid, such as whether it is a conductor, semiconductor, or insulator.

The conductivity of a semiconductor that does not contain impurities generally increases exponentially with temperature. From a reliability standpoint, an exponential dependence of conductivity on temperature is undesirable for electronic devices that generate heat as they operate. Thus, semiconductors are doped (i.e., impurities are intentionally added) in order to control the conductivity of semiconductors with extreme precision. We will learn how **dopants** change the band structure of a semiconductor so that the electrical conductivity can be tailored for particular applications.

Metals, semiconductors, and insulators are all critical components of integrated circuits. Some features of integrated circuits are now approaching atomic-scale dimensions, and the fabrication of integrated circuits is arguably the most sophisticated manufacturing process in existence. It involves simultaneously fabricating hundreds of millions, and even billions, of devices on a single microchip and represents a fundamentally different manufacturing paradigm from that of any other process. We will learn about some of the steps involved in fabricating integrated circuits, including the process of depositing thin films (films on the order of 10 Å to 1 μm in thickness).

We will examine some of the properties of insulating materials. Insulators are used in microelectronic devices to electrically isolate active regions from one another. Insulators are also used in capacitors due to their dielectric properties. Finally, we will consider piezoelectric materials, which change their shape in response to an applied voltage or vice versa. Such materials are used as actuators in a variety of applications.

Superconductors comprise a special class of electronic materials. Superconductors are materials that exhibit zero electrical resistance under certain conditions (which usually includes a very low temperature on the order of 135 K or less) and that completely expel a magnetic field. A discussion of superconductivity is beyond the scope of this text.

19-1   Ohm's Law and Electrical Conductivity

Most of us are familiar with the common form of Ohm's law,

$$V = IR \tag{19-1}$$

where V is the voltage (volts, V), I is the current (amperes or amps, A), and R is the resistance (ohms, Ω) to the current flow. This law is applicable to most but not all materials. The resistance (R) of a resistor is a characteristic of the size, shape, and properties of the material according to

$$R = \rho \frac{l}{A} = \frac{l}{\sigma A} \tag{19-2}$$

where l is the length (cm) of the resistor, A is the cross-sectional area (cm^2) of the resistor, ρ is the electrical resistivity (ohm $\cdot$ cm or $\Omega \cdot$ cm), and σ, which is the reciprocal of ρ, is the electrical conductivity (ohm$^{-1} \cdot$ cm^{-1}). The magnitude of the resistance depends upon the dimensions of the resistor. The resistivity or conductivity does not depend on the dimensions of the material. Thus, resistivity or conductivity allows us to compare different materials. For example, silver is a better conductor than copper. Resistivity is a **microstructure-sensitive property**, similar to yield strength. The resistivity of pure copper is much less than that of commercially pure copper, because impurities in commercially pure copper scatter electrons and contribute to increased resistivity. Similarly, the resistivity of annealed, pure copper is slightly lower than that of cold-worked, pure copper because of the scattering effect associated with dislocations.

In components designed to conduct electrical energy, minimizing power losses is important, not only to conserve energy, but also to minimize heating. The electrical power P (in watts, W) lost when a current flows through a resistance is given by

$$P = VI = I^2R \tag{19-3}$$

A high resistance R results in larger power losses. These electrical losses are known as Joule heating losses.

A second form of Ohm's law is obtained if we combine Equations 19-1 and 19-2 to give

$$\frac{I}{A} = \sigma \frac{V}{l}$$

If we define I/A as the **current density** J (A/cm^2) and V/l as the **electric field** E (V/cm), then

$$J = \sigma E \tag{19-4}$$

The current density J is also given by

$$J = nq\bar{v}$$

where n is the number of charge carriers (carriers/cm^3), q is the charge on each carrier (1.6×10^{-19} C), and $\bar{v}$ is the average **drift velocity** (cm/s) at which the charge carriers move [Figure 19-1(a)]. Thus,

$$\sigma E = nq\bar{v} \text{ or } \sigma = nq\frac{\bar{v}}{E}$$

Diffusion occurs as a result of temperature and concentration gradients, and drift occurs as a result of an applied electric or magnetic field. Conduction may occur as a result of diffusion, drift, or both, but drift is the dominant mechanism in electrical conduction.

The term $\bar{v}/E$ is called the **mobility** $\mu \left(\dfrac{\text{cm}^2}{\text{V} \cdot \text{s}}\right)$ of the carriers (which in the case of metals is the mobility of electrons):

$$\mu = \frac{\bar{v}}{E}$$

Finally,

$$\sigma = nq\mu \tag{19-5a}$$

The charge q is a constant; from inspection of Equation 19-5a, we find that we can control the electrical conductivity of materials by (1) controlling the number of charge carriers in the material or (2) controlling the mobility—or ease of movement—of the charge carriers. The mobility is particularly important in metals, whereas the number of carriers is more important in semiconductors and insulators.

Electrons are the charge carriers in metals [Figure 19-1(b)]. Electrons are, of course, negatively charged. In semiconductors, electrons conduct charge as do positively charged carriers known as **holes** [Figure 19-1(c)]. We will learn more about holes in Section 19-4. In semiconductors, electrons and holes flow in opposite directions in response to an applied electric field, but in so doing, they both contribute to the net current. Thus, Equation 19-5a can be modified as follows for expressing the conductivity of semiconductors:

$$\sigma = nq\mu_n + pq\mu_p \tag{19-5b}$$

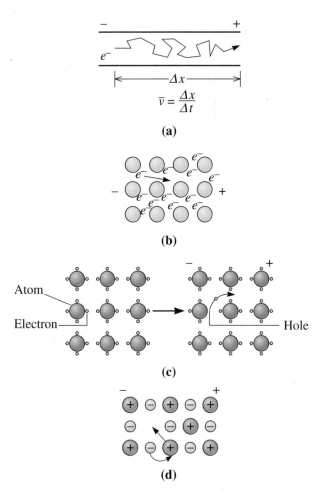

Figure 19-1
(a) Charge carriers, such as electrons, are deflected by atoms or defects and take an irregular path through a conductor. The average rate at which the carriers move is the drift velocity $\bar{v}$.
(b) Valence electrons in metals move easily. (c) Covalent bonds must be broken in semiconductors and insulators that do not contain impurities for an electron to be able to move. (d) Entire ions may diffuse via a vacancy mechanism to carry charge in many ionically bonded materials.

In this equation, μ_n and μ_p are the mobilities of electrons and holes, respectively. The terms n and p represent the concentrations of free electrons and holes in the semiconductor.

In ceramics, when conduction does occur, it can be the result of electrons that "hop" from one defect to another or the movement of ions [Figure 19-1(d)]. The mobility depends on atomic bonding, imperfections, microstructure, and, in ionic compounds, diffusion rates.

Because of these effects, the electrical conductivity of materials varies tremendously, as illustrated in Table 19-1. These values are approximate and are for high-purity materials at 300 K (unless noted otherwise). Note that the values of conductivity for metals and semiconductors depend very strongly on temperature. Table 19-2 includes some useful units and relationships.

Electronic materials can be classified as (a) superconductors, (b) conductors, (c) semiconductors, and (d) dielectrics or insulators, depending upon the magnitude of their electrical conductivity. Materials with conductivity less than $10^{-12}\ \Omega^{-1} \cdot cm^{-1}$, or resistivity greater than $10^{12}\ \Omega \cdot cm$, are considered insulating or dielectric. Materials with conductivity less than $10^3\ \Omega^{-1} \cdot cm^{-1}$ but greater than $10^{-12}\ \Omega^{-1} \cdot cm^{-1}$ are considered semiconductors. Materials with conductivity greater than $10^3\ \Omega^{-1} \cdot cm^{-1}$, or resistivity less than $10^{-3}\ \Omega^{-1} \cdot cm^{-1}$, are considered conductors. (These are approximate ranges of values.)

TABLE 19-1 ■ *Electrical conductivity of selected materials at T = 300 K**

Material	Conductivity $(ohm^{-1} \cdot cm^{-1})$	Material	Conductivity $(ohm^{-1} \cdot cm^{-1})$
Superconductors		**Semiconductors**	
Hg, Nb_3Sn		Group 4B elements	
$YBa_2Cu_3O_{7-x}$	Infinite (under certain conditions	Si	4×10^{-6}
MgB_2	such as low temperatures)	Ge	0.02
Metals		Compound semiconductors	
Alkali metals		GaAs	2.5×10^{-9}
Na	2.13×10^5	AlAs	0.1
K	1.64×10^5	SiC	10^{-10}
Alkali earth metals		**Ionic Conductors**	
Mg	2.25×10^5	Indium tin oxide (ITO)	
Ca	3.16×10^5	Yttria-stabilized zirconia (YSZ)	
Group 3B metals		**Insulators, Linear, and Nonlinear Dielectrics**	
Al	3.77×10^5	Polymers	
Ga	0.66×10^5	Polyethylene	10^{-15}
Transition metals		Polytetrafluoroethylene	10^{-18}
Fe	1.00×10^5	Polystyrene	10^{-17} to 10^{-19}
Ni	1.46×10^5	Epoxy	10^{-12} to 10^{-17}
Group 1B metals		Ceramics	
Cu	5.98×10^5	Alumina (Al_2O_3)	10^{-14}
Ag	6.80×10^5	Silicate glasses	10^{-17}
Au	4.26×10^5	Boron nitride (BN)	10^{-13}
		Barium titanate ($BaTiO_3$)	10^{-14}
		C (diamond)	$< 10^{-18}$

Unless specified otherwise, assumes high-purity material.

TABLE 19-2 ■ *Some useful relationships, constants, and units*

Electron volt = 1 eV = 1.6×10^{-19} Joule = 1.6×10^{-12} erg
1 amp = 1 coulomb/second
1 volt = 1 amp·ohm
$k_B T$ at room temperature (300 K) = 0.0259 eV
c = speed of light 2.998×10^8 m/s
ε_0 = permittivity of free space = 8.85×10^{-12} F/m
q = charge on electron = 1.6×10^{-19} C
Avogadro constant N_A = 6.022×10^{23}
k_B = Boltzmann constant = 8.63×10^{-5} eV/K = 1.38×10^{-23} J/K
h = Planck's constant 6.63×10^{-34} J·s = 4.14×10^{-15} eV·s

We use the term "dielectric" for materials that are used in applications where the dielectric constant is important. The **dielectric constant (k)** of a material is a microstructure-sensitive property related to the material's ability to store an electrical charge. We use the term "insulator" to describe the ability of a material to stop the flow of DC or AC current, as opposed to its ability to store a charge. A measure of the effectiveness of an insulator is the maximum electric field it can support without an electrical breakdown.

Example 19-1 *Design of a Transmission Line*

Design an electrical transmission line 1500 m long that will carry a current of 50 A with no more than 5×10^5 W loss in power. The electrical conductivity of several materials is included in Table 19-1.

SOLUTION

Electrical power is given by the product of the voltage and current or

$$P = VI = I^2R = (50)^2R = 5 \times 10^5 \, \text{W}$$
$$R = 200 \, \text{ohms}$$

From Equation 19-2,

$$A = \frac{l}{R \cdot \sigma} = \frac{(1500 \, \text{m})(100 \, \text{cm/m})}{(200 \, \text{ohms})\sigma} = \frac{750}{\sigma}$$

Let's consider three metals—aluminum, copper, and silver—that have excellent electrical conductivity. The table below includes appropriate data and some characteristics of the transmission line for each metal.

	σ (ohm$^{-1} \cdot$ cm^{-1})	A (cm^2)	Diameter (cm)
Aluminum	3.77×10^5	0.00199	0.050
Copper	5.98×10^5	0.00125	0.040
Silver	6.80×10^5	0.00110	0.037

Any of the three metals will work, but cost is a factor as well. Aluminum will likely be the most economical choice (Chapter 14), even though the wire has the largest diameter. Other factors, such as whether the wire can support itself between transmission poles, also contribute to the final choice.

Example 19-2 *Drift Velocity of Electrons in Copper*

Assuming that all of the valence electrons contribute to current flow, (a) calculate the mobility of an electron in copper and (b) calculate the average drift velocity for electrons in a 100 cm copper wire when 10 V are applied.

SOLUTION

(a) The valence of copper is one; therefore, the number of valence electrons equals the number of copper atoms in the material. The lattice parameter of copper is 3.6151×10^{-8} cm and, since copper is FCC, there are four atoms/unit cell. From Table 19-1, the conductivity $\sigma = 5.98 \times 10^5 \, \Omega^{-1} \cdot$ cm^{-1}

$$n = \frac{(4 \, \text{atoms/cell})(1 \, \text{electron/atom})}{(3.6151 \times 10^{-8} \, \text{cm})^3} = 8.466 \times 10^{22} \, \text{electrons/cm}^3$$
$$q = 1.6 \times 10^{-19} \, \text{C}$$

$$\mu = \frac{\sigma}{nq} = \frac{5.98 \times 10^5}{(8.466 \times 10^{22})(1.6 \times 10^{-19})}$$

$$= 44.1 \frac{cm^2}{\Omega \cdot C} = 44.1 \frac{cm^2}{V \cdot s}$$

(b) The electric field is

$$E = \frac{V}{l} = \frac{10}{100} = 0.1 \, V/cm$$

The mobility is 44.1 cm^2/(V · s); therefore,

$$\bar{v} = \mu E = (44.1)(0.1) = 4.41 \, cm/s$$

19-2   Band Structure of Solids

As we saw in Chapter 2, the electrons of atoms in isolation occupy fixed and discrete energy levels. The Pauli exclusion principle is satisfied for each atom because only two electrons, at most, occupy each energy level, or orbital. When N atoms come together to form a solid, the Pauli exclusion principle still requires that no more than two electrons in the solid have the same energy. As two atoms approach each other in order to form a bond, the Pauli exclusion principle would be violated if the energy levels of the electrons did not change. Thus, the energy levels of the electrons "split" in order to form new energy levels.

Figure 19-2 schematically illustrates this concept. Consider two atoms approaching each other to form a bond. The orbitals that contain the valence electrons are located (on average) farther from the nucleus than the orbitals that contain the "core" or innermost electrons. The orbitals that contain the valence electrons of one atom thus interact with the orbitals that contain the valence electrons of the other atom first. Since the orbitals of these electrons have the same energy when the atoms are in isolation, the orbitals shift in energy or "split" so that the Pauli exclusion principle is satisfied. As shown in Figure 19-2, when considering two atoms, each with one orbital of interest, one of the orbitals shifts to a higher energy level while the other orbital shifts to a lower energy level. The electrons of the atoms will occupy these new orbitals by first filling the lowest energy levels. As the number of atoms increases, so does the number of energy levels. A new orbital with its own energy is formed for each orbital of each atom, and as the number of atoms in the solid increases, the separation in energy between orbitals becomes finer, ultimately forming what is called an energy band. For example, when N atoms come together to form a solid, the $2s$ energy band contains N discrete energy levels, one for each atom in the solid since each atom contributes one orbital.

In order for charge carriers to conduct, the carriers must be able to accelerate and increase in energy. The energy of the carriers can increase only if there are available energy states to which the carriers can be promoted. Thus, the particular distribution of energy states in the band structure of a solid has critical implications for its electrical and optical properties.

Depending on the type of material involved (metal, semiconductor, insulator), there may or may not be a sizable energy gap between the energy levels of the orbitals that shifted to a lower energy state and the energy levels of the orbitals that shifted to a

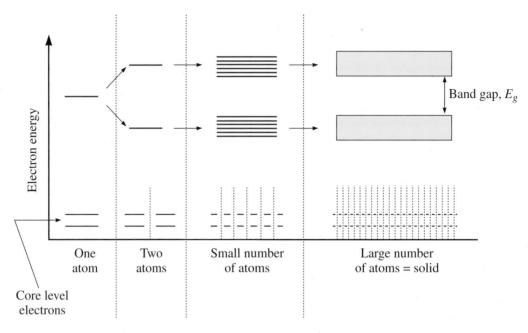

Figure 19-2   The energy levels broaden into bands as the number of electrons grouped together increases. (*Courtesy of John Bravman.*)

higher energy state. This energy gap, if it exists, is known as the bandgap. We will discuss the bandgap in more detail later.

Band Structure of Sodium

Sodium is a metal and has the electronic structure $1s^2 2s^2 2p^6 3s^1$. Figure 19-3 shows a schematic diagram of the band structure of sodium as a function of the interatomic separation. (Note that Figure 19-2 shows a

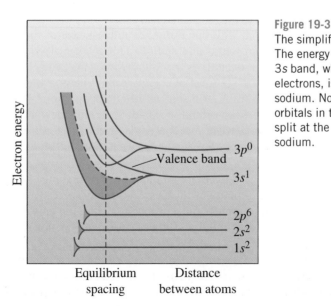

Figure 19-3
The simplified band structure for sodium. The energy levels broaden into bands. The $3s$ band, which is only half filled with electrons, is responsible for conduction in sodium. Note that the energy levels of the orbitals in the $1s$, $2s$, and $2p$ levels do not split at the equilibrium spacing for sodium.

general band diagram for a fixed interatomic separation.) The energies within the bands depend on the spacing between the atoms; the vertical line represents the equilibrium interatomic spacing of the atoms in solid sodium.

Sodium and other alkali metals in column 1A of the periodic table have only one electron in the outermost s level. The $3s$ valence band in sodium is half filled and, at absolute zero, only the lowest energy levels are occupied. The **Fermi energy** (E_f) is the energy level at which half of the possible energy levels in the band are occupied by electrons. It is the energy level where the probability of finding an electron is 1/2. When electrons gain energy, they are excited into the empty higher energy levels. The promotion of carriers to higher energy levels enables electrical conduction.

Band Structure of Magnesium and Other Metals

Magnesium and other metals in column 2A of the periodic table have two electrons in their outermost s band. These metals have a high conductivity because the p band overlaps the s band at the equilibrium interatomic spacing. This overlap permits electrons to be excited into the large number of unoccupied energy levels in the combined $3s$ and $3p$ band. Overlapping $3s$ and $3p$ bands in aluminum and other metals in column 3B provide a similar effect.

In the transition metals, including scandium through nickel, an unfilled $3d$ band overlaps the $4s$ band. This overlap provides energy levels into which electrons can be excited; however, complex interactions between the bands prevent the conductivity from being as high as in some of the better conductors. In copper, the inner $3d$ band is full, and the atom core tightly holds these electrons. Consequently, there is little interaction between the electrons in the $4s$ and $3d$ bands, and copper has a high conductivity. A similar situation is found for silver and gold.

Band Structure of Semiconductors and Insulators

The elements in Group 4—carbon (diamond), silicon, germanium, and tin—contain two electrons in their outer p shell and have a valence of four. Based on our discussion in the previous section, we might expect these elements to have a high conductivity due to the unfilled p band, but this behavior is not observed!

These elements are covalently bonded; consequently, the electrons in the outer s and p bands are rigidly bound to the atoms. The covalent bonding produces a complex change in the band structure. The $2s$ and $2p$ levels of the carbon atoms in diamond can contain up to eight electrons, but there are only four valence electrons available. When carbon atoms are brought together to form solid diamond, the $2s$ and $2p$ levels interact and produce two bands (Figure 19-4). Each hybrid band can contain $4N$ electrons. Since there are only $4N$ electrons available, the lower (or **valence**) band is completely full, whereas the upper (or **conduction**) band is empty.

A large **energy gap** or **bandgap** (E_g) separates the valence band from the conduction band in diamond ($E_g \sim 5.5$ eV). Few electrons possess sufficient energy to jump the forbidden zone to the conduction band. Consequently, diamond has an electrical conductivity of less than 10^{-18} ohm$^{-1} \cdot$ cm^{-1}. Other covalently and ionically bonded materials have a similar band structure and, like diamond, are poor conductors of electricity. Increasing the temperature supplies the energy required for electrons to overcome the energy gap. For example, the electrical conductivity of boron nitride increases from about 10^{-13} at room temperature to 10^{-4} ohm$^{-1} \cdot$ cm^{-1} at 800°C.

Figure 19-5 shows a schematic of the band structure of typical metals, semiconductors, and insulators. Thus, an important distinction between metals and semiconductors is that the conductivity of semiconductors *increases* with temperature, as more and

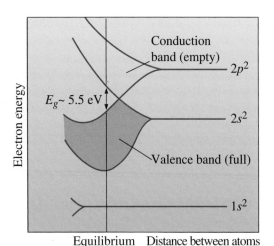

Figure 19-4
The band structure of carbon in the diamond form. The 2s and 2p levels combine to form two hybrid bands separated by an energy gap, E_g.

more electrons are promoted to the conduction band from the valence band. In other words, an increasing number of electrons from the covalent bonds in a semiconductor is freed and becomes available for conduction. The conductivity of most metals, on the other hand, *decreases* with increasing temperature. This is because the number of electrons that are already available begin to scatter more (i.e., increasing temperature reduces mobility).

Although germanium (Ge), silicon (Si), and α-Sn have the same crystal structure and band structure as diamond, their energy gaps are smaller. In fact, the energy gap (E_g) in α-Sn is so small ($E_g = 0.1$ eV) that α-Sn behaves as a metal. The energy gap is somewhat larger in silicon ($E_g = 1.1$ eV) and germanium ($E_g = 0.67$ eV)—these elements behave as semiconductors. Typically, we consider materials with a bandgap greater than 4.0 eV as insulators, dielectrics, or nonconductors; materials with a bandgap less than 4.0 eV are considered semiconductors.

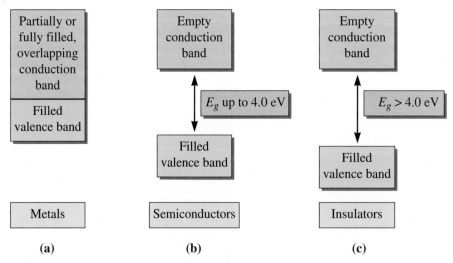

Figure 19-5   Schematic of band structures for (a) metals, (b) semiconductors, and (c) dielectrics or insulators. (Temperature is assumed to be 0 K.)

19-3   Conductivity of Metals and Alloys

The conductivity of a pure, defect-free metal is determined by the electronic structure of the atoms, but we can change the conductivity by influencing the mobility, μ, of the carriers. Recall that the mobility is proportional to the average drift velocity, $\bar{v}$. The average drift velocity is the velocity with which charge carriers move in the direction dictated by the applied field. The paths of electrons are influenced by internal fields due to atoms in the solid and imperfections in the lattice. When these internal fields influence the path of an electron, the drift velocity (and thus the mobility of the charge carriers) decreases. The **mean free path** (λ_e) of electrons is defined as

$$\lambda_e = \tau\bar{v} \qquad (19\text{-}6)$$

The average time between collisions is τ. The mean free path defines the average distance between collisions; a longer mean free path permits higher mobilities and higher conductivities.

Temperature Effect

When the temperature of a metal increases, thermal energy causes the amplitudes of vibration of the atoms to increase (Figure 19-6). This increases the *scattering cross section* of atoms or defects in the lattice. Essentially, the atoms and defects act as larger targets for interactions with electrons, and interactions occur more frequently. Thus, the mean free path decreases, the mobility of electrons is reduced, and the resistivity increases. The change in resistivity of a pure metal as a function of temperature can be estimated according to

$$\rho = \rho_{RT}(1 + \alpha_R\Delta T) \qquad (19\text{-}7)$$

where ρ is the resistivity at any temperature T, ρ_{RT} is the resistivity at room temperature (i.e., 25°C), $\Delta T = (T - T_{RT})$ is the difference between the temperature of interest and room temperature, and α_R is the *temperature resistivity coefficient*. The relationship between resistivity and temperature is linear over a wide temperature range (Figure 19-7). Examples of the temperature resistivity coefficient are given in Table 19-3.

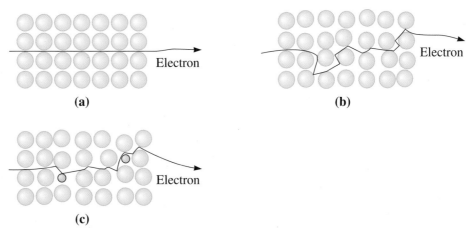

Figure 19-6   Movement of an electron through (a) a perfect crystal, (b) a crystal heated to a high temperature, and (c) a crystal containing atomic level defects. Scattering of the electrons reduces the mobility and conductivity.

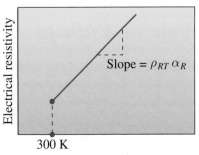

Figure 19-7
The effect of temperature on the electrical resistivity of a metal with a perfect crystal structure.

TABLE 19-3 ■ *The temperature resistivity coefficient α_R for selected metals*

Metal	Room Temperature Resistivity (ohm · cm)	Temperature Resistivity Coefficient (α_R) [ohm/(ohm · °C)]
Be	4.0×10^{-6}	0.0250
Mg	4.45×10^{-6}	0.0037
Ca	3.91×10^{-6}	0.0042
Al	2.65×10^{-6}	0.0043
Cr	12.90×10^{-6} (0°C)	0.0030
Fe	9.71×10^{-6}	0.0065
Co	6.24×10^{-6}	0.0053
Ni	6.84×10^{-6}	0.0069
Cu	1.67×10^{-6}	0.0043
Ag	1.59×10^{-6}	0.0041
Au	2.35×10^{-6}	0.0035
Pd	10.8×10^{-6}	0.0037
W	5.3×10^{-6} (27°C)	0.0045
Pt	9.85×10^{-6}	0.0039

(*HANDBOOK OF ELECTROMAGNETIC MATERIALS: MONOLITHIC AND COMPOSITE VERSIONS AND THEIR APPLICATIONS by P. S. Neelkanta. Copyright 1995 by TAYLOR & FRANCIS GROUP LLC - BOOKS. Reproduced with permission of TAYLOR & FRANCIS GROUP LLC - BOOKS in the format Textbook via Copyright Clearance Center.*)

The following example illustrates how the resistivity of pure copper can be calculated.

Example 19-3   *Resistivity of Pure Copper*

Calculate the electrical conductivity of pure copper at (a) 400°C and (b) −100°C.

SOLUTION

The resistivity of copper at room temperature is 1.67×10^{-6} ohm · cm, and the temperature resistivity coefficient is 0.0043 ohm/(ohm · °C). (See Table 19-3.)

(a) At 400°C:

$$\rho = \rho_{RT} (1 + \alpha_R \Delta T) = (1.67 \times 10^{-6})[1 + 0.0043(400 - 25)]$$

$$\rho = 4.363 \times 10^{-6} \, \text{ohm} \cdot \text{cm}$$

$$\sigma = 1/\rho = 2.29 \times 10^5 \, \text{ohm}^{-1} \cdot \text{cm}^{-1}$$

(b) At $-100°C$:

$$\rho = (1.67 \times 10^{-6})[1 + 0.0043(-100 - 25)] = 7.724 \times 10^{-7} \, \text{ohm} \cdot \text{cm}$$

$$\sigma = 12.9 \times 10^{5} \, \text{ohm}^{-1} \cdot \text{cm}^{-1}$$

Effect of Atomic Level Defects

Imperfections in crystal structures scatter electrons, reducing the mobility and conductivity of the metal [Figure 19-6(c)]. For example, the increase in the resistivity due to solid solution atoms for dilute solutions is

$$\rho_d = b(1 - x)x \tag{19-8}$$

where ρ_d is the increase in resistivity due to the defects, x is the atomic fraction of the impurity or solid solution atoms present, and b is the defect resistivity coefficient. In a similar manner, vacancies, dislocations, and grain boundaries reduce the conductivity of the metal. Each defect contributes to an increase in the resistivity of the metal. Thus, the overall resistivity is

$$\rho = \rho_T + \rho_d \tag{19-9}$$

where ρ_d equals the contributions from all of the defects. Equation 19-9 is known as **Matthiessen's rule**. The effect of the defects is *independent* of temperature (Figure 19-8)

Effect of Processing and Strengthening

Strengthening mechanisms and metal processing techniques affect the electrical properties of a metal in different ways (Table 19-4). Solid-solution strengthening is *not* a good way to obtain high strength in metals intended to have high conductivities. The mean free paths are short due to the random distribution of the interstitial or substitutional atoms. Figure 19-9 shows the effect of zinc and other alloying elements on the conductivity of copper; as the amount of alloying element increases, the conductivity decreases substantially.

Age hardening and dispersion strengthening reduce the conductivity to an extent that is less than solid-solution strengthening, since there is a longer mean free path between precipitates, as compared with the path between point defects. Strain hardening and grain-size control have even less effect on conductivity (Figure 19-9 and Table 19-4). Since dislocations and grain boundaries are further apart than solid-solution atoms, there are large volumes of metal that have a long mean free path. Consequently, cold working is an effective way to increase the strength of a metallic conductor without seriously impairing the electrical properties of the material. In addition, the effects of cold working on conductivity can be eliminated by the low-temperature recovery heat treatment in which good conductivity is restored while the strength is retained.

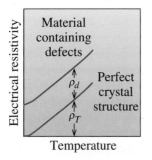

Figure 19-8
The electrical resistivity of a metal is due to a constant defect contribution ρ_d and a variable temperature contribution ρ_T.

TABLE 19-4 ■ *The effect of alloying, strengthening, and processing on the electrical conductivity of copper and its alloys*

Alloy	$\dfrac{\sigma_{alloy}}{\sigma_{Cu}} \times 100$	Remarks
Pure annealed copper	100	Few defects to scatter electrons; the mean free path is long.
Pure copper deformed 80%	98	Many dislocations, but because of the tangled nature of the dislocation networks, the mean free path is still long.
Dispersion-strengthened Cu-0.7% Al$_2$O$_3$	85	The dispersed phase is not as closely spaced as solid-solution atoms, nor is it coherent, as in age hardening. Thus, the effect on conductivity is small.
Solution-treated Cu-2% Be	18	The alloy is single phase; however, the small amount of solid-solution strengthening from the supersaturated beryllium greatly decreases conductivity.
Aged Cu-2% Be	23	During aging, the beryllium leaves the copper lattice to produce a coherent precipitate. The precipitate does not interfere with conductivity as much as the solid-solution atoms.
Cu-35% Zn	28	This alloy is solid-solution strengthened by zinc, which has an atomic radius near that of copper. The conductivity is low, but not as low as when beryllium is present.

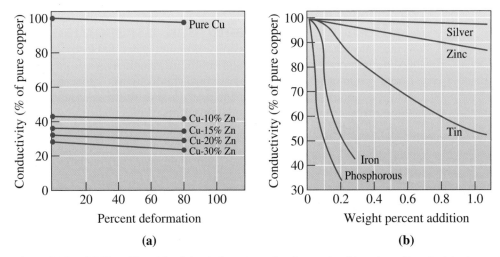

Figure 19-9 (a) The effect of solid-solution strengthening and cold work on the electrical conductivity of copper, and (b) the effect of the addition of selected elements on the electrical conductivity of copper.

Conductivity of Alloys

Alloys typically have higher resistivities than pure metals because of the scattering of electrons due to the alloying additions. For example, the resistivity of pure Cu at room temperature is $\sim 1.67 \times 10^{-6}\ \Omega \cdot$ cm and that of pure gold is $\sim 2.35 \times 10^{-6}\ \Omega \cdot$ cm. The resistivity of a 35% Au-65% Cu alloy at room temperature is much higher, $\sim 12 \times 10^{-6}\ \Omega \cdot$ cm. Ordering of atoms in alloys by heat treatment can decrease their resistivity. Compared to pure metals, the resistivity of alloys tends to be stable in regards to temperature variation. Relatively high-resistance alloys such as nichrome ($\sim 80\%$ Ni-20% Cr) can be used as heating elements. Certain alloys of Bi-Sn-Pb-Cd are used to make electrical fuses due to their low melting temperatures.

19-4   Semiconductors

Elemental semiconductors are found in Group 4B of the periodic table and include germanium and silicon. Compound semiconductors are formed from elements in Groups 2B and 6B of the periodic table (e.g., CdS, CdSe, CdTe, HgCdTe, etc.) and are known as II–VI (two–six) semiconductors. They also can be formed by combining elements from Groups 3B and 5B of the periodic table (e.g., GaN, GaAs, AlAs, AlP, InP, etc.). These are known as III–V (three–five) semiconductors.

An **intrinsic semiconductor** is one with properties that are not controlled by impurities. An **extrinsic semiconductor** (*n*- or *p*-type) is preferred for devices, since its properties are stable with temperature and can be controlled using ion implantation or diffusion of impurities known as dopants. Semiconductor materials, including silicon and germanium, provide the building blocks for many electronic devices. These materials have an easily controlled electrical conductivity and, when properly combined, can act as switches, amplifiers, or information storage devices. The properties of some of the commonly encountered semiconductors are included in Table 19-5.

As we learned in Section 19-2, as the atoms of a semiconductor come together to form a solid, two energy bands are formed [Figure 19-5(b)]. At 0 K, the energy levels of the valence band are completely full, as these are the lowest energy states for the electrons. The valence band is separated from the conduction band by a bandgap. At 0 K, the conduction band is empty.

The energy gap E_g between the valence and conduction bands in semiconductors is relatively small (Figure 19-5). As a result, as temperature increases, some electrons possess enough thermal energy to be promoted from the valence band to the conduction band. The excited electrons leave behind unoccupied energy levels, or holes, in the valence band. When an electron moves to fill a hole, another hole is created; consequently, the holes appear to act as positively charged electrons and carry an electrical charge. When a voltage is applied to the material, the electrons in the conduction band accelerate toward the positive terminal, while holes in the valence band move toward the negative terminal (Figure 19-10). Current is, therefore, conducted by the movement of both electrons and holes.

TABLE 19-5 ■ *Properties of commonly encountered semiconductors at room temperature*

Semiconductor	Bandgap (eV)	Mobility of Electrons (μ_n) $\left(\frac{cm^2}{V \cdot s}\right)$	Mobility of Holes (μ_p) $\left(\frac{cm^2}{V \cdot s}\right)$	Dielectric Constant (k)	Resistivity ($\Omega \cdot cm$)	Density $\left(\frac{g}{cm^2}\right)$	Melting Temperature (°C)
Silicon (Si)	1.11	1350	480	11.8	2.5×10^5	2.33	1415
Amorphous Silicon (a:Si:H)	1.70	1	10^{-2}	~11.8	10^{10}	~2.30	—
Germanium (Ge)	0.67	3900	1900	16.0	43	5.32	936
SiC (α)	2.86	500		10.2	10^{10}	3.21	2830
Gallium Arsenide (GaAs)	1.43	8500	400	13.2	4×10^8	5.31	1238
Diamond	~5.50	1800	1500	5.7	$> 10^{18}$	3.52	~3550

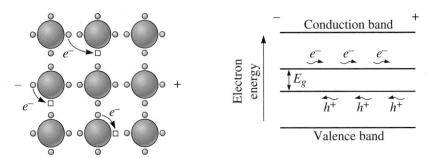

Figure 19-10   When a voltage is applied to a semiconductor, the electrons move through the conduction band, while the holes move through the valence band in the opposite direction.

The conductivity is determined by the number of electrons and holes according to

$$\sigma = nq\mu_n + pq\mu_p \tag{19-10}$$

where n is the concentration of electrons in the conduction band, p is the concentration of holes in the valence band, and μ_n and μ_p are the mobilities of electrons and holes, respectively (Table 19-5). This equation is the same as Equation 19-5b.

In intrinsic semiconductors, for every electron promoted to the conduction band, there is a hole left in the valence band, such that

$$n_i = p_i$$

where n_i and p_i are the concentrations of electrons and holes, respectively, in an intrinsic semiconductor. Therefore, the conductivity of an intrinsic semiconductor is

$$\sigma = qn_i(\mu_n + \mu_p) \tag{19-11}$$

In intrinsic semiconductors, we control the number of charge carriers and, hence, the electrical conductivity by controlling the temperature. At absolute zero temperature, all of the electrons are in the valence band, whereas all of the levels in the conduction band are unoccupied [Figure 19-11(a)]. As the temperature increases, there is a greater probability that an energy level in the conduction band is occupied (and an equal probability that a level in the valence band is unoccupied, or that a hole is present) [Figure 19-11(b)]. The number of electrons in the conduction band, which is equal to the number of holes in the valence band, is given by

$$n = n_i = p_i = n_0 \exp\left(\frac{-E_g}{2k_B T}\right) \tag{19-12a}$$

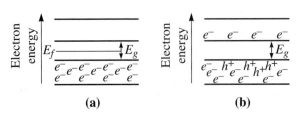

Figure 19-11
The distribution of electrons and holes in the valence and conduction bands (a) at absolute zero and (b) at an elevated temperature.

where n_0 is given by

$$n_0 = 2\left(\frac{2\pi k_B T}{h^2} \right)^{3/2} (m_n^* m_p^*)^{3/4} \tag{19-12b}$$

In these equations, k and h are the Boltzmann and Planck's constants and m_n^* and m_p^* are the effective masses of electrons and holes in the semiconductor, respectively. The effective masses account for the effects of the internal forces that alter the acceleration of electrons in a solid relative to electrons in a vacuum. For Ge, Si, and GaAs, the room temperature values of n_i are 2.5×10^{13}, 1.5×10^{10}, and 2×10^6 electrons/cm³, respectively. The $n_i p_i$ product remains constant at any given temperature for a given semiconductor. This allows us to calculate n_i or p_i values at different temperatures.

Higher temperatures permit more electrons to cross the forbidden zone and, hence, the conductivity increases:

$$\sigma = n_0 q(\mu_n + \mu_p) \exp\left(\frac{-E_g}{2k_B T} \right) \tag{19-13}$$

Note that both n_i and σ are related to temperature by an Arrhenius equation, rate $= A \exp\left(\frac{-Q}{RT} \right)$. As the temperature increases, the conductivity of a semiconductor also increases because more charge carriers are available for conduction. Note that as for metals, the mobilities of the carriers decrease at high temperatures, but this is a much weaker dependence than the exponential increase in the number of charge carriers. The increase in conductivity with temperature in semiconductors sharply contrasts with the decrease in conductivity of metals with increasing temperature (Figure 19-12). Even at high temperatures, however, the conductivity of a metal is orders of magnitudes higher than the conductivity of a semiconductor. The example that follows shows the calculation for carrier concentration in an intrinsic semiconductor.

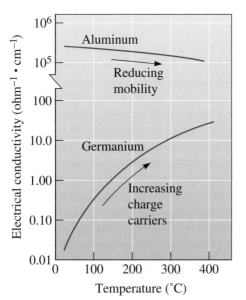

Figure 19-12
The electrical conductivity versus temperature for intrinsic semiconductors compared with metals. Note the break in the vertical axis scale.

Example 19-4   *Carrier Concentrations in Intrinsic Ge*

For germanium at 25°C, estimate (a) the number of charge carriers, (b) the fraction of the total electrons in the valence band that are excited into the conduction band, and (c) the constant n_0 in Equation 19-12a.

SOLUTION

From Table 19-5, $\rho = 43\ \Omega \cdot cm, \therefore \sigma = 0.0233\ \Omega^{-1} \cdot cm^{-1}$
Also from Table 19-5,

$$E_g = 0.67\ eV,\ \mu_n = 3900\ \frac{cm^2}{V \cdot s},\ \mu_p = 1900\ \frac{cm^2}{V \cdot s}$$

$$2k_B T = (2)(8.63 \times 10^{-5}\ eV/K)(273 + 25) = 0.05143\ eV\ \text{at}\ T = 25°C$$

(a) From Equation 19-10,

$$n = \frac{\sigma}{q(\mu_n + \mu_p)} = \frac{0.0233}{(1.6 \times 10^{-19})(3900 + 1900)} = 2.51 \times 10^{13}\ \frac{\text{electrons}}{cm^3}$$

There are 2.51×10^{13} electrons/cm^3 and 2.51×10^{13} holes/cm^3 conducting charge in germanium at room temperature.

(b) The lattice parameter of diamond cubic germanium is 5.6575×10^{-8} cm.
The total number of electrons in the valence band of germanium at 0 K is

$$\text{Total electrons} = \frac{(8\ \text{atoms/cell})(4\ \text{electrons/atom})}{(5.6575 \times 10^{-8}\ cm)^3}$$

$$= 1.77 \times 10^{23}$$

$$\text{Fraction excited} = \frac{2.51 \times 10^{13}}{1.77 \times 10^{23}} = 1.42 \times 10^{-10}$$

(c) From Equation 19-12a,

$$n_0 = \frac{n}{\exp(-E_g/2k_B T)} = \frac{2.51 \times 10^{13}}{\exp(-0.67/0.05143)}$$

$$= 1.14 \times 10^{19}\ \text{carriers/cm}^3$$

Extrinsic Semiconductors   The temperature dependence of conductivity in intrinsic semiconductors is nearly exponential, but this is not useful for practical applications. We cannot accurately control the behavior of an intrinsic semiconductor because slight variations in temperature can significantly change the conductivity. By intentionally adding a small number of impurity atoms to the material (called doping), we can produce an extrinsic semiconductor. The conductivity of the extrinsic semiconductor depends primarily on the number of impurity, or dopant, atoms and in a certain temperature range is independent of temperature. This ability to have a tunable yet temperature independent conductivity is the reason why we almost always use extrinsic semiconductors to make devices.

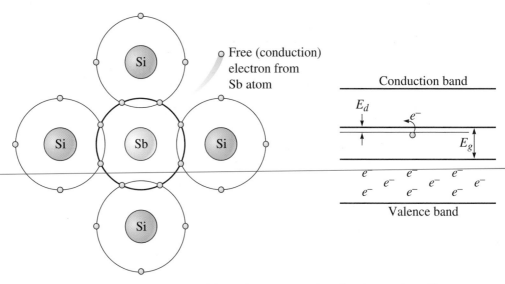

Figure 19-13 When a dopant atom with a valence greater than four is added to silicon, an extra electron is introduced and a donor energy state is created. Now electrons are more easily excited into the conduction band.

n-Type Semiconductors

Suppose we add an impurity atom such as antimony (which has a valence of five) to silicon or germanium. Four of the electrons from the antimony atom participate in the covalent-bonding process, while the extra electron enters an energy level just below the conduction band (Figure 19-13). Since the extra electron is not tightly bound to the atoms, only a small increase in energy, E_d, is required for the electron to enter the conduction band. This energy level just below the conduction band is called a donor state. An *n*-type dopant "donates" a free electron for each impurity added. The energy gap controlling conductivity is now E_d rather than E_g (Table 19-6). No corresponding holes are created when the donor electrons enter the conduction band. It is still the case that electron-hole pairs are created when thermal energy causes electrons to be promoted to the conduction band from the valence band; however, the number of electron-hole pairs is significant only at high temperatures.

p-Type Semiconductors

When we add an impurity such as gallium or boron, which has a valence of three, to Si or Ge, there are not enough electrons to complete the covalent bonding process. A hole is created in the valence band that can be filled by electrons from other locations in the band (Figure 19-14). The holes act as "acceptors" of electrons. These hole sites have a somewhat higher than normal energy and create an acceptor level of possible electron energies just above the valence band (Table 19-6). An electron must gain an energy of only E_a in order to create a hole in the valence band. The hole then carries charge. This is known as a *p*-type semiconductor.

Charge Neutrality

In an extrinsic semiconductor, there has to be overall electrical neutrality. Thus, the sum of the number of donor atoms (N_d) and holes per unit volume (p_{ext}) (both are positively charged) is equal to the number of acceptor atoms (N_a) and electrons per unit volume (n_{ext}) (both are negatively charged):

$$p_{ext} + N_d = n_{ext} + N_a$$

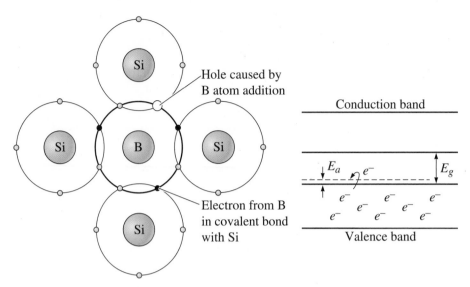

Figure 19-14 When a dopant atom with a valence of less than four is substituted into the silicon structure, a hole is introduced in the structure and an acceptor energy level is created just above the valence band.

TABLE 19-6 ■ *The donor and acceptor energy levels (in electron volts) when silicon and germanium semiconductors are doped*

	Silicon		Germanium	
Dopant	E_d	E_a	E_d	E_a
P	0.045		0.0120	
As	0.049		0.0127	
Sb	0.039		0.0096	
B		0.045		0.0104
Al		0.057		0.0102
Ga		0.065		0.0108
In		0.160		0.0112

In this equation, n_{ext} and p_{ext} are the concentrations of electrons and holes in an extrinsic semiconductor.

If the extrinsic semiconductor is heavily *n*-type doped (i.e., $N_d \gg n_i$), then $n_{ext} \sim N_d$. Similarly, if there is a heavily acceptor-doped (*p*-type) semiconductor, then $N_a \gg p_i$ and hence $p_{ext} \sim N_a$. This is important, since this says that by adding a considerable amount of dopant, we can dominate the conductivity of a semiconductor by controlling the dopant concentration. This is why extrinsic semiconductors are most useful for making controllable devices such as transistors.

The changes in carrier concentration with temperature are shown in Figure 19-15. From this, the approximate conductivity changes in an extrinsic semiconductor are easy to follow. When the temperature is too low, the donor or acceptor atoms are not ionized and hence the conductivity is very small. As temperature begins to increase, electrons (or holes) contributed by the donors (or acceptors) become available for conduction. At sufficiently high temperatures, the conductivity is nearly independent of temperature

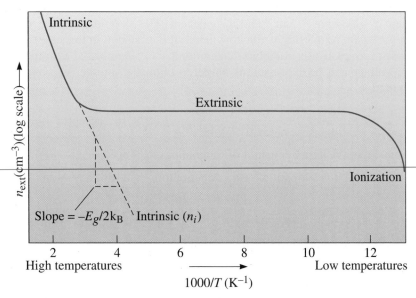

Figure 19-15 The effect of temperature on the carrier concentration of an *n*-type semiconductor. At low temperatures, the donor or acceptor atoms are not ionized. As temperature increases, the ionization process is complete, and the carrier concentration increases to a level that is dictated by the level of doping. The conductivity then essentially remains unchanged until the temperature becomes too high and the thermally generated carriers begin to dominate. The effect of dopants is lost at very high temperatures, and the semiconductor essentially shows "intrinsic" behavior.

(region labeled as extrinsic). The value of conductivity at which the plateau occurs depends on the level of doping. When temperatures become too high, the behavior approaches that of an intrinsic semiconductor since the effect of dopants essentially is lost. In this analysis, we have not accounted for the effects of dopants concentration on the mobility of electrons and holes and the temperature dependence of the bandgap. At very high temperatures (not shown in Figure 19-15), the conductivity *decreases* again as scattering of carriers dominates.

Example 19-5   *Design of a Semiconductor*

Design a *p*-type semiconductor based on silicon, which provides a constant conductivity of 100 ohm^{-1} · cm^{-1} over a range of temperatures. Compare the required concentration of acceptor atoms in Si with the concentration of Si atoms.

SOLUTION

In order to obtain the desired conductivity, we must dope the silicon with atoms having a valence of +3, adding enough dopant to provide the required number of charge carriers. If we assume that the number of intrinsic carriers is small compared to the dopant concentration, then

$$\sigma = N_a q \mu_p$$

where $\sigma = 100$ ohm$^{-1} \cdot$ cm^{-1} and $\mu_p = 480$ cm$^2/$(V $\cdot$ s). Note that electron and hole mobilities are properties of the host material (i.e., silicon in this case) and not the dopant species. If we remember that coulomb can be expressed as ampere-seconds and voltage can be expressed as ampere $\cdot$ ohm, the number of charge carriers required is

$$N_a = \frac{\sigma}{q\mu_p} = \frac{100}{(1.6 \times 10^{-19})(480)} = 1.30 \times 10^{18} \text{ acceptor atoms/cm}^3$$

Assume that the lattice constant of Si remains unchanged as a result of doping:

$$N_a = \frac{(1 \text{ hole/dopant atom})(x \text{ dopant atom/Si atom})(8 \text{ Si atoms/unit cell})}{(5.4307 \times 10^{-8} \text{ cm})^3/\text{unit cell}}$$

$$x = (1.30 \times 10^{18})(5.4307 \times 10^{-8})^3/8 = 26 \times 10^{-6} \text{ dopant atom/Si atom}$$

or 26 dopant atoms/10^6 Si atoms

Possible dopants include boron, aluminum, gallium, and indium. High-purity chemicals and clean room conditions are essential for processing since we need 26 dopant atoms in a million silicon atoms.

Many other materials that are normally insulating (because the bandgap is too large) can be made semiconducting by doping. Examples of this include $BaTiO_3$, ZnO, TiO_2, and many other oxides. Thus, the concept of n- and p-type dopants is not limited to Si, Ge, GaAs, etc. We can dope $BaTiO_3$, for example, and make n- or p-type $BaTiO_3$. Such materials are useful for many sensor applications such as **thermistors**.

Direct and Indirect Bandgap Semiconductors   In a direct bandgap semiconductor, an electron can be promoted from the conduction band to the valence band without changing the momentum of the electron. An example of a direct bandgap semiconductor is GaAs. When the excited electron falls back into the valence band, electrons and holes combine to produce light. This is known as **radiative recombination**. Thus, direct bandgap materials such as GaAs and solid solutions of these (e.g., GaAs-AlAs, etc.) are used to make light-emitting diodes (LEDs) of different colors. The bandgap of semiconductors can be tuned using solid solutions. The change in bandgap produces a change in the wavelength (i.e., the frequency of the color (v) is related to the bandgap E_g as $E_g = hv$, where h is Planck's constant). Since an optical effect is obtained using an electronic material, often the direct bandgap materials are known as optoelectronic materials (Chapter 21). Many lasers and LEDs have been developed using these materials. LEDs that emit light in the infrared range are used in optical-fiber communication systems to convert light waves into electrical pulses. Different colored lasers, such as the blue laser using GaN, have been developed using direct bandgap materials.

In an indirect bandgap semiconductor (e.g., Si, Ge, and GaP), the electrons cannot be promoted to the valence band without a change in momentum. As a result, in materials that have an indirect bandgap (e.g., silicon), we cannot get light emission. Instead, electrons and holes combine to produce heat that is dissipated within the material. This is known as **nonradiative recombination**. Note that both direct and indirect bandgap materials can be doped to form n- or p-type semiconductors.

19-5   Applications of Semiconductors

We fabricate diodes, transistors, lasers, and LEDs using semiconductors. The **p-n junction** is used in many of these devices, such as transistors. Creating an *n*-type region in a *p*-type semiconductor (or vice versa) forms a *p-n* junction [Figure 19-16(a)]. The *n*-type region contains a relatively large number of free electrons, whereas the *p*-type region contains a relatively large number of free holes. This concentration gradient causes diffusion of electrons from the *n*-type material to the *p*-type material and diffusion of holes from the *p*-type material to the *n*-type material. At the junction where the *p*- and *n*-regions meet, free electrons in the *n*-type material recombine with holes in the *p*-type material. This creates a *depleted region* at the junction where the number of available charge carriers is low, and thus, the resistivity is high. Consequently, an electric field develops due to the distribution of exposed positive ions on the *n*-side of the junction and the exposed negative ions on the *p*- side of the junction. The electric field counteracts further diffusion.

Electrically, the *p-n* junction is conducting when the *p*-side is connected to a positive voltage. This **forward bias** condition is shown in Figure 19-16(a). The applied voltage directly counteracts the electric field at the depleted region, making it possible for electrons from the *n*-side to diffuse across the depleted region to the *p*-side and holes from the *p*-side to diffuse across the depleted region to the *n*-side. When a negative bias is applied to the *p*-side of a *p-n* junction (**reverse bias**), the *p-n* junction does not permit much current to flow. The depleted region simply becomes larger because it is further depleted of carriers. When no bias is applied, there is no current flowing through the *p-n* junction. The forward current can be as large as a few milli-amperes, while the reverse-bias current is a few nano-amperes.

The current–voltage (*I–V*) characteristics of a *p-n* junction are shown in Figure 19-16(b). Because the *p-n* junction permits current to flow in only one direction, it passes only half of an alternating current, therefore converting the alternating current to direct current [Figure 19-6(c)]. These junctions are called **rectifier diodes**.

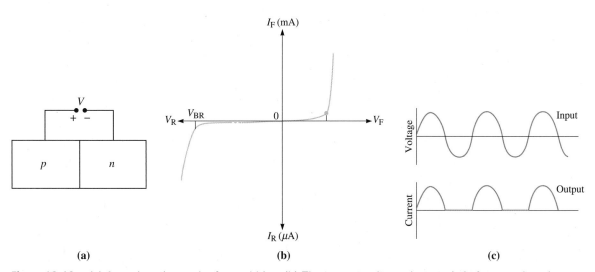

Figure 19-16   (a) A *p-n* junction under forward bias. (b) The current–voltage characteristic for a *p-n* junction. Note the different scales in the first and third quadrants. At sufficiently high reverse bias voltages, "breakdown" occurs, and large currents can flow. Typically this destroys the devices. (c) If an alternating signal is applied, rectification occurs, and only half of the input signal passes the rectifier. (*From Floyd,* Thomas L., ELECTRONIC DEVICES (CONVENTIONAL FLOW VERSION), *6th, © 2002. Electronically reproduced by permission of Pearson Education, Inc., Upper Saddle River, New Jersey.*)

Bipolar Junction Transistors

There are two types of **transistors** based on *p-n* junctions. The term transistor is derived from two words, "transfer" and "resistor." A transistor can be used as a switch or an amplifier. One type of transistor is the *bipolar junction transistor* (BJT). In the era of mainframe computers, bipolar junction transistors often were used in central processing units. A bipolar junction transistor is a sandwich of either *n-p-n* or *p-n-p* semiconductor materials, as shown in Figure 19-17(a). There are three zones in the transistor: the emitter, the base, and the collector. As in the *p-n* junction, electrons are initially concentrated in the *n*-type material, and holes are concentrated in the *p*-type material.

Figure 19-17(b) shows a schematic diagram of an *n-p-n* transistor and its electrical circuit. The electrical signal to be amplified is connected between the base and the emitter, with a small voltage between these two zones. The output from the transistor, or the amplified signal, is connected between the emitter and the collector and operates at a higher voltage. The circuit is connected so that a forward bias is produced between the emitter and the base (the positive voltage is at the *p*-type base), while a reverse bias is produced between the base and the collector (with the positive voltage at the *n*-type collector). The forward bias causes electrons to leave the emitter and enter the base.

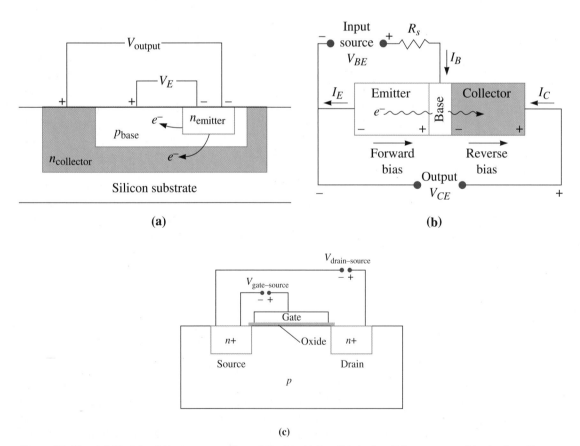

(a)   **(b)**

(c)

Figure 19-17   (a) Sketch of the cross-section of the transistor. (b) A circuit for an *n-p-n* bipolar junction transistor. The input creates a forward and reverse bias that causes electrons to move from the emitter, through the base, and into the collector, creating an amplified output. (c) Sketch of the cross-section of a metal oxide semiconductor field effect transistor.

Electrons and holes attempt to recombine in the base; however, if the base is exceptionally thin and lightly doped, or if the recombination time τ is long, almost all of the electrons pass through the base and enter the collector. The reverse bias between the base and collector accelerates the electrons through the collector, the circuit is completed, and an output signal is produced. The current through the collector (I_c) is given by

$$I_c = I_0 \, \exp\left(\frac{V_E}{B}\right) \qquad (19\text{-}14)$$

where I_0 and B are constants and V_E is the voltage between the emitter and the base. If the input voltage V_E is increased, a very large current I_c is produced.

Field Effect Transistors   A second type of transistor, which is almost universally used today for data storage and processing, is the *field effect transistor* (FET). A metal oxide semiconductor (MOS) field effect transistor (or MOSFET) consists of two highly doped *n*-type regions (*n*+) in a *p*-type substrate or two highly doped *p*-type regions in an *n*-type substrate. (The manufacturing processes by which a device such as this is formed will be discussed in Section 19-6.) Consider a MOSFET that consists of two highly doped *n*-type regions in a *p*-type substrate. One of the *n*-type regions is called the source; the second is called the drain. A potential is applied between the source and the drain with the drain region being positive, but in the absence of a third component of the transistor (a conductor called the gate), electrons cannot flow from the source to the drain under the action of the electric field through the low conductivity *p*-type region. The gate is separated from the semiconductor by a thin insulating layer of oxide and spans the distance between the two *n*-type regions. In advanced device structures, the insulator is only several atomic layers thick and comprises materials other than pure silica.

A potential is applied between the gate and the source with the gate being positive. The potential draws electrons to the vicinity of the gate (and repels holes), but the electrons cannot enter the gate because of the silica. The concentration of electrons beneath the gate makes this region (known as the channel) more conductive, so that a large potential between the source and drain permits electrons to flow from the source to the drain, producing an amplified signal ("on" state). By changing the input voltage between the gate and the source, the number of electrons in the conductive path changes, thus also changing the output signal. When no voltage is applied to the gate, no electrons are attracted to the region between the source and the drain, and there is no current flow from the source to the drain ("off" state).

19-6   General Overview of Integrated Circuit Processing

Integrated circuits (ICs, also known as microchips) comprise large numbers of electronic components that have been fabricated on the surface of a substrate material in the form of a thin, circular wafer less than 1 mm thick and as large as 300 mm in diameter. Two particularly important components found on ICs are transistors, which can serve as electrical switches, as discussed in Section 19-5, and **capacitors**, which can store data in a digital format. Each wafer may contain several hundred chips. Intel's well-known Xeon™ microprocessor is an example of an individual chip.

When first developed in the early 1960s, an integrated circuit comprised just a few electrical components, whereas modern ICs may include several billion components, all in the area of a postage stamp. The smallest dimensions of IC components (or "devices") now approach the atomic scale. This increase in complexity and sophistication, achieved simultaneously with a dramatic decrease in the cost per component, has enabled the entire information technology era in which we live. Without these achievements, mobile phones, the Internet, desktop computers, medical imaging devices, and portable music systems—to name just a few icons of contemporary life—could not exist. It has been estimated that humans now produce more transistors per year than grains of rice.

Since the inception of modern integrated circuit manufacturing, a reduction in size of the individual components that comprise ICs has been a goal of researchers and technologists working in this field. A common expression of this trend is known as "Moore's Law," named after Gordon Moore, author of a seminal paper published in 1965. In that paper, Moore, who would go on to co-found Intel Corporation, predicted that the rapid growth in the number of components fabricated on a chip represented a trend that would continue far into the future. He was correct, and the general trends he predicted are still in evidence four decades later.

As a result, for instance, we see that the number of transistors on a microprocessor has grown from a few thousand to several hundred million, while dynamic random access memory (DRAM) chips have passed the billion-transistor milestone. This has led to enormous advances in the capability of electronic systems, especially on a per-dollar cost basis. During this time, various dimensions of every component on a chip have shrunk, often by orders of magnitude, with some dimensions now best measured in nanometers. This scaling, as it is called, drove and continues to drive Moore's Law and plays out in almost every aspect of IC design and fabrication. Maintaining this progress has required the commitment of enormous resources, both financial and human, as IC processing tools (and the physical environments in which they reside) have been serially developed to meet the challenges of reliably producing ever-smaller features.

Fabrication of integrated circuits involves several hundred individual processing steps and may require several weeks to effect. In many instances, the same types of processing step are repeated again and again, with some variations and perhaps with other processing steps interposed, to create the integrated circuit. These so-called "unit processes" include methods to deposit thin layers of materials onto a substrate, means to define and create intricate patterns within a layer of material, and methods to introduce precise quantities of dopants into layers or the surface of the wafer. The length scales involved with some of these processes are approaching atomic dimensions.

The equipment used for these unit processes includes some of the most sophisticated and expensive instruments ever devised, many of which must be maintained in "clean rooms" that are characterized by levels of dust and contamination orders of magnitude lower than that found in a surgical suite. A modern IC fabrication facility may require several billion dollars of capital expenditure to construct and a thousand or more people to operate.

Silicon wafers most often are grown using the Czochralski growth technique [Figure 19-18(a)]. A small seed crystal is used to grow very large silicon single crystals. The seed crystal is slowly rotated, inserted into, and then pulled from a bath of molten silicon. Silicon atoms attach to the seed crystal in the desired orientation as the seed crystal is retracted. Float zone and liquid encapsulated Czochralski techniques also are used. Single crystals are preferred, because the electrical properties of uniformly doped and essentially dislocation-free single crystals are better defined than those of polycrystalline silicon.

Following the production of silicon wafers, which itself requires considerable expense and expertise, there are four major classes of IC fabrication procedures. The first,

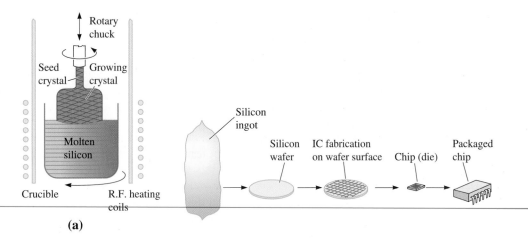

(a)

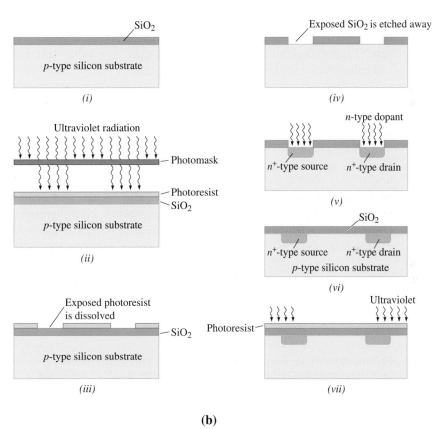

(b)

Figure 19-18 (a) Czochralski growth technique for growing single crystals of silicon. (*From Microchip Fabrication, Third Edition, by P. VanZant, Fig. 3-7. Copyright © 1997 The McGraw-Hill Companies. Reprinted by permission The McGraw-Hill Companies.*) (b) Overall steps encountered in the processing of semiconductors. Production of a FET semiconductor device: (i) A *p*-type silicon substrate is oxidized. (ii) In a process known as photolithography, ultraviolet radiation passes through a photomask (which is much like a stencil), thereby exposing a photosensitive material known as photoresist that was previously deposited on the surface. (iii) The exposed photoresist is dissolved. (iv) The exposed silica is removed by etching. (v) An *n*-type dopant is introduced to produce the source and drain. (vi) The silicon is again oxidized. (vii) Photolithography is repeated to introduce other components, including electrical connections, for the device. (*From Fundamentals of Modern Manufacturing, by M.P. Groover, p. 849, Fig. 34-3. © John Wiley & Sons, Inc. Reproduced with permission of John Wiley & Sons, Inc.*)

known as "front end" processing, comprises the steps in which the electrical components (e.g., transistors) are created in the uppermost surface regions of a semiconductor wafer. It is important to note that most of the thickness of the wafer exists merely as a mechanical support; the electrically active components are formed on the surface and typically extend only a few thousandths of a millimeter into the wafer. Front-end processing may include a hundred or more steps. A schematic diagram of some exemplar front-end processing steps to produce a field-effect transistor are shown in Figure 19-18(b).

"Back end" processing entails the formation of a network of "interconnections" on and just above the surface of the wafer. Interconnections are formed in thin films of material deposited on top of the wafer that are patterned into precise networks; these serve as three-dimensional conductive pathways that allow electrical signals to pass between the individual electronic components, as required for the IC to operate and perform mathematical and logical operations or to store and retrieve data. Back-end processing culminates with protective layers of materials applied to the wafers that prevent mechanical and environmental damage. One feature of IC fabrication is that large numbers of wafers—each comprising several hundred to several thousand ICs, which in turn may each include several million to several billion individual components—are often fabricated at the same time.

Once back-end processing has been completed, wafers are subjected to a number of testing procedures to evaluate both the wafer as a whole and the individual chips. As the number of components per chip has increased and the size of the components has diminished, testing procedures have themselves become ever more complex and specialized. Wafers with too small a fraction of properly functioning chips are discarded.

The last steps in producing functioning ICs are collectively known as "packaging," during which the wafers are cut apart to produce individual, physically distinct chips. To protect the chips from damage, corrosion, and the like, and to allow electrical signals to pass into and out of the chips, they are placed in special, hermetically sealed containers, often only slightly larger than the chip itself. A computer powered by a single microprocessor contains many other chips for many other functions.

19-7   Deposition of Thin Films

As noted in Section 19-6, integrated circuit fabrication depends in part on the deposition of **thin films** of materials onto a substrate. This is similarly true for many technologies that employ films, coatings, or other thin layers of materials, such as wear-resistant coatings on cutting tools, anti-reflective coatings on optical components, and magnetic layers deposited onto aluminum discs for data storage. Thin films may display very different microstructures and physical properties than their bulk counterparts, features that may be exploited in a number of ways. Creating, studying, and using thin films represents a tremendously broad area of materials science and engineering that has enormous impact on modern technology.

Thin films are, as the name implies, very small in one dimension—especially in comparison to their extent in the other two dimensions. There is no well-defined upper bound on what constitutes "thin," but many modern technologies routinely employ thicknesses of several microns down to just a few atomic dimensions. There are myriad ways by which thin films can be deposited, but in general, any technique involves both a *source* of the material to be deposited and a means to *transport* the material from the source to the workpiece surface upon which it is to be deposited. Many deposition techniques require that the source and workpiece be maintained in a vacuum system, while others place the workpiece in a liquid environment.

Physical vapor deposition (PVD) is one very important category of thin-film growth techniques. PVD takes places in a vacuum chamber, and by one means or another creates a low-pressure vapor of the material to be deposited. Some of this vapor will condense on the workpiece and thereby start to deposit as a thin film. Simply melting a material in vacuum, depending on its vapor pressure, may sometimes produce a useful deposit of material.

Sputtering is an example of physical vapor deposition and is the most important PVD method for integrated circuit manufacturing. The interconnections that carry electrical signals from one electronic device to another on an IC chip typically have been made from aluminum alloys that have been sputter deposited. Sputtering can be used to deposit both conducting and insulating materials.

As shown in Figure 19-19, in a sputtering chamber, argon or other atoms in a gas are first ionized and then accelerated by an electric field towards a source of material to be deposited, sometimes called a "target." These ions dislodge and eject atoms from the surface of the source material, some of which drift across a gap towards the workpiece; those that condense on its surface are said to be deposited. Depending on the how long the process continues, it is possible to sputter deposit films that are many microns thick.

Chemical vapor deposition (CVD) represents another set of techniques that is widely employed in the IC industry. In CVD, the source of the material to be deposited exists in gaseous form. The source gas and other gases are introduced into a heated vacuum chamber where they undergo a chemical reaction that creates the desired material as a product. This product condenses on the workpiece (as in PVD processes) creating, over time, a layer of the material. In some CVD processes, the chemical reaction may take place preferentially on the workpiece itself. Thin films of polycrystalline silicon, tungsten, and

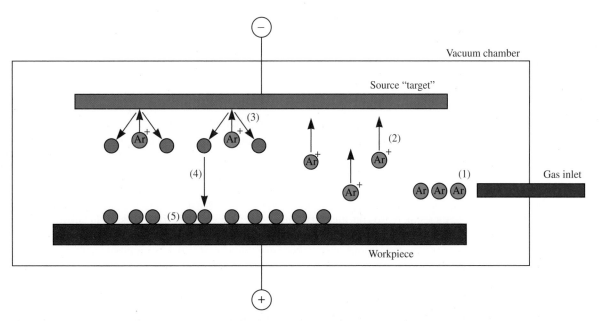

Figure 19-19   Schematic illustration of sputtering. The workpiece and sputter target are placed in a vacuum chamber. (1) An inlet allows a gas such as argon to enter at low pressure. In the presence of the electric field across the target and workpiece, some of the argon atoms are ionized (2) and then accelerated towards the target. By momentum transfer, atoms of the target are ejected (3), drift across the gap towards the workpiece (4), and condense on the workpiece (5), thereby depositing target atoms and eventually forming a film. *(Courtesy of John Bravman.)*

titanium nitride are commonly deposited by CVD as part of IC manufacturing. Nanowire growth, which was discussed in Chapter 11, also often proceeds via a CVD process.

Electrodeposition is a third method for creating thin films on a workpiece. Although this is a very old technology, it has recently been adopted for use in IC manufacturing, especially for depositing the copper films that are replacing aluminum films in most advanced integrated circuits. In electrodeposition, the source and workpiece are both immersed in a liquid electrolyte and also are connected by an external electrical circuit. When a voltage is applied between the source and workpiece, ions of the source material dissolve in the electrolyte, drift under the influence of the field towards the workpiece, and chemically bond on its surface. Over time, a thin film is thus deposited. In some circumstances, an external electric field may not be required; this is called electroless deposition. Electrodeposition and electroless deposition are sometimes referred to as "plating," and the deposited film is sometimes said to be "plated out" on the workpiece.

19-8 Conductivity in Other Materials

Electrical conductivity in most ceramics and polymers is low; however, special materials provide limited or even good conduction. In Chapter 4, we saw how the Kröger-Vink notation can be used to explain defect chemistry in ceramic materials. Using dopants, it is possible to convert many ceramics (e.g., $BaTiO_3$, TiO_2, ZrO_2) that are normally insulating into conductive oxides. The conduction in these materials can occur as a result of movement of ions or electrons and holes.

Conduction in Ionic Materials
Conduction in ionic materials often occurs by movement of entire ions, since the energy gap is too large for electrons to enter the conduction band. Therefore, most ionic materials behave as insulators.

In ionic materials, the mobility of the charge carriers, or ions, is

$$\mu = \frac{ZqD}{k_B T} \tag{19-15}$$

where D is the diffusion coefficient, k_B is the Boltzmann constant, T is the absolute temperature, q is the electronic charge, and Z is the charge on the ion. The mobility is many orders of magnitude lower than the mobility of electrons; hence, the conductivity is very small:

$$\sigma = nZq\mu \tag{19-16}$$

For ionic materials, n is the concentration of ions contributing to conduction. Impurities and vacancies increase conductivity. Vacancies are necessary for diffusion in substitutional types of crystal structures, and impurities can diffuse and help carry the current. High temperatures increase conductivity because the rate of diffusion increases. The following example illustrates the estimation of mobility and conductivity in MgO.

Example 19-6 *Ionic Conduction in MgO*

Suppose that the electrical conductivity of MgO is determined primarily by the diffusion of the Mg^{2+} ions. Estimate the mobility of the Mg^{2+} ions and calculate the electrical conductivity of MgO at 1800°C. The diffusion coefficient of Mg^{2+} ions in MgO at 1800°C is 10^{-10} cm^2/s.

SOLUTION

For MgO, $Z = 2/\text{ion}$, $q = 1.6 \times 10^{-19}$ C, $k_B = 1.38 \times 10^{-23}$ J/K, and $T = 2073$ K:

$$\mu = \frac{ZqD}{k_BT} = \frac{(2)(1.6 \times 10^{-19})(10^{-10})}{(1.38 \times 10^{-23})(2073)} = 1.12 \times 10^{-9}\,\text{C}\cdot\text{cm}^2/(\text{J}\cdot\text{s})$$

Since one coulomb is equivalent to one ampere · second, and one Joule is equivalent to one ampere · second · volt:

$$\mu = 1.12 \times 10^{-9}\,\text{cm}^2/(\text{V}\cdot\text{s})$$

MgO has the NaCl structure with four magnesium ions per unit cell. The lattice parameter is 3.96×10^{-8} cm, so the number of Mg^{2+} ions per cubic centimeter is

$$n = \frac{4\,Mg^{2+}\,\text{ions/cell}}{(3.96 \times 10^{-8}\,\text{cm})^3} = 6.4 \times 10^{22}\,\text{ions/cm}^3$$

$$\sigma = nZq\mu = (6.4 \times 10^{22})(2)(1.6 \times 10^{-9})(1.12 \times 10^{-9})$$

$$= 23 \times 10^{-6}\,\text{C}\cdot\text{cm}^2/(\text{cm}^3\cdot\text{V}\cdot\text{s})$$

Since one coulomb is equivalent to one ampere · second (A · s) and one volt is equivalent to one ampere · ohm (A · Ω),

$$\sigma = 2.3 \times 10^{-5}\,\text{ohm}^{-1}\cdot\text{cm}^{-1}$$

Applications of Ionically Conductive Oxides

The most widely used conductive and transparent oxide is indium tin oxide (ITO), used as a transparent conductive coating on plate glass. Other applications of (ITO) include touch screen displays for computers and devices such as automated teller machines. Other conductive oxides include ytrria-stabilized zirconia (YSZ), which is used as a solid electrolyte in solid oxide fuel cells. Lithium cobalt oxide is used as a solid electrolyte in lithium ion batteries. It is important to remember that, although most ceramic materials behave as electrical insulators, by properly engineering the point defects in ceramics, it is possible to convert many of them into semiconductors.

Conduction in Polymers

Because their valence electrons are involved in covalent bonding, polymers have a band structure with a large energy gap, leading to low-electrical conductivity. Polymers are frequently used in applications that require electrical insulation to prevent short circuits, arcing, and safety hazards. Table 19-1 includes the conductivity of four common polymers. In some cases, however, the low conductivity is a hindrance. For example, if lightning strikes the polymer-matrix composite wing of an airplane, severe damage can occur. We can solve these problems by two approaches: (1) introducing an additive to the polymer to improve the conductivity, and (2) creating polymers that inherently have good conductivity.

The introduction of electrically conductive additives can improve conductivity. For example, polymer-matrix composites containing carbon or nickel-plated carbon fibers combine high stiffness with improved conductivity; hybrid composites containing metal fibers, along with normal carbon, glass, or aramid fibers, also produce lightning-safe aircraft skins. Figure 19-20 shows that when enough carbon fibers are introduced to nylon

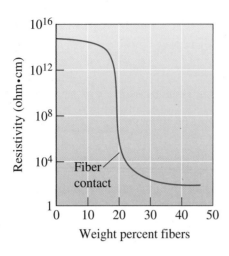

Figure 19-20
Effect of carbon fibers on the electrical resistivity of nylon.

in order to ensure fiber-to-fiber contact, the resistivity is reduced by nearly 13 orders of magnitude. Conductive fillers and fibers are also used to produce polymers that shield against electromagnetic radiation.

Some polymers inherently have good conductivity as a result of doping or processing techniques. When acetal polymers are doped with agents such as arsenic penta-fluoride, electrons or holes are able to jump freely from one atom to another along the backbone of the chain, increasing the conductivity to near that of metals. Some polymers, such as polyphthalocyanine, can be cross-linked by special curing processes to raise the conductivity to as high as 10^2 ohm$^{-1} \cdot$ cm^{-1}, a process that permits the polymer to behave as a semiconductor. Because of the cross-linking, electrons can move more easily from one chain to another. Organic light-emitting diodes are fabricated from semiconducting polymers including polyanilines.

19-9   Insulators and Dielectric Properties

Materials used to insulate an electric field from its surroundings are required in a large number of electrical and electronic applications. Electrical insulators obviously must have a very low conductivity, or high resistivity, to prevent the flow of current. Insulators must also be able to withstand intense electric fields. Insulators are produced from ceramic and polymeric materials in which there is a large energy gap between the valence and conduction bands; however, the high-electrical resistivity of these materials is not always sufficient. At high voltages, a catastrophic breakdown of the insulator may occur, and current may flow. For example, the electrons may have kinetic energies sufficient to ionize the atoms of the insulator, thereby creating free electrons and generating a current at high voltages. In order to select an insulating material properly, we must understand how the material stores, as well as conducts, electrical charge. Porcelain, alumina, cordierite, mica, and some glasses and plastics are used as insulators. The resistivity of most of these is $> 10^{14} \, \Omega \cdot$ cm, and the breakdown electric fields are ~5 to 15 kV/mm.

19-10   Polarization in Dielectrics

When we apply stress to a material, some level of strain develops. Similarly, when we subject materials to an electric field, the atoms, molecules, or ions respond to the applied electric field (E). Thus, the material is said to be polarized. A dipole is a pair of opposite charges separated by a certain distance. If one charge of $+q$ is separated from another charge of $-q$ (q is the electronic charge) and d is the distance between these charges, the dipole moment is $q \times d$. The magnitude of polarization is given by $P = zqd$, where z is the number of charge centers that are displaced per cubic meter.

Any separation of charges (e.g., between the nucleus and electron cloud) or any mechanism that leads to a change in the separation of charges that are already present (e.g., movement or vibration of ions in an ionic material) causes **polarization**. There are four primary mechanisms causing polarization: (1) electronic polarization, (2) ionic polarization, (3) molecular polarization, and (4) space charge (Figure 19-21). Their occurrence depends upon the electrical frequency of the applied field, just like the mechanical behavior of materials depends on the strain rate (Chapters 6 and 8). If we apply a very rapid rate of strain, certain mechanisms of plastic deformation are not activated. Similarly, if we apply a rapidly alternating electric field, some polarization mechanisms may be unable to induce polarization in the material.

Polarization mechanisms play two important roles. First, if we make a **capacitor** from a material, the polarization mechanisms allow charge to be stored, since the dipoles created in the material (as a result of polarization) can bind a certain portion of the charge on the electrodes of the capacitor. Thus, the higher the dielectric polarization, the higher the dielectric constant (k) of the material. The dielectric constant is defined as the ratio of capacitance between a capacitor filled with dielectric material and one with vacuum between its electrodes. This charge storage, in some ways, is similar to the elastic strain in a material subjected to stress. The second important role played by polarization mechanisms is that when polarization sets in, charges move (ions or electron clouds are displaced). If the electric field oscillates, the charges move back and forth. These displacements are extremely small (typically < 1 Å); however, they cause **dielectric losses**. This energy is lost as heat. The dielectric loss is similar to the viscous deformation of a material. If we want to store a charge, as in a capacitor, dielectric loss is not good; however, if we want to use microwaves to heat up our food, dielectric losses that occur in water contained in the food are great! The dielectric losses are often measured by a parameter known as *tan* δ. When we are interested in extremely low loss materials, such as those used in microwave communications, we refer to a parameter known as the dielectric *quality factor* ($Q_d \sim 1/\tan \delta$). The dielectric constant and dielectric losses depend strongly on electrical frequency and temperature.

Electronic polarization is omnipresent since all materials contain atoms. The electron cloud gets displaced from the nucleus in response to the field seen by the atoms. The separation of charges creates a dipole moment [Figure 19-21(a)]. This mechanism can survive at the highest electrical frequencies ($\sim 10^{15}$ Hz) since an electron cloud can be displaced rapidly, back and forth, as the electrical field switches. Larger atoms and ions have higher electronic polarizability (tendency to undergo polarization), since the electron cloud is farther away from the nucleus and held less tightly. This polarization mechanism is also linked closely to the refractive index of materials, since light is an electromagnetic wave for which the electric field oscillates at a very high frequency ($\sim 10^{14} - 10^{16}$ Hz). The higher the electronic polarizability, the higher the refractive index. We use this mechanism in making "lead crystal," which is really an amorphous glass that contains up to 30% PbO.

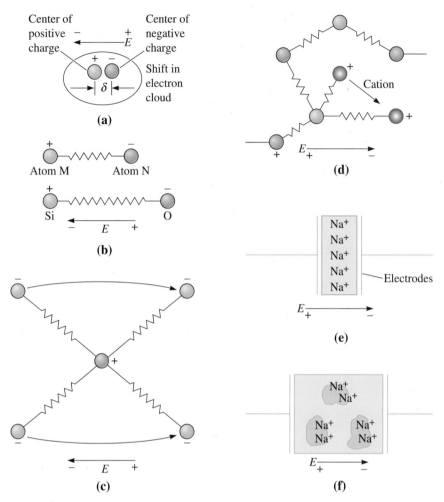

Figure 19-21   Polarization mechanisms in materials: (a) electronic, (b) atomic or ionic, (c) high-frequency dipolar or orientation (present in ferroelectrics), (d) low-frequency dipolar (present in linear dielectrics and glasses), (e) interfacial-space charge at electrodes, and (f) interfacial-space charge at heterogeneities such as grain boundaries. (*From* Principles of Electronic Ceramics, *L.L. Hench and J.K. West, p. 188, Fig. 5-2. Copyright © 1990 Wiley Interscience. Reprinted by permission. This material is used by permission of John Wiley & Sons, Inc.)*

The large lead ions (Pb^{+2}) are highly polarizable due to the electronic polarization mechanisms and provide a high-refractive index when high enough concentrations of lead oxide are present in the glass.

Example 19-7   *Electronic Polarization in Copper*

Suppose that the average displacement of the electrons relative to the nucleus in a copper atom is 10^{-8} Å when an electric field is imposed on a copper plate. Calculate the electronic polarization.

SOLUTION

The atomic number of copper is 29, so there are 29 electrons in each copper atom. The lattice parameter of copper is 3.6151 Å. Thus,

$$z = \frac{(4\,\text{atoms/cell})(29\ \text{electrons/atom})}{(3.6151 \times 10^{-10}\,\text{m})^3} = 2.455 \times 10^{30}\,\text{electrons/m}^3$$

$$P = zqd = \left(2.455 \times 10^{30}\,\frac{\text{electrons}}{\text{m}^3}\right)\left(1.6 \times 10^{-19}\,\frac{\text{C}}{\text{electron}}\right)(10^{-8}L)(10^{-10}\,\text{m/Å})$$

$$= 3.93 \times 10^{-7}\,\text{C/m}^2$$

Frequency and Temperature Dependence of the Dielectric Constant and Dielectric Losses

A capacitor is a device that is capable of storing electrical charge. It typically consists of two electrodes with a dielectric material situated between them. The dielectric may or may not be a solid; even an air gap or vacuum can serve as a dielectric. Two parallel, flat-plate electrodes represent the simplest configuration for a capacitor.

Capacitance C is the ability to store charge and is defined as

$$C = \frac{Q}{V} \tag{19-17}$$

where Q is the charge on the electrode plates of a capacitor and V is the applied voltage [Figure 19-22(a)]. Note that a voltage must be applied to create the charge on the electrodes, but that the charge is "stored" in the absence of the voltage until an external circuit allows it to dissipate. In microelectronic devices, this is the basis for digital data

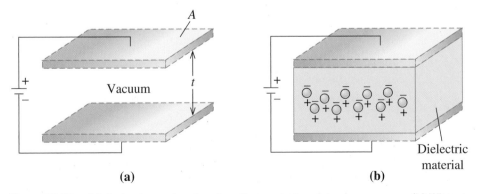

(a)   **(b)**

Figure 19-22   (a) A charge can be stored on the conductor plates in a vacuum. (b) When a dielectric is placed between the plates, the dielectric polarizes, and additional charge is stored.

storage. If the space between two parallel plates (with surface area A and separated by a distance t) is filled with a material, then the dielectric constant k, also known as the relative permittivity ε_r, is determined according to

$$C = \frac{k\varepsilon_0 A}{t} \qquad (19\text{-}18)$$

The constant ε_0 is the **permittivity** of a vacuum and is 8.85×10^{-12} F/m. As the material undergoes polarization, it can bind a certain amount of charge on the electrodes, as shown in Figure 19-22(b). The greater the polarization, the higher the dielectric constant, and therefore, the greater the bound charge on the electrodes.

The dielectric constant is the measure of how susceptible the material is to the applied electric field. The dielectric constant depends on the composition, microstructure, electrical frequency, and temperature. The capacitance depends on the dielectric constant, area of the electrodes, and the separation between the electrodes. Capacitors in parallel provide added capacitance (just like resistances add in series). This is the reason why multi-layer capacitors consist of 100 or more layers connected in parallel. These are typically based on $BaTiO_3$ formulations and are prepared using a tape-casting process. Silver-palladium or nickel is used as electrode layers.

For electrical insulation, the **dielectric strength** (i.e., the electric field value that can be supported prior to electrical breakdown) is important. The dielectric properties of some materials are shown in Table 19-7

Linear and Nonlinear Dielectrics

The dielectric constant, as expected, is related to the polarization that can be achieved in the material. We can show that the dielectric polarization induced in a material depends upon the applied electric field and the dielectric constant according to

$$P = (k - 1)\varepsilon_0 E \text{ (for linear dielectrics)} \qquad (19\text{-}19)$$

TABLE 19-7 ■ *Properties of selected dielectric materials*

Material	Dielectric Constant (at 60 Hz)	Dielectric Constant (at 10^6 Hz)	Dielectric Strength (10^6 V/m)	tan δ (at 10^6 Hz)	Resistivity (ohm · cm)
Polyethylene	2.3	2.3	20	0.00010	$> 10^{16}$
Teflon	2.1	2.1	20	0.00007	10^{18}
Polystyrene	2.5	2.5	20	0.00020	10^{18}
PVC	3.5	3.2	40	0.05000	10^{12}
Nylon	4.0	3.6	20	0.04000	10^{15}
Rubber	4.0	3.2	24		
Phenolic	7.0	4.9	12	0.05000	10^{12}
Epoxy	4.0	3.6	18		10^{15}
Paraffin wax		2.3	10		10^{13}–10^{19}
Fused silica	3.8	3.8	10	0.00004	10^{11}–10^{12}
Soda-lime glass	7.0	7.0	10	0.00900	10^{15}
Al_2O_3	9.0	6.5	6	0.00100	10^{11}–10^{13}
TiO_2		14–110	8	0.00020	10^{13}–10^{18}
Mica		7.0	40		10^{13}
$BaTiO_3$		2000–5000	12	~0.0001	10^8–10^{15}
Water		78.3			10^{14}

where E is the strength of the electric field (V/m). For materials that polarize easily, both the dielectric constant and the capacitance are large and, in turn, a large quantity of charge can be stored. In addition, Equation 19-19 suggests that polarization increases, at least until all of the dipoles are aligned, as the voltage (expressed by the strength of the electric field) increases. The quantity (k−1) is known as dielectric susceptibility (χ_e). The dielectric constant of vacuum is one, or the dielectric susceptibility is zero. This makes sense since a vacuum does not contain any atoms or molecules.

In **linear dielectrics**, P is linearly related to E and k is constant. This is similar to how stress and strain are linearly related by Hooke's law. In linear dielectrics, k (or χ_e) remains constant with changing E. In materials such as $BaTiO_3$, the dielectric constant changes with E, and hence, Equation 19-19 cannot be used. These materials in which P and E are not related by a straight line are known as **nonlinear dielectrics** or ferroelectrics. These materials are similar to elastomers for which stress and strain are not linearly related and a unique value of the Young's modulus cannot be assigned.

19-11 Electrostriction, Piezoelectricity, and Ferroelectricity

When any material undergoes polarization, its ions and electron clouds are displaced, causing the development of a mechanical strain in the material. This effect is seen in all materials subjected to an electric field and is known as **electrostriction.**

Of the total 32 crystal classes, eleven have a center of symmetry. This means that if we apply a mechanical stress, there is no dipole moment generated since ionic movements are symmetric. Of the 21 that remain, 20 point groups, which lack a center of symmetry, exhibit the development of dielectric polarization when subjected to stress. These materials are known as **piezoelectric.** (The word *piezo* means pressure.) When these materials are stressed, they develop a voltage. This development of a voltage upon the application of stress is known as the *direct* or *motor piezoelectric effect* (Figure 19-23). This effect helps us make devices such as spark igniters, which are often made using lead zirconium titanate (PZT). This effect is also used, for example, in detecting submarines and other objects under water.

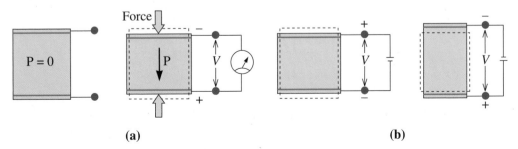

(a)   (b)

Figure 19-23   The (a) direct and (b) converse piezoelectric effect. In the direct piezoelectric effect, applied stress causes a voltage to appear. In the converse effect (b), an applied voltage leads to the development of strain.

Conversely, when an electrical voltage is applied, a piezoelectric material shows the development of strain. This is known as the *converse* or *generator piezoelectric effect*. This effect is used in making actuators. For example, this movement can be used to generate ultrasonic waves that are used in medical imaging, as well as such applications as ultrasonic cleaners or toothbrushes. Sonic energy can also be created using piezoelectrics to make the high-fidelity "tweeter" found in most speakers. In addition to $Pb(Zr_xTi_{1-x})O_3$ (PZT), other piezoelectrics include SiO_2 (for making quartz crystal oscillators), ZnO, and polyvinylidene fluoride (PVDF). Many naturally occurring materials such as bone and silk are also piezoelectric.

The "*d*" constant for a piezoelectric is defined as the ratio of strain (ε) to electric field:

$$\varepsilon = d \cdot E \tag{19-20}$$

The "*g*" constant for a piezoelectric is defined as the ratio of the electric field generated to the stress applied (X):

$$E = g \cdot X \tag{19-21}$$

The *d* and *g* piezoelectric coefficients are related by the dielectric constant as:

$$g = \frac{d}{k\varepsilon_0} \tag{19-22}$$

We define **ferroelectrics** as materials that show the development of a spontaneous and reversible dielectric polarization (P_s). Examples include the tetragonal polymorph of barium titanate. Lead zirconium titanate is both ferroelectric and piezoelectric. Ferroelectric materials show a **hysteresis loop** (i.e., the induced polarization is not linearly related to the applied electric field) as seen in Figure 19-24.

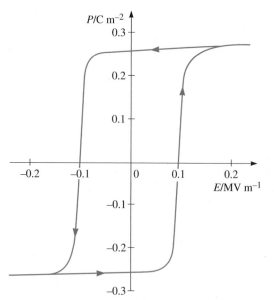

Figure 19-24
The ferroelectric hysteresis loop for a single-domain single crystal of $BaTiO_3$. (*From* Electroceramics: Material, Properties, Applications, *by A.J. Moulson and J.M. Herbert, p. 76, Fig. 2-46. Copyright © 1990 Chapman and Hall. Reprinted with kind permission of Kluwer Academic Publishers and the author.*)

Figure 19-25
Ferroelectric domains
can be seen in the
microstructure of
polycrystalline BaTiO₃.
(*Courtesy of Dr. Rodney
Roseman, University of
Cincinnati.*)

Ferroelectric materials exhibit ferroelectric domains in which the region (or domain) has uniform polarization (Figure 19-25.) Certain ferroelectrics, such as PZT, exhibit a strong piezoelectric effect, but in order to maximize the piezoelectric effect (e.g., the development of strain or voltage), piezoelectric materials are deliberately "poled" using an electric field to align all domains in one direction. The electric field is applied at high temperature and maintained while the material is cooled.

The dielectric constant of ferroelectrics reaches a maximum near a temperature known as the **Curie temperature**. At this temperature, the crystal structure acquires a center of symmetry and thus is no longer piezoelectric. Even at these high temperatures, however, the dielectric constant of ferroelectrics remains high. $BaTiO_3$ exhibits this behavior, and this is the reason why $BaTiO_3$ is used to make single and multi-layer capacitors. In this state, vibrations and shocks do not generate spurious voltages due to the piezoelectric effect. Since the Curie transition occurs at a high temperature, use of additives in $BaTiO_3$ helps to shift the Curie transition temperature to lower temperatures. Additives also can be used to broaden the Curie transition. Materials such as $Pb(Mg_{1/3}Nb_{2/3})O_3$ or PMN are known as *relaxor ferroelectrics*. These materials show very high dielectric constants (up to 20,000) and good piezoelectric behavior, so they are used to make capacitors and piezoelectric devices.

Example 19-8   *Design of a Multi-Layer Capacitor*

A multi-layer capacitor is to be designed using a $BaTiO_3$ formulation containing $SrTiO_3$. The dielectric constant of the material is 3000. (a) Calculate the capacitance of a multi-layer capacitor consisting of 100 layers connected in parallel using nicklel electrodes. The area of each layer is 10 mm × 5 mm, and the thickness of each layer is 10 μm.

SOLUTION

(a) The capacitance of a parallel plate capacitor is given by

$$C = \frac{k\varepsilon_0 A}{t}$$

Thus, the capacitance per layer will be

$$C_{\text{layer}} = \frac{(3000)(8.85 \times 10^{-12}\,\text{F/m})(10 \times 10^{-3}\,\text{m})(5 \times 10^{-3}\,\text{m})}{10 \times 10^{-6}\,\text{m}}$$

$$C_{\text{layer}} = 13.28 \times 10^{-8}$$

We have 100 layers connected in parallel. Capacitances add in this arrangement. All layers have the same geometric dimensions in this case.

$$C_{\text{total}} = (\text{number of layers}) \cdot (\text{capacitance per layer})$$

$$C_{\text{total}} = (100)(13.28 \times 10^{-8}\,\text{F}) = 13.28\,\mu\text{F}$$

Summary

- Electronic materials include insulators, dielectrics, conductors, semiconductors, and superconductors. These materials can be classified according to their band structures. Electronic materials have enabled many technologies ranging from high-voltage line insulators to solar cells, computer chips, and many sensors and actuators.

- Important properties of conductors include the conductivity and the temperature dependence of conductivity. In pure metals, the resistivity increases with temperature. Resistivity is sensitive to impurities and microstructural defects such as grain boundaries. Resistivity of alloys is typically higher than that of pure metals.

- Semiconductors have conductivities between insulators and conductors and are much poorer conductors than metals. The conductivities of semiconductors can be altered by orders of magnitude by minute quantities of certain dopants. Semiconductors can be classified as elemental (Si, Ge) or compound (SiC, GaAs). Both of these can be intrinsic or extrinsic (*n*- or *p*-type). Some semiconductors have direct bandgaps (e.g., GaAs), while others have indirect bandgaps (e.g., Si).

- Creating an *n*-type region in a *p*-type semiconductor (or vice versa) forms a *p-n* junction. The *p-n* junction is used to make diodes and transistors.

- Microelectronics fabrication involves hundreds of precision processes that can produce hundreds of millions and even a billion transistors on a single microchip.

- Thin films are integral components of microelectronic devices and also are used for wear-resistant and anti-reflective coatings. Thin films can be deposited using a variety of techniques, including physical vapor deposition, chemical vapor deposition, and electrodeposition.

- Ionic materials conduct electricity via the movement of ions or electrons and holes.

- Dielectrics have large bandgaps and do not conduct electricity. With insulators, the focus is on breakdown voltage or field. With dielectrics, the emphasis is on the dielectric constant, frequency, and temperature dependence. Polarization mechanisms in materials dictate this dependence.

- In piezoelectrics, the application of stress results in the development of a voltage; the application of a voltage causes strain.

- Ferroelectrics are materials that show a reversible and spontaneous polarization. $BaTiO_3$, PZT, and PVDF are examples of ferroelectrics. Ferroelectrics exhibit a large dielectric constant and are often used to make capacitors.

Glossary

Bandgap (E_g) The energy between the top of the valence band and the bottom of the conduction band.

Band structure The band structure consists of the array of energy levels that are available to or forbidden for electrons to occupy and determines the electronic behavior of a solid, such as whether it is a conductor, semiconductor, or insulator.

Capacitor A device that is capable of storing electrical charge. It typically consists of two electrodes with a dielectric material situated between them, but even an air gap can serve as a dielectric. A capacitor can be a single layer or multi-layer device.

Chemical Vapor Deposition (CVD) A thin-film growth process in which gases undergo a reaction in a heated vacuum chamber to create the desired product on a substrate.

Conduction band The unfilled energy levels into which electrons are excited in order to conduct.

Curie temperature The temperature above which a ferroelectric is no longer piezoelectric.

Current density Current per unit cross-sectional area.

Dielectric constant (k) The ratio of the permittivity of a material to the permittivity of vacuum, thus describing the relative ability of a material to polarize and store a charge; the same as relative permittivity.

Dielectric loss A measure of how much electrical energy is lost due to motion of charge entities that respond to an electric field via different polarization mechanisms. This energy appears as heat.

Dielectric strength The maximum electric field that can be maintained between two conductor plates without causing a breakdown.

Doping Deliberate addition of controlled amounts of other elements to increase the number of charge carriers in a semiconductor.

Drift velocity The average rate at which electrons or other charge carriers move through a material under the influence of an electric or magnetic field.

Electric field The voltage gradient or volts per unit length.

Electrodeposition A method for depositing materials in which a source and workpiece are connected electrically and immersed in an electrolyte. A voltage is applied between the source and workpiece, and ions from the source dissolve in the electrolyte, drift to the workpiece, and gradually deposit a thin film on its surface.

Electrostriction The dimensional change that occurs in any material when an electric field acts on it.

Energy gap (E_g) (Bandgap) The energy between the top of the valence band and the bottom of the conduction band.

Extrinsic semiconductor A semiconductor prepared by adding dopants, which determine the number and type of charge carriers. Extrinsic behavior can also be seen due to impurities.

Fermi energy The energy level at which the probability of finding an electron is 1/2.

Ferroelectric A material that shows spontaneous and reversible dielectric polarization.

Forward bias Connecting a *p-n* junction device so that the *p*-side is connected to a positive terminal, thereby enabling current to flow.

Holes Unfilled energy levels in the valence band. Because electrons move to fill these holes, the holes produce a current.

Hysteresis loop The loop traced out by the nonlinear polarization in a ferroelectric material as the electric field is cycled. A similar loop occurs in certain magnetic materials.

Integrated circuit An electronic package that comprises large numbers of electronic devices fabricated on a single chip.

Intrinsic semiconductor A semiconductor in which properties are controlled by the element or compound that is the semiconductor and not by dopants or impurities.

Linear dielectrics Materials in which the dielectric polarization is linearly related to the electric field; the dielectric constant is not dependent on the electric field.

Matthiessen's rule The resistivity of a metallic material is given by the addition of a base resistivity that accounts for the effect of temperature (ρ_T) and a temperature independent term that reflects the effect of atomic level defects, including solutes forming solid solutions (ρ_d).

Mean free path (λ_e) The average distance that electrons move without being scattered by other atoms or lattice defects.

Microstructure-sensitive property Properties that depend on the microstructure of a material (e.g., conductivity, dielectric constant, or yield strength).

Mobility The ease with which a charge carrier moves through a material.

Nonlinear dielectrics Materials in which dielectric polarization is not linearly related to the electric field (e.g., ferroelectric). These have a field-dependent dielectric constant.

Nonradiative recombination The generation of heat when an electron loses energy and falls from the conduction band to the valence band to occupy a hole; this occurs mainly in indirect bandgap materials such as Si.

p-n junction A device made by creating an *n*-type region in a *p*-type material (or vice versa). A *p-n* junction behaves as a diode and multiple *p-n* junctions function as transistors. It is also the basis of LEDs and solar cells.

Permittivity The ability of a material to polarize and store a charge within it.

Physical Vapor Deposition (PVD) A thin-film growth process in which a low-pressure vapor supplies the material to be deposited on a substrate. Sputtering is one example of PVD.

Piezoelectrics Materials that develop voltage upon the application of a stress and develop strain when an electric field is applied.

Polarization Movement of charged entities (i.e., electron cloud, ions, dipoles, and molecules) in response to an electric field.

Radiative recombination The emission of light when an electron loses energy and falls from the conduction band to the valence band to occupy a hole; this occurs in direct bandgap materials such as GaAs.

Rectifier A *p-n* junction device that permits current to flow in only one direction in a circuit.

Reverse bias Connecting a junction device so that the *p*-side is connected to a negative terminal; very little current flows through a *p-n* junction under reverse bias.

Sputtering A thin-film growth process by which gas atoms are ionized and then accelerated by an electric field towards the source, or "target," of material to be deposited. These ions eject atoms from the target surface, some of which are then deposited on a substrate. Sputtering is one type of a physical vapor deposition process.

Superconductor A material that exhibits zero electrical resistance under certain conditions.

Thermistor A semiconductor device that is particularly sensitive to changes in temperature, permitting it to serve as an accurate measure of temperature.

Thin film A coating or layer that is small or thin in one dimension. Typical thicknesses range from 10 Å to a few microns depending on the application.

Transistor A semiconductor device that amplifies or switches electrical signals.

Valence band The energy levels filled by electrons in their lowest energy states.

Problems

Section 19-1 Ohm's Law and Electrical Conductivity

Section 19-2 Band Structure of Solids

Section 19-3 Conductivity of Metals and Alloys

19-1 A current of 10 A is passed through a 1-mm diameter wire 1000 m long. Calculate the power loss if the wire is made from

(a) aluminum and
(b) silicon (see Table 19-1).

19-2 A 0.5-mm-diameter fiber, 1 cm in length, made of boron nitride is placed in a 120 V circuit. Using Table 19-1, calculate

(a) the current flowing in the circuit and
(b) the number of electrons passing through the boron nitride fiber per second.
(c) What would the current and number of electrons be if the fiber were made of magnesium instead of boron nitride?

19-3 The power lost in a 2-mm-diameter copper wire is to be less than 250 W when a 5 A current is flowing in the circuit. What is the maximum length of the wire?

19-4 A current density of 100,000 A/cm^2 is applied to a gold wire 50 m in length. The resistance of the wire is found to be 2 ohms.

Calculate the diameter of the wire and the voltage applied to the wire.

19-5 We would like to produce a 5000-ohm resistor from boron-carbide fiber having a diameter of 0.1 mm. What is the required length of the fiber?

19-6 Ag has an electrical conductivity of $6.80 \times 10^5 \, \Omega^{-1} \cdot cm^{-1}$. Au has an electrical conductivity of $4.26 \times 10^5 \, \Omega^{-1} \cdot cm^{-1}$. Calculate the number of charge carriers per unit volume and the electron mobility in each in order to account for this difference in electrical conductivity. Comment on your findings.

19-7 A current density of 5000 A/cm^2 is applied to a magnesium wire. If half of the valence electrons serve as charge carriers, calculate the average drift velocity of the electrons.

19-8 We apply 10 V to an aluminum wire 2 mm in diameter and 20 m long. If 10% of the valence electrons carry the electrical charge, calculate the average drift velocity of the electrons in km/h.

19-9 In a welding process, a current of 400 A flows through the arc when the voltage is 35 V. The length of the arc is about 0.25 cm, and the average diameter of the arc is about 0.45 cm. Calculate the current density in the arc, the electric field across the arc, and the electrical conductivity of the hot gases in the arc during welding.

19-10 Draw a schematic of the band structures of an insulator, a semiconductor, and a metallic material. Use this to explain why the conductivity of pure metals decreases with increasing temperature, while the opposite is true for semiconductors and insulators.

19-11 A typical thickness for a copper conductor (known as an interconnect) in an integrated circuit is 250 nm. The mean free path of electrons in pure, annealed copper is about 40 nm. As the thickness of copper interconnects approaches the mean free path, how do you expect conduction in the interconnect is affected? Explain.

19-12 Calculate the electrical conductivity of platinum at −200°C.

19-13 Calculate the electrical conductivity of nickel at −50°C and at +500°C.

19-14 The electrical resistivity of pure chromium is found to be 18×10^{-6} ohm·cm. Estimate the temperature at which the resistivity measurement was made.

19-15 After finding the electrical conductivity of cobalt at 0°C, we decide we would like to double that conductivity. To what temperature must we cool the metal?

19-16 From Figure 19-9 (b), estimate the defect resistivity coefficient for tin in copper.

19-17 (a) Copper and nickel form a complete solid solution. Draw a schematic diagram illustrating the resistivity of a copper and nickel alloy as a function of the atomic percent nickel. Comment on why the curve has the shape that it does.

(b) Copper and gold do not form a complete solid solution. At the compositions of 25 and 50 atomic percent gold, the ordered phases Cu_3Au and $CuAu$ form, respectively. Do you expect that a plot of the resistivity of a copper and gold alloy as a function of the atomic percent gold will have a shape similar to the plot in part (a)? Explain.

19-18 The electrical resistivity of a beryllium alloy containing 5 at% of an alloying element is found to be 50×10^{-6} ohm·cm at 400°C. Determine the contributions to resistivity due to temperature and due to impurities by finding the expected resistivity of pure beryllium at 400°C, the resistivity due to impurities, and the defect resistivity coefficient. What would be the electrical resistivity if the beryllium contained 10 at% of the alloying element at 200°C?

Section 19-4 Semiconductors

Section 19-5 Applications of Semiconductors

Section 19-6 General Overview of Integrated Circuit Processing

Section 19-7 Deposition of Thin Films

19-19 Explain the following terms: semiconductor, intrinsic semiconductor, extrinsic semiconductor, elemental semiconductor, compound semiconductor, direct bandgap semiconductor, and indirect bandgap semiconductor.

19-20 What is radiative and nonradiative recombination? What types of materials are used to make LEDs?

19-21 For germanium and silicon, compare, at 25°C, the number of charge carriers per cubic centimeter, the fraction of the total

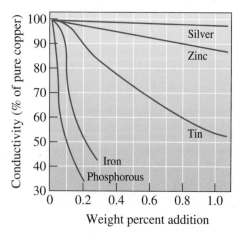

Figure 19-9 (Repeated for Problem 19-16.) (b) The effect of selected elements on the electrical conductivity of copper.

electrons in the valence band that are excited into the conduction band, and the constant n_0.

19-22 For germanium and silicon, compare the temperature required to double the electrical conductivity from the room temperature value.

19-23 Determine the electrical conductivity of silicon when 0.0001 at% antimony is added as a dopant and compare it to the electrical conductivity when 0.0001 at% indium is added.

19-24 We would like to produce an extrinsic germanium semiconductor having an electrical conductivity of 2000 ohm^{-1} · cm^{-1}. Determine the amount of phosphorous and the amount of gallium required to make *n*- and *p*-type semiconductors, respectively.

19-25 Estimate the electrical conductivity of silicon doped with 0.0002 at% arsenic at 600°C, which is above the plateau in the conductivity-temperature curve.

19-26 Determine the amount of arsenic that must be combined with 1 kg of gallium to produce a *p*-type semiconductor with an electrical conductivity of 500 ohm^{-1} · cm^{-1} at 25°C. The lattice parameter of GaAs is about 5.65 Å, and GaAs has the zinc blende structure.

19-27 Calculate the intrinsic carrier concentration for GaAs at room temperature. Given that the effective mass of electrons in GaAs is $0.067m_e$, where m_e is the mass of the electron, calculate the effective mass of the holes.

19-28 Calculate the electrical conductivity of silicon doped with 10^{18} cm^{-3} boron at room temperature. Compare the intrinsic carrier concentration to the dopant concentration.

19-29 At room temperature, will the conductivity of silicon doped with 10^{17} cm^{-3} of arsenic be greater than, about equal to, or less than the conductivity of silicon doped with 10^{17} cm^{-3} of phosphorus?

19-30 When a voltage of 5 mV is applied to the emitter of a transistor, a current of 2 mA is produced. When the voltage is increased to 8 mV, the current through the collector rises to 6 mA. By what percentage will the collector current increase when the emitter voltage is doubled from 9 mV to 18 mV?

19-31 Design a light-emitting diode that will emit at 1.12 micrometers. Is this wavelength in the visible range? What is a potential application for this type of LED?

19-32 How can we make LEDs that emit white light (i.e., light that looks like sunlight)?

19-33 Investigate the scaling relationship known as Moore's Law. Is it expected that this trend will continue to be followed in the future using established microelectronics fabrication techniques? If not, what are some of the alternatives currently being considered? Provide a list of the references or websites that you used.

19-34 Silicon is the material of choice for the substrate for integrated circuits. Explain why silicon is preferred over germanium, even though the electron and hole mobilities are much higher and the bandgap is much smaller for germanium than for silicon. Provide a list of the references or websites that you used.

Section 19-8 Conductivity in Other Materials

19-35 Calculate the electrical conductivity of a fiber-reinforced polyethylene part that is reinforced with 20 vol% of continuous, aligned nickel fibers.

19-36 What are ionic conductors? What are their applications?

19-37 How do the touch screen displays on some computers work?

19-38 Can polymers be semiconducting? What would be the advantages in using these instead of silicon?

Section 19-9 Insulators and Dielectric Properties

Section 19-10 Polarization in Dielectrics

19-39 With respect to mechanical behavior, we have seen that stress (a cause) produces strain (an effect). What is the electrical analog of this?

19-40 With respect to mechanical behavior, elastic modulus represents the elastic energy stored, and viscous dissipation represents the mechanical energy lost in deformation. What is the electrical analog for this?

19-41 Calculate the displacement of the electrons or ions for the following conditions:

(a) electronic polarization in nickel of 2×10^{-7} C/m^2;

(b) electronic polarization in aluminum of 2×10^{-8} C/m^2;

(c) ionic polarization in NaCl of 4.3×10^{-8} C/m^2; and

(d) ionic polarization in ZnS of 5×10^{-8} C/m^2.

19-42 A 2-mm-thick alumina dielectric is used in a 60 Hz circuit. Calculate the voltage required to produce a polarization of 5×10^{-7} C/m^2.

19-43 Suppose we are able to produce a polarization of 5×10^{-8} C/m^2 in a cube (5 mm side) of barium titanate. Assume a dielectric constant of 3000. What voltage is produced?

19-44 What polarization mechanism will be present in (a) alumina, (b) copper, (c) silicon, and (d) barium titanate?

Section 19-11 Electrostriction, Piezoelectricity, and Ferroelectricity

19-45 Define the following terms: electrostriction, piezoelectricity (define both its direct and converse effects), and ferroelectricity.

19-46 Calculate the capacitance of a parallel-plate capacitor containing five layers of mica for which each mica sheet is 1 cm $\times$ 2 cm $\times$ 0.005 cm. The layers are connected in parallel.

19-47 A multi-layer capacitor is to be designed using a relaxor ferroelectric formulation based on lead magnesium niobate (PMN). The apparent dielectric constant of the material is 20,000. Calculate the capacitance of a multi-layer capacitor consisting of ten layers connected in parallel using Ni electrodes. The area of the capacitor is 10 mm $\times$ 10 mm, and the thickness of each layer is 20 μm.

19-48 A force of 90 N is applied to the face of a 0.5 cm $\times$ 0.5 cm $\times$ 0.1 cm thickness of quartz crystal. Determine the voltage produced by the force. The modulus of elasticity of quartz is 72×10^3 MPa.

 Design Problems

19-49 We would like to produce a 100-ohm resistor using a thin wire of a material. Design such a device.

19-50 Design a capacitor that is capable of storing 1 μF when 100 V is applied.

19-51 Design an epoxy-matrix composite that has a modulus of elasticity of at least 240×10^3 MPa and an electrical conductivity of at least 1×10^5 ohm$^{-1} \cdot$ cm^{-1}.

 Computer Problems

19-52 *Design of Multi-layer Capacitors.* Write a computer program that can be used to calculate the capacitance of a multi-layer capacitor. The program, for example, should ask the user to provide values of the dielectric constant and the dimensions of the layer. The program should also be flexible in that if the user provides an intended value of capacitance and other dimensions, the program should provide the required dielectric constant.

Ⓚ Knovel® **Problems**

K19-1 Calculate the resistivity of pure iridium at 673 K using its temperature resistivity coefficient.

K19-2 Electrical conductivity is sometimes given in the units of %IACS. What does IACS stand for? Define the unit using the information found.

K19-3 Can organic materials such as polymers and carbon nanotubes be semiconductors? If they are, what determines their semiconducting properties?

A magnetic hard drive is the heart of personal and laptop computers. These disks use magnetic materials that are unique in that information can be easily written to them, but information cannot be easily erased. The hard drive system is complex in that it makes use of nanostructured and nanoscale thin-film magnetic materials for information storage. (*PhotoDisc Green/Getty Images*)

Chapter 20

Magnetic Materials

Have You Ever Wondered?

- *What materials are used to make audio and video cassettes?*
- *What affects the "lifting strength" of a magnet?*
- *What are "soft" and "hard" magnetic materials?*
- *Are there "nonmagnetic" materials?*
- *Are there materials that develop mechanical strain upon the application of a magnetic field?*

Every material in the world responds to the presence of a magnetic field. Magnetic materials are used to operate such things as electrical motors, generators, and transformers. Much of data storage technology (computer hard disks, computer disks, video and audio cassettes, and the like) is based on magnetic particles. Magnetic materials are also used in loudspeakers, telephones, CD players, telephones, televisions, and video recorders. Superconductors can also be viewed as magnetic materials. Magnetic materials, such as iron oxide (Fe_3O_4) particles, are used to make exotic compositions of "liquid magnets" or ferrofluids. The same iron oxide particles are also used to bind DNA molecules, cells, and proteins.

In this chapter, we look at the fundamental basis for responses of certain materials to the presence of magnetic fields. We will also examine the properties and applications of different types of magnetic materials.

20-1   Classification of Magnetic Materials

Strictly speaking, there is no such thing as a "nonmagnetic" material. Every material consists of atoms; atoms consist of electrons spinning around them, similar to a current-carrying loop that generates a magnetic field. Thus, every material responds to a magnetic field. The manner in which this response of electrons and atoms in a material is scaled determines whether a material will be strongly or weakly magnetic. Examples of **ferromagnetic** materials are materials such as Fe, Ni, Co, and some of their alloys. Examples of **ferrimagnetic** materials include many ceramic materials such as nickel zinc ferrite and manganese zinc ferrite. The term "nonmagnetic," usually means that the material is neither ferromagnetic nor ferrimagnetic. These "nonmagnetic" materials are further classified as **diamagnetic** (e.g., superconductors) or **paramagnetic**. In some cases, we also encounter materials that are **antiferromagnetic** or **superparamagnetic**. We will discuss these different classes of materials and their applications later in the chapter. Ferromagnetic and ferrimagnetic materials are usually further classified as either soft or hard magnetic materials. High-purity iron or plain carbon steels are examples of a magnetically soft material as they can become magnetized, but when the magnetizing source is removed, these materials lose their magnet-like behavior.

Permanent magnets or **hard magnetic materials** retain their magnetization. These are permanent "magnets." Many ceramic ferrites are used to make inexpensive refrigerator magnets. A hard magnetic material does not lose its magnetic behavior easily.

20-2   Magnetic Dipoles and Magnetic Moments

The magnetic behavior of materials can be traced to the structure of atoms. The orbital motion of the electron around the nucleus and the spin of the electron about its own axis (Figure 20-1) cause separate magnetic moments. These two motions (i.e., spin and orbital) contribute to the magnetic behavior of materials. When the electron spins, there is a magnetic moment associated with that motion. The **magnetic moment** of an electron due to its spin is known as the **Bohr magneton** (μ_B). This is a fundamental constant and is defined as

$$\mu_B = \text{Bohr magneton} = \frac{qh}{4\pi m_e} = 9.274 \times 10^{-24} \text{A} \cdot \text{m}^2 \qquad (20\text{-}1)$$

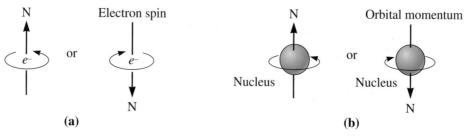

Figure 20-1   Origin of magnetic dipoles: (a) The spin of the electron produces a magnetic field with a direction dependent on the quantum number m_s. (b) Electrons orbiting around the nucleus create a magnetic field around the atom.

where q is the charge on the electron, h is Planck's constant, and m_e is the mass of the electron. This moment is directed along the axis of electron spin.

The nucleus of the atom consists of protons and neutrons. These also have a spin; however, the overall magnetic moment due to their spin is much smaller than that for electrons. We normally do not encounter the effects of a magnetic moment of a nucleus with the exception of such applications as nuclear magnetic resonance (NMR).

We can view electrons in materials as small elementary magnets. If the magnetic moments due to electrons in materials could line up in the same direction, the world would be a magnetic place! However this, as you know, is not the case. Thus, there must be some mechanism by which the magnetic moments associated with electron spin and their orbital motion get canceled in most materials, leaving behind only a few materials that are "magnetic." There are two effects that, fortunately, make most materials in the world not "magnetic."

First, we must consider the magnetic moment of atoms. According to the Pauli exclusion principle, two electrons within the same orbital must have opposite spins. This means their electron spin derived magnetic moments have opposite signs (one can be considered "up ↑" and the other one "down ↓") and cancel. The second effect is that the orbital moments of electrons also cancel each other. Thus, in a completely filled shell, all electron spin and orbital moments cancel. This is why atoms of most elements do not have a net magnetic moment. Some elements, such as transition elements ($3d$, $4d$, $5d$ partially filled), the lanthanides ($4f$ partially filled), and actinides ($5f$ partially filled), have a net magnetic moment due to an unpaired electron.

Certain elements, such as the transition metals, have an inner energy level that is not completely filled. The elements scandium (Sc) through copper (Cu), the electronic structures of which are shown in Table 20-1, are typical. Except for chromium and copper, the valence electrons in the $4s$ level are paired; the unpaired electrons in chromium and copper are canceled by interactions with other atoms. Copper also has a completely filled $3d$ shell and thus does not display a net magnetic moment.

The electrons in the $3d$ level of the remaining transition elements do not enter the shells in pairs. Instead, as in manganese, the first five electrons have the same spin. Only after half of the $3d$ level is filled do pairs with opposing spins form. Therefore, each atom in a transition metal has a permanent magnetic moment, which is related to the number of unpaired electrons. Each atom behaves as a magnetic dipole.

In many elements, these magnetic moments exist for free individual atoms, however, when the atoms form crystalline materials, these moments are "quenched" or canceled out. Thus, a number of materials made from elements with atoms that have a net magnetic moment do not exhibit magnetic behavior. For example, the Fe^{+2} ion has a net magnetic moment of $4\mu_B$ (four times the magnetic moment of an electron); however, $FeCl_2$ crystals are not magnetic.

TABLE 20-1 ■ *The electron spins in the 3d energy level in transition metals with arrows indicating the direction of spin*

Metal	3d					4s
Sc	↑					↑↓
Ti	↑	↑				↑↓
V	↑	↑	↑			↑↓
Cr	↑	↑	↑	↑	↑	↑
Mn	↑	↑	↑	↑	↑	↑↓
Fe	↑↓	↑	↑	↑	↑	↑↓
Co	↑↓	↑↓	↑	↑	↑	↑↓
Ni	↑↓	↑↓	↑↓	↑	↑	↑↓
Cu	↑↓	↑↓	↑↓	↑↓	↑↓	↑

The response of the atom to an applied magnetic field depends on how the magnetic dipoles of each atom react to the field. Most of the transition elements (e.g., Cu, Ti) react in such a way that the sum of the individual atoms' magnetic moments is zero. The atoms in nickel (Ni), iron (Fe), and cobalt (Co), however, undergo an exchange interaction, whereby the orientation of the dipole in one atom influences the surrounding atoms to have the same dipole orientation, producing a desirable amplification of the effect of the magnetic field. In the case of Fe, Ni, and Co, the magnetic moments of the atoms line up in the same directions, and these materials are known as ferromagnetic.

In certain materials, such as BCC chromium (Cr), the magnetic moments of atoms at the center of the unit cell are opposite in direction to those of the atoms at the corners of the unit cell; thus, the net moment is zero. Materials in which there is a complete cancellation of the magnetic moments of atoms or ions are known as anti-ferromagnetic.

Materials in which magnetic moments of different atoms or ions do not completely cancel out are known as ferrimagnetic materials. We will discuss these materials in a later section.

20-3   Magnetization, Permeability, and the Magnetic Field

Let's examine the relationship between the magnetic field and magnetization. Figure 20-2 depicts a coil having n turns. When an electric current is passed through the coil, a magnetic field H is produced, with the strength of the field given by

$$H = \frac{nI}{l} \qquad (20\text{-}2)$$

where n is the number of turns, l is the length of the coil (m), and I is the current (A). The units of H are therefore ampere turn/m, or simply A/m. An alternate unit for magnetic field is the oersted, obtained by multiplying A/m by $4\pi \times 10^{-3}$ (see Table 20-2).

When a magnetic field is applied in a vacuum, lines of magnetic flux are induced. The number of lines of flux, called the flux density, or *inductance B*, is related to the applied field by

$$B = \mu_0 H \qquad (20\text{-}3)$$

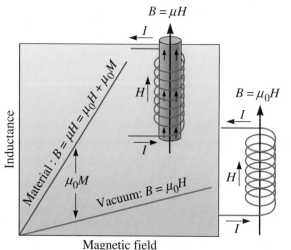

Figure 20-2
A current passing through a coil sets up a magnetic field H with a flux density B. The flux density is higher when a magnetic core is placed within the coil.

TABLE 20-2 ■ *Units, conversions, and values for magnetic materials*

	Gaussian and cgs emu (Electromagnetic Units)	SI Units	Conversion
Inductance or magnetic flux density (B)	gauss (G)	Tesla [or weber (Wb)/m^2]	1 tesla = 10^4 G, Wb/m^2
Magnetic flux (ϕ)	maxwell (Mx), G · cm^2	Wb, volt · second	1 Wb = 10^8 G-cm^2
Magnetic potential difference or magnetic electromotive force (U, F)	gilbert (Gb)	ampere (A)	1 A = $4\pi \times 10^{-1}$ Gb
Magnetic field strength, magnetizing force (H)	oersted (Oe), gilbert (Gb)/cm	A/m	1 A/m = $4\pi \times 10^{-3}$ Oe
(Volume) magnetization (M)	emu/cm^3	A/m	1 A/m = 10^{-3} emu/cm^3
(Volume) magnetization ($4\pi M$)	G	A/m	1 A/m = $4\pi \times 10^{-3}$ G
Magnetic polarization or intensity of magnetization (J or I)	emu/cm^3	T, Wb/m^2	1 tesla = $(1/4\pi) \times 10^4$ emu/cm^3
(Mass) magnetization (σ, M)	emu/g	A · m^2/kg Wb-m/kg	1 A·m^2/kg = 1 emu/g 1 Wb·m/kg = $(1/4\pi) \times 10^7$ emu/g
Magnetic moment (m)	emu, erg/G	A · m^2, Joules per tesla (J/T)	1 J/T = 10^3 emu
Magnetic dipole moment (j)	emu, erg/G	Wb · m	1 Wb·m = $(1/4\pi) \times 10^{10}$ emu
Magnetic permeability (μ)	Dimensionless	Wb/(A · m) [henry (H)/m]	1 Wb/(A·m) = $(1/4\pi) \times 10^7$
Magnetic permeability of free space (μ_0)	1 gauss/oersted	$\mu_0 = 4\pi \times 10^{-7}$ H/m	
Relative permeability (μ_r)	Not defined	Dimensionless	
(Volume) energy density, energy product (W)	erg/cm^3	J/m^3	1 J/m^3 = 10 erg/cm^3

where B is the inductance, H is the magnetic field, and μ_0 is a constant called the **magnetic permeability of vacuum**. If H is expressed in units of oersted, then B is in gauss and μ_0 is 1 gauss/oersted. In an alternate set of units, H is in A/m, B is in tesla (also called weber/m^2), and μ_0 is $4\pi \times 10^{-7}$ weber/(A · m) (also called henry/m).

When we place a material within the magnetic field, the magnetic-flux density is determined by the manner in which induced and permanent magnetic dipoles interact with the field. The flux density now is

$$B = \mu H \tag{20-4}$$

where μ is the permeability of the material in the field. If the magnetic moments reinforce the applied field, then $\mu > \mu_0$, a greater number of lines of flux that can accomplish work are created, and the magnetic field is magnified. If the magnetic moments oppose the field, however, $\mu < \mu_0$.

We can describe the influence of the magnetic material by the relative permeability μ_r, where

$$\mu_r = \frac{\mu}{\mu_0} \tag{20-5}$$

A large relative permeability means that the material amplifies the effect of the magnetic field. Thus, the relative permeability has the same importance that conductivity has in dielectrics. A material with higher magnetic permeability (e.g., iron) will carry magnetic flux more readily. We will learn later that the permeability of ferromagnetic or ferrimagnetic materials is not constant and depends on the value of the applied magnetic field (H).

The **magnetization** M represents the increase in the inductance due to the core material, so we can rewrite the equation for inductance as

$$B = \mu_0 H + \mu_0 M \tag{20-6}$$

The first part of this equation is simply the effect of the applied magnetic field. The second part is the effect of the magnetic material that is present. This is similar to our discussion on dielectric polarization and the mechanical behavior of materials. In materials, stress causes strain, electric field (E) induces dielectric polarization (P), and a magnetic field (H) causes magnetization ($\mu_0 M$) that contributes to the total flux density B.

The **magnetic susceptibility** χ_m, which is the ratio between magnetization and the applied field, gives the amplification produced by the material:

$$\chi_m = \frac{M}{H} \tag{20-7}$$

Both μ_r and χ_m refer to the degree to which the material enhances the magnetic field and are therefore related by

$$\mu_r = 1 + \chi_m \tag{20-8}$$

As noted before, the μ_r and, therefore, the χ_m values for ferromagnetic and ferrimagnetic materials depend on the applied field (H). For ferromagnetic and ferrimagnetic materials, the term $\mu_0 M \gg \mu_0 H$. Thus, for these materials,

$$B \cong \mu_0 M \tag{20-9}$$

We sometimes interchangeably refer to either inductance or magnetization. Normally, we are interested in producing a high inductance B or magnetization M. This is accomplished by selecting materials that have a high relative permeability or magnetic susceptibility.

The following example shows how these concepts can be applied for comparing actual and theoretical magnetizations in pure iron.

Example 20-1   *Theoretical and Actual Saturation Magnetization in Fe*

Calculate the maximum, or saturation, magnetization that we expect in iron. The lattice parameter of BCC iron is 2.866 Å. Compare this value with 2.1 tesla (a value of saturation flux density experimentally observed for pure Fe).

SOLUTION

Based on the unpaired electronic spins, we expect each iron atom to have four electrons that act as magnetic dipoles. The number of atoms per m^3 in BCC iron is

$$\text{Number of Fe atoms/m}^3 = \frac{2 \text{ atoms/cell}}{(2.866 \times 10^{-10} \text{m})^3} = 8.496 \times 10^{28}$$

The maximum volume magnetization (M_{sat}) is the total magnetic moment per unit volume:

$$M_{sat} = \left(8.496 \times 10^{28} \frac{\text{atoms}}{\text{m}^3} \right)(9.274 \times 10^{-24} \text{ A} \cdot \text{m}^2)\left(4 \frac{\text{Bohr magnetons}}{\text{atom}} \right)$$

$$M_{sat} = 3.15 \times 10^6 \frac{\text{A}}{\text{m}}$$

To convert the value of saturation magnetization M into saturation flux density B in tesla, we need the value of $\mu_0 M$. In ferromagnetic materials $\mu_0 M \gg \mu_0 H$ and therefore, $B \cong \mu_0 M$.

Thus, the saturation induction or saturation flux density in tesla is given by $B_{sat} = \mu_0 M_{sat}$.

$$B_{sat} = \left(4\pi \times 10^{-7} \frac{\text{Wb}}{\text{A} \cdot \text{m}} \right)\left(3.15 \times 10^6 \frac{\text{A}}{\text{m}} \right)$$

$$B_{sat} = 3.96 \frac{\text{Wb}}{\text{m}^2} = 3.96 \text{ tesla}$$

This is almost two times the experimentally observed value of 2.1 tesla. Reversing our calculations, we can show that the each iron atom contributes only about 2.1 Bohr magneton and not 4. This is the difference between behavior of individual atoms and their behavior in a crystalline solid. It can be shown that in the case of iron, the difference is due to the $3d$ electron orbital moment being quenched in the crystal.

20-4 Diamagnetic, Paramagnetic, Ferromagnetic, Ferrimagnetic, and Superparamagnetic Materials

As mentioned before, there is no such thing as a "nonmagnetic" material. All materials respond to magnetic fields. When a magnetic field is applied to a material, several types of behavior are observed (Figure 20-3).

Diamagnetic Behavior A magnetic field acting on any atom induces a magnetic dipole for the entire atom by influencing the magnetic moment caused by the orbiting electrons. These dipoles oppose the magnetic field, causing the magnetization to be less than zero. This behavior, called **diamagnetism**, gives a relative permeability of about 0.99995 (or a negative susceptibility approximately -10^{-6}, note the negative sign). Materials such as copper, silver, silicon, gold, and alumina are diamagnetic at room temperature. Superconductors are perfect diamagnets ($\chi_m = -1$); they lose their superconductivity at higher temperatures or in the presence of a magnetic field. In a diamagnetic material, the magnetization (M) direction is opposite to the direction of applied field (H).

Paramagnetism When materials have unpaired electrons, a net magnetic moment due to electron spin is associated with each atom. When a magnetic field is applied,

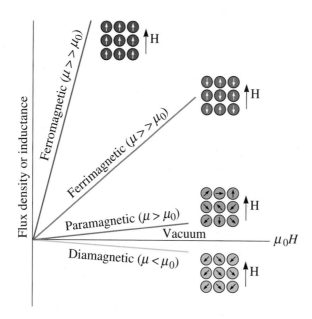

Figure 20-3
The effect of the core material on the flux density. The magnetic moment opposes the field in diamagnetic materials. Progressively stronger moments are present in paramagnetic, ferrimagnetic, and ferromagnetic materials for the same applied field.

the dipoles align with the field, causing a positive magnetization. Because the dipoles do not interact, extremely large magnetic fields are required to align all of the dipoles. In addition, the effect is lost as soon as the magnetic field is removed. This effect, called **paramagnetism**, is found in metals such as aluminum, titanium, and alloys of copper. The magnetic susceptibility (χ_m) of paramagnetic materials is positive and lies between 10^{-4} and 10^{-5}. Ferromagnetic and ferrimagnetic materials above the Curie temperature also exhibit paramagnetic behavior.

Ferromagnetism   Ferromagnetic behavior is caused by the unfilled energy levels in the $3d$ level of iron, nickel, and cobalt. Similar behavior is found in a few other materials, including gadolinium (Gd). In ferromagnetic materials, the permanent unpaired dipoles easily line up with the imposed magnetic field due to the exchange interaction, or mutual reinforcement of the dipoles. Large magnetizations are obtained even for small magnetic fields, giving large susceptibilities approaching 10^6. Similar to ferroelectrics, the susceptibility of ferromagnetic materials depends upon the intensity of the applied magnetic field. This is similar to the mechanical behavior of elastomers with the modulus of elasticity depending upon the level of strain. Above the Curie temperature, ferromagnetic materials behave as paramagnetic materials and their susceptibility is given by the following equation, known as the Curie-Weiss law:

$$\chi_m = \frac{C}{(T - T_c)} \tag{20-10}$$

In this equation, C is a constant that depends upon the material, T_c is the Curie temperature, and T is the temperature above T_c. Essentially, the same equation also describes the change in dielectric permittivity above the Curie temperature of ferroelectrics. Similar to ferroelectrics, ferromagnetic materials show the formation of hystereis loop domains and magnetic domains. These materials will be discussed in the next section.

Antiferromagnetism   In materials such as manganese, chromium, MnO, and NiO, the magnetic moments produced in neighboring dipoles line up in

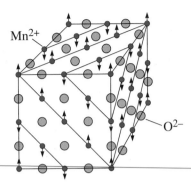

Figure 20-4
The crystal structure of MnO consists of alternating layers of {111} type planes of oxygen and manganese ions. The magnetic moments of the manganese ions in every other (111) plane are oppositely aligned. Consequently, MnO is antiferromagnetic.

opposition to one another in the magnetic field, even though the strength of each dipole is very high. This effect is illustrated for MnO in Figure 20-4. These materials are **antiferromagnetic** and have zero magnetization. The magnetic susceptibility is positive and small. In addition, CoO and $MnCl_2$ are examples of antiferromagnetic materials.

Ferrimagnetism

In ceramic materials, different ions have different magnetic moments. In a magnetic field, the dipoles of cation A may line up with the field, while dipoles of cation B oppose the field. Because the strength or number of dipoles is not equal, a net magnetization results. The **ferrimagnetic** materials can provide good amplification of the imposed field. We will look at a group of ceramics called ferrites that display this behavior in a later section. These materials show a large, magnetic-field dependent magnetic susceptibility similar to ferromagnetic materials. They also show Curie-Weiss behavior (similar to ferromagnetic materials) at temperatures above the Curie temperature. Most ferrimagnetic materials are ceramics and are good insulators of electricity. Thus, in these materials, electrical losses (known as eddy current losses) are much smaller compared to those in metallic ferromagnetic materials. Therefore, ferrites are used in many high-frequency applications.

Superparamagnetism

When the grain size of ferromagnetic and ferrimagnetic materials falls below a certain critical size, these materials behave as if they are paramagnetic. The magnetic dipole energy of each particle becomes comparable to the thermal energy. This small magnetic moment changes its direction randomly (as a result of the thermal energy). Thus, the material behaves as if it has no net magnetic moment. This is known as superparamagnetism. Thus, if we produce iron oxide (Fe_3O_4) particles in a 3 to 5 nm size, they behave as superparamagnetic materials. Such iron-oxide superparamagnetic particles are used to form dispersions in aqueous or organic carrier phases or to form "liquid magnets" or ferrofluids. The particles in the fluid move in response to a gradient in the magnetic field. Since the particles form a stable sol, the entire dispersion moves and, hence, the material behaves as a liquid magnet. Such materials are used as seals in computer hard drives and in loudspeakers as heat transfer (cooling) media. The permanent magnet used in the loudspeaker holds the liquid magnets in place. Superparamagnetic particles of iron oxide (Fe_3O_4) also can be coated with different chemicals and used to separate DNA molecules, proteins, and cells from other molecules.

The following example illustrates how to select a material for a given application.

Example 20-2   *Design/Materials Selection for a Solenoid*

We want to produce a solenoid coil that produces an inductance of at least 2000 gauss when a 10 mA current flows through the conductor. Due to space limitations, the coil should be composed of 10 turns over a 1 cm length. Select a core material for the coil. Refer to Table 20-4.

SOLUTION

First, we can determine the magnetic field H produced by the coil. From Equation 20-2,

$$H = \frac{nI}{l} = \frac{(10)(0.01\ \text{A})}{0.01\ \text{m}} = 10\ \text{A/m}$$

$$H = (10\ \text{A/m})[4\pi \times 10^{-3}\ \text{oersted/(A/m)}] = 0.12566\ \text{oersted}$$

If the inductance B must be at least 2000 gauss, then the permeability of the core material must be

$$\mu = \frac{B}{H} = \frac{2000}{0.12566} = 15{,}916\ \text{gauss/oersted}$$

The relative permeability of the core material must be at least

$$\mu_r = \frac{\mu}{\mu_0} = \frac{15{,}916}{1} = 15{,}916$$

If we examine the magnetic materials listed in Table 20-4, we find that 4750 alloy has a maximum relative permeability of 80,000 and might be a good selection for the core material.

20-5   Domain Structure and the Hysteresis Loop

From a phenomenological viewpoint, ferromagnetic materials are similar to ferroelectrics. A single crystal of iron or a polycrystalline piece of low-carbon steel is ferromagnetic; however, these materials ordinarily do not show a net magnetization. Within the single crystal or polycrystalline structure of a ferromagnetic or ferrimagnetic material, a substructure composed of magnetic domains is produced, even in the absence of an external field. This spontaneously happens because the presence of many domains in the material, arranged so that the net magnetization is zero, minimizes the magnetostatic energy. **Domains** are regions in the material in which all of the dipoles are aligned in a certain direction. In a material that has never been exposed to a magnetic field, the individual domains have a random orientation. Because of this, the net magnetization in the virgin ferromagnetic or ferrimagnetic material as a whole is zero [Figure 20-5(a)]. Similar to ferroelectrics, application of a magnetic field (poling) will coerce many of the magnetic domains to align with the magnetic field direction.

Boundaries, called **Bloch walls**, separate the individual magnetic domains. The Bloch walls are narrow zones in which the direction of the magnetic moment gradually and continuously changes from that of one domain to that of the next [Figure 20-5(b)]. The domains are typically very small, about 0.005 cm or less, while the Bloch walls are about 100 nm thick.

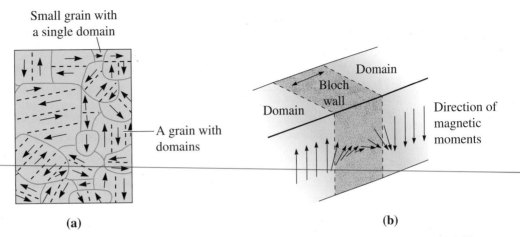

Small grain with
a single domain

A grain with
domains

Domain

Bloch
wall

Domain

Direction of
magnetic
moments

(a)

(b)

Figure 20-5 (a) A qualitative sketch of magnetic domains in a polycrystalline material. The
dashed lines show demarcation between different magnetic domains; the dark curves show the
grain boundaries. (b) The magnetic moments change direction continuously across the
boundary between domains.

Movement of Domains in a Magnetic Field

When a
magnetic field is imposed on the material, domains that are nearly lined up with the
field grow at the expense of unaligned domains. In order for the domains to grow, the
Bloch walls must move; the field provides the force required for this movement.
Initially, the domains grow with difficulty, and relatively large increases in the field are
required to produce even a little magnetization. This condition is indicated in Figure
20-6 by a shallow slope, which is the initial permeability of the material. As the field
increases in strength, favorably oriented domains grow more easily, with permeability
increasing as well. A maximum permeability can be defined as shown in the figure.
Eventually, the unfavorably oriented domains disappear, and rotation completes the
alignment of the domains with the field. The **saturation magnetization**, produced when all

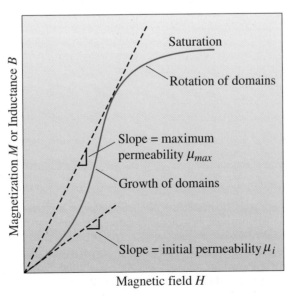

Saturation

Rotation of domains

Slope = maximum
permeability μ_{max}

Growth of domains

Slope = initial permeability μ_i

Magnetization M or Inductance B

Magnetic field H

Figure 20-6
When a magnetic field is first applied
to a magnetic material,
magnetization initially increases
slowly, then more rapidly as the
domains begin to grow. Later,
magnetization slows, as domains
must eventually rotate to reach
saturation. Notice the permeability
values depend upon the magnitude
of H.

of the domains are oriented along with the magnetic field, is the greatest amount of magnetization that the material can obtain. Under these conditions, the permeability of these materials becomes quite small.

Effect of Removing the Field

When the field is removed, the resistance offered by the domain walls prevents regrowth of the domains into random orientations. As a result, many of the domains remain oriented near the direction of the original field and a residual magnetization, known as the **remanance** (M_r) is present in the material. The value of B_r (usually in Tesla) is known as the retentivity of the magnetic material. The material acts as a permanent magnet. Figure 20-7(a) shows this effect in the magnetization-field curve. Notice that the M-H loop shows saturation, but the B-H loop does not. The magnetic field needed to bring the induced magnetization to zero is the **coercivity** of the material. This is a microstructure-sensitive property.

For magnetic recording materials, Fe, γ-Fe_2O_3, Fe_3O_4, and needle-shaped CrO_2 particles are used. The elongated shape of magnetic particles leads to higher coercivity (H_c). The dependence of coercivity on the shape of a particle or grain is known as **magnetic shape anisotropy**. The coercivity of recording materials needs to be smaller than that for permanent magnets since data written onto a magnetic data storage medium should be erasable. On the other hand, the coercivity values should be higher than soft magnetic materials since we want to retain the information stored. Such materials are described as magnetically semi-hard.

Effect of Reversing the Field

If we now apply a field in the reverse direction, the domains grow with an alignment in the opposite direction. A coercive field H_c (or coercivity) is required to force the domains to be randomly oriented and cancel

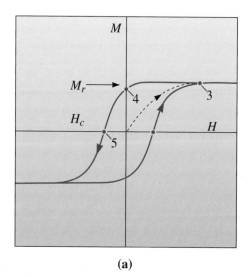
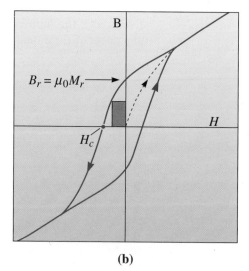

(a)   (b)

Figure 20-7   (a) The ferromagnetic hysteresis M-H loop showing the effect of the magnetic field on inductance or magnetization. The dipole alignment leads to saturation magnetization (point 3), a remanance (point 4), and a coercive field (point 5). (b) The corresponding B-H loop. Notice the B value does not saturate since $B = \mu_0 H + \mu_0 M$. (*Adapted from Permanent Magnetism, by R. Skomski and J.M.D. Coey, p. 3, Fig. 1-1. Edited by J.M.D. Coey and D.R. Tilley. Copyright © 1999 Institute of Physics Publishing. Adapted by permission.*)

one another's effect. Further increases in the strength of the field eventually align the domains to saturation in the opposite direction.

As the field continually alternates, the magnetization versus field relationship traces out a **hysteresis loop**. The hysteresis loop is shown as both *B-H* and *M-H* plots. The area contained within the hysteresis loop is related to the energy consumed during one cycle of the alternating field. The shaded area shown in Figure 20-7(b) is the largest *B-H* product and is known as the power of the magnetic material.

20-6   The Curie Temperature

When the temperature of a ferromagnetic or ferrimagnetic material is increased, the added thermal energy increases the mobility of the domains, making it easier for them to become aligned, but also preventing them from remaining aligned when the field is removed. Consequently, saturation magnetization, remanance, and the coercive field are all reduced at high temperatures (Figure 20-8). If the temperature exceeds the **Curie temperature** (T_c), ferromagnetic or ferrimagnetic behavior is no longer observed. Instead, the material behaves as a paramagnetic material. The Curie temperature (Table 20-3), which depends on the material, can be changed by alloying elements. French scientists Marie and Pierre Curie (the only husband and wife to win a Nobel prize; Marie Curie actually won two Nobel prizes) performed research on magnets, and the Curie temperature refers to their name. The dipoles still can be aligned in a magnetic field above the Curie temperature, but they become randomly aligned when the field is removed.

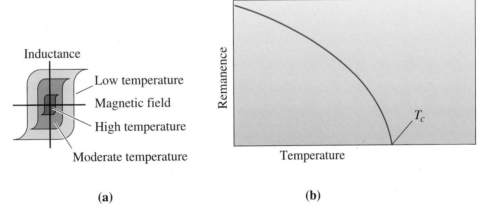

(a)   **(b)**

Figure 20-8   The effect of temperature on (a) the hysteresis loop and (b) the remanance. Ferromagnetic behavior disappears above the Curie temperature.

TABLE 20-3 ■ *Curie temperatures for selected materials*

Material	Curie Temperature (°C)	Material	Curie Temperature (°C)
Gadolinium	16	Iron	771
$Nd_2Fe_{12}B$	312	Alnico 1	780
Nickel	358	Cunico	855
$BaO \cdot 6Fe_2O_3$	469	Alnico 5	900
Co_5Sm	747	Cobalt	1117

20-7   Applications of Magnetic Materials

Ferromagnetic and ferrimagnetic materials are classified as magnetically soft or magnetically hard depending upon the shape of the hysteresis loop [Figure 20-9(a)]. Generally, if the coercivity value is $\sim > 10^4 \, A \cdot m^{-1}$, we consider the material as magnetically hard. If the coercivity values are less than $10^3 \, A \cdot m^{-1}$, we consider the materials as magnetically soft. Figure 20-9(b) shows classification of different commercially important magnetic materials. Note that while the coercivity is a strongly *microstructure-sensitive* property, the saturation magnetization is constant (i.e., it is not microstructure dependent) for a material of a given composition. This is similar to the way the yield strength of metallic materials is strongly dependent on the

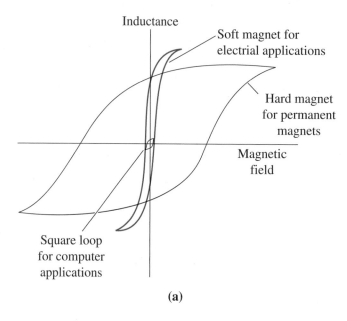

(a)

Figure 20-9
(a) Comparison of the hysteresis loops for three applications of ferromagnetic and ferrimagnetic materials. (b) Saturation magnetization and coercivity values for different magnetic materials. (*Adapted from "Magnetic Materials: An Overview, Basic Concepts, Magnetic Measurements, Magnetostrictive Materials," by G.Y. Chin et al. In D. Bloor, M. Flemings, and S. Mahajan (Eds.), Encyclopedia of Advanced Materials, Vol. 1, 1994, p. 1424, Table 1. Copyright © 1994 Pergamon Press. Reprinted with permission of the editor.*)

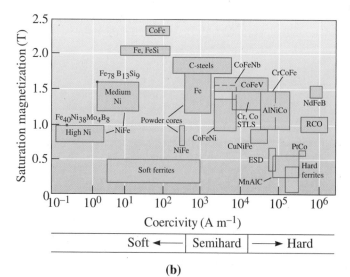

(b)

microstructure, while the Young's modulus is not. Many factors, such as the structure of grain boundaries and the presence of pores or surface layers on particles, affect the coercivity values. The coercivity of single crystals depends strongly on crystallographic directions. There are certain directions along which it is easy to align the magnetic domains. There are other directions along which the coercivity is much higher. Coercivity of magnetic particles also depends upon shape of the particles. This is why in magnetic recording media we use acicular and not spherical particles. This effect is also used in Fe-Si steels, which are textured or grain oriented so as to minimize energy losses during the operation of an electrical transformer.

Let's look at some applications for magnetic materials.

Soft Magnetic Materials

Ferromagnetic materials are often used to enhance the magnetic flux density (B) produced when an electric current is passed through the material. The magnetic field is then expected to do work. Applications include cores for electromagnets, electric motors, transformers, generators, and other electrical equipment. Because these devices utilize an alternating field, the core material is continually cycled through the hysteresis loop. Table 20-4 shows the properties of selected soft, magnetic materials. *Note that in these materials the value of relative magnetic permeability depends strongly on the strength of the applied field* (Figure 20-6).

These materials often have the following characteristics:

1. High-saturation magnetization.
2. High permeability.
3. Small coercive field.

TABLE 20-4 ■ *Properties of selected soft magnetic materials*

Name	Composition	Permeability μ_r Initial	Permeability μ_r Maximum	Coercivity H_c(A · m^{-1})	Retentivity B_r (T)	B_{max} (T)	Resistivity ($\mu\Omega$ · m)
Ingot Iron	99.8% Fe	150	5000	80	0.77	2.14	0.10
Low-carbon steel	99.5% Fe	200	4000	100		2.14	1.12
Silicon iron, unoriented	Fe-3% Si	270	8000	60		2.01	0.47
Silicon iron, grain-oriented	Fe-3% Si	1400	50,000	7	1.20	2.01	0.50
4750 alloy	Fe-48% Ni	11,000	80,000	2		1.55	0.48
4-79 permalloy	Fe-4% Mo-79% Ni	40,000	200,000	1		0.80	0.58
Superalloy	Fe-5% Mo-80% Ni	80,000	450,000	0.4		0.78	0.65
2V-Permendur	Fe-2% V-49% Co	800	450,000	0.4		0.78	0.65
Supermendur	Fe-2% V-49% Co		100,000	16	2.00	2.30	0.40
Metglas[a] 2650SC	$Fe_{81}B_{13.5}Si_{3.5}C_2$		300,000	3	1.46	1.61	1.35
Metglas[a] 2650S-2	$Be_{78}B_{13}S_9$		600,000	2	1.35	1.56	1.37
MnZn Ferrite	H5C2[b]	10,000		7	0.09	0.40	1.5×10^5
MnZn Ferrite	H5E[b]	18,000		3	0.12	0.44	5×10^4
NiZn Ferrite	K5[b]	290		80	0.25	0.33	2×10^{12}

[a]*Allied Corporation trademark.*
[b]*TDK ferrite code.*
(*Adapted from "Magnetic Materials: An Overview, Basic Concepts, Magnetic Measurements, Magnetostrictive Materials," by G.Y. Chin et al. In R. Bloor, M. Flemings, and S. Mahajan (Eds.),* Encyclopedia of Advanced Materials, *Vol. 1, 1994, p. 1424, Table 1. Copyright © 1994 Pergamon Press. Reprinted with permission of the editor.*)

4. Small remanance.
5. Small hysteresis loop.
6. Rapid response to high-frequency magnetic fields.
7. High electrical resistivity.

High saturation magnetization permits a material to do work, while high permeability permits saturation magnetization to be obtained with small imposed magnetic fields. A small coercive field also indicates that domains can be reoriented with small magnetic fields. A small remanance is desired so that almost no magnetization remains when the external field is removed. These characteristics also lead to a small hysteresis loop, therefore minimizing energy losses during operation.

If the frequency of the applied field is so high that the domains cannot be realigned in each cycle, the device may heat due to dipole friction. In addition, higher frequencies naturally produce more heating because the material cycles through the hysteresis loop more often, losing energy during each cycle. For high frequency applications, materials must permit the dipoles to be aligned at exceptionally rapid rates.

Energy can also be lost by heating if eddy currents are produced. During operation, electrical currents can be induced into the magnetic material. These currents produce power losses and Joule, or I^2R, heating. Eddy current losses are particularly severe when the material operates at high frequencies. If the electrical resistivity is high, eddy current losses can be held to a minimum. Soft magnets produced from ferrimagnetic ceramic materials have a high resistivity and therefore are less likely to heat than metallic ferromagnetic materials. Recently, a class of smart materials, known as magnetorheological or MR fluids based on soft magnetic carbonyl iron (Fe) particles, has been introduced in various applications related to vibration control, such as Delphi's MagneRide™ system. These materials are like magnetic paints and can be made to absorb energy from shocks and vibrations by turning on a magnetic field. The stiffening of MR fluids is controllable and reversible. Some of the models of Cadillac and Corvette offer a suspension based on these smart materials.

Data Storage Materials

Magnetic materials are used for data storage. Memory is stored by magnetizing the material in a certain direction. For example, if the "north" pole is up, the bit of information stored is 1. If the "north" pole is down, then a 0 is stored.

For this application, materials with a square hysteresis loop, a low remanance, a low saturation magnetization, and a low coercive field are preferable. Hard ferrites based on Ba, CrO_2, acicular iron particles, and γ-Fe_2O_3 satisfy these requirements. The stripe on credit cards and bank machine cards are made using γ-Fe_2O_3 or Fe_3O_4 particles. The square loop ensures that a bit of information placed in the material by a field remains stored; a steep and abrupt change in magnetization is required to remove the information from storage in the ferromagnet. Furthermore, the magnetization produced by small external fields keeps the coercive field (H_c), saturation magnetization, and remanance (B_r) low.

The B_r and H_c values of some typical magnetic recording materials are shown in Table 20-5.

Many new alloys based on Co-Pt-Ta-Cr have been developed for the manufacture of hard disks. Computer hard disks are made using sputtered thin films of these materials. As discussed in earlier chapters, many different alloys, such as those based on nanostructured Fe-Pt and Fe-Pd, are being developed for data storage applications. More recently, a technology known as *spintronics* (*spin-based electronics*) has evolved. In spintronics, the main idea is to make use of the spin of electrons as a way of affecting the flow of electrical

TABLE 20-5 ■ *Properties of typical magnetic recording materials in a powder form*

	Particle Length (μm)	Aspect Ratio	Magnetization B_r (Wb/m^2)	(emu/cm^3)	Coercivity H_c (kA/m)	(Oe)	Surface Area (m^2/g)	Curie Temp. T_c(°C)
γ-Fe$_2$O$_3$	0.20	5:1	0.44	350	22–34	420	15–30	600
Co-γ-Fe$_2$O$_3$	0.20	6:1	0.48	380	30–75	940	20–35	700
CrO$_2$	0.20	10:1	0.50	400	30–75	950	18–55	125
Fe	0.15	10:1	1.40[a]	1100[a]	56–176	2200	20–60	770
Barium Ferrite	0.05	0.02 μm thick	0.40	320	56–240	3000	20–25	350

[a]*For overcoated, stable particles use only 50 to 80% of these values due to reduced magnetic particle volume (From* The Complete Handbook of Magnetic Recording, Fourth Edition, *by F. Jorgensen, p. 324, Table 11-1. Copyright © 1996. The McGraw-Hill Companies. Reprinted by permission of The McGraw-Hill Companies.)*

current (known as spin-polarized current) to make devices such as field effect transistors (FET). The spin of the electrons (up or down) is also being considered as a way of storing information. A very successful example of a real-world spintronic-based device is a giant magnetoresistance (GMR) sensor that is used for reading information from computer hard disks.

Permanent Magnets

Finally, magnetic materials are used to make strong permanent magnets (Table 20-6). Strong permanent magnets, often called hard magnets, require the following:

1. High remanance (stable domains).
2. High permeability.
3. High coercive field.
4. Large hysteresis loop.
5. High power (or BH product).

The *record* for any energy product is obtained for Nd$_2$Fe$_{14}$B magnets with an energy product of ~445 kJ · m^{-3} [~56 Mega-Gauss-Oersteds (MGOe)]. These magnets are made in the form of a powder by the rapid solidification of a molten alloy. Powders are either bonded in a polymer matrix or by hot pressing, producing bulk materials. The

TABLE 20-6 ■ *Properties of selected hard, or permanent, or magnetic materials*

Material	Common Name	$\mu_0 M_r$ (T)	$\mu_0 H_c$ (T)	$(BH)_{max}$ (kJ · m^{-3})	T_c (°C)
Fe-Co	Co-steel	1.07	0.02	6	887
Fe-Co-Al-Ni	Alnico-5	1.05	0.06	44	880
BaFe$_{12}$O$_{19}$	Ferrite	0.42	0.31	34	469
SmCo$_5$	Sm-Co	0.87	0.80	144	723
Nd$_2$Fe$_{14}$B	Nd-Fe-B	1.23	1.21	290–445	312

(Adapted from Permanent Magnetism, *by R. Skomski and J.M.D. Coey, p. 23, Table 1-2. Edited by J.M.D. Coey and D.R. Tilley. Copyright © 1999 Institute of Physics Publishing. Adapted by permission.)*

energy product increases when the sintered magnet is "oriented" or poled. Corrosion resistance, brittleness, and a relatively low Curie temperature of ~312°C are some of the limiting factors of this extraordinary material.

The **power** of the magnet is related to the size of the hysteresis loop, or the maximum product of *B* and *H*. The area of the largest rectangle that can be drawn in the second or fourth quadrants of the *B-H* curve is related to the energy required to demagnetize the magnet [Figure 20-10(a) and Figure 20-10(b)]. For the product to be large, both the remanance and the coercive field should be large.

In many applications, we need to calculate the lifting power of a permanent magnet. The magnetic force obtainable using a permanent magnet is given by

$$F = \frac{\mu_0 M^2 A}{2} \tag{20-11}$$

In this equation *A* is the cross-sectional area of the magnet, *M* is the magnetization, and μ_0 is the magnetic permeability of free space.

One of the most successful examples of the contributions by materials scientists and engineers in this area is the development of strong rare earth magnets. The progress made in the development of strong permanent magnets is illustrated in Figure 20-10(b). Permanent magnets are used in many applications including loudspeakers, motors, generators, holding magnets, mineral separation, and bearings. Typically, they offer a nonuniform magnetic field; however, it is possible to use geometric arrangements known as Halbach arrays to produce relatively uniform magnetic fields. The following examples illustrate applications of some of these concepts related to permanent magnetic materials.

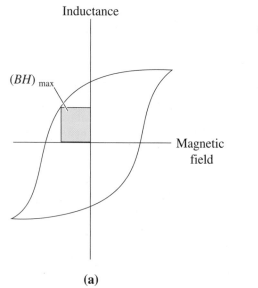

(a)

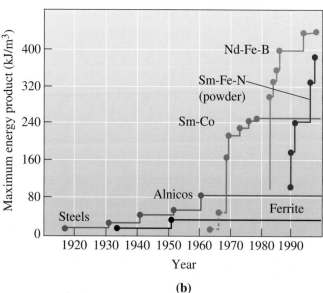

(b)

Figure 20-10   (a) The largest rectangle drawn in the second or fourth quadrant of the *B-H* curve gives the maximum *BH* product. (*BH*)$_{max}$ is related to the power, or energy, required to demagnetize the permanent magnet. (b) Development of permanent magnet materials. The maximum energy product is shown on the vertical axis. (*Adapted from* Permanent Magnetism, *by R. Skomski and J.M.D. Coey, p. 25, Fig. 1-15. Edited by J.M.D. Coey and D.R. Tilley. Copyright © 1999 Institute of Physics Publishing. Adapted by permission.*)

| **Example 20-3** | *Energy Product for Permanent Magnets* |

Determine the power, or *BH* product, for the magnetic material with the properties shown in Figure 20-11.

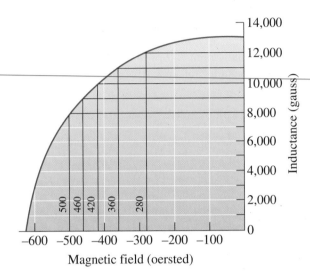

Figure 20-11
The fourth quadrant of the *B-H* curve for a permanent magnetic material (for Example 20-3).

Magnetic field (oersted)

SOLUTION

Several rectangles have been drawn in the fourth quadrant of the *B-H* curve. The *BH* product in each is

$$BH_1 = (12,000)(280) = 3.4 \times 10^6 \text{ gauss} \cdot \text{oersted}$$
$$BH_2 = (11,000)(360) = 4.0 \times 10^6 \text{ gauss} \cdot \text{oersted}$$
$$BH_3 = (10,000)(420) = 4.2 \times 10^6 \text{ gauss} \cdot \text{oersted} = \text{maximum}$$
$$BH_4 = (9,000)(460) = 4.1 \times 10^6 \text{ gauss} \cdot \text{oersted}$$
$$BH_5 = (8,000)(500) = 4.0 \times 10^6 \text{ gauss} \cdot \text{oersted}$$

Thus, the power is about 4.2×10^6 gauss $\cdot$ oersted.

| **Example 20-4** | *Design/Selection of Magnetic Materials* |

Select an appropriate magnetic material for the following applications: a high electrical-efficiency motor, a magnetic device to keep cupboard doors closed, a magnet used in an ammeter or voltmeter, and magnetic resonance imaging.

SOLUTION

High electrical-efficiency motor: To minimize hysteresis losses, we might use an oriented silicon iron, taking advantage of its anisotropic behavior and its small hysteresis loop. Since the iron-silicon alloy is electrically conductive, we would produce a laminated structure with thin sheets of the silicon iron sandwiched between a nonconducting dielectric material. Sheets thinner than about 0.5 mm might be recommended.

Magnet for cupboard doors: The magnetic latches used to fasten cupboard doors must be permanent magnets; however, low cost is a more important design feature than high power. An inexpensive ferritic steel or a low-cost ferrite would be recommended.

Magnets for an ammeter or voltmeter: For these applications, alnico alloys are particularly effective. We find that these alloys are among the least sensitive to changes in temperature, ensuring accurate current or voltage readings over a range of temperatures.

Magnetic resonance imaging: One of the applications for MRI is in medical diagnostics. In this case, we want a very powerful magnet. A $Nd_2Fe_{12}B$ magnetic material, which has an exceptionally high *BH* product, might be recommended for this application. We can also make use of very strong electromagnets fabricated from superconductors.

The example that follows shows how the lifting power of a permanent magnet can be calculated.

Example 20-5   *Lifting Power of a Magnet*

Calculate the force in kN for one square meter area of a permanent magnet with a saturation magnetization of 1.61 tesla.

SOLUTION

As noted before, the attractive force from a permanent magnet is given by

$$F = \frac{\mu_0 M^2 A}{2}$$

We have been given the value of $\mu_0 M = 1.61$ tesla. We can rewrite the equation that provides the force due to a permanent magnet as follows:

$$F = \frac{\mu_0 M^2 A}{2} = \frac{(\mu_0 M)^2 A}{2\mu_0}$$

$$\therefore \frac{F}{A} = \frac{(1.61 \text{ T})^2}{2\left(4\pi \times 10^{-7} \dfrac{\text{H}}{\text{m}}\right)} = 1031.4 \frac{\text{kN}}{\text{m}^2}$$

Note that the force in this case will be 1031 kN since the area (*A*) has been specified as 1 m².

20-8   Metallic and Ceramic Magnetic Materials

Let's look at typical alloys and ceramic materials used in magnetic applications and discuss how their properties and behavior can be enhanced. Some polymeric materials have shown magnetic activity; however, the Curie temperatures of these materials are too low compared to those for metallic and ceramic magnetic materials.

Magnetic Alloys   Pure iron, nickel, and cobalt are not usually used for electrical applications because they have high electrical conductivities and relatively large hysteresis loops, leading to excessive power loss. They are relatively poor permanent magnets; the domains are easily reoriented and both the remanance and the *BH* product are small compared with those of more complex alloys. Some change in the magnetic properties is obtained by introducing defects into the structure. Dislocations, grain boundaries, boundaries between multiple phases, and point defects help pin the domain boundaries, therefore keeping the domains aligned when the original magnetizing field is removed.

Iron-Nickel Alloys.   Some iron-nickel alloys, such as Permalloy, have high permeabilities, making them useful as soft magnets. One example of an application for these magnets is the "head" that stores or reads information on a computer disk (Figure 20-12). As the disk rotates beneath the head, a current produces a magnetic field in the head. The magnetic field in the head, in turn, magnetizes a portion of the disk. The direction of the field produced in the head determines the orientation of the magnetic particles embedded in the disk and, consequently, stores information. The information can be retrieved by again spinning the disk beneath the head. The magnetized region in the disk induces a current in the head; the direction of the current depends on the direction of the magnetic field in the disk.

Silicon Iron.   Silicon irons are processed into grain-oriented steels. Introduction of 3 to 5% Si into iron produces an alloy that, after proper processing, is useful in electrical applications such as motors and generators. We take advantage of the anisotropic magnetic behavior of silicon iron to obtain the best performance. As a result of rolling and subsequent annealing, a sheet texture is formed in which the $\langle 100 \rangle$ directions in each grain are aligned. Because the silicon iron is most easily magnetized in $\langle 100 \rangle$ directions, the field

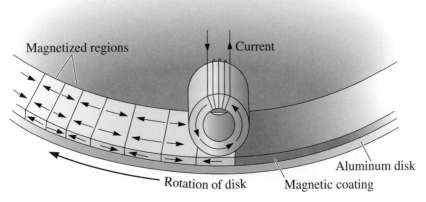

Figure 20-12   Information can be stored or retrieved from a magnetic disk by use of an electromagnetic head. A current in the head magnetizes domains in the disk during storage; the domains in the disk induce a current in the head during retrieval.

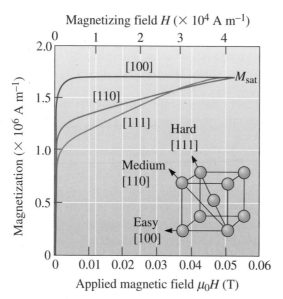

Magnetizing field H ($\times 10^4$ A m^{-1})

M_{sat}

[100]
[110]
[111]
Hard [111]
Medium [110]
Easy [100]

Magnetization ($\times 10^6$ A m^{-1})

Applied magnetic field $\mu_0 H$ (T)

Figure 20-13
The initial magnetization curve for iron is highly anisotropic; magnetization is easiest when the $\langle 100 \rangle$ directions are aligned with the field and hardest along [111]. (*From* Principles of Electrical Engineering Materials and Devices, *by S.O. Kasap, p. 623, Fig. 8-24. Copyright © 1997 Irwin. Reprinted by permission of The McGraw-Hill Companies.*)

required to give saturation magnetization is very small, and both a small hysteresis loop and a small remanance are observed (Figure 20-13). This type of anisotropy is known as **magnetocrystalline anisotropy**.

Composite Magnets. Composite magnets are used to reduce eddy current losses. Thin sheets of silicon iron are laminated with sheets of a dielectric material. The laminated layers are then built up to the desired overall thickness. The laminate increases the resistivity of the composite magnets and makes them successful at low and intermediate frequencies.

At very high frequencies, losses are more significant because the domains do not have time to realign. In this case, a composite material containing domain-sized magnetic particles in a polymer matrix may be used. The particles, or domains, rotate easily, while eddy current losses are minimized because of the high resistivity of the polymer.

Data Storage Materials. Magnetic materials for information storage must have a square loop and a low coercive field, permitting very rapid transmission of information. Magnetic tape for audio or video applications is produced by evaporating, sputtering, or plating particles of a magnetic material such as γ-Fe_2O_3 or CrO_2 onto a polyester tape.

Hard disks for computer data storage are produced in a similar manner. In a hard disk, magnetic particles are embedded in a polymer film on a flat aluminum substrate. Because of the polymer matrix and the small particles, the domains can rotate quickly in response to a magnetic field. These materials are summarized in Table 20-5.

Complex Metallic Alloys for Permanent Magnets. Improved permanent magnets are produced by making the grain size so small that only one domain is present in each grain. Now the boundaries between domains are grain boundaries rather than Bloch walls. The domains can change their orientation only by rotating, which requires greater energy than domain growth. Two techniques are used to produce these magnetic materials: phase transformations and powder metallurgy. Alnico, one of the most common of the complex metallic alloys, has a single-phase BCC structure at high temperatures, but when alnico slowly cools below 800°C, a second BCC phase rich in iron

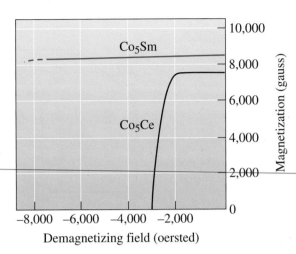

Figure 20-14
Demagnetizing curves for Co_5Sm and Co_5Ce, representing a portion of the hysteresis loop.

and cobalt precipitates. This second phase is so fine that each precipitate particle is a single domain, producing a very high remanance, coercive field, and power. Often the alloys are permitted to cool and transform while in a magnetic field to align the domains as they form.

A second technique—powder metallurgy—is used for a group of rare earth metal alloys, including samarium-cobalt. A composition giving Co_5Sm, an intermetallic compound, has a high *BH* product (Figure 20-14) due to unpaired magnetic spins in the 4*f* electrons of samarium. The brittle intermetallic is crushed and ground to produce a fine powder in which each particle is a domain. The powder is then compacted while in an imposed magnetic field to align the powder domains. Careful sintering to avoid growth of the particles produces a solid-powder metallurgy magnet. Another rare earth magnet based on neodymium, iron, and boron has a *BH* product of 45 mega-gauss-oersted (MGOe). In these materials, a fine-grained intermetallic compound, $Nd_2Fe_{14}B$, provides the domains, and a fine HfB_2 precipitate prevents movement of the domain walls.

Ferrimagnetic Ceramic Materials

Common magnetic ceramics are the ferrites, which have a spinel crystal structure (Figure 20-15). These ferrites have nothing to do with the *ferrite phase* we encountered in studying the Fe-C phase diagram (Chapters 12 and 13). Ferrites are used in wireless communications and in microelectronics in such applications as inductors. Ferrite powders are made using ceramic processing techniques.

We can understand the behavior of these ceramic magnets by looking at magnetite, Fe_3O_4. Magnetite contains two different iron ions, Fe^{2+} and Fe^{3+}, so we could rewrite the formula for magnetite as $Fe^{2+}Fe_2^{3+}O_4^{2-}$. The magnetite, or spinel, crystal structure is based on an FCC arrangement of oxygen ions, with iron ions occupying selected interstitial sites. Although the spinel unit cell actually contains eight of the FCC arrangements, we need examine only one of the FCC subcells:

1. Four oxygen ions are in the FCC positions of the subcell.
2. Octahedral sites, which are surrounded by six oxygen ions, are present at each edge and the center of the subcell. One Fe^{2+} and one Fe^{3+} ion occupy octahedral sites.

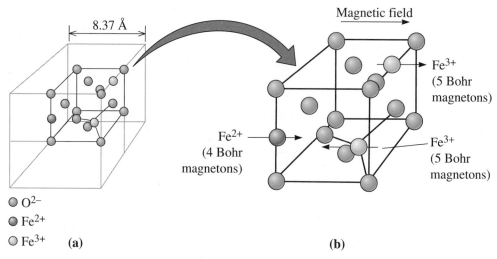

Figure 20-15 (a) The structure of magnetite, Fe_3O_4. (b) The subcell of magnetite. The magnetic moments of ions in the octahedral sites line up with the magnetic field, but the magnetic moments of ions in tetrahedral sites oppose the field. A net magnetic moment is produced by this ionic arrangement.

3. Tetrahedral sites have positions in the subcell such as (1/4, 1/4, 1/4). One Fe^{3+} ion occupies one of the tetrahedral sites.

4. When Fe^{2+} ions form, the two $4s$ electrons of iron are removed, but all of the $3d$ electrons remain. Because there are four unpaired electrons in the $3d$ level of iron, the magnetic strength of the Fe^{2+} dipole is four Bohr magnetons. When Fe^{3+} forms, both $4s$ electrons and one of the $3d$ electrons are removed. The Fe^{3+} ion has five unpaired electrons in the $3d$ level and, thus, has a strength of five Bohr magnetons.

5. The ions in the tetrahedral sites of the magnetite line up so that their magnetic moments oppose the applied magnetic field, but the ions in the octahedral sites reinforce the field [Figure 20-15(b)]. Consequently, the Fe^{3+} ion in the tetrahedral site neutralizes the Fe^{3+} ion in the octahedral site (the Fe^{3+} ion coupling is antiferromagnetic). The Fe^{2+} ion in the octahedral site is not opposed by any other ion, and it therefore reinforces the magnetic field. The following example shows how we can calculate the magnetization in Fe_3O_4, which is one of the ferrites.

Example 20-6   *Magnetization in Magnetite (Fe_3O_4)*

Calculate the total magnetic moment per cubic centimeter in magnetite. Calculate the value of the saturation flux density (B_{sat}) for this material.

SOLUTION

In the subcell [Figure 20-15(b)], the total magnetic moment is four Bohr magnetons obtained from the Fe^{2+} ion, since the magnetic moments from the two Fe^{3+} ions located at tetrahedral and octahedral sites are canceled by each other.

In the unit cell overall, there are eight subcells, so the total magnetic moment is 32 Bohr magnetons per cell.

The size of the unit cell, with a lattice parameter of 8.37×10^{-8} cm is

$$V_{cell} = (8.37 \times 10^{-8})^3 = 5.86 \times 10^{-22} \, cm^3$$

The magnetic moment per cubic centimeter is

$$\text{Total moment} = \frac{32 \text{ Bohr magnetons/cell}}{5.86 \times 10^{-22} \, cm^3/cell} = 5.46 \times 10^{22} \, \text{magnetons/cm}^3$$

$$= (5.46 \times 10^{22})(9.274 \times 10^{-24} \, A \cdot m^2/\text{magneton})$$

$$= 0.51 \, A \cdot m^2/cm^3 = 5.1 \times 10^5 \, A \cdot m^2/m^3 = 5.1 \times 10^5 A/m$$

This expression represents the magnetization M at saturation (M_{sat}). The value of $B_{sat} \simeq \mu_0 M_{sat}$ will be $= (4\pi \times 10^{-7})(5.1 \times 10^5) = 0.64$ Tesla.

When ions are substituted for Fe^{2+} ions in the spinel structure, the magnetic behavior may be changed. Ions that may not produce ferromagnetism in a pure metal may contribute to ferrimagnetism in the spinels, as shown by the magnetic moments in Table 20-7. Soft magnets are obtained when the Fe^{2+} ion is replaced by various mixtures of manganese, zinc, nickel, and copper. The nickel and manganese ions have magnetic moments that partly cancel the effect of the two iron ions, but a net ferrimagnetic behavior, with a small hysteresis loop, is obtained. The high electrical resistivity of these ceramic compounds helps minimize eddy currents and permits the materials to operate at high frequencies. Ferrites used in computer applications may contain additions of manganese, magnesium, or cobalt to produce a square hysteresis loop behavior.

Another group of soft ceramic magnets is based on garnets, which include yttria iron garnet, $Y_3Fe_5O_{12}$ (YIG). These complex oxides, which may be modified by substituting aluminum or chromium for iron or by replacing yttrium with lanthanum or praseodymium, behave much like the ferrites. Another garnet, based on gadolinium and gallium, can be produced in the form of a thin film. Tiny magnetic domains can be produced in the garnet film; these domains, or *magnetic bubbles*, can then serve as storage units for computers. Once magnetized, the domains do not lose their memory in case of a sudden power loss.

Hard ceramic magnets used as permanent magnets include another complex oxide family, the hexagonal ferrites. The hexagonal ferrites include $SrFe_{12}O_{19}$ and $BaFe_{12}O_{19}$.

The example that follows highlights materials selection for a ceramic magnet.

TABLE 20-7 ■ *Magnetic moments for ions in the spinel structure*

Ion	Bohr Magnetons	Ion	Bohr Magnetons
Fe^{3+}	5	Co^{2+}	3
Mn^{2+}	5	Ni^{2+}	2
Fe^{2+}	4	Cu^{2+}	1
		Zn^{2+}	0

Example 20-7 *Design/Materials Selection for a Ceramic Magnet*

Design a cubic ferrite magnet that has a total magnetic moment per cubic meter of $5.5 \times 10^5 A/m$.

SOLUTION

We found in Example 20-6 that the magnetic moment per cubic meter for Fe_3O_4 is $5.1 \times 10^5 A/m$. To obtain a higher saturation magnetization, we must replace Fe^{2+} ions with ions having more Bohr magnetons per atom. One such possibility (Table 20-7) is Mn^{2+}, which has five Bohr magnetons.

Assuming that the addition of Mn ions does not appreciably affect the size of the unit cell, we find from Example 20-6 that

$$V_{cell} = 5.86 \times 10^{-22}\,cm^3 = 5.86 \times 10^{-28}\,m^3$$

Let x be the fraction of Mn^{2+} ions that have replaced the Fe^{2+} ions, which have now been reduced to $1 - x$. Then, the total magnetic moment is

Total moment

$$= \frac{(8\ \text{subcells})[(x)(5\ \text{magnetons}) + (1 - x)(4\ \text{magnetons})](9.274 \times 10^{-24}A \cdot m^2)}{5.86 \times 10^{-28}\,m^3}$$

$$= \frac{(8)(5x + 4 - 4x)(9.274 \times 10^{-24})}{5.86 \times 10^{-28}} = 5.5 \times 10^5$$

$$x = 0.344$$

Therefore we need to replace 34.4 at % of the Fe^{2+} ions with Mn^{2+} ions to obtain the desired magnetization.

Magnetostriction

Certain materials can develop strain when their magnetic state is changed. This effect is used in actuators. The magnetostrictive effect can be seen either by changing the magnetic field or by changing the temperature. Iron, nickel, Fe_3O_4, $TbFe_2$, DyFe, and $SmFe_2$ are examples of some materials that show this effect. Terfenol-D, which is named after its constituents terbium (Tb), iron (Fe), and dysprosium (Dy) and its developer, the Naval Ordnance Laboratory (NOL), is one of the best known magnetostrictive materials. Its composition is $\sim Tb_xDy_{1-x}Fe_y$ $(0.27 < x < 0.30, 1.9 < y < 2)$. The magnetostriction phenomenon is analogous to electrostriction. Recently, some ferromagnetic alloys that also show magnetostriction have been developed.

Summary

- All materials interact with magnetic fields. The magnetic properties of materials are related to the interaction of magnetic dipoles with a magnetic field. The magnetic dipoles originate with the electronic structure of the atom, causing several types of behavior.

- Magnetic materials have enabled numerous technologies that range from high intensity superconducting magnets for MRI; semi-hard materials used in magnetic data storage;

permanent magnets used in loud speakers, motors, and generators; to superparamagnetic materials used to make ferrofluids and for magnetic separation of DNA molecules and cells.

- In diamagnetic materials, the magnetic dipoles oppose the applied magnetic field.

- In paramagnetic materials, the magnetic dipoles weakly reinforce the applied magnetic field, increasing the net magnetization or inductance.

- Ferromagnetic and ferrimagnetic materials are magnetically nonlinear. Their permeability depends strongly on the applied magnetic field. In ferromagnetic materials (such as iron, nickel, and cobalt), the magnetic dipoles strongly reinforce the applied magnetic field, producing large net magnetization or inductance. In ferrimagnetic materials, some magnetic dipoles reinforce the field, whereas others oppose the field. A net increase in magnetization or inductance occurs. Magnetization may remain even after the magnetic field is removed. Increasing the temperature above the Curie temperature destroys the ferromagnetic or ferrimagnetic behavior.

- The structure of ferromagnetic and ferrimagnetic materials includes domains, within which all of the magnetic dipoles are aligned. When a magnetic field is applied, the dipoles become aligned with the field, increasing the magnetization to its maximum, or saturation, value. When the field is removed, some alignment of the domains may remain, giving a remanant magnetization.

- For soft magnetic materials, little remanance exists, only a small coercive field is required to remove any alignment of the domains, and little energy is consumed in reorienting the domains when an alternating magnetic field is applied.

- For hard, or permanent, magnetic materials, the domains remain almost completely aligned when the field is removed, large coercive fields are required to randomize the domains, and a large hysteresis loop is observed. This condition provides the magnet with a high power.

- Magnetostriction is the development of strain in response to an applied magnetic field or a temperature change that induces a magnetic transformation. Terfenol type magnetostrictive materials have been developed for actuator applications.

Glossary

Antiferromagnetism Arrangement of magnetic moments such that the magnetic moments of atoms or ions cancel out causing zero net magnetization.

Bloch walls The boundaries between magnetic domains.

Bohr magneton The strength of a magnetic moment of an electron (μ_B) due to electron spin.

Coercivity The magnetic field needed to force the domains in a direction opposite to the magnetization direction. This is a microstructure-sensitive property.

Curie temperature The temperature above which ferromagnetic or ferrimagnetic materials become paramagnetic.

Diamagnetism The effect caused by the magnetic moment due to the orbiting electrons, which produces a slight opposition to the imposed magnetic field.

Domains Small regions within a single or polycrystalline material in which all of the magnetization directions are aligned.

Ferrimagnetism Magnetic behavior obtained when ions in a material have their magnetic moments aligned in an antiparallel arrangement such that the moments do not completely cancel out and a net magnetization remains.

Ferromagnetism Alignment of the magnetic moments of atoms in the same direction so that a net magnetization remains after the magnetic field is removed.

Hard magnet Ferromagnetic or ferrimagnetic material that has a coercivity $> 10^4 \, \text{A} \cdot \text{m}^{-1}$. This is the same as a permanent magnet.

Hysteresis loop The loop traced out by magnetization in a ferromagnetic or ferrimagnetic material as the magnetic field is cycled.

Magnetic moment The strength of the magnetic field associated with a magnetic dipole.

Magnetic permeability The ratio between inductance or magnetization and magnetic field. It is a measure of the ease with which magnetic flux lines can "flow" through a material.

Magnetic susceptibility The ratio between magnetization and the applied field.

Magnetization The total magnetic moment per unit volume.

Magnetocrystalline anisotropy In single crystals, the coercivity depends upon crystallographic direction creating easy and hard axes of magnetization.

Paramagnetism The net magnetic moment caused by the alignment of the electron spins when a magnetic field is applied.

Permanent magnet A hard magnetic material.

Power The strength of a permanent magnet as expressed by the maximum product of the inductance and magnetic field.

Remanance The polarization or magnetization that remains in a material after it has been removed from a magnetic field. The remanance is due to the permanent alignment of the dipoles.

Saturation magnetization When all of the dipoles have been aligned by the field, producing the maximum magnetization.

Shape anisotropy The dependence of coercivity on the shape of magnetic particles.

Soft magnet Ferromagnetic or ferrimagnetic material that has a coercivity $\leq 10^3 \, \text{A} \cdot \text{m}^{-1}$.

Superparamagnetism In the nanoscale regime, materials that are ferromagnetic or ferrimagnetic but behave in a paramagnetic manner (because of their nano-sized grains or particles).

Problems

Section 20-1 Classification of Magnetic Materials

Section 20-2 Magnetic Dipoles and Magnetic Moments

20-1 State any four real-world applications of different magnetic materials

20-2 Explain the following statement "Strictly speaking, there is no such thing as a non-magnetic material."

20-3 Normally we disregard the magnetic moment of the nucleus. In what application does the nuclear magnetic moment become important?

20-4 What two motions of electrons are important in determining the magnetic properties of materials?

20-5 Explain why only a handful of solids exhibit ferromagnetic or ferrimagnetic behavior.

20-6 Calculate and compare the maximum magnetization we would expect in iron, nickel, cobalt, and gadolinium. There are seven electrons in the $4f$ level of gadolinium. Compare the calculated values with the experimentally observed values.

Section 20-3 Magnetization, Permeability, and the Magnetic Field

Section 20-4 Diamagnetic, Paramagnetic, Ferromagnetic, Ferrimagnetic, and Superparamagnetic Materials

Section 20-5 Domain Structure and Hysteresis Loop

20-7 Define the following terms: magnetic induction, magnetic field, magnetic susceptibility, and magnetic permeability.

20-8 Define the following terms: ferromagnetic, ferrimagnetic, diamagnetic, paramagnetic, superparamagnetic, and antiferromagnetic materials.

20-9 What is a ferromagnetic material? What is a ferrimagnetic material? Explain and provide examples of each type of material.

20-10 How does the permeability of ferromagnetic and ferrimagnetic materials change with temperature when the temperature is greater than the Curie temperature?

20-11 Derive the equation $\mu_r = 1 + \chi_m$ using Equations 20-4 through 20-7.

20-12 A 4-79 permalloy solenoid coil needs to produce a minimum inductance of 1.5 Wb/m^2. If the maximum allowed current is 5 mA, how many turns are required in a wire 1 m long?

20-13 An alloy of nickel and cobalt is to be produced to give a magnetization of 2×10^6 A/m. The crystal structure of the alloy is FCC with a lattice parameter of 0.3544 nm. Determine the atomic percent cobalt required, assuming no interaction between the nickel and cobalt.

20-14 Estimate the magnetization that might be produced in an alloy containing nickel and 70 at% copper, assuming that no interaction occurs.

20-15 An Fe-80% Ni alloy has a maximum permeability of 300,000 when an inductance of 3500 gauss is obtained. The alloy is placed in a 20-turn coil that is 2 cm in length. What current must flow through the conductor coil to obtain this field?

20-16 An Fe-49% Ni alloy has a maximum permeability of 64,000 when a magnetic field of 0.125 oersted is applied. What inductance is obtained and what current is needed to obtain this inductance in a 200-turn, 3-cm-long coil?

20-17 Draw a schematic of the *B-H* and *M-H* loops for a typical ferromagnetic material. What is the difference between these two loops?

20-18 Is the magnetic permeability of ferromagnetic or ferrimagnetic materials constant? Explain.

20-19 From a phenomenological viewpoint, what are the similarities between elastomers, ferromagnetic and ferrimagnetic materials, and ferroelectrics?

20-20 What are the major differences between ferromagnetic and ferrimagnetic materials?

20-21 Compare the electrical resistivities of ferromagnetic metals and ferrimagnetic ceramics.

20-22 Why are eddy current losses important design factors in ferromagnetic materials but less important in ferrimagnetic materials?

20-23 Which element has the highest saturation magnetization? What alloys have the highest saturation magnetization of all materials?

20-24 What material has the highest energy product of all magnetic materials?

20-25 Is coercivity of a material a microstructure sensitive property? Is remanance a microstructure sensitive property? Explain.

20-26 Is saturation magnetization of a material a microstructure sensitive property? Explain.

20-27 Can the same material have different hysteresis loops? Explain.

20-28 The following data describe the effect of the magnetic field on the inductance in a silicon steel. Calculate the initial permeability and the maximum permeability for the material.

H (A/m)	B (tesla)
0.00	0
20	0.08
40	0.30
60	0.65
80	0.85
100	0.95
150	1.10
250	1.25

20-29 A magnetic material has a coercive field of 167 A/m, a saturation magnetization of 0.616 tesla, and a residual inductance of 0.3 tesla. Sketch the hysteresis loop for the material.

20-30 A magnetic material has a coercive field of 10.74 A/m, a saturation magnetization of 2.158 tesla, and a remanance induction of 1.183 tesla. Sketch the hysteresis loop for the material.

20-31 Using Figure 20-16, determine the following properties of the magnetic material: remanance, saturation magnetization, coercive field, initial permeability, maximum permeability, and power (maximum *BH* product).

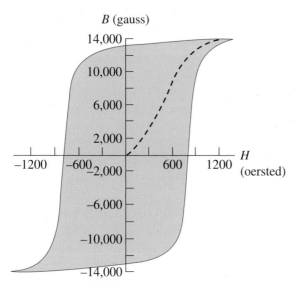

Figure 20-16 Hysteresis curve for a hard magnetic material (for Problem 20-31).

20-32 Using Figure 20-17, determine the following properties of the magnetic material: remanance, saturation magnetization,

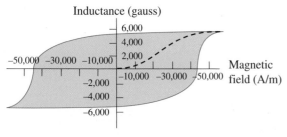

Figure 20-17 Hysteresis curve for a hard magnetic material (for Problem 20-32).

coercive field, initial permeability, maximum permeability, and power (maximum *BH* product).

Section 20-6 The Curie Temperature

Section 20-7 Applications of Magnetic Materials

Section 20-8 Metallic and Ceramic Magnetic Materials

20-33 Sketch the *M-H* loop for Fe at 300 K, 500 K, and 1000 K.

20-34 Define the terms soft and hard magnetic materials. Draw a typical *M-H* loop for each material.

20-35 What important characteristics are associated with soft magnetic materials?

20-36 Are materials used for magnetic data storage magnetically hard or soft? Explain.

20-37 Give examples of materials used in magnetic recording.

20-38 What are the advantages of using Fe-Nd-B magnets? What are some of their disadvantages?

20-39 Estimate the power of the Co_5Ce material shown in Figure 20-14.

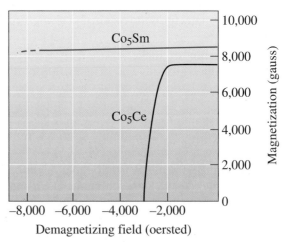

Figure 20-14 (Repeated for Problem 20-39.) Demagnetizing curves for Co_5Sm and Co_5Ce, representing a portion of the hysteresis loop.

20-40 What advantages does the Fe-3% Si material have compared with permalloy for use in electric motors?

20-41 The coercive field for pure iron is related to the grain size of the iron by the relationship $H_c = 1.83 + 4.14/\sqrt{A}$, where A is the area of the grain in two dimensions (mm^2) and H_c has units of A/m. If only the grain size influences the 99.95% iron (coercivity 0.9 oersted), estimate the size of the grains in the material. What happens to the coercivity value when the iron is annealed to increase the grain size?

20-42 Calculate the attractive force per square meter from a permanent magnet with a saturation magnetization of 1.0 tesla.

20-43 Suppose we replace 10% of the Fe^{2+} ions in magnetite with Cu^{2+} ions. Determine the total magnetic moment per cubic centimeter.

20-44 Suppose that the total magnetic moment per cubic meter in a spinel structure in which Ni^{2+} ions have replaced a portion of the Fe^{2+} ions is 4.6×10^5 A/m. Calculate the fraction of the Fe^{2+} ions that have been replaced and the wt% Ni present in the spinel.

20-45 What is magnetostriction? How is this similar to electrostriction? How is it different from the piezoelectric effect?

20-46 State examples of materials that show the magnetostriction effect.

20-47 What is spintronics? Give an example of a spintronics-based device used in personal and laptop computers.

Design Problems

20-48 Design a solenoid no longer than 1 cm that will produce an inductance of 3000 gauss.

20-49 Design a permanent magnet that will have a remanance of at least 5000 gauss, that will not be demagnetized if exposed to a temperature of 400°C or to a magnetic field of 1000 oersted, and that has good magnetic power.

20-50 Design a spinel-structure ferrite that will produce a total magnetic moment per cubic meter of 5.6×10^5 A/m.

20-51 Design a spinel-structure ferrite that will produce a total magnetic moment per cubic meter of 4.1×10^5 A/m.

20-52 Design a permanent magnet to lift a 1000 kg maximum load under operating temperatures as high as 750°C. Which material(s) listed in Table 20-6 will meet the above requirement?

Computer Problems

20-53 *Converting Magnetic Units.* Write a computer program that will convert magnetic units from the cgs or Gaussian system to the SI system. For example, if the user provides a value of flux density in Gauss, the program should provide a value in Wb/m^2 or tesla.

Optical materials play a critical role in the infrastructure of our communications and information technology systems. There are millions of kilometers of optical fiber, such as shown in the image above, installed world-wide. Optical fibers for communications and medical applications, lasers for medical and manufacturing applications, micromachined mirror arrays, light-emitting diodes, and solar cells have enabled a wide range of new technologies. (*PhotoDisc Blue/Getty Images.*)

21

Photonic Materials

Have You Ever Wondered?

- *Why does the sky appear blue?*

- *How does an optical fiber work?*

- *What factors control the transmission and absorption of light in different materials?*

- *What does the acronym LASER stand for?*

- *What is a ruby laser made from?*

- *Does the operation of a fluorescent tube light involve phosphorescence?*

- *How did the invention of blue lasers enable high definition DVDs?*

Photonic or optical materials have had a significant impact on the development of the communications infrastructure and information technology. Photonic materials have also played a key role in many other technologies related to medicine, manufacturing, and astronomy, just to name a few. Today, millions of kilometers of optical fiber have been installed worldwide. The term "optoelectronics" refers to the science and technology that combine electronic and optical materials. Examples include light-emitting diodes (LEDs), solar cells, and semiconductor lasers. Starting with simple mirrors, prisms, and lenses to the latest photonic band gap materials, the field of optical materials and devices has advanced at a very rapid pace. The goal of this chapter is to present a summary of fundamental principles that have guided applications of optical materials.

Optical properties of materials are related to the interaction of a material with electromagnetic radiation in the form of waves or particles of energy called photons. This radiation may have characteristics that fall in the visible light spectrum or may be invisible to the human eye. In this chapter, we explore two avenues by which we can use the optical properties of materials: emission of photons from materials and interaction of photons with materials.

21-1   The Electromagnetic Spectrum

Light is energy, or radiation, in the form of waves or particles called **photons** that can be emitted from a material. The important characteristics of the photons—their energy E, wavelength λ, and frequency ν—are related by the equation

$$E = \text{h}\nu = \frac{\text{h}c}{\lambda} \tag{21-1}$$

where c is the speed of light (in vacuum, the speed c_0 is 3×10^{10} cm/s), and h is Planck's constant (6.626×10^{-34} J · s). Since there are 1.6×10^{-19} J per electron volt (eV), the value of "h" is also given by 4.14×10^{-15} eV · s. This equation permits us to consider the photon either as a particle of energy E or as a wave with a characteristic wavelength and frequency.

The spectrum of electromagnetic radiation is shown in Figure 21-1. Gamma and x-rays have short wavelengths, or high frequencies, and possess high energies; microwaves and radio waves possess low energies; and visible light represents only a very narrow portion of the electromagnetic spectrum. Figure 21-1 also shows the response of the human eye to different colors. Bandgaps (E_g) of semiconductors (in eV) and corresponding wavelengths of light are also shown. As discussed in Chapter 19, these relationships are used to make LEDs of different colors.

21-2   Refraction, Reflection, Absorption, and Transmission

All materials interact in some way with light. Photons cause a number of optical phenomena when they interact with the electronic or crystal structure of a material (Figure 21-2). If incoming photons interact with valence electrons, several things may happen. The photons may give up their energy to the material, in which case *absorption* occurs. Or the photons may give up their energy, but photons of identical energy are immediately emitted by the material; in this case, *reflection* occurs. Finally, the photons may not interact with the electronic structure of the material; in this case, *transmission* occurs. Even in transmission, however, photons are changed in velocity, and *refraction* occurs. A small fraction of the incident light may be scattered with a slightly different frequency (Raman scattering).

As Figure 21-2 illustrates, an incident beam of intensity I_0 may be partly reflected, partly absorbed, and partly transmitted. The intensity of the incident beam I_0 therefore can be expressed as

$$I_0 = I_r + I_a + I_t \tag{21-2}$$

where I_r is the portion of the beam that is reflected, I_a is the portion that is absorbed, and I_t is the portion finally transmitted through the material. Reflection may occur at both the front and back surfaces of the material. Figure 21-2(a) shows reflection only at the front surface. Also, reflection occurs at a certain angle with respect to the normal of the surface (specular reflection) and also in many other directions [diffuse reflection, not shown in Figure 21-2(a)]. Several factors are important in determining the behavior of the photon, with the energy required to excite an electron to a higher energy state being of particular importance.

Let's examine each of these four phenomena. We begin with refraction, since it is related to reflection and transmission.

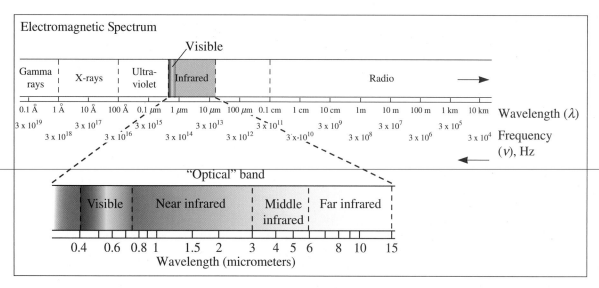

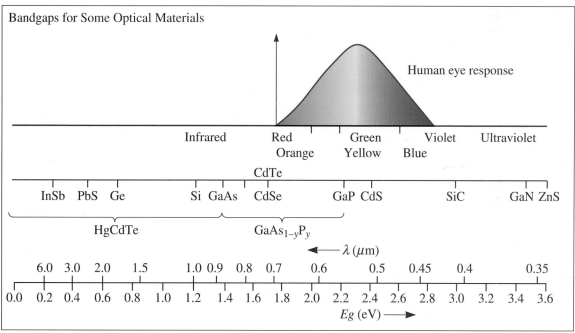

Figure 21-1   The electromagnetic spectrum of radiation; the bandgaps and cutoff frequencies for some optical materials are also shown. (*From Optoelectronics: An Introduction to Materials and Devices, by J. Singh. Copyright © 1996 The McGraw-Hill Companies. Reprinted by permission of The McGraw-Hill Companies.*)

Refraction   Even when a photon is transmitted, the photon causes polarization of the electrons in the material and, by interacting with the polarized material, loses some of its energy. The speed of light (c) can be related to the ease with which a material polarizes both electrically (permittivity ε) and magnetically (permeability μ) through

$$c = \frac{1}{\sqrt{\mu\varepsilon}}$$

(21-3)

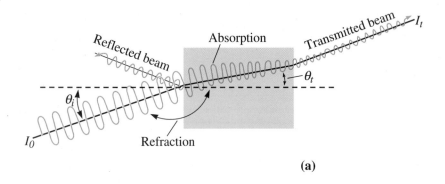

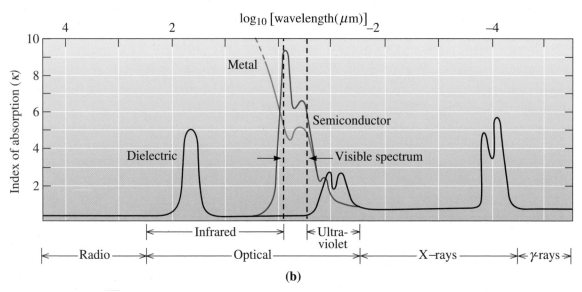

Figure 21-2   (a) Interaction of photons with a material. In addition to reflection, absorption, and transmission, the beam changes direction, or is refracted. The change in direction is given by the index of refraction n. (b) The absorption index (κ) as a function of wavelength.

Generally, optical materials are not magnetic, and the permeability can be neglected. Because the speed of the photons decreases, the beam of photons changes direction when it enters the material [Figure 21-2(a)]. Suppose photons traveling in a vacuum impinge on a material. If θ_i and θ_t respectively, are the angles that the incident and refracted beams make with the normal of the surface of the material, then

$$n = \frac{c_0}{c} = \frac{\lambda_{\text{vacuum}}}{\lambda} = \frac{\sin \theta_i}{\sin \theta_t} \tag{21-4}$$

The ratio n is the **index of refraction**, c_0 is the speed of light in a vacuum (3×10^8 m/s), and c is the speed of light in the material. The frequency of light does not change as it is refracted. Typical values of the index of refraction for several materials are listed in Table 21-1.

We can also define a complex refractive index (n^*). This includes κ, a parameter known as the absorption index:

$$n^* = n(1 - i\kappa) \tag{21-5}$$

TABLE 21-1 ■ *Index of refraction of selected materials for photons of wavelength 5890 Å*

Material	Index of Refraction (n)	Material	Index of Refraction (n)
Air	1.00	Polystyrene	1.60
Ice	1.309	TiO_2	1.74
Water	1.333	Sapphire (Al_2O_3)	1.8
Teflon™	1.35	Leaded glasses (crystal)	2.50
SiO_2 (glass)	1.46	Rutile (TiO_2)	2.6
Polymethyl methacrylate	1.49	Diamond	2.417
Typical silicate glasses	~1.50	Silicon	3.49
Polyethylene	1.52	Gallium arsenide	3.35
Sodium chloride (NaCl)	1.54	Indium phosphide	3.21
SiO_2 (quartz)	1.55	Germanium	4.0
Epoxy	1.58		

In Equation 21-5, $i = \sqrt{-1}$ is the imaginary number. The absorption index is defined as

$$\kappa = \frac{\alpha\lambda}{4\pi n} \tag{21-6}$$

In Equation 21-6, α is the **linear absorption coefficient** (see Equation 21-12), λ is the wavelength of light, and n is the refractive index. Figure 21-2(b) shows the variation in index of absorption with the frequency of electromagnetic waves. Drawing an analogy, the refractive index is similar to the dielectric constant of materials, and the absorption index is similar to the dielectric loss factor.

If the photons are traveling in Material 1, instead of in a vacuum, and then pass into Material 2, the velocities of the incident and refracted beams depend on the ratio between their indices of refraction, again causing the beam to change direction:

$$\frac{c_1}{c_2} = \frac{n_2}{n_1} = \frac{\sin\theta_i}{\sin\theta_t} \tag{21-7}$$

Equation 21-7 is also known as Snell's law.

When a ray of light enters from a material with refractive index (n_1) into a material of refractive index (n_2), and if $n_1 > n_2$, the ray is bent away from the normal and toward the boundary surface [Figure 21-3(a)]. A beam traveling through Material 1 is reflected rather than transmitted if the angle θ_t becomes 90°.

More interaction of the photons with the electronic structure of the material occurs when the material is easily polarized. We saw different dielectric polarization mechanisms in Chapter 19. Among these, the electronic polarization (i.e., displacement of the electron cloud around the atoms and ions) is the one that controls the refractive index of materials. Consequently, we find a relationship between the index of refraction n and the high-frequency dielectric constant k_∞ of the material. From Equations 21-3 and 21-4 and for nonferromagnetic or nonferrimagnetic materials,

$$n = \frac{c_0}{c} = \sqrt{\frac{\mu\varepsilon}{\mu_0\varepsilon_0}} \cong \sqrt{\frac{\varepsilon}{\varepsilon_0}} = \sqrt{k_\infty} \tag{21-8}$$

In Equation 21-8, c_0 and c are the speed of light in a vacuum and in the material, respectively, μ_0 and μ are the magnetic permeabilities of the vacuum and material,

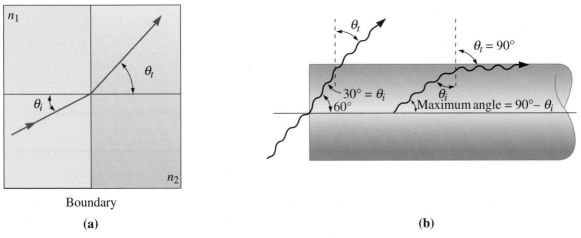

Boundary

(a)

(b)

Figure 21-3 (a) When a ray of light enters from Material 1 into Material 2, if the refractive index of Material 1 (n_1) is greater than that of Material 2 (n_2), the ray bends away from the normal and toward the boundary surface. (b) Diagram of a light beam in a glass fiber for Example 21-1.

respectively, and ε_0 and ε are the dielectric permittivities of the vacuum and material, respectively. As we discussed in Chapter 19, the material known as "lead crystal," which is actually amorphous silicate glass with up to ~30% lead oxide, has a high index of refraction ($n \sim 2.4$), since Pb^{2+} ions have high electronic polarizability. We use a similar strategy to dope silica fibers to enhance the refractive index of the core of optical fibers as compared to the outer cladding region. This helps keep the light (and hence information) in the core of the optical fiber. The difference in the high-frequency dielectric constant and the low-frequency dielectric constant is a measure of the other polarization mechanisms that are contributing to the dielectric constant.

The refractive index n is not a constant for a particular material. The frequency, or wavelength, of the photons affects the index of refraction. **Dispersion** of a material is defined as the variation of the refractive index with wavelength:

$$(\text{Dispersion})_\lambda = \frac{dn}{d\lambda} \tag{21-9}$$

This dependence of the refractive index on wavelength is nonlinear. The dispersion within a material means light pulses of different wavelengths, starting at the same time at the end of an optical fiber, will arrive at different times at the other end. Thus, material dispersion plays an important role in fiber optics. This is one of the reasons why we prefer to use a single wavelength source of light for fiber-optic communications. Dispersion also causes chromatic aberration in optical lenses.

Since dielectric polarization (P) is equal to the dipole moment per unit volume, and the high-frequency dielectric constant is related to the refractive index, we expect that (for the same material), a denser form or polymorph will have a higher refractive index (compare the refractive indices of ice and water or glass and quartz).

The following example illustrates how an optical fiber is designed to minimize optical losses during insertion. It is followed by an example calculating the index of refraction.

Example 21-1 *Design of a Fiber Optic System*

Optical fibers are commonly made from high-purity silicate glasses. They consist of a core with a refractive index that is higher than the refractive index of the coating on the fiber (the coating is called the cladding). Even a simple glass fiber in air can serve as an optical fiber because the fiber has a refractive index that is greater than that of air.

Consider a beam of photons that is introduced from a laser into a glass fiber with an index of refraction of 1.5. Choose the angle of introduction of the beam with respect to the fiber axis that will result in a minimum of leakage of the beam from the fiber. Also, consider how this angle might change if the fiber is immersed in water.

SOLUTION

To prevent leakage of the beam, we need total internal reflection, and thus the angle θ_t must be at least 90°. Suppose that the photons enter at a 60° angle to the axis of the fiber. From Figure 21-3(b), we find that $\theta_i = 90 - 60 = 30°$. If we let the glass be Material 1 and if the glass fiber is in air ($n = 1.0$), then from Equation 21-7:

$$\frac{n_2}{n_1} = \frac{\sin \theta_i}{\sin \theta_t} \quad \text{or} \quad \frac{1}{1.5} = \frac{\sin 30°}{\sin \theta_t}$$

$$\sin \theta_t = 1.5 \sin 30° = 1.5(0.50) = 0.75 \quad \text{or} \quad \theta_t = 48.6°$$

Because θ_t is less than 90°, photons escape from the fiber. To prevent transmission, we must introduce the photons at a shallower angle, giving $\theta_t = 90°$.

$$\frac{1}{1.5} = \frac{\sin \theta_i}{\sin \theta_t} = \frac{\sin \theta_i}{\sin 90°} = \sin \theta_i$$

$$\sin \theta_i = 0.6667 \quad \text{or} \quad \theta_i = 41.8°$$

If the angle between the beam and the axis of the fiber is $90 - 41.8 = 48.2°$ or less, the beam is reflected.

If the fiber were immersed in water ($n = 1.333$), then

$$\frac{1.333}{1.5} = \frac{\sin \theta_i}{\sin \theta_t} = \frac{\sin \theta_i}{\sin 90°} = \sin \theta_i$$

$$\sin \theta_i = 0.8887 \quad \text{or} \quad \theta_i = 62.7°$$

In water, the photons would have to be introduced at an angle of less than $90 - 62.7 = 27.3°$ in order to prevent transmission.

Example 21-2 *Light Transmission in Polyethylene*

Suppose a beam of photons in a vacuum strikes a sheet of polyethylene at an angle of 10° to the normal of the surface of the polymer. Polyethylene has a high-frequency dielectric constant of $k_\infty = 2.3$. Calculate the index of refraction of polyethylene, and find the angle between the incident beam and the beam as it passes through the polymer.

SOLUTION

The index of refraction is related to the high-frequency dielectric constant:

$$n = \sqrt{k_\infty} = \sqrt{2.3} = 1.52$$

The angle θ_t is

$$n = \frac{\sin \theta_i}{\sin \theta_t}$$

$$\sin \theta_t = \frac{\sin \theta_i}{n} = \frac{\sin 10°}{1.52} = \frac{0.174}{1.52} = 0.1145$$

$$\theta_t = 6.57°$$

Reflection   When a beam of photons strikes a material, the photons interact with the valence electrons and give up their energy. In metals, the radiation of almost any wavelength excites the electrons into higher energy levels. One might expect that, if the photons are totally absorbed, no light would be reflected and the metal would appear black. In aluminum or silver, however, photons of almost identical wavelength are immediately re-emitted as the excited electrons return to their lower energy levels—that is, reflection occurs. Since virtually the entire visible spectrum is reflected, these metals have a white, or silvery color.

The **reflectivity** R gives the fraction of the incident beam that is reflected and is related to the index of refraction. If the material is in a vacuum or in air:

$$R = \left(\frac{n - 1}{n + 1} \right)^2 \tag{21-10}$$

If the material is in some other medium with an index of refraction of n_i, then

$$R = \left(\frac{n - n_i}{n + n_i} \right)^2 \tag{21-11}$$

These equations apply to reflection from a single surface and assume a normal (perpendicular to the surface) incidence. The value of R depends upon the angle of incidence. Materials with a high index of refraction have a higher reflectivity than materials with a low index. Because the index of refraction varies with the wavelength of the photons, so does the reflectivity.

In metals, the reflectivity is typically on the order of 0.9 to 0.95, whereas the reflectivity of typical glasses is nearer to 0.05. The high reflectivity of metals is one reason that they are *opaque*.

There are many applications for which we want materials to have very good reflectivity. Examples include mirrors and certain types of coatings on glasses. In fact, these coatings must also be designed such that much of a certain part of the electromagnetic spectrum (e.g., infrared, the part that produces heat) must be reflected. Many such coatings have been developed for glasses. There are also many applications for which the reflectivity must be extremely limited. Such coatings are known as antireflective (AR) coatings. These coatings are used for glasses, in automobile rear view mirrors, on windows, or for the glass of picture frames so that you see through the glass without seeing your own reflection.

Absorption

That portion of the incident beam that is not reflected by the material is either absorbed or transmitted through the material. The fraction of the beam that is absorbed is related to the thickness of the material and the manner in which the photons interact with the material's structure. The intensity of the beam after passing through the material is given by

$$I = I_0 \exp(-\alpha x) \tag{21-12}$$

where x is the path through which the photons move (usually the thickness of the material), α is the linear absorption coefficient of the material for the photons, I_0 is the intensity of the beam after reflection at the front surface, and I is the intensity of the beam when it reaches the back surface. Equation 21-12 is also known as Bouguer's law, or the Beer–Lambert law. Figure 21-4 shows the linear absorption coefficient as a function of wavelength for several metals.

Absorption in materials occurs by several mechanisms. In *Rayleigh scattering*, photons interact with the electrons orbiting an atom and are deflected without any change in photon energy; this is an example of "elastic" scattering. Rayleigh scattering is more significant for higher photon energies and is responsible for the color of the sky. Since the blue (high energy) end of the visible spectrum scatters most efficiently, the sunlight scattered by molecules and atmospheric particles that reaches our eyes is mostly blue. The effect is quite strong as the intensity of scattering is proportional to the energy of the photons raised to the fourth power. Rayleigh scattering also depends on the particle size and is most efficient for particles that are much smaller than the wavelength of the light. Scattering from particles much larger than the wavelength of light occurs because of the *Tyndall effect*. This is why clouds, consisting of water droplets, look white. *Compton scattering* also occurs when a photon interacts with an electron, but because the incoming photon loses some of its energy to the electron, the wavelength of the light increases; this is an example of "inelastic" scattering.

The *photoelectric effect* occurs when the energy of a photon is fully absorbed, resulting in the ejection of an electron from an atom. The atom thus becomes ionized. No electrons are ejected when the incoming photons have energies less than the electron binding energy, regardless of the intensity of the light. When the photons have energies greater than the binding energy, electrons are ejected with a kinetic energy equal to the photon energy minus the binding energy. As the energy of the photon increases (or the wavelength

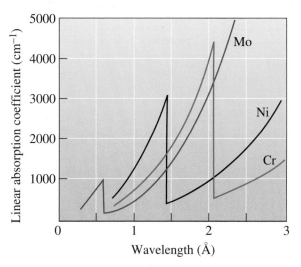

Figure 21-4
The linear absorption coefficient relative to wavelengths for several metals. Note the sudden decrease in the absorption coefficient for wavelengths greater than the absorption edge.

decreases, see Figure 21-4), less absorption occurs until the photon has an energy equal to that of the binding energy. At this energy, the absorption coefficient increases abruptly because electrons may now be ejected, thereby providing an efficient means for photons to be absorbed. The energy or wavelength at which this occurs is referred to as the **absorption edge** because of the shape of the peaks in Figure 21-4. The abrupt change in the absorption coefficient corresponds to the energy required to remove an electron from the atom; this absorption edge is important to certain x-ray analytical techniques. Albert Einstein received the Nobel Prize in Physics in 1921 for explaining the photoelectric effect.

In some cases, the effect of scattering can be written as

$$I = I_0 \exp\left[(-\alpha_i + \alpha_s)x\right] \tag{21-13}$$

In Equation 21-13, α_i is what we previously termed α, the intrinsic absorption coefficient, and α_s is the scattering coefficient.

Examples of a portion of the characteristic spectra for several elements are included in Table 21-2. The K_α, K_β, and L_α lines correspond to the wavelengths of radiation emitted from transitions of electrons between shells, as discussed later in this chapter.

Transmission
The fraction of the beam that is not reflected or absorbed is transmitted through the material. Using the following steps, we can determine the fraction of the beam that is transmitted (see Figure 21-5).

1. If the incident intensity is I_0, then the loss due to reflection at the front face of the material is RI_0. The fraction of the incident beam that actually enters the material is $I_0 - RI_0 = (1 - R)I_0$:

$$I_{\text{reflected at front surface}} = RI_0$$
$$I_{\text{after reflection}} = (1 - R)I_0$$

2. A portion of the beam that enters the material is lost by absorption. The intensity of the beam after passing through a material having a thickness x is

$$I_{\text{after absorption}} = (1 - R)I_0 \exp(-\alpha x)$$

3. Before the partially absorbed beam exits the material, reflection occurs at the back surface. The fraction of the beam that reaches the back surface and is reflected is

$$I_{\text{reflected at back surface}} = R(1 - R)I_0 \exp(-\alpha x)$$

TABLE 21-2 ■ *Characteristic emission lines and absorption edges for selected elements*

Metal	K_α (Å)	K_β (Å)	L_α (Å)	Absorption Edge (Å)
Al	8.337	7.981	—	7.951
Si	7.125	6.768	—	6.745
S	5.372	5.032	—	5.018
Cr	2.291	2.084	—	2.070
Mn	2.104	1.910	—	1.896
Fe	1.937	1.757	—	1.743
Co	1.790	1.621	—	1.608
Ni	1.660	1.500	—	1.488
Cu	1.542	1.392	13.357	1.380
Mo	0.711	0.632	5.724	0.620
W	0.211	0.184	1.476	0.178

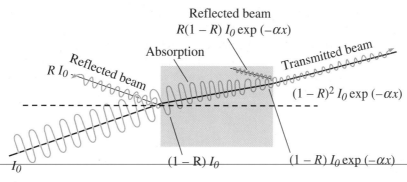

Reflected beam
$R(1 - R) I_0 \exp(-\alpha x)$

Absorption

Transmitted beam

$R I_0$ Reflected beam

$(1 - R)^2 I_0 \exp(-\alpha x)$

I_0

$(1 - R) I_0$

$(1 - R) I_0 \exp(-\alpha x)$

Figure 21-5 Fractions of the original beam that are reflected, absorbed, and transmitted.

4. Consequently, the fraction of the beam that is completely transmitted through the material is

$$I_{transmitted} = I_{after\ absorption} - I_{reflected\ at\ back}$$
$$= (1 - R)I_0 \exp(-\alpha x) - R(1 - R)I_0 \exp(-\alpha x)$$
$$= (1 - R)(1 - R)I_0 \exp(-\alpha x) \tag{21-14}$$
$$I_t = (1 - R)^2 I_0 \exp(-\alpha x)$$

The intensity of the transmitted beam may depend on the wavelength of the photons in the beam. In metals, because there is no energy gap, virtually any photon has sufficient energy to excite an electron into a higher energy level, thus absorbing the energy of the excited photon [Figure 21-6(a)]. As a result, even extremely thin samples of metals are opaque. Dielectrics, on the other hand, possess a large energy gap between the valence and conduction bands. If the energy of the incident photons is less than the energy gap, no electrons gain enough energy to escape the valence band and, therefore, absorption does not occur [Figure 21-6(b)]. In intrinsic semiconductors, the energy gap is smaller

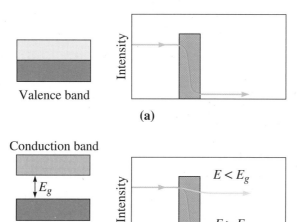

Figure 21-6
Relationships between absorption and the energy gap: (a) metals and (b) dielectrics and intrinsic semiconductors. The diagram on the left represents the band structure of the material under consideration. The diagram on the right represents the intensity of light as it passes from air into the material and back into air.

than that for insulators, and absorption occurs when the photons have energies exceeding the energy gap E_g, whereas transmission occurs for less energetic photons [Figure 21-6(b) also applies to semiconductors]. In other words, only the light below a certain wavelength is absorbed. Extrinsic semiconductors include donor or acceptor energy levels within the bandgap that provide additional energy levels for absorption. Therefore, semiconductors are opaque to short wavelength radiation but transparent to long wavelength photons (see Example 21-3). For example, silicon and germanium appear opaque to visible light, but they are transparent to longer wavelength infrared radiation. Many of the narrow bandgap semiconductors (e.g., HgCdTe) are used for detection of infrared radiation. These detector materials have to be cooled to low temperatures (e.g., using liquid nitrogen) since the thermal energy of electrons at room temperature is otherwise enough to saturate the conduction band.

The intensity of the transmitted beam also depends on microstructural features. Porosity in ceramics scatters photons; even a small amount of porosity (less than 1 volume percent) may make a ceramic opaque. For example, alumina that has relatively low mass density (owing to porosity) is opaque, while high density alumina is optically transparent. High density alumina is often used in the manufacture of lightbulbs. Crystalline precipitates, particularly those that have a much different index of refraction than the matrix material, also cause scattering. These crystalline *opacifiers* cause a glass that normally may have excellent transparency to become translucent or even opaque. Typically, smaller pores or precipitates cause a greater reduction in the transmission of light.

Photoconduction occurs in semiconducting materials if the semiconductor is part of an electrical circuit. If the energy of an incoming photon is sufficient, an electron is excited into the conduction band from the valence band, thereby creating a hole in the valence band. The electron and hole then carry charge through the circuit [Figure 21-7(a)]. The maximum

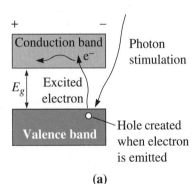

(a)

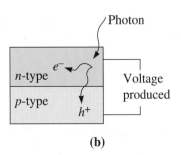

(b)

Figure 21-7
(a) Photoconduction in semiconductors involves the absorption of a stimulus by exciting electrons from the valence band to the conduction band. Rather than dropping back to the valence band to cause emission, the excited electrons carry a charge through an electrical circuit. (b) A solar cell takes advantage of this effect.

wavelength of the incoming photon that will produce photoconduction is related to the energy gap in the semiconducting material:

$$\lambda_{max} = \frac{hc}{E_g} \tag{21-15}$$

We can use this principle for photodetectors or "electric eyes" that open or close doors or switches when a beam of light focused on a semiconducting material is interrupted.

Solar cells also use the absorption of light to generate voltage [Figure 21-7(b)]. Essentially the electron-hole pairs generated by optical absorption are separated, and this leads to the development of a voltage. This voltage causes a current flow in an external circuit. Solar cells are *p-n* junctions designed so that photons excite electrons into the conduction band. The electrons move to the *n*-side of the junction, while holes move to the *p*-side of the junction. This movement produces a contact voltage due to the charge imbalance. If the junction *p-n* is connected to an electric circuit, the junction acts as a battery to power the circuit. Solar cells make use of antireflective coatings so that maximum key elements of the solar spectrum are captured.

LEDs   As discussed in Chapter 19, the light that is absorbed by a direct band gap semiconductor causes electrons to be promoted to the conduction band. When these electrons fall back into the valence band, they combine with the holes and cause emission of light. Many semiconductor solid solutions can be tailored to have particular bandgaps, producing LEDs of different colors. This phenomenon is also used in semiconductor lasers (Figure 21-8). The design of LEDs is discussed later in this chapter.

The following examples illustrate applications of many of these concepts related to the absorption and transmission of light.

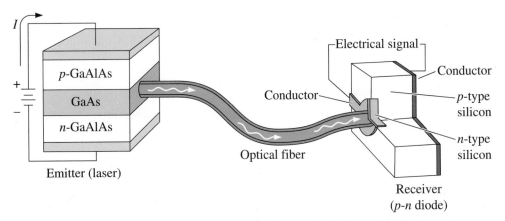

Figure 21-8   Elements of a photonic system for transmitting information involves a laser or LED to generate photons from an electrical signal, optical fibers to transmit the beam of photons efficiently, and an LED receiver to convert the photons back into an electrical signal.

Example 21-3 | *Determining Critical Energy Gaps*

Determine the critical energy gaps for a semiconductor that provide complete transmission and complete absorption of photons in the visible spectrum.

SOLUTION

In order for complete transmission to occur, the bandgap of the semiconductor must be larger than the energies of all photons in the visible spectrum. The visible light spectrum varies from 4×10^{-5} cm to 7×10^{-5} cm. The photons with shorter wavelengths have higher energies. Thus, the minimum band gap energy E_g required to ensure that no photons in the visible spectrum are absorbed (and all photons are transmitted) is

$$E_g = \frac{hc}{\lambda} = \frac{(6.626 \times 10^{-34}\,\text{J}\cdot\text{s})(3 \times 10^{10}\,\text{cm/s})}{(4 \times 10^{-5}\,\text{cm})(1.6 \times 10^{-19}\,\text{J/eV})} = 3.1\,\text{eV}$$

If the semiconductor band gap is 3.1 eV or larger, all photons in the visible spectrum will be transmitted.

In order for a photon to be absorbed, the energy gap of the semiconductor must be less than the photon energy. The photons with longer wavelengths have lower energies. Thus, the maximum bandgap energy that will allow for complete absorption of all wavelengths of the visible spectrum is

$$E_g = \frac{hc}{\lambda} = \frac{(6.626 \times 10^{-34}\,\text{J}\cdot\text{s})(3 \times 10^{10}\,\text{cm/s})}{(7 \times 10^{-5}\,\text{cm})(1.6 \times 10^{-19}\,\text{J/eV})} = 1.8\,\text{eV}$$

If the bandgap is 1.8 eV or smaller, all photons in the visible spectrum will be absorbed. For semiconductors with an E_g between 1.8 eV and 3.1 eV, a portion of the photons in the visible spectrum will be absorbed.

Example 21-4 | *Design of a Radiation Shield*

A material has a reflectivity of 0.15 and an absorption coefficient (α) of 100 cm^{-1}. Design a shield that will permit only 1% of the incident radiation to be transmitted through the material.

SOLUTION

From Equation 21-14, the fraction of the incident intensity that will be transmitted is

$$\frac{I_t}{I_0} = (1 - R)^2 \exp(-\alpha x)$$

and the required thickness of the shield can be determined:

$$0.01 = (1 - 0.15)^2 \exp(-100x)$$

$$\frac{0.01}{(0.85)^2} = 0.01384 = \exp(-100x)$$

$$\ln(0.01384) = -4.28 = -100x$$

$$x = 0.0428 \text{ cm}$$

The material should have a thickness of 0.0428 cm in order to transmit 1% of the incident radiation.

If we wished, we could determine the amount of radiation lost in each step:

$$\text{Reflection at the front face: } I_r = RI_0 = 0.15I_0$$
$$\text{Intensity after reflection: } I = I_0 - 0.15I_0 = 0.85I_0$$
$$\text{Intensity after absorption: } I_a = (0.85)I_0 \exp[(-100)(0.0428)] = 0.0118I_0$$
$$\text{Absorbed Intensity: } 0.85I_0 - 0.0118I_0 = 0.838I_0$$
$$\text{Reflection at the back face: } I_r = R(1 - R)I_0 \exp(-\alpha x)$$
$$= (0.15)(1 - 0.15)I_0 \exp[-(100)(0.0428)] = 0.0018I_0$$

21-3   Selective Absorption, Transmission, or Reflection

Unusual optical behavior is observed when photons are selectively absorbed, transmitted, or reflected. We have already seen that semiconductors transmit long wavelength photons but absorb short wavelength radiation. There are a variety of other cases in which similar selectivity produces unusual optical properties.

In certain materials, replacement of normal ions by transition or rare earth elements produces a *crystal field*, which creates new energy levels within the structure. This phenomenon occurs when Cr^{3+} ions replace Al^{3+} ions in Al_2O_3. The new energy levels absorb visible light in the violet and green-yellow portions of the spectrum. Red wavelengths are transmitted, giving the reddish color in ruby. In addition, the chromium ion replacement creates an energy level that permits luminescence (discussed later) to occur when the electrons are excited by a stimulus. Lasers made from chromium-doped ruby produce a characteristic red beam because of this.

Glasses can also be doped with ions that produce selective absorption and transmission (Table 21-3). Similarly, electrons or hole traps called *F-centers*, can be present in crystals. When fluorite (CaF_2) is formed with excess calcium, a fluoride ion vacancy is produced. To maintain electrical neutrality, an electron is trapped in the vacancy, producing energy levels that absorb all visible photons—with the exception of purple.

Polymers—particularly those containing an aromatic ring in the backbone—can have complex covalent bonds that produce an energy level structure which causes selective absorption. For this reason, chlorophyll in plants appears green, and hemoglobin in blood appears red.

TABLE 21-3 ■ *Effect of ions on colors produced in glasses*

Ion	Color	Ion	Color
Cr^{2+}	Blue	Mn^{2+}	Orange
Cr^{3+}	Green	Fe^{2+}	Blue-green
Cu^{2+}	Blue-green	U^{6+}	Yellow

21-4   Examples and Use of Emission Phenomena

Let's look at some particular examples of emission phenomena which, by themselves, provide some familiar and important functions.

Gamma Rays—Nuclear Interactions

Gamma rays, which are very high-energy photons, are emitted during the radioactive decay of unstable nuclei of certain atoms. Therefore, the energy of the gamma rays depends on the structure of the atom nucleus and varies for different materials. The gamma rays produced from a material have fixed wavelengths. For example, when cobalt 60 decays, gamma rays having energies of 1.17×10^6 and 1.34×10^6 eV (or wavelengths of 1.06×10^{-10} cm and 0.93×10^{-10} cm) are emitted. The gamma rays can be used as a radiation source to detect defects in a material (a nondestructive test).

X-rays—Inner Electron Shell Interactions

X-rays, which have somewhat lower energy than gamma rays, are produced when electrons in the inner shells of an atom are stimulated. The stimulus could be high-energy electrons or other x-rays. When stimulation occurs, x-rays of a wide range of energies are emitted. Both a continuous and a characteristic spectrum of x-rays are produced.

Suppose that a high-energy electron strikes a material. As the electron decelerates, energy is given up and emitted as photons. Each time the electron strikes an atom, more of its energy is given up. Each interaction, however, may be more or less severe, so the electron gives up a different fraction of its energy each time and produces photons of different wavelengths (Figure 21-9). A **continuous spectrum** is produced (the smooth portion of the curves in Figure 21-10). If the electron were to lose all of its energy in one impact, the minimum wavelength of the emitted photons would correspond to the

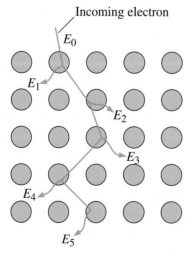

$$E_1 + E_2 + E_3 + E_4 + E_5 = E_0$$

Figure 21-9
When an accelerated electron strikes and interacts with a material, its energy may be reduced in a series of steps. In the process, several photons of different energies E_1 to E_5 are emitted, each with a unique wavelength.

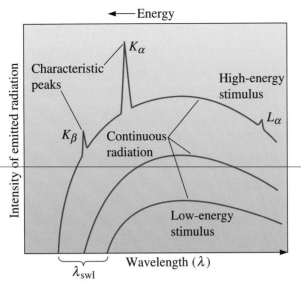

Figure 21-10
The continuous and characteristic spectra of radiation emitted from a material. Low-energy stimuli produce a continuous spectrum of low-energy, long wavelength photons. A more intense, higher energy spectrum is emitted when the stimulus is more energetic until, eventually, characteristic radiation is observed.

original energy of the stimulus. The minimum wavelength of x-rays produced is called the **short wavelength limit** λ_{swl}. The short wavelength limit decreases, and the number and energy of the emitted photons increase, when the energy of the stimulus increases.

The incoming stimulus may also have sufficient energy to excite an electron from an inner energy level to an outer energy level. The excited electron is not stable and, to restore equilibrium, electrons from a higher level fill the empty inner level. This process leads to the emission of a **characteristic spectrum** of x-rays that is different for each type of atom.

The characteristic spectrum is produced because there are discrete energy differences between any two energy levels. When an electron drops from one level to a second level, a photon having the corresponding energy and wavelength is emitted. This effect is illustrated in Figure 21-11. We typically refer to the energy levels by the *K, L, M, . . .* designation, as described in Chapter 2. If an electron is excited from the *K* shell, electrons may fill that vacancy from an outer shell. Normally, electrons in the closest shells fill the vacancies. Thus, photons with energy $\Delta E = E_K - E_L$ (K_α x-rays) or $\Delta E = E_K - E_M$ (K_β x-rays) are emitted. When an electron from the *M* shell fills the *L* shell, a photon with energy $\Delta E = E_L - E_M$ (L_α x-rays) is emitted; it has a long wavelength, or low energy. Note that we need a more energetic stimulus to produce K_α x-rays than that required for L_α x-rays.

As a consequence of the emission of photons having a characteristic wavelength, a series of peaks is superimposed on the continuous spectrum (Figure 21-10). The wavelengths at which these peaks occur are unique to each type of atom (Table 21-2). Thus, each element produces a different characteristic spectrum, which serves as a "finger-print" for that type of atom. If we match the emitted characteristic wavelengths with those expected for various elements, the identity of the material can be determined. We can also measure the intensity of the characteristic peaks. By comparing measured intensities with standard intensities, we can estimate the percentage of each type of atom in the material and, hence, we can estimate the composition of the material. The energy (or wavelength) of the x-rays emitted when an electron beam impacts a sample (such as that in a scanning or transmission electron microscope) can be analyzed to get chemical information about

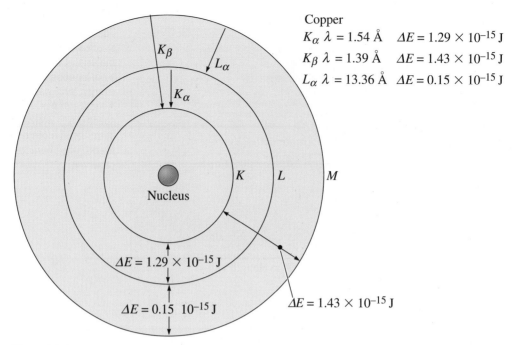

Figure 21-11 Characteristic x-rays are produced when electrons transition from one energy level to a lower energy level, as illustrated here for copper. The energy and wavelength of the x-rays are fixed by the energy differences between the energy levels.

a sample. This technique is known as energy dispersive x-ray analysis (EDXA). The examples that follow illustrate the application of x-ray emission as used in x-ray diffraction (XRD) and EDXA analytical techniques.

| **Example 21-5** | *Design/Materials Selection for an X-ray Filter* |

Design a filter that preferentially absorbs K_β x-rays from the nickel spectrum but permits K_α x-rays to pass with little absorption. This type of filter is used in x-ray diffraction (XRD) analysis of materials.

SOLUTION

When determining a crystal structure or identifying unknown materials using various x-ray diffraction techniques, we prefer to use x-rays of a single wavelength. If both K_α and K_β characteristic peaks are present and interact with the material, analysis becomes much more difficult.

To avoid this difficulty, we can use selective absorption to isolate the K_α peak. Table 21-2 includes the information that we need. If a filter material is selected such that the absorption edge lies between the K_α and K_β wavelengths, then the K_β is almost completely absorbed, whereas the K_α is almost completely transmitted. In nickel, $K_\alpha = 1.660$ Å and $K_\beta = 1.500$ Å. A filter with an absorption edge between these characteristic peaks will work. Cobalt, with an absorption edge of 1.608 Å, would be our choice. Figure 21-12 shows how this filtering process occurs.

Absorption edge

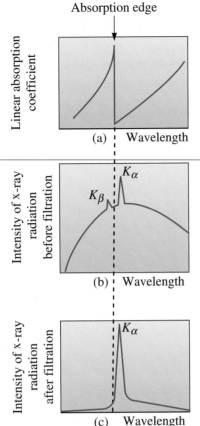

(a) Wavelength

(b) Wavelength

(c) Wavelength

Figure 21-12
Elements have a selective lack of absorption of certain wavelengths. If a filter is selected with an absorption edge between the K_α and K_β peaks of an x-ray spectrum, all x-rays except K_α are absorbed (for Example 21-5). (a) The linear absorption coefficient of a filter material as a function of wavelength; (b) the intensity of the x-ray radiation before filtration; and (c) the intensity of the x-ray radiation after filtration.

Example 21-6   *Design of an X-ray Filter*

Design a filter to transmit at least 95% of the energy of a beam composed of zinc K_α x-rays, using aluminum as the shielding material. (The aluminum has a linear absorption coefficient of 108 cm^{-1}.) Assume no loss to reflection.

SOLUTION

Assuming that no losses are caused by the reflection of x-rays from the aluminum, we simply need to choose the thickness of the aluminum required to transmit 95% of the incident intensity. The final intensity will therefore be $0.95I_0$. Thus, from Equation 21-12,

$$\ln\left(\frac{0.95I_0}{I_0}\right) = -(108)(x)$$

$$\ln(0.95) = -0.051 = -108x$$

$$x = \frac{-0.051}{-108} = 0.00047 \text{ cm}$$

We would like to roll the aluminum to a thickness of 0.00047 cm or less. The filter could be thicker if a material were selected that has a lower linear absorption coefficient for zinc K_α x-rays.

Example 21-7   *Generation of X-rays for X-ray Diffraction (XRD)*

Suppose an electron accelerated at 5000 V strikes a copper target. Will K_α, K_β, or L_α x-rays be emitted from the copper target?

SOLUTION

The electron must possess enough energy to excite an electron to a higher level, or its wavelength must be less than that corresponding to the energy difference between the shells:

$$E = (5000\,\text{eV})(1.6 \times 10^{-19}\,\text{J/eV}) = 8 \times 10^{-16}\,\text{J}$$

$$\lambda = \frac{hc}{E} = \frac{(6.626 \times 10^{-34}\,\text{J}\cdot\text{s})(3 \times 10^{10}\,\text{cm/s})}{8 \times 10^{-16}\,\text{J}}$$

$$= 2.48 \times 10^{-8}\,\text{cm} = 2.48\,\text{Å}$$

Note that one electron volt (eV) is the kinetic energy acquired by an electron moving through a potential difference of one volt.

For copper, K_α is 1.542 Å, K_β is 1.392 Å and L_α is 13.357 Å (Table 21-2). Therefore, the L_α peak may be produced, but K_α and K_β will not.

Example 21-8   *Energy Dispersive X-ray Analysis (EDXA)*

The micrograph in Figure 21-13 was obtained using a scanning electron microscope at a magnification of 1000. The beam of electrons in the SEM was directed at the three different phases, creating x-rays and producing the characteristic peaks. From the energy spectra, determine the probable composition of each phase. Assume each region represents a different phase.

SOLUTION

All three phases have an energy peak of about 1.5 keV = 1500 eV, which corresponds to a wavelength of

$$\lambda = \frac{hc}{E} = \frac{(6.626 \times 10^{-34}\,\text{J}\cdot\text{s})(3 \times 10^{10}\,\text{cm/s})}{(1500\,\text{eV})(1.6 \times 10^{-19}\,\text{J/eV})(10^{-8}\,\text{cm/Å})} = 8.283\,\text{Å}$$

In a similar manner, energies and wavelengths can be found for the other peaks. These wavelengths are compared with those in Table 21-2, and the identity of the elements in each phase can be found, as summarized in the table.

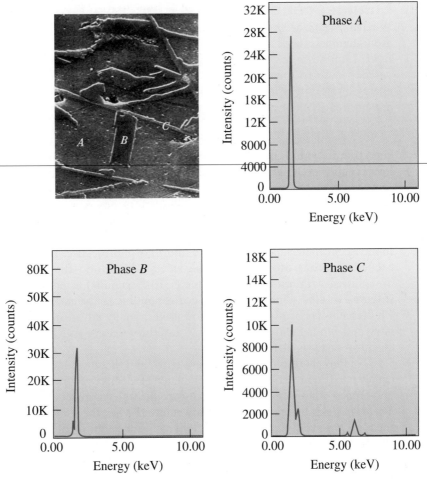

Figure 21-13 Scanning electron micrograph of a multiple phase material. The energy distributions of emitted radiation from the three phases marked *A*, *B*, and *C* are shown. The identity of each phase is determined in Example 21-8. (*Reprinted courtesy of Don Askeland.*)

Phase	Peak Energy	λ	λ (Table 21-2)	Line
A	1.5 keV	8.283 Å	8.337 Å	K_αAl
B	1.5 keV	8.283 Å	8.337 Å	K_αAl
	1.7 keV	7.308 Å	7.125 Å	K_αSi
C	1.5 keV	8.283 Å	8.337 Å	K_αAl
	1.7 keV	7.308 Å	7.125 Å	K_αSi
	5.8 keV	2.142 Å	2.104 Å	K_αMn
	6.4 keV	1.941 Å	1.937 Å	K_αFe
	7.1 keV	1.750 Å	1.757 Å	K_βFe

Thus, Phase *A* appears to be an aluminum matrix, Phase *B* appears to be a silicon needle (perhaps containing some aluminum), and Phase *C* appears to be an Al-Si-Mn-Fe compound. Actually, this is an aluminum-silicon alloy. The stable phases are aluminum and silicon with inclusions forming due to the presence of manganese and iron as impurities.

Luminescence—Outer Electron Shell Interactions

Whereas x-rays are produced by electron transitions in the inner energy levels of an atom, **luminescence** is the conversion of radiation or other forms of energy to visible light. Luminescence occurs when the incident radiation excites electrons from the valence band to the conduction band. The excited electrons remain in the higher energy levels only briefly. When the electrons drop back to the valence band, photons are emitted. If the wavelength of these photons is in the visible light range, luminescence occurs.

Luminescence does not occur in metals. Electrons are merely excited into higher energy levels within the unfilled valence band. When the excited electron returns to the lower energy level, the photon that is produced has a very small energy and a wavelength longer than that of visible light [Figure 21-14(a)].

In certain ceramics and semiconductors, however, the energy gap between the valence and conduction bands is such that an electron dropping through this gap produces a photon in the visible range. Two different effects are observed in these luminescent materials: fluorescence and phosphorescence. In **fluorescence,** all of the excited electrons drop back to the valence band and the corresponding photons are emitted within a very short time ($\sim10^{-8}$ seconds) after the stimulus is removed [Figure 21-14(b)]. One wavelength, corresponding to the energy gap E_g, predominates. Fluorescent dyes and microscopy are used in many advanced techniques in biochemistry and biomedical engineering. X-ray fluorescence (XRF) is widely used for the chemical analysis of materials.

Phosphorescent materials have impurities that introduce a donor level within the energy gap [Figure 21-14(c)]. The stimulated electrons first drop into the donor level and are trapped. The electrons must then escape the trap before returning to the valence band. There is a delay before the photons are emitted. When the source is removed, electrons in the traps gradually escape and emit light over some additional period of time. The intensity of the luminescence is given by

$$\ln\left(\frac{I}{I_0}\right) = -\frac{t}{j} \tag{21-16}$$

where τ is the **relaxation time**, a constant for the material. After time t following removal of the source, the intensity of the luminescence is reduced from I_0 to I. Phosphorescent

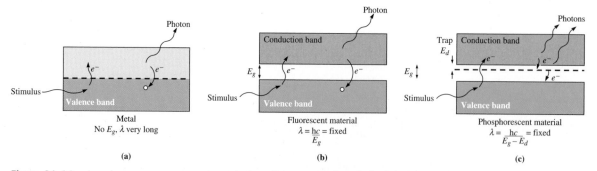

Figure 21-14 Luminescence occurs when photons have a wavelength in the visible spectrum. (a) In metals, there is no energy gap, so luminescence does not occur. (b) Fluorescence may occur if there is an energy gap. (c) Phosphorescence occurs when the photons are emitted over a period of time due to donor traps in the energy gap.

materials are very important in the operation of television screens. In this case, the relaxation time must not be too long or the images begin to overlap. In color television, three types of phosphorescent materials are used; the energy gaps are engineered so that red, green, and blue colors are produced. Oscilloscope and radar screens rely on the same principle. Fluorescent lamps contain mercury vapor. The mercury vapor, in the presence of an electric arc, fluoresces and produces ultraviolet light. The inside of the glass of these lamps is coated with a phosphorescent material. The role of this material is to convert the small wavelength ultraviolet radiation into visible light. The relaxation times range from 5×10^{-9} seconds to about 2 seconds. The following example illustrates the selection of a phosphor for a television screen.

Example 21-9   *Design/Materials Selection for a Television Screen*

Select a phosphor material that will produce a blue image on a television screen.

SOLUTION

Photons having energies that correspond to the color blue have wavelengths of about 4.5×10^{-5} cm (Figure 21-1). The energy of the emitted photons therefore is

$$E = \frac{hc}{\lambda} = \frac{(4.14 \times 10^{-15} \, \text{eV} \cdot \text{s})(3 \times 10^{10} \, \text{cm/s})}{4.5 \times 10^{-5} \, \text{cm}}$$

$$= 2.76 \, \text{eV}$$

Figure 21-1 includes energy gaps for a variety of materials. None of the materials listed has an E_g of 2.76 eV, but ZnS has an E_g of 3.54 eV. If a suitable dopant were introduced to provide a trap $3.54 - 2.76 = 0.78$ eV below the conduction band, phosphorescence would occur.

We would also need information concerning the relaxation time to ensure that phosphorescence would not persist long enough to distort the image. Typical phosphorescent materials for television screens might include $CaWO_4$, which produces photons with a wavelength of 4.3×10^{-5} cm (blue). This material has a relaxation time of 4×10^{-6} s. ZnO doped with excess zinc produces photons with a wavelength of 5.1×10^{-5} cm (green), whereas $Zn_3(PO_4)_2$ doped with manganese gives photons with a wavelength of 6.45×10^{-5} cm (red).

Light-Emitting Diodes—Electroluminescence

Luminescence can be used to advantage in creating **light-emitting diodes** (LEDs). LEDs are used to provide the display for watches, clocks, calculators, and other electronic devices. The stimulus for these devices is an externally applied voltage, which causes electron transitions and **electroluminescence**. LEDs are *p-n* junction devices engineered so that the E_g is in the visible spectrum (often red). A voltage applied to the diode in the forward-bias direction causes holes and electrons to recombine at the junction and emit photons (Figure 21-15). GaAs, GaP, GaAlAs, and GaAsP are typical materials for LEDs.

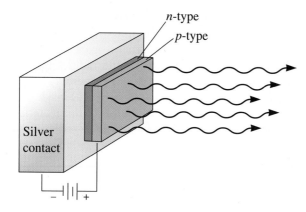

n-type
p-type

Silver
contact

− +

Figure 21-15
Diagram of a light-emitting
diode (LED). A forward-bias
voltage across the *p-n*
junction produces photons.

Lasers—Amplification of Luminescence

The **laser** (*l*ight *a*mplification by *s*timulated *e*mission of *r*adiation) is another example of a special application of luminescence. In certain materials, electrons excited by a stimulus produce photons which, in turn, excite additional photons of identical wavelength. Consequently, a large amplification of the photons emitted in the material occurs. By proper choice of stimulant and material, the wavelength of the photons can be in the visible range. The output of the laser is a beam of photons that are parallel, of the same wavelength, and coherent. In a *coherent* beam, the wavelike nature of the photons is in phase, so that destructive interference does not occur. Lasers are useful in heat treating and melting of metals, welding, surgery, and transmission and processing of information. They are also useful in a variety of other applications including reading of compact discs and DVDs. Blu-ray^{TM} technology has enabled high-resolution DVDs to be commercialized. The shorter wavelength of the blue–violet light from a laser can read finer pits that encode the digital information than longer wavelength light. This has enabled a nearly six-fold increase in data storage capability compared to conventional DVDs, making it possible to store a two hour, high-definition movie on a single disc.

A variety of materials are used to produce lasers. Ruby, which is single crystal Al_2O_3 doped with a small amount of Cr_2O_3 (emits at 6943 Å) and yttrium aluminum garnet ($Y_3Al_5O_{12}$ YAG) doped with neodymium (Nd) (emits at 1.06 μm) are two common solid-state lasers. Other lasers are based on CO_2 gas.

Semiconductor lasers, such as those based on GaAs solid solutions which have an energy gap corresponding to a wavelength in the visible range, are also used (Figure 21-16).

Thermal Emission

When a material is heated, electrons are thermally excited to higher energy levels, particularly in the outer energy levels where the electrons are less tightly bound to the nucleus. The electrons immediately drop back to their normal levels and release photons, an event known as **thermal emission**.

As the temperature increases, thermal agitation increases, and the maximum energy of the emitted photons increases. A continuous spectrum of radiation is emitted, with a minimum wavelength and an intensity distribution dependent on the temperature. The photons may include wavelengths in the visible spectrum; consequently, the color of the material changes with temperature. At low temperatures, the wavelength of the radiation is too long to be visible. As the temperature increases, emitted photons have shorter wavelengths. At 700°C, we begin to see a reddish tint; at 1500°C, the orange and red wavelengths are emitted (Figure 21-17). Higher temperatures produce all wavelengths in the

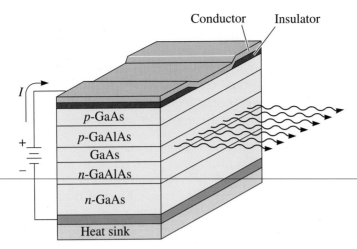

Figure 21-16
Schematic cross-section of a GaAs laser. Because the surrounding p- and n-type GaAlAs layers have a higher energy gap and a lower index of refraction than GaAs, the photons are trapped in the active GaAs layer.

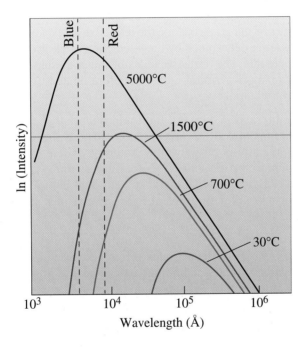

Figure 21-17
Intensity in relation to wavelengths of photons emitted thermally from a material. As the temperature increases, more photons are emitted in the visible spectrum.

visible range, and the emitted spectrum is white light. By measuring the intensity of a narrow band of the emitted wavelengths with a pyrometer, we can estimate the temperature of the material.

21-5 Fiber-Optic Communication System

In 1880, Alexander Graham Bell invented a light-based communication system known as the photophone, and William Wheeler was granted a patent (U.S. Patent 247,229) in 1881 for a system that used pipes to light distant rooms. These two inventions were the predecessors of the fiber-optic communications systems that exist today. The other key invention that helped commercialize fiber-optics technologies was the invention of the laser in 1960. The laser provided a monochromatic source of light so fiber optics could be used effectively. Another advancement came when high-purity silica glass fibers became available. These fibers provide very small optical losses and are essential for carrying information over longer distances without the need for equipment to boost the signal. Optical fibers are also free from electromagnetic interference (EMI) since they carry signals as light, not radio waves.

A fiber-optic system transmits a light signal generated from some other source, such as an electrical signal. The fiber-optic system transmits the light to a receiver using an optical fiber, processes the data received, and converts the data to a usable form. Photonic materials are required for this process. Most of the principles and materials presently used in photonic systems have already been introduced in the previous sections.

Summary

- The optical properties of materials include the refractive index, absorption coefficient, and dispersion. These are determined by the interaction of electromagnetic radiation, or photons, with materials. The refractive index of materials depends primarily upon the extent of electronic polarization and is therefore related to the high-frequency dielectric constant of materials.

- As a result of the interaction between light and materials, refraction, reflection, transmission, scattering, and diffraction can occur. These phenomena are used in a wide variety of applications of photonic materials. These applications include fiber optics for communication and lasers for surgery and welding. Devices that use optoelectronic effects include LEDs, solar cells, and photodiodes. Other applications include phosphors for fluorescent lights, televisions, and many analytical techniques.

- Emission of photons occurs by electron transitions or nuclear decay within an atom. Fluorescence, phosphorescence, electroluminescence (used in light-emitting diodes), and lasers are examples of luminescence. Photons are emitted by thermal excitation, with photons in the visible portion of the spectrum produced when the temperature is sufficiently high. X-ray emission from materials is used in EDXA and XRF analysis.

Glossary

Absorption edge The wavelength at which the absorption characteristics of a material abruptly change.

Characteristic spectrum The spectrum of radiation emitted from a material. It shows peaks at fixed wavelengths corresponding to particular electron transitions within an atom. Every element has a unique characteristic spectrum.

Continuous spectrum Radiation emitted from a material having all wavelengths longer than a critical short wavelength limit.

Dispersion Frequency dependence of the refractive index.

Electroluminescence Use of an applied electrical signal to stimulate photons from a material.

Fluorescence Emission of light obtained typically within $\sim 10^{-8}$ seconds of stimulation.

Index of refraction Relates the change in velocity and propagation direction of radiation as it passes through a transparent medium (also known as the refractive index).

Laser The acronym stands for light amplification by stimulated emission of radiation. A beam of monochromatic coherent radiation produced by the controlled emission of photons.

Light-emitting diodes (LEDs) Electronic *p-n* junction devices that convert an electrical signal into visible light.

Linear absorption coefficient Describes the ability of a material to absorb radiation.

Luminescence Conversion of radiation to visible light.

Phosphorescence Emission of radiation from a material after the stimulus is removed.

Photoconduction Production of a voltage due to the stimulation of electrons into the conduction band by radiation.

Photons Energy or radiation produced from atomic, electronic, or nuclear sources that can be treated as particles or waves.

Reflectivity The percentage of incident radiation that is reflected.

Refractive index See Index of refraction.

Relaxation time The time required for $1/e$ of the electrons to drop from the conduction band to the valence band in luminescence.

Short wavelength limit The shortest wavelength or highest energy radiation emitted from a material under particular conditions.

Solar cell A *p-n* junction device that creates a voltage due to excitation by photons.

Thermal emission Emission of photons from a material due to excitation of the material by heat.

X-rays Electromagnetic radiation in the wavelength range ~ 0.1 to 100 Å.

Problems

Section 21-1 The Electromagnetic Spectrum

Section 21-2 Refraction, Reflection, Absorption, and Transmission

21-1 State the definitions of refractive index and absorption coefficient. Compare these with the definitions of dielectric constant, loss factor, Young's modulus, and viscous deformation.

21-2 What is Snell's law? Illustrate using a diagram.

21-3 Upon what does the index of refraction of a material depend?

21-4 What is "lead crystal?" What makes the refractive index of this material so much higher than that of ordinary silicate glass?

21-5 What polarization mechanism affects the refractive index?

21-6 Why is the refractive index of ice smaller than that of water?

21-7 What is dispersion? What is the importance of dispersion in fiber-optic systems?

21-8 What factors limit the transmission of light through dielectric materials?

21-9 What is the principle by which LEDs and solar cells work?

21-10 A beam of photons strikes a material at an angle of 25° to the normal of the surface. Which, if any, of the materials listed in Table 21-1 could cause the beam of photons to continue at an angle of 18 to 20° from the normal of the material's surface?

21-11 A laser beam passing through air strikes a 5-cm-thick polystyrene block at a 20° angle to the normal of the block. By what distance is the beam displaced from its original path when the beam reaches the opposite side of the block?

21-12 A length of 6000 km of fiber-optic cable is laid to connect New York to London. If the core of the cable has a refractive index of 1.48 and the cladding has a refractive index of 1.45, what is the time needed for a beam of photons introduced at 0° in New York to reach London? Assume that dispersion effects can be neglected for this calculation. What is the maximum angle of incidence at which there is no leakage of light from the core?

21-13 A block of glass 10 cm thick with $n = 1.5$ transmits 90% of light incident on it. Determine the linear absorption coefficient (α) for this material. If this block is placed in water, what fraction of the incident light will be transmitted through it?

21-14 A beam of photons passes through air and strikes a soda-lime glass that is part of an aquarium containing water. What fraction of the beam is reflected by the front face of the glass? What fraction of the remaining beam is reflected by the back face of the glass?

21-15 We find that 20% of the original intensity of a beam of photons is transmitted from air through a 1-cm-thick material having a dielectric constant of 2.3 and back into air. Determine the fraction of the beam that is
(a) reflected at the front surface,
(b) absorbed in the material, and
(c) reflected at the back surface.
Determine the linear absorption coefficient of the photons in the material.

21-16 A beam of photons in air strikes a composite material consisting of a 1-cm-thick sheet of polyethylene and a 2-cm-thick sheet of soda-lime glass. The incident beam is 10° from the normal of the composite. Determine the angle of the beam with respect to the normal of the composite as it
(a) passes through the polyethylene,
(b) passes through the glass, and
(c) passes through air on the opposite side of the composite.
By what distance is the beam displaced from its original path when it emerges from the composite?

21-17 A glass fiber ($n = 1.5$) is coated with Teflon. Calculate the maximum angle that a beam of light can deviate from the axis of the fiber without escaping from the inner portion of the fiber.

21-18 A material has a linear absorption coefficient of 591 cm^{-1} for photons of a particular wavelength. Determine the thickness of the material required to absorb 99.9% of the photons.

Section 21-3 Selective Absorption, Transmission, or Reflection

21-19 What is a photochromic glass?

21-20 How are colored glasses produced?

21-21 What is ruby crystal made from?

Section 21-4 Examples and Uses Of Emission Phenomena

21-22 What is the principle of energy dispersive x-ray analysis (EDXA)?

21-23 What is fluorescence? What is phosphorescence?

21-24 What is XRF?

21-25 How does a fluorescent lamp work?

21-26 What is electroluminescence?

21-27 A scanning electron microscope has three settings for the acceleration voltage (a) 5 keV, (b) 10 keV, and (c) 20 keV. Determine the minimum voltage setting needed to produce K_α peaks for the materials listed in Table 21-2.

21-28 The relaxation time of a phosphor used for a TV screen is 5×10^{-2} seconds. If the refresh frequency is 60 Hz, then what is the reduction in intensity of the luminescence before it is reset to 100% by the refresh?

21-29 Calcium tungstate ($CaWO_4$) has a relaxation time of 4×10^{-6} s. Determine the time required for the intensity of this phosphorescent material to decrease to 1% of the original intensity after the stimulus is removed.

21-30 The intensity of radiation from a phosphorescent material is reduced to 90% of its original intensity after 1.95×10^{-7} s. Determine the time required for the intensity to decrease to 1% of its original intensity.

21-31 A phosphor material with a bandgap of 3.5 eV with appropriate doping will be used to produce blue (475 nm) and green (510 nm) colors. Determine the energy level of the donor traps with respect to the conduction band in each case.

21-32 What is a laser?

21-33 Determine the wavelength of photons produced when electrons excited into the conduction band of indium-doped silicon

(a) drop from the conduction band to the acceptor band and

(b) then drop from the acceptor band to the valence band (See Chapter 19).

21-34 Which, if any, of the semiconducting compounds listed in Chapters 19 and 21 are capable of producing an infrared laser beam?

21-35 What type of electromagnetic radiation (ultraviolet, infrared, visible) is produced when an electron recombines with a hole in pure germanium and what is its wavelength?

21-36 Which, if any, of the dielectric materials listed in Chapter 19 would reduce the speed of light in air from 3×10^{10} cm/s to less than 0.5×10^{10} cm/s?

21-37 What filter material would you use to isolate the K_α peak of the following x-rays: iron, manganese, or nickel? Explain your answer.

21-38 What voltage must be applied to a tungsten filament to produce a continuous spectrum of x-rays having a minimum wavelength of 0.09 nm?

21-39 A tungsten filament is heated with a 12,400 V power supply. What is

(a) the wavelength and

(b) the frequency of the highest energy x-rays that are produced?

21-40 What is the minimum accelerating voltage required to produce K_α x-rays in nickel?

21-41 Based on the characteristic x-rays that are emitted, determine the difference in energy between electrons in tungsten for

(a) the K and L shells,

(b) the K and M shells, and

(c) the L and M shells.

21-42 Figure 21-18 shows the results of an x-ray fluorescent analysis in which the intensity of x-rays emitted from a material is plotted relative to the wavelength of the x-rays. Determine

(a) the accelerating voltage used to produce the exciting x-rays and

(b) the identity of the elements in the sample.

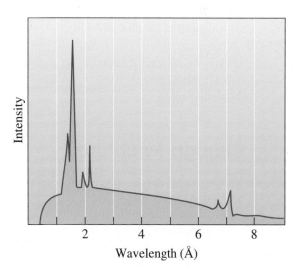

Figure 21-18 Results from an x-ray fluorescence analysis of an unknown metal sample (for Problem 21-42).

21-43 Figure 21-19 shows the intensity as a function of energy of x-rays produced from an energy-dispersive analysis of radiation emitted from a specimen in a scanning electron microscope. Determine the identity of the elements in the sample.

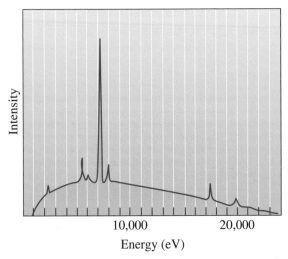

Figure 21-19   X-ray emission spectrum (for Problem 21-43).

Section 21-5 Fiber Optic Communication System

21-44 What is the principle by which information is transmitted via an optical fiber?

21-45 What material is used to make most optical fibers? From what material is the cladding made? What material is used to enhance the refractive index of the core?

Design Problems

21-46 Nickel x-rays are to be generated inside a container, with the x-rays being emitted from the container through only a small slot. Design a container that will ensure that no more than 0.01% of the K_α nickel x-rays escape through the rest of the container walls, yet 95% of the K_α nickel x-rays pass through a thin window covering the slot. The following data give the mass absorption coefficients of several metals for nickel K_α x-rays. The mass absorption coefficient α_m is α/ρ, where α is the linear mass absorption coefficient and ρ is the density of the filter material.

Material	α_m (cm²/g)
Be	1.8
Al	58.4
Ti	247.0
Fe	354.0
Co	54.4
Cu	65.0
Sn	322.0
Ta	200.0
Pb	294.0

21-47 Design a method by which a photoconductive material might be used to measure the temperature of a material from the material's thermal emission.

21-48 Design a method, based on a material's refractive characteristics, that will cause a beam of photons originally at a 2° angle to the normal of the material to be displaced from its original path by 2 cm at a distance of 50 cm from the material.

21-49 Amorphous selenium is used in photocopiers. Conduct a literature search and find out how amorphous selenium works in this application.

Computer Problems

21-50 *Calculating Power in Decibels.* In an optical communications system or electrical power transmission system, the power or signal often is transferred between several components. The decibel (dB) is a convenient unit to measure the relative power levels. If the input power to a device is P_1 and the output power is P_2, then P_2/P_1 is the ratio of power transmitted, thus representing efficiency. This ratio in decibels is written as

$$dB = 10 \log \frac{P_2}{P_1}$$

Power must be expressed in similar units. Write a computer program that will calculate the dB value for the transmission of power between two components of a fiber-optic system (e.g., light source to fiber). Then, extend this calculation to three components (e.g., light transmitted from source to fiber and then fiber to a detector).

Ⓚ Knovel® **Problem**

K21-1 A beam of light passes through benzene to glass (silicon dioxide). The angle of incidence of the light is 30°. What is the angle of the refracted light?

K21-2 Calculate the reflectivity of mercury in air from the index of refraction n.

Thermal properties of materials play a very important role in many technologies. From refractories used in the production of metallic materials to thermal barrier coatings (TBCs) for turbine blades, many applications require materials that minimize heat transfer. Thermal management is of paramount importance in the microelectronics industry. Some computer memory chips now have more than a billion transistors; the heat that is produced must be dissipated effectively for devices to function properly.

The thermal expansion of materials also plays a key role in many situations; such expansion can lead to the development of stresses that lead to material failure. Materials scientists and engineers have created many novel materials that have a negative or near zero thermal expansion coefficient. This is an important characteristic in optical materials such as those used in telescope mirrors because the focus of the mirrors changes depending upon thermal expansion and contraction. This photograph shows an artist's rendering of the Chandra x-ray telescope. The mirror substrates of NASA's Chandra and Hubble telescopes have been made from a glass-ceramic material known as Zerodur™, manufactured by Schott Glass Technologies. Notice the solar panels as well. Zerodur™ is just one of the many examples of how the science and engineering of materials has had an impact on other scientific disciplines, creating knowledge and benefits to society. (*Image Courtesy of NASA.*)

Chapter 22

Thermal Properties of Materials

Have You Ever Wondered?

- *What material has the highest thermal conductivity?*
- *Are there materials that have zero or negative thermal-expansion coefficients?*
- *What material is used to make the Chandra x-ray telescope mirror substrate?*
- *What materials are used to protect the space shuttle against extreme high and low temperatures?*

In previous chapters, we described how a material's properties can change with temperature. In many cases, we found that a material's mechanical and physical properties depend on the service or processing temperature. An appreciation of the thermal properties of materials is helpful to understanding the mechanical failure of materials when the temperature changes; in designing processes in which materials must be heated; or in selecting materials to transfer heat rapidly.

Thermal management has become a very important issue in electronic packaging materials. Some computer memory chips now have more than a billion transistors; the heat that is produced must be dissipated effectively for devices to function properly.

In metallic materials, electrons transfer heat. In ceramic materials, the conduction of heat involves phonons. In certain other applications, such as thermal barrier coatings or space shuttle tiles, we want to minimize the heat transfer through the material. Heat transfer is also important in many applications ranging from, for example, polystyrene foam cups used for hot beverages to sophisticated coatings on glasses to make energy efficient buildings. In this chapter, we will discuss heat capacity, thermal expansion properties, and the thermal conductivity of materials.

22-1   Heat Capacity and Specific Heat

In Chapter 21, we noted that optical behavior depends on how photons are produced and interact with a material. The photon is treated as a particle with a particular energy or as electromagnetic radiation having a particular wavelength or frequency. Some of the thermal properties of materials can also be characterized in the same dual manner; however, these properties are determined by the behavior of **phonons**, rather than photons.

At absolute zero, the atoms in a material have a minimum energy. When heat is supplied, the atoms gain thermal energy and vibrate at a particular amplitude and frequency. The vibration of each atom is transferred to the surrounding atoms and produces an elastic wave called a phonon. The energy of the phonon E can be expressed in terms of the wavelength where h is Planck's constant and c is the speed of light or frequency v, just as in Equation 21-1:

$$E = \frac{hc}{\lambda} = hv \tag{22-1}$$

The energy required to change the temperature of the material one degree is the **heat capacity** or **specific heat**.

The heat capacity is the energy required to raise the temperature of one mole of a material by one degree. The specific heat is defined as the energy needed to increase the temperature of one gram of a material by 1°C. The heat capacity can be expressed either at constant pressure, C_p, or at a constant volume, C_v. At high temperatures, the heat capacity for a given volume of material approaches

$$C_p = 3R \simeq 6 \frac{cal}{mol \cdot K} \tag{22-2}$$

where R is the gas constant (1.987 cal/mol); however, as shown in Figure 22-1, heat capacity is not a constant. The heat capacity of metals approaches $6 \frac{cal}{mol \cdot K}$ near room temperature, but this value is not reached in ceramics until near 1000 K.

The relationship between specific heat and heat capacity is

$$\text{Specific heat} = C_p = \frac{\text{heat capacity}}{\text{atomic weight}} \tag{22-3}$$

In most engineering calculations, specific heat is used more conveniently than heat capacity. The specific heat of typical materials is given in Table 22-1. Neither the heat capacity

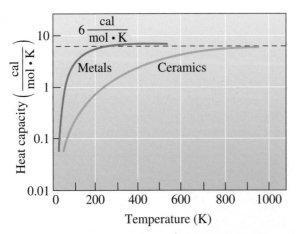

Figure 22-1
Heat capacity as a function of temperature for metals and ceramics.

TABLE 22-1 ■ *The specific heat of selected materials at 300 K*

Material	Specific Heat $\left(\dfrac{cal}{g \cdot K}\right)$	Material	Specific Heat $\left(\dfrac{cal}{g \cdot K}\right)$
Metals		Ceramics	
Al	0.215	Al_2O_3	0.200
Cu	0.092	Diamond	0.124
B	0.245	SiC	0.250
Fe	0.106	Si_3N_4	0.170
Pb	0.038	SiO_2 (silica)	0.265
Mg	0.243	Polymers	
Ni	0.106	High-density polyethylene	0.440
Si	0.168	Low-density polyethylene	0.550
Ti	0.125	6,6-nylon	0.400
W	0.032	Polystyrene	0.280
Zn	0.093	Other	
		Water	1.000
		Nitrogen	0.249

Note: $1 \dfrac{cal}{g \cdot K} = 4184 \dfrac{J}{kg \cdot K}$

nor the specific heat depends significantly on the structure of the material; thus, changes in dislocation density, grain size, or vacancies have little effect.

The most important factor affecting specific heat is the lattice vibrations or phonons; however, other factors affect the heat capacity. One striking example occurs in ferromagnetic materials such as iron (Figure 22-2). An abnormally high heat capacity is observed in iron at the Curie temperature, where the normally aligned magnetic moments of the iron atoms are randomized and the iron becomes paramagnetic. Heat capacity also depends on the crystal structure, as shown in Figure 22-2 for iron.

The following examples illustrate the use of specific heat.

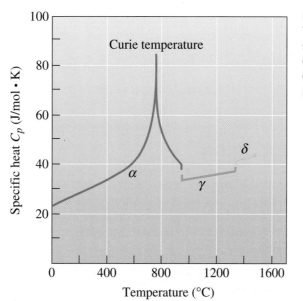

Figure 22-2
The effect of temperature on the specific heat of iron. Both the change in crystal structure and the change from ferromagnetic to paramagnetic behavior are indicated.

Example 22-1   *Specific Heat of Tungsten*

How much heat must be supplied to 250 g of tungsten to raise its temperature from 25°C to 650°C?

SOLUTION

The specific heat of tungsten is $0.032 \dfrac{\text{cal}}{\text{g} \cdot \text{K}}$. Thus,

$$\text{Heat required} = (\text{specific heat})(\text{mass})(\Delta T)$$
$$= (0.032 \, \text{cal/g} \cdot \text{K})(250 \, \text{g})(650 - 25)$$
$$= 5000 \, \text{cal}$$

If no losses occur, 5000 cal (or 20,920 J) must be supplied to the tungsten. A variety of processes might be used to heat the metal. We could use a gas torch, we could place the tungsten in an induction coil to induce eddy currents, we could pass an electrical current through the metal, or we could place the metal into an oven heated by SiC resistors.

Example 22-2   *Specific Heat of Niobium*

Suppose the temperature of 50 g of niobium increases 75°C when heated. Estimate the specific heat and determine the heat in calories required.

SOLUTION

The atomic weight of niobium is 92.91 g/mol. We can use Equation 22-3 to estimate the heat required to raise the temperature of one gram by one °C:

$$C_p \approx \frac{6}{92.91} = 0.0646 \, \text{cal/(g} \cdot °\text{C)}$$

Thus, the total heat required is

$$\text{Heat} = \left(0.0646 \, \frac{\text{cal}}{\text{g} \cdot °\text{C}} \right)(50 \, \text{g})(75°\text{C}) = 242 \, \text{cal}$$

Note that a temperature change of 1°C is equivalent to a temperature change of 1 K.

22-2   Thermal Expansion

An atom that gains thermal energy and begins to vibrate behaves as though it has a larger atomic radius. The average distance between the atoms and therefore the overall dimensions of the material increase. The change in the dimensions of the material Δl per unit length is given by the **linear coefficient of thermal expansion** α:

$$\alpha = \frac{l_f - l_0}{l_0(T_f - T_0)} = \frac{\Delta l}{l_0 \Delta T} \qquad (22\text{-}4)$$

TABLE 22-2 ■ *The linear coefficient of thermal expansion at room temperature for selected materials*

Material	Linear Coefficient of Thermal Expansion (α) ($\times 10^{-8}/°C$)	Material	Linear Coefficient of Thermal Expansion (α) ($\times 10^{-6}/°C$)
Al	25.0	6,6-nylon	80.0
Cu	16.6	6,6-nylon—33% glass fiber	20.0
Fe	12.0	Polyethylene	100.0
Ni	13.0	Polyethylene—30% glass fiber	48.0
Pb	29.0	Polystyrene	70.0
Si	3.0	Al_2O_3	6.7
W	4.5	Fused silica	0.55
1020 steel	12.0	Partially stabilized ZrO_2	10.6
3003 aluminum alloy	23.2	SiC	4.3
Gray iron	12.0	Si_3N_4	3.3
Invar (Fe-36% Ni)	1.54	Soda-lime glass	9.0
Stainless steel	17.3		
Yellow brass	18.9		
Epoxy	55.0		

where T_0 and T_f are the initial and final temperatures and l_0 and l_f are the initial and final dimensions of the material. A *volume* coefficient of thermal expansion (α_v) also can be defined to describe the change in volume when the temperature of the material is changed. If the material is isotropic, $\alpha_v = 3\alpha$. An instrument known as a dilatometer is used to measure the thermal-expansion coefficient. It is also possible to trace thermal expansion using x-ray diffraction (**XRD**). Coefficients of thermal expansion for several materials are included in Table 22-2.

The coefficient of thermal expansion of a material is related to the atomic bonding. In order for the atoms to vibrate about their equilibrium positions, energy must be supplied to the material. If the potential well is asymmetric, the atoms separate to a larger extent with increased energy compared to a symmetric well, and the material has a high thermal-expansion coefficient (Chapter 2). The thermal-expansion coefficient also tends to be inversely proportional to the depth of the potential well, i.e., materials with high melting temperatures have low coefficients of thermal expansion (Figure 22-3).

Consequently, lead (Pb) has a much larger coefficient than high melting point metals such as tungsten (W). Most ceramics, which have strong ionic or covalent bonds, have low coefficients compared with metals. Certain glasses, such as fused silica, also have a poor packing factor, which helps accommodate thermal energy with little dimensional change. Although bonding within the chains of polymers is covalent, the secondary bonds holding the chains together are weak, leading to high coefficients. Polymers that contain strong cross-linking typically have lower coefficients than linear polymers such as polyethylene.

Several precautions must be taken when calculating dimensional changes in materials:

1. The expansion characteristics of some materials, particularly single crystals or materials having a preferred orientation, are anisotropic.

2. Allotropic materials have abrupt changes in their dimensions when the phase transformation occurs (Figure 22-4). These abrupt changes contribute to the cracking of refractories on heating or cooling and quench cracks in steels.

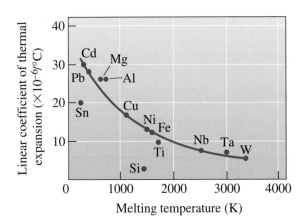

Figure 22-3
The relationship between the linear coefficient of thermal expansion and the melting temperature in metals at 25°C. Higher melting point metals tend to expand to a lesser degree.

3. The linear coefficient of expansion continually changes with temperature. Normally, α either is listed in handbooks as a complicated temperature-dependent function or is given as a constant for only a particular temperature range.

4. Interaction of the material with electric or magnetic fields produced by magnetic domains may prevent normal expansion until temperatures above the Curie temperature are reached. This is the case for Invar, an Fe-36% Ni alloy, which undergoes practically no dimensional changes at temperatures below the Curie temperature (about 200°C). This makes Invar attractive as a material for bimetallics (Figure 22-4).

The thermal expansion of engineered materials can be tailored using multi-phase materials. Upon heating, one phase can show thermal expansion while the other phase can show thermal contraction. Thus, the overall material can show a zero or negative thermal-expansion coefficient. Zerodur™ is a glass-ceramic material that can be controlled to have zero or slightly negative thermal expansion. It was developed by Schott Glass Technologies. It consists of a ~ 70 to 80 wt% crystalline phase. The remainder is a glassy phase. The negative thermal expansion coefficient of the glassy phase and the positive thermal expansion of the crystalline phase cancel, leading to a zero thermal-expansion

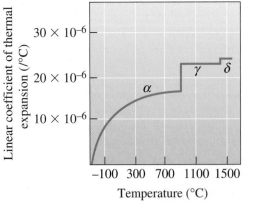

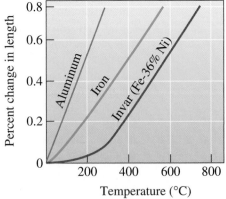

Figure 22-4   (a) The linear coefficient of thermal expansion of iron changes abruptly at temperatures where an allotropic transformation occurs. (b) The expansion of Invar is very low due to the magnetic properties of the material at low temperatures.

material. Zerodur™ has been used as the mirror substrate on the Hubble telescope and Chandra x-ray telescope. A dense, optically transparent, and zero thermal-expansion material is necessary in these applications, since any changes in dimensions as a result of changes in temperature in space will make it difficult to focus the telescopes properly. Zerodur™ is one example of how engineered materials have helped astronomers and society learn about far away galaxies. Many ceramic materials based on sodium zirconium phosphate (NZP) that have a near-zero thermal-expansion coefficient also have been developed by materials scientists.

The following examples show the use of the linear coefficients of thermal expansion.

Example 22-3 *Bonding and Thermal Expansion*

Explain why, in Figure 22-3, the linear coefficients of thermal expansion for silicon and tin do not fall on the curve. How would you expect germanium to fit into this figure?

SOLUTION

Both silicon and tin are covalently bonded. The strong covalent bonds are more difficult to stretch than the metallic bonds (a deeper trough in the energy separation curve), so these elements have a lower coefficient. Since germanium also is covalently bonded, its thermal expansion should be less than that predicted by Figure 22-3.

Example 22-4 *Design of a Pattern for a Casting Process*

Design the dimensions for a pattern that will be used to produce a rectangular shaped aluminum casting having dimensions at 25°C of 25 cm × 25 cm × 3 cm.

SOLUTION

To produce a casting having particular final dimensions, the mold cavity into which the liquid aluminum is to be poured must be oversized. After the liquid solidifies, which occurs at 660°C for pure aluminum, the solid casting contracts as it cools to room temperature. If we calculate the amount of contraction expected, we can make the original pattern used to produce the mold cavity that much larger.

The linear coefficient of thermal expansion for aluminum is $25 \times 10^{-6}\,°C^{-1}$. The temperature change from the freezing temperature to 25°C is $660 - 25 = 635°C$. The change in any dimension is given by

$$\Delta l = l_0 - l_f = \alpha l_0 \Delta T$$

For the 25 cm dimensions, $l_f = 25$ cm. We wish to find l_0:

$$l_0 - 25 = (25 \times 10^{-6})(l_0)(635)$$
$$l_0 - 25 = 0.015875 l_0$$
$$0.984 l_0 = 25$$
$$l_0 = 25.40\,cm$$

For the 3 cm dimension, $l_f = 3$ cm.

$$l_0 - 3 = (25 \times 10^{-6})(l_0)(635)$$
$$l_0 - 3 = 0.015875l_0$$
$$0.984l_0 = 3$$
$$l_0 = 3.05 \, \text{cm}$$

If we design the pattern to the dimensions 25.40 cm $\times$ 25.40 cm $\times$ 3.05 cm, the casting should contract to the required dimensions.

When an isotropic material is slowly and uniformly heated, the material expands uniformly without creating any residual stress. If, however, the material is restrained, the dimensional changes may not be possible and, instead, stresses develop. These **thermal stresses** are related to the coefficient of thermal expansion, the modulus of elasticity E of the material, and the temperature change ΔT:

$$\sigma_{\text{thermal}} = \alpha E \Delta T \qquad \qquad (22\text{-}5)$$

Thermal stresses can arise from a variety of sources. In large rigid structures such as bridges, restraints may develop as a result of the design. Some bridges are designed in sections with steel plates between sections, so that the sections move relative to one another during seasonal temperature changes.

When materials are joined—for example, coating cast iron bathtubs with a ceramic enamel or coating superalloy turbine blades with a yttria stabilized zirconia (YSZ) thermal barrier—changes in temperature cause different amounts of contraction or expansion in the different materials. This disparity leads to thermal stresses that may cause the protective coating to spall off. Careful matching of the thermal properties of the coating to those of the substrate material is necessary to prevent coating cracking (if the coefficient of the coating is less than that of the underlying substrate) or spalling (flaking of the coating due to a high expansion coefficient).

A similar situation may occur in composite materials. Brittle fibers that have a lower coefficient than the matrix may be stretched to the breaking point when the temperature of the composite increases.

Thermal stresses may even develop in a nonrigid, isotropic material if the temperature is not uniform. In producing tempered glass (Chapter 15), the surface is cooled more rapidly than the center, permitting the surface to initially contract. When the center cools later, its contraction is restrained by the rigid surface, placing compressive residual stresses on the surface.

Example 22-5   *Design of a Protective Coating*

A ceramic enamel is to be applied to a 1020 steel plate. The ceramic has a fracture strength of 28 MPa, a modulus of elasticity of 103×10^3 MPa, and a coefficient of thermal expansion of $10 \times 10^{-6} \, °\text{C}^{-1}$. Design the maximum temperature change that can be allowed without cracking the ceramic.

SOLUTION

Because the enamel is bonded to the 1020 steel, it is essentially restrained. If only the enamel was heated (and the steel remained at a constant temperature), the maximum temperature change would be

$$\sigma_{thermal} = \alpha E \Delta T = \sigma_{fracture}$$

$$(10 \times 10^{-6}{}^{\circ}C^{-1})(103 \times 10^3 \, \text{MPa})\Delta T = 28 \, \text{MPa}$$

$$\Delta T = 26.7{}^{\circ}C$$

The steel also expands, thereby permitting a larger temperature increase prior to fracture. Its coefficient of thermal expansion (Table 22-2) is $12 \times 10^{-6}{}^{\circ}C^{-1}$, and its modulus of elasticity is 207×10^3 MPa. Since the steel expands more than the enamel, a stress is still introduced into the enamel. The net coefficient of expansion is

$$\Delta \alpha = 12 \times 10^{-6} - 10 \times 10^{-6} = 2 \times 10^{-6}{}^{\circ}C^{-1}$$

$$\sigma = (2 \times 10^{-6})(103 \times 10^3 \, \text{MPa})\Delta T = 28 \, \text{MPa}$$

$$\Delta T = 136{}^{\circ}C$$

In order to permit greater temperature variations, we might select an enamel that has a higher coefficient of thermal expansion, an enamel that has a lower modulus of elasticity (so that greater strains can be permitted before the stress reaches the fracture stress), or an enamel that has a higher strength.

22-3 Thermal Conductivity

The **thermal conductivity** k is a measure of the rate at which heat is transferred through a material. The treatment of thermal conductivity is similar to that of diffusion (Chapter 5). *Thermal conductivity, similar to the diffusion coefficient, is a microstructure sensitive property.* The conductivity relates the heat Q transferred across a given plane of area A per second when a temperature gradient $\Delta T / \Delta x$ exists (Figure 22-5):

$$\frac{Q}{A} = k \frac{\Delta T}{\Delta x} \tag{22-6}$$

Note that the thermal conductivity k plays the same role in heat transfer that the diffusion coefficient D does in mass transfer. Among all metals, silver (Ag) has the highest thermal conductivity at room temperature ($430 \, \text{W} \cdot \text{m}^{-1} \cdot \text{K}^{-1}$). Copper is next with a thermal conductivity of $400 \, \text{W} \cdot \text{m}^{-1} \cdot \text{K}^{-1}$. In general, metals have higher thermal conductivity than ceramics; however, diamond, a ceramic material, has a very high thermal conductivity of $2000 \, \text{W} \cdot \text{m}^{-1} \cdot \text{K}^{-1}$. Values for the thermal conductivity of some materials are included in Table 22-3.

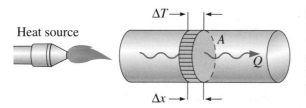

Heat source

Figure 22-5
When one end of a bar is heated, a heat flux Q/A flows toward the cold end at a rate determined by the temperature gradient produced in the bar.

TABLE 22-3 ■ *Typical values of room temperature thermal conductivity of selected materials*

Material	Thermal Conductivity (k) $(W \cdot m^{-1} \cdot K^{-1})$	Material	Thermal Conductivity (k) $(W \cdot m^{-1} \cdot K^{-1})$
Pure Metals		Ceramics	
Ag	430	Al_2O_3	16–40
Al	238	Carbon (diamond)	2000
Cu	400	Carbon (graphite)	335
Fe	79	Fireclay	0.26
Mg	100	Silicon carbide	up to 270
Ni	90	AIN	up to 270
Pb	35	Si_3N_4	up to 150
Si	150	Soda-lime glass	0.96–1.7
Ti	22	Vitreous silica	1.4
W	171	Vycor™ glass	12.5
Zn	117	XrO_2	4.2
Zr	23	Polymers	
Alloys		6,6-nylon	0.25
1020 steel	100	Polyethylene	0.33
3003 aluminum alloy	280	Polyimide	0.21
304 stainless steel	30	Polystyrene	0.13
Cementite	50	Polystyrene foam	0.029
Cu-30% Ni	50	Teflon	0.25
Ferrite	75		
Gray iron	79.5		
Yellow brass	221		

Note: $1 \dfrac{cal}{s} cm^{-1} K^{-1} = 418.4 \ W m^{-1} K^{-1}$

Thermal energy is transferred by two important mechanisms: transfer of free electrons and lattice vibrations (or phonons). Valence electrons gain energy, move toward the colder areas of the material, and transfer their energy to other atoms. The amount of energy transferred depends on the number of excited electrons and their mobility; these, in turn, depend on the type of material, lattice imperfections, and temperature. In addition, thermally induced vibrations of the atoms transfer energy through the material.

Metals   Because the valence band is not completely filled in metals, electrons require little thermal excitation in order to move and contribute to the transfer of heat. Since the thermal conductivity of metals is due primarily to the electronic contribution, we expect a relationship between thermal and electrical conductivities:

$$\frac{k}{\sigma T} = L = 5.5 \times 10^{-9} \frac{cal \cdot ohm}{s \cdot K^2} \tag{22-7}$$

where L is the **Lorenz number**. This relationship is followed to a limited extent in many metals.

When the temperature of the material increases, two competing factors affect thermal conductivity. Higher temperatures are expected to increase the energy of the electrons, creating more "carriers" and increasing the contribution from lattice vibrations; these effects increase the thermal conductivity. At the same time, the increased lattice vibrations scatter the electrons, reducing their mobility, and therefore tend to decrease the thermal conductivity. The combined effect of these factors leads to very different behavior for different metals. For iron, the thermal conductivity initially decreases with increasing temperature (due to the lower mobility of the electrons), then increases slightly (due to increased lattice vibrations). The conductivity *decreases* continuously when aluminum is heated but *increases* continuously when platinum is heated (Figure 22-6).

Thermal conductivity in metals also depends on crystal structure defects, microstructure, and processing. Thus, cold-worked metals, solid-solution-strengthened metals, and two-phase alloys might display lower conductivities compared with their defect-free counterparts.

Ceramics

The energy gap in ceramics is too large for many electrons to be excited into the conduction band except at very high temperatures. Thus, the transfer of heat in ceramics occurs primarily by lattice vibrations (or phonons). Since the electronic contribution is absent, the thermal conductivity of most ceramics is much lower than that of metals. The main reason why the experimentally observed conductivity of ceramics is low, however, is the level of porosity. Porosity increases scattering. The best insulating brick, for example, contains a large porosity fraction. Effective sintering reduces porosity (and therefore increases thermal conductivity).

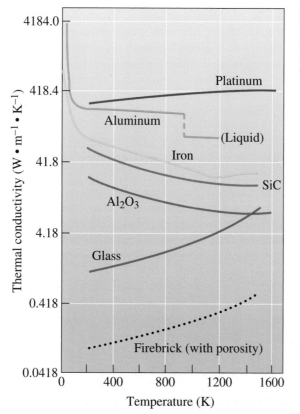

Figure 22-6
The effect of temperature on the thermal conductivity of selected materials. Note the log scale on the vertical axis.

Glasses have low thermal conductivity. The amorphous, loosely packed structure minimizes the points at which silicate chains contact one another, making it more difficult for the phonons to be transferred. The thermal conductivity increases as temperature increases; higher temperatures produce more energetic phonons and more rapid transfer of heat. In some applications, such as window glasses, we use double panes of glass and separate them using a gas (e.g., Ar) to provide better thermal insulation. We can also make use of different coatings on glass to make buildings and cars more energy efficient.

The more ordered structures of the crystalline ceramics, as well as glass-ceramics that contain large amounts of crystalline precipitates, cause less scattering of phonons. Compared with glasses, these materials have a higher thermal conductivity. As the temperature increases, however, scattering becomes more pronounced, and the thermal conductivity decreases (as shown for alumina and silicon carbide in Figure 22-6). At still higher temperatures, heat transfer by radiation becomes significant, and the conductivity may increase. The thermal conductivity of polycrystalline ceramic materials is typically lower than that of single crystals.

Some ceramics such as diamond have very high thermal conductivities. Materials with a close-packed structure and high modulus of elasticity produce high energy phonons that encourage high thermal conductivities. It has been shown that many ceramics with a diamond-like crystal structure (e.g., AlN, SiC, BeO, BP, GaN, Si, AlP) have high thermal conductivities. For example, thermal conductivities of $\sim 270\,\mathrm{W\,m^{-1}K^{-1}}$ have been reported for SiC and AlN. Although SiC and AlN are good thermal conductors, they are also electrical insulators; therefore, these materials are good candidates for use in electronic packaging substrates where heat dissipation is needed.

Semiconductors

Heat is conducted in semiconductors by both phonons and electrons. At low temperatures, phonons are the principal carriers of energy, but at higher temperatures, electrons are excited through the small energy gap into the conduction band, and thermal conductivity increases significantly.

Polymers

The thermal conductivity of polymers is very low—even in comparison with silicate glasses. Vibration and movement of the molecular polymer chains transfer energy. Increasing the degree of polymerization, increasing the crystallinity, minimizing branching, and providing extensive cross-linking all produce a more rigid structure and provide for higher thermal conductivity.

The thermal conductivity of many engineered materials depends upon the volume fractions of different phases and their connectivity. Silver-filled epoxies are used in many heat-transfer applications related to microelectronics. Unusually good thermal insulation is obtained using polymer foams, often produced from polystyrene or polyurethane. Styrofoam™ coffee cups are a typical product. The next example illustrates the design of a window glass for thermal conductivity.

Example 22-6 *Design of a Window Glass*

Design a glass window 120 cm × 120 cm square that separates a room at 25°C from the outside at 40°C and allows no more than 5×10^6 cal of heat to enter the room each day.

Assume that the thermal conductivity of glass is $0.96\,\mathrm{W} \cdot \mathrm{m}^{-1} \cdot \mathrm{K}^{-1}$ or $0.0023\,\dfrac{\mathrm{cal}}{\mathrm{cm} \cdot \mathrm{s} \cdot \mathrm{K}}$.

SOLUTION

From Equation 22-6,

$$\frac{Q}{A} = k \frac{\Delta T}{\Delta x}$$

where Q/A is the heat transferred per second through the window.

$$1 \text{ day} = (24 \text{ h/day})(3600 \text{ s/h}) = 8.64 \times 10^4 \text{ s}$$

$$A = (120 \text{ cm})^2 = 1.44 \times 10^4 \text{ cm}^2$$

$$Q = \frac{(5 \times 10^6 \text{ cal/day})}{8.64 \times 10^4 \text{ s/day}} = 57.87 \text{ cal/s}$$

$$\frac{Q}{A} = \frac{57.87 \text{ cal/s}}{1.44 \times 10^4 \text{ cm}^2} = 0.00402 \frac{\text{cal}}{\text{cm}^2 \cdot \text{s}}$$

$$\frac{Q}{A} = 0.00402 \frac{\text{cal}}{\text{cm}^2 \cdot \text{s}} = \left(0.0023 \frac{\text{cal}}{\text{cm} \cdot \text{s} \cdot \text{K}} \right)(40 - 25°\text{C})/\Delta x$$

$$\Delta x = 8.58 \text{ cm} = \text{thickness}$$

The glass would have to be exceptionally thick to prevent the desired maximum heat flux. We might do several things to reduce the heat flux. Although all of the silicate glasses have similar thermal conductivities, we might use instead a transparent polymer material (such as polymethyl methacrylate). The polymers have thermal conductivities approximately one order of magnitude smaller than the ceramic glasses. We could also use a double-paned glass, with the glass panels separated either by a gas (air or Ar have very low thermal conductivities) or a sheet of transparent polymer.

22-4 Thermal Shock

Stresses leading to the fracture of brittle materials can be introduced thermally as well as mechanically. When a piece of material is cooled quickly, a temperature gradient is produced. This gradient can lead to different amounts of contraction in different areas. If residual tensile stresses become high enough, flaws may propagate and cause failure. Similar behavior can occur if a material is heated rapidly. This failure of a material caused by stresses induced by sudden changes in temperature is known as **thermal shock**. Thermal shock behavior is affected by several factors:

1. *Coefficient of thermal expansion*: A low coefficient minimizes dimensional changes and reduces the ability to withstand thermal shock.

2. *Thermal conductivity*: The magnitude of the temperature gradient is determined partly by the thermal conductivity of the material. A high thermal conductivity helps to transfer heat and reduce temperature differences quickly in the material.

3. *Modulus of elasticity*: A low modulus of elasticity permits large amounts of strain before the stress reaches the critical level required to cause fracture.

4. *Fracture stress*: A high stress required for fracture permits larger strains.

5. *Phase transformations*: Additional dimensional changes can be caused by phase transformations. Transformation of silica from quartz to cristobalite, for example, introduces residual stresses and increases problems with thermal shock. Similarly, we cannot use pure $PbTiO_3$ ceramics in environments subjected to large, sudden temperature changes, since the stresses induced during the cubic to tetragonal transformation will cause the ceramic to fracture.

One method for measuring the resistance to thermal shock is to determine the maximum temperature difference that can be tolerated during a quench without affecting the mechanical properties of the material. Pure (fused) silica glass has a thermal shock resistance of about 3000°C. Figure 22-7 shows the effect of quenching temperature difference on the modulus of rupture in sialon ($Si_3Al_3O_3N_5$) after quenching; no cracks and therefore no change in the properties of the ceramic are evident until the quenching temperature difference approaches 950°C. Other ceramics have poorer resistance. Shock resistance for partially stabilized zirconia (PSZ) and Si_3N_4 is about 500°C; for SiC 350°C; and for Al_2O_3 and ordinary glass, about 200°C.

Another way to evaluate the resistance of a material to thermal shock is by the thermal shock parameter (R' or R)

$$R' = \text{Thermal shock parameter} = \frac{\sigma_f k(1 - v)}{E \cdot \alpha} \tag{22-8a}$$

or

$$R = \frac{\sigma_f(1 - v)}{E \cdot \alpha} \tag{22-8b}$$

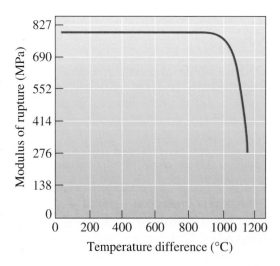

Figure 22-7
The effect of quenching temperature difference on the modulus of rupture of sialon. The thermal shock resistance of the ceramic decreases at about 950°C.

where σ_f is the fracture stress of the material, v is the Poisson's ratio, k is the thermal conductivity, E is the modulus of elasticity, and α is the linear coefficient of thermal expansion. Equation 22-8b is used in situations where the heat transfer rate is essentially infinite. A higher value of the thermal shock parameter means better resistance to thermal shock. The thermal shock parameter represents the maximum temperature change that can occur without fracturing the material.

Thermal shock is usually not a problem in most metals because metals normally have sufficient ductility to permit deformation rather than fracture. As mentioned before, many ceramic compositions developed as zero thermal-expansion ceramics (based on sodium zirconium phosphates) and many glass-ceramics have excellent thermal shock resistance.

Summary

- The thermal properties of materials can be at least partly explained by the movement of electrons and phonons.

- Heat capacity and specific heat represent the quantity of energy required to raise the temperature of a given amount of material by one degree, and they are influenced by temperature, crystal structure, and bonding.

- The coefficient of thermal expansion describes the dimensional changes that occur in a material when its temperature changes. Strong bonding leads to a low coefficient of expansion. High melting-point metals and ceramics have low coefficients, whereas low melting-point metals and polymers have high coefficients.

- The thermal expansion of many engineered materials can be made to be near zero by properly controlling the combination of different phases that show positive and negative thermal expansion coefficients. Zerodur™ glass-ceramics and NZP ceramics are examples of such materials.

- Because of thermal expansion, stresses develop in a material when the temperature changes. Care in design, processing, or materials selection is required to prevent failure due to thermal stresses.

- Heat is transferred in materials by both phonons and electrons. Thermal conductivity depends on the relative contributions of each of these mechanisms, as well as on the microstructure and temperature.

- Phonons play an important role in the thermal properties of ceramics, semiconductors, and polymers. Disordered structures, such as ceramic glasses or amorphous polymers, scatter phonons and have low conductivity.

- Thermal conductivity is sensitive to microstructure. Most commonly encountered ceramics have a low thermal conductivity because of porosity and defects.

- Crystalline ceramics and polymers have higher conductivities than their glassy or amorphous counterparts.

- Electronic contributions to thermal conductivity are most important in metals; consequently, lattice imperfections that scatter electrons reduce conductivity. Increasing temperature increases phonon energy but also increases the scattering of both phonons and electrons.

Glossary

Heat capacity The energy required to raise the temperature of one mole of a material by one degree.

Linear coefficient of thermal expansion Describes the amount by which each unit length of a material changes when the temperature of the material changes by one degree.

Lorenz number The constant that relates electrical and thermal conductivity.

Phonon A packet of elastic waves. It is characterized by its energy, wavelength, or frequency, which transfers energy through a material.

Specific heat The energy required to raise the temperature of one gram of a material by one degree.

Thermal conductivity (k) A microstructure-sensitive property that measures the rate at which heat is transferred through a material.

Thermal shock Failure of a material caused by stresses introduced by sudden changes in temperature.

Thermal stresses Stresses introduced into a material due to differences in the amount of expansion or contraction that occur because of a temperature change.

Problems

Section 22-1 Heat Capacity and Specific Heat

22-1 State any two applications for which a high thermal conductivity is desirable.

22-2 State any two applications for which a very low thermal conductivity of a material is desirable.

22-3 Calculate the increase in temperature for 1 kg samples of the following metals:
(a) Al,
(b) Cu,
(c) Fe, and
(d) Ni
when 20,000 Joules of heat is supplied to the metal at 25°C.

22-4 Calculate the heat (in calories and joules) required to raise the temperature of 1 kg of the following materials by 50°C:
(a) lead,
(b) nickel,
(c) Si_3N_4, and
(d) 6,6-nylon.

22-5 Calculate the temperature of a 100 g sample of the following materials (originally at 25°C) when 3000 calories are introduced:
(a) tungsten,
(b) titanium,
(c) Al_2O_3, and
(d) low-density polyethylene.

22-6 An alumina insulator for an electrical device is also to serve as a heat sink. A 10°C temperature rise in an alumina insulator 1 cm × 1 cm × 0.02 cm is observed during use. Determine the thickness of a high-density polyethylene insulator that would be needed to provide the same performance as a heat sink. The density of alumina is 3.96 g/cm^3.

22-7 A 200 g sample of aluminum is heated to 400°C and is then quenched into 2000 cm^3 of water at 20°C. Calculate the temperature of the water after the aluminum and water reach equilibrium. Assume no heat loss from the system.

Section 22-2 Thermal Expansion

22-8 A 2-m-long soda-lime glass sheet is produced at 1400°C. Determine its length after it cools to 25°C.

22-9 A copper casting requires the final dimensions to be 2.5 cm × 50 cm × 10 cm. Determine the size of the pattern that must be used to make the mold into which the liquid copper is poured during the manufacturing process.

22-10 A copper casting is to be produced having the final dimensions of 2.5 cm × 30 cm × 60 cm. Determine the size of the pattern that must be used to make the mold into which the liquid copper is poured during the manufacturing process.

22-11 An aluminum casting is made using the permanent mold process. In this process, the liquid aluminum is poured into a gray cast iron mold that is heated to 350°C. We wish to produce an aluminum casting that is 38 cm long at 25°C. Calculate the length of the cavity that must be machined into the gray cast iron mold.

22-12 A Ti-alloy strip and stainless steel strip are roll bonded at 760°C. If the total length of the bimetallic strip is 1 m long when hot rolled, calculate the length of each metal at 25°C. What is the nature of the stress (compressive or tensile) just below the surface on the Ti-alloy side and on the stainless steel side?

22-13 We coat a 100-cm-long, 2-mm-diameter copper wire with a 0.5-mm-thick epoxy insulation coating. Determine the length of the copper and the coating when their temperature increases from 25°C to 250°C. What is likely to happen to the epoxy coating as a result of this heating?

22-14 We produce a 25-cm-long bimetallic composite material composed of a strip of yellow brass bonded to a strip of Invar. Determine the length to which each material would expand when the temperature increases from 20°C to 150°C. Draw a sketch showing what will happen to the shape of the bimetallic strip.

22-15 Give examples of materials that have negative or near-zero thermal expansion coefficients.

22-16 What is Zerodur™? What are some of the properties and applications of this material?

22-17 A nickel engine part is coated with SiC to provide corrosion resistance at high temperatures. If no residual stresses are present in the part at 20°C, determine the thermal stresses that develop when the part is heated to 1000°C during use. (See Table 15-3 and Table 22-2.)

22-18 Alumina fibers 2 cm long are incorporated into an aluminum matrix. Assuming good bonding between the ceramic fibers and the aluminum, estimate the thermal stresses acting on the fiber when the temperature of the composite increases 250°C. Are the stresses on the fiber tensile or compressive? (See Table 15-3 and Table 22-2.)

22-19 A 60-cm-long copper bar with a yield strength of 207 MPa is heated to 120°C and immediately fastened securely to a rigid framework. Will the copper deform plastically during cooling to 25°C? How much will the bar deform if it is released from the framework after cooling?

22-20 Repeat Problem 22-19, but using a silicon carbide rod rather than a copper rod. (See Table 15-3 and Table 22-2.)

Section 22-3 Thermal Conductivity

Section 22-4 Thermal Shock

22-21 Define the terms thermal conductivity and thermal shock of materials.

22-22 Thermal conductivity of most ceramics is low. True or False? Explain.

22-23 Is the thermal conductivity of materials a microstructure sensitive property? Explain.

22-24 If a 1-m long silver bar is heated at one end to 300°C and the temperature measured at the other end is 100°C, calculate the heat transferred per unit area.

22-25 A 3-cm-plate of silicon carbide separates liquid aluminum (held at 700°C) from a water-cooled steel shell maintained at 20°C. Calculate the heat Q transferred to the steel per cm^2 of silicon carbide each second.

22-26 A sheet of 0.025 cm polyethylene is sandwiched between two 90 cm × 90 cm × 0.3 cm sheets of soda-lime glass to produce a window. The thermal conductivity of polyethylene is Calculate (a) the heat lost through the window each day when the room temperature is 25°C and the outside air is 0°C and (b) the heat entering through the window each day when the room temperature is 25°C and the outside air is 40°C.

22-27 We would like to build a heat-deflection plate that permits heat to be transferred rapidly parallel to the sheet but very slowly perpendicular to the sheet. Consequently, we incorporate 1 kg of unidirectional copper wires, each 0.1 cm in diameter, into 5 kg of a polyimide polymer matrix. Estimate the thermal conductivity parallel and perpendicular to the wires.

22-28 An exothermic reaction at a battery electrode releases 80 MW of power per square meter. The electrode consists of a 5-μm-thick graphite layer on a 20-μm- thick copper foil; the reaction occurs at the graphite surface; and heat needs to be transferred out through the graphite to the copper foil. The temperature on the outer side of the copper foil is maintained at 30°C. What is the temperature on the inner surface of the electrode?

22-29 Suppose we just dip a 1-cm-diameter, 10-cm-long rod of aluminum into one liter of water at 20°C. The other end of the rod is in contact with a heat source operating at 400°C. Determine the length of time required to heat the water to 25°C if 75% of the heat is lost by radiation from the bar.

22-30 Write down the equations that define the thermal shock resistance (TSR) parameter. Based on this, what can you say about materials that show a near-zero thermal-expansion coefficient?

22-31 Determine the thermal shock parameter for silicon nitride, hot-pressed silicon carbide, and alumina. Compare it with the thermal-shock resistance as defined by the maximum quenching temperature difference. (See Table 15-3.)

22-32 Gray cast iron has a higher thermal conductivity than ductile or malleable cast iron. Review Chapter 13 and explain why this difference in conductivity might be expected.

 Design Problems

22-33 A chemical-reaction vessel contains liquids at a temperature of 680°C. The wall of the vessel must be constructed so that the outside wall operates at a temperature of 35°C or less. Design the vessel wall and appropriate materials if the maximum heat transfer is 6000 cal/s.

22-34 Design a metal panel coated with glass enamel capable of thermal cycling between 20°C and 150°C. The glasses generally available are expected to have a tensile strength of 34 MPa and a compressive strength of 345 MPa.

22-35 What design constraints exist in selecting materials for a turbine blade for a jet engine that is capable of operating at high temperatures?

22-36 Consider the requirements of a low dielectric constant, low dielectric loss, good mechanical strength, and high thermal conductivity for electronic packaging substrates. What materials would you consider?

 Computer Problems

22-37 *Thermal Shock Resistance Parameters.* Write a computer program or use spreadsheet software that will provide the value of thermal shock-resistance parameters (TSR) when the values of the fracture stress, thermal conductivity, Young's modulus, Poisson's ratio, and thermal-expansion coefficient are provided.

Ⓚ Knovel® **Problems**

K22-1 The molar heat capacity is the energy required to increase the temperature of one mole of a material by one degree. How much heat is required to bring 50 moles of ethanol stored at 25°C to a boil? (Note that ethanol is also known as ethyl alcohol.)

K22-2 The thermal-expansion coefficient is the change of length of a material per unit length per degree of temperature change. A gold bar is initially 1 m long. What is the length of the gold bar after it has been heated from 298 to 473 K?

K22-3 Three heat sinks are identical except that they are made from different materials: gold, silver, and copper. Which heat sink will extract heat the fastest?

Most developed nations spend about 6% of their total gross domestic product in addressing corrosion-related issues. In the United States, this amounts to about $550 billion annually. The process of corrosion affects many important areas of technology, including the construction and manufacturing of bridges, airplanes, ships, and microelectronics; the food industry; space exploration; fiber optic networks; and nuclear and other power utilities, just to name a few. Corrosion can be prevented or contained using different techniques that include the use of coatings and anodic or cathodic protection.

Since steel is the most important structural material and may be highly susceptible to corrosion depending on composition, the corrosion of steel is a significant engineering issue. The photo above shows rusting of steel bar used to reinforce concrete (known as "rebar") when it is exposed to the environment. The rust that is commonly observed on steel is iron oxide that forms as a result of an electrochemical reaction known as galvanic corrosion. In this process, steel is consumed to form the brittle oxide, which may later flake off, thereby degrading the bar (© *Glenn Volkman/Alamy*).

Corrosion and Wear

Have You Ever Wondered?

- *Why does iron rust?*

- *What does the acronym "WD-40^{TM}" stand for?*

- *Is the process of corrosion ever useful?*

- *Why do household water-heater tanks contain Mg rods?*

- *Do metals like aluminum and titanium undergo corrosion?*

- *How does wear affect the useful life of different components such as crankshafts?*

- *What process makes use of mechanical erosion and chemical corrosion in the manufacture of semiconductor chips?*

The composition and physical integrity of a solid material is altered in a corrosive environment. In chemical corrosion, a corrosive liquid dissolves the material. In electrochemical corrosion, metal atoms are removed from the solid material as the result of an electric circuit that is produced. Metals and certain ceramics react with a gaseous environment, usually at elevated temperatures, and the material may be destroyed by the formation of oxides or other compounds. Polymers degrade when exposed to oxygen at elevated temperatures. Materials may be altered when exposed to radiation or even bacteria. Finally, a variety of wear and wear-corrosion mechanisms alter the shape of materials. According to a study concluded in 2001, in the United States, combating corrosion costs about 6% of the gross domestic product (GDP). This amount, which includes direct and indirect costs, was approximately $550 billion in 1998.

The corrosion process occurs in order to lower the free energy of a system. The corrosion process occurs over a period of time and can occur either at high or low temperatures. Chemical corrosion is an important consideration in many sectors including transportation (bridges, pipelines, cars, airplanes, trains, and ships), utilities (electrical, water, telecommunications, and nuclear power plants), and production and manufacturing (food industry, microelectronics, and petroleum refining).

In some applications, corrosion or oxidation is useful. The processes of chemical corrosion and erosion are used to make ultra-flat surfaces of silicon wafers for computer chips. Similarly, degradation and dissolution of certain biopolymers is useful in some medical applications, such as dissolvable sutures.

The goal of this chapter is to introduce the principles and mechanisms by which corrosion and wear occur under different conditions. This includes the aqueous corrosion of metals, the oxidation of metals, the corrosion of ceramics, and the degradation of polymers. We will offer a summary of different technologies that are used to prevent or minimize corrosion and associated problems.

23-1   Chemical Corrosion

In **chemical corrosion**, or direct dissolution, a material dissolves in a corrosive liquid medium. The material continues to dissolve until either it is consumed or the liquid is saturated. An example is the development of a green patina on the surface of copper-based alloys. This is due to the formation of copper carbonate and copper hydroxides and is why, for example, the Statue of Liberty looks greenish. The chemical corrosion of copper, tantalum, silicon, silicon dioxide, and other materials can be achieved under extremely well-controlled conditions. In the processing of silicon wafers, for example, a process known as chemical mechanical polishing uses a corrosive silica-based slurry to provide mechanical erosion. This process creates extremely flat surfaces that are suitable for the processing of silicon wafers. Chemical corrosion also occurs in nature. For example, the chemical corrosion of rocks by carbonic acid (H_2CO_3) and the mechanical erosion of wind and water play an important role in the formation of canyons and caverns.

Liquid Metal Attack   Liquid metals first attack a solid at high-energy locations such as grain boundaries. If these regions continue to be attacked preferentially, cracks eventually grow (Figure 23-1). Often this form of corrosion is complicated by the presence of fluxes that accelerate the attack or by electrochemical corrosion. Aggressive metals such as liquid lithium can also attack ceramics.

Figure 23-1
Molten lead is held in thick steel pots during refining. In this case, the molten lead has attacked a weld in a steel plate, and cracks have developed. Eventually, the cracks propagate through the steel, and molten lead leaks from the pot. (*Reprinted courtesy of Don Askeland.*)

Figure 23-2
Micrograph of a copper deposit in brass, showing the effect of dezincification ($\times$ 50). (*Reprinted courtesy of Don Askeland.*)

Selective Leaching

One particular element in an alloy may be selectively dissolved, or leached, from the solid. **Dezincification** occurs in brass containing more than 15% Zn. Both copper and zinc are dissolved by aqueous solutions at elevated temperatures; the zinc ions remain in solution while the copper ions are replated onto the brass (Figure 23-2). Eventually, the brass becomes porous and weak.

Graphitic corrosion of gray cast iron occurs when iron is selectively dissolved in water or soil, leaving behind interconnected graphite flakes and a corrosion product. Localized graphitic corrosion often causes leakage or failure of buried gray iron gas lines, sometimes leading to explosions.

Dissolution and Oxidation of Ceramics

Ceramic refractories used to contain molten metal during melting or refining may be dissolved by the slags that are produced on the metal surface. For example, an acid (high SiO_2) refractory is rapidly attacked by a basic (high CaO or MgO) slag. A glass produced from SiO_2 and Na_2O is rapidly attacked by water; CaO must be added to the glass to minimize this attack. Nitric acid may selectively leach iron or silica from some ceramics, reducing their strength and density. As noted in Chapters 7 and 15, the strength of silicate glasses depends on flaws that are often created by corrosive interactions with water.

Chemical Attack on Polymers

Compared to metals and oxide ceramics, plastics are considered corrosion resistant. Teflon^{TM} and Viton^{TM} are some of the most corrosion-resistant materials and are used in many applications, including the chemical processing industry. These and other polymeric materials can withstand the presence of many acids, bases, and organic liquids. Aggressive solvents do, however, often diffuse into low-molecular-weight thermoplastic polymers. As the solvent is incorporated into the polymer, the smaller solvent molecules force apart the chains, causing swelling. The strength of the bonds between the chains decreases. This leads to softer, lower-strength polymers with low glass-transition temperatures. In extreme cases, the swelling leads to stress cracking.

Thermoplastics may also be dissolved in a solvent. Prolonged exposure causes a loss of material and weakening of the polymer part. This process occurs most easily when the temperature is high and when the polymer has a low molecular weight, is highly branched and amorphous, and is not cross-linked. The structure of the monomer is also important; the CH_3 groups on the polymer chain in polypropylene are more easily removed from the chain than are chloride or fluoride ions in polyvinyl chloride (PVC) or polytetrafluoroethylene (Teflon^{TM}). Teflon has exceptional resistance to chemical attack by almost all solvents.

23-2   Electrochemical Corrosion

Electrochemical corrosion, the most common form of attack of metals, occurs when metal atoms lose electrons and become ions. As the metal is gradually consumed by this process, a byproduct of the corrosion process is typically formed. Electrochemical corrosion occurs most frequently in an aqueous medium, in which ions are present in water, soil, or moist air. In this process, an electric circuit is created, and the system is called an **electrochemical cell**. Corrosion of a steel pipe or a steel automobile panel, creating holes in the steel and rust as the byproduct, are examples of this reaction.

Although responsible for corrosion, electrochemical cells may also be useful. By deliberately creating an electric circuit, we can *electroplate* protective or decorative coatings onto materials. In some cases, electrochemical corrosion is even desired. For example, in etching a polished metal surface with an appropriate acid, various features in the microstructure are selectively attacked, permitting them to be observed. In fact, most of the photographs of metal and alloy microstructures in this text were obtained in this way, thus enabling, for example, the observation of pearlite in steel or grain boundaries in copper.

Components of an Electrochemical Cell

There are four components of an electrochemical cell (Figure 23-3):

1. The **anode** gives up electrons to the circuit and corrodes.
2. The **cathode** receives electrons from the circuit by means of a chemical, or cathode, reaction. Ions that combine with the electrons produce a byproduct at the cathode.
3. The anode and cathode must be electrically connected, usually by physical contact, to permit the electrons to flow from the anode to the cathode and continue the reaction.
4. A liquid **electrolyte** must be in contact with both the anode and the cathode. The electrolyte is conductive, thus completing the circuit. It provides the means by which metallic ions leave the anode surface and move to the cathode to accept the electrons.

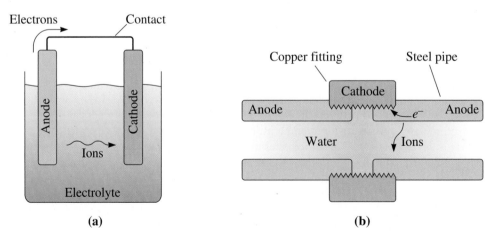

(a)   **(b)**

Figure 23-3 The components in an electrochemical cell: (a) a simple electrochemical cell and (b) a corrosion cell between a steel water pipe and a copper fitting.

This description of an electrochemical cell defines electrochemical corrosion. If metal ions are deposited onto the cathode, *electroplating* occurs.

Anode Reaction

The anode, which is a metal, undergoes an **oxidation reaction** by which metal atoms are ionized. The metal ions enter the electrolytic solution, while the electrons leave the anode through the electrical connection:

$$M \rightarrow M^{n+} + ne^- \tag{23-1}$$

Because metal ions leave the anode, the anode corrodes, or oxidizes.

Cathode Reaction in Electroplating

In electroplating, a cathodic **reduction reaction,** which is the reverse of the anode reaction, occurs at the cathode:

$$M^{n+} + ne^- \rightarrow M \tag{23-2}$$

The metal ions, either intentionally added to the electrolyte or formed by the anode reaction, combine with electrons at the cathode. The metal then plates out and covers the cathode surface.

Cathode Reactions in Corrosion

Except in unusual conditions, plating of a metal does not occur during electrochemical corrosion. Instead, the reduction reaction forms a gas, solid, or liquid byproduct at the cathode (Figure 23-4).

1. *The hydrogen electrode*: In oxygen-free liquids, such as hydrochloric acid (HCl) or stagnant water, hydrogen gas may be evolved at the cathode:

$$2H^+ + 2e^- \rightarrow H_2 \uparrow \text{ (gas)} \tag{23-3}$$

If zinc were placed in such an environment, we would find that the overall reaction is

$$Zn \rightarrow Zn^{2+} + 2e^- \text{ (anode reaction)}$$
$$2H^+ + 2e^- \rightarrow H_2 \uparrow \text{ (cathode reaction)} \tag{23-4}$$
$$Zn + 2H^+ \rightarrow Zn^{2+} + H_2 \uparrow \text{ (overall reaction)}$$

The zinc anode gradually dissolves, and hydrogen bubbles continue to evolve at the cathode.

2. *The oxygen electrode*: In aerated water, oxygen is available to the cathode, and hydroxyl, or $(OH)^-$, ions form:

$$\frac{1}{2}O_2 + H_2O + 2e^- \rightarrow 2(OH)^- \tag{23-5}$$

The oxygen electrode enriches the electrolyte in $(OH)^-$ ions. These ions react with positively charged metallic ions and produce a solid product. In the case of rusting of iron:

$$Fe \rightarrow Fe^{2+} + 2e^- \text{ (anode reaction)}$$

$$\left.\begin{array}{l} \dfrac{1}{2}O_2 + H_2O + 2e^- \rightarrow 2(OH)^- \\[2mm] Fe^{2+} + 2(OH)^- \rightarrow Fe(OH)_2 \end{array}\right\} \text{ (cathode reactions)} \tag{23-6}$$

$$Fe + \frac{1}{2}O_2 + H_2O \rightarrow Fe(OH)_2 \text{ (overall reaction)}$$

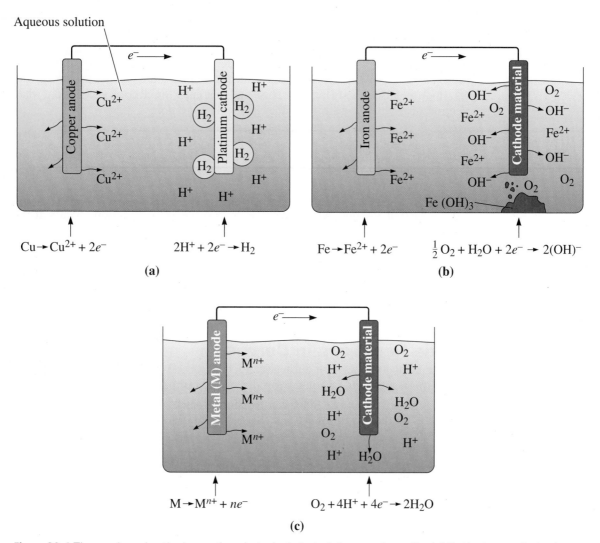

Figure 23-4 The anode and cathode reactions in typical electrolytic corrosion cells: (a) the hydrogen electrode, (b) the oxygen electrode, and (c) the water electrode.

The reaction continues as the $Fe(OH)_2$ reacts with more oxygen and water:

$$2Fe(OH)_2 + \frac{1}{2}O_2 + H_2O \rightarrow 2Fe(OH)_3 \qquad (23\text{-}7)$$

$Fe(OH)_3$ is commonly known as *rust.*

    3. *The water electrode*: In oxidizing acids, the cathode reaction produces water as a byproduct:

$$O_2 + 4H^+ + 4e^- \rightarrow 2H_2O \qquad (23\text{-}8)$$

If a continuous supply of both oxygen and hydrogen is available, the water electrode produces neither a buildup of solid rust nor a high concentration or dilution of ions at the cathode.

23-3   The Electrode Potential in Electrochemical Cells

In corrosion, a potential naturally develops when a material is placed in a solution. Let's see how the potential required to drive the corrosion reaction develops.

Electrode Potential

When a pure metal is placed in an electrolyte, an **electrode potential** develops that is related to the tendency of the material to give up its electrons; however, the driving force for the oxidation reaction is offset by an equal but opposite driving force for the reduction reaction. No net corrosion occurs. Consequently, we cannot measure the electrode potential for a single electrode material.

Electromotive Force Series

To determine the tendency of a metal to give up its electrons, we measure the potential difference between the metal and a standard electrode using a half-cell (Figure 23-5). The metal electrode to be tested is placed in a 1 molar (M) solution of its ions. A reference electrode is also placed in a 1 M solution of ions. The reference electrode is typically an inert metal that conducts electrons but does not react with the electrolyte. The use of a 1 M solution of hydrogen (H^+) ions with the reference electrode is common. The reaction that occurs at this hydrogen electrode is $2H^+ + 2e^- = H_2$. H_2 gas is supplied to the hydrogen electrode. An electrochemical cell such as this is known as a standard cell when the measurements are made at 25°C and atmospheric pressure with 1 M electrolyte concentrations. The two electrolytes are in electrical contact but are not permitted to mix with one another. Each electrode establishes its own electrode potential. By measuring the voltage between the two electrodes when the circuit is open, we obtain the potential difference. The potential of the hydrogen electrode is arbitrarily set equal to zero volts. If the metal has a greater tendency to give up electrons than the hydrogen electrode, then the potential of the metal is negative—the metal is anodic with respect to the hydrogen electrode. If the potential of the metal is positive, the metal is cathodic with respect to the hydrogen electrode.

The **electromotive force** (or **emf**) **series** shown in Table 23-1 compares the standard electrode potential E^0 for each metal with that of the hydrogen electrode under standard

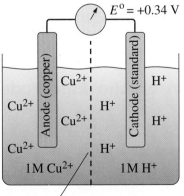

Figure 23-5
The half-cell used to measure the electrode potential of copper under standard conditions. The electrode potential of copper is the potential difference between it and the standard hydrogen electrode in an open circuit. Since E^0 is greater than zero, copper is cathodic compared to the hydrogen electrode.

$E^0 = +0.34$ V

Screen that permits transfer of charge but not mixing of electrolytes

TABLE 23-1 ■ *The standard reduction potentials for selected elements and reactions*

	Metal	Electrode Potential E^0 (Volts)
Anodic ↑	$Li^+ + e^- \rightarrow Li$	−3.05
	$Mg^{2+} + 2e^- \rightarrow Mg$	−2.37
	$Al^{3+} + 3e^- \rightarrow Al$	−1.66
	$Ti^{2+} + 2e^- \rightarrow Ti$	−1.63
	$Mn^{2+} + 2e^- \rightarrow Mn$	−1.63
	$Zn^{2+} + 2e^- \rightarrow Zn$	−0.76
	$Cr^{3+} + 3e^- \rightarrow Cr$	−0.74
	$Fe^{2+} + 2e^- \rightarrow Fe$	−0.44
	$Ni^{2+} + 2e^- \rightarrow Ni$	−0.25
	$Sn^{2+} + 2e^- \rightarrow Sn$	−0.14
	$Pb^{2+} + 2e^- \rightarrow Pb$	−0.13
	$2H^+ + 2e^- \rightarrow H_2$	0.00 — (defined)
	$Cu^{2+} + 2e^- \rightarrow Cu$	+0.34
	$O_2 + 2H_2O + 4e^- \rightarrow 4OH^-$	+0.40
	$Ag^+ + e^- \rightarrow Ag$	+0.80
	$Pt^{4+} + 4e^- \rightarrow Pt$	+1.20
	$O_2 + 4H^+ + 4e^- \rightarrow 2H_2O$	+1.23
Cathodic ↓	$Au^{3+} + 3e^- \rightarrow Au$	+1.50

conditions of 25°C and a 1 M solution of ions in the electrolyte. Note that the measurement of the potential is made when the electric circuit is open. The voltage difference begins to change as soon as the circuit is closed.

The more negative the value of potential for the oxidation of metal, the more electropositive is the metal; this means the metal will have a higher tendency to undergo an oxidation reaction. For example, alkali and alkaline earth metals (e.g., Li, K, Ba, Sr, and Mg) are so reactive that they have to be kept under conditions that prevent any contact with oxygen. On the other hand, metals that are toward the bottom of the chart (e.g., Ag, Au, and Pt) will not tend to react with oxygen. This is why we call them "noble metals." Metals such as Fe, Cu, and Ni have intermediate reactivities; however, this is not the only consideration. Aluminum, for example, has a strongly negative standard electrode potential and does react easily with oxygen to form aluminum oxide; it also reacts easily with fluoride to form aluminum fluoride. Both of these compounds form a tenacious and impervious layer that helps stop further corrosion. Titanium also reacts readily with oxygen. The quickly formed titanium oxide creates a barrier that prevents the further diffusion of species and, thus, avoids further oxidation. This is why both aluminum and titanium are highly reactive, but can resist corrosion exceptionally well. Note that the emf series tells us about the thermodynamic feasibility and driving force, it does *not* tell us about the kinetics of the reaction.

Effect of Concentration on the Electrode Potential

The electrode potential depends on the concentration of the electrolyte. At 25°C, the **Nernst equation** gives the electrode potential in nonstandard solutions:

$$E = E^0 + \frac{0.0592}{n} \log (C_{ion}) \tag{23-9}$$

where E is the electrode potential in a solution containing a concentration C_{ion} of the metal in molar units, n is the charge on the metallic ion, and E^0 is the standard electrode potential in a 1 M solution. Note that when $C_{ion} = 1$, $E = E^0$. The example that follows illustrates the calculation of the electrode potential.

Example 23-1   *Half-Cell Potential for Copper*

Suppose 1 g of copper as Cu^{2+} is dissolved in 1000 g of water to produce an electrolyte. Calculate the electrode potential of the copper half-cell in this electrolyte. Assume $T = 25°C$.

SOLUTION

From chemistry, we know that a standard 1 M solution of Cu^{2+} is obtained when we add 1 mol of Cu^{2+} (an amount equal to the atomic mass of copper) to 1000 g of water. The atomic mass of copper is 63.54 g/mol. The concentration of the solution when only 1 g of copper is added must be

$$C_{ion} = \frac{1}{63.54} = 0.0157\,M$$

From the Nernst equation, with $n = 2$ and $E^0 = +0.34$ V,

$$E = E^0 + \frac{0.0592}{n} \log(C_{ion}) = 0.34 + \frac{0.0592}{2} \log(0.0157)$$

$$= 0.34 + (0.0296)(-1.8) = 0.29\,V$$

Rate of Corrosion or Plating   The amount of metal plated on the cathode in electroplating, or removed from the anode by corrosion, can be determined from **Faraday's equation**,

$$w = \frac{ItM}{nF} \tag{23-10}$$

where w is the weight plated or corroded (g), I is the current (A), M is the atomic mass of the metal, n is the charge on the metal ion, t is the time (s), and F is Faraday's constant (96,500 C). This law basically states that one gram equivalent of a metal will be deposited by 96,500 C of charge. Often the current is expressed in terms of current density, $i = I/A$, so Equation 23-10 becomes

$$w = \frac{iAtM}{nF} \tag{23-11}$$

where the area A (cm^2) is the surface area of the anode or cathode.

The following examples illustrate the use of Faraday's equation to calculate the current density.

Example 23-2   *Design of a Copper Plating Process*

Design a process to electroplate a 0.1-cm-thick layer of copper onto a 1 cm $\times$ 1 cm cathode surface.

SOLUTION

In order for us to produce a 0.1-cm-thick layer on a 1 cm^2 surface area, the mass of copper must be

$$\rho_{Cu} = 8.93\ \text{g/cm}^3 \quad A = 1\ cm^2$$

$$\text{Volume of copper} = (1 \text{ cm}^2)(0.1 \text{ cm}) = 0.1 \text{ cm}^3$$
$$\text{Mass of copper} = (8.93 \text{ g/cm}^3)(0.1 \text{ cm}^3) = 0.893 \text{ g}$$

From Faraday's equation, where $M_{Cu} = 63.54$ g/mol and $n = 2$:

$$It = \frac{wn\text{F}}{M} = \frac{(0.893)(2)(96,500)}{63.54} = 2712 \text{ A} \cdot \text{s}$$

Therefore, we might use several different combinations of current and time to produce the copper plate:

Current	Time
0.1 A	27,124 s = 7.5 h
1.0 A	2,712 s = 45.2 min
10.0 A	271.2 s = 4.5 min
100.0 A	27.12 s = 0.45 min

Our choice of the exact combination of current and time might be made on the basis of the rate of production and quality of the plated copper. Low currents require very long plating times, perhaps making the process economically unsound. High currents, however, may reduce plating efficiencies. The plating effectiveness may depend on the composition of the electrolyte containing the copper ions, as well as on any impurities or additives that are present. Currents such as 10 A or 100 A are too high—they can initiate other side reactions that are not desired. The deposit may also not be uniform and smooth. Additional background or experimentation may be needed to obtain the most economical and efficient plating process. A current of ~1 A and a time of ~45 minutes are not uncommon in electroplating operations.

Example 23-3   *Corrosion of Iron*

An iron container 10 cm × 10 cm at its base is filled to a height of 20 cm with a corrosive liquid. A current is produced as a result of an electrolytic cell, and after four weeks, the container has decreased in weight by 70 g. Calculate (a) the current and (b) the current density involved in the corrosion of the iron.

SOLUTION

(a) The total exposure time is

$$t = (4 \text{ wk})(7 \text{ d/wk})(24 \text{ h/d})(3600 \text{ s/h}) = 2.42 \times 10^6 \text{ s}$$

From Faraday's equation, using $n = 2$ and $M = 55.847$ g/mol,

$$I = \frac{wn\text{F}}{tM} = \frac{(70)(2)(96,500)}{(2.42 \times 10^6)(55.847)}$$

$$= 0.1 \text{ A}$$

(b) The total surface area of iron in contact with the corrosive liquid and the current density are

$$A = (4 \text{ sides})(10 \times 20) + (1 \text{ bottom})(10 \times 10) = 900 \text{ cm}^2$$

$$i = \frac{I}{A} = \frac{0.1}{900} = 1.11 \times 10^{-4} \text{ A/cm}^2$$

Example 23-4 *Copper-Zinc Corrosion Cell*

Suppose that in a corrosion cell composed of copper and zinc, the current density at the copper cathode is 0.05 A/cm². The area of both the copper and zinc electrodes is 100 cm². Calculate (a) the corrosion current, (b) the current density at the zinc anode, and (c) the zinc loss per hour.

SOLUTION

(a) The corrosion current is

$$I = i_{Cu} A_{Cu} = (0.05 \text{ A/cm}^2)(100 \text{ cm}^2) = 5 \text{ A}$$

(b) The current in the cell is the same everywhere. Thus,

$$i_{Zn} = \frac{I}{A_{Zn}} = \frac{5}{100} = 0.05 \text{ A/cm}^2$$

(c) The atomic mass of zinc is 65.38 g/mol. From Faraday's equation:

$$w_{\text{zinc loss}} = \frac{ItM}{nF} = \frac{\left(5 \, \frac{A}{cm^2}\right)(3600 \text{ s/h})(65.38 \text{ g/mol})}{(2)(96,500 \text{ C})}$$

$$= 6.1 \text{ g/h}$$

23-4   The Corrosion Current and Polarization

To protect metals from corrosion, we wish to make the current as small as possible. Unfortunately, the corrosion current is very difficult to measure, control, or predict. Part of this difficulty can be attributed to various changes that occur during operation of the corrosion cell. A change in the potential of an anode or cathode, which in turn affects the current in the cell, is called **polarization**. There are three important kinds of polarization: (1) activation, (2) concentration, and (3) resistance polarization.

Activation Polarization

This kind of polarization is related to the energy required to cause the anode or cathode reactions to occur. If we can increase the degree of polarization, these reactions occur with greater difficulty, and the rate of corrosion is reduced. Small differences in composition and structure in the anode and cathode materials dramatically change the activation polarization. Segregation effects in the electrodes cause the activation polarization to vary from one location to another. These factors make it difficult to predict the corrosion current.

Concentration Polarization

After corrosion begins, the concentration of ions at the anode or cathode surface may change. For example, a higher concentration of metal ions may exist at the anode if the ions are unable to diffuse rapidly into the electrolyte. Hydrogen ions may be depleted at the cathode in a hydrogen electrode, or a high $(OH)^-$ concentration may develop at the cathode in an oxygen electrode. When this situation occurs, either the anode or cathode reaction is stifled because fewer electrons are released at the anode or accepted at the cathode.

In any of these examples, the current density, and thus the rate of corrosion, decreases because of concentration polarization. Normally, the polarization is less pronounced when the electrolyte is highly concentrated, the temperature is increased, or the electrolyte is vigorously agitated. Each of these factors increases the current density and encourages electrochemical corrosion.

Resistance Polarization

This type of polarization is caused by the electrical resistivity of the electrolyte. If a greater resistance to the flow of the current is offered, the rate of corrosion is reduced. Again, the degree of resistance polarization may change as the composition of the electrolyte changes during the corrosion process.

23-5   Types of Electrochemical Corrosion

In this section, we will look at some of the more common forms taken by electrochemical corrosion. First, there is *uniform attack.* When a metal is placed in an electrolyte, some regions are anodic to other regions; however, the location of these regions moves and even reverses from time to time. Since the anode and cathode regions continually shift, the metal corrodes uniformly.

Galvanic attack occurs when certain areas always act as anodes, whereas other areas always act as cathodes. These electrochemical cells are called galvanic cells and can be separated into three types: **composition cells, stress cells,** and **concentration cells.**

Composition Cells

Composition cells, or *dissimilar metal corrosion*, develop when two metals or alloys, such as copper and iron, form an electrolytic cell. Because of the effect of alloying elements and electrolyte concentrations on polarization, the emf series may not tell us which regions corrode and which are protected. Instead, we use a **galvanic series**, in which the different alloys are ranked according to their anodic or cathodic tendencies in a particular environment (Table 23-2). We may find a different galvanic series for seawater, freshwater, and industrial atmospheres.

TABLE 23-2 ■ *The galvanic series in seawater*

Active, Anodic End	Magnesium and Mg alloys
	Zinc
	Galvanized steel
	5052 aluminum
	3003 aluminum
	1100 aluminum
	Alclad
	Cadmium
	2024 aluminum
	Low-carbon steel
	Cast iron
	50% Pb-50% Sn solder
	316 stainless steel (active)
	Lead
	Tin
	Cu-40% Zn brass
	Nickel-based alloys (active)
	Copper
	Cu-30% Ni alloy
	Nickel-based alloys (passive)
	Stainless steels (passive)
	Silver
	Titanium
	Graphite
	Gold
	Platinum
	Noble, Cathodic End

(*After ASM* Metals Handbook, *Vol. 10, 8th Ed., Copyright 1975 ASM International.*)

The following example is an illustration of the galvanic series.

Example 23-5   *Corrosion of a Soldered Brass Fitting*

A brass fitting used in a marine application is joined by soldering with lead-tin solder. Will the brass or the solder corrode?

SOLUTION

From the galvanic series, we find that all of the copper-based alloys are more cathodic than a 50% Pb-50% Sn solder. Thus, the solder is the anode and corrodes. In a similar manner, the corrosion of solder can contaminate water in freshwater plumbing systems with lead.

Composition cells also develop in two-phase alloys, where one phase is more anodic than the other. Since ferrite is anodic to cementite in steel, small microcells cause steel to galvanically corrode (Figure 23-6). Almost always, a two-phase alloy has less resistance to corrosion than a single-phase alloy of a similar composition.

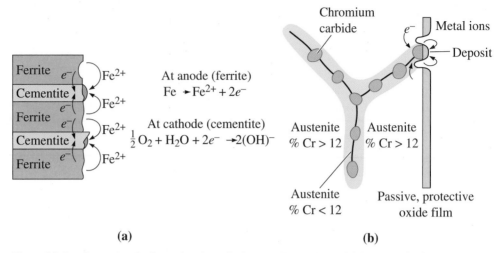

At anode (ferrite)

$$Fe \rightarrow Fe^{2+} + 2e^-$$

At cathode (cementite)

$$\tfrac{1}{2}O_2 + H_2O + 2e^- \rightarrow 2(OH)^-$$

(a)   (b)

Figure 23-6   Example of microgalvanic cells in two-phase alloys: (a) In steel, ferrite is anodic to cementite. (b) In austenitic stainless steel, precipitation of chromium carbide makes the low Cr austenite in the grain boundaries anodic.

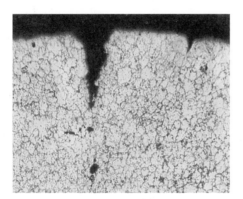

Figure 23-7
Micrograph of intergranular corrosion in a zinc die casting. Segregation of impurities to the grain boundaries produces microgalvanic corrosion cells ($\times$ 50). (*Reprinted courtesy of Don Askeland.*)

Intergranular corrosion occurs when the precipitation of a second phase or segregation at grain boundaries produces a galvanic cell. In zinc alloys, for example, impurities such as cadmium, tin, and lead segregate at the grain boundaries during solidification. The grain boundaries are anodic compared to the remainder of the grains, and corrosion of the grain boundary metal occurs (Figure 23-7). In austenitic stainless steels, chromium carbides can precipitate at grain boundaries [Figure 23-6(b)]. The formation of the carbides removes chromium from the austenite adjacent to the boundaries. The low-chromium (<12% Cr) austenite at the grain boundaries is anodic to the remainder of the grain and corrosion occurs at the grain boundaries. In certain cold-worked aluminum alloys, grain boundaries corrode rapidly due to the presence of detrimental precipitates. This causes the grains of aluminum to peel back like the pages of a book or leaves. This is known as exfoliation.

Stress Cells   Stress cells develop when a metal contains regions with different local stresses. The most highly stressed or high-energy regions act as anodes to the less-stressed cathodic areas (Figure 23-8). Regions with a finer grain size, or a higher

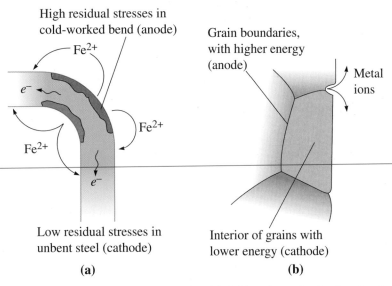

Figure 23-8   Examples of stress cells. (a) Cold work required to bend a steel bar introduces high residual stresses at the bend, which then is anodic and corrodes. (b) Because grain boundaries have high energy, they are anodic and corrode.

density of grain boundaries, are anodic to coarse-grained regions of the same material. Highly cold-worked areas are anodic to less cold-worked areas.

Stress corrosion occurs by galvanic action, but other mechanisms, such as the adsorption of impurities at the tip of an existing crack, may also occur. Failure occurs as a result of corrosion and an applied stress. Higher applied stresses reduce the time required for failure.

Fatigue failures are also initiated or accelerated when corrosion occurs. *Corrosion fatigue* can reduce fatigue properties by initiating cracks (perhaps by producing pits or crevices) and by increasing the rate at which the cracks propagate.

Example 23-6   *Corrosion of Cold-Drawn Steel*

A cold-drawn steel wire is formed into a nail by additional deformation, producing the point at one end and the head at the other. Where will the most severe corrosion of the nail occur?

SOLUTION

Since the head and point have been cold-worked an additional amount compared with the shank of the nail, the head and point serve as anodes and corrode most rapidly.

Concentration Cells   Concentration cells develop due to differences in the concentration of the electrolyte (Figure 23-9). According to the Nernst equation, a difference in metal ion concentration causes a difference in electrode potential. The metal in

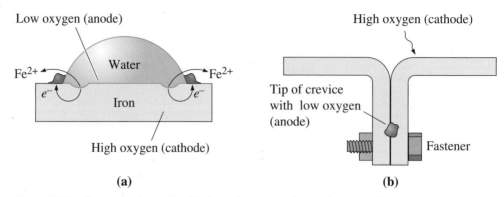

Figure 23-9   Concentration cells: (a) Corrosion occurs beneath a water droplet on a steel plate due to low oxygen concentration in the water. (b) Corrosion occurs at the tip of a crevice because of limited access to oxygen.

contact with the most concentrated solution is the cathode; the metal in contact with the dilute solution is the anode.

The oxygen concentration cell (often referred to as **oxygen starvation**) occurs when the cathode reaction is the oxygen electrode, $H_2O + \frac{1}{2}O_2 + 2e^- \rightarrow 2(OH)^-$. Electrons flow from the low oxygen region, which serves as the anode, to the high oxygen region, which serves as the cathode.

Deposits, such as rust or water droplets, shield the underlying metal from oxygen. Consequently, the metal under the deposit is the anode and corrodes. This causes one form of pitting corrosion. Waterline corrosion is similar. Metal above the waterline is exposed to oxygen, while metal beneath the waterline is deprived of oxygen; hence, the metal underwater corrodes. Normally, the metal far below the surface corrodes more slowly than metal just below the waterline due to differences in the distance that electrons must travel. Because cracks and crevices have a lower oxygen concentration than the surrounding base metal, the tip of a crack or crevice is the anode, causing **crevice corrosion.**

Pipe buried in soil may corrode because of differences in the composition of the soil. Velocity differences may cause concentration differences. Stagnant water contains low oxygen concentrations, whereas fast-moving, aerated water contains higher oxygen concentrations. Metal near stagnant water is anodic and corrodes. The following is an example of corrosion due to water.

Example 23-7   *Corrosion of Crimped Steel*

Two pieces of steel are joined mechanically by crimping the edges. Why would this be a bad idea if the steel is then exposed to water? If the water contains salt, would corrosion be affected?

SOLUTION

By crimping the steel edges, we produce a crevice. The region in the crevice is exposed to less air and moisture, so it behaves as the anode in a concentration cell. The steel in the crevice corrodes.

Salt in the water increases the conductivity of the water, permitting electrical charge to be transferred at a more rapid rate. This causes a higher current density and, thus, faster corrosion due to less resistance polarization.

Microbial Corrosion

Various microbes, such as fungi and bacteria, create conditions that encourage electrochemical corrosion. Particularly in aqueous environments, these organisms grow on metallic surfaces. The organisms typically form colonies that are not continuous. The presence of the colonies and the byproducts of the growth of the organisms produce changes in the environment and, hence, the rate at which corrosion occurs.

Some bacteria reduce sulfates in the environment, producing sulfuric acid, which in turn attacks metal. The bacteria may be either aerobic (which thrive when oxygen is available) or anaerobic (which do not need oxygen to grow). Such bacteria cause attacks on a variety of metals, including steels, stainless steels, aluminum, and copper, as well as some ceramics and concrete. A common example occurs in aluminum fuel tanks for aircraft. When the fuel, typically kerosene, is contaminated with moisture, bacteria grow and excrete acids. The acids attack the aluminum, eventually causing the fuel tank to leak.

The growth of colonies of organisms on a metal surface leads to the development of oxygen concentration cells (Figure 23-10). Areas beneath the colonies are anodic, whereas unaffected areas are cathodic. In addition, the colonies of organisms reduce the rate of diffusion of oxygen to the metal, and—even if the oxygen does diffuse into the colony—the organisms tend to consume the oxygen. The concentration cell produces pitting beneath the regions covered with the organisms. Growth of the organisms, which may include products of the corrosion of the metal, produces accumulations (called **tubercules**) that may plug pipes or clog water-cooling systems in nuclear reactors, submarines, or chemical reactors.

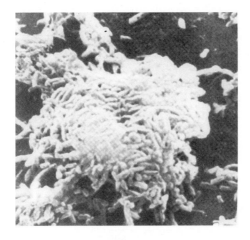

(a)

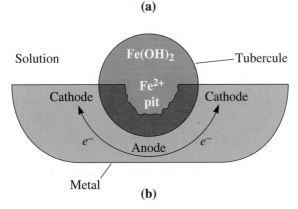

(b)

Figure 23-10
(a) Bacterial cells growing in a colony ($\times 2700$). (b) Formation of a tubercule and a pit under a biological colony. (Micrograph *reprinted courtesy of Don Askeland.*)

23-6   Protection Against Electrochemical Corrosion

A number of techniques are used to combat corrosion, including proper design and materials selection and the use of coatings, inhibitors, cathodic protection, and passivation.

Design   Proper design of metal structures can slow or even avoid corrosion. Some of the steps that should be taken to combat corrosion are as follows:

1. Prevent the formation of galvanic cells. This can be achieved by using similar metals or alloys. For example, steel pipe is frequently connected to brass plumbing fixtures, producing a galvanic cell that causes the steel to corrode. By using intermediate plastic fittings to electrically insulate the steel and brass, this problem can be minimized.

2. Make the anode area much larger than the cathode area. For example, copper rivets can be used to fasten steel sheet. Because of the small area of the copper rivets, a limited cathode reaction occurs. The copper accepts few electrons, and the steel anode reaction proceeds slowly. If, on the other hand, steel rivets are used for joining copper sheet, the small steel anode area gives up many electrons, which are accepted by the large copper cathode area; corrosion of the steel rivets is then very rapid. This is illustrated in the following example.

Example 23-8   *Effect of Areas on Corrosion Rate for Copper-Zinc Couple*

Consider a copper-zinc corrosion couple. If the current density at the copper cathode is 0.05 A/cm^2, calculate the weight loss of zinc per hour if (1) the copper cathode area is 100 cm^2 and the zinc anode area is 1 cm^2 and (2) the copper cathode area is 1 cm^2 and the zinc anode area is 100 cm^2.

SOLUTION

For the small zinc anode area,

$$I = i_{Cu}A_{Cu} = (0.05\,\text{A/cm}^2)(100\,\text{cm}^2) = 5\,\text{A}$$

$$w_{Zn} = \frac{ItM}{nF} = \frac{(5)(3600)(65.38)}{(2)(96,500)} = 6.1\,\text{g/h}$$

For the large zinc anode area,

$$I = i_{Cu}A_{Cu} = (0.05\,\text{A/cm}^2)(1\,\text{cm}^2) = 0.05\,\text{A}$$

$$w_{Zn} = \frac{ItM}{nF} = \frac{(0.05\,\text{A/cm}^2)(3600\,\text{s})\left(65.38\,\dfrac{\text{g}}{\text{mol}}\right)}{(2)(96,500\,\text{C})} = 0.061\,\text{g/h}$$

The rate of corrosion of the zinc is reduced significantly when the zinc anode is much larger than the cathode.

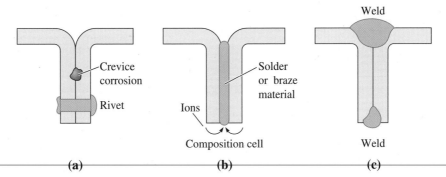

Figure 23-11   Alternative methods for joining two pieces of steel: (a) Fasteners may produce a concentration cell, (b) brazing or soldering may produce a composition cell, and (c) welding with a filler metal that matches the base metal may avoid the formation of galvanic cells.

3. Design components so that fluid systems are closed, rather than open, and so that stagnant pools of liquid do not collect. Partly filled tanks undergo waterline corrosion. Open systems continuously dissolve gas, providing ions that participate in the cathode reaction, and encourage concentration cells.

4. Avoid crevices between assembled or joined materials (Figure 23-11). Welding may be a better joining technique than brazing, soldering, or mechanical fastening. Galvanic cells develop in brazing or soldering since the filler metals have a different composition from the metal being joined. Mechanical fasteners produce crevices that lead to concentration cells. If the filler metal is closely matched to the base metal, welding may prevent these cells from developing.

5. In some cases, the rate of corrosion cannot be reduced to a level that will not interfere with the expected lifetime of the component. In such cases, the assembly should be designed in such a manner that the corroded part can be easily and economically replaced.

Coatings   Coatings are used to isolate the anode and cathode regions. Coatings also prevent diffusion of oxygen or water vapor that initiates corrosion or oxidation. Temporary coatings, such as grease or oil, provide some protection but are easily disrupted. Organic coatings, such as paint, or ceramic coatings, such as enamel or glass, provide better protection; however, if the coating is disrupted, a small anodic site is exposed that undergoes rapid, localized corrosion.

Metallic coatings include tin-plated and hot-dip galvanized (zinc-plated) steel (Figure 23-12). This was discussed in earlier chapters on ferrous materials. A continuous coating of either metal isolates the steel from the electrolyte; however, when the coating is scratched, exposing the underlying steel, the zinc continues to be effective, because zinc is anodic to steel. Since the area of the exposed steel cathode is small, the zinc coating corrodes at a very slow rate and the steel remains protected. In contrast, steel is anodic to tin, so a small steel anode is created when the tin is scratched, and rapid corrosion of the steel subsequently occurs.

Chemical conversion coatings are produced by a chemical reaction with the surface. Liquids such as zinc acid orthophosphate solutions form an adherent phosphate layer on the metal surface. The phosphate layer is, however, rather porous and is more often used to improve paint adherence. Stable, adherent, nonporous, non-conducting

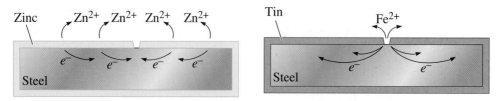

Figure 23-12   Zinc-plated steel and tin-plated steel are protected differently. Zinc protects steel even when the coating is scratched since zinc is anodic to steel. Tin does not protect steel when the coating is disrupted since steel is anodic with respect to tin.

oxide layers form on the surface of aluminum, chromium, and stainless steel. These oxides exclude the electrolyte and prevent the formation of galvanic cells. Components such as reaction vessels can also be lined with corrosion-resistant Teflon™ or other plastics.

Inhibitors   When added to the electrolyte, some chemicals migrate preferentially to the anode or cathode surface and produce concentration or resistance polarization— that is, they are **inhibitors**. Chromate salts perform this function in automobile radiators. A variety of chromates, phosphates, molybdates, and nitrites produce protective films on anodes or cathodes in power plants and heat exchangers, thus stifling the electrochemical cell. Although the exact contents and the mechanism by which it functions are not clear, the popular lubricant WD-40™ works from the action of inhibitors. The name "WD-40™" was designated water displacement (WD) that worked on the fortieth experimental attempt!

Cathodic Protection   We can protect against corrosion by supplying the metal with electrons and forcing the metal to be a cathode (Figure 23-13). Cathodic protection can use a sacrificial anode or an impressed voltage.

A **sacrificial anode** is attached to the material to be protected, forming an electrochemical circuit. The sacrificial anode corrodes, supplies electrons to the metal, and thereby prevents an anode reaction at the metal. The sacrificial anode, typically zinc or magnesium, is consumed and must eventually be replaced. Applications include preventing the

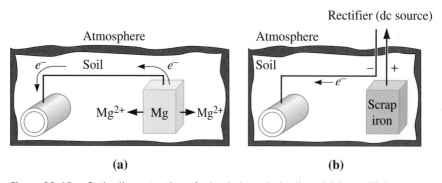

Figure 23-13   Cathodic protection of a buried steel pipeline: (a) A sacrificial magnesium anode ensures that the galvanic cell makes the pipeline the cathode. (b) An impressed voltage between a scrap iron auxiliary anode and the pipeline ensures that the pipeline is the cathode.

corrosion of buried pipelines, ships, off-shore drilling platforms, and water heaters. A magnesium rod is used in many water heaters. The Mg serves as an anode and undergoes dissolution, thus protecting the steel from corroding.

An **impressed voltage** is obtained from a direct current source connected between an auxiliary anode and the metal to be protected. Essentially, we have connected a battery so that electrons flow to the metal, causing the metal to be the cathode. The auxiliary anode, such as scrap iron, corrodes.

Passivation or Anodic Protection

Metals near the anodic end of the galvanic series are active and serve as anodes in most electrolytic cells; however, if these metals are made passive or more cathodic, they corrode at slower rates than normal. **Passivation** is accomplished by producing strong anodic polarization, preventing the normal anode reaction; thus the term anodic protection.

We cause passivation by exposing the metal to highly concentrated oxidizing solutions. If iron is dipped in very concentrated nitric acid, the iron rapidly and uniformly corrodes to form a thin, protective iron hydroxide coating. The coating protects the iron from subsequent corrosion in nitric acid.

We can also cause passivation by increasing the potential on the anode above a critical level. A passive film forms on the metal surface, causing strong anodic polarization, and the current decreases to a very low level. Passivation of aluminum is called **anodizing**, and a thick oxide coating is produced. This oxide layer can be dyed to produce attractive colors. The Ta_2O_5 oxide layer formed on tantalum wires is used to make capacitors.

Materials Selection and Treatment

Corrosion can be prevented or minimized by selecting appropriate materials and heat treatments. In castings, for example, segregation causes tiny, localized galvanic cells that accelerate corrosion. We can improve corrosion resistance with a homogenization heat treatment. When metals are formed into finished shapes by bending, differences in the amount of cold work and residual stresses cause local stress cells. These may be minimized by a stress-relief anneal or a full recrystallization anneal.

The heat treatment is particularly important in austenitic stainless steels (Figure 23-14). When the steel cools slowly from 870°C to 425°C, chromium carbides precipitate at the grain boundaries. Consequently the austenite at the grain boundaries may contain less than 12% chromium, which is the minimum required to produce a passive oxide layer. The steel is **sensitized**. Because the grain boundary regions are small and highly anodic, rapid corrosion of the austenite at the grain boundaries occurs. We can minimize the problem by several techniques.

1. If the steel contains less than 0.03% C, chromium carbides do not form.
2. If the percent chromium is very high, the austenite may not be depleted to below 12% Cr, even if chromium carbides form.
3. Addition of titanium or niobium ties up the carbon as TiC or NbC, preventing the formation of chromium carbide. The steel is said to be **stabilized**.
4. The sensitization temperature range—425 to 870°C—should be avoided during manufacture and service.
5. In a **quench anneal** heat treatment, the stainless steel is heated above 870°C, causing the chromium carbides to dissolve. The structure, now containing 100% austenite, is rapidly quenched to prevent formation of carbides.

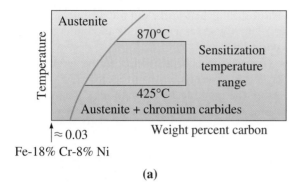

(a)

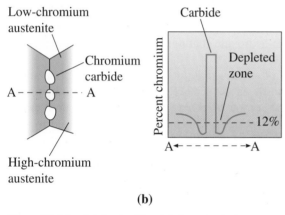

(b)

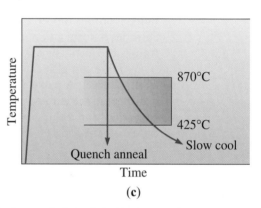

(c)

Figure 23-14   (a) Austenitic stainless steels become sensitized when cooled slowly in the temperature range between 870°C and 425°C. (b) Slow cooling permits chromium carbides to precipitate at grain boundaries. The local chromium content is depleted. (c) A quench anneal to dissolve the carbides may prevent intergranular corrosion.

The following examples illustrate the design of a corrosion protection system.

Example 23-9	*Design of a Corrosion Protection System*

Steel troughs are located in a field to provide drinking water for a herd of cattle. The troughs frequently rust through and must be replaced. Design a system to prevent or delay this problem.

SOLUTION

The troughs are likely a low-carbon, unalloyed steel containing ferrite and cementite, producing a composition cell. The waterline in the trough, which is partially filled with water, provides a concentration cell. The trough is also exposed to the environment, and the water is contaminated with impurities. Consequently, corrosion of the unprotected steel tank is to be expected.

Several approaches might be used to prevent or delay corrosion. We might, for example, fabricate the trough using stainless steel or aluminum. Either would provide better corrosion resistance than the plain carbon steel, but both are considerably more expensive than the current material.

We might suggest using cathodic protection; a small magnesium anode could be attached to the inside of the trough. The anode corrodes sacrificially and prevents corrosion of the steel. This would require that the farm operator regularly check the tank to be sure that the anode is not completely consumed. We also want to be sure that magnesium ions introduced into the water are not a health hazard.

Another approach would be to protect the steel trough using a suitable coating. Painting the steel (that is, introducing a protective polymer coating) or using a tin-plated steel provides protection as long as the coating is not disrupted.

The most likely approach is to use a galvanized steel, taking advantage of the protective coating and the sacrificial behavior of the zinc. Corrosion is very slow due to the large anode area, even if the coating is disrupted. Furthermore, the galvanized steel is relatively inexpensive, readily available, and does not require frequent inspection.

Example 23-10   *Design of a Stainless Steel Weldment*

A piping system used to transport a corrosive liquid is fabricated from 304 stainless steel. Welding of the pipes is required to assemble the system. Unfortunately, corrosion occurs, and the corrosive liquid leaks from the pipes near the weld. Identify the problem and design a system to prevent corrosion in the future.

SOLUTION

Table 13-4 shows that 304 stainless steel contains 0.08% C, causing the steel to be sensitized if it is improperly heated or cooled during welding. Figure 23-15 shows the maximum temperatures reached in the fusion and heat-affected zones during welding. A portion of the pipe in the HAZ heats into the sensitization temperature

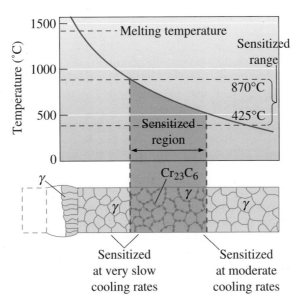

Figure 23-15
The peak temperature surrounding a stainless steel weld and the sensitized structure produced when the weld slowly cools (for Example 23-10).

range, permitting chromium carbides to precipitate. If the cooling rate of the weld is very slow, the fusion zone and other areas of the heat-affected zone may also be affected. Sensitization of the weld area, therefore, is the likely reason for corrosion of the pipe.

Several solutions to the problem may be considered. We might use a welding process that provides very rapid rates of heat input, causing the weld to heat and cool very quickly. If the steel is exposed to the sensitization temperature range for only a brief time, chromium carbides may not precipitate. Joining processes such as laser welding or electron-beam welding are high-rate-of-heat-input processes, but they are expensive. In addition, electron beam welding requires the use of a vacuum, and it may not be feasible to assemble the piping system in a vacuum chamber.

We might heat treat the assembly after the weld is made. By performing a quench anneal, any precipitated carbides are re-dissolved during the anneal and do not reform during quenching; however, it may be impossible to perform this treatment on a large assembly.

We might check the original welding procedure to determine if the pipe was preheated before joining in order to minimize the development of stresses due to the welding process. If the pipe were preheated, sensitization would be more likely to occur. We would recommend that any preheat procedure be suspended.

Perhaps our best design is to use a stainless steel that is not subject to sensitization. For example, carbides do not precipitate in a 304L stainless steel, which contains less than 0.03% C. The low-carbon stainless steels are more expensive than the normal 304 steel; however, the extra cost does prevent corrosion and still permits us to use conventional joining techniques.

23-7   Microbial Degradation and Biodegradable Polymers

Attack by a variety of insects and microbes is one form of "corrosion" of polymers. Relatively simple polymers (such as polyethylene, polypropylene, and polystyrene), high-molecular-weight polymers, crystalline polymers, and thermosets are relatively immune to attack.

Some polymers—including polyesters, polyurethanes, cellulosics, and plasticized polyvinyl chloride (which contains additives that reduce the degree of polymerization)—are particularly vulnerable to microbial degradation. These polymers can be broken into low-molecular-weight molecules by radiation or chemical attack until they are small enough to be ingested by the microbes.

We take advantage of microbial attack by producing *biodegradable* polymers, thus helping to remove the material from the waste stream. Biodegradation requires the complete conversion of the polymer to carbon dioxide, water, inorganic salts, and other small byproducts produced by the ingestion of the material by bacteria. Polymers such as cellulosics easily can be broken into molecules with low molecular weights and are therefore biodegradable. In addition, special polymers are produced to degrade rapidly; a copolymer of polyethylene and starch is one example. Bacteria attack the starch portion of the polymer and reduce the molecular weight of the remaining polyethylene. Certain other polymers such as polycaprolactone (PCL), polylactic acid (PLA), and polylactic glycolic acid (PLGA) are useful in a number of biomedical applications such as sutures and scaffolds.

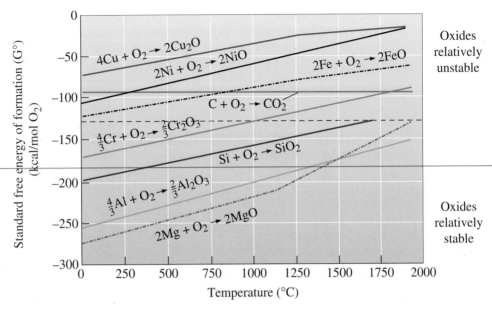

Figure 23-16   The standard free energy of formation of selected oxides as a function of temperature. A large negative free energy indicates a more stable oxide.

23-8   Oxidation and Other Gas Reactions

Materials of all types may react with oxygen and other gases. These reactions can, like corrosion, alter the composition, properties, or integrity of a material. As mentioned before, metals such as Al and Ti react with oxygen very readily.

Oxidation of Metals
Metals may react with oxygen to produce an oxide at the surface. We are interested in three aspects of this reaction: the ease with which the metal oxidizes, the nature of the oxide film that forms, and the rate at which **oxidation** occurs.

The ease with which oxidation occurs is given by the standard free energy of formation for the oxide. The standard free energy of formation for an oxide can be determined from an Ellingham diagram, such as the one shown in Figure 23-16. Figure 23-16 shows that there is a large driving force for the oxidation of magnesium and aluminum compared to the oxidation of nickel or copper. This is illustrated in the following example.

Example 23-11   *Chromium-Based Steel Alloys*

Explain why we should not add alloying elements such as chromium to pig iron before the pig iron is converted to steel in a basic oxygen furnace at 1700°C.

SOLUTION

In a basic oxygen furnace, we lower the carbon content of the metal from about 4% to much less than 1% by blowing pure oxygen through the molten metal. If chromium were already present before the steel making began, chromium would oxidize before the carbon (Figure 23-16), since chromium oxide has a lower free energy of formation (or is more stable) than carbon dioxide (CO_2). Thus, any expensive chromium added would be lost before the carbon was removed from the pig iron.

The type of oxide film influences the rate at which oxidation occurs (Figure 23-17). For the **oxidation reaction**

$$n\text{M} + m\text{O}_2 \rightarrow \text{M}_n\text{O}_{2m} \tag{23-12}$$

the **Pilling-Bedworth (P-B) ratio** is

$$\text{P-B ratio} = \frac{\text{oxide volume per metal atom}}{\text{metal volume per metal atom}} = \frac{(M_{\text{oxide}})(\rho_{\text{metal}})}{n(M_{\text{metal}})(\rho_{\text{oxide}})} \tag{23-13}$$

where M is the atomic or molecular mass, ρ is the density, and n is the number of metal atoms in the oxide, as defined in Equation 23-12.

If the Pilling-Bedworth ratio is less than one, the oxide occupies a smaller volume than the metal from which it formed; the coating is therefore porous and oxidation continues rapidly—typical of metals such as magnesium. If the ratio is one to two, the volumes of the oxide and metal are similar and an adherent, non-porous, protective film

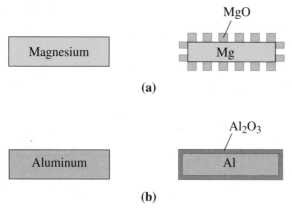

(a)

(b)

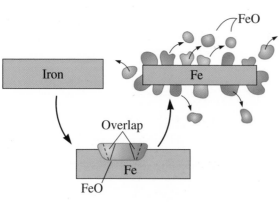

(c)

Figure 23-17
Three types of oxides may form, depending on the volume ratio between the metal and the oxide: (a) magnesium produces a porous oxide film, (b) aluminum forms a protective, adherent, nonporous oxide film, and (c) iron forms an oxide film that spalls off the surface and provides poor protection.

forms—typical of aluminum and titanium. If the ratio exceeds two, the oxide occupies a large volume and may flake from the surface, exposing fresh metal that continues to oxidize—typical of iron. Although the Pilling-Bedworth equation historically has been used to characterize oxide behavior, many exceptions to this behavior are observed. The use of the P-B ratio equation is illustrated in the following example.

Example 23-12 *Pilling-Bedworth Ratio*

The density of aluminum is 2.7 g/cm³ and that of Al_2O_3 is about 4 g/cm³. Describe the characteristics of the aluminum-oxide film. Compare these with the oxide film that forms on tungsten. The density of tungsten is 19.254 g/cm³ and that of WO_3 is 7.3 g/cm³.

SOLUTION

For $2Al + 3/2 O_2 \rightarrow Al_2O_3$, the molecular weight of Al_2O_3 is 101.96 g/mol and that of aluminum is 26.981 g/mol.

$$\text{P-B} = \frac{M_{Al_2O_3}\rho_{Al}}{nM_{Al}\rho_{Al_2O_3}} = \frac{\left(101.96 \frac{g}{mol}\right)(2.7 \text{ g/cm}^3)}{(2)\left(26.981 \frac{g}{mol}\right)(4 \text{ g/cm}^3)} = 1.28$$

For tungsten, $W + 3/2 O_2 \rightarrow WO_3$. The molecular weight of WO_3 is 231.85 g/mol and that of tungsten is 183.85 g/mol.

$$\text{P-B} = \frac{M_{WO_3}\rho_W}{nM_W\rho_{WO_3}} = \frac{\left(231.85 \frac{g}{mol}\right)(19.254 \text{ g/cm}^3)}{(1)\left(183.85 \frac{g}{mol}\right)(7.3 \text{ g/cm}^3)} = 3.33$$

Since P-B $\simeq$ 1 for aluminum, the Al_2O_3 film is nonporous and adherent, providing protection to the underlying aluminum. Since P-B $>$ 2 for tungsten, the WO_3 should be nonadherent and nonprotective.

The rate at which oxidation occurs depends on the access of oxygen to the metal atoms. A linear rate of oxidation occurs when the oxide is porous (as in magnesium) and oxygen has continued access to the metal surface:

$$y = kt \tag{23-14}$$

where y is the thickness of the oxide, t is the time, and k is a constant that depends on the metal and temperature.

A parabolic relationship is observed when diffusion of ions or electrons through a nonporous oxide layer is the controlling factor. This relationship is observed in iron, copper, and nickel:

$$y = \sqrt{kt} \tag{23-15}$$

Finally, a logarithmic relationship is observed for the growth of thin-oxide films that are particularly protective, as for aluminum and possibly chromium:

$$y = k \ln(ct + 1) \qquad (23\text{-}16)$$

where k and c are constants for a particular temperature, environment, and composition. The example that follows shows the calculation for the time required for a nickel sheet to oxidize completely.

Example 23-13   *Parabolic Oxidation Curve for Nickel*

At 1000°C, pure nickel follows a parabolic oxidation curve given by the constant $k = 3.9 \times 10^{-12}$ cm²/s in an oxygen atmosphere. If this relationship is not affected by the thickness of the oxide film, calculate the time required for a 0.1-cm nickel sheet to oxidize completely.

SOLUTION

Assuming that the sheet oxidizes from both sides:

$$y = \sqrt{kt} = \sqrt{\left(3.9 \times 10^{-12} \, \frac{cm^2}{s}\right)(t)} = \frac{0.1 \, cm}{2 \, sides} = 0.05 \, cm$$

$$t = \frac{(0.05 \, cm)^2}{3.9 \times 10^{-12} \, cm^2/s} = 6.4 \times 10^8 \, s = 20.3 \, years$$

Temperature also affects the rate of oxidation. In many metals, the rate of oxidation is controlled by the rate of diffusion of oxygen or metal ions through the oxide. If oxygen diffusion is more rapid, oxidation occurs between the oxide and the metal; if the metal ion diffusion is more rapid, oxidation occurs at the oxide-atmosphere interface. Consequently, we expect oxidation rates to follow an Arrhenius relationship, increasing exponentially as the temperature increases.

Oxidation and Thermal Degradation of Polymers

Polymers degrade when heated and/or exposed to oxygen. A polymer chain may be ruptured, producing two macroradicals. In rigid thermosets, the macroradicals may instantly recombine (a process called the *cage* effect), resulting in no net change in the polymer. In the more flexible thermoplastics—particularly for amorphous rather than crystalline polymers—recombination does not occur, and the result is a decrease in the molecular weight, viscosity, and mechanical properties of the polymer. Depolymerization continues as the polymer is exposed to high temperatures. Polymer chains can also *unzip*. In this case, individual monomers are removed one after another from the ends of the chain, gradually reducing the molecular weight of the remaining chains. As the degree of polymerization decreases, the remaining chains become more heavily branched or cyclization may occur. In *cyclization*, the two ends of the same chain may be bonded together to form a ring.

Polymers also degrade by the loss of side groups on the chain. Chloride ions [in polyvinyl chloride (PVC)] and benzene rings (in polystyrene) are lost from the chain, forming byproducts. For example, as polyvinyl chloride is degraded, hydrochloric acid (HCl) is produced. Hydrogen atoms are bonded more strongly to the chains; thus, polyethylene does not degrade as easily as PVC or polystyrene. Fluoride ions (in Teflon^{TM}) are more difficult to remove than hydrogen atoms, providing Teflon^{TM} with its high temperature resistance. As mentioned before, degradation of polymers, such as PCL, PLGA, and PLA, actually can be useful for certain biomedical applications (e.g., sutures that dissolve) and also in the development of environmentally friendly products.

23-9   Wear and Erosion

Wear and erosion remove material from a component by mechanical attack of solids or liquids. Corrosion and mechanical failure also contribute to this type of attack.

Adhesive Wear

Adhesive wear—also known as scoring, galling, or seizing—occurs when two solid surfaces slide over one another under pressure. Surface projections, or asperities, are plastically deformed and eventually welded together by the high local pressures (Figure 23-18). As sliding continues, these bonds are broken, producing cavities on one surface, projections on the second surface, and frequently tiny, abrasive particles—all of which contribute to further wear of the surfaces.

Many factors may be considered in trying to improve the wear resistance of materials. Designing components so that loads are small, surfaces are smooth, and continual lubrication is possible helps prevent adhesions that cause the loss of material.

The properties and microstructure of the material are also important. Normally, if both surfaces have high hardnesses, the wear rate is low. High strength, to help resist the applied loads, and good toughness and ductility, which prevent the tearing of material from the surface, may be beneficial. Ceramic materials, with their exceptional hardness, are expected to provide good adhesive wear resistance.

Bond

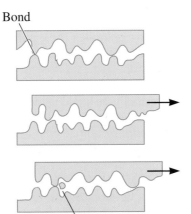

Debris

Figure 23-18
The asperities on two rough surfaces may initially be bonded. A sufficient force breaks the bonds, and the surfaces slide. As they slide, asperities may be fractured, wearing away the surfaces and producing debris.

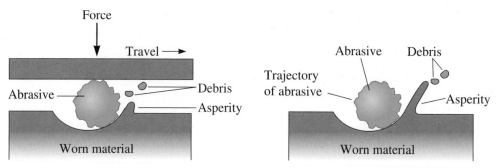

Figure 23-19 Abrasive wear, caused by either trapped or free-flying abrasives, produces troughs in the material, piling up asperities that may fracture into debris.

Wear resistance of polymers can be improved if the coefficient of friction is reduced by the addition of polytetrafluoroethylene (Teflon™) or if the polymer is strengthened by the introduction of reinforcing fibers such as glass, carbon, or aramid.

Abrasive Wear

When material is removed from a surface by contact with hard particles, **abrasive wear** occurs. The particles either may be present at the surface of a second material or may exist as loose particles between two surfaces (Figure 23-19). This type of wear is common in machinery such as plows, scraper blades, crushers, and grinders used to handle abrasive materials and may also occur when hard particles are unintentionally introduced into moving parts of machinery. Abrasive wear is also used for grinding operations to remove material intentionally. In many automotive applications (e.g., dampers, gears, pistons, and cylinders), abrasive wear behavior is a major concern.

Materials with a high hardness, good toughness, and high hot strength are most resistant to abrasive wear. Typical materials used for abrasive-wear applications include quenched and tempered steels; carburized or surface-hardened steels; cobalt alloys such as Stellite; composite materials, including tungsten carbide cermets; white cast irons; and hard surfaces produced by welding. Most ceramic materials also resist wear effectively because of their high hardness; however, their brittleness may sometimes limit their usefulness in abrasive wear conditions.

Liquid Erosion

The integrity of a material may be destroyed by erosion caused by high pressures associated with a moving liquid. The liquid causes strain hardening of the metal surface, leading to localized deformation, cracking, and loss of material. Two types of liquid erosion deserve mention.

Cavitation occurs when a liquid containing a dissolved gas enters a low pressure region. Gas bubbles, which precipitate and grow in the liquid in the low pressure environment, collapse when the pressure subsequently increases. The high-pressure, local shock wave that is produced by the collapse may exert a pressure of thousands of atmospheres against the surrounding material. Cavitation is frequently encountered in propellers, dams and spillways, and hydraulic pumps.

Liquid impingement occurs when liquid droplets carried in a rapidly moving gas strike a metal surface. High localized pressures develop because of the initial impact and the rapid lateral movement of the droplets from the impact point along the metal surface. Water droplets carried by steam may erode turbine blades in steam generators and nuclear power plants.

Liquid erosion can be minimized by proper materials selection and design. Minimizing the liquid velocity, ensuring that the liquid is deaerated, selection of hard, tough materials to absorb the impact of the droplets, and coating the material with an energy-absorbing elastomer all may help minimize erosion.

Summary

- About 6% of the gross domestic product in the United States is used to combat corrosion. Corrosion causes deterioration of all types of materials. Designers and engineers must know how corrosion occurs in order to consider suitable designs, materials selection, or protective measures.

- In chemical corrosion, a material is dissolved in a solvent, resulting in the loss of material. All materials—metals, ceramics, polymers, and composites—are subject to this form of attack. The choice of appropriate materials having a low solubility in a given solvent or use of inert coatings on materials helps avoid or reduce chemical corrosion.

- Electrochemical corrosion requires that a complete electric circuit develop. The anode corrodes and a byproduct such as rust forms at the cathode. Because an electric circuit is required, this form of corrosion is most serious in metals and alloys.

- Composition cells are formed by the presence of two different metals, two different phases within a single alloy, or even segregation within a single phase.

- Stress cells form when the level of residual or applied stresses varies within the metal; the regions subjected to the highest stress are anodic and, consequently, they corrode. Stress corrosion cracking and corrosion fatigue are examples of stress cells.

- Concentration cells form when a metal is exposed to a nonuniform electrolyte; for example, the portion of a metal exposed to the lowest oxygen content corrodes. Pitting corrosion, waterline corrosion, and crevice corrosion are examples of these cells. Microbial corrosion, in which colonies of organisms such as bacteria grow on the metal surface, is another example of a concentration cell.

- Electrochemical corrosion can be minimized or prevented by using electrical insulators to break the electric circuit, designing and manufacturing assemblies without crevices, designing assemblies so that the anode area is large compared to that of the cathode, using protective and even sacrificial coatings, inhibiting the action of the electrolyte, supplying electrons to the metal by means of an impressed voltage, using heat treatments that reduce residual stresses or segregation, and a host of other actions.

- Oxidation degrades most materials. While an oxide coating provides protection for some metals such as aluminum, most materials are attacked by oxygen. Diffusion of oxygen and metallic atoms is often important; therefore oxidation occurs most rapidly at elevated temperatures.

- Other factors, such as attack by microbes, damage caused by radiation, and wear or erosion of a material, may also cause a material to deteriorate.

Glossary

Abrasive wear Removal of material from surfaces by the cutting action of particles.

Adhesive wear Removal of material from moving surfaces by momentary local bonding and then bond fracture at the surfaces.

Anode The location at which corrosion occurs as electrons and ions are given up in an electrochemical cell.

Anodizing An anodic protection technique in which a thick oxide layer is deliberately produced on a metal surface.

Cathode The location at which electrons are accepted and a byproduct is produced during electrochemical corrosion.

Cavitation Erosion of a material surface by the pressures produced when a gas bubble collapses within a moving liquid.

Chemical corrosion Removal of atoms from a material by virtue of the solubility or chemical reaction between the material and the surrounding liquid.

Composition cells Electrochemical corrosion cells produced between two materials having different compositions. Also known as galvanic cells.

Concentration cells Electrochemical corrosion cells produced between two locations on a material where the composition of the electrolyte is different.

Crevice corrosion A special concentration cell in which corrosion occurs in crevices because of the low concentration of oxygen.

Dezincification A special chemical corrosion process by which both zinc and copper atoms are removed from brass, but the copper is replated back onto the metal.

Electrochemical cell A cell in which electrons and ions can flow by separate paths between two materials, producing a current which, in turn, leads to corrosion or plating.

Electrochemical corrosion Corrosion produced by the development of a current that removes ions from the material.

Electrode potential Related to the tendency of a material to corrode. The potential is the voltage produced between the material and a standard electrode.

Electrolyte The conductive medium through which ions move to carry current in an electrochemical cell.

Electromotive force (emf) series The arrangement of elements according to their electrode potential, or their tendency to corrode.

Faraday's equation The relationship that describes the rate at which corrosion or plating occurs in an electrochemical cell.

Galvanic series The arrangement of alloys according to their tendency to corrode in a particular environment.

Graphitic corrosion A special chemical corrosion process by which iron is leached from cast iron, leaving behind a weak, spongy mass of graphite.

Impressed voltage A cathodic protection technique by which a direct current is introduced into the material to be protected, thus preventing the anode reaction.

Inhibitors Additions to the electrolyte that preferentially migrate to the anode or cathode, cause polarization, and reduce the rate of corrosion.

Intergranular corrosion Corrosion at grain boundaries because grain boundary segregation or precipitation produces local galvanic cells.

Liquid impingement Erosion of a material caused by the impact of liquid droplets carried by a gas stream.

Nernst equation The relationship that describes the effect of electrolyte concentration on the electrode potential in an electrochemical cell.

Oxidation Reaction of a metal with oxygen to produce a metallic oxide. This normally occurs most rapidly at high temperatures.

Oxidation reaction The anode reaction by which electrons are given up to the electrochemical cell.

Oxygen starvation In the concentration cell, low-oxygen regions of the electrolyte cause the underlying material to behave as the anode and to corrode.

Passivation Formation of a protective coating on the anode surface that interrupts the electric circuit.

Pilling-Bedworth ratio Describes the morphology of the oxide film that forms on a metal surface during oxidation.

Polarization Changing the voltage between the anode and cathode to reduce the rate of corrosion. *Activation* polarization is related to the energy required to cause the anode or cathode reaction; *concentration* polarization is related to changes in the composition of the electrolyte; and *resistance* polarization is related to the electrical resistivity of the electrolyte.

Quench anneal The heat treatment used to dissolve carbides and prevent intergranular corrosion in stainless steels.

Reduction reaction The cathode reaction by which electrons are accepted from the electrochemical cell.

Sacrificial anode Cathodic protection by which a more anodic material is connected electrically to the material to be protected. The anode corrodes to protect the desired material.

Sensitization Precipitation of chromium carbides at the grain boundaries in stainless steels, making the steel sensitive to intergranular corrosion.

Stabilization Addition of titanium or niobium to a stainless steel to prevent intergranular corrosion.

Stress cells Electrochemical corrosion cells produced by differences in imposed or residual stresses at different locations in the material.

Stress corrosion Deterioration of a material in which an applied stress accelerates the rate of corrosion.

Tubercule Accumulations of microbial organisms and corrosion byproducts on the surface of a material.

Problems

Section 23-1 Chemical Corrosion

Section 23-2 Electrochemical Corrosion

Section 23-3 The Electrode Potential in Electrochemical Cells

Section 23-4 The Corrosion Current and Polarization

23-1 Explain how it is possible to form a simple electrochemical cell using a lemon, a penny, and a galvanized steel nail.

23-2 Silver can be polished by placing it in an aluminum pan containing a solution of baking soda, salt, and water. Speculate as to how the polishing process works.

23-3 A gray cast iron pipe is used in the natural gas distribution system for a city. The pipe fails and leaks, even though no corrosion noticeable to the naked eye has occurred. Offer an explanation for why the pipe failed.

23-4 A brass plumbing fitting produced from a Cu-30% Zn alloy operates in the hot water system of a large office building. After some period of use, cracking and leaking occur. On visual examination, no metal appears to have been corroded. Offer an explanation for why the fitting failed.

23-5 Plot the electronegativity of the elements listed in Table 23-1 as a function of electrode potential. (Refer to Chapter 2 for electronegativity values.) What is the trend that you observe in this data? Using an example, explain how the electronic structure of a metal is consistent with its anodic or cathodic tendency.

23-6 Suppose 10 g of Sn^{2+} are dissolved in 1000 ml of water to produce an electrolyte. Calculate the electrode potential of the tin half-cell.

23-7 A half-cell produced by dissolving copper in water produces an electrode potential of +0.32 V. Calculate the amount of copper that must have been added to 1000 ml of water to produce this potential.

23-8 An electrode potential in a platinum half-cell is 1.10 V. Determine the concentration of Pt^{4+} ions in the electrolyte.

23-9 A current density of 0.05 A/cm^2 is applied to a 150 cm^2 cathode. What period of time is required to plate out a 1-mm-thick coating of silver onto the cathode?

23-10 We wish to produce 100 g of platinum per hour on a 1000 cm^2 cathode by electroplating. What plating current density is required? Determine the current required.

23-11 A 1-m-square steel plate is coated on both sides with a 0.005-cm-thick layer of zinc. A current density of 0.02 A/cm^2 is applied to the plate in an aqueous solution. Assuming that the zinc corrodes uniformly, determine the length of time required before the steel is exposed.

23-12 A 5-cm-inside-diameter, 360-cm-long copper distribution pipe in a plumbing system is accidentally connected to the power system of a manufacturing plant, causing a current of 0.05 A to flow through the pipe. If the wall thickness of the pipe is 0.313 cm, estimate the time required before the pipe begins to leak, assuming a uniform rate of corrosion.

23-13 A steel surface 10 cm $\times$ 100 cm is coated with a 0.002-cm-thick layer of chromium. After one year of exposure to an electrolytic cell, the chromium layer is completely removed. Calculate the current density required to accomplish this removal.

23-14 A corrosion cell is composed of a 300 cm^2 copper sheet and a 20 cm^2 iron sheet, with a current density of 0.6 A/cm^2 applied to the copper. Which material is the anode? What is the rate of loss of metal from the anode per hour?

23-15 A corrosion cell is composed of a 20 cm^2 copper sheet and a 400 cm^2 iron sheet, with a current density of 0.7 A/cm^2 applied to the copper. Which material is the anode? What is the rate of loss of metal from the anode per hour?

23-16 Provide at least two reasons as to why steel alone in water with no other metal present will corrode.

Section 23-6 Protection Against Electrochemical Corrosion

Section 23-7 Microbial Degradation and Biodegradable Polymers

23-17 Explain why titanium is a material that is often used in biomedical implants even though it is one of the most anodic materials listed in Table 23-1.

23-18 Alclad is a laminar composite composed of two sheets of commercially pure aluminum (alloy 1100) sandwiched around a core of 2024 aluminum alloy. Discuss the corrosion resistance of the composite. Suppose that a portion of one of the 1100 layers was machined off, exposing a small patch of the 2024 alloy. How would this affect the corrosion resistance? Explain.

Would there be a difference in behavior if the core material were 3003 aluminum? Explain.

23-19 The leaf springs for an automobile are formed from a high-carbon steel. For best corrosion resistance, should the springs be formed by hot working or cold working? Explain. Would corrosion still occur even if you use the most desirable forming process? Explain.

23-20 Several types of metallic coatings are used to protect steel, including zinc, lead, tin, aluminum, and nickel. In which of these cases will the coating provide protection even when the coating is locally disrupted? Explain.

23-21 An austenitic stainless steel corrodes in all of the heat-affected zone (HAZ) surrounding the fusion zone of a weld. Explain why corrosion occurs and discuss the type of welding process or procedure that might have been used. What might you do to prevent corrosion in this region?

23-22 A steel nut is securely tightened onto a bolt in an industrial environment. After several months, the nut is found to contain numerous cracks. Explain why cracking might have occurred.

23-23 The shaft for a propeller on a ship is carefully designed so that the applied stresses are well below the endurance limit for the material. Yet after several months, the shaft cracks and fails. Offer an explanation for why failure might have occurred under these conditions.

23-24 An aircraft wing composed of carbon fiber-reinforced epoxy is connected to a titanium forging on the fuselage. Will the anode for a corrosion cell be the carbon fiber, the titanium, or the epoxy? Which will most likely be the cathode? Explain.

23-25 The inside surface of a cast iron pipe is covered with tar, which provides a protective coating. Acetone in a chemical laboratory is drained through the pipe on a regular basis. Explain why, after several weeks, the pipe begins to corrode.

23-26 A cold-worked copper tube is soldered, using a lead-tin alloy, into a steel connector. What types of electrochemical cells might develop due to this connection? Which of the materials would you expect to serve as the anode and suffer the most extensive damage due to corrosion? Explain.

23-27 Pure tin is used to provide a solder connection for copper in many electrical uses. Which metal will most likely act as the anode?

23-28 Sheets of annealed nickel, cold-worked nickel, and recrystallized nickel are placed into an electrolyte. Which is the most likely to corrode? Which is the least likely to corrode? Explain.

23-29 A pipeline carrying liquid fertilizer crosses a small creek. A large tree washes down the creek and is wedged against the steel pipe. After some time, a hole is produced in the pipe at the point where the tree touches the pipe, with the diameter of the hole larger on the outside of the pipe than on the inside of the pipe. The pipe then leaks fertilizer into the creek. Offer an explanation for why the pipe corroded.

23-30 Two sheets of a 1040 steel are joined together with an aluminum rivet (Figure 23-20). Discuss the possible corrosion cells that might be created as a result of this joining process. Recommend a joining process that might minimize corrosion for these cells.

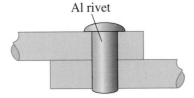

Al rivet

Figure 23-20 Two steel sheets joined by an aluminum rivet (for Problem 23-30).

23-31 Figure 23-21 shows a cross-section through an epoxy-encapsulated integrated circuit, including a microgap between the copper lead frame and the epoxy polymer. Suppose chloride ions from the manufacturing process penetrate the package. What types of corrosion cells might develop? What portions of the integrated circuit are most likely to corrode?

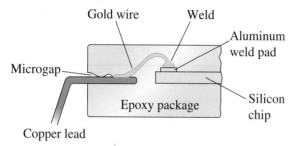

Figure 23-21 Cross-section through an integrated circuit showing the external lead connection to the chip (for Problem 23-31).

23-32 A current density of 0.1 A/cm² is applied to the iron in an iron-zinc corrosion cell. Calculate the weight loss of zinc per hour
(a) if the zinc has a surface area of 10 cm² and the iron has a surface area of 100 cm² and
(b) if the zinc has a surface area of 100 cm² and the iron has a surface area of 10 cm².

Section 23-8 Oxidation and Other Gas Reactions

Section 23-9 Wear and Erosion

23-33 Determine the Pilling-Bedworth ratio for the following metals and predict the behavior of the oxide that forms on the surface. Is the oxide protective, does it flake off the metal, or is it permeable? (See Appendix A for the metal density.)

	Oxide Density (g/cm³)
Mg-MgO	3.60
Na-Na₂O	2.27
Ti-TiO₂	5.10
Fe-Fe₂O₃	5.30
Ce-Ce₂O₃	6.86
Nb-Nb₂O₅	4.47
W-WO₃	7.30

23-34 Oxidation of most ceramics is not considered to be a problem. Explain.

23-35 A sheet of copper is exposed to oxygen at 1000°C. After 100 h, 0.246 g of copper are lost per cm² of surface area; after 250 h, 0.388 g/cm² are lost; and after 500 h, 0.550 g/cm² are lost. Determine whether oxidation is parabolic, linear, or logarithmic, then determine the time required for a 0.75-cm sheet of copper to be completely oxidized. The sheet of copper is oxidized from both sides.

23-36 At 800°C, iron oxidizes at a rate of 0.014 g/cm² per hour; at 1000°C, iron oxidizes at a rate of 0.0656 g/cm² per hour. Assuming a parabolic oxidation rate, determine the maximum temperature at which iron can be held if the oxidation rate is to be less than 0.005 g/cm² per hour.

Design Problems

23-37 A cylindrical steel tank 90 cm in diameter and 240 cm long is filled with water. We find that a current density of 0.015 A/cm² acting on the steel is required to prevent corrosion. Design a sacrificial anode system that will protect the tank.

23-38 The drilling platforms for offshore oil rigs are supported on large steel columns resting on the bottom of the ocean. Design an approach to ensure that corrosion of the supporting steel columns does not occur.

23-39 A storage building is to be produced using steel sheet for the siding and roof. Design a corrosion protection system for the steel.

23-40 Design the materials for the scraper blade for a piece of earthmoving equipment.

Computer Problems

23-41 Write a computer program that will calculate the thickness of a coating using different inputs provided by the user (e.g., density and valence of the metal, area or dimensions of the part being plated, desired time of plating, and current).

Ⓚ Knovel® Problems

K23-1 Describe the mechanism of stray current corrosion.

K23-2 Describe how the corrosion rate of copper changes with increasing solution velocity at different temperatures. Explain the mechanism for this change.

K23-3 Five grams of zirconium are dissolved in 1000 g of water producing an electrolyte of Zr^{4+} ions. Calculate the electrode potential of the copper half-cell in this electrolyte at standard temperature and pressure using the Nernst equation.

Appendix A: Selected Physical Properties of Metals

Metal		Atomic Number	Crystal Structure	Lattice Parameters (Å)	Atomic Mass (g/mol)	Density (g/cm³)	Melting Temperature (°C)
Aluminum	Al	13	FCC	4.04958	26.981	2.699	660.4
Antimony	Sb	51	hex	$a = 4.307$	121.75	6.697	630.7
				$c = 11.273$			
Arsenic	As	33	hex	$a = 3.760$	74.9216	5.778	816
				$c = 10.548$			
Barium	Ba	56	BCC	5.025	137.3	3.5	729
Beryllium	Be	4	HCP	$a = 2.2858$	9.01	1.848	1290
				$c = 3.5842$			
Bismuth	Bi	83	mono	$a = 6.674$	208.98	9.808	271.4
				$b = 6.117$			
				$c = 3.304$			
				$\beta = 110.3°$			
Boron	B	5	rhomb	$a = 10.12$	10.81	2.36	2076
				$\alpha = 65.5°$			
Cadmium	Cd	48	HCP	$a = 2.9793$	112.4	8.642	321.1
				$c = 5.6181$			
Calcium	Ca	20	FCC	5.588	40.08	1.55	839
Cerium	Ce	58	HCP	$a = 3.681$	140.12	6.6893	798
				$c = 5.99$			
Cesium	Cs	55	BCC	6.13	132.91	1.892	28.6
Chromium	Cr	24	BCC	2.8844	51.996	7.19	1907
Cobalt	Co	27	HCP	$a = 2.5071$	58.93	8.832	1495
				$c = 4.0686$			
Copper	Cu	29	FCC	3.6151	63.54	8.93	1084.9
Gadolinium	Gd	64	HCP	$a = 3.6336$	157.25	7.901	1313
				$c = 5.7810$			
Gallium	Ga	31	ortho	$a = 4.5258$	69.72	5.904	29.8
				$b = 4.5186$			
				$c = 7.6570$			
Germanium	Ge	32	DC	5.6575	72.59	5.324	937.4
Gold	Au	79	FCC	4.0786	196.97	19.302	1064.4
Hafnium	Hf	72	HCP	$a = 3.1883$	178.49	13.31	2227
				$c = 5.0422$			

Metal		Atomic Number	Crystal Structure	Lattice Parameters (Å)	Atomic Mass (g/mol)	Density (g/cm³)	Melting Temperature (°C)
Indium	In	49	tetra	$a = 3.2517$ $c = 4.9459$	114.82	7.286	156.6
Iridium	Ir	77	FCC	3.84	192.9	22.65	2447
Iron	Fe	26	BCC	2.866	55.847	7.87	1538
			FCC	3.589	(>912°C)		
			BCC		(>1394°C)		
Lanthanum	La	57	HCP	$a = 3.774$ $c = 12.17$	138.91	6.146	918
Lead	Pb	82	FCC	4.9489	207.19	11.36	327.4
Lithium	Li	3	BCC	3.5089	6.94	0.534	180.7
Magnesium	Mg	12	HCP	$a = 3.2087$ $c = 5.209$	24.312	1.738	650
Manganese	Mn	25	cubic	8.931	54.938	7.47	1244
Mercury	Hg	80	rhomb		200.59	13.546	−38.9
Molybdenum	Mo	42	BCC	3.1468	95.94	10.22	2623
Nickel	Ni	28	FCC	3.5167	58.71	8.902	1453
Niobium	Nb	41	BCC	3.294	92.91	8.57	2468
Osmium	Os	76	HCP	$a = 2.7341$ $c = 4.3197$	190.2	22.57	3033
Palladium	Pd	46	FCC	3.8902	106.4	12.02	1552
Platinum	Pt	78	FCC	3.9231	195.09	21.45	1769
Potassium	K	19	BCC	5.344	39.09	0.855	63.2
Rhenium	Re	75	HCP	$a = 2.760$ $c = 4.458$	186.21	21.04	3186
Rhodium	Rh	45	FCC	3.796	102.99	12.41	1963
Rubidium	Rb	37	BCC	5.7	85.467	1.532	38.9
Ruthenium	Ru	44	HCP	$a = 2.6987$ $c = 4.2728$	101.07	12.37	2334
Selenium	Se	34	mono	$a = 9.054$ $b = 9.083$ $c = 11.60$ $\beta = 90.8°$	78.96	4.809	217
Silicon	Si	14	DC	5.4307	28.08	2.33	1410
Silver	Ag	47	FCC	4.0862	107.868	10.49	961.9
Sodium	Na	11	BCC	4.2906	22.99	0.967	97.8
Strontium	Sr	38	FCC	6.0849	87.62	2.6	777
			BCC	4.84			(>557°C)
Tantalum	Ta	73	BCC	3.3026	180.95	16.6	2996
Technetium	Tc	43	HCP	$a = 2.735$ $c = 4.388$	98.9062	11.5	2157
Tellurium	Te	52	hex	$a = 4.4565$ $c = 5.9268$	127.6	6.24	449.5
Thorium	Th	90	FCC	5.086	232	11.72	1775
Tin	Sn	50	tetra	$a = 5.832$ $c = 3.182$	118.69	5.765	231.9
			DC	6.4912			
Titanium	Ti	22	HCP	$a = 2.9503$ $c = 4.6831$	47.9	4.507	1668
			BCC	3.32			(>882°C)

(Continued)

Metal		Atomic Number	Crystal Structure	Lattice Parameters (Å)	Atomic Mass (g/mol)	Density (g/cm^3)	Melting Temperature (°C)
Tungsten	W	74	BCC	3.1652	183.85	19.254	3422
Uranium	U	92	ortho	$a = 2.854$	238.03	19.05	1133
				$b = 5.869$			
				$c = 4.955$			
Vanadium	V	23	BCC	3.0278	50.941	6.1	1910
Yttrium	Y	39	HCP	$a = 3.648$	88.91	4.469	1522
				$c = 5.732$			
Zinc	Zn	30	HCP	$a = 2.6648$	65.38	7.133	420
				$c = 4.9470$			
Zirconium	Zr	40	HCP	$a = 3.2312$	91.22	6.505	1852
				$c = 5.1477$			
			BCC	3.6090			(>862°C)

Appendix B: The Atomic and Ionic Radii of Selected Elements

Element	Atomic Radius (Å)	Valence	Ionic Radius (Å)
Aluminum	1.432	+3	0.51
Antimony	1.45	+5	0.62
Arsenic	1.15	+5	2.22
Barium	2.176	+2	1.34
Beryllium	1.143	+2	0.35
Bismuth	1.60	+5	0.74
Boron	0.46	+3	0.23
Bromine	1.19	−1	1.96
Cadmium	1.49	+2	0.97
Calcium	1.976	+2	0.99
Carbon	0.77	+4	0.16
Cerium	1.84	+3	1.034
Cesium	2.65	+1	1.67
Chlorine	0.905	−1	1.81
Chromium	1.249	+3	0.63
Cobalt	1.253	+2	0.72
Copper	1.278	+1	0.96
Fluorine	0.6	−1	1.33
Gallium	1.218	+3	0.62
Germanium	1.225	+4	0.53
Gold	1.442	+1	1.37
Hafnium	1.55	+4	0.78
Hydrogen	0.46	+1	1.54
Indium	1.570	+3	0.81
Iodine	1.35	−1	2.20
Iron	1.241 (BCC)	+2	0.74
	1.269 (FCC)	+3	0.64
Lanthanum	1.887	+3	1.15
Lead	1.75	+4	0.84
Lithium	1.519	+1	0.68
Magnesium	1.604	+2	0.66
Manganese	1.12	+2	0.80
		+3	0.66
Mercury	1.55	+2	1.10
Molybdenum	1.363	+4	0.70

(Continued)

Element	Atomic Radius (Å)	Valence	Ionic Radius (Å)
Nickel	1.243	+2	0.69
Niobium	1.426	+4	0.74
Nitrogen	0.71	+5	0.15
Oxygen	0.60	−2	1.32
Palladium	1.375	+4	0.65
Phosphorus	1.10	+5	0.35
Platinum	1.387	+2	0.80
Potassium	2.314	+1	1.33
Rubidium	2.468	+1	1.48
Selenium	1.15	−2	1.91
Silicon	1.176	+4	0.42
Silver	1.445	+1	1.26
Sodium	1.858	+1	0.97
Strontium	2.151	+2	1.12
Sulfur	1.06	−2	1.84
Tantalum	1.43	+5	0.68
Tellurium	1.40	−2	2.11
Thorium	1.798	+4	1.02
Tin	1.405	+4	0.71
Titanium	1.475	+4	0.68
Tungsten	1.371	+4	0.70
Uranium	1.38	+4	0.97
Vanadium	1.311	+3	0.74
Yttrium	1.824	+3	0.89
Zinc	1.332	+2	0.74
Zirconium	1.616	+4	0.79

Note that 1 Å = 10^{-8} cm = 0.1 nanometer (nm)

Answers to Selected Problems

CHAPTER 2

2-6   (i) 3.30×10^{22} atoms/cm^3. (ii) 4.63×10^{22} atoms/cm^3

2-7   (b) 4.7 cm^3

2-8   (a) 5.99×10^{23} atoms. (b) 0.994 mol

2-24   MgO, MgO has ionic bonds.

2-25   Si, Si has covalent bonds.

CHAPTER 3

3-13   (a) 1.426×10^{-8} cm. (b) 1.4447×10^{-8} cm.

3-15   (a) 5.3349 Å. (b) 2.3101 Å.

3-17   FCC.

3-19   BCT.

3-21   (a) 8 atoms/cell. (b) 0.387.

3-31   0.6% contraction.

3-41 A: $[00\bar{1}]$. B: $[1\bar{2}0]$. C: $[\bar{1}11]$. D: $[2\bar{1}\,\bar{1}]$

3-43 A: $(1\bar{1}1)$ B: (030). C: $(10\bar{2})$.

3-45 A: $[1\bar{1}0]$ or $[1\bar{1}00]$. B: $[11\bar{1}]$ or $[11\bar{2}\,3]$. C: $[011]$ or $[\bar{1}2\bar{1}3]$.

3-47 A: $(1\bar{1}01)$. B: (0003). C: $(1\bar{1}00)$.

3-53 $[\bar{1}10]$, $[1\bar{1}0]$, $[101]$, $[\bar{1}0\bar{1}]$, $[011]$, $[0\bar{1}\,\bar{1}]$.

3-55   Tetragonal—4; orthorhombic—2; cubic—12.

3-57   (a) (111). (b) (210). (c) $(0\bar{1}2)$. (d) (218).

3-59   [100]: 0.35089 nm, 2.85 nm^{-1}, 0.866. [110]: 0.496 nm, 2.015 nm^{-1}, 0.612. [111]: 0.3039 nm^{-1}, 3.291 nm^{-1}, 1. The [111] is close packed.

3-61   (100): 1.617×10^{15}/cm^2, packing fraction 0.7854. (110): 1.144×10^{15}/cm^2, packing fraction 0.555. (111): 1.867×10^{15}/cm^2, 0.907. The (111) is close packed.

3-63   4,563,000.

3-66   (a) 0.2978 Å. (b) 0.6290 Å.

3-69   (a) 6. (c) 8. (e) 4. (h) 6.

3-72   Fluorite. (a) 5.2885 Å. (b) 12.13 g/cm^3. (c) 0.624.

3-74   Cesium chloride. (a) 4.1916 Å. (b) 4.8 g/cm^3. (c) 0.693.

3-76   (111): 1.473×10^{15}/cm^2, 0.202 (Mg^{+2}). (222): 1.473×10^{15}/cm^2, 0.806 (O^{-2}).

3-81   0.40497 nm.

3-83   (a) BCC. (c) 0.2327 nm.

CHAPTER 4

4-2 4.98×10^{19} vacancies/cm^3.

4-4   (a) 0.00204. (b) 1.39×10^{20} vacancies/cm^3.

4-6   (a) 1.157×10^{20} vacancies/cm^3. (b) 0.532 g/cm^3.

4-11   8.262 g/cm^3.

4-13   (a) 0.0449. (b) one H atom per 22.3 unit cells.

4-15   (a) 0.0522 defects/unit cell. (b) 2.47×10^{20} defects/cm^3.

4-19   (a) $[0\bar{1}1], [01\bar{1}], [\bar{1}10], [1\bar{1}0], [\bar{1}01], [10\bar{1}]$.
             (b) $[1\bar{1}1], [\bar{1}1\bar{1}], [\bar{1}\bar{1}1], [11\bar{1}]$.

4-21 $(1\bar{1}0), (\bar{1}10), (0\bar{1}1), (01\bar{1}), (10\bar{1}), (\bar{1}01)$.

4-23 $(111)[1\bar{1}0]$: b = 2.863 Å, d = 2.338 Å. $(110)[1\bar{1}1]$:
             b = 7.014 Å, d = 2.863 Å. Ratio = 0.44.

4-52   284 Å.

CHAPTER 5

5-9 1.08×10^9 jumps/s.

5-16 $D_H = 1.07 \times 10^{-4}$ cm^2/s versus $D_N = 3.9 \times 10^{-9}$ cm^2/s. Smaller H atoms diffuse more rapidly.

5-18   (a) 59,390 cal/mol. (b) 0.057 cm^2/s.

5-24   (a) -0.02495 at% Sb/cm. (b) -1.246×10^{19} Sb/(cm$^3 \cdot$ cm).

5-26   (a) -1.969×10^{11} H atoms/(cm$^3 \cdot$ cm).
             (b) 3.3×10^7 H atoms/(cm$^2 \cdot$ s).

5-28 1.25×10^{-3} g/h.

5-30 -198°C.

5-42 $D_0 = 3.47 \times 10^{-16}$ cm^2/s versus $D_{Al} = 2.48 \times 10^{-13}$ cm^2/s. It is easier
             for the smaller Al ions to diffuse.

5-43   0.01 cm: 0.87% C. 0.05 cm: 0.43% C. 0.10 cm: 0.16% C.

5-45   907°C.

5-47   0.53% C.

5-49   190 s.

5-51   1184 s.

5-53   667°C.

5-61   50,488 cal/mol; yes.

CHAPTER 6

6-21   (a) Deforms. (b) Does not neck.

6-22   (b) 4891 N.

6-34   (a) 274 MPa. (b) 417 MPa. (c) 172 GPa. (d) 18.55%. (e) 15.8%. (f) 397.9 MPa. (g) 473 MPa.
             (h) 0.17 MPa.

6-42   (a) 41 mm; will not fracture.

6-50   29.8 kg/mm^2.

6-61   Not notch-sensitive; poor toughness.

CHAPTER 7

7-4   0.99 MPa$\sqrt{m}$.

7-5   No; test will not be sensitive enough.

7-26   22 MPa; max = $+22$ MPa, min = -22 MPa, mean = 0 MPa; a higher frequency will reduce
             fatigue strength due to heating.

7-28   (a) 2.5 mm. (b) 0.0039 mm.

7-32   C = 2.061×10^{-3}; n = 3.01.

7-43 101,329 h.

7-45 $n = 6.82$. $m = -5.7$.

7-47 29 days.

CHAPTER 8

8-5 (a) 1 – the slope is steeper. (b) 1 – it is stronger. (c) 2 – it is more ductile and less strong.

8-7 $n = 0.12$; BCC.

8-11 $n = 0.15$.

8-12 0.56.

8-45 (b) Will not break.

8-58 (a) 550°C, 750°C, 950°C. (b) 700°C. (c) 900°C. (d) 2285°C.

8-68 Slope $= 0.4$. Yes.

CHAPTER 9

9-11 (a) 6.65 Å. (b) 109 atoms.

9-13 1.136×10^6 atoms.

9-30 (a) 0.0333. (b) 0.333. (c) All.

9-31 1265°C.

9-35 31.15 s.

9-37 $B = 300$ s/cm^2, $n = 1.6$.

9-40 (a) ~4.16×10^{-3} cm. (b) 90 s.

9-42 $c = 0.003$ s, $m = 0.35$.

9-44 0.03 s.

9-50 (a) 900°C. (b) 430°C. (c) 470°C. (d) ~250°C/min. (e) 9.7 min.
(f) 8.1 min. (g) 60°C. (h) Zinc. (i) 87.3 min/in^2.

9-60 V/A (riser) $= 0.68$, V/A (thick) $= 1.13$, V/A (thin) $= 0.89$; not effective.

9-62 $D_{Cu} = 1.48$ in. $D_{Fe} = 1.30$ in.

9-64 (a) 46 cm^3. (b) 4.1%.

9-69 23.04 cm.

9-72 0.046 cm^3/100 g Al.

CHAPTER 10

10-23 (a) Yes. (c) No. (e) No. (g) No.

10-26 Cd should give the smallest decrease in conductivity; none should give unlimited solid solubility.

10-33 (a) 2330°C, 2150°C, 180°C. (c) 2570°C, 2380°C, 190°C.

10-35 (a) 100% L containing 30% MgO. (b) 70.8% L containing 38% MgO, 29.2% S containing 62% MgO. (c) 8.3% L containing 38% MgO, 91.7% S containing 62% MgO. (d) 100% S containing 85% MgO.

10-38 (a) L: 15 mol% MgO or 8.69 wt% MgO. S: 38 mol% MgO or 24.85 wt% MgO. (b) L: 78.26 mol% or 80.1 wt%; S: 21.74 mol% or 19.9 wt% MgO. (c) 78.1 vol% L, 21.9 vol% S.

10-40 750 g Ni, Ni/Cu $= 1.62$.

10-42 331 g MgO.

10-44 64.1 wt% FeO.

10-46 (a) 49 wt% W in L, 70 wt% W in α. (b) Not possible.

10-50 Ni dissolves; when the liquid reaches 10 wt% Ni, the bath begins to freeze.

10-54 (a) 2900°C, 2690°C, 210°C. (b) 60% L containing 49% W, 40% α containing 70% W.

10-56 (a) 55% W. (b) 18% W.

10-60 (a) 2900°C. (b) 2710°C. (c) 190°C. (d) 2990°C. (e) 90°C. (f) 300 s. (g) 340 s. (h) 60% W.

10-63 (a) 2000°C. (b) 1450°C. (c) 550°C. (d) 40% FeO. (e) 92% FeO.
(f) 65.5% L containing 75% FeO, 34.5% S containing 46% FeO.
(g) 30.3% L containing 88% FeO, 69.7% S containing 55% FeO.

10-64 (a) 3100°C. (b) 2720°C. (c) 380°C. (d) 90% W. (e) 40% W.
(f) 44.4% L containing 70% W, 55.6% α containing 88% W.
(g) 9.1% L containing 50% W, 90.9% α containing 83% W.

CHAPTER 11

11-7 (a) θ, nonstoichiometric. (b) $\alpha, \beta, \gamma, \eta$. B is allotropic, existing in three different forms at different temperatures. (c) 1100°C: peritectic. 900°C: monotectic. 690°C: eutectic. 600°C: peritectoid. 300°C: eutectoid.

11-10 (c) $SnCu_3$.

11-11 $SiCu_4$.

11-13 (a) 2.5% Mg. (b) 600°C, 470°C, 400°C, 130°C. (c) 74% α containing 7% Mg, 26% L containing 26% Mg. (d) 100% α containing 12% Mg. (e) 67% α containing 1% Mg, 33% β containing 34% Mg.

11-15 (a) Hypereutectic. (b) 98% Sn. (c) 22.8% β containing 97.5% Sn, 77.2% L containing 61.9% Sn. (d) 35% α containing 19% Sn, 65% β containing 97.5% Sn. (e) 22.8% primary β containing 97.5% Sn, 77.2% eutectic containing 61.9% Sn. (f) 30% α containing 2% Sn, 70% β containing 100% Sn.

11-19 (a) Hypoeutectic. (b) 1% Si. (c) 78.5% α containing 1.65% Si, 21.5% L containing 12.6% Si (d) 97.6% α containing 1.65% Si, 2.4% β containing 99.83% Si. (e) 78.5% primary α containing 1.65% Si, 21.5% eutectic containing 12.6% Si. (e) 21.5% eutectic containing 12.6% Si. (f) 96% α containing 0% Si, 4% β containing 100% Si.

11-21 56% Sn, hypoeutectic.

11-23 52% Sn.

11-25 17.5% Li, hypereutectic.

11-27 60.5% α, 39.5% β; 27.7% primary α, 72.3% eutectic; 0.54.

11-29 (a) 1150°C. (b) 150°C. (c) 1000°C. (d) 577°C. (e) 423°C.
(f) 10.5 min. (g) 11.5 min. (h) 45% Si.

CHAPTER 12

12-3 $c = 8.9 \times 10^{-6}, n = 2.81$.

12-24 (a) For Al – 4% Mg: solution treat between 210 and 451°C, quench, and age below 210°C. For Al – 6% Mg: solution treat between 280 and 451°C, quench, and age below 280°C. For Al – 12% Mg: solution treat between 390 and 451°C, quench, and age below 390°C. (c) The precipitates are not coherent.

12-36 (a) Solution treat between 290 and 400°C, quench, and age below 290°C. (c) Not a good candidate. (e) Not a good candidate.

12-49 (a) 795°C. (b) Primary ferrite. (c) 56.1% ferrite containing 0.0218% C and 43.9% austenite containing 0.77% C. (d) 95.1% ferrite containing 0.0218% C and 4.9% cementite containing 6.67% C. (e) 56.1% primary ferrite containing 0.0218% C and 43.9% pearlite containing 0.77% C.

12-51 0.53% C, hypoeutectoid.

12-53 0.156% C, hypoeutectoid.

12-55 0.281% C.

12-57 760°C, 0.212% C.

12-61 (a) 900°C; 12% CaO in tetragonal, 3% CaO in monoclinic, 16% CaO in cubic; 30.8% monoclinic, 69.2% cubic. (c) 250°C; 47% Zn in β', 36% Zn in α, 59% Zn in γ; 52.2% α, 47.8% γ.

12-71 (a) 615°C. (b) 1.67×10^{-5} cm.

12-73 Bainite with HRC 47.

12-75 Martensite with HRC 66.

12-84 (a) 37.2% martensite with 0.77% C and HRC 65.
(c) 84.8% martensite with 0.35% C and HRC 58.

12-86 (a) 750°C. (b) 0.455% C.

12-88 ~3% expansion.

12-90 Austenitize at 750°C, quench, and temper above 330°C.

CHAPTER 13

13-3 (a) 97.8% ferrite, 2.2% cementite, 82.9% primary ferrite, 17.1% pearlite.
(c) 85.8% ferrite, 14.2% cementite, 3.1% primary cementite, 96.9% pearlite.

13-8 For 1035: $A_1 = 727$°C; $A_3 = 790$°C; anneal = 820°C; normalize = 845°C; process anneal = 557 – 647°C; not usually spheroidized.

13-12 (a) Ferrite and pearlite. (c) Martensite. (e) Ferrite and bainite.
(g) Tempered martensite.

13-14 (a) Austenitize at 820°C, hold at 600°C for 10 s, cool.
(c) Austenitize at 780°C, hold at 600°C for 10 s, cool.
(e) Austenitize at 900°C, hold at 320°C for 5000 s, cool.

13-20 0.48% C in martensite; austenitized at 770°C; should austenitize at 860°C.

13-22 1080: fine pearlite. 4340: martensite.

13-26 May become hypereutectoid with grain boundary cementite.

13-28 (a) Not applicable. (c) 8 to 10°C/s. (e) 32 to 36°C/s.

13-30 (a) 16°C/s. (b) Pearlite with HRC 38.

13-32 (a) Pearlite with HRC 36. (c) Pearlite and martensite with HRC 46.

13-37 0.30 h.

13-41 0.05 mm: pearlite and martensite with HRC 53. 0.15 mm: medium pearlite with HRC 38.

13-44 δ-ferrite; nonequilibrium freezing; anneal.

13-49 2.4% Si.

CHAPTER 14

14-9 27% β versus 2.2% β.

14-17 Al: 340%. Mg: 31%. Cu: 1004%.

14-19 Lead may melt during hot working.

14-21 γ' because it is smaller and more numerous than the carbide.

14-24 Ti-15% V: 100% β transforms to 100% α', which then transforms to 24% β precipitate in an α matrix. Ti-35% V: 100% β transforms to 100% β_{ss}, which then transforms to 27% α precipitate in a β matrix.

14-28 Spalls off; cracks.

CHAPTER 15

15-22 $B = 2.4$; true porosity = 22.58%; fraction = 0.557.

15-28 1.276 kg BaO; 0.249 kg Li_2O.

CHAPTER 16

16-16 Polybutadiene and silicone; the T_g must be lower than $-78°C$.

16-18 Polyethylene and polypropylene.

16-27 1.055×10^5 cal/mol.

CHAPTER 17

17-8 7.8×10^{13} per cm^3.

17-10 2.48%.

17-14 9.41 g/cm^3.

17-17 (a) 0.507. (b) 0.507. (c) 7.77 g/cm^3.

17-19 11.2 to 22.2 kg.

17-26 (a) 2.56 g/cm^3. (b) 29×10^6 psi. (c) 15.2×10^6 psi.

17-29 188 MPa.

17-31 For $d = 20$ μm, $l_c = 0.30$ cm, $l_c/d = 150$. For $d = 1$ μm, $l_c = 0.15$ cm, $l_c/d = 1500$.

17-39 Sizing improves strength.

17-41 Pyrolize at 2500°C.

17-49 0.417 g/cm^3; 20.0 kg versus 129.6 kg.

CHAPTER 18

18-2 (b) 0.77 g/cm^3.

CHAPTER 19

19-1 (a) 3377 W. (b) 3.183×10^{14} W.

19-4 $d = 0.0864$ cm; 1174 V.

19-8 0.0234 km/h.

19-13 3.03×10^5 $ohm^{-1} \cdot cm^{-1}$ at $-50°C$; 0.34×10^5 $ohm^{-1} \cdot cm^{-1}$ at 500°C.

19-15 $-81.8°C$.

19-18 At 400°C, $\rho = 41.5 \times 10^{-6}$ ohm $\cdot$ cm; $\rho_d = 8.5 \times 10^{-6}$ ohm $\cdot$ cm; $b = 1.79 \times 10^{-4}$ ohm $\cdot$ cm. At 200°C, $\rho = 3.8 \times 10^{-5}$ ohm $\cdot$ cm.

19-21 (a) $n(Ge) = 2.51 \times 10^{13}$ per cm^3. (b) $f(Ge) = 1.42 \times 10^{-10}$. (c) $n_0(Ge) = 1.14 \times 10^{19}$ per cm^3.

19-23 Sb: 10.8 $ohm^{-1} \cdot cm^{-1}$. In: 3.8 $ohm^{-1} \cdot cm^{-1}$.

19-30 2600%.

19-41 (a) 4.85×10^{-19} m. (c) 1.15×10^{-17} m.

19-43 9.4 V.

19–46 0.001239 μF.

CHAPTER 20

20-6 Fe: 39,604 Oe. Co: 31,573 Oe.

20-14 6068 Oe.

20-16 8000 G; 1.49 mA.

20-31 (a) 13,000 G. (b) 14,000 G. (c) 800 Oe. (d) 5.8 G/Oe. (e) 15.6 G/Oe. (f) 6.3×10^6 G Oe.

20-39 15×10^6 G $\cdot$ Oe.

20-40 High saturation inductance.

20-43 4.68×10^5 A/m.

CHAPTER 21

21-10 Ice, water, Teflon.
21-14 4%; 0.35%.
21-16 (a) 6.56°. (b) 6.65°. (c) 10°. (d) 0.18 cm.
21-18 0.0117 cm.
21-29 1.84×10^{-5} s.
21-33 (a) 13.1×10^{-5} cm. (b) 77.6×10^{-5} cm.
21-35 1.85×10^{-4} cm; infrared.
21-38 13,800 V.
21-40 7480 V.
21-42 (a) 24,825 V. (b) Cu, Mn, Si.

CHAPTER 22

22-4 (a) 7950 J. (c) 35,564 J.
22-6 0.0375 cm.
22-8 1.975 m.
22-11 15.182 m.
22-26 (a) 78.5×10^{6} cal/day. (b) 47.09×10^{6} cal/day.
22-29 19.6 min.
22-32 Interconnected graphite flakes in gray iron.

CHAPTER 23

23-3 Graphitic corrosion.
23-6 -0.172 V.
23-8 0.000034 g/1000 mL.
23-10 0.055 A/cm^2; 55 A.
23-12 34 years.
23-14 187.5 g Fe lost/h.
23-18 1100 alloy is anode and continues to protect 2024; 1100 alloy is cathode and the 3003 will corrode.
23-20 Al, Zn, Cd.
23-22 Stress corrosion cracking.
23-24 Ti is the anode; carbon is the cathode.
23-28 Cold worked nickel will corrode most rapidly; annealed nickel will corrode most slowly.
23-32 (a) 12.2 g/h. (b) 1.22 g/h.
23-34 Most ceramics are already oxides.
23-36 698°C.

Index

PRINCIPAL UNITS USED IN MECHANICS

Quantity	International System (SI)			U.S. Customary System (USCS)		
	Unit	Symbol	Formula	Unit	Symbol	Formula
Acceleration (angular)	radian per second squared		rad/s^2	radian per second squared		rad/s^2
Acceleration (linear)	meter per second squared		m/s^2	foot per second squared		ft/s^2
Area	square meter		m^2	square foot		ft^2
Density (mass) (Specific mass)	kilogram per cubic meter		kg/m^3	slug per cubic foot		$slug/ft^3$
Density (weight) (Specific weight)	newton per cubic meter		N/m^3	pound per cubic foot	pcf	lb/ft^3
Energy; work	joule	J	$N{\cdot}m$	foot-pound		ft-lb
Force	newton	N	$kg{\cdot}m/s^2$	pound	lb	(base unit)
Force per unit length (Intensity of force)	newton per meter		N/m	pound per foot		lb/ft
Frequency	hertz	Hz	s^{-1}	hertz	Hz	s^{-1}
Length	meter	m	(base unit)	foot	ft	(base unit)
Mass	kilogram	kg	(base unit)	slug		$lb\text{-}s^2/ft$
Moment of a force; torque	newton meter		$N{\cdot}m$	pound-foot		lb-ft
Moment of inertia (area)	meter to fourth power		m^4	inch to fourth power		$in.^4$
Moment of inertia (mass)	kilogram meter squared		$kg{\cdot}m^2$	slug foot squared		$slug\text{-}ft^2$
Power	watt	W	J/s ($N{\cdot}m/s$)	foot-pound per second		ft-lb/s
Pressure	pascal	Pa	N/m^2	pound per square foot	psf	lb/ft^2
Section modulus	meter to third power		m^3	inch to third power		$in.^3$
Stress	pascal	Pa	N/m^2	pound per square inch	psi	$lb/in.^2$
Time	second	s	(base unit)	second	s	(base unit)
Velocity (angular)	radian per second		rad/s	radian per second		rad/s
Velocity (linear)	meter per second		m/s	foot per second	fps	ft/s
Volume (liquids)	liter	L	$10^{-3}\ m^3$	gallon	gal.	$231\ in.^3$
Volume (solids)	cubic meter		m^3	cubic foot	cf	ft^3

Principles of Construction Management
Third Edition

PRINCIPLES OF CONSTRUCTION MANAGEMENT Third Edition

Principles of Construction Management, third edition, is a new title in the **McGraw-Hill International Series in Civil Engineering**

Principles of Construction Management
Third Edition

Roy Pilcher

Formerly A. J. Clark Chair Professor in Construction Engineering and Management
University of Maryland

McGRAW-HILL BOOK COMPANY

London · New York · St Louis · San Francisco · Auckland · Bogotá · Caracas ·
Hamburg · Lisbon · Madrid · Mexico · Milan · Montreal · New Delhi · Panama · Paris ·
San Juan · São Paulo · Singapore · Sydney · Tokyo · Toronto

Published by
McGraw-Hill Book Company Europe
Shoppenhangers Road, Maidenhead, Berkshire, SL6 2QL, England
Telephone 0628 23432
Fax 0628 770224

British Library Cataloguing in Publication Data
Pilcher, Roy, *1926–*
   Principles of construction management.—3rd ed.
   I. Title
   624.0684
   ISBN 0-07-707236-7

Library of Congress Cataloging-in-Publication Data
Pilcher, Roy.
   Principles of construction management/Roy Pilcher.—3rd ed.
     p. cm. — (McGraw-Hill International series in civil engineering)
   Includes bibliographical references and index.
   ISBN 0-07-707236-7: £19.95
   1. Construction industry—Management. I. Title. II. Series.
   TH438.P47 1992
   690′.068—dc20   91-44893 CIP

1234 9432
Typeset by Interprint Ltd, Malta.
Printed and bound in Great Britain by
BPCC Hazells Ltd
Member of BPCC Ltd

Contents

Preface

This third edition of *Principles of Construction Management* comes some fifteen years after the second edition and twenty-four after the first. Because the book is primarily intended to present a basic course of principles, the general approach in this edition has not changed significantly. It remains a course text for teaching, and therefore continues to concern itself with those principles that can form the basis for future thought and development. This must not be confused with a detailed practical approach to day-to-day administrative matters that are best dealt with in other types of books which, nevertheless, make a comparable contribution to the development of the general subject area.

As to the content and arrangement of the text, the first chapter has largely been rewritten to focus more directly on *organizational behaviour*. Clearly, this is an introductory chapter acting as a signpost to wider studies. It is none the less an increasingly important area of study that should not be omitted from construction management studies. Chapters 2–5 inclusive are concerned with managerial economics. They have also been rewritten and expanded to deal more fully with one of the prime considerations of management, that of money and its effective use.

Chapter 6 now recognizes the fundamental importance of *productivity* as an area warranting close study and examination. Much of the subject matter concerning work study is included in this chapter with slightly more material on work measurement than hitherto. Chapter 7 starts a series of five chapters concerned with construction planning and the principles underlying it. Chapter 7 itself contains some new material on estimating work durations and also on learning and forgetting curves—a subject rarely taken into account in planning. Chapters 8, 9, and 10 deal with network analysis techniques and resource allocation. This material is essentially the same as that presented in the earlier edition. Precedence diagrams and line-of-balance methods form the content of Chapter 11. Precedence diagramming has been expanded in the light of its popularity in use in construction. Line-of-balance methods have still to achieve their full potential as a planning tool for repetitive work. They are often not used because the repetitiveness of some work in construction is not recognized as such at the outset of the work. The treatment of line-of-balance methods has also been expanded.

A new chapter has been included as Chapter 12. This is concerned with *financial management*, and the elements of balance sheets and their interpretation is included. The importance of understanding the financial workings of an enterprise is stressed and ratio analysis is included. It is increasingly necessary that construction managers (and they are not alone in this respect) should have an awareness of financial matters and how the work that they carry out, as construction managers, affects the finances of clients, contractors, consultants, and so on. They also need to have an appreciation of what their accountant and financial manager colleagues are talking about and doing. The objective of this additional chapter is to stir an interest in such matters, if it is not already there, and to stimulate readers to take the matter further in the abundant professional literature that is available on the subject.

The chapter on financial management leads on to the subject of site cost control and how that interrelates with financial provision. To this work has been added an introduction to standard costs and variance analysis. Some very small part of the latter (as did an example used later in decision trees) first appeared in the author's *Project Cost Control in Construction* published by Collins (London, 1985). Variance analysis has the important property of assigning a reason to overspending/underspending—something that many cost control systems in practice are unable to do.

The final two chapters are devoted to an introduction to operational research techniques. To Chapter 15 has been added the important subject of *decision analysis*. This has general usefulness in that it allows a problem to be formulated, and thus aired, before attempting a solution. Related to, and part of it, are decision trees and utility theory, both of which condition our thinking about the structure of problems.

Wherever possible, additional examples and problems for solution have been added throughout the revised text.

Sums of money that necessarily appear in such a textbook, for various reasons, are rarely the current market value of an operation, hire rate, initial capital cost or cost of many other factors. Authors do their best in this respect, and so they should. However, time goes by and market values change. Construction generally is not an area in which precise market values are universally adopted for one reason or another. This book is concerned with *principles*; as such it does not rely on accurate market values in every calculation, though it would be nice if this were possible. Its purpose is to provide a basis that will give an appropriate answer no matter what variation takes place in the initial data used (unless the calculation is qualified in this respect) and to get readers used to thinking in disciplined ways about problems.

Further reading appears at the end of most chapters. This is not intended to be an exhaustive bibliography in the relevant subject area because that would not usefully serve the purpose. The lists of reading material are intended to supplement the text and to provide the enquiring reader with a starting point for gaining additional breadth of the subject. They are all references that either my students or I have found useful from time to time.

The subject of construction management and related areas is always expanding and some related areas of interest have obviously not been covered. They include information technology, knowledge-based expert systems, robotics, automation, data collection, quality control, communication, risk anaysis and bidding theory. All are areas that could either have been included or, where there is some inclusion, could legitimately have been extended. Perhaps this will be done at another time.

I have avoided, in the book, detailed reference to computer programs as aids in solving problems that are raised in this text. Of course there are many available in all shapes and sizes. Many are transient and pass quickly through their useful phase only to disappear for good. All have some deficiency in terms of the personal likes and dislikes of users. Many are very useful. As far as the subject matter of this book is concerned there is almost nothing that cannot be usefully assisted by the intelligent and skilful use of a good spreadsheet and I commend that as a first step when the need for computation facilities arises.

Finally, once more, I come to my acknowledgements. I am indebted to many, many individuals impossible to name, both in my family and in my work. I am sure however that each will recognize the part they have played should they have the inclination to read this book. But there is one to whom the debt surpasses that to all the others and who is unlikely to read it—my wife. For so long she has needed to live with both me and this book—her competitor—and, for once, at the end of the day will almost certainly be glad to see its passing!

Roy Pilcher

PART ONE: Theory and Practice

1. Management and organizational behaviour

1.1 Introduction

This chapter is concerned with the interaction of management and *organizational behaviour*—the study of individuals and groups in organizations. It is appropriate that a textbook such as this should begin with an introduction to a subject area that has considerable impact on managers and their interactions with the performance of people. Much of this book is concerned with providing a framework and explaining techniques for the analysis and solution of construction problems. Frequently, little or no regard is paid to the human factors involved in the problems or how the outcome of the decisions will be influenced by the behaviour of the people concerned with the solutions. Construction is essentially concerned with the behaviour of people within an organization. Organizational behaviour provides the background for managers to develop a systematic method of taking into account the influence of human work behaviour in their problem-solving techniques. The two facets, techniques and organizational behaviour, need to be considered together. By the end of this chapter the reader should be familiar with an outline of organizational systems, know something about organizational design, managing people in organizations, and the behaviour of groups in organizations, and, more importantly, will have an adequate basis for making a further, much more detailed study of such important matters.

1.2 Historical

The early days of management as a subject for study in Britain were related to the Industrial Revolution. Though the Industrial Revolution is generally associated with a period between 1750 and 1850 or thereabouts, it would be more accurate to say that management's association relates to the latter half of this period.

In the first instance, attention was paid to improving the methods necessary for an increase in the production of goods by the use of mechanical power. The celebrated James Watt, of steam-engine fame, became one of the earliest pioneers of such developments in his Soho Foundry. The Foundry was laid out so that the flow of materials through the various processes was logically and thoughtfully arranged.

3

From these early beginnings the techniques of management have tended to find their support and application in the field of production engineering, rather than in construction. This, in spite of the fact that one of the early pioneers in the field of motion study, Gilbreth, started work as a bricklayer and subsequently became a contractor in construction. Such thought as was given to the social scientific aspects of management in the early nineteenth century often proved to be too much for contemporary industry, and pioneers in personnel administration, of which Robert Owen was the leading exponent, lived many years before their principles became generally accepted.

If the formation of a body of scientific knowledge is measured by the amount of original material on that subject which is published, then the history of the formation of the science of management jumps to the end of the nineteenth century. In the intervening years, developments in Britain had not been at a standstill, for the growth of the trade union system in the mid-nineteenth century meant that considerable attention was paid to wages and working conditions. Simultaneously, the rapid expansion of industry meant that more thought had to be paid to the means of providing capital. With the principle of limited liability being accepted legally and commercially, there arose a need to reassure those stockholders providing the necessary capital that it was not being misused. Attention, therefore, had to be paid to proper financial accounting, maintenance of the appropriate books of account, and the periodic preparation of a balance sheet for the benefit of the interested parties.

By the end of the nineteenth century, two of the major subjects of management, both of that time and of today, had been drawn together and their importance recognized. Over the next few years a number of books on the combined subject area appeared, among them *The Commercial Organization of Factories* by J. Slater-Lewis—a historic textbook study of the highest class in organization and management. It is particularly striking that such a large number of books published on management topics in the early days of the development of the subject should remain for many years as standard works of current practice.

During the 1880s in the USA, the 'Father of Scientific Management', Frederick Winslow Taylor (1856–1915), commenced his researches in the Midvale Steel Works of the Bethlehem Steel Company where he was employed as a chargehand supervising lathe operators. Soon, with his colleagues, Henry Lawrence Gantt and Frank Bunker Gilbreth, he was to found the movement that bore the title of 'Scientific Management' and in 1911, he published the first version of his book on the subject.[1]

One of Taylor's main preoccupations was to create a mental revolution among both men and management in industry, believing as he did that scientific analysis of work should be undertaken by managers. In general, his activities were oriented towards practical studies with an engineering emphasis. He believed that both sides of industry were too concerned with how the surplus moneys of a business were to be divided, when they should be more concerned with how to increase the extent of the surplus. In that way he saw a means to increase the earnings of both employees and employers. As a means to increasing the surplus he advocated better planning and better motivation to work, with proper and adequate use of an incentive and bonus system. As well as being concerned with proposals for a mental revolution, Taylor, Gantt, and Gilbreth spent a considerable amount of time and effort on the elaboration of methods such as the specification of job responsibilities, time and motion study, planning schedules, and other aids to the adequate planning and control of production. Unfortunately, Taylor was misunderstood in many circles, particularly among the leaders of organized labour, and the methods he proposed and developed were harshly criticized and treated with deep suspicion. While his philosophies have survived his death and are now accepted by the majority of students of management, some of the suspicions associated with the methods in application have lingered to this day, and are met frequently in the dealings between employees and managers.

While Taylor and his colleagues were working in the field of scientific management in the USA, Henri Fayol read a paper in France, *Administration Industrielle et Générale*, the basis of which is still relevant today in describing the processes of management. Fayol was the general manager of a large French iron and steel combine which had mining as well as metallurgical interests. In 1908 he presented his paper to a congress of one of the metallurgical societies and in it attempted to categorize the processes he understood to be involved in his day-to-day practice as a chief executive. The basis of his analysis was that the management process consisted of five areas—planning, organizing, commanding, coordinating, and controlling. Fayol was the first man to advocate what, at that time, was considered to be a somewhat revolutionary thought, namely that management principles could and should be taught. His paper was not published by the sponsoring society until 1916, and did not appear in English until 1926. Fayol did not follow up his intention to dissect the management process even further, possibly because of the apparent lack of real enthusiasm with which his paper was received, but also because he was in the twilight of his career when it first appeared. He was well over eighty when he died in 1925.

Simultaneously with Fayol's display of interest in the principles of management in France, Mary Follett,[2] in the USA, was working on the many social and industrial problems of the time. She continued to gain experience in this field for the first quarter of the twentieth century before publicly displaying her interest in the human principles of management by presenting a series of papers on the subject. These made one of the outstanding contributions to the contemporary literature of management.

From the 1920s onwards, systematic study of the many branches of management has led to the promulgation of a multitude of theories. Many textbooks have been written and research in universities and business schools has come to be accepted as one part of the necessary education and training of competent managers. A body of knowledge has been built up concerning the principles of management and is now well established.

1.3 The nature of management

Management is often misinterpreted as meaning either the board of directors and the senior executives of a company collectively, or alternatively their function within a company. Management is in fact a process. It involves the five formal activities of *planning, organizing, staffing, directing,* and *controlling.* A manager is responsible for selecting, obtaining, distributing, organizing, and putting to use all of those resources that are necessary to pursue and achieve an organization's objectives. Not least of these resources will be the employees of the organization. In construction, for example, within a project organization, management is a function of all the participants down to and including the foreman level. Below this level the function tends to be supervisory.

Such terms as *the science of management, scientific management,* and *management science* are sometimes used freely, but incorrectly, with little distinction, again incorrectly, drawn between them. The common use of these terms implies that management is a science and not an art. In fact, successful management must be a blend of the two. The organized body of knowledge forming the science consists of principles from which are developed a large number of tools commonly used in management—techniques and methods, usually with a mathematical basis—to assist in the decision-making aspects of the process. The skill necessary for the proper and efficient organization of human beings for production is frequently the art.

Scientific management has for its basis a rational or systematic approach. Firstly, it relies upon the premise that problems cannot be solved properly if assumptions are made about any of the relevant facts. The problem situation must first be studied rigorously and factors lending themselves to quantitative evaluation must be measured before analysis takes place. Secondly, scientific management is based upon prescribing a standardized set of thoughtfully designed and relevant procedures together with the necessary control functions. Thirdly, it requires the study and analysis of all operations in order to make them as efficient as possible. This implies that they will be designed to achieve objectives at the minimum cost that is compatible with the minimum duration or time and with the appropriate quality. There is often a fourth premise based upon the belief that reward in the form of wages is the prime motivator for employees. It will be shown shortly that the truth of this is limited in practice and it is not necessarily always the case.

Management science, while derived from the same conceptual basis as scientific management, should not be confused with it. Management science is alternatively named *operational research*—a collection of mainly mathematically supported approaches to problem solving that involves, among other sciences, statistics, the decision sciences, and quantitative methods. Such methods can be used to deal effectively with problems such as programming, resource allocation, scheduling, inventory control, and queueing, but they have found relatively little application in the behavourial aspects of management.

Of all the techniques and methods available to assist managers, there is no single one that can be applied successfully in every situation. A manager must therefore assess each situation, note in particular its differences from other situations, and then select the most appropriate approach to reaching a solution.

Because of the complex construction projects that are undertaken today, there is a great need for efficient managers in the construction industry. We live in the age of organization. A manager must be able to organize not only technologies but also human resources. The manager must view the management activity as a whole and not from the bias of a particular professional education and training, which may have been as an engineer, an accountant, a lawyer, or as a member of any of numerous other callings. In the organization of human resources it is necessary for a manager to realize that improvements in a worker's mental, physical, and social conditions, together with sharing a sense of participation in the enterprise, are prime movers of effective performance.

In order to be able to lead or organize human beings effectively it is essential to have a knowledge of the way in which they think. No two people have identical reactions to a given situation; no two people have the identical ability to absorb instructions and carry out work with the same efficiency. Therefore, each individual must be treated according to his personal characteristics if optimum effectiveness is to be achieved. The behavioural sciences—psychology, sociology, and anthropology—are comparatively recent intellectual and academic disciplines. They exist because of the need to better understand human behaviour over the many facets of everyday life. However, the behavioural sciences are not yet developed to the extent that they can be used to predict with certainty the behaviour of human beings under specified circumstances. Success comes to an organization only as the result of sound decisions being made by managers, both in the technological and commercial sense. Such decisions cannot be made after consideration of the material factors alone; human factors must always be taken into account.

1.4 Leadership

Wherever a group of human beings collects in order to work or play as a team then, unless the group is very small or the circumstances quite abnormal, the situation requires that one of the group becomes the leader. Groups are usually formed in order to achieve a common objective. Someone must organize the group, make all the arrangements necessary to achieve the objective, and supervise the group or team in order to ensure that such arrangements are implemented.

Leadership is not synonymous with management but is an aspect of it. The leader of a group is appointed to that position because of superior wisdom, knowledge, ability, strength, or cunning, and acquires such a position through the ability to inspire others by both personality and example. The leader must have authority for the purpose of coordinating and motivating the team for the appropriate action, and it is undesirable that such authority should rely on the subordinate members of the group holding the leader in fear. The task of coordinating the varying types of human resources, amongst the people working or playing in the team, will be all the easier if the team's members are inspired by personality and example.

A leader must be capable of the following:

1. Establishing good communications. People can only be carried along when they are kept informed both of current and future events within the organization and of known possible developments in the foreseeable future. The leader must engender a sense of participation in all the team members and a belief that they play an essential and very important role.
2. Evoking loyalty from the team members by being fair and impartial in all dealings with them concerning matters of rates of pay, bonus payments, promotion, discipline, and work allocation.
3. Selecting suitable subordinates who in turn will display adequate qualities of leadership.
4. Fostering self-discipline within the team by encouraging its members to seek responsibility in running the group's affairs.

In order that one person is not overwhelmed by having to deal with too much detail at all levels in an organization there must be *delegation*. Delegation results in specialization since it is the process of conferring the authority to carry out certain functions on individuals at a lower level of the organization. The jobs best delegated are those that one is able to do least well oneself. It is a fault of many managers that they are unable to delegate authority to others and, because of it, work less efficiently themselves. Inability to delegate may well hamper the expansion of a thriving business. When duties are delegated within an organization, the terms and limitations of the authority and responsibility that are being delegated must be clearly and precisely delineated.

As in other areas of management, there appears to be no one best way to lead. Researchers have failed so far to determine those personal qualities that contribute to making a 'good leader' as exhibited by the ability of one person to be a better leader than others. However, there has been some success in identifying personality traits that detract from good leadership. These include, among others, insensitivity to other individuals, excessive ambition, inflexibility, betrayal of confidences, arrogance, aloofness, and playing politics. Tannenbaum and Schmidt[3] have developed a theory that concerns a continuum of leadership styles. This ranges from entirely autocratic, where a manager makes decisions that exercise managerial authority alone with no contribution from employees, to participative, where subordinates have full freedom to make decisions within authority limits defined by the manager. In between is an infinite number of varying proportions of the two components.

1.5 Organizations and their managers

The prime function of a manager within an organization is to solve problems. Before this can be done, that a problem actually exists needs to be established, as does the exact definition of what that problem is. This is the *diagnosis* stage of problem solving and, in the context of human resources at work, there are usually many indicators of the existence of a problem. Some of these indicators are poor productivity, indifferent quality of work, high labour turnover (that is, a high proportion of employees spending only a relatively short time in the employment of the company before leaving), absenteeism, and a general attitude of indifference to work. Once the problem has been diagnosed and defined, a manager must develop a solution and then implement it. A knowledge of organizational behaviour can be used to help in making this process successful.

Organizations exist because no one individual can successfully cope, either mentally or physically, with all the various demands for skills, experience, knowledge, and ability that are required when, for example, a large construction project is to be undertaken. An organization therefore results from a combination of people who work within the boundaries of their particular strengths and skills. They contribute one part of the total collective expertise that is required to achieve a common objective. This process of breaking down the work to be carried out into smaller tasks, then to be carried out by groups of people or by individuals who have specialist skills and knowledge, is known as the *division of labour*.

In the context of the systems approach to organization theory, an organization may be seen to be an *open system*. A *system* is an interrelated set of elements functioning as a whole wherein a change in one part necessarily has an effect on the others. An open system is one in which the system has free interaction with the *environment* within which it exists. It must also be capable of maintaining its stability to act throughout, regardless of any changes that may take place in the environment. The environment is generally very wide-ranging and will consist of all those influences that can act on an organization. It will include legal requirements, technological constraints, political structure, sociological considerations, educational limitations, and so on. In the case of a construction project, any aspect of the environment in which the project is being constructed that can affect the total process is included. The interaction with its environment may take the form of an exchange of information and/or materials in order to create some form of finished product as a result of the transformation or conversion process being undertaken by an organization. Figure 1.1 illustrates diagrammatically a construction organization as an open system. Feedback at many points in the process results from control action and serves to enable adjustments to be made to the transformation process and its interaction with its environment. Within such an open system a manager can seek resources, probably in competition with others, from the outside (the environment) but nevertheless, because of the control functions, the freedom to act indiscriminately is contained.

A *closed system* is one that operates in a specified way, with a given output from a specified input, either under conditions in which it cannot be influenced by a changing environment or within a strictly specified and constant environment which is taken into account in the design of the system. The components of the system are designed to achieve the transformation process and all of the

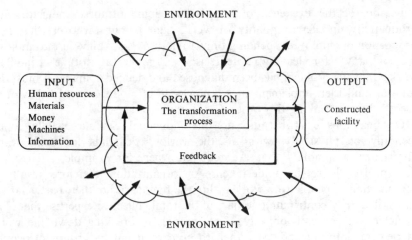

Figure 1.1

operating conditions are anticipated. In the management sense a closed system would be one in which a manager would be able to direct all the components of the system without any influence being exerted from the environment. Many mechanical and electrical systems are closed systems because they are designed to operate only when all the conditions related to them, including the environment, are in position. In practice, organizational systems tend not to be entirely open or entirely closed but to lean towards either one or other end of the range between them.

The construction project open system model of Fig. 1.1 accepts its input of human resources, materials, money, machines, and information from the environment and tranforms them into a constructed facility for which sufficent reward (usually financial) should be received to compensate the members of the organization for their effort as well as to pay for the other resources used. A surplus is required to ensure, where it is desirable, the continued existence of the organization. Viewing organizations as open systems has the important attribute of stressing the dependence of the organization, with its related human resources, on its ability to provide a desired service to the environment in order to satisfy an established environmental demand.

Managers in organizations are required to coordinate both human and other resources so as to achieve a particular objective. A manager not only has to achieve an objective in terms of an output of work of the required quantity and the appropriate quality, but also has to take steps to ensure that sufficient human resources are recruited and retained so as to transform the material resources available into that output. In order to do so, managers need to have knowledge of, and use in the appropriate context, a wide variety of techniques that will facilitate these processes.

Robert L. Katz[4] considered that a manager needs skills that can be categorized under three headings:

1. *Technical skill.* This usually follows some formal education and training in the skill and enables the recipient to exercise expertise related to the procedures or methods of the organization in a proficient manner. Civil engineers are expected to exercise technical skills, as are all other kinds of engineers in management positions, as well as accountants, lawyers, and business graduates. Many managers will have developed their technical skills through training and experience with less formal education. Tradition has been a strong influence (now much diminished) in some professions on the method of achieving professional competence.
2. *Human relations skill.* This is the ability to get on with other people and to work in cooperation with them. If interpersonal relationships are to be strong and healthy, a manager needs to be self-aware, understanding, and sensitive to the feelings and thoughts of others.
3. *Conceptual skill.* This is the ability to see problems from a broad range of views and to provide solutions for the benefit of the whole rather than a small part of an organization's objectives.

These three categories of skill need to be present in different proportions, depending on the relative position of a manager in the organizational hierarchy. At lower levels, technical skills will be more important and in greater demand than at higher levels. In the case of conceptual skills, the reverse will be the case. If any skill is likely to be in constant demand throughout the hierarchy it will be human relations skill, though even this is likely to vary in the nature and in the extent of its use at different levels.

1.6 Job satisfaction

Increasingly the *quality of life* for everyone is a matter for concern. A major contributor to the quality of life is quality of work, since work occupies a major part of most individuals' lives. People need to work in order to have the means of obtaining goods which they cannot provide for themselves. If work is valuable to an organization, a reward will be paid for it and, one hopes, both participants, the worker and the organization, will be satisfied with the transaction. The aim of a manager should be to create an environment in which the objectives of the organization can be most economically and satisfactorily achieved while at the same time good working conditions are provided for the human resources involved. *Job satisfaction* is the extent to which a person feels satisfied or dissatisfied with the work they have to carry out, with their place in the hierarchy in relation to the colleagues with whom they work and with the general environment in which they carry out their work.

Many researchers have attempted to establish useful measures of job satisfaction, but the need to use qualitative responses such as *good*, *very good*, *poor*, etc., to questions about a person's job satisfaction or dissatisfaction leads to a lack of precision in the answers. It is difficult to compare the response 'good' from one person with a similar response from another. One person's perception of the meaning of a descriptive term may be very different from that of another. This aspect of measurement can, at first, seem quite unsatisfactory to someone from an engineering, accounting, or other numerate discipline, in which extraordinary faith is often put in calculated numbers. However, there are ways in which questioning can be scientifically designed so that these deficiencies are overcome and a considerable degree of precision established in the results.

High job satisfaction will result in employees spending more time at their work, increasing their effort and raising productivity, enjoying good relations with their co-employees, seeking and accepting advice of their supervisors, and being punctual with a minimum amount of absenteeism. The outcome of low job satisfaction is usually the reverse. An employee will complain, argue, seek transfer, lower effort, be absent, and ultimately quit.

Federick Herzberg[5, 6, 7] a psychologist, has undertaken considerable research into the attitudes of employees towards their work, and among other things has suggested that job satisfaction is a fundamental influence on work performance. Herzberg's original research was based upon an extensive survey of engineers and accountants in a cross-section of industry. Workers were asked (1) what aspects of their work they felt created an increase in their job satisfaction and (2) what aspects caused a reduction in their job satisfaction. As a result, *Herzberg's two-factor theory* was developed. This consisted of two sets of factors, one of which acted as determinants for job satisfaction and the other for dissatisfaction. The theory resulted in the identification of a number of factors, five of which appeared to be strong determinants of job satisfaction: *achievement*, *recognition*, *the work itself*, *responsibility*, and *advancement*. Few of those questioned put any of these five factors as a cause of job dissatisfaction. The last three of the five were identified as factors which were most likely to have a lasting, long-term influence on the degree of job satisfaction. In similar fashion, factors were identified that tended to cause job dissatisfaction (mostly in the short term). The major factors in this category were *company policy and administration*, *supervision*, *salary*, *interpersonal relations*, and *working conditions*.

Factors that appeared to promote job dissatisfaction generally had a link to the job environment or job context, rather than to the work itself. Herzberg labelled these *hygiene factors*. The job satisfiers he labelled *motivators*. The motivation–hygiene concept formed the basis of the *two-factor theory*. Herzberg saw the two sets of factors as being quite distinct. The hygiene factors influenced the degree of dissatisfaction felt by an employee, and attention to those factors resulting in improved conditions, salary for example, and acted to lessen job dissatisfaction rather than contribute to job satisfaction. In order to improve job satisfaction and enhance performance, managers must concentrate on the motivation factors.

It must be said that Herzberg's two-factor theory has not received universal acceptance. Critics have regarded it as acceptable only if the original research methodology used by Herzberg is repeated. For general acceptance a scientific theory requires confirmation by the use of different research methodologies.

1.7 Motivation theories

An employee in an organization responds to instructions in order to satisfy a personal need. Such needs are developed as a result of a personality, often not fully understood by the individual concerned. They do, however, greatly affect a person's performance. Needs may be for a number of things such as for money, security, friendship, respect, pride, proving one's ability at work, and so on. *Motivation* to work refers to which one or more of these factors dictate the extent to which an employee will work satisfactorily, or which direction he will take, or the persistence he will show at work. A manager's task is to identify the personal needs of each employee and then to design a job in such a way that they can be met. The more accurately a manager can identify the needs, the more readily can the appropriate conditions be provided and the employee be properly motivated. It is important in promoting motivation that the objectives of the organization are also met, since only by effectively meeting these as well will an employee become truly productive.

Abraham Maslow,[8, 9] in his *hierarchy of needs theory*, averred that people perform in the light of the needs they feel at a particular time. These are likely to change from time to time because some will eventually be satisfied. For example, an employee's needs are likely to change over a period of time after a pay raise is awarded. The pressing need before the raise will not necessarily be present to the same degree after the raise has been awarded; another need may then have become first priority. This is known as the *deficit principle*—once a need is satisfied there is no longer a satisfaction 'deficit'. Maslow identified a hierarchy of needs consisting of five levels. In descending order of importance these are: *self-actualization, esteem, social, safety or security,* and *physiological* needs; see Fig. 1.2. The *progression principle* is implicit in the theory, which is formulated as a strict hierarchy. The principle holds that a need arises at one particular level only if the one immediately below it in the hierarchy has been satisfied.

Maslow's two principles of deficit and progression have failed to find universal support among other researchers, who have advanced alternative theories. It has been found, however, that in Maslow's hierarchy the first two lower-level needs tend to be satisfied by providing satisfactory financial rewards such as good wages, supplementary benefits like pensions and health insurance, sick pay, holiday pay, and some protection from dismissal. The three higher levels tend to be satisfied by job attributes such as independence of action, increased responsibility, recognition and public endorsement of success, a challenging job, creative task demands, a high status job title, and freedom in decision making.

An alternative to Maslow's theory is Clayton Alderfer's modification in the form of his *existence–relatedness–growth (ERG) theory*.[10, 11] Alderfer reduced Maslow's

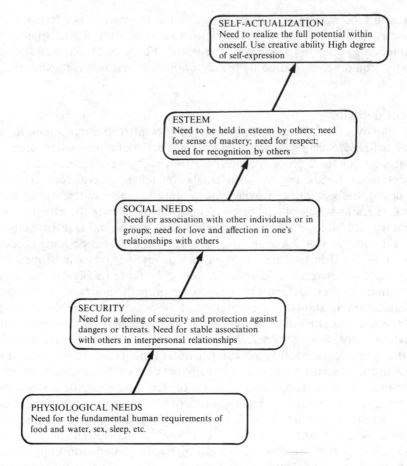

Figure 1.2

five hierarchical categories of need to three. Firstly, the existence needs took care of an individual's physiological and material needs—broadly, Maslow's two lowest levels. Secondly, relatedness needs encompassed Maslow's social needs that involved relations with other people. Thirdly, growth needs included the upper two levels concerned with improvements in one's job and one's work environment. ERG theory adopts a *frustration–regression* principle in respect of progress up through the hierarchy of needs in opposition to the satisfaction–progress concept of Maslow. In other words, Alderfer argued against the Maslow concept that an individual only moves up one level in the hierarchy when satisfaction is reached with the current level, but suggested that individuals regressed to a once-satisfied lower-level need when frustrated by not being able to satisfy a higher-level need. ERG theory also contends that an individual may well have more than one active need at a time, as opposed to the Maslow theory that only one is active at a time.

The two motivation theories briefly described above are two out of a number of *content theories* that exist and are in current use. Content theories of motivation are attempts to categorize the different needs that motivate the behaviour of individuals. The purpose of doing this is to assist a manager to understand the relative value to an individual of various job rewards. In addition to content theories there is also a set of *process theories*. Process theories are claimed to be more dynamic, seeking an understanding of the thought processes of individuals that influence their behaviour. Each type of theory has its place in offering some useful indicators of motivation to a manager who must identify and understand the needs of individuals.

1.8 Groups in organizations

The consideration of groups as collections of employees is important to a manager because the relationship between groups is quite different from that between a manager and an individual. Not only are the relationships different but they are more difficult to establish and to sustain in such a way that they create lasting benefit to an organization.

A group, in the context of management organization, is a collection of people that is recognized as a group by others, has interdependent relationships with other groups as well as the interrelationships between its own members (who have different roles to play, both from the group's point of view and from that of each individual), and has the objective of the attainment of common aims that are acceptable to all the members of the group. A group, because of these attributes, takes on a character of its own. The essence of the definition is that there is member interaction, common objective achievement by combination, and lasting establishment.

There are two forms of group, as follows:

1. *Formal.* These are created by authority in the organization for a specific purpose and with well-defined relationships with others. The task or tasks of each member of the group will also be well defined. A formal group may be *permanent* or *temporary*. Examples of permanent groups are the departments, usually small, of a company organization, which have a well-defined place in the organizational structure of the company. Size may vary from group to group, but in each case the objectives of the group will be to make a contribution to the organization's aims and objectives. Other examples of formal groups are working parties, project task forces, and committees. A group may well be temporary, set up to achieve an objective over a short period of time and then disbanded. The responsibility for managing a formal group in an organization will normally be that of a manager, who ranks above the other group members.
2. *Informal.* Such groups are created without formal authority from within an organization. They may or may not exist within formal groups. The common purpose behind the formation is most frequently a shared interest. Informal groups are rarely work-orientated, although common job interests or job positions are often a prime motivation to create an informal group. People with

similar personalities and/or outside common interests frequently form an informal group. Even though such groups are not work-orientated, they can
nevertheless have considerable influence within the organization.

There are many tasks in organizations that would not be achieved without group
effort because they are beyond the endeavours of a single individual. Research[12] has
shown that groups can often achieve more than the sum of each individual group
member's capability. This phenomenon is known as *synergy*, defined as 'the creation
of a whole that is greater than the sum of the parts'. Among other things, the
research has demonstrated that groups are frequently more successful than individuals when a problem responds to a collection of individuals' skills and the sharing of
information; a group containing no outstanding expert tends to come to better
conclusions than would an average individual. Groups also tend to be more creative
and innovative than individuals because they have a tendency to make decisions
with less regard for the risks involved.

On the other hand, it seems that members of a group may not work as hard as
they would if acting alone. This is in line with the *Ringlemann effect*, named after a
German psychologist who experimented with people pulling on a rope, both
individually at first and then as a team. He found that if he asked a group of people
by themselves to pull on a rope as hard as they were able and then to join the others
pulling on a rope as a team, the average pull for the team effort dropped below that
for the individuals. The conclusion is that people may not work as hard in groups as
they do individually because slacking cannot be detected so easily and because they
have a tendency to like to see others doing the bulk of the work.

Leavitt[13] is a strong supporter of groups and sees them as an essential, if
somewhat neglected, component of organizations. The advantages of human resource groups he believes to be as follows:

1. Groups frequently make better decisions than do individuals and demonstrate
    greater creativity and innovation; decisions made by groups seem to bear the
    commitment of the group membership when implementation takes place.
2. Groups, which often develop spontaneously, even without encouragement, tend
    to compensate for the disadvantages found in organizations when they grow
    considerably in size.
3. Groups appear to have beneficial effects on individuals and seem to be able to
    keep track of and control their members in circumstances where it is sometimes
    otherwise difficult to do so. By providing for social interactions a group has the
    ability to satisfy some of a member's needs in this direction.

Leavitt's findings encourage the use of groups in organizations and it would seem
that a manager should view them as an essential part of his general strategy. The
manager, however, must take steps to determine, for his own organization, those
matters regarding which groups can make a positive contribution and where, and
how the groups can be most effectively placed in the organizational structure.

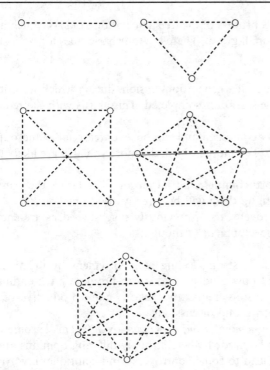

Figure 1.3

Group size is clearly important. The larger the group membership the greater the number of personal relationships that can be formed. The number in fact increases in a geometric progression. For two members there is one mutual relationship; for three, three; for four, six; for five, ten, and for six, fifteen. (See Fig. 1.3.)

It has been shown[14] that if there are less than five members in a group then there tend to be more personal discussions, with more complete participation by members. There are, however, fewer members to participate in the work responsibilities than in a larger group. A group with more than seven members has the disadvantages of providing fewer opportunities for each member to participate, possible domination of the group by a member using more aggressive tactics, a tendency for a breakdown of membership to occur by the formation of subgroups, and more member inhibitions. This postulates an optimum group size of from five to seven. A manager needs to weigh the advantage of having a larger group, namely that there are more members to participate, against the increasing difficulties of communication and coordination that come with more members.

It is useful to review some of the strategies that have evolved in order to assist groups to make their decisions and also to enhance the creativity of the process. The first strategy is that of *brainstorming*, the idea of Alex Osborn, an advertising executive. This is probably the most used and most effective method of generating ideas in groups. It relies on a process of the free association of ideas, which are freely and openly exchanged in a small group. There are four basic rules for brainstorming, as follows:

1. First, the group holds an idea generation session, during which no criticism of ideas is allowed until the session is completed. This assists with the reduction of members' inhibitions.
2. Ideas are not in any way contained and can be as unreasonable and outrageous as the contributors wish. The more radical, the more valuable are likely to be the ideas.
3. The number of ideas forthcoming is all important. The larger the number, the more likely it is that among them will be a really good one.
4. Combinations and improvements of previously suggested ideas are encouraged so as to stimulate the development of thought.

Brainstorming can lead to a strong feeling of involvement by members of the group and to an uninhibited flow of ideas, and it can promote great enthusiasm in the participants. All these reasons frequently make it more productive of creative thinking than the use of open group discussions.

The second strategy is the *nominal group technique* (*NGT*). This is somewhat like brainstorming, but it might be needed where strongly differing opinions among the members of the group may lead to some controversy and emotional involvement in the decision-making situation. Hence NGT can be used instead of having an open meeting of the group or a brainstorming session. A typical approach to NGT starts with a silent session, in which people work as individuals rather than in a group and ideas are generated and written down. There follows a round-robin read-aloud session, without criticism or discussion, wherein ideas are presented to the group and written down on a blackboard so as to be on display to the whole group. Thirdly, the ideas are discussed to clarify information only. Finally, a written voting procedure is followed leading to a rank ordering of the alternatives.

The third strategy is the *Delphi technique*. This was developed by the Rand Corporation and was particularly designed to avoid members of a group having to meet face to face. It allows group members to participate in the decision-making process, even though they may be physically dispersed. The technique involves using a questionnaire to solicit problem solutions. A coordinator collates and summarizes the answers and returns the result to the group members. The process is repeated until step by step a consensus is reached and a clear decision emerges.

Using a strategy such as the three described above in order to undertake group decision making tends to produce a better performance than an open, face-to-face meeting with a more prolific flow of ideas. It is important for a manager who is designing groups and structuring their decision-making processes to take these possibilities into account. People like to be involved, particularly in the successful outcomes of the decision-making processes. The reward is commonly the satisfaction of taking part in a positive process.

1.9 Organization structures

So far, there has been frequent reference to the aims, goals, or objectives of an organization. In simple terms, each one of these is a statement about what it is that an organization, whether a company, group, division, or individual, seeks to achieve in the future. Without them an organization cannot be structured or managed with confidence and authority. Rarely can the goals be defined in a simple statement such as 'Our goal is to build beautiful buildings.' While this statement may be true, an organization's goals are likely to have many facets and, not infrequently, some of them will appear to be in contradiction to others.

Goals can be extremely varied. For example, it is often deemed desirable for an organization to have *societal* goals, and a construction company may well put building beautiful buildings into that category. Such a goal tends to be an expression of intent from an organization in order to justify its existence and activities within the scope of society as a whole, and to state how its function will meet some of the needs of society. In this way, an organization attempts to achieve an implicit acceptance by society, and, perhaps more particularly, at the same time a specific group within society. Often the societal goals of an organization will be directed towards, for example, the company's shareholders and/or the company's employees as an expression of how it intends to look after these particular groups. Overall goals are normally formulated and set by the top managers in an organization. They are usually couched in very general terms and, in spite of the expression of intent, no specific criteria are generally available by which the extent of achievement of the societal goals can be measured. These goals must be structured so that they are capable of achievement as a result of the organization's operations. This can be done by establishing the goals in the form of a hierarchy, providing at each level the structure and resources needed for the goals' achievement.

In large and complex organizations, there will be several levels at which goals are set for achievement and thus several different categories of goals are required. At the top of the hierarchy, broad levels of *strategic goals* are set, designed so as to cover the organization's interaction with its environment and its future achievements in this respect. At the level below the top will be *coordinative goals*. These are an interpretation of the strategic goals into a form that is necessary for their operational achievement. This set of goals determines what it is necessary to achieve at each level of the structure in order to meet the requirements of the strategic goals. Clearly, this hierarchy will have particular implications for the way in which the organization is structured. The lowest level is that of the operating activities. This is the level at which the actual work is carried out. *Operating goals* will be detailed, specific, usually quantitative, and as a rule concerned with the relatively short term. In the case of production units they will normally state specific outputs that are required.

Organizations have a planned *formal structure* or skeleton which provides a basis for managers' actions. It shows the layout of the various parts of an organization and how they relate to each other. It provides a description for the jobs in the structure and how the positions related to those jobs are coordinated. The hierarchical relationships of authority are shown linking the relevant parts of the structure. The structure is planned in order to meet the goals of the organization. While the formal structure is deliberately planned, in practice there is also an *informal structure*. Informal relationships are often used where the formal system tends to be too slow and ponderous to deal with situations that demand speedier responses.

Organization charts are diagrams that illustrate the formal structure of an organization and are generally used to communicate the details of the structure to those who need to know. Charts are usually drawn depicting the hierarchical structure in such a way that they show the direct relationships between managers and their subordinates. Supporting information will be needed to supplement the brief job descriptions superimposed on the diagram and to detail the processes and work tasks with which each is involved. Organization charts become too complicated if they include every component and job for all but the smallest and simplest organizations. A chart must necessarily be simplified, and thus often does not fully detail lines of authority, staff relationships, detailed jobs carried out by each individual in a group, informal relationships, lateral links showing cooperative action between various positions in the structure, and those situations where informalities have become the norm. The two predominant aspects of most organization charts are those depicting *vertical* and *horizontal specialization*. Vertical specialization refers to the hierarchical structure of authority; horizontal specialization is the differentiation of functions within a level of the hierarchy. For example, at one particular level may be shown managers of departments each having approximately equal importance in the hierarchy, though they each have quite different functions. A typical but simplified construction company organization chart is shown in Fig. 1.4.

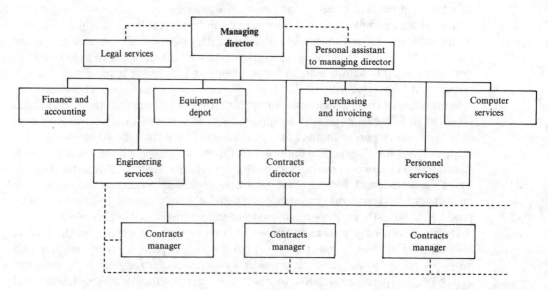

Figure 1.4

There are a number of long-standing, traditional principles underlying the design of organization structures. Many of these were developed early in the twentieth century and suited well the requirements of those times. These traditional principles were based on the need for rigid organization structures viewed as closed systems, that is, having no interrelationship with the environment in which they existed. Today this concept is criticized for its inability to take account of changing technology and, in many cases, different organizational requirements. However, the principles are still likely to apply to the design of structures for organizations that are small, or that exist in a stable environment and use technology that is not advancing at a significant pace. For this reason the traditional principles are now described briefly.

Initially, the structure of an organization was all-important. Traditionally, individuals were *not* the essence of an organization; a structure was established in a way that was thought to result in the most efficient, if rigid, organization. In the structure authority was vested in the job position rather than in the person who occupied it. The first principle is known as the *scalar principle*. It requires that authority flows vertically from the top of the structure, through intervening levels, to the lowest level. It is the basis of the hierarchical structure of an organization and results in the allocation of authority and responsibility to the various stages in the vertical path.

The *division of labour* is also a basic traditional concept in designing organization structures. The work that needs to be accomplished to achieve an organization's objectives is broken down into tasks that have been identified by their component of specialist skills. Departments, sections, or groups are then formed and fitted together in a logical structure. The application of the scalar principle then sets levels of authority and responsibility. Authority can be assumed on the basis of the position of a job in the hierarchy. (Authority and responsibility must be dealt with as one—if an individual is given the responsibility of carrying out a task, then that individual must also be given the necessary authority.) Having been allocated the responsibility and authority, the individual must then accept that he will have to be accountable for resulting actions in line with the vertical flow of both through the structure. In this way the individuals in an organization can be integrated efficiently so as to achieve its objectives.

The concept of *span of control* became important in traditional management theory because structures based on a vertical flow of authority and responsibility depend on a manager–subordinate relationship. Span of control relates to the question of the number of subordinates that a manager can supervise effectively, in other words, the number he can integrate efficiently. There appears to be little or no clear direction as to how broad or narrow a span of control should be. It has been suggested that it should be as small as five, yet many organizations operate quite successfully with as many as fifteen subordinates reporting to one manager. The concept is a difficult one to resolve and much depends on the interpretation of the extent and depth of the control exercised. Traditionally the span of control has been seen to be quite limited in the light of a true vertical flow of authority. However, other methods and techniques are available to distribute authority and to deal with specialization, since the concept of span of control is not as direct and simple as it may seem. One of these other methods is the use of *line and staff units*.

Line positions in an organizational structure are those that have formal authority to make decisions; *staff* positions are filled by individuals, usually with specialist attributes, who advise those in line positions. Individuals in staff positions are sometimes referred to as having *functional* authority. In Fig. 1.4, line relationships are shown by full lines and staff relationships by broken lines. In construction, line positions for a site organization will be those that are directly concerned with the basic production process.

1.10 Design of organizational systems

Managers have to design organizational structures so that each organization can best achieve its objectives. There is an infinity of different types, shapes, and forms of structure and the final selection must be one of best fit for the purpose, provided it remains as flexible as possible and is such that adjustments to meet new conditions and demands can be made quickly and easily. Although there is a large number of such structures they tend to fall into broadly defined categories such as those described above. In order to progress in the matter of designing a structure, it is necessary to have in mind the general theoretical and ideal concepts of form and, taking all relevant information into account, to work forward from this position.

A useful starting point is provided by Burns and Stalker (see Further Reading at the end of this chapter) with their concept of *mechanistic* and *organic* organizations. Theirs was not an attempt to establish an idealized structure for an organization that would meet all the demands on it, whether they came from the technology of the company or from its environment. Rather, they were looking in the reverse direction and were asserting the need to adopt a structure built according to the particular type of technology employed by the organization, and the kind of environment encountered by it.

Mechanistic organizations are considered to have strong vertical specialization and control. They have well-documented policies and procedures of all kinds, for both decision making and control. The hierarchy of control is centralized, with a well-defined division of labour such as is found in the functional departmentation model (see Fig. 1.4). Most aspects of the system are formal and impersonal. Staff are centralized. *Organic organizations* represent the opposite end of the spectrum from that of mechanistic structures and are typified by a decentralized hierarchy of authority with horizontal specialization, few policies, procedures, and rulebooks, and with informal coordination and control. Between the two ends of the spectrum is an infinite number of structures with varying proportions of the characteristics of each.

Reference has already been made to the traditional concept of an organization as a rigid, closed system with no interaction with its environment, and also to the limitations of such an approach. Few organizations can exist successfully if they are designed strictly in accordance with traditional theories. Organizations must be designed as open systems having strong links with the environment. They must be seen to be, and designed to be, capable of changing and developing as time passes, adapting to the environment's changing demands.

Managers design organization structures to interact with the environment *as they see it*. They make judgements about the subsystems that are required in the organization structure in the light of all the relevant factors. The more uncertain and changing the environment, the more complex is the organization structure likely to be. Managers use the following three main determinants to guide them in the design of organization structures:[15]

1. Environmental interactions;
2. The technology used by the organization;
3. Psychosocial influences.

Environmental interactions deal with the ways in which an organization will communicate with the environment through input and output (see Fig. 1.1). Each input or output requires the action of some one individual, or a group of individuals, to respond to each input or generate each output. For example, a construction company has a demand from the environment for the input to the organization of a wide range of professionally qualified staff. It therefore needs a personnel department to attract applications from the environment and to deal with interviewing, recruitment, negotiating conditions, and so on. Successful staff are an input from the environment. In similar fashion, bids are required from specialist subcontractors, who will input their services to the organization. A purchasing section or department will therefore need to be placed in a strong position in the organization. The company organization will output finished product in the form of houses, schools, dams, roadways, etc. It needs a subsystem in the form of a group, department, division, etc., to carry out this work. There are many environmental interactions of which the above are but a few simple examples. The manager, therefore, needs to study the interactions of his organization with the environment and then to determine a suitable organization structure accordingly.

Technology utilization also has an influence on the internal structure of an organization. Clearly, the technology of the production process employed by the organization will have the greatest effect on the structure of that part of the organization structure that is directly responsible for transforming the input of all types of resources and outputting the finished product. In a construction company one example of this is the organization structure adopted for a project site organization, the site organization being part of a larger structure for the company as a whole.

The third determinant is the *psychosocial* system. The psychosocial system is the province of the behavioural sciences. It is the basis for one of two major thrusts within the study of organizations during the last forty years. The other comes from management scientists, economists, engineers, mathematicians, and others who tend to view an organization from the point of view of achieving its goals with maximum efficiency. This has led to considerable advances in methods to aid decision making by managers, and many numerate and normative models have been developed. The psychosocial approach by behavioural scientists, on the other hand, has led to the study of human factors in organizations and of the manner in which people behave within them. Recognition must be made of the expectations of the individuals who are employed by the organization and the way in which they respond to varying degrees of control, or to different structural hierarchies of authority and responsibility. A major factor in this determinant is likely to be the educational levels of the individuals concerned, together with the generally accepted standards of autonomy that are associated with relevant positions in their experience. This is likely to vary considerably over the range of employees' jobs, but in each case is likely to influence the organization structure.

Having studied and considered these determinants, a manager, with specialist help where need be, will make the decisions concerning the principal features of the organization structure.

Vertical specialization has been referred to earlier as hierarchical organization established to enable authority to flow down through the structure and to define where and how decisions are made. A manager designing an organization structure also needs to give consideration to *horizontal specialization* in which individuals carrying out similar processes or work tasks are grouped together in departments or similar subsystems within an organization structure. The division of labour is thus carried out on a horizontal basis. Horizontal specialization can take one of three forms:

1. functional;
2. divisional;
3. matrix.

Functional departmentation results in a structure as illustrated in Fig. 1.4. Like functional skills and types of work task are grouped together. This form of structure is very commonly used. Its advantages are that, with like personnel grouped together, individuals can learn from each other and hence job training is simplified and becomes more effective. Also, the structures are readily understandable by employees and the specialization leads to ease in specifying task assignments. On the other hand, specialization sometimes makes it difficult to avoid boring and tedious jobs and may render horizontal communication between groups difficult.

Divisional departmentation results in horizontal grouping by product, geographical location, customer, or service. A construction company might be divided into civil engineering, small works, piling, etc., as set out in Fig. 1.5. An international company might wish to use divisional departmentation on a geographical basis. A company that has expanded by the acquisition of another company may use divisional departmentation to retain the latter's staff and other resources in a self-contained and identifiable unit within its overall organizational structure. The use of divisional departmentation has the main advantages that competition between divisions, each of which is seen as a separate business, can be encouraged and attention can be focused on the success or failure of those units. Such organizations tend to be adaptable and to give greater flexibility in meeting new external demands. On the negative side, it is easier for the objectives of one division to be given priority over those of others; the duplication of effort in divisions may be difficult to eradicate; the

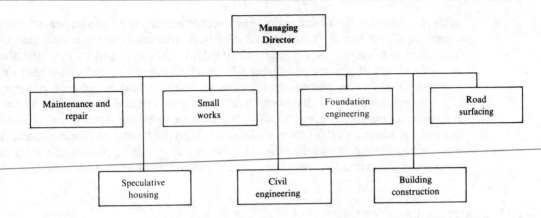

Figure 1.5

training of experts is not so easy and effective as with functional departmentation; and lastly, conflict may occur in joint endeavours between divisions when resources need to be interchanged.

Matrix departmentation is an organizational structure that attempts to place equal emphasis on technical and product development and is a major departure from traditional structures. This is achieved by having a structure in which reponsibility for reporting from within is to two authorites. It is used for undertaking project-orientated activities where the traditional functional or departmentalized structures would be inappropriate. This form of structure is frequently used as a temporary expedient. In it each group of specialists reports both to a functional manager of their particular group expertise and to a project manager (Fig. 1.6). The project

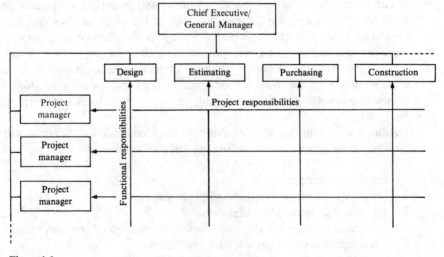

Figure 1.6

manager is responsible for achieving the project objectives by building a project team consisting of members of the functional groups. A structure of this type does tend to form new but sometimes complex relationships, although it does combine the strengths of both functional and divisional departmentalization. It also combines the business skills of a commercial organization with the technical demands of achieving a highly complex project. It has been found that a matrix organization is difficult and expensive to manage, largely because its structure does not have a well-defined hierarchical structure of authority and responsibility with unity of command, and it is altogether less structured than the other forms. Frequently employees operating in a matrix organization find the principles difficult to accept and understand.

1.11 Contractual relationships

An organization with which construction managers need to be familiar is that for the purchase of a constructed facility because it forms the basis of and influences so much that is carried out under construction contracts. In general, to produce a constructed facility, a number of parties have to be brought together to work within a contractual relationship. The technical and/or financial responsibilities of each party need to be defined and the interrelationships between the parties need to be established. Fortunately, many of these relevant organizational matters have been resolved in practice and standard documents and model procedures which are accepted by the parties have been used for a long time. Further, the considerable experience gained in their use has resulted in modifications and improvements being made to the systems employed from time to time. In addition, quite different new concepts have occasionally been proposed, and a variety of types of organization have been developed to carry out a wide range of construction work smoothly and efficiently.

The party that is the customer of the construction industry, and proposes to purchase either a constructed facility or one of the other services that are offered by the industry, is variously known as the *client*, *owner*, *promoter*, or *employer*. One or more of these terms may have specific legal significance, depending on the context in which they are used. For the purpose of this general discussion of contractual organization the term *client* will be used. A client may be an individual, a group of people, a partnership, a limited liability company, or a local or central government authority.

Among other things, a client not skilled in construction practice will probably need to obtain expert advice on one or more of the following:

1. feasibility studies;
2. the design of the works that are proposed;
3. specialist equipment installations;
4. the preparation of the contract documents and other contract procedures;
5. tendering procedures and tender evaluation;
6. construction programming and scheduling;

7. the supervision of the construction of the works;
8. the certification of completed work for payment;
9. dealing with variation orders and claims for additional payments.

The advisers may be from the client's own in-house staff, or they may be appointed from outside organizations. The organizations to be drawn on may be those whose members have professional skills in engineering—whether civil, structural, mechanical, building services, etc.—or in architecture, quantity surveying, project management, etc.; also, with certain forms of contract, contractors may be employed who have suitable experience in design, construction, and/or construction management. The generic term of *consultants* will be used here to describe these advisers in general, unless the context requires the description of one particular professional category so as to be more precise.

A *contractor* is an individual or a company (usually protected by limited liability) that contracts to carry out the construction *works*. If only one contractor is appointed he may be known as the *main contractor*. It is likely that a contractor will subcontract or sublet some of the work to *subcontractors* who have specialist skills, experience and equipment to deal with specialized aspects of the work.

The *contractual relationship* between the parties, and indeed the professional and commercial skills of the parties involved, will depend upon the type of organization that a client choses in order to obtain the construction of the work. It must be borne in mind that organizations are fluid and must change to suit changes in the functions they are required to fulfil. The general principles of construction contractual organizations are discussed here, but it should not be assumed that their use must be so rigid as to prevent change or modification, or that they cover all the ways in which work can be carried out. Organizational contractual relationships can be classified within three groups, as follows:

1. traditional;
2. design and construction;
3. management.

The *traditional* contractual organization (see Fig. 1.7) is one in which a client has a direct contract with consultants to carry out the design of the works and also probably the supervision of the construction, with a quantity surveyor as one of the consultants. The quantity surveyor will give advice on a range of matters relating to the cost of the work as well as preparing some of the contract documents and measuring the work completed for valuation and variation purposes together with the preparation of a final account. Consultants are normally in independent professional practice, with no ties to construction or property development commercial undertakings. The client also has a direct contract with a contractor. The latter is likely to be in contract with suppliers of materials of all kinds and with subcontractors for carrying out specialist works and equipment installations. Some of the suppliers and/or subcontractors may be *nominated* by the client or on his behalf by

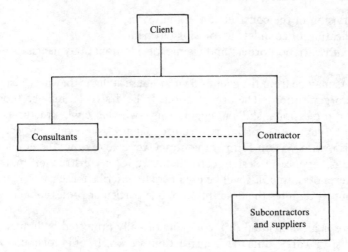

Figure 1.7

one of the consultants. Such subcontractors will normally be selected after submission of their tender to the client, and the contractor is then instructed to enter into a contract with the nominated subcontractor in terms that are specified by the client or the consultant. Other subcontracts, those arranged by the contractor, are known as *non-nominated* or *domestic* subcontracts and are subject to the approval of the engineering or architectural consultant.

The organization for a *design and construct* method involves a client having a contractual relationship with a design and construct contractor (see Fig. 1.8). In this relationship the contract is for the contractor to design the proposed constructed facility and to build it. If the client does not have the necessary in-house skills to arrange for tenders for the work to be submitted and then for their evaluation and the selection of a suitable contractor, a consultant may be appointed to act on behalf of the client and to advise the client accordingly. In such an arrangement, the contractor may wish to arrange a contract with a consultant for design services where the technical skills are not available to him from his own organization.

Management contractual organizations are generally formed to provide one of two types of service. The first is that of *management contracting*; the other is *construction management*. For management contracting (Fig. 1.9) a client has a contractual relationship with a contractor who acts as a *management contractor*. It is normal practice for the management contractor to be precluded from undertaking any of the construction and to provide purely management services. A client also contracts directly with consultants to provide design and cost consultancy services. The management contractor then contracts directly with other contractors to carry out the construction work.

A *construction management* organization (Fig. 1.10) is one in which a client enters into direct contracts with a professional construction manager, design and cost

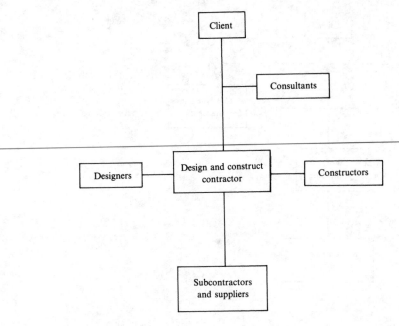

Figure 1.8

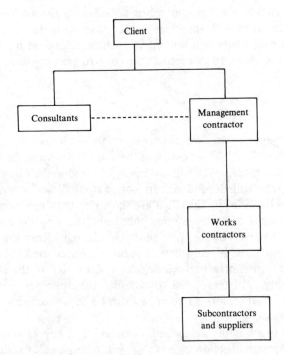

Figure 1.9

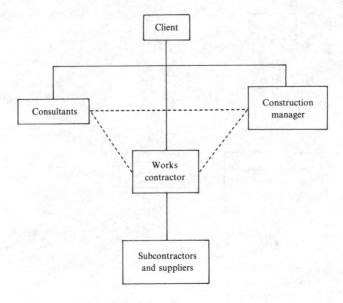

Figure 1.10

consultants, and a works contractor. The contractor undertaking the work is then in a direct contractual relationship with the client rather than with the construction manager. The construction manager will undertake such management functions as are delegated directly by the client. In this respect the construction manager may act as the agent of the client.

1.12 Types of contract

Figure 1.11 shows a very general sequence of operations that will take place when a client seeks construction works. While almost all of the actions shown in the diagram are needed to complete the process, they will not always be undertaken by the same party in each case, but that will depend on the contractual organization that is selected for use. In Fig. 1.11 actions by the client are shown by the single rectangular boxes, those by the consultants in double-framed boxes, and those by the contractor in boxes with rounded corners. The diagram most closely represents the flow of actions where a traditional form of contractual organization is used. The client, having decided on the form of contractual relationships best suited to the construction of the required facility, will then need to consider the appropriate types of contract that are available. Not all forms of contract can be implemented with every contractual organization.

A *construction contract* is a binding agreement, enforceable in law, containing the conditions under which the construction of a facility will take place. It results from an undertaking made by one party to another, for a consideration, to construct the

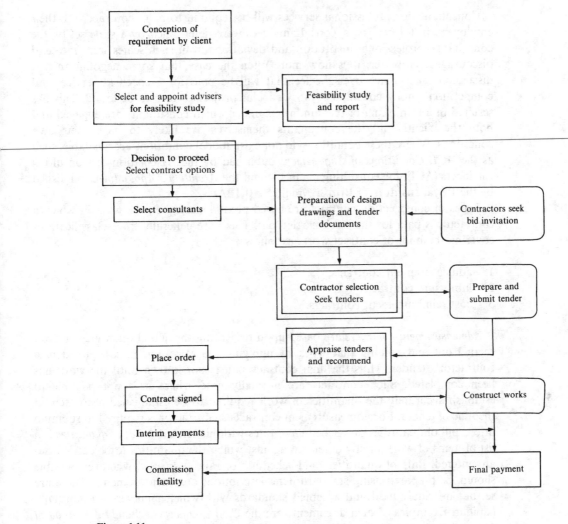

Figure 1.11

works that are the subject of the contract. The offer in construction is normally in the form of a *tender* and, when full and complete agreement about the *conditions* and the *consideration* (usually payment) has been reached, the acceptance can be formalized. There are a number of essential general conditions for a valid contract to be formed, not all of which will necessarily apply to a construction contract. The principal requirements are that the parties to the contract have the legal capacity to be so, that their objectives are legal, that the parties genuinely agree to be parties to the contract, i.e., they have the intention to create a legally binding association between them, and that, in the case of simple contracts, something of value passes in both directions.

Consultants for professional services will be required to be in contract with their employers. In former years, consultants' fees were determined by a scale set by the consultants' professional institution and deviations from such scales were rare and discouraged. However, it is now more often the case that some negotiation and discussion takes place over the fee that will be paid for a specified service, and competition among consultants for particular projects is also to be found. This has resulted in a rethinking of the conditions under which consultants are engaged and both the clients and the consultants themselves are likely to have their own conditions. Where these conditions do not exist, model conditions are available, such as the *ACE Conditions of Engagement* published by the Association of Consulting Engineers (ACE) for consulting engineers and the *Architect's Appointment* published by the Royal Institute of British Architects (RIBA).

Types of construction contract are based predominantly on the ways in which a contractor is paid for the work carried out. There are generally three classifications covering the bulk of contract work, as follows:

1.   admeasurement contracts;
2.   lump sum contracts;
3.   cost reimbursement contracts.

Admeasurement contracts are based upon measuring the actual quantities of work carried out and valuing that work by applying the rates and prices quoted in a contractor's tender. Thus the final contract sum is not known until the work has been completed. Such contracts are normally used where it is not possible to establish accurately the quantities of work required to be established accurately at the time of tender. For admeasurement contracts a contractor is required to submit a priced bill of quantities or a schedule of rates with his tender. A *bill of quantities* is a list of items of work briefly described against which the quantities to be carried out are entered. Bills of quantities (and schedules of rates) can, and wherever possible should, be prepared using standard item descriptions and measurements. These are set out in widely used and accepted standards with which tenderers are normally familiar. Examples of such documents are the *Civil Engineering Standard Method of Measurement, 2nd edition* (CESMM2), for civil engineering works and the *Standard Method of Measurement of Building Works, 7th edition* (SMM7), for building works. A contractor in preparing his tender is required to enter a unit rate or a price against each item. The bill of quantities is often prepared for work that has been only partly designed and detailed, and for which all the detailed drawings are not yet available. Hence the quantities in the bill are approximate. Similarly, a *schedule of rates* is a list of categories of work believed to be required in the works (but with no quantities given) against which the tenderer is expected to write in the rate required for carrying out one unit of measurement of the item. An alternative method to the tenderer quoting a rate is for the consultant who prepares the schedule of rates to insert suitable rates and to require the tenderer to quote a percentage by which the tenderer wishes to raise or lower them if awarded the contract.

The main advantage of using admeasurement contracts stems from the fact that clients can often gain an advantage in the project programme, since the work can be put out to tender before the fine detail of the design and the drawings are finalized. This advantage, however, is offset by the increased risk that arises from the uncertainty of not knowing the exact contract sum before work is commenced.

A *lump sum* contract is one in which the *contract sum* is fixed and agreed before construction work commences and can be used in traditional design-and-construct contracts, and for the works contractors shown in Figs 1.9 and 1.10. Where lump sum contracts are to be used a client must have a clear idea of exactly what is required. There is a wide spectrum of different ways in which this information can be conveyed to a contractor. At one end the specification of requirements can be performance-oriented, such as 'a factory to produce 150 cars per week'. Of course, more detail would be required about the performance, but the essence of such a contract is that a client hands over a lump sum in exchange for a factory capable of performing at a specified rate. At the other end of the spectrum a lump sum tender is requested for a project that is fully designed and detailed, complete with specifications for quality of workmanship and materials and fully detailed drawings. Because of its nature, it is not expected that there will be any variations to the lump sum basis once the contract is agreed and signed, unless the requirements of the client are altered during the course of the construction. Payment to a contractor, in all but the smallest contracts, is normally phased throughout the various stages of the work, the method and sequence of payment being part of the contract terms.

Lump sum contracts have the advantages to clients that the overall price for the work is known at the outset, that the client is not too involved in the construction process itself, and that it is readily possible to arrange for competitive tenders at the pre-contract stage. The evaluation of lump sum tenders is relatively straightforward. An additional advantage to a client is that a considerable amount of the risk and responsibility for the outcome of the project can be transferred to the contractor.

Some disadvantages of a lump sum contract to a client are, firstly, that the overall project programme is usually longer than by employing some of the other methods available because of the need to provide precise details of the works; also if changes are made to the scope of the work, or unforeseen difficulties arise, then disputes over payment will occur and the lump sum price will be driven up. To assist with the resolution of disputes, in some instances a contractor will be required to quote rates for some items of work delineated in the tender documents so that they provide a basis for re-evaluating work that arises from new or changed conditions.

One form of lump sum contract that has been used increasingly in recent years is the *all-in contract*. It follows from the use of the performance-oriented specification noted above. The client prepares a brief of what is required. This is then sent out to contractors for tendering. Each tenderer prepares a scheme to provide the facility required by the specification and submits to the client the relevant design drawings, specifications, etc., and lump sum price with a completion date. The client is therefore in a contractual relationship with one party, the successful contractor, although he may require to take professional advice during the preparation of the brief and in the assessment of the tenders. The all-in contract method has the advantage that the client pays a sum, established before construction starts, for the facility that the client requires when the client wants it, and has no problems of coordinating the various services. On the other hand, the client is unable to make changes in the design and construction processes unless they conform with the original specification, because this is likely to give rise to considerable additional costs.

Cost reimbursement contracts involve recording the total actual costs of materials, plant and labour, known as the *allowable* or *prime costs*, incurred in order to carry out the works, and then adding to them a previously agreed fee to cover profit and head office overheads. Such a contract is used for work where it is not possible to prepare accurate definitions of the extent and nature of the works involved prior to commencing construction work. It follows, therefore, that a tenderer would find it difficult, if not impossible, to prepare a realistic price, in either lump sum or admeasurement terms. An example might be where a building has collapsed and urgent work needs to be carried out to render it safe and perhaps watertight.

Cost reimbursement contracts have a great deal of flexibility built into them, so that the extent of the works to be carried out can be varied easily, as can the overall programme and the total duration involved. On the other hand, the method has the disadvantages that it is sometimes difficult to impose a ceiling on the amount of money to be spent and that an unscrupulous contractor may abuse the system unless closely supervised. The method also demands considerable administrative input in supervising and recording expenditure, in giving approvals for work to be undertaken and in generally controlling costs.

Where a *prime cost plus percentage fee* contract is used, that is where the fee consists of a fixed and previously agreed percentage of the total incurred prime costs, a contractor has little incentive to reduce total costs. So as to provide a greater incentive for a contractor to pursue the work as efficiently as possible, a number of variations to the prime cost plus percentage fee for overheads and profit arrangement, as described above, have been developed. Not all are particularly efficacious in producing the desired result. One such alternative is to have a *cost plus fluctuating fee* contract where the percentage fee is on a reducing sliding scale as the total prime cost increases. The greater the prime cost, the lower the percentage for calculating the fee. This method, however, will not in all circumstances necessarily encourage a contractor to reduce the level of his prime costs.

Another arrangement, more likely to encourage economy of prime costs, is that of *cost plus fixed fee*. The fixed fee is either tendered by the contractor as a lump sum or alternatively, it may be negotiated. The fixed amount is then added to the prime costs, when established, to give the total costs. The fixed fee may refer to a stated range of total prime costs and different fixed fees would then apply to different ranges.

One more alternative method, *target cost plus fee*, which is believed to be effective in economy of cost, is to agree a target for the prime cost of the work before any work is carried out. This is commonly effected by using a priced bill of quantities for the prime costs. Such a bill is useful in valuing any variations in the scope of work as well as dealing with fluctuations of cost as a result of inflation. It is often not easy to set this target accurately because of the nature of some types of work and there needs to be an awareness that setting it too high or too low may adversely affect any incentive to be efficient. The fee ultimately payable to a contractor is then established by adjusting the agreed basic fee (which is usually a percentage of the agreed target estimate). The actual fee payable is arrived at by increasing the basic fee, in accordance with a previously agreed scale, where the actual total prime cost is less than the target cost, and decreasing it if the target cost is exceeded.

1.13 Tender documents

The prime function of the tender documents is to provide would-be contractors who are tendering with sufficient information about the works to enable them to prepare competitive and well-informed tenders. Contractors tendering need to know everything that will influence their estimation of the costs for the works to be constructed, the form in which the tender should be prepared and how it should be submitted, the conditions governing the way in which the work must be carried out, the mode of payment and factors that will affect it, and their contractual responsibilities and those of the other parties concerned with the ownership, engineering, design and construction.

The required tender documents will vary with the type of contract being prepared, depending on its detailed requirements. For example, a bill of quantities will be required with an admeasurement contract but almost certainly not with a lump sum contract. However, the following is a list of the documents that are required for use with an admeasurement contract:

1. form of tender;
2. general conditions of contract;
3. specification;
4. drawings;
5. bill of quantities;
6. form of agreement.

In addition, though not usually considered to be a formal contract document, *instructions to tenderers* are issued with the contract documents in order to assist each tenderer with the preparation of their submission. The instructions will set out the what, why, when, where, and how of the submission, and will in most cases give instructions as to how tenderers may visit the site of the proposed works.

The *form of tender* normally follows a standard form and is used by a tenderer to make his offer to carry out the works that are the subject of the tender documents. It acknowledges the other relevant contract documents and will state whether the time required for completion is as required by the client. Otherwise an alternative project duration may be submitted.

The *general conditions of contract* define the terms under which the contract will be carried out. The conditions will include the definitions of the terms used in the conditions, so that there will be no doubt as to the identity of the parties. In particular, the conditions set out the powers and responsibilities and obligations of the parties involved and the relationship between the client and the contractor. The conditions include many clauses as to what procedures are to be used in the case of specific events occurring, so as to avoid ambiguity or difference of opinion if the events do occur. Matters such as extensions of time for completion of the works, liquidated damages, defects liability, payment, default or failure of the contractor, damage to public property, and so on are all important matters dealt with in the conditions.

It is always advisable, wherever possible, to use model conditions of contract that have been agreed by representative bodies of the construction industry, including the professional institutions. Such conditions will normally have been tried and tested in practice and modifications to them should not be made without the appropriate legal advice. For civil engineering works there are the *ICE Conditions of Contract*; for building, the Joint Contracts Tribunal (JCT) *Standard Form of Contract with Quantities* (other forms are available where the contract in use does not incorporate bills of quantities in the contract documents); for central government capital projects, the *General conditions of government contracts for building and civil engineering works*, Forms GC/Works/1 and 2. These are examples from a wider range of standard conditions both for general and for some more specific purposes. Model conditions, for example, exist for simple building works (JCT), admeasurement contracts with approximate quantities (JCT), cost reimbursement contracts (JCT), management contracts (JCT) and for the erection of mechanical and process plant (Institution of Mechanical Engineers; Institution of Chemical Engineers).

The *specification* has two main functions. Firstly, it describes the nature, class, and quality of materials and workmanship that must be used in the works. Secondly, it covers any specific conditions of contract that do not appear in the general conditions. As a rule, these will be conditions that have particular reference to the contract in hand and may, for example, be concerned with detailing where particular construction methods will be used and in what sequence. As with conditions of contract, a number of standard specifications are in existence and reference is often made to these rather than drawing up new documents on each tendering occasion.

The *drawings* of the works are self-explanatory. They are extremely important in the context of tendering, and as much information as possible should be given at the tender stage so that tenderers have at least sufficient information to prepare a sensible and competitive proposal. In particular, the drawings should show the general layout of the work site and the areas that will be available to the contractor for his various installations of facilities and equipment. Soil investigation information should also be indicated on the drawings, giving the location and detail of boreholes with the log of the material recorded and/or sampled.

Reference has already been made to the *bill of quantities* under admeasurement contracts in Section 1.12.

The *form of agreement* may or may not be used. It is the formal legal undertaking that is completed when a contractor's tender is accepted to carry out the works specified in the contract documents. Frequently, such acceptance is effected in writing to the contractor by the client, in which case a form of agreement is not necessarily required.

1.14 Tendering procedures

Should a client wish to enter into a contract with a contractor in order to obtain the construction of a facility, the client has first to arrange to obtain an offer or tender from the contractor. The client may well wish to have several such offers from which he may select the one best suited to his purposes. Before the tendering process can proceed, the client, in association with any consultants that may have been engaged as advisers, will need to decide on the contractual organization and the type of contract to be used. Tenders are invited in one of three ways, namely:

1. open tendering;
2. selective tendering;
3. negotiated tenders.

Open tendering is a procedure that allows practically any contractor to submit a tender for the works. The procedure usually involves a client (or their advisers) placing a public advertisement in the national and/or technical press, giving a brief description of the works and inviting contractors to apply to the client or to their consultant for the contract documents prior to making a bid. Sometimes the client will require a cash deposit when the contract documents are requested, the deposit being returned in the event of a bona fide tender being received subsequently. This condition is used to discourage contractors and others seeking the documents out of idle curiosity. Such a method of tendering appears to have the advantage that it is likely to attract the maximum number of tenders and hence be the most competitive. However, through using this method, tenders may be received from contractors who are ill-equipped to carry out the work either financially or technically. They may also lack sufficient experience of the most appropriate kind. While it is not always a requirement of the client to accept the lowest tender, there is often some reluctance not to do so and it is particularly difficult not to where public money is involved. In addition, if large numbers of contractors are to be invited to tender, an unnecessary burden is placed on the industry by way of the unrewarded expense for time and effort that is necessarily put into unsuccessful bids. The use of this method has been declining for some time.

The disadvantages of open tendering can be overcome by *selective tendering*. Selective tendering consists of drawing up a short list of contractors that are known to have the appropriate qualifications to carry out the work satisfactorily. Such a list is drawn up either from the experience of the client and their advisers or from advertising for contractors who wish to have their names included on the list. Those contractors who seek to be listed are then asked for further details concerning their technical competence, their financial standing, the resources that they have at their disposal and their relevant experience. This process is known as *prequalifying*. The list of contractors invited to tender using this method will usually vary from three to eight or so.

Negotiated tendering is used in several different contexts, but the essence of the system is that an acceptable tender is arrived at by discussion between a client, consultants and a single contractor. Thus a negotiated tender is arrived at without necessarily obtaining competitive tenders from other contractors for the works being constructed. The circumstances in which a negotiated tender may be obtained are where the magnitude of the work is not fully established at the outset, where an early start is of great importance, and for work which is extremely difficult to undertake. Many contracts negotiated with single contractors are of the prime cost type—either cost plus, cost plus fixed fee, or target cost with fluctuating fee. An alternative use of negotiated tendering takes place where the negotiations with a single contractor are undertaken after the contractor has been identified by an initial stage of competitive tendering. This process is known as *two-stage tendering*. The parties involved, the one making an offer and the other accepting it, move progressively towards a proposal based on the initial tender which is unconditional and acceptable to both parties. Neither is committed throughout the negotiations until a tender proposal is reached that is considered to be in such a state that it is open for acceptance or rejection.

Negotiation can be used in extending the scope of the works beyond a first-stage contract. The *extension contract* may be established as a result of a round of competitive tendering, but more usually it results from negotiations based on the tender submitted for the original contract. *Serial tendering* is another form of extension contract, but the intention to have an *extension* is normally expressed at the time of the initial invitation to tender. In this case, when the contract documents for the first tender are prepared, the intention to allow further subsequent contracts based upon the tender for the first is incorporated. Conditions that will apply to subsequent contracts are stated where possible, and a negotiation will take place at the appropriate time.

1.15 Selecting a successful contractor

Once competitive tenders have been submitted in accordance with the instructions to tenderers, they should not be opened until the time and date that have been set out in the instructions have arrived. In the meantime the unopened tenders should be kept in a secure place. The submissions should then be opened by at least two representatives of the client and the total price of each tender should be recorded. The consultant, if one is appointed, dealing with the tenders should then report their findings to the client, taking into account any matters, such as the contractor's proposed construction methods, and any qualifications the contractor may have added in a covering letter from the tenderer. The latter must be considered as a legally binding part of the tender.

Before a tender based on a priced bill of quantities is accepted there is a need to check the bill for arithmetical accuracy. The extended prices, that is the quantities times the unit rates that have been inserted by the tenderer, must be checked. Any inaccuracies found in this check must be corrected before the tenders are compared, as will need to be the additions of the extended prices. As part of this check the client/consultant will need to scrutinize the rates to make sure that they are reasonable and fall within the range that might be expected from normal historic standards. Where significant under- or overpricing of the rates occurs, so long as the variation is not common to several tenders, the differences may be due to *front-end loading* or *unbalancing* the tender. This process is one in which the rates for operations, such as excavation, which are scheduled to be undertaken very early in the life of the project, are enhanced at the expense of rates for items scheduled later. In doing this a contractor can receive more income in the early phases of the work so as to finance the establishment costs and avoid the high cost of borrowing that money. An investigation of such practices should be made in order to establish their likely effect on the total tendered price, especially if some of the higher priced items have estimated quantities against them which are likely to increase, and vice versa. In the event that the apparent overpricing of some items is common to several tenders, the cause may be due to the contract documents not being clear in some respects. If this is the case, unless it is a minor difficulty to put the matter right, it may be necessary to adjust the offending documents and reinvite tenders.

When the client/consultant believes that all the necessary checks and adjustments have been made—it may have been necessary to discuss some of the adjustments with the appropriate tenderers if they have put in a low bid that has a chance of being accepted—one tender will be recommended for acceptance. This may or may not be the lowest bid, though where selective tendering has been used it may be difficult to find a basis for not doing so. At this stage there may be a need to agree with the successful tenderer a number of adjustments to the tender, thus delaying the issue of a formal letter of acceptance. If it is somewhat urgent to get the succcessful tenderer preparing to start work then a *letter of intent* may be sent to them. This will state the intention of the client to enter into a contract with the tenderer and set out the steps that the contractor may take before formal acceptance is made. These steps should have financial limits placed on them as to the amount of expense that will be reimbursed and some specific steps, such as ordering materials or setting up site accommodation. When the modifications to the tender are completed and agreed by both parties, the formal letter of acceptance can be issued to the contractor. The unsuccessful tenderers, if they have not already received a letter of rejection of their tender, can now be notified.

If the client proposes to use a *form of agreement* this can now be executed by the client and the successful contractor, unless the client wishes the contractor to execute a *performance bond* beforehand. A performance bond is a document by which the contractor provides a formal assurance to the client that the company will complete all of its obligations under the contract. The *guarantor* or *surety* is usually a bank, insurance company or other financial institution that provides the bond on payment of a fee or premium by the contractor. In the event of the contractor defaulting in their performance then the surety is required to reimburse the client for any loss by such default. Bonds are not always required by clients.

Summary

This chapter has sought to outline some of the influences that human behaviour has on organizations and their structure. It began by looking at the early history of management theories and the ways in which exponents of the time influenced managerial matters. Some emphasis was placed on management as a science and why it is important for there to be a rational and systematic approach to management problems. Next, the personal qualities and qualifications for leadership were examined, since leadership is an important aspect of a manager's role in coordinating the activities of the human resources that are seeking to achieve the goals of the organization. It was here that it was concluded that there is neither a *best* way to lead nor a set of personal qualities that ensure good leadership will result.

The organization as a system and a manager's role in it was examined. The importance of seeing the organization as an open system, interacting with its environment, was stressed, as opposed to the traditional concept, from other times, of its being a closed system. The skills that a manager needs in an organization were examined. Job satisfaction was the next topic investigated, since a manager must have a clear idea of what motivates people to work well, so as to promote the concepts as part of the management process. This led us to the work of Herzberg and his two-factor theory. Pursuing motivation to work, Maslow's hierarchy of needs was examined and Alderfer's subsequent modification to it resulting in the ERG theory. Both are important in their general concepts.

Groups are important units in all organizations, and a manager needs to understand how and why they work in order to make full use of them in both designing and operating his organization. The attributes of groups were therefore discussed, as were strategies such as brainstorming, NGT, and the Delphi technique which facilitate their decision making. The goals of an organization were discussed in the context of what it has to achieve and how is it structured in order to do it in the best, most expeditious way. This led on to organization charts as a means of depicting and communicating the details of an organization and the traditional, somewhat rigid, principles underlying the design of organization structures were discussed. Finally, the points a manager needs to know in order to design a sound organizational structure were developed. The ways in which environmental interactions, technology and psychosocial influences affect the design of a structure were examined, as were the various descriptive forms of organizational structures such as the functional, divisional and matrix arrangements.

The remainder of the chapter dealt with some of the organizations that are necessary in order to get construction works built. It dealt with the various contractual relationships between the client, consultants and contractors and the ways in which their responsibilities differ. The procedures for obtaining tenders and then evaluating them were described.

Problems

Problem 1.1   What is a system? Describe the differences between an open and a closed system in the context of organizational structure.

Problem 1.2   Contrast organic and mechanistic organizations using possible examples from the construction industry.

Problem 1.3   What are the objectives of an organization? Describe four objectives that you think a construction company might reasonably adopt. It is quite possible that different functional departments will have conflicting objectives. Explain how this might happen and how the matter should be rectified.

Problem 1.4   Distinguish between a formal and an informal organization. What methods are available for illustrating organizational structure? Give the advantages and disadvantages of each method you describe.

Problem 1.5   Draw an organizational chart for an organization to which you have access. Comment on both its vertical and its horizontal integration and distinguish between line and staff positions. Analyse the organization for span of control.

Problem 1.6   Describe what is meant by the motivation–hygiene concept. Make a comparison of this concept with that of the hierarchy of needs, explaining in detail how you view the philosophical differences between the two.

Problem 1.7   Why is it necessary to develop a concept of teamwork in an organization? What are the benefits of doing so and where may conflicts arise within the organizational structure as a result of being inflexible about its implementation?

Problem 1.8   Describe how the demands on the skills of a manager will change as they move up the hierarchical ladder in a construction company. Give examples of the revised demands.

Problem 1.9   Explain why it is important for a manager to encourage the use of groups, both formal and informal, in an organizational structure. Give both the advantages and the disadvantages of doing so.

Problem 1.10   Describe three strategies that could be adopted in assisting groups to come to a decision about the matter under discussion. Give examples of the type of situation that lends itself to the use of each strategy and why.

Problem 1.11   Describe the detail of the steps that should be taken in the process of designing an organizational structure. Link each part of the process to a specific feature of the structure that will be established.

Problem 1.12   Draw up a table in the form of a matrix setting out the types of contract that are feasible for the various forms of contractual organization that a client may adopt.

Problem 1.13   A client wishes to install a new, large-diameter water pipe under a factory workshop floor. It is esssential that the work be carried out during a works holiday of one week so that normal production is not interrupted. The client is advised that it would be best if a form of cost reimbursement contract is adopted. The client's adviser prepares a note for the client showing the likely costs for four alternative forms of the contract—cost plus, cost plus fixed fee, cost plus fluctuating fee and target cost plus fee.

   The adviser suggests that, in the case of the cost plus contract, 20, 12, 5 and 12 per cent should be added to the actual labour, plant, materials, and subcontract costs, respectively. In the case of the cost plus fixed fee contract the fixed fee should be £18 000. In the case of the cost plus fluctuating fee contract, the fee should be 16, 14 and 12 per cent for resulting total costs in the ranges of £120 000 to £140 000, £140 000 to £150 000 and £150 000 to £170 000, respectively. For the target cost contract the fee should be £20 000 less 25 per cent of any actual cost above the target cost of £140 000.

   The expected actual costs of the work are shown in Table 1.1.

Table 1.1

	Cost-plus, £	Cost-plus, fixed-fee, £	Cost-plus, fluct-fee, £	Target cost, £
Labour	62 000	61 000	59 000	59 000
Plant	4 500	4 500	4 500	4 500
Materials	60 000	60 000	59 000	58 000
Subcontracts	21 000	21 000	21 000	21 000

Calculate the expected total cost to the client, including the fee, in each case.

(£165 960; 164 500; 163 590; 161 875)

References

1. Taylor, F. W.: *The Principles of Scientific Management*, Norton, New York, 1967.
2. Metcalfe, H. C. and L. Urwick, eds: *Dynamic Administration: The Collected Papers of Mary Parker Follett*, Harper, New York, 1940.
3. Tannenbaum, R. and W. H. Schmidt: How to choose a leadership pattern, *Harvard Business Review*, vol. 51, May–June 1973, pp. 178–179.
4. Katz, R. L.: Skills of an effective administrator, *Harvard Business Review*, vol. 52, Sept.–Oct. 1974, p. 94.
5. Herzberg, F.: *Work and the Nature of Man*, World Publishing, Cleveland, 1966.
6. Herzberg, F., B. Mausner and B. B. Synderman: *The Motivation to Work*, 2nd edn, Wiley, New York, 1967.

7. Herzberg, F.: One more time: how do you motivate employees?, *Harvard Business Review*, vol. 46, Jan.–Feb. 1968, pp. 53–62.
8. Maslow A. H.: A theory of human motivation, *Psychological Review*, vol. 50, 1943, pp. 370–396.
9. Maslow, A. H.: *Motivation and Personality*, 2nd edn, Harper and Row, New York, 1970.
10. Alderfer, C. P.: An empirical test of a new theory of human needs, *Organizational Behavior and Human Performance*, vol. 4, 1969, pp. 142–175.
11. Alderfer, C. P.: Group and intergroup relations, in *Improving Life at Work*, eds J. R. Hackman and J. Lloyd Suttle, Goodyear Publishing, Santa Monica, 1977, p. 230.
12. Shaw, M. E.: *Group Dynamics: The Psychology of Small Group Behavior*, 2nd edn, McGraw-Hill, New York, 1976.
13. Leavitt, H. J.: Suppose we took groups seriously, in E. L. Cass and F. G. Zimmer, eds, *Man and Work in Society*, Van Nostrand Reinhold, New York, 1975, pp. 67–77.
14. Thomas, E. J. and C. F. Fink: Effects of group size, in L. L. Cummings and W. E. Scott, eds, *Readings in Organizational and Human Performance*, Irwin, Homewood, 1969, pp. 394–408.
15. Kast, F. E. and J. E. Rosenzweig, *Organization and Management—A Systems and Contingency Approach*, 4th edn, McGraw-Hill, New York, 1985, pp. 240–244.

Further reading

Beer, S.: *The Brain of the Firm*, Allen Lane The Penguin Press, London, 1972.
Burns, T. and G. M. Stalker: *The Management of Innovation*, 2nd edn, Tavistock, London, 1966.
Cyert, R. M. and J. G. March: *A Behavioral Theory of the Firm*, Prentice-Hall, Englewood Cliffs, NJ, 1963.
Galbraith, J.: *Designing Complex Organizations*, Addison-Wesley, Reading, Mass., 1973.
Kast, F. E. and J. E. Rosenzweig: *Organization and Management—A Systems and Contingency Approach*, 4th edn, McGraw-Hill, New York, 1985.
Likert, R.: *The Human Organization*, McGraw-Hill, New York, 1967.
Mintzberg, H.: *The Nature of Managerial Work*, Harper and Row, New York, 1973.
Mintzberg, H.: *The Structuring of Organizations—A Synthesis of Research*, Prentice-Hall, Englewood Cliffs, NJ, 1979.
Maloney, W. F. and J. M. McFillen: Motivational implications of construction work, *Journal of Construction Engineering and Management*, ASCE, vol. 112, no. 1, pp. 137–151, March 1986.
Maloney, W. F. and J. M. McFillen: Motivational impact of work crews, *Journal of Construction Engineering and Management*, ASCE, vol. 113, no. 2, pp. 208–221, June, 1987.
McFillen, J. M. and W. F. Maloney: Human resource data in the construction industry, *Journal of Construction Engineering and Management*, ASCE, vol. 112, no. 1, March 1986.
Mitchell, T. R., *People in Organizations*, McGraw-Hill, New York, 1982.
Oglesby, C. H. Parker and G. Howell: *Productivity Improvement in Construction*, McGraw-Hill, New York, 1989, particularly Chapters 9 and 10.
Schermerhorn, J. R. Jr, J. G. Hunt and R. N. Osborn: *Managing Organizational Behavior*, 2nd edn, Wiley, New York, 1985.
Woodward, J.: *Industrial Organization: Theory and Practice*, Oxford University Press, London, 1965.
Woodward, J.: *Industrial Organization: Behaviour and Control*, Oxford University Press, London, 1970.

Construction procedures

Abrahamson, M. W.: *Engineering Law and the ICE Contracts*, 4th edn, Applied Science, London, 1975.

Barrie, D. S. and B. C. Paulson: *Professional Construction Management*, 2nd edn, McGraw-Hill, London, 1984.

Horgan, M. O.: *Competitive Tendering for Engineering Contracts*, Spon, London, 1984.

Institution of Civil Engineers: *Civil Engineering Procedure*, 4th edn, Telford, London, 1986.

Institution of Civil Engineers: *Civil Engineering Standard Method of Measurement CESMM2*, 2nd edn, Institution of Civil Engineers and Federation of Civil Engineering Contractors, 1985.

Institution of Civil Engineers: *ICE Conditions of contract*, 5th edn, Institution of Civil Engineers, London, 1986.

McCaffer, R. and A. N. Baldwin: *Estimating and Tendering for Civil Engineering Works*, 2nd edn, Blackwell, Oxford, 1991.

Royal Institution of Chartered Surveyors: *Standard Method of Measurement for Building Works SMM7*, 7th edn, RICS and Building Employers Federation.

Turner, A.: *Building Procurement*, Macmillan, London, 1990.

Uff, J. F.: *Construction Law*, 4th edn, Sweet and Maxwell, London, 1985.

2. Engineering economics— the principles

2.1 Introduction

This chapter presents an introduction to engineering economy studies. It sets out the factors that must be taken into account. It deals with the mathematical methods of calculating simple and compound interest and equivalence as a basis for making engineering economy studies.

Economy studies are concerned with making comparisons between a number of alternative opportunities for investing resources with a view to selecting the opportunity that will give the optimal future return for an investment. Invariably, an enterprise will have a limit on the funds available to be invested in a given period and a choice must be made between competing investment opportunities. In the case of the design of a facility there is normally a cost budget, frequently set by the owner, constraining the expenditures within which a best economic design must be effected. Because an investment decision is of paramount importance to the future financial health and profitability of an enterprise, it is necessary to establish a rational and satisfactory means of evaluation that will ensure the efficient utilization of capital when taking into account a wide variety of criteria. Examples of economic criteria are rate of return on investment, initial capital investment, and operating and maintenance costs; intangible criteria include safety, quality, and public acceptability. Note that an important alternative is almost always that of not investing at all if none of the other options is attractive.

Where evaluation demands the support of technical knowledge, it must apply the principles of *engineering economics* to the various available alternatives. An engineer has responsibility for the practicability and technical efficiency of the various schemes under consideration. However, it is important that engineers should be able to appreciate the value of money and to interpret its use in a way similar to that in which they interpret all the other materials and resources in which they endeavour to design economically.

A decision to proceed with investment is sometimes made on the basis of guesswork—it is then a *feel* or *hunch* decision. Sometimes managers wrongly pride themselves on a flair for taking the hunch decision. As often as not, few if any alternatives are considered except superficially. When and if they are, the trouble is not taken to evaluate carefully and logically their respective worths. While the hunch decision does not always result in failure, there is no logical background of evaluation to support it or to form a basis for any subsequent enquiry into what has gone wrong should failure occur.

A construction engineer—more than any other—is concerned with the assessment and creation of works having a considerable life and, as a rule, having a need for relatively high capital investments. In addition to the initial investment, the costs of operating and maintaining the works will therefore be significant in the evaluation of the investment. It is not always possible to evaluate in monetary terms the benefits that will arise from a constructed facility, say a highway bridge. A facility for evaluating a benefit of this type is therefore necessary. In addition, there is often a long period of waiting after the initial investment before returns from a constructed facility become available. It is important, therefore, that a construction engineer should have a sound knowledge of the principles of engineering economics and their application. It is not only at the higher administrative/management levels that an engineer needs such a knowledge in order to make decisions in relation to large projects. The principles should also be applied to the design of relatively simple works and structures. By carefully considering, for example, the arrangement of the various components of a structure together with their design sizes and the cost of construction, considerable economies may be made within the structure.

An engineering economy study is desirably carried out before any commitments are made or any money is spent. Any commitments already made are likely to influence a decision in favour of certain alternatives and adversely bias the approach to other, perhaps more suitable, proposals. In such a situation it can be said that a proper evaluation will at least make the best use of commitments already made, and will sometimes justify a decision to abandon or modify them without fear of rational criticism.

It is also necessary that proper technical consideration be given to the alternative ways in which a project may be carried out. An engineering economy study is used to evaluate the differences between proposed alternatives. While the choice of the best alternative can be made by the use of these principles, it is possible that incomplete recognition of the options available may lead to an indifferent solution.

2.2 Irreducible factors

In comparing alternative arrangements for an engineering scheme it is not always possible to convert all the factors under consideration into monetary terms. The influence of the human factor on the appraisal must be evaluated to establish possible effects. A scheme for the installation of mechanical equipment to distribute mixed concrete in a precast concrete factory may be under consideration. One of the factors noted in carrying out the engineering economy study is that the labour required in the works will be reduced by 20 per cent if the equipment is installed. The future wage-bill can be compared with the present wage-bill in terms of money, but reducing the labour requirements may have many other far-reaching effects that cannot be measured in such terms.

If the company is a long-established firm in a small town then it may have its local reputation in the labour market to consider. It may enjoy the reputation of being a good and fair employer, retaining labour in less prosperous times when other companies may take steps to pay off their unwanted or underemployed resources. As a result, the company probably finds that there is little difficulty in obtaining labour at times when expanding business makes it both desirable and necessary. Such attitudes, if carefully controlled, are likely to lead to a low turnover rate and a more highly skilled and experienced workforce, resulting in higher production through greater efficiency. The inevitable costs of training new labour and disposing of the existing can be avoided in meeting the fluctuating cycles of business.

This loss of reputation is an *irreducible*, *intangible*, or *judgement* factor, which, though it cannot be evaluated in strict monetary terms, must be recorded as a factor to be considered in the final assessment. There is a possibility, for instance, that work can be found elsewhere for redundant labour—a position that can as a rule be assessed only by the senior executives taking an overall view of the company's activities. It is not often that the individual making the detailed engineering economy study is called upon to make the final decision between alternatives, and therefore it becomes very important that all known factors, irreducible or otherwise, are recorded for consideration at the time of the study.

Another illustration of a problem that produces an irreducible factor, or a judgement factor, is that involved in the purchase of mechanical equipment for construction work. A company having bought mechanical excavators from a single manufacturer in the past has probably done so, in part, to rationalize its problems concerning spare parts and maintenance procedures. The company probably bought originally on the basis of efficiency, reliability, and safety in service. Though a manufacturer of another machine may now offer reduced operating costs it is very difficult to assess, in terms of money, the value of a unified spares system, the proven safety of the established machine, and the goodwill that may have been established by previous transactions. Decisions in such cases become decisions of policy.

2.3 Interest calculations

The following symbols and definitions are used throughout this text:

P the principal, a sum of money invested in the initial year or a present sum of money;

i the interest rate per unit of time expressed as a decimal;

n time, the number of units of time over which interest accumulates;

I simple interest; the total sum paid for the use of the money at simple interest;

F a compound amount; a sum of money at the end of n units of time at interest i—made up of the principal plus the interest payable;

A uniform series end-of-period payment or receipt that extends for n periods;

S the salvage or resale value at the end of n years.

Interest may be defined as the time value of money—it is the money or consideration that is paid by the borrower for the use of money provided by the lender. In wider terms it may be thought of as the return that may be obtained when capital is invested in a productive concern. It is payment for the use of the money under circumstances that may involve risk to the lender. The effect of interest is to increase the value of invested capital over a period of time.

Interest rates are normally quoted as a percentage for a period of 1 year or less. If the length of the time period is not quoted then it is accepted by convention that the unit of time is 1 year. An interest rate of 5 per cent therefore means 5 per cent *per year*.

2.4 Simple interest

In the calculation of *simple interest* the elements have the following relationship:

$$\text{interest} = \text{principal} \times \text{interest rate} \times \text{time}$$

or, symbolically, $I = Pin$

EXAMPLE 2.1

If the amount to be deposited in the bank is £10 000 and the bank is offering 2.5 per cent per year simple interest, then at the end of the first year the interest payable by the bank will be

$$I = 10\,000 \times 0.025 \times 1 = \underline{£250}$$

The interest for the second and subsequent years will also be £250, unless the interest rate changes. Therefore, at the end of the second year the total amount in the account will be £10 500.

Simple interest is seldom used in engineering economy studies except where very short periods of time are involved.

2.5 Cash flow diagrams

The solution of engineering economics problems that involve evaluation and/or comparison of cash flows is facilitated by preparing diagrams representing the movement of cash over time. The objective of drawing a cash flow diagram is to portray, in a simple and concise fashion, all the data that are available about the problem so that the required calculations can be formulated. A horizontal line is used as the time scale and vertical arrows are used to indicate the direction in which the cash flows. Arrows pointing downwards are used conventionally to indicate negative cash flows or cash going out (payments); capital outlays, wages and salaries, and costs of raw materials are examples of negative cash flows. Arrows pointing upwards indicate positive cash flows or cash coming in (receipts) and will be such items as salvage values, sales of goods produced and sale of equipment or buildings.

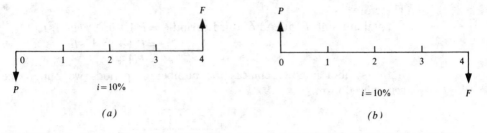

Figure 2.1

Figure 2.1a is drawn from the point of view of the depositor and represents a deposit of P (a negative cash flow) in a bank and its subsequent withdrawal, F (a positive cash flow), including the interest of 10 per cent, four years later. Figure 2.1b represents the cash flow diagram from the point of view of the banker. To the banker the deposit represents a positive cash flow and subsequently the withdrawal represents a negative cash flow. The length of the arrow is not usually to scale.

Cash flows are often referred to as *net cash flows* since they are usually the result of subtracting one or more negative from one or more positive cash flows—all taking place at what is assumed to be the same point in time.

It should be noted that interest is not shown at every year end, since it is not an actual cash flow and it is left to accumulate in the account, being withdrawn with the initial deposit at the end of the fourth year. Therefore no cash flow takes place at intermediate points.

2.6 Compound interest

If a sum of £1000 is invested at 5 per cent per year interest rate, then at the end of the first year interest of £50 will be added to the initial sum of £1000. If the interest is now allowed to remain invested with the capital at the same rate of interest, it will itself earn interest, and at the end of the second year the interest due will be 5 per cent of £1050 or £52.50. The process of paying interest on interest, as well as on the initial investment, can go on from year to year. Such a process is known as *compounding* interest. Mathematically it is represented as follows:

If an initial sum, P, is invested at an interest rate, i, then the interest earned by the end of the first period amounts to Pi. At the end of the first period the total amount is equal to $P + Pi$ or $P(1 + i)$.

Interest for second period	$= P(1 + i)i$
Total amount, F, at end of second period	$= P(1 + i) + P(1 + i)i$
	$= P(1 + i)(1 + i)$
	$= P(1 + i)^2$
Interest for third period	$= P(1 + i)^2 i$

Therefore,

Total amount, F, at end of third period
$$\begin{aligned} &= P(1+i)^2 + P(1+i)^2 i \\ &= P(1+i)^2(1+i) \\ &= P(1+i)^3 \end{aligned}$$

If the symbol n now replaces the number of periods we obtain the following generalized form:

$$\boxed{F = P(1+i)^n} \tag{2.1}$$

The factor $(1+i)^n$ is known as the *single payment compound amount factor*. It is convenient to represent such a factor in symbolic form using a standard notation rather than to repeat continually the factors in mathematical form. For example, the single payment compound amount factor, $(1+i)^n$, will be denoted by $(F|P, i\%, n)$, where the symbols in the brackets have the meaning already assigned to them. (This notation was one of those suggested by the Engineering Economy Division of the American Society for Engineering Education.) The symbolic notation above is read as 'factor to find F given P at an interest rate of i per cent for n periods'. All other such factors will be given similar notation. The full equation for single-payment compound interest will thus be

$$\boxed{F = P(1+i)^n = P(F|P, i\%, n)} \tag{2.2}$$

Each such formula will also be represented by a cash flow diagram. For formula (2.2) it is as shown in Fig. 2.2.

EXAMPLE 2.2

£150 is deposited in a bank for 5 years at a compound interest rate of 6 per cent per year. What is the amount in the account at the end of the 5 years? (See Fig. 2.3.)

$P = £150$; $i = 0.06$; $n = 5$; $F = ?$

$F = P(F|P, i\%, n) = 150(1 + 0.06)^5 = \underline{£220.73}$

In order to facilitate the solution of problems that involve the use of compounding and discounting formulae, tables have been compiled listing values of the various

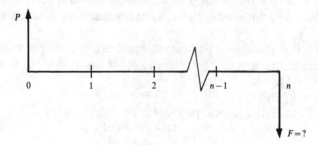

Figure 2.2

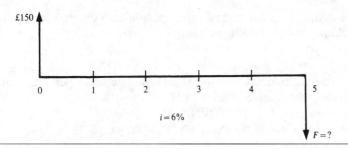

Figure 2.3

factors for a range of values of n and i. A series of such tables is given in Appendix B for a maximum number of periods up to 100. Thus the values for the factor above, $(F|P, i\%, n)$, and those factors elsewhere in this text, can be drawn from the tables directly. Each column of factors in a table carries the relevant symbolic representation at its head. It is important in practice, however, to have an understanding of the principles that underly the function and derivation of the factors.

2.7 Nominal and effective interest rates

The significance of the terms *period* and *interest* in connection with compound interest as expressed in formula (2.1) is that n refers to a number of periods for each of which the interest rate, i, applies.

If an interest rate of 10 per cent per year is *compounded quarterly*, the period or unit of time over which interest is calculated becomes three months and i, the interest rate, becomes 2.5 per cent for that period. If a total period of three years is being considered, n becomes 12 and for every £100 of the initial capital and

$$F = F(F|P, i\%, n) = F(F|P, 2\tfrac{1}{2}\%, 12) = 100(1.3448) = \underline{£134.48}$$

If the calculation had been carried out by compounding annually rather than quarterly then F would have been

$$F = F(F|P, 10\%, 3) = 100(1.3310) = \underline{£133.10}$$

Thus compounding quarterly results in a larger amount of interest accruing than does using the same annual interest rate but compounding annually. Where an annual rate is quoted, but compounding is carried out for a period other than one year, the annual rate is known as a *nominal annual interest rate*; the actual annual ;interest rate which results from compounding over periods of less than one year is known as an *effective annual interest rate*.

The effective annual interest rate, I_{eff}, can be calculated from

$$I_{\text{eff}} = \left(1 + \frac{r}{m}\right)^m - 1 \qquad (2.3)$$

where r is the nominal annual interest rate expressed as a decimal, and m denotes the number of compounding periods per year.

EXAMPLE 2.3

If a loan of £1000 is made at a nominal interest rate of 10 per cent compounded quarterly, what is the effective interest rate?

Effective interest rate, $I_{eff} = (1 + 0.10/4)^4 - 1 = 0.1038$.

The annual interest would then be $= 1000(0.1038) = £103.80$.

2.8 Present worth

If formula (2.1) is rearranged so as to express P in terms of F, i and n, then

$$P = F\left[\frac{1}{(1+i)^n}\right] = F(P|F, i\%, n) \qquad (2.4)$$

(See Fig. 2.4.)

Formula (2.4) establishes the principal and/or the initial capital, P, which must be placed for n periods at an interest rate, i, in order to generate an amount F at the end of that time. In other words, the *present worth* of F in n periods of time, for the given rate of interest. The sum F is being *discounted* to an equivalent value at a previous date.

The factor $1/(1+i)^n$ or $(P|F, i\%, n)$ is referred to as the *single-payment present worth factor*.

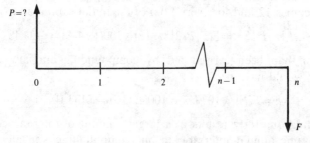

Figure 2.4

EXAMPLE 2.4

An investor's bank statement shows a credit of £345 as a result of a small investment made 10 years previously. Interest over this period has been 2.5 per cent. What was the original investment? (See Fig. 2.5.)

$i = 0.025$; $n = 10$; $F = £345$

From formula (2.4), $P = F(P|F, 2.5\%, 10) = 345(0.7812) = £269.51$

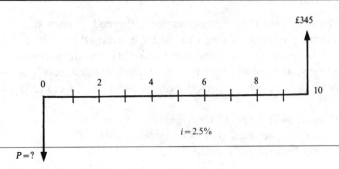

Figure 2.5

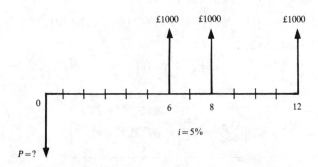

Figure 2.6

EXAMPLE 2.5

How much must a family invest now to provide a lump sum of £1000 for school fees at the end of each of 6 years, 8 years, and 12 years from now if interest is at 5 per cent (See Fig. 2.6.)

Present worth of £1000 in 6 years' time $= 1000(P|F, 5\%, 6) = 1000(0.742) =$ £746.20
Present worth of £1000 in 8 years' time $= 1000(P|F, 5\%, 8) = 1000(0.6768) =$ £676.80
Present worth of £1000 in 12 years' time $= 1000(P|F, 5\%, 12) = 1000(0.5568) =$ £556.80

Total present worth to be invested now $=$ £1979.80

2.9 Uniform series of payments

There are four ways in which a principal sum P, or a compound amount F, and a number of periods of time n, can be linked together with a uniform series of end-of-period payments A, for any given interest rate i. A series of equal payments made at the end of equal periods is known as an *annuity*. It should be noted that the period of time need not be confined to one year, so long as the interest rate is linked to that period of time.

In the first case, a payment A can be made at the end of each of n periods at interest rate i and the payments allowed to gather interest when a final sum, F, will be accumulated. F is the sum of the compound amounts of each individual payment.

An amount invested at the end of the first period will earn interest for $(n-1)$ years and therefore its final value after n periods will be $A(1+i)^{n-1}$, from (2.1).

Compound amount of 2nd period's payment $\quad = A(1+i)^{n-2}$
Compound amount of the third period's payment $= A(1+i)^{n-3}$
Final payment which earns no interest $\quad\quad\quad\quad = A$

$$F = A(1+i)^{n-1} + A(1+i)^{n-2} + A(1+i)^{n-3} + \cdots + A$$
$$= A[1 + (1+i) + (1+i)^2 + (1+i)^3 + \cdots + (1+i)^{n-1}] \tag{2.5A}$$

Multiplying both sides of the equation by $(1+i)$,

$$(1+i)F = A[(1+i) + (1+i)^2 + (1+i)^3 + \cdots + (1+i)^n] \tag{2.5B}$$

Subtract Eq. (2.5A) from (2.5B):

$$iF = A[(1+i)^n - 1]$$

and

$$F = A\left[\frac{(1+i)^n - 1}{i}\right] = A(F\,|\,A,\,i\%,\,n) \tag{2.5C}$$

The expression $(F\,|\,A,\,i\%,\,n)$ is known as the *uniform series compound amount factor*. (See Fig. 2.7.)

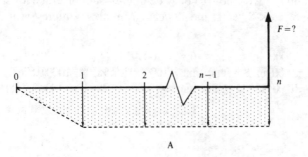

F=?

A

Figure 2.7

EXAMPLE 2.6
Given an interest rate of 5 per cent per year, what sum would be accumulated after 6 years if £200 were invested at the end of each year for the 6 years? (See Fig. 2.8.)

$$F = 200(F\,|\,A,\,5\%,\,6) = 200(6.8019) = \underline{£1360.38}$$

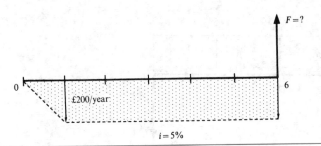

Figure 2.8

EXAMPLE 2.7

A speculator buys a site near the fringe of an industrial area in a large city for £1 000 000. Annual outgoings on the site for maintenance, fencing, watching, etc., amount to £45 000. It is estimated that the site will not be sold for 8 years, at which time that area is due for development. For what minimum price must the site be sold at that time so as to break even on the costs if the original purchase price and the annual outgoings could have been alternatively invested at 12 per cent per year? (See Fig. 2.9.)

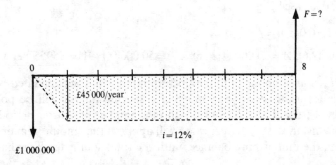

Figure 2.9

Future value of capital sum = 1 000 000($F|P$, 12%, 8) = 1 000 000(2.4760) = £2 476 000

Future value of annual costs = 45 000($F|A$, 12%, 8) = 45 000(12.2997)   =   £553 487

Minimum selling price for site in 8 years' time = £2 476 000 + 553 487   = £3 029 487

See Fig. 2.10. The algebraic expression (2.5) can be restated as

$$A = F\left[\frac{i}{(1+i)^n - 1}\right] = F(A|F, i\%, n) \tag{2.6}$$

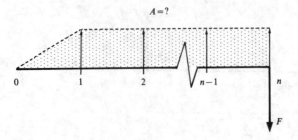

Figure 2.10

Thus the uniform amount, A, to be invested at the end of each period in order to produce a specific amount at the end of n periods, can be calculated directly for a given interest rate. A fund into which such payments are made is known as a *sinking fund*, and the expression within the outer brackets of formula (2.6) is known as the *uniform series sinking fund deposit factor*.

EXAMPLE 2.8

A uniform annual investment is to be made into a sinking fund with a view to providing the capital at the end of 7 years for the replacement of a tractor. An interest rate of 6 per cent is available. What is the annual investment needed to provide for £50 000? (See Fig. 2.11.)

$F = £50\,000; \quad i = 0.06; \quad n = 7$

From formula (2.6), $A = 50\,000(A|F, 6\%, 7) = 50\,000(0.1191) = \underline{£5955 \text{ per year}}$

The principal, P, is an amount of money before interest has accumulated with it. If an amount P is set aside at the beginning of a specific time it will be possible to withdraw a sum A from it at the end of each of a number of shorter periods within that time. In this instance, the interest rate i will apply to the amount remaining after each withdrawal, the diminishing balance, until the total sum is finally exhausted by

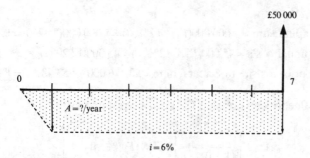

Figure 2.11

the withdrawals at the end of n periods. The amount P required to give a set number of withdrawals under these conditions can be calculated or alternatively the equation can be restated in such a form that the magnitude of withdrawal can be calculated directly, given the remaining data.

Formula (2.6):

$$A = F\left[\frac{i}{(1+i)^n - 1}\right]$$

By substituting $F = P(1+i)^n$ (formula (2.1)),

$$A = P(1+i)^n\left[\frac{i}{(1+i)^n - 1}\right]$$

therefore

$$A = P\left[\frac{i(1+i)^n}{(1+i)^n - 1}\right] = P(A\,|\,P, i\%, n) \qquad (2.7)$$

See Fig. 2.12. The bracketed factor in formula (2.7) is known as the *uniform series capital recovery factor*. It is frequently used and has special significance in engineering economics. It should be noted that it is equal to the sinking fund factor plus the interest rate, and P is the uniform series end-of-period payment necessary to repay a debt in n periods if the interest rate is i.

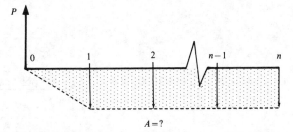

$A = ?$

Figure 2.12

EXAMPLE 2.9
With an interest rate of 6 per cent, what uniform end-of-period payment must be made for 10 years to repay an initial debt of £2000? (See Fig. 2.13.)

$P = £2000;\ i = 0.06;\ n = 10$

$A = 2000P(A\,|\,P, 6\%, 10) = 2000(0.1359) = \underline{£271.80}$

EXAMPLE 2.10
A fund is set up with an initial investment of £1000 to provide a uniform year-end payment each year for 6 years. How much will the payment amount to if the

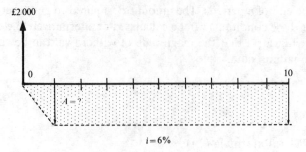

Figure 2.13

investment is made at an interest rate of 3 per cent and the fund is to be completely exhausted by the sixth payment? (See Fig. 2.14.)

$P = £1000$; $i\ 0.03$; $n = 6$

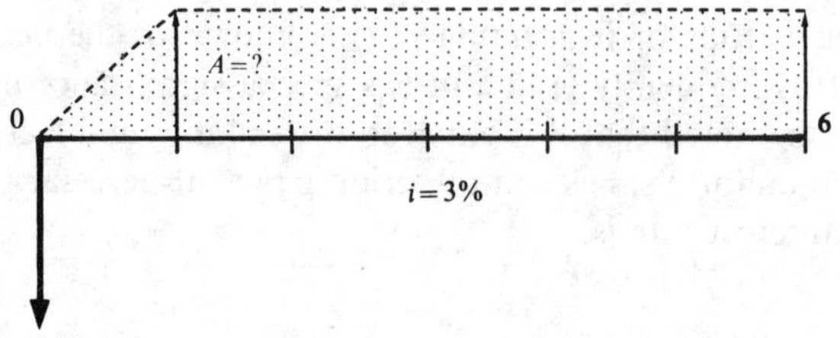

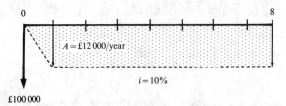

Figure 2.14

From formula (2.7), $A = 1000(A \mid P, 3\%, 6) = 1000(0.1846) = £184.60$ per year

EXAMPLE 2.11

A unit of mechanical equipment has an initial cost of £100 000 and annual maintenance expenditure is expected to average £12 000 for its 8 years of life. If interest is at 10 per cent and the equipment has no salvage value, what is its equivalent annual cost, excluding labour, fuels, etc.? (See Fig. 2.15.)

Figure 2.15

To convert the capital sum to an equivalent uniform annual series, use formula (2.7) to find the capital recovery:

$A = P(A|P, 10\%, 8) = 100\,000(0.1874) = £18\,740$

Maintenance expenditure is already at annual cost, therefore

Total equivalent annual cost $= £18\,740 + 12\,000 = \underline{£30\,740}$

By rearranging (2.7),

$$P = A\left[\frac{(1+i)^n - 1}{i(1+i)^n}\right] = A(A|P, i\%, n) \tag{2.8}$$

See Fig. 2.16. The bracketed factor in formula (2.8) is known as the *uniform series present worth factor*.

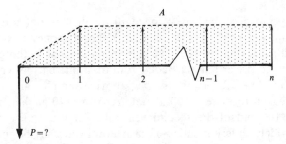

Figure 2.16

EXAMPLE 2.12

What sum of money should be deposited in a bank in order to provide five equal annual withdrawals of £1000, the first of which will be made one year after the deposit? The fund pays 9 per cent (See Fig. 2.17.)

$A = £1000; \ i = 0.09; \ n = 5$ years

$P = 1000(A|P, 9\%, 5) = 1000(3.8897) = \underline{£3889.70}$

The above formulae can be applied to all cases where there is a regular pattern of payments or withdrawals of uniform amounts. Where the payments or with-

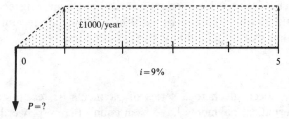

Figure 2.17

drawals vary from period to period then each amount must be dealt with separately.

2.10 Equivalence

The concept of *equivalence* underlies the methods used to compare different series of cash flows, one with another. Taking the simplest case first, that of a single payment, it has been demonstrated that a sum of money now, say £100, compounded at a rate of interest of 10 per cent per year, will become £110 at the end of the first year, £121 at the end of the second, and £133.10 at the end of the third. The compounding may continue in perpetuity. The sum of £110 at the end of the first year is said to be *equivalent* to £100 at time zero or £133.10 at the end of the third year—*given that the interest rate is 10 per cent per year.* If the interest rate is different then the sums of money (except for the original £100) will be different.

When sums of money are described as being equivalent, it is not intended to mean that they are *equal*, but that, taking the time-value of money into account and given risk-free conditions (that is, that there is certainty that nothing will go wrong), there is no reason to prefer one particular sum to a different but equivalent one at another point in time. Referring to the above example, given no-risk conditions, there is no reason to prefer £100 at time zero to £110 at the end of the first year or £121.00 at the end of the second year, etc. They are all equivalent. Comparisons between widely differing series of cash flows can thus be made at a given interest rate.

EXAMPLE 2.13

£10 000 at the end of 10 years from now is equivalent to £X now if the interest rate is 5 per cent. What is X?

X is the present worth of £10 000 with $i = 0.05$

therefore $X = 10\,000(P|F, 5\%, 10) = 10\,000(0.6139) = £6139$

Therefore, £6139 is equivalent to £10 000 at the end of 10 years from now with an interest rate of 5 per cent.

Similarly, £6139 is equivalent to $P(1 + 0.05)^5 = 6139(1.276) = \underline{£7833.40}$

at the end of 5 years from now at an interest rate of 5 per cent. The importance of the concept of equivalence cannot be overemphasized in engineering economics, since this is the basis upon which comparisons of expenditure or receipts over a period of years, or for that matter indefinitely, can be compared.

2.11 Uniform gradient series

Consideration has so far been given to a series of payments or receipts which is uniform—that is, the periodical payments have been equal. Before proceeding with the various methods that facilitate the economic comparison of alternatives it is desirable to formulate a method of finding the uniform series equivalent of a number of payments that are increasing each year by a similar amount. In the case of most items of mechanical equipment, for example, experience shows that maintenance costs have a tendency to increase each year. While the increases may not be constant each year, it is often possible to convert them to an equivalent approximate uniform gradient, and thus facilitate their conversion to a uniform series, in the following manner.

Let the uniform increment at the end of each year (or period, if other than a year is being considered) be G. Therefore the payment or receipt at the end of the second period is G greater than that at the end of the first period.

Similarly, the payment or receipt at the end of the third year is G greater than at the end of the second. Thus at the end of $(n-1)$ years the payment or receipt will be $(n-2)G$ greater than that at the end of the first year. At the end of the nth year it will be $(n-1)G$ greater.

Each increment G that is added at successive year ends is the start of a new series of instalments, A, which are carried on until the end of n periods. Each of these series can be totalled to establish its final compound amount.

The first increment will have a compound amount $= G\left[\dfrac{(1+i)^{n-1}-1}{i}\right]$

and the second increment will have a compound amount $= G\left[\dfrac{(1+i)^{n-2}-1}{i}\right]$

All these compound amounts can now be totalled, giving the sum of the compound amounts:

$$F = G\left[\frac{(1+i)^{n-1}-1}{i}+\frac{(1+i)^{n-2}-1}{i}\cdots+\frac{(1+i)^2-1}{i}+\frac{(1+i)-1}{i}\right]$$

$$=\frac{G}{i}[(1+i)^{n-1}+(1+i)^{n-2}\cdots+(1+i)^2+(1+i)-(n-1)]$$

$$=\frac{G}{i}[(1+i)^{n-1}+(1+i)^{n-2}\cdots+(1+i)^2+(1+i)+1]-\frac{nG}{i}$$

$$=\frac{G}{i}\left[\frac{(1+i)^n-1}{i}\right]-\frac{nG}{i}$$

The final equation above is obtained by substituting the uniform series compound amount factor for the expression in the square brackets. In order to convert this sum into an equivalent uniform period payment over n periods, it is necessary to substitute the above sum for F in Eq. (2.6), the sinking fund factor, giving

$$A = \frac{G}{i}\left[\frac{(1+i)^n - 1}{i}\right]\left[\frac{i}{(1+i)^n - 1}\right] - \frac{nG}{i}\left[\frac{i}{(1+i)^n - 1}\right]$$

$$= \frac{G}{i} - \frac{nG}{i}\left[\frac{i}{(1+i)^n - 1}\right]$$

$$A = \frac{G}{i} - \frac{nG}{i}(A\,|\,F, i\%, n)] = G(A\,|\,G, i\%, n) \tag{2.9}$$

$(A\,|\,G, i\%, n)$ is known as the *arithmetic gradient conversion factor*.

EXAMPLE 2.13
If the maintenance cost of a bulldozer amounts to £2000 by the end of the first year of its service, £2500 by the end of the second and £3000, £3500, and £4000 by the end of the third, fourth, and fifth years respectively, find the equivalent uniform series cost each year over a period of 5 years. Interest is at 5 per cent. (See Fig. 2.18.)

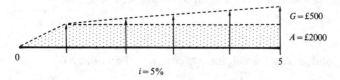

Figure 2.18

Annual equivalent of increment $= G(A\,|\,G, 5\%, 5) = 500(1.90) = £950.00$
Therefore, uniform series equivalent annual cost of maintenance
$= 2000 + 950 = \underline{£2950 \text{ for each of five years}}$

Summary

An introduction to engineering economics is dealt with in this chapter. It starts by putting engineering economy studies into context and looking at some of the factors concerned—both quantitative and qualitative. The main thrust of this chapter is the way in which interest calculations to establish the time value of money for engineering economy studies can be carried out.

Simple interest is dealt with first. Cash flow diagrams as pictorial representations of the way in which money flows in and out of an investment situation are described as an aid to problem solving. Calculations for compound interest of lump sum, regular uniform or regularly varying payments are derived from first principles and illustrated with typical solved problems. Equivalence, the basis for comparing investments and their returns, is described. Finally, the various equations for calculating the time value of money are summarized in Table 2.1.

Table 2.1 Discrete compound interest factors

Factor	Expressions		Find	Given
A. Single payment				
1. Compound amount	$F = P[1+i]^n$	$(F\|P, i\%, n)$	F	P
2. Present worth	$P = F\left[\dfrac{1}{(1+i)^n}\right]$	$(P\|F, i\%, n)$	P	F
B. Uniform series				
3. Compound amount	$F = A\left[\dfrac{(1+i)^n - 1}{i}\right]$	$(F\|A, i\%, n)$	F	A
4. Sinking fund	$A = F\left[\dfrac{i}{(1+i)^n - 1}\right]$	$(A\|F, i\%, n)$	A	F
5. Present worth	$P = A\left[\dfrac{(1+i)^n - 1}{i(1+i)^n}\right]$	$(P\|A, i\%, n)$	P	A
6. Capital recovery	$A = P\left[\dfrac{i(1+i)^n}{(1+i)^n - 1}\right]$	$(A\|P, i\%, n)$	A	P
C. Arithmetic gradient				
7. Uniform series equivalent	$A = G\left\{\dfrac{1}{i} - \dfrac{n}{i}\left[\dfrac{i}{(1+i)^n - 1}\right]\right\}$	$(A\|G, i\%, n)$	A	G

Problems

Problem 2.1   What is the equivalent payment now to one of £1200 in 8 years' time if interest is at 6 per cent?   (£752.88)

Problem 2.2   If £6000 is invested at an interest rate of 8 per cent compounded annually, what will be the value of the investment in 8 years? What will the total interest amount to at the end of this period?   (£11 105; £5105)

Problem 2.3   What is the present worth of a future sum of £2450 in 10 years' time if interest is at 10 per cent?   (£945)

Problem 2.4   An investor wishes to accumulate a sum of £5000 by investing a lump sum in a bank account which pays interest of 5 per cent. How much should be invested if the investment is continued for 5 years?   (£3918)

Problem 2.5   What maximum uniform end-of-year payment can be withdrawn from an invested sum of £5000 at 5 per cent for a period of 6 years?   (£985)

Problem 2.6   What total amount will accumulate in the bank if a payment of £100 is made at the end of each quarter for 6 years with interest of 8 per cent compounded quarterly?   (£3042)

Problem 2.7   Deposits of £600, £1000, and £2000 are made at the end of years 1, 2 and 3 respectively. Each payment is compounded at 10 per cent. How much will be accumulated in the account at the end of year 5?   (£4631)

Problem 2.8   A plot of land is to be purchased for an initial sum of £40 000. It is agreed with the bank financing the purchase that a series of uniform annual payments will be made for the land over a period of 10 years. What will the annual payment be if interest is at 10 per cent?   (£6510)

Problem 2.9   An investor places £5000 in an account which bears interest at 9.5 per cent compounded monthly. At the end of the first year the investor decides to deposit, at the end of each month, a sum of £50 for the next 4 years. If the interest rate is constant throughout, calculate how much the investor will have in the account at the end of the fourth year.   (£9374)

Problem 2.10   A person wishes to borrow £8000 now to be paid back over 8 years at 15 per cent compounded annually. What will be the annual payment for the 8-year period?   (£1783)

Problem 2.11   A man and wife buy a house and take out a mortgage of £60 000 to meet part of the cost. They agree to pay off the mortgage over 25 years making monthly payments. Interest on the mortgage is 10.5 per cent. To what will the monthly payment amount? What amount of the original debt of £60 000 will remain after they have made 250 payments?   (£566; £22 903)

Problem 2.12   A credit card bears a nominal interest rate of 18 per cent per year compounded monthly. What is the effective annual rate?   (19.56 per cent)

Problem 2.13   £275 is available for investment now in order to provide for a lump sum of £380 in 10 years' time. What interest rate must be obtained in order to achieve this increase in capital?   (3.2724 per cent)

Problem 2.14   A man deposits £1000 every 3 months in a bank account that bears an interest rate of 12 per cent compounded monthly. What is the effective quarterly interest rate? What will be the balance in the account at the end of 3 years?   (3.03 per cent; £14 216)

Problem 2.15   If a proposal for the installation of equipment in a factory requires a capital investment now of £10 000, what saving per year must be shown over the next ten years to justify the expenditure at an interest rate of 5 per cent? (£1295)

Problem 2.16   £1200 is deposited in a special savings account at the end of year 1. The account bears an interest rate of 10 per cent. At the end of each succeeding year a further sum is deposited but the amount is increased by £200 per year. How much will be in the account after 5 years?   (£9536)

Problem 2.17   Four different banks are offering interest rates on their savings account as follows:
(a) 9.0 per cent compounded annually;
(b) 8.8 per cent compounded semiannually;
(c) 8.7 per cent compounded quarterly;
(d) 8.5 per cent compounded monthly.
Which represents the highest interest rate for a customer?
                    (9 per cent; 8.9936 per cent; 8.9879 per cent; 8.3981 per cent)

Problem 2.18   How many years will it take, at an interest rate of 12 per cent compounded annually, for an initial sum to become doubled?   (6.11 years)

Problem 2.19   A debtor repays £3000, £5000, £4000, £4000, and £2000 at the end of years 1, 2, 3, 4, and 5 respectively for a debt incurred at time 0. If with these payments the debt is cleared when interest of 10 per cent compounded annually is accumulated, what is the size of the initial debt?   (£13 862.8)

Problem 2.20   The purchaser of an automobile is paying for it at the rate of £600 per half-year, having agreed to make 10 such payments, but after 2 years, when the fourth payment becomes due, decides to make a lump sum payment to settle the account. With an interest rate of 10 per cent, how much will be needed to do this if there is no rebate of the interest to be charged for the whole of the 5 years?   (£3645.42)

Problem 2.21   It is estimated that in the case of an industrial building an increased expenditure of £2000 per year over the next 8 years will avoid a lump sum replacement cost of £19 000 at the end of that time. Which is the cheaper scheme with interest at 6 per cent?   (Cheaper to pay £19 000)

Problem 2.22   What is the worth today of a series of 5 annual payments starting at £300 and increasing by £100 for each successive year, the first payment to be received 6 years hence?   (£1537.32)

Problem 2.23   If a mill building is constructed of reinforced concrete, it will have an estimated initial cost of £200 000 and no maintenance costs for the first 10 years. A building to serve a similar purpose but erected in structural steelwork and clad in plastic coated metal sheets has an initial cost of £160 000 but the steelwork needs to be painted every 2 years at a cost of £14 000. With interest at 10 per cent, which is the cheaper investment considered over the first 10 years of the building's life?   (Reinforced concrete is cheaper)

Problem 2.24   A bond is issued for which the interest rate varies with other market rates. Over a 10-year period the interest paid is 9.5 per cent for the first 3 years, 8.5 per cent for years 4 and 5, 9.75 per cent for years 6 to 9 and 10.5 per cent for the final year. If the bond is issued for £10 000, what is it worth after 10 years? What is the equivalent uniform rate of interest over the 10 years?   (£24 723; 9.474 per cent)

Further reading

Couper, J. R. and W. H. Radar: *Applied Finance and Economic Analysis for Scientists and Engineers*, Van Nostrand Reinhold, New York, 1986, Chapter 5.

Dixon, R.: *Investment Appraisal—a Guide for Managers*, Kogan Page in association with the Chartered Institute of Management Accountants, London, 1988. (This slim but useful book gives less detail about the underlying principles but covers the subject of this chapter briefly in its Chapter 3.)

Grant, E. L., W. G. Ireson and R. S. Leavenworth: *Principles of Engineering Economy*, 8th edn, Wiley, New York, 1990, Chapters 1–3.

Pilcher, R.: *Project Cost Control in Construction*, Collins, London, 1985, Chapters 1–3.

Riggs, J. L. and T. M. West: *Engineering Economics*, 3rd edn, McGraw-Hill, New York, 1986, Chapters 1–3.

White, J. A., M. H. Agee and K. E. Case: *Principles of Engineering Economic Analysis*, 3rd edn, Wiley, New York, 1989, Chapter 3.

3. Economic comparisons

3.1 Economic feasibility

This chapter is concerned with the ways in which decisions concerning economic choices between alternative investments, methods and materials can be made. It contains the details of methods that employ the basic concepts of the time value of money, as developed in Chapter 2. Two less rigorous methods are also included which, nevertheless, are popular among many commercial and industrial organiz-ations. In many cases, the less rigorous methods will have a place in the evaluation process so as to establish order of magnitude studies before detailed analysis takes place.

It is advantageous at this stage in the evaluation process to keep the economic feasibility analysis separate from considerations of how a project that meets the economic requirements will be financed. The financing methods that will be used subsequently, and the feasibility analysis, must be fitted together before a project is finally approved. However, in the *economic evaluation* process the con-cern is primarily about cash flows that will arise as a result of investments and operations both during the facility's construction and when it is commissioned into service. In the *financial evaluation* process the concern is primarily with the positive and negative cash flows that will arise from the financing plan and the time schedule of the total funding that will be required. The latter is dealt with in Chapter 13.

3.2 Economy study methods

Economy studies are made to evaluate two aspects of the decision problem. The first is to decide whether a single investment should be made at all—the decision alternatives being two in number: either invest or don't invest. The second situation occurs where it is necessary to make a choice between several alternatives that are available for investment. This latter situation can then be subdivided into a further two categories. The first is where one is to be selected from a list of projects, all the projects on the list being *mutually exclusive*, i.e., if one is chosen, the remainder are excluded. The second is a category in which projects are being selected from a list where the total finance available for investment may be limited. In this case not all projects can necessarily be implemented, even though all of them may achieve the minimum required rate of return or greater. Generally speaking, the methods used for the evaluation in the first situation are, with some modest modification, equally applicable in the second.

The projects for investment can also be cross-classified into those investments to reduce cost and those to expand income. However, those projects in the category to reduce cost may lead ultimately to an expansion of income and vice versa. Investments to reduce cost can be typified by those that propose the replacement of manual labour and/or obsolescent equipment by modern, automated, more efficient machinery. The *income* from the investment in the new equipment is usually the saving in outgoings that is made as a result of modernization and, where relevant, increased production as well as enhanced productivity. The category to expand income can be typified by the purchase of additional assets so as to increase the total income generated. A study is carried out to determine whether the additional investment achieves a satisfactory rate of return.

An economic appraisal will normally follow a well-established sequence of three phases, as follows:

1. estimation;
2. calculation;
3. evaluation.

Estimation is the phase during which the various cash flows associated with the project are established, usually in the common denominator of money. The *negative cash flows* or payments will include invested capital for buildings, machinery, plant, raw materials, production costs of labour, fuel, overheads, tax, and so on. The *positive cash flows* or receipts will be the estimated returns that will flow back to the project as a result of the investment, such as income from sales, tax allowances, sales of patent rights, and salvage or resale values. Some influential factors, like equipment life, market size, and the state of the national economy, will also need to be estimated, though it may be difficult, at this stage, to express all of them accurately in terms of money.

Calculation is the phase in which the positive and negative cash flows for the project are put into an appropriate sequence and their relative time scheduling is established. This enables the rate of return for the project to be calculated and the economic merit of the investment to be reviewed. In addition, where necessary, a *sensitivity study* is carried out to establish the sensitivity of a project's outcome to possible changes in the values of some of the principal variables of the project, such as size of investment, inflationary effects, positive cash flow, timing of cash flows, etc.

Finally, *evaluation* takes place. Frequently, the numerical analysis of the cash flows of a project will not provide the only basis on which the project can be evaluated. There may be risks associated with the project that cannot be expressed in quantitative terms. Judgement will need to be exercised, not only in this but in the many other intangible areas surrounding the investment, such as those concerned with the social obligations of an employer to his employees and the local community.

3.3 Traditional methods of appraisal

It is commonplace for the merits of investment in one or other projects to be assessed using what might be called *traditional* methods, taking little or no account of the time value of money in accordance with the principles already described.

One traditional method is known as the *rate of return on capital*, the basis of which is to calculate the rate of return by dividing the annual net profit by the capital invested. There are many variants of this particular method, depending on the interpretation of *annual net profit*, both in the sense of what is deducted from the gross profit and whether or not it is treated as an average profit over a period of years. In its favour it is often said that the outcome of the rate of return method is well understood by managers and can be applied readily to evaluate the worth of a single project or to confirm the choice of one from a number of alternatives.

Against the method—and the disadvantages heavily outweigh the advantages—is that it takes no cognizance of the time value of payments and receipts and therefore ignores the fact, which has already been illustrated, that a payment at an early date is worth more than the same payment at a distant date. In methods where a single year's profits or estimated profits are used to calculate the rate of return, a very distorted picture may be obtained, especially where the profit figure chosen is not truly representative of the future years' earnings. At best, the rate of return method can be considered to be a crude approximation of some of the more refined methods.

A method that is probably more common, but belongs to the same category as the rate of return on capital method, is that of *payback* or *payout*. While it is very simple to use—and therefore tends to be popular—it too suffers from a number of deficiencies in practice. The principle of the method is to determine how quickly the gross capital invested in a project can be recovered by the net cash flowing in as a result of the project's operations. It is deficient because it considers the money-earning capacity of a project only during the period when the investment is being paid back. No account is taken of the potential profitability after the payback period. Again, no account is taken either of the time value of money or of the timing of cash flows during capital recovery.

In its favour it can be said that payback has uses in situations where alternatives are being assessed under highly uncertain or risky conditions, in which it may be very desirable to recover one's investment as quickly as possible. Another area in which it has its uses is where relatively small investments, obviously having a high profitability, are being considered.

The following example illustrates some of the defects of using a rule-of-thumb system.

EXAMPLE 3.1

In Table 3.1 are listed, over the first 6 years of operation, the profits as assessed for three projected mutually exclusive investments, Alpha, Beta, and Gamma.

It is required to select one for investment from the three alternatives. The capital outlay of £20 000 for each of the three projects is the same. If the three projects are now considered in the light of the return on capital method, it would appear that

Table 3.1

| Year | Project | | |
	Alpha	Beta	Gamma
0	−£20 000	−£20 000	−£20 000
1	3 000	6 500	7 000
2	4 000	6 500	6 000
3	6 000	6 500	7 000
4	8 000	6 500	5 000
5	8 000	—	3 000
6	8 000	—	1 000
	£37 000	£26 000	£29 000
Average annual profits	£6 170	£6 500	£4 833
Return on capital	30.85	32.50	24.17
Payback	3.13 years	3.08 years	3.00 years

Project Beta is the most favourable, having an average annual net profit of £6500. This is 32.5 per cent of the initial investment. Alternatively, on examining Project Alpha, support might be generated for it because, having settled down in years 4, 5, and 6 with a net cash flow at the rate of £8000 a year, it is then returning 40 per cent per year on the initial investment. There is, of course, little justification for this calculation. If the return on capital method is to be used, there seems to be no stronger case than that for using the average net cash inflow or profit over the project life for the calculation of the rate of return on capital. It is 30.85 per cent in this case.

The total net cash flow over the 6 years in the case of Project Beta is considerably less than that for the other projects even though, on the basis of return on capital, it should be selected. It shows a total net cash flow over the 6 years of £26 000, some £11 000 less than project Alpha and some £3000 less than Project Gamma.

Should the payback method be used to determine which of the three projects should go ahead, then Project Gamma clearly becomes the selection with the repayment of the initial investment in the first 3 years.

The timing of the net cash flow has not so far been considered in the assessment for any of the three cases, except in the case of Project Gamma, where high cash flows in the early years gave it favourable consideration under the payback method.

Failure to consider the time element in the case of the return on capital method allows the selection of Project Beta, but few companies, from a visual inspection of the figures, would not be attracted to Project Alpha with its total positive cash flow of £37 000. The consideration by the return on capital method disregards the fact that the high average net cash flow of Project Beta does not extend beyond the fourth year.

In order to overcome some of the deficiencies of these traditional methods in the calculation of the rate of return on an investment, a means is required of assessing investments which takes into account the initial capital outlay, the cash flow of money into or out of the project for its subsequent life and the points of time during the life of the project when the payments and receipts will be made. In addition, it must be easy to use and to interpret.

3.4 The equivalent annual cost method

Engineering economy comparisons are made with a view to aiding the decision as to which alternative should be chosen for investment or whether an investment should be made at all. They are properly made, therefore, before the capital is invested. The decision on whether the investment in the project should be made at all demands a knowledge of the magnitude or rate of return which could be obtained by investing the capital elsewhere.

In using the *equivalent annual cost* method, as well as other methods of comparison using the time value of money, account is taken of both the capital and recurrent investment that is made over the full period of assessment. Such expenses as salaries, maintenance charges, fuels, etc., are allocated in the study on an annual basis. While the expenses may be accumulated day-by-day through-out the year, an *end-of-year* convention is used. All such recurrent charges are assumed to occur at the end of the year in which they are incurred. No difficulty arises out of making this assumption, though more detailed and frequent allocations could be made if required. No invalidation of the methods used is caused by so doing.

In using the equivalent annual cost method for the purposes of comparison, all payments and receipts, no matter how diverse in quantity or timing, are converted to their *equivalent uniform annual costs*. It is necessary to make an assumption about the required rate of return before it is possible to convert variable cash flows to a uniform series of payments over the life of an investment proposal. The subject of establishing the required rate of return is addressed further in Section 4.5. For the purposes of illustrating the use of methods of comparison, suitable rates of return will be assumed without justification. In such calculations the term *minimum required rate of return* is freely interchangeable with that of the *interest rate, i.* The following example illustrates the application of the equivalent annual cost technique for a simple case.

EXAMPLE 3.2

At a long-term strip-mining coal site it is proposed to maintain the main temporary haulage roads serving the excavation by using hand labour. The annual wage bill is estimated to be £78 000. With other associated expenses, the total cost of labour to the contractor will be £108 000 per year. The production of coal on the site is expected to last for 6 years, and alternative methods of constructing and maintaining haulage roads need to be investigated.

The first alternative is to buy a motor grader for £70 000 and, as a con-sequence, reduce the labour force. Maintenance of the grader is estimated to average £3000 per year for the 6 years, after which it will have a salvage or resale value of £15 000. The labour costs associated with the use of the grader amount to £60 000 per year.

The second alternative is to lay more substantial roads in the first instance, extending these after 2 years and again after 4 years. Initial costs are then £60 000, with further investments of £30 000 and £28 000 after 2 and 4 years respectively. Total labour costs in this scheme amount to £48 000 per year.

If the return of at least 10 per cent is desirable on the capital invested, which is the most economic scheme?

Scheme 1
See Fig. 3.1.

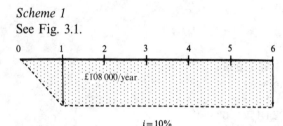

Figure 3.1

Annual cost of labour $=£108\,000$

This is the sole annual outgoing and requires no conversion to annual payments.

Scheme 2
See Fig. 3.2.

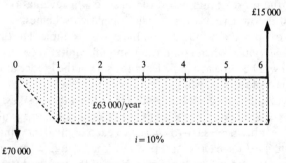

Figure 3.2

Annual capital recovery cost of the motor grader (where S is the salvage value of the grader) $= (P-S)(A|P, 10\%, 6) + Si$

$=(70\,000 - 15\,000)(0.2296) + 15\,000(0.10)$	$=£14\,128$
Annual maintenance costs of grader	$=\quad 3\,000$
Annual labour costs	$=\quad 60\,000$
Total equivalent annual cost	$=£77\,128$

Scheme 3
See Fig. 3.3.

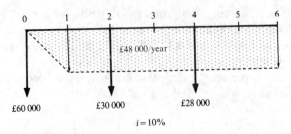

Figure 3.3

Annual capital recovery of initial cost
  $= 60\,000(A|P, 10\%, 6) = 60\,000(0.2296)$ $= £13\,776$
  Annual capital recovery for capital cost at end of 2 years
    $= 30\,000(P|F, 10\%, 2)(A|P, 10\%, 6)$
    $= 30\,000(0.8265)(0.2296)$ $=\quad 5\,693$
  Annual capital recovery for capital cost at end of 4 years
    $= 28\,000(P|F, 10\%, 4)(A|P, 10\%, 6)$
    $= 28\,000(0.6830)(0.2296)$ $=\quad 4\,391$
  Annual labour costs $=\quad \underline{48\,000}$
  Total equivalent annual cost $= \underline{\underline{£71\,860}}$

Scheme 3 is therefore the most economic on the basis of this evaluation because its equivalent annual cost is lower than those of the other two schemes.

There are a number of points to be noted from the above example. The first concerns the treatment of salvage values when computing annual capital recovery costs. The salvage value (£15 000) will become available from the sale of the grader at the end of 6 years. Therefore, the part of the cost which is invested over the 6 years of the grader's useful life, and which will not be recoverable as salvage, is the initial cost *less* the salvage value (£70 000 − £15 000 = £55 000). Since the salvage value will become available again at the end of 6 years it is only necessary to *charge* to each equivalent annual cost the interest on that amount. Treating each year separately, the salvage value can be looked on as being locked up or loaned for the initial purchase of the grader during each year and it is therefore not possible to earn interest or profit by investing the money elsewhere. Account is taken of this in the calculation.

In Scheme 3, each of the payments is converted to present worth before being converted to an equivalent uniform series of payments over the 6 years of the comparison.

Finally, the one overriding assumption is that each of the three schemes considered will either give equally good service if put into operation and/or at least will provide the minimum service required. In making an economic choice between alternatives, it is assumed that the technical merit of each alternative has been examined and found to be satisfactory. The only considerations that may now affect the ultimate decision are the *irreducible* factors.

One example of an irreducible factor might be that there is an ample supply of skilled labour in an area where unemployment is high. It therefore becomes a social obligation of the contractor to act as beneficially as he is able towards the local community. There may, for the contractor, be other spinoffs in doing that, which though irreducible in themselves, create a better climate in which to work—a benefit that may well outweigh some of the other considerations.

In the above example, the comparison between the schemes was made on the basis that each of them represented the annual cost for 6 years. The equivalent annual costs were therefore comparable because the lives of the alternatives were assumed to be the same. This may not always be the case, particularly where the construction of more permanent installations is under consideration.

EXAMPLE 3.3
The board of directors of a manufacturing company has under consideration the erection of a building for the storage of finished products. They are advised that the two technically acceptable alternatives are for a reinforced concrete shell roof structure having an initial cost of £2 000 000 and for a steel-framed structure with brick cladding for an initial cost of £1 350 000. The life of the concrete building is estimated to be 60 years and, while there will be no maintenance costs for this building during the first 10 years, there will thereafter be an annual maintenance cost of £25 000. The life of the other building is estimated to be 20 years with, an equivalent annual maintenance cost from completion of construction of £30 000. The salvage value of the concrete building is estimated at £60 000 and that of the steel-framed building at £20 000. An acceptable rate of return is assessed at 10 per cent. Which is the better economic proposition?

Reinforced concrete building
See Fig. 3.4.

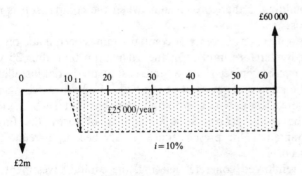

Figure 3.4

Capital recovery $= (P-S)(A\,|\,P, 10\%, 60) + Si$
$ = (2\,000\,000 - 60\,000)(0.1003) + 60\,000(0.10)$
$ = £200\,582$ per year

The sum of money at the end of year 10 equivalent to £25 000 per year from years 11 to $60 = 25\,000(P\,|\,A, 10\%, 50) = 25\,000(9.9148) = £247\,870$
Present worth of £247 870 at year $0 = 247\,870(P\,|\,F, 10\%, 10)$
$ = 247\,870(0.3856) = £95\,579$

Therefore, equivalent annual cost over 60 years of £25 000 a year from years 11 to $60 = 95\,579(A\,|\,P, 10\%, 60) = 95\,579(0.1003) = £9587$
Therefore, total equivalent annual cost $= 200\,582 + 9587 = \underline{£210\,169}$

Steel-framed building
See Fig. 3.5.

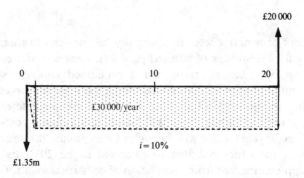

Figure 3.5

Capital recovery $= (1\,350\,000 - 20\,000)(A\,|\,P, 10\%, 20) + 20\,000(i)$
$ = 1\,330\,000(0.1175) + 20\,000(0.10) = £158\,275$ a year
Therefore total equivalent annual cost $= 158\,275 + 30\,000 = \underline{£188\,275}$

The steel-framed building is therefore cheaper when the comparison is made on the basis of annual cost.

This example raises a number of points. A comparison has been made on the basis of annual cost and it is therefore implicit in the calculation that after 20 years the steel-framed building can be replaced at the same cost as the initial installation and that replacement will continue at this cost at intervals of 20 years. Rising costs are inevitable in this context, though it is not unreasonable that such a method of comparison should be used because, in the majority of cases, the future cost increases, when discounted to the present time, quickly become a relatively small proportion of present costs.

In the case of the reinforced concrete building, the capital investment is being made now, and therefore no question of increased costs in the replacement situation arises. If the replacement cost of the steel-framed building in 20 years' time is increased by 50 per cent over the present-day cost, that is, it becomes £2 025 000, then the present worth of the increase in cost under similar conditions of interest amounts to £675 000(0.1486)=£10 305. If the second replacement cost in 40 years' time increases by 50 per cent over the first replacement cost, that is, it becomes £3 037 500, the present worth of the increase amounts to £1 012 500(0.022 09)= £22 366. These two sums produce an equivalent uniform annual cost of £32 671(0.100 32)=£3278 over the total life of the 60 years under consideration. This is a small amount when considered in relation to the other annual costs already calculated, though it should be considered when making the selection, because on this basis the steel-framed building is no longer the cheaper.

Quite apart from the financial aspects of the economic appraisal, there may be considerable advantages within many businesses from constructing buildings with a shorter life. New developments in products and building materials may enable such a company to replace the building in 20 years' time with one that gives improved performance. Replacement may well take place at a cost comparable to that of the original building or investment because of technical improvements. With the long-life building in such a situation it may be difficult to make good use of it in changed circumstances unless money is spent on its rehabilitation. This aspect becomes an irreducible factor in such a situation.

Alternatively, future costs can be estimated only by the interpretation of historic trends. Since, historically, costs have always risen continuously and steadily (with few exceptions), it seems likely that they will continue to do so. A longer-life investment is clearly advantageous in this circumstance.

EXAMPLE 3.4

In an economic assessment concerned with the alignment of a new road, one of the alternatives to be evaluated on the basis of equivalent annual cost consists of a bridge at an estimated cost of £1 000 000, an embankment costing £160 000, and other earthworks at an estimated cost of £28 000. Maintenance on the earthworks and the embankment is estimated to reach an annual cost of £20 000 over the first 4 years of its service and then to drop to £10 000 for every year thereafter. Maintenance on the bridge is expected to remain constant throughout its life at a figure of £50 000 a year. What is the total equivalent uniform annual cost of this alternative if the life of the bridge is estimated at 60 years, the life of the earthworks and the embankments is in perpetuity and the interest rate to be used is 15 per cent?

Capital recovery for the bridge
$$= 1\ 000\ 000(A\,|\,P, 15\%, 60) = 1\ 000\ 000(0.1500) \qquad\qquad = £150\ 000$$

Interest for embankment and earthworks $= 188\ 000(0.15) \qquad\qquad = \ \ 28\ 200$

Annual maintenance of the bridge $\qquad\qquad\qquad\qquad\qquad = \ \ 50\ 000$

Basic annual maintenance on the embankment and earthworks $\qquad = \ \ 10\ 000$

Equivalent annual cost of excess maintenance over first 4 years

$= 10\,000(P|A, 15\%, 4)(0.15)$

$= 10\,000(2.8550)(0.15)$ $= \underline{\qquad 4\,283}$

Total equivalent annual cost $= \underline{\underline{£242\,483}}$

The treatment of the more extensive maintenance over the first 4 years should be noted. This can be dealt with most readily by assuming a basic maintenance of £10 000 a year in perpetuity and then converting the uniform series of £10 000 a year for the first 4 years to a present worth lump sum. The interest rate can then be applied as in the case of the capital recovery for the bridge.

3.5 The present worth method

This second method of appraising alternative capital investment projects is long established and well tried. Assuming that the interest or acceptable rate of return used in both the annual cost and the present worth methods is the same, both methods of appraisal will lead to the selection, on economic grounds, of the same alternative. The present worth method is alternatively known as the *net present worth, present value* or *net present value (NPV)* method. This method is sometimes more readily applied than the equivalent annual cost method to situations where widely varying sums of money are paid out or received over a period of time. On the other hand, annual cost methods of comparison may appear to be more straightforward and may, by virtue of their method of presentation, be more easily interpreted.

The basis of the present worth method is that all future receipts or payments concerned with an investment project are converted to present worth, using an interest rate which, as previously, represents the cost of the money involved or the acceptable rate of return for that money.

EXAMPLE 3.5

The application of the present worth method to the situation of Example 3.2 is as follows:

Scheme 1

Present worth of annual labour cost over 6 years

$= 108\,000(P|A, 10\%, 6) = 108\,000(4.3552)$ $= \underline{£470\,362}$

Scheme 2

Initial cost of motor grader $= £70\,000$

Present worth of maintenance and labour costs

$= 60\,000(P|A, 10\%, 6) = 60\,000(4.3552)$ $= \underline{\quad 261\,312}$

$331\,312$

Less: Present worth of salvage value

$= 15\,000(P|F, 10\%, 6) = 15\,000(0.564\,48)$ $= \underline{\qquad 8\,467}$

Present worth of total costs $= \underline{\underline{£322\,845}}$

Scheme 3

Initial cost of first section of road	=	£60 000
Present worth of second investment		
$= 30\,000(P\,|\,F, 10\%, 2) = 30\,000(0.826\,45)$	=	24 794
Present worth of third investment		
$= 28\,000(P\,|\,F, 10\%, 4) = 28\,000(0.683\,02)$	=	19 125
Present worth of annual labour costs		
$= 48\,000(P\,|\,A, 10\%, 6) = 48\,000(4.3552)$	=	209 050
Present worth to total costs	=	£312 969

Therefore, on the basis of the above present worth evaluation the economic appraisal comes out in favour of Scheme 3 since, in effect, with the given interest rate, the whole scheme can be financed with a smaller lump sum than the other two.

In the case of Scheme 3, where there are several staged investments over the period under consideration, it will be noted that one step in the computation has been saved in considering present worth rather than equivalent annual cost methods for comparison purposes. On the other hand, all the payments for labour, for example, that are already in the convenient form for annual costs, need to be converted to a lump-sum present worth.

EXAMPLE 3.6
The present worth method can be applied to Example 3.3.

Reinforced concrete building

Initial cost of building		= £2 000 000
Less: Present worth of salvage value $= 60\,000(P\,|\,F, 10\%, 60)$		
$= 60\,000(0.0033)$	=	198
		£1 999 802

Equivalent capital value at the end of year 10 of annual
maintenance of £25 000 from year 11 to year 60
$= 25\,000(P\,|\,A, 10\%, 50) = 25\,000(9.9148) = £247\,870$
Present worth of £247 870 at the end of year 10

$= 247\,870(P\,|\,F, 10\%, 10) = 247\,870(0.3856)$	=	95 579
Present worth of total payments over 60 years		£2 095 381

Steel-framed building

Initial cost of building		= £1 350 000
Present worth of maintenance costs $= 30\,000(P\,|\,A, 10\%, 60)$		
$= 30\,000(9.9671)$	=	299 013
Present worth of renewal cost less salvage value at the		
end of 20 years $= (1\,350\,000 - 20\,000)(P\,|\,F, 10\%, 20)$		
$= 1\,330\,000(0.14865)$	=	197 705

Present worth of renewal cost less salvage value at the
end of 40 years $= 1\,330\,000(P\,|\,F, 10\%, 40)$
$= 1\,330\,000(0.022\,10)$

$$= \quad \underline{29\,393}$$
$$\underline{£1\,876\,111}$$

Less: Present worth of £20 000$(P\,|\,F, 10\%, 40)$
$= 20\,000(0.003\,28)$

$$= \quad \underline{66}$$

Present worth of total payments over 60 years $\quad = \underline{£1\,876\,045}$

This confirms the result of the analysis made by the equivalent uniform annual cost method.

In the above example, using the present worth method where the buildings have different lives, it should be noted that the comparison has to be made over a period of time that is the lowest common multiplier of the lives of the alternatives. It is therefore necessary in the case of the steel building to consider the replacement costs at the end of 20 and 40 years, together with salvage values at the end of 20, 40, and 60 years.

The present worth of the series of maintenance payments for the concrete building could have been calculated in a different way. The payments did not commence until year 11 and they continued until the end of year 60. If the factor for conversion of an annual payment to present worth for the first 10 years is subtracted from the similar factor for a 60-year period and is then multiplied by the annual amount, the same result will obtain (Note the small arithmetical error due to the rounding of the factors.)

Present worth of payments for years 11–60
$$= 25\,000[(P\,|\,A, 10\%, 60) - (P\,|\,A, 10\%, 10)]$$
$$= 25\,000(9.9671 - 6.1445) = \underline{£95\,566}.$$

Having obtained either total equivalent annual costs or total present worths, then either of these amounts can readily be converted into the other. For example, the total payments at total present worth for the concrete building can be converted to total equivalent annual cost as follows:

Equivalent annual cost $= 2\,095\,381(A\,|\,P, 10\%, 60)$
$$= 2\,095\,381(0.1003) = \underline{£210\,162}.$$

3.6 Capitalized costs

In construction works, the precise life of an asset may be very difficult to assess with accuracy. This is particularly so for such works as rail or road cuttings where an initial capital investment must be made in order to shape the natural ground and the life of the work when completed may stretch far into the future—in fact, forever. In such cases the computation of capital recovery takes a similar form to the computation of simple interest.

The capital recovery formula is as follows:

$$A = P\left[\frac{i(1+i)^n}{(1+i)^n - 1}\right]$$

It will be seen that as the value of n increases, so $(1+i)^n/[(1+i)^n-1]$ approaches 1, so that the factor by which P is multiplied approaches equality with the interest rate i. If specific values are assigned to n, such as 50, 80, and 100, then the values of the factor $(A|P, i\%, n)$ for an interest of 10 per cent are 0.100 85, 0.100 04, and 0.100 01 respectively. It will be seen that these values tend toward that of the interest rate (expressed as a decimal). Therefore, if the life of an asset is considered to be 100 years and not in perpetuity, there will be only a very small difference in the resulting calculation between using the appropriate capital recovery factor itself and using the relevant interest rate.

The term *capitalized cost* is commonly used by engineers in cases where comparisons of cost are made over periods of time in perpetuity and annual costs are assumed to be incurred on a perpetual basis. The cost of the renewal of certain assets having a definite life can be converted into a uniform series of payments as between renewals. If it is assumed therefore that the asset is renewed on each occasion that it comes to the end of its useful life, a uniform perpetual series of payments can be established.

It has already been shown above that as n approaches infinity so A approaches Pi. Capitalized cost in this example is represented by P and equals the initial cost plus the perpetual annual cost divided by the interest rate.

Where the capitalized cost of a road scheme is required and the total equivalent uniform annual cost has been calculated as £100 000, using a 10 per cent interest rate, capitalized cost can be found by dividing £100 000 by 0.10. The capitalized cost therefore becomes £1 000 000.

The capitalized cost is the present sum which will finance the initial construction costs plus an annual cost at a fixed rate of interest in perpetuity. Capitalized costs can be used as a basis for making economy studies and are considered to be a particular case of the present worth study method.

$$\boxed{\text{Capitalized cost} = P + A/i} \qquad (3.1)$$

The capitalized cost method was used extensively in construction works because many of the projects that were undertaken, such as railways, dams, sewers, etc., were considered to have perpetual life. It has been shown that there is little difference in economic studies between 100 years and perpetual life, and therefore there would appear to be some justification for this premise. Capitalized costs are still favoured in some local and central government offices because the nature of long-lived assets so often places them in the public domain. However, they have generally given way to present worth methods in view of the more accurate and realistic estimation of structural lives that is now possible.

3.7 The choice between annual cost and present worth methods

In both the equivalent annual cost and present worth methods of investment appraisal the calculations are based upon the selection of a suitable interest rate, i. The interest rate selected must equate to the rate of return required from an investment in order to justify it being made. It can be set by comparison with the return that will accrue if investment is made in a alternative project. Difficulty is often experienced in selecting a rate that gives a fair reflection of the cost of capital to an investor. In both methods an interest rate is selected for use in comparing initial capital investment and future cash flows one with the other.

It is often considered more meaningful to calculate the interest rate which equates a given or estimated set of future cash flows over the life of the project to the capital invested, and to make comparisons on this basis. The methods for doing this are discussed later in this chapter.

Equivalent uniform annual cost as a method of comparison is more readily understood and interpreted than present worth, and it particularly lends itself to the comparison or assessment of schemes involving more or less regular annual costs. Where annual costs are irregular, the method necessitates their conversion into regular cash flows by first converting to present worth. Alternatively, the irregularities may be ignored or an approximation of their uniform annual equivalent may be made. With modern computing facilities this disadvantage has all but disappeared.

Present worth methods make for ease of computation where cash flows are very irregular but, generally, the implications of the result obtained are not so readily understood. They involve the calculation of a large sum of money, especially if the studies are concerned with a long period of time. Such large sums of money can be relatively meaningless in conducting a study except for comparative purposes.

Annual cost methods are often more commonly used in very large, public-oriented industries, where future budgeted cash flows tend to be reasonably uniform, and, to a lesser extent, in the analysis of property investments.

However, the correct use of either one or the other method will lead to the same conclusion.

EXAMPLE 3.7

A civil engineering contractor operates a fleet of dumpers, and from past experience has found that a dumper normally has a useful working life of 5 years. Such a machine has an initial capital cost of £10 000 and at the end of the 5-year period has a salvage value of £800. The cost of maintenance of each dumper amounts to £3000 for the first year, and increases by £500 for each succeeding year. If the current interest rate is 12 per cent, what is the equivalent annual cost of owning and maintaining each dumper? If the contractor can sell the dumpers for £1200 each at the end of the fourth year, should he be advised to do so? (See Fig. 3.6.)

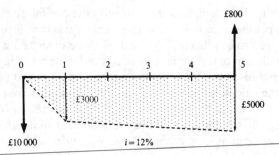

Figure 3.6

Equivalent annual cost for 5-year life:
Capital recovery $= (10\,000 - 800)(A\,|\,P, 12\%, 5) + 800(i)$
$$= 9200(0.277\,41) + 800(0.12) = 2552 + 96 = £2648 \text{ per year}$$
Equivalent annual maintenance cost $= 3000 + 500(A\,|\,G, 12\%, 5)$
$$= 3000 + 500(1.7745) = 3000 + 887 \qquad = £3887 \text{ per year}$$
Therefore, total equivalent annual cost $= 2648 + 3887 \qquad = \underline{£6535}$
See Fig. 3.7.

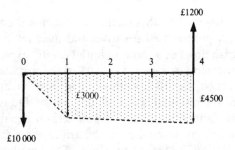

Figure 3.7

Considering a 4-year life:
Capital recovery $= (10\,000 - 1200)(A\,|\,P, 12\%, 4) + 1200(0.12)$
$$= 8800(0.329\,24) + 144 \qquad\qquad = £3014 \text{ per year}$$
Equivalent annual maintenance cost $= 3000 + 500(A\,|\,G, 12\%, 4)$
$$= 3000 + 500(1.3588) \qquad\qquad = 3679$$
Therefore, total equivalent annual cost $= 3041 + 3679 \qquad = \underline{£6720}$
It will therefore be to the contractor's benefit to keep his machines for 5 years.

3.8 Internal rate of return method (IRR)

In both the equivalent annual cost and the present worth methods of using the time value of money to make comparisons, it is necessary to assume a required rate of return before commencing the calculation that converts the capital investments and future cash flows to a comparable basis, and to establish whether a proposal will indeed bring a positive return at all. The next method to be described is one that is concerned with the calculation of the actual rate of return for a proposal, rather than commencing from the premise of an acceptable minimum required rate of return to be used as a basis for the calculations. One well-established method of doing this is by the use of what is known as the *internal rate of return method (IRR)*, which employs discounted cash flow techniques. There are a number of alternative names in common use for these methods of appraisal. Some may differ slightly in the detail of application, but all share the same underlying fundamental principle. Other names are the *investors' method*, the *profitability index*, the *interest rate of return*, the *marginal efficiency of capital*, the *yield method*, and the *discounted cash flow method*.

In Chapter 2 the mathematics of compound interest and present worth were explained. The underlying assumption in the compounding of money is that a sum of money at any time in the future will amount to more than its present value given a positive interest rate. Conversely, if one requires to exchange a sum of money that will either be paid out or received in the future, for one at the present day, then, if the payment of interest is involved, the present sum would be smaller. This latter process is known as *discounting*. Discounting is the process whereby a future sum of money is evaluated prior to the future date—it is the calculation of a *present value* or *present worth*.

Internal rate of return, in common with other methods of economic comparison, is concerned with *cash flows*, both positive and negative, and their relation to a time scale. There may be negative cash flows as a result of initial capital expenditure or as a result of the replacement, extension or renewal of part of the installation, during the course of its life. By definition, the IRR method of investment appraisal is a means of arriving at a rate of interest that will discount all the future cash flows associated with an investment, both positive and negative, into equality with the initial capital investment. Account must be taken of the arithmetic sign of the cash flows in undertaking the calculation. The basic principle of the method is that the future positive cash flows are regarded as being payments that reduce the outstanding capital outlay together with paying the interest that is accruing on the outstanding capital.

A simple example will illustrate the basic principle of IRR. If an undertaking invests the sum of £1000 today and it receives back a payment of £500 at the end of the first year, followed by a final payment of £500 at the end of the second year, then the initial investment is recovered but at a zero rate of return. If no further receipt of cash occurs as a result of this investment, then the capital has earned no return during the 2 years in which it has been locked up in the investment.

If the same company invests the sum of £1000 in another project and at the end of the first year receives a positive cash flow of £553, together with a similar sum at the end of the second year, it can be said that the investment has made a return or paid interest. More money has been received than was originally invested, although at this stage it is not possible to put a precise figure on the actual rate of return. Let it be assumed, for the time being, that the rate of return for the investment equals 7 per cent. (A means of calculating this rate will be developed later.) By the end of the first year the interest payable on the outstanding capital of £1000 amounts to £70. This amount, therefore, must be deducted from the receipt of £553 to cover the interest at the end of the first year. A net sum of £483 remains from the first year's income to pay off part of the initial capital investment, leaving an outstanding debt of £517. It is now necessary to calculate the interest that will be due at the end of the second year on the outstanding amount of £517. At 7 per cent this amounts to £36. After deducting this from the second annual payment of £553 there is a balance left of £517 which just clears the outstanding capital debt.

This example shows that the rate of return of 7 per cent is calculated only on the outstanding balance of the capital remaining invested, after making allowance for income to be set against the initial investment. This is a reasonable basis for calculation, since the capital that has been recovered from the positive cash flow can now be invested elsewhere for other purposes.

The principle by which the interest on the outstanding balance of the capital sum is calculated using the IRR method can be further demonstrated by assuming that the receipts total the same amount as before, that is £1106, but that they are received in instalments of different size. If it is assumed that a positive cash flow of £900 is received at the end of the first year, and that at the end of the second year £206 is received, then the rate of return amounts to 9 per cent rather than 7 per cent. This would not seem to be a deviation from the general principle of the time value of money because, from the first receipt of £900, a far greater proportion of the capital investment of £1000 is returned than in the first example. At the end of the first year the interest that has accrued to the capital investment of £1000 is £90, leaving £810 available to pay off part of the outstanding capital investment. The outstanding capital for the second year now amounts to £190, and the interest on this amount together with the outstanding capital sum can just be repaid by the positive net cash flow of £206 at the end of the second year.

3.9 The calculation of IRR

The following example illustrates the principles of the calculations in the IRR method applied to the cash flows for a single investment. It will subsequently be extended to deal with choices between a number of alternatives.

EXAMPLE 3.8

A company purchases a small computer for the sum of £15 000. For 5 years the computer is hired out to clients and the gross receipts are those appearing in column 2 of Table 3.2. (Receipts, or positive cash flows, are listed without a sign; negative cash flows are listed in brackets.) After 5 years, the company decides to recondition the computer at a cost of £500 and to sell it for £2000. The resale value is shown in column 2 of the table, and for clarity it is kept separate from the annual cash flows.

Table 3.2

Year	Receipts	Payments	Net cash flow
0	—	(£15 000)	(£15 000)
1	£6 000	(2 000)	4 000
2	7 500	(3 500)	4 000
3	6 500	(2 500)	4 000
4	8 000	(3 000)	5 000
5	5 000	(2 500)	2 500
6	2 000	(500)	1 500
Total	£35 000	(£29 000)	£6 000

The payments or negative cash flows which have taken place over the 5-year period are listed in column 3 of the table. The first entry in column 3, at the commencement of the period under study, is the initial cost of the computer. The five subsequent entries are concerned with payments arising from the operation of the computer, such as overheads, insurance, maintenance, repairs, etc., and the final sum in the column, £500, is the cost of the reconditioning of the computer prior to its sale. In column 4 of the table is shown the net cash flow, the result of subtracting the payments from the receipts.

The totals of the three columns indicate that the positive cash flow over the 5-year period, including the amount received for the resale of the computer, exceeds the negative cash flow by £6000. It is required to establish the IRR on this investment by the hire company.

In order to find the IRR, i, it is necessary to establish the value of the interest rate that will just equate the present worth of all of the future cash flows, both positive and negative, considered over the full period of 5 years, to the initial capital investment. The discounting of the future cash flows will need to take into account the arithmetic sign representing the direction in which the flows are taking place. Therefore, in this example, the interest rate to be established will just equate the present worth of each of the positive cash flows shown in column 4 of Table 3.2 to the initial investment of £15 000, a negative cash flow.

The IRR can only be determined by using *trial and error*. This is achieved by estimating a value of i, the required IRR. The present worth of the future net cash flows is then calculated using this value. (It obviously saves time and calculation if the initial guess is reasonably close to the true rate of return.) If the resulting calculation gives a positive answer, that is, if the discounted future cash flows sum to a larger amount than the negative initial investment, then a second, higher interest rate is chosen. This will give smaller discount factors and hence a smaller total discounted cash flow. If the net answer is now negative, the actual rate of return can be calculated by interpolation between the two assumed rates. If the second answer is still positive, a further assumption of a higher discount rate needs to be made and the calculation repeated until two interest rates are reached that give one positive net summation and one negative net summation.

Table 3.3

Year	Net cash flow	$(P\mid F, 11\%, n)$	Present worth of net cash flow	$(P\mid F, 13\%, n)$	Present worth of net cash flow
0	(£15 000)	1.0000	(£15 000)	1.0000	(£15 000)
1	4 000	0.9009	3 604	0.8850	3 540
2	4 000	0.8116	3 246	0.7831	3 132
3	4 000	0.7312	2 925	0.6931	2 772
4	5 000	0.6587	3 294	0.6133	3 067
5	4 000	0.5935	2 374	0.5428	2 171
	£6 000		£443		(£318)

In Table 3.3 the information to allow the calculation of the present worths in order to establish the rate of return is listed. In this example the single-payment present worth factors are also included for the sake of explanation.

Each cash flow is multiplied by the present worth factors relating to the particular year in which the flow is estimated to take place. Separate columns of results are created for each of the two estimated interest rates and the discounted cash flows are totalled in each. At this time it will be possible to tell from the initial totals whether the chosen interest rates are appropriate. In Table 3.3 interest rates of 11 per cent and 13 per cent have been chosen. In the case of the former, the total present worth of the future cash flows (they all happen to be positive but that is not necessarily always the case) amounts to £15 443, and in the case of an interest rate of 13 per cent the figure amounts to £14 682. One total is above the initial capital investment of £15 000; the other is below it. It is therefore possible to make a linear interpolation between the two in order to arrive at the appropriate IRR.

$$\text{Rate of return (per cent)} = 11 + \left[\frac{15\,443 - 15\,000}{15\,443 - 14\,682}\right] \quad (2)$$

$$= 11 + \frac{(443)(2)}{761} = 11 + 1.17 = \underline{12.17 \text{ per cent}}$$

This interpolation indicates that the IRR for the capital investment amounts to 12.17 per cent.

Figure 3.8 illustrates graphically the process by which the IRR is established. The curve is drawn for a series of incremental interest rates. The net present worth of the net cash flows of Table 3.2 is calculated for a range of interest rates. For interest rates less than the IRR, the net present worths are positive; where the interest rates are greater, the net present worths are negative. Where the curve intersects the zero net present worth line, the IRR can be read off on the x-axis. For minimum required

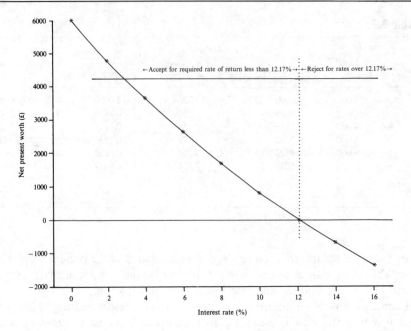

Figure 3.8

rates of return less than the calculated IRR, that is over the range where the discounted net cash flow is positive, the project is acceptable; for minimum required rates of return to the right, in excess of 12.17 per cent, and where the discounted net cash flow is negative, the investment would be unacceptable.

In the above example, the tables of cash flows could refer either to a historic transaction or to an estimated forecast for the future. The computer could have been bought 5 years ago and the cash flows could then result from having kept accounts during the course of the time period over which the study has been made. If the question is whether to purchase the computer in the first place, some estimate of the cash flows will be required before the prospective IRR can be assessed. In the case of assessing the success or otherwise of a past investment, the precise IRR is a matter of general interest only, since, from the control point of view, the opportunity of influencing profitability has gone. However, the analysis should provide meaningful data to support future purchasing decisions of a like nature.

Returning to the example of the three alternative projects, Alpha, Beta, and Gamma, portrayed by the figures in Table 3.1, between which it was required to make a choice for the investment of £20 000, the IRRs are 17.50 per cent for Project Alpha, 11.33 per cent for Project Beta, and 14.00 per cent for project Gamma. Project Alpha would therefore be selected on the basis of IRR, a different selection from that made on the basis of either the payback or return on capital selection.

Table 3.1 shows that the IRR as now calculated is considerably smaller than the rate calculated by the rule-of-thumb method of return on capital. This is because the figures listed in Table 3.1 do not make any allowance for depreciation of the asset being purchased, that is, for paying back the cost of the initial investment of £20 000. Using Project Alpha as an example, if the initial investment is deducted from the £37 000, which is the total positive cash flow from the investment over a period of 6 years, a net sum of £17 000, or an average of £2833 per year, is received. This is the net gain for the investment not taking into account any time value of money. If the average annual profit of £2833 per year is now calculated as a percentage of the initial investment, it amounts to slightly in excess of 14 per cent. On the other hand, the IRR method of calculation automatically allows for depreciation since it allows for the initial investment to be paid off from the positive cash flows over the life of the project. The interest is then calculated only on the outstanding amount of the investment. The IRR method, therefore, avoids the necessity of separately depreciating the assets throughout their lives at this, the evaluation stage. However, depreciation methods are an important influence in determining cash flows when tax is taken into account. This matter is discussed further in Chapter 4.

There are several ways to aid the initial selection of suitable interest rates, between which interpolation will take place, so as to avoid unnecessary recalculation. If the future cash flows are reasonably uniform over the life of the project, the uniform series capital recovery factor, $(A|P, i\%, n)$, can be obtained by dividing the uniform cash flow, A, by the initial investment, P. Interest tables for the capital recovery factor can then be searched for the appropriate factor at varying interest rates, along the rows for n, the life of the project, until a close approximation is obtained. If a close value of the capital recovery factor is not obtainable in the tables, two values—one larger and one smaller—can be found and interpolation between their associated interest rates can take place.

Where the future cash flows for a project are irregular, it is frequently helpful to find the average net cash flow and then follow the above procedure. Allowances can be made for concentrations of higher net cash flows in the early years of a project when the rate of return will tend to be higher (and the capital recovery factors will be larger) than that calculated for the average situation and vice versa.

Interpolation between interest rates is carried out as though the relationship between the interest factors and present worth is linear in form. This is not in fact the case and therefore a small but not overly significant error occurs in the final interest rate calculated by linear interpolation. The closer are the two interest rates between which interpolation will take place, the smaller will tend to be the error.

3.10 Calculation of yield on bonds by IRR

One method by which a company can raise money to finance future capital investment is by issuing a *bond* or *debenture*. In issuing a bond the borrower agrees to make to the lender a number of fixed-size interest payments at a stated interval. The intervals are usually regular time intervals of 6 or 12 months. In addition to making the fixed-interest payments, the borrower also contracts in the bond to repay its face value at a stated future date. If, for example, a company issues a 30-year bond with a face value of £100 at an interest rate of 6 per cent payable annually, it contracts to pay to the purchaser of the bond a sum of £6 at the end of every year for 30 years, together with the sum of £100 at the end of year 30 in repayment of the loan. In this case the *par* or *face value* of the bond is £100 and the interest rate quoted, that is, 6 per cent, is the *bond rate*. The bond rate is calculated for the payment of interest on the face value of the bond.

If the bond is sold at other than its face value, either when issued initially or at some time before redemption, the purchaser's actual yield on the investment in the bond will not be the same as the bond rate. The actual yield on the bond can be calculated by using the IRR method.

EXAMPLE 3.9

A small company wishes to raise the sum of £100 000 by issuing 25-year debentures at 5 per cent. To make the issue more attractive to the public, each £100 debenture is to be offered for sale at a price of £95. Interest on the debentures will be payable at the end of each 6 months, until the contract terminates with the repayment to the purchaser of the face value of the debenture. A purchaser of some of the debentures requires to know the yields that will be purchased in terms of a nominal interest rate.

Since the debentures are not due for repayment until the expiry of 25 years, the number of 6-monthly periods over which interest will be payable amounts to 50. For each bond of £100, £2.50 will be payable to the lender at the end of each 6-month period. The cash flow diagram over the period is a simple one. Initially there is an investment of £95, that is, a negative £95 for the purchase of one bond; there will be a positive cash flow of £2.50 for 50 periods, the final figure being coupled with one of £100, for the redemption of the bond at the end 25 years. The calculation of the present worths is facilitated if it is completed separately for the series of payments of £2.50 from the one final payment of £100. (See Fig. 3.9.)

Solve the following equation for i:
Present worth $= 95 = 2.50(P|A, i\%, 50) + 100(P|F, i\%, 50)$
Try 2.5 per cent;
Present worth $= 2.50(28.361) + 100(0.290\ 95) = 70.90 + 22.81 = £100$
Try 3 per cent:
Present worth $= 2.50(25.729) + 100(0.228\ 12) = 64.32 + 22.81 = £87.13$
By interpolation:
Interest rate $= 2.5 + 5/12.87(0.5) = \underline{2.69\ \text{per cent per 6 months}}$

The nominal annual interest rate will therefore amount to $2(2.69) = 5.38$ per cent and the effective annual rate will be $I_{\text{eff}} = (1 + 0.0269)^2 - 1 = 0.545$ or 5.45 per cent.

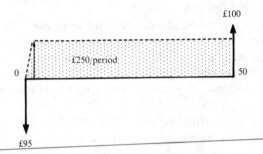

Figure 3.9

EXAMPLE 3.10
In Example 3.9 the company issues the debentures for a total face value of £100 000.
In doing so it incurs certain legal fees, commission, etc., to the extent of £5000. What
is the true cost to the company of borrowing the money in terms of an annual
interest rate? (See Fig. 3.10.)

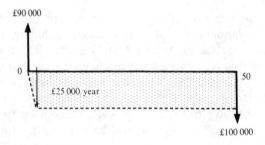

Figure 3.10

Solve the following equation for i:
Present worth $= 90\,000 = 2500(P\,|\,A, i\%,\,50) + 100\,000(P\,|\,F, i\%,\,50)$
Try 2.5 per cent:
Present worth $= 2500(28.361) + 100\,000(0.290\,95)$ $= 70\,902.50 + 29\,095$
$= £99\,997.50$

Try 3 per cent:
Present worth $= 2500(25.729) + 100\,000(0.228\,12) = 64\,322.50 + 22\,812$
$= £87\,134.50$

By interpolation:
$2.50 + 9997.50/12\,863(0.5) = 2.889$ per cent
Effective interest rate $= (1 + 0.028\,89)^2 - 1 = \underline{5.86\ \text{per cent}}$

3.11 Selection from multiple alternatives

Multiple-alternative investments may be classified as being either *mutually exclusive* or *independent*. When seeking a best way of carrying out a particular operation it is usual to look at a number of alternative solutions from which one will be selected. One best solution is sought. Mutually exclusive alternatives are those from which one only is to be selected; selecting one means the remaining alternatives are automatically rejected.

An independent alternative is selected in its own right and on its own merit, as a result of its meeting the minimum requirements that are set for selection. From a set of independent alternatives, each one that meets the minimum requirements will be implemented so long as capital is available to fund them.

With the exception of the example with the three projects, Alpha, Beta, and Gamma (Table 3.1), which was used to illustrate the principles underlying some of the rule-of-thumb methods of comparison and subsequently the IRR method, the examples of the IRR method given so far have been concerned with single projects. IRR methods can be used for determining in which one of a number of mutually exclusive projects investment should be made, rather than for the straightforward establishment of the rate of return for single investment.

EXAMPLE 3.11
Consider the case for the reorganization of a precast concrete production factory. As the factory is currently operating, the total cost of labour, including overheads, insurances, holidays with pay, etc., amounts to £200 000 a year. The manager of the factory produces two alternative schemes for mechanizing the handling and placing of the concrete into the moulds. The first scheme involves an initial investment in equipment of £100 000. This will reduce the labour cost for producing the same quantity of units to £170 170 a year. The revised labour cost takes into account the additional insurance and power that will be required for running the new scheme.

The second alternative is to invest rather more capital, £200 000, in new equipment, reducing the inclusive labour figure to £144 520 a year. The three schemes are to be compared over a period of 5 years, since this is the estimated life of the mechanical equipment. The factory manager has decided that an interest rate of 11 per cent represents a necessary minimum rate of return for any investment which is made at that particular time. The cash flows for the three schemes are listed in Table 3.4. In evaluating mutually exclusive multiple alternatives, the proposals are set out in order of increasing capital investment from left to right. Note that all cash flows in Table 3.4 are negative, that is, they are payments.

The calculation of the IRR for each condition is shown in Table 3.5.

In the first instance, Alternative B, the one with the initial investment of £100 000, can be compared with existing conditions, Alternative A. The difference in net cash flows between these two, calculated by subtracting those of B from those of A, are shown in column 2 of Table 3.5. In effect, £100 000 is being invested to save £29 830 a year for five years. The IRR amounts to 15 per cent and therefore, since this exceeds

Table 3.4

Year	Existing arrangement A	Alternative B	Alternative C
	£	£	£
0	—	100 000	200 000
1	200 000	170 170	144 520
2	200 000	170 170	144 520
3	200 000	170 170	144 520
4	200 000	170 170	144 520
5	200 000	170 170	144 520

Table 3.5

Year	Compare B with A	Compare C with B	Compare C with A
	£	£	£
0	(100 000)	(100 000)	(200 000)
1	29 830	25 650	55 480
2	29 830	25 650	55 480
3	29 830	25 650	55 480
4	29 830	25 650	55 480
5	29 830	25 650	55 480
	IRR 15%	9%	12%

the attractive rate of return required by the manager, Alternative B would be acceptable. In column 3 of Table 3.5 Alternative C is compared with Alternative B. There is an additional initial investment of £100 000 in Alternative C, with a further reduction in the annual cost of £25 650. This additional investment of £100 000 brings an IRR of slightly less than 9 per cent. This means that it is not an acceptable proposition to proceed with Alternative C, because the additional investment over and above that for Alternative B should be invested elsewhere to provide a more attractive rate of return, namely, a minimum of 11 per cent.

If Alternative C is now compared with Alternative A, for which the figures of net cash flow are listed in column 3 of Table 3.5, it will be seen that the investment of £200 000 brings a reduction in annual cost of £55 480. On computing the IRR it is found to be 12 per cent. Since this is in excess of the acceptable rate of return of 11 per cent, it would appear that there are conflicting views as to whether the company should proceed with Alternative C or not.

It can be argued that if Alternative B had not been presented, then C would automatically have been compared with existing conditions under A and the scheme of investment would have been approved. However, since Alternative B is available, it must preclude the investment of capital in Alternative C, since investing an additional £100 000 of capital to bring a IRR of 9 per cent, less than the minimum

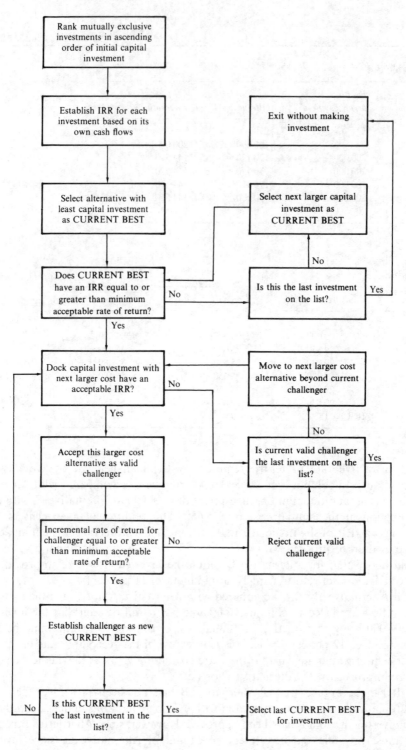

Figure 3.11

required, is not acceptable. This discussion also necessarily revolves around the objectives of the undertaking that is making the investment. It has been presumed that an undertaking's objective is to maximize the return for any investment made. However, there may be special reasons, in the context of the undertaking's circumstances, that make it desirable to invest up to a specific amount of capital, even though some elements are at less than the minimum rate of return. In this case the undertaking may wish to adjust its procedures accordingly in the knowledge that it is not, in this example, maximizing its rate of return by so doing.

Where mutually exclusive multiple alternatives are being considered, it is essential to compare them in pairs on an incremental basis. Should the increment in capital cost between one project and another not provide the minimum attractive rate of return, then the project with the higher capital investment should be rejected in favour of the alternative with the lower capital investment. Such a process of elimination finally results in the optimal investment for the capital. Figure 3.11 shows a flow diagram of the procedure to be followed in selecting one alternative from any number of mutually exclusive proposals.

EXAMPLE 3.12

Five mutually exclusive investment projects are available from which one must be chosen. The projects are V, W, X, Y, and Z. Table 3.6 lists the relevant details for each of these projects, which are ranked from left to right in increasing order of capital outlay. Line 2 of the table shows the increment in initial capital investment that is proposed as between one project and its predecessor in the table. Project W, therefore, involves capital investment of £30 000 more than Project V. Line 3 shows the uniform annual net receipts, considered over a period of 20 years, for each of the projects, and the fifth line indicates the increment in receipts as between projects. The IRR is calculated for each of the projects from the information concerning annual net receipts and the total initial capital outlay. It will be noted that the IRR for each individual project, which is shown in line 4, ranges from 12 to 14.5 per cent.

Table 3.6

	Project				
	V	W	X	Y	Z
Total capital outlay, £	150 000	180 000	200 000	230 000	270 000
Incremental capital outlay over preceeding project, £	—	30 000	20 000	30 000	40 000
Annual net receipts, £	20 000	25 000	31 000	36 000	39 000
IRR, %	12	12.5	14.5	14.25	13.75
Increase in net receipts, £	—	5 000	6 000	5 000	3 000
IRR on incremental outlay, %	—	15.75	29.8	15.75	4.0

Note: Attractive rate of return = 13 per cent.

Line 6 of Table 3.6 gives the results of calculating the IRR for the incremental annual net receipts resulting from the additional initial capital outlay of line 2. For example, the effect of an additional investment of £20 000 in Project X, over and above Project W, is to bring about an increase in annual net receipts of £6000—an IRR of almost 30 per cent when calculated on the incremental basis. For this particular example it is assumed that 13 per cent is the minimum required rate of return. This means that, if the company decides not to invest money in any of these projects, they have other sources of investment at equal risk which will return a minimum of 13 per cent.

Following the procedure as set out in Fig. 3.11, a start is made with the project having the lowest total capital outlay and it is compared with the project having the next largest capital outlay. Such a comparison is based, not only on the IRR for each individual project in its own right, but also on the IRR for the incremental capital which is invested. From the example it will be seen that neither Project V nor Project W, individually, achieves an IRR in line 4 that is equal to or greater than the desired rate of return of 13 per cent. Neither of these projects is therefore acceptable for investment.

Project X has an IRR of 14.5 per cent when considered on the merits of the initial investment and the returns which it brings, and is therefore acceptable overall. It becomes the *current best* against which subsequent *challengers* are measured. Project Y, the fourth in order of precedence, has an IRR of 14.5 per cent, and therefore it too is an acceptable investment in its own right. The additional capital investment for Project Y over Project X amounts to £30 000 and results in incremental annual net receipts of £5000 per year for the 20 years over which the study is made. The IRR for this increase in annual net receipt, considered in the light of the additional capital outlay, amounts to 15.75 per cent and, therefore, by the set standard of minimum required rate of return, it is an acceptable and worthwhile investment. Project Y becomes the current best. It should be noted that the investment of an additional £30 000 is acceptable in spite of the fact that the overall IRR (line 4) is reduced from 14.5 to 14.25 per cent. It is acceptable because 13 per cent is the minimum required rate of return and the additional £30 000 can be invested to bring in an IRR in excess of 13 per cent, namely 15.75 per cent.

Having concluded that Project Y is a better investment, on the whole, than Project X, the new defender, Project Y, becomes the standard against which to measure the value of the project having the next higher capital outlay. Project Z must now be compared with Project Y. Project Z involves the investment of an additional £40 000 of capital, which will bring in an increase in net receipts of £3000 per year for the 20-year life of the projects. The IRR for Project Z is 13.75 per cent (from line 4 of Table 3.6), and as this is in excess of the minimum required rate of return of 13 per cent, it would appear on its face value that Project Z is one worthy of investment.

If we turn now to the examination of the IRR on an incremental basis between Projects Y and Z, the rate of return has to be calculated on the basis of an increase of capital outlay of £40 000, together with an increase in net receipts of £3000 per year. This produces a rate of return on increased outlay of only 4 per cent when considered over the 20-year period. In spite of the overall rate of return of 13.75 per cent, it is obviously better to invest in Project Y and invest elsewhere the additional £40 000 which would have been required for Project Z.

Unless the question of multiple alternatives is tackled on a systematic, incremental basis rather than from an inspection of the individual returns on projects only, errors will arise in investment decisions. In the first place, by looking along line 4, it would seem that Project X provides the best investment. From the previous calculations it is known that this is not so when the minimum required rate of return is 13 per cent. If the rate of return in line 4 is the only figure which is inspected in order to make the choice, then, if the sum invested is to be maximized, Project Z would be accepted because its overall rate of return is in excess of 13 per cent. Investigation on an incremental basis shows this to be a wrong decision.

3.12 Comparison of methods for deciding between alternatives

The three principal methods of making economic comparison studies using the time value of money are the annual cost, present worth, and IRR methods. The latter two methods have much in common in their methodology. The principal difference between the rate of return method and the annual cost and present worth methods is that by using the first-named it is possible to calculate a rate of return as a percentage and then to compare it with a minimum required rate of return. With the other two methods it is necessary to assume a minimum required rate of return before commencing the calculation, without which the calculation cannot be completed.

This feature in itself leads to one of the main advantages of using rate of return methods. In attempting to assess the relative merit of investment as between one proposal and another, or between many proposals, it may not be possible beforehand to fix precisely what rate of return will be required. In the case of the rate of return method this difficulty is obviated. With the other methods it may be necessary to carry through the calculations based upon a range of interest rates to gauge what effect a relatively small change in the minimum required rate of return will have upon the overall assessment of the possible investments. In the case of a comparison between many complex alternatives, this may lead to the production of a large number and variety of cost figures and rates, which in turn may tend to complicate the overall assessment.

Further, it may be that for some reason or other, during the period in which the possible investment is being considered, the minimum required rate of return for a particular investor may change. The IRR method does not require a reassessment of the investments for this cause. Both of the other methods described require a recalculation to be made on the basis of a new minimum required rate of return, always assuming that this rate now falls outside the range of interest rates that were examined in the first assessment.

Another advantage of the IRR method over other methods is that, at the end of an investigation, a firm percentage of the likely yield is obtained. This is usually more meaningful than the results of the other calculations. A rate of return as a percentage figure needs little interpretation by a manager as compared with the assessment in present worth terms and the meaning of the final outcome taken in conjunction with its chosen interest rate.

One of the disadvantages of the rate of return method is that it does not provide a ready means of listing the projects in the order of their best return for investment. Returning to Table 3.6, we see that the best rate of return in line 4 amounts to 14 per cent for Project X. Subsequently it was shown that this was not the most satisfactory project in which the money should be invested if the minimum required rate of return was 13 per cent. It is necessary, therefore, with this method of assessing multiple alternatives to use the principles of incremental gain, and assess changes in capital investment in conjunction with the changes in net annual revenue that it brings about.

The annual cost method of comparison lends itself to comparisons where the forecast of annual expenditures and receipts will tend to be fairly uniform. If this is not the case with a proposal, more calculation is involved in using the annual cost method because of the need to convert the irregularities in the cash flows into a uniform series of flows. However, if one of the many computer programs is to be used for processing the cash flows then this is unlikely to be a serious consideration in the choice of method. Ease with which the data can be input to the program is likely to be a more significant factor.

The IRR method does have some technical disadvantages, especially when a series of future cash flows includes large negative values toward the end of the life of the project. However, situations in which they occur can readily be recognized and precautions can then be taken against obtaining a false result. The disadvantages are discussed further in Chapter 4.

In making present worth appraisals between alternative projects, the present worth values to be compared may change considerably with only a very small change in the interest rate. If, for example, the interest rate used in calculating the present worth is lower than the attractive rate of return required for an investment then the present worth method will tend to favour the alternative that has a higher initial investment. There is a tendency to use present worth methods of comparison where the investment of capital outlay in a project occurs in two or three steps at different times, since the annual cost methods makes it rather difficult to interpret the results of such a case.

Summary

The general theme of Chapter 3 has been that of how economic comparisons between engineering alternatives are made. The process of comparison is described and two traditional methods of comparison—neither involving the time value of money—are explained. Three more sophisticated methods, equivalent annual cost, present worth and internal rate of return (IRR), each using time value of money concepts, are then explained and demonstrated with examples. The capitalized cost derivative for the special case of asset lives in perpetuity is included. The merits and demerits of each method are discussed. The IRR method is applied to various aspects of raising finance by selling bonds. The importance of using incremental analysis for IRR selection of mutually exclusive projects is investigated. The chapter is rounded off by presenting a block flow diagram for incremental analysis comparison of the situations in which each method of analysis gives good results.

Problems

Problem 3.1   An excavator needs repairing at an estimated cost of £4000. If the repair is not carried out it is estimated that the operating expenses for the excavator will involve an increase of £1400 per year for each of of the next 3 years. If the minimum acceptable rate of return is 10 per cent, compare the two alternatives using the net present worth method.

(Carry out the repair; NPWs £4000 versus £6753)

Problem 3.2   A specialized piling rig is purchased by a contractor for one project only. The duration of the project is two years. The economic life of the rig is 10 years, but if it is sold at the end of the project, that is after 2 years, then the contractor will be able to get half the purchase price. If the rig costs £50 000 and the required rate of return is 10 per cent, what is the annual cost of the rig to the contractor if operating expenses are ignored?

(£16 905)

Problem 3.3   A small office building can be purchased for £120 000 and it is expected to generate an annual rental income of £15 000. General expenses in running the office building are estimated at £7000 per year. The office is said to have a resale value at the end of 5 years of £90 000. What rate of return can be expected if the office building is purchased as an investment?

(1.85 per cent)

Problem 3.4   Five projects, A, B, C, D, and E, in ascending order of investment magnitude, are mutually exclusive and are being evaluated to determine the one to be implemented. In using incremental analysis, Project D involves an additional investment of £50 000 over Project C for an increase of cash flow of £5873 per year for 20 years. If the rate of return required is 8 per cent, determine whether Project D should be implemented if, considering its own related cash flows it has a return of 12 per cent.

(Implement Project D)

Problem 3.5   A government bond has a £10 000 face value and pays interest of 5 per cent semiannually. The bond is due to mature in 20 years time. The first payment of interest is due in 6 months time. What price should be paid for this bond if it is to have a nominal yield of 10 per cent compounded semiannually?

(£5710)

Problem 3.6   A new-issue bond bearing an interest rate of 12 per cent paid semiannually, and which matures in 10 years' time, is purchased at its face-value of £5000. After the sixth dividend is received, the bond is sold to a new owner for a price sufficient to yield a total return on the original purchase price that equates to 10 per cent compounded semiannually. What was the selling price?

If the new owner redeems the bond for £5000 at its maturity, what approximate annual effective yield rate will be received?

(£4660; 14.5)

Problem 3.7   A small office building is designed to provide 10 000 square feet of usable space. If completely constructed in one phase, its estimated cost of construction is £350 000. However, the design allows the building to be constructed in two phases—the first of 6000 square feet (at an estimated cost of £230 000) and the second for the remainder (at an estimated cost of £170 000). The developer, concerned only with the first 6 years of the building's life, decides to evaluate the consequences of building the whole building initially against that of building the first phase initially and completing the second phase at the end of year 3. The required rate of return is 10 per cent. Table 3.7 shows the net revenues during this period of time. Carry out the evaluation of the two proposals and state clearly the decision that you would recommend, together with your reasons.

(Construct building in two phases)

Table 3.7

Year	Revenue—complete building	Revenue—two phases
1	£80 000	£30 000
2	90 000	50 000
3	120 000	40 000
4	120 000	100 000
5	110 000	90 000
6	100 000	110 000

Problem 3.8   A hydroelectric project, if completely developed now, will cost £70 000 000. Annual operation and maintenance charges will amount to £3 500 000 per year. Alternatively, £40 000 000 may be invested in the project now and the remainder of the work carried out in 12 years' time at a cost of £39 000 000. In this alternative case annual operation and maintenance charges will be £2 400 000 per year for the first 12 years and £4 000 000 per year thereafter. Both schemes are assumed to have perpetual life. Compare their equivalent annual costs with interest at 12 per cent.

(£11.9m; £8.812m)

Problem 3.9   In order to provide access to a factory site, a designer is faced with the problem of crossing a main railway track, using either a bridge or an underpass. The estimated costs and lives concerning both schemes are shown in Table 3.8. Compare the equivalent annual costs of each scheme with a rate of return of 10 per cent.

Check the computation for annual costs by the use of the present worth method.

(£140 198 a year; £196 130 a year)

Table 3.8

		Bridge	Underpass
1	Initial cost of structure	£600 000	£500 000
2	Initial cost of earthworks	400 000	300 000
3	Maintenance per year	40 000	100 000
4	Annual cost of pumping surface water	0	15 000
5	Life of structure	60 years	40 years
6	Life of associated works	Perpetual	Perpetual

Problem 3.10 A reinforced concrete road pavement, including the base, is laid for £7.5 per square foot. A flexible pavement to give the same service is laid for £6.75 per square foot.

The flexible pavement has major maintenance every 5 years, which costs the equivalent of 25 pence per square foot every year. The concrete pavement has a first life of 40 years, after which it is resurfaced with asphalt costing £2.35 per square foot. Thereafter it is maintained at the same cost as a flexible pavement. In addition, both types of road require annual maintenance estimated to amount to 5 pence per square foot.

On the basis of both roads giving perpetual service, compare the capitalized costs of 2000 square yards of road at an interest rate of 12 per cent.

(£145 922)

Problem 3.11 In diverting river water for an irrigation project, two alternative schemes are prepared, as follows:

Scheme 1 Open ditch and tunnel with a capital cost of £2 500 000 and an annual maintenance cost of £40 000 per year.

Scheme 2 Pipework and open flume with a capital cost of £1 750 000 and a maintenance cost of £80 000 per year, with a major replacement cost of £120 000 every 10 years.

Either of the above schemes will provide the service required. If the current interest rate is 12 per cent, compare the two schemes on the basis of capitalized cost.

(£2.833m against £2.474m)

Problem 3.12 It is required to replace the pumps that are pumping waste water from a manufacturing process to the water treatment plant. Based upon previous experience, three schemes are prepared with the objective of providing the pumping capacity for the estimated future life of the manufacturing process, which is 12 years.

Table 3.9

	Scheme 1	Scheme 2	Scheme 3
Life, years	6	3	4
Capital cost, £	25 000	70 000	100 000
Annual cost, £	12 000	2 400	1 500

The investment details as estimated are shown in Table 3.9. Evaluate each of the three schemes, using firstly net present worth and then equivalent annual cost, with a required rate of return of 10 per cent.

(Present worths: £120 875; £208 146; £225 173)

Problem 3.13   A plant to produce additives for use with high-grade concretes will cost £2 000 000 and net revenues from the plant are expected to average £300 000 per year.

In planning the plant, account is taken of a possible expansion. If the expansion is built with the original plant then its capital cost will amount to an additional £3 000 000, but as a result net revenue will increase to an average of £400 000 per year for the first 4 years of operation and will then increase further at the rate of £100 000 per year for the next 15 years before remaining constant for the foreseeable future.

The future expansion of the plant can be delayed for 5 years, but important items of equipment need to be installed at the time of construction because of restricted access. It is estimated that these items will cost £500 000. After 5 years, when the extension is also in production, net revenues are expected to increase by £150 000 per year and this gradient increase is expected to continue until the end of the the year 5, after the expansion is put into production. Thereafter, until the end of the year 20, the gradient increase is expected to reduce to £100 000 per year.

Analyse the above proposals (using a rate of return of 10 per cent where required) and make a recommendation for a future course of action, giving the reasons why.

(Postpone the investment)

Problem 3.14   A decision has to be made with regard to the installation of automatic control equipment on a concrete batching plant installed at the construction site of a power station. Quotations for the equipment show its cost to be £210 000, but its installation will have the effect of reducing annual labour costs from an estimated £100 000 to £30 000. Maintenance of the automatic plant is expected to amount to £4200 per year more than the manually controlled plant and only this excess cost need be considered in the analysis. The automatic equipment, if installed, will have a salvage value of £20 000 irrespective of the length of time it is in use. The company carrying out the work state their rate of return on capital to be 10 per cent. Will the selection of the automatic equipment for the contract with a duration of $3\frac{1}{2}$ years be justified, and what is the minimum contract period that will do this?

(No; 3.72 years)

Problem 3.15 An appraisal of three alternative, mutually exclusive projects, A, B, and C, is being made for a company that requires a return of at least 10 per cent on its invested capital. The estimated details of the investments are shown in Table 3.10.

Which investment should be recommended and why? Support your recommendation and reasoning by calculation.   (B is acceptable)

Table 3.10

	Project A	Project B	Project C
Initial cost, £	100 000	160 000	280 000
Scrap value, £	nil	nil	40 000
Net annual receipts, £	18 400	30 600	42 300
Life, years	8	8	10

Further reading

Couper, J. R. and W. H. Radar: *Applied Finance and Economic Analysis for Scientists and Engineers*, Van Nostrand Reinhold, New York, 1986, Chapter 5.

Dixon, R.: *Investment Appraisal—a Guide for Managers*, Kogan Page in association with the Chartered Institute of Management Accountants, London, 1988.

Grant, E. L., W. G. Ireson and R. S. Leavenworth: *Principles of Engineering Economy*, 8th edn, Wiley, New York, 1990, Chapters 4–7.

Pilcher, R.: *Project Cost Control in Construction*, Collins, London, 1985, Chapter 3.

Riggs, J. L. and T. M. West: *Engineering Economics*, 3rd edn, McGraw-Hill, New York, 1986, Chapters 3–7.

White, J. A., M. H. Agee and K. E. Case: *Principles of Engineering Economic Analysis*, 3rd edn, Wiley, New York, 1989, Chapters 4 and 5.

4. Influences on economic analysis

4.1 IRR—apparent technical irregularities

In Section 3.12, in reviewing the advantages and disadvantages of the various methods for the appraisal of proposed investments, reference was made to some technical disadvantages of the IRR method. There are typically two related situations in which these can possibly occur and, in doing so, may appear to contradict the findings obtained through the use of either annual cost or present worth methods of comparison, unless steps are taken to avoid misinterpretation of the results. Both situations result from particular patterns of cash flows and their sequences. Too much emphasis is often placed on these technical problems, and when this occurs the comparatively simple explanations of how to deal with them are sometimes ignored.

EXAMPLE 4.1

Two proposals for investment, Projects A and B, are mutually exclusive and have the series of cash flows over a period of 5 years and the incremental cash flows of A over B shown in Table 4.1. The net present worths of the two projects for a range of interest rates from 0–20 per cent are shown in Table 4.2 and these values are plotted in Fig. 4.1.

The IRR for Project A, that is the interest rate at which the future cash flows equate to the initial investment, is 14.14 per cent, while for Project B it is 10.73 per cent. To be able to decide between the two, since they are mutually exclusive, needs

Table 4.1

Year	Project A	Project B	Incremental cash flows
0	(£11 000)	(£10 000)	−£1 000
1	8 000	1 000	+7 000
2	2 500	1 500	+1 000
3	1 000	3 000	−2 000
4	1 500	4 000	−2 500
5	1 000	5 000	−4 000

Table 4.2

Interest rate, %	Net present worth of A	Net present worth of B
0	£3000	£4500
2	2480	3473
4	1997	2544
6	1547	1702
8	1128	936
10	736	239
12	368	−397
14	24	−979
16	−300	−1511
18	−605	−2001
20	−893	−2450

an incremental analysis as described in the previous chapter. From the graphs of net present worth against interest rate shown in Fig. 4.1, it will be seen that, for the first part of the range of interest rates (up to 6.86 per cent), Project B has the greater net present worth, and for the remainder of the range Project A has the greater net present worth. Thus, if net present worth is used as the criterion for selecting between the projects then the choice would be Project B if a minimum acceptable rate of return less than 6.86 per cent is required and Project A if a greater return is required.

The point at which the changeover takes place is that where the curves intersect at an interest rate of 6.86 per cent. This is the rate of return that would be calculated for the incremental cash flows between Projects A and B. At the same time it is the rate of return which discounts the cash flows of both projects in Example 4.1 into equality of net present worth one with the other. At this crossover point there is nothing to choose between Projects A and B from the point of view of the return.

If IRR is used as the criterion for selecting between two or more mutually exclusive projects, without using incremental analysis on the differences between successive projects, an incorrect selection may be made. In Example 4.1, without incremental analysis, Project A with an IRR of 14.14 per cent will be chosen before Project B, even though the latter has higher net present values, with minimum acceptable rates of return of less than 6.86 per cent. Incremental analysis enables this misinterpretation to be avoided and results in the same choice being made between mutually exclusive alternatives as would be made by using net present worth or annual cost. Clearly this adjustment is only required where the projects are mutually exclusive. For projects where this is not the case, selection of those that are suitable can be made on consideration of their IRR as compared with the minimum acceptable rate of return.

The second area of technical deficiency is that of *multiple returns*. The calculation of IRR for an investment proposal may result in more than one rate of return that

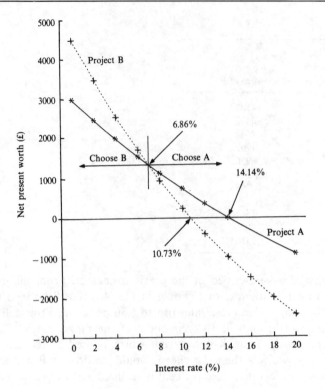

Figure 4.1

equates the future cash flows into equality with the initial investment if its cumulative cash flow switches from negative to positive or vice versa on more than one occasion through the project duration. The cumulative cash flow at any point in time of the project duration is defined as the arithmetic summation of the positive and negative cash flows, taking due regard of the signs, commencing from time zero until that point of time. When this situation occurs (though it is relatively infrequent in practice) there may be more than one IRR that discounts the cash flows to a net present worth of zero. In this case it becomes difficult to rank the investment with others having a single IRR.

A project has the cash flows illustrated in Table 4.3. Such a pattern is not uncommon, for example, in surface mining situations where there are initial costs in setting up the operation prior to mining, to be followed by costs for the reinstatement of the landscape when the mine is depleted.

The cumulative cash flow of Table 4.3 illustrates a double change of arithmetic sign from negative to positive and back to negative again. This sequence of switching signs typically results in a concave downwards curve. Figure 4.2 shows a plot of the net present worths of the cash flows for a range of interest rates. It will be noted that

Table 4.3

Year	Cash flow, £000	Cumulative cash flow, £000
0	(320)	(320)
1	250	(70)
2	250	180
3	350	530
4	300	830
5	300	1130
6	(1400)	(270)

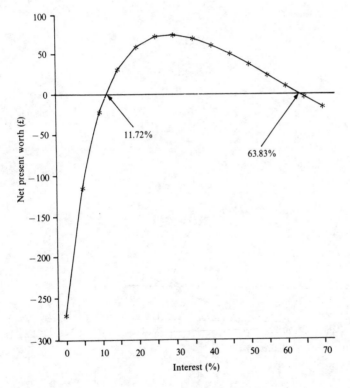

Figure 4.2

the curve passes through the zero net present worth axis twice, giving two IRRs, 11.72 per cent and 63.83 per cent.

Table 4.4 and Fig. 4.3 illustrate another series of project cash flows. In this example the switching of sign, which is the reverse of the previous example, is from positive to negative and back to positive again. The curve of net present worth

Table 4.4

Year	Cash flow, £000	Cumulative cash flow, £000
0	200	200
1	200	400
2	70	470
3	(2000)	(1530)
4	500	(1030)
5	500	(530)
6	500	(30)
7	400	370

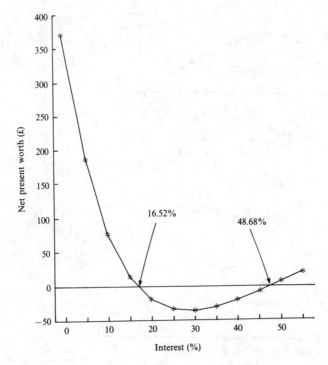

Figure 4.3

against interest rate is now concave upwards and also cuts the interest rate axis in two places.

Given the two reversals of sign for the cumulative cash flows, there will be either two points at which the curves cross the zero net present worth line or none. In the latter case the curve either peaks or bottoms out before it reaches the zero net present worth horizontal axis. The important conclusion to be drawn is the possibility of recognizing beforehand the likely existence of a situation which does not respond conventionally to the IRR method of evaluation.

4.2 External rate of return

The *external rate of return method* (ERR) has been developed in order to overcome the difficulty arising from the possibility that multiple rates of return will occur when the internal rate of return method is used. The ERR method makes the assumption that all receipts resulting from the investment are explicitly reinvested at what is the generally accepted and available rate of return. This rate of return will normally be assumed to be the minimum acceptable rate of return.

The ERR can be calculated by equating the future worth of all receipts (or positive cash flows) compounded at a reinvestment interest rate, taken to be the minimum acceptable rate of return, to the future worth of all payments (or negative cash flows) compounded at the ERR. If the interest rate for reinvested funds is i, and the ERR is i', then

$$\boxed{\begin{array}{c} \text{future worth of receipts } (+\text{ive} \\ \text{cash flows) compounded at } i \end{array} = \begin{array}{c} \text{future worth of payments } (-\text{ive} \\ \text{cash flows) compounded at } i' \end{array}} \quad (4.1)$$

There is only one external rate of return that can be found as a result of solving the above equation for a specific set of cash flows. Where on solution i' is greater than i, and i is taken to be the minimum acceptable rate of return, then the investment is attractive. Note that the value of the ERR will always fall between the values of the calculated IRR (where this is possible) and the minimum acceptable rate of return, thus resulting in the same economic choice of project whether IRR or ERR is used.

EXAMPLE 4.2

Using the cash flows set out in Table 4.3 and a minimum acceptable rate of return of 10 per cent,

$$250(F|P, 10\%, 5) + 250(F|P, 10\%, 4) + 350(F|P, 10\%, 3) + 300(F|A, 10\%, 2)$$

$$= 320(F|P, i'\%, 6) + 1400$$

$$250(1.6105) \quad + 250(1.4641) + 350(1.3310) + 300(2.0999) = 1864.47$$

$$= 320(F|P, i'\%, 6) + 1400$$

$$(1864.47 - 1400)/320 = (F|P, i'\%, 6) = 1.4515$$

Therefore,

$$i' = (1.4515 - 1.4185)/(1.5007 - 1.4185) + 6 = 6.40\%$$

and   ERR = 6.40 per cent, so the investment is not acceptable

4.3   Asset life

Brief reference was made in Chapter 3 to the economic comparison of alternatives having unequal lives. Where competing investments have different lives it is necessary to decide how to allow for this inequality in the economic analysis. There are several methods that can be adopted, but whichever one is used the underlying principle that *alternatives must be compared on an equivalent basis* must not be impaired.

The following are commonly used bases for dealing with different lives:

1. The *least common multiple method*—the least common multiple of the lives of the set of alternatives is used as the analysis period. If three assets have lives of 2, 5, and 10 years, then the study period would be 10 years—the life of the longest-lived asset. The implication of this choice of study period is that the asset with a 2-year life will be replaced 4 times and that with a 5-year life will be replaced once during the analysis period. Each replacement will result in a repeat of the cash flow characteristics determined for each asset in its first period of service. This is not an unreasonable assumption if the asset can be replaced for a cost similar to the original investment—a situation which is often enabled by improved methods of manufacture or construction. Replacement of the short-lived assets can take advantage of technological advances that have taken place since the previous purchase, thus enabling cheaper future purchase, lower annual costs and enhanced production.

This is the most commonly adopted method.

2. The *shortest life among competing alternatives*—this method insures against technological obsolescence, but it necessitates the estimation of the value of the unused portion of any asset that has a longer life than the shortest in the group under consideration. Using the example in (1) above, it is necessary to estimate the salvage or resale value of the 10-year life asset at the end of 2 years. This is clearly advantageous, however, in situations where least common multiple periods become very long because of the lives of the competing assets. In the case of two assets with lives of 7 and 11 years, a study period of 77 years would be necessary. It is likely to be more realistic to estimate the residual value of the 11-year life asset at 7 years.

3. The *longest life among competing alternatives*—this variant is seldom adopted because of the difficulty in specifying the cash flows that will take place between the end of lives of the shorter-lived and the longer-lived assets.

4. The *study period method*—if the life of the facility that is required or the length of time that it is expected to provide a service is known with reasonable accuracy, this period is adopted for the analysis. It does, however, suffer from the need to estimate salvage or resale values when the facility's useful period of service expires, and also when components of the main asset being purchased have a longer life than the period of use required of the total asset.

It is not a sound practice to assume that the use of the equivalent annual cost method precludes the need to consider the individual lives of the set of alternatives by presuming that the calculated equivalent annual cost continues for the life of the investment. If this is the case it assumes implicitly that salvage or resale values do not affect the annual worth and/or that the least common multiple method is the appropriate one to use. It is essential in the analysis stage that all features of each alternative are consciously considered in the light of the study period chosen.

4.4   Inflation

At one time inflation exerted a small and relatively insignificant effect on economic conditions; in modern economies it has become a much more telling influence and cannot be ignored in economic appraisal of investments in all but the simplest of proposals. The causes of inflation are subject to much discussion and theorizing—both economic and political. The postulated cures are many, both simple and complicated, but in themselves appear to be effective only intermittently, with no great long-term stability being established.

In simple terms, inflation is caused by an increase in the stock of money that is available for spending while the quantity of goods available for purchase does not increase by a proportionate amount. Assuming that the whole of the increase in the stock of money is exchanged for the goods that are available, the price of these goods must rise. Similarly, if the stock of money is not increased under the same circumstances, but less of it is saved, then prices will again rise. A simple example explains the theory. A construction company buys 100 000 tonnes of cement a year. If the price of that cement remains constant from year to year, the construction company (everything else being equal) has no need to raise its own prices in order to maintain its purchasing programme. If, however, the price of cement is increased by 5 per cent per year, the company has either to reduce its consumption of cement, or raise the additional funds by raising its own prices, with reference to cement content, by approximately 5 per cent. In turn, the owner of the facilities that are constructed by the company has to find the additional funding to pay the extra cost. These consequential increases are passed on by all who are subjected to them without genuinely contributing to increased wealth creation or profitability. The increased prices and costs are the result of what is known as *inflation*. Another term used in this context is that of *escalation*. Escalation covers a wider spectrum of cost changes than does inflation and includes inflation within it. Other factors covered by the term escalation include cost changes due to those of supply and demand, technological demands, and changes brought about by legislation concerning, for example, safety issues, labour employment conditions, building control regulations, etc.

The greatest difficulty of dealing with inflationary effects in economic appraisals is being able to arrive at a realistic measure of current inflation and being able to forecast what it is likely to be over the study period for the proposal. The difficulties are compounded by the fact that not all goods increase or decrease in price by a similar amount and certainly not simultaneously. Inflation rates can change very quickly. Some of these difficulties can be overcome by the use of general indices compiled from data collected by statisticians. Many of these major indices emanate from government departments. Indices are constructed from historical data and therefore are a record of what has happened in the past. They can, however, help with predicting what will happen in the future, since trends can be extrapolated from them by examining the structure of past movements in the index. Projecting costs into the future can often be a highly speculative process and historic trends can only be a very approximate guide to what will happen next.

EXAMPLE 4.3

If an index representing the price of cement increases from 231 to 287 over a period of 3 years, the annual price trend, f, is found from

$$231(1+f)^3 = 287$$
$$(1+f)^3 = 287/231 = 1.2424$$

Therefore

$$f = (1.2424)^{1/3} - 1 = \underline{0.075 \text{ or } 7.5 \text{ per cent}}$$

The effect of inflation in investment appraisal can be established in one of two ways, both of which, if used properly, will lead to the same result.

1. By establishing all cash flows in *constant*, *base-year*, *current*, or *real* pounds. Thus the effect of inflation on purchasing ability is eliminated and the *real discount rate* without a component for inflation can be used.
2. By using *future*, *then-current*, *inflated*, or *actual* pounds. Thus all the future cash flows are estimated in terms of the number of pounds actually being used, or needed, at the time of spending. In this case, since estimated actual cash flows are being used, a component for inflation must be incorporated into the minimum required rate of return when used for discounting.

Method 2 has the advantage that it tends to be easier to understand, and hence to use, than the real pound approach of method 1. The most important aspect is that whichever alternative method is chosen, it must be used consistently throughout an analysis, and real cash flows must be used only with a real discount rate or actual cash flows only with an inflation-loaded rate of return. In this text the terms *real* and *actual* will be used from the selection of terms that may be encountered in practice and are listed in (1) and (2) above.

If a government bond is purchased with, say, an interest rate of 12 per cent per year based upon the capital value of the bond, then every year the government promises to pay the 12 per cent to its legal owner or beneficiary. Assume that the bond's principal is £1000, though it may well be possible to purchase it for more or less depending on its market value at the time. The interest rate of 12 per cent is quoted as a *nominal* rate to be applied to the principal sum, rather than a *real* rate based on the varying market value. If the current inflation rate is 7 per cent per year, the real interest rate will be less than the nominal rate of 12 per cent and the real capital value of the bond will become less than the principal of £1000 as time goes by.

Suppose that the bond is purchased for £1000 at the beginning of the year, and interest of £120 together with the principal of £1000 is received back at the end of the year on the sale of the bond. Owing to inflation the £1120 received will no longer purchase the same quantity of goods at the end of the year as it would have done at the beginning. Owing to inflation, £1.07 will be required at the end of the year to provide the same purchasing power as £1.00 at the beginning of the year. The *real* value of £1120 at the end of the year is therefore £1120/1.07 = £1046.73. The original £1000 has only grown to £1046.73 in real terms and hence the real rate of return is only 46.73/1000, or approximately 4.67 per cent rather than the 12 per cent nominal rate.

Thus, because £1046.73 = 1120/1.07, it follows that

$$1000(1+d) = 1000(1+i)/(1+j)$$

where d = real interest rate
i = combined interest rate
j = rate of inflation

Hence
$$(1+d) = (1+i)/(1+j)$$

or
$$\boxed{d = [(1+i)/(1+j)] - 1} \tag{4.2}$$

and
$$i = d + j + dj$$

If all cash flows concerned with an investment are expressed in actual terms then it is necessary to use a combined discount/interest rate that includes an inflation factor when discounting. The cash flows will take into account inflationary trends and price movements for all the various components of the cash flows such as labour costs, material prices, depreciation rates, etc. Many of these components will inflate at different rates. Alternatively, real discount or interest rates can be used with cash flows that are in real terms as paragraph 1 above, though this is not the method most commonly adopted.

EXAMPLE 4.4
An investment under review has the cash flows shown in Table 4.5, estimated in real terms. The real discount rate to be used is 12 per cent and inflation is estimated to be 10 per cent per year. Compare the net present values of the project as calculated using both inflated and real terms.

(a) All cash flows in real terms:

$$NPV = -50 + 15/(1.12) + 30/(1.12)^2 + 20/(1.12)^3$$
$$= -50 + 13.39 + 23.92 + 14.24$$
$$= +£1550 \text{—acceptable}$$

(b) All cash flows converted to actual terms by multiplying real cash flows by the inflation factor, as shown in Table 4.6.

Table 4.5

Year	£000
0	− 50
1	+ 15
2	+ 30
3	+ 20

Table 4.6

Year	Real £000	Actual £000
0	− 50	− 50
1	15	16.50
2	30	36.30
3	20	26.62

Then, discount cash flows by combined interest rate,

$$i = d + j + dj = 0.12 + 0.10 + 0.12 = 0.232$$

$$\text{NPV} = -50 + 16.50/(1.232) + 36.30/(1.232)^2 + 26.62/(1.232)^3$$
$$= -50 + 0.12 + 13.39 + 23.92 + 14.24$$
$$= +£1150\text{—acceptable}$$

EXAMPLE 4.5
A machine is purchased for £2000 and the expected cash flow in actual pounds is as follows:

Year	0	1	2	3
Cash flow, £	− 2000	750	1350	950

The inflation rate is forecast to be 7 per cent over the period of time involved. The purchaser of the machine expects a minimum rate of return of 10 per cent.

Determine the combined interest rate of the purchaser and then calculate the present worth of the investment.

Establish the real cash flows to produce the same present worth when discounted at 10 per cent and show that they do so.

Combined interest rate $i = d + j + dj$ (from 4.2)

$$= 0.10 + 0.07 + (0.10 \times 0.07)$$
$$= 0.1707 \text{ or } 17.07 \text{ per cent}$$

Present worth $= -2000 + 750/(1.1707) + 1350/(1.1707)^2 + 950(1.1707)^3$
$$= -2000 + 640.64 + 985.01 + 592$$
$$= £217.65$$

$$\text{Real cash flows} = -2000 \times 1.00 = -2000$$
$$750 \times 1/1.07 = 700.93$$
$$1350 \times 1/1.07^2 = 1179.93$$
$$950 \times 1/1.07^3 = 775.48$$

$$\text{Present worth} = -2000 + 700.93/1.10 + 1179.14/1.10^2 + 775.48/1.10^3$$
$$= -2000 + 637.20 + 985.01 + 592$$
$$= \text{£217.65}$$

4.5 The cost of capital

In the use of various methods of investment appraisal, one of the elements common to all such methods has been the use of an *interest* or *discount* rate that has played a crucial role in the evaluation and decision process. The rate of return for an investment project has either been calculated directly as in the IRR method, or it has been used in the calculation to arrive at a sum of money as in the net present worth method, by which the relative merit of an investment can be judged. The correct rate of return is therefore an extremely important parameter, and the assumption of different rates of return may well cause different choices to be made from among the available alternatives.

The *cost of capital* is the cost of funds to the firm. If a company is to be successful in attracting funds then it must ensure that it provides opportunities for investors to receive returns that are at least equal to those of equivalent risk that are available to them from other sources. Almost all companies will have a mixture of funding sources and most of these sources will have a different cost and will involve different degrees of risk to an investor. Funds may be drawn from the issue of ordinary shares, borrowing from a bank, issuing loan stock, earnings from profitable operations of the company retained in the business, or sundry other less regularly available sources such as selling company assets and reducing the working capital required in the day-to-day operations. Not only do these various funding sources have different costs but the costs will vary over time and sometimes quite quickly and erratically, from no cause that can be attributed directly to the company. So too will the mix of sources tend to change, which also will have an effect on the overall cost of capital. One method of allowing for most of these factors and arriving at a company's cost of capital is now explained.

If someone is just starting up in business they may be able to borrow the initial capital that is required from their parents at a rate of interest of, say, 10 per cent per year. It can be said, ignoring for the time being the effects of taxation, that the cost of capital to the newly established entrepreneur is 10 per cent. Clearly, if the new business is going to be able to service the debt, that is, pay the interest on it when if falls due, it must earn at least that return.

A second, more sophisticated, situation might occur if the business prospers. The owner requires to raise £20 000, having no other debts since these have been repaid. The owner decides to raise three-quarters of it by selling stock in the company and taking the remainder in the form of a bank loan. In selling the stock, a dividend of 12 per cent has to be promised to potential stockholders in order to make an attractive investment; the bank is charging 16 per cent interest on its loan. The required return to pay both the dividends and interest can be calculated as shown in Table 4.7.

Table 4.7

Source	Cost		Proportion	Weighted return
Stock	0.12	×	0.75	= 0.09
Debt	0.16	×	0.25	= 0.04
	Overall return			= 0.13

The minimum overall return on the capital which must be made by the company is therefore 13 per cent. This return is referred to as the *weighted average cost of capital*. Each source of funds is *weighted* by its proportion of the total sources of funds. This principle can be extended to the total funding of a company in order to establish its cost of capital, but first it is necessary to look at the estimation of cost for various types of funding.

Ordinary shares or *common stock* are the ownership interest in a limited company and are issued by a company up to a maximum number specified in the articles of association of the company and known as *authorized share capital*. Holders of ordinary shares bear the greatest uncertainty as to the payment of dividends, since the holders of debentures and preference shareholders have prior claim to their interest (in that order). However, ordinary shareholders can exercise a greater degree of control over the company as a whole. The ordinary share dividend has no upper limit, whereas the other shares, such as debentures and preference shares, are generally issued at a fixed interest.

The return required by holders of ordinary shares is closely related to the returns for similar investments elsewhere. Ordinary shareholders generally look for rewards in two categories—one the dividend yield and the other capital appreciation in the form of an increase in price of the share.

Several models have been suggested to determine the return including the *dividend growth valuation model*:

$$P_0 = \frac{D_1}{r-g}$$

(4.3)

where P_0 = present worth of the stream of future dividends or current price of
          share
     r = required rate of return
     g = expected annual growth in dividends
     D_1 = annual dividend for next year

Formula (4.3) can be transposed to give directly the value of r:

$$r = \frac{D_1}{P_0} + g$$

(4.4)

and D_1/P_0 is known as the *dividend yield*.

It is clear that formula (4.3) can be used only if r is greater than g.

The dividend yield for groups of shares can be obtained from *The Financial Times* or other similar publications. If the value of dividend yield is 5.5 per cent and dividends are expected to grow at a rate of 12 per cent per year then the after-tax cost of capital is $5.5 + 12$ per cent = 17.5 per cent.

While the annual dividend is expected to grow at a constant annual rate, 12 per cent in this case, it may well not happen exactly like this. However, the calculation is not invalidated since the estimate of 12 per cent is intended to be a general trend and small perturbations from the trend over the years are to be expected.

There are a number of other influences on the cost of capital when it takes the form of ordinary shares. Firstly, when such capital is created by issuing the shares for subscription there will be the costs of the issue to be taken into account. This results in less capital being received by the company than the sum that is actually subscribed. These costs will generally amount to between 3 and 4 per cent of the total issued amount. Second, and more important, is the question of tax from the company's point of view. There is a crucial difference between the way that interest on debt finance and dividends on ordinary shares are treated from the point of view of the company's tax payments. Interest that is paid on debt is charged to the profit and loss account before the amount of profit is determined. Dividends that are paid on ordinary shares are not so treated and are paid out of profits after tax has been deducted. This has the effect of creating the net cost of a loan to the company at its interest rate, less corporation tax. (Corporation tax is the tax levied in the United Kingdom on the profits of companies.) If the interest on a loan is 12 per cent and corporation tax is at 35 per cent, the net-of-tax interest rate becomes $12 \times 0.65 = 7.8$ per cent. If a dividend on an ordinary share is paid at a rate of 12 per cent then the cost to the company remains at 12 per cent, because the dividend is paid out of after-tax profits.

Preferred shares are another source of funding that has much in common with loan stocks. Such stock has a nominal value on which a fixed rate of dividend is paid. The stock is *preferred* because it ranks before ordinary stock for the payment of dividends and no dividend can be paid on a company's ordinary stock until the preferred dividend has been paid in full. However, there is no obligation on the part of the directors of a company to pay a dividend on preferred stock if they consider it unjustified by the circumstances in which the company finds itself.

The dividend on preference shares is paid out of after-tax profits and therefore its current cost can be calculated as the ratio of dividend per share to the market price per share, or

$$K_p = \frac{D}{P} \qquad (4.5)$$

where K_p = cost of preferred stock
         D = annual dividends per share
         P = market price per share

Debt will be treated in general and will refer to long-term debt such as debenture stocks, loan stocks, bonds, and notes. When a long-term loan is raised, one alternative method is to state a fixed rate of interest that will be paid and the date at which repayment of the loan will be made. The given interest rate is also known as a *coupon* or *face rate* and is stated as a percentage of the *principal*, *nominal*, or *par value*. The loan stock or bond can be issued at its par value and redeemed at the same value, though other arrangements are possible. If loan stock is issued at its par value then the yield on the loan is the same as the interest rate. If, however, the loan is issued and redeemed at a *premium* (greater than par) or at a *discount* (less than par) then the effective yield needs to be calculated, as already demonstrated in Section 3.10. Alternatively, an *approximate* method of determining the yield to maturity is by using the following formula:

$$Y_m = \frac{I + (P - P_m)/n}{(P + P_m)/2} \qquad (4.6)$$

where Y_m = yield to maturity for existing loan
         I = annual interest paid, £
         P = par value of loan stock, £
         P_m = current market value of stock, £
         n = number of years to maturity

If debt capital is raised by a debenture then a fixed period of time after which the debenture will be redeemed may or may not be stated. If it is then the yield can be determined in the same way as for a bond, for example. If no date of redemption is fixed, then payments of interest will continue in perpetuity. A person wishing to purchase the debenture would do so at a market price and thus continue to receive the interest payments until the debenture is sold to another buyer. The cost of capital for such loan capital of this kind amounts to:

$$K_d = \frac{i}{P_d} \qquad (4.7)$$

where K_d = cost of debt capital
    i = interest paid
    P_d = current market price of debt capital after payment of interest

Because the interest paid on debt capital is deductible against tax, the cost of debt is usually lower than capital raised by some other means, such as ordinary or preference shares. However, much debt needs to be *secured* by assets of a company, whether defined as specific assets or assets in general. In the latter case they are known as *floating* debentures. A company that defaults against the terms of such debt is taking the risk that the providers of the debt will require the company's assets to be sold in order to repay their loans. A company must therefore spread its sources of capital over various types in order to avoid too much emphasis on such risks. When taking tax into account, formula (4.7) becomes:

$$K_d = \frac{i}{P_d}(1-t)$$

(4.8)

where K_d = cost of debt capital after tax
    i = interest paid
    P_d = current market price of debt capital after payment of interest
    t = relevant taxation rate

EXAMPLE 4.6

Hopeful Limited has a capital structure as follows:

Ordinary shares
Current market value: £500 000
Current market price per share: £40
Next year's dividend: £4
Expected annual growth in dividends: 8 per cent

Preferred stock
Current market value: £300 000
Current market price per share: £60
Annual dividend per share: £6

Debt
Current market value: £1 000 000
Interest rate: 9.5 per cent
Market price per bond: £100
Years till maturity: 18

The company's tax rate on profits is 35 per cent. Calculate the weighted average cost of capital after tax for Hopeful Limited.

Cost of ordinary shares: from (4.4), $r = 4/40 + 0.08 = 18\%$
Cost of preferred stock: from (4.5), $K_p = 6/60 = 10\%$
Cost of debt before tax: from (4.6),

$$Y_m = \frac{9.5 + (100 - 80)/18}{(100 + 80)/2} = 11.79\%$$

after tax:

$$K_d = 11.79(1 - 0.35) = 7.66\%$$

See also Table 4.8.

The cost of capital to the company is therefore 10.92 per cent. However, this must be viewed as being the minimum rate of return required, since it makes no allowance for risks taken as a result of investing. A premium needs to be added for such risks, and also an allowance for anticipated inflation.

Table 4.8

Source	Market value	After tax cost	Proportion	Weighted cost
Ordinary shares	£500 000	0.18	0.278	0.500
Preferred stock	300 000	0.10	0.167	0.0167
Debt	1 000 000	0.0766	0.555	0.0425
	£1 800 000		1.000	0.1092

4.6 Capital rationing

It has so far been assumed, in evaluating investment proposals, that capital is freely available and that a proposal has only to have a rate of return (however measured) in excess of the minimum acceptable rate of return in order to be approved and to be put in hand. However, if the availability of capital for investment has a fixed upper limit placed on it, and the sum total of the capital in the proposals put forward is in excess of the constraint level, then rationing has to take place. Only those proposals with the best returns can be funded (this does not of course apply to proposals of high social and/or environmental value, because these must be judged by other standards).

One way of illustrating this requirement is to order all the available capital investment proposals in descending order of prospective rate of return after taxes. The cumulative total of the capital investment that is required can then be established as in Table 4.9.

If the minimum acceptable rate of return for the company is 12 per cent and capital is freely available, then Projects A to G inclusive would be funded at a total cost of £6 220 000. However, if only £4 000 000 is available for investment, only Projects A to D inclusive can be funded and the minimum rate of return is then increased to 19 per cent.

This concept can alternatively be illustrated graphically, by plotting a curve of rate of return (per cent) against cumulative investment amount. This is shown in Fig. 4.4.

Table 4.9

Project	Capital investment, £000	Estimated rate of return, %	Cumulative total capital, £000
A	1000	42	1000
B	900	32	1900
C	1200	22	3100
D	800	19	3900
E	350	17	4250
F	670	16	4920
G	1300	16	6220
H	524	10	6744

The curve labelled 'Rate of return' represents prospective returns from subsequent amounts of investment where they are arranged in decreasing order of return from left to right. The second curve, 'Minimum acceptable rate of return', represents the cost of the capital available. Often when large and increasing amounts of capital are required, its cost begins to increase above its starting level as the total amount increases, so as to compensate the lender for the greater risk of a borrowing company being overloaded with debt. The intersection of the two graphs represents a cutoff point for investment capital.

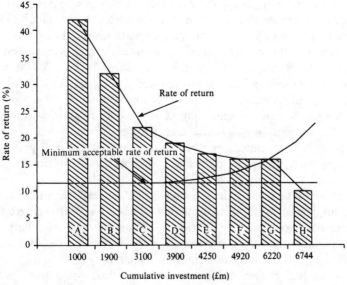

Figure 4.4

4.7 Depreciation

Depreciation, in the accounting or engineering economics sense, refers to the decrease in value of an asset (machinery, equipment, buildings and structures) which results from deterioration, wear and tear, obsolescence arising from improvements in the design and construction of new buildings and equipment, or the depletion of resources such as the mining of stone for construction purposes from a quarry. *Accounting* is a function which is concerned with the historical financial records of an enterprise. It is carried out, among other reasons, to satisfy legal requirements, including those of the taxation authority. Managers of an enterprise also require historical information in the course of their activities. In the accounting process, all expenditures and revenues of an enterprise are classified and recorded and are periodically summarized so as to present information in report form to a wide variety of recipients. From the information available in an enterprise's accounts, its overall trading position can be established at any given time. It is an important function of accounting to ensure that the cost of an asset is recovered over its production life. This is effected by apportioning the cost of the asset to the production costs of the goods produced, thus reflecting the depreciation of the capital assets in monetary terms in the prices charged for the goods when they are sold. The sales of the goods give rise to a positive cash flow. The following is a simple example to illustrate the application of these principles.

The FJK Contracting Company purchases, for £20 000, a number of steel moulds to be used for casting standard concrete beams. The company owner, F. J. King, is able, with limited facilities, to produce 1000 beams a year, which sell for £100 each. The concrete materials, labour, and other accessories cost £70 for each beam and, as a result, the available surplus is computed to be £30 per beam, leaving the owner to draw from the business a salary of £30 000 per year. After 2 years of continuous use, the steel moulds are no longer fit for use and the contractor is without any capital to replace them, having taken the whole of the immediate surplus of £30 000 a year salary. The £20 000 forming the original capital investment has been swallowed up by the use of the moulds and no depreciation allowance in the price of the beams has been made for its recovery. In fact, it can be said that £20 000/2000, or £10, was consumed by every use of a mould, and thus the true surplus was only £20 for each beam. The moulds had depreciated by their full value over the 2-year period and the contractor had made no allowance for this in the running costs of the business.

If the contractor had set aside £10 (neglecting interest) each time a mould was used, sufficient capital would have accumulated after 2 years to replace the moulds. Had this cost been considered in relation to the market and the selling price of the beams then possibly the selling price could have been raised to cover all or part of the depreciation costs, leaving the balance for salary.

If F. J. King had drawn up proper accounts this would have produced the following annual *balance sheet* and *profit and loss statement* for the first year of operations, which show clearly the effect of depreciation on the state of the King finances:

Balance sheet—FJK Contracting Company—end of year 1

Liabilities		Assets		
Capital	£20 000	Cash		£30 000
Retained earnings	20 000	Moulds	£20 000	
		less: Depreciation allowance	10 000	10 000
	£40 000			£40 000

Profit and loss statement—FJK Contracting Company—end of year 1

Receipts from sales		£100 000
less: Operating expenses		
Materials and labour	£70 000	
Depreciation of moulds	10 000	80 000
Profit from first year's operations		£20 000

It is difficult to estimate exactly the extent and rate of depreciation of an asset, since it is necessary for an asset to run for its full useful life before a true figure can be calculated. Capital costs must be paid, as a rule, before an asset can be put into production, and, in this respect, they are different from costs arising out of the use of materials and labour in production—costs that are incurred as production proceeds. However, the estimated depreciation must be considered in the same way as the many other costs which will need to be estimated before an investment in an asset is made and before economic analysis can take place.

Depreciation allowances are simply entries in the books of an undertaking and are not in themselves cash flows in the same way as the expenses of overheads, raw materials and fuel (in which money and goods or services are exchanged) are cash flows. However, taxation systems generally allow some cash allowances to be set off against income in recognition of the depreciation of capital assets, thus reducing the amount of tax to be paid where a surplus is available for this purpose. The payment of taxes does create a negative cash flow, and it is important that any thorough study in engineering economics takes tax charges and their timing into account when the relevant cash flows are established. Tax payments are normally delayed beyond the incidence of the income on which they are levied, and therefore the pattern of cash flows for an investment project, taking tax into account, can vary considerably from one in which the cash flows are considered to be gross of tax. The delays in payments of tax confer considerable benefits on the rate of return for a project when this delay in timing is taken into account.

In making allowances for depreciation in engineering economy studies, the objective is for the cash flows to reflect the benefit of any tax relief on depreciation allowances over the economic life of an asset. Indeed, there is no need to include a specific allowance for depreciation in the cash flows for the time value calculations, as described in Chapter 3, since these calculations already distribute the capital investment in each case in such a way as to represent its full repayment. The process of arriving at the appropriate tax relief through depreciation allowances involves an estimation of the useful life of an asset in order that the allowances may be distributed in a systematic and acceptable fashion. Depreciation allowances are not usually calculated by attempting to value an asset at the end of each year of its life and then deducting this value from that calculated at the end of the previous year. For simplicity, depreciation allowances are calculated in one of a number of empirical ways, so that the process becomes one of allocation rather than valuation.

Empirical methods of calculating depreciation allowances are usually adopted in accounting in order to determine the influence, through taxation, of depreciation on future cash flows. They can be divided into three broad groups. The first includes the methods that are based on writing off a fixed similar value for every year of the life of the asset. These methods are simple, both to calculate and to operate, though in practice they may not be particularly realistic in terms of the actual rate of depreciation. The second group of methods is that used to write off a greater proportion of an asset's value in its early years, with a gradually decreasing amount in the later years of its life. The third and last group represents the reverse of the procedure in the second group; the basis is to write off a smaller amount in the early years of the asset's life, compared with that written off in the later years.

The first group reflects a constant annual depreciation cost, and for simplicity can be applied to any asset depreciation. The purchase of a motor-car in private life might give rise to the use of the second form of depreciation accounting, where in the early years of a car's life the value depreciates very quickly and the amount of devaluation gradually decreases as the years go by. In the third group might be the case of the industrial building of a specialist nature which, when new, has a small depreciation; in subsequent years, when the structure deteriorates more rapidly, the depreciation increases at a greater rate.

4.8   Straight-line depreciation

Straight-line depreciation is sometimes alternatively known as the *fixed-instalment* method. The basis of the method is that the difference between the initial capital cost of an asset and the estimated salvage value at the end of its life is divided by the estimated useful life in years, to arrive at the uniform instalment of depreciation that is written off every year. The total current depreciation of the asset at any point in its life is, therefore, assumed to be directly proportional to its age. This may be expressed as follows:

if P = initial capital cost of asset
         k = number of years that the asset has been depreciated
         L = asset salvage or resale value
         n = estimated life of asset in years

$$\text{annual straight-line depreciation} = \frac{\text{initial cost} - \text{estimated salvage value}}{\text{estimated life in years}}$$

$$= (P - L)/n \qquad\qquad (4.9)$$

A concrete vibrator having an initial cost of £10 000 and an estimated salvage value of £2000 after a life of 8 years has a depreciation of (£10 000 − £2000)/8 = £1000 a year, as calculated by the straight-line depreciation method. It is often the case that the cost of an asset is to be written off completely, that is, it has no salvage value after its estimated useful life. In this case the initial capital cost is divided by the estimated useful life in years. In the case of this machine, the straight-line depreciation would be £10 000/8 = £1250 a year.

Straight-line depreciation may be calculated in terms other than those of money. It is often stated as a percentage and the depreciation can be calculated in terms of machine hours of operation. If it is estimated that a machine will work for 12 000 hours during its useful life, after which it will have to be scrapped, and its service life is estimated to be 10 years, then the depreciation can be calculated in terms of machine hours. This modification of the straight-line depreciation method (sometimes known as the *working hours method*) is not entirely satisfactory when applied to machinery used in construction work, which may have a greatly varying rate of utilization. An item of specialist construction equipment may be used continuously for a period of 6 months and then spend the next 15 months comparatively idle in a contractor's plant yard. At the end of this period the machine will probably be in good condition, but it will have a relatively small market value because of its specialist nature. However, the depreciation recorded by the working hours method will be proportionate to the comparatively small amount of time that the machine has spent working.

The example of the concrete vibrator above is used to illustrate the working hours method. If the expected useful life of the vibrator is 12 000 hours, its depreciable cost of (£10 000 − £2000) = £8000 represents an hourly depreciation rate of £8000/12 000 = £0.667/hour. If the machine operates for 2000 hours in its first year of life then the depreciation will be (2000 × £0.667) = £1334; in the second year if the machine works for 500 hours then the depreciation allowance will be £333.

It is difficult with this method to forecast the annual use of the machine and hence the depreciation charges.

Book value is a term which is used to represent the current amount of the initial cost of an asset which remains, and thus still appears on the undertaking's accounts, after deducting from the initial capital cost the total of the depreciation allowances for the life of an asset up to date. It is also referred to as *unamortized cost*, the *amortized cost* being that part of the initial cost of an asset which has already been written off in using depreciation methods during its past life.

For straight-line depreciation, book value is calculated as follows:

$$\boxed{\text{book value} = P - k[(P - L)/n]} \tag{4.10}$$

In the case of the vibrator, the book value after 4 years of use is calculated as follows:

$$\text{book value} = 10\,000 - 4[(10\,000 - 2000)/8)] = £6000$$

The advantages of straight-line depreciation are its simplicity and the fact that the depreciation instalments can be used to write off completely the initial capital cost of an asset (or the initial capital cost less any salvage or resale value) during its estimated useful life. Many of the figures to be used in calculating depreciation are estimated and, therefore, more sophisticated methods will not necessarily guarantee a more accurate forecast of what will happen. This is often quoted as being an advantage of the method.

4.9   Declining-balance depreciation

The declining-balance method is one of those methods that are used to write off the initial cost of assets at a greater rate in the early years of their life than later. This pattern of depreciation tends to follow the pattern of an asset's ability to contribute to income arising from its use. In the *declining or diminishing balance method* (also referred to as the *Matheson formula* or *constant percentage method*), a fixed percentage of the book value of an asset is written off annually. The ratio of the amount of depreciation for any year to the book value of the asset at the beginning of that year is therefore constant for every year of the asset's life.

If the declining-balance method of depreciation accounting is applied to the example used to illustrate the straight-line method, and the depreciation rate is set at 20 per cent, then the depreciation in the first year amounts to 20 per cent of £10 000 = £2000. In the second year the depreciation charge will be 20 per cent of (£10 000 − 2000) = £1600. In the third year it becomes 20 per cent of £6400 and so on until the expiry of the life of the asset.

For an asset with a known initial cost, an estimated salvage value, and an estimated service life, it is possible to calculate the percentage depreciation that will be necessary to depreciate the asset fully during its lifetime. If d = declining balance percentage expressed as a decimal, then

$$\boxed{\text{depreciation in } k\text{th year} = Pd(1-d)^{k-1}} \tag{4.11}$$

$$\boxed{\text{book value of asset after } k \text{ years} = P(1-d)^k} \tag{4.12}$$

At the end of an asset's life, the book value must be equal to the salvage or resale value L of the asset. Therefore,

$$L = P(1-d)^n$$

or

$$\boxed{d = 1 - \sqrt[n]{(L/P)}} \tag{4.13}$$

Formula (4.13) shows that the declining-balance method of depreciation cannot be used with an asset that has zero salvage value at the end of its estimated life. It is rarely needed in practice in order to set a declining balance rate. Such a rate is normally set, having a knowledge of the industry and the asset on which the depreciation is to be calculated. For example, in the depreciation of construction plant the percentage may well range from about 22 to 35 per cent. At the lower end of this range would be equipment having a relatively long life, such as concrete mixers, hoists, light cranes, etc., and at the upper end of the range would appear those units of equipment that experience fairly heavy wear during their working life, such as tracked excavators, bulldozers, shovel loaders, etc. Note that the percentage rate selected is likely to be two to three times the one that would depreciate an asset from a similar capital value to the same salvage value over the same life as by the straight-line method.

In using this method where the lives of assets are likely to be fairly long, and thus the information concerning their possible salvage values not very accurate, the calculation of a percentage rate in accordance with formula (4.13) may depend very much on the salvage or resale value that is estimated. A small variation in this estimate can lead to a large difference in the declining balance rate, as the following example illustrates.

EXAMPLE 4.7
The pumps to be installed in a sewage pumping station have an initial cost of £40 000. It is required to depreciate these pumps by the use of the declining-balance method but there is some doubt about the salvage value of the pumps at the end of their 10-year life. What difference will it make to the percentage depreciation rate if a salvage value of £1000 is used instead of £4000?

Case A
Initial cost = £40 000; salvage value = £1000; life = 10 years

$$d = 1 - (1000/40\,000)^{1/10} = 1 - 0.6915 = 0.3085 = \underline{30.85\%}$$

Case B
Initial cost = £40 000; salvage value = £4000; life = 10 years

$$d = 1 - (4000/40\,000)^{1/10} = 1 - 0.7943 = 0.2057 = \underline{20.57\%}$$

From the relatively small difference in the estimated salvage values of the two assets with otherwise identical information, it will be seen in Case A that the declining-balance rate amounts to 30.85 per cent, but in Case B it decreases sharply to 20.57 per cent.

One particular form of the declining balance method is named the *double declining-balance method*. This is an alternative from which the depreciation rate d can be determined. The formula for doing so is

$$d = \frac{200 \text{ per cent}}{\text{estimated life in years}} \tag{4.14}$$

Note that the rate d is double that for the straight-line method where an asset has no salvage or resale value. If there is a salvage or resale value it is usually disregarded in arriving at the value of the depreciation rate.

4.10 Sum-of-the-years-digits (SOYD) depreciation

This is another method that enables a greater proportion of the initial cost of an asset to be depreciated in the early years of its life. Each year's depreciation for an asset is calculated as a fraction of the total cost where there is no salvage value, or alternatively of the initial cost less the salvage value. The denominator of the fraction is formed by finding the sum of all the digits corresponding to the sequential numbers of each year of the life of the asset. For example, if the life of an asset is estimated to be 5 years, then the denominator of the fraction becomes $1+2+3+4+5=15$. Alternatively, the denominator may be established by the use of the algebraic progression equation, $\text{Sum}=n[(n+1)/2]$, where n is the life of the asset. The depreciation for each year is then calculated by using the sequential year numbers in reverse order as the numerator of the fraction. In the above example, five-fifteenths of the total value will be depreciated during the first year; for the second year it becomes four-fifteenths, for the third, three-fifteenths, and so on. In the case of the earlier example, where the cost to be depreciated was £8000 over 8 years, the first year's depreciation would be $\frac{8}{36}$ of £8000 = £1778. By similar calculation, in the second year the depreciation would amount to $\frac{7}{36}$ of £8000 = £1555.

The depreciation for any one particular year, k, of an asset's life may be found from the following expression:

$$D_k = \frac{n-(k-1)}{n(n+1)/2}(P-L) \qquad (4.15)$$

where D_k = the depreciation cost required
P = initial capital cost
L = salvage or resale value
n = total life of asset
k = sequential year number for which depreciation is required

The book value BV at the end of each year k is given by

$$BV_k = (P-L)\frac{(n-k)(n-k+1)}{n(n+1)} + L \qquad (4.16)$$

This method does enable an asset to be depreciated to zero value and it is rather easier to use than the declining-balance method, while having some of its principal advantages.

4.11 Sinking fund depreciation

There are a number of variations of the basic sinking fund depreciation method, all of which at times carry that title. As its name implies, this method involves the assumed deposit of a uniform series of end-of-year payments into a notional sinking fund at a given rate of interest. The amount of the annual deposit is calculated so that the accumulated sum at the end of the estimated life of the asset, and at the stated interest rate, will just equal that part of the value of the asset that is being depreciated. This value, as before, will be either the initial capital cost of the asset itself, or, if the asset has a salvage value, the initial capital cost less its salvage value.

The amount of the depreciation written off in any one year is the uniform end-of-year payment plus the interest charge on the amount previously accumulated in the sinking fund up to that point in time. Reverting to the original example of an asset having an initial capital cost of £10 000 and a salvage value after 8 years of £2000, if it is assumed that the interest rate is 10 per cent, then the uniform end-of-year payment into the imaginary sinking fund amounts to

$$8000(A|F, 10\%, 8) = 8000(0.0874) = £699.20 \text{ per year}$$

The amount of the uniform end-of-year payment will be the payment into the sinking fund for depreciation in the first year. In succeeding years, however, the amount of annual depreciation will increase because of the fixed percentage interest on an ever-increasing accumulation in the sinking fund to be added to each uniform end-of-year sum. Interest on the first payment will increase the depreciation set against the asset in the second year to £699.20 plus 10 per cent of the sum paid in the first year, bringing it to £769.12.

An interest payment, based upon the total sum (£699.20 + £769.12) now resting in the imaginary sinking fund, must be added to the uniform payment in order to arrive at the depreciation accounted for in the third year. In the third year it will be $[699.20 + 0.10(699.20 + 769.12)] = £846.03$. The basic assumption being made by such a method is that, if this amount of money is actually put on one side in a sinking fund, then it can be invested elsewhere so as to bring in interest at the established rate. If this is so, then the uniform end-of-year payment multiplied by the estimated life of the asset will always amount to less than the sum which has to be taken into account as depreciation. A quick check on the above example will show this to be the case. The total amount of accumulated payments plus interest at the end of any given year, k, can be found from $(P - L)(A|F, i\%, n)(F|A, i\%, k)$, where i is the interest rate for the fund.

The asset's book value at any time can be readily calculated by finding, at the given interest rate, the amount that will have accumulated in the imaginary sinking fund (including interest) and subtracting this value from the initial capital value of the asset.

EXAMPLE 4.8

Using the sinking fund method, determine the annual depreciation cost and the book value at the end of the fifth year for a truck that has an initial cost of £10 000, an estimated life of 8 years and a resale value of £2000. Interest is at 10 per cent.

Annual sinking fund deposit $= (10\,000 - 2000)(A|F, 10\%, 8)$

$$= 8000(0.0874) = £699.20$$

Sinking fund at end of year $5 = £699.20(F|A, 10\%, 5)$

$$= 699.20(6.1051) = £4268.69$$

Therefore,

Book value at end of 5 years $= £10\,000 - 4268.69$

$$= \underline{£5731.32}$$

The sinking fund method is based upon sound principles of compounding or discounting, but it does give rise to the objection that the asset's book value in the early stages of its life may well be in excess of its market value. Although the nominal amount of each year's depreciation is the same, because interest is then taken into account it appears to increase with time. For these reasons, the method is used only infrequently.

4.12 Graphical comparison of depreciation methods

Figure 4.5 illustrates, in the form of four curves, a comparison of the effect of using different depreciation methods on the book value of an asset. The four methods described above are applied to the example of an asset having an initial cost of £10 000 and a salvage value of £2000 after a life of 8 years.

The graphs clearly show the effect of each method on book value. In the case of the sinking fund method, where an interest rate of 8 per cent has been assumed, the book value is always higher than that calculated by using the straight-line method. The straight-line method can be said to be the limiting case of the sinking fund method, wherein the interest rate is zero. There is very little difference between the book values as computed by the sum-of-the-years-digits and the declining-balance methods.

Figure 4.6 illustrates the annual rate of depreciation for each of the four methods plotted against the age, in years, of the asset. This figure confirms the conclusions to be drawn from the previous graphs, inasmuch as the annual depreciation with the sinking fund method increases (if interest is taken into account) as time goes by, whereas, when both the declining balance and sum-of-the-years-digits methods are used, a comparatively large depreciation is indicated in the asset's life as well as decreasing depreciation with increasing age.

EXAMPLE 4.9

A warehouse building is constructed for an initial cost of £500 000. It is expected to have no salvage value at the end of its estimated life of 50 years. In order to handle the goods in the warehouse, mechanical equipment at the initial capital cost of

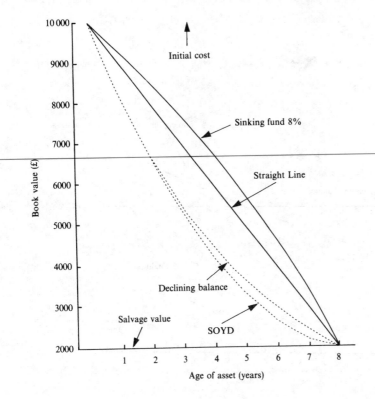

Figure 4.5

£150 000 is installed in the first instance. The estimated life of the mechanical equipment is 15 years, after which it is estimated that it will have a salvage value of £10 000. Using the straight-line, the declining-balance with a depreciation of 25 per cent, the sum-of-the-years-digits and the sinking fund methods for the calculation of depreciation, calculate the total annual depreciation charges by the end of the fifth year after construction, and the total book value for both the building and equipment at the end of the fifth year. Assume that the interest rate on a sinking fund is 10 per cent.

Straight-line method

	£
Warehouse buildings:	
Annual depreciation charge = £5000 000/50	10 000
Mechanical equipment:	
Annual depreciation charge = (150 000 − 10 000)/15	9 333
Total annual depreciation charge	£19 333

Book value after 5 years = (500 000 + 150 000) − 5(19 333)

= 650 000 − 96 665 = £533 335

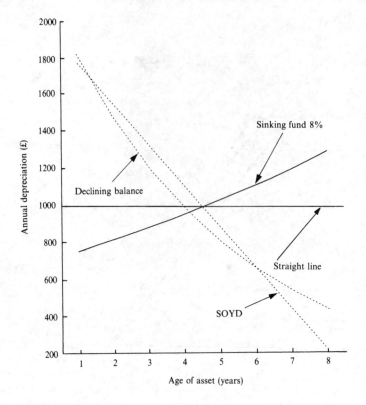

Figure 4.6

Sinking fund method

Warehouse building:

Depreciation in year $1 = £500\,000(A|F, 10\%, 50)$

$$= 500\,000(0.0009) = £450$$

At the end of year 4 the sinking fund will amount to:

$$£450(F|A, 10\%, 4) = 450(4.6410) = £2088.45$$

At the end of year 5 the sinking fund will amount to:

$$£450(F|A, 10\%, 5) = 450(6.1051) = £2747.30$$

Therefore, the amount to be charged in year 5

$$= 2747.30 - 2088.45 = \underline{£658.85}$$

Similarly, for the mechanical equipment:

Amount charged in year 5

$$= £140\,000(A\,|\,F, 10\%, 15)[(F\,|\,A, 10\%, 5) - (F\,|\,A, 10\%, 4)]$$
$$= 140\,000(0.0315)(6.1051 - 4.6410)$$
$$= £6456.68$$

Total depreciation charged in year $5 = £658.85 + 6456.68 = £7115.53$

Book value of warehouse after 5 years

$$= £500\,000 - 2747.30 = £497\,252.70$$

Book value of mechanical equipment

$$= £150\,000 - 140\,000(A\,|\,F, 10\%, 15)(F\,|\,A, 10\%, 5)$$
$$= £150\,000 - 140\,000(0.0315)(6.1051)$$
$$= £123\,076.51$$

Total book value at the end of 5 years $= £497\,252.70 + 123\,076.51$
$$= £620\,329.21$$

4.13 Taxation and its effect on cash flows

Reference has been made, in the previous sections on depreciation, to the influence of the taxation structure of a country on the cash flows for an investment project. It can happen that a decision taken as the result of an investment appraisal that ignores the effects of taxation will be different if taxation is considered as part of the analysis. While the structure of national and local taxation systems is frequently complex and the application of the systems to situations that are not routine requires expert interpretation, it is essential that the important implications of taxes on investment appraisals are understood in broad, general terms. It is not intended here to do other than treat taxation in those terms. The particular details of taxation systems in any one country change from year to year, placing the emphasis on different aspects of the economy. However, the general principles tend to remain much the same. In addition, many applications of investment appraisal will be international, and therefore subject to differing tax structures. Their application will thus require detailed knowledge of the relevant countries' tax systems.

The first aspect of taxation to be considered is the tax on profits, whether it be for an individual, for a partnership or for a corporation. In the case of the first two, the tax is known as *income* tax; in the case of a corporation or company, it is known as *corporation* tax. The following notes will deal with the influence of taxation principles on corporate investment appraisals. Tax is payable on the profits that a company makes. Profit is derived from income after certain charges and expenses are set against it and these are set out in a company's profit and loss account. The payment of this tax will have the effect of reducing the cash flows that will arise as a result of a profitable investment being made, since the company will not be able to use these tax payments for further investment.

Since corporation taxes are not normally paid until some 9 months or more after the end of a company's accounting period (currently 9 months and 1 day after the financial year end), there will be a delay between the cash flows that give rise to profits or losses and the payment of taxes where due. This is an important consideration in calculating a rate of return because it makes the relevant cash flows subject to the time value of money.

The second important way in which taxation influences investment appraisal is concerned with the *allowances* and *incentives* that are normally used to encourage capital investment. The first significant allowance is that for the depreciation of capital assets. It is essential to understand that the depreciation costs, as established by the undertaking's management (and the method chosen for its calculation), will not necessarily have any influence on the way that tax on profits is calculated. The company's managers are required to value the company's productive assets on the books at a fair value and to ensure that any loss of current value in those assets due to their use is charged against the current profits. In the profit and loss account for a company, depreciation (as calculated on a basis chosen by the managers and usually by one of the methods described earlier in this chapter), is deducted from the gross profit, together with other expenses, before arriving at the net profit. If depreciation were not deducted in this way, the profitability of the undertaking could be overstated. However, before tax is calculated on profits, this depreciation must be added back into the net profit, and it is on the figure then resulting that the tax calculation takes place. (A number of other items, such as *political contributions*, are similarly treated but depreciation is usually by far the most significant.) Depreciation shown in the profit and loss account is disallowed by the Inland Revenue and added back into the taxable profit because of the wide variety of methods and bases that can be adopted as already shown. This leads to little consistency between companies and a danger that manipulation of the calculations may take place.

To offset this disallowment of depreciation items in the profit and loss account, the Inland Revenue sets off, against profit before tax, a variety of standard capital allowances. These allowances are known as *writing-down* allowances and are stated as a percentage per year for different types of capital assets. For example, plant and machinery are currently (1991/2) written down at 25 per cent per year and industrial buildings at 4 per cent per year. The percentages are calculated on a declining balance basis. The following example illustrates the general principles.

EXAMPLE 4.10

A contractor purchases a new scraper on 1 January of year 1 for £150 000. His profits (shown in the profit and loss account) for years 1, 2, and 3 are anticipated to be £100 000, £200 000 and £220 000 respectively after charging straight-line depreciation on the assumption that the life of the scraper is 5 years and its salvage value at the end of that time will be £30 000. If the contractor pays corporation tax at 35 per cent, the writing-down allowance for the equipment is 25 per cent per year on a reducing-balance basis, and his financial accounting year ends on 31 December, estimate his tax payments, their timing, and the profits after tax arising from the purchase of the scraper, over the 3 years in question.

The capital investment for the machine is £150 000 ((a) in Table 4.10), and the annual straight-line depreciation used by the company (from formula (4.1)) is (150 000 − 30 000)/5 = £24 000 (d). With straight line depreciation it will remain at that level throughout the machine's life. Since the annual depreciation is disallowed by the Internal Revenue, it must be added back into the profit as shown in the balance sheet and therefore 'Profit + depreciation' = £100 000 + 24 000 = £124 000 (f). The writing-down allowance is calculated on the written-down value of the machine, which in the first year is its full cost, in this case £150 000. The writing-down allowance in the first year, therefore, comes to £150 000 × 0.25 = £37 500 (h) and this sum may be set off against profits. The taxable profits now become £124 000 − 37 500 = £86 500 (i) on which tax at 35 per cent amounts to £30 275 (j). The profit after tax is then £86 500 − 30 275 = £56 225 (k). The calculations for years 2 and 3 are similar. It should be noted that, since the writing down allowance for that year, is calculated on the written down value, the written down value for year 2 will be £150 000 − 37 500 = £112 500 and the writing down allowance is therefore £112 500 × 0.25 = £28 125.

Table 4.10

	Year 1 £	Year 2 £	Year 3 £
Capital value of machine	(a) 150 000		
Salvage value of machine	(b) 30 000		
Capital value less salvage	(c) 120 000		
Accounting depreciation	(d) 24 000	24 000	24 000
Accounting profit	(e) 100 000	200 000	220 000
Profit plus depreciation	(f) 124 000	224 000	244 000
Written-down value of machine	(g) 150 000	112 500	84 375
Writing-down allowance	(h) 37 500	28 125	21 094
Taxable profits	(i) 86 500	195 875	222 906
Tax	(j) 30 275	68 556	78 017
Profit after tax	(k) £56 225	£127 319	£144 889

When an item of capital plant, equipment or building is sold, there has to be a balancing-up process with the Inland Revenue, which results in a *balancing charge*. If the machine in the above example is sold at the end of year 3 for £50 000, this sum is less than the written down value of £84 375 − 21 094 = £63 281. The company's total tax relief resulting from its capital allowance claim would then be increased by £13 281 × 0.25 = £3320 in order to bring the total tax relief from the standard writing down allowances to the actual level based upon the re-sale value. Alternatively, if the machine had been sold for £70 000, the total relief resulting from its capital allowance claim would have been reduced by (£70 000 − 63 281)0.25 = £1680.

As far as the timing of the tax payments is concerned, and this is important when considering the cash flow diagram, the tax calculated on year 1 profits (£30 275) will not be paid until 1 October of year 2. It will therefore, on a year end basis, appear as a negative cash flow at the end of year 2. Similarly with tax payments for other years.

EXAMPLE 4.11

The Drillen Company is about to launch a new machine for its bored pile division. The machine has been under development for some time and it is now proposed to manufacture an operational version for £100 000. The new machine is expected to produce a real annual net income before tax of £20 000 in year 1, £35 000 in year 2, £40 000 in year 3, £30 000 in year 4 and £30 000 in year 5. At the end of year 5 the machine is expected to have no salvage value. The company has an after-tax minimum acceptable rate of return of 9 per cent and inflation is expected to be 5 per cent for the first 3 years and 7 per cent thereafter. Drillen depreciates all its equipment by the straight-line method and corporation tax is currently at the rate of 45 per cent.

Determine whether the investment in the machine is worth while.

The first column of Table 4.11 gives the time scale of the cash flows. The second column shows the expected real cash flows as listed above. In order to convert the real to the actual cash flows of column 3, each (except that at year 0) is compounded by the inflation rate of 5 per cent for each of the first 3 years and 7 per cent for the remaining 2 years where applicable. For example:

Table 4.11

Year	Before tax real cash flows, £	Before tax actual cash flows, £	Actual depreciation, £	Before tax actual after depreciation, £	Actual income tax, £	After tax actual cash flows, £	After tax real cash flows, £	Net present worth, £
1	2	3	4	5	6	7	8	9
0	− 100 000	− 100 000				− 100 000	− 100 000	− 100 000
1	20 000	21 000	20 000	1 000	450	20 550	19 571	17 955
2	35 000	38 588	20 000	18 588	8 364	30 223	27 413	23 073
3	40 000	46 305	20 000	26 305	11 837	34 468	29 775	22 991
4	30 000	37 160	20 000	17 160	7 722	29 438	23 766	16 836
5	30 000	39 761	20 000	19 761	8 892	30 869	23 291	15 137
								− 4 006

Actual cash flow, year 4 = (real cash flow, year 4)$(1.05)^3(1.07)$
$$= 30\,000(1.05)^3(1.07)$$
$$= \text{£}37\,160$$

Annual straight-line depreciation is £100 000/5 = £20 000 (as in column 4) and this is then deducted from the actual cash flows of column 3 for each year in turn. It should be noted that, for tax purposes, depreciation is calculated on the *original purchase price* of the equipment irrespective of the timing of its calculation.

Since the depreciation as calculated is an allowance, it is set against the 'before tax actual cash flows' before the amount of tax to be paid is assessed. The annual amount of depreciation in column 4 is then deducted from the equivalent annual cash flow in column 3. The result is the series of cash flows in column 5. Tax is calculated at 45 per cent of each cash flow and listed in column 6. Deducting the annual taxes of column 6 from the equivalent actual cash flows of column 3 gives rise to the series of actual cash flows in column 7. These are the cash flows in actual terms that are expected to arise after tax has been paid and from which the net present worth may be calculated. Alternatively, the cash flows can be brought back to real terms by discounting each by the appropriate single-payment present worth factor, as has been done to arrive at the figures in column 8. For example:

After-tax real cash flow, year $3 = \text{£}34\,468/(1.05)^3 = \underline{\text{£}29\,775}$

and

After-tax real cash flow, year $5 = \text{£}30\,869/(1.05)^3/(1.07)^2 = \underline{\text{£}23\,291}$

If the net present worth is to be calculated on the actual cash flows of column 7, a combined discount rate must be used:

For years 1–3, the combined interest rate, $i = 0.09 + 0.05 + (0.09)(0.05)$
$$= \underline{0.1445 \text{ or } 14.45 \text{ per cent}}$$
For years 4–5, $i = 0.09 + 0.07 + (0.09)(0.07) = \underline{0.1663 \text{ or } 16.63 \text{ per cent}}$

Therefore, in the case of year 2 the discounted cash flow of column 9 will be
$$30\,223/(1.1445)^2 = \text{£}23\,073$$

and in the case of year 5 it will be
$$\text{£}30\,869/(1.1445)^3/(1.1663)^2 = \underline{\text{£}15\,137}$$

Since the net present worth is negative, −£4006, the investment should not be made.

In the case of the after-tax real cash flows of column 8, the discount rate for calculating net present worth does not need to include a component for inflation. It will therefore be 9 per cent. Discounting the cash flows of column 8 by 9 per cent will produce the same cash flows as those listed in column 9, giving rise to the same decision as that above.

Two points related to the environment in which the problem is set should be noted from Table 4.11. Firstly, if the company is not profitable for some time during the life of the equipment, the assessment will be different. If advantage is to be taken of tax allowances there need to be net positive cash flows. Secondly, depending on the timing of the accounting and tax year of the company, some tax may be paid in the year after the related net positive cash flows are established. This should be allowed for in the table and will then result in a greater net present worth for the project.

Summary

In Chapter 3 the IRR method of investment appraisal was presented. Chapter 4 opens with a discussion of the apparent technical difficulties that may arise in the use of these methods where there are certain patterns of cash flows. Situations in which those difficulties are likely to arise are defined and explained. The *external rate of return* method is outlined as one that overcomes the possibility of obtaining multiple rates of return. Different methods of incorporating the asset life of a project as an essential component of the analysis are raised and discussed. The influence of inflation is rarely absent from matters dealing with future cash flows and the effects of this are presented giving the essential formulae and their explanation. A method of determining the cost of capital raised to finance project investments is set out, dealing with it from the point of view of the various forms of financial instruments that are used in practice to raise capital monies; this is followed by a discussion of the changes that may have to be made to acceptable rates of return in the light of a shortage of capital for investment purposes. Depreciation is an important facet to be taken into account in the establishment of cash flows prior to economic analysis taking place. Four commonly used methods of allowing for depreciation are described and this is followed by a brief description of the effect of taxation and inflation on the same cash flows.

Problems

Problem 4.1 An investment proposal has the following cash flows:

Year	0	1	2	3	4	5	6	7	8
Cash flows £000	(50)	10	15	20	18	(22)	16	10	(21)

Explain, by sketching a graph of net present worth (*y*-axis) against interest/rate of return (*x*-axis), the nature of the results you might obtain if you were to analyse the cash flows to determine the IRR. The sketch needs to be diagrammatic only; that is, it needs to be accurate in shape but calculated values are not required.

Problem 4.2 Two mutually exclusive projects are being analysed using IRR as a criterion. Explain fully, using sketches wherever possible, *why* an incorrect choice may be made if an incremental analysis is not used.

Problem 4.3 Calculate:
(a) The book value of an item of equipment at the end of the ninth year of its life if its initial cost was £10 000 and it is estimated to have a useful life of 20 years, with a salvage value of £1000. Use the straight-line method of depreciation.

(b) The depreciation in year 6 of the life of the same equipment as in (a) above by the use of the sum-of-the-years method of depreciation.

(c) The percentage depreciation and the amount of depreciation in year 14 of the life of the above asset.

(d) The book value at the end of year 10 of the above equipment's life if the sinking fund method of depreciation is used and an interest rate of 3 per cent is assumed.   (£5959; £643; 10.87 per cent; £244; £6162)

Problem 4.4   An excavator having an initial cost of £25 000 is expected to have a useful life of 5 years, after which it will have a salvage value of £5000. Calculate the percentage depreciation each year if the declining-balance method of depreciation is used.   (27.52 per cent)

Problem 4.5   A company buys a building to serve as a plant repair depot for a sum of £720 000. The remaining life of the building is expected to be 20 years. What sum must be set aside each year, with interest compounded at 12 per cent per year, in order to replace the present cost of the building at the end of its life if it is not expected to have a salvage value at that time?   (£10 008)

Problem 4.6   A machine is purchased for £100 000 and is sold after 4 years for £35 000. Calculate the book value of the machine after 3 years of its life using the following depreciation methods:

(a) straight line;
(b) sum-of-the-years digits;
(c) declining balance   (£51 250; £41 500; £45 511)

Problem 4.7   What is the purpose of using depreciation methods looked at through the eyes of an economist? Why is it unlikely that the straight-line method of depreciation will give realistic results?

Problem 4.8   A scheme for the exploitation of a small tin deposit allows for the constant recovery of the tin over a period of 8 years. If the mine originally cost £1 324 000 and a rate of return of 8 per cent on invested capital is required, what must be the annual charge plus profit allowed against the depletion of the investment?   (£230 376)

Problem 4.9   The Jimbo Hydroelectric Company is to construct a hydroelectric plant at Lakeside. The plant is expected to have a net send-out generation of 300 million kW h per year. The estimated capital and production costs are as shown in Table 4.12.

The complete scheme will be financed by borrowing £38m at an interest rate of 16 per cent per year. £10m will be borrowed at the start of construction, a further £10m one year later and the final instalment of £18m will be borrowed at the end of the second year. Interest on the loan will be calculated annually in arrears and it will be allowed to accrue during the project construction period. The loan and the

Table 4.12

Capital	£
Land and land rights	10 000 000
Structure	3 000 000
Reservoirs and dams	13 500 000
Equipment	9 500 000
Roads, railways and bridges	2 000 000
Production	
Operation supervision	86 000/year
Generation expenses	350 000/year
Maintenance and engineering	150 000/year

interest thereon will ultimately be repaid by 20 equal annual instalments, the first of which is to be paid one year after the start-up of the plant.

Construction of the project is scheduled to commence now and it is anticipated that the start-up date will be $3\frac{1}{2}$ years from now. Ignoring the effects of inflation, determine the average rate that must be charged per kW h in order that the loan may be repaid as required.

If inflation of 3 per cent per year is expected on the production costs, what does the new average charge per kW h become?   (£0.0323; £0.0331)

Problem 4.10   The directors of a company propose to invest £10 000 in modern equipment in order to bring about an estimated saving of £4000 per year over the first three years of the new equipment's life. The managing director of the company estimates that the company's present cost of capital is 12 per cent. The effect of inflation is expected to be at a rate of 3 per cent per year in the future. Explain why the investment should not be made and justify your explanation by calculation.

Problem 4.11   Inflation is at 10 per cent; the required rate of return is 10 per cent; a machine is purchased for £10 000. The real income from the machine's operations is expected to increase by 15 per cent every year from an initial level of £2500. The real operating costs for the machine are estimated to be £1200 in the first year and to increase over the life of the machine, which is 6 years, in line with the rate of inflation. The salvage value of the machine will be £2000 at the end of its useful life. Show that the machine is a good investment when taking inflation into account.

Problem 4.12   Calculate the weighted-average cost of capital for a company that has the existing capital structure shown in Table 4.13.

Note: For 'Debt' use an accurate calculation but check your answer by the approximate method.   (10.33 per cent)

Problem 4.13   In the annual capital budgeting situation it often happens that not all of the proposals for investment can be funded at the company's usual cost of capital; some may have to be funded at a higher cost of capital. Additionally, where capital is rationed, that is where there is insufficient capital to fund all of the

Table 4.13

Debt	
Market value of total bonds	£7 830 000
Market value of single bond	£900
Annual interest paid	8 per cent on principal
Years to maturity	20
Principal value of one bond	£1000
Preferred stock	
Market value of preferred stock	£500 000
Market value of one share	£85
Annual dividend per share	£7
Common stock	
Market price of stock	£18/share
Dividends expected next year	£2/share
Dividend growth expected to be	5 per cent/year
Total common stock	100 000 shares

projects with acceptable forecast rates of return, a higher average rate of return for the projects that are funded is likely to be achieved.

Explain these two situations with diagrams. Annotate the diagrams fully with a complete explanation of the diagrams you draw.

Problem 4.14 Inflation is at 10 per cent; the required rate of return is 10 per cent; a machine is purchased for £10 000. The real income from the machine's operations is expected to increase by 15 per cent every year from an initial level of £2500. The real operating costs for the machine are estimated to be £1200 in the first year and to increase over the life of the machine, which is 6 years, in line with the rate of inflation. The salvage value of the machine will be £2000 at the end of its useful life.

State the combined rate of return, taking inflation into account.

Calculate whether the machine is a good investment, taking inflation into account. (21 per cent; no)

Problem 4.15 Shaky Structures Inc is subject to 40 per cent income tax and a 10 per cent cost of capital. The company is about to buy a grader at a cost of £100 000 which, it is estimated, will reduce outgoings by £25 000 a year. The grader is expected to have a life of 10 years and, at the end of that time, will have zero salvage value.

Assume straight-line depreciation will be used and
(a) identify the relevant cash flows;
(b) compute the net present worth for the proposal;
(c) advise if this is an attractive project? ((b) £16 745.50; (c) yes)

Problem 4.16 Shaky Structures Inc has a capital structure as set out in Table 4.14. Calculate the before-tax and the after-tax cost of debt, the after-tax cost of preferred stock, and the after-tax cost of common stock, given that the company has a 50 per cent tax rate on regular income.

Table 4.14

Debt	
Market value	£2 000 000
Face interest rate	7.1 per cent
Market price per bond	£800
Principal value per bond	£1000
Years till maturity	20
Preferred stock	
Market value	£1 000 000
Market price per share	£100
Annual dividend per share	£9
Common stock	
Market value	£1 000 000
Market price per share	£80
Dividends for the coming year	£4
Expected continuous annual growth rate in dividends	7 per cent

Compute the after-tax weighted average cost of capital for Shaky Structures.

(7.5 per cent)

Problem 4.17   The WWA Corporation is preparing to purchase a new machine for £75 000. It estimates that it will have no salvage value at the end of its life, believed to be 5 years. WWA have a required minimum acceptable rate of return of 10 per cent, which includes an allowance for inflation. The company pays tax at 35 per cent and it estimates that the future annual inflation rate is expected to be 4 per cent during the next 5 years. The machine is expected to contribute £24 000 before tax per year, at today's prices, for each of the next 5 years. Should the company go ahead with the purchase?   (No)

Problem 4.18   A mining company has been awarded the mining rights for an area of forest land on condition that it replants the area with trees on completion of the mining in approximately 6 years' time. The cash flows for the project as estimated by the company are as shown in Table 4.15.

Table 4.15

Year	Cash flow, £000
0	(160)
1	90
2	120
3	170
4	150
5	100
6	(600)

Explain the likely shape of the net present worths of the cash flows for a range of interest rates to illustrate one of the irregularities that might occur in this situation.

Calculate the rate of return for the cash flows by the ERR method if funds arising from the project can be reinvested at an interest rate of 10 per cent.

(6.89 per cent)

Further reading

Couper, J. R. and W. H. Radar: *Applied Finance and Economic Analysis for Scientists and Engineers*, Van Nostrand Reinhold, New York, 1986, Chapters 3, 6 and 12.

Dixon, R.: *Investment Appraisal—a Guide for Managers*, Kogan Page in association with the Chartered Institute of Management Accountants, London, 1988.

Grant, E. L., W. G. Ireson and R. S. Leavenworth: *Principles of Engineering Economy*, 8th edn, Wiley, New York, 1990, Chapters 7–9 and 14.

Pilcher, R.: *Project Cost Control in Construction*, Collins, London, 1985, Chapter 3.

Riggs, J. L. and T. M. West: *Engineering Economics*, 3rd edn, McGraw-Hill, New York, 1986, Chapters 6, 9–11 and 13.

White, J. A., M. H. Agee and K. E. Case: *Principles of Engineering Economic Analysis*, 3rd edn, Wiley, New York, 1989, Chapters 5–7.

Merrett, A. J. and A. Sykes: *Capital Budgeting and Company Finance*, 2nd edn, Longmans, London, 1973.

Merrett, A. J. and A. Sykes: *The Finance and Analysis of Capital Projects*, Longmans, London, 1973.

Taylor, G. A.: *Managerial and Engineering Economy*, 6th edn, Van Nostrand Reinhold, New York, 1980.

5. Other economic analyses and risk

5.1 Benefit–cost analysis

In the investment appraisals described hitherto it has been assumed that the investments are being made by companies or individuals in the private sector and hence, in the majority of cases, the objective has been the maximization of return. In the public sector it is also necessary to evaluate the costs and benefits that arise from investments in monetary terms and often the analytical methods used for the private sector can be applied with only slight modification. However, many public-sector projects have different characteristics in themselves from those in the private sector.

Public-sector projects cover a particularly wide spectrum of applications and include those concerned with economic services such as the transportation infrastructure, education, cultural aspects of life such as museums, the preservation of historic aspects of the country, recreation, the protection of natural resources, and so on. These are but a few of the areas in which there is a continuous need for major investments after a thorough evaluation of their merit. While the principles of analytical methods similar to those in the private sector are usually appropriate for part of the evaluation, such projects do tend to have some characteristics which differentiate them from those of the private sector. Frequently, the projects will have a *social* cost, some examples of which are increasing the overall demand for labour, facilitating movement from one geographical location to another, creating a better living environment, adding to the cultural environment of the populace, and providing better facilities for health protection. Such benefits are generally available to a much wider section of population than those provided by the private sector.

Public-sector projects are often much larger and involve much greater investments than those of the private sector. When the investment is for a constructed facility, it tends to be for a longer life—often in perpetuity. Another facet of such investments is that neither are the benefits always immediate, nor do they accrue even in the short-term future. The total benefits, when they are realized, are not always readily evaluated in terms of success or failure in the way that a rate of return can so readily be judged by placing it on a scale of acceptability/rejection.

Benefit–cost analysis has a long history and it is used to evaluate public projects in which the *benefits* are the positive or good effects that the investments will have

142

for the public. The *disbenefits* are also evaluated, and wherever possible both are measured in monetary terms, though this is often not an easy process. Another difficulty is the establishment of a satisfactory minimum required rate of return. The general procedure for benefit–cost analysis follows very closely that used in the evaluation of private sector projects except that it will usually take a much wider view of the benefits and their associated costs. When all of the aspects of the appraisal are determined, conversion of the data enables total equivalent annual costs and benefits or the present worths of costs and benefits to be assessed. The *benefit–cost ratio* (*B/C*) or sometimes a measure of benefits less costs (*B − C*) is established. The benefit–cost ratio can be calculated by dividing the present worth of the monetary value of the benefits accruing from the project by the present worth of the total costs associated with the project. The calculation of (*B − C*) produces, in effect, a net present worth. A simple example will be used to illustrate the method, though the satisfactory use of benefit–cost analysis requires considerable experience in real life projects.

EXAMPLE 5.1

Consideration is being given to the construction of a road bypass to the village of Havitall. Presently, the main road passes through the centre of Havitall, causing serious traffic jams with their consequent delays to all traffic. Repairs are urgently needed to the main road in Havitall. A bypass will skirt the village and will require 8 miles of new road at an initial cost of £12 000 000, with renovation of the surface every 5 years at a cost of £600 000. The renovation of the existing main road through Havitall will cost £2 000 000 and subsequent annual repairs and maintenance are estimated to cost £150 000. The length of main road through Havitall affected by the scheme is 5.5 miles. Assume a period of 20 years for the evaluation and a rate of return of 10 per cent.

Using equivalent annual cost as a means of comparison:

New Bypass

$$[12\,000\,000 + 600\,000(P|F, 10\%, 5) + 600\,000(P|F, 10\%, 10)$$
$$+ 600\,000(P|F, 10\%, 15) + 600\,000(P|F, 10\%, 20)](A|P, 10\%, 20)$$
$$= [12\,000\,000 + 600\,000(0.6209 + 0.3855 + 0.2394 + 0.1486)](0.1175)$$
$$= 12\,836\,640(0.1175) = \underline{£1\,508\,305}$$

Existing

$$[2\,000\,000 + 150\,000(P|A, 10\%, 20)](0.1175)$$
$$= [2\,000\,000 + 150\,000(8.5136)]0.1175 = (3\,277\,040)0.1175$$
$$= \underline{£385\,052/\text{year}}$$

To refurbish the existing road is clearly the cheaper option on the basis of equivalent annual cost, but other factors such as benefits to the public now have to be taken into account.

Table 5.1

Class of vehicle	Number per day	Average cost £/mile
Heavy lorries	195	1.50
Medium lorries	240	1.25
Light vans	320	1.10
Motor cars	620	0.60
Motor cycles	50	0.20

A traffic survey for the existing route reveals the densities shown in Table 5.1 for various kinds of vehicles on a daily basis. The estimated cost per mile of each class of vehicle is also featured in Table 5.1.

The vehicle costs for travelling through Havitall or on the bypass can now be calculated.

Through Havitall

$$\text{Cost to public/year} = [195(1.50) + 240(1.25) + 320(1.10) + 620(0.60) + 50(0.20)]5.5(365)$$

$$= (1326.50)(5.5)(365) = \underline{£2\,662\,949}$$

For Bypass

$$\text{Cost to public/year} = (1326.50)(8)(365) = \underline{£3\,873\,380}$$

There will be a saving of journey time by using the bypass. It is estimated that commercial traffic is valued at £12.5 per hour saved and non-commercial traffic at £3 per hour saved. The estimated saving in time for different categories of traffic is shown in Table 5.2.

Table 5.2

Class of vehicle	Average saving in time, min.
Heavy lorries	20
Medium lorries	25
Light vans	25
Cars	30
Motorcycles	15

Table 5.3

	Existing	Bypass
Equivalent annual cost for 20 years	£385 052	£1 508 305
add: Public travelling costs	2 662 949	3 873 380
	3 048 001	5 381 685
Saving in cost of time	—	2 073 154
	£3 048 001	£3 308 531

Assuming that all of the trucks and lorries, together with a third of the cars but none of the motorcycles are commercial, the saving in cost by the public using the bypass is:

$$\text{Daily saving} = [195(20/60) + 240(25/60) + 320(25/60) + 207(30/60)]12.50$$
$$+ [413(30/60) + 50(15/60)]3.00$$

Saving per year $= (5022.875 + 657)365 = £2\,073\,154$

Therefore, costs of the two proposals can be summarized as shown in Table 5.3.

The annual benefit to the public may be assumed to be the difference in the cost of them using the Havitall route as compared with that for the bypass route, that is £2 662 949 − 3 873 380 + 2 073 154 = £862 723. The public will benefit by £862 723 a year by having the bypass. The incremental annual cost of the bypass amounts to £1 508 305 − 385 052 = £1 123 253.

The benefit–cost ratio = 862 723/1 123 253 = 0.768 and since this is less than 1.00, on the basis of benefit–cost ratio the bypass scheme would not be implemented.

The above example is clearly oversimplified for the sake of demonstration. Many other factors could be taken into account, particularly on the benefit side of the equation. For example, in small towns there is a serious problem of damage from the vibrations caused by large vehicles, particularly to older properties close to the line of the road. There are also problems of dust and dirt, and dangers to pedestrians, who find difficulty crossing the roads and generally using the town's facilities. Safety is also likely to be improved for traffic using a bypass with the consequent saving of lives and expense.

5.2 Break-even cost analysis

Break-even cost analysis is a form of sensitivity analysis that is very useful as a managerial production planning tool. It can, however, be used in engineering economics for making cost comparisons between two or more alternatives. The basis of the analysis concerns the variation in total cost y, as it is affected by a single common variable, x, such as the number of units of production, number of hours of use per year, amount of sales per time period, time, etc. The costs of the alternatives are frequently assumed to have a linear relationship with the chosen variable, though this is not an essential requirement of the method. Essentially, each alternative has a *fixed* cost plus one or more *variable* or *semi-variable* elements of cost. Fixed costs are those that remain at a reasonably constant level no matter what the value or percentage of the variable x. Figure 5.1 illustrates the general principles of the method. Fixed costs are alternatively known as *indirect* costs and can be typified by examples of rent or mortgage repayments for buildings, insurance, depreciation of equipment and buildings, management salaries, technical support, and the costs of general supervision. They are indirect costs because it is difficult to apportion them accurately (or directly) to a specific operation. Fixed costs are incurred over the course of time whether production or income is high, low, or non-existent, and usually it is not possible to make reductions in them that will have an immediate and quick-acting effect.

Variable or *direct* costs fluctuate with production levels or other relevant factors, and are those costs that can be directly attributed to a specific production operation or cost centre. They can include direct labour wages, materials, equipment purchased or hired for a specific operation, and direct supervision. Semi-variable costs are those that consist of two or more components of cost. These include one such as a fixed

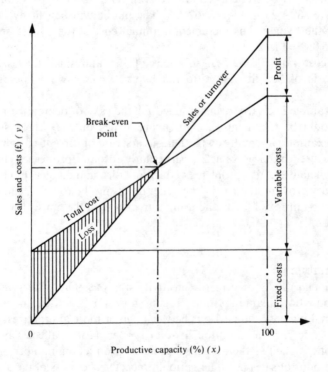

Figure 5.1

(or relatively fixed) component that remains constant and thus contributes a fixed amount to each unit of x, say production, and another varying component. An example may be that of purchasing fuel. An incentive to purchase diesel fuel from a single supplier may be promoted on the basis of paying a fixed standard charge for a minimum quantity to be purchased during a fixed period of time. The fixed standard charge becomes payable whether any or all of the minimum quantity of fuel is drawn. Thereafter the price per gallon of fuel varies on a sliding scale depending on the total amount purchased in the agreed period. The unit price paid for the fuel, therefore, becomes a semi-variable cost.

The break-even analysis displays, either mathematically or graphically, the sensitivity of the total cost to changes in the variable cost component. The value of the variable on the x-axis for the point where the cost of each alternative or pair of alternatives is the same is also established. The point at which the alternatives have the same values of x and y is known as the *break-even point*. One of the simplest and most commonly illustrated break-even charts is that relating income in the form of sales or turnover at a range of production outputs to fixed and variable costs. It shows the profit or loss for each volume of sales. Such a chart is shown in Fig. 5.1 and illustrates the variation of both sales (or turnover) and total cost of production for all percentages of the productive capacity that are achieved.

Figure 5.2 shows an alternative form of a break-even chart to that in Fig. 5.1. In Fig. 5.2 the relative positions of the fixed and variable costs are reversed. This is useful in illustrating the extent to which each level of production contributes to the sum of the fixed costs and profit—the combination of which is known as the *contribution*. Alternatively, contribution may be defined as the cost difference

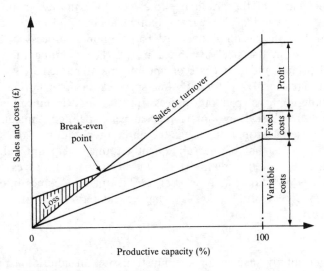

Figure 5.2

between sales or turnover and the total variable costs for a specific percentage output of total capacity.

Break-even analysis is useful when one alternative solution to a problem is more acceptable under a given set of conditions, for example, a certain range of values for the variable on the x-axis, than another. The choice of solution might well change again for yet another different set of conditions. Where the suitability of one alternative changes to that of another is the break-even point, at which both of the alternatives are equally acceptable from the point of view of the parameter being used for comparison—usually cost. This type of economic analysis is often very useful in considering preliminary engineering designs, for example, the arrangement and spacing of the structural components, such as reinforced concrete colums and beams of a building.

The solution of problems by break-even methods can be carried out either mathematically or graphically. However, where the relationships between the value of x and, say, costs are relatively simple and straightforward, a mathematical solution is often easier. The method is illustrated in Examples 5.2, 5.3, and 5.4 below. Alternatively, there will be many cases where it is not easy to relate the variables by a mathematical equation because the relationship is neither continuous nor a relatively simple one. In such cases it is usually quicker and more reliable to use a graphical method of presentation and solution. In addition, there is always a danger that mathematical formulation may be interpolated by persons who, for one reason or another, do not know or fully understand the original concepts and qualifications behind its development. Graphical solutions are almost always easier to understand and to interpret.

An example of a graphical break-even cost analysis is illustrated by Fig. 5.3. It concerns the comparison of cost between two types of painting for structural steelwork. On the y-axis of the graph the total cost of painting, in pounds per square metre, over a period of 35 years is plotted against time in years on the x-axis. Treatment 1 is the accumulative cost based upon a process of removing the loose mill-scale from the steel by abrasion and then painting the steelwork with one coat of primer and two coats of finishing paint immediately and every 5 years thereafter. Treatment 2 is for the same specification for painting, but with the steelwork undergoing a pickling process in the first instance, instead of the removal of loose mill-scale by abrasion, then being painted immediately and every 8 years. The fixed cost in each case is the initial preparation of the structural steel; the variable cost is that of the subsequent treatment. Up until about year 17 after commencement of the treatment routine, Treatment 1 proves to be the cheaper on the basis of total cost. After year 17 Treatment 2 comes into its own and if a period in excess of 17 years is being considered it will always be the cheaper in total cost.

EXAMPLE 5.2

As part of a water supply scheme for an industrial site, an additional pipeline is required. It will be 750 m long. If a pipe of 200 mm diameter is installed at a capital

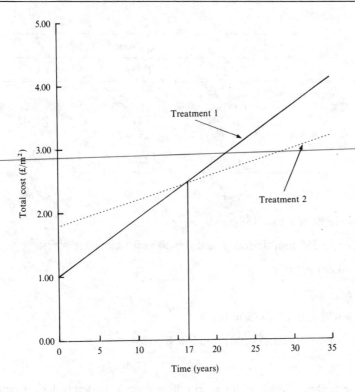

Figure 5.3

cost of £65 000, the cost of pumping water will be £0.52 per hour. If a 250-mm-diameter pipe is used the pumping cost will fall to £0.26 per hour in addition to a capital cost of £70 000. A 300-mm-diameter pipe will cost £75 000 and have a pumping cost of £0.10 per hour. The useful life of the pipeline is expected to be 20 years, after which there will be no salvage value. Interest is at a rate of 12 per cent per year. Determine the most economical size of pipe when it is expected that 5000 hours of pumping will be required each year. Also, determine what length of pumping time each year would make the annual cost of other sizes of pipe comparable with that of the 300-mm-diameter pipe. Draw the break-even chart for the investigation and determine during what periods each pipe will prove the most economical installation with up to 5000 hours of pumping each year.

See Table 5.4: the 300-mm-diameter pipe will be cheapest for 5000 hours per year pumping.

Let the hours of pumping for equal cost with the 300-mm-diameter pipe = x.

(a) Compare the 200-mm-diameter and the 300-mm-diameter pipes:

$$8704 + 0.52x = 10\,043 + 0.10x$$

Table 5.4

	Diameter of pipe		
	200 mm	250 mm	300 mm
Annual capital recovery cost of pipeline for $i = 12\%$ and $n = 20$	0.1339(65 000) =£8 704	0.1339(70 000) =£9 373	0.1339(75 000) =£10 043
Cost of 5000 hours' pumping	£2 600	£1 300	£500
Total annual cost for 5000 hours' pumping	£11 304	£10 673	£10 543

therefore

$0.42x = 1339$ and $x = 3188$ hours

(b) Compare the 250-mm-diameter and the 300-mm-diameter pipes:

$9373 + 0.26x = 10 043 + 0.10x$

therefore

$0.16x = 670$ and $x = 4188$ hours

The break-even chart is illustrated in Fig. 5.4.

EXAMPLE 5.3

A contractor builds a small plant to produce precast concrete facing panels. The plant is set up with a view to selling concrete products to the open market. The following facts are relevant:

   Annual capacity of plant: 1000 panels
   Selling price of each panel: £300
   Initial cost of plant: £90 000
   Fixed operating expenses: £50 000 per year
   Cost of labour per unit product: £120
   Cost of materials per unit product: £100

The plant is the subject of a mortgage to be repaid over 20 years at a fixed rate of interest of 15 per cent (this is not included in the fixed costs above). Draw a break-even chart to relate the number of panel sales to annual profitability.

Annual cost of mortgage $= £90\ 000(A \,|\, P, 15\%, 20)$

$= 90\ 000(0.1703)$

$= £15\ 327$

Total annual fixed costs $= £15\ 327 + 50\ 000$

$= £65\ 327$

Variable cost per panel $= £120 + 100$

$= \underline{£220}$

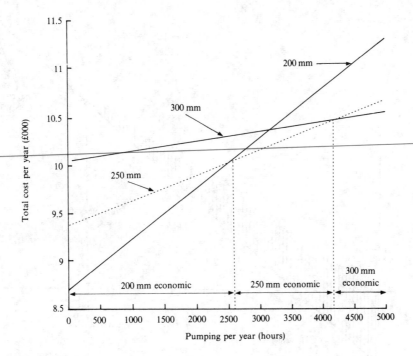

Figure 5.4

Figure 5.5 shows the standard form of the break-even chart. The break-even point occurs at the intersection of the total (fixed + variable) cost and the revenue graphs. The significance of this point is that it shows the number of panels that need to be sold for the operation to break even between cost and income. Up to this point the vertical distance between the two curves shows the extent of the loss being made; to the right of the break-even point the distance between the two curves shows the profit being made for each production output up to the maximum of 1000.

Output at break-even = revenue − total cost = 0

= n(revenue per panel − variable cost per panel) − fixed cost

where n = number of panels produced.

Therefore, at break-even

$$0 = n(300 - 200) - 65\,327 \qquad \text{or} \qquad n = 65\,327/80 = \underline{817 \text{ panels}}$$

To achieve a break-even point at as low a level of units of output as possible is highly desirable because it means that the company will meet its contribution target at a lower production of panels. The alternative break-even chart is illustrated in Fig. 5.6.

By reducing the fixed cost by £10 000 (Fig. 5.6), the total cost is reduced by a similar amount. For the same revenue derived from sales the break-even point falls

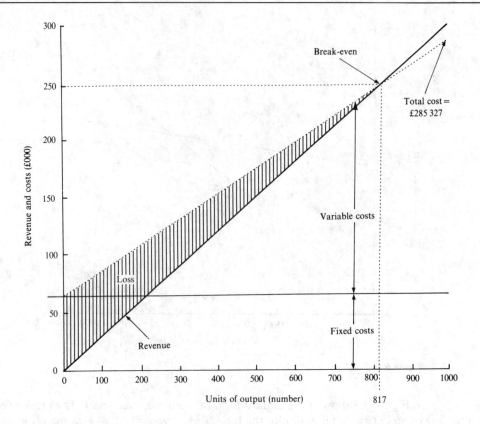

Figure 5.5

back from A to B and the break-even unit production reduces from 817 to 696. Similar effects can be noted by holding the other two variables constant and varying variable costs and revenue in Figs 5.5 and 5.6.

EXAMPLE 5.4

A firm of civil engineering contractors owns 15 trucks. Their demand for trucks depends upon the types of project that they are carrying out, their stage of completion, each project's locality and the weather. They can buy such trucks, having a useful life of 4 years, for £30 000 and the operating and maintenance costs for each vehicle amount to £6000 per year. On the other hand they can hire trucks from a local plant hire company for £280 per day including all operating and maintenance costs. The contractors need to supply and pay a maintenance mechanic in either case. Assuming that the trucks, if bought, have no salvage value at the end of their life, calculate the number of days per year when the contractors' daily demand has to exceed 15 to make it worth their while to purchase an additional truck. Their required rate of return is 15 per cent.

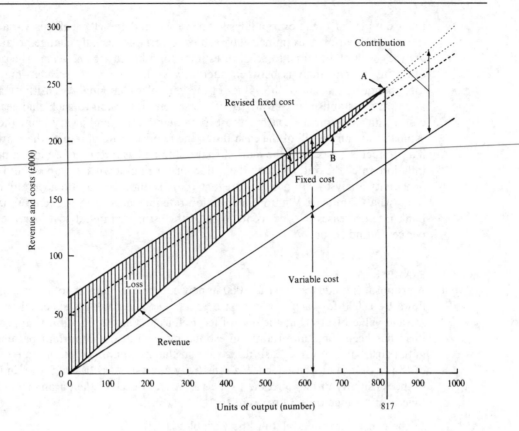

Figure 5.6

Annual cost of purchasing a truck $= £30\,000(A|P, 15\%, 4) + 6000$
$$= 30\,000(0.3587) + 600 = £16\,761$$

Let the number of days/year for hiring a truck $= D$
Annual cost of renting truck $= £280D$

Therefore,

At break-even point $280D = £16\,761$ and $D = 59.86$, say, 60 days

In this case, if demand for an additional truck is likely to exceed 15 on more than 60 days per year, an additional truck should be bought.

5.3 Sensitivity analysis

Break-even analysis is one particular form of a *sensitivity analysis*. The latter is an analysis in which one or more of the variables, the cost in the case of the illustrative examples of break-even analysis, are varied and the effect of this variation on the whole situation is examined. The values of variables usually change because they are the result of an initial forecast or estimate and hence are *uncertain*. The outcome of an investment analysis between alternatives—the *selection decision*—may therefore change as a result of the values of the variables changing. A sensitivity analysis attempts to establish the sensitivity of the selection decision to such changes. In the case of an investment decision, those parameters that might vary from the initial estimates are any or all of the cash flows, the required rate of return, the market size if a product is concerned, the market share obtained as a result of an investment, the inflation rate, the rate of tax, the value of writing-down allowances and so on. Sensitivity analyses fall into the category of those models that are generally referred to as 'what-if' models. What if the inflation rate increases to 15 per cent? What if the bank rate decreases to 5 per cent? What if the estimated initial cost increases by 25 per cent? And so on.

EXAMPLE 5.5

A proposal is made to invest £20 000 in new equipment in order to increase net cash flows by £6000 for each of the next 6 years. The equipment is expected to have a salvage value of £1000 at the end of its useful life. On examination it appears that there has been some uncertainty in estimating each of the important parameters— initial cost, salvage value, cash flow, life and the minimum acceptable rate of return of 18 per cent. As a consequence, a sensitivity analysis is requested for an indication of the sensitivity of the outcome to the variation of each of the parameters one at a time over a range of $+30$ per cent to -30 per cent.

The general formula relating the variables is

$$\text{NPV} = -(\text{initial cost}) + (\text{salvage value})(P|F, i\%, n) + (\text{net cash flows})(P|A, i\%, n)$$

The analysis is set out in Fig. 5.7.

EXAMPLE 5.6

An excavator is purchased for a capital cost of £100 000 and it is expected to have a useful life of 12 years. At the end of the excavator's life it is estimated to have a salvage value of £10 000. Examine the sensitivity of the annual capital cost of the excavator over variations of cost of capital from 9 per cent to 19 per cent and the excavator's life from 4 to 20 years.

Draw curves for i per cent from 9 to 19 per cent in steps of 2 per cent on axes of life (x) against annual cost (y). These curves are drawn in Fig. 5.8.

EXAMPLE 5.7

A plant hire company investigates the possibility of establishing the first phase of a new regional depot for the hire, maintenance, and repair of mechanical equipment. The depot is estimated to have an initial capital cost of £2 500 000, including all the

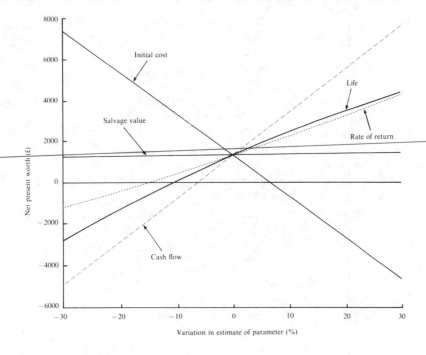

Figure 5.7

necessary equipment. The company's policy is to use a building and equipment life of 20 years in the appraisal of such facilities. The operating and maintenance costs for the depot (including labour costs and overheads) are estimated to be £325 000 per year. Savings elsewhere due to the establishment of the depot, together with additional revenue generated because it serves a new area, are expected to produce a revenue amounting to £825 000 per year.

Some doubt is expressed by the company about the estimates of initial capital cost and the costs of operating and maintaining the depot once it is established. Make an analysis on a before-tax basis showing the sensitivity of the outcome to these two parameters. The company requires a rate of return of 18 per cent on its investments.

Let x represent a percentage variation (expressed as a decimal fraction) in the estimated initial capital cost, calculated as

$$\text{(actual cost} - \text{estimated cost)/estimated cost}$$

and let y represent a similar percentage variation in the estimated operating and maintenance costs.

$$\text{Present worth} = -£2\,500\,000(1+x) + [825\,000 - 325\,000(1+y)](P|A, 18\%, 20) \geqslant 0$$
$$= -2\,500\,000 - 2\,500\,000x + (825\,000 - 325\,000$$
$$- 325\,000y)(5.3527) \geqslant 0$$

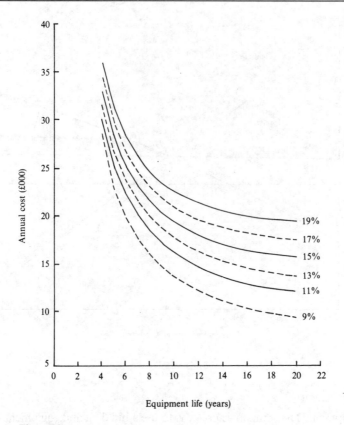

Figure 5.8

Therefore,

$y = 0.101\,37 - 1.437\,09x$ is a straight line of zero present worth, giving a boundary between areas where present worth $\geqslant 0$ and present worth $\leqslant 0$.

Figure 5.9 shows in a graphical form the implications of variations in the estimated figures. The straight line of $y = 0.101\,37 - 1.437\,09x$ has negative slope and represents the dividing line between net present worth being positive, downwards and to the left, and being negative, upwards and to the right. If the initial capital investment has been overestimated, then the variation will be negative and the equivalent point on the x-axis will be to the left of zero error. If it has been underestimated, then its point on the x-axis will be to the right of zero error. The same reasoning can be applied to the variation in the operating and maintenance costs, which are also negative cash flows and are subtracted from the income in the present worth equation. This means that an overestimation of the operating and maintenance costs at the *appraisal* stage will lead ultimately to a greater net revenue at the *actual* stage and therefore a move towards a larger net present worth. It will

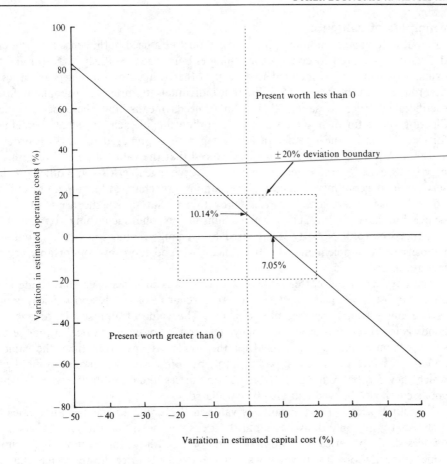

Figure 5.9

be seen from the slope of the graph that variations of underestimation of the capital cost need a compensatory overestimation in the operating and maintenance costs so as to maintain a positive net present worth and vice versa.

If there is no variation in the estimated capital cost of the project then it is possible to contemplate an increase of 10.14 per cent in the actual operating and maintenance costs over those estimated before the net present worth becomes negative. If the operating and maintenance costs turn out to have been correctly estimated, then the capital cost can have been underestimated by 7.05 per cent before a similar effect occurs.

A large proportion of the area of the ±20 per cent zone falls within the acceptable area of the diagram, thus indicating that the project is not particularly sensitive to variations of this magnitude at the estimating stage. The slope of the indifference boundary line is approximately 45°, indicating that the sensitivity of present worth to variations of both the x and y values is much the same.

5.4 The principles of valuation

For a need to have economic significance, it must be a need in the sense that whoever desires it is prepared to give up something else in order to satisfy it. Naturally, the extent to which one is inclined to give up other things for this purpose varies in accordance with the desire. That different individuals are prepared to make different sacrifices in order to satisfy their similar needs means that these persons have differing values for them. One way in which the value of such a need can be defined, for example, is by the number of hours a person is prepared to work in order to satisfy it. A general statement may be deduced that the value of a good or a service may be assessed by the extent to which it is exchangeable for some other good or service. To an economist, something that cannot be exchanged for something else has no real value. There are, however, many things that satisfy the needs of human beings but have no monetary value. Sunshine is one such thing. It cannot be exchanged or sold. Nobody is prepared to exchange money for sunshine unless it happens to be supplemented by the beach, by good food, by comfortable accommodation, and so on.

The days when the barter system of trading was in general use have long since gone in most developed countries, and the exchange of goods and services is now carried out through the medium of *money*. The value of almost all exchangeable goods is interpreted in terms of money. The goods are given a price so that certain sums of money may be exchanged for the goods. At any given time, the value of goods and services can be compared simply by comparing the amount of money for which they can be exchanged. By comparing prices, the quantitative rates at which goods can be exchanged can also be calculated.

Note that price is not the same as value in this context. The exchange rates for money may vary on a daily basis, but this does not mean that the value of goods or services similarly varies. The value of a loaf of bread now, for example, is much the same to most people as its value was 50 years ago. Its price, however, has changed considerably.

Where the value of a capital asset is concerned, there are two main values to be considered. The first is the *market value*. This is the price which the owner of a house will obtain if it is sold on the open market. It is the price paid by a willing buyer to a willing seller, where neither is under any compulsion to buy or sell. The value to the houseowner, however, may be very different, and value can vary greatly if one considers other assets of a specialist nature. A machine for taking deep soil samples, for instance, has considerable value to someone engaged as a soil mechanics specialist. That person almost certainly cannot conduct their business without it. To someone who is not concerned with soil sampling or is not associated with a soil mechanics laboratory, the machine has a market value of almost nothing; in fact, it has probably only the equivalent of its *scrap value* (unless it has *antique* value). Therefore, the market value and the value to the owner, or *use value* as it is sometimes called, can show a wide discrepancy. This discussion must include the likely value to the prospective owner. If another individual chooses to set up as an expert in soil mechanics and requires a machine to take soil samples then the market value of such a machine to that person will immediately rise to an owner's value, or use value. In general, the value of an asset to its owner does not exceed the replacement cost of that asset, making due allowance for any variation there may be in the quality of the replacement. On the other hand, the value to the owner is not less than the market value, being the price that the owner can obtain by selling the asset. To define more specifically the value of an asset to its owner, it can be said to be the compensation, and no more, in terms of money, which is required by the owner on being deprived of the asset.

A value that has already been used is a *resale* or *salvage value*. This is the price that can be obtained when an asset is sold in the second-hand market before its useful life is completed. Scrap value is the price that will be obtained by selling the asset when its useful life is completed, and its value is little or no more than the value of the materials of which it is made when broken down.

The valuation of an asset is needed in a number of situations during the consideration of capital budgeting problems. A valuation is often required by an engineer when considering depreciation policy or the costs of using existing machinery in a new project, when using an existing building to house a new plant, when deciding the best life of an asset, when determining the insurable value of an asset, and for many other purposes.

The definition of value to an owner of an asset in terms of maximum and minimum prices needs amplification. That an owner's valuation should not exceed the cost of a replacement, making due allowance for any variation in quality or capacity of the replacement from the original, is only strictly true if the replacement can be made available immediately. In the example of a scraper owned by a plant hire company that, for some reason or other, loses the use of this asset and cannot replace it immediately, the company also loses the revenue-earning capability of the scraper. Therefore, if replacement cannot take place for 3 months, the owner's valuation of the scraper can reasonably be assumed to be the replacement cost plus the net income for that 3 months. This difference between replacement cost and an owner's value is sometimes known as the *revenue factor*. Replacement cost plus the revenue factor is equivalent to the maximum value of the machine to the owner.

An important means of valuing an asset is to consider its ability to earn money or produce revenue over its expected future life. The asset valuation is achieved by discounting the future net cash flows in ways that have already been described in the previous chapters. One of the difficulties with such a means of valuation is obviously that of predicting exactly the future net cash flows that are directly attributable to the asset itself. However, in some instances, this method may be the only means of valuing an asset, as for example in the case of an item of equipment that is no longer manufactured and therefore cannot be replaced. In practice, unless the discounted future net cash flows are in excess of the asset's resale value, the asset is not being operated as an economic proposition.

In enlightened practice, the discounted net cash flow technique will have been used to determine whether the purchase of the asset should have been made in the first instance. If the discounted net cash flows were originally in excess of the purchase price then the original purchase will have been an economic one to make. At any point in the life of an asset, a check can be made on the estimated future cash flows arising from it, and for the initial period of its life it is likely that, when discounted, they will exceed the replacement value. As the asset's life becomes extended, so will its estimated future life shorten and therefore its ability to earn revenue. Ultimately the value of the estimated net cash flows, when discounted, will fall below the replacement value of the asset. For example, if the discounted net cash flows for a trenching machine amount to £40 000 and the replacement price of that machine amounts to £45 000 then its owner's valuation is now less than its replacement cost. As time progresses the next stage of economic assessment is reached, when the discounted net cash flows are less than the resale value of the machine. The resale value of the machine may be £10 000, and the estimated future life of the trencher will now be so short that the discounted net cash flows for the future may have dropped to £8000. In this situation there is no financially satisfactory option other than for the machine to be sold, because from the principles of present worth, it becomes evident that the future value of the asset at today's price is less than the amount of money which would be obtained today if it were sold.

It will be appreciated that to forecast accurately the net cash flows over a future period is an extremely difficult task. The analysis of previous records of similar situations may well help in this forecasting procedure. In construction work such records are not always readily available, and if they are then their accuracy is often open to question. Because of this, there may be an area in which it is difficult to make a clear-cut decision. In general, however, if the discounted value either greatly exceeds or is very significantly lower than the resale value, then the decision should be to either retain or dispose of the asset respectively.

5.5 A valuation procedure

The valuation of an asset already in a company's possession necessitates, in the first instance, the establishment of the net cash flows that will arise from it, taking into account the need for it to be maintained and replaced at regular intervals. These net cash flows must then be compared with those that will result from the action the company will have to take if it is assumed that it does not possess the asset, but takes immediate steps to acquire one of comparable performance. For example, if a company owns a bulldozer that is already some years old and now wishes to value it, the company can initially arrive at the estimated net cash flows that will arise from the possession of the existing machine. If it is assumed that the company is not in possession of a bulldozer but purchases a new one at the same time as the valuation is being made, then the acquisition of the new bulldozer will give rise to a different set of net cash flows. The value of the existing asset in both this and the general case is then the difference between the present worths of the two series of net cash flows.

For the sake of simplicity in the following example, it will be assumed that the existing asset is perpetually replaced at regular intervals of time at a capital cost that will be similar for each period under consideration. In addition, it will be assumed that the capital and operating costs of the replacement assets will be similar to each other in every case of replacement, but may be different from those of the existing asset for which the valuation is being carried out.

EXAMPLE 5.8

The XY Construction Co. Limited owns a bulldozer which it has been using for the past 3 years. The economic life of this bulldozer has been calculated to be 8 years, after which time it has an estimated resale value of £10 000. The capital cost of replacement by a similar bulldozer amounts to £60 000, and it is estimated that any replacement made in the future will have a resale value similar to the existing model after the same length of life. The operating costs (excluding the cost of the driver) for the existing machine amount to £5000 per year, but in the case of a new replacement these costs are estimated to be reduced to £3000 per year. The cost of capital to the company is 10 per cent per year. What is the value of the existing bulldozer to the company at the present time?

The net cash flows for the existing bulldozer and for the hypothetical case, where the company does not possess a bulldozer but purchases one at today's date, are listed in Table 5.5. The assumption is made that the bulldozer will be replaced, after 8 years in both cases, by one at the capital cost of £60 000 and the withdrawal of the existing bulldozer will result in a receipt of £10 000 from its sale.

In the case of the hypothetical new bulldozer the operating costs can be converted to a single payment at the commencement of the economic life of the bulldozer to which they refer:

$$P = A(P|A, 10\%, 8) = -3000(5.3349) = -£16\,005$$

Similarly, the discounted resale value for a life period of 8 years

$$= S(P|F, 10\%, 8) = 10\,000(0.4665) = £4665$$

Table 5.5

Year	Existing bulldozer	New bulldozer
0	—	−£60 000
1	−£5 000	−3 000
2	−5 000	−3 000
3	−5 000	−3 000
4	−5 000	−3 000
5	−5 000 − 60 000 + 10 000	−3 000
6	−3 000	−3 000
7	−3 000	−3 000
8	−3 000	−3 000 − 60 000 + 10 000
9	−3 000	−3 000
	etc.	etc.

The series of payments for the new bulldozer in perpetuity

$$= -60\,000 - 16\,005 + 4665 = -£71\,340 \text{ every 8 years}$$

In general terms, if a payment A is made every t years, then the present worth of such a series in perpetuity, commencing in t years' time, amounts to

$$P = \frac{A}{(1+i)^t} + \frac{A}{(1+i)^{2t}} + \frac{A}{(1+i)^{3t}} + \cdots + \frac{A}{(1+i)^{\infty}}$$

If both sides of the equation are multiplied by $(1-i)^t$, then

$$P(1+i)^t = A + \frac{A}{(1+i)^t} + \frac{A}{(1+i)^{2t}} + \cdots + \frac{A}{(1+i)^{\infty}}$$

Subtracting the first from the second equation, then

$$P(1+i)^t - P = A = P[(1+i)^t - 1]$$

or

$$\boxed{P = \frac{A}{(1+i)^t - 1}} \qquad (5.1)$$

In this case therefore, the present worth of the payment of £71 340 every 8 years, commencing with an initial payment of £71 340 at year 0,

$$= \frac{-71\,340}{(1+0.10)^8 - 1} - 71\,340 = -63\,382 - 71\,340 = -£133\,722$$

For the existing bulldozer the series of payments and receipts from the end of year 6 onwards will be similar to those for the hypothetical asset, including the capital cost of £60 000 at the end of year 5.
Present worth of this series

$$= -133\,722(P|F, 10\%, 5) = -133\,722(0.6209) = -£83\,028$$

Present worth of the operating costs of the existing asset from years 0–5

$$= -5000(P|A, 10\%, 5) = -5000(3.7908) = -£18\,954$$

Present worth of resale value of 5 years' time

$$= 10\,000(P|F, 10\%, 5) = 10\,000(0.6209) = £6209$$

Therefore,
Present worth of net cash flows of existing asset

$$= -£83\,028 - 18\,954 + 6209 = -£95\,773$$

and the present value of the existing asset $= £133\,722 - 95\,773 = \underline{£37\,949}$

This example has been worked through with simplified and uniform cash flows. It will be appreciated, however, that more variable cash flows can be dealt with using similar principles.

5.6 Retirements and replacements

The stage is reached in the life of every asset when its owner has to consider its retirement from service. This may be due to a number of reasons, among which may be the increasing cost of the asset's maintenance and repairs, obsolescence due to the availability of new and improved alternatives, inadequacy because of increased demands that must be made on it, or inadequacy because of mechanical or structural deficiencies. An asset may be retired for one or more of these reasons, and it may be retired by one owner, only to be bought and put back into service by another.

The replacement of an asset may come about because it is retired and a new one is now needed to take its place. Unless an asset fails completely during performance, in such a way that it is impossible to repair so that it may continue to provide its service, then at some time in its life a decision must be made as to whether it should be retired and replaced. This point in time usually occurs when increasing maintenance and repair costs make the asset uneconomic. It is highly desirable to have a means of establishing, as precisely as possible, when the retirement and replacement decision should be put into effect as a result of economic justification. A knowledge of valuation and depreciation procedures is necessary in order to make this decision on rational grounds.

Table 5.6 presents the results of a series of calculations made at the end of each year for the valuation of the asset. These yearly valuations, shown in column 10, may

Table 5.6

Year	Cash flows	Estimated resale values	Present worth factors at 10%	Present worth of operating costs	Cumulative total of column 5	Present worth of resale value	Present worth of total cash flows	$1-(1+i)^{-t}$	Present worth of perpetual series of payments
				(2×4)		(3×4)	$(6+7)$		$(8/9)$
1	2	3	4	5	6	7	8	9	10
	£	£		£	£	£	£		£
0	(10 000)	10 000	1	(10 000)	(10 000)	10 000			
1	(300)	8 000	0.909 09	(273)	(10 273)	7 273	(3 000)	0.090 91	(33 000)
2	(400)	6 000	0.826 45	(331)	(10 603)	4 959	(5 645)	0.173 55	(32 524)
3	(500)	4 000	0.751 31	(376)	(10 979)	3 005	(7 974)	0.248 69	(32 063)
4	(1 000)	3 000	0·683 01	(683)	(11 662)	2 049	(9 613)	0.316 99	(30 326)
5	(1 500)	2 500	0.620 92	(931)	(12 593)	1 552	(11 041)	0.379 08	(29 126)
6	(3 000)	2 000	0.564 47	(1 693)	(14 287)	1 129	(13 158)	0.435 53	(30 211)
7	(4 000)	1 500	0.513 16	(2 053)	(16 339)	770	(15 570)	0.486 84	(31 981)
8	(4 000)	1 000	0.466 51	(1 866)	(18 205)	467	(17 739)	0.533 49	(33 251)

be plotted in graph form for clarity, as in Fig. 5.10. The valuations are carried out on the basis already described in Section 5.4. The asset should be replaced at that time when the present worth valuation of column 10, as calculated, is at a minimum.

In the operation of an item of equipment, whether it be a concrete mixer, a scraper, a tractor, a computer or any asset of a nature that involves the expenditure of money on maintenance and repair throughout its life, these costs will, generally speaking, increase as the life of the asset increases. There will come a time when it is obviously better, from an economic point of view, to replace the asset by another machine capable of performing similar duties, and a decision must be taken to determine just when this particular moment has arrived. An assessment of the appropriate time can be made if the relevant information concerning operating costs and the resale market value of the asset is known or can be estimated. In column 2 of Table 5.6 are listed the cash flows for such an item of construction equipment over a period of 8 years. For the sake of simplicity, it is assumed that there will be a continuing demand for the work carried out by this equipment and that it will be replaced by another item capable of providing a similar service when the optimal time for its retirement comes along. In column 2 of Table 5.6 the initial capital cost of the machine is shown as

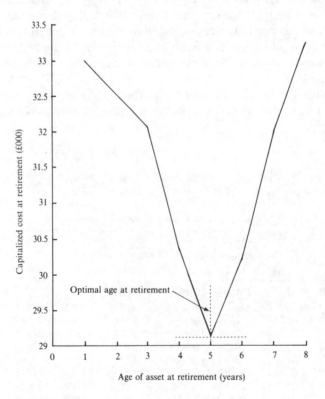

Figure 5.10

£10 000 at the outset of the study period. For the years 1 to 8, cash flows representing the repair and maintenance costs are listed. It will be seen that in the initial stages these are fairly low but they increase at a higher rate towards the end of the 8-year period. These figures take account of any reduction in the efficiency of the plant as a producing unit as well as the repair and maintenance costs associated with it. Column 3 contains the estimated resale values or valuations of the machine at the end of each year. For example, if the row of figures for year 1 is examined, the resale value of the machine is seen to be £8000 at the end of that year; at the end of year 7 the resale value has dropped to £1500, and it finally drops to £1000 at the end of year 8.

Firstly, it is necessary to find the present worth of the initial cost, the operating costs and the resale value on the assumption that the asset's life will end at each year consecutively. That is, the present worth of the relevant cash flows is firstly found as though the asset's life finished at the end of year 1, then a separate similar calculation is made up to the end of year 2, and so on until, in this case, the end of year 8 is reached. For simplicity, this calculation is made on the assumption that the asset could be sold at the end of each year. While for the purposes of illustration the valuations have been worked out in this particular example for the end of each of the 8 years, in practice, by inspecting the figures of cash flow, it is often possible to determine the general area in which the minimum valuation will fall. By calculating the valuations at intervals of, say, two years over this area, it is possible to narrow down the precise point still further without having to investigate the whole range of the asset's life. Alternatively, such tables can be created very quickly by the use of a suitable microcomputer spreadsheet package.

Column 4 of Table 5.6 lists the present worth factors for an interest rate of 10 per cent (which is assumed to be the attractive rate of return for the company at this time) related to the end of the years shown on the left-hand side of the table. The present worth of the initial cost will always be £10 000 in the table, but it is necessary to calculate the present worths of the operating costs, and, for illustration, the results of these calculations are shown in column 5. Column 5 is the arithmetical result of multiplying together the figures in columns 2 and 4. Column 6 is the progressive cumulative total of the present worths of the initial cost and operating costs for each year shown in column 5. At this point, for example, the initial cost plus the present worth of the total operating costs up until the end of year 4 amount to £11 662. At the end of year 8 the equivalent present worth of the initial cost and the whole of the series of maintenance costs for the 8 years life of the machine is shown as £18 205.

The next step is to find the present worth of the resale values, and these are listed in column 7. They are shown as positive cash flows because, by the convention used in this table, the outgoings are shown as negative sums. Column 7 is the arithmetical result of multiplying together the figures in columns 3 and 4. The present worth of all the payments and receipts in a period representing the life of the machine can now be totalled for a range of lives from 0 to 8 years. These figures are listed in column 8. Against year 4, for example, the total of £9613 is the result of adding the present worths of all the maintenance costs up to the end of year 4 to the initial cost of

£10 000, making a total of −£11 662, and then adding to that figure the present worth of the resale value at that time, that is, £2049. This leaves a net figure of £9613. If necessary, the cash flow diagrams for each year can be drawn.

If, therefore, it is assumed that the machine will be replaced at the end of year 4 by a machine of a similar capacity and at a similar initial cost, then a perpetual series of 4-yearly payments of £9613 will be created. It is necessary to find the present worth of such as series of payments and this can be done, as has already been described, by dividing the total figure in column 8 of Table 5.6 by $[(1 + i)^t − 1]$, from formula (5.1), where t is the interval at which the payments will be made. To be added to it is the initial present worth of the series at the beginning of year 1 which is not taken into account by the above formulation.

Alternatively, the calculation can be made as follows: if A is the uniform periodic payment, the total present worth of the series plus the initial payment

$$= A + \frac{A}{(1+i)^t - 1} \quad \text{(from (5.1))}$$

$$= \frac{A((1+i)^t - 1)}{(1+i)^t - 1} + \frac{A}{(1+i)^t - 1}$$

$$\boxed{= \frac{A(1+i)^t}{(1+i)^t - 1} = \frac{A}{1 - (1+i)^{-t}}} \tag{5.2}$$

The values of the denominator of formula (5.2) for an interest rate of 10 per cent and values of t from 1 to 8 are listed in column 9 of Table 5.3 and column 10 results from dividing the present worth of the total cash flows up to the end of each year (shown in column 8) by these factors.

Note that the minimum figure of column 10 is at the end of year 5, and as a consequence for this particular set of figures the best time to replace the asset appears to be then. The time to retire the machine is obviously when its operation costs reach a disproportionately high level and it loses production because it is continually needing repair. By inspecting the figures of Table 5.3, it can be seen that between years 5 and 6 the cash flow for maintenance expenses increases more rapidly than previously, so that we might expect to find the optimal time for retirement within this area.

5.7 Lease or buy?

If a contractor (or other organization carrying out a similar function) wishes to acquire the use of mechanical equipment then there are various courses of action open to them. The four most common methods are:

1. outright purchase;
2. rental from a plant hire company;

3. hire purchase;
4. finance leasing.

Typical factors that will influence the choice of method include the nature and the extent of the work for which the equipment is required, whether the machine will become obsolete in the near future, and the financial status of the company requiring the equipment. The methods of outright purchase and rental from a plant hire company need no further explanation. The first has already been dealt with as an investment appraisal; the second results in a negative periodic cash flow to be balanced against the income arising from productive activities of the plant.

Hire purchase is a means of hiring the equipment for an agreed period, with an option to purchase at the end of the period. The hirer is required to put down a reasonable deposit (the size of which is sometimes controlled by government regulation) at the commencement of the hire and then to make regular payments for the remainder of the hire period. If the hirer exercises his option to purchase then the equipment becomes wholly owned by the hirer and the company financing the deal has no further interest in it. Hire purchase payments in this context are generally allowed against tax.

With a *financial lease* the lessee, i.e., the user, is usually responsible for the service and maintenance costs of the equipment. However, it is possible to have an *operating lease* under which the lessor, i.e., the owner, is responsible for those services. In this case the lease is usually for a shorter life than the equipment's economic life and the lessor may lease it to others at the end of the initial or subsequent agreement. The advantages of leasing are many. Firstly they include the one mentioned above, of tax deduction under certain conditions. There must, of course, be a tax liability against which the payments can be set. The financial status of the company is therefore important. Leasing conserves liquid capital that might otherwise be absorbed outright by a series of large purchases or alternatively it is useful where a firm lacks sufficient cash resources. Leasing can be viewed as a protection against the risks of equipment becoming obsolescent and, in a similar fashion, against the possible effects of inflation. Leases frequently impose fewer financial restrictions than loans for the purpose of buying such equipment and do not affect a firm's future capacity to borrow by debt financing, as may happen with an excess of existing loan capital.

On the other hand there are disadvantages. In effect, leasing large equipment reduces the magnitude of the fixed assets of the company and hence possibly its future debt-raising capacity. No indication is given in a balance sheet as to the nature or magnitude of assets that are leased. The lease payments are not identified as such in the profit and loss account of the company, though they do effectively reduce profit since they are an expense, which may subsequently reduce distributions by way of dividends and interest.

Financial leases are mostly arranged for new assets. A potential lessee will identify an item of equipment that is needed in his company and then engage a leasing company to purchase it and to arrange a *direct lease* between them. In order to raise cash a company may sell one or more of its assets to a leasing company and then lease the asset back in order to continue to have its use. This is known as having a *sale and lease back* agreement. It is commonly applied to buildings such as those used for offices and factories but may also be applied to aeroplanes, ships and other equipment. Under such an arrangement, the lessee continues to have the right to use the buildings, etc., though the leasing company becomes the legal owner of the property.

EXAMPLE 5.9

A survey company decides to consider a leasing alternative for a diving bell to be used for inspecting underwater structures. The bell has an initial cost of £180 000 and a life of 6 years, after which it is expected to have no significant resale or salvage value. Since this diving bell will be one of several owned by the company, the lease excludes the costs of operating, maintenance, and insurance expenses that can be undertaken by the existing organization. The manufacturer of the diving bell offers a lease to the company at an annual payment of £26 000, payable in advance.

Draw up a table of the annual cash flows that will arise as a consequence of using leased equipment, assuming that the company is allowed, for tax purposes, to depreciate the bell by the declining-balance method at a rate of 25 per cent a year. The company is profitable to the extent that it can take advantage of all the available tax reliefs and it pays corporation tax at 35 per cent.

The company requires a minimum rate of return on its investments of 12 per cent.

The results of the calculations establishing the cash flows are shown in Table 5.7. In the case of the lease, the company does not have an initial capital expenditure of £180 000, so there is the equivalent of a positive net cash flow of this amount at the outset of the lease, representing the saving of the investment. To be set against this is the fact that a lessee cannot depreciate an asset that does not belong to it for tax purposes, so in effect it loses that benefit and this is shown as a negative cash flow

Table 5.7

	Year 0 £	1 £	2 £	3 £	4 £	5 £	6 £	*Net present worth,* £
(1) Capital cost	180 000							
(2) Depreciation	45 000	33 750	25 313	18 984	14 238	10 679	32 036	
(3) Less tax on depreciation	(15 750)	(11 813)	(8 860)	(6 644)	(4 983)	(3 738)	(11 213)	
(4) Lease payment	(26 000)	(26 000)	(26 000)	(26 000)	(26 000)	(26 000)		
(5) Tax relief on lease payment	9 100	9 100	9 100	9 100	9 100	9 100		
(6) Net cash flow	147 350	(28 713)	(25 760)	(23 544)	(21 883)	(20 638)	(11 213)	53 133

in line 3. The actual annual amount of depreciation is shown in the shaded line 2. (It has been assumed that the company pays its tax in the same year as the depreciation accrues.) Because declining balance is used as the depreciation method, the diving bell cannot be depreciated to a zero salvage or resale value at the end of its life. The residual book value at year 6 amounts to £32 036 and the benefit of the tax relief on this would have accrued at the end of the bell's life. The balancing charge of £32036(0.35) = £11 213 is shown as a benefit lost under year 6.

The payments under the lease, shown in line 4, are tax deductible, and therefore a credit is shown in line 5 against these payments.

The net cash flows arising from taking the lease rather than purchasing the bell are shown in line 6 of Table 5.7. In effect, these cash flows are the incremental cash flows between the outright purchase of the diving bell and the lease. If the incremental rate of return exceeds 12 per cent, the minimum required rate of return of the company, then the lease will be the better financial opportunity for the company. In this case the net present worth at 12 per cent is positive, so the company should take the leasing alternative.

5.8 Risk analysis in investment appraisal

Investment appraisal almost invariably deals with events that will occur in the future. Cash flows of many kinds, arising from a wide variety of causes, need to be predicted in order that potential returns may be evaluated. A major risk here is that, in practice, the chances are that things may not turn out as expected. There may be difficulties in forecasting that lead to an uncertainty about the outcome of taking a decision between two or more alternatives. Methods of *risk analysis* have been developed to provide a variety of ways of examining problems that involve uncertainty. These provide a clearer and more detailed understanding of the risks that are involved in decision making, that is making the choice between alternatives. Risk is concerned with the chances or probability of a loss being made. The loss may not necessarily be directly financial; the risk of physical injury (or death), collapse of a building structure, having a car accident, falling into a river, loosing a golf ball off the tee, etc., are all situations where someone or something is at risk—probably as a result of making a decision, whether after due consideration or not, to take some specific action.

Prominent writers have drawn a distinction in the past between *risk* and *uncertainty*. In essence, with modest variations, the distinction has been based upon a concept that *risk* is concerned with situations in which statistical data are available concerning the variability of the environment in which a decision problem exists, and *uncertainty* is concerned with situations in which no such data are available. In the former case it is assumed to be possible to postulate, for example, a distribution showing the variation in the probability of particular outcomes resulting; in the latter case there are insufficient or no data available to define such probabilities. Where such data are not available, decision makers need to resort to subjective probability assessments.

While much emphasis has been placed on risk analysis as a tool to assist with decision making concerning financial investments, it must not be overlooked that the theories involved are capable of application to most decision problems, whether they involve alternative financial investments or not. For example, in construction, it is likely that risk analysis can have beneficial results from its application in special areas of insurance where unusual risks are being undertaken on new or difficult methods, in assessing situations where opportunities for alternative contractual arrangements are available involving different allocations of risk and reward among the construction organizations involved, in assisting with the deliberations on the circumstances in which competitive or other types of bids should be made or not as the case may be, and for generally enhancing the processes of design and construction as a result of assessing the risk to achieving rapid, accurate, and true construction. In this chapter risk will be dealt with in the context of financial appraisal.

Until this point, with the exception of the section dealing with sensitivity analysis, it has been assumed that there has been no uncertainty associated with the forecasts of cash flows, salvage or resale values, capital investments, etc., and that a single estimate in each case represents the best that can be done to arrive at a true forecast. In dealing with sensitivity analysis it was demonstrated how consideration could be given to the variations in outcomes that may arise as a result of variations in inputs, particularly of errors in estimating, and the results of those variations on the outcome as a whole. Other than these analyses it has been assumed that an estimated positive cash flow of £10 000 for a resale value at year 10 of the cash flow diagram would actually be that value when the time came, without investigating the consequences of what may have happened in practice. However, in the real world, *variability* is a fact of life and risk analysis attempts to acknowledge that variations in estimated quantities will occur. Risk analysis is based upon applying probability and other theories to cash flow models so that a broader, more complete, treatment can be undertaken of an economic evaluation by providing a quantitative expression of the chances or probabilities that certain outcomes will be achieved given the data concerning the problems.

5.9 Introduction to probability concepts

As long ago as the seventeenth century, gamblers were curious as to their chances of winning or losing in a specific play. Mathematicians such as Fermat and Pascal became involved, at the invitation of their friends who gambled, in determining the probabilities of their winning while playing a variety of games of chance. It was this curiosity that led to the beginning of the development of what have become many theories of probability. Axioms were developed early in this century and modern probability theory has increased in importance to the point that it is now commonly used in practically every academic discipline and its application.

If the outcome of a process or experiment is always the same, given that the same input is used on each occasion, the process is said to be *deterministic*. For example, determining the roots of a quadratic equation by use of the relevant formula will always yield the same result if the constants entered in the formula remain the same. Given that the process is error-free, it is then possible to *forecast* the result for the next occasion on which the process is put in hand. When an experiment is conducted, such as predicting by whatever means the number of cubic metres an excavator will dig during the first hour of work on the following day, there is little certainty as to what the number will be in spite of its historic performance. The outcome then becomes a *random event*.

The gambling antecedents of modern probability theory are reflected in the dice, coins, and playing cards still used today to illustrate and explain its simple concepts. The probability of an event happening (or not happening as the case may be) is measured on a scale from zero to unity. The zero end of the scale represents the situation of no chance at all that an event will take place, and it is written as $p = 0$. The unity end of the scale, $p = 1$, represents absolute certainty that an event will take place. There is an infinite number of points between 0 and 1 which represent intervening probabilities.

A fair coin has two sides, a head and a tail. If it is tossed into the air and lands on a flat, unimpeded surface, either one or the other side uppermost, the set of all possible outcomes, S, is called the *sample space*. An element in S, that is a specified outcome, is called a *sample point* or a *sample*. A reasonable person will not disagree with the proposition that the outcome of a head (H) and that of a tail (T) are equally likely and therefore on the scale of probability $p(H) = 0.5$ and $p(T) = 0.5$. Since there is absolute certainty that it will be either one or the other (unless given some major phenomenon in contradiction of the law of gravity), $p(H \text{ or } T) = 1$. In the case of rolling a die, the sample space consists of all the possible outcomes, a 1, 2, 3, 4, 5, or 6. If the die is fair, each of its faces is equally likely to be uppermost when it comes to rest and, since the probability of getting one of the numbers on the top is one, the probability of getting some particular number, $p(\text{some number}) = \frac{1}{6}$. The probability of getting either 1, 3, or 5 then becomes $\frac{3}{6}$ or 0.5.

Establishing probabilities as above, by reasonable experience and consideration of the nature of the event, produces *subjective* or *prior* (also called *a priori*) probabilities. The probability of such an event can be determined before it takes place. Such prior knowledge of probabilities is possible only infrequently and in relatively few applications. Where, as an alternative, the establishment of probabilities is based upon an analysis of historical records of relevant events (taking care to ensure that some records do not unwittingly influence others), they are known as *objective* or *a posteriori* (after the fact) probabilities. Such probabilities provide a useful basis for application in practice but must necessarily be considered as approximate.

If a fair coin is tossed 500 times, it would be *expected* that 250 times it would fall with the head uppermost and 250 times with the tail uppermost. However, 'expectation' needs some further examination. It is difficult to believe even with this large number of tosses of a coin that it would fall head uppermost for *exactly* 250 times. What is really expected is that it will fall head uppermost *approximately* 250 times. However, it is also not unreasonable to expect that the longer the series of tosses that takes place, the more accurate will be the series as a true indicator of the probability of a head or a tail appearing. The *empirical, objective,* or *a posteriori* probability of an event may be expressed as follows:

$$\text{probability} = \frac{\text{number of occurrences of the event}}{\text{total number of trials}}$$

Thus, if the record shows that 25 out of 400 concrete test cubes fail, $p(\text{failure}) = 25/400 = 0.0625$.

Events can be classified statistically as *dependent* or *independent*. In the former case the event can be influenced by, or is dependent upon, the occurrence of some other event. If an event is *mutually exclusive*, that is if it occurs in one way and that occurrence excludes any possibility that it will occur in another way, the probabilities of those events can be added.

In rolling a die on a single occasion it is equally likely that a 1, 2, 3, 4, 5, or 6 will appear on the upper face. The probability of a 5 showing as a result of a single roll is therefore

$$p(5) = \tfrac{1}{6}$$

Since the possible outcomes in the case of rolling a die are mutually exclusive—if a 4 shows, for example, then that excludes any of the other even numbers from showing—the probability of one of the 3 even numbers showing becomes the sum of the probabilities of each of the eligible alternatives. This leads to the probability that none of the even numbers will show being

$$p(\text{odd numbers}) = 1 - p(\text{even numbers})$$
$$= 1 - \tfrac{3}{6} = \tfrac{3}{6}$$

The *addition law* may be stated as follows:

Where an event can occur in one of several mutually exclusive different ways, its probability of occurrence may be calculated as the sum of the probabilities of the occurrence of each of the several different ways.

EXAMPLE 5.10

Twenty-five stainless steel bolts are collected in a box. It is subsequently discovered that 10 of the bolts were made of steel which did not meet the specification (classified as B bolts) and that 7 bolts had metallurgical faults that were not visible to the naked eye (classified as C bolts). The remaining 8 bolts were good (classified A). What are:

(a) The probability of making a random selection of a bolt from the box and it being good?
(b) The probability of getting either a class A or a class B bolt with a single selection?
(c) The probability of not getting a class A bolt, $p(\overline{\text{class A}})$, with a single selection?

(a) $p(\text{class A}) = 8/25 = 0.32$
(b) $p(\text{class A or class B}) = 8/25 + 10/25 = 18/25 = 0.72$
(c) $p(\overline{\text{class A}}) = p(\text{class B or C}) = 10/25 + 7/25 = 17/25 = 0.68$

or

$$= 1 - p(\text{class A}) = 1 - 8/25 = 17/25 = 0.68$$

The addition law of probability is also known as the *first law of probability*. The *second law of probability* is alternatively known as the *multiplication law:*

The probability that two or more uncertain and independent events will occur, whether both together or in succession, is the product of the probabilities that each individual event will occur.

EXAMPLE 5.11

Using the box of 25 bolts of Example 5.10:

(a) Determine the probability of getting a class B bolt and then a class C bolt in that order in making two successive selections from the box without replacing the first before drawing the second.
(b) What is the probability of obtaining the two bolts of (a) above, but in any order?

(a) $p(\text{1st bolt in class B}) = 10/25$
$p(\text{2nd bolt in class C}) = 7/24$
$p(\text{class B, class C}) = (10/25)(7/24) = 7/60$

(b) In addition to the order of (a) above:
$p(\text{1st bolt in class C}) = 7/25$
$p(\text{2nd bolt is class B}) = 10/24$
$p(\text{class C, class B}) = 7/60$
$p(\text{C and B, any order}) = 7/60 + 7/60 = 7/30$

5.10 Expected monetary value

Probability theory leads to the concept of *expected value* in engineering economy and decision analysis. Expected value uses weighted averages in order to make allowance for risk in the results obtained from making decisions. It results from the multiplication of a probability and an outcome. The latter is usually, though not exclusively, an associated monetary value. In general, expectation can be defined as follows:

If the amounts A_1, A_2, $A_3, \ldots,$ A_k can be obtained with the probabilities p_1, p_2, $p_3, \ldots, p_k$, then the expected value, *EV*, or the expected monetary value, *EMV*, amounts to:

$$\boxed{EV = A_1 p_1 + A_2 p_2 + A_3 p_3 + \cdots + A_k p_k} \qquad (5.3)$$

A simple example of the calculation of an expected monetary value arises in a gamble involving the toss of a coin.

Suppose that an offer is made that a fair coin will be tossed and if it comes down as a head a prize of £1 will be given. If it comes down as a tail a penalty of 75 pence will be required of the gambler. It is agreed that 50 tosses of the coin will be made under these conditions.

The probability of getting a head at each toss is 0.5, with the similar probability of getting a tail. The expected monetary value of the winnings from the gamble then is:

$$EMV = p(\text{H})(£1)50 + p(\text{T})(-£0.75)50$$
$$= 0.5(£1)50 + 0.5(-£0.75)50 = £25 - 18.75 = \underline{£6.25}$$

It will be seen that the average expectation for each toss of the coin amounts to $6.25/50 = 12\frac{1}{2}$ pence. The calculation takes into account the probability of certain events and the values associated with the possible outcomes of those events. Over 50 tosses (or on average over one) the gamble is seen to be profitable. It was the view of early protagonists of expected value that any gamble showing an expected profit should be accepted. However, it must be noted that a single toss of the coin in the above example will result either in the winning of £1 or the loss of 75 pence. On no occasion will the outcome of a single toss of the coin be $12\frac{1}{2}$ pence. The evaluation is undertaken on the basis of a long-term, repetitive gamble, but it does incorporate a measure of the risk that is being taken. However, in choosing between alternatives in the face of uncertainty, it is desirable to make decisions that are consistent with rationality and long-term objectives. Such objectives are likely to be those of minimizing expected costs or maximizing expected profits. If, therefore, the concern is for making comparisons of risk in the long-term or for repetitive situations then it seems likely that expected monetary value has merit.

EXAMPLE 5.12

A firm of contractors have opportunities to bid for one of three projects. They estimate that the first has probabilities of 0.2, 0.4, and 0.4 of making a profit of £500 000, a profit of £250 000, and a loss of £50 000 respectively. The second has probabilities of 0.2, 0.3, and 0.5 of making a profit of £600 000, one of £400 000, or a loss of £100 000 respectively; a third has probabilities of 0.3, 0.4, 0.1, and 0.2 of making a profit of £400 000, £300 000, $-£20 000$, and $-£200 000$ respectively. For which project should they bid?

$$EMV(\text{Project } 1) = 0.2(£500 000) + 0.4(£250 000) + 0.4(-£50 000) = \underline{£180 000}$$

$$EMV(\text{Project } 2) = 0.2(£600 000) + 0.3(£400 000) + 0.5(-£100 000) = \underline{£190 000}$$

$$EMV(\text{Project } 3) = 0.3(£400 000) + 0.4(£300 000) + 0.1(-£20 000) + 0.2(-£200 000)$$
$$= \underline{£198 000}$$

Therefore the contractor should bid for either Project 2 or Project 3. If the possibility of loosing as much as £200 000 is daunting and would create difficulties of, say, cash flow, then bid for Project 2.

5.11 Investment risk reviews

In making a single estimate of a cost, a cash flow, an asset life, future inflation rates, etc., there is an implication that the estimated figure is known with some certainty. While the actual figure may vary in practice, all the estimator's faith is being put in a single figure. However, if the estimator is approached and asked if there is not a reasonable chance (a risk) that the number presented can be 10 per cent higher, the response will almost certainly be that there is, while further pressure will probably produce an estimate of the chances, such as 1 in 4 or perhaps 15 per cent probability, or whatever is felt intuitively. In considering alternative formats for the establishment of the return on an investment, especially in looking at the probability that the significant factors in the calculation will vary with specific probabilities, the different possibilities become many and their evaluation becomes more complex. One graphical aid to assist with overcoming this problem is the *investment risk profile*. The investment risk profile is a graphical summary of the probabilities of specific returns being made in an investment project over a range of alternatives and statements of the probabilities of occurrence. An example will now be used to demonstrate the method.

EXAMPLE 5.13

A company proposes to invest £350 000 in major production equipment and anticipates receiving back an after-tax net income of £110 000 for each of the next 5 years, when it is anticipated that the project's life will come to an end. The company has a requirement for an after-tax return of 10 per cent.

The finance director, reviewing the proposals, asks for a full analysis of the risks associated with the project. The risk analyst comes up with the possibility that the annual after-tax net income is likely to take one of four values, £70 000, £90 000, £110 000, or £120 000, with a probability of achievement of 0.1, 0.3, 0.4, and 0.2 respectively. The analyst is satisfied that the initial capital cost is unlikely to vary from £350 000. In addition, the analyst suggests that the income for the project will vary in its duration in accordance with the probabilities set out in Table 5.8.

The salvage value of the equipment varies with the life of the project, being £20 000, £15 000, and £10 000 in each case where the project life is 4, 5, or 6 years respectively. In each case the salvage value is regarded as certain, with a probability, therefore, of one.

The finance director asks for a simple, clear-cut summary of the analyst's findings. The analyst prepares an investment risk profile.

The tree structure of the investment risk profile is shown in Fig. 5.11. This diagram summarizes all of the information provided by the analyst. The introductory trunk

Table 5.8 Probability of annual after-tax net income

Project life	Annual after-tax net income			
	£70 000	£90 000	£110 000	£120 000
4 years	0.1	0.2	0.3	0.6
5 years	0.5	0.5	0.5	0.4
6 years	0.4	0.3	0.2	—

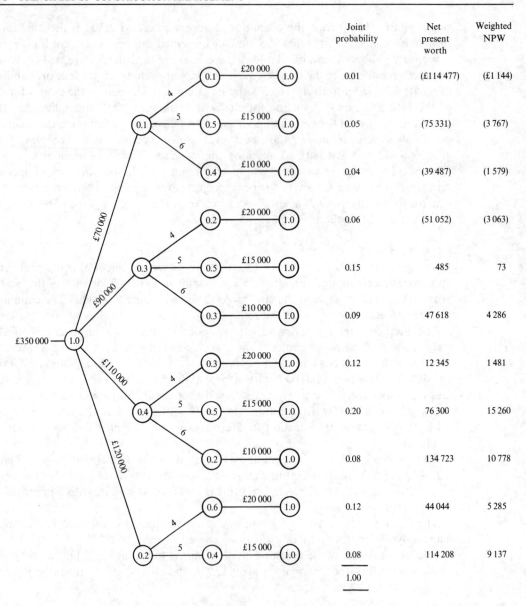

	Joint probability	Net present worth	Weighted NPW
	0.01	(£114 477)	(£1 144)
	0.05	(75 331)	(3 767)
	0.04	(39 487)	(1 579)
	0.06	(51 052)	(3 063)
	0.15	485	73
	0.09	47 618	4 286
	0.12	12 345	1 481
	0.20	76 300	15 260
	0.08	134 723	10 778
	0.12	44 044	5 285
	0.08	114 208	9 137
	1.00		

Figure 5.11

of the tree shows the initial capital investment and the probability of its achievement is shown in the circular node at its right-hand end. In this case one single certain estimate has been made, and therefore its probability is 1.0. If there had been more than one estimate, there could have been more than one introductory trunk to the tree.

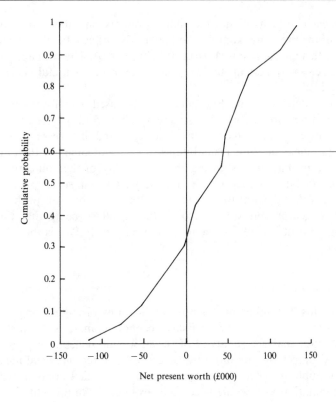

Figure 5.12

The first branches from the 'estimated capital cost trunk' represent the alternative annual after-tax net incomes, and from each of these spring the branches representing the projected length of each alternative's life, together with the probability of each occurring placed in its end node. Finally, the certain salvage value for each projected situation is added. Ignoring for the time being the question of the probabilities on the tree structure, it can be seen that the uppermost alternative outcome for the project is for it to have a capital investment of £350 000, an annual after-tax net income of £70 000 for each of 4 years and a resale value of £20 000. The net present worth of such a project would be:

$$NPW = -£350\,000 + 70\,000(P|A, 10\%, 4) + 20\,000(P|F, 10\%, 4)$$
$$= -350\,000 + 70\,000(3.1699) + 20\,000(0.6830)$$
$$= \underline{-£114\,477}$$

This equivalent calculation is carried out for each of the alternative outcomes and the results are listed at the end of each relevant complete branch in Fig. 5.11.

The joint probability for each complete branch is now calculated and added to the figure. The joint probability is the result of multiplying the individual independent probabilities for each branch of the tree that represents one alternative outcome. Thus in the case of the previous example, the uppermost branch, the joint probability of the outcome of this alternative is $(1.0)(0.1)(0.1)(1.0) = 0.01$. The probabilities of all outcomes must necessarily add up to 1.0. When they do so a useful check on the arithmetic is provided.

The information derived from Fig. 5.11 can be used to plot a cumulative probability curve of net present worth, as shown in Fig. 5.12. In the first instance the net present worths of the outcomes must be reordered, if necessary, in ascending order of magnitude. The joint probabilities of each outcome are then added to become cumulative probabilities showing the probability of that outcome or less. The cumulative probability is then plotted on the y-axis against net present worth on the x-axis. It will be seen from Fig. 5.12 that the overall probability of making a negative net present worth is 0.31, the corollary being that the probability of making a profit is 0.69. The sum of the weighted NPWs is the EV for the proposal.

Summary

This chapter contains a number of different applications of engineering economic principles as well as an introduction to risk in economic analysis. It starts with a glance at benefit–cost analysis—analysis of the economic benefits of projects that cannot be assessed simply by rate of return or any other direct financial measure. It is used for many public projects in the government and allied areas in which the benefits to the population as a whole need to be estimated. An introduction is then given to break-even cost analysis, which is one beginning of how to measure the sensitivity of projects to variations in the assumptions on which are based the original calculations. Examples are worked through on a number of applications to production and purchase. There follows a more formal examination of sensitivity analyses. Next comes valuation. Valuation is an important tool in business, since assets in particular have a value and it is often necessary to have a good idea of what that is. A procedure is set out for calculating value by discounting the cash flows that evolve from being with or without assets in order to assess their value. It is highly desirable to replace or retire an asset at the point in its life when it is about to become more costly than an alternative. One method for using this thesis is set out and followed by a treatment of the lease-or-buy problem. The chapter closes with a brief review of some aspects of risk analysis including an introduction to probability theory—the basis of much risk analysis.

Problems

Problem 5.1 It is proposed to construct an underpass to ease the congestion at a very busy city centre street junction. The underpass will have the benefits of improving safety and saving the time of users. It is expected to cost £20 million and to involve an additional annual operating cost of £700 000.

It is estimated that each of the 8000 vehicles making daily use of the underpass will save 15 minutes at an estimated cost saving of £12 per hour. The congestion now occurs on approximately 270 days in the year. It is expected that better safety provisions of the the underpass will result in an average of two less accidents per day each of which may be considered to cost £650.

Using a benefit–cost analysis, a discount rate of 7 per cent and an asset life of 20 years, determine whether the underpass should be built.

Problem 5.2   A company designs a new device for providing ventilation at the eaves of domestic houses. In considering its production and sales plans it is decided to sell the device for £18.75 per unit. The company estimates that it will need to install £300 000 worth of new equipment to produce the units. The labour cost for 1000 units amounts to £3750 and the cost of overheads is £75 000 per year. The company considers only the first 5 years of such ventures in deciding whether they are worthwhile, after which it assumes that the equipment has no value. If the company has a minimum acceptable rate of return of 14 per cent, determine how many units per year it needs to make and sell to break even. Draw the break-even chart.   (10 826 units)

Problem 5.3   A manufacturer produces mobile compressors for the construction industry at a selling price of £35 000 each, of which £21 000 is a variable cost. The manufacturer's maximum output is 100 units a year. Overheads amount to £650 000 for a year. At the present time 65 units are made and sold each year, but the manufacturer believes that reducing the selling price to £33 000 and spending £150 000 on marketing can raise sales to 95 units a year. Reducing the selling price means making some modifications to the compressors, which will increase the variable cost by £1000 per machine. Assess the effects of carrying out the new policies and use a break-even chart to illustrate the analysis.

(Original break-even = 47 units; new break-even = 73 units; original profit = £260 000; new profit = £245 000)

Problem 5.4   The contractor building a large office complex considers installing a central furnace to provide hot air so that bricklaying and concreting can go ahead in cold weather. The furnace will cost £65 000, have a life of 10 years, and have no resale or salvage value at the end of its useful life. The maintenance and operating costs of the equipment will amount to £14 000 per year independent of the extent of use. The managers of an adjacent building offer to provide the necessary heat from their standby boiler at a cost of £295 per day. If the contractor's cost of capital equals 12 per cent, calculate by break-even analysis the number of days per year on which heat would have to be provided to make the external source of heat a cheaper alternative. What would the contractor's cost of capital have to be for the two heat sources to be of equal cost assuming use on 80 days of the year?   (87 days; 7.8 per cent)

Problem 5.5   The manager of a plant producing ready-mix concrete wishes to expand his turnover by 25 per cent. The present turnover of the plant, which is at

100 per cent effective capacity, is £1 000 000 per year. The fixed overhead is currently £300 000 per year and the variable costs are £500 000.

One possibility for expanding the turnover is to work longer hours in the plant. An investigation of this situation shows that the fixed overhead will remain at £300 000 per year but that the variable overhead will increase at a faster rate than currently, owing to the necessity of paying overtime rates to the labour employed. In fact, the estimate shows that from £500 000 at what is currently 10 per cent effective turnover, the variable costs will increase to £700 000 by the time that the turnover is increased by 15 per cent, and to £1 000 000 by the time that the full expansion of 25 per cent is achieved.

An alternative proposal is to extend the plant by adding new equipment. This, it is estimated, will increase the fixed overhead to £450 000 annually and the variable costs will increase linearly from zero to £750 000 at the new maximum turnover.

Compare the two schemes using a break-even chart and comment on the relative merits of each proposal. Frame recommendations to the board of directors of the company that owns the plant concerning expansion of turnover by the methods proposed.

Problem 5.6 A proposal comes to the board of directors for the investment of £150 000 in a plant modification with a life of 10 years, after which the plant is expected to have a salvage value of £60 000. The modified plant will give rise to additional operating costs of £42 000 per year. Check the sensitivity of the proposal to changes in initial cost, salvage value, economic life, and operating costs by establishing the amount of after-tax revenue required in each circumstance to achieve a rate of return of 14 per cent. Examine the sensitivity due to changes of ± 30 per cent in each of the variables. Assume that the company pays tax at the rate of 35 per cent and that the plant will be depreciated by the straight-line method.

Comment on the outcome of the analysis.

Problem 5.7 Two years ago the ABC Plant Hiring Company bought a fleet of dumptrucks for their own use. The life of the trucks was estimated to be 5 years. Their initial cost was £14 350 each and the salvage value of each at the end of its useful life was believed to be £2000. Operating costs have proved so far to be £1300 per year and this rate is expected to be maintained for the next 3 years. The initial price of the dumper has remained unchanged, though some improvements in design have been effected which will now reduce the operating costs of a dumper to an expected £1000 per year. If an attractive rate of return for the company is 11 per cent, what is an appropriate valuation for each of the existing dumpers at the present time?

State the assumptions that you have made in arriving at the valuation.

Problem 5.8 What would the valuation of the trucks of Problem 5.7 amount to if for the first year the existing trucks had an operating cost of £1300, which then increased by £100 for each succeeding year and the new trucks had a uniform cost of operation of £1500 per year?

Problem 5.9 The same company wishes to carry out a valuation of other equipment that they own and that has been in operation for some years. One particular item of equipment has been in use for 2 years of its estimated economic life of 9 years. It is estimated that it will have a resale value of £10 000 at the end of its economic life. A new replacement for this machine at today's price will cost £55 000 and will have a similar economic life and resale price to that of the existing machine. The annual operating and maintenance costs of the existing machine are estimated to be £15 000 for the third year of service and thereafter to increase by £1000 per year until the time comes to dispose of it. A new replacement machine to provide the same service has the advantage that its annual operating and maintenance costs are estimated to be lower than those of the existing machine and to start at £10 000 for the first year, increasing by £1200 per year throughout its life. If the cost of capital to the company is 11 per cent, what is the value of the machine to the company at the present time?

Problem 5.10 A company intends to purchase a new computer for £75 000. The company's records show the data set out in Table 5.9.

Table 5.9

Resale year	Resale value	Operating costs
1	£65 000	£0
2	55 000	6 500
3	35 000	9 000
4	30 000	15 000
5	20 000	17 000

If the company's cost of capital is 14 per cent, what is the expected economic life of the computer and the present worth of its total cash flows at that life?

If tax is at 35 per cent and the computer is depreciated by 100 per cent in its first year, what uniform annual revenue is required to justify the purchase and subsequent costs?   (3 years; £109 700)

Problem 5.11 Examine the sensitivity of the economic life of the equipment that is the subject of Table 5.6 to establish to what the interest rate needs to be increased before the economic life changes.

Problem 5.12 A company wishes to buy cars for three of its senior executives at an initial cost of £90 000. The cars are expected to have a four-year life and will be depreciated by the straight-line method to an expected resale value of £30 000 at the end of that life. The company is alternatively offered a lease for the three cars at £14 000 per year payable in advance. The operating, maintenance, and insurance expenses will be undertaken by the company using the cars. If the company pays corporation tax at 35 per cent, should it take up the lease with a required rate of return of at least 13 per cent?

Problem 5.13 A leasing organization wishes to offer a competitive lease to a company that wants to purchase a number of machine tools. The leasing organization knows that the prospective purchaser has a minimum required rate of return of 11 per cent and pays tax at 32 per cent. It has also found out that it proposes to pay £120 000 for the machine tools, which will have a useful life of 7 years (with no resale value) and that they will be depreciated in the company's books by the declining-balance method.

At what annual value should the leasing company offer the lease so that it makes it competitive with outright purchase?

Problem 5.14 When rolling a single die, what is the probability of throwing a number less than three with one throw?

What is the probability of getting each of the even numbers in any order with three consecutive throws?

What is the probability of getting each of the even numbers in ascending numerical order with three consecutive throws of the die? $(\frac{1}{3}; \frac{1}{2}; \frac{1}{216})$

Problem 5.15 Suppose that two dice, each of which can be identified, are rolled at the same time. What is the probability of the throw resulting in a total of seven showing on the two faces?

What is the probability of showing a total of six on the two faces? $(\frac{1}{6}; \frac{5}{36})$

Problem 5.16 A box contains three black, three white, and five red balls. Calculate the probability of two successive draws from the box giving: (a) a red then a white ball; (b) two black balls; (c) no white balls.

Assume in each case that the ball from the first draw is replaced.

For the same events find the probabilities when the ball from the first draw is not replaced. $(\frac{2}{27}; \frac{16}{81}; \frac{25}{81}. \frac{1}{12}; \frac{1}{6}; \frac{20}{72})$

Problem 5.17 A contractor has the financial resources to bid for either an earthmoving project or the manufacture of precast concrete beams, but not both. In each case working capital of £100 000 will be required.

In the case of the earthmoving contract, the contractor estimates that if the weather is good the project should return 12 per cent on his investment; if it is bad, it may only reach $1\frac{1}{2}$ per cent. Alternatively, the precast concrete beam project will bring a certain return of 8 per cent.

What will the probability of bad weather need to be if the two projects are to yield the same expected monetary value? (0.381)

Problem 5.18 It costs a firm of contractors £100 000 to prepare a bid for a construction project. For initial consideration within the firm, four bids are prepared—one of £5 000 000 which would make them £200 000 profit, one of £4 500 000, the profit on which would be £180 000, one of £3 900 000 for a profit if

£160 000, and one of £3 250 000 for a profit of £120 000. The firm estimates the probability of winning each of these four bids as 0.05, 0.25, 0.30 and 0.40 respectively.

Which bid should the contractors make?   (Choose 2 or 4 to minimize loss)

Problem 5.19 A radio mast is to be constructed on the fringe of an earthquake zone. Seismologists estimate that the probability of severe earthquake damage to the mast is 1 in 9 in any one year. It is estimated that repair costs for the mast will amount to £200 000 if an earthquake does occur. The economic life of the mast is estimated to be 20 years. If the owner's cost of capital is 12 per cent, would it be better to add £250 000 to the cost of the original structure in order to design it to earthquake-resistant standards in the first instance?   (No)

Problem 5.20 A project manager has three mutually exclusive projects from which one must be selected. Project 1 has a probability of 0.3, 0.1, 0.2, and 0.4 of achieving a return of −5 per cent, 0 per cent, 10 per cent, or 20 per cent respectively. In the case of Project 2, the probabilities of the same returns are 0.0, 0.3, 0.5, and 0.2; for Project 3 they are 0.15, 0.15, 0.4, and 0.3.

Determine the project to be recommended for investment if the criterion to be used is that of maximum return with the most probability of achievement.

Determine the expected value of the investments and then, from these values, select the optimum project.   (1; 2)

Further reading

Couper, J. R. and W. H. Radar: *Applied Finance and Economic Analysis for Scientists and Engineers*, Van Nostrand Reinhold, New York, 1986, Chapter 9.

Dixon, R.: *Investment Appraisal—a Guide for Managers*, Kogan Page in association with the Chartered Institute of Management Accountants, London, 1988.

Grant, E. L., W. G. Ireson and R. S. Leavenworth: *Principles of Engineering Economy*, 8th edn, Wiley, New York, 1990, Chapters 14–16.

Lindley, D. V.: *Making Decisions*, Wiley, London, 1971.

Pilcher, R.: *Project Cost Control in Construction*, Collins, London, 1985, Chapters 4, 6 and 7.

Raiffa, H.: *Decision Analysis—Introductory Lectures on Choices Under Uncertainty*, Addison-Wesley, Reading, Mass., 1968.

Riggs, J. L. and T. M. West: *Engineering Economics*, 3rd edn, McGraw-Hill, New York, 1986, Chapters 8, 10 and 13–16.

White, J. A., M. H. Agee and K. E. Case: *Principles of Engineering Economic Analysis*, 3rd edn, Wiley, New York, 1989, Chapters 5, 7 and 8.

Taylor, G. A.: *Managerial and Engineering Economy*, 6th edn, Van Nostrand Reinhold, New York, 1980.

6. Construction productivity improvement

6.1 Productivity

Productivity is the rate of producing. A satisfactory comprehensive and universal definition of productivity does not exist for construction. In general terms it can be, and is, defined as *output divided by input*. While input is normally relatively easy both to define and then to establish in quantitative terms, output has successfully defied definition by investigators in precise and truly meaningful terms. This is largely because of the widely varying output of construction, in which a *unit*, except when measured at the lowest levels of physical size or cost, varies immensely in production content. Figures are frequently quoted, for example, about the labour content in the unit of 'a house'. It is said that construction productivity in country A must be greater than in country B, because less man-hours go into building a house in country A. Unfortunately there is no such thing as a standard house. Even if each house provides the same floor area and the same accommodation, houses are built to different standards, have to meet different design requirements, have to cope with different climatic conditions, can incorporate materials of widely varying qualities, need to satisfy different building regulations and so on. Even if seemingly well-defined units, such as a cubic metre of reinforced concrete, are used for productivity studies, the multitude of ill-defined influences on production times, or costs that cannot be ascribed to that unit, make the answer obtained unreliable.

Cost proves to be no more accurate a measure of output than physical quantity or description—even within a fairly close geographical locality. The majority of construction work is undertaken on a competitive bid basis and hence the true cost of a unit of work is difficult to establish with reasonable accuracy. In many instances construction contractors are unable to determine the accurate cost of a unit of work so as to form a basis of comparison with or to create an estimate for the next similar work. Such figures, if available, often have a large indeterminate variation within them—usually the subject of circumstances beyond the direct control of the contractor.

The fact that productivity is difficult to measure accurately should not, however, be used as a satisfactory excuse for neglecting all efforts to keep it to a high level and constantly to try to improve it. The three principal components in the basic direct cost of carrying out construction work at a site are those for *labour, equipment,* and *materials.* Sometimes *capital* is included in this context, but, important though it is, it is rarely possible to have direct control over the use of capital, as such, at a job site. To the above three components must be added the *site overheads* as they affect *productive work.* Many site overheads such as the cost of providing offices, telephones, ancillary buildings such as stores, canteens, toilets, temporary roads, etc., do not contribute directly to specific productive work and therefore cannot be incorporated into direct costs. However, management, supervision and technical support are very important to productive work and can and do have a direct influence on it. In this respect it is informative to review the breakdown given by R. L. Tucker in the paper listed under Further Reading at the end of this chapter. In reviewing the type of development and research needed for construction Tucker wrote about the way in which a typical skilled operative on a job site has their productive time reduced by the effects of indirect requirements. The breakdown in Table 5, on page 167 of Tucker's paper, is shown here in Table 6.1. This, while clearly not universal (and never intended to be so), gives an indication of where a manager's attention to productivity improvement needs to be focused. If it was not impossible, it can be seen that the complete elimination of administrative delays and inefficient work methods would double the productive time of an operative.

An essential step to improving job-site productivity is to undertake a careful and thorough investigation of all the relevant factors before job site production commences. Such a process is known as *preplanning.* It involves deciding in detail what is the extent of the task under consideration, how work will be carried out and by whom, what resources need to be made available, what specialized equipment, if any, will be needed and when temporary works, such as formwork manufacture or temporary support for adjacent structures, will be required, the sequence and requirements for delivery and locating materials at site, and indications of special safety and inspection requirements. These matters are referred to again in Chapter 7.

The remainder of this chapter is devoted to productivity improvement studies and the work study techniques that can be used in order to make job site operations more efficient both before and after they have started. Such studies have the added function of providing feedback to assist with the task of ensuring that the most efficient ways of undertaking work in the future are available.

Table 6.1

Productive time	40%
Unproductive time:	
Administrative delays	20%
Inefficient work methods	20%
Work restrictions	
jurisdictions	15%
Personal	5%

6.2 The scope of work study

Work study is defined in BS 3138 (1979) as

The systematic examination of activities in order to improve the effective use of human and other material resources.

The natural division of work study falls between *method study* and *work measurement*. The structure of the subject is outlined in Fig. 6.1. Method study encompasses a number of techniques that are concerned with the critical review of the ways in which work is done, or is proposed to be done. Its objectives are to reduce the cost of the work and to make it easier to carry out and to be more effective. An intensive study is made of the processes involved; these are recorded and then critically examined with a view to making changes if necessary so as to improve as many aspects of the work as is possible. Work measurement is concerned with the establishment, by a variety of methods, of yardsticks for human effort and, as such, involves the measurement of the time that is required to carry out a specific job under specified conditions on the basis that it is carried out by a qualified worker. Each of these broad divisions of work study can be applied to problem solving without reference to the other, though the best results are obtained most often by a carefully planned combination of the two. Method study, a creative process, when properly applied should result in higher productivity through the improved layout of the production facility, a better and safer environment of work, the reduction of fatigue, improved quality of work, and better plant and equipment design. All of these factors will contribute to the more economical utilization of productive facilities. Work measurement facilitates improved planning and control of production and costs, reduced costs of resources by providing established yardsticks that allow

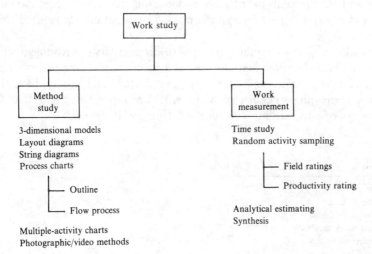

Figure 6.1

ready comparison of alternative job methods, and improved estimating of the cost of work; it also provides a sound basis for incentive schemes and scheduling processes.

Work study can be applied to all situations where human resources are used in carrying out work. It is clearly desirable that the benefits that result from its application are enumerated and quantified in terms of cost savings, increased productivity, or shorter production times. On the other hand, improvements may be justified by way of better motivation of operatives, better industrial relations, or greater job satisfaction, all of which defy precise or immediate quantitative measurement. Whatever the form of benefit—quantitative or qualitative—work study is not important for its own sake. Its justification is that it will assist with the achievement of the objectives set out above.

Before a decision is taken as to the depth of a work study to be undertaken or as to the particular techniques that should be selected for use as being best suited to the project in question, a preliminary economic appraisal of the likely benefits to arise should be considered. Where possible, savings should be expressed in terms of cost, whether as an absolute quantity or in costs per unit of production such as the cost per cubic metre of concrete placed. Savings are likely to accrue not only in direct labour costs but also in overheads. Frequently, the better organization of work will result in more efficient handling of materials from both the point of view of design and that of standardization. Care must be taken that ill-chosen or poorly designed incentive schemes do not lead to wastage or excessive use of materials as a result of sacrifices made by operatives seeking to achieve difficult targets. Otherwise some or all of the savings that may have been made will be dissipated.

The apt choice of an appropriate work study technique is important to the economic success of a project. This is particularly so in construction, where many situations are of a one-off nature and the opportunity for method improvement and a possible saving occurs at one time only. Quick and satisfactory answers are required, either before the work is commenced or during its early stages, so that the maximum benefit may be obtained. There are repetitive situations in construction, but usually the duration of a particular repetitive operation or set of operations is unlikely to be as extended as a long production run in an engineering production shop. Because this is the nature of construction work, not all work study techniques are as appropriate for use as those applied in engineering production. Emphasis can be placed on such techniques as *activity sampling*, which is very useful in providing a rapid indication of the utilization of labour and mechanical equipment. It also may provide a guide as to the critical areas for further and more detailed investigation, particularly those areas that may yield the greatest savings. *Flow process charting* and *multiple activity charts* are other techniques that can yield good results and are appropriate to many construction activities. On the other hand, *predetermined motion time systems* are likely to offer too detailed an analysis of a work situation for all but very few construction situations.

6.3 Work measurement

Work measurement is defined in BS 3138 (1979) as

The application of techniques designed to establish the time for a qualified worker to carry out a task at a defined rate of working.

It is not possible to separate work measurement from method study completely, except perhaps in describing the various techniques that are used in carrying out each. It is not the intention here to treat work measurement exhaustively but rather to draw attention to some of the limitations and possibilities for its use in construction by describing techniques that have useful applications. It is widely recognized, even in production engineering, where manufacturing takes place in a more stable environment than in construction, that work measurement cannot be an exact science. In site construction its full use is limited to a relatively small number of situations, because circumstances change so radically from task to task. One of the prime difficulties of applying many of the work measurement techniques to site construction stems from the fact it is not always easy to define small elements of the work that is being undertaken. This frequently necessitates the study of larger and hence coarser elements of work, and therefore a broad brush approach rather than a large number of short, detailed measurements of time. In addition, the difficulty is accentuated by the widely varying skills and expertise of the operatives and the ever-changing conditions under which they work. It is affected, too, by the variety of ways in which an operative's fatigue can influence a study and the speed at which an operative can work. In spite of these difficulties, many techniques of work measurement have been used successfully in site construction work, though its application has tended to follow fluctuating patterns of fashion rather than being consistent. An understanding of the principles of work measurement, however, is important, so as to condition a disciplined approach to thinking about estimating the cost, the planning, and the scheduling of production problems. One most useful group of techniques of work measurement which has found increasing use in recent years is that of *work sampling*. Work sampling in various forms is used as a means of collecting data by observation about how work is currently being carried out with a view to effecting improvements in methods and costs.

6.4 Work sampling—statistical principles

Work sampling, also known as activity sampling, employs techniques of statistical analysis in order to take measurements of the state of activity (and hence also inactivity) of the study operation by examining small but statistically representative samples. It thus allows further quantitative analysis to be made about the effectiveness of the way in which the work is managed. Statistical sampling is well known for its widespread use in many areas from quality control to political polling. In contrast to time study, work sampling is concerned with making a controlled number of random observations so as to develop a quick and approximate analysis of the subject of study rather than making a more costly continuous and detailed time study. The approximate study (or experiment), however, if properly designed, will be within prescribed statistical limits of accuracy.

Sampling is a statistical method whereby valid generalizations are made about a *population* on the basis of samples drawn from it. (The reader should refer to a standard textbook on statistics for a more rigorous treatment of sampling theory. Here it is intended only to indicate general directions so as to enable a reader to follow the ideas underlying the theory used in applications.) *Statistical inference* is the name given to the processes of drawing inferences about a population from samples that have been derived from it and the use of probability theory to indicate the accuracy of those inferences. A population can be *finite* or *infinite*—the one has a fixed or known number of *elements* and the other, hypothetically, has no limit to the elements it contains. The elements can be objects, observations, measurements, items, members, costs, etc. Samples are drawn on the basis either of a specified *attribute* or *characteristic* of each element or of a variable or characteristic having a specified quantity or measurement. For example, in taking a sample to establish the proportion of operatives actually working productively at a given time, it is necessary to distinguish between the attributes of 'working' and 'not working'. In sampling to check the correctness of quantities of packaged bricks delivered to a construction site it is necessary to check the variable 'quantity', so that if occasional pallets of bricks are being checked the specified quantity may be 'number of bricks per pallet', or perhaps for very large quantities, 'number of pallets'. In order to employ correct sampling procedures, samples must be drawn that are representative of the population as a whole. Work sampling is concerned with *random sampling*. Random sampling may be defined as taking place when each element of the population has an equal chance of being selected or drawn. Random sampling is one of a number of methods whereby representative samples may be obtained. Other methods include *stratified* and *sequential* sampling.

One simple method of obtaining a random sample (where the population is small) is to write out a series of numbers on similar pieces of paper, and at the same time assign each of those numbers to one element of the population so that each bears a different number. The pieces of paper are then placed in a hat, box, drum, or other suitable receptacle, and thoroughly mixed together. A piece of paper is drawn from the receptacle and the number on it thus represents a particular element of the population. Before each further draw, the pieces of paper must be thoroughly mixed again. An alternative method of randomly drawing elements is to assign a *random number* to each. Tables of random numbers are published for this purpose or can readily be generated by computer. After drawing a sample from a population it can either be replaced, so allowing the chance of drawing it again in the future, or not replaced, thus excluding it from subsequent drawings. These alternatives are known as *sampling with replacement* and *sampling without replacement* respectively. In theory, if replacement takes place, a finite population may be considered to be infinite, since there is no limit to the number of samples that may be withdrawn without reducing the population to zero elements.

A population as a whole can be described in terms of *parameters* such as its mean, standard deviation, median, etc. After taking a random sample of elements from a population and calculating what are known as its *statistics*—its mean, standard deviation, etc.—these statistics may be used to estimate the value of the population's parameters. If for each sample drawn from a population a specific statistic is calculated (and it may reasonably be expected that the statistic will vary from sample to sample) then, after sufficient samples have been drawn, a distribution of each or any specific statistic can be compiled. Such a distribution is called a *sampling distribution*, and for a specific statistic such as, for example, a mean it is referred to as the *sampling distribution of the mean*. To go one step further, sampling distributions themselves have measurable statistics which may be computed.

The standard deviation of a sampling distribution, because of its importance as a measure of the precision of an estimate, is called its *standard error*. For example, the *standard error of the mean* is the standard deviation of a sampling distribution of the sample means. The use of the word 'error' in this context is not to be confused with the more general use of the word in the context of measurement or calculation. It is used here with a technical connotation.

If a situation is considered in which samples of size n are drawn from an infinite population (or from a finite population on which sampling with replacement is carried out) then the mean and standard deviation of the sampling distribution of the means are as follows:

$$\mu_s = \mu \quad \text{and} \quad \sigma_s = \frac{\sigma}{\sqrt{n}} \tag{6.1}$$

where μ_s = mean of sampling distribution
μ = mean of population
σ_s = standard error of the mean
σ = standard deviation of the population

Where n is large, that is greater than 30, the sampling distribution of the means can be shown to be approximately normal.

If the probability of the occurrence of a particular event (a success) is p then the non-occurrence of that event (a failure) becomes q or $(1-p)$. In the case of tossing a coin, for example, getting a head may be seen to be a success and a tail failure. If both the toss and the coin are fair, the probability of success, p, will be 0.5 and that of failure, q, will be $(1-0.5)=0.5$. If samples are now taken $(n>30)$ and each is examined in such a way as to determine the proportion of successes, a sampling distribution of proportions can be compiled. In this example, the number of heads achieved divided by the number of tosses in the sample would be the proportion sought. The sampling distribution of proportions so derived would be very close to a normal distribution and it should be noted that the population as a whole is binomially distributed. The mean and the standard deviation of the distribution can be calculated from (6.1) by substituting $\mu = p$ and $\sigma = (pq)^{1/2}$:

$$\mu_p = p \quad \text{and} \quad \sigma_p = \left(\frac{pq}{n}\right)^{1/2} = \left(\frac{p(1-p)}{n}\right)^{1/2} \tag{6.2}$$

where μ_p and σ_p are the mean and the standard error of the sampling distribution of proportions.

Since the sampling distributions using large samples are either normally distributed or nearly so, it is possible to use the properties of a normal distribution to derive a measure of confidence that a particular estimate of a statistic, such as a mean, will fall between two given values. For a normal distribution 68.27 per cent of the values fall between ± 1 standard deviation from the mean, 95.45 per cent between ± 2 standard deviations from the mean, and 99.73 per cent between ± 3 standard deviations from the mean. These values are called the 68.27 per cent, the 95.45 per cent, and the 99.73 per cent *confidence limits*. An alternative way of looking at the confidence measures is that 95 per cent of the values will fall between ± 1.96 standard deviations from the mean and 99 per cent will fall between ± 2.58 standard deviations. Other values for different confidence limits can be found by using normal curve area tables.

In the formula (6.2) for the standard error of the sampling distribution of proportions, n can be assumed to be the total number of observations (each being considered as a sample). The formula can be rearranged to enable the number of observations to be read directly:

$$n = \frac{p(1-p)}{\sigma_p^{\,2}} \tag{6.3}$$

$\sigma_p^{\,2}$ is the standard error of the sampling distribution of the proportions. Let it be assumed that the confidence limits for a particular situation are set and it is decided the allowable limit of error is ± 5 per cent either side of the true value. The *limit of error*, k, is the assessment of the accuracy that is required in the result to be obtained by sampling. It is the percentage variation on either side of the value obtained using the sampling process within which the results can be anticipated to fall. It may alternatively be described as the tolerance allowable in the result. A statement that 'the accuracy of the result will be ± 2 per cent at a 95 per cent confidence level' means that the result obtained will fall within ± 2 per cent of the value required on 95 occasions out of 100, or that it will only fall outside those limits on 5 occasions out of 100. It should be noted particularly that the limit of error is calculated on the total value being sought and not on the value of the proportion, p. The limit of error and the confidence limits allow a numerical value to be ascribed to σ_p in formula (6.3). The confidence limits determine the number of standard errors either side of the true value within which the result will fall and the tolerance or limit of error prescribes the same limits but in terms of proportions. Therefore,

$$z\sigma_p = k \tag{6.4}$$

where z = number of standard errors for confidence limits

σ_p = standard error of proportions

k = limit of error

Thus, if confidence limits are set at 95 per cent with a limit of error of ± 3 per cent, the values obtained must fall within ± 1.96 standard errors of the true value, and therefore

$$1.96\sigma_p = 0.03$$

and $\quad \sigma_p = 0.03/1.96 = \underline{0.0153}$

or, if percentages are used throughout instead of decimal proportions,

$$\sigma_p = 3/1.96 = \underline{1.53}$$

Returning to formula (6.3) and substituting from (6.4) for

$$\frac{1}{\sigma_p{}^2}$$

the formula for the number of samples becomes:

$$n = \frac{z^2 p(1-p)}{k^2} \tag{6.5}$$

EXAMPLE 6.1

Establish the number of observations that are required to determine the proportion of 100 operatives who are 'active' as opposed to 'inactive' on a construction site. The result is required within confidence limits of 95 per cent and within a tolerance of ± 2.5 per cent.

Firstly, it is necessary to make an approximate estimate of p in formula (6.5). This should be for the larger proportion rather than the smaller and may, if necessary, be determined from a relatively short preliminary study. Let it be assumed to be 0.75. z, for 95 per cent confidence limits, is 1.96—usually rounded to 2—and k will be 0.025.

$$n = (2)^2 0.75(1-0.75)/0.025^2 = \underline{1200 \text{ observations}}$$

6.5 Activity sampling

Activity sampling has been found to be very effective in enabling supervisors and managers to identify inefficiencies in the field which are reducing overall productivity. It enables field observations to be made in a planned and systematic fashion. If the programme of observations has been properly designed, the data collected can then be analysed in accordance with well-established statistical procedures, as briefly described in Section 6.4. Such programmes are designed to give specific degrees of confidence in the results, assuming of course that the fundamental precautions necessary to the proper collection of data are observed. Studies of this kind have considerable value where it is required to obtain an estimate of the utilization of plant or labour and/or the various proportions of time that the subjects under study are occupied by specific activities. Work sampling can be carried out by observation, by interview, or by a written questionnaire.

A main advantage of using work sampling is that it allows a larger number of machines or men to be studied at one time than can be managed using a continuous time study. This often leads to a broader picture of the efficiency of a particular operation than that obtained from a more concentrated but continuous study on a smaller group. Work sampling can be used in order to detect areas that may subsequently respond to a study in greater detail using other techniques. Clearly, the more observations that are recorded about one situation, the greater will be the accuracy of the reported outcome. However, data are usually expensive to accumulate and a balance must be struck between increased and necessary accuracy and additional cost. It is of considerable importance that work sampling can be undertaken by personnel less well trained in work study than can a detailed time study.

In setting up an activity sampling exercise, the first step is to decide on the various categories of activity that are possible and that are believed to be in need of examination, together with the reasons for such. Clearly, the simplest division is a two-category one, for example, between 'activity' and 'idleness'. Even this division in a study of an excavator with its attendant trucks carrying excavated material away can provide high-quality information on plant utilization and the balancing of the lorry numbers against the excavator performance. A more detailed classification for the observations, using the same example, may be as shown in Table 6.2.

A typical observation record sheet for an activity sampling study is reproduced in Fig. 6.2. It has been completed with sample data for the simple exercise in observing the excavator given the four activities and the four reasons for inactivity set out in Table 6.2.

In arriving at the number of observations that need to be taken, it is necessary to define the confidence limits and the limit of error that are desirable. In construction work it is generally accepted that 95 per cent confidence limits with a limit of error of ± 5 per cent will give satisfactory results that can assist in making a real contribution to increasing effectiveness. The value of p, the proportion of the total operation being observed, for construction work usually falls within the limits of 0.40 to 0.60. The maximum number of observations required for a sampling task, however many

Table 6.2

Excavator	
Activity	*Inactivity due to:*
Digging	breakdown
Loading	waiting for trucks
Turning	servicing and refuelling
Travelling	operator absent

ACTIVITY SAMPLING OBSERVATION RECORD

1. Digging	5. Breakdown	CONTRACT: Downshire	OBSERVER:
2. Loading	6. Waiting	STUDY 3 m³ excavation	DATE:
3. Turning	7. Servicing		REFERENCE: 2673
4. Travelling	8. No operator		

No.	TIME	Active					Inactive				
		1	2	3	4		5	6	7	8	
1	0809									✓	
2	0817									✓	
3	0829								✓		
4	0837	✓									
5	0847	✓									
6	0859		✓								
7	0912				✓						
8	0925		✓								
9	0939		✓								
10	0950				✓						
11	1015			✓							
12	1032	✓									
13	1042							✓			
14	1057							✓			
15											
16											
17											
18											
19											
20											
21											
22											
23											
24											
25											
26											
27											
28											
29											
30											

Figure 6.2

operatives or machines are involved, occurs when $p = 0.50$. At 95 per cent confidence limits and ± 5 per cent limit of error, the number of observations required is as follows:

$$n = 1.96^2(0.50)(1 - 0.50)/(0.50)^2 = \underline{384}$$

Therefore, for construction work, 384 observations is normally accepted as a minimum satisfactory number. It will allow statistically significant results to be obtained whatever proportion of the activity turns out to be the case.

There are no limitations, other than physical ones, to restrict the size of the task that can be subjected to a work sampling observation. If 384 observations are required, however, and 13 operatives are performing the work under examination, then it will be necessary to sample the 13 men on 30 different occasions to obtain the statistically necessary data. Alternatively, if there are 25 operatives, each will need to be sampled 16 times in order to get a satisfactory result. Table 6.3 lists the sample sizes for confidence limits of 90, 95, and 99 per cent, given various category proportions and limits of error. Alternatively, the nomograph set out in Fig. 6.3 may be used. Note that the values of k are expressed as percentages and, therefore, the values of p must also be percentages in the nomograph as opposed to decimals in previous example.

Table 6.3

Category proportion, %	Limits of error, %			
	1	2.5	5	10
99% confidence limits				
50, 50	16 641	2 663	666	166
40, 60	15 975	2 556	639	160
30, 70	13 978	2 237	559	140
20, 80	10 650	1 704	426	107
10, 90	5 991	959	240	60
95% confidence limits				
50, 50	9 604	1 537	384	96
40, 60	9 220	1 475	369	92
30, 70	8 067	1 291	323	81
20, 80	6 147	983	246	61
10, 90	3 457	553	138	35
90% confidence limits				
50, 50	6 765	1 082	271	68
40, 60	6 494	1 039	260	65
30, 70	5 683	909	227	57
20, 80	4 330	693	173	43
10, 90	2 435	390	97	24

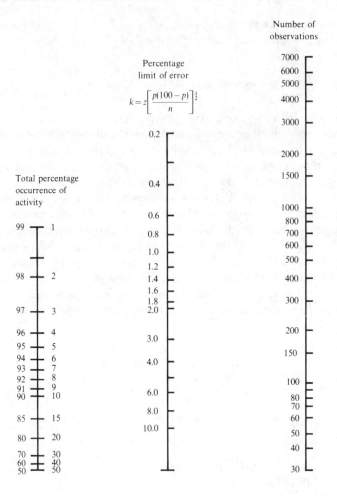

Figure 6.3

In determining the pattern of the observations that are to be made as part of an activity sampling study, a number of factors need to be taken into account. The degree to which randomness needs to be introduced into the schedule of observations often receives too much emphasis. Randomness is desirable where the activity under study consists of a cycle of operations over a short period of time—one hour, for example. Rarely will this occur in construction practice, but when it does then randomness can be introduced by the use of a table of random numbers such as those commonly found in the appendices of statistics textbooks.

The pattern of observations will all be affected by the variability in the pattern of the subject to be studied. If it follows a similar pattern over a period of, say, one day then the observations can be carried out during one day. This will be influenced, of course, by practical considerations as well, but the validity of the observations as part of one particular study will pertain only if the work situation remains substantially the same over the course of the study.

Care must be taken in carrying out a study that it is not biased by introducing subjective observations. Care must also be taken to see that observations result in instantaneous judgements of what the observed subject is doing at that moment. The observations become of considerably less value when the observer speculates as to what has just happened or is just about to happen and allows that to influence the record. For example, an operative under study may be observed standing with his shovel resting on the ground. He is clearly inactive. However, if a fraction of a second after observation commences he lifts his shovel ready to dig, there may be a temptation to record him as being active. This should be resisted. At the same time, observations should be made as discreetly as possible so as to allow the study subject no time in which to react to the fact that he is under observation.

Activity sampling encompasses a number of different approaches to suit different situations and to give an insight into differing aspects of work. Two variations to be covered in the subsequent sections are:

1. field ratings;
2. productivity ratings.

6.6 Field ratings

A field rating is the simplest of the activity sampling techniques in practical use. Much of the discussion up to this point in relation to activity sampling has encompassed field rating, in which observations are made with a view to categorizing them into two divisions, 'working' or 'not working', 'active' or 'inactive', 'in attendance' or 'absent', and so on. The value of such a simple analysis hinges very much on the validity of the observations. It is frequently difficult to distinguish between 'working' and 'being productive'. A tradesman talking to a foreman may be receiving instructions without which he cannot make any further progress with his work or he may be asking for the name of the favourite in the 3.30 p.m. race at Ascot. With snap observations it is difficult to differentiate realistically between the two.

Making such a study requires the full-time attention of the observer. The study should not be carried out by somebody who also has another task to perform at the same time. The recording of data is best carried out by using two mechanical counters—one of which is used to record the total number of personnel being observed and the other to record the number of personnel who come within one of the two categories being used as a classification. As well as complying with the general conditions for activity sampling as set out in Sections 6.4 and 6.5, wherever possible field ratings should cover as large a percentage as possible of the personnel who are being observed and for this purpose ideally would be included in the survey. If all of the relevant personnel cannot be covered then an attempt should be made to include at least 75 per cent of them.

For the most realistic results, field rating studies should not commence immediately work starts, either at the commencement of the day or after a meal break, unless the survey is concerned with highlighting the general timekeeping standards for the project. Time should be allowed for the personnel to settle into their particular tasks, and wherever possible to develop a rhythm in carrying out the relevant operations.

6.7 Productivity ratings

Productivity ratings involve a further stage of sophistication beyond the two-division field ratings discussed in the previous section. They are used in a planned and systematic way in order to highlight inefficiencies in construction activities by classifying the effectiveness (in a production sense) of the work being carried out. While it has been seen in the example of the excavator and its associated trucks that it is possible to break down each of the two categories of activity that become the subject of the investigation, in the case of a productivity rating the division is based upon the effectiveness of the work. One such division postulated by Oglesby, Parker, and Howell is to have three divisions, namely *effective work*, *essential contributory work*, and *ineffective work*.

Effective work is defined as work that directly contributes to a physical addition to the unit of the construction. (In some cases, where dismantling or demolition is required, there may be a physical removal.) The effective work must necessarily include some movement in the work area, such as that associated with hand excavation in a trench, that used to spread concrete in a paving slab or that of a loader driving into a spoil heap for removal.

Essential contributory work is that which cannot obviously be identified immediately with a particular unit of construction, but without which the work that is the subject of the investigation could not proceed. Examples are the erection of a staging from which bricklayers may work in elevated positions; setting up and adjusting timber guides prior to driving steel sheet piles; heating and preparing mastic asphalt before applying tanking to concrete walls; extending the outflow pipe to a concrete pump during concrete placement.

Ineffective work falls into the category of inactivity; the ineffective worker may be active in the sense of walking or driving about but is doing so without purpose and is not directly contributing to the physical addition of the unit under construction. Having to rework units of construction after failure to comply with the drawings or the quality requirement will fall into this category.

While it would be useful if every observer for the purpose of productivity sampling were to place every observation of similar work in the same one of the three categories, this is almost impossible. It is not possible to make an exhaustive list of what is to be classified as effective, ineffective, and essential contributory work and then to ensure that the classification is perfectly applied. This means that it is often useful to have the observer carry out the analysis of the data and be a partner in any discussions which go on concerning improvements which may be made to the work methods. Oglesby, Parker, and Howell quote series of typical productivity ratings that represent good performance and were collected by a large construction firm over a period of years. A few of these values to give an indication of what may be expected are, bricklayer: 42/33/25; carpenter: 29/38/33; labourer: 44/26/30; painter: 46/26/28. The numbers are percentages and reflect the proportions of effective, essential contributory, and ineffective work respectively.

Labour utilization factors can be established with the results of a work sampling study. Where a productivity rating study has been carried out there arises the question of the extent to which essential contributory work should be included in the calculations. Experience with both including and excluding the essential contributory work will be the best guide. Without any inclusion, the labour utilization factor will be:

$$\text{labour utilization factor} = \frac{\text{effective work observations}}{\text{total observations}} \qquad (6.6)$$

A compromise concerning the essential contributory work frequently leads to a formula acknowledging its existence but only in part:

$$\text{labour utilization factor} = \frac{(\text{effective work}) + \frac{1}{4}(\text{essential contributory})}{\text{total observations}} \qquad (6.7)$$

6.8 Time study

Time study is a principal technique of work measurement. It is concerned with data collection in the form of recording times and rates of working. As a result of analysing the data, calculations are made of the standard times and/or standard production outputs for carrying out a specified job at defined levels of performance. Generally speaking, a time study will be the most satisfactory method of obtaining standard times for construction operations. The subject of a study may be an individual operative or a gang. The process is flexible in that it can be made on a very detailed basis or to a much coarser degree of detail, depending on the objective of the study, the nature of the results required and the subject of the study. In order to produce accurate and reliable time studies, a practitioner needs intensive training in the laboratory and practical training on the job.

The usual phases of a time study consist of the following:

1. familiarization with the work to be studied;
2. collection of data by timing;
3. rating by assessing performance against accepted standards;
4. adding allowances for relaxation to obtain standard times.

The first phase of the process of time study, that of familiarization, is one in which a practitioner makes himself totally conversant with the objectives of the study—why it is being carried out—together with its location, the study subjects, the general environmental conditions surrounding the study, the details of the numbers of operatives involved, and the plant, equipment, and materials that may be used. Sketches should be made of the layout of the workplace at which the study operation is to be carried out and all the detail collected should be recorded for future reference. In studying the operation at this stage, thought should be given to the way in which it should be broken down into component *elements*. When a conclusion is reached the elements should be listed, with a complete description of each of them. It should be remembered in preparing the breakdown that each element needs to be timed and that the *breakpoints* between elements of work should be readily distinguishable. It is not practical to use elements that have a duration of less than approximately 0.10 to 0.15 minutes. It is also important to keep separate those elements that involve different levels of effort. For example, walking on a horizontal surface, climbing a ladder, and walking up a slope wheeling a loaded barrow demand widely differing efforts on the part of an operative, and should therefore be timed as separate elements. Again, where machines are in use, if they have fixed cycles, running speeds, or rates of working then these should be kept distinct from the activities of an associated operative whose pace of working may vary from hour to hour and from day to day.

The second phase involves the collection of data. A practitioner will need to equip himself with a stopwatch or other suitable timing device, a clipboard and prepared observation sheets, and a supply of pencils. The essential use or not of a stopwatch will depend on the nature of the job to be observed. If the elements to be studied have estimated durations of about a minute or less, a stopwatch will have advantages. Where the minimum units of time are likely to be in hours and minutes with an occasional requirement for a breakdown into seconds, a watch with a second hand is probably quite satisfactory. Such an example may be for the driving of steel sheet piles, where the elements will tend to take hours and minutes rather than seconds. If the units of time are larger still, such as for the erection of a large but repetitive structural steel transmission pylon on prepared foundations, the overall process is likely to take several days and almost any timepiece measuring hours and minutes is suitable.

The measurement of time does not consist solely of taking the start and finish times of the work task under observation, but the start and finish times of each element of the task while the work proceeds. The sum of the time for the elements will be equal to the total time for the task. Where the elements are of such a size that a stopwatch is called for, the use of *flyback* timing is extremely useful. A flyback stopwatch for time study is normally calibrated in minutes, and decimal parts of a minute, leading to simplicity of calculation. It has, in effect, two dials—an outer larger dial recording decimal parts of a minute, with one revolution of the hand being one minute, and a smaller, minor dial recording the number of elapsed minutes. Pressing and releasing the winding button causes both hands revert to zero. The stopwatch can be started or stopped by sliding a small catch on the top of the watch. In timing the elements of a work task, the watch is started at the commencement of the task. When the first element is completed the observer quickly reads the number of minutes and seconds that have elapsed, then presses the flyback button (the winding button); at this, the hands revert to zero and, without any pause, the timing of the second element begins. This goes on until the study is finished.

At the end of the study the finish time is recorded on the observation sheet. A typical time study observation sheet is shown in Fig. 6.4. The observed times are recorded in minutes and decimals thereof in the third column of the sheet, continued if necessary through to the sixth column. If a flyback stopwatch is not available to the observer, *cumulative* timing may be used for the study, recording the cumulative time expired to each breakpoint. This does mean slightly more calculation to find the times for each element, but in construction work, where there are frequently few elements to a work task, it should not prove to be a real disadvantage. Wherever possible, a series of abbreviations of element names should be developed for a study in order to make the recording of a task more simple and less prone to errors as a result of being distracted by writing as timing proceeds.

Few operatives carry out work at the same tempo as any of their colleagues. Not only can the rate of working for an individual vary from that of others, but it is also likely to vary for the individual from time to time during a day. After a break in work, either first thing in the morning or in the afternoon after lunch, it takes some time for an individual's tempo of work to increase to a peak, and there is then a tendency for it to fall away towards the onset of the next break. Clearly this variation in the rate of doing work reflects on the length of time it takes to complete an operation. When an observer is making a time study it is necessary to make allowances for this variation by a process known as *rating*. The *standard rating* for an activity is difficult to define, but an observer needs to establish one in his own mind and to rate consistently the tempo of an operative against it. Standard rating is assigned a value of 100 in British Standards terminology. At the same time as an observer records the time that an element takes, the rating should also be recorded alongside it (in column 2 of the form shown in Fig. 6.2). For a tempo less than standard, a rate of 85 or 90 may be assigned for example; for a rate above standard, 105 or more may be assigned.

There are two influences on an operative that will cause him to vary his production rate: firstly if he alters the speed at which he carries out tasks, and secondly if he makes a change in the method he is using to carry out the work. Rating is used to assess the first of these variables only, since it is very difficult to make allowance for the second. An observer should attempt to assess the speed at which the observed operative is working and make a mental comparison with what he believes would be the speed of the work on the same task if it was carried out at the standard rating. Rating is an operation that requires considerable experience and skill and which, even in observers with both of these attributes, requires periodic practice and testing sessions to ensure consistency and accuracy of assessment.

On completion of the study, after checking that the sum of the observed times agrees broadly with the independent check start and finish times on the observation sheet, a ring should be placed round any observation that is not typical of normal

TIME STUDY OBSERVATION SHEET

Study by:		Date:	Started:		Finished:		Study No	Sheet	of
Element			Rating	OT	Element			Rating	OT
Opening check time				0915	B/F				24.67
					Drill timber			105	2.15
Collect tools and walk 50 m to site			90	6.70					
					Fix screws			100	6.78
Clean out			90	1.80					
Measure			95	0.40					
Talk to foreman			100	1.20					
Measure			95	0.73					
Cut timber			105	1.25					
Re-check measure			100	0.25					
Fetch timber			80	3.21					
Adjust bandage on thumb			IT	2.21					
Move to adjacent site			90	3.56					
Measure			100	0.75					
Cut timber			105	2.61					
					Final check time				0943
Total C/F				24.67	Total				33.60

Figure 6.4

conditions so that it can be ignored in the calculations that follow. The observed times may then be adjusted according to the given rating to arrive at a *basic time* shown in the extended observation sheet in Fig. 6.5. The basic time may be defined as that time in which an element would be completed if it had been undertaken at a standard rating. It is calculated as follows:

$$\text{basic time} = \frac{\text{observed time} \times \text{observed rating}}{\text{standard rating}} \qquad (6.8)$$

In observing and recording the various times of the elements of the studied work task, the observer needs to ignore any elements of relaxation on the part of the operatives. *Relaxation allowances* consist of two components. One is an allowance for the *personal needs* of an operative; the other is an allowance for *fatigue*. Both allowances are normally made by adding a percentage to the basic time. While the allowance for personal needs can be assessed with some reasonable degree of certainty, that for fatigue is usually very difficult to gauge. When making observations, the observer has to make sure that his or her rating does not include the effect of fatigue. This can be significant in a task that lasts several hours or more. Not only that, but different individuals are affected in different ways and to different extents by fatigue. Fatigue may be influenced, for example, by the familiarity that the operative has with the nature of the work and the conditions under which it is carried out. Many companies and industries make their own recommendations for the allowances to be used. A typical set of allowances is set out in Table 6.4.

In Table 6.4 is shown an allowance for personal needs which covers such activities as going to the toilet, getting a drink of water, fetching something from a top-coat pocket, and so on. This allowance may need some adjustment, depending on the location and size of the particular site involved. A fatigue allowance needs to be set for the operative's posture at his workplace. This will cover whether he is standing, sitting, bending, crouching, etc. Clearly, an operative excavating in a tunnel of about 1.75 m diameter requires to have some allowance over and above an operative excavating in the same material but in an open excavation or trench. The operative in the tunnel would be likely to have an allowance at the top end of the range.

Fatigue will also be influenced by the degree of attention that an operative needs to focus on his work. Open excavation may not warrant any allowance, but a tower crane driver, working at considerable heights above the ground and lifting large items or those difficult to position precisely to the upper floors of a building and then manoeuvring them into position before carefully lowering them on holding down bolts, will need an upper-range allowance.

The allowance for environmental conditions, other than climatic ones, is made for work that may be carried out in badly ventilated areas or areas in which there may be extremes of temperature, such as boiler rooms, steel furnace shops, or cold stores for food. Another common factor that makes this allowance very necessary is a dusty atmosphere, which is typical of a demolition operation.

Operation Description:									Date:		
Cut and fix timber supports to formwork									Study no:		
									Sheet of		

Notes:

Elements	Basic times	% Relaxation						% Cont^y	Total %	Standard times
		PN	P	A	C	E	M			
Collect tools	6.03	6	3	0	2	10	0	5	26	7.60
Clean out	1.62	6	4	1	2	5	0	5	23	1.99
Measure	1.82	6	2	1	2	5	0	5	21	2.22
Discussion	1.20	6	2	1	2	0	0	5	16	1.39
Cut timber	4.05	6	3	1	2	10	0	5	27	5.14
Recheck measure	0.25	6	2	1	2	0	0	5	16	0.29
Fetch timber	2.57	6	4	1	2	20	0	5	38	3.55
Move	3.20	6	4	0	2	10	0	5	27	4.06
Drill timber	2.26	6	3	1	2	10	0	5	27	2.57
Fix supports	6.78	6	4	1	2	15	0	5	33	9.02

Figure 6.5

Table 6.4 Table of relaxation allowances

Category	Range of allowance	Comments
Personal needs	6–12%	The minimum to be added to all activities. Large sites with longer walking distances to facilities may be 8%; 12% for tall buildings and sites remote from facilities such as small tunnels, jetty works, etc.
Posture	2–10%	Standing allowance: 2%. Walking with light loads and bending and stretching: 3–4%. Constant bending and stretching: 5–7%. Awkward positions in cramped conditions—restricted movement: 8–10%.
Attention	0–6%	Few jobs require more than 2–3%. Operating moving mechanical equipment needs 4–6% where operation is to close limits or in difficult surroundings.
Conditions	0–15%	Up to 10% for working under normal conditions. Light, airy conditions inside a building should require little or no allowance. Underground in dark, dusty, badly lit conditions, up to 15%.
Effort	4–50%	Work in the sitting postion requires little allowance. Wheeling a loaded barrow: 20%. Shovelling wet concrete: 30%. Carrying bags of cement: 50%
Monotony	0–8%	Most normal jobs require very little allowance. Unloading materials by hand: 6–8%. Repetitive cutting holes in concrete by hand: 8%.

One of the most obvious influences of fatigue is that of the effort that is necessary to perform the work. Lifting, pulling, and pushing weight has a variable influence on fatigue, depending on the weight involved. The degree to which an operative is trained to lift and generally deal with weights properly will have an influence on the extent to which an allowance needs to be made.

The last category of fatigue allowances is that resulting from the monotony of the work task. This means monotony of physical rather than mental effort, though in certain circumstances an allowance may be given for the effects of both. Physical monotony comes with a wide range of repetitive operations; a common example is using a jack-hammer to break up a concrete slab where there is a continual need to lift the hammer, guide it into the slab, and split it. The vibration and noise concerned will also have a considerable effect on the operative's fatigue.

The final allowance that is made on top of the basic time is one for contingency. This is a general allowance to cover many of the small items that cannot be covered by a specific allowance, such as sharpening and setting tools, adjusting small machinery like pump and vibrators, refuelling small equipment, allowing inspection of work and receiving modified orders from a foreman. When decisions have been taken about the size of the allowances to be made they are then totalled in percentage terms and added to the basic times that have been derived from the observed times and the ratings. The *standard time* for an element of work is the time taken to carry out the element at a rating of 100 and with the addition of the relaxation and contingency allowances for that particular task. The standard time or output can then be used as the output that may be produced in carrying out that task by an average operative. Figure 6.5 shows a typical sheet used for adding relaxation allowances and develops the basic times arrived at in the example used for the observations recorded in Fig. 6.4.

While time study has been discussed so far in terms of observing a single operative, it is possible to extend it to a *gang study* where the gang includes several operatives. It is important in a gang study that the observer has every operative in view throughout the study and that the work task being undertaken is not broken down into too many small elements.

6.9 Synthesis and analytical estimating

Synthesis is a process that results from time studies and enables a time for a task to be built up from the element times of other studies collected previously. The level of performance for the new composite task will be known. In construction there are many elemental activities that have a part or parts in common with other operations, even though the total operation may be quite different. Spreading concrete in slabs, fixing formwork, laying bricks, and earthmoving are such activities. Spreading concrete may be part of pouring concrete for the fifth floor of an office building, concreting a road pavement, or placing concrete in the suspended roof of a basement garage. Similarly, earthmoving by scraper has an element of cutting the soil and, assuming that the soil types are comparable, the time it takes to complete an earthmoving cycle of cut, run to tip, discharge, and run to cut is dependent on the distance of the tip from the cut. Other elements of the process are likely to be similar in the times that are taken for many other similar activities. Where relaxation allowances are estimated to be similar, standard times may be used, but in the case of the likelihood of wide variation, basic times can be used and the percentage allowances can be added afterwards. It is, of course, essential that the relaxation percentages are stored with the original data. Data collected from time studies need, therefore, to be referenced and entered into a database so that they are available for future use. Data stored in this way are known as *synthetic data*. Synthetic data need not be stored only in the form in which they are collected in the field; often, after sufficient data have been gathered, a formula can be derived combining a number of the variables that influence the element times. This is another method of using synthetic data.

Analytical estimating is used in order to fill gaps in the process of synthesis, where not all of the required synthetics are known or available. If the standard times of one or more of the elements in the new situation are not known because a study of that work has not previously been carried out, it may be necessary to make an estimate of the missing element times from previous data on similar activities. Often handbooks, such as those published by manufacturers of earthmoving equipment or estimating manuals, can be used to help in this process.

6.10 Method study

Method study is defined in BS 3138 (1979) as:

> The systematic recording and critical examination of ways of doing things in order to make improvements.

Its essence is in a systematic approach, which should become an attitude of mind. It is a flexible technique which has universal application. Though practitioners and writers vary in the number of steps they assign to the process, the general approach is similar in each case. The steps are:

1. Select the work to be studied.
2. Identify the problem.
3. Record the facts.
4. Examine the recorded facts.
5. Develop an improved or new method.
6. Install the new method.
7. Maintain the new method and check the results.

1. *Select the work to be studied*   The emphasis in the selection of the work to be studied is usually on the greatest potential for cost reduction. It may not be an obvious direct cost reduction in, say, that a bottleneck may need to be smoothed out so as to effect improvements elsewhere. Clearly, the greater the potential cost savings in work, the higher is likely to be the priority for the selection of that work for study. The speed with which a benefit may be achieved is also likely to be an influence in the selection stage. The importance of improving conditions, particularly safety, under which operatives undertake their work, should not be overlooked as an indicator for the priority of selection, in spite of the fact that such improvements may not lead directly to cost savings. Other factors that may have a bearing on the selection of work are those concerned with a reduction of waste materials, reducing idle time, relieving fatigue and monotony, smoothing the flow of materials, eliminating unnecessary movement of all resources by improved layout, improving the quality of production and reducing the working capital required to finance work in progress.

2. *Identify the problem*   At this stage it is necessary to identify the problem boundaries and to ensure that there is a clear understanding by the construction management as to exactly what problem is being tackled. It is necessary, on the other hand, for an individual carrying out a method study to have the confirmation from the construction management as to the detail of the problem that they believe requires attention and to which a solution will be offered.

3. *Record the facts*   Collecting the facts as well as recording them is part of this stage in the procedure. An incomplete story can be extremely misleading, so it is

essential to obtain all the facts concerning the project under examination. If it is possible to obtain the facts by measurement and in numerical form then this is to be preferred. The various recording techniques will be discussed later. Graphical recording methods should be used wherever possible.

4. *Examine the recorded facts* The recorded facts must be critically and systematically examined, using the critical examination procedure of *What?*, *Where?*, *When?*, *Who?*, and *How?* These five questions form the basis of what are known as the *primary questions*. The question of *Why?* is then asked of each of these primary questions. Table 6.5 illustrates a useful form of critical examination chart to place the questioning procedure in a logical framework.

Table 6.5

	Primary questions		*Secondary questions*	*Alternatives*
Purpose	What is achieved?	Is it necessary? Why?	What is the alternative?	What should be done?
Place	Where is it done?	Why that place?	Where else is possible?	Where should it be done?
Sequence	When is it done?	Why then?	When else could it be done?	When should it be done?
Person	Who does it?	Why that person?	Who else could do it?	Who should do it?
Means	How is it done?	Why that way?	How else could it be done?	How should it be done?

5. *Develop an improved or new method* In carrying out the examination phase, improved methods or new methods will be revealed. Development will almost inevitably necessitate the testing of new ideas. Pilot schemes can be implemented where conditions permit. Where possible several individuals may be used at this stage—preferably from different backgrounds and with varying experience.

6. *Install the new method* This may cause disturbance, especially if new materials and/or different methods are to be used. Retraining of personnel may be necessary and care may be required in gaining the cooperation and assistance of those operatives who are affected by the changes. Early participation in the study by those affected is likely to smooth the path of installation.

7. *Maintain the new method and check the results* Maintaining the new method involves the monitoring and modifying required to keep the method operational. Site management must take over responsibility at this stage, bringing work study advisers into discussions where difficulties are experienced. Work study personnel must check

at this stage the success of the changes which have been made so that economic benefits can be established.

6.11 Process charts

Process charts provide diagrammatic means for recording the sequence of activities in an existing method under study. In this way it is possible to have a visual aid to the overall conception of the method and a sound basis on which to effect improvements.

Process charts may be of several kinds (see below), but those of all types are constructed by using only five symbols. The ASME symbols are most commonly used today and are recommended in BS 3138 (1979). The ASME symbols were derived by a committee of the American Society of Mechanical Engineers from those originally used by Gilbreth. The symbols and their interpretations are shown in Table 6.6.

In a process chart the symbols are linked together to represent the sequence of individual events or activities in the total operation under study. The total operation must be clearly defined before the study commences, as must be the points where it commences and finishes. When using the ASME or alternative symbols for the purposes of recording an existing process, the purpose of the recording should constantly be kept in mind. This may help in the classification of activities where some doubt arises or there is some contention as to the precise classification. Recording a process so as to facilitate subsequent analysis is more important than doctrinaire attitudes towards the means of doing it.

Table 6.6

Activity	Symbol	Interpretation
Operation	◯	A definable step in a process, method, or procedure. Some change usually takes place, for example, a hole is drilled, timber is cut to length, a crane is loaded.
Transport	⇦	Any movement of operatives, material or plant, however small. For example, bricks are carried, concrete is transported by mixer truck, a hoist rises.
Permanent storage	▽	Storage that is planned and so creates a delay that was anticipated; authorized storage that is controlled as to place and duration, for example, ballast at a mixer, banded bricks in a stockyard, bent reinforcement at the steel-fixers' shop.
Temporary storage or delay	D	An unintentional delay, usually caused by one part of the process not fitting in to the timing of another, for example, material awaiting processing, operatives awaiting material, an excavator awaiting a lorry to load.
Inspection	☐	An inspection for quality and/or quantity, for example, checking a measurement, checking the weight of ballast at the concrete mixer, checking the vertical height of courses in brickwork.

Process charts can be very useful in assisting with the establishment of an economic layout for a workshop or yard where a repetitive process is to be undertaken, for example a concrete precasting yard, an area in which steel reinforcement will be stored, cut, and bent, or a stores compound such as is often provided for some of the more valuable building materials. Process charts can be drawn at varying levels of detail, depending on the scope of the process to be recorded. An *outline process chart* may be used to record a broad and general picture of a process, often at the initiation of an investigation. In such cases it is usually only necessary to employ the operation and inspection symbols. A *flow process chart* is used to show greater detail. Many of the component operations of the outline chart can now be expanded. All five symbols will be used and the chart should be used to record the flow of material through a process, the sequence of activities of an operative or the sequence of activities of an item of equipment or plant. These are the three types of flow process chart, which are known as material-type, man-type, and equipment-type respectively. Whatever the subject the chart is initiated for, it must subsequently remain consistent throughout and must not switch from operatives' activities to material flow, for example. The third form of process chart is the *two-handed process chart*. This chart is designed to portray an even greater degree of detail in an operation, to the extent of recording the relative movements of each of an operative's hands. The analysis of construction operations to this degree of detail on a site of work is rarely warranted, since few tasks are repeated often enough to yield sufficient benefit from such analysis. Charts of the first two categories offer the greatest scope for improvement in construction methods.

6.12 The outline process chart

An example of an outline process chart is illustrated in Fig. 6.6. The operation is that of unloading a baulk of timber from a truck, measuring its length, cutting it to the required length, and placing and fixing it into position as timber supporting an excavation. The whole process is carried out five times. This chart introduces a convention for recording simultaneous operations, such as measuring a length and marking the position of cut required. It also introduces a convention that saves repetition when recording. Since the operations are repeated for each baulk of timber, this can be recorded in a shorthand fashion. In this example the start of the operation can clearly be defined as 'timber delivered by truck and inspected' and the completion of the operation recorded can be defined as 'final check for position'.

Note that each symbol bears a number for convenience and identification. The system used is to number similar symbols sequentially. The numbering system enables a summary of the various types of activity to be prepared on completion of recording. When work is repetitive, due allowance should be made in the symbol numbering for the activities which are covered under the umbrella convention. For example, in Fig. 6.7, if there were to be another operation beyond the 'repeat four times' it would bear the number 31.

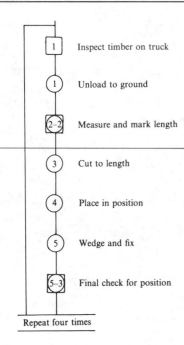

1	Inspect timber on truck
1	Unload to ground
2–2	Measure and mark length
3	Cut to length
4	Place in position
5	Wedge and fix
5–3	Final check for position

Repeat four times

Figure 6.6

One stage more detail in this particular outline process chart can be introduced by including the cutting and supply of the wedges—operations that are out of the mainstream of activity but are necessary to the completion of the total job. Figure 6.7 illustrates the recording conventions for doing so, as well as the usual way of recording the description of materials that are introduced into the process.

The outline process charts illustrated in Figs 6.6 and 6.7 record the way in which the work is being carried out. It is only when this stage has been reached that a further critical examination of the process can take place with a view to effecting some improvement in the method currently in use. In the case of the operations recorded in the example, it may well be the immediate reaction that all the timber should be offloaded before any other operations are undertaken. The outline process chart would take a different form if this were to be the case.

6.13 Flow process chart

A flow process chart represents the next level of detail greater than that of an outline process chart. It may result from the expansion of an outline process chart or a more detailed study of some operation that forms part of an outline process chart. All five of the ASME symbols are used in its preparation.

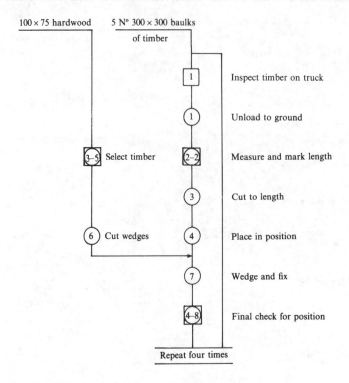

Figure 6.7

An example of a man-type flow process chart is illustrated in Fig. 6.8. Sometimes it is advantageous to the analysis of a problem if the flow process charts for an operative or a machine are drawn with the related material-type chart alongside. The various operations and movements can then be interrelated. If it is thought likely that the distance to be travelled or through which some object or material is to be moved will be of significance in the subsequent analysis, it can be recorded against each appropriate transport symbol on one side of the chart. Similarly, where the time that it takes to carry out an operation may affect the outcome of an analysis, this may also be recorded alongside the relevant operation symbols.

While a flow chart as described displays only the sequence of events in a process, it can be adapted in such a way as to provide a pictorial view of the various paths of movement. This is effected by superimposing a flow process sequence on a plan of the area that shows the facilities involved in the operations. This adaptation is known as a *flow diagram* and in depicting the movement of people, materials, or machines within a general working area it assists with the detection of congested areas, paths of excessive length, and unnecessary journeys. A flow diagram tends to complement a flow process chart. Examination

Unload timber from truck, cut to length, position and wedge in excavation timbering

Chart begins: Operative walking to truck
Chart ends: Final inspection of timbering

<u>Distance</u>

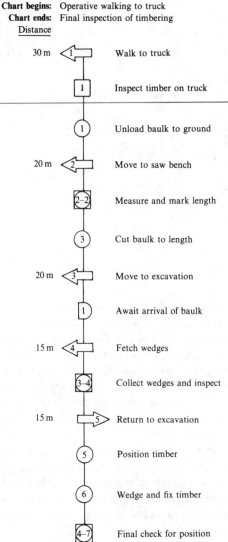

30 m	Walk to truck
	Inspect timber on truck
	Unload baulk to ground
20 m	Move to saw bench
	Measure and mark length
	Cut baulk to length
20 m	Move to excavation
	Await arrival of baulk
15 m	Fetch wedges
	Collect wedges and inspect
15 m	Return to excavation
	Position timber
	Wedge and fix timber
	Final check for position

Figure 6.8

of a problem is generally facilitated by examining both records side by side. It is useful to use lines of different colours to represent the movements of different people or plant in a flow diagram. Figure 6.9 illustrates a flow diagram for the flow process chart of Fig. 6.8.

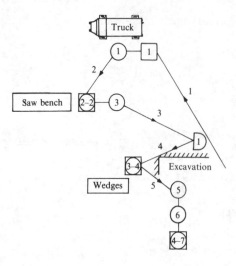

Figure 6.9

6.14 Multiple-activity charts

In contrast to the flow charts, a *multiple-activity chart* incorporates a time scale against which various activities are plotted. It is used to illustrate the interrelationships between the activities of two or more objects. For example, it can be used to examine the interrelationship between an operator and his machine or between each of a number of operatives in a gang engaged in concreting, bricklaying, or unloading, where the activity is a team activity. The use of a multiple-activity chart facilitates balancing the various activities of the members of a gang by highlighting idle time and thus any weak link in a team. It is an invaluable method in construction studies, where very few operations are not team efforts.

Time data must be collected before a multiple-activity chart can be constructed. This can usually be done by direct observation, using a stop watch if necessary. Rarely, however, is it necessary to use anything more accurate than an observation made from an ordinary watch having a second hand. Where large groups of men or machines are involved, it may be necessary to use several observers and to develop a carefully coordinated plan of study.

Where it is necessary to study a large team in operation, perhaps on a fundamental and repetitive operation, the expense of filming may be justified as a means of collecting the relevant information. A multiple-activity chart is then constructed from watching the film. To lend itself to filming, the operation involved must clearly be such that it can be seen in its entirety through the lens of a fixed cine camera. The use of *time-lapse photography* does effect some economies in film. This is alternatively known as *memo-motion photography* and involves exposing single frames at intervals while taking care not to destroy the continuity of the action. Intervals between frames of somewhere between half a second and four seconds are usual.

The following example illustrates the use of a multiple-activity chart.

EXAMPLE 6.2

An examination and study of the mixing and distribution of concrete to the moulds in a small, site precasting yard results in the data given in Table 6.7.

Table 6.7

Operation identification	Description	Standard time, standard minutes
A	Load mixer drum and mix one batch of concrete by mixer gang	5.2
B	Discharge concrete into dumper by mixer gang	0.7
C	Deliver concrete to moulds by dumper	1.7
D	Pour concrete into moulds by dumper and concretor gang	3.2
E	Vibrate and finish off concrete in moulds by concretor gang	1.4
F	Dumper returns to mixer	1.5
G	Place reinforcement and boxes in moulds prior to concreting by concretor gang	4.5

There are three components to the concreting team. One mixer with its mixer gang, the dumper with its driver, and the concretor gang. The standard times above are expressed in minutes and tenths of minutes. The multiple-activity chart can be constructed with either horizontal bars or with vertical bars, as is the case in Fig. 6.10. The multiple-activity chart needs to be constructed through at least two or three cycles of an operation from its commencement before utilization factors are calculated, so as to allow the operations to settle down. If this is not the case, the initial waiting of the dumper and the concretor gang at the beginning of the chart has a biasing influence on the assessments. The utilization of the various components of the team can be calculated as follows:

$$\text{Utilization} = \frac{\text{working time per cycle}}{\text{cycle time}} \times 100 \text{ per cent}$$

$$\text{Utilization of mixer} = \frac{5.9}{9.1} \times 100 = 64.8 \text{ per cent}$$

$$\text{Utilization of dumper} = \frac{7.1}{9.1} \times 100 = 78 \text{ per cent}$$

$$\text{Utilization of concretors} = \frac{9.1}{9.1} \times 100 = 100 \text{ per cent}$$

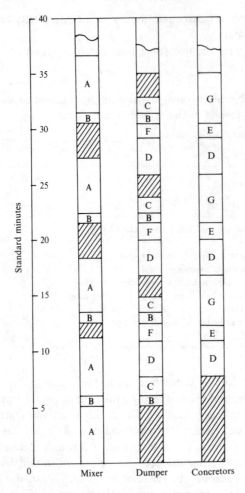

Figure 6.10

Comment: the concretor gang have a utilization of 100 per cent; therefore, if delays to the use of the other resources are to be reduced, the concretor gang needs to work more efficiently or it needs to be reorganized, perhaps adding an additional man. Clearly the addition of another dumper would, in the first instance, eliminate delays at the mixer, but very quickly the concretor gang would lengthen the cycles of both dumpers and hence the mixer. The construction of a further multiple-activity chart will illustrate this.

6.15 String diagram

A string diagram provides a charting device that is similar in purpose to a flow diagram. However, it does provide a facility for recording and analysing more complex situations than with the former, and more than one subject can be charted on a single diagram. It is an ideal means by which a delivery or transporting activity can be analysed, particularly where the movement takes place in or through a congested area and there is likely to be an overlapping between one transport resource and another. In addition, a string diagram can be used in order to compare a new, and, it is to be hoped, improved, situation with an existing one. A string diagram is prepared by using a scaled plan of the area (which may be a site, a precast concrete manufacturing yard, or a joiner's shop, for example) into which pins are stuck at every key position or change of direction in the movement of the transport resources under study. Stout thread, which can be of different colours to indicate different resources, can then be stretched between the pins following the route of travel. The finished record illustrates the extent to which paths of different vehicles or men will cross each other during the course of the operation and the length of each route can be determined by the length of thread that is needed to cover it.

6.16 Travel chart

A travel chart can be used for two principal purposes: firstly, as an aid to the collection of facts during observation so that a flow or string diagram can be constructed; secondly, so as to provide a general picture of the pattern of movement within, say, a workshop or a warehouse, highlighting the locations between which operatives or materials most frequently travel. As with string diagrams, more than one subject can, and usually is, being studied. Figure 6.11 provides an example of a travel chart outline. Stations 1–7 would each be identified. Station 1 might be a store, 2 an outside delivery point, 3 a paint shop—all within a workshop under study. From the study it can be seen, for example, that during the period of the

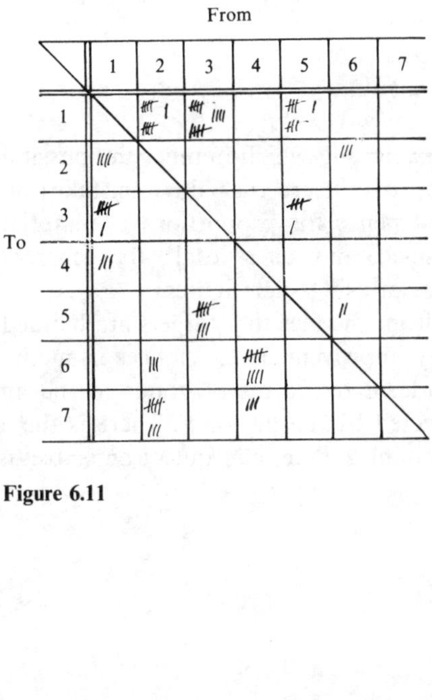

Figure 6.11

observation, there were 14 movements from the paint shop to the store and 11 movements from the outside delivery point to the store.

6.17 Incentive schemes

Incentive schemes are alternatively known as *bonus schemes* or *payment-by-results schemes*. They play an important part in the makeup of the gross wage of a construction worker, and payments may frequently pretend to be such as a matter of convenience, although they are not directly linked to results as they should be. Work study has long been associated with payment-by-results schemes since the histories of the two are closely intertwined. An incentive scheme is usually designed to supplement a basic hourly wage rate and, in such cases, the supplement takes the form of an addition or bonus, the extent of which depends on the performance achieved by the operative or the group in which the operative works. Incentives need not always be financial, but it is intended here to describe the principal types of scheme and how they work rather than to indulge in controversial discussion of their effectiveness and when and whether they should be installed.

Before a payment-by-results scheme can be designed, a decision must be made as to the standard performance that will form the basis for the scheme and the level of earnings that will be related to it. In addition, it is necessary to determine the level of performance below which bonus payments will not be made and, related to it, the rate at which such paymets will be made for performance in excess of the standard. It may be necessary to determine whether there will be any modification to payments made where performance does not reach the standard. In designing an incentive scheme all agreements negotiated within the industry must be taken into account.

The standards upon which an incentive scheme is based are, in construction, usually derived from one of three sources:

1. a priced bill of quantities for the work;
2. negotiated targets based on experience;
3. targets based on work measurement.

Of these, (1) can be full of pitfalls, among which number the possibility of prices being loaded or unbalanced to improve future cash flow, mistakes at the time of estimating the prices, methods, and hence the proportions of material, plant, and labour within the rate being changed, and, as a result, the general difficulty in convincing operatives of the fairness of using such figures.

The second source often suffers from the fact that targets are handed down from job to job without reference to any improvement or changes in method. It is also difficult to reconcile payments made in the forms of bonus to the amount to be received for the work done as reflected by the bill rates. There is the added doubt that targets are produced by bargaining rather than out of objective assessment of the work involved.

The last of the above three sources depends on the measurement of the work content of a task by using work study techniques. It results in an accurate assessment of this work content. Changes in content due to the use of different methods can readily be accommodated. Bargaining has little part to play in these processes, but the techniques of work measurement need to be accepted by both sides.

6.18 Incentive schemes and their characteristics

The basic scheme from which variants have been developed, and with which variants are often compared, is that of the *strictly proportional* scheme. This is illustrated graphically by line X in Fig. 6.12. Earnings are plotted on the vertical axis and performance based on work measurement is plotted horizontally. Operatives' performance has been plotted on the British Standards scale wherein standard performance is assumed to be 100. The vertical scale is plotted in terms of *bonus ratio*—the ratio of gross wage including the bonus payment to the basic wage without bonus. Earnings, as will be seen, are in direct proportion to an operative's performance. At the standard performance of 100 it is generally agreed that an operative should receive 33.3 per cent bonus as a proportion of the basic wage. The aim, therefore, is to set standard performance such that at that performance, 33.3 per cent of the basic rate will be paid as a bonus. The minimum labour cost of the bonused operation is therefore fixed and once an operative achieves a performance of 75 or better, any savings accruing from the better performance will be paid in the form of a bonus. Once the standard performance is exceeded, operatives can earn high rates of bonus, and therefore it is essential that accurate standard times can be and are set. It is also

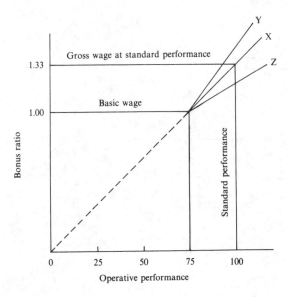

Figure 6.12

vital that a system exists whereby accurate time records can be kept of an operative's involvement. The strictly proportional scheme does not allow for a reduction in pay for performance less than 75.

Similar to the strictly proportional bonus system is that of *straight proportional*. It is also known as a *geared* incentive scheme. This system is illustrated in Fig. 6.13. The system is based on starting bonus payments at a lower operative performance, say 50, but retaining the 1.33 bonus ratio for a performance of 100. With this system an operative will always have a higher bonus ratio compared with a strictly proportional scheme until a performance of 100 is achieved, but thereafter the lower rate of bonus will result in less bonus than with the strictly proportional scheme. Such a system has its application in areas in which it is difficult, owing to the nature of the work, to establish accurate work standards. Establishing a bonus line with a flatter gradient, but starting at a lower performance, tends to reduce fluctuations in earnings for work on which there will be fluctuating performance.

Sometimes it may be advisable to increase or decrease the slope of the bonus line beyond the intersection of the unit bonus ratio line and the operative's performance 75 line. These two conditions are represented by lines Y and Z respectively in Fig. 6.12. Where the bonus line is flattened, variations in bonus earnings will be *less than proportional* to changes in operative performance. Such a system is suitable for situations in which it may be necessary to pay a higher basic wage, which should consequently require less supplementation by incentives.

Bonus line Y in Fig. 6.12 represents the general case where the bonus rate increases at a *greater than proportional* rate than the operative's performance. Such

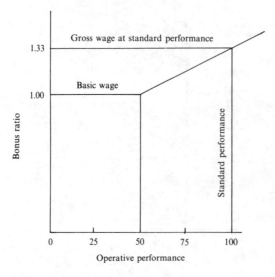

Figure 6.13

schemes need very careful and accurate control if gross earnings are not to get out of hand. They can be applied only in situations where standards of performance can be set with great accuracy and confidence. They may be required in situations where it is vital to reach high operative performance but where a feature of the work involved automatically prevents bonus payments beyond a reasonable level.

Payments by results schemes are usually installed so as to provide an incentive for increasing production—but not at the expense of quality and sound workmanship. Good supervision is therefore required to ensure that quality is maintained. It may be necessary, also, to embody in an incentive scheme some supplementary provision in relation to the quality achieved. Such provisions are difficult, but not impossible, to establish, and should be based on a method of calculation that is readily understood by the operatives involved.

EXAMPLE 6.3

The basic rate for an operative is £3.90 per hour. A gang of such operatives take 475 hours to complete an operation that is work-measured to take 445 hours. Calculate the bonus rate of the work if (a) a strictly proportional scheme is in operation where bonus starts at 75 performance and rises to $33\frac{1}{3}$ per cent at standard performance and continues to rise at that rate; (b) a 50 per cent geared scheme is used.

(a) *Strictly proportional scheme*

$$\text{Performance for gang} = \frac{\text{total standard hours for work done}}{\text{total actual hours worked}} \times 100$$

$$= \frac{445}{475} \times 100 = 93.7$$

At a performance of 100,

$$\text{Bonus} = \frac{33\frac{1}{3}}{100} \times 3.90 = £1.30 \text{ per hour}$$

Therefore,

Bonus per unit increase in performance	$= 1.30/25 = £0.052$
Actual performance	$= 93.7$
Increase in performance above basic rate	$= 93.7 - 75 = 18.7$ units

Therefore,

Bonus earned by each operative	$= £0.97 \text{ per hour}$
Gross rate (basic plus bonus)	$= £4.87 \text{ per hour}$

(b) *50 per cent geared scheme*

Bonus per unit increase in performance	$= 1.30/50 = £0.026$
Increase in performance above basic rate	$= 93.7 - 5 = 43.7$

Therefore,

Bonus earned by each operative	$= 43.7 \times 0.026 = £1.136 \text{ per hour}$
Gross rate	$= £5.036 \text{ per hour}$

The problem is illustrated graphically in Fig. 6.14.

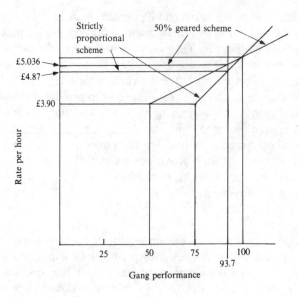

Figure 6.14

Summary

This chapter is concerned with describing techniques for improving and rewarding construction productivity. The techniques are constituent parts of work study. Work study is treated in its two main divisions of work measurement and method study. The use of activity sampling is emphasized and the underlying statistical principles are described. Activity sampling is seen to be an important technique for use in construction, and therefore some emphasis is placed on it. Two variations of activity sampling—field ratings and productivity ratings—are described. Time study, a fundamental facet of work measurement but not one of overriding importance to construction, is explored and the basic aspects of the process for the collection and analysis of data are set out. Method study, the other main division of work study, has strong applications in construction. Process charts of various types, multiple-activity charts, string diagrams, and travel charts, all for improving the methods used, are described and applied. The remainder of the chapter is devoted to a description of the principles of designing incentive schemes based upon improvements in productivity.

Problems

Problem 6.1   Make a random drawing of 32 samples consisting of two numbers from a population of seven numbers, 1, 2, 3, 4, 5, 8, and 9. The draw should be made with replacement. Calculate the following:

(a) the population mean;
(b) the standard deviation of the population;

(c) the mean of the sampling distribution of means;
(d) the standard error of the means.

Problem 6.2 If a fair coin is tossed 95 times, what is the probability that the outcome of a head will occur between 45 per cent and 55 per cent of the times?
(0.6806)

Problem 6.3 A series of 100 sample observations of operatives working indicated that 55 per cent of them were active. Find 95 per cent and then 99 per cent confidence limits for the proportion of all the operatives being active.
$(0.55 \pm 0.10; \ 0.55 \pm 0.13)$

Problem 6.4 Explain why it is that an activity sampling exercise on a construction site is almost certain not to give a completely accurate indication of the actual position. Your explanation should include a note about the statistical aspects of your answer.

Activity sampling observations are taken on a total of 40 operatives excavating the foundations for a large building. Assume the appropriate limits of error and confidence level for the work, assuming that the study is to establish the working/not working conditions for the operatives. Design the sampling experiment on the basis that the proportion of working operatives is 70 per cent. Given an indication in the programme of how the observations will be taken.

Problem 6.5 A preliminary study has been made to establish that a particular site manager spends 65 per cent of working time actually on the site and 35 per cent in the site office. How many observations would be needed for an activity sampling study based on the preliminary work in order to give a 95 per cent confidence limit with a limit of error of ± 3.5 per cent?

Of the work outside the office 485 observations were eventually made out of a total of 785: 95 were taken when the manager was inspecting the quality and correctness of the work, 73 in discussions, 89 in measuring work, 56 in walking, 93 in generally organizing operatives, and 79 in checking and discussing the work with the clerk of works. Check which items come within the required limits of accuracy and confidence level.

Problem 6.6 In using activity sampling for a routine check on the performance of a dumptruck which is hauling the spoil arising from an excavator, 322 observations were made: 123 of the observations showed the truck to be moving, either full or empty, 65 showed it to be under the excavator receiving its load, 63 showed it to be dumping the spoil and on 71 occasions it was idle. Determine the proportion of its time that the truck was idle and the degree of accuracy of the answer. How would the study have to be adjusted if the degree of accuracy were required within ± 5 per cent?

How would you timetable the observations to be made in an activity sampling study? How would you record the results of the observations so as to provide a continuous, up-to-date picture of a study involving approximately 2000 observations?

Problem 6.7 Outline the advantages of work sampling as a work measurement technique.

Design a work sampling analysis form listing, among the other relevant details, an element classification suitable for a work sampling study on carpenters and labourers engaged on construction work. Give the reasons for your choice of layout and headings.

Problem 6.8 The operations listed in Table 6.8 refer to the cycle of events in the process of concreting a 1.500 m lift on a long retaining wall. The wall is 9 m high. A gang of operatives are assisted by a tower crane. The complete gang is involved on jobs 1, 2, 5, 7, and 8. On the other jobs the gang is split into two groups—one does jobs 4 and 6 and the other does job 3.

It is decided to construct two similar sections of the wall concurrently using the same crane and a second similar gang. The work on the two sections will have the same gang and crane times, and hence the same cycle times, but they may be out of phase one with the other. Use a multiple-activity chart to examine the problem of the crane utilization.

Table 6.8

Job	Gang time, standard hours	Crane time, standard hours
1 Remove fixing bolts, props, spacers, etc., to formwork	6	nil
2 Remove formwork to ground	4	4
3 Clean and oil formwork and bolts, etc.	3	nil
4 Prepare supports for next lift of concrete	2	nil
5 Lift formwork into position and support for next lift	8	8
6 Plumb formwork and finally fix	4	nil
7 Prepare platform for concreting	4	4
8 Concrete wall lift	5	5

Problem 6.9 Describe the five ASME symbols used in method study process charts and what each symbol represents. Give examples from civil engineering of an activity that would be represented by each one of the five symbols.

A crane is engaged in hoisting small bundles of reinforcement to the roof level of an 8-storey building. Two operatives load the crane at ground level and another two unload and position the reinforcement at roof level. The whole operation has been measured in a work study exercise and the information showed in Table 6.9 results. Draw a multiple-activity chart to represent the whole operation and calculate the labour and plant utilization if the process is a continuous one.

(Operatives at ground level 69.6 per cent; crane, 86.9 per cent; operatives at roof level, 69.6 per cent)

Table 6.9

	Element	Standard time, standard minutes
1	Pick up and carry reinforcement 40 m to crane at ground level	1.10
2	Place slings round reinforcement and attach to crane hook	0.20
3	Raise crane hook with reinforcement 30 m	1.20
4	Slew crane jib and land material	0.16
5	Release reinforcement from crane hook	0.15
6	Lower crane hook to ground level	0.60
7	Pick up and carry reinforcement from crane to required point on roof and return to crane	1.50
8	Return from crane to reinforcement stock at ground level	0.30

Problem 6.10   A crane is used to lift structural sections to the upper levels of a building where they are then bolted into position. One gang of operatives is employed in this fixing and another is working at ground level, bringing the sections from a storage area to the crane loading area. A work study exercise yields the data shown in Table 6.10. Construct a multiple-activity chart to represent the total operation. Find the crane utilization and that of each gang after the cycle has continued long enough to overcome starting-up effects.

Comment on the efficiency of the operations and make suggestions for improving the utilizations.

(Lower gang, 100 per cent; crane, 76.4 per cent; upper gang, 57 per cent)

Table 6.10

	Element	Standard time, standard minutes
1	Load and transport sections from storage area to crane at ground level	6.10
2	Load crane at ground level	0.60
3	Raise load by crane	2.30
4	Slew crane to work position	0.60
5	Hold crane in position for temporary bolting	2.10
6	Release crane hook from positioned section	0.20
7	Lower crane hook empty	1.50
8	Complete bolting at work position	4.30
9	Return from crane loading area to storage area for more sections	2.80

Problem 6.11   The multiple-activity chart in Fig. 6.15 depicts the present sequence of operations followed by two operatives handling loose bricks. They share the use of a brick barrow which will accommodate 20 bricks and a mobile hoist to lift the bricks to a height of 3.5 m. The power unit on the hoist is

controlled by the operative on the ground and the operative on the scaffold has to release the safety catch on the cage to remove the load on arrival at the upper level.

(a) Devise an improved sequence of operations, stating any assumptions made.
(b) Compare the existing and the improved values for the cycle time; manpower utilization; machine utilization; the output in terms of bricks stacked at the upper level for the first hour of operation.
(c) Enumerate any additional modifications to the method that you envisage.

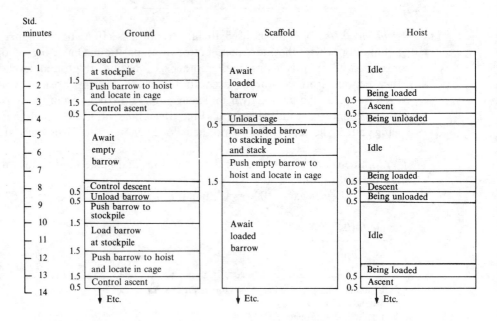

Figure 6.15

Problem 6.12 Differentiate between direct and geared incentives for construction work. What are the main features of both methods?

Fifty shallow inspection chambers of almost uniform dimensions are required as part of a housing estate drainage system. It is considered by the site manager that one bricklayer should reasonably take 12 working hours to build a chamber and that doing so will entitle the bricklayer to a 15 per cent bonus on basic wage. Compare the direct scheme and a 60 per cent geared scheme for this work, indicating the relationship between the actual time taken and the bonus earned by the bricklayer. Show how the cost of the work to the employer on one chamber (in man-hours) varies if the time taken to complete a chamber varies between 2 and 18 hours.

Problem 6.13   You are required to prepare a draft report for your company's senior management, explaining the various types of incentive scheme that can be implemented on construction sites. Set out such a report and explain the advantages and disadvantages of each system using the following information as the basis for your explanation:

(a) A gang of carpenters takes 1 man-hour to fix and strike $0.35\,\mathrm{m^2}$ of horizontal formwork.
(b) A concrete gang takes 8 man-hours to place $1\,\mathrm{m^3}$ of mass concrete.
(c) A gang of bricklayers takes 1 man-hour to lay $1.5\,\mathrm{m^2}$ of brickwork 1 brick thick.

Problem 6.14   The following is a statement from a construction company's policy document:

This company proposes to maintain a direct productive incentive scheme such that employees may receive 30 per cent of their basic earnings for achieving standard performance.

Given that the basic rate for labour is £4.00 per hour, discuss the implications of the above statement, using a one-standard-hour job as your basis and ignoring indirect costs.

If the allowed time for fixing shuttering at standard performance is 1.50 man-hours/$\mathrm{m^2}$, and 4 joiners of the above company working 8 hours per day fix $1000\,\mathrm{m^2}$ in a 5-day week, how much bonus have they earned and what is their rating?

Further reading

Advisory Service for the Building Industry, *The Principles of Incentives for the Construction Industry*, London, 1981.

Currie, R. M.: *Work Study*, Pitman, London, 1959.

International Labour Office, *Introduction to Work Study*, ILO, Geneva, 1969.

International Labour Office, *Payment by Results*, ILO, Geneva, 1951.

Liou, F. S. and J. D. Borcherding: Work sampling can predict unit rate productivity, *Journal of Construction Engineering and Management*, ASCE, vol. 112, no. 1, March 1986, pp. 90–103.

Moroney, M. J.: *Facts from Figures*, 3rd edn, Penguin, London, 1956.

Oglesby, C. H., H. W. Parker and G. A. Howell: *Productivity Improvement in Construction*, McGraw-Hill, New York, 1989.

Olson, R. C.: Planning, scheduling and communicating effects on crew productivity, *Journal of the Construction Division*, ASCE, vol. 108, no. CO1, March 1982, pp. 121–128.

Richardson, W. J.: *Cost Improvement, Work Sampling and Short Interval Scheduling*, Reston Publishing, Reston, Virginia, 1976.

Thomas, H. R.: Can work sampling lower construction costs? *Journal of the Construction Division*, ASCE, vol. 107, no. CO2, June 1981, pp. 263–278.

Thomas, H. R. and J. Daily: Crew performance measurement via activity sampling, *Journal of Construction Engineering and Management, ASCE*, vol. 109, no. 3, September 1983, pp. 309–320.

Thomas, H. R., J. M. Guevara and G. T. Gustenhoven: Improving productivity estimates by work sampling, *Journal of Construction Engineering and Management, ASCE*, vol. 110, no. 2, June 1984, pp. 178–188.

Thomas, H. R. and M. P. Holland: Union challenges to methods improvement programs, *Journal of the Construction Division, ASCE*, vol. 106, no. CO4, December 1980, pp. 455–468.

Thomas, H. R., M. P. Holland and C. T. Gustenhoven: Games people play with work sampling, *Journal of the Construction Division, ASCE*, vol. 108, no. CO1, March 1982, pp. 13–22.

Thomas, H. R.: Can work sampling lower construction costs?, *Journal of the Construction Division, ASCE*, vol. 107, no. CO2, June 1981, pp. 263–278.

Tucker, R. L.: Perfection of the buggy whip, *ASCE Journal of Construction Engineering and Management* vol. 114, no. 2, June 1988, pp. 158–171.

Turner, G. J. and K. R. Elliott: *Project Planning and Control in the Construction Industry*, Cassell, London, 1964.

7. Estimating and construction planning

7.1 Estimating

Many different types of cost estimate are used in construction. Which is employed in any particular instance will depend on the stage in the construction process at which a project currently stands (and therefore the purpose for which it is being prepared) and the amount of information concerning the project that is available to the estimator. A project will generally involve three parties: a client (otherwise known as a promoter, employer or owner), a designer/consultant, and a contractor. While the detail of their organization and interrelationship may vary (see Sections 1.11–1.15), these three parties, or the functions they carry out, are necessary to the majority of construction projects. Each needs to estimate the cost of the project for their own specific purposes.

Initially, a client will need to have a *rough estimate* of the amount of fixed investment that will be required for the constructed facility that is in mind. Such an estimate will result from little formal calculation and will probably be extrapolated using the *feel* of experts associated with previous similar works. The function of such an estimate is to provide a measure of the further interest in the likely profitability of pursuing the venture. Such estimates of cost are frequently referred to as *guesstimates* or *ballpark figures*.

If the rough estimate generates further interest in the scheme then the client will usually appoint professional advisers to prepare preliminary designs and estimates of cost for the work. Which advisers are selected will vary with the nature of the work to be undertaken, but in most cases they are likely to consist partly or entirely of engineers, architects, and quantity surveyors. The group of advisers will undertake a *feasibility study* for the project in which the designers of the facility will examine design alternatives. It will be necessary to estimate the relative costs of each and hence propose an economical design that will meet the client's aesthetic and functional requirements and at the same time fall within the total cost that the client is prepared to pay. Such *preliminary estimates* are usually derived by using one of such methods as the following:

1. the unit method;
2. the cube method;
3. the square method.

The *unit method* is based upon the cost for a functional unit, for example, the cost of storage facilities per tonne of material stored, per million litres of water retained in a reservoir, per pupil accommodated in a school, per occupant of an office building, etc. The estimated cost of the work is then the capacity, in unit terms, multiplied by the cost per unit.

The *cube method* is based upon using historical costs, updated when necessary, of one cubic metre volume of a structure or building. The volume of the proposed building is calculated in cubic metres and then multiplied by the cost of one cubic metre.

The *square metre method* is similar to the cube method, but it is based on the superficial floor area of the proposed building.

In order to obtain a reasonable and reliable preliminary estimate of the probable cost of a project for budgeting purposes, considerable experience and judgement needs to be brought into play. Only very rarely are any two construction projects the same, and therefore due allowance must be made in unit costs for differences in quality, workmanship, material, location, availability of resources, etc. Such estimates are useful, however, for the process of the evaluation of alternative designs and arrangements and for the budgeting purposes of a client. They are not, of course, sufficiently detailed or accurate for the purpose of a contractor establishing a competitive tender. It is the preparation of estimates of cost for such purposes that is primarily the objective of the estimating work included in this chapter.

7.2 The tendering process

It will be assumed throughout that the *traditional* system of procurement is being used, i.e., a system in which the client appoints consultants to carry out the design work and the cost control, to prepare contract documentation and invite tenders for the work, subsequently recommending acceptance of one or other tender to the client, and to administer the construction. On the approval by the client of the recommendation by a consultant, one or several contractors will be appointed to carry out the work. Modifications to this general process inevitably occur to a greater or a lesser extent. However, whatever these modifications are, if an accurate cost of the work depicted in the contract documents is required prior to the commencement of work on site, a detailed estimate of cost must be made. It is this detailed estimate of cost that is the subject of this section of the book.

The outline of the process of construction bidding, for a contract of traditional type, from the point of view of a contractor, is shown in Fig. 7.1. This process is one whereby a contractor can produce a detailed estimate of the cost for carrying out a construction project by taking into account the cost of all the necessary resources to be used such as those of labour, materials, machines, and finance, together with the cost of subcontract work, overheads, and profit.

The first decision to be made by a contractor when invited to submit a tender is whether a bid will be submitted or not. An invitation to tender may have come as a result of a number of the various actions referred to in Section 1.14. A senior

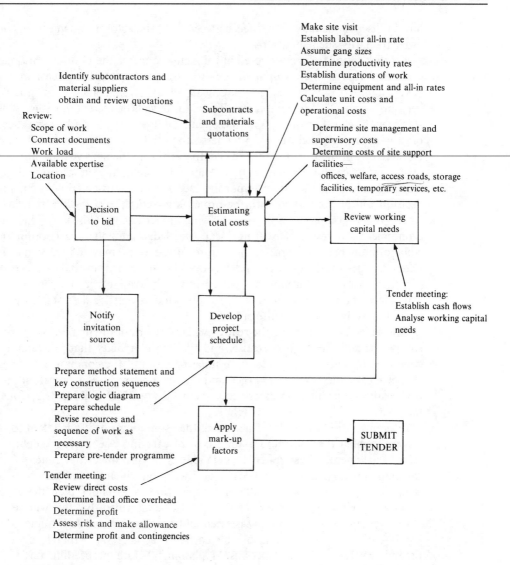

Figure 7.1

manager, probably at director level for all but relatively small projects, will generally make the decision as to whether the company will submit a bid or not. The decision will follow a detailed examination of the *contract documents* (see Section 1.13) to see if there are any unusual requirements that will involve greater than normal risks of higher expense, or are particularly onerous in other ways. If these occur they may be offset by an increased estimate of cost or if they are not feasible, for example by calling for an insurance providing unlimited indemnity against an accident, then further discussions may be needed between the contractor and owner/consultants in order to clarify such situations.

Another factor to be considered at this stage concerns the present workload of the company. This is often a difficult area to assess because a contractor may already have submitted, say, 20 tenders that await adjudication, some of which may be for very large projects. The company will perhaps have little idea at this stage of where it stands with regard to the bulk of these tenders. It perhaps has a historic success rate of gaining one contract out of 15 tenders submitted. There is considerable uncertainty, therefore, in the workload for the company during the next year or two. A sudden excess of successful tenders may overload the company in respect of technical resources and finances. Other important factors to be considered before making this decision will be concerned with the project's geographical location. Is it in an area where other company projects are being undertaken? This may lead to some economies as a result of existing local knowledge and possible coordination and collaboration over resources. The nature of the work may influence the decision. Often it is desirable that a company should bid for a project that, if successful, will provide experience of a kind not currently widely available in the company. If such experience could lead to more work of this kind, it may be desirable to pursue the project actively and competitively.

Other factors will include the ready availability or not of any necessary specialized equipment and technical expertise, and the bond capacity that is available to the company. In addition, consideration must be given to the company's business plan, its need to achieve a sales target and its need to receive a contribution to the overheads and profit from its projects, the details of which will have previously been targeted.

If the invitation to tender is accepted, the estimating process is taken to the next stage, in which the contractor's total cost of carrying out the work defined in the tender documents is estimated in detail. It is in this stage that much of the cost calculation is carried out and a cost is derived upon which the bid may be based. If it is not intended to proceed with the preparation of a tender then the office from which the invitation to tender emanated must be informed that this will be the case.

In preparing a cost estimate it is essential that planning and scheduling go hand in hand with the development of the costs. It is necessary, therefore, to take a preliminary look at the processes of planning before proceeding further with a detailed description of the cost estimating system.

7.3 Planning generally

Planning is an administrative process. As such, it is necessary so that instructions may be issued in order to instigate action to achieve a specific objective. Planning is interpreted in different ways by different people. To some, the word planning is synonymous with programing; to others it brings to mind the work of an office etc. bearing the label 'Planning Department', which in itself may have duties varying widely as between one company and another. Not only is the planning process seen as being one which varies widely in its scope, but also different people have different views as to both the requirements of planning and the detail which should be undertaken in varying circumstances. Planning, however, is the most important of the management processes and without it the proper and successful running of a company, a project, or a private life, must be very much a matter of chance.

There are two broad divisions of planning by management, both of which are necessary for the successful operation of a construction company. The first area is concerned with the operation of the company as a commercial undertaking; the second is the planning associated with the company's technical processes. The former is policy planning and is concerned with the overall means of achieving the company's objectives as defined by the board of directors. The second area concerns the various methods of arranging and employing money, materials, people, and plant (i.e., resources) to carry out the day-to-day operations of the company, and, as such, is dealt with here.

Planning, in this context, is the deliberate consideration of all the circumstances concerned with a project in order to evolve the best method of achieving a stated objective. Without it, wasteful, unproductive time is unavoidable. In planning, a logical attempt is being made to foresee all of those events that are likely to prevent or defer attainment of the stated objective. The objective in construction work is usually that of completing a prescribed amount of work within a fixed duration and at a previously estimated cost. Alternatively, it may be described as an attempt to establish the length of time that a given amount of work will take having known resources at one's disposal. Time is the unit that is most frequently planned. Cost is a unit that receives less attention but to which, for greater efficiency, the planning of time should be very closely linked.

The executive process that is complementary to the administrative process of planning is *controlling*. Without a proper and adequate plan the controlling process cannot take place, since it is essential to have some means by which the progress of the work—both financially and by time—can be measured against the planned requirements. Controlling, therefore, must be preceded by planning and as a process cannot function without it. It is important to emphasize that, where practicable, it is vitally important that the man who is going to control work should have a hand in its planning.

The controlling process may lead to *replanning*. As work progresses and performance is checked against the plan, adjustments to it are frequently necessitated as a result of the deviations from it. In the event of major deviations, the effect they will have on the remainder of the plan must be established, and every attempt must be made to adjust or remodel the plan so that the original objective can still be achieved. Adjustments and remodelling of the plan must be based upon rational calculations and scientific method. There is no point in reducing, on paper, the time span of a programme to an unrealistic period, since failure to achieve this will have its effect on many aspects of the work ranging beyond the operations of immediate interest. Every plan must be simple and flexible in order to facilitate any necessary adjustments.

A graphical schedule known as a *programme* forms the basis for effective planning. The programme must show sufficient detail to enable proper consideration to be given to the timing and durations of all the operations concerned, material and equipment delivery dates, manpower requirements, and subcontractors' visits to site. In this way, it should bring to light likely problems or delays.

The purpose and aims of a good programme may be summarized as follows:

1. It must expose difficulties likely to occur in the future and facilitate reorganization to overcome them.
2. It must enable the unproductive time of both humans and machines to be minimized.
3. It must be suitable for use as a control tool against which progress can be measured.
4. It must be sufficiently accurate to enable its use for forecasting material, manpower, machine, and money requirements.
5. It must aid the establishment of a work method which is efficient but at optimal cost, bearing in mind the availability of resources.

7.4 Preliminary planning processes

It is desirable to approach the development of a programme for construction work under two headings. The first is *construction planning*, which principally includes the determination of the technology and the methods to be used to carry out the work so as to fulfil the requirements of the facility design and specification. Second is the *work scheduling phase*, in which the resources for carrying out the methods are considered, together with their sequencing, cost, and likely productivity, resulting in a time and resource usage schedule for the project.

The scheduling of the work cannot be commenced until the first phase of the preparation is virtually complete. This part of the process is usually carried out in the early stages of a project in order for a contractor to be able to establish the cost of the work and to bid for a project, although the accepted methods may well be changed as work proceeds. In settling the technology and methods, although the process tends to be somewhat qualitative, it is frequently the case that several alternatives are conceived and costed; the availability and cost of the resources required for each are reviewed and preliminary plans and schedules showing the implications of the respective schemes are almost certainly prepared. Each method should be evaluated for quality, safety, and risk, as well as cost. Risk is included here as a separate heading, even though it pervades each of the other areas. Risk in this instance refers to the chances that the work, the subject of the planning and scheduling, will not be completed within the programme set for it. Quality and safety are not too difficult to evaluate because they are constrained by specifications for materials, work, and performance, as well as having to comply with legal requirements and regulations. Generally, construction methods will either comply or not comply; elimination of some alternatives can be made on this basis.

Decisions will be required, for example, on whether to use a sheet piling or open-cut for excavation, ready-mix or site-mixed concrete, pumping or tower cranes and buckets for concrete distribution, tracked or rubber-tyred excavators or loaders, etc. Before this stage can be reached it will be desirable to visit the site on which the work is to be constructed to obtain local information concerning many aspects of the project. Typical information collated in a pre-tender report will include the following:

1. A brief description of the project to be carried out and its precise location. Information concerning the client, the engineer, and the architect, their addresses and telephone numbers, and named individuals in their offices with whom direct contact can be made.

2. The name and address of the local authority engineer in whose district the work is to be carried out, together with such other information concerning the local authority as is considered necessary.

3. The access to the site of the works, both by public transport and for transport that will be conveying materials, heavy items of plant, etc. Under this item might also be included the details of any temporary access roads that may have to be constructed to provide the proper access for the duration of the contract.

4. Details of services already available at the site, for example, water, gas, electricity, and telephone, and the location of points to which connection can be made. In the absence of such services, firm estimates of the distances over which they will have to be carried to the construction site will be required. Contact with the various authorities responsible for the supply of the services will be necessary.

5. General details of the geography of the area, with any likely uses to which the site has been put during the past; details of any boreholes, groundwater levels, or local knowledge that are available in this respect.

6. The location of tips, if these are required, for the disposal of surplus excavation material and other rubbish. It is often a useful exercise in this connection to locate existing quarries and the suppliers of hardcore, ballast, etc. The two investigations often have a common solution.

7. The local availability of labour. This is one of the key points to discuss and investigate on a site visit. The local Employment Office is an important focal point when collecting information of this type. In the event of the supply of local labour being insufficient, then some estimate needs to be made of the areas from which suitable labour can be drawn and the likely cost of importation. Important too, in this context, is the question whether other similar types of work will be starting in the area at about the same time as the contract under review and whether they are likely to offer serious competition for the employment of the labour that is available. On the other hand, it may happen that a local contract will be closing or running down at about this time, with the result that there will be an adequate supply of labour available, ready, willing, and experienced to carry out the work.

8. The type of weather that will occur over an extended period. It should be possible to predict this, very approximately, from the local weather records. The amount of rainfall, for example, will not, on the whole, vary tremendously from one year to another, and statistically it should be possible to build up a picture of the probable extent of rainfall at one particular period of the year.
9. Details of possible sub-contractors who already work in the area. This information can be collected when the site area is visited. In addition, a talk to some of the individuals concerned will often produce valuable results relevant to the local labour situations.

All the information collected must be prepared in the form of a report and set down in a clear and concise manner so that the team of people undertaking the planning and preparing the estimate can have it available for ready reference. If the information has been collected properly then the planner will have a good idea of the resources that can be used and the estimator, among other things, will be in a position to establish comprehensive labour rates for labourers and craftsmen, allowing for all the items over and above the basic hourly rate of pay.

7.5 Pre-tender planning

Before pre-tender planning can commence a *method statement* must be prepared detailing the method of carrying out every operation of consequence and the combinations of equipment and manpower established as being most suitable for the work. For example, the method statement might suggest that a certain type of concrete mixing installation is required to give the required outputs. The installation can range from a central batching plant to a small tilting drum mixer or a combination of several sizes and types of mixers for different purposes in different locations. The method statement will, where required, list the plant to be used for excavation, since this is often an important feature of a large construction project. A great deal of excavation is involved on certain types of contract, for example, those for motorways.

At this stage in the planning procedure, the contracting firm usually has a programme period set by the owner during which all of the work must be completed, or they will have the responsibility of calculating, according to the methods to be used, the length of time this will require. The owner is almost invariably only too anxious to get the work started at the earliest possible moment and for it to be completed in the shortest possible time. A keen contractor will gain nothing by stating that the work can be completed in a given time when it is known from the pre-tender planning exercises that it will be extremely difficult, if not impossible, to do so.

As important as preparing the method statement for planning purposes is breaking down the work to be programmed into work tasks. A *work task*, sometimes called a *work activity*, is a specific item of work that can be clearly identified and delineated in such a way that its commencement, content and completion can readily be recognized. A work task uses resources including that of time. The definition of work tasks for a specific project is very dependent upon the experience of the planner because there are no formal methods of defining such tasks precisely. The definition of work tasks is essential to the process of *scheduling*, that is, the process of establishing the sequence of work on a time base. *Planning* is normally considered to be the process of devising the work tasks and the order in which they are to be undertaken, whereas scheduling puts the plans into the time dimension.

An aid to the establishment of suitable work tasks, their interrelationship and the future schedule, is the *work breakdown structure*. This structure results from dividing the tasks into a number of groups, each of which represents a major division of the work to be accomplished. Such a structure links together the tasks and illustrates the relationships between them.

The work breakdown structure shows the hierarchy of a project. At the highest level is shown the facility to be achieved as a result of the construction activity; at the lowest level, as a result of successively breaking down the work at each level, is the work task or activity. Each work task is therefore a component of the project facility and each level in the hierarchy above it is a summary of the levels below. Groups of work tasks can be seen to be components of the unit from which they derive at the next level above and so on through each level up to the project facility. A work task will be defined as a result of the complexity of the project, the detail required for the planning process and, if relevant to the managerial processes, the cost or value of the work. Each work task will have its own scheduled commencement and completion date and, in every way, will be clearly distinguishable from all the other project work tasks. A typical work breakdown structure for a simple facility is illustrated in Fig. 7.2.

A work breakdown structure has other uses beyond those described above. One important facet is that of relating the work and its management or supervision to the organizational structure of the contractor carrying out the work. If the contractor's organization is interrelated with the work breakdown structure then functional responsibilities for the work tasks can readily be allocated to individuals, the means of control by monitoring performance being established by the schedule, cost, and work content of each work task.

At the pre-tender planning stage it is necessary to consider the type of staff organization structure that will be used if the tender is successful. It may even be necessary for specific individuals to be considered for key posts in the organization structure, since, in the case of specialist operations particularly, provisions for obtaining the right type of staff may have to be made at an early date. The drafting of an organization structure enables an estimate to be made of the amount of time for which key men and other staff will be required on the site. The estimator can then allow in his tender for the salaries and oncosts that will be associated with such personnel. The staff requirements are best portrayed in the form of a simple schedule or bar chart so that the continuity of supervision can be checked and the staff requirements are not either over- or underestimated. The bar chart, or Gantt chart (after Henry Gantt; see Section 1.2), is a simple schedule listing the operations to be

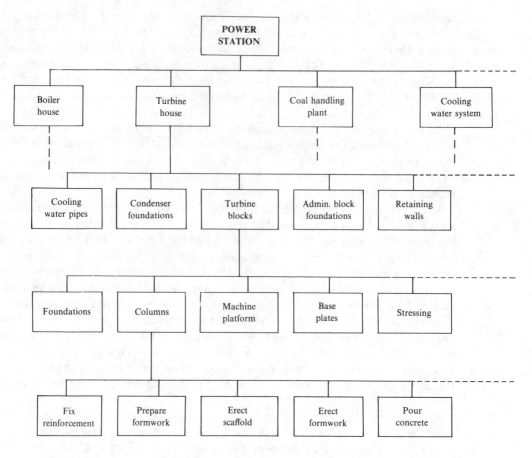

Figure 7.2

carried out on the left-hand side of the chart, one beneath the other, and carrying a time schedule along the horizontal scale. Figure 7.3 illustrates a typical arrangement for a staff requirement chart.

As early as possible in the pre-tender planning stage it is desirable to prepare a draft overall programme which at this time can conveniently be in bar-chart or Gantt chart form. Such a programme gives a general idea of the length of time that it will take to complete the work, on the basis of estimated outputs for the major items of plant and equipment. It need only be prepared for the key items of work in the contract that will have a major influence on the estimated price—particular attention being paid to those items that will be subcontracted. Figure 7.4 is an illustration of a draft pre-tender programme as it might be drawn up for a small bridge.

The pre-tender programme will enable the estimator to check whether there are any gaps in his overall assessment of the work or whether major items of work have

Staff Requirement Schedule															
Enquiry Nº 1634		Location Allhalls							Prepared by						
Staff	Month	1	2	3	4	5	6	7	8	9	10	11	12	13	14
Project Manager															
Deputy P M															
Chief Engineer															
Section Engineer A															
—do— B															
—do— C															
Quality Surveyor															
Asst Q S															
—do—															
Office Manager															
General Foreman															
Section Foreman 1															
—do— 2															

Figure 7.3

Pre-Tender Programme															
Enquiry Nº 1637		Location Towson							Prepared by						
Operation	Month	1	2	3	4	5	6	7	8	9	10	11	12	13	14
Preliminaries															
Site clearance															
Foundation piling															
West abutment															
East abutment															
Centre support portal															
Concrete deck															
Handrails															
Road surface finish															
Site works															

Figure 7.4

been omitted from the deliberations. Allowance will have been made for the use of specific mechanical equipment for carrying out particular operations. The pre-tender programme enables the estimator to coordinate the periods during which this equipment will be on the site. Consequently, it will enable some allowance to be made in the tender price for those periods when it is not possible to remove the plant but during which, of necessity, it must stand idle or work at a reduced rate of output. Figure 7.5 illustrates a simple bar chart representing the preliminary assessment of the requirements for major plant items during the construction of an underground pumphouse. At this stage the periods of use of equipment must be reconciled. A

Operation	Plant	Month	May	June	July	Aug	Sept	Oct	Nov	Dec
		Plant Schedule — Location				Date — Prepared by				
			1	2	3	4	5	6	7	8
Site clearance	D8 Bulldozer		▭							
"	Drott skid shovel		▭							
Excavate drains	$\frac{1}{2}$ m^3 Excavator		▭							
Pumphouse Exc^n	$1\frac{1}{4}$ m^3 Excavator		▭							
"	D4 Bulldozer		▭							
Concrete in Pumphouse	1 m^3 Mixer	Erection				▭				
"	10 T Derrick						▭			
Concrete in Culverts	$\frac{1}{2}$ m^3 Crane						▭			
Backfilling	D8 Bulldozer									▭

Figure 7.5

decision has to be taken, for instance, as to whether the $\frac{1}{2}$-m^3 excavator will be kept on site doing nothing during July and then converted into a crane for use in placing concrete, or whether it would be cheaper to send it away to another site, if there is a demand, for that period. Alternatively, by some adjustment of the excavation methods it may be possible to utilize it elsewhere on the site during that month. Yet again, the programme as a whole might be shuffled so as to bring about a higher utilization of the plant. The portrayal of these preliminary assessments, in the form of the simple bar chart, facilitates the optimization of the tender as a whole, though this is not necessarily the best way of tackling the problem.

By preparing a number of alternative plans based upon different method statements it may be possible in some degree to optimize the cost of carrying out the work. Comparisons should be made in terms of money, based upon various method statements. Network analysis, which will be described later in this book, is an excellent tool for the purpose of pre-tender planning. It does enable fallacies of logic to be highlighted at the pre-tender stage, whereas a bar chart tends to be somewhat general in its application and leaves much to be assumed by the user.

As far as the unit of time is concerned for pre-tender programmes, sufficient detail can generally be displayed by using periods of one month. There may be some virtue in the illustration of a programme in units of one week, where more detail is essential or where the programme is of short duration.

When the estimator has completed his basic procedures then the whole of the planning at this stage can be checked to make sure that there are no outstanding errors of judgement. In most organizations it is the responsibility of a senior member of the staff, if not a director, to check the tender in principle before it is submitted to the client in competition with other contractors' bids. Should the contractor be awarded the contract, then the pre-tender planning forms a ready-made basis for the work, which he can commence to carry out at the earliest possible time.

7.6 Estimating work durations

The time it takes to carry out a particular operation or work task is known as its *duration*. Before a series of work tasks (or larger sections of work consisting of more than one work task) can be scheduled, it is necessary to define at least two of its parameters:

1. its duration;
2. its relationship to its predecessor activities.

The duration of an activity is usually estimated, in the first instance, for *normal* conditions. Its normal duration is the one that results from an examination of one or more methods of carrying out the work on the basis of minimum cost. Most construction managers would normally seek to carry out the work at minimum cost and use the duration of the work that is associated with it. Minimum costs and minimum duration rarely coincide. The cost of work depends largely on the extent of the resources that are selected to carry out the work and how the different resources are combined and balanced to achieve the chosen construction method. Adding more resources to a work task increases the cost but does not necessarily result in a proportionate decrease in activity duration. Carrying out work with less than normal resources does not necessarily extend its duration proportionately.

Figure 7.6 shows a theoretical relationship between cost and time that often occurs in practice. The flat part of the curve can be supported by a simple illustration. If it is required to excavate by hand a length of trench in reasonable ground, point A on the curve may represent the point at which 10 operatives take 5 days; point B may represent the point at which 5 operatives take 10 days. If only the direct cost is considered (that is, excluding overheads and oncosts) then the costs of the excavation will be the same. However, if it is required to excavate the trench in a duration of less than 5 days then either addition of operatives to the trench may be physically impossible or the balance of the smooth working may be upset, or both. Another alternative may be to work overtime at enhanced rates of pay, causing an increase in the unit cost of the excavation. Beyond point B, the gang of operatives will have to be reduced below five, in which case the work may become inefficient because, again, the balance of the gang may be upset. Production will therefore fall and costs would rise. This simple example illustrates what is usually well established, that there will be a range of durations for most work tasks, but outside certain limits of duration, costs will tend to rise. Normal conditions, in the sense of the previous discussion, occur along the flat part of the curve.

Experience of similar work will play a significant part in the duration estimation process, but many factors will affect the calculation of the duration and each

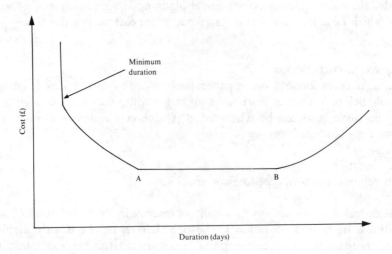

Figure 7.6

should be given consideration before a critical duration is settled. Some of the factors are:

1. weather;
2. availability, quality, and training of operatives;
3. familiarity with the work;
4. quality of workmanship specified;
5. quality of management/supervision;
6. size and completion date of project;
7. length and incidence of holidays;
8. repetitiveness leading to learning advantages;
9. physical constraints of site such as access, size, initial condition, location relative to adjacent buildings, storage space, etc.

Making allowances for the weather is an important decision to be reached in respect of activity durations. It is particularly critical for excavation and earth-moving activities, which are completely exposed to the weather. It may not be quite as critical for building work, especially when the structure has reached the stage of being almost completely closed in. There are generally two approaches to the problem. Firstly, each activity has an allowance added to it to take account of possible delays due to the weather. This has the advantage that each activity can be assessed on its merits, its overall place in the seasons and its duration. Sometimes a fixed percentage is added to each activity for this purpose, though this can produce difficulties with activities of long duration, particularly if they are not too sensitive to the weather.

The second method is to make a single allowance at the end of the project. This method may work best where all work tasks on the project have more or less the same sensitivity to weather and the weather does not vary significantly from period to period.

As a starting point for estimating the duration of a work task the following formula may be used:

$$\text{duration} = \frac{\text{quantity of work to be carried out}}{(\text{average productivity per operative})(\text{number of operatives})} \qquad (7.1)$$

For example, where $200\,\text{m}^3$ of excavation is to be carried out by 5 operatives each with an average productivity of $2\,\text{m}^3$ per working day in this kind of material,

$$\text{duration} = \frac{200}{(2)(5)} = 20 \text{ working days}$$

If the excavation needs to be timbered as it proceeds, since it cannot commence until some excavation has been carried out, say $10\,\text{m}^3$, then a start time can be determined for the timbering relative to its predecessor work task. Timbering can start $10/(2)(5)=1$ working day after the start of excavation. If each of the two activities are represented by bars on a time scale they will look like Fig. 7.7.

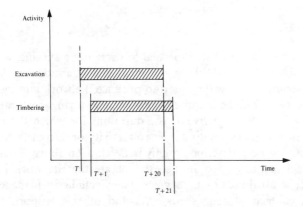

Figure 7.7

7.7 Estimating and tendering

A tender price is made up of the cost categories shown in Fig. 7.8. The estimated cost to a contractor of carrying out the work is known as the *construction cost* and is composed of the *direct cost* of carrying out the work to which are added the *site overheads* (*oncosts*). A direct cost consists of the cost of the resources—materials,

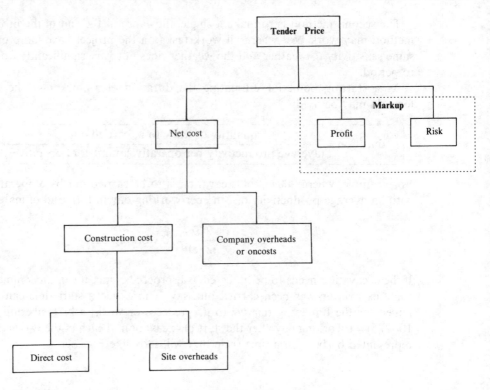

Figure 7.8

labour, equipment, and subcontractors—needed to carry out a specific, well-defined item of work. It is found by establishing the measured quantities in the item of work, defining the resources that will be used to produce the work and the duration of time over which each will be required, and then applying cost rates to the quantities and times. In order to arrive at the durations for which the productive resources are required it is necessary to use a productivity or output factor for each resource. This process has been described briefly in Section 7.6 above. Site overheads or oncosts include all of those costs needed to operate the site work production activities that cannot be attributed to direct costs. They include site management and supervision, offices, canteen, storage sheds, cars and other transport, temporary roads and services, and general labour not assigned to production. Most companies have a standard checklist for oncosts, which acts as an *aide-mémoire* for estimate preparation.

The construction cost then forms the basis for determining the *net cost* for a contract. Every contract within a company incurs overheads with regard to the company's head office. These are usually termed *general* or *company overheads*. Such overheads cover the cost of the company management and administration and often design services for temporary or permanent works, computer processing of wage-sheets, cost and bugetary control, personnel services, training, and so forth. The total relevant overheads for a company are usually distributed over the contracts as a percentage of the construction cost completed. For example, if the total overhead cost for a company amounts to £500 000 and the projected turnover for the company is £12 000 000 then the percentage to be added to a contract turnover is $500 000 \times 100/12 000 000 = 4.17$. If a contract for £1 000 000 is obtained then the total contribution to general overheads will be £41 700. This is the sum that needs to be added to the construction cost in order to arrive at the net cost.

The estimated net cost of the project having been determined, the *estimating process* is completed and the *tendering process* begins. Tendering is the process whereby a contractor, given the net cost, converts this to the sum what will actually be submitted to the client, together with any qualifications that are seen to be required. At this stage the principal discussions are concerned with the profit and the risk, together known as the *margin* or the *markup*. In order to determine the margin, an assessment of the possibilities of over- or underestimation of the costs is made. Where, for example, a tunnel is to be driven through waterlogged ground containing boulders, a relatively high allowance for risk will be required because of the possibility of disruptions and delays to the work. For building a standard, two-storey, detached house, on the other hand, the risk of experiencing construction difficulties is very low. It is at the tendering stage that judgements of this kind need to be made.

Another factor to be taken into account at the tendering stage is the financial effect that taking on the contract may have. Payment for work completed is almost always in arrears, and therefore a contractor who has yet to make sufficient profit on a contract to meet those arrears (and this may not be until the later stages of the work, if at all on some occasions) must provide sufficient *working capital* to keep the work going until that stage is reached. This can be calculated (see Chapter 13) by assessing the *net cash flows* and their timing. Working capital has either to be borrowed at the cost of the interest rate or else be provided from company resources at the cost of not having the money available for alternative uses.

7.8 The all-in labour and equipment rate

The calculation of the cost of labour for a construction activity is complex. This is due to the many elements that go to make up the total cost of employing an operative, including the payments that must be made as a result of the various agreements between employers and unions, and also as a result of national legislation. It would be uneconomic to calculate the wages of all types of operative with various skills that may be employed on a single operation. Because of this, an *all-in rate* is established, which includes all aspects of payments reduced to a rate for one hour; thus in civil engineering, for example, for the sake of simplicity, an all-in rate is normally established for a typical labourer, a craftsman, and a plant operator only. An all-in rate for other craft operatives is calculated only where it is thought to be necessary because of the beneficial effect it may have on the estimated costs. The components of the all-in rate tend to change from time to time, since negotiations are almost continuously in progress about some aspect or other of the total wage package. In civil engineering, the current Working Rule Agreement of the Civil Engineering Construction Conciliation Board for Great Britain should be consulted for details of operative's wages; for building work, the National Working Rules for the Building Industry of the National Joint Council for the Building Industry is the equivalent document. In both cases these documents, and all their supplements, set out comprehensively the detailed rules that apply to the wages of operatives in the construction industry. The following calculation gives the principles of calculating an all-in rate for a civil engineering craftsman. Clearly, the rates included in the example are for illustrative purposes only and should not be assumed to be the current rates and/or deductions or additions for practical purposes.

Assume a regular working week to be from 8.00 a.m. to 5.15 p.m. (less half an hour for lunch) from Monday to Friday. With a normal week of 39 hours, this timetable gives $(8\frac{3}{4} \times 5) = 43\frac{3}{4}$ hours; that is, 39 hours plus $4\frac{3}{4}$ hours overtime.

Since overtime is paid at $1\frac{1}{2}$ times the standard rate, the *non-productive overtime*, that is the additional rate paid for the overtime over and above normal time, amounts to $4\frac{3}{4} \times \frac{1}{2} = 2\frac{3}{8}$ hours.

Therefore, total time paid $= 39 + 4\frac{3}{4} + 2\frac{3}{8} = 46\frac{1}{8}$ hours

It is now necessary to make allowance for time, in working days, that will not be worked during the year:

Total time worked in 1 year	$= 52(43\frac{3}{4})$	$= 2275$ hours
less: Holiday of 30 working days	$= 30(8\frac{3}{4})$	$= 263\frac{1}{2}$
Total hours worked in 1 year		$= 2011\frac{1}{2}$
Assume 8 days lost for inclement weather		70
		$1941\frac{1}{2}$
Assume 10 days of $8\frac{3}{4}$ hours lost for sick leave		$87\frac{1}{2}$
Actual hours worked in 1 year		$= 1854$

The total non-productive time in one year will be affected by the time assumed lost to inclement weather and sick leave:

Total non-productive time	$= 52(2\frac{3}{8})$ hours	$= 123\frac{1}{2}$ hours
less: Time for inclement weather and sick leave		
	$= 18(4\frac{3}{4})/5 = $ (say)	17
Total non-productive time for year		$= 106\frac{1}{2}$ hours

Annual cost of labour. Assume:

(a) A basic wage rate of £2.90 per hour.
(b) Guaranteed minimum bonus of £16.00 for a 39-hour week.
(c) An allowance of 3 per cent on the *total basic wage* for plus rates (that is, the additional hourly rates paid over the basic rate for conditions, skills, etc.).
(d) 11 per cent of *total wage paid* for National Insurance.

(e) 2 per cent of *total wage paid* for training levy.
(f) Annual holiday pay and death benefit of £14.45 per week.
(g) Small tool and clothing allowance of 4 per cent of total wage paid.
(h) Travel allowance of £5.00 per week.
(i) Allowance of 6.5 per cent *total wage paid + allowances* for insurance and severance pay.
(k) Allowance for sick pay—3 per cent of *total basic wage.*
(l) 8 days public holidays at 8 hours per day.

From Table 7.1, we see that the total annual cost of the operative in the above example is then £9896.32, and since the operative is expected to work for 1854 hours during the year, the all-in rate becomes £9896.32/1854 or £5.34/hour. This is the total cost to hire an operative, and therefore the cost that must be incorporated into calculations for developing the estimated cost of construction work.

Table 7.1

	Rate, £	Hours	Weeks	Cost, £	Summation, £
Basic wage[a]	2.90	1854			5376.60
Allowance for inclement weather	2.90	70			203.00
Total basic wage					5579.60
Guaranteed minimum bonus[b]	16.00		47	752.00	
Non-productive overtime	2.90	106.5		308.85	
Public holidays[l]	2.90	64		185.60	
Sick pay allowance[k]				167.39	
Plus rates[c]				167.39	1581.23
Total wage paid					7160.83
National Insurance[d]				787.69	
Training levy[e]				143.22	
Travel allowance[h]	5.00		47	235.00	
Small tool allowance[g]				286.43	
Annual holidays and death benefit[f]	14.45		47	679.15	2131.49
Total paid + allowances					9292.32
Allowance for severance pay and insurance[i]				604.00	604.00
Total annual cost					£9896.32

All-in rates for mechanical equipment and plant can be calculated in a similar fashion. Reference should be made to Section 14.12 in which this subject is discussed. When a contractor purchases construction equipment it is with the view to using it within the company to make a profit. It is therefore very important that the total cost of owning, maintaining, repairing, and operating each item of equipment is established, and that that cost is then incorporated in the price for any item of work for which the contractor uses it. Equipment may alternatively be hired. Establishing the cost of such equipment is, of course, more straightforward, but care needs to be taken to establish exactly what is included in the hire rate and what will need to be provided by a contractor.

7.9 Preparing the estimate

When an invitation to tender for construction works is received in a contractor's office, as soon as a decision is made to bid for the contract, several processes can go forward in parallel with one another. One such action will be to obtain quotations for the supply of materials. Before the quotations are received, because of the need to save time, an estimator may have to assume the prices of the materials and incorporate them in the calculations of the estimates. Later, when the actual prices are received, the calculations will have to be reviewed and adjusted. Because time is usually short for preparing detailed estimates, when enquiries are distributed for materials, it is essential that all the relevant details such as quantities, grades, specifications, location, delivery requirements, and so on are made available to the potential suppliers. When the quotations are received it is important to check to ensure that all the requirements are being met and that the discounts being offered on the prices are in accordance with what should be expected.

Subcontracts will be of two kinds—either *nominated* or *domestic*, to which reference has already been made. The nominated subcontractors will be detailed in the contract documents, but the contractor must make a decision as to what work that is not the subject of a nomination will be subcontracted. Work subcontracted is often of a specialized nature, requiring special equipment and skilled labour. Many processes concerned with excavation, for example, come into this category. Diaphragm walling, grouting, freezing, consolidation, and piling are some of those processes. On large civil engineering works with extensive excavation, the basic earthmoving is frequently subcontracted to companies having a fleet of earthmoving equipment in their plant inventory. Subcontractors' quotations are normally obtained in the terms of the contractor's contract conditions and the quotations, when received, will be in the form required by the contractor to insert in the contract documents. A contractor will wish to discuss with the subcontractor, before the subcontract is placed, the proposals for the programme and the specification of the work. The contractor must be satisfied that the subcontractor will perform as required and will need to add a sum of money to the subcontractor's quotation for attendance, including accommodation and temporary works, and supervision as well as profit and risk.

Two other general items, which are included in a bill of quantities forming part of the contract documents that come into the estimator's consideration, are *provisional sums* and *prime cost sums*. A provisional sum of money is inserted into a bill at the time of tender, as a provision for work to be carried out by the contractor, the precise nature of which is undefined at the time of tendering. A prime cost sum is a reservation of money inserted in the bill, either for work to be carried out by a nominated contractor, which includes a statutory authority (diverting a main water supply pipe, for example) or a public undertaking, or else for materials or goods to be obtained from a nominated supplier.

Items in bills of quantities for nominated subcontractors, nominated suppliers, and prime costs will normally have associated items for the contractor's profit, for general attendance, where required. These need to be dealt with by an estimator before the bill is completed with tendered rates.

7.10 Estimating direct costs

Before the direct costs of construction can be estimated, the pre-tender planning, as described in Section 7.5, must be brought to such a stage that it can form a suitable basis for estimating. This, with the cost of the resources, such as the labour, plant and materials, can be combined to generate direct cost estimates. It is important that the estimator joins with the planners when the programmes and method statement are being prepared, so that the estimator may have a complete understanding of what is being done and provide some input from his experience. When the method statement and the associated programmes are being prepared, the estimator can help with the evaluation of alternative methods that may be available and produce cost evaluations for comparison purposes.

If the tender documents include a bill of quantities, the estimator has eventually to be able to arrive at rates to insert in it for each item. This has to be borne in mind when direct costs for operations are being built up. There are two ways in which estimates may be prepared—*unit rate estimating* and *operational estimating*. If a direct cost is calculated by using an established production output rate for a specified quantity of work of a particular description, say, as set out in a bill of quantities, then it is described as unit rate estimating. If the direct cost of an operation, activity, or work package is calculated on the basis of the time it will take to complete then it is known as operational estimating. Four simple examples are given below to illustrate the major features of each type.

EXAMPLE 7.1
Estimate the direct cost of excavating by hand a trench 1 m wide, 1.25 m deep, and 4 m long in stiff clay. Strutting to the sides is not required. Spoil is to be removed by wheelbarrow to not more than 50 m from the trench. Labour rate is £6.00/hour.

Using unit rate estimating
   Volume of soil $= 1 \times 1.25 \times 4 = 5 \text{ m}^3$
The stiff clay will need loosening with a pick—allow 2.5 hours/m^3
   Loosening soil $= 5 \times 2.5 = 12.5$ hours
Shovelling soil from trench: allow 0.8 m/hour (bank measure)*
   Shovelling soil $= 5/0.8 = 6.25$ hours

Bank measure is the measurement of the volume in its original state in the ground—in other words, before it has been disturbed by the excavation process. *Loose measure* applies to the volume after it has bulked by being removed from its original state. For this soil one might expect the loose measure to be some 20–40 per cent greater in volume than the bank measure, depending on the exact nature of the soil.

Filling wheelbarrow, capacity $0.1 \, \text{m}^3$ bank measure—allow 3 minutes. Wheel barrow 50 m: allow $2\frac{1}{2}$ minutes each way

Fill and wheel barrow and return — 8 minutes, i.e., $60/8 \times 0.1 = 0.75 \, \text{m}^3/\text{hour}$

Soil disposal $= 5/0.75 = 6.67$ hours

Therefore,

Total labour hours $= 12.5 + 6.25 + 6.67 = 25.42$ hours

Total labour cost $= 25.42 \times 6.00$ $= £152.52$

Add 10 per cent for small tools, barrow $=$ $\underline{15.25}$

Total cost $= \underline{\underline{£167.77}}$

That is, $167.77/5 = £33.55/\text{m}^3$

The unit rates for production, such as 2.5 hours of labour to loosen $1 \, \text{m}^3$ of soil, are based either on the experience of the estimator on previous works, or on a company database of performance figures. Such figures are also available in the textbooks or pricing guides that deal with estimating, though these should only be used with caution, because many such figures do not clearly specify the exact conditions under which the work was carried out.

EXAMPLE 7.2

A crane is erected over a deep excavation for a pumphouse for the sole purpose of transporting concrete in skips to the reinforced concrete foundations. There are $12\,000 \, \text{m}^3$ of concrete to be placed over a period of 12 weeks. Calculate the direct costs per m^3 of cranage for placing concrete that should be included in the rate for concrete placed in position.

	£
Site levelling and preparation for crane	$= 2\,500$
Cost of crane, £250 × 12 weeks	$= 3\,000$
Cost of skips, £40 × 12 weeks	$=\ \ 480$
Erection and dismantling of crane	$= 2\,000$
Transport of crane to site and return	$=\ \ 500$
Cost of operator, £75 × 5 days × 12 weeks	$= 4\,500$
Cost of banksman, £48 × 5 × 12	$= 2\,880$
Cost of mechanic £65 × $\frac{1}{2}$ day × 12 weeks	$=\ \ \underline{390}$
Total	$\underline{\underline{£16\,250}}$

The cranage cost per m^3 of concrete is therefore £16 250/12 000, or £1.354.

This example illustrates the operational estimating method. The concrete is to be placed over a period of 12 weeks, no matter what the production capacity of the crane. The programme dictates the need for the crane to be available. The cranage cost can then, if required, be reduced to one for a unit quantity of concrete, and that cost (£1.354) then needs to be incorporated into the cost of every cubic metre of concrete placed in the foundations by this method. It will be noted that in this example of the operational estimating method no production rates are used.

EXAMPLE 7.3
Determine the total cost and the cost per cubic metre of an excavation, the bottom of which is 7 m below ground level and measures 30 m by 60 m with sides battered back at 45° to the horizontal. The soil is sand and gravel and the excavated material is run to a tip 500 m distant.

$$\text{Volume of excavation} = 10\,192 \text{ m}^3 \text{ (bank measure)}$$
$$= 10\,192 \times 1.15 = 11\,721 \text{ m}^3 \text{ (loose)}$$

Use a dragline for excavation with a $1\frac{1}{4}$-m³ bucket having an estimated output in sand and gravel of 150 m³ (loose) per 60-minute hour under ideal conditions. Since the maximum depth of excavation is 7 metres, which is deeper than that at which optimum output could be obtained, use a correction factor of 0.75 on output. Assume that the dragline will be working for 45 minutes in every hour.

$$\text{Adjusted output} = 0.75 \times 45/60 \times 150 = 84.4 \text{ m}^3/\text{hour}$$
$$\text{Time required to excavate sand and gravel} = 11\,721/84.4 = 139 \text{ hours}$$

The cost of the excavation will be:

	£
Dragline 139 hours at £45.00	= 6 255
Operator 139 hours at £8.50	= 1 181.50
Mechanic 139 hours at £6.50	= 903.50
Banksman 139 hours at £6.00	= 834
Labour for trimming excavation: 139 hours at £6.00	= 834
Total direct cost	£10 008.00

Cost/m³, 10 008/10 192 = £0.982

The cost of hauling spoil will be:
Use 20-tonne tippers loaded with 16 m³ (loose) per trip
Assume a haulage speed of 7 km/h

Travel time to tip $= 500/7000 \times 60 =$	4.29 minutes
Return time	= 4.29
Loading time $= 16/84.4 \times 60$	= 11.37
Tipping time	= 2.00
Cycle time	= 21.95 minutes

Use 5 trucks at £25 = 25 × 5 × 139 = £17 375
Cost/m³, 17 375/10 192 = £1.705
Therefore, total cost/m³ = 0.982 + 1.705 = £2.687

EXAMPLE 7.4

Estimate the cost of making, fixing, and stripping the formwork for constructing a reinforced concrete wall 28 m long by 2.25 m high by 250 mm thick. The wall will require formwork to both sides. The general arrangement of the formwork is shown in Fig. 7.9.

The wall will be concreted to its full height in one pour, using formwork panels made up of standard sheets of plywood measuring 2.44 m × 1.22 mm (8 ft 0 in × 4 ft 0 in). The area of one standard sheet is 2.98 m^2; use 3.00 m^2.

Number of sheets in vertical direction = 2.25/1.22 = 1.84; use 2
Number of sheets in horizontal direction = 28.00/2.44 = 11.48; use 12
If panels are made up of 3 sheets in the horizontal and 2 in the vertical direction, then the composite formwork panel can be used a total of 12/3 = 4 times.
Each panel made up of 6 standard sheets will have a surface area of 6 × 3 = 18 m^2, but a concrete wall contact area of, for measurement purposes, 28/4 × 2.25 = 15.75 m^2

Materials for two composite formwork panels:
Area of 24 mm plywood required = 2 × 6 × 3   = 36 m^2
Allow 20 per cent for wastage   = 7
                        Total   43 m^2

Timber in twin walings = 2 × 6 × 7.50 × 0.175 × 0.050 = 0.788 m^3
Timber in studs = 2 × 11 × 3.75 × 0.125 × 0.044   = 0.454
                                     1.242
Allow 20 per cent for wastage   0.248
                        Total   1.490 m^3
Cost of 24-mm plywood = £8.50/m^2
Cost of timber   = £275/m^3
Therefore,
Total material cost for panels = (43 × 8.50) + (1.49 × 275) = £775.25

Labour cost for making panels:
Assume 2.5 labour hours to make 10 m^2 of formwork panel
Total hours required = 36/10 × 2.5 = 9.00 hours
Assume that panels are made by 1 carpenter at £6.00 all-in/hour and 1 labourer at £5.30 all-in/hour. This is a composite rate of (6.00 + 5.30)/2 = £5.65/hour
Labour cost to make 2 panels = 9.00 × 5.65 = £50.85

Labour cost to fix and strike 2 panels:
Assume 12 labour hours per 10 m^2 to fix and strike
Total hours required = 36/10 × 12 × 2 = 86.4 labour hours
Labour cost = 86.4 × 5.65 = £488.16
Allow for 4 uses of each panel = 488.16 × 4 = £1952.64

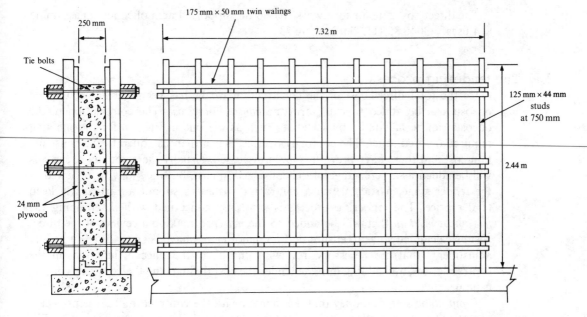

Figure 7.9

Labour cost for cleaning and oiling formwork:
 Assume $\frac{1}{2}$ hour per $10\,m^2 = (36 \times 2)/(10 \times 2) = 3.6$ labour hours
 Labour cost $= 3.6 \times 5.65 = £120.34$

Materials for oiling and cleaning formwork:
 Allow 5.0 litres of mould oil at £2.00/litre per $10\,m^2$ of formwork
 Cost of mould oil $= 2(36 \times 2)/10 = £14.4$

Other materials:
 2×33 tie bolts at £1.25   = £82.50
 Allow 25 per cent wastage =   20.88
               Total = £103.38
 Nails and screws for panels say, £15.00 per panel $= 2 \times 15.00 = £30.00$

Summary of costs:

		£	£
Materials:	timber for panels	775.25	
	mould oil	14.40	
	tie bolts	103.38	
	nails and screws	30.00	923.03
Labour:	making panels	50.85	
	fixing and striking panels	1952.64	
	cleaning and oiling	120.34	2123.83
			£3046.86

The direct cost rate for formwork (based on supported area of concrete) assuming 4 uses $= 3046.86/(15.75 \times 2) = £96.73$

7.11 The tendering process

Once the total direct cost of the construction works has been calculated, the site oncosts can be added to it to form the *construction cost*. The compilation of the oncosts can be facilitated by schedules such as the one in Fig. 7.4 for the site staff required on a contract. The company overheads can then be added to the construction cost and these have previously been discussed. Usually the company overheads will be added as a percentage of the construction cost, but they may be reviewed at the tender stage, particularly if a contractor wishes to submit a particularly keen tender price. The tendering process is primarily concerned with determining the *margin* or *markup* that will be added to the *net cost*, making a review of the direct cost as calculated so as to ensure that no major errors have been made effecting any adjustments that are necessary, reviewing the likely influence of inflation on the contract prices, and adding into the tender the cost of financing the contract working capital.

Profit is the sum of money that will remain with the contractor after the project is completed and once all of the costs of carrying out the works have been paid. The percentage profit to be added has already been referred to in Section 7.7. An important influence on the percentage of profit added is the evaluation of *risk* in the project. This is because *risk* and *cost* tend to be synonymous. A failure by a contractor to manage successfully the risk in a project will certainly increase the costs; successful management of risk reduces cost. The main reason why risk exists in construction projects is because tenders are submitted ahead of construction taking place. Construction may not start for some time after a bid is compiled and then it may continue for some years into the future. All of this delay increases the uncertainty of the project conditions and hence increases the risk.

The apportionment of risk between the parties to a construction contract is in the hands of the client when the decision to accept a particular form of contract and all the other conditions established in the contract documents is made. One means of reducing the risk of uncertainty to all the parties is for the client to specify exactly what is required of the other parties by way of performance, quality, quantity, and the conditions under which it will be provided, etc., so that there are few if any doubts in a tenderer's mind as to what is being priced and the responsibilities that will be borne. The client by so doing then reduces most of the uncertainties concerning the project.

There will be some risks that a contractor assumes as a matter of course. Among these are managing the project so that the productivity of resources allowed for in the estimate is achieved, bad weather and the interference with production and the programme that it may cause, the availability of materials for incorporation in the work, delays due to industrial disputes, the financial stability of the client, and the performance of equipment. Generally speaking, compensation is not expected for failure that is due to these causes and a contractor will make allowances in preparing the tender to cover the risk resulting from them. (The client, of course, could take on these risks if it were so wished. All, including the first, by employing a cost-reimbursement-plus-percentage-fee contract, for example. Otherwise compensation could be allowed for delays and disruption resulting from them.) Another important area where a contractor is at risk from the unexpected is where site conditions, particularly subsurface ones, are found to be other than were expected. While the additional cost of the changed works will be reimbursed, it is often difficult to substantiate and thus cover all the costs of delays and reprogramming, which may not be immediately apparent. This applies to unexpected changes of scope to the contract as a whole. One of the serious aspects that results from changes being made to a contract is the delay that frequently results in getting the variations agreed and paid for. A contractor sometimes tends to act as a financing institution for the client, and, having committed resources to additional work—especially if there is a difference of opinion about its extent and/or value—finds that prompt payment is not necessarily forthcoming. In most of these matters of risk, the contractor uses intuition, judgement, and experience to assess the compensation to be included in the markup. The contractor must also bear in mind, if the tender is to be competitive, that others are also bidding competitively. Also to be assessed is the contractor's own workload, the availability of resources, and the need to keep good specialist resources fully and continuously employed.

Inflation is another source of risk to be taken into account at the tendering stage where a tenderer is expected to bear the cost increases of materials, plant, and labour. For projects of an expected duration in excess of 12 months, a formula that is generally adopted by the construction industry can be used. The formula method is based upon dividing the work into *elements* or *work categories* and using an index published monthly to measure the reimburseable inflation cost. The methods are available, one for building and the other for civil engineering works. The method for building is described in *Price Adjustment Formulae for Building Contracts* and that for civil engineering in the *Report of Price Adjustment Formulae for Civil Engineering Contractors* (see Further Reading at the end of this chapter). The monthly indices are collated and published by the Department of the Environment. With the use of the formula method the risk to be allowed for at the tendering stage is considerably reduced if not avoided altogether.

7.12 Learning and forgetting curves

It is a common experience that repeating a job operation a number of times enables it to be carried out in successively shorter times. A point may be reached when the time taken cannot be further shortened and then levels off. In the repetition process a conscientious operative becomes increasingly familiar with the nature of the work, and as a result usually seeks to develop more efficient methods and perhaps better equipment and tools in order to carry out the work more efficiently. Repetitive work can also lead to a better coordination and organization in carrying it out, since it tends to draw more attention from managers and supervisors. This is due principally to the lengthy production runs required. Any inefficiency in a single repetitive process runs the risk of being multiplied many times if early diagnosis of the cause is not made, nor steps taken immediately to deal with it.

Learning curves (sometimes alternatively referred to as *experience curves*) have existed on an industry-wide basis for 50 years or more. They are mathematical and graphical representations of the likely improvements in an operation being repeated over and over again because of the ongoing enhancement of skill and experience. Learning curves can be used to predict what will happen when future activities of a similar nature are undertaken. As such they can have a considerable influence on the design, planning, and scheduling of work and thus also the estimating of its cost. Learning curves tend to show the greatest benefit when applied to situations in which the repetitive operations are labour-intensive. Some literature sources distinguish unnecessarily between learning and experience. However, mathematical formulation to represent many situations can, at best, reflect human behaviour only approximately and should not be incorporated in inflexible rules and instructions that are to be used without the benefit of experience and judgement.

Forgetting curves are learning curves that have been modified to take account of delays or interruptions to repetitive job operations. On interrupting the repetitive work, some of the enhanced skills and methods gained by the experience up to the point of interruption begin to decay. When the repetitive work is resumed, it is usually found that the first unit to be undertaken takes longer to complete than the one completed immediately prior to the interruption. However, from the resumption, the learning experience once more results in the reduction of the time taken to complete each subsequent unit. Nevertheless, a penalty is paid for the interruption in terms of both cost and time. The theory and practice of forgetting curves are less well understood than those of learning curves and relatively few attempts to gather field data to support the theories have been recorded.

T. P. Wright contributed a paper to the *Journal of Aeronautical Science* in February 1936, proposing an equation that became the starting point of much of the subsequent work on learning curves. Wright's equation is:

$$T_{an} = T_1 n^r \tag{7.2}$$

where T_{an} = the cumulative average man-hours or average cost per unit after the construction of n units

T_1 = man-hours or cost per unit for the first unit

n = unit serial number: 1, 2, 3, ..., etc.

r = *index of learning* or *improvement function* (where $r < 0$)

Because when it is plotted on a log–log scale, Eq. (7.2), in logarithmic form, $\log T_{an} = \log T_1 + r \log n$, reduces to a straight line, it is sometimes referred to as the *straight-line model*.

Studies have demonstrated that improvement in performance in similar conditions can be expressed as a fixed percentage each time the production of units doubles. If the production of a unit for the first time takes 60 minutes and the production of the second unit takes 36 minutes then the cumulative average duration $T_{an} = (60 + 36)/2 = 48$ minutes. The operative producing the units is then said to have a learning rate of 80 per cent (48/60). Unit 4 then has a cumulative average duration of $48(0.8) = 38.4$ minutes, Unit 8 of $38.4(0.8) = 30.72$ minutes, and so on.

For this situation, where improvement is a fixed percentage each time production doubles, the improvement function is

$$r = \log \text{(learning rate}/100)/\log 2 \qquad (7.3)$$

Using Eq. (7.3), if the learning rate = 80 per cent, the improvement function = $\log 0.8/\log 2 = -0.322$.

In this case, $T_{a2} = 60(2)^{-0.322} = 48$ minutes, and
$$T_{a4} = 60(4)^{-0.322} = 38.4 \text{ minutes}$$

Other improvement functions for various learning rates are given in Table 7.2.

Table 7.2

Learning rate, %	Improvement function, r	$(1 + r)$
100	0.0	1.0
95	−0.074	0.926
90	−0.152	0.848
85	−0.234	0.766
80	−0.322	0.678
75	−0.415	0.585
70	−0.515	0.485
65	−0.621	0.379

A typical learning curve for the production of 50 units, with a man-hour content for the first unit of 150 and a learning rate of 85 per cent, is shown in terms of both *man-hours per unit* and *cumulative average man-hours* in Fig. 7.10.

In construction, learning rates fall commonly between 80 per cent and 95 per cent, depending on the work conditions, the environment in which the work is carried out, and the nature of the work. Relatively simple labour-intensive elements of work with few operations uncomplicated by external constraints and carried out in a workshop environment might lead to learning rates of 80 per cent. More complicated large operations involving multi-trade and complex highly skilled functions, such as occur in the finishes of high-quality building, might merit a learning rate of 90–95 per cent. In the latter cases the opportunity for learning benefits will often come through enhanced and enlightened management and supervisory activities.

Clearly, there will be some operations that do not benefit, or benefit very little, from the effects of learning. These will generally fall into two categories. The first is that where the skills, experience, and motivation to learn of the operatives are already highly developed. The second category contains activities that are controlled.

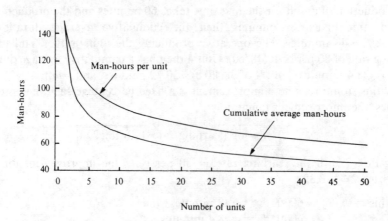

Figure 7.10

by the rate of a machine working and having a fixed output, so that no amount of experience or learning can influence productivity. In this last category the learning rate will be 100 per cent or close to it.

The above formulations have concentrated on cumulative average durations for each unit, or at any point in the overall progress of a project. However, the direct labour cost of an operation is usually directly proportional to the duration of that activity. It is important to be able to calculate this duration for any unit in the repetitive process in order to obtain a meaningful measure of progress or cost, as shown in Example 7.5.

EXAMPLE 7.5

A repetitive process is estimated to have a learning rate of 85 per cent. The initial unit under construction takes 150 man-hours. Determine the time that will be taken by the fifteenth unit.

The time taken by the nth unit, T_n, can be found from

$$T_n = nT_{an} - (n-1)T_{a(n-1)}$$

Therefore, from Eq. (7.2),

$$\boxed{\begin{aligned} T_n &= nT_1 n^r - (n-1)T_1(n-1)^r \\ &= T_1[n^{r+1} - (n-1)^{r+1}] \end{aligned}} \tag{7.4}$$

when $r = \log(85/100)/\log 2 = -0.234$
$$T_n = 150(15^{0.766} - 14^{0.766})$$
$$= 150(7.96 - 7.55) = 61.50 \text{ man-hours}$$

This can be compared with the cumulative average duration of unit 15:

$$T_{a15} = 150(15)^{-0.234} = 79.60 \text{ man-hours}$$

The use of learning curves can be important in determining schedules for progress control or in determining the direct costs of repetitive construction. If such effects are not taken into account then time and cost estimates can vary widely from what is likely to be achieved in practice. The following example illustrates the advantages of using a learning curve.

EXAMPLE 7.6

An estimate of cost is required for the direct labour costs in the production of 75 precast concrete beams. The labour wage rate is £5 per hour, with an addition of 50 per cent for other costs and benefits. In the past the steady-state production rate of beams has been 100 man-hours per beam. If the learning rate is believed to be 90 per cent, calculate the additional direct labour cost that is to be added for learning, and the total direct labour cost of producing the beams if the steady-state production rate is reached after producing 50 beams.

Labour wage rate $= 1.5(5) = £7.50$ per hour
Steady-state labour input $= 100$ man-hours after 50 beams
r for 90 per cent learning rate $= -0.152$

From Eq. (7.4)

$$100 = T_1(50^{0.848} - 49^{0.848})$$
$$= T_1(27.59 - 27.12) = 0.47T_1$$

Therefore,

$$T_1 = 100/0.47 = 212.77 \text{ man-hours}$$

Cumulative average man-hours during learning process $= T_{an}$

and
$$T_{an} = T_1 n^r = 212.77(50)^{-0.152}$$
$$= 117.41 \text{ manhours}$$

Total direct labour cost for making 75 beams

$$= 7.50\{[50(17.41)] + [75(100)]\} = 7.50(8370.50)$$
$$= £62\,778.75$$

Additional cost for learning $= [50(117.41 - 100)]7.5 = £6528.75$.

This additional cost is shown as the shaded area in Fig. 7.11.

Assuming that repetitive construction continues without interference causing a delay or a break, the full benefits of learning are likely to be achieved.

However, when the production system is subjected to an interruption after a number of units have been produced, whether for a comparatively short time of a few days or for a longer period, *forgetting* is likely to take place. The extent to which it occurs will depend generally on the rate at which learning was taking

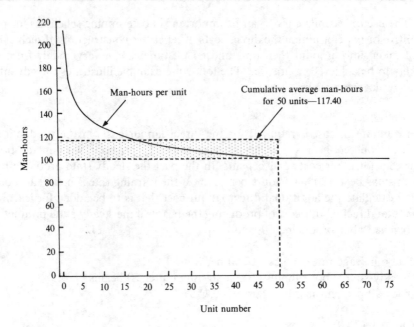

Figure 7.11

place prior to the break, the break's duration and the number of units completed up until the time of the interruption. When the break takes place, the forgetting curve tends to proceed back along the learning curve in the reverse direction as the forgetting takes place. Although the forgetting curve will not necessarily regress at exactly the same reverse slope as the learning curve, it is usually assumed that it will do so. Once the break is concluded, the learning process resumes, though the rate at which it does will depend on a number of factors including the extent of the break, the number of units already produced and the number of operatives who are available at the time of resumption who also benefited in the original learning process.

In general terms, the forgetting process can be described by Fig. 7.12. The example used is for the production of 25 units with a learning rate of 85 per cent, the first unit taking 150 man-hours. Production time for successive units is plotted against the serial sequence of unit numbers, as shown by the full-line curve up until the point of interruption and thereafter by a dotted line. Production proceeds until after the fifth unit is produced, when an interruption to the process occurs. During the break, the forgetting process takes place and the time taken for the sixth unit, the first unit produced after the interruption, is given by the y-coordinate of point A. The distance travelled back up the learning curve cannot be established with accuracy at the present time and it will depend upon many factors. From point A, the resumed learning curve will then take the form of the original curve as though it were resuming from point B on that curve. In other words, the penalty

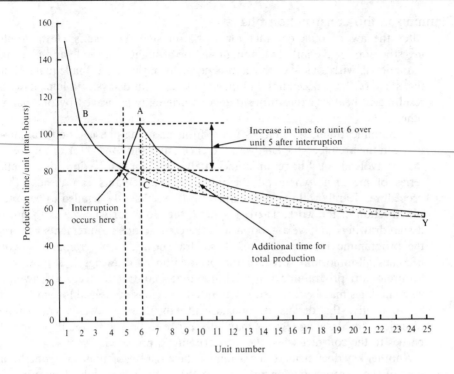

Figure 7.12

to be paid for the interruption in the time taken for the sixth unit will be *AC*. The total penalty to be paid for the interruption over the production of the rest of the units is illustrated by the shaded area between the two curves and on this example amounts to 111.70 man-hours. It can be found by calculating the cumulative average of the displaced curve and subtracting from it the cumulative average of the original curve (between units 6 and 25 for both curves). The result is multiplied by the number of units for 6–25 inclusive ($(62.439 - 56.560)19 = 111.70$ man-hours). This illustration reinforces the importance of keeping interruptions in repetitive work to an absolute minimum.

Learning and forgetting curves are by no means freely used within the construction industry and they are by no means free of their detractors. Insufficient evidence is as yet available to establish the theories once and for all. That learning can take place on the majority of repetitive operations *given the right conditions* is almost certain, but the conditions for it to foster are many and varied. However, what is important is that an awareness of the processes is always present, and that steps are taken to ensure that proper and thorough planning and scheduling techniques are employed, together with the optimum mix of equipment and labour in order to enhance production on repetitive work. Designers should design so as to avoid breaks in repetitive construction sequences caused by coping with too many changed details from one unit to another.

7.13 Planning in the construction phase

After the award of a contract for work on site, the results of the preliminary investigation for the initial planning must be brought up to date and investigated in more detail, with a view to the commencement of the work. Inadequate planning at this stage of the proceedings will inevitably result in delays at a later date, possibly resulting in heavier expenditure than was originally envisaged by one party or the other.

Communication is the first essential of this planning phase. Good communication between the architect, the engineer, the client, and the contractor or any other party that is involved, must be set up immediately. This can best be done in a meeting, or a series of meetings, where all the interested participants can discuss the future procedures. Rarely will the tender drawings be sufficiently detailed to be used for the construction of the work. There is a need, therefore, for whoever is to produce the design drawings to have a good idea of the contractor's requirements with regard to the programme for the issue of these drawings. The programme of work then becomes all-important. Because the production of drawings is a most important feature of any programme, it is advisable that a contractor keeps a drawing register wherein dates may be recorded when initial drawings are issued for work on the site. Subsequently, the details of amendments to those drawings can similarly be recorded. The drawing register may later prove to be a very useful document for all parties to the contract when the final account is prepared.

Another key document at this planning stage is a bill of quantities, whether it forms part of the contract documents or not. While the priced bill of quantities cannot necessarily be considered as a document representing the final quantity of work to be done, it has considerable value inasmuch as it can now be broken down to give a broad indication of the quantities of materials that will be required during the course of the contract, the number of specialist subcontractors and suppliers who have to be brought into the picture, and the probable amount of money that will be outstanding in the form of provisional sums. The latter may or may not be utilized during the course of the contract. Lists of all these items can be drawn up so that the initial routine work in this connection can be put in hand. The prices in the final tender bill of quantities prior to its submission to the client for consideration must now be broken down into rates for labour, plant, material, overheads, and profit. This financial information will form the basis for much of the planning during this stage.

One of the next essential planning tasks to be undertaken is the preparation of a *labour requirement chart*. This is particularly important with regard to the work that is to be carried out from the contractor's own resources. The approximate quantity of work can be assessed from the bill of quantities. The labour man-hours associated with that work can be assessed by using the labour element of the billed rate for that work item. For example, if the total rate for one square metre of formwork appears in the bill of quantities as £25.00 (excluding the margin for profit and overheads), this rate can be broken down into those for labour, plant, and material costs. These three rates, for the sake of the example, might be £10.00, £10.00 and £5.00 respectively. If the total quantity of formwork to be erected amounts to 475 square metres and the labour rate used in building up the figure of £10.00 per square metre amounts to £4.00 per hour, then the total time available for the whole of this one item of formwork amounts to $475 \times 10.00/4 = 1187.5$ man-hours. Performed by a gang of 6 operatives, the work should take approximately 198 hours or, in round figures, 25 working days.

By relating the numbers of operatives required at various stages in the work to the draft programme it is possible to see whether there will be peaks to the labour requirements curve that are unrealistic. If this is so, they need to be smoothed out in order to spread the number of operatives on the site more evenly throughout the contract period. Having calculated the approximate amount of labour that will be required for the various stages of the contract, it is then possible to define the type and amount of supervision by key personnel that will be required. Figure 7.13 shows a typical labour requirement chart that has been developed in the pre-contract planning period. It will be noted that the total labour requirement rises very rapidly within the first two or three months to a peak and then tails away over rather a long period. It would be possible at this stage to adjust the programme so that the maximum numbers to be employed were reduced but remained on the site for a longer period, rising fairly quickly at the beginning of the contract and tailing away in the same manner. It is desirable to aim at a level labour requirement for as long as possible during the contract period. The broken line on the graph of Fig. 7.13 is a better distribution of labour.

Contract Nº 375 D	Labour schedule Location Green Hoe									Date Prepared by				
Labour \ Month	Aug	Sept	1976 Oct	Nov	Dec	Jan	Feb	1977 Mar	Apr	May	Jun			
Labourers – General	20	40	48	50	42	30	22	20	15	15				
–do– –Carpenters	2	10	13	16	16	14	6	6	2	2				
–do– –Bricklayers	–	5	10	14	13	13	10	10	5	2				
Carpenters	3	15	25	25	26	20	15	12	10	8				
Bricklayers	–	10	20	20	21	21	17	14	10	8				
Steelfixers	–	–	10	10	12	12	12	8	8	–				
Totals	25	80	126	135	130	110	82	70	50	35	0			

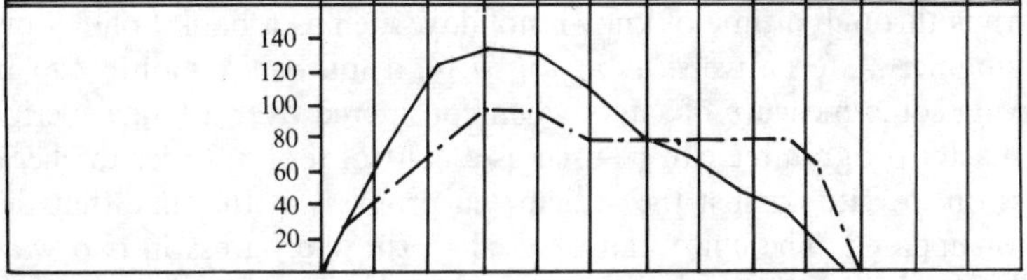

Figure 7.13

The period of time before work is commenced on site provides an opportunity for the critical re-examination of the methods that will be used to carry out the work. The pre-tender method statement can now be examined by those of the staff who will be engaged on the control of the work, and alternative methods can be examined to see if a cheaper and better alternative is available. When a scheme is developed that is to the satisfaction of all who are involved, then it should be set down for record purposes and future reference.

Attention can now be directed towards the preparation of a master programme for carrying out the work. Such a programme will be incomplete unless it gives, directly, information concerning the required dates of issue of drawings, the latest date by which orders need to be placed for materials to be delivered at or before the appropriate time, subcontracts, etc., and the requirements of any special plant that needs to be brought in for limited periods of work on the contract. It must show clearly the sequence of operations that is necessary to complete the works in the appropriate time period. Where possible, it should show the interdependence of one activity upon another, especially where key labour and plant are required to move from the early activity to a subsequent one. The programme must be agreed with the clients and their advisers, for the programme may well be important to the clients who have to make provision for the necessary finance to be available to meet the contractor's certificates, as well as to make the preliminary arrangements to put the subject of the contract to use on its completion.

Having made the necessary calculations to determine the elements of time and quantity in the programme, it is necessary to put this information into a compact form in which it can be successfully communicated, with ease and accuracy, to the people who are going to supervise and control the work in accordance with the programme. Such a composite chart is called the *master programme*. There are a number of ways of doing this and one of the most common of the pictorial methods of presentation is that of the bar chart. An example of a bar chart for a master programe is shown in Fig. 7.14. Note the column provided giving the approximate quantities of the work to be carried out under each item, the provision for the name of any subcontractor who will be employed upon the particular operation against which their name appears, and the use of standard symbols to show clearly the various requirements for drawings, materials, and schedules.

Such a chart, used as a master programme, should show at this stage the whole of the contract period. A week is a suitable time unit on its horizontal scale. Appropriate allowances must be made for holidays, especially where a contract period runs through a time of known holiday, such as a bank holiday or the middle of the summer. Alternatively, as in Fig. 7.14, a four-week month can be assumed, thus giving some measure of safety when considered over a longer period.

The master programme can be used as a control tool in order to check the actual progress on the site against the anticipated progress at the time that the chart was drawn or updated. The chart can be used to show progress in two ways. The first method is to show the actual percentage of work completed. The second is to relate how the work is actually carried out in relation to the calendar dates. The latter can

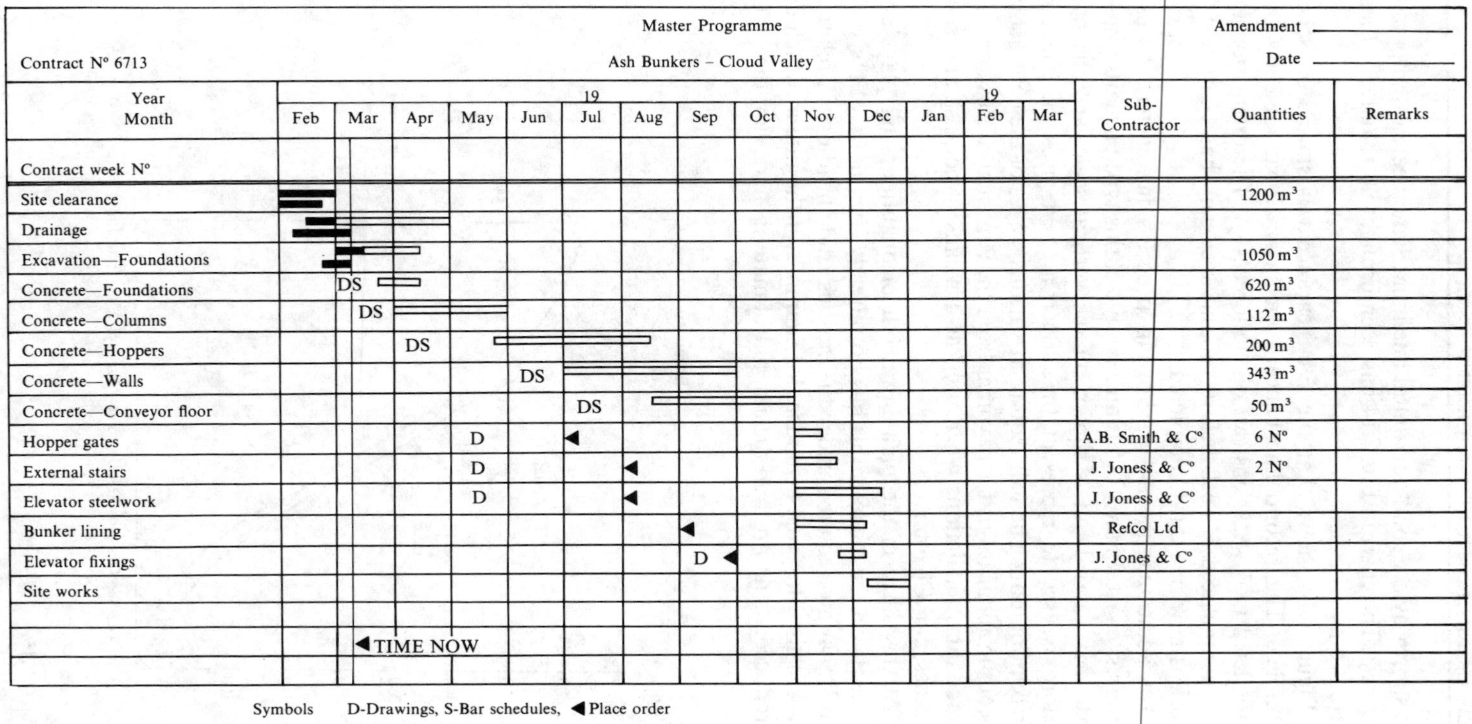

Master Programme
Ash Bunkers – Cloud Valley

Contract Nº 6713

Amendment _____
Date _____

Year Month	Sub-Contractor	Quantities	Remarks
Site clearance		1200 m³	
Drainage			
Excavation—Foundations		1050 m³	
Concrete—Foundations		620 m³	
Concrete—Columns		112 m³	
Concrete—Hoppers		200 m³	
Concrete—Walls		343 m³	
Concrete—Conveyor floor		50 m³	
Hopper gates	A.B. Smith & Cº	6 Nº	
External stairs	J. Jones & Cº	2 Nº	
Elevator steelwork	J. Jones & Cº		
Bunker lining	Refco Ltd		
Elevator fixings	J. Jones & Cº		
Site works			

Month columns: Feb Mar (19__) Apr May Jun Jul Aug Sep Oct Nov Dec Jan Feb Mar (19__)

Symbols   D-Drawings, S-Bar schedules, ▼ Place order

▼ TIME NOW

Figure 7.14

265

be shown by a second horizontal line immediately below and parallel to the original forecast. The chart in Fig. 7.14 has been brought up to date for the first few weeks of operation.

Figure 7.14 shows that the site clearance was started on schedule and is now complete. The second line beneath the original bar indicates that the work was completed in 3 weeks instead of the forecast 4. The excavation for the foundation was commenced 1 week before programmed and has continued for 2 weeks. The work is now one-third completed.

An additional means of showing progress in different terms is the financial progress chart. Figure 7.15 is an example of this type of chart, where the valuation of the work carried out is shown on the vertical scale, and the horizontal scale, as before, is used to represent time. Such a chart gives the client some idea of the amount of money he will require at various stages of the contract period in order to meet the applications for payment from the contractor. This type of chart can lead to serious miscalculations when used as a means of control, and such aspects will be discussed in Chapter 13.

At this stage of the planning a bill of quantities still plays an important part. Extracts can be made, if this has not already been done, of the various quantities of work that are required under the various trade headings in relation to the contract plan. This will give a firm idea, for example, of the rate at which concrete has to be produced, its nature, and the amount of material and plant that will be required over

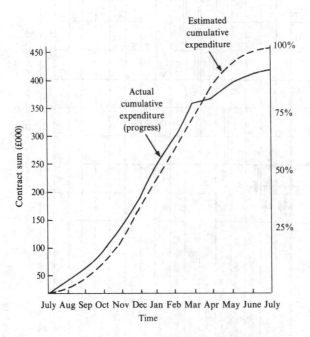

Figure 7.15

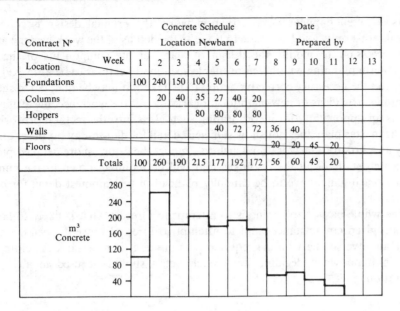

Contract Nº	Concrete Schedule Location Newbarn								Date Prepared by				
Location \ Week	1	2	3	4	5	6	7	8	9	10	11	12	13
Foundations	100	240	150	100	30								
Columns		20	40	35	27	40	20						
Hoppers				80	80	80	80						
Walls					40	72	72	36	40				
Floors								20	20	45	20		
Totals	100	260	190	215	177	192	172	56	60	45	20		

Figure 7.16

any given period. Figure 7.16 shows an example of this sort of summary. Smoothing action can be taken in order to get rid of peaks of one sort of work which result in excessive demands for either plant or labour or both.

Similarly, details of the labour requirements (as at the pre-tender stage) can be extracted from the various detailed programmes and brought together to make a master schedule. Smoothing action can also be taken with this schedule.

7.14 Control

Once a number of programmes have been set up, all within the contract plan, the work commences on site and the question of controlling its progress now arises. The planning process has been carried out in order to give every opportunity for the work to proceed on certain directed lines and within certain estimated costs. Control is the process of measuring the actual progress made against these standards and adjusting the use of resources to meet a divergence from the original intention. Unfortunately, very few projects, for one reason or another, proceed along the well-defined and intended lines of the programme. In the first place, in order to make effective use of the programme, short-term planning must be put into effect. Short-term planning involves the frequent checking of the progress to date for all the operations envisaged in the master programme. It involves taking a look ahead for perhaps three weeks in order both to foresee any immediate detailed requirements and to attempt to anticipate any imminent delays or causes of delay.

As so often happens in construction work, the original design is conceived in broad terms only, and from time to time, as the details of the work become available, it is necessary to modify the design of many of the components of the project. Short-term planning will therefore enable such modifications to be absorbed into the plan and the necessary corrective action to be taken. It may be that, in the short-term planning, new ideas or new methods for carrying out the work become available, and this is an opportunity for incorporating such ideas into the method of working.

The master plan does not show sufficient detail for day-to-day purposes and serves only as a broad framework on which to support the future, more detailed plans. The intricate details of all the resources that will be required, down to the numbers of men in each gang, have to be carefully planned in the minutest detail for optimum operations.

The whole basis of planning is to attempt to forecast what is likely to happen in certain given circumstances and to attempt to forestall those happenings that will have an adverse affect on the required proramme. It is necessary, therefore, to bear this in mind when deciding how much detail should be used in any planning operation.

Summary

This chapter has introduced the important closely linked subjects of construction estimating and planning. The essence of planning is to coordinate all the relevant information, involving all the parties in the construction of a project, to formulate a method and technology for carrying out the work, to present the summary of the analysis in such a way that it is meaningful to those who need to use it, and then to use the final proposal for the purposes of control. Estimating is concerned with the establishment of a cost of carrying out the works. Emphasis has been placed on the information that needs to be collected at the outset for the planning that needs to be undertaken in the pre-tender stages in order to begin to prepare the estimate. Typical bar charts that may be used at this stage are described. The estimating and tendering processes are described and examples are given of the derivation of direct costs for construction activities. The basis of a programme is the determination of the durations for each item of work that needs to be carried as part of the project. A method for doing this is dealt with, as is the subject of learning and forgetting. This latter aspect of planning is particularly important in work that has repetitive or near-repetitive elements. It aids the identification of activity durations that will benefit from learning and those that will be affected by an interruption in the smooth progress of the work. Methods of taking these factors into account are described. The component costs that go to make up an estimate are described, together with how they are put together by a contractor to form a construction cost. The final stage of tendering is then described when a contractor adds the necessary margin to the direct costs so that recompense for the risk that is being taken is included in the tender. Finally in this chapter descriptions of some of the programmes and control charts that may be used for the actual construction are presented.

Problems

Problem 7.1 It is decided to use a dragline with a $1\frac{1}{2}$-m^3 bucket to excavate a trench with a trapezoidal cross-section. The ground is wet, sticky clay. The trench is 15 m wide at the top, 5 m wide at the bottom, 3 m deep, and 2000 m long. The excavated material will be cast on the side of the trench. A reasonable output for the dragline is estimated to be 115 m^3 per 60-minute hour.

Assume that the dragline will actually be working a 50-minute hour and that its transport to and from the site costs £2500. The cost per hour for the dragline is £75.00, for an operator £9.00, for a mechanic £7.00, and for a labourer £6.00.

Estimate the cost per cubic metre, bank measure, for excavating the trench.

Problem 7.2 Ordinary earth is excavated to build up a fill area for a road embankment. The earth is removed from a borrow area approximately 700 m from the embankment site by using twin-engined scrapers having a capacity, for the prevailing conditions, of 25 m^3, loose measure, per load. Three scrapers are available to carry out this work. The scrapers will use a level road between the site of the borrow pit and the embankment and they are expected to average 20 km/h when running full and 30 km/h when running empty. The time it takes to load the scraper, discharge it at the embankment, and any waiting time that it may incur in the process, is estimated to be 4 minutes per trip. Assume that the scrapers work an effective 50-minute hour.

If the scrapers cost £102.00 per hour, the drivers £8.50 per hour, and the two operators required £6.00 per hour, estimate the cost of moving 1 m^3 (bank measure). It may be assumed that the excavated material swells by 20 per cent on removal from the borrow pit. Estimate the length of time it will take to move 10 000 m^3 (bank measure) from the borrow pit to the embankment.

Problem 7.3 A reinforced concrete structure consists of columns and beam and slab floors—all placed *in situ*. The columns are 400 mm × 400 mm and the reinforcement amounts to 5 per cent of the cross-sectional area of each. The columns have a clear height of 3500 mm. There are 60 such columns and 10 sets of column formwork need to be made. Estimate the total direct cost of making the formwork, erecting it, and stripping it after use, as well as placing the reinforcement. Assume a labour cost of £7.50 for all trades. Obtain costs for materials and output figures from current texts, journals, or local firms.

Problem 7.4 Sketch a learning curve to show

(a) the cumulative average man-hours for successive units, and
(b) the man-hours per unit for successive units

for a repetitive process consisting of the construction of 120 bays of concrete road pavement, where the learning rate is expected to be 90 per cent. The first bay takes 200 man-hours. Read off from your graph the number of man-hours it is expected to take to construct bay number 56 and the cumulative average hours up until that point.

(92.11; 108.47)

Problem 7.5 An engineer is preparing a schedule for the construction of a series of 200 identical bridge beams. Past experience indicates that the steady-state duration for the construction of each beam amounts to 5 man-days and that when 125 such beams have been constructed there is negligible further improvement in the time taken.

Calculate the overall duration that should be scheduled for the beams taking learning into account and compare this with the duration based upon average performance.

What beam in the sequence of 200 should have the same construction time as the cumulative average for the first 125 of 200 beams? Assume a learning rate of 90 per cent.

(1111.6 man-days against 1000; 42/43 beam)

Problem 7.6 An operative, in order to be fully trained to carry out a routine repetitive process and to reach a regular steady-state output, needs to gain experience of the job for a period of 60 working days. After this period of experience the process takes 10 man-hours. The learning rate for the operation is 90 per cent.

The operative falls ill and is away for 35 working days. The job is temporarily taken by another operative with no previous experience of the task involved. Estimate the loss of production that arises from the illness of the operative.

(62.78 per cent)

Problem 7.7 A conctractor is tunnelling by hand excavation for a large-diameter sewer. The tunnel is lined with precast concrete rings. A ring is placed and fixed for every metre advancement of the excavation. A total of 200 metres of such tunnel is to be excavated and lined. The learning rate of the tunnelling crew is approximately 90 per cent for the first 100 rings and thereafter the installation is carried out in a constant duration. After ring number 30 is fixed, the design engineer for the tunnel stops the work for 15 days in order to check the design for a fault in the current ground conditions. The initial cycle for excavation and fixing a ring took 4 days. If the interruption of the work causes forgetting to the extent that the first ring fixed after the break takes the same duration as ring number 20, and improvement goes on for the next 100 rings, estimate the total duration for carrying out the work from start to finish.

What would the duration have been without the break?

(387; 368)

Further Reading

Planning

Harris, F. and R. McCaffer: *Modern Construction Management*, 3rd edn, BSP Professional Books, Oxford, 1989.

Neale, R. H. and D. E. Neale: *Construction Planning*, Thomas Telford, London, 1989.

O'Brien, J. J. (ed.): *Scheduling Handbook*, McGraw-Hill, New York, 1969.

Peurifoy, R. L.: *Construction Planning, Equipment and Methods*, 2nd edn, McGraw-Hill, New York, 1970.

Turner, G. and K. Elliott: *Project Planning and Control in the Construction Industry*, Cassell, London, 1964.

Willis, E. M.: *Scheduling Construction Projects*, Wiley, New York, 1986.

Learning curves

Bailey, C. D.: Forgetting and the learning curve: a laboratory study, *Management Science*, vol. 35, 1989, pp. 340–352.

Cockran, E. B.: *Planning Production Costs: Using the Improvement Curve*, Chandler, New York, 1968.

Conley, P.: Experience curves as a planning tool, IEEE Spectrum, vol. 7, 1970, pp. 63–68.

Gates, M. and A. Scarpa: Learning and experience curves, *Journal of the Construction Division, ASCE*, vol. 98, no. CO1, March 1972, pp. 79–101 and errata, *ibid.* vol. 102, no. CO4, December 1976, p. 689.

Peck, G. M.: Significant factors in cost efficient design and construction with particular reference to segmental bridges, *Proc. Concrete Bridge Conference*, Singapore, August 1982.

Thomas, H. R., C. T. Mathews and J. G. Ward.: Learning curve models of construction productivity, *Journal of Construction Engineering and Management, ASCE*, vol. 112, no. 2, June 1986.

Wright, T. P.: Factors affecting the cost of airplanes, *Journal of Aeronautical Science*, vol. 3, February 1936, pp. 122–128.

Estimating

Bentley, J.: *Construction Estimating and Tendering*, Spon, London, 1987.

Economic Development Committee for Civil Engineering, *Report on Price Adjustment Formula for Civil Engineering Contractors* NEDO, 1971.

McCaffer, R. and A. N. Baldwin: *Estimating and Tendering for Civil Engineering Works*, 2nd edn, BSP Professional Books, Oxford, 1991.

Mudd, R. D.: *Estimating and Tendering for Construction Work*, Butterworths, London, 1984.

Peurifoy, R. L. and G. D. Oberlender: *Estimating Construction Costs*, 4th edn, McGraw-Hill, New York, 1989.

Property Services Agency, *Price Adjustment Formulae for Building Contracts*, HMSO, 1977.

8. Planning—basic network analysis

8.1 Introduction to critical path methods

The Gantt chart or bar chart has already been introduced as a means of the visual presentation of a construction programme. It has been used since the turn of the century and represents planning and control in a somewhat static dimension. It was developed as a means of scheduling the output of fixed production equipment. In those circumstances, the need for a means of dealing with detailed changes of programme and resources did not occur very frequently. The bar chart is a simple and readily understood means of scheduling but, nevertheless, it is a static means of illustrating what should be a dynamic situation. It can be more readily applied in planning, for example, the outputs of lathes (for which it was originally designed) than the very many different types of construction production.

For construction work, the bar chart, though it is frequently used, does not in fact allow a programme to be devised on a scientific basis. If the overall requirement is for a contract period of 2 years, the horizontal lines in the bar chart can readily be juggled so as to fit them within the 2-year period. There is no interdependency shown between one bar and the next and the really important aspects of the programme must remain in the mind of whoever draws the chart.

Network analysis methods of scheduling had their beginning in two separate spheres. One of these was known as *CPM* and the other as *PERT*. The former is an acronym for *critical path method*, although alternative expressions such as *critical path scheduling* (CPS) or *critical path analysis* (CPA) are in fairly common use. *PERT* started life as the *programme evaluation research task* but ultimately became *programme evaluation review technique*. Since the beginnings of the use of these methods for scheduling purposes they have grown very much together, with the result that the two basic expressions are now frequently interchanged and the different situations for which they were designed in their early days do not generally apply now to any great degree. The general procedures, for scheduling and programming operations with a network, are often described under the general heading of *network analysis* or *critical path methods*. Both methods relied upon the basic network plan as a graphical portrayal of the way in which a project should be carried out and the logical sequence of operations.

272

Critical path method (or CPM, as it will henceforth be called) had its beginnings in early 1957. In those days, Morgan R. Walker, of the Du Pont Engineering Sevices, collaborated with James E. Kelley, Jun., then of Remington Rand, in order to provide a more precise and dynamic model for scheduling purposes. They finally settled on the graphical network diagram, which was to prove a great departure from the accepted traditional methods of planning, largely represented by the bar chart. Walker and Kelley produced a scheduling system which demanded the use of simple, straightforward arithmetical processes. Calculations could be carried out using no more than addition and subtraction for the basic scheduling process. As a result of such calculations, it was possible to show that every project to be programmed has at least one sequence of operations, jobs, or activities, however they are described, that is critical to the completion of the task in hand. The result of the calculation is the definition of a critical path through the network diagram, and any of the activities on the critical path that are not completed in at least their estimated duration will inevitably cause the overall project length to be extended.

Kelley pursued his enquiries still further, and in May 1957 produced a solution to the question of optimizing the total length of time for a project so that the cost was at a minimum. By this time, both the basic critical path method and the addition of time–cost optimization were ready for test, and certain pilot projects at Du Pont's various installations were operated on the basis of control by the critical path method. By 1958 a plant maintenance shut-down was scheduled by this method, and in March 1959, at Du Pont's Works in Louisville, Kentucky, it was claimed that the shut-down period for one of the chemical plants was reduced from a normal 125 hours to 93 hours by the use of such models.

In 1959 both Kelley and Walker joined Mauchly Associates, and continued to devote their energies to the development of critical path methods in fields of resources planning, costing, etc.

PERT was developed by June 1958 as a result of the collaboration of Willard Fazar, of the Special Projects Office of the United States Navy, with D. G. Malcolm, J. H. Roseboom, and C. E. Clark, of Booz, Allen and Hamilton of Chicago, Illinois. It had become necessary to produce a method of programme evaluation for the development of the Fleet Ballistic Missile Weapon. Owing to the very wide range of activities that had to be considered in the development programme, which ranged far into the future, it was considered that conventional methods of planning and programme evaluation were of little use. Inadequate traditional methods might well have led to serious slippages of programme, which in turn, where such weapon systems are concerned, could well have had drastic results on the defence policies of many countries. As was ultimately proved to be the case, some 3000 contractors and agencies were employed in the development of the Polaris missile. In carrying out research and development work of this nature, the question of forecasting the possible durations of individual activities becomes exceedingly difficult, particularly of those activities that are to be projected many years into the future. The PERT method made allowances for probable errors in the estimation of durations. So effective did the PERT method of scheduling prove to be, that the US Navy credits

PERT with making a considerable contribution to the completion of the Polaris missile programme ahead of time. This is especially significant when one takes into account that, in the field of weapon systems development and construction in the United States, the average excess of contract time over that estimated had been of the order of 36 per cent.

The basis of both CPM and PERT systems is the *network diagram*, and the way in which the two avenues of approach have finally come together and overlapped is the logical consequence of basically similar thinking.

8.2 The fundamentals of network analysis

The network is really a project graph—a means of representing a plan so that it displays clearly the series of operations that must follow in order to complete the project. It also shows clearly their interrelationship and interdependence. The graph, without giving, at this stage, consideration as to the durations of each operation, is concerned only with the planning phase of the programming operation. It is one of the major advantages of using such a method that the planning may be divorced entirely from the scheduling or the estimation of time durations for the project.

There are three basic methods of preparing a diagram. The first is the *activity-on-the-arrow* system. It is this one that will be used to explain the basic construction and use of the network diagram. The second system is the *activity-on-the-node* system now made popular by a derivative, *precedence diagramming*. Thirdly there is the *event* or *milestone* system, which is largely represented by the PERT method.

The principal component of a network diagram in the activity-on-the-arrow system is an *arrow*. An arrow is used to represent an *activity* which in itself is a time-consuming element of the programme. All projects can be broken down into a number of necessary activities. The detail required will decide the number of activities into which the project is subdivided. Each of these activities that go to make up the whole project is then represented by an arrow in the diagram. Not only will an arrow represent the consumption of time but it can also represent the consumption of certain resources such as labour, money, or the use of plant or materials. The length of the arrow as drawn in the diagram bears no relationship to the time the activity takes nor to the resources the activity consumes. The direction of the arrow simply indicates the direction of work flow. The tail of the arrow represents the starting point for a particular activity and the head represents the completion point. Typical arrows are shown in Fig. 8.1, each one representing a specific activity. It should be noted that at some time in the process of planning or scheduling, a brief description of what the activity represents as part of the task is written against the arrow. There is no reason why the arrow should be drawn as a straight line; it can well be bent or curved to suit the construction of the diagram.

At either end of the arrow must appear an *event*. This is a milestone or a point at which an operation, the activity, is completed or another can start. It is in theory an instantaneous point in time. An event may also be known as a *node* or *connector*. Figure 8.2 shows a simple relationship of event to arrow. In this figure, the activity

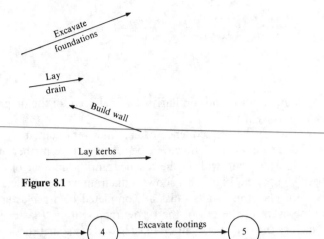

Figure 8.1

Figure 8.2

lies between the two events (in this particular case, excavating footings is the time-consuming element). The circle at the head of the arrow represents the event, or the milestone, or the point in time when it can be said that the footing excavation is completed. The circle at the tail of the arrow is the similar event when it can be said that the excavation of the footings can commence. An event may be represented by a circle, a square, a triangle, or any geometric figure. Very often, especially in the initial stages of preparing a diagram, a symbol is not used at all. The event can be an intersection of two or more activities and need not be a point, as in Fig. 8.2, of the connection between the head of a single arrow and the tail of one other.

All networks are constructed logically on the principle of *dependency*. Dependency can be illustrated by the simple sequence of events as shown in Fig. 8.3. No event can be reached in a project before the activity that immediately precedes it is completed. Similarly, no activity can be started until the event that immediately precedes it has been reached. In the case of Fig. 8.3 it can be inferred from the sequence of activities that the concrete cannot be poured until the formwork is completely erected. In like fashion, it is obviously necessary for the concrete to have been poured before curing can proceed, and stripping of the formwork cannot proceed until the curing is completed. This simple sequence illustrates the principle of dependency—one of the basic fundamentals of the logic of a network diagram.

It should now be apparent that the representation of activities and events in the fashion of Fig. 8.3 provides a pictorial representation of the sequence in which the work must be carried out if successful completion of the whole project is to be achieved. It should also be apparent that each activity must be preceded and succeeded by an event and, conversely, that each event must have an activity before it and an activity following it. There will be two exceptions to this latter basic rule, inasmuch as the starting event of a network or arrow diagram representing a project

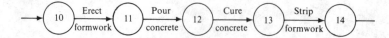

Figure 8.3

will have no activity leading into it and, similarly, the final event of the diagram will have no activity leading away from it.

In Fig. 8.3 is shown a very simple sequence of activities represented by a linear flow. The principle of dependency may, however, be taken a stage further, and it is possible that an event in a diagram will not be reached until a number of activities preceding it have been completed. Figure 8.4 shows a diagram where all the activities preceding the activity of plastering walls must be completed before plastering can proceed and before the event at the tail of the arrow representing 'Plaster walls' is achieved. An event such as the one at the tail of 'Plaster walls' is sometimes described as a *merge* event because a number of arrows merge into it. If the situation arises that a number of arrows emerge from an event, this is sometimes known as a *burst* event.

If the logic of Fig. 8.4 is examined, it becomes clear that items, such as 'Rough plumbing', 'Run electric conduit', 'Joiners' first fixings', and 'Fix window frames', can all proceed at the same time or with the same overall duration. The only restriction which has been imposed upon this diagram is that *all* these activities should be completed before the walls are plastered.

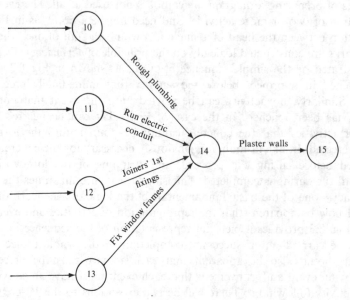

Figure 8.4

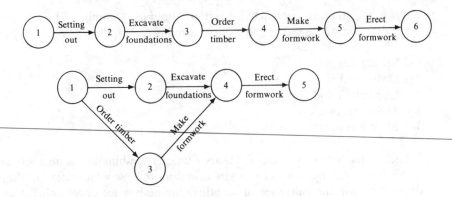

Figure 8.5

The upper part of Fig. 8.5 illustrates a sequence of five activities that go to make up part of a simple project. In setting them out one after the other in this fashion, it is not possible to fault the logic of the diagram, for each of these activities could well take place in this sequence. However, if time is considered at this point, it is clear from the diagram that the overall time required to carry out these five operations is simply the sum of the durations estimated for each of the individual activities. Further examination of Fig. 8.5 shows that some of these activities can readily be carried out concurrently with others. An alternative diagram, as portrayed in the lower part of Fig. 8.5, can be prepared. The diagram illustrates that while the setting out and the excavation of the foundation are taking place, timber for the formwork can be ordered, and on its receipt the formwork can be made. However, erection of formwork cannot take place until two conditions have been satisfied. In the first place, the foundations have to be excavated and, secondly, the formwork has to be made. By rearranging the sequence of operations, that is, by making some concurrent with others, the overall time for completing all the work will obviously be shortened. It is extremely important at the planning stage that as much of the work as possible is arranged to be carried out concurrently if the shortest possible programme period is required. Other considerations will arise at a later date that will further influence the concurrence of a number of activities.

8.3 Activity identification

In order to avoid confusion among the activities in a network, it is necessary to produce a unique method of identification for each activity forming a part of the diagram. As part of this identification procedure all the events are numbered and the *i–j convention* is used. By this convention the number of the event at the tail of the arrow is known as the *i-number* and the number at the event at the head of the arrow is known as the *j-number*. The activity can therefore be identified by its *i–j* number and it is essential that each activity carries a unique *i–j* number. The activities in the lower diagram of Fig. 8.5 are identified as follows:

1–2 Setting out
1–3 Order timber
2–4 Excavate foundations
3–4 Make formwork
4–5 Erect formwork

Each of the activities in this list bears a unique combination of numbers and there is no need, having numbered each event in this way, to refer to the activity description for the purposes of identification unless for some other reason it is desirable to do so.

Numbers can be assigned to each event in a network in one of three ways. Firstly, there is the *forward numbering* method, whereby the first event in the network is numbered 1 (alternatively it is often 0), the second 2, and so on, in such a way that the event number at the tail of each activity arrow is always smaller than the event number at the head of that activity arrow. This implies that the final event of the network will be the one bearing the largest number and it will always be the number of nodes in the network.

Secondly, there is the *backward numbering* method, which is simply the reverse of the first method, entailing the numbering of the final event with 1 (or 0) and working backwards through the network, so that the initial event will bear the highest number. Both of these methods are known as *serial numbering* methods and both impose a strict discipline in the way that the numbers are allocated to the events. Serial numbering was demanded by many of the early computer programs, but in more recent years computers have been programmed to deal with the third type of numbering, that of *random* or *non-consecutive numbering*.

In the random numbering system the event numbers, as the title implies, are set to the events in any order so that *i* may be greater or less than *j* for each activity. It is important however that no two events should bear the same number. With random numbering it is very necessary to keep a note of all the numbers already used in a network so as to ensure that a number is not repeated. Random numbering makes it very much easier to add an activity arrow into the network at a late stage without having to go back through the previous or subsequent part of the network and renumber the events in order to maintain the consecutive succession. Serial numbering has the advantage that one can tell from the identification numbers of the activity or events approximately where in the overall project they appear.

Which numbering system is adopted is largely a matter of the preference of the originator of the network. It is normally more practicable to omit the numbering from the network altogether until the diagram has been completed in its preliminary form. In this way, activities and events may be added or deleted as required during the preliminary planning period without disrupting the flow of numbers from either end.

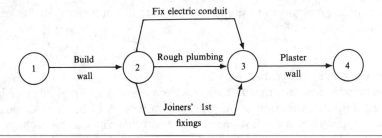

Figure 8.6

Figure 8.6 illustrates an example in which each event shown in this simple sequence of activities bears a different number. If the activities are now identified by their *i–j* notation it will be seen that three of the activities bear the same *i–j* configuration. Thus the identification of the activities is not unique.

In order to correct this situation, logical restraints called *dummies* are added to the diagram. The dummy in this situation is used only to maintain the unique numbering system and represents no consumption of time or resource. It is conventionally shown in a diagram by the use of a broken line.

Figure 8.7 illustrates the same programme as that in Fig. 8.6, but with dummy activities inserted between events 3 and 5 and 4 and 5 so that each activity bears a unique identification. The dummy events could well have been inserted between events 2 and 3 and 2 and 4, the two activities becoming 3–5 and 4–5 respectively. Conventionally, the dummy is added after the activity but it can happen that, in the construction of a diagram, it is more useful to place the dummy before the activity for the purposes of graphical portrayal. The logic of the diagram in Fig. 8.7 does not differ in any way from that shown in Fig. 8.6. The same dependencies are illustrated and the same succession of activities and work to be completed is maintained.

Dummy activities can have applications in networks other than those concerned directly with the maintenance of a unique numbering system. They can be used to

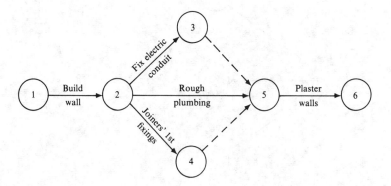

Figure 8.7

maintain the logic of a certain situation where the use of activity arrows would not allow this.

8.4 The logic of network diagrams

A network or arrow diagram is built up from a detailed knowledge of just how the work forming the subject of the programme will be carried out. If the activities of the network follow one after the other in a serial fashion, as was illustrated in Figs 8.3 and 8.5, the logic is perfectly simple and the arrows form a straight line. In a case such as this there is little point in drawing a network diagram, since the overall period of the programme will be the summation of the time durations for each activity, and there are no complex dependencies to be illustrated.

If the next situation to arise is one in which it is known that activities R, S, and T can all follow on from an activity P, then logically that part of the diagram will appear as Fig. 8.8. These are simple examples and will rarely give cause for an error in network logic.

If it is now required that activity R follows P and activity S follows T, more care must be exercised in the preparation of the network. The left-hand diagram of Fig. 8.9 represents one arrangement which, as far as the initial requirement that R follows P and S follows T is concerned, is perfectly correct. On examination it will be noted that there is an additional dependency arising in the diagram that was not called for in the method statement. Not only does R follow P and S follow T, but S is shown as being dependent on both P and T, and R is similarly shown. This was not required. The right-hand diagram of Fig. 8.9 is the logically correct diagram for the stated requirement.

Figure 8.10 illustrates part of a diagram in which it has been necessary to use a dummy in order to maintain the logic. The statement of the work method is that

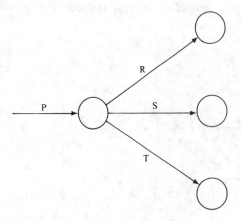

Figure 8.8

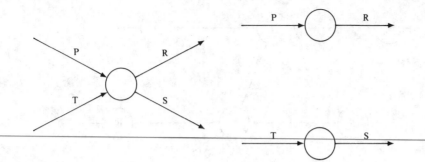

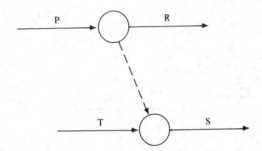

Figure 8.9

Figure 8.10

activity S is dependent upon both activity P and activity T, but activity R is dependent upon activity P only. Since we are unable to split activities the dummy arrow is required in order to maintain this logic. The dummy arrow in the activity-on-the-arrow method is almost indispensable.

Other situations frequently arise in which the use of dummy arrows is the only solution in order that the logic may be maintained. In Fig. 8.11, if activity 8–11 is neglected, a situation arises that demands the use of a single dummy between two parallel flows of work. The dummy has the identification of 7–8. This particular series of activities may well be part of a larger diagram that is under review. On reviewing the situation, it may be decided that a particular activity has been omitted and that this activity should logically follow that of 6–8. Care is needed not to add it in the way that 8–11 has been added to Fig. 8.11, for this represents dependence of 8–11 not only upon activity 6–8 but also upon activity 5–7. Such a situation demands partial redrawing of the network, and Fig. 8.12 illustrates the logical way of representing such a process of working and has entailed the addition of a further dummy and an additional event.

Dummies have other uses in maintaining the logic of the diagram. For example, where a network is being used to control a project and the work is under way, if all the completed activity arrows are removed from the diagram then there may be a

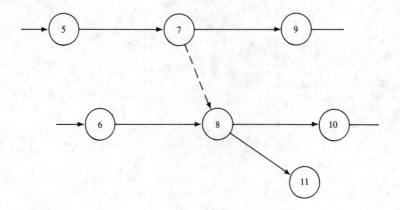

Figure 8.11

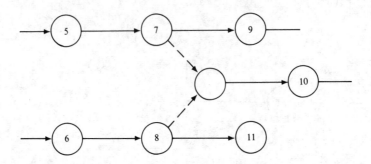

Figure 8.12

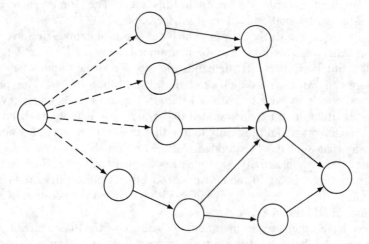

Figure 8.13

series of loose ends that are derived from work yet to be carried out. While such a situation can be handled manually, it is often necessary, when a computer is being used, to deal with a network that commences with a single activity or event. Dummies can be used as in Fig. 8.13 to close up the open end of such a diagram, and, having no resource usage, they do not alter the logic of the diagram. The loose events of the left-hand end of the diagram, before the addition of the dummies, are commonly known as *dangling* events.

8.5 Network construction

The network is used to portray, in graphical form, the logic of the construction plan, that is, it is used as a sequence plan at this stage because time duration of various activities has not yet been considered. The network, therefore, should be drawn in as neat a fashion as possible so that it can readily be understood by those persons to whom the information is to be communicated. The first thing to do, before drawing up the network, is to define the programme's objective. The objective will be the last event in the network to which all of the various branches of the network must eventually converge.

It is argued in many cases that a network diagram should have but one initial activity and but one final activity, passing from the initial and to the final event respectively. While this situation is sometimes demanded by the requirements of a particular computer program, the protagonists of such a system would argue that it is logical that the commencement of any contract springs from but one initial activity. The antagonists of such an argument, that is, those people who believe that dangling events at the commencement of a network are no disadvantage, would argue that it is possible for a project to commence simultaneously in several departments within an organization.

In the event of a single activity being required at the commencement of a project, there may be a need for an artificial activity to be generated. For example, one may call this a *lead time* or the *lead in*. In other circumstances it may be that a single activity offers itself readily, such as *place order*, *seek approval* or some similar action. The same can be said of the final activity in a network, which may well be *clean up*, or, in the event of an artificial activity having to be generated, *end time*. A further advantage of using a single activity at the commencement of a network can well be that of facility in updating. The updating process is that of recalculation of the end time, etc., once the project is under way and the programme is partially completed. The length of time expended on the construction activities already completed can be assigned in total to the 'lead in' activity only. This method does result in the saving of a certain amount of arithmetical computation and thus of planning time.

In developing the arrow diagram, three questions must be asked of each operation:

1. What activity must immediately precede this operation?
2. What activity can immediately follow this operation?
3. What activities can be taking place concurrently with this operation?

If all the activities to be shown in the network are listed, it can be of considerable assistance, before drawing the network, if they can be manipulated into a reasonably logical work sequence before being committed to the network. Answering the three questions posed above for each activity will enable this manipulation to be carried out more easily.

Drawing a network diagram imposes a much stricter discipline on the planner than that of drawing the equivalent bar chart. There can be no overlapping of operations in drawing up the diagram like there may well be in the bar chart preparation. If one operation, as conceived in the initial stages, overlaps another operation on the list of activities in the actual performance of the job, then it is necessary to break down such activities in such a way that they can be represented by a linear flow of arrows head to tail. An example of this might occur in a wall-concreting sequence. In a small contract the formwork would be erected and, on completion, would be followed by the concreting. This would be depicted in a network by a serial work flow as shown in Fig. 8.14.

Figure 8.14

If, however, the concrete wall is of considerable length, it may be desirable to break it down into a number of lengths, which will be concreted one after the other. If the situation now arises that a restriction on the amount of available labour has to be imposed and it becomes clear that one gang of men will be erecting the formwork and another gang laying the concrete, then the logic that such restrictions impose on the network can be illustrated by Fig. 8.15. It illustrates the breakdown that is necessary in the various operations in order that such a succession of work can be illustrated by an arrow diagram.

The initial attempt at drawing the network diagram may result in a somewhat rough and ready representation, and may involve curved arrows, and arrows going from right to left on the page, instead of in the conventional direction, in order to illustrate the logic for the first time. After the network has received a high degree of consideration and appears to be in its almost final form, for the time being, there is considerable

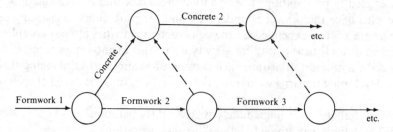

Figure 8.15

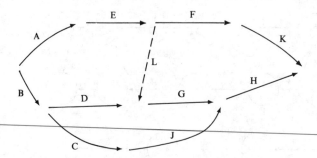

Figure 8.16

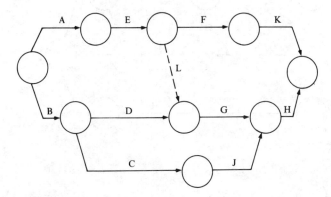

Figure 8.17

advantage in tidying the diagram for clarity of presentation. An attempt should be made to portray the majority of the arrows on a horizontal plane working from left to right throughout the diagram. Where possible, it is desirable to give each arrow at least part of its length horizontally, as this will make it easier to add the description of each activity and the result will be easier to read. Figures 8.16 and 8.17 give a comparison of two networks describing the same work. Figure 8.16 represents the initial attempt at drawing the network and Fig. 8.17 shows an attempt at a more polished arrangement.

It is inevitable in the majority of networks that the logic will demand that in a number of instances one activity arrow crosses another. By arranging the diagram carefully it is possible to avoid too many such situations, and Fig. 8.18 illustrates how this may be achieved by careful planning. Where crossovers are unavoidable, there are a variety of ways of showing them: Fig. 8.19 shows three such methods which are in common use.

8.6 Planning example

Figure 8.20 illustrates a simple reinforced concrete culvert which is to be constructed through an embankment during the diversion of a stream. The planning of the

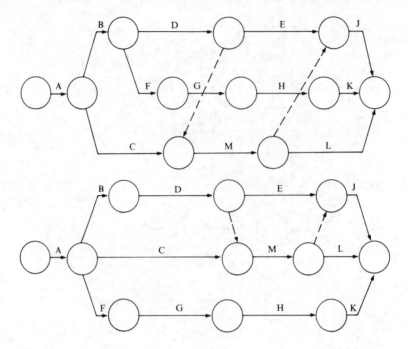

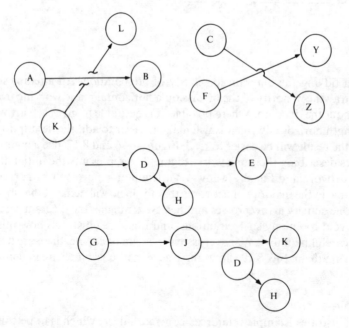

Figure 8.18

Figure 8.19

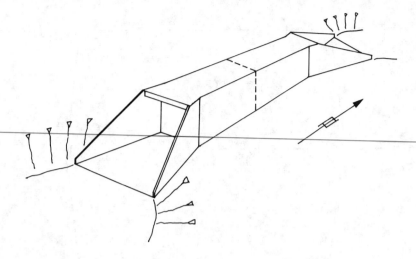

Figure 8.20

example is confined to the reinforced concrete work in the culvert, together with the grading of the approaches to the culvert on either side.

The list of operations to be considered in the plan is as follows:

Construct base slab
Construct north apron slab
Construct south apron slab
Construct north section of side walls
Construct south section of side walls
Construct north wing walls
Construct south wing walls
Construct north section of roof
Construct south section of roof
Grade north approaches
Grade south approaches

From this list of activities, it will be clear that certain restrictions have been placed on the method of working. These restrictions are made in the light of the availability of certain items of plant, materials, and labour. They are known as *restraints*, and they will dictate to a large extent the order in which the work can be carried out.

In the example under consideration, the method statement, that is, the way in which the work will be carried out, is as follows:

The base slab will be constructed in one piece. The apron slabs must be constructed before the grading of the approaches is commenced. The aprons will be constructed after the pouring of the concrete in the base slab. In order to economize

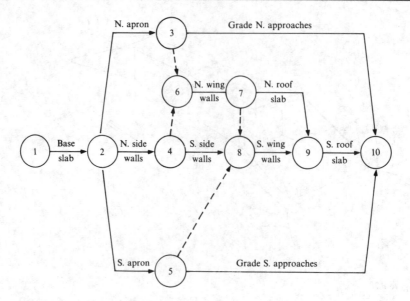

Figure 8.21

with formwork, the parallel walls to the sides of the tunnel are to be constructed in two sections—both side walls in the north half and then both side walls in the south half. The walls will be constructed in one lift up to the underside of roof level. One set of formwork will be made up so that both wing walls to one apron can be constructed at the same time. The north wing walls will be constructed first and will be followed by the construction of the south wing walls. The roof slab will similarly be constructed in two sections, but it cannot be poured until both the side walls and the wing walls for that particular section have been completed.

The fact that the formwork has been restricted for this particular work has placed a restraint on the free planning of the method of construction. For example, if it is decided that unlimited formwork could be made up and used, it is possible that the complete length of both parallel walls of the tunnel and the wing walls on both sides would be poured at one and the same time.

The network for carrying out this work is shown in Fig. 8.21. A single activity has been used for the commencement of the diagram and this has been represented by the pouring of the concrete for the base slab. A single activity has not been used in this simple example as leading into the final event but, had it been desirable to do so, then an activity labelled 'Clean up site' could readily have been inserted.

Having completed the diagram and numbered the events a list may be made of the activities with their descriptions in numerical order as follows:

1–2   Construct base slab
2–3   Construct north apron slab

2–4   Construct north section of side walls
2–5   Construct south apron slab
3–6   Dummy
3–10 Grade north approaches
4–6   Dummy
4–8   Construct south section of side walls
5–8   Dummy
5–10 Grade south approaches
6–7   Construct north wing walls
7–8   Dummy
7–9   Construct roof on north side
8–9   Construct south wing walls
9–10 Construct roof on south side

At this stage, the network diagram represents clearly and concisely the plan for doing the work. No time duration has been considered for any of the activities. The great disadvantage of a bar chart is that these two operations are inseparable, and time and sequence have to be considered at one and the same time.

When the diagram has been drawn, it must be critically examined to ensure that it represents the required and desirable method of carrying out the work, that the logic is correct, and that the principle of dependency applies logically throughout the diagram.

In Fig. 8.21 the network events have been numbered from left to right, that is, by the forward numbering system, and each number is consecutive. Every activity to meet this condition has an *i*-number which is lower than its *j*-number.

Summary

This, the second chapter to deal with aspects of construction planning and scheduling, is devoted to introducing the basic aspects of critical path analysis and, in particular, the arrow-on-the-activity method of portraying programmes. The fundamentals are introduced by way of defining the various components of a network and the ways in which they are used in composition in order to establish the logical sequence of events and activities having decided on the construction technology and the methods that will be used in order to construct a facility. Various ways are detailed in order to make sure that the logic of networks is clearly set out in such a way as to avoid mistakes of communication.

Problems

Problem 8.1   Draw the following network:
Activities   B and C both follow A
Activity   D follows B
Activity   E follows C
Activities   D and E precede F

Problem 8.2   Draw the following network:
Activity   H follows G
Activities   J and K both follow H
Activity   O follows K only
Activities   L and P both follow J
Activity   M follows L
Activity   Q follows P and O
Activity   N follows M
Activity   R follows Q
Activity   S follows both R and N

Problem 8.3   Draw the following network:
Activities   G and H both follow F
Activity   D follows both B and C
Activity   C precedes both F and D
Activity   J follows E only and precedes L
Activity   K follows D and precedes L
Activity   L follows both J and K and merges with M and N at the objective event
Activity   M follows G
Activity   A commences the network
Activity   N follows H
Activities   B and C follow A
Activity   E follows B and precedes J
Activity   F follows C and precedes G and H

Problem 8.4   Number the events in the network shown in Fig. 8.18 by the

(a) forward method—serial numbering;
(b) backward method—serial numbering;
(c) random method.

Problem 8.5 Draw a network of not less than 20 activities to illustrate the organization of a dance.

Further reading

Ahuja, H. N.: *Project Management: Techniques in Planning and Controlling Construction Projects*, Wiley, New York, 1984.

Antill, J. M. and R. W. Woodhead: *Critical Path Methods in Construction Practice*, 2nd edn, Wiley–Interscience, New York, 1970.

Harris, R. B.: *Precedence and Arrow Networking Techniques for Construction*, Wiley, New York, 1978.

Moder, J. J., C. R. Phillips and E. W. Davis: *Project Management with CPM, PERT and Precedence Diagramming*, 3rd edn, Van Nostrand Reinhold, New York, 1983.

O'Brien, J. J.: *CPM in Construction Management*, 3rd edn, McGraw-Hill, New York, 1984.

9. Planning—scheduling for arrow networks

9.1 Time and a network

After considering and reconsidering a network to ensure that the planning logic is sound, the duration of each activity must be estimated so that the *scheduling* process may commence. In estimating the duration and considering each activity on its own merits it may be necessary again to refine the network in the light of this procedure. The term *time* as applied to network analysis is generally used to indicate a particular event time, a point in time at which one of the milestones will occur. The estimated length of time required to carry out an activity, part of a project, or the project as a whole, is referred to as its *duration*.

Any convenient unit of time can be used in the scheduling process—the hour, the working day, the calendar day, the week, or the month. It is often necessary that the unit of time chosen should be a shift or half-shift, if the type of work involved is carried out on this basis. The most important point related to the use of units of time for estimating durations is that they should be consistent throughout the whole of the network. The most suitable units of time for construction work are the *working day* or *working week*. In this context, it is very necessary to distinguish between the working day and the *calendar day*. Deliveries, for example, will almost certainly be quoted in terms of calendar days, and if the working day is used as a unit throughout the network then conversion from calendar days to working days must be made before applying the duration to that particular activity.

In arriving at a first estimate of the duration for an activity, it must be assumed that normal and reasonable circumstances will apply to the work. This means that the estimator must assume that normal hours will be worked, that a normal amount of plant will be used and will be available as required, and that the labour gangs employed will be of the size that the estimator would normally expect on work of that nature. In carrying out this first estimate, no safety factors and no contingencies or interference outside of the control of the site management should be allowed for in estimating durations. Under the latter item the estimate would exclude such influences as bad weather, fires, or strikes. If all durations are estimated on the basis of normal activity, certain adjustments can be made at a later date when the whole of the network is reviewed. It is very important at this stage that the estimated

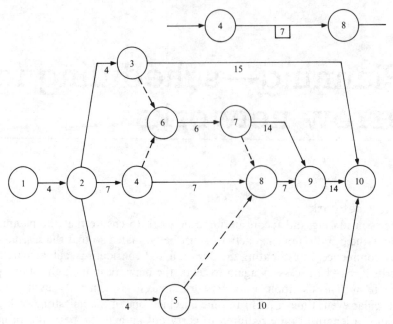

Figure 9.1

duration should be accurate and realistic, and, particularly, that the person who is going to be responsible for carrying out the work should have a hand in its calculation.

The estimated activity duration is shown as a figure below the relevant activity arrow in the diagram, and in Fig. 9.1 the estimated durations for all activities in working days have been added to the network example set out in Section 8.6 for the construction of the reinforced concrete culvert. Alternatively, the estimated duration for the activity can be placed in a small box below the relevant activity arrow, and an example of this method is also shown, inset to Fig. 9.1, using activity 4–8.

Consideration of the likelihood of weather delays in the programme for a project can be made in two ways, as far as representation on the network diagram is concerned. In the first case an activity can be inserted at the beginning or end of the project as a single activity representing the estimated period by which the work will be delayed during the overall programme. In such a case the activity durations, individually, will have no allowance added for such delays. The second method is to allow for the expected weather delay in individual activity durations, so that the total allowance throughout the job would amount to the total figure allowed in the first method.

9.2 Earliest event time (T_E)

The computation process, having estimated the activity durations, involves the use of simple arithmetical operations only. The processes are simple step-by-step operations involving the addition and subtraction of the activity durations. These operations are normally carried out using the network diagram as a worksheet, although there are other methods such as using a *work table* or a *matrix* for carrying out this process. This chapter is concerned with using the network diagram and work table only.

The first calculation, the forward pass, establishes, for each event, the earliest event time (T_E). The earliest event time is the earliest time by which the event under consideration can be achieved, that is, by which all the preceding activities merging into that event have been completed. Figure 9.2 shows a series of three activities, A, B, and C, with durations of 5, 7, and 3 working days respectively. If the tail of activity A represents the commencement of these three activities at zero time, then it can be seen that event number 2 will be reached by the end of day 5. The earliest event time for event 2 is then 5 and this is indicated in a box adjacent to the event number. A square is frequently used to enclose the earliest event time, but this can vary depending upon the individual concerned. If activity 2–3 is now examined it will be seen that its duration is 7 working days. Calculating the earliest event time for event number 3 gives $5+7=12$ days and therefore the figure 12 is placed in a box adjacent to event number 3. In similar fashion 15 is placed in a box adjacent to event number 4.

It has therefore been established that if these three activities are to be carried out in accordance with their estimated durations it must be expected that the earliest event time for event 4, the final event in this simple programme, will be 15. This means that the earliest time by which it is possible to arrive at event 4, in view of the information which is given, is the end of day 15.

Figure 9.2 represents an extremely simple programme, and, since many activities are carried out concurrently in a normal project, the earliest event time for any particular event must be the result of calculating through several paths of the network up to the point under consideration. Figure 9.3 illustrates the network for the reinforced concrete culvert with the earliest event times superimposed upon it. Work is started at time zero, and event 2 can be achieved by the end of day 4. The consideration of earliest event times for events 3, 4, and 5 presents no problems, and these become 8, 11, and 8 respectively. However, when the earliest event time is considered for event 6, it is seen that this event may be approached through two paths or sequences of activities in the diagram. The first path is traced through events 1–2–3–6. The second path is traced through events 1–2–4–6. Because activities 3–6 and 4–6 are dummies, it should be borne in mind that these bear no duration against them since they do not represent the consumption

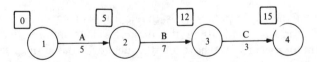

Figure 9.2

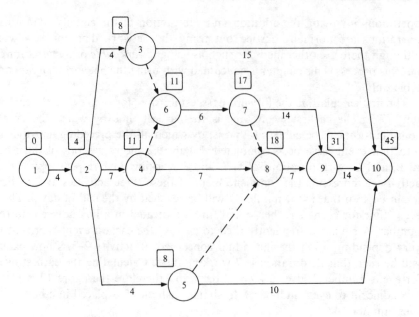

Figure 9.3

of time. Nevertheless, they must be considered as part of a possible chain of activities through the network. Through the first path an earliest event time of 8 is arrived at for event 3, to which is added zero for the dummy. Tracing the durations through this path, it would be expected that the earliest event time for event 6 would be 8 days. Following the second of the two available paths, an earliest event time at event 6 is 11. Reverting to the definition of an earliest event time, that is, the earliest time by which all the activities preceding that event must be completed, it is clear that, of the two available figures 8 and 11, the smaller must be rejected. Event 6 will not be achieved until the end of day 11. Therefore, the 11 is superimposed upon the diagram adjacent to event 6. The remainder of the diagram can now be analysed in a similar way. In considering event 8, it will be seen that there are three arrows merging into this event and therefore there are three immediately adjacent paths through which the earliest event time under consideration will be 8 via dummy 5–8, 17 via the dummy 7–8, and 18 via activity 4–8. As before, all but the largest of these numbers are rejected and the earliest event time for event number 8 is therefore the end of day 18.

The final result of these computations is that a total of 45 days is placed as the earliest event time for event 10 in the diagram. This implies that, on the basis of the information which we have on the network, the project cannot be completed in less than 45 days. The earliest event time is 45 for the ultimate event in the network.

9.3 Latest event time (T_L)

The second computation is to establish the latest event times (T_L) for each event in the network. The latest event time may be defined as that time by which a particular event must be achieved if there is to be no delay to the completion of the project in the set overall duration. This phase of the computation is known as the backward pass.

The latest event time for the objective event is arbitrarily allocated at this stage of the scheduling procedures to be equal to the earliest event time established in the forward pass computation. In Fig. 9.4, the network diagram for the reinforced concrete culvert, event 10 is assigned the figure 45—placed in a triangle adjacent to the earliest event time—because this is the largest figure arrived at in the forward pass computation.

The procedure for the establishment of the latest event times is simply a reversal of that for calculating the earliest event times. On working back from event 10 to event 9, it will be noted that there is only one activity, 9–10, emanating from event 9. The duration of activity 9–10 has been estimated at 14 days, so that, if the programme duration is to be maintained and the whole of the work completed in the computed 45 days, event 9 must be achieved in $45-14=31$ days. Event 9 can be dealt with in this straightforward manner because of the single activity between itself and event 10. Similarly, event 8 has but one activity emanating from it, and therefore the latest event time for this event will be $31-7=24$ days. This means that

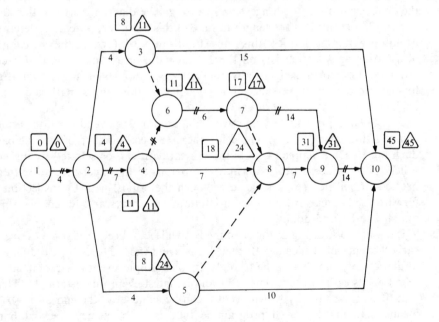

Figure 9.4

if the time for the objective of 45 days is to be achieved then event 8 must be completed on or before the end of day 24. Any delay beyond day 24 means that inevitably there will be a delay to the programme completion as a whole if the durations estimated for activities 8–9 and 9–10 are maintained.

On moving to event 7, it can be seen that there are two merging paths through which the backward pass must be made. The first is directly through event 9 to 7, and the second through event 8 to 7. Checking back through activities 9–7, $T_L = 31 - 14 = 17$, and through dummy 7–8, $T_L = 24 - 0 = 24$. From our definition of latest event time it is clear that, if event 7 is not completed by the end of day 17 then the programme as a whole cannot be achieved in 45 days. The selection between the alternatives for the latest event times is made on exactly the opposite basis from that of the earliest event times, and thus the lowest figure is chosen. In this case the lower figure is 17. Figure 9.4 has all the lastest event times added to it, working back to event 1. If the arithmetic has been correct throughout the diagram, and the latest event time for the end event is the same as the earliest event time, excepting compensating errors, then the figure for the latest event time at event 1 must necessarily be the same as the figure for the earliest event time, or start date, for event 1.

9.4 The critical path

The forward pass computation shows that the overall time for the project (in the case of the example for the reinforced concrete culvert it is 45 days) is established by the determination of the longest path through the diagram between the start event and the end event. It is the length in time of this path, therefore, that determines the overall programme and it is this longest path that is known as the *critical path*. Any delay to an activity that lies on the critical path, or any duration that exceeds the estimated time for a critical individual activity, must result in the overall schedule for the contract being extended. Each activity on this longest path is critical to the achievement of the project target.

Every plan represented by a network diagram must have at least one critical path running through it. It is possible for there to be a number of critical paths in a diagram. In some simple projects the critical path can be selected by inspection; in others, parts of the critical path may be obvious to the experienced eye. In complex projects, it will not be possible to establish the critical path by inspection and the forward and backward pass computations are necessary to enable this to be established beyond doubt.

If the network of Fig. 9.4 is examined, it will be noticed that some events have an earliest event time which equals their latest event time. Others have a different figure in the square from that in the triangle. If, for example, event 6 is examined it will be seen that its earliest event time is 11 and its latest event time is also 11. This means that the earliest time by which event 6 can be achieved is the end of day 11. It also means that, if the overall programme time is to be achieved, event 6 must be accomplished not later than the end of day 11. Event 6 is therefore critical, because there must be no delay between its achievement and the start of the following activity. It is a *critical event*, as are all the other events in the network bearing an earliest event time which equals the latest event time.

The critical events have therefore been established by the forward and backward pass computations, and the critical paths will obviously pass through them. Figure 9.5 illustrates a very simple network of four activities. The forward and backward passes have been made. Judged by the criterion already established—that the critical path passes through critical events—it is not clear from this diagram whether the critical path follows the path through events 1–3–4 or whether it will pass through events 1–2–3–4. These are the only two alternative paths in this network. A second criterion has to be applied, therefore, in order to distinguish between situations similar to the one presented by this diagram.

The second test that can establish whether an activity is critical or not is to subtract, from the latest event time at the head of the activity arrow in question, the earliest event time at the tail of that arrow, and then to subtract from the result the duration assigned to the arrow. If the result is zero, then the arrow is on the critical path. Reverting to Fig. 9.5 and applying the test in this case for activity 1–3, the latest event time at the arrow head, that is, at event 3, is 5, from which must be subtracted the earliest event time at its tail, that is, event 1. The activity duration of 4 is then subtracted from the result, in this case $5-0-4=1$. Therefore, activity 1–3 does not lie on the critical path. Examining in a similar way activity 1–2, the computation results in $2-0-2=0$; activities 1–2 and 2–3, therefore, must be critical.

Figure 9.4 does not present such a difficulty and the critical path can be established as being through events 1–2–4–6–7–9–10.

The critical path can be indicated on a diagram in a number of ways. It can be shown either as a thick line, identified in some such way as showing two dashes across the critical activity arrow, or by similarly marking across it with another symbol. In the reinforced concrete culvert example, the critical path runs through a dummy arrow. This is quite legitimate, bearing in mind that although a dummy

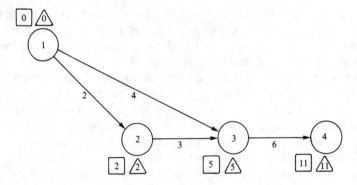

Figure 9.5

arrow may be critical it only shows dependency and does not require the use of time or any other resource.

To summarize, the two criteria for finding the critical path are:

1. $T_E = T_L$ for an event on the critical path,
2. $T_L j - T_E i - y = 0$ for a critical activity,

where $y =$ the estimated duration for the activity.

The use of the diagram as a basis for the simple computations, examples of which have been given above, reveals activities which are critical to performance if the project programme as a whole is to be achieved. These critical activities assume a new importance, whereas, up to this point, they may have been looked upon as being quite trivial when considered as part of the overall project. On the other hand, many activities which may have been thought to be very important now assume a position of relative unimportance as far as the schedule is concerned. In typical networks for large and complex projects, the critical jobs revealed by such an analysis rarely number more than 10 per cent of the total. In smaller networks this percentage may well increase, since there must always be at least one critical path through any network.

9.5 Ladder construction

Considering the construction of a simple, but long, reinforced concrete retaining wall, it may be possible to employ only one gang of steel-fixers, one gang of carpenters, and one gang of concretors. This may be necessary because the wall forms part of a much larger contract; but there may be many other reasons. It may be desirable to divide the wall up into a number of lengths so as to provide, as far as possible, continuity of working for the gangs selected to do the work, as well as to break the work down into areas that can be dealt with more easily. A network diagram to illustrate how the wall will be constructed if it is divided into four such sections is shown in Fig. 9.6.

In the diagram, it will be seen that initially the reinforcement for the first section is fixed (S_1) and is described by activity 1–2. The duration of this activity is 2 working days. On completion of the fixing of the reinforcement, the formwork for the first section of wall can be erected, and the gang of steel-fixers can be moved to the second section of the wall. The second operation of the steel-fixers is represented by activity 2–3 (S_2). The activity of erecting the formwork emerges from event 2 and is represented by arrow 2–4, having a duration of 4 days (J_1). Similarly, the pouring of concrete for the first section can start when event 4 is completed, and the carpenters employed upon erection of the formwork for the first section can now be moved to erection of formwork for the second section, arrow 5–6.

The logic of the diagram described so far is not complete, inasmuch as no indication has been given that the erection of the formwork for the second section

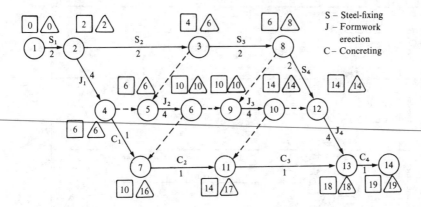

Figure 9.6

depends upon the completion of the fixing of the reinforcement for that section. A dummy arrow is therefore inserted between events 3 and 5 to illustrate this dependency. The whole of the work by the three trades is represented over the four sections of wall by the complete diagram Fig. 9.6, and it will be noted that the overall duration for the project is 19 days. Once it is decided that this is a suitable target at which to aim, the backward pass computation can be made in order to determine the critical path. This follows the path 1–2–4–5–6–9–10–12–13–14, and, as one would expect, it passes through the carpenters' work for most of its path, the carpenters taking 4 days to erect the formwork on a particular section of wall, and therefore longer than the time required by any other trade.

This network diagram has been constructed for a simple project and only three trades have been considered in the light of splitting the wall into four sections. It can be envisaged that a diagram for a 20-storey block of flats involving perhaps as many as 15 or so trades would become an extremely complex diagram to understand and considerable repetition would be involved. A simplification can be made involving the use of *lead* and *lag* arrows.

Figure 9.7 represents the simplification in diagrammatic form. The repetitive work of one trade is now represented by a single horizontal arrow. Since there are three trades involved, there are three horizontal arrows, 50–51, 52–53, and 54–55. The duration shown against each of these arrows is the summation of the durations for individual activities within that trade of Fig. 9.6. In the case of the steel-fixing the total time taken equals 4×2 days $= 8$ days duration, and similarly with the other two trades involved.

In order to maintain the logic of the diagram, allowance must be made for the fact that the erection of the formwork on the first section is not commenced until the fixing of the reinforcement is completed. This can be represented by a lead arrow, that is, an arrow which connects the leading events of two activities. These are dummy arrows with a duration allocated to them and are represented on the diagram by a chain-dotted line. In the case of Fig. 9.7, formwork erection can

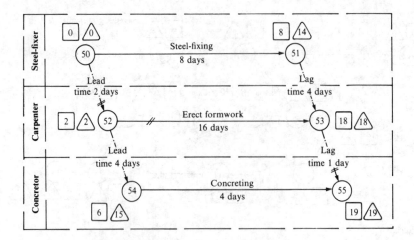

Figure 9.7

commence 2 days after the start on the wall as a whole, so that the lead arrow 50–52 bears a duration of 2 days. Similarly, the pouring of concrete in the first section cannot start until at least 4 days after the commencement of the erection of formwork—assuming that this work goes in accordance with the plan—so that the lead arrow, in this case 52–54, bears a duration of 4 days.

The end events of each of the arrows concerned are joined by lag arrows. Activity 51–53 is such a lag arrow and bears the duration of 4 days because the erection of the formwork to the last section of wall cannot be completed until 4 days after the completion of reinforcement fixing to the last section. Similarly, arrow 53–55, another lag arrow, bears a duration of 1 day.

By carrying out the backward and forward passes through the ladder diagram, the completion time for the project is calculated as 19 days, being the same as that of Fig. 9.6. However, it should be noted that, in using such a technique, some sacrifice of internal logic is made when compared with the detailed diagram of Fig. 9.6. If some of the activities, for example, 5–6 or 9–10, were to become unbalanced in the duration of time required for their completion then it would be possible for the critical path to run through activities other than those shown. Also, if the overall duration of the completion for all the similar activities within their master arrow of the ladder diagram remained the same then this changed path would not become obvious with the use of the ladder technique.

Lead arrows may be used in another context within the network diagram. Certain activities can be allocated, for one reason or another, a far greater duration than is absolutely necessary for their completion; in other words, there is considerable leeway at one end or the other of the arrow. It is better for the activity to be defined in duration and time more closely, and this will be particularly advantageous if the activity is best carried out at one particular time of the year. For example, in the grass seeding of the banks of a motorway cutting or embankment it may be desirable to sow the seed during September or October or during April or May. If the network diagram indicates that there is considerable leeway for this operation as a whole because it is not critical, without indicating time against a calendar, then it is possible to absorb part of the activity duration for sowing seed in such a way that the timing of the activity is more closely defined. This time may readily be absorbed with a lead arrow or a lag arrow. In effect, the activity arrow is being divided into two arrows—one a lead or lag arrow and the other the activity.

One of the advantages of using the ladder technique arises from the fact that trades or departments can be banded through the diagram. In Fig. 9.7 horizontal lines have been drawn across the diagram dividing activity 50–51 from activity 52–53, and again dividing activity 52–53 from 54–55. This means that if the plan is to be communicated to, say, the foreman carpenter then they need only to look within the band that is labelled with their trade on the left-hand side to see the work in which they or their operatives are involved. This technique can be used in many other fields, particularly where the design processes, as they should be, are shown on the same diagram as those of the construction. For example, the work to be carried out by the architect or the engineer is readily portrayed in this way, the work of the contractor is highlighted, and it is often desirable to illustrate the work of various subcontractors so that they, too, may be readily informed of their part in the complete process.

9.6 Float

If Fig. 9.5 is now reconsidered, and activity 1–3 is examined in particular, it will be seen that the latter's duration is estimated to be 4 days, whereas the durations of the critical activities 1–2–3 have an estimated total duration of 5 days. This means that activity 1–3 has spare time or leeway, known as *float*. Because activity 1–3 has float, it means that it can either be started at the beginning of day 0 and completed by the end of day 4 or it can be started at the beginning of day 1 and completed by the end of day 5. When shown in bar chart form it looks like Fig. 9.8. Whichever alternative is adopted, the overall project duration of 11 units of time will still be maintained because by the basic rule previously laid down, if each event is achieved by its latest event time then the overall project duration will be achieved. Activities on the critical path have no spare time or leeway; that is, their float is 0.

Each event (except the first and the last) in a network must have at least one activity preceding and succeeding it. After the backward and forward passes have been computed through the network, each event has two figures associated with it, described as the earliest and the latest event times. The earliest event time associated with the event at the tail of any activity arrow is the earliest time by which that activity can commence. It is usually referred to as the *early start* (*ES*). If the activity proceeds from its early start time and takes the estimated duration against the arrow, it will have an *early finish* (*EF*), and

$$\text{early finish} = ES + \text{duration} = ES + y$$

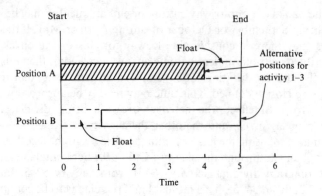

Figure 9.8

In a similar way, if the event at the head of the arrow is examined it will be seen that the *late finish* (*LF*) is the same as the latest event time for that event. The *late start* may be derived from the late finish by subtracting the duration from it:

$$\text{late start} = LF - y$$

A table of activities forming any network may now be drawn up and the early start, early finish, late start, and late finish can be tabulated in such a way as to make the calculation of float easier. If the early start and late finish columns of the table are completed from the network diagram after the computations have taken place, then the early finish and the late start may be calculated by the addition or the subtraction respectively of the duration from these figures. Table 9.1 is such a table

Table 9.1

Activity	Description	Estimated duration	Early		Late	
			Start	Finish	Start	Finish
1–2	Base slab	4	0	4	0	4
2–3	N. apron	4	4	8	7	11
2–4	N. side walls	7	4	11	4	11
2–5	S. apron	4	4	8	20	24
3–6	Dummy	0	8	8	11	11
3–10	N. approaches	15	8	23	30	45
4–6	Dummy	0	11	11	11	11
4–8	S. side walls	7	11	18	17	24
5–8	Dummy	0	8	8	24	24
5–10	S. approaches	10	8	18	35	45
6–7	N. wing walls	6	11	17	11	17
7–8	Dummy	0	17	17	24	24
7–9	N. roof	14	17	31	17	31
8–9	S. wing walls	7	18	25	24	31
9–10	S. roof	14	31	45	31	45

for the diagram relating to the reinforced concrete culvert shown in Fig. 8.20. This table shows that the activities on the critical path of the diagram have similar early and late start times and similar early and late finish times.

Reverting now to the consideration of float, if the early start time for any activity is subtracted from its late finish time, an indication is given of the total overall time available for carrying out the activity. If now the duration is subtracted from the answer so obtained, it will be clear that this is the *total float* available for this particular activity. This is the leeway that the activity has and can be used without affecting the overall duration of the project. The calculation of total float may be stated as follows:

$$\text{total float} = T_\text{L}j - T_\text{E}i - y$$

or

$$= LF - ES - y$$

A variety of types of float can be derived, but total float is by far the most important and useful. The following paragraphs briefly describe three other types of float that have been used but have mainly proved to be of only academic interest in the development of the use of networks.

Free float is the difference for any activity between its early finish time and the early start time of the succeeding activities. Free float becomes significant if the activity under consideration commences at its early start time, since it then gives an indication of the float that is not shared by any of its succeeding activities.

Total float is often related to more than one activity and the same quantity of total float can be calculated as referring to each of a number of activities on a subcritical path. If the whole or part of the total float in any one of these activities is used up, it will affect float on any of the other activities forming part of the same subcritical path. It may make critical other activities forming part of that path, and therefore the use of total float requires considerable care. If, for example, in Fig. 9.4, activities 4–8 and 8–9 are considered, there is a quantity of total float common to both of them. By definition the total float for activity 4–8 will be $24 - 11 - 7 = 6$. The total float for activity 8–9 will be $31 - 18 - 7 = 6$. If activity 4–8 does not start until its latest possible start date, that is, by the end of day 17, this activity will be completed by the end of day 24. The calculation of total float for activity 8–9 now becomes $31 - 24 - 7 = 0$, and activity 8–9 has become critical. The six days' float calculated for these activities is thus common to both of them, and if it is utilized by one activity then it must be denied to the other. In such cases, where part or all of the float is common to two activities, the common float is known as *shared float*.

Another variety is *independent float*. Activity 5–10 of Fig. 9.4 provides an illustration of independent float. The minimum time available for carrying out this activity is 21 days. Because the activity is scheduled to have a duration of only 10 days, then 11 days of the time available must be float. As it is impossible for any other activity to use this float, it is completely irreducible and thus independent. Figure 9.9 illustrates graphically the extent of float for any linear activity in series. Remember that $T_\text{E}i$ and $T_\text{L}i$ for the start event and $T_\text{E}j$ and $T_\text{L}j$ for the finish event

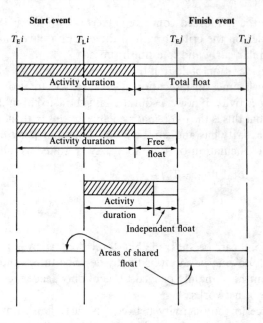

Figure 9.9

of Fig. 9.9 may depend on a number of activities at a merge or a burst event, and that a single linear activity may not necessarily be considered. In the latter case, if the activity commences at $T_E i$ it will finish at $T_E j$. It must be stressed again that total float is the one of greatest importance in the work of critical path methods of planning.

Table 9.2

		Estimated duration	Early		Late		Float			
Activity	Description		Start	Finish	Start	Finish	Total	Independent	Free	CP
1–2	Base slab	4	0	4	0	4	0	0	0	*
2–3	N. apron	4	4	8	7	11	3	0	0	
2–4	N. side walls	7	4	11	4	11	0	0	0	*
2–5	S. apron	4	4	8	20	24	16	0	0	
3–6	Dummy	0	8	8	11	11	3	0	3	
3–10	N. approaches	15	8	23	30	45	22	19	22	
4–6	Dummy	0	11	11	11	11	0	0	0	*
4–8	S. side walls	7	11	18	17	24	6	0	0	
5–8	Dummy	0	8	8	24	24	16	0	10	
5–10	S. approaches	10	8	18	35	45	27	11	27	
6–7	N. wing walls	6	11	17	11	17	0	0	0	*
7–8	Dummy	0	17	17	24	24	7	0	1	
7–9	N. roof	14	17	31	17	31	0	0	0	*
8–9	S. wing walls	7	18	25	24	31	6	0	6	
9–10	S. roof	14	31	45	31	45	0	0	0	*

In a complex network diagram it is sometimes inconvenient and even difficult to calculate the floats for various activities, and these may more easily and more tidily be calculated from the table of early and late start and finish times. Table 9.2 includes total, independent, and free float for all the activities in the example of the reinforced concrete culvert. In the final column of this table are indicated those activities that lie upon the critical path.

9.7 The presentation of network diagrams

The detail with which a network needs to be presented is one of the more important points to be considered in the use of such a technique for programming and scheduling purposes. It may be found in practice that a minimum amount of detail is necessary, in the first instance, in order to arrive at a completion date because of the logical interdependencies between one activity and another. It may have been the purpose of the programmer of Fig. 9.10 that a mechanical excavator would be used for the first 21 days of activity 10–11 and that the cleaning-up of the excavation would be carried out by hand labour. Further, it may have been intended that the excavator would move from activity 10–11 to enable activity 41–42 to begin, when it had completed its 21 days' work on the excavations for the foundations.

Since it is not possible to commence a new activity arrow at any other point than at an event in the diagram, it is not possible to draw an arrow from an intermediate point along activity arrow 10–11 to join event 41. It is necessary, therefore, to introduce an additional event and to split the one activity 10–11 into two subsidiary activities, as shown in Fig. 9.11. A linking dummy between new event 71 and event 41 indicates that the excavation to the culverts cannot commence until the foundation excavations have been taken out by the mechanical excavator. The working breakdown of activities will, in this way, often dictate the extent of the detail to be carried out for a particular network.

On the other hand, the upper levels of management of an organization do not routinely require to know a vast amount of detail about the project but will require to know general progress against key dates, and the overall picture. It must therefore be possible to reduce the detail of a network in its operational form so that it

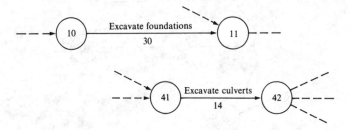

Figure 9.10

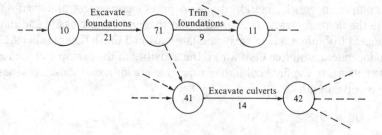

Figure 9.11

presents a simple overall picture. Figure 9.12 illustrates how this may be done without disturbing the logic of the network. Where a serial run of activities that is not dependent upon other activities occurs, it may be condensed and represented by a single activity. It is important in carrying out such condensation that all relationships of dependency between one set of activities and the other should be maintained. In Fig. 9.12, 15 activities and 13 events have been reduced to a total of 7 activities and 5 events in the condensed network.

There is little point in using the network diagram as a communication tool between the planner or management and the operatives on the site unless these operatives have themselves been trained in the use of network analysis techniques. Where operatives and supervisory personnel are familiar with the use of the bar chart as a means of pictorially representing a programme, there is little reason why

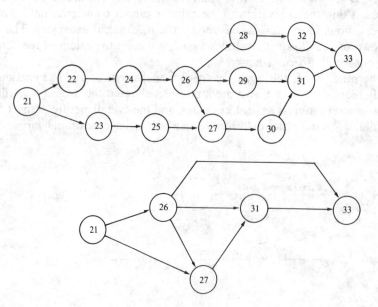

Figure 9.12

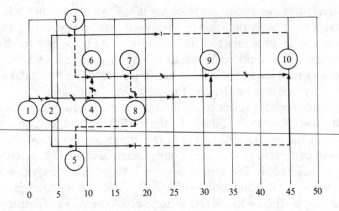

Figure 9.13

the network diagram should not be translated into a bar chart for the purposes of communication. Some printouts of a network analysis from a computer can be directly in the form of a bar chart. Care must be taken, in the conversion to a bar chart, that total float and its implications are understood by the person receiving and interpreting the programme.

As an alternative to the bar chart, it is possible to draw a network diagram to a horizontal time scale. Figure 9.13 shows the example of the culvert network of Fig. 9.4 drawn to a time scale. Activity durations are shown by a solid arrow on the diagram, and float is shown by a broken line. Connections between arrows in a vertical direction have no time significance and are provided in the diagram only as a means of continuity.

9.8 PERT

The PERT method of network analysis was devised to take account of the difficulty of estimating the durations of activities that realistically cannot be established from past experience. In the case of construction work, excavation, for example, has been carried out under many varied conditions and in all types of soil. The experience of conducting this operation is, therefore, very broad, and, in estimating durations for carrying out future work, there is little doubt about the estimated normal durations if the facts about the excavation conditions are known.

Where it is necessary to estimate the possible duration of such work as research or development, or work that is being carried out for the first time, then the basis for estimation is by no means as sound as that for construction works. It is for this type of work that the PERT statistical approach was developed. Statisticians have relatively simple means for the determination of uncertainty in quantitative terms and such methods, though they have been challenged in this context, are the basis for the PERT approach to network analysis. PERT methods do take account of the activity that is likely to have a wide range of durations. An activity in CPM may be estimated to have a most likely duration of 10 days. It may be that, due to the nature of the operations, the possible range of the duration could be from 2 to 18 days or, for a different operation, it could be from 8 to 15 days. No distinction between these two types of activity is made in CPM; the duration is stated to be 10 days and the probable uncertainty is not considered in the subsequent calculations.

Other than the different approach to estimating the durations of activities, PERT is known as an *event-orientated* method. In the PERT method, certain events are selected throughout the programme to act as milestones for the programme. These are events that are important in the overall programme and against which progress in general can be measured. The milestones are labelled with a specific title, for example, 'Completion of all foundation work'. With CPM, the tendency is to refer to the activity and not to the event. When a number of events or milestones in the PERT network has been established, they are linked together by arrows in the same way as the CPM network. The expenditure of resource is assumed to occur on the activity arrow in PERT in the same way as in CPM.

The method of estimating durations in the PERT technique uses elementary probability theory to measure the probability of each event being achieved as predicted. In simple terms, probability theory is a means of putting a figure to such statements as *most likely, very probable, unlikely*, etc. A brief introduction to probability theory was given in Section 5.9.

A frequency distribution is a means of setting out the number of occurrences of a specific value in a set of readings. For example, if a well-defined activity is observed on a large number of occasions, and its duration on each occasion is noted, then a table of the readings can be prepared listing the values of the duration in a chosen numerical order and giving the number of times each value has occurred in the set of observations. This table represents a frequency distribution of observed durations. The frequency distribution can then be plotted on a base of observed values (all in the same units of time) against a vertical axis of frequency of occurrence, as in Fig. 9.14. If it is preferred, the vertical axis can be in terms of relative frequency, either parts or percentages. In this case, each value of observation is expressed as a ratio of the total number of observations. The frequency data can be drawn as a histogram (in which the area of each vertical bar is proportional to the number of observations) or as a smooth curve. The latter allows the shape of the curve to be seen more clearly. From Fig. 9.14 it can be seen that the frequency of occurrence of any duration can be read off from the curve (or histogram)—for example, there were 100 observations of a duration of 17 days. The distribution provides a great deal of information about the duration for which it is drawn. Among this information is its variability, that is, whether there is a wide range between the highest and the lowest value showing large variability, or whether the distribution has a narrow base, illustrating small variability. Another factor is whether the mean of the set of values is central to the range as a whole or whether it is offset or skewed. When PERT was developed, it was assumed that a *beta distribution* was a close fit to activity duration distribution. Being able to find a standard distribution to which the data provide a

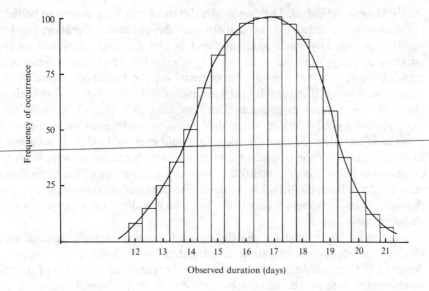

Figure 9.14

close fit is advantageous in that the characteristics and parameters of the standard distribution are already determined and thus, for any frequency distribution of that form, they can be readily calculated.

Having accepted that a beta distribution is a satisfactory model for the estimation of a duration, the expected mean duration for an activity, t_e, and the standard deviation, s, of the distribution can be found from

$$t_e = \frac{a + 4m + b}{6} \qquad (9.1)$$

and

$$s = \frac{(b - a)}{6} \qquad (9.2)$$

where m = the *most likely* duration of the activity
  a = the *optimistic* time, that is, the shortest time that could be anticipated for the activity
  b = the *pessimistic* estimate of the time, that is, the duration of the activity assuming that everything goes at its worst.

The three estimates of the activity duration can be obtained in a number of ways. The first is by a review of historic performance in the same way as durations for CPM are established. The pessimistic duration may be assumed to be the longest duration ever recorded for the activity and the optimistic duration would be the shortest. Statistically, each of these shold be the durations achieved on only one occasion in every hundred. A second way of establishing the durations would be to question those persons who are experienced on the type and nature of the work concerned. Asking 'If the conditions are adverse and everything that is likely to affect the work goes against its progress then how long do you think it would take?' and then prompting and questioning the replies would result in a figure for the pessimistic duration. Similar questioning would eventually elicit a reasoned answer to the questions for the optimistic and most likely durations as well. With the three durations in place it is then possible to arrive at an estimate of the mean duration by substitution in formula (9.1). The value of t_e so obtained will then take account of the chances of the duration varying from that which is obtained simply as a result of a single estimate.

If the most likely time for the duration of a certain activity is estimated to be 10 days, the optimistic duration is estimated to be 3 days, and the pessimistic duration of time is taken to be 15 days, then these three estimates of duration can be reduced to one for the activity by substitution in the formula above:

$$t_e = \frac{3 + (4 \times 10) + 15}{6} = 9.7 \text{ days}$$

It is usual in calculating durations for a PERT network to allow one decimal place. In the above example the time to be shown against the activity would be 9.7 days.

EXAMPLE 9.1

The arrow diagram in Fig. 9.15 represents the logical sequence of activities in a research and development programme for the design and construction of a re-inforced concrete nuclear pressure vessel. Because of the uncertain nature of the programme it is decided to analyse the arrow diagram using three estimates of duration for each activity. The three duration estimates (in weeks) are shown against

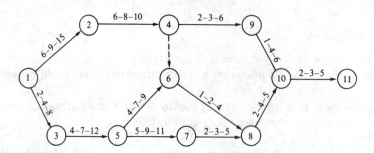

Figure 9.15

each arrow in the figure. Thus for arrow 1–2 the estimated optimistic, most likely and pessimistic durations are 6, 9, and 15 weeks respectively.

What is the probability that the programme will be completed in 33 weeks or less?

The first step in the analysis is to calculate the expected time, t_e, for each activity from the general formula stated above. It should be noted that the mean of the three given durations will be equal to the value of t_e only where the optimistic and the pessimistic durations are symmetrically placed about the most likely duration. Activity 2–4 in Fig. 9.15 is an example. The values of t_e give a measure of the central tendency for the beta distribution of the activity durations. In order to measure the spread or range of a distribution, a *standard deviation* is used. The result of calculating values of the mean, t_e, and the standard deviation, s, for each activity in the network are summarized in Table 9.3. The values of t_e and s describe the distribution for each activity. From the values of t_e, the critical path through the network can be found in the same way as in a CPM network, by carrying out the forward and backward passes and applying the two criteria for critical activities to each activity. The critical path passes through events 1–3–5–7–8–10–11.

The central limit theorem is next invoked. As applied to a PERT network, it can be stated that where a series of sequential independent activities lie on the critical path of a network then the sum of the individual activity durations will be distributed in approximately normal fashion regardless of the way in which the individual activity durations are themselves distributed. The mean of the distribution of the sum of the activity durations will be the sum of the means of the individual activities and its variance will be the sum of the variances.

Using the central limit theorem allows the calculation of the standard deviation for the distribution of the total project duration. The project duration is the sum of the values of t_e for the length of the critical path. From the central limit theorem the duration will be normally distributed and will have a *variance* that will be the sum of the variances of each activity on the critical path.

Table 9.3

Activity		Estimated duration			Mean	Standard deviation
i	j	a	m	b	t_e	s
1	2	6	9	15	9.5	1.50
1	3	2	4	8	4.3	1.00
2	4	6	8	10	8.0	0.67
3	5	4	7	12	7.3	1.33
4	9	2	3	6	3.3	0.67
5	6	4	7	9	6.8	0.83
5	7	5	9	11	8.7	1.00
6	8	1	2	4	2.2	0.50
7	8	2	3	5	3.2	0.50
8	10	2	4	5	3.8	0.50
9	10	1	4	6	3.8	0.83
10	11	2	3	5	3.2	0.50

The variance for a distribution is the square of its standard deviation. The standard deviation for the normally distributed duration of the network of Fig. 9.15, therefore, becomes the sum of the variances for each activity on the critical path:

$$\text{standard deviation} = (1^2 + 1.33^2 + 1^2 + 0.5^2 + 0.5^2 + 0.5^2)^{1/2}$$
$$= \underline{2.13 \text{ weeks}}$$

The sum of the activity durations on the critical path amounts to 30.5 weeks. The completion time for the project can therefore be said to be normally distributed, having a mean of 30.5 weeks and a standard deviation of 2.13 weeks. This distribution is illustrated in Fig. 9.16.

A normal distribution follows a mathematical formula and is a unimodal, symmetrical frequency distribution. Theoretically a normal distribution ranges from minus infinity to plus infinity, but only a very small proportion of values lie very far from the mean. No matter what the mean and/or standard deviation, normal curves always assume the same shape. In Section 6.4 some properties of the normal distribution were discussed in the context of random sampling. Those properties are again used here.

Table 9.4 is a simplified extract from those tables describing a normal distribution that are published in sets of statistical tables or as an appendix to other textbooks. This table gives the proportion of the total area under the normal curve that is bounded by the vertical line at a distance of x standard deviations above the mean and the curve itself. As an example, if a vertical line is drawn through the curve to the horizontal axis, at the value of the distribution's mean, t_e, on the horizontal axis, that

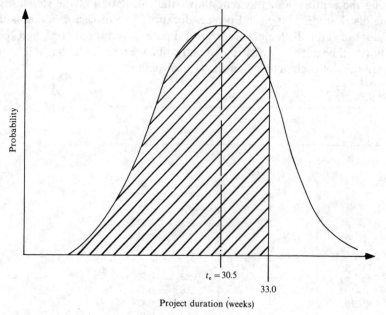

Project duration (weeks)

Figure 9.16

Table 9.4

Standard deviations above mean	Proportion of area to right	Standard deviations above mean	Proportion of area to right
0.0	0.500	1.5	0.067
0.1	0.460	1.6	0.055
0.2	0.421	1.7	0.045
0.3	0.382	1.8	0.036
0.4	0.345	1.9	0.029
0.5	0.309	2.0	0.023
0.6	0.274	2.1	0.018
0.7	0.242	2.2	0.014
0.8	0.212	2.3	0.011
0.9	0.184	2.4	0.008
1.0	0.159	2.5	0.006
1.1	0.136	2.6	0.005
1.2	0.115	2.7	0.004
1.3	0.097	2.8	0.003
1.4	0.081	2.9	0.002

is at 0.0 standard deviations above the mean, then 0.50, or one-half, of the area under the total curve will be to the right of this line. If the vertical line is drawn at 1.3 standard deviations above the mean, then 0.097 of the area under the curve will be to the right of this line. The first result is, of course, what would be expected for a symmetrical curve divided at its mean. The same argument is true for standard deviations below the mean (which are usually expressed as negative standard deviations). Where the area under the curve is assumed to be 1.0, as is the case with a probability curve, then the total area under the curve to the left of the moveable vertical line of known standard deviations from the mean will give the probability of achieving that value at the vertical line or below.

The problem calls for the probability that the programme will be completed in 33 weeks or less. This probability is represented by the shaded area of the normal distribution curve shown in Fig. 9.16. The mean to the distribution is 30.5 weeks, and therefore the duration called for, 33 weeks, is above the mean. In order to make use of the normal distribution table (Table 9.4) it is necessary to state the difference between the mean and the variable value ($33.00 - 30.50 = 2.50$ weeks) in terms of standard deviations. This is carried out by *standardizing* the variable by dividing it by the standard deviation, giving $2.5/2.13 = 1.17$ standard deviations. Therefore 33 weeks is 1.17 standard deviations above the mean. A reading of 1.17 standard deviations gives an area of 0.121 (by interpolation). But this is the area above the variable (not hatched in Fig. 9.16). The probability of completing the programme in 33 weeks or less is therefore $(1 - 0.121) = 0.879$ or 87.9 per cent. The probability of taking longer than 33 weeks is therefore 0.121, or 12.1 per cent.

In analysing a network in order to determine its critical path it has been assumed that there will be only one such path through the network. However, if there is more than one then the path having the largest variance should be used in the calculations.

Summary

Once the logic of the network has been decided and the network to represent the sequence of activities has been drawn, attention can be concentrated on the addition of durations so as to schedule the activities on a time scale. It is this phase of network analysis that is dealt with in this chapter. In order to arrive at a point where it is known at what points in time each activity should take place, a forward and a backward pass need to be carried out. This establishes the earliest and latest times at which activities can start and finish. A means of determining which is the critical path through the network is described. Repetitive work can cause networks to become very complicated, with many activity arrows and complex calculations—the result is usually considerable expense in their processing. Ladder networks that will overcome some of these difficulties are utilized. Not every activity is a critical one. Many, if not most, activities have some leeway within the times that are established for carrying them out. Calculations for float of several different kinds are carried out. The final section of the chapter is devoted to a description and an example of PERT, using probabilistic durations for each arrow to take into account the uncertainty that inevitably arises in making estimations of activity durations, particularly when operations that have not previously been experienced are used.

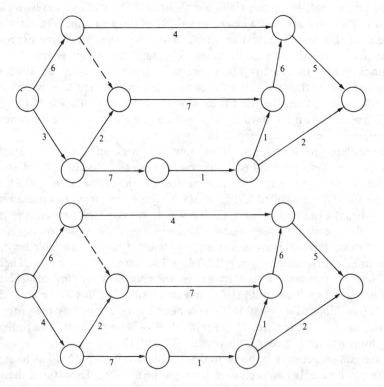

Figure 9.17

Problems

Problem 9.1 Carry out the forward and backward passes and then establish the critical paths for the networks in Fig. 9.17.

Problem 9.2 What would be the effect on the overall programme if, because of unforeseen circumstances, operations S_2 and S_4 only took a $\frac{1}{2}$-day each, operation S_3 took 5 days, operations J_1 and J_2 took 2 days each, and J_3 took 8 days in Fig. 9.6? All other activity durations remain the same. Compare the result with the ladder diagram of Fig. 9.7.

Problem 9.3 Prepare a table of independent, free, and total float for the upper network in Fig. 9.17, having used a forward serial numbering system for the events.

Problem 9.4 Draw the upper network of Fig. 9.17 as a time-scaled diagram and check the results obtained from it against the results obtained by the backward and forward passes.

Problem 9.5 The newly graduated son of the Chairman of Movem (Earthworks) Corporation Limited dangled a network diagram under the nose of Bullit, the long-serving, painstaking, and mathematically minded contracts manager.

'We can do this job in 36 weeks and here is the evidence to prove it,' he said as he laid the network (Fig. 9.18) and information concerning the activity durations (Table 9.5) before Bullit. Bullit scribbled a few numbers on the back of an envelope.

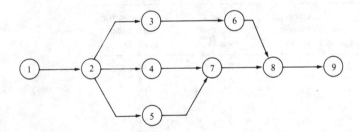

Figure 9.18

'I am confident, given these figures, that there is a 99:1 chance of completing the work in the duration that I have in mind,' said Bullit at last. What contract duration should Bullit have had in his mind?

Problem 9.6 The Nuclear Bombastic Missile Co. Ltd are striving to develop their Captain Scarlet rocket system before the end of the current financial year which is now 28 weeks away. If development can be completed before this target date then the company will benefit from a large bonus payment to be made by the sponsors of the research and development programme.

Table 9.5

i–j	Optimistic duration, weeks	Most likely duration, weeks	Pessimistic duration, weeks
1–2	2	7	8
2–3	5	6	9
2–4	3	7	8
2–5	4	5	6
3–6	6	9	10
4–7	6	8	11
5–7	4	6	7
6–8	5	7	10
7–8	4	8	9
8–9	4	7	9

The decision is taken by the chief engineer of the company to use PERT techniques in order to plan, schedule, and control the programme of development work yet to be completed. A network is drawn representing the 11 major activities yet to be carried out and 3 estimates of duration for each activity are established. Table 9.6 lists the activities and their estimated durations.

What is the probability that the programme will be completed by the end of the current financial year?

Table 9.6

i	j	Optimum duration, weeks	Most likely duration, weeks	Pessimistic duration, weeks
1	2	3	8	13
1	3	2	5	8
1	4	3	8	10
2	5	4	7	10
3	6	2	3	8
4	6	7	9	10
4	8	4	6	9
5	7	6	9	12
6	9	5	7	11
7	9	2	5	14
8	9	4	6	9

Further reading

Ahuja, H. N.: *Project Management: Techniques in Planning and Controlling Construction Projects*, Wiley, New York, 1984.

Antill, J. M. and R. W. Woodhead: *Critical Path Methods in Construction Practice*, 2nd edn, Wiley–Interscience, New York, 1970.

Harris, R. B.: *Precedence and Arrow Networking Techniques for Construction*, Wiley, New York, 1978.

Moder, J. J., C. R. Phillips and E. W. Davis: *Project Management with CPM, PERT and Precedence Diagramming*, 3rd edn, Van Nostrand Reinhold, New York, 1983.

O'Brien, J. J.: *CPM in Construction Management*, 3rd edn, McGraw-Hill, New York, 1984.

10. The allocation of resources

10.1 Introduction

So far, in the preceding chapters, network analysis has been considered using one resource only, that is, it has been a *time-only* network. The assumption implicit in the use of a time-only network is that all the resources required to carry out the job are available as and when required. Such an assumption implies, for example, that on a construction site not only will all the labour be available as required but that it will be available in each category of skill needed at any one specific time in the programme. It implies that the necessary plant will be available as required. A time-only network, for instance, may assume that three large mechanial excavators will be on the site for two or three weeks, only to send them away for three weeks and bring them back again for another short period subsequently. This is obviously uneconomic, and some consideration must be given to the high utilization of the available resources in planning a project.

On any project it is undesirable that there should be peaks of labour employment at a number of different points in the programme. Not only is it bad for the morale of the labour force that there should be sackings one week and additional operatives taken on a few weeks later, but the very action of getting rid of employees and re-engaging them is an expensive one. In addition, new workers are bound to be less efficient during the period in which they are becoming accustomed to their new surroundings, and their supervisors need time in order to assess their capabilities. Contracts offering security of employment for long periods have a distinct advantage when attracting labour which may be in short supply.

From the above points of view it is desirable that there should be an as even as possible demand for labour over the whole of the contract period, with a smooth increase in the numbers required at the beginning of the contract and a smooth tailing off at the end. Thus the first problem of resource allocation is that of smoothing or levelling the demand for labour within the specified project time. The smoothing or levelling that will be achieved will almost never be perfect, but a solution is sought as near the optimum as possible.

The nature of the other problem that can be tackled in resource allocation is that of optimizing the project duration, knowing that there are certain constraints as to the quantity of resources that will be available during the project. A very critical situation of this nature arises more often in the factory type of production, where casual labour is not employed to such a great extent as in construction work, and perhaps in a maintenance group there will be a more or less fixed number of key tradesmen. Such a situation might well occur, however, in construction work in, for example, an under-developed country where the key labour has to be imported from far afield. If, for example, 12 carpenters are brought in to do a particular job then the project must be scheduled on the basis that a maximum of 12 will be available, since it may take some weeks to obtain additional ones at the site if required.

There are many sophisticated resource allocation computer programs available and the use of a computer is the only rational way in which to tackle such a problem to any depth. Even with the use of a computer there must always be an element of human judgement, in the end, since a computer will never be able to smooth to the last degree or to work within resource restraints on every occasion, and always give an acceptable programme schedule. Even without computer programs, there are many methods of allocating resources with different degrees of sophistication, from the brief visual check to some form of systematic tabular approach to the problem.

10.2 Resource aggregation

Resource aggregation is simply a summation on a time-period basis, for example, day to day or week to week, of the resources that are used in order to carry out the programme. Figure 10.1 illustrates a simple bar chart to which has been added the total labour requirements on a weekly basis for each individual operation. Along the bottom of the bar chart the totals are added up, forming a resource aggregation. These totals may be plotted in graphic form as indicated below the bar chart. The variable demand for labour will be seen by the fluctuating level of the chart and, obviously, some attention needs to be given to the programme if this is to be smoothed or levelled satisfactorily. In comparing this bar chart with a network diagram, it becomes clear that the former is an attempt in precise terms to set the activities between fixed dates. This is not the case with a network diagram, where only the critical activities are tied to performance at fixed times and the remaining activities in the network have float and thus can be varied in time. It will be appreciated, therefore, that whereas the bar chart as it is shown lends itself to the formation of a unique resource aggregation graph, the network diagram has a certain amount of flexibility. For instance, if the bar for the 'Floors' in the bar chart, which may have float in the network, can be moved to weeks 5 and 6 then this is obviously going to affect the smoothness of the resource aggregation.

For a programme represented by a network diagram there will be a number of resource aggregation graphs, depending on where in time the activities with float are actually carried out. Nevertheless, having carried out a network analysis for a project and having arrived at an optimal duration in the view of limited resources or alternatively having settled on a fixed programme duration, the resources can then be aggregated to advantage. If a further visual inspection is made of the bar chart in

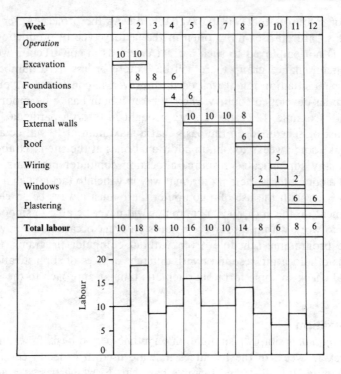

Week	1	2	3	4	5	6	7	8	9	10	11	12
Operation												
Excavation	10	10										
Foundations		8	8	6								
Floors				4	6							
External walls					10	10	10	8				
Roof								6	6			
Wiring										5		
Windows									2	1	2	
Plastering											6	6
Total labour	10	18	8	10	16	10	10	14	8	6	8	6

Figure 10.1

Fig. 10.1, then clearly, with the peak of labour as it is in the second week, it is desirable either to extend the excavation into week 3 and reduce the labour demand over the 3 weeks or to delay the commencement of the work on the foundations for a few days in order to reduce the immediate demand of labour for this work. Alternatively, it is possible to make a combination of both these adjustments in order to get rid of the peak of 18 operatives in week number 2. Similarly, in weeks 5 and 8 it is desirable to make some adjustment to the programme to smooth resource demand at these points. This is a simple programme and the smoothing and levelling can be carried out by eye and a manual calculation.

10.3 Priorities and sorts

Figure 10.2 shows a simple example network comprising 11 events and 15 activities. The network has been numbered in serial fashion from left to right and the duration in weeks of each activity has been placed against the arrows. The forward and backward passes have been carried out and the duration of the programme is established at 34 weeks. Against each arrow, in a box, is shown the resource requirement of each activity. This requirement is the minimum amount of resource required to do the work. For example, against activity 1–2 is shown 3X, which means

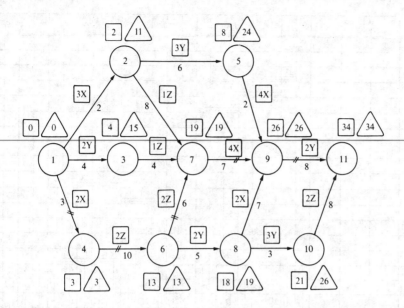

Figure 10.2

that 3 units of resource X are required throughout the duration of the activity 1–2. For simplicity X has been chosen to represent a particular type of tradesman. Of the three types of resource required, X, Y, and Z, it is assumed that not more than four, three, and two units respectively of the resource are available at any one time. It is required to establish a programme for carrying out the project represented by the network in such a way that demand for these resources will never be exceeded during the project duration.

First of all, the activities making up the network must be listed in order of their priority of resource allocation. There must therefore be some means of sorting the initial sequence. One thing that has been established by the network diagram is the logical sequence of activities, giving a clear indication of those activities which must precede others. The order in which the activities are listed must, if they are to be dealt with from the top of the list to the bottom, reflect the dependency of some activities on the preceding completion of others. The process of arranging the activities in a list to certain specified rules and conditions is known as a *sort*.

Figure 10.3 shows the activities of the diagram of Fig. 10.2 sorted in order of the activity *j*-numbers. Activity 1–2 therefore commences at the head of the list, followed by 1–3 and 1–4. It will be noted that this does not necessarily result in a list arranged with *i*-numbers ascending in numerical order.

Sorting in this fashion illustrates one of the reasons for using a serial numbering system. If the number at the head of an activity arrow is always greater than that of the tail, and the activities are sorted in the order of precedence as indicated on the left-hand side of Fig. 10.3, by ascending numerical order of the *j*-number, it means

i–j	Duration	Total float	X	Resource Y	Z	Loading (0 → 40 Weeks)
1–2	2	9	3	–	–	3X ②
1–3	4	11	–	2	–	2Y 2Y ③
1–4	3	0	2	–	–	2X ④
2–5	6	16	–	3	–	3Y 3Y 3Y ⑤
4–6	10	0	–	–	2	2Z 2Z 2Z 2Z 2Z 2Z ⑥
6–7	6	0	–	–	2	2Z 2Z 2Z 2Z ⑦
2–7	8	9	–	–	1	1Z 1Z … 1Z 1Z 1Z ⑦
3–7	4	11	–	–	1	1Z … 1Z 1Z ⑦
6–8	5	1	–	2	–	2Y 2Y 2Y ⑧
7–9	7	0	4	–	–	4X 4X 4X 4X ⑨
8–9	7	1	2	–	–	2X 2X 2X … 2X 9
5–9	2	16	4	–	–	4X
8–10	3	5	–	3	–	3Y 3Y ⑩
9–11	8	0	–	2	–	2Y 2Y 2Y 2Y ⑪
10–11	8	5	–	–	2	2Z 2Z 2Z 2Z
			X			3 3 2 2 2 – – – – – 4 4 – – – – – – – – 2 2 2 2 2 2 4 4 4 4 4 4 2 – – – – – – –
			Y			2 2 2 2 3 3 3 3 3 3 3 – – – – 2 2 2 2 2 3 3 3 – – – – – – – – – – – 2 2 2 2 2 2 2
			Z			1 1 2 2 2 2 2 2 2 2 2 2 2 2 2 2 2 2 2 2 2 1 1 2 2 2 2 2 2 2
			Total			5 5 5 5 7 5 5 5 5 5 6 6 2 2 2 4 4 4 4 7 7 7 4 3 3 6 6 6 6 6 6 6 4 2 2 2 2 2 2 2

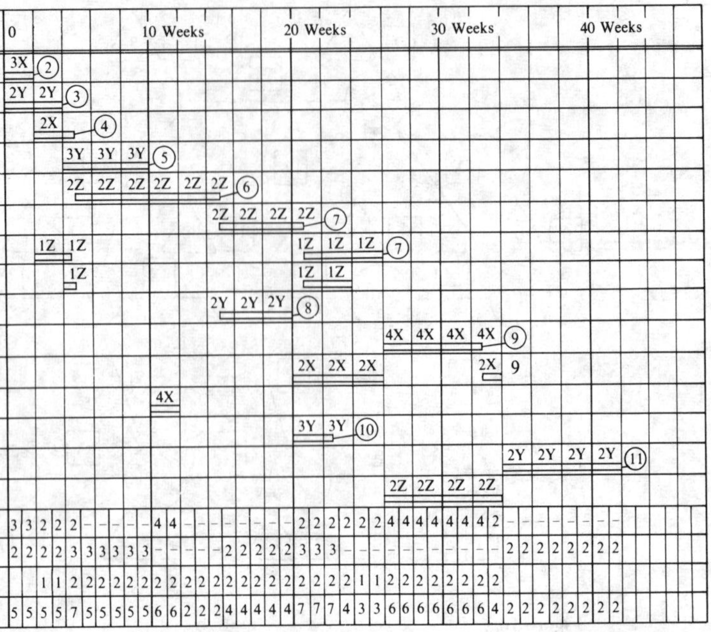

Figure 10.3

that the list will have the important property that those activities above any particular activity in the list will always be those of preceding activities. Conversely, those below the particular activity will always be the activities succeeding that chosen. Thus, in the case of activity 6–8, all those activities below it in the list can be carried on independently or in succession to activity 6–8. The sort of the table also ensures that all those activities, upon which the start of 6–8 depends, will appear in the list above activity 6–8.

If this sorting of activities is examined more closely, it becomes apparent that it is by no means a unique listing of the activities and depends to a large extent on the way in which the network diagram is numbered. For example, in Fig. 10.2 the logic of the diagram would not have been disturbed if, instead of placing a 4 where it is on the diagram, it had been placed where now appears event number 2 or 3. This would have necessarily meant that a different order of precedence of activities would have been established in the sort. Attention is drawn to this feature, since, if the resource allocation is a particularly critical or difficult one, it may be advisable to renumber the network and carry through a process of resource allocation again in order to see if a better solution can be obtained.

The sort so far has been carried out on the basis of an ascending numerical order of the *j*-number. Where an event is a merge event there will be more than one activity bearing the same *j*-number, as in Fig. 10.3 there are activities 2–7, 3–7, and 6–7. It is necessary, therefore, to have a secondary rule whereby an order of pecedence is given to activities that cannot be sorted conclusively using the first rule. The primary rule for sorting is known as the *major sort* and the secondary rule is known as the *minor sort*. In this example, the minor sort has been made in accordance with the magnitude of total float, an activity having a small total float taking precedence over one with a larger total float. If one is going to consider each activity in accordance with the order in which it appears in the sorted list, this means that priority of resource allocation is being given to an activity of low total float. The reason for this will be apparent.

Sorting can be carried out on the basis of many priorities related to the network diagram. The activities can be sorted in order of ascending numerical order of *i*-number or *j*-number and the minor sort can be in terms of duration, total float, free float, resource requirement, and many others.

The resource requirements for each activity are stated against the activities in Fig. 10.2 and the resources can then be allocated manually to each activity in turn within the restraints imposed. A bar chart can be drawn out as the allocation proceeds and a note made at the time of the allocation of each resource. In Fig. 10.3 the first activity to be dealt with is 1–2, with a duration of 2 weeks, and this is filled in as a bar 2 weeks long, showing a demand for resource 3X during each of those weeks. The next activity to be dealt with is 1–3, which has a duration of 4 weeks and uses a resource of 2Y for each of these weeks. It is useful at this stage, for quick identification subsequently, to make a note of the *j*-number at the end of each bar as it is drawn. In the two activities so far plotted, these are 2 and 3 respectively.

Since no activity can proceed unless its preceding event or its *i*-number event has been achieved, care must be taken in the allocation process to ensure that no bar is plotted before it may proceed in accordance with the logic of the original network diagram of Fig. 10.2. When activity 1–4 is considered, it too can start at zero date in accordance with the logic of the original network diagram of Fig. 10.2, but its demand of resource type X is for two units and three units have already been allocated for the first two weeks to activity 1–2. It is therefore necessary to delay activity 1–4, until the resource becomes available from the maximum number of 4. Activity 1–4 cannot be scheduled to commence until the beginning of week 3. This means that since it takes 3 weeks to complete that activity, 4–6 cannot start now until the beginning of week 6. Clearly this is a case where human judgement can be used to improve the allocation, because activity 1–4 is critical and it would have been better to deal with this before activity 1–2, which has 9 days' total float.

When activity 2–7 is considered, an important decision has to be made. This activity can follow on from activity 1–2, being the sole activity upon which it depends. One unit of resource Z is available at this time but owing to the prior allocation of resources to activities 4–6 and 6–7 to the full extent of their availability, activity 2–7 can proceed for only 3 weeks in the first instance. The decision has to be taken whether activity 2–7 is a *splittable* activity, in other words, whether its operation can be divided over two or more periods of time. In construction work there are many activities which cannot be treated in this way. For example, a large pour of concrete in a foundation, which is specified as a continuous pour and may be programmed in terms of hours on a detailed programme, is a non-splittable activity. Other jobs in construction, such as the spreading of topsoil on the verges of roadways, tend to become the sort of jobs which are used to fill in gaps in the times between periods of full employment of the labour elsewhere. Under these conditions they can be considered as splittable jobs. The chart in Fig. 10.3 has been prepared on the basis that, where necessary, activities can be split.

The process is carried out by working through the list of activities in the order of precedence, and a final date is now achieved, showing an overall programme of 42 weeks to carry out the work. This extension of the programme over the 34 weeks shown by the network is a direct result of having to work within the restraints on the number of available resources and the sorting of the activities in the list in accordance with the initial priority rules. In carrying out the resource allocation, a new total restraint has been put on the project which does not appear in a time-only network of the type shown in Fig. 10.2. The concept of float has been completely removed from the network and, under the conditions previously stated, each activity now has a clearly defined time within which it must be carried out in order that it does not produce conflicting resource demands. It is important to realize that activities no longer have float of any description, if one is to comply with the allocation of resources in a network.

The resources may now be aggregated as in Fig. 10.3 under the various categories in which they are listed.

10.4 Early start or late start?

Figure 10.4 is a similar diagram to Fig. 10.2, having similar activity durations and similar resource demands, with the exception that no distinction is made between any of the types of resource. They are all assumed to be units of one operative of a similar skill. Table 10.1 has been drawn up from the information contained in Fig. 10.4 and the activities have been listed by a major sort of early start time and a minor sort of total float. Early start is the earliest time by which an activity may be commenced. By comparing the column under early start with the figures in the square boxes against each event in Fig. 10.4, it will be appreciated that the early start information can readily be obtainable from the network diagram, after the backward and forward passes have been completed. Where two activities have equal early starts, such as 2–7 and 2–5, then total float is used as the secondary sort and priority is given to the activity of lower total float. Total float and the number of units of resource that are to be used are also included in Table 10.1.

On the basis of the early start data, a resource aggregation can now be made, and this is plotted in the upper graph of Fig. 10.5.

Table 10.2 now shows a listing of activities for the same network, but the major sort has been made by late start and the minor sort is again by total float. The diagram in the lower part of Fig. 10.5 now represents the resource aggregation based on late start and the precedence of activities listed in Table 10.2. It will be noted from the two diagrams that there is considerable variation in the demand for

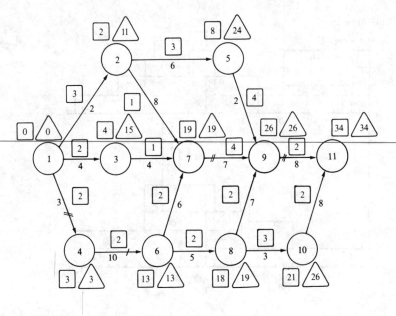

Figure 10.4

resources, depending upon whether the activities are sorted by late start or by early start. Neither of the resource aggregation curves gives an ideal or even a near-ideal situation. In the case of early start, there are two predominant peaks but the major demand for resource does not exceed 9 operatives, whereas in the diagram for the late start there is a single peak but it will be noted that the maximum demand rises to 16 in the extreme case.

Table 10.1

Activity	Duration	Early start	Total float	Resource units
1–4	3	0	0	2
1–2	2	0	9	3
1–3	4	0	11	2
2–7	8	2	9	1
2–5	6	2	16	3
4–6	10	3	0	2
3–7	4	4	11	1
5–9	2	8	16	4
6–7	6	13	0	2
6–8	5	13	1	2
8–9	7	18	1	2
8–10	3	18	5	3
7–9	7	19	0	4
10–11	8	21	5	2
9–11	8	26	0	2

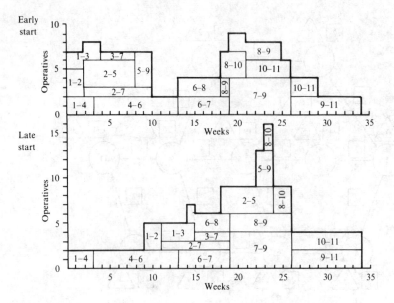

Figure 10.5

These two resource aggregation diagrams represent two extremes, inasmuch as the early start aggregation represents the requirements of resource for a situation in which every activity in the project commenced at its earliest possible date. If the logic and timing of the network diagram is not to be upset, any smoothing process must take place between the two extremes already plotted.

Table 10.2

Activity	Duration	Late start	Total float	Resource units
1–4	3	0	0	2
4–6	10	3	0	2
1–2	2	9	9	3
2–7	8	11	9	1
1–3	4	11	11	2
6–7	6	13	0	2
6–8	5	14	1	2
3–7	4	15	11	1
2–5	6	18	16	3
7–9	7	19	0	4
8–9	7	19	1	2
5–9	2	22	16	4
8–10	3	23	5	3
9–11	8	26	0	2
10–11	8	26	5	2

Figure 10.6 is a diagram that is a modification of both early start and late start resource aggregation curves. Part of the area is common to both curves, but the cross-hatched area is the excess of the early start over the late start resource aggregation, and the vertical hatching denotes the excess of the late start over the early start aggregation. Since the areas under the curves are representative of operative-weeks, the area of the cross-hatched portion must equal the sum of the areas of the vertical-lined portion since the durations and the resource requirements are the same in both cases. If smoothing is to be carried out within the overall duration of 34 weeks, it must be carried out in an area between the two extremes as plotted. In carrying out any smoothing or levelling operation, care must be taken not to upset the logic of the original network diagram. An inspection of the combined diagram shows that the peak brought about in weeks 22 and 23 is largely attributable to activities 5–9, 8–10, 8–9, 2–5, and 7–9. Since 7–9 is a critical activity, nothing can be done about it without affecting the end date of the programme. Activity 5–9, however, has a total float of 16 days and may therefore be moved in such a way that it helps to fill the gap between the two peaks. It will be seen that the manipulation of the diagrams becomes easier if the resource demands for the critical activities are kept to the lower part of the chart, since, unless the contract duration is to be extended, these cannot be adjusted.

The peak of the late start diagram can be removed by bringing forward activity 5–9 to commence at week 16 and also bringing forward activity 8–10 to its earliest start date. Neither of these changes affects the logic or the overall duration of the original network. Figure 10.6 shows the new positions of activities 5–9 and 8–10 by a broken outline and when considered in conjunction with the late start diagram results in a curve with but one major peak. This is just one solution to the problem, as there must be a great many solutions between the limits of early and late start. It is often desirable to tend towards the resource aggregation curve illustrated by early start, because this means that, if there are to be labour difficulties to be overcome, they will occur in the early part

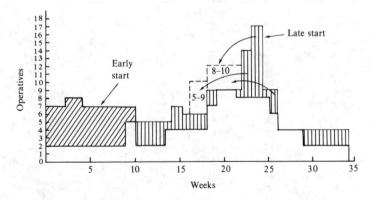

Figure 10.6

of the contract, leaving as long a time possible for the difficulties to be resolved. Leaving peaks of resource requirement until late in the contract does mean that, should something go wrong at this stage when there are heavy demands on the resource, unless they can be righted immediately there will be a permanent effect on the contract duration.

10.5 Allocation within resource restraints

The previous method of aggregating resources has been carried out within a fixed project duration, the prime object being to total the amount of resource used at any particular period of the project but to do this so as to maintain the given duration. The next method to be described is one for the allocation of resources within known limitations of availability. Such a method may result in a project duration which is longer than the minimum.

Use is made of Fig. 10.2 with three types of resource, X, Y, and Z. Two levels of resource restraint will be used—a normal level of availability at four units of X, three of Y, and two of Z, and a maximum level of resource availability. The latter will be of X, five units, of Y, five units, and of Z, three units. The activities are again sorted by late start as the major sort and by total float as the minor sort and the requirement of units and type of resource are shown at the right-hand side of Table 10.3. In addition it is necessary to compile a list of events showing the earliest date by which they will be achieved, and a column headed 'Scheduled date.' The list for this example is shown in Table 10.4.

The only scheduled date known at the beginning of the analysis is for Event 1. This is zero. No specific sorting method need be used on the event numbers, and they can be listed in ascending numerical order, since the only facility

Table 10.3

Activity	Duration	Late start	Total float	Resource	Units of resource
1–4	3	0	0	X	2
4–6	10	3	0	Z	2
1–2	2	9	9	X	3
2–7	8	11	9	Z	1
1–3	4	11	11	Y	2
6–7	6	13	0	Z	2
6–8	5	14	1	Y	2
3–7	4	15	11	Z	1
2–5	6	18	16	Y	3
7–9	7	19	0	X	4
8–9	7	19	1	X	2
5–9	2	22	16	X	4
8–10	3	23	5	Y	3
9–11	8	26	0	Y	2
10–11	8	26	5	Z	2

Table 10.4

Event number	Earliest date	Scheduled date
1	0	0
2	2	5
3	4	4
4	3	3
5	8	11
6	13	13
7	19	19
8	18	18
9	26	33
10	21	21
11	34	41

required of them is that they should be easily found at the time of calculation.

To facilitate the allocation of resources during the procedure it is necessary to compile a loading chart for each category of resource. Three loading charts are shown in Fig. 10.7. The lower of the thick horizontal lines shows the normal availability of resource, and the horizontal line immediately above this indicates the maximum possible availability of resource. For the purpose of this example, it will be assumed that no activity can be split for its operation but must be performed as a continuous operation.

The first on the list in Table 10.3 is activity 1–4, which requires 2 units of resource X for a duration of 3 weeks, commencing at day zero. This resource demand is then loaded into the chart, occupying an area of 2 units of resource by 3 weeks,

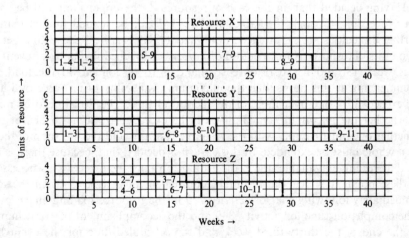

Figure 10.7

commencing at an earliest date of zero, the figure for the scheduled date on the event list. Since the resource requirement is less than the availability of the resource, this demand can be satisfied. If the activity is carried out at this point, event 4, as far as can be seen at the time of calculation, will be reached at the end of week 3. The scheduled date in the event list against the event number 4 should therefore be listed as 3.

The next activity on the list of Table 10.3 is 4–6. The earliest date for event 4, the *i*-event of activity 4–6, is 3 and the scheduled date is 3. Therefore, the activity demand for resource of 2 units of Z for a duration of 10 weeks is loaded into the resource loading chart of Z to cover this demand and the time required. There is no difficulty with loading this particular resource and since it does not overload the availability, the scheduled date for event 6 is noted as being 13 and added to the event list. It is important to remember in this systematic procedure that, having successfully loaded a demand for a resource in a loading chart, a note must be made of the time at which the end event will be achieved for this particular activity. This is necessary in order to maintain the logic of the diagram. It can be seen that even after loading activity 4–6, event 6 will not be achieved until at least the end of week 13. There may be other activities running into event 6 that have not yet been considered, but so far no activity with an *i*-number of 6 can commence before this scheduled date.

Activity 1–2 can be loaded with no difficulty. When activity 2–7 is considered, having a demand of one unit of resource Z for 8 days, and a scheduled date of 5, it will be seen that the demand for resource, when added to that for activity 4–6, exceeds the normal availability. It is necessary to make a decision at this point as to whether the normal availability can be exceeded or whether the project's duration will be extended to keep the resource demand within the normal. In this case it has been decided that the normal can be exceeded, so the resource demand for activity 2–7 is loaded and the scheduled date of the end of week 13 is recorded.

Having decided that, in the case of resource Z, the upper limit will be used, then on proceeding through the table, activity 3–7 can be loaded in this manner. No further difficulties are experienced until activity 8–9 is considered with a demand for two units of resource X for a period of 7 weeks. The scheduled date for event number 8 is week 18, and if the resource demand is to be loaded on the loading chart commencing at this point then the maximum availability of resource of 5 units will be exceeded by one over most of this period of time. The decision to be made now is whether the maximum availability can be exceeded by some means or other, or whether one must comply with the resource restraint and the duration of the project as a whole must be extended. In this case, it is decided that, because there is no more of that resource available, the project duration must be extended as necessary. The earliest point at which the resource requirement of activity 8–9 can be loaded is immediately following that of activity 7–9 in the resource loading chart for unit X. The completion time for activity 8–9 and the accomplishment of event number 9 is at the end of the thirty-third week and the scheduled date for the event list must therefore be 33.

The next activity on the list, 5–9 has an earliest date of 8 and a scheduled date of 11 and can readily be loaded into the loading chart for resource X, beginning at the scheduled date of 11. Activity 8–10 can be dealt with in normal circumstances, but activity 9–11 will take the completion of the programme beyond the duration already calculated at 34 weeks. The new duration for the project is 41 weeks. Activity 10–11, with its demand for 2 units of resource Z, can readily be fitted into the resource loading chart for resource Z, beginning at the earliest date as shown on the event list.

The effect of the restraint on the quantity of resource available is again shown, inasmuch as a maximum number of 5 units of resource X is insufficient to cope with the peak demand for that resource. It is necessary to lengthen the duration of the project unless an additional unit of resource X can be found over a period of 6 weeks during the course of the project. It is as well to remember at this point that this is just one solution of the resource allocation problem and, by considering now a table ranked in order of early start as the major sort and a minor sort of total float with the resource requirements as indicated in Table 10.5, a different solution is obtained.

The allocation procedure is carried out as before and results in resource loading charts of Fig. 10.8, together with an event list as Table 10.6. By using this alternative method of sorting, the overall duration of the project is reduced by one week and, among other things, now dictates a slightly different sequence in which the activities must be carried out. This latter variation may be desirable in certain circumstances since, if there is a choice as to when a particularly critical operation can be carried out, it should be done at the earliest possible moment on the premise that has already been stated.

Table 10.5

Activity	Duration	Early start	Total float	Resource	Units of resource
1–4	3	0	0	X	2
1–2	2	0	9	X	3
1–3	4	0	11	Y	2
2–7	8	2	9	Z	1
2–5	6	2	16	Y	3
4–6	10	3	0	Z	2
3–7	4	4	11	Z	1
5–9	2	8	16	X	4
6–7	6	13	0	Z	2
6–8	5	13	1	Y	2
8–9	7	18	1	X	2
8–10	3	18	5	Y	3
7–9	7	19	0	X	4
10–11	8	21	5	Z	2
9–11	8	26	0	Y	2

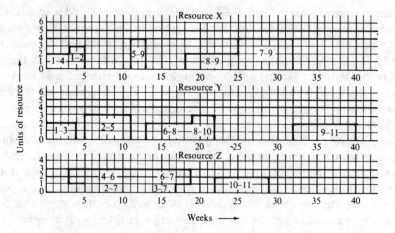

Figure 10.8

The above is an example of the *serial method* of the allocation of resources. In such a method, the general principle is that the activities are sorted into a list so that any activity in the list has those preceding it above it and those succeeding it below it. The principle of allocation is such that one activity is dealt with at a time, working in sequence down the list. As the activity is considered an allocation of resource is made to it, and so on until each activity in the list has received its respective allocation. From the method of working it will be seen that the method can easily be used for more than one project using common resources simply by listing all the activities properly sorted.

There is a further general approach to the problem of resource allocation and it is termed the *parallel method* of allocation. Here, each period of time is considered in turn rather than each activity. The available resources in any one time period are allocated on the basis of some criteria that set up priorities. For example, it may be

Table 10.6

Event number	Latest date	Scheduled date
1	0	3
2	11	5
3	15	4
4	3	3
5	24	11
6	13	13
7	19	19
8	19	18
9	26	32
10	36	21
11	34	40

that if two or three or more activities compete for the available resources then the resource will be allocated to that activity with the least total float, the total float being used as a measure of the criticality of any activity. Having allocated, in the parallel method, the resources available for a particular time period, allocation moves forward to the next time period and the activities that perhaps have not previously received an allocation of resource are then considered. Because consideration is made on one time period later in the programme, the criticality of those activities, which have already been considered but rejected in the light of other more critical competition for the resources, has increased. For example, if an activity cannot obtain an allocation of resource that has a total float of 10 weeks at one particular time period, then, in considering the next time period of 1 week, the total float of the activity brought forward will be 9 weeks and so on, until it is in such a position that its total float enables it to obtain priority over all other activities that are brought forward.

In such a system a situation may arise where three critical activities are competing for the resources that are sufficient only to satisfy two of them. In this case one of the critical activities will have to be put back for consideration in the next time period and inevitably, since the activity is critical, the overall duration of the project will be extended.

10.6 Judgement and resource allocation

As has been shown in the above examples, it is necessary to adopt some sort of priority rules in the allocation of resources, in order to tackle the resource allocation problem in a systematic manner and, at the same time, to give priority to activities on a rational basis, especially when there is competition for limited resource. It has been illustrated that the choice of priority will give different results for similar examples, and therefore the decision as to which priority rules will be used must be one of judgement based upon experience.

Not only is the choice or priority rule a matter of judgement but, having levelled and allocated the resources for any project in a systematic manner, it is then a question of judgement as to how the results of the process can be moulded to suit the situation in hand. No systematic process of resource allocation, whether carried out manually or by the use of computer, will necessarily give the optimum solution for the levelling of resources. It is a combination of the judgement of the planner with the experienced and intelligent use of a computer that can combine to give the best solution.

The above examples have been used to illustrate some of the principles of resource allocation and have involved the use of manual methods only. When problems are only slightly more complicated than those illustrated, the manual allocation of resources becomes out of the question. It is a tedious and complicated process that will lead to mistakes on the part of the allocator. It is in this field that the computer comes into its own; use of its facility to calculate quickly enables not only a quick review of allocation in accordance with a specific priority rule, but also, if satisfactory results are not obtained, a look at the allocation in the light of several different priority rules.

10.7 The advantages and disadvantages of critical path methods

One of the prime advantages that critical path methods have had, particularly in the construction field or the field of the one-off project, is that for the first time there has had to be some logical thinking about the sequence of events that must take place on a site. The bar chart of the Gantt chart does not require the discipline of thought that is required by the construction of a network with its interdependence of activities. It could well be that this particular advantage is the most important of them all. On the other hand, it must not be thought that someone who implements the use of critical path techniques has the solution to all their problems, since the critical path method can be used only as a tool to assist in making decisions—it will not make decisions by itself.

Critical path methods not only set up a discipline in the planning and scheduling of a project but they also enable the control of a project to be carried out within strictly defined limits. In addition, they encourage the preparation of long-term plans for projects where the use of traditional planning methods have been somewhat woolly in their application. The logic of the network cannot be neglected in the situation where it is desirable to reduce a project duration. For example, if it is required to reduce the duration of a project by 4 weeks, this can be done only by considering the critical activities first, followed by the subcritical activities. If the logic of the diagram is accepted as being correct then the saving in overall time must come about by a saving in individual activities. It is not possible to overlap activities in the network. With the bar chart such a discipline is not apparent, since the only thing that has to be done in order to plan a job and to make a bar chart conform to a reduced project duration is to squeeze up the bars and make them overlap a little more. This is a easy enough operation and does not require any consideration of the interdependence of one activity with another, so far as the drawing and presentation of the bar chart are concerned.

Management by exception can be practised when using network analysis as a control tool. The use enables the attention to be focused on a few critical activities that are likely to have the most effect upon the overall duration of the project. The remaining activities, so long as they all carry float, can be ignored by the top manager looking at the general picture. It is only when such activities tend to become critical, or near critical, that attention must be concentrated on them.

One of the original reasons for developing network analysis was in order to harness the speed and accuracy of the computer. Certainly, this is one of the advantages of using such methods on a large scale, since revisions to the plan and periodic updating can be carried out by computer in very little time.

The attention to planning in construction is usually placed on the contractor's part in the scheme of operation. With the use of network analysis techniques, there is no reason why the architect and/or the engineer, and even the client, should not allow the operations coming within their control to be scheduled on the network diagram together with those of the contractor. Certain milestones should be agreed between the client and the contractor, such as the issue of key drawings, and these should be recorded within the network so that the contractor and the client both have a clear understanding of what is required of the other. The network diagram, therefore, becomes a clear and certain means of communication and coordination between all the parties bound together by a contract to carry out work. The subsequent use of the network diagram as a basis for contract sum adjustments by either the contractor or the client is a fair and equitable one. If the contractor's work has been held up by the non-issue of certain essential drawings and additional expense has been incurred by the delay then it is only right and proper that the contractor should be recompensed. The use of a network diagram can show conclusively what the effect is of non-delivery of items such as drawings on the whole of the project duration. The responsibility for delays of this kind can be recorded and pinpointed with great clarity.

The very process of drawing a network diagram can itself be a profitable procedure for the site manager. This person is not required, and should not be required, to keep the network diagram up to date as the work proceeds, but if they draw a simple network for the whole of the project before it commences then they will have in mind a general idea of the sequence of events and the relationship of one activity with another.

There are disadvantages of using such techniques, but these are similar to the disadvantages that occur when any new technique is used. There is scope in such a situation for empires to be built and a lot of paper to be produced that will be ineffective and, in fact, will have a negative effect on the progress of events. There is a tendency for such methods to be looked upon as the panacea for all the ills that have gone before; but let us repeat—this is a tool to be used in combination with personal judgement and it will not in itself resolve difficulties or make decisions.

The use of critical path methods, generally speaking, is a more expensive process of planning than the use of conventional methods. It is not easy to evaluate the benefits of its use in terms of money. There is the assurance however that, if a diagram has been completed, the project has been planned. There is no similar assurance from using traditional methods of planning. This aspect of critical path planning may well have its effect through a planning organization, because it will surely expose the man who has ostensibly been planning for many years but, in fact, has been quickly and lightly sketching the bars of a chart to arrive at a likely figure for the overall duration of the project.

Many people who have used critical path methods find it to have definite advantages over other types of planning. It is necessary, however, in using such methods to train all the personnel who will come in contact with the diagram as a means of communication and control. With more sophisticated methods, extensions, and derivatives of the basic critical path method, more and more use will be made of such techniques in the construction industry.

Summary

This chapter addresses the problem of aggregating and smoothing resources with regard to a network diagram. The two problems of resource allocation are either to fix the amount of a resource that is available and to determine the best schedule for the work within this constraint or to fix the project duration and then determine the optimum arrangement of resources to meet it. Both problems are addressed within the established premise that there is no analytical method available to meet exactly the optimization of resource allocation. A discussion is included of the advantages and disadvantages of using critical path methods for the planning and scheduling of construction projects.

Problems

Problem 10.1   What is the effect on the project duration if the absolute maximum available number of units of resource Z in Fig. 10.7 is two?

Problem 10.2   If the network in Fig. 10.4 represents a small project in which a road roller is required continuously for activities 1–4, 5–9, 2–7, 8–10, and 9–11, is it possible to schedule the programme for 34 weeks and use only one roller?

Problem 10.3   Draw the loading charts for the network shown in Fig. 10.4, assuming an upper limit on the resource availability of 8 operatives together with one roller on the activities listed in Problem 10.2.

Problem 10.4   What are the maximum resource requirements at any one time for the network in Fig. 10.2 if the duration of the project must not exceed 34 days?

Problem 10.5   The bar chart in Fig. 10.9 represents the programme for the construction of a small garage together with the necessary approach drive. The garage has brick walls and a timber roof. Each bar in the chart is shown in the

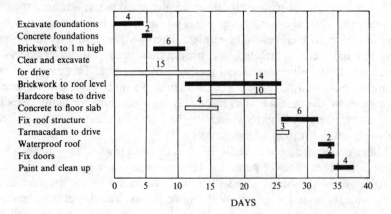

Figure 10.9

earliest position that the operation can take up and the solid lines represent activities which lie on the critical path of the programme. Of the non-critical activities, 'Concrete to floor slab' depends only on the completion of 'Brickwork to 1 m high', 'Tarmacadam to drive' depends on the completion of both 'Concrete to floor slab' and 'Hardcore base to drive'. Of the critical activities, 'Fix doors' depends on both 'Fix roof of structure' and 'Concrete to floor slab'. 'Clear and excavate for drive' must precede 'Hardcore base to drive'. 'Tarmacadam to drive' must be completed before 'Paint and clean up'.

Draw the network diagram, with events serially numbered, which represents the above programme of work. Using the durations, in days, above each bar, establish the overall duration for the work together with the float that is available for non-critical activities. If two operatives are required in order to carry out each of the activities in the stated durations, and four operatives (each capable of carrying out all varieties of work) are available, establish whether the programme is still feasible. What will be the minimum overall duration of the work using these resources only and assuming that each activity once started must be continued until it is completed? Illustrate your answers to those questions concerning resources by providing a self-explanatory, time-scaled network diagram together with a resource aggregation in the form of an histogram.

Problem 10.6   A bridge pier foundation is to be constructed in a cofferdam. The work is programmed by the use of the critical path method and the network is illustrated in Fig. 10.10. Each activity duration is initially estimated on the basis that unlimited resources are available to carry out the work and these durations (in working weeks) are shown against the appropriate arrow in Fig. 10.10. On further examination of the programme it is found that each activity in the network necessitates the continuous use of a derrick crane. Because of space restrictions, however, only two such cranes can be made available for the work. What is the minimum overall duration of the work, assuming that once an activity is started it must be completed without a break?

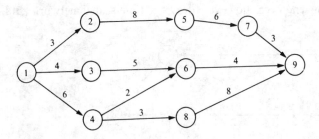

Figure 10.10

Problem 10.7   Each activity in the network diagram shown in Fig. 10.11 requires the continuous use of a mechanical excavator throughout its duration. The

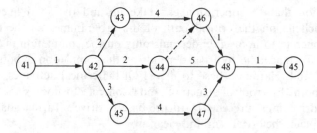

Figure 10.11

The estimated duration (in weeks) for each activity is shown against the appropriate arrow.

Establish the overall contract duration for the work represented by the network diagram if it is known that three excavators of the type required will be available throughout the work. What will be the effect on the contract duration if no more than two excavators can be made available for the work, and if it is assumed that having started an activity it must be completed without a break?

Problem 10.8 The Popdiggers Quarrying Co. Ltd are desperate to have the preliminary earthworks to their new quarrying venture completed in a period of not more than 12 weeks. All of their own excavators are in use and a decision is taken by their managing director to employ Hurryon Excavation and Mining Company to carry out the work on their behalf. Hurryon pride themselves in their use of sophisticated planning techniques and quickly draw up a network diagram (reproduced in Fig. 10.12), taking account of the restraints to the free planning of the project. Hurryon estimate the durations and the number of similar machines that will be required for each activity in order to achieve these target durations (detailed in Table 10.7). The managing director of Hurryon is deeply concerned by the knowledge that they have only four suitable machines available. The firm's chief planning engineer, however, draws a time-scaled network and a resource

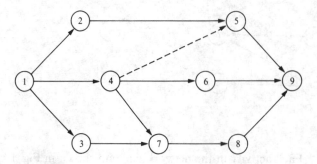

Figure 10.12

Table 10.7

Activity	Duration, weeks	Number of machines
1–2	4	1
1–3	5	2
1–4	4	1
2–5	2	2
3–7	1	1
4–6	5	1
4–7	4	1
5–9	4	1
6–9	3	1
7–8	3	2
8–9	1	3

aggregation diagram, both based on the network diagram, and convinces the managing director that the work can be carried out and to time. Hurryon take the contract for completion in a duration of 12 weeks.

Reproduce the final time-scaled network and resource aggregation diagram that convinced Hurryon's managing director that four machines were sufficient to carry out the work satisfactorily.

Further reading

Antill, J. M. and R. W. Woodhead: *Critical Path Methods in Construction Practice*, 2nd edn, Wiley-Interscience, New York, 1970.

Harris, R. B.: *Precedence and Arrow Networking Techniques for Construction*, Wiley, New York, 1978.

Moder, J. J., C. R. Phillips and E. W. Davis: *Project Management with CPM, PERT and Precedence Diagramming*, 3rd edn, Van Nostrand Reinhold, New York, 1983.

O'Brien, J. J.: *CPM in Construction Management*, 3rd edn, McGraw-Hill, New York, 1984.

11. Planning—precedence diagrams and line of balance

11.1 Activity-on-the-node and precedence diagrams

The emphasis in the two previous chapters has been placed on activity on the arrow networks wherein the consumption of resources—mainly time up until this point—has been described by activity designations on the arrows. In order to reflect the logic of a network, the arrows (including dummies) have been linked together by nodes, usually drawn as circles. The nodes represent points in time, or milestones, in the programme.

However, other general arrangements for networks have been developed, the most prominent of which is that in which the activities are represented by nodes and these nodes are then connected together by arrows (often without an arrow head) forming links between the nodes to represent the network logic. One distinct advantage of such a scheme is that dummy arrows are not required to maintain explicit logic. The networks are thus easier to process. John Fondahl at Stanford University developed the original activity-on-the-node scheme and referred to it as the 'circle and connecting node' network. A simple activity-on-the-node network is illustrated in Figure 11.1. This method has formed the basis of a further development from activity-on-the-node diagrams which have become increasingly popular in construction. It is known as *precedence planning* and, in using it, activities are represented by rectangles rather than arrows. The rectangles are then linked together to show the sequence and precedence of the activities and a *precedence diagram* is the result. Precedence diagrams have been found to have certain advantages over activity-on-the-arrow networks. Not only need dummies not be used, but time delays can be introduced into precedence diagrams without resort to the detail that may be required to represent the same logic in the alternative. It is claimed that precedence diagrams can be constructed more quickly and more easily and that they do not have the incidence of logic errors that can occur with arrow-on-the-activity networks.

At one time it was claimed that the huge investment in computer programs for the activity-on-the-arrow logic would prove a stumbling block to the rapid introduction

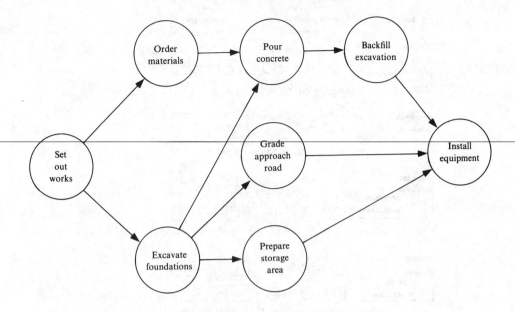

Figure 11.1

of precedence diagramming. However, this has not been the case and many major computer control system packages have been developed based on precedence diagramming. As a rule the printout from use of these packages can be in either notation.

One of the disadvantages of a precedence diagram is that the direct link between an arrow network and its related bar chart is apparently lost, as is the facility for readily drawing the activities against a time scale.

In Fig. 11.2 some of the commonly occurring conventions are compared for bar charts, activity-on-the-arrow networks, and precedence diagrams. The planning example used in Section 8.6 and illustrated using an activity-on-the-arrow network in Fig. 8.21, is now shown in Fig. 11.3 as a precedence diagram. An additional activity of 'Clear site' has been added to complete the diagram. *Burst* and *merge* activities also occur in precedence diagramming much as do burst and merge events in CPM. A burst activity is one that has two or more subsequent activities dependent on it. A merge activity is one that depends upon two or more previous activities. In Fig. 11.3 'Clear site' and 'South apron' are examples of a merge activity and a burst activity respectively.

The activities shown in Fig. 11.3 are all such that, where relevant, they cannot be started until the previous dependent activities have been wholly completed and, in themselves, they have to be wholly completed before passing to the subsequent dependent activities. As such, the dependency arrows in the diagram have all conventionally emanated from the extreme right-hand side of an activity box and have then entered the extreme left-hand side of a subsequent activity box. It is

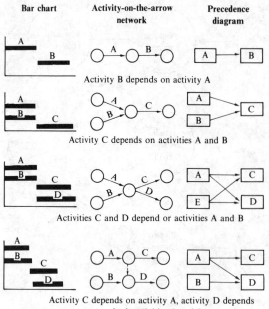

Figure 11.2

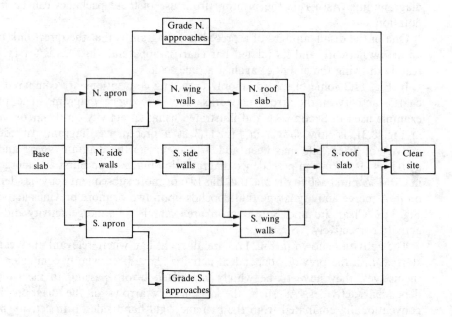

Figure 11.3

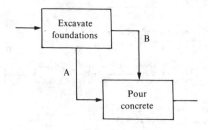

Figure 11.4

possible in precedence diagramming to represent overlapping activities as in Fig. 11.4. If the activity 'Excavate foundations' is extensive, it will probably be possible to 'Pour concrete' before its completion. The positioning of the start of dependency arrow A indicates that 'Excavate foundations' is not complete before 'Pour concrete' begins. The positioning of the dependency arrow B indicates that 'Excavate foundations' must finish before 'Pour concrete' can be completed.

11.2 Time and precedence networks

So far precedence diagramming has been explained in terms of planning the sequence and precedence of work only. Time can now be taken into account. Unlike in activity-on-the-arrow networks, the subscript to a dependency arrow does not refer to an activity duration. It refers to a delay that may occur between either the start or finish of one activity and the appropriate point in another activity in sequence. The activity duration in units of time to suit the subject of the diagram is usually shown in the bottom right-hand corner of each activity box.

Figure 11.5 illustrates some of the connections in precedence diagramming which are time-dependent and compares them with comparable situations shown in bar-charting and activity-on-the-arrow networks. Note that not all situations in precedence diagramming have an exact equivalent in activity-on-the-arrow networks and that the strict logic of networks is not always present in the bar chart equivalent.

In Fig. 11.5, the first relationship in the diagram is simply concerned with notation. The second concerns a simple serial relationship between two activities, the only one allowed in the activity-on-the-arrow method, wherein the second activity starts d units of time after the first one finishes. If $d=0$, the start of activity B can proceed immediately activity A is finished. This relationship is described as a *finish-to-start* (FS) constraint and the *lag* time can be placed on the link connecting the finish of A with the start of B or as $FS=d$ on the link between the two activities. This relationship is quite common in construction. An example is that of pouring concrete into formwork and then having to wait for the curing or lag time before the formwork can be removed. If the lag time is, for example, 2 days, d equals 2.

The next relationship is shown in two forms in the third and fourth diagrams of Fig. 11.5. This is the *start to start* (SS) constraint in which two or more activities

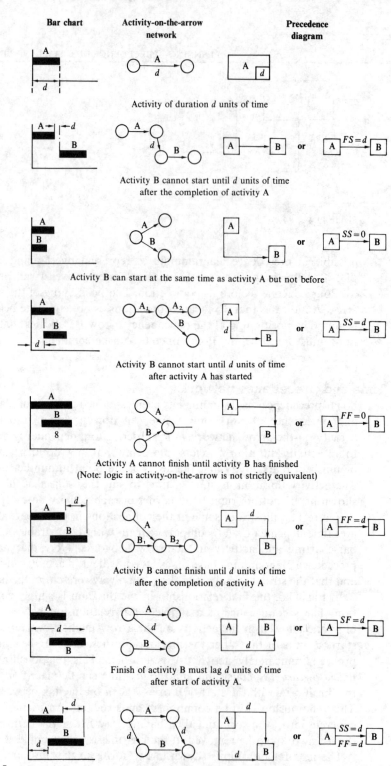

Activity of duration d units of time

Activity B cannot start until d units of time
after the completion of activity A

Activity B can start at the same time as activity A but not before

Activity B cannot start until d units of time
after activity A has started

Activity A cannot finish until activity B has finished
(Note: logic in activity-on-the-arrow is not strictly equivalent)

Activity B cannot finish until d units of time
after the completion of activity A

Finish of activity B must lag d units of time
after start of activity A.

Figure 11.5

Start of activity B must lag d units of time
after start of activity A; finish of activity
B must lag d units of time after finish
of activity A

can start at the same time or activity B can start d units of time after the start of activity A. Again, as in the case of all of these relationships, there are two alternative forms of logic diagram. The connecting link arrow can be drawn so that it starts and finishes at the appropriate position for the relationship, or the relationship and its numerical value can be shown on the link between the two activities. In general, the second option will be adopted here because it is often easier to draw the network diagram without having to position the connecting links accurately—particularly when drawing draft networks. This relationship, like the previous one, is commonly used in construction. In excavating a trench, for example, the bottoming-up process, or the timbering, cannot start until the excavation has proceeded for some way ahead of the other processes.

The *finish to finish* (*FF*) constraint is complementary to the previous one insomuch as the end of the follow-up activities must always lag behind the completion of the initial activity.

Of the last two relationships the penultimate, the *start to finish* (*SF*) constraint, does not have frequent use. However, it might be used in a situation where activity A represents the design and manufacture of some complex formwork for, say, a bifurcated cooling water duct where it entered a pumphouse at constantly changing angles and cross-sections. The major and critical part of the formwork will become available for fixing after 25 days of a total manufacturing period of 35 days. After the formwork is delivered to the site of work for setting and fixing into position, the fixing of reinforcement and other formwork, activity B, is estimated to take 12 days. This would mean that activity B, with a total duration of 17 days, would be completed $(25+12)=37$ days after the start of activity A. Therefore, $SF = 37$ days.

The last relationship is a composite one, which represents the ladder of the CPM system. It is applied to a situation wherein a series of successive activities are staggered by the lag times in the way described in Section 9.5.

It is clear from Fig. 11.5 that the relationships in precedence diagrams allow the overlapping of activities much more readily than with activity-on-the-arrow networks. In activity-on-the-arrow networks, it is frequently necessary to break activities into several parts in order to obtain true overlapping logic.

The analysis of time in precedence diagrams is carried out in a similar fashion to that in CPM networks using a backward and forward pass. The scheduling computations are facilitated by using a suitably designed node diagram such as that shown in Fig. 11.6. To keep the diagram simple for demonstration purposes, the node shown in this figure includes total float only, though other forms of float may be added if required. The activity being on the node does mean, however, that earliest and late start times and earliest and late finish times need to be included in the diagram. The precedence diagram shown in Fig. 11.7 is in overall sequential logic identical to that of Fig. 11.4, the culvert example. However, in order to demonstrate some precedence relationships not readily available to the CPM method, several have been introduced where suitable. These are marked on the precedence diagram.

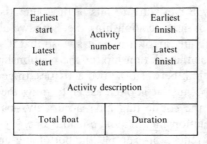

Figure 11.6

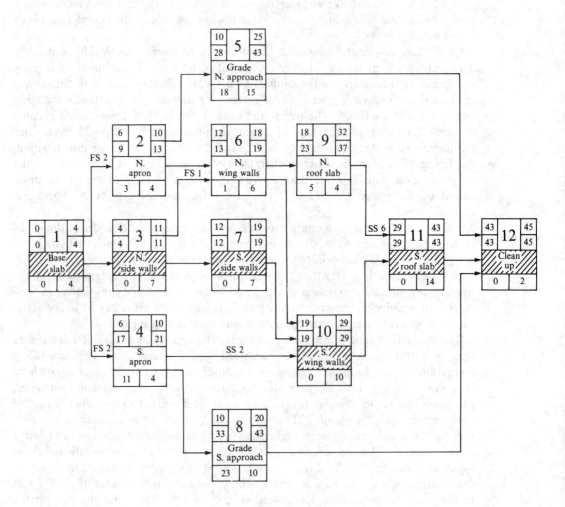

Figure 11.7

The forward pass can start at activity 1 and the earliest start time will be at zero time, giving an earliest finish of $0+4=4$ days. In moving to nodes 2 and 4, *finish-to-start* constraints are encountered, both of 2 days' duration. This means that nodes 2 and 4 must have an earliest start of 2 days after the earlist finish of the preceding activity. In both cases the earliest start becomes $4+2=6$ and the earliest finish therefore becomes $6+4$ in both cases. The forward pass can continue in exactly the same fashion as with CPM, but a difference will occur with the *start-to-start* relationships between nodes 4 and 10, and 9 and 11. In the first case this means that the earliest start of activity 10 (in respect of path 4–10 only) will be $6+2=8$ days. However, other constraints merging into node 10 affect the situation. In the case of node 7, the earliest finish is 19, and in the case of node 6 it is 18. The choice becomes the largest value of 8, 18, and 19. Turning to node 11, the earliest start through path 9–11 will be $18+6=24$ and through 10–11 it will be 29. 29 therefore becomes the earliest start for node 11.

It is often useful to resort to the bar chart representation of the precedence relationship when carrying out the forward and backward passes because it facilitates the calculation. In the case of path 9–11, the bar chart representation is as Fig. 11.8. Where the node is a merge node, other paths into the node must not, however, be forgotten.

The backward pass can now be carried out as previously and the total floats can be calculated from

$$\text{latest finish time} - \text{earliest start time} - \text{activity duration}$$

The critical path for the diagram in Fig. 11.7 is indicated by light shading of the relevant node descriptions.

The use of precedence diagrams greatly simplifies the construction of a ladder over that of an activity-on-the-arrow network. The ladder diagram illustrated in Figs 9.6 and 9.7 is reproduced as a precedence diagram in Fig. 11.9. This precedence diagram incorporates two of the connections of Fig. 11.5—a lead and a lag arrow. In making the forward pass to establish the earliest start time, it will be appreciated that activity 'Erect formwork' cannot start before the lead delay of 2 units of time has occurred. Likewise, 'Concreting' will have an earliest start time after 6 units of time have elapsed. Clearly the earliest time at which 'Erect formwork' can be finished is $2+16=18$ units and therefore 'Concreting' cannot finish before $18+1(\text{lag})=19$ units or $6+4=10$, whichever is the longer. For completion by 19 (the one selected), the

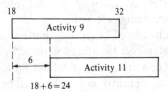

Figure 11.8

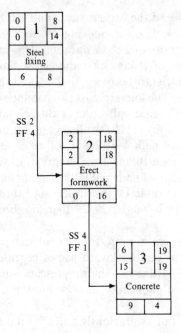

Figure 11.9

latest start time for 'Concreting' must be $19-4=15$. Similarly, for 'Erect form-work' the lastest start time must be $19-1$(for the lag)-16(for the duration)$=2$. The latest start time for 'Steel-fixing' must be $19-1$(for one lag)-4(for the other lag)-8(for its duration)$=6$. 'Erect formwork' with zero float will be on the critical path.

11.3 Line-of-balance methods

The origins of the line-of-balance method of programming, as applied to construction, are not entirely clear. The original method, used in production engineering, was developed during the late 1950s. As a planning and scheduling technique, line of balance has generally remained in the shadow of other methods. This is possibly because of its prime application to situations in which there are predominantly repetitive elements—situations that are often overlooked or not identified or recognized clearly when planning methods are being selected.

Typical construction applications for which line of balance is particularly suited include multi-house development projects, high-rise buildings with repetitive floor construction, tunnels, roadway pavement construction, pipelines, and bridges. For many of these applications, the logic of detailed networks can be complex both to draw and to maintain if they are to include the details of all the repetitive units involved. One of the consequences is that networks can be expensive to use in these circumstances. By expressing the basic construction programming logic of one repetitive unit in the form of a network known as a *production diagram*, and then using this diagram in the line-of-balance method as a basis for producing target and progress charts, an integrated technique is available. This technique is somewhat simpler than network analysis methods in respect of calculation, evaluation, and presentation. As a direct result, it is more efficient, meaningful, and economic in use.

Initially, the line-of-balance method requires the establishment of a delivery schedule for complete units of repetitive work, for example, houses from a batch of like houses. This delivery schedule will normally be expressed as a *rate of delivery* or a *handover* rate for the finished product and is plotted as the planned number of units produced against time. It is known as an *objective diagram*. For the purposes of programming, delivery will normally be at a constant rate (though not necessarily so), such as two per week. The graph of quantity against time will be straight and will obey the following linear relationship:

$$Q = mt + C \qquad\qquad (11.1)$$

where Q = line-of-balance quantity of units produced
m = required rate of delivery (number per unit time)
t = time
C = a constant—the value of Q at the intercept of the graph on the y-axis

Such a curve is illustrated in Fig. 11.10.

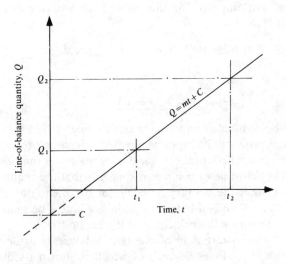

Figure 11.10

If the line of balance quantity completed, Q_1, is required at time t_1 then it will be equal to

$$Q_1 = mt_1 + C \qquad (11.2)$$

The constant C, for a given curve, may be eliminated by taking two points on the curve:

$$Q_1 = mt_1 + C \text{ (a)}$$

and

$$Q_2 = mt_2 + C \text{ (b)}$$

Subtracting (a) from (b) gives

$$Q_2 - Q_1 = m(t_2 - t_1)$$

from which

$$Q_2 = m(t_2 - t_1) + Q_1 \qquad (11.3)$$

or

$$t_2 = \frac{(Q_2 - Q_1)}{m} + t_1 \qquad (11.4)$$

EXAMPLE 11.1

A project for the construction of 120 similar houses is programmed to have a handover rate of completed houses of 1 per 4 working days. The fifth house is to be completed at the end of working day 25; after how many working days will the fortieth unit be completed?

$$Q_1 = 5 \qquad t_1 = 25 \qquad Q_2 = 40 \qquad m = \tfrac{1}{4} \text{ per day}$$

Therefore,

$$t_2 = (40 - 5)4 + 25 = \underline{165}$$

The fortieth unit will be completed on working day number 165.

From the objective and production diagrams, a schedule can be prepared for the delivery, manufacture, or construction of the various components that go to make up one complete unit. The schedule of completed elements that are required at any point in time, in order to meet the delivery schedule for completion as planned, is known as the *line of balance*. A line-of-balance schedule will now be built up using a number of examples to illustrate the principles of the method.

Let it be assumed that the construction of one unit, a house, is made up of the activities shown in Table 11.1, alongside each of which is shown its duration in weeks.

Table 11.1

Operation	Weeks
Foundations	2
Brickwork	4
Roof	1
Window frames, etc.	1
Plumbing	3
Plaster	1
Joinery	1
Electrics	1
Decorate	2
Clean out	1

In the line-of-balance method, the target emphasis is to maintain a predetermined delivery schedule as opposed to the emphasis being on the appropriate start of activities. The time analysis of the activities for one unit of housing is, therefore, made with relation to their completion dates in accordance with the delivery schedule, rather than their commencement dates. Figure 11.11 shows the network that is the production diagram for one housing unit. It is drawn on a time base and within the known resource constraints for this work. The time scale is reversed, with the final Event 10 being at time zero. Working back from Event 10, the *lead times* for each of the other events in the network are calculated. The lead time for an event is the number of time periods (weeks in this case) by which that event must precede the end event if delivery is eventually to be made on time. For example, 'Foundations' must be completed 10 weeks before the house is scheduled to be completed. It will be noted that there is no float in the production diagram. A schedule of lead

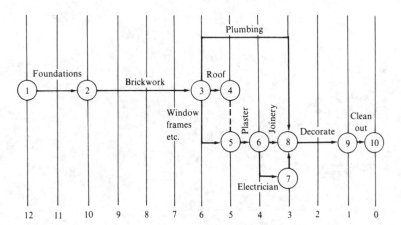

Figure 11.11

Table 11.2

Control points	Lead times, weeks
1 Commence	12
2 Complete foundations	10
3 Complete brickwork	6
4 Complete roof	5
5 Complete window frames, etc.	5
6 Complete plaster	4
7 Complete electrics	3
8 Complete plumbing and joinery	3
9 Complete decoration	1
10 Clean out	0

times can be drawn up as in Table 11.2. Because of this, the emphasis in the table is placed on completion events rather than starting events.

The delivery of houses from the schedule may be stated in two ways. Either it may be a straightforward, week-by-week schedule giving the number of units to be delivered during each separate, successive week. Or it may be represented diagrammatically as the *objective diagram*, illustrated in Fig. 11.12. This example illustrates a

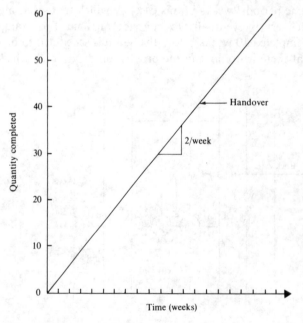

Figure 11.12

handover rate of 2 units per week from a batch of 60 houses. The ordinate to the graph represents the number of houses that should be completed after any period of time in the programme. The line of balance can be determined for each control point (event) in the production diagram constructed for a single unit of housing. In Fig. 11.13 a delivery schedule is shown for the handover of 70 houses in 55 weeks. At any point in time from the schedule shown on the x-axis, the necessary information for control regarding progress compared with each of the unit's control points established in the production diagram of Fig. 11.11 can be established. At the point marked 'Today', at the end of week 35, the number of house foundations that should be completed at this point in time can be read off against the graph of completed deliveries by going forward in time by the lead time for this activity of 10 weeks. If the required number of foundations for sufficient houses, as established in this way, are not completed by 'Today' then delivery at the handover rate will not take place in 10 weeks' time unless some corrective action is taken. Similar information can be established for each control event, and the control diagram for week 35, or for any other point in time, can be established as illustrated in Fig. 11.14. Figure 11.14 can then be used for monitoring progress. Actual performance can be plotted against the *line of balance*, that is, the line displaying the numbers of components that must be completed in order to be on the required programme, by making a visual check of the work on the various unit activities completed so far.

Alternatively, the line of balance for each of the individual activities within a repetitive unit can be portrayed graphically by drawing parallel lines displayed to the left of the handover line. This is known as *parallel scheduling*. The parallel lines are offset to the left by the lead times calculated from the production diagram of the repetitive basic unit. Using the unit production diagram in Fig. 11.11 and the lead times of Table 11.2, the line-of-balance chart for the construction and handover of 70 houses, started and completed within a period of 55 weeks, is shown in Fig. 11.15.

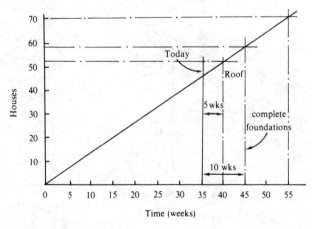

Figure 11.13

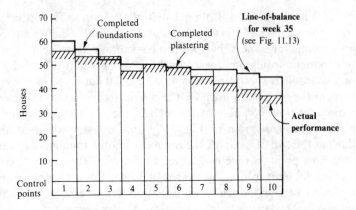

Figure 11.14

Each activity is scheduled for every unit of housing between two parallel lines which are identified by their *i–j* event numbers from the basic unit network and which are shown at the top of the chart in Fig. 11.15. For example, activity 2–3, 'Brickwork', for the tenth house is scheduled to take place as shown on the

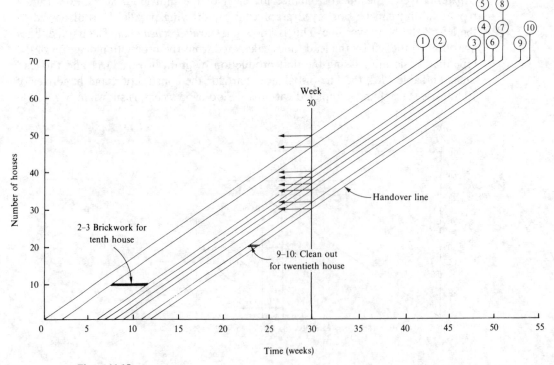

Figure 11.15

diagram—the scheduled times of start and finish can be read off from the x-axis. Similarly, the schedule for 'Clean out' of the twentieth house is shown on the same chart.

By drawing a vertical line at, for example, week 30 of the overall schedule, the scheduled numbers of starts and completions for each activity that should have occurred can be read off on the y-axis. In this case, where the vertical line cuts the event 1 line, the projection on the y-axis will provide the number of houses that should have been started; where it cuts the event 2 line it will indicate the number of house foundations that should have been completed and the number of houses in which the brickwork should now have commenced. Actual progress at week 30 can be plotted by drawing a horizontal bar against each house number, as illustrated in Fig. 11.15.

11.4 Resource utilization in parallel scheduling

An important aspect of scheduling is the consideration of the quantity of resources that are required in order to carry out the work in accordance with the programme. Resources may be labour, equipment, materials, and/or money. Each is important in the context of a construction programme. By considering a simple example programme, the implications of a schedule on the utilization of labour as a resource will be examined. The principles that are derived may then be used in considering the utilization of resources of other different categories. Parallel scheduling is one method of drawing up a schedule in which the handover rate is imposed upon the various activities, and the consequences of doing so are then examined. In particular, it is necessary to examine the utilization of resources in the light of the required handover rate.

EXAMPLE 11.2

For the purpose of programming, the replacement of a length of road base and its pavement it is divided into eight sections, each of approximately equal length. The activities for each section with the estimated durations for each are shown in Table 11.3.

Each activity follows on from the previous one on serial fashion from section to section.

A handover rate of one section per 5 working days is required.

Table 11.3

		Working days
1–2	Excavate existing road	6
2–3	Lay sub-base	3
3–4	Lay base	5
4–5	Lay road pavement	7
5–6	Clean up	4

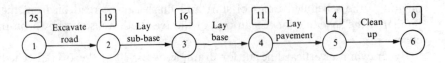

Figure 11.16

The production diagram showing the sequence of activities for each section of road is illustrated in Fig. 11.16. The lead times are shown in boxes at each event.

The line-of-balance programme for carrying out the road replacement, using parallel scheduling, is shown in Fig. 11.17. The horizontal spacing between the start and finish schedule lines represents the duration for each activity as set out in the production diagram for one section of work. All of the lines of balance are drawn parallel one with another and hence parallel to the handover line. The handover line coincides with the completion of clean up, event 6.

The first of the eight sections of the road, Section 1, will be started at time 0 and be completed at day 25. The intermediate stages will be completed as shown in its

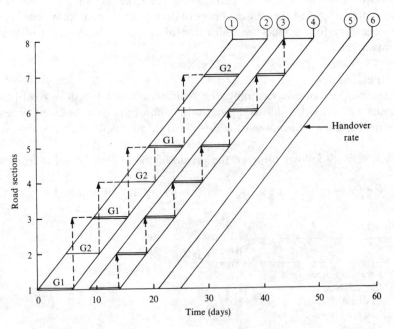

Figure 11.17

production diagram. The slope of the handover line of balance can be determined from the following:

$$t_2 = \frac{(Q_2 - Q_1)}{m} + t_1$$

where $t_1 = 25$, $Q_1 = 1$, $Q_2 = 8$, and $m = \frac{1}{5}$
Therefore,

$$t_2 = (8 - 1)5 + 25 = 60 \text{ days}$$

After compiling the line-of-balance programme to meet the requirements of constructing eight sections of road at a handover rate of one every 5 working days, it is possible to consider the influence of resource use on the programme and vice versa. Let it be assumed that all the equipment requirements are readily available and that one gang of operatives is started on the excavation of the existing road, activity 1–2. The gang should complete this work in 6 days and then move on to the next unit of excavation. However, Section 2 excavation should have started at day 5, and if Gang 1 (G1) is to be used then the start of the second section will have to be delayed by 1 day, thus delaying the completion programme as a consequence. In order to maintain the programme, it is necessary to introduce an additional gang to carry out the excavation for the second road section. These two gangs will then leapfrog one another, excavating for alternate sections until the excavation for all 8 units is complete. (It could be argued that the successive alternate activity 1–2 could be started as soon as the appropriate gang is available, that is, could be brought forward by 4 days. This could be done because the activity is the first of the production diagram, but in principle it contravenes the requirement of the scheduling and may well not be acceptable for other reasons.)

With the need for the employment of two gangs on activity 1–2, it is clear that an inefficient use of resource is introduced. When Gang 1 has completed the excavation for Section 1 and then moves to Section 3, Section 3 is not yet available for starting and Gang 1 has to wait for 4 days before being able to commence work. Unless the gang can be employed elsewhere, Gang 1 (and also Gang 2 after they have completed the excavation for Section 2) will incur unproductive time. This will increase the unit cost of the excavation activity above that for a smooth changeover of resources. This inefficiency will arise in each of the activities of the line-of-balance programme as depicted in Fig. 11.17, except for activity 3–4, laying the road base. In this case one gang can move from one unit to the next with no unproductive time because, coincidentally, the activity duration is the same as the interval between the handover of successive activities. For the programme shown in Fig. 11.16, two gangs are required for activities 1–2 and 4–5 and one gang will be sufficient for each of the other activities, though with varying degrees of unproductive time.

In the case of activities 1–2 and 4–5, the *resource unit factor*, F, is said to be 2; in the case of the other activities it is 1. The *resource unit* for this example is one gang. The resource unit factor is the number of units of the resource that are required in order to achieve the rate of working necessary to meet the handover programme. It

Table 11.4

Activity	Utilization, %
1–2	60
2–3	60
3–4	100
4–5	40
5–6	80

is calculated here on the assumption that each activity requires the employment of a different gang(s). This may not necessarily be true in practice.

The degree to which each gang is being utilized can be calculated from:

$$\text{utilization} = \frac{\text{activity duration}}{\text{gang waiting time} + \text{activity duration}} \times 100 \text{ per cent} \tag{11.5}$$

In the case of activity 1–2, utilization $= 6/(4+6) \times 100 = \underline{60\%}$

Similarly, the utilization for other gangs is shown in Table 11.4. There is a need to be able to assess the sum of the underutilization that occurs throughout a total project programme when line-of-balance methods are used. In order to do so it is first necessary to determine how many gangs will be needed for each activity in order to maintain not less than the handover rate. This is calculated by multiplying the rate of handover, m, by the duration of the activity, d, and, if the product is not an integer then the number obtained is rounded to the next largest integer. Care must be taken to ensure that the units of time are the same for both variables—working days, weeks, etc. A handover rate of 2 units per week for an activity with a duration of 1.7 weeks will therefore require 4 gangs to maintain at least the handover rate.

By using the first activity in the production diagram for Example 11.2, 'Excavate existing road', with a duration of 6 working days, it is possible to develop an expression to calculate the non-productive time for each repetition of the activity after the first occurrence.

Let I = non-productive time for each occurrence of one repetitive activity
  F = resource unit factor
  m = rate of handover
  d = activity duration

Then, referring to Fig. 11.18,

$$Im = F - md$$

and

$$I = \frac{(F - md)}{m} = \frac{F}{m} - d \tag{11.6}$$

Therefore, in Example 11.2, activity 1–2, $md = 6/5 = 1.2$

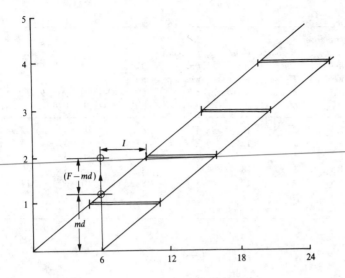

Figure 11.18

that is

$$F = 2 \text{ since 2 gangs are required}$$

and

$$I = 2 - (1/5 \times 6)5 = \underline{4 \text{ working days}}$$

No gang will incur non-productive time on the first occasion on which it is engaged on the activity because the gang will become available only at the start of the first activity; therefore, if Q = the total number of units to be constructed then the total number of activities incurring non-productive time will equal $(Q - F)$.

If T = total units of time required to complete Q repetitive units then

$$T = [(Q - F)(d + I)] + Fd \qquad (11.7)$$

The minimum time required to complete the activities, T_0, assuming that there is no non-productive time, is

$$T_0 = Qd \qquad (11.8)$$

These times can be converted to man-hours, etc. by multiplying the number of operatives in a gang by either T or T_0.

For activity 1–2, therefore,

$$T = [(8 - 2)(6 + 4)] + 2 \times 6$$
$$= 60 + 12 = \underline{72 \text{ working days}}$$

Table 11.5

Activity (1)	Duration d, working days (2)	Handover rate, units/ working day (3)	Resource unit F, gangs (4)	Gang size R, operatives (5)	Non-productive time I, working days (6)	Total units Q (7)	Q−F (8)	d+I (9)	(8)×(9) (10)	Fd (11)	T (12)	T_0 (13)
1–2	6	1/5	2	10	4	8	6	10	60	12	72	48
2–3	3	1/5	1	12	2	8	7	5	35	3	38	24
3–4	5	1/5	1	6	0	8	7	5	35	5	40	40
4–5	2	1/5	1	8	3	8	7	5	35	2	37	16
5–6	7	1/5	2	10	3	8	6	10	60	14	74	56
											261	184

and

$$T_0 = 8 \times 6 = \underline{48 \text{ working days}}$$

If there are 10 members in the gang, the total idle time for activity 1–2 is equal to $(72-48)10 = \underline{240 \text{ man-days}}$, or the overall utilization of the gang equals $48/72 \times 100 = 66.6$ per cent.

Calculations of this type, if they are to be carried out manually, can usefully be tabulated for simplicity. The full calculations for Example 11.2 are set out in Table 11.5. Total non-productive time for the project, in terms of man-days, is equal to:

$$10(78-48) + 12(38-24) + 8(37-16) + 10(74-56) = \underline{756 \text{ man-days}}$$

11.5 Resource scheduling

In Section 11.4, the resource aspect of line of balance was examined using parallel scheduling. In that case, the emphasis was placed on the schedule, and in particular on the handover rate. All activities were assumed to be carried out so as to be completed for each production unit at the handover rate. As was seen in the chosen example, parallel scheduling can lead to considerable non-productive time where gangs continually have to wait between finishing one unit activity and starting on the same activity for another unit. An alternative method of taking resources into account as part of the scheduling process will now be examined. This is usually known as *resource scheduling* for line of balance.

EXAMPLE 11.3
The site for a project consisting of 120 similar houses will become available for construction to commence on 1 January next. A handover rate of six houses

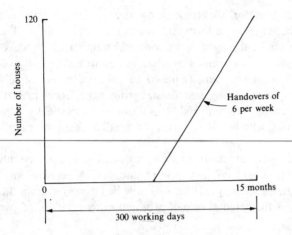

Figure 11.19

per week is required, the project to be completed by 31 March of the following year.

The contractor will be working a 5-day week at 42 hours per week, which, taking holidays, etc., into account enables the programme to be based on a total of 300 working days. Figure 11.19 shows the base programme. Table 11.6 can now be drawn up after consideration of the resources that will be used. The sequence of activities will be that adopted in Fig. 11.11. Against each activity is placed, in Table 11.6, the estimated number of man-hours that are required to complete it. In column 3 can then be put the total number of operatives, G, that will be required to complete six similar activities per 42-hour week:

$$G = M \times 6/42 = M/7$$

Table 11.6

Operation	Man-hours per house M	Gang size G, operatives	Men per house H	Actual gang size S, operatives	Revised rate of house production R, per week	Time for one house D, day	Slope of line for 120 houses T, days
Foundations	100	14.29	3	15	6.30	3.97	94.44
Brickwork	220	31.43	6	30	5.73	4.37	103.84
Window frames, etc.	50	7.14	1	6	5.04	5.95	118.06
Roof	200	28.57	4	28	5.88	5.95	101.19
Plumbing	150	21.43	3	21	5.88	5.95	101.19
Plaster	40	5.71	3	6	6.30	1.59	94.44
Joinery	120	17.14	4	20	7.00	3.57	85.00
Electrician	90	12.86	3	12	5.60	3.57	106.25
Decorate	130	18.57	4	20	6.46	3.87	92.11
Clean up	20	2.86	2	2	4.20	1.19	141.67
Total	1120						

Rarely does the calculation for the total gang size, G, result in a round and convenient number; therefore an adjustment has to be made to it. The adjustment must also take place in the light of the optimal number of workers that can be employed in one gang at a time. For the 'Foundations' activity, G as calculated results in 14.29 workers being required to produce 6 sets of foundations in one week and the optimum gang size (found from experience) is 3. Therefore, the actual number of operatives required, S, should be divisible by 3 *and* be as near to 14.29 as possible, whether it is larger or smaller than the exact number. Fifteen is chosen.

Because of the rounding-up or -down process, production will not be quite at the same rate as if the gang size of column 2 were adopted. A revised rate of house components per week, R, for each operation, can now be calculated from the formula $R = R_1 \times S/G$, where R_1 is the original rate of production required, i.e., 6 per week in this example.

The duration in working days for the completion of each operation in one house, D, can now be calculated on the assumption that the 42 working hours per week are equally spread over each working day as 8.4 hours per day:

$$D = M(H \times 8.4)$$

These figures are listed in column 7 of Table 11.6. In column 8 are listed the durations for the completion of each operation in the total number of houses for the project. If N is the number of houses then the completion time will therefore span a duration of $N-1$ construction periods. The duration for each operation over all houses, T, is then

$$T = \frac{(N-1) \times 5}{R} \text{ working days}$$

Enough information is now available to enable a programme chart to be produced. To explain the principles of the line-of-balance chart it will be assumed, initially, that each of the operations will be sequential rather than having some concurrent activities as in the diagram for the unit of housing. The activities will be assumed to be in the sequence of the list in Table 11.6, reading from top to bottom. Against a horizontal axis of working days and a vertical axis of numbers of houses the chart of Fig. 11.20 is plotted.

The line-of-balance programme shows a series of bars, sloping from left to right since they represent construction progress, each of which displays the schedule for one particular activity, such as 'Foundations', for each of the 120 units to be constructed. If any bar crosses or impinges on another then the logic of the diagram is destroyed. For a smooth succession of activities there must be clear space between activity bars, though in an extreme case of parallel scheduling the completion of one activity could be coincident with the start of its successor throughout its length. Each bar has a horizontal width which represents the time to undertake the particular activity or operation for one unit. For example, in the case of 'Foundations', the

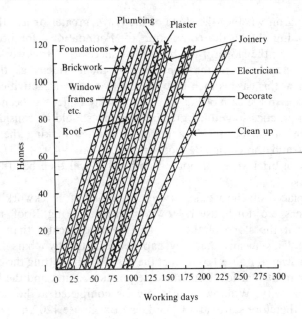

Figure 11.20

width is 3.97 days (from Table 11.6), and in the case of 'Joinery' it is 3.57 days. (Note that it is impracticable to use durations which are expressed to two decimal places in planning and controlling actual work; this is only done here so as to maintain reasonably correct arithmetic and to make the argument easier to follow.) These widths are a function of the total man-hours per unit involved for the activity and the size of the gang carrying out the work. The *slope* of the bar (also obtained from Table 11.6) is a function of the revised rate of house production that has been calculated following the determination of the actual gang size. If a gang size of 12 had been chosen for 'Foundations' instead of one of 15, then the revised rate of production of houses, R, would have been $6(12)/14.29 = 5.04$ houses per week instead of 6.30 and the slope of the bar would have been $5(120-1)/5.04 = 118.06$ instead of 94.44. In other words, it would mean that the completion of foundations for house 1 would have been at $+3.97$ days and for house 120 would have been at $+3.97 + 118.60 = 122.57$ days instead of 98.41 days as before. The slope of the bar from the final column of Table 11.6 is in units of days and represents the span of time from the start (or completion) of unit 1 to that of unit 120. The smaller the number of days, the steeper the slope of the bar.

It is normally impracticable to commence a different activity on a repetitive unit immediately its predecessor finishes, if for no other reason than that there must be some contingency in the programme. Subcontractors do not always arrive as planned; the weather is worse or better than expected; operatives fall ill; etc. A line-of-balance chart, therefore, has *buffers* introduced between its bars at one end or the other, depending on whether the slope of the bar is greater or less than that of its predecessor. In Fig. 11.20, the completion of 'Foundations' for house 1 is scheduled for $+3.97$ days; the slope of the bar for its successor, 'Brickwork', is less than its own slope, and therefore the gap between them widens as the work progresses. The buffer in this case is then between the two relevant activities for the first house on the x-axis. In the case of Fig. 11.20, a one-day buffer has been allowed in all cases, though in practice something greater than this should be considered. At house 1, 'Foundations' finish at $+3.97$, a buffer is inserted making the start of 'Brickwork' at $+4.97$ and the completion of 'Brickwork' at $+4.97+4.37 = +9.34$ days. The completion of brickwork for house number 120 will then be at $+9.34+103.84 = +113.18$ days.

The buffer is also placed on the x-axis between the finish of 'Brickwork' and the start of 'Window frames, etc.' for house 1. However, in considering 'Roof', it can be seen in Table 11.6 that the slope of the bar for 'Roof' is greater than that for 'Window frames, etc.' This means that any gap between the two bars for these activities must be narrower at house 120 than at the first. When plotting the diagram, the position of the *top* of the bar for 'Roof' must be positioned first and the bar then drawn from it. The activity 'Window frames, etc.' is completed at house 120 in $+135.35$ days. 'Roof' therefore starts at 136.35 days from house 120 and will finish at $+136.35+5.95 = +142.30$ days. 'Roof' will start for the first house at $+41.11-5.95 = +35.16$ days. The gap between the finish of 'window frames, etc.' and the start of 'Roof' will therefore be 17.87 days at house 1 and 1 day at house 120.

The complete line-of-balance chart having been drawn showing the programme for all activities for 120 houses, it can now be examined to see if there are any ways of making the programme more efficient. One obvious anomaly in Fig. 11.20 is the last activity, 'Clean up', which does not finish for house 120 until $+238.85$ days, whereas its predecessor finishes at $+187.10$ days. One way to improve this situation is to increase the resource used to carry out this activity. This can be done from the start, at house 1, or at some point between that and the last house. If it were decided to increase the gang size to 4 from the start of the work then the revised rate of production, R, becomes $6(4)/2.86 = 8.39$ houses per week and T, the slope of the line for 120 houses, becomes $119(5)/8.39 = 70.92$. The slope of the line is now greater than that of its preceding activity and the buffer between them will need to be at the top. The work will be completed $+187.10+1+1.19 = +189.29$ days instead of $+238.85$ days. While 'Clean up' is a trivial activity, this calculation does emphasize the improvement in efficiency that may be gained by analysing such charts.

An alternative method of improvement would be to put six operatives on the work from, say, house 61. The slope of the line to house 60 would then be the same as originally but R would become $6(6)/2.86 = 12.59$ days for the second half. The average slope for the 'Clean up' bar then becomes $+141.67/2+24.43 = +95.27$ days. 'Clean up' will then finish at $+94.99+1.00+95.27 = +192.26$ days. The $+94.99$ days is the completion of the activity of 'Decorate' for house 1. The diagram then looks like Fig. 11.21. Other activities may then be examined in like manner.

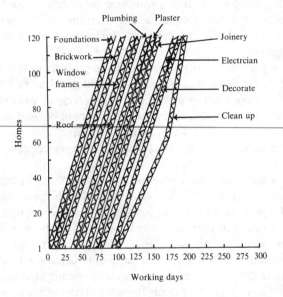

Figure 11.21

If there are concurrent operations to be considered such as occur in the typical network for a unit of housing in Fig. 11.11 then care must be taken that all necessary preceding operations have been scheduled before scheduling the one in hand. Operations which are in a straight-line sequence can be scheduled as they were in Example 11.1. Concurrent operations can be scheduled as such and a buffer band parallel to the activity to be projected in sequence can be used.

11.6 The linear scheduling method

Among the many planning models available for use with construction activity, some, such as bar charts, have general application on most projects. Some, however, are more suitable for use on projects having particular features of project organization and/or the nature of the construction method and sequence. One such specialized category of project is that of the *linear project* typified by pipelines, tunnels, highways, and bridges. The overriding characteristic of such construction is that progress is made continuously along the physical length of the project, or part of it, since there can be more than one starting point. The construction is not made up of discrete activities that are readily broken down for representation by an arrow in a network. Progress can be measured by the rate of completion, usually for both the completed construction and for the component activities that make up the whole. In the case of a pipeline, the component activities selected for planning purposes might be 'Clear site' and 'Remove vegetation', 'Remove topsoil', 'Excavate trench', 'String out pipes', 'Lay pipes', 'Weld pipes', 'Backfill', and 'Reinstate surface'. Each of these component activities can be undertaken continuously from the starting point of the project to its physical end, though few of them can be carried out simultaneously or even in an overlapping fashion.

One method of planning designed to be used with the projects in this format is the linear scheduling method (LSM), the origins of which are unclear except that its roots were almost certainly developed in production engineering. This is not to say that other methods of planning, such as network analysis, cannot be used for projects of this type, but the LSM can often be quicker, simpler, cheaper, and more effective to use since it has features designed for ready application to projects that are linear by nature.

The elementary principles of the method are similar to those of line of balance and can be established by examining Fig. 11.22, which illustrates the planning for the excavation and laying of a pipeline six sections long, each of which is one-quarter of a mile. The scale on the vertical axis represents the total length of the pipeline (it will be assumed that installation of the pipeline commences at one end and proceeds towards the other). A point on that axis refers to a particular location on the pipeline at a specific distance from its origin. The horizontal axis represents time or duration, and time will normally progress from start-up at the intersection of the axes towards the right. The units in which time is measured will depend on the length of the project, what part of it is portrayed on the diagram, and what level of detail is to be shown on the diagram. In the case of Fig. 11.22, the unit of one working day is used.

The first inclined line on the diagram, labelled 'Remove topsoil', illustrates that excavation commences at the beginning of the second working day of the schedule and is completed for the whole 6-mile stretch in a duration of 15 working days. The line representing the programme is linear, and therefore progress throughout the six

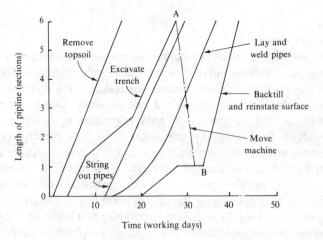

Figure 11.22

sections is planned to be constant. The progress expected at any location on the pipeline can be read off from the diagram. It will be seen that it is expected to have completed excavation for two sections of the pipeline by the end of the week 2 of the schedule. Alternatively, if it is required to know when excavation for the pipeline will reach the end of the second section, the process can be reversed.

The second sloping line, for 'Excavate trench', introduces two changes in the production rate when measured as a function of the length of trench excavated against time. Parts of sections 2 and 3 have much deeper trenches than the remainder of the pipeline. The breakpoint in the sloping line shows the commencement of the deep section, the lower rate of production when measured by the length of trench completed, and the return to the previous rate partway through section 3. The third sloping line, for 'String out pipes', assumes constant progress throughout the $1\frac{1}{2}$ miles of the pipeline, but the fourth line, 'Lay and weld pipes', introduces a new concept. The curve is one that starts out at a modest slope, gradually increasing its slope as it progresses through the pipeline. It does this in order to make allowance for *learning*. As the operatives laying the pipes, and the welders jointing them, get more used to the conditions, the learning effect enables the production rate to be increased.

Lastly, the line to represent 'Backfill and reinstate surface' introduces two further actions. Firstly, the backfilling and reinstating start and progress to the end of section one before a break in the work takes place. This is because there is a waiting period prior to a machine becoming available from the 'Excavate trench' activity in order to proceed. The constraint imposed by the wait for the machine is represented by the broken line. The machine needs a few days for servicing and transporting from A to B in Fig. 11.22 and then, after setting up, backfilling and reinstatement proceeds. The whole of the work is completed in 42 working days.

The LSM can be extended to recognize the different types of activity that may be carried out on linear projects. Figure 11.23 shows a chart that illustrates a typical roadworks project and incorporates a number of activities having different planning attributes. It represents the earthmoving, grading part of the drainage and construction of the abutments and piers for a bridge. The diagram has been kept simple in order to illustrate the principles. Firstly, the single line is used to relate location or distance with time. Examples in Fig. 11.23 are for the removal of topsoil and the installation of surface water drainage. Secondly, the earthmoving is represented by a series of rectangles showing the location and time period for the removal, shaded rectangles indicating the location and timing for the filling by the arising spoil. Thirdly, the construction of the bridge abutments and the piers is illustrated by a long rectangle, of notional width with regard to location and with its length representing the period of time over which the construction takes place. While progress in the first two activities, removal of topsoil and excavation, is indicated as movement along the relevant line or series of rectangles, the rectangle for the bridge foundations shows static construction and how it integrates with the construction of the remaining works.

Monitoring progress by using the LSM can be carried out by drawing a line perpendicular to the x-axis at the current point in time and seeing where this line

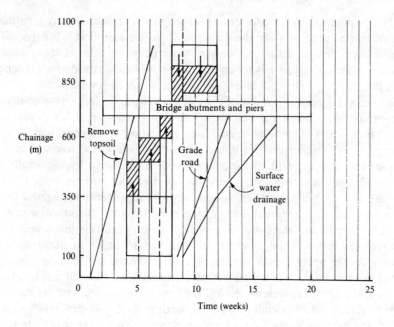

Figure 11.23

crosses the various bars or lines. This can then be related to the chainage and progress on site can be compared with the result.

11.7 Which method of planning?

The main criterion to be used in selecting a scheduling system will depend to a large extent on the amount of detail that is already available, or is required from the process, and who will be undertaking and using the chosen method. Taking the second point first, the choice of system will often depend on the training of the personnel who will be carrying out the scheduling process and thereafter the training of the individuals who will need to use its output in the field to install, monitor, and control the construction works.

Of the methods described in this book, there is little doubt that the bar or Gantt chart is the simplest to use. It employs a graphical plot of a horizontal bar, representing the use of resources, against a time scale. The various activities or operations are usually plotted in order of occurrence with the first to the left side of the diagram and in the upper part of the list. The system lacks the rigour of having to interrelate one activity with others and the overlapping of activities can take place without too much consideration of the detailed content of the relevant operations or their phasing. Progress can be monitored with the bar chart by drawing another bar below that shown on the plan. This second bar should show the actual dates between which work is carried out and the actual proportion of the work represented by the original bar that has been completed during the work period. Where work is not planned to be completed at a uniform rate of progress throughout the bar, either by cost or quantity, the percentages completed should, where possible, take this into account.

Networks take account of the interrelationship of activities and display clearly the various sequential paths through a project. The logical interdependencies are set in such a way that float or leeway can be calculated and attention can be focused on the critical path(s) on which the overall duration of the project depends. For large and/or repetitive projects, networks can become very complex and without editing, or by conversion to other methods for visual presentation, such as the bar chart, difficulties of communication with the users in the field can arise.

Linear scheduling methods are designed to cope with specialized forms of projects. Scheduling linear projects by the use of networks is not satisfactory, through a combination of network analysis and linear scheduling can prove to be effective. Linear scheduling methods can result in good visual charts that facilitate monitoring and control. Users can see the progress that should have been made and readily compare with it the progress actually made. The slopes of the line provide good visual information on the rate of progress required.

Summary

An alternative method of planning and scheduling to that of CPM is detailed. Known as precedence programming, it is an activity on the node method which deals particularly well with overlapping activities. A different type of method is then introduced—that of line of balance—which is particularly well suited for use with projects on which there is a considerable amount of repetitive work. The influence that resource allocation has on the line-of-balance approach is examined for parallel scheduling, also where the slope of the bars can be varied in order to use resources more efficiently. Linear scheduling methods are examined for use with linear projects such as roads.

Problems

Problem 11.1   Draw the precedence networks for the logic set out in Problems 8.1, 8.2, and 8.3.

Problem 11.2   Draw the upper network in Fig. 9.17 as a precedence diagram and complete the time analysis.

Problem 11.3   Use the precedence diagram for the construction of a culvert shown in Fig. 11.7. After 15 days the progress report shown on Table 11.7 is received.

Table 11.7

Activity number	Actual start date	Duration so far	Percentage completion
1	0	4	100
2	7	5	100
3	4	8	100
4	6	5	100
5	13	10	75
6	14	18	40
7	15	6	60
8	13	6	40
(No other activity started.)			

Establish:

1. the time required to complete the project;
2. the revised project duration;
3. the critical path for the remaining activities;
4. the revised total and free floats for the remaining activities.

Problem 11.4 In terms of line-of-balance programming, define the following terms using sketches wherever possible to illustrate your answer:

1. Objective diagram.
2. Production diagram.
3. The line of balance.
4. Lead times.

Problem 11.5 For the purposes of planning and scheduling, the construction of a highway pavement is divided into 85 approximately equal sections. Handover of completed sections is required at the rate of 2 per week. The first 15 sections are completed to programme by the end of working day 43 after the commencement of the work. There are 5 working days in one week. When is section 80 scheduled to be completed?

Draw an accurate graph, clearly labelling each axis with title and units, showing the completion rate for the pavement extending it from the point where it crosses the vertical y-axis (state the value for this point on the graph) to the point where all 85 sections are complete.

State the changes that would need to take place in the programming if the project is to be completed in 180 working days.

Problem 11.6 The production diagram for the construction of a domestic house shows a total duration of 12 weeks from commencement to completion. Work is programmed to commence on a project for the construction of 105 such houses at the start of week 1. A handover rate of 2 houses a week is required. Draw the objective diagram for the project.

Calculate:

1. how many houses will be completed by the end of week 22;
2. the time at which houses 55 and 101 will be completed;
3. after the project has been in progress for 18 weeks, it is decided to increase the handover rate to 3 per week. Recalculate the time at which houses 55 and 100 will now be completed.

Problem 11.7   Table 11.8 lists the lead times (in weeks) for each operation in a typical housing unit. The operations may be assumed to be in series. If 50 such units are to be constructed having a handover rate of 5 per week, draw the line-of-balance control chart for week 10 of the programme.

Table 11.8

1	Substructure	14
2	Brickwork	12
3	Roof (carpenter)	9
4	Roof tiler	7
5	Joinery	6
6	Electrician	5
7	Plumber	3
8	Decorator	2

Problem 11.8   If the 50-house project of Problem 11.7 above commences on 1 January and is required to be completed by 31 December of the same year, select suitable resources for each operation and draw for it a line-of-balance chart similar to the one featured in Fig. 11.20. Increase the resources for the brickwork operation and examine the ultimate interference effect of one operation on another. Propose a solution to this problem.

Problem 11.9   Using line-of-balance parallel scheduling, draw a full scheduling diagram for all activities in a project which has three serial activities, A, B, and C. The durations of these activities are 0.7, 1.3, and 0.8 weeks respectively. 12 units are to be built at a rate of 2 per week.

Mark clearly on your diagram the time, accurately calculated, at which the units 1 and 12 are completed.

For your diagram, calculate how many activities A, B, and C should be completed by the end of week 7.

Problem 11.10   Using the details contained in Problem 11.9, draw a line-of-balance diagram in which one gang is used for each activity.

By drawing a second diagram, illustrate how the overall duration of the total project for the construction of all 12 units can be reduced. In both cases above, state clearly on your diagram the completion times of each activity for both the first and the twelfth units.

Problem 11.11 A reinforced concrete jetty is supported from the river bed by driven concrete piles. The deck of the jetty is a concrete slab supported on concrete beams spanning between the pile caps. The jetty can be divided into 24 repetitive sections, which must be handed over at a rate of 1 every 2 weeks. Using parallel scheduling line of balance, determine the overall duration of the total project and then draw a line-of-balance chart for control purposes.

The sequence of activities is serial and the durations are as shown in Table 11.9

Construct a line-of-balance chart showing what should have been achieved by the end of week 30.

Calculate the non-productive time of the gangs constructing the jetty, assuming that a different gang carries out each activity.

Table 11.9

	Operation	Duration, weeks
A	Cap piles	1.0
B	Construct beams	1.5
C	Construct slab	3.0
D	Install pipework	2.0
E	Paint pipework	0.5

Problem 11.12 For the jetty described in Problem 11.11, the man-hour content per section and the desirable size of each gang is set out in Table 11.10.

Using this information, draw up a line-of-balance chart for the construction of the jetty minimizing the non-productive time. Use a 40-hour week in your calculations.

Table 11.10

	Operation	Man-hours per section	Operatives per section
A	Cap piles	375	3
B	Construct beams	600	6
C	Construct slab	840	5
D	Install pipework	920	2
E	Paint pipework	120	2

Further reading

Precedence diagramming

Fondahl, J.: *A Non-computer Approach to the Critical Path Method for the Construction Industry*, Technical Report no. 9, The Construction Institute, Department of Civil Engineering, Stanford University, Stanford, California, 1962.

Fondahl, J.: *Methods for Extending the Range of Non-computer Critical Path Applications*, Technical Report no. 47, The Construction Institute, Department of Civil Engineering, Stanford University, Stanford, California, 1964.

Harris, R. B.: *Precedence and Arrow Networking Techniques for Construction*, Wiley, New York, 1978.

Moder, J. J., C. R. Phillips and E. W. Davis: *Project Management with CPM, PERT and Precedence Diagramming*, 3rd edn, Van Nostrand Reinhold, New York, 1983.

Neale, R. H. and D. E. Neale: *Construction Planning*, Telford, London, 1989.

O'Brien, J. J.: *CPM in Construction Management*, 3rd edn, McGraw-Hill, New York, 1984.

Line-of-balance and linear scheduling

Arditi, D. and M. Z. Albulak: Line of balance scheduling in pavement construction, *Journal of Construction Engineering and Management*, ASCE, vol. 112, no. 3, September 1986, pp. 411–424.

Arditi, D. and M. Z. Albulak: Comparison of network analysis with line of balance in a linear repetitive construction project, *Proceedings of the Sixth Internet Congress*, vol. 2, Garmisch-Partenkirchen, Germany, September 1979, pp. 13–25.

Ashley, D. B.: Simulation of repetitive unit construction, *Journal of the Construction Division*, ASCE, vol. 106, no. CO2, June 1980, pp. 185–194.

Carr, R. I. and W. L. Meyer: Planning construction of repetitive building units, *Journal of the Construction Division*, ASCE, vol. 100, no. CO3, September 1974, pp. 149–158.

Chrzanowski, E. N. and D. W. Johnston: Application of linear scheduling, *Journal of Construction Engineering and Management*, ASCE, vol. 112, no. 4, December 1986, pp. 476–491.

Johnson, D. W.: Linear scheduling method for highway construction, *Journal of the Construction Division*, ASCE, vol. 107, no. CO2, June 1981, pp. 247–261.

Lumsden, P.: *The Line of Balance Method*, Pergamon, London, 1968.

O'Brien, J. J.: VPM scheduling for high rise buildings, *Journal of the Construction Division*, ASCE, vol. 101, no. CO4, December 1975, pp. 895–905.

12. Financial management

12.1 Introduction

It is desirable that a construction engineer has an understanding of basic accounting and financial practices inasmuch as accounting is a process of collecting and classifying financial data concerning the activities of a business, which are then presented, in various forms, for the benefit of the managers of those activities. The subsequent analysis of that information is known as *financial analysis*, and the ways in which it is then interpreted so as to become one basis for making management decisions is known as *financial management*. This chapter will introduce the basic principles of an important accounting statement and its interpretation, since a construction engineeer will need an appreciation of these practices in order to seek data that will assist him in managing and controlling both individual projects and business units. Accounts are used as a means to assess the planning, progress, and financial efficiency of a business or a sector of it, such as a construction project, quite apart from other aspects such as legal requirements, etc.

While it is not essential for a construction engineer to become familiar with the detailed techniques employed in accounting, it will quickly be appreciated that the capacity to read and understand the information presented in various accounting reports will lead to being able to:

1. compare one company's financial performance with that of another in the same industry;
2. establish past trends of markets, sales, costs, performance, etc;
3. understand the relationships between the allocation of resources and profitability;
4. prepare budgets for future activities of the business;
5. understand the means by which a company, or a subdivision of a company, can compare performance with their plan or budget and assign reasons to the differences;
6. evaluate the performance of an organization's managers.

There are two broad classifications of accounting, *financial* and *management*. *Financial accounting* is concerned with the ordering of information, after the collection, recording, and classification of financial data concerning the revenue and expenditure incurred by the organization, so that the financial or trading position of a company may be established at any point in time. The two principal summary reports so produced, for external as well as internal purposes, are a *balance sheet* and a *profit and loss account*. These reports are required by law to present a *true and fair view* of the financial state of affairs of a company, mainly, but not exclusively, for the benefit of the shareholders and prospective investors. While a specific definition of 'a true and fair view' has not been included in successive Companies Acts, it is now generally accepted to mean that such accounts will be objective, free from falsehood or bias, and will be prepared in accordance with established good accounting practices and standards. Such financial accounts for companies are required to be published annually and presented to the shareholders, together with a directors' report and an auditor's report on the accounts. The Companies Act 1981 specified a number of formats to use for the presentation of annual accounts. Two options were established for balance sheets and one of four different formats can be selected for profit and loss accounts. This requirement came about under the influence of the European Community (EC) countries, such as France and Germany, which already had legal requirements for fixed formats. As a result of changes brought about by successive Companies Acts (mainly those of 1948, 1967, and 1981), financial statements have moved away from those in which emphasis was placed on stewardship of resources to those concerned with appraising investment performance.

Management accounting is a system designed to assist managers of an organization in planning and budgeting, both short and long term, and the subsequent control of the activities of their organization. A management accountant's main task is to provide reports to managers, often on an *ad hoc* basis, in order to facilitate their decision making, particularly concerning the internal operations of the organization. This area of activity clearly covers such functions as budgetary control, standard costing, financial management, and the application of theories and practices from other branches of knowledge such as economics, operational research, and statistics.

The definition of limited companies in the context of accounts is important. The Companies Act 1980, containing the first legislation on company law as a result of influence from the EC, introduced a new classification of companies. Up until this Act, all limited companies were assumed to be public unless they met the somewhat narrower definition of private companies. With the implementation of this Act, this situation was reversed. A public limited company, as defined below, now has 'Public Limited Company' (PLC) in its title and a private limited company has 'Limited' (Ltd) at the end of its title. For example 'Conman Public Limited Company' or 'Conman PLC' and 'Conman Company Limited' identify public and private limited companies respectively.

A *private limited company* has

1. at least two members;
2. a share capital and is registered as a private limited company;
3. a memorandum (in which the corporate trading objectives of the company are set out) which includes a statement that the company is a private company;
4. a name which ends in 'Limited'.

A private *limited* company, that is one wherein the liability of the owners (the shareholders) is limited, usually to the amount of their shares, should the company become insolvent, is forbidden to raise capital from the public by issuing shares and debentures.

A *public limited company* has

1. at least two members;
2. a limit by shares and a share capital;
3. a memorandum that includes a statement that the company is a public company;
4. a name which ends in either 'Public Limited Company' or 'PLC';
5. *allotted* share capital with a nominal value of not less than £50 000.

A *public limited company* has powers, under stringent regulations, to raise capital by issuing shares and/or debentures.

12.2 The balance sheet

A *balance sheet* is a fundamental statement, in money terms, of what a company *owns* and what it *owes* at a given date. What the company owns are called its *assets* and what it owes are called its *liabilities*. In addition, a balance sheet shows, as a liability, the capital that has been subscribed by its shareholders in return for an ownership interest. An example of a balance sheet is shown in Fig. 12.1.

The balance sheets shown in Figs 12.1 and 12.2 are laid out in one of the formats that are prescribed in the Companies Act 1981 for the publication of balance sheets of companies registered in the UK. In Fig. 12.3 the balance sheet of Fig. 12.2 is alternatively presented in *narrative* form. The narrative form is an alternative to the double-sided presentation wherein the liabilities are shown on the left half of a sheet of paper and the assets on the right. The narrative form of presentation gives the same degree of information as that of the prescribed format used but it is arranged in a fashion that makes it rather more useful for a manager who is using it for financial analysis and management.

Before a company can acquire assets, buy raw materials, pay staff, etc., money must be obtained from outside sources. The sources are varied, for example, money may be borrowed from one or more banks, it may be raised from shareholders who subscribe for company shares, it may come from private persons or from one of many other places. Once a business is successful, with a good profit-making record and sufficient fixed assets in ownership, then other sources become available, such as the issue of debentures, wider share ownership and profits ploughed back into the business. When obtaining this additional finance, the business must have as its objective to use it in the most productive way.

Let it be assumed that Conman PLC, a fictitious company the balance sheet of which is shown in Fig. 12.1, has used various sources of finance to fund its business. The company uses two different kinds of assets in the course of its activities. The

The Conman PLC
Balance Sheet as at 31 March 1991

	£000	£000	£000
Fixed assets			
Intangible assets			
Development costs			450
Tangible assets			
Land and buildings			12 000
Plant and machinery			3 000
Fixtures, fittings, tools, and equipment			200
			15 650
Investments			
Current assets			
Stocks			
Raw materials		2 310	
Work in progress		1 900	
Finished goods		1 500	
		5 710	
Debtors			
Trade debtors		4 850	
Prepayments and accrued income		220	
Cash at bank and in hand		2 100	
		12 880	
Current liabilities			
Creditors: amounts falling due within one year			
Bank and other loans	2 500		
Trade creditors	4 100		
Taxation and social security	2 500	9 100	
Net current assets			3 780
Total assets less *current liabilities*			£19 430
Creditors: amounts falling due after more than one year			
15% mortgage debenture repayable 2011			3 000
Bank and other loans			2 130
Capital and *reserves*			
Called up share capital			13 200
General reserve		1 000	
Profit and loss account		100	1 100
			£19 430

Figure 12.1

first are *fixed assets*. These can be of two kinds and if they have a physical presence they are known as *tangible assets*. They can consist of land, buildings, plant or machinery, etc., and they are considered to be permanent or semi-permanent acquisitions by the company as a result of capital investment. They are purchased to be used in its productive processes and not for resale in its day-to-day business. The

The Conman PLC
Balance Sheet as at 31 March 1990

	£000	£000	£000
Fixed assets			
Intangible assets			
Development costs			450
Tangible assets			
Land and buildings			11 308
Plant and machinery			1 800
Fixtures, fittings, tools, and equipment			180
			13 738
Investments			
Current assets			
Stocks			
Raw materials		1 670	
Work in progress		1 600	
Finished goods		1 200	
		4 470	
Debtors			
Trade debtors		4 920	
Prepayments and accrued income		100	
Cash at bank and in hand		1 600	
		11 090	
Current liabilities			
Creditors: amounts falling due within one year			
Bank and other loans	2 100		
Trade creditors	3 970		
Taxation and social security	2 000	8 070	
Net current assets			3 020
Total assets less *current liabilities*			£16 758
Creditors: amounts falling due after more than one year			
Bank and other loans			2 558
Capital and *reserves*			
Called up share capital			13 200
General reserve		800	
Profit and loss account		200	1 000
			£16 758

Figure 12.2

assets will be required to generate income for the business over a long period of time. It will be seen from Fig. 12.1 that, at the date of its balance sheet, Conman PLC had in its ownership £15 200 000 worth of tangible fixed assets.

Fixed assets can also be *intangible*. That is, they do not take a physical form such as buildings or equipment. Intangible fixed assets may consist of patents, copyrights, trademarks, goodwill, chemical process formulae, etc. There is little

The Conman PLC
Balance Sheet as at 31 March 1990

	£000	£000	£000
Fixed assets			
Development costs			450
Land and buildings		11 500	
Less depreciation		192	11 308
Plant and machinery		2 000	
less: Depreciation		200	1 800
Fixtures, fittings, tools & equipment		200	
less: Depreciation		20	180
Total fixed assets			13 738
Current assets			
Raw materials		1 670	
Work in progress		1 600	
Finished goods		1 200	
Debtors		4 920	
Prepayments and accruals		100	
Cash		1 600	
TOTAL CURRENT ASSETS		11 090	
less: Current liabilities			
Bank loans and overdraft	2 100		
Creditors	3 970		
Taxation	2 000		
TOTAL CURRENT LIABILITIES		8 070	
NET CURRENT ASSETS (WORKING CAPITAL)			3 020
NET CAPITAL EMPLOYED			16 758
less: Loan capital			2 558
NET WORTH OF COMPANY			£14 200
The above net worth has been financed by:			
Subscribed capital			
13 200 Ordinary shares of £1 each			13 200
Reserves			
General reserve			800
Profit and loss account			200
TOTAL SHAREHOLDERS' INTEREST			£14 200

Figure 12.3

doubt that intangible though these assets are, they exist and have a value because a buyer would pay a seller for them in the anticipation of making profits from the purchase. Conman PLC has valued its intangible fixed assets at £450 000, which, together with those that are tangible, make a total of £15 650 000 worth of fixed assets.

An item for *investments* appears on the balance sheet. Conman PLC does not have a financial entry against this heading. An investment item under fixed assets on a balance sheet refers to an investment in another company and is usually itemized at the original cost. If there were investments in companies that are *unlisted* on the Stock Exchange then there would need to be a separate heading for these as opposed to those that are *listed*. If investments are held by a company as a way of employing cash that is surplus to the immediate needs of the business, such investments do not count as fixed assets but as current assets. They would then appear under that heading.

The second category of assets a business owns and uses are *current assets*. These are alternatively called *circulating assets* because it is highly desirable that they should be employed as actively as possible in the normal course of business and will circulate round the production cycle. They are converted from cash to raw materials, then to work in progress, and on to finished product, being converted by the production process, and finally to sales, when the assets are turned back again to cash when payment is received. In the normal course of events it is expected that current assets, whether in the form of cash, stocks, debtors, marketable securities, or prepaid expenses will be put to use in the business operating cycle and within one year from the date of the balance sheet. The operating cycle for a manufacturing or production company is the period between buying the raw materials and receiving the cash for the sale of the finished product.

The first subheading under *current assets* on Conman's balance sheet is *raw materials*. Raw materials, in money terms, represent the cost of the constituent materials awaiting to be converted to finished product in the production process. *Work in progress*, the second heading, represents the value of that work which falls between raw materials on the one hand and finished product on the other. Not all the operations to produce the finished product may have been completed and such work, particularly in construction, may neither be in the form, nor make up part of a recognizable unit, that can be invoiced to a client. The valuation of work in progress, particularly in construction, is a difficult operation and care must be taken not to overvalue it, since this will increase the total value of the assets of the business on paper only and tend to mislead a reader of the balance sheet.

Further current assets are shown under *debtors*. *Trade debtors* arise from goods that are sold to customers. The customers have received the goods on credit and their accounts are likely to be paid within the course of a month or so. Thus their accounts will then be converted into cash. In the event that it is expected that a customer of consequence will default on the payment of his account, some early adjustment may be required to trade debtors. *Prepayments* arise when payments for certain general expenses, such as rent, insurance, electricity, gas, water, etc., are made during the period before the date of the balance sheet but which will refer to expenses that will not actually be incurred until after that date. Rent, for example, may be paid 12 months in advance, but the rental period may have 6 months to run at the date of the balance sheet. Only half the cost of the rent then becomes a current asset, the remainder being a prepayment. *Accrued income* is income that can arise in matching the expenditures up to the date of the balance sheet with their associated incomes up to that date, whether the expenditures have actually been made or the income has actually been received in cash. This is required in order to make the balance sheet a truly fair and accurate statement of the assets and liabilities.

Cash at the bank and in hand needs no explanation as a current asset.

A company's balance sheet in the conventional form already described literally has two sides to it. The side described above sets out how the funds that have been received into the business have been applied—on what they have been spent, for example, raw materials, and how the temporarily unused funds are held. The detail is relatively broad, but a balance sheet is intended to be a summary only and at a particular point in time. It is now necessary to turn to its second side, which lists the sources of the finance used in the business. This second aspect is a statement concerning the *liabilities* of an enterprise and what it owes.

The first group of liabilities listed in Fig. 12.1 are known as *current liabilities*. These have to be met within the short term, usually within the period of one year. In Fig. 12.1 are listed three types of current liabilities applicable to the business. *Bank and other loans* are the first item because they can normally be recalled at relatively short notice. *Trade creditors* are those businesses that supply many materials and services to the company, such as raw materials, stationery, machinery repairs, etc. and as such need to be paid reasonably promptly so as to ensure prompt future service as required. *Taxation* and social security payments not yet made need to be established and set on one side for payment during the 12 months. *Accruals* are the expenses, such as those unbilled, that must be matched to the income that has been earned from the act of expending them, whether either has been paid in cash or not. In the balance sheet it will be noted that the total current liabilities amount to £9 100 000.

The liabilities shown in Fig. 12.1 are not intended to form an exclusive list (this comment also applies to assets) but rather an indication of the several types of liabilities that companies have. Other items that may be seen include *payments received on account, amounts owed to group and/or related companies, interest on fixed liabilities*, and so on.

The second group of liabilities is sometimes called *fixed liabilities*, which are the company's long-term liabilities that include all those means of financing the company that do not have to be paid back to their source during the next 12 months. Conman's first such item is a 15 per cent mortgage debenture. A mortgage debenture is secured by mortgaging one or more of the fixed assets of the company, probably in this case a particular building or buildings owned by the company. It is repayable in the year 2011. If the company fails then the building or buildings will be sold and the first claim on the proceeds will be used to pay off the debenture loan of £3 000 000. Debentures are loan capital and the interest payable on them constitutes a debt, the non-payment of which may cause the company to be put into liquidation. This is to be contrasted with the dividends payable on ordinary shares, which are at the discretion of the board of directors. If it is decided not to pay a dividend no such drastic action can be taken.

Conman has *called-up share capital* of £13 200 000. This money has been received as a result of issuing ordinary shares to the public. When the shares were sold, probably not all at the same time, the asking price of the shares was paid to the company. This is shown as a liability because the company must account for the money received. When it was received it would be noted as a liability and, at the same time, the cash obtained would become a current asset under that heading. The cash could then be used in the business.

The entry under *profit and loss account* is the balance from that account that is not used or distributed. This is alternatively referred to as *retained earnings* and subsequently becomes part of the *general reserve* and is an investment in the business to support its financial requirements for future operations.

Turning to the narrative form of balance sheet presentation shown in Fig. 12.3, a number of general features of a balance sheet can be described. Firstly, the fixed assets of the company are set out at the top of the balance sheet. It will be noted that the depreciation (see Chapter 4 for a detailed discussion of depreciation) for the year up to the date of the balance sheet is included. In each of the three cases cited, for simplicity, the straight-line method with zero residual value has been used. The lines of the three varieties of fixed assets have been assumed as 60 years for the 'land and buildings' and 10 years for the 'plant and machinery' and 'fixtures, fittings, tools and equipment', respectively

Next below are listed the *current assets*. These are assets which can be converted into cash without too much difficulty, and in a short time if need be. Some of the current assets are already in the form of *cash* but others are potentially cash, such as those for *debtors*. They become cash when the debtors have settled their accounts. Since the *current liabilities* are assumed to be paid during the next 12 months, it is more than likely that they will be settled from the current assets. In the case of Fig. 12.3, the current assets exceed the current liabilities by £3 020 000, so the settlement, if it is required, can take place. It is important that a business should always have a reasonable margin by which current assets exceed current liabilities. The margin between the two is known as *working capital*. The working capital plus the capital of fixed assets is known as the *net capital employed* in the business. In this case it amounts to £16 758 000. The *net worth* of the company is the value of the total assets, both fixed and current, less current liabilities and long-term loans. The net worth of the company is alternatively its *capital*, also known as the *total shareholders' interest*. The net worth of Conman PLC is £14 200 000 at the end of the year of the balance sheet in Fig. 12.3.

12.3 Funds flow from a balance sheet

Balance sheets are static reports detailing the financial state of affairs in a business at a particular point in time. Certain information is not available from a balance sheet, such as how the profit was actually made or how the income of the business was used to increase the wealth of the company. The first of these two points is settled by reference to a profit and loss account, the second by reference to a *statement of funds flow*. It is the latter statement that is now described.

A statement of funds flow shows the additional sources of funding for the business over a specific period of time and how those funds were applied in the business. The period of time is frequently one year and is the period between successive balance sheets. A statement of funds flow is now constructed for the period of time

Balance sheet variations

	1990 (£000)	1991 (£000)	+ (£000)	− (£000)
Liabilities				
Ordinary shares	13 200	13 200	—	—
Reserves	1 000	1 100	100	—
Debentures	—	3 000	3 000	—
Creditors	3 970	4 100	130	—
Taxation	2 000	2 500	500	—
Bank loans, etc,	4 658	4 630	—	28
	£24 828	£28 530	£3 730	£28
Assets				
Development costs	450	450	—	—
Land and buildings	11 308	12 000	692	—
Plant and machinery	1 800	3 000	1 200	—
Fixtures and fittings	180	200	20	—
Raw materials	1 670	2 310	640	—
Work in progress	1 600	1 900	300	—
Finished goods	1 200	1 500	300	—
Trade debtors	4 920	4 850	—	70
Prepayments, etc.	100	220	120	—
Cash at bank/in hand	1 600	2 100	500	—
	£24 828	£28 530	£3 772	£70

Figure 12.4

intervening between the balance sheets of Figs 12.2 and 12.1, the balance sheet of Fig. 12.1 being 12 months later than that of Fig. 12.2.

Firstly, it is necessary to construct tables showing the variation between the *liabilities* and the *assets* of the two balance sheets. These are set out in Fig. 12.4. There has been no change in the ordinary shares issued and thus no variation. The reserves of the first balance sheet have increased by £100 000 in the second and raising a debenture for £3 000 000 when none existed at the time of the first has increased the long-term capital by that amount in the period under examination. Similar variations can be established for the other liabilities and it should be noted that the only one with a negative variation, a decrease, is that for bank loans. The bank loans have been decreased by £28 000. The assets are dealt with in similar fashion. All have shown an increase with the exception of trade debtors, which have decreased by £70 000.

There is an adjustment to be made to the figures in the variation table in respect of capital assets and reserves before transferring them to the statement of funds flow. In the case of capital assets, the balance sheets show the current historic cost of the fixed assets *after depreciation has been deducted*. This is more easily demonstrated in the case of the earlier balance sheet of Fig. 12.2 by referring to Fig. 12.3, the same balance sheet but in a different format. The historic cost value of the 'land and buildings', at the beginning of the year concerned, amounted to £11 500 000. Depreciation of £192 000 has been deducted for the depreciation during the year, leaving an investment of £11 308 000 appearing on the balance sheet in both Figs 12.2 and 12.3. However, in Fig. 12.1, for 'land & buildings' the value of £12 000 000 appears. This must therefore have increased as a result of the investment of an additional amount of capital in that item over and above the earlier value of the investment. The amount of depreciation deducted, however, is not evident. Using the basis of assuming a life of 60 years with no residual value as defined previously, the depreciation will amount to £11 308 000$(\frac{1}{60})$ = £188 460. The new written-down value is therefore £11 308 000 − 188 460 = £11 119 540, say £11 120 000 for the purposes of the balance sheet. Similar calculations for the next two items of fixed assets arrive at written-down values of £1 620 000 and £128 000 for the 'plant and machinery' and the 'fixtures, fittings, tools, and equipment' respectively. For the purposes of the fund flows, because the depreciation must be put into reserve to preserve the value of the assets to the business, the expenditure on additional fixed assets for the year in question must be increased by the amount of the depreciation and the reserves must similarly be increased by the same amount. In this case, the total depreciation amounts to £400 000 and therefore the reserves must be increased by this amount. Additional investment in 'land and buildings' was actually £692 000 + 188 000 = £880 000; for 'plant and machinery' it was £1 200 000 + 180 000 = £1 380 000 and for 'fixtures and fittings' it was £20 000 + 32 000 = £52 000.

The funds flow statement can now be prepared (Fig. 12.5). The first section lists the sources of additional funds used in the business during the year between the two balance sheets. The addition of the depreciation to the reserves now makes the

Funds flow statement

	£000
Sources	
Reserves	500
Debentures	3 000
Creditors	130
Taxation due	500
Reduction in trade debtors	70
	£4 200
Applications	
Land and buildings	880
Plant and machinery	1 380
Fixtures and fittings	52
Raw materials	640
Work in progress	300
Finished goods	300
Prepayments, etc.	120
Cash at bank/in hand	500
Reduction in bank loans	28
	£4 200

Figure 12.5

reserves a total of £500 000. The big item concerns the new debt financing, a mortgage debenture of £3 000 000. To be included in this list as a source is the reduction in trade debtors of £70 000, shown as a negative variation in the summary of balance sheet variations. The second part of the statement shows to what the additional funds have been applied. A major item is for additional plant and machinery and there have been increases in all of the stock items of raw materials, work in progress, and finished goods.

It is important to analyse the flow of funds into the company to see where the finance has been obtained to promote the growth of the business. Of the total of £4 200 000 used in the funds flow statement, £500 000 or 11.9 per cent has come from the business itself, £3 000 000 or 71.4 per cent has come from long-term debt or loan capital and the remainder or 16.7 per cent has come from short-term creditors. The increase in long-term debt is a large proportion of the funds obtained during the year, and although this need not be a concern at this point in time, the proportion of long-term debt in the company as a whole needs to be watched because of the vulnerability caused by the need to pay relatively high interest charges over a long period of time. The use of funds for which a fixed return is paid is known as *financial leverage*. A company making a return on its operations in excess of the percentage paid for its debt finance can increase the return paid on its *equity* or ordinary shares. If the financial leverage becomes too large the risk to the company is increased since interest must be paid by a company even in years when the return may be low.

12.4 Analysing the balance sheet

A company balance sheet, with a small amount of other information, enables the *liquidity* and *solvency* and the *performance* or *activity* and the *profitability* of a company to be assessed. The first two areas, liquidity and solvency, are concerned with testing whether a company can meet its liabilities and to what extent. The profitability and activity of a company can be measured in a number of different ways, but whatever the measures used they are the ultimate verdict on the way in which a company is being managed. In assessing a balance sheet for these four characteristics, it is useful to compare performance over a number of years so as to establish relative performance between one year and another. It also enables the establishment of any trends that may be appearing, either adverse or beneficial. The examination here, where relevant, will look at the two successive balance sheets for Conman PLC for the years to 31 March 1990 (the 1990 year) and to 31 March 1991 (the 1991 year), which are set out in Figs 12.2 and 12.1 respectively, and a hybrid statement that includes some manufacturing information and some that would appear in a profit and loss account for the year to 31 March 1991, in Fig. 12.6. (The figures on the relevant sheets are in thousands of pounds and for the sake of clarity the figures used in the analysis are in the same units and the £ sign has been omitted.)

It is difficult to compare broad statements concerning the financial performance of different companies, or about the same company in different years, without having

Manufacturing, profit and loss account for year to 31 March 1991

	(£000)	(£000)
Materials consumed:		
Opening stock	1 670	
Purchases	20 650	
	22 320	
less: Closing stock	2 310	20 010
Wages		14 320
		34 330
Overheads	4 620	
Depreciation	150	4 770
		39 100
add: Initial work in progress	1 600	
less: Final work in progress	1 900	(300)
Production costs		£38 800
Opening stock— finished goods		1 200
Cost of goods made		38 800
		40 000
less: Closing stock— finished goods		1 500
Cost of sales		38 500
Gross profit		8 396
Sales		£46 896
Administration expenses		2 325
Selling expenses		2 216
Research and development		2 413
Provision for bad debts		520
Depreciation		50
Provision for taxation		218
		7 742
Net profit		654
Gross profit		£8 396

Figure 12.6

the presented data converted to a suitable form. *Ratio analysis* is one method of converting such data for the analysis of financial statements. However, where ratios are used, there is a need to compare them with standard ratios, with ratios for different years, and with those of other similar companies. When ratios are used for analysis emphasis can be placed on one or more different aspects of the analysis, depending on why the analysis is taking place. Someone wishing to establish whether a short-term loan should be offered to a business may place more emphasis on the company's liquidity, whereas someone considering making a long-term loan may wish to see good earning power and operating efficiency. Investors in the ordinary shares, unless speculators, will probably be interested in the long-term position, too. The following ratios are some of those in more common use.

Liquidity ratios:

$$\text{working capital } or \text{ current ratio} = \frac{\text{current assets}}{\text{current liabilities}} \tag{12.1}$$

$$\text{liquid capital ratio} = \frac{\text{current assets} - \text{stock}}{\text{current liabilities}} \tag{12.2}$$

Activity ratios:

$$\text{stock turnover} = \frac{\text{cost of materials used}}{\text{average stock of raw materials}} \tag{12.3}$$

$$\text{debtors' ratio} = \frac{\text{debtors}}{\text{average daily sales}} \tag{12.4}$$

$$\text{net asset turnover} = \frac{\text{sales}}{\text{total assets } less \text{ current liabilities}} \tag{12.5}$$

$$\text{debt ratio} = \frac{\text{long-term debt}}{\text{total assets } less \text{ current liabilities}} \tag{12.6}$$

Profitability ratios:

$$\text{return on equity} = \frac{\text{net profit}}{\text{ordinary shareholders funds}} \tag{12.7}$$

$$\text{return on net assets} = \frac{\text{profit before interest and tax}}{\text{Total assets } less \text{ current liabilities}} \tag{12.8}$$

$$\text{profit margin} = \frac{\text{profit before interest and tax}}{\text{sales}} \tag{12.9}$$

Liquidity ratios are used to measure the ability of a business to pay its short-term debts in the near future. *Liquid assets* are those components of current assets that are available as cash or near-cash. They are alternatively known as *quick assets*. In the case of a balance sheet these consist of *cash, debtors,* and *short-term investments and securities (marketable securities)*. All of these assets can be converted into cash relatively quickly. A business that has total assets that will meet its outside liabilities, that is the liabilities that are owed to persons outside the business, is said to be *solvent*. The outside liabilities are current liabilities plus long-term debt. If a company can meet its current liabilities out of its current assets it is said to be *liquid*. For the year to 1990 it will be seen that Conman PLC had current assets (in thousands of pounds) of 11 090, current liabilities of 8 070, total assets of 24 828, and outside liabilities of 10 628. Conman is therefore both liquid and solvent by these measures.

It is desirable that a company has a working capital ratio (12.1) in excess of 1.00. For 1990, Conman had a working capital ratio of $11\,090/8070 = 1.37$, which is satisfactory. For the following year its ratio was $12\,880/9100 = 1.41$, so that it had marginally improved and remained satisfactory. These ratios are at the bottom of the scale that would be expected of a solid company.

As far as the liquid capital ratio is concerned, Conman had the following:

1990 $(11\,090 - 4470)/8070 = 0.82$
1991 $(12\,880 - 5710)/9100 = 0.79$

This ratio is alternatively known as the *acid test ratio* and is a critical test of solvency. It is usually considered to be a better test than the working capital ratio, because it places particular emphasis on stocks as a part of the current assets of a company. Its desirable value is 1.00 or more, and the two values above indicate that Conman will have to rely on disposing of its stocks in order to meet its current liabilities if the need arises. This may not be very easy to carry out in the short term. A firm having such ratios is weak in respect of liquid assets. In addition, Conman's ratio has deteriorated slightly over the intervening year.

A business should never endanger its supply of working capital at the expense of purchasing too many fixed assets, thus resulting in *over-capitalization*. Such a course inevitably necessitates high-interest bank loans to supplement working capital, and may consequently add to the difficulties of making a reasonable level of profit. Another danger is to have stock at too high a percentage of current assets. In 1991, Conman had let this percentage rise to nearly 45 from 40 the year before. The higher the percentage stock, the less flexibility in the use of working capital. Conman needs to pay attention to reducing its stocks of raw materials and finished goods. Another danger of a shortage of working capital, as reflected by liquidity ratios, is that of *overtrading*, and the company may need to do something about the trade debtors as well as the stock.

Another key factor in the analysis of the liquidity of a business is the length of time that an item of stock remains on the books. The speed of selling inventory can be measured by the *stock turnover* ratio (12.3). Clearly, with a high rate of turnover there is a reduced risk that stock will change in value while a part of the inventory. The cost of materials used can be obtained from the profit and loss account, and in this case it amounts to 15 010. The cost of the average stock of raw materials is $(1670 + 2310)/2 = 2825$. The stock turnover therefore amounts to $15\,010/2825 = 5.31$. The average time the materials are in stock is therefore $3655/5.31 = 68.7$ days.

Another relevant ratio is the *debtors' ratio* (12.4). This ratio represents the average number of days that accounts remain unpaid after making a sale. In the case of Conman for 1991, the debtors' ratio $= 4850(365)/46\,896 = 37.7$ days. This length of time is pertinent to the state of liquidity of the business in that the longer the debtor takes before the account is paid, the longer can the money be used in the debtor's own business, and the longer the period of time that Conman has to finance it. The figure of 37.7 days is difficult to judge, since much depends on the terms of sale offered to the debtors. If Conman sells on the basis of 60 days' credit, 37.7 is excellent; if the credit is 30 days, it is poor. Irrespective of the terms offered, the shorter the collection period the better, and the length of time for successive years should be established, compared, and examined for trends. This analysis needs to be supplemented by a list of customers and the age of their accounts before being paid. The age can be classified into periods of 0–10 days, 11–20, 21–30, etc. with the number of customers in each. It may show that a few very good payers are reducing the average disproportionately and that there are some very bad payers who need particular attention.

Net asset turnover (12.5) summarizes the efficiency in use of the business assets. The 1991 ratio is equal to 46 896/19 430 = 2.41. This should be compared with an industry ratio. A decline in this ratio from year to year is an indication that the business is utilizing its capacity in the form of assets to a lesser extent in each year by achieving less and less annual sales (or by achieving the same sales with increasing assets). The company is therefore not generating sufficient volume of business for the extent of the assets employed and should either sell some of its assets or generate more sales or both.

The *debt ratio* is a financial ratio which concerns the capital structure of a business by indicating the proportion of the assets that are funded by long-term debt. For Conman in 1990, the debt ratio was 2558/16 758 = 15.26 per cent and in 1991 it has risen to 5130/19 430 = 26.40 per cent. There has been a considerable increase in this ratio over the intervening year between balance sheets due to raising the 15 per cent mortgage debenture. Even so, the ratio for 1991 is still unlikely to cause a problem. The significance of the ratio is that it indicates how much the assets of the business could be allowed to fall below their book value, before they fail to realize sufficient funds to pay off the long-term debt, in the event of failure of the business. The higher the value of the fixed assets, the easier it is to raise long-term debt against them, hence a move to revalue assets prior to raising debt capital. The credibility of the ratio depends, of course, on a realistic valuation of the company's assets. Owners may wish to see high debt ratios because raising new equity almost inevitably means giving up some measure of control over the business.

The final three ratios included in the list above are all measures of profitability. The *return on equity* (12.7) is a measure of the efficiency with which shareholders' equity is employed in the business. The *ordinary shareholders' funds* include the *ordinary share capital*, the *general reserve*, and the *retained earnings*. For 1991, the net profit is 654 from Fig. 12.6. The return on equity ratio is 654/14 300 = 4.57 per cent. This is a low return on capital compared with that to be obtained elsewhere and the ordinary shareholders cannot be satisfied with it. It is probably due to either falling profit margins on sales or a low rate of asset turnover or both. Both need to be investigated and improved.

Return on net assets measures the financial efficiency with which capital is being used in the business. Conman's ratio for 1991 is 872/19 430 = 4.48 per cent—not vastly different from the *return on equity ratio* for that year. Comments are similar to those for the above ratio.

Finally, the *profit margin ratio* for 1991 is 872/46 896 = 1.86 per cent. This ratio attempts to answer why the *return on equity* should be so low by looking at one of the two probable causes cited above. This low profit margin ratio indicates that the sales prices charged by the business are relatively low or that the production costs are relatively high, or both. The trend over past years should be established to see if this is a falling margin or not.

Typical comments about Conman, made without prescribing a particular industry or working environment, may be that its liquidity is just all right, although its inventory of materials and finished product is badly in need of reduction. The company does not have a high debt ratio, and it could almost certainly be increased if debt finance is needed. The net asset turnover is probably on the low side and that compared with the return obtained on assets may lead to consideration that some assets should be sold. The production costs are almost certainly too high, sales are weak and profit margins are poor.

12.5 Construction financing

It can be seen from the previous sections of this chapter how important is the question of investing in fixed assets for a business. Care must be taken that the assets that are purchased are the best for the purpose that the business needs at that particular time and for the foreseeable future. It is necessary that economic feasibility studies are carried out in order to ensure that the financial implementation of the purchase is such that an acceptable rate of return will result from the investment. When the feasibility phase is completed and the decision to proceed has been taken, the business has to find the finance in order to fund the purchase, whether from self-generated funds or from outside. There are a number of ways in which this can be done, and some of these have already been touched upon in previous discussions. However, at this stage, particularly if the project proposed is large, it is necessary to investigate the timing and the amounts of finance that will be required, and the implications of these on the project construction and the ultimate return of the investment when the asset is commissioned.

The importance of working capital to a business has been stressed in reviewing the implications of a balance sheet and in using ratio analysis. This situation applies equally when a construction project is put in hand. On the one hand the owner wishes to pay the least cost for the project compatible with getting an end result that is satisfactory for the purpose, and will wish to have the project constructed and commissioned in the least possible time so that production can begin, followed by a positive cash flow; also, during the construction period the owner will wish to delay the payment of the project accounts as long as is reasonably possible so as to keep interest charges on borrowed finance to as low a level as possible. On the other hand, the contractors appointed to construct the works look forward to prompt payment by the owner and a construction programme that will reduce the working capital that they have to find during the construction period to an absolute minimum. The following example examines some of these problems.

EXAMPLE 12.1

An owner wishes to construct a new factory at an estimated capital cost of £5 million. The capital sum will be borrowed at an interest rate of 14 per cent. A commitment charge to reserve the funds is payable half-yearly on the whole of the sum to be borrowed until the total sum has been received, at a rate of $\frac{1}{2}$ per cent per year. The limit of the borrowing is to be £5 million and any excess, whether capital or interest, is to be provided by the owner at a cost of 14 per cent per year.

A tentative construction programme over 4 years is prepared together with a cash flow statement. At this stage, expenditure in any one year is assumed to occur at mid-year. The first form to be examined is for a 4-year programme and this is detailed in Table 12.1. The time scale has been reversed in that the completion of construction is at year 0 and the construction programme starts at year −4. It will be seen that this programme of construction results in interest and commitment charges of £1 306 856 in addition to the capital investment in the asset.

The second programme to be examined reduces the overall construction programme to 3 years and results in a reduced charge of £1 074 938—a reduction of £231 918. This alternative is detailed in Table 12.2. Two further alternatives are examined. The first is one in which as much as possible of the construction work (in cost terms) is brought forward to the early part of the programme in a 3-year period, as in Table 12.3, when the total interest charges reduce to £512 298. The second alternative is to look at the possibility that interest rates will increase to 18 per cent before the construction work commences. This is detailed in Table 12.4 and shows, for the same 3-year programme as that in Table 12.2, that

Table 12.1

Year	Capital outlay, £	Commitment charge $\frac{1}{2}$% p.a., £	Interest at 14% p.a. on cumulative outlay, £	Cumulative outlay, £
−4	—	12 500	—	12 500
−3½	500 000	12 500	875	525 875
−3	—	12 500	36 811	575 186
−2½	1 000 000	12 500	40 263	1 627 949
−2	—	12 500	113 956	1 754 405
−1½	1 500 000	12 500	122 808	3 389 713
−1	—	12 500	237 280	3 639 493
−½	2 000 000	—	254 765	5 894 258
0	—	—	412 598	6 306 856
Totals	5 000 000	87 500	1 219 356	6 306 856

Table 12.2

Year	Capital outlay, £	Commitment charge ½% p.a., £	Interest at 14% p.a. on cumulative outlay, £	Cumulative outlay, £
−3	—	12 500	—	12 500
−2½	1 000 000	12 500	875	1 025 875
−2	—	12 500	71 811	1 110 186
−1½	2 000 000	12 500	77 713	3 200 399
−1	—	12 500	224 028	3 436 927
−½	2 000 000	—	240 585	5 677 512
0	—	—	397 426	6 074 938
Totals	5 000 000	62 500	1 012 438	6 074 938

Table 12.3

Year	Capital outlay, £	Commitment charge ½% p.a., £	Interest at 14% p.a. on cumulative outlay, £	Cumulative outlay, £
−3	—	12 500	—	12 500
−2½	500 000	12 500	875	525 875
−2	—	12 500	36 811	575 186
−1½	500 000	12 500	40 263	1 127 949
−1	—	12 500	78 956	1 219 405
−½	1 500 000	12 500	84 308	2 816 213
0	2 500 000	—	196 085	5 512 298
Totals	5 000 000	75 000	437 298	5 512 298

Table 12.4

Year	Capital outlay, £	Commitment charge ½% p.a., £	Interest at 18% p.a. on cumulative outlay, £	Cumulative outlay, £
−3	—	12 500	—	12 500
−2½	1 000 000	12 500	1 125	1 026 125
−2	—	12 500	92 351	1 130 976
−1½	2 000 000	12 500	101 788	3 245 264
−1	—	12 500	292 074	3 549 838
−½	2 000 000	—	319 485	5 869 323
0	—	—	528 239	6 397 562
Totals	5 000 000	62 500	1 335 062	6 397 562

the interest charges will rise to £1 397 562, putting £322 624, or approximately 6½ per cent, on the total interest charges of the Table 12.2 programme.

If it is assumed that, when the construction programme is completed, the factory commences production and in the first year of production results in a surplus of £400 000 and in each subsequent year a surplus of £1 200 000 is available. Figure 12.7 shows the way in which capital is recovered. It is assumed that interest at the rate during the construction period continues to be charged on the outstanding balance of the initial capital investment. The full line graph of Fig. 12.7 shows the 4-year construction programme of Table 12.1 and the broken line the 3-year programme of Table 12.2. In the case of the 4-year construction programme the project breaks even at year 13; in the case of the 3-year project it breaks even at year 12. This example, even though the returns from the production when it starts are generous indeed, illustrates the dramatic effect of high interest rates, both on the size of interest payments during construction periods and on subsequent capital recovery after construction is completed. Not only are the these two factors important but so is that of risk. The shorter construction programme illustrates that the return

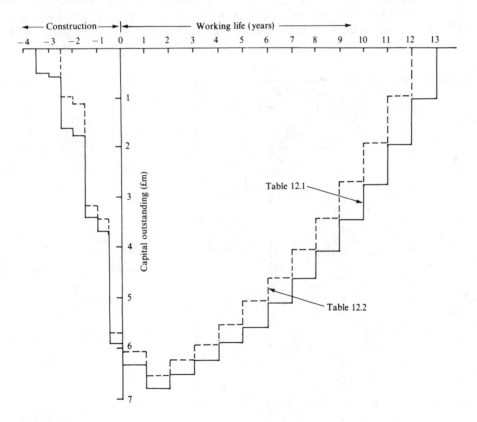

Figure 12.7

of capital and interest is effected 2 years earlier than in the longer programme. This means that the investment, as a profitable venture, is at risk for a shorter period of time.

12.6 Expenditure forecasting

Having decided to go ahead with the creation of a capital asset, an owner will require to know how the finance must be made available to pay for the cost of the work and for all the professionals who are involved in the project as the work proceeds. There will be fees to engineering designers, architects, construction supervisors, etc. One very useful and basic tool for establishing this phasing of the finance is an *S-curve*. This name stems from the fact that typical S-curves usually take the form of an old-fashioned, long 'S'. The S-curve is a graphical statement of the cumulative cost or value that is involved in a project displayed over time. Figure 12.8 shows a simple preliminary curve that is in sufficient detail to produce data as a first approximation. It will be seen that monthly costs are distributed over the bars of a bar chart, each bar representing an activity in a construction of the project. The costs for each successive month are then summed vertically and are plotted to produce a curve that represents the cumulative cost at any point in time of the project.

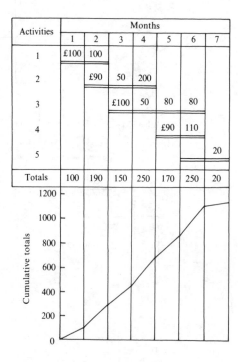

Activities	Months						
	1	2	3	4	5	6	7
1	£100	100					
2		£90	50	200			
3			£100	50	80	80	
4					£90	110	
5							20
Totals	100	190	150	250	170	250	20

Figure 12.8

Sometimes it is possible to use an empirical curve that predicts the pattern of expenditure. One such alternative assumes that a quarter of the cost will be expended in the first third of the project, half will be expended in the middle third and the remaining quarter will be expended in the last third of the time. A second simple model has been proposed by the US Corps of Engineers for forecasting the earnings of a contractor against the expenditure of project time. The model is

$$y = \sin^2(90x)$$

where y is the decimal fraction of the contractor's earnings and x equals the decimal fraction of the total project time completed.

More sophisticated models have been produced, one example of which is that produced by Schlomo Peer in developing a cash flow planning system for housing construction. The curve obtained was:

$$y = [0.009 + 0.2731(W/T) - 1.0584(W/T)^2 + 5.4643(W/T)^3 - 3.6778(W/T)^4]C$$

where y = cumulative cost
   C = total cost
   W = number of time units passed
   T = total number of time units, i.e., total construction duration

(W/T), therefore, is the relative fraction of total time. y is the cumulative cost that an owner will need to disburse when the work has proceeded for W units of time out of a total of T units. Care must be taken, of course, to ensure that due account is taken of payments that are subject to retention, disputed items and so on.

12.7 Optimizing programme duration

We have seen that the length of the construction programme is a strong influence on the financial outcome of the construction of a fixed asset. Clearly there is a need to give careful consideration to trading off extra interest payments to those payments that are often needed to reduce programme length. When a contractor has produced an estimate of the price an owner needs to pay to get a project constructed, it is usually at what is known as the *normal cost*. That is, the price is the one that a contractor believes will result from carrying out the work at minimum cost to the contractor and by the best and most economic and efficient methods. If the programme is then to be accelerated, additional costs are likely to be incurred by such items as premium payments for overtime and bonus and by additional resources.

The owner can undertake an exercise to establish the amount of interest for finance that will be saved for different lengths of construction programme and for different phasing of the work elements. The result of both the increased costs for programme reduction and the decreased costs for interest saving can be plotted

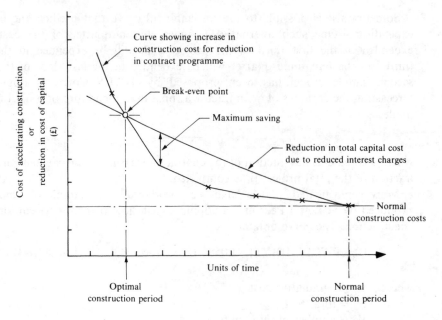

Figure 12.9

as two graphs, as in Fig. 12.9. The straighter, shallower curve of Fig. 12.9 shows the reduction in total capital cost due to reduced interest charges and, as expected, the graph has negative slope downwards from left to right. The other curve shows the increase in construction cost for reductions in the contract programme. The two curves start off from the same point on the right-hand side of the graph at normal construction cost. They ultimately cross at the break-even point where interest saved just equals increased construction cost. At this point an owner would obtain a reduced construction programme at no extra cost above that for the normal cost. The vertical cut-off between the two curves, assuming that the interest reduction curve is the upper one of the two, represents the saving to be obtained for various construction periods and the longest cut-off is identified as the maximum saving.

Summary

Understanding the wider implications of financial matters is an important feature of a construction manager's job. This entails understanding how a company is financed and what part an individual project plays in the company finances as whole. This chapter looks at a few fundamentals of financial management and starts off with an introduction to what a company is when considered as a legal entity. The chapter goes on to discuss in some detail what a company balance sheet is and the type of information that is included in it and where it comes from. Next it deals with the funds flow in a company. It details the sources of funds for a particular year using an example balance sheet and goes on to explain how it can be established where those funds have been put to use in the business during the previous year. In order to tell whether a business is properly funded and managed, it is necessary to examine successive balance sheets to see what changes in performance and results are being achieved and whether financial success is improving or otherwise. Ratio analysis is the chosen method for doing this and the commonly used ratios are stated, described and put to use in examples. Finally, the question of how a manager should manage the cash flows of a project to create fixed assets for the business is introduced, with a comment on the cost trade-off between interest saving and programme reduction.

Problems

Problem 12.1 The XYZ Company Limited borrows £5 000 000 on 1 January at an interest rate of 18 per cent per year, payable half-yearly, in order to construct a factory. In order not to have the money idle it is all deposited in a bank account which brings a return of 12 per cent each half-year. Funds are withdrawn from this deposit account as they are required in order to pay the contractors who are building the factory and supplying the production equipment.

The factory takes 3 years to complete to the stage where it can be used for production purposes. The payments to be made to the contractors from the account are estimated to be £500 000, £1 000 000, £2 500 000, and £1 000 000 each on 30 June at one-yearly intervals, starting in the same year as the money is borrowed.

During the year ended 31 December of year 4 after the money was borrowed, the company expects to make a net profit of £200 000 on production from the new factory. In each subsequent year it is expected that the net profit will be £700 000 per year.

How soon will the firm be able to pay back the loan and the interest thereon, assuming that it devotes the whole of the net profits as they arise each year, to doing so? Illustrate your answer graphically.

Problem 12.2 A company is in need of approximately £750 000 of capital. The board of directors considers a number of ways of raising this amount of money, the first of which is to issue more equity stock. It is estimated that capital raised in this way will cost the company 8.25 per cent.

The only other viable alternative at this time is considered to be the sale of the company's head office building on the basis that it can be leased back from the purchaser for the remaining 30 years of its estimated useful life. This situation is investigated further and an offer of £750 000 is received for the building.

The arrangement for leaseback rental is offered in two alternative forms:
(a) Rent for each of the first 10 years is to be £100 000. The rent is then to be increased to take into account inflation in the previous 10 years at the rate of 3 per cent per year; the adjusted rent is to remain in force until the end of year 20, when a further increase will be applied, again taking account of inflation at

3 per cent per year for each of the previous 10 years. This third rate of rental is then to remain in force for the rest of the building's 30-year life.

OR

(b) A flat rate rental of £120 000 per year for the whole of the 30 years.

The company, if it sells and leases back, is then allowed the rental of the building against profits for taxation purposes. It is anticipated that there will always be sufficient profit to take full advantage of this. A corporation tax at the rate of 35 per cent of profits is payable 1 year in arrears. It is estimated that it will remain at this rate for the future duration of the lease.

If the rent is payable on an annual basis in advance, which is the cheapest source of capital for the company?

Problem 12.3   There are three categories of financial ratios commonly in use.
(a) Explain the form of each ratio in the three groups and its function to a company's financial manager.
(b) Explain why the use of ratios may have a different significance to different managers, depending on their type of employment.

Problem 12.4   The following data derive from financial statements of the PQR Construction Company Limited:

Balance sheet

Cash	£12 300
Receivables	15 500
Inventory	46 300
Total current assets	74 100
Land and buildings (less depreciation)	60 500
Total assets	£134 600
Trade creditors	9 000
Bank loans	22 000
Taxation	5 250
Total current liabilities	36 250
Long-term debt (12 per cent)	15 000
Net worth	83 350
Total claims on assets	£134 600

Income statement

Sales		£142 000
Cost of goods sold		
Materials	£51 300	
Wages	30 500	

Energy	8 500	
Depreciation (15 per cent)	9 075	99 375
Gross profit		42 625
Selling expenses	£14 200	
Administration expenses	12 000	26 200
Operating profit		16 425
less: Interest		1 800
Net profit before tax		14 625
less: Tax		4 388
Net profit		£10 237

Calculate the ratios for PQR and comment on the outcome.

Problem 12.5 The following is the balance sheet of the AJX Co. Ltd, a family concern in which two families hold all of the issued share capital. The company is controlled by two directors.

Balance sheet

	Year 21		Year 22	
Fixed assets				
Land and buildings		£25 500		£25 500
Plant and machinery	28 750		37 700	
Depreciation	10 000	18 750	12 000	25 700
		44 250		51 200
Current assets				
Stock and work in progress		35 850		46 700
Trade debtors		20 000		34 000
Cash		4 700		1 000
		£104 800		£132 900
Authorized and issued share capital				
30 000 ordinary shares				
of £1.00 each		30 000		30 000
Profit and loss account		12 200		18 300
10 per cent mortgage debenture				
repayable 2010		10 000		15 000
Current liabilities				
Trade creditors	51 600		67 100	
Taxation	1 000	52 600	2 500	69 600
		£104 800		£132 900

Sales	£250 000	£275 000
Cost of sales	220 000	235 000
Directors' fees	8 000	20 000
Depreciation	5 000	2 000

Determine whether you would be prepared to lend money to the company, giving your reasons and saying under what conditions.

Problem 12.6　From the following information prepare a statement showing the monthly bank balance for the 6 months ended 31 March.

At 1 October the opening bank balance is £10 000. For the next 6 months the sales are expected to be £160 000, 155 000, 165 000, 165 000, 190 000, and 155 000 respectively. The credit terms are 1 month. The company is likely to have sundry expenses during each month of £9 000.

Wages are paid in the month in which they are earned and are estimated to be £30 000 per month, rising to £38 000 in and after December. Trade creditors are paid at the rate of £50 000 per month up to December, when they rise to £75 000 for December, only reverting to £60 000 thereafter. Tax is paid in December and will amount to £25 000.

A project for the construction of a factory extension will start on 1 November, having a capital cost of £100 000. A mobilization payment of £15 000 has to be paid to the contractor on 1 November, with the balance of the cost paid in equal instalments, the last being paid on 1 April, 6 months later.

Further reading

Drake, B. E.: A mathematical model for expenditure forecasting post contract, *Proceedings CIB W-65 Second Symposium on Organisation and Management of Construction*, Haifa, 31 October–2 November, 1978, pp. II–163 to II–183.

Foster, G.: *Financial Statement Analysis*, 2nd edn, Prentice-Hall, London, 1986.

Horngren, C. T. and G. L. Sundem: *Introduction to Financial Accounting*, 3rd edn, Prentice-Hall, London, 1987.

Merrett, A. J. and A. Sykes.: *The Finance and Analysis of Capital Projects*, 2nd edn, Longmans, London, 1973.

Paish, F. W.: *Business Finance*, 6 edn, Pitman, London, 1982.

Peer, S.: Application of cost-flow forecasting models, *Journal of the Construction Division*, ASCE, vol. 108, no. CO2, June 1982, pp. 226–232.

Rockley, L. E.: *Finance for the Non-accountant*, 4 edn, London, 1984.

13. The classification and distribution of costs

13.1 Introduction

Cost control is a process that should be carried out throughout the life of a project, from the inception of an idea in the client's mind to the final completion of the project and the final payment to the contractor who has constructed the work at the site. As a subject in the field of construction, it can be divided into two major areas:

1. The control of cost during the design stages, so that the proposals under design fall within the original estimates for the scheme.
2. The control of cost, principally by the contractor, once the work of construction has commenced. This is an attempt by the contractor to keep the cost to them of carrying out the work within the moneys that will be paid to them by the client as a result of valuing the completed work in accordance with prearranged rates and prices.

This and the next chapter are concerned with the second area, that of controlling cost during the execution of work.

Traditionally, cost control in the construction industry has not received the attention that it has in other industries, for example, the production engineering industry. The reason for this is undoubtedly the ever-changing environment of construction work as opposed to the more precisely defined environment of production engineering. Many of the less technically-minded construction companies, usually the smaller ones, are content to wait until the termination of a contract before they know whether there is a positive or negative difference between the price for which they offered to do the work and the cost of it to themselves. Some firms rely on making good a deficiency in this respect by submitting claims to the client for additional moneys. Not always will the claim have a totally fair and legitimate basis. While in many instances there are legitimate claims to be made because of the circumstances of contract work, it is unfortunate that some firms will gauge the size of their claim by the extent of the deficiency and the profit margin that they consider to be appropriate.

401

In the construction field, cost control must be the responsibility of the construction engineer. This is the person with the feeling for the value of the work and who is most likely to be intimate with the very detail of the work. The accountant, who might also be expected to make a contribution in this field, has responsibilities that are largely historical as far as the money is involved. Historical information from accounts is of little use in construction cost control, because it is invariably available at too late a date to be effective as a control tool.

13.2 Definitions

The following are definitions of the terms commonly used in connection with cost control procedures:

Accounting includes book-keeping and is the proper recording of information about a company's trading with its clients, subcontractors, suppliers, and employees. The accounts of a company facilitate the preparation of profit or loss statements on the trading of the company and also on its financial position at any time. There are certain minimum legal requirements with regard to the maintenance of accounts.

A *budget* is a plan for the future against which actual results can be measured.

Cost is the amount a purchaser will have to pay for goods or a service. The total amount a client has to pay for a building, for example, is the cost of that building to the client.

Cost analysis is the subdivision of cost under various elements of the whole contract or building. It is normally used in those aspects of cost control concerned with design, and a heading might be related to the various structural elements, for example, of a building.

Cost control is the whole process of controlling the expenditure of cost on a project, from the inception of the idea in the client's mind to the completion and final payment on site.

Costing is the analysis of all expenditure so that it can be allocated to various contracts, processes, or services, for the purpose of ascertaining cost.

Price is the value of goods or services measured in terms of money. Price is what is received by a vendor in a sale and would be the total amount received by the contractor for carrying out a development.

Spot cost is a cost for a particular operation based upon a limited observation and the measurement of associated cost for a short period.

Standard costs are the costs of standard outputs or consumption for plant or employees under specified conditions of environment.

Unit costing is evaluating the cost per unit, whether this is a cubic metre of concrete, a square metre of formwork, or a cubic metre of excavation.

13.3 The purpose of cost control

The most important day-to-day use of a cost control system is that of drawing immediate attention to any operation that is being pursued on a contract and is proving to be uneconomic to the contractor. If a particular operation or process is being carried out inefficiently, immediate warning must be given to the contract management so that action can be taken forthwith to put the matter right. There is little point in having sound historical information when the operation is completed, because nothing can be done at that stage but accept that the operation was carried out inefficiently from a cost point of view, if it is proved that a loss was incurred. In addition to providing a control facility over very short operational periods for current work, the cost control system should be such that overall financial control can be exercised and the relationship between the monthly valuation for work completed and the cost thereof to the contractor can be assessed at regular intervals. The different purposes of these two areas of control are that the day-to-day warning is required by the immediate site management and the overall financial control on, say, a monthly basis is for management at a higher level.

The second function of a cost control system is to provide feedback to the estimator who was responsible for pricing the tender in the first instance and will be responsible for pricing more tenders in the future. The value of such feedback is often of a limited nature, owing to the fact that, in construction work, conditions tend to vary very widely from one contract to another. Any feedback of this nature provided by a cost control system must be accompanied by a full and complete description of the conditions pertaining to the particular costs. Such feedback tends to be most valuable when concerned with the output of machines, for example, excavators. Such outputs as will be fed back are best obtained when the machine's operation is in full swing on part of a contract where it will be working more or less continuously for a reasonably long period such as two or three weeks. This period is long enough for the output to be averaged in such a way that it takes account of the normal day-to-day ups and downs of operation.

Thirdly, the cost control system must provide data for the valuation of variations that may occur during the course of the contract. Frequently, during a construction contract, rates have to be calculated for operations that differ in one way or another from those originally conceived. The maintenance of proper cost records enables the contractor to strengthen the case for the buildup of a new rate and, although this is not all-convincing, since the contractor's operation may not be very efficient, at least there is a basis for the ensuing discussions.

It will be evident that, for a system of control, the first of the above purposes is the most important. The other two functions tend to make historic use of information that is collected. In addition to having the above functions, it is necessary that a cost control system is simple and easy to install. Its accurate operation will depend upon a large number of individuals in the organization playing their part, and many of them will be individuals who have not been trained to make their living by the allocation of costs. It is important, too, that any system that is installed should be readily reconcilable with standard forms used within the company for other purposes. Examples are paysheets, the plant hire returns, the company's plant returns, and the system of preparing valuations for payment.

Cost control for a construction contract is normally limited to the cost of labour and plant. This is because the labour and plant are the two areas in which there is most likely to be inefficient working. In addition, the interaction of the use of mechanical equipment with labour on the site is highly important. The best combination of mechanization and labour must be achieved with the object of obtaining the optimal cost for the work. It is possible, too, to arrive at a quick check of the allocations of labour and plant from the wage-sheets and the plant returns. The total of the field allocations of labour should agree with the totals recorded on the weekly payroll.

The control of the cost of materials used on a site is obviously necessary, but it is normally quite satisfactory to carry out checks at approximately monthly intervals. One of the principal difficulties of carrying out such a check is the accurate assessment of the amount of materials on the site, in other words, the difference between the materials on delivery and those incorporated in the work. A system for doing this must be sufficiently flexible to allow for adjustment, since clearly, in certain cases, especially where a large quantity of high-priced material is being used, there will be a need to carry out a check at more frequent intervals. Cement is a material that should be checked on a weekly basis, especially where it is being used through a large batching unit with a high output every week. A small error in the weighing mechanism can easily waste large quantities of cement.

The estimate prepared for any contract must be the basis for the cost control of that contract, if the tender is successful. It is desirable, therefore, that the figures prepared by the estimator before submitting the tender should be in a form suitable for use in cost control at a subsequent date. This does not mean that a cost control system should be tied to the rates in a bill of quantities. While most estimating methods allow for the ultimate pricing of a bill of quantities, the rates are rarely built up in strict accordance with the descriptions of the bill. On the other hand, there is little point in the estimator receiving feedback on costing from the site if it is not able to compare this information with that assembled in the preparation of the tender.

In preparing a competitively priced bill of quantities with a tender, the prime objective of the company concerned is to win the award of the contract. The object of pricing the bill of quantities is to ensure that the company submitting the tender will be repaid the cost of the work to them plus a margin for their overheads, profit, and risk, even though the bill may not be a part of the contract. It assumes that the work to be carried out within the contract will be carried out at the level of efficiency that was assumed at the time of the estimate. The estimate of the cost of the work, therefore, that is used in the control process will not necessarily coincide with the estimate as it was submitted in the tender. Certain facts may have come to light since the award of the contract that will result in the making of a re-estimate and new thinking about the way in which the work will be carried out. One important piece of information that is required from the estimate for the purposes of control is a precise statement showing the cost of every operation involved in the work. It may not be possible to relate this too closely to the items in the bill of quantities because there will be, on many construction jobs, a large amount of temporary works, some possibly at very high cost. The construction of a cofferdam for a bridge pier, for example, may be a major item of expenditure in a bridge foundation contract, but no specific item for it may appear in the bill of quantities. It is important, too, that the estimate used for cost control purposes should be related very closely to the contract programme.

In making a comparison of the costs incurred with the priced bill of quantities or the estimator's analysis, a *cost standard* is being used. It must be remembered that this cost standard is one that is set by the estimator at the time of preparation of the estimate and one that falls within the estimator's judgement. In the light of further information which becomes available it may be necessary to use other judgements of a better-informed nature for the purposes of control.

13.4 Classification of cost control methods

Prior to the operation of cost control for a contract, it is necessary to decide upon the extent to which control is required and upon the amount of detail that will be entered into when carrying out the work. Many methods of cost control have been tried in the past in different companies. Many methods have not survived with time and others have little support from those who should be operating them. One of the prime difficulties in operating a detailed cost control system is that of cost itself. To carry out a comprehensive and detailed cost control system can be an extremely expensive operation in the time of cost clerks, etc., for a large contract. The following are a number of categories in which costing systems tend to fall:

1. *By comparison with a cost standard.* The comparison with the standard set up by the estimator at the time of the tender has been discussed above. These are not the only standards by which efficiency can be judged; others include those set up by the work study department of a company, previous 'record' outputs that have been achieved within the company or within the knowledge of the company's employees in the past, and standards that have been published in books primarily for the use of estimators, giving data on recommended outputs for labour and plant.

2. *By subdivision by detail* The method involving the least detail is the one whereby the contractor, usually the small builder, waits until the end of the contract before he compares the amount of money that he has coming in with the amount of money he has had to pay out to complete the contract. This is an inexpensive but extremely risky operation, which involves little or no control of cost.

The next stage entails the comparison of costs at the valuation period for the contract. This means that, normally at a monthly intervals, the valuation or the amount of money claimed by the contractor is compared with the amount of money that he has expended for carrying out the work over a similar period. Such a method lends itself to inaccuracies, since it is well known that many surveyors tend to keep some amount of cost up their sleeve for the 'rainy day' and contentious items are often omitted from a valuation, even though it is reasonably certain that some payment will result from them. It is, however, a better system than the previous one, since at least it is carried out regularly and for a shorter interval of time.

The system with the greatest amount of detailed costing is that where costing is carried out for each operation as it is performed on the site. This, as has been stated before, can be an expensive way of costing, particularly as many operations are unimportant in the context of the whole contract.

3. *By detail of system*   It may be decided that the most important feature of the costing for the contract should be that of keeping costs on particular parts of the contract that seem to be vulnerable to inefficiency. Such costs might be kept, for instance, on particular trades or over particular periods of the work (to check with the duration of the costing period). Under this heading might also be included the detail as to whether the costing is going to be kept for labour only, for labour, plant and materials, or for any other suitable combination.

4. *By integration with other functions*   It may well be that the cost control system will not be operated as a separate entity but will be combined with some other necessary operation in the administration of a contract. For example, it may be combined with the organization of an incentive or bonusing scheme, since in many of the profit-sharing types of incentive schemes it is necessary to know the cost efficiency of the work before a bonus can be allocated. Alternatively, such a control system may be combined with a labour utilization scheme, where control is kept on the optimal utilization of the labour employed.

13.5 The budget

The *budget* is a financial plan for the contract as a whole. It is used in construction work to determine the amount of liquid cash that will be required over the various periods of a contract, and as a yardstick against which actual progress can be measured. The budget is a financial version of the programme—a financial forecast. It is an important tool of management, inasmuch as the trading position of a construction company can be established by having budgets for all of the work on hand. It is a means of insuring against the risk of *overtrading*. (Overtrading arises when the current liabilities of a company exceed the current assets, even though the business is solvent.) In such a situation, a company requires more working capital. Considerable amounts can be tied up in construction work, particularly if a high rate of working is achieved on a contract where the client does not meet the valuations promptly.

The financial budget may be shown in graphical form and a typical example is shown in Fig. 13.1. The schedule shows the starting and completion date for the contract, and a curve is plotted of the estimated progress of the work in terms of money. The estimated progress is usually shown as a smooth curve and will be based upon the experience of the particular company carrying out that type of construction work. The curve, as will be seen in Fig. 13.1, represents a slow acceleration of the rate of working in the first few months, a steady but faster rate during the middle period, and a gradual tailing-off of the rate of working towards

—

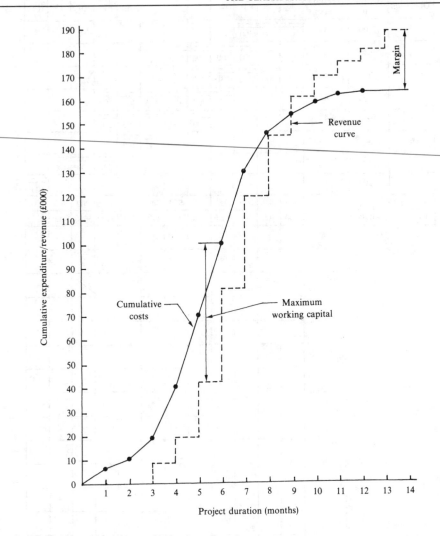

Figure 13.1

the end of the contract. The empirical rule of carrying out 50 per cent of the work in the middle third of the time and the remainder of the work equally divided between the first and third thirds of the duration is often used and represents a fairly realistic estimate, as discussed in the previous chapter.

Alternatively, the budget curve may be derived from an initial distribution of cost against a bar chart, as in Fig. 13.2. It is as well to remember that the total cost of work may be split into a number of constituents and, from a contractor's point of view, each may have a different phasing when considering payment. For example, most labour costs will be paid on a weekly basis; materials, plant, and subcontractors

Operation	Direct cost(£)	Duration (weeks)	1	2	3	4	5	6	7	8	9	10	11	12
1	1200	3	200	500	500									
2	650	4			100	100	200	250						
3	700	3			200	250	250							
4	1100	6				150	200	350	250	300	100			
5	1000	5					200	250	250	200	100			
6	300	2							200	100	100			
7	500	3									200	200	100	
8	200	2											100	100
Weekly totals (£)			200	500	800	500	850	850	700	600	400	200	200	100
Cumulative totals (£)			200	700	1500	2000	2850	3700	4400	5000	5400	5600	5800	5900

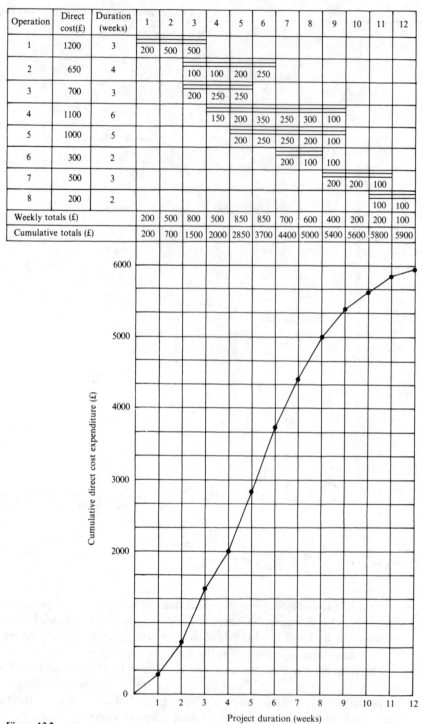

Figure 13.2

will be paid for on a monthly basis unless other arrangements are negotiated. Hence the budget cost curve for a contractor may need to take these factors into account. The individual curves and the cumulative curve for these items are shown in Fig. 13.3.

A second line must be plotted against the estimated progress line, representing the forecast of money that will be received by the contractor for work already completed. Since this money will be paid as a result of monthly valuations the line will be stepped, and, because of the normal procedure whereby the client retains a percentage of the valuation, it will commence on the lower side of the estimated progress line, as illustrated in Fig. 13.1. At the budget stage it is estimated that a profit will be made on the contract and, ultimately, the stepped line should cross the estimated progress line, the difference between the final points on both lines being the anticipated profit to be gained. From this diagram it can be calculated, by taking the difference on the vertical ordinate at any point in time, how much money the contractor will have outstanding for any period of the contract. It will also locate the time during the contract when the maximum amount of money will be outstanding. This is important in taking the company view because, with a number of contracts operating, it will be important to attempt to phase them together so that the outstanding capital is within the limits of the working capital of the company. Progress in the field of cost can be plotted against the budget line as another line on the budget graph.

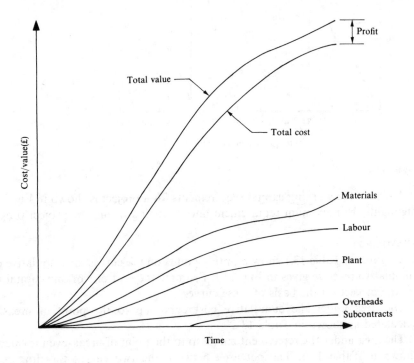

Figure 13.3

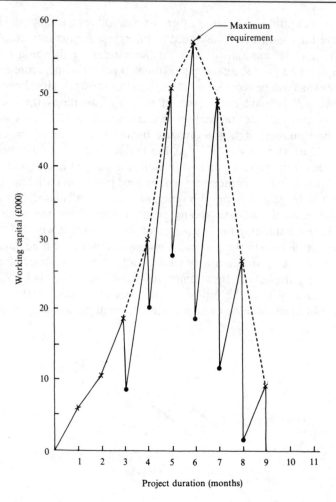

Figure 13.4

A plot of the working capital requirements for a project is shown in Fig. 13.4. This highlights the maximum requirement and its timing within the project programme.

EXAMPLE 13.1

The figures in Table 13.1 show monthly points on the curves for cumulative cost and cumulative revenue gives in Fig. 13.1. Estimate the cost of working capital required for the project on the basis of these curves.

The area under the cumulative cost curve up to the point of break-even is calculated as shown in Table 13.2.

The area under the repayment curve up to the point of break-even is calculated as shown in Table 13.3. The net area between the two curves therefore represents $(596\,000 - 413\,000) = 183\,000$ £-months.

Table 13.1

End of month	Cumulative cost, £	Cumulative revenue, £
0	0	0
1	6 000	0
2	10 000	0
3	18 000	9 000
4	40 000	19 000
5	70 000	42 000
6	100 000	81 000
7	130 000	118 000
8	146 000	144 000
9	153 000	161 000

Table 13.2

Month	Calculation	Area, £-months
1	$\frac{1}{2} \times 6\,000$	3 000
2	$6\,000 + (\frac{1}{2} \times 4\,000)$	8 000
3	$10\,000 + (\frac{1}{2} \times 8\,000)$	14 000
4	$18\,000 + (\frac{1}{2} \times 22\,000)$	29 000
5	$40\,000 + (\frac{1}{2} \times 30\,000)$	55 000
6	$70\,000 + (\frac{1}{2} \times 30\,000)$	85 000
7	$100\,000 + (\frac{1}{2} \times 30\,000)$	115 000
8	$130\,000 + (\frac{1}{2} \times 16\,000)$	138 000
9	$146\,000 + (\frac{1}{2} \times 7\,000)$	149 500
	Total	596 500

Table 13.3

Month	Calculation	Area, £-months
1	—	0
2	—	0
3	—	0
4	$9\,000 \times 1$	9 000
5	$19\,000 \times 1$	19 000
6	$42\,000 \times 1$	42 000
7	$81\,000 \times 1$	81 000
8	$118\,000 \times 1$	118 000
9	$144\,000 \times 1$	144 000
	Total	413 000

If the cost of capital is 15 per cent, then the cost of providing the working capital for this project alone amounts to

$$\frac{183\,000}{12} \times \frac{15}{100} = \underline{£2288}$$

This sum is $\frac{2288}{163\,000} \times 100 = 1.40$ per cent of the total contract cost.

Clearly, one of the significant factors that affects the cost of the working capital is the timing of the payment from the owner. As payment from the owner is delayed, so the revenue curve in Fig. 13.1 moves to the right, the break-even point in the contract comes later (if at all) and the area between the two curves gets larger. If, for example, the owner does not pay until two months after the end of the month at which the valuation is made then the area under the revenue curve to be subtracted will remain the same. The area under the cumulative cost curve, however, increases by $(153\,000 \times 1) + (5000 \times 1)/2 = 155\,500$ £-months (where the cost ordinate at end of month 10 is £158 000). This would increase the interest charge by $[155\,500/(12 \times 10)]1.5 = £1938$, making a total charge of £4226 or 2.59 per cent of the total contract cost. There is, therefore, scope by adjusting the programme, by adjusting credit terms on both sides and by negotiating payment terms, to optimize generally the working capital that will be required at any one stage of the company's activities.

Figure 13.1 deals with one particular contract only, but it may be desirable, and certainly it will be useful, to portray graphically the current amount of work on hand at any one time within the whole company. This can be done by drawing a *work load diagram* such as that illustrated in Fig. 13.5, where the full lines represent the outstanding budgeted value of work on a number of different contracts that are being carried out over a given period of time. A diagram such as this is clearly carried out on a 'running' basis, because it will rarely be possible to forecast the commencement of a further contract in 15 months' time.

The chain-dotted line represents the total of the outstanding work, which is arrived at by summing vertically the amounts shown by the contract curves for one particular date. It illustrates the peaks and valleys that occur when new contracts are awarded and others are nearing completion. For instance, in Fig. 13.5 at the beginning of October on the calendar scale and before the award of contracts B and C, the outstanding work has fallen to a value of £14 000. On the award of the two additional contracts the peak value of £176 000 is achieved. Planning work can be greatly facilitated by this type of picture.

Another feature of the diagram is that it can be used to plan the utilization of key personnel and equipment. In the case illustrated, a large excavator, of which the particular company owns only one, is required for three of the contracts. By plotting the requirements on the diagram it can be seen whether the

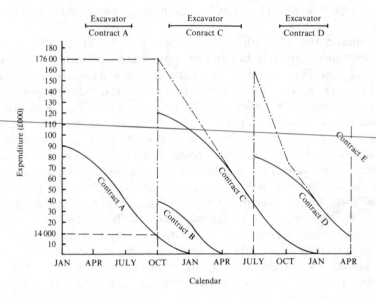

Figure 13.5

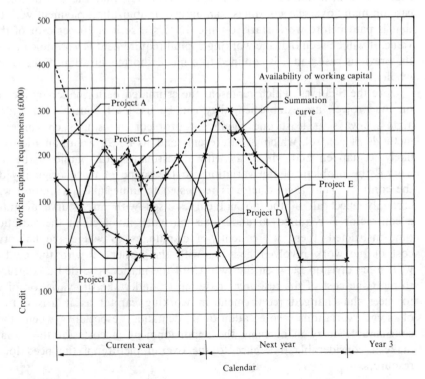

Figure 13.6

demands overlap or not. Demands for site managers and others can be similarly scheduled.

An alternative form of graph for this purpose is shown in Fig. 13.6.

The importance of the schedule and an appropriate repayment curve is reinforced in Fig. 13.7 and 13.8. The first represents the cash flows for a typical project where the repayments for work done start only when sufficient work has been carried out to qualify the payment. The working capital that is incurred as the work proceeds is shown in Fig. 13.7 by the black area under the horizontal axis. The black area represents working capital actually in use at any single point in time. Eventually, if and when the project breaks down, the cumulative working capital curve will cross the horizontal axis and working capital is then financed out of surpluses made on the project. It should be noted that this budget curve has three forms. The first is that for value, which is amount paid to the contractor by the owner on an agreed basis, usually in arrears. Secondly, there is a cost curve, which represents the amount of money that the contractor expects to have to pay for the work—the difference between this curve and the value curve is the contractor's margin. Finally, the third payment curve is the expenditure curve, which is the curve representing the actual rate at which the contractor makes payment, taking into account the delays for paying bills, etc. Figure 13.8 now shows the same details assuming that the contractor, as in the case on many overseas contracts, can agree a mobilization payment with the owner that is paid before work starts on site. In this case it is 10 per cent of the project value. It will be seen that the working capital requirement is very much reduced over the previous case and it is confined to a much shorter duration of the contract as a whole.

Once the S-curves, which represent the estimated performance by the contractor on the project, are established, they can be used for control. However, there is almost always at least one significant problem in using such curves or other methods of control, which in any event tend to be an approximation of what is likely to happen and which, as with all estimates, are bound to be inaccurate. This is the problem of the variations to the scope of work originally defined in the contract. That a contractor is requested to undertake a variation will almost certainly affect all of the normal curves used with an S-curve diagram. While a contractor is required to undertake the work involved in the variation at the time instructed, there is frequently a dispute as to where the liability for the cost of the work involved lies, i.e., with the owner or with the contractor, or with some division between the two. Until the matter is settled a contractor needs to meet the cost of the resources required to undertake the variation of work and to meet the changed requirement in working capital as it arises. In the case of some variations these additional demands, especially when there is a considerable delay in agreeing the apportionment of costs for the variation, can make a considerable difference to the overall picture of the need for financial resources.

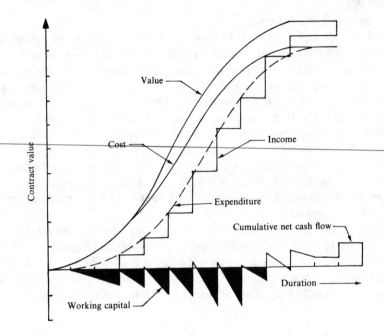

Figure 13.7

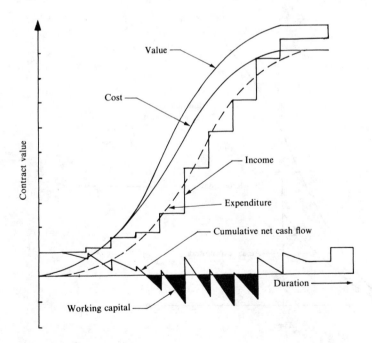

Figure 13.8

13.6 Budget curves used for control

The traditional form of the financial status chart in use is that of the budget S-curve, the construction of which has already been discussed. The S-curve can be drawn from experience of previous similar operations in the company or, at the estimating stage, it can be prepared on an arbitrary basis. The chart is frequently used as a means of control by plotting the actual expenditure curve against the budget curve, as shown in Fig. 13.9. It is sometimes wrongly assumed that whenever the *actual expenditure* curve keeps below the budget curve for value, the cost of the work is being controlled effectively and that the work is being carried out efficiently and profitably. In fact, from the control point of view, this combination of information is almost meaningless. As with the work schedule chart, the actual expenditure curve is not related to the physical quantity of the work that has been carried out, nor to the efficiency of the operations.

The inadequacy of such a chart may be demonstrated by the use of a simple example. Figure 13.10 illustrates the budget curve established from the simple outline bar chart for a project estimated to have a total cost of £670 000. A uniform linear rate of expenditure over each activity is assumed. If the activities are appropriately chosen then this need not be a false assumption. While actual expenditure was at one

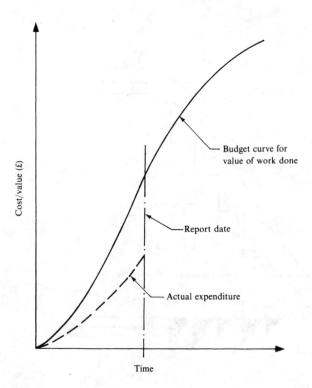

Figure 13.9

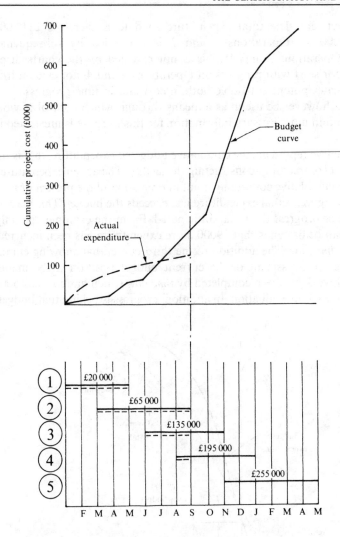

Figure 13.10

point in the project proceeding at a rate in excess of that budgeted, the cumulative total at the end of September has fallen below the target. Reference to the updating on the bar chart (by means of a broken line) indicates that progress is apparently in accordance with the programme.

The actual situation, however, may be a long way from being satisfactory. For example, Operation 1 may have been completed at a cost of £23 000, that is £3000 over the budget figure; Operation 2 may have been completed for £80 000, that is £15 000 over the budget figure. Operations 3 and 4 evidently started on programme, but owing to poorly planned progress may have resulted in the expenditure of £3000 and £4000 respectively. The total expenditure to date is therefore £110 000. An overrun of £18 000 on Operations 1 and 2 is concealed by subsequent poor performance of Operations 3 and 4. This is not revealed by the charts displayed. Moreover, the poor start with progress on Operations 3 and 4, not evident from the programme, indicates potential trouble both in cost and in time progress.

An S-curve can however be useful as a means through which control of work can be effected. The minimum necessary information for this purpose is illustrated in Fig. 13.11.

The full-line S-curve represents the cumulative budgeted expenditure, that is the cost to be incurred by a contractor in constructing the facility. The actual expenditure curve shows the current cumulative cost at any point in time up until the present or *time now*. At *time now* in the figure, actual expenditure just exceeds the budget. This may or may not represent better progress than that anticipated. From the example, the only firm conclusion that can be drawn is that £9000 more expenditure has been incurred than was budgeted at *time now*. The addition of the third curve, that showing cumulative *valuation*, enables an assessment of the current budget position to be made. The assessment is that work has been completed by *time now* which has cost £62 500 more than will be paid for it—the valuation. In addition, progress is behind that budgeted by

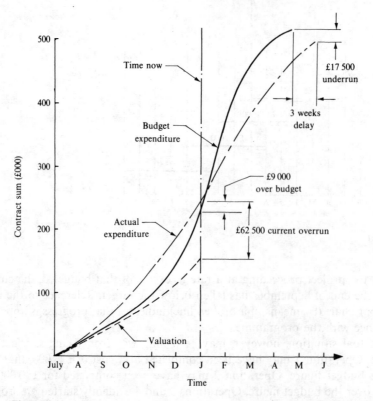

Figure 13.11

£62 500 − £9000 = £53 500. Cost is being expended at a faster rate than budgeted and may cause cash flow problems. Attention overall, therefore, needs to be paid to the efficiency of the work methods, productivity and financing.

From Fig. 13.11 it will be seen that a forecast of expenditure to complete the work shows a likely final underrun of £17 500, though this will cause a delay of 3 weeks in the completion of the project.

In the use of an S-curve in conjunction with monitoring and controlling a project, reference has most frequently been made in the terms of the vertical axis of the graph bearing a scale expressed in units of money. This can equally usefully be expressed in terms of work content, for example, cubic metres of excavation, tonnes of structural steel erected, or man-hours of work content for a particular operation or a group of activities. For a project in which there is a number of major activities of some magnitude, it is useful to use an S-curve for each one, so as to ensure that progress advances as required. In addition to using units of work on the vertical scale, it is feasible and also useful to use a scale of percentages, the whole of the particular work under consideration being expressed as 100 per cent. Such a scale obviates the need to change the total quantity of work involved when variations to its scope occur.

An elaboration of the S-curve may be used to provide a broader picture of the implications in making a decision about the future progress and cost. Figure 13.12

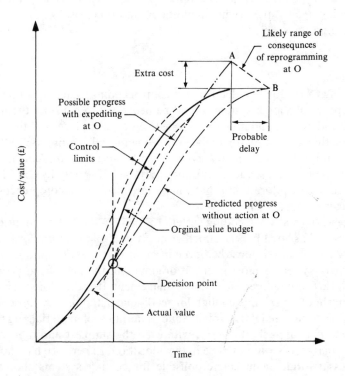

Figure 13.12

illustrates an adaption of the normal S-curve for cost against time for a project (though again it could well be of any other units). It is possible to provide an envelope of control giving limits, perhaps ± 10 per cent on either axis, within which a project can proceed without major adjustment being made to the resources in use. If the actual progress as recorded on the chart breaks through the envelope of control limits at either side then a decision must be made about future resource utilization. In the light of the project time progress and/or cost progress, it must be decided either whether more or less resources must now be committed to the project in order to put it on the original path again so as to attain the original completion target, or whether some modified target will be calculated. In other words, a decision point has been reached when some fundamental change has to be made to the project programme involving additional expenditure, if the original targets are to be achieved or bettered. The two extremes of action that can be taken are to reprogramme the work so as to speed it up (in the case of the example illustrated), which almost inevitably will result in increased costs, or to take no action at this point. The latter course will allow the future progress to continue in the trend already established, resulting in an extension of the project duration. Between these two extremes is a large number of alternatives, the end results of which can be represented by the curve which joins points A and B on Fig. 13.12. The most economic answer will be arrived at as a result of balancing the cost of the delay in project completion against the saving in additional resource costs within the context of the existing situation.

13.7 Cost control codes

Before attempting to set down in detail a cost control coding system it is necessary to review the purposes of such a system. It will be necessary to consider the nature of the work to be undertaken and how this is scheduled to take place, and the nature of the information that the estimator or head office staff will require as feedback for future use. Possibly also the client may have some special requirement for cost information, perhaps for subsequent insurance purposes. It is only by considering all the potential uses of the information that will be recorded that a comprehensive and useful system can be organized.

It is not necessary to ally the coding system to the relevant bill of quantities. The bill of quantities, as has already been stated, is rarely in a suitable form for the proper control of site costs. Again, it must be borne in mind that the information being fed into a cost control system is most likely to originate, in the field, from people who have received little training in construction management. It is therefore essential that a coding system should be simple enough for ready understanding by those who will be allocating the hours of labour, plant, and quantities of materials against the individual codes. It is unlikely, however, with even the simplest and most straightforward system that errors will not arise in the allocation. Experience has shown that it is very necessary that engineers responsible for costing systems should check allocations very closely with their own knowledge of the work involved.

There is a wide school of thought that supports the opinion that a cost control coding system should be uniform throughout a company. Undoubtedly there are advantages of such a system, especially for the staff who are permanently employed and will be moving from one contract to another. There are also doubts as to the desirability of such a system, since no two contracts are similar in many respects, and most construction companies carry out a very wide range of work, from small buildings to large, heavy civil engineering contracts.

The policy with regard to the detail of the coding system must be settled as early as possible. It must be borne in mind that a system that is too detailed will lead to difficulties in the allocation of expenses, particularly in the field. If insufficient detail is provided then the costing system as a control system will be ineffective, since it will not be possible to locate the most inefficient of the operations that have been included under a single, all-embracing code.

It is desirable, therefore, that a universal coding system is designed to suit all the contracts under operation. The system must be clear and precise in its terms and it is necessary that it should be flexible, since, by the very nature of construction work, additional items and operations are likely to arise throughout the progress of the work. There will undoubtedly be occasions when codes have to be introduced for new sections of the work or for new operations that were not envisaged at the commencement of the work. It may also be that the amount of information required during the course of the contract will itself alter. The original assessment of what is required may be inaccurate and, therefore, flexibility of the system is also required in this area.

A further point to be borne in mind, when setting out a cost control coding system, is that there is little point in subdividing the system in such a way that one arrives at a number of subheadings for which allocations cannot be made. For example, a bulldozer may be used to excavate material on the site of a road to a depth of half a metre or so, and, in pushing this material forward, it is pushing it into a fill area. In other words, the bulldozer is being used for both excavation and filling, in one operation. It will not be possible to allocate accurately costs against the filling and excavating operations, and there is therefore little point in having a separate code for the two operations in these circumstances. Similarly, other operations may well be combined—for example, where a contiguous wall and column in reinforced concrete are being constructed, it will be almost impossible to separate the allocation of time between the column and the wall for the erection of formwork, concreting, and stripping the formwork on completion.

If the information to be obtained from the cost control system is going to provide feedback for the estimator, great care is necessary to ensure that a distinction is made between certain operations, which may on the surface appear to be similar but in practice be quite different. An example of this may be the concreting of a reinforced concrete column at a height of 30 metres on a bunker, as compared to concreting a similar column at ground level. Concreting a beam 600 millimetres deep by 300 millimetres wide is quite a different proposition from concreting a beam of four times those dimensions. Some provision must be made in the classification system for dayworks and for additional work that may well become the subject of a subsequent claim.

When the cost control code has been set out for a contract, it must be distributed for information to all those employees who will be expected to use it during the course of their work. These will include the estimators, buyers, accountants, clerks, timekeepers, engineers, and supervisory staff in the field, such as foremen and gangers. With the cost code must be a clear explanation of why it has been adopted and what are its purposes. It is only by such an explanation that the staff can be carried with the system and that it can be operated with success. For any cost control system to be a success, the cooperation of the site supervisors is absolutely necessary. There are many methods of classification, which are represented by combinations of numbers and letters, both upper and lower case, full stops, colons, hyphens, etc., all of which have taken their place in the system. Two types of classification for widely varying work are explained below as being fairly representative of the type of coding systems in common use.

Coding systems for particular projects may be linked directly to the work breakdown structure for planning purposes described in Section 7.3 *et seq.*

13.8 A simple coding system

The first system to be described is a simple one for use with a small company and is primarily concerned with direct labour costs only. For such a company, labour costs are the biggest variable, the overheads being relatively fixed, and materials not requiring the same degree of supervision as labour and plant. The use of plant by such a contractor is comparatively limited and special provision in a coding system, therefore, need not be made. The following system is based largely upon alphabetical notation, using where possible the initial letter of the operation under consideration. Such a system quickly becomes familiar to the employees allocating the labour.

Alphabetical trade code

B–Bricklaying
C–Concreting
D–Drainlaying
E–Excavation
F–Formwork
M–Mechanical equipment
O–Preliminaries
P–Piling
R–Roadworks
S–Structural steel

Example of the subdivision of the code:

B—Bricklaying

BBZ – laying 215-mm blocks
BC   – laying common bricks

BEM – laying engineering bricks in manhole
BF   – laying facing bricks
BFZ – laying facing bricks in 215-mm wall
BFY – laying facing bricks in 255-mm wall
BP   – pointing

C—Concreting
CB   – concreting beams
CC   – concreting columns
CF   – concreting floors
CFX – concreting floors 150 mm thick
CFO – concreting footings or foundations
CW   – concreting in walls
CDB – concreting to drain bed

D—Drainlaying
DC3 – laying 300-mm-diameter concrete pipes
DE2 – laying 225-mm-diameter earthenware pipes
DJ   – jointing pipes
DT   – laying drains in tunnel

The above system is broken down into major divisions under trades. The initial letter of any particular trade is used as the first part of the code; for example, in the case of bricklaying the first letter of all the codes is the letter B, and under concreting the first letter is C. The second letter of the code represents the operation being carried out within the trade. For example, in bricklaying BF represents laying facing bricks and BC represents laying common bricks.

Clearly, in a cost sense, there is a need to differentiate between laying these two types of brick. Under other trades the second letter of the code is used to indicate a location for the work which is being carried out. For example, under concreting, CC is the code for allocating labour expended on the concreting of columns. Under drainlaying, DT is the code for laying drains in a tunnel. A further refinement can be made to such a system by appending either an additional letter or a number to the code to give further detail. Examples of this appear in drainlaying, for instance, where DC3 represents laying 300-mm concrete pipes and CDB represents concreting to drains in the bed of the drain. Such a system is suitable for the small contract, where the foreman, who will be allocating the costs, will be familiar with the names of subordinates and will be aware of the area of the contracts in which they are working.

13.9 The decimal classification of work

Decimal classification has been widely adopted in costing systems and has the advantage that it is extremely flexible in its use. As will be seen from the following examples, the coding system can be expanded or reduced to meet all conditions and to meet any level of detail required in the contract. An additional advantage of such a coding system is that, if it is chosen to divide the code functionally, for example in the case of a welfare building, an administration building, and a control room, by using different key numbers to represent each of the three buildings, then costs of similar operations in each of the three buildings may later be collected either by hand, by machine accounting, or by computer.

The following is a cost coding system in the decimal classification which might be used for a large contract such as a power station. The first division is on a functional basis and the boiler house is represented by 10, the turbine room by 20, the water-cooling system by 30, the control room by 40, etc. In inspecting a cost allocation, it is possible, by looking at the first digit of coding, to identify the area in the contract to which the allocation is being made. The second figure of the code represents that part of the functional area which is being considered. The following example is limited in scope, but has been taken far enough to indicate the method of procedure.

Functional codes:

    10.   Boiler house
    20.   Turbine room
    30.   Water-cooling system
    40.   Control room etc.
    01.   Equipment
    02.   Superstructure
    03.   Culverts
    04.   Cable trenches etc.

Therefore 12, represents the superstructure of the boiler house.

.1	Clearing site	.01	Excavation
.2	Foundations	.02	Concreting
.3	Substructure	.03	Piling
.4	Floors	.04	Backfilling etc.
.5	Roadworks		

Therefore 42.4 represents the floors to the superstructure of the control room, and 21.21 represents excavation to the foundations for equipment in the turbine room.

Having subdivided by trades, the various operations within the trade must be represented. For example, for excavation (.01) the following subdivision can be made:

    .011   Excavation by machine
    .012   Excavation by hand

For concreting (.02) more subdivisions may be required:

    .021   Formwork
    .022   Reinforcement
    .023   Inserts
    .024   Placing etc.

A fourth figure after the decimal point can be used for further subdivision, and theoretically this process can go on indefinitely in order to satisfy the requirements of the contract under consideration.

A fourth decimal figure under concreting might be used as follows:

.0211   Erecting formwork to foundations
.0212   Erecting formwork to straight walls
.0213   Erecting formwork to curved walls
.0214   Erecting formwork to floor soffits
.0215   Erecting formwork to beam sides etc.

22.2211 would therefore represent the erection of formwork to the foundations of the turbine room superstructure.

As the code has been set out above, it will be seen that each subdivision is limited to 9 digits (unless the figure 0 is used as one of the subdivisions). This can be overcome: for example, if it is felt that under concreting (which is .02 in the code) there will be more than 9 or 10 subdivisions—and there may well be where there is considerable reinforced concrete works—then the first subdivision could be .0201, and this would be formwork. Reinforcement would be .0202, inserts .0203, and so on, giving considerable flexibility, inasmuch as 99 divisions are available.

Such a code as this is very comprehensive and can be expanded and reduced as the need arises. For instance, if the contract is not large enough to be divided functionally, then the first digit can be omitted altogether.

If consideration is given to machine accounting or to the use of a computer for sorting costs, there is little point in using the decimal point to break up the figure sequence. However, if the decimal point is omitted, care should be taken that each allocation has the correct number of digits, even where some of the figure spaces are not required.

The above classification has dealt with the operational side of contracting, but site on costs, margin, and preliminaries have not been included. These can be allocated to a separate account, and a number, within the system as described, should be allocated to them in accordance with the breakdown of the preliminaries in the original estimate.

The question of head office expenses and margin need not arise in the cost control system as operated at a contract site. These will be figures that will be dealt with at the contract manager or director level and should not in any way influence the manner in which the work is carried out or the control which is placed upon it.

13.10 The allocation of costs

The coding of the allocation of costs in the field is the job of a cost engineer or a cost clerk. In order to be able to code allocations correctly, the person dealing with this matter must be familiar with the programme of work for the contract; they must be familiar, too, with the terms used in construction work. It is also important that whoever is coding costs should be familiar with the materials and the way in which they are used in the work. This makes it very important that the supervisor of such operations should be a construction engineer, not only with regard to the site cost control system, but also because the engineer responsible for the work will also have a better idea of the use to which feedback will be put. In this instance it is clearly better that the controlling engineer should have a sound knowledge of the estimating processes.

It should be continually borne in mind, in setting out a coding system and subsequently allocating the costs, that the accuracy of the system will be in direct proportion to the accuracy of the original allocation of costs in the field. In addition to the point already made concerning the undesirability of having a cost code for which the costs cannot be allocated, it should be remembered that there are a number of external influences that can in the first place affect the allocation of cost. Not the least influential among these is the effect on the allocation of a bonusing system. The costs may tend to be allocated in the field in order to give a healthy picture to work that is being bonused on an output and time basis. This means, of necessity, that if the overall allocations of time are to agree with those on the wage-sheets in total then costs must be allocated elsewhere in order to meet the deficit. It is desirable, therefore, that, in such a system, general accounts, accounts for miscellaneous work, etc., should be kept to an absolute minimum so that the abuse will be at a minimum too.

It is important to distinguish in the allocation procedure between *direct* and *indirect costs*. Direct labour costs are those labour costs that can be allocated to a specific task in the field. Generally, the performance of this task will make a direct contribution to the progress of the job and will probably be related to a specific item in the bill of quantities. Similarly, material, and direct expenses are assigned to specific tasks or parts of the work. Indirect labour, materials, and expenses all form part of the oncost for the project and cannot, and must not, be allocated to a specific task. For example, the cost of supervision, the foreman, the engineers, the site manager, will all be part of the indirect cost of carrying out the work. The foreman may well supervise ten different jobs during the day and it will be impossible to allocate the time so spent to any one specific task; similarly with the site manager and the site engineers, who will be supervising the contract on a broad basis. The cost of the site offices and other similar facilities is a further example of indirect costs, which must be spread over the contract as a whole and cannot be allocated to a specific task on the site. These indirect costs will generally appear in the preliminaries of the bill of quantities whether a formal contract document or not, and will be allocated in the original estimate quite apart from some of the operational costs related to items in the bill.

Summary

After analysing the merits of projects before the decision is taken to invest capital in them, there is a need to make sure that they cost no more than was expected of them at the estimating stage. This applies equally well to the potential owner, as well as to the contractor who undertakes the work. The viability of the businesses of both parties to the work is vulnerable to a breakdown in the proper control of cost. This chapter begins a dialogue on collecting, classifying and controlling the costs that arise in a construction project. The purposes of cost control are discussed, as are the various methods of carrying it out. A budget is an important forerunner of the control process and the principles of drawing up a cost budget are explained, particularly using an S-curve. This leads on to the use of working capital both on a single project and in the multi-project company. An essential aspect of a comprehensive cost control system is a sound cost coding. Two diverse methods are described and form the basis for others to be developed.

Problems

Problem 13.1   Draw a workload diagram for a company view of its work in hand over a period of 12 months, assuming that the following contracts are running or will be obtained and that the work is carried out at the rate given by the empirical rule in the text:

(a) Three months before the period under construction a contract for £90 000 of 3 months' duration is started.
(b) At the commencement of the period a contract for £15 000 with a duration of 3 months is started.
(c) Four months after the commencement of the period a contract for £120 000 (15 months) is obtained.
(d) Seven months after the commencement of the period a contract for £60 000 (10 months) is obtained.

Problem 13.2   Set out a six-figure decimal coding system for a large contract with which you are familiar.

Problem 13.3   Having prepared a tender for six reinforced concrete bunkers, there only remains to calculate the cost of providing the necessary working capital. It is estimated that the total value (except the cost of finance) of the work will amount to £300 000 and that it will be completed during a contract duration of 16 months. A budget is drawn up based upon the anticipated programme of work and this is set down in Table 13.4. The contract documents provide for the contractor to be paid at intervals of 2 months from the commencement of the work. Payment will be made on the basis of 90 per cent of the cumulative value of the work completed by the end of each payment period. At the end of the contract, half of the 10 per cent retention will be paid and the remaining 5 per cent will be paid promptly 6 months after the end of the contract.

If the contractor estimates that actual costs will be 90 per cent of the value of the work completed and that money for work completed will be received $1\frac{1}{2}$ months after it is due, i.e., $1\frac{1}{2}$ months after the end of each period of 2 months, what will the

Table 13.4

End of month	Cumulative value of work completed
2	£30 000
4	60 000
6	85 000
8	140 000
10	200 000
12	260 000
14	280 000
16	300 000

working capital cost the contractor, over the duration of the contract only, at an interest rate of 10 per cent?

Problem 13.4 Define the term working capital.

Table 13.5 lists the cumulative monthly costs incurred by a contractor and the corresponding monthly payments that are received from the promoter of a project.

Calculate the cost to the contractor of borrowing the working capital necessary to finance the project if the annual rate of interest on borrowed money is 12 per cent. If the promoter makes all the payments one month later than anticipated in the table, by what percentage will be cost of working capital increase?

Table 13.5

End of month	Cumulative cost, £000	Cumulative revenue, £000
0	0	0
1	12	0
2	20	0
3	54	0
4	90	14
5	130	40
6	180	100
7	220	130
8	240	190
9	260	210
10	290	300
11	290	320
12	290	340

Problem 13.5 Explain what is meant by the term budgetary control.

A contracting firm is invited to tender for a small hydroelectric scheme in a developing country overseas. The firm calculates its actual costs to be £5 000 000 and this figure is accepted by the client subject to a condition regarding payment of the money for completed work. The condition is that £1 500 000 will be paid at the end of each of the third, fourth, fifth, and sixth years after construction is commenced, to cover the actual costs of the contractor plus his requirement for profit.

The contractor programmes the work to cover a 3 year period and estimates that the costs to be incurred will amount to £1 000 000 by the end of the first year, £3 000 000 in total by the end of the second year, and £5 000 000 in total by completion of the work at the end of the third year. The contractor estimates that all the money to finance the work will have to be found from sources outside the firm, borrowed at an interest rate of 10 per cent per annum. Produce a budget statement for the contractor, year by year, assuming that all payments and receipts are on a year-end basis. If inflation is assumed to be at the rate of 3 per cent p.a. establish the true profit or loss for the contract reckoned in terms of the value of money at the commencement of the contract.

Problem 13.6   During the preparation of a cost estimate for a small reinforced concrete jetty, a contractor wishes to assess as accurately as possible the interest charges on the working capital requirements.

The contractor estimates the duration of the contract to be 12 months and the total tender price to be £250 000, and budgets on carrying out the work to the cumulative value shown in Table 13.6 after successive periods of 2 months. The contract provides for valuations to be submitted by the contractor every 2 months (the first to be submitted at the end of the second month of the contract duration) and for progress payments to be made by the client up to the limit of 90 per cent of each of the first 5 cumulative valuations. The client will then pay in the usual way 95 per cent of the sixth valuation and the balance of 5 per cent exactly 6 months after the completion of the contract unless there are any claims outstanding against the contractor.

The contractor estimates that the client's cheque for work carried out will not arrive until 1 month after the date of each valuation and that this cost of capital will be 8 per cent per year. What interest charges will the contractor have to pay on working capital if the cumulative expenditure is estimated to be as shown in Table 13.7? If the contractor can invest surplus income over expenditure for this contract at 5 per cent per year interest, will this balance outgoings against

Table 13.6

End of month	Cumulative value of work completed
2	£20 000
4	54 000
6	120 000
8	200 000
10	240 000
12	250 000

Table 13.7

End of month	Cumulative expenditure
2	£5 000
4	10 000
6	80 000
8	120 000
10	180 000
12	195 000
14	205 000
16	225 000

receipts as far as interest charges are concerned? What difference is there between the two amounts?

Further reading

American Society of Civil Engineers, *Construction Cost Control*, ASCE Manuals and Reports of Engineering Practice No. 65, revised edn, New York, 1985.

Antill, J. M. and R. W. Woodhead: *Critical Path Methods in Construction Practice*, 3rd edn, Wiley, New York, 1982.

Batty, J.: *Standard Costing*, 3rd edn, McDonald and Evans, London, 1970.

Gobourne, J.: *Site Cost Control in the Construction Industry*, Butterworths, London, 1982.

Kharbanda, O. P., E. A. Stallworthy and L. F. Williams: *Project Cost Control in Action*, Gower, Farnborough, England, 1980.

Moder, J. J., C. R. Phillips and E. W. Davis: *Project Management with CPM, PERT and Precedence Diagramming*, 3rd edn, Van Nostrand Reinhold, New York, 1983.

Owler, L. J.: *Wheldon's Cost Accounting and Costing Methods*, 15th edn, MacDonald and Evans, London, 1984.

Park, W. R., *The Strategy of Contracting for Profit*, Prentice-Hall, Englewood Cliffs, NJ, 1966.

Walker, T. M.: *Understanding Standard Costing*, Gee, London, 1980.

14. Controlling costs

14.1 The use of cost control

The previous chapter has been concerned with describing some of the mechanics of cost control and how the necessary data can be collected and sorted. After collecting the data, it is necessary to arrange them so that ready use can be made of them by those managers who need them for control action. Consideration must be given to the amount of detail required in reports, which will vary according to the level of management for which they are prepared. The director in charge of a contract will not require to know the details of cost down to every single operation, but will require an overall picture of the situation, together with some idea of the trends being established. At the contracts manager level a little more detail will be required, but still not to the depth of each operation that the site manager will need. The time interval between reports may also vary with the level of management that receives them. For instance, at site manager level, it may be necessary to receive the cost report on a weekly basis. Above this level in the organization structure, it may be necessary to have a regular report at monthly intervals, with the possibility of additional reports at shorter intervals of time, should there be indications that things are not going right in one particular area.

As well as preparing reports based upon the cost data already collected for a contract, it is necessary to project the costs into the future and to estimate or re-estimate the cost of the work yet to be completed. It is only in this way that the ultimate profit or loss for a contract can be estimated. The re-estimate must take into account any new information that has come to light since the commencement of the contract. For example, if excavation is found to be losing money and no amount of method study or reorganization, the use of new equipment, or different methods can achieve the estimated output at the required cost, all the future excavation of this nature in the contract must be costed at the new rate in the light of these conditions. An adequate reporting system is then an essential part of a cost control system.

Often an estimator will find, too late, that an estimated rate for a particular operation has been based on some cost data that was received in the past and about which the full details are not known. The cost data may have been accompanied by a certain amount of detail, but important conditions which affected the output or the cost of the operation have been omitted. Quite naturally, the estimator becomes more reluctant in the future to use cost data that have been fed back from the site.

In addition, it may take a considerable amount of an estimator's time to ascertain sufficient detail about the data and the conditions under which they were obtained, so much in fact that the estimator is disinclined to carry out this exercise. The only time at which the estimator might show some interest in a return of data from a contract is likely to be during the pricing of similar work for another tender. Two contracts of a similar work content will enable this straightforward transference and comparison of rates. There is a good case, therefore, for taking the advice of the estimator when setting up a costing system, so that some of the data at least will be of direct use in the costing work. It may mean that the cost data collected on site will have to be translated into operations covered by bill items, and full details of the operation, describing the method by which it was carried out, the plant used, the size of the gangs, the weather conditions, the ground conditions, etc., must accompany each item.

There are two areas within which the cost control reporting system can indicate efficiency or inefficiency. The first is a check on the profitability of the work. This must mean, in general terms, that the value of the work that will be returned must be compared with the cost of the work, which has been carried out, to the contractor. If no detailed cost control system is used then this must be a straightforward comparison between the valuation figure and the total expenditure, giving a single profit or loss. In such a simple method as this, there is only one true figure that can be accepted as being reasonably accurate, and this is the final figure for the contract. The second area is covered by a check on efficiency, carried out against the standard of the output rates that were used by the estimator in compiling the estimate. Efficiency is the concern of the site management and supervisory staff; profitability is the direct concern of the contract management level and above.

Upper levels of management often use the test of profitability to decide whether their intervention in the running of a project is necessary. A project that is showing, as a result of periodical checks, a good profitability will receive little attention from the contracts management and director levels of the organization; attention will be paid to the contracts that show minimum or negative profitability. It is probably in these contracts, showing little or negative profitability, that the upper levels of management will require to investigate the costs in more detail, in order to locate the particular operations that, through inefficiency or underpricing, are the cause of the trouble.

14.2 Unit costs

One of the most significant ways of using cost information is by the *unit cost*. The unit cost for any item or operation is the direct cost for one unit of measurement for that item. For example, in excavation the unit cost can be the cost of excavating one cubic metre of material under those given conditions; in other circumstances, it might be the cost of one square metre at a stated average depth.

In order that unit costs can be calculated and used for control purposes, it is necessary to ensure that the measurement of the actual work carried out relates directly and accurately to the costs incurred in carrying it out over the operational period. It is necessary, too, that the measurement of the quantity of actual work carried out should be very accurately made for the operation under consideration. Care must be taken that a 'give and take' method of interim measurement does not clash with the calculation of unit rates and result in a temporarily high or low unit cost for a particular item, thus warranting an investigation that is needless if the true situation is recorded in the first instance. It is also important that in calculating unit rates the costs are related solely to the work under consideration. Care must be taken to see that the supervisory staff do not allocate the hours when their operatives are not properly and efficiently employed on productive work, to some item or operation that will inevitably result in a poor unit cost for that work.

Opinion is usually divided as to whether unit costs should be stored in terms of pounds and pence or whether they should be stored as man-hours. The argument in favour of using man-hours is, of course, that the historical use of such data is more meaningful. It is always a risky and approximate business to bring up to date costs in pounds and pence, whereas outputs measured in terms of man-hours will be more likely to remain constant, given similar conditions, over the years. On the other hand, in these days, a considerable number of units of mechanical equipment are employed and it is not easy to correlate the working of a machine having a particular size, under specific circumstances, for one hour with one or more man-hours of labour.

Since cost control has the prime purpose of measuring the efficiency of an operation in progress at the time when the data are collected, money often means more than do man-hours. Money is likely to make a greater impact than man-hours upon the managers involved in controlling the costs. There is no reason, of course, why data in terms of both money and man-hours should not be collected, the former for immediate use and the latter as historical data and feedback for use in future estimating work. In using either unit, it should be borne in mind that it will rarely be necessary for details of every single operation involved in a contract to be fed back to the estimating department. For many operations, they will already have received a considerable amount of data, and many of the other operations will be of such little significance, in comparison with the contract as a whole, that there is no point in collecting the relevant information. The '80/20 rule' is often quoted with regard to civil engineering bills of quantities, meaning that 80 per cent of the value of the work is often in 20 per cent of the items in the bill. Obviously, this 20 per cent must receive a great deal of consideration both in the tendering procedure and in subsequent cost control.

14.3 Reporting systems

One of the basic reports for the control of costs on site is the *weekly cost record*. This gives a complete record of the quantities of work which have been carried out in the week previous to the date of the report, together with the lump sum total costs of the labour and plant that have been incurred in respect of those operations. Where applicable, it also lists the unit costs for the operation. Each of the figures of quantity, cost, and unit cost are prepared on the basis of the work carried out during the week, the total amount of work carried out in previous weeks, the total work carried out to date, and an estimated total of the work that has to be carried out in order to complete that operation. The relevant unit costs for each of these four totals are similarly recorded. Figure 14.1 shows an example of a typical weekly cost record with two sample operations filled in.

In this particular example all the costs are collected under the code number given in the left-hand column of the form. The example quoted is for the excavation and blinding concrete to the pumphouse. On looking at the unit costs for excavation, it will appear that the work during this week has been carried out at a lower unit cost than the average of the work previously. The unit cost to date is consequently reduced below the previous unit cost, indicating a trend towards the reduction in the overall unit cost for the work. In the fourth column under 'Unit cost' is shown the unit cost of the estimate. Such a record is not an attempt to indicate the time progress of the work, although a comparison of the total quantities of the work carried out to date with the estimated total to be carried out gives an accurate figure for the percentage completion on the particular operation.

Such a cost record would not be prepared for every operation currently being carried out on a site during the week. For the weekly cost record, only the major

XYZ Contracting Ltd			**Weekly Cost Record**									Sheet 1 of 1			
Contract *Bigend Power Station*			Contract No H 4321									Week ending 13/10/			
												Week of programme 5			
Code	Operation	Unit	Quantity				Cost (£)				Unit cost (£)				
			This week	Previous total	Total to date	Est. Overall total	This week	Previous total	Total to date	Est. Overall total	This week	Previous	To date	Estimate
35.3114	Excavation to pumphouse	m³	750	1200	1950	7640	200	350	550	1910	0.27	0.29	0.28	0.25
35.3247	Blinding concrete	m²	56	130	186	780	21	68	89	390	0.38	0.52	0.48	0.50

REMARKS:–

INITIALS _____

Figure 14.1

operations involving large quantities and considerable cost would be chosen, since the preparation of cost records of this nature involves a great deal of time and cost. If they are to be of use to the staff controlling the contract, such records must be issued promptly at the end of the week to which they refer and regularly, to the extent that their issue can be forecast and relied upon by the site manager. Items that are showing a unit cost in excess of the estimate will require further investigation into the plant and labour costs involved. In the normal course of events, checking costs for a period of less than one week tends to be far too expensive an operation, unless an operation on site is being carried out at an extremely fast rate and costs and quantities can be measured accurately. It is not usual to include the cost of materials used in a weekly cost record.

The decimal system of classification, which has been described previously, facilitates the collection of cost data for almost any operation or particular aspect of an operation. Should, for example, the operation on the weekly cost record be one of concreting and the unit costs show up poorly against the estimated cost, then, by selecting costs under further subdivisions of the decimal classification, unit costs can be compared for the formwork, the reinforcing steel, the mixing, the transporting, or the placing quite readily and quickly, in order to locate the area of inefficient or unprofitable working.

It is important to stress at this stage that, in the production of a weekly cost record, which will be seen by the site staff at comparatively short intervals, the object is to provide information in as simple, concise, and accurate a manner as possible. The provision of too much extraneous information will encourage the controlling staff to ignore the cost record sheets. What is needed is a statement that makes a quick but lasting impression and draws attention to the salient features of the situation.

An alternative method of presenting the cost information on a weekly basis or for a longer period, without the use of unit costs, is illustrated by the typical form shown in Fig. 14.2. The figures inserted in the form in this figure are the same as those used in the weekly cost record of Fig. 14.1. The information on this form is restricted to the total quantity of work carried out in the week, the total to date, the value of the total to date measured by the estimated figure, and the total cost that has been incurred in carrying out the work. Column G shows the excess or deficiency on the total operation to date—in this particular case the figure is negative, indicating a loss so far of £62.00. Column H shows the last reported estimate and indicates a loss at that stage of £50. While this type of weekly cost record does not give nearly as much information as the previous record, it does indicate that the operation is still being carried out unprofitably, although without further calculation it is not possible to identify the trend. However, the indication remains that some further investigation is required into this operation. It will be noted that the preparation of the form in Fig. 14.2, while apparently much simpler than that of Fig. 14.1, requires the collection of a similar amount of data. As the collection of data is one of the most expensive operations in any cost control system, the record of Fig. 14.1 is likely to be favoured as long as the quantity of information presented does not become

Code	Operation	Unit	C Quantity this week	D Total to date	E Total value	F Cost	G Profit (+) Loss (−)	H Prev. P or L
	Weekly Cost Record					Sheet 1 of 1		
XYZ Contracting Ltd						Week ending 13/10/		
Contract *Bigend P. Station*		Contract No. H 4321				Week of programme		5
35.3114	Excavation	m³	750	1950	£488	£550	−£62	−£50
35.3247	Blinding concrete	m²	56	106	£93	£89	+£4	+£2

Remarks:–

Initials _____

Figure 14.2

overwhelming. The effect of both of these reports is to concentrate attention on operations which are proving to be unprofitable, and probably therefore inefficient.

Yet a further variation of the presentation as illustrated by Fig. 14.2 is the division of the total value of column E between plant and labour costs and carrying out the same procedure with column F, which is the total cost to date. The excess or deficit of columns G and H is also divided into plant and labour costs, giving a closer indication of the unprofitable areas.

The information presented in Figs 14.1 and 14.2, and the variations on this presentation, can be presented in the form of graphs. Graphs give a good indication of trends and also, as with the comparison of previous and present unit costs, can indicate whether action by the management in controlling cost has been effective. For example, if excavation costs are clearly too high during one week of the programme and some attempt is made to modify the methods employed or the machines used, then, provided the rearrangement is effective, it would be expected that the unit cost would fall week by week thereafter. In the graphical presentation of this information, it is possible to chart the time when specific action is taken and relate to it the progress made in the reduction of costs. Figure 14.3 is an example of a report, in graphical form, in respect of excavation work.

To be complete, a report of this type needs to present two graphs, as in Fig. 14.3. In the first, the average cost of excavation is plotted on a weekly basis and is the result of dividing the total cost of the excavation during that week by the quantity

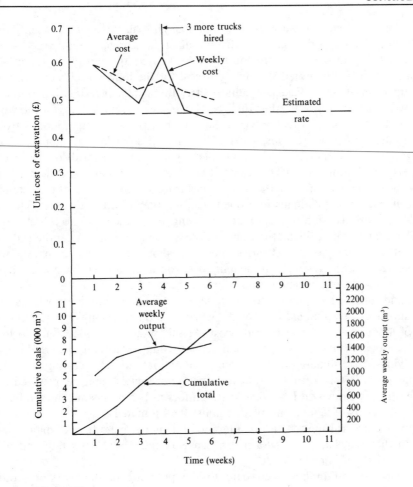

Figure 14.3

of excavation carried out in the same time. This is plotted as the full line in the upper graph of Fig. 14.3. To indicate the trend of the unit cost per week, it is necessary to plot the average cost over the period to date on the same chart and to the same vertical and horizontal scales. This is indicated in Fig. 14.3 by the broken line in the upper graph. Care must be taken in such a case, since there will be, in all probability, a variable quantity of excavation carried out in each week. Cumulative costs must be divided by the cumulative total, and the average must *not* result from adding the unit weekly costs of the excavation together and dividing by the number of weeks.

Adjacent to the graph of unit cost, it is useful to plot, as in the lower part of Fig. 14.3, a graph of the cumulative total week by week, in order to get some indication of the amount of work in the operation being carried out and whether this is being carried out at a uniform rate or whether there is some variation. Against the graph cumulative total, a graph of average output each week can also be plotted, thus indicating the trend. In the particular example chosen, the average weekly output has been plotted to a different scale from that of the cumulative output.

On receiving, at the end of the fourth week, the relevant part of the report shown in Fig. 14.3, it will be noted that the unit cost for the fourth week's excavation has shot up from £0.49 to 0.63 per cubic metre. Immediate investigation is therefore required in order to see whether this was due to inefficient planning of the work. It may be due to a major breakdown in the equipment, bad weather, labour troubles, or a number of other reasons, most of which will soon become apparent on investigation. Inefficiency in working may be highlighted as a result of the investigation. As an example, it may be found that a face shovel has insufficient trucks to carry away the excavated material at a fast enough rate, causing it to stand idle between discharging its loads into one truck and the arrival of the next. The action of adding more trucks to the pool already being fed by the machine will clearly help to solve a problem of this nature, and a note of the action taken can be put on the chart. If the action taken is effective, the result should be similar to that shown for weeks 5 and 6 in the graph, with the unit cost dropping considerably and the average cost showing a trend, which also falls over the job's duration.

To add further information to the report, a horizontal line is drawn, as in Fig. 14.3, showing the estimated rate for the work, including, so long as it is defined, the cost of labour and plant in the unit cost. At this stage it is unnecessary to include site oncosts and margin. A further refinement would be to add another horizontal line dividing the estimated rate between plant and labour, and plotting the unit and average costs in this form, the sum total or the ordinates being represented by the graphs already drawn on Fig. 14.3. A straightforward comparison can then be made, setting the costs of labour and plant against the estimated costs.

Similar graphical reports can be prepared for many operations of major importance in the contract, such as formwork, costs of mixing, transporting and placing concrete, costs of brickwork, etc.

If proper attention has been paid to the planning of the work in both the pre-tender and the contract stages then a labour schedule and graph will have been prepared of the type shown in Fig. 7.10. It is now desirable, both from a control point of view and for feedback for estimating future works, to keep a graph of the numbers of operatives employed on the work for each week, plotting, on the same diagram, a cumulative total. Such a chart can be plotted against a vertical scale of numbers or money. Comparison of this chart with the one produced in the planning stages will give an indication whether the numbers of operatives employed on the contract are approximately the same as was originally thought. If it is wildly astray then some investigation is required to establish whether the rate of working is in excess of that planned in the initial stages, or whether the working is less efficient than was assumed.

Charts such as those described and illustrated above tend to make easier and more interesting reading than tabulated figures like Figs 14.1 and 14.2. Trends become far more obvious and easier to spot in reports graphically presented. No reporting system, whether in cost control, planning, or any other area of management control, will be effective unless it is read and understood by the individual for whom it is intended—the one who is controlling, or having a hand in the control of, the work. Charts and graphs that are made simple and easy to read will achieve this purpose more easily than tabulated figures.

14.4 Monthly reporting

At monthly intervals, it is desirable to prepare a cost statement in somewhat more detail than the weekly control statements. In addition to information regarding the costs to date and the value of the work carried out, it is highly desirable that an estimate be prepared of the cost of the work yet to be completed. This may well vary from the original estimate, and should be prepared in the light of information that has been gained on the work so far. For example, if laying drains has so far proved more expensive than was originally estimated and every step has been taken to improve the efficiency of operations without success, some account of the known facts must be incorporated in the estimate for work yet to be performed. A suitable arrangement of a cost report for the monthly interval is shown in Fig. 14.4. Since this type of report is prepared with more accuracy than the intermediate weekly reports and is more an account than a controlling record, there is an opportunity of inserting more detailed information than on the

XYZ Contracting Ltd.			**Monthly Cost Summary**							Sheet 1 of 1				
Contract *Snooker Bridge*			Contract No. *B 761*							Week of programme 10				
Code number	Operation	Quantity					Cost (£)							
		Unit	Initial estimate	To date	Estimated final total	To complete	Initial estimate	Cost to date	Valuation to date	± to date	Estimated final cost	Final valuation	Final ±
A	B	C	D	E	F	G	H	J	K	L	M	N	O
67.8112	Temporary cofferdam	item	100%	100%			3.500	3.750	3.500	−250	3.750	3.500	−250
67.2117	Excavation	m³	1.000	500	1.200	700	500	300	250	−50	650	600	−50
67.2371	Driving pile-cases	No.	750	400	750	350	37.500	19.900	20.000	+1.000	35.000	37.500	2.500
67.2211	Formwork	m²	1.000	100	1.250	1.150	125	150	125	−25	1.563	1.563	

Remarks:

Initials _____

Figure 14.4

previous types of record discussed. In this particular example, quantities are separated from cost, so that comparisons can be made between these different units. A statement is made in column D of the quantity that was initially estimated, E states the amount of work carried out to date, F, the estimated final amount of work to be carried out in the light of the known conditions, and G gives the quantity of work yet to be carried in order to complete the contract. This information gives some idea whether the scope of the work has increased or decreased, how the work is currently progressing, and whether the operation is nearly complete or in its initial stages. Both of these situations must be known, since many operations are expensive in the starting-up stages but their unit costs become less as production gets under way. It should be noted under the first operation in Fig. 14.4, that of a major temporary work, namely a cofferdam, that this is a single item. It is therefore difficult to give any idea of the progress of the item in the columns under quantities. One method of overcoming this is to refer to it in terms of percentage completion.

The columns under 'Cost' follow similar lines, except that it is possible to make a statement here whether the work is being carried out profitably or not. H and J simply give the cost of the work in accordance with the initial estimate and the cost of the work to date, respectively. K gives the valuation of the work carried out to date in accordance with the estimated rates, or with those rates taken from the bill of quantities. The difference between the valuation and the actual cost appears in L, which also shows the profit or loss on the particular operation as it has so far proceeded. In order to be able to see the true position of cost for an operation, an estimate of the total cost to completion of that operation must be made. In the case of Fig. 14.4, this is placed in M and can be compared with N, which is the valuation in accordance with the tendered rates for the estimated final quantities of work in the operation. These figures will give rise to O, in the form of a final profit or loss figure for the complete operation.

The view is sometimes expressed that operations that have been completed should be omitted from this type of monthly report. This enables the historical information presented to be reduced in quantity. It is necessary, however, to indicate on the report those operations that have been done by underlining some of the key figures. It is desirable, on balance, that the report should contain full details of all the operations, whether they have been completed or not. Very often, where similar operations are concerned in different parts of the work, it is of value to compare the costs of one operation with the costs of a similar operation elsewhere. Not only is this so, but figures that are completed are firm and can be treated in a different light from those which represent estimated costs to completion. In looking at Fig. 14.4, it will be seen that the temporary cofferdam has been installed, incurring a total loss of £250. This is a figure which is irretrievably lost on this particular operation and, if the overall scheme is to break even, at least £250 must be picked up on another operation. In the case of driving the pile-cases, the work is only a little over 50 per cent complete, but the estimated profit on the operation (and, if this is a labour and plant cost report, the term profit refers only to these two resources and does not include oncosts, overheads, or margin) amounts to £2500. This £2500 is estimated at this stage on a considerable amount of work yet to be carried out. It, therefore, cannot be treated as a firm figure like the −£250, but must be treated with reserve until it becomes a final figure. At best, it is an exception and it would be improper to take it into full account in balancing operations, although it would appear that there may well be a figure to take into account as profit from this operation.

14.5 Control of materials

As has been mentioned previously, the question of control on materials does not receive a great deal of consideration in existing cost control systems. This is largely due to the fact that labour and plant are the areas in which greater variations can result, and in which it is most likely that money will be made or lost over the original estimated costs. Control systems for materials consumption should not be carried out in terms of cost, but in terms of their unit measurement. In obvious cases of deficiency, one cannot be sure, if cost is the basis, whether there is a high wastage of material bought at prices lower than in the estimate, or whether the materials are being used economically, with little wastage, but have been bought at prices in excess of the estimate. In addition, if a cost comparison alone is made for materials, it reveals no true quantity of the rate of usage for the materials.

One of the simplest and most widely accepted methods of controlling materials is to draw a graph of the amounts of the various materials that should have been used, calculated from the current measurement of the quantities of work carried out. Such a graph would be drawn for the major items of materials in use, such as bricks, concrete aggregates, cement, etc., depending upon the nature of the work. Against this graph for each individual material can be drawn a second graph, which indicates the quantities of material that have been delivered to the site. This information is readily obtainable from the invoices received and a running total can quite easily be maintained. The intercept on the vertical ordinate between the two graphs should be equal in quantity to the amount of materials on site. It is, therefore, necessary to check the stock on the site in order to make the comparison effective. The desirable interval for such a check is a month but, in the event of large discrepancies, it may be necessary to carry out checks in specific materials more frequently.

14.6 Standard costs and variances

It is not possible to provide meaningful analyses of cost data that have been collected in the field unless there is available a *norm* or *standard* to which they can be compared. Standards in construction are usually set by estimating the possible outcome from historic performance and experience or, in special cases, from the use of work measurement techniques. Initial budgets for work are established from estimated costs and the variances between actual and estimated is calculated. If the actual cost is greater than the standard then the variance is negative or *unfavourable*. In the reverse situation the variance is *favourable*.

Variances occur for one or both of two reasons:

1. The price actually paid for resources is greater or less than that estimated in the standard.
2. The quantity of resource actually used is either greater or less than that estimated in the standard.

If work is to be carried out at its estimated or budgeted cost, then attention must be paid to the above two factors. In using variance analysis, inefficiencies are high-lighted in terms of cost, and attention can be concentrated on limited areas where these inefficiencies exist. It also draws a distinction between those variances that occur because of price differences and those due to quantity differences. The following generalizations are true:

Actual cost (AC)   = actual quantity (AQ) × actual price (AP)
Standard cost (SC) = standard quantity (SQ) × standard price (SP)
Total cost variance = standard cost (SC) − actual cost (AC)
                    = (from above) (SQ × SP) − (AQ × AP)

Variance analysis will first of all be applied to the purchase and use of materials. The outcome of material costs will be due to variances either in price or in quantity from those estimated—there will be either a *material price variance* or a *material usage variance*. The first will amount to the difference between the standard and actual prices for the quantity of materials used; the second will be the difference between the standard and actual quantities of materials used.

Therefore,

Material price variance = AQ(SP − AP)
Material usage variance = SP(SQ − AQ)
Material cost variance   = SC − AC = (SQ × SP) − (AQ × AP)

The calculation of material usage variance follows a convention adopted in the United Kingdom, that variations in quantities actually used against standard quantities are priced at the standard price, since this buying price is very frequently outside the control of individuals responsible for using the materials.

EXAMPLE 14.1
When a civil engineering contractor bid for a construction project, the amount of cement in a cubic metre of one of the mixes of concrete was calculated to be 280 kg. The price of cement at the time of the estimate was £1.00 per 50 kg and this was

adopted as the standard price (SP). 400 m^3 of this concrete were mixed and placed on one particular day and a series of checks on the cement storage facility revealed that 120 000 kg of cement had been used. The actual price (AC) of the cement paid to the supplier was £0.98 per 50 kg.

Calculate the material price and usage variances.

Material price variance $= (120\,000/50)(1.00-0.98) = £48.00$ (favourable).
Material usage variance $= 1.00[(280 \times 400)/50) - (120\,000/50)]$
$$= -£160.00 \text{ (unfavourable)}$$

Therefore,

Material cost variance $=$ standard cost $-$ actual cost
$$= [280(400)1.00/50] - [120\,000/50(0.98)]$$
$$= -£112 \text{ (unfavourable)}$$
$$= \text{price variance} + \text{usage variance}$$
$$= 48 - 160 = -£112 \text{ (unfavourable)}$$

This may be interpreted as meaning that the purchase of cement on this particular day and at the rates given above was favourable, the rate being less than that assumed at the time of setting the standard. However, more cement appears to have been used than was expected and has resulted in an adverse variance. Overall, taking both aspects into account, the balance of cost proves to be unfavourable. The extent of the adverse material usage variance would indicate that an investigation is needed into the use of cement.

14.7 Labour variances

The two important variances for labour costs are the *labour rate* and the *labour efficiency* variances. They are defined as follows:

1. The labour rate variance arises from the difference between the standard wage rate and the actual wage rate paid; that is,
    Labour rate variance $=$ actual time worked (standard rate $-$ actual rate)
    $$= AH(SR - AR)$$
2. The labour efficiency variance arises from the difference between the actual time and the standard time to do a job, measured at the standard rate.
    Efficiency variance $=$ standard rate (standard time $-$ actual time)
    $$= SR(SH - AH)$$
    As a result,
    Labour cost variance $=$ standard cost $-$ actual cost
    $$= SC - AC$$
    $$= \text{labour rate variance} + \text{labour efficiency variance}$$
    $$= AH(SR - AR) + SR(SH - AH)$$
    $$= (\text{standard hours} \times \text{standard rate}) - (\text{actual hours} \times \text{actual rate})$$

EXAMPLE 14.2

An average and uniform wage rate is calculated for the purposes of estimating and it amounts to £4.00 per hour. In undertaking some excavation, 30 operatives are employed. Ten of them are paid an average of £3.75 per hour and the remaining 20 are paid £4.10 per hour. The standard output per man-hour is set at $0.18\,m^3$ and a total of $250\,m^3$ of excavation are achieved in one particular 40-hour week.

$$\text{Labour rate variance} = (10)40(4.10 - 3.75) + 20(40)(4.00 - 4.10)$$
$$= 100 - 80 = \underline{£20} \text{ (favourable)}$$

$$\text{Labour efficiency variance} = 4.00[(250/0.18) - 1200]$$
$$= \underline{£755.55} \text{ (favourable)}$$

$$\text{Labour cost variance} = 20.00 + 755.55 = \underline{£775.55}$$

This may be checked as follows:

$$\text{Standard labour cost of production (estimated cost)} = 250/0.18(4.00)$$
$$= £5555.55$$

$$\text{Actual cost of hours worked} = (10)(40)3.75 + (20)(40)4.10$$
$$= £4780$$

$$\text{Labour cost variation} = \text{standard cost} - \text{actual cost}$$
$$= 5555.55 - 4780 = \underline{£775.55}$$

The labour rate variance is positive, that is it is favourable, which is interpreted as meaning that the labour rates used for this particular operation have been less than was expected at the estimating stage. The labour efficiency variance can be interpreted in a similar fashion; this work has been carried our more efficiently than was expected in terms of the labour content. The total benefit in cost terms for the operation, that is for both the labour cost and the efficiency, is the labour cost variance, which amounts to £775.55.

14.8 Overhead variances

An *overhead cost variance* is a variance obtained as a result of calculating the difference between the overhead actually incurred on a specific operation or activity and the standard overhead cost set for that operation related to the production budgeted to be achieved. The units upon which the budgets are based, so as to establish standards, can vary. They can be units of production or of time. Units of production might be 100 cubic metres of concrete, 50 cubic metres of excavation, 1 square metre of topsoil removal, or 100 square metres of brickwork of a particular kind. The unit of time can be in hours, days, weeks, months, etc. It is important that the units are selected to suit the nature of work being carried out—factory-type work may necessitate the choice of different units from construction work on a site—and that they are chosen for convenience and economy of calculation and data collection. The choice will also be influenced by the purpose of the analysis and the accuracy and the form in which the ultimate information is required.

In establishing overheads there is often a need to recognize two components that have different characteristics. These are *fixed* and *variable costs.* Fixed costs are those that do not fluctuate with changing levels of production, although they may change after a certain level of production has been reached. A simple example is that of the overhead for site supervision. One foreman can be used for a wide variety of levels of production and numbers of operatives. If, however, a certain level of production is reached, or the numbers of operatives exceed a maximum number, a second foreman may be needed. The cost of the first foreman is fixed over a wide range of activity until the second foreman is needed. Variable costs vary as the level of activity fluctuates. For example, if fuel is included as an overhead on a machine-intensive operation then the consumption will vary almost directly with the level of activity of the machines.

An *overhead volume variance* occurs when there is a difference in the recovery of the related overhead, because the budgeted production or the time taken to produce the budgeted quantity is greater or less than the actual production or the production time. An unfavourable volume variance will occur when a fixed overhead is not fully utilized. For example, a precast concrete production plant is not utilized for the number of production hours for which it is available; the fixed overhead must then necessarily be spread over fewer hours of production. Conversely, when production time bearing a fixed overhead is in excess of that budgeted then a favourable volume variance will be the result, because more overhead will be recovered from the greater volume of product than was budgeted, though this variance may indicate that overtime is being worked at enhanced labour wage rates.

An *overhead budget variance* is the difference between the actual cost of overhead incurred and the overhead cost budgeted for that level of production. An *overhead efficiency variance* is the difference between the hours actually worked in order to achieve a specific production target and the standard hours budgeted to be used to achieve that production target.

EXAMPLE 14.3

A contractor who undertakes earthmoving, tenders a sum of £105 000 for moving 300 000 m^3 of material for a road embankment. The tender is based on using 3 rubber-tyred scrapers, each having an average output of 100 m^3 per working hour. Of the tendered sum, 5 per cent is profit, 5 per cent is for variable overheads such as site repair operatives, workshop facilities, etc., paid for in relation to the hours worked by the machines, and 10 per cent is for fixed overheads such as site office erection, head office overheads, etc. The work is abandoned for the winter because of extremely bad weather, when 240 000 m^3 of the material have been moved and a total of 2000 scraper-hours have been used. The total overheads incurred amount to £12 000. Calculate the overhead variances at this stage in the work.

Analysis of tender:

		£
Direct costs	(80%)	84 000
Fixed overhead	(10%)	10 500
Variable overhead	(5%)	5 250
Profit	(5%)	5 250
		£105 000

Total standard machine-hours for production $= \dfrac{300\,000}{100} = 3000$

Standard machine-hours for production achieved at abandonment $= \dfrac{24\,000}{100} = 2400$

Actual machine-hours worked at abandonment $= 2000$

Budgeted fixed overhead for complete contract $= £10\,500$

Pro rata basis for partially completed work $= \dfrac{24\,000}{30\,000} \times 10\,500 = £8400$

Standard *fixed* overhead cost per hour $= \dfrac{10\,500}{3000} = £3.50$

Budgeted variable overhead $= £5250$

Standard *variable* overhead cost per hour $= \dfrac{5250}{3000} = £1.75$

Therefore,
Total standad overhead cost
cost per hour $= £3.50 + 1.75$ $= £5.25$

Actual overhead incurred at abandonment $= £12\,000$

Total standard cost of overhead flexible budget for actual hours worked $= \left(\dfrac{2400}{3000} \times 10\,500 + 2000 \times 1.75 \right)$

$= £11\,900$

Standard overhead cost for actual production at abandonment $= 2400 \times 5.25$ $= £12\,600$

1. Overhead budget
   variance $= 12\,000 - 11\,900$ $= £100$ (unfavourable)
2. Overhead volume
   variance $= 11\,900 - 5.25\,(2000)$ $= £1400$ (unfavourable)
3. Overhead efficiency
   variance $= (2000 - 2400)\,5.25$ $= -£2100$ (favourable)
4. Overhead cost variance $=$ budget
   variance $+$ volume variance $+$
   efficiency variance $= 100 + 1400 - 2100 = \underline{-£600}$ (favourable)

The first assumption to be made is that the fixed overhead, that is £10 500, will be incurred on a *pro rata* basis according to the volume of material moved. Since the evaluation is being carried out using average outputs for the scrapers, this is the same thing as saying that the overheads will be incurred on a time or 'hours' basis, using standard machine-hours. Note that there may well be some additional overhead costs owing to the abandonment of the work, such as personnel move-ments, temporary buildings not in use, etc., that will not come into this analysis but must be kept in mind for some future one.

In the analysis the total standard machine-hours for production are established at 3000. This is the estimated total time in machine-hours that will be taken to do the work. When the work was abandoned, then, on a proportionate basis, 2400 standard hours should have been used for the work completed at that time. In fact, 2000 hours had actually been used. Again on a proportionate basis, £8400 of fixed costs was budgeted for the work completed at abandonment, to which must be added £(2000)1.75 for the variable overhead, making a total of £11 900. The standard overhead cost for actual production at abandonment can be established by using the hourly rate of £5.25, which includes both fixed and variable costs. This costs amounts to £12 600.

Firstly, the overhead budget variance is calculated by substracting the standard overhead cost from the actual cost of overhead. The actual cost is greater and therefore the variance is unfavourable. £100 more has been spent than was originally allowed for this production. Secondly, the overhead volume variance is established by subtracting standard overhead cost per hour times the actual production hours from the total standard overhead flexible budget (that is, fixed plus variable costs) for the actual hours worked. This again gives an unfavourable variance, this time of £1400. Thirdly, the overhead efficiency variance can be found by subtracting the standard overhead for the production actually achieved from the standard overhead costs of hours actually worked. This is a measure of the efficiency of output per unit time. This variance is favourable to the extent of £2100.

The overhead volume variance can be explained by the fact that the actual production of $240\,000\,\text{m}^3$ was achieved in 2000 hours rather than the 2400 standard machine-hours. Therefore, because fixed overheads have to be recovered at the actual rate of production, that is in 2000 hours in this case, when 2400 hours were allowed in estimating, there would be a shortfall in the outcome or an unfavourable variance. However, the machines have worked far more efficiently than it was estimated they would, having moved the material, so far, in less time than was estimated, thus producing a favourable variance.

The three variances are added together to give the overhead cost variance, which gives a result of £600 favourable variance. Thus the efficiency variance has more than made up for the deficiencies in the other two. This figure can be checked by subtracting the standard overhead for the actual production at abandonment from the actual overhead incurred at abandonment, or $12\,000 - 12\,600 = -£600$ (favourable). This calculation provides a useful and simple check on the accuracy of the variance analysis. However, such a calculation does not enable the reasons for the various variances to be ascertained and good performance, for example, in the efficiency variance as in this example, may well conceal poor performance elsewhere.

14.9 Time–cost optimization with network analysis

Up to this point, in considering the duration of an activity within a network, only one estimate of duration has been considered (except in the application of PERT, though this, too, ultimately results in one time estimate). Initially, in estimating the activity duration, normal resources and hours of working are considered. In adjusting such durations to remodel the schedule, or even in the original estimation of the duration under normal conditions, little consideration has been given to the cost of an activity so carried out.

When considering the optimization of cost through the use of a network it is necessary to consider the two main (though not exclusive) types of cost associated with a construction project. The first is *direct costs*, which include the cost to the contractor of the labour, plant, and materials to carry out the activity. Alternatively, direct costs may entail the cost to the main contractor of employing a subcontractor, if work is carried out in this manner. *Indirect costs* generally consist of costs incurred in direct proportion to the length of time that the contract takes—for example, the wages of the site staff or the head office expenses. These will tend to increase directly as the length of the contract increases. Other indirect costs that may have considerable influence on the overall cost of a project can be those incurred by a *penalty* or *bonus* clause in the terms of the contract.

Traditionally, if construction work must be speeded up, this is done by putting all the operatives, who are currently working, on an overtime basis and by increasing the amount of plant available. The use of the critical path method demonstrates that such a measure can be extremely expensive in two respects. In the first respect, overtime is paid and plant is made available to complete many jobs that have float attached to them and their activity in the network, and therefore they are being expedited unnecessarily. The second point is that it is usually towards the middle or in the second half of the programme when the need for expedition is apparent and it is at this stage in the contract that the maximum resources in the form of labour and plant are being used.

The aim of a contractor is usually to minimize the direct costs for a project, for the important reason that this will induce the best chance of securing the contract in the first instance. The minimum cost associated with construction work rarely coincides with the minimum duration necessary to complete the work. The motives regarding minimization of cost with respect to time, therefore, clash when considering direct and indirect costs since, in the former case, minimizing duration will almost certainly increase cost, and in the latter, increasing duration increases total indirect costs. Some balance between the two must be struck.

The normal duration is the one that is usually estimated for the initial preparation of a construction network. The *normal time* for *time–cost optimization* is arrived at by calculation on the basis of the minimum direct cost for the activity. A *normal time* is thereby established as to the minimum cost or *normal cost* for the activity.

Estimating the cost of carrying out an activity in the least time possible is relatively easy. This way of executing an activity involves a great deal of overtime, increased sizes of gang, additional equipment, or a combination of all these three. There will be a minimum duration below which the activity cannot be reduced. If, for example, there is a basement to be excavated, there is a limit to the number of machines or operatives that can be placed in the excavation. If this limit is exceeded, it results in extremely inefficient working, and in the limit would result in very little work being carried out. The minimum duration that can be associated with an activity is known as the *crash time*. It becomes clear that, if it is required to crash the duration of an activity, it is also inevitable that the cost of this activity will be increased a great deal, owing to the general inefficiencies that will result from crowding the operation with labour and plant, and also the increase in the non-productive element of the overtime that is paid

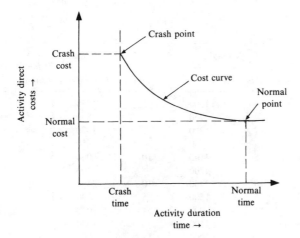

Figure 14.5

for in carrying out the job. The cost that is associated with the crash time is called the *crash cost*.

In considering the activity time–cost relationship it may be represented in the form of Fig. 14.5. The exact nature of the curve, drawn on the vertical axis of cost and horizontal axis of duration, will vary from case to case. Figure 14.5 shows a typical relationship between cost and duration for a construction activity. On this curve have been plotted the two points that have been discussed concerning the activity, that is, the crash cost and the crash time associated with it, and the normal cost and the normal time associated with it. It often happens (and this has been illustrated in the graph) that the cost of an operation may well increase if it allowed to drag out beyond the normal time.

14.10 The procedure for optimization

For every activity in a network, therefore, crash and normal costs and time can be estimated. Table 14.1 lists the activities, together with their crash and normal times and costs, for the previous example of the reinforced concrete culvert. It will be noted from Table 14.1 that some of the activities have not been crashed. The roof slab is such a example, where the activity of curing concrete is specified to take a certain length of time. If it is assumed that rapid-hardening cement was to be used in the first place for such a project, then there are no practicable possibilities for the reduction in curing time, and the crash and normal times must be similar. A network can now be drawn for the project, placing the revised durations under crash conditions against the activities instead of those of the normal durations. The backward and forward passes can be made, and, as can be seen from Fig. 14.6, a new project time, the crash time, is arrived at. In this case it amounts to 41 days, a reduction of 4 days over the normal duration for the project. In this example the critical path

Table 14.1

Operation	Description	Normal		Crash		Range of duration	Rate, £/day
		Time	Cost	Time	Cost		
1–2	Base slab	4	£1500	3	£1750	1	250
2–3	N. apron	4	750	3	850	1	100
2–4	N. side walls	7	1200	5	1400	2	100
2–5	S. apron	4	750	3	850	1	100
3–10	N. approaches	15	4100	10	6000	5	380
4–8	S. side walls	7	1200	5	1400	2	100
5–10	S. approaches	10	2400	10	2400	—	—
6–7	N. wing walls	6	1500	5	1700	1	200
7–9	N. roof	14	2000	14	2000	—	—
8–9	S. wing walls	7	1600	6	1900	1	300
9–10	S. roof	14	2000	14	2000	—	—
			£19 000		£22 250		

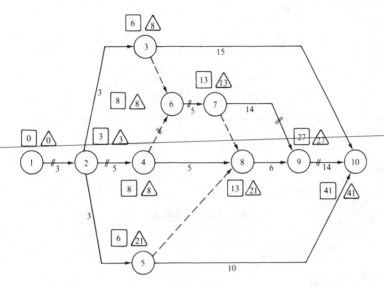

Figure 14.6

runs through the same events as for the normal conditions, but it is important to check this point. It is a matter of coincidence that it does.

By adding up the cost of each activity in Table 14.1, a total direct cost for the contract can be calculated. The normal direct cost for the culvert amounts to £19 000 and the crash direct cost, that is, the cost of crashing all the activities involved in the construction of the network, amounts to £22 250. In effect, an additional cost of £3250 would be paid in order to crash all the activities and secure a reduction of 4 days in the overall project programme. The object of time–cost optimization procedures is to find the minimum direct cost for which the reduction of 4 days in the project schedule can be achieved and, therefore, the minimum total direct cost associated with the new schedule. It also enables the direct cost for reductions of less than 4 days to be calculated.

For each of the network activities there are now two points on the time–cost curve for the activity and Fig. 14.7 illustrates, by way of example, these two points plotted for activity 3–10. It is necessary to assume in the following procedure that there is a linear relationship between cost and duration and that the two points in Fig. 14.7 are thus joined by a straight line. While this assumption may not be strictly true, it is not an unreasonable one, owing to some variations in the exactitude of the estimating procedures and the fact that during the course of the optimization procedures certain errors will inevitably cancel each other out. In addition, there are frequently cases where either the crash or the normal cost as calculated is used in the procedure, in which cases there is no need to use even the linear relationship for calculation purposes.

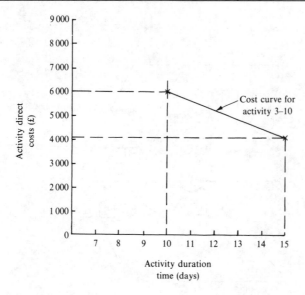

Figure 14.7

It will be seen that the graph in Fig. 14.7 implies that a reduction of activity duration from 15 to 10 days, i.e., of five days, can be achieved at a cost of £6000 − 4100 = £1900. Because of the linear relationship it may be assumed that the activity duration can be reduced for an additional cost of £380 per day. The cost of carrying out the activity for any duration between 10 and 15 days can therefore be calculated from this relationship. The final two columns of Table 14.1, illustrating the range of the durations for each activity and the cost of reducing the duration by one day, may now be completed.

It becomes clear from examination of the rates in the final column of Table 14.1 that some activities can be expedited at less cost than others, and therefore the general principle can be established that where a choice of activity is presented for expedition the one that is cheapest is selected in the first instance. In crashing all the activities making up the network, many will be crashed that would not have affected the overall duration of the project, even if they had been left as normal durations. In order to optimize the cost at which this reduction in time can now be made, it is necessary to distinguish between those activities that must be reduced in duration and those that need not be reduced. It is necessary, if the overall duration of a programme is to be reduced, that attention be paid in the first instance to those activities on the critical path.

In carrying out the optimization procedure it is possible to work from the network, which shows the normal durations, by crashing the optimal activities for the purpose of reducing the overall duration. Alternatively, it is possible to work from the network, which shows all of the crash durations, by expanding the duration of those activities that have float, thus allowing expansion to take place without affecting the overall time schedule. If this philosophy is applied to the culvert example under review, it will be noted from Table 14.1 that activity 2–4 can be expedited or slowed down for a rate of £100 per day to the total extent of 2 days. It also lies on the critical path. By reducing the duration of an activity on the critical path by 2 days, the overall time schedule for the project will almost certainly be reduced by the same amount. It is, however, necessary to check that, in reducing the duration of one of these critical activities, another path does not become critical and the normal critical path is therefore not diverted through another series of events. In the example of the culvert, working from the normal duration network of Fig. 9.4, the most likely direction along which the critical path will be diverted in this instance is through event number 3. Since this path has a float of 3 days and a reduction of 2 days is proposed, there is no danger of such a diversion in this case. The first line of Table 14.2 may now be filled in, relating to the reduction in duration of activity 2–4 by 2 days at a total cost of £200. The total cost of the programme as dictated by the new schedule of 43 days amounts to £19 000 + 200 = £19 200. This amount, the revised total, is shown in the penultimate column of the table.

Having reduced the duration of activity 2–4 to its fullest extent, that is, from its normal to its crash time, attention must be paid to other activities, and preferably to those lying on the critical path of the network. In selecting the next cheapest critical activity by which the duration may be reduced, attention is drawn to activity 6–7, which can be reduced in duration for a total of £200 per day. Examination of the network (Fig. 9.4) indicates that expediting activity 6–7 by one week will not divert the critical path and, therefore, line 2 of Table 14.2 may be completed. A cost of £19 400 is now established for carrying out the project in 42 days.

By reducing the duration of activity 1–2 by 1 week at a cost of £250, the crash duration of the schedule, that is, 41 days, can be achieved. This has been achieved for a total increase in cost of £650, making the new total cost for the project £19 650. This is to be compared with the figure of £22 250 that was calculated as a result of crashing every one of the activities. The cost of the network activities making up the project has been optimized for various overall durations. The optimum tradeoff of time against cost has been made. Many of the activities that would otherwise have been crashed, causing the unnecessary expenditure of money, are kept at their normal durations and attention has been paid to the three key activities that provide the maximum reduction in schedule time at least cost.

Table 14.2

Activity	Rate, £/week	Duration reduced by	Cost of reduction	New total cost	New schedule
2–4	10	2 days	£200	£19 200	43 days
6–7	20	1 day	200	19 400	42 days
1–2	25	1 day	250	19 650	41 days

In the example just quoted, reductions have been possible on the critical path without affecting other paths through the network. In the event of a reduction of duration in a critical activity causing a new critical path, a joint reduction in the duration of two activities is necessitated in order to achieve a reduced overall time schedule. Care, in such circumstances, must be taken to select the lowest joint cost if a number of alternatives present themselves. For example, a critical activity may present itself as a possibility for reduction in duration at a cost of £1000 per day. This may be rejected in favour of an alternative activity that can be reduced in time at a cost of £750 per day. On examination of the network, it may become apparent that, if the second activity is reduced in its duration, it will also necessitate the reduction in time of a parallel activity at a cost of £400 per day. In order to achieve a reduction in the overall schedule it will be necessary to take into account the cost of reducing both of these activities at a total cost of £1150 per day compared with the single activity of £1000 per day.

A final network can now be drawn, using a combination of both normal and expedited durations. In the culvert example all activities will bear normal durations except those of activities 2–4, 6–7, and 1–2, which will bear the shortened durations of five, five, and three days respectively. The backward and forward computations can be made once more to confirm that the overall project duration of 41 days will be achieved, and a new table of floats for subcritical activities can be presented.

So far only direct costs have been considered in optimizing the project duration, and it now becomes necessary to consider these costs in association with the indirect costs, so that the duration of the network can be optimized from the point of view of the two variables. From the figures that have been tabulated in Table 14.2 in carrying out the analysis so far, a graph can be drawn of total project indirect cost against overall project duration. Figure 14.8 shows the graph for the culvert example. Such a graph is of immense value to a contractor, but the detail work involved in its preparation is normally beyond the scope of most contractor's staffs because of the time involved in its preparation. Such an analysis clearly lends itself to the use of the computer and a number of programs have been written for this work.

The direct and indirect costs can be combined graphically, as in Fig. 14.9, by summing the curves for direct cost and indirect costs on the same graph. Most companies have a general formula for calculating their indirect costs and this may often be represented by a straight line. The low point of the curve of total cost has a project duration time associated with it and it is this time that will result in the minimum cost of the project being achieved. At this point in the calculation it may be very desirable to add in the indirect costs in the form of a penalty or bonus, which may well be shown as a separate line to total project cost curves.

By adjusting the time of the project schedule to coincide with the optimal time obtained from the combined graphs, it is possible to carry out the work in the optimal overall duration.

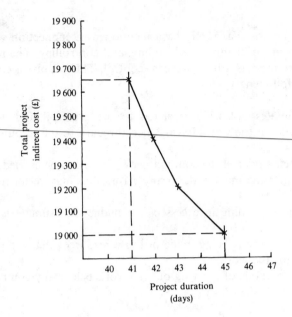

Figure 14.8

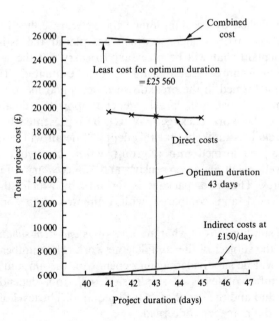

Figure 14.9

14.11 Costs on a network

Other than in this chapter, cost has rarely been mentioned in connection with the network diagram as a means of planning, scheduling, and controlling. The network diagram, however, can be used as a basis for cost control. This enables a construction manager to do the following:

1. Ensure that all estimates of cost are related to the practical and accepted proposals for carrying out the work, having special regard to the resources that will be used.
2. Check whether the remainder of the contract programme can be carried out for the estimated costs and the magnitude of any errors that may result from the work.
3. Locate areas in which spending in excess of the budget or estimate is taking place.
4. Check whether the best use is being made of those resources that are planned and made available for the work.
5. Evaluate alternative methods of carrying out the work before any commitment is made to any one method.

Before such a system can be implemented, a network plan and schedule must be prepared, illustrating the logic of the activity sequences and the estimated durations of each activity. It is this scheduled network that provides a basis for the costing procedures.

Work packages are established by breaking down the network either into single activities or into groups of activities, against which are placed the estimates of materials, employees, and plant that will be necessary to carry out the work. The cost associated with these resource requirements is then estimated. The work package is the smallest unit formed in the breakdown process and its size, whether in terms of calendar time or cost, will largely depend upon the detail that is required. In small projects, the work package may well be represented by a single activity but in large projects it is unlikely that this depth of detail will be required, and several activities may be formed into a group within one work package. Too much detail may well result in an expensive and unnecessary effort going into the costing procedures. The work package is the unit by which all the cost data are ultimately collected and compared with estimated data for control purposes.

The work packages are subdivisions of what are known as *end item subdivisions—the project end item* being the subject of the complete network. The number of times that the project end item will be subdivided again depends on the amount of detail that is required. The number of levels of subdivision will, too, depend on the magnitude of the task in hand and the detail of control required. The level of detail is a matter of judgement by the manager concerned.

Figure 14.10 illustrates a straightforward breakdown structure for the civil engineering works of a power station. The project end item is the power station

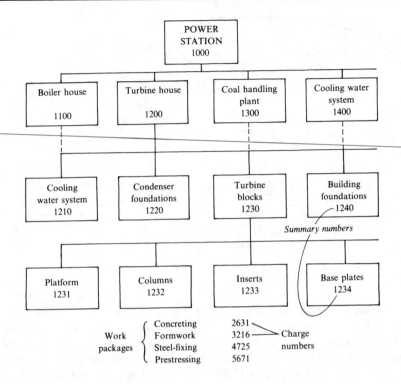

Figure 14.10

itself. The end item subdivisions occur in row 4 of the structure, being the last row of subdivision. The work packages are listed under each subdivision and can be themselves broken down, if so desired, into smaller work packages.

Figure 14.11 represents that part of a network which includes the activities making up the end item subdivision. Every activity within the loop contributes to the end item subdivision (in this case the column) and each activity, sometimes two, contributes towards a work package.

A simple example will be used to illustrate the principles of using a network in order to control cost. Figure 14.12 shows an activity-on-the-arrow network representing the schedule for the construction of the foundations of an oil storage tank, the site preparation and access roads, and the erection of the tank itself. Also included is an arrow for *overheads and supervision*. Note that this latter arrow flows from the first node, node 1, to the last node, node 13, without connecting to any intermediate nodes. This is known as a *hammock activity* and as such is used to represent indirect costs in general because of the difficulty, for example, of allocating such items as overheads to individual activities. Hammock activities do not necessarily have to span the whole network as a single activity; they may span a group of activities to which they particularly

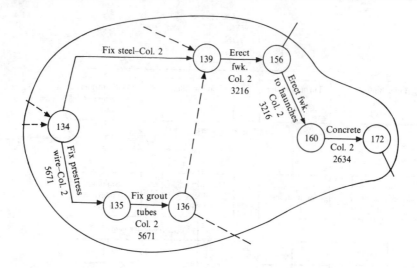

Figure 14.11

refer and they may represent any resource that is to be allocated in this way. In the case of each activity, in addition to the description of the operation and its duration, here is stated a money value of the work to be carried out. The overall duration of the schedule is 20 weeks and this duration will automatically be adopted by the hammock while, unless other distributions are given, the indirect costs will be distributed uniformly throughout the 20 weeks.

Table 14.3 shows the breakdown of the network into work packages. Although with such a small network the scope for the breakdown is limited, for the purposes of control five work packages are arranged. The table shows the number of activities to each package, ranging from four in package 0102 to single activities for the indirect costs and fees,. The work packages should be compiled from a cost organizational point of view. Clearly, the foundations will make one package from the control point of view and may well be undertaken by a single contractor. That contractor may also undertake the site preparation and the access roads, which are work of a similar nature. It is unlikely, however, that the erection of the tank will be carried out by the civil engineering contractor unless a subcontractor is employed. The tank therefore makes another convenient package. Unless other provisions are made to the contrary, it is convenient to assume that expenditure will be incurred on a linear basis throughout a work package. With small packages and short durations this will cause no problems. However, it can be seen in the case of the tank that it has a long delivery (10 weeks) and a short erection period of 1 week. If a linear distribution of cost is used for this work package, it will tend to give a false picture. The possibility of dividing the 'deliver' activity into 'manufacture' and 'deliver' activities with the major cost being on the latter, might be considered. Otherwise the bulk of the cost could be redistributed to 'erect tank'.

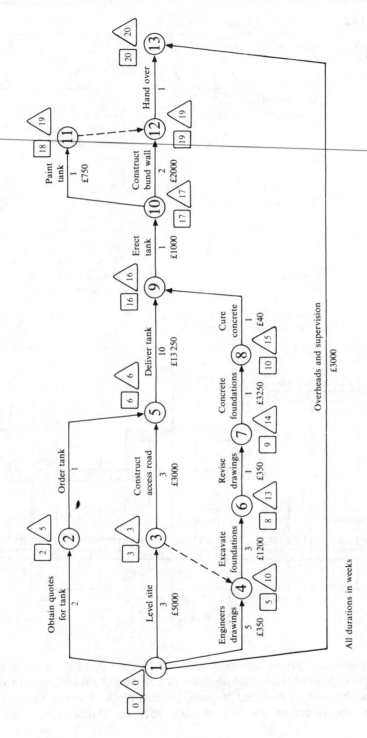

All durations in weeks

Figure 14.12

459

Table 14.3

Work package code number	Work package description	Estimated total cost, £	i	j	Activities in work package	Estimated activity cost, £
0101	Site preparation access roads	8 000	1	3	Level site	5 000
			3	5	Construct access road	3 000
0102	Foundations	6 490	4	6	Excavation	1 200
			7	8	Concrete	3 250
			8	9	Curing	40
			10	12	Bund wall	2 000
0103	Storage tank	15 000	5	9	Deliver tank	13 250
			9	10	Erect tank	1 000
			10	11	Paint tank	750
0198	Fees	700	1	4	Engineer's drawings	350
			6	7	Revise drawings	350
0199	Overheads and supervision	3 000	1	13	Overheads and supervision	3 000
	Total project estimated cost	£33 190				£33 190

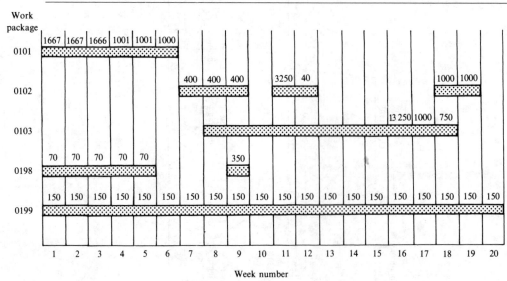

Figure 14.13

The associated bar chart at early start is shown in Fig. 14.13. With the exception of work package 0103, linear distributions of cost have been assumed. In this case the cost for the tank is allocated to week 16. Figure 14.14 shows the S-curve for the project expenditure on the basis of early start, the choice made in the bar chart.

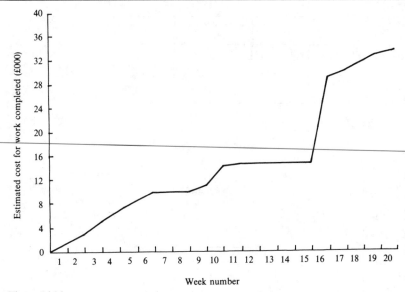

Figure 14.14

Figure 14.15 illustrates the S-curve of cost expenditure for work package 4–12, coded 0102. This is illustrated at early start and the horizontal portions of the S-curve in both cases represent periods where no cost is incurred for this work package. Much of the upper horizontal portion of the curve is due to the early start situation and could be eliminated by the use of total float to move some of the early activities later in the programme. The linear interpolation of the expenditure is superimposed on the curve. On the face of it, some considerable deviations from the actual curve are highlighted. The use of float would tend to bring the linear curve and the actual curve closer together.

A more meaningful monitoring device than the single S-curve is the S-curve envelope formed by the early-start and the late-start curves. The cost envelope for this illustration is shown in Fig. 14.16. A narrow envelope, with little horizontal space between the two curves, illustrates a project that requires very close attention to its progress because there is little float to give leeway. If the envelopes are further apart then control is not required to be as stringent. If the envelope is being used for control purposes then the *actual cost* curve and the *value of work done* curve can be superimposed on the envelope as the work proceeds. The *actual cost* curve represents the amount of cost actually incurred in doing the work. The *value of work done* curve represents the actual value, that is the payment due, for the work completed at the cost shown in the other curve. If the 'value' curve includes a margin for profit, risk, etc., then the 'value' curve should always be above the 'cost' curve if the work is proceeding profitably. If the envelope curves are drawn for cost then the 'cost' curve should start off and remain within the envelope provided the work is proceeding to programme. If the curves are for value then it should be the 'value' curve that follows this

Activity	Duration (weeks)	Week Nos.
4–6	3	6, 7, 8
7–8	1	10
8–9	1	11
10–12	2	18 & 19

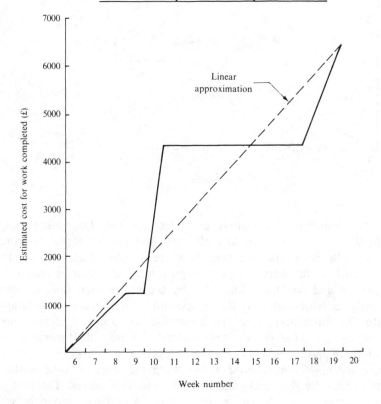

Figure 14.15

path. Assuming that the two curves forming the envelope are cost curves, if the 'actual cost' curve breaks through the upper, early start, curve, then the project is ahead of programme; if it breaks through the lower, late start, curve, then all the float for the work completed so far has probably been used up and the project is now behind programme. In the latter case some action needs to be taken to bring the project back on course and this may involve additional expense. In both of these cases the relationship of the 'cost' and 'value' curves will indicate whether the work is profitable or not. In Fig. 14.16 the curves are drawn for cost. The project is ahead of programme but more cost has been incurred than will be repaid by the value to be received. Probably the increased cost is due to more resource of one sort or another being used to speed up the programme unnecessarily.

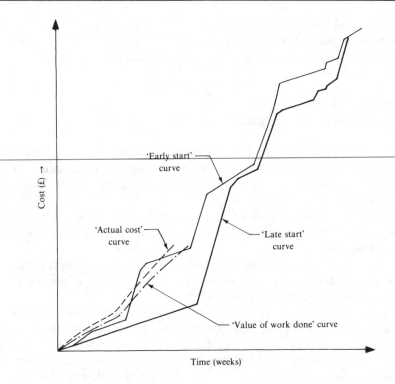

Cost (£) →

'Early start' curve

'Actual cost' curve

'Late start' curve

'Value of work done' curve

Time (weeks)

Figure 14.16

14.12 Control of mechanical plant costs

If the cost of owning and operating mechanical equipment is to be controlled effectively, account must be taken of all the individual influences on those costs. The total costs must be recovered from the valuation of the work being carried out, and, in addition, allowance must be made for the times when plant is idle. The method of achieving these aims is by deriving a comprehensive hire rate per unit of time, which is applied whether the equipment is used internally or externally to the owning company.

The following factors influence the cost of owning and operating an item of mechanical plant in the construction industry:

1. Capital cost.
2. Residual or resale value or scrap value.
3. The useful life of the equipment and obsolescence.
4. Tax allowances.
5. Required return on capital employed.
6. Cost of capital.
7. Depot repair and overhaul costs including overheads.

8. Insurance and licence where payable.
9. Site maintenance and overhead costs.
10. Grease and oils for lubrication and other consumables.
11. Operators' wages.
12. Fuel.
13. Transport to and from site of work.
14. Erection and dismantling, where applicable.

Items 1 to 6 inclusive are all intimately concerned with the rate and the extent to which capital is recovered. Some factors can be established with certainty—1 above, for example—and some are very much estimates at the time a hire rate is to be calculated initially—2 and 3, for example.

Item 7 is a cost associated with the central organization that is set up to hold, store, repair, and overhaul plant and equipment or to carry out some of these functions as required.

Item 8 is a straight charge to each item of plant or equipment as it arises.

The remaining factors, 12 to 14 inclusive, are all costs that are incurred when the plant is operating or on hire to a contract.

There can be no question about the actual capital cost of plant—the price paid is well established and documented and must be recovered through the hire rate over a period not greater than the useful life of the equipment. Its full recovery and the rate of recovery, however, are very dependent on other factors, which cannot be established with any such certainty at the time that the hire rate needs to be calculated. Two of the predominant factors in the establishment of the proper recovery of capital are the anticipated life and the estimated resale value of the equipment. To a very much lesser extent the assumed cost of capital, i.e., the general rate at which money can be made available, influences the economic hire rate to be charged.

The cost of capital, in this context, must be distinguished from the rate of return on capital employed. The latter is the ratio of profit/capital employed and can be calculated in a wide variety of ways. This latter return is that which should accrue as a result of using an item of plant in a production process. The term 'cost of capital' used in this section is the interest rate payable on capital that can be made available to purchase plant.

In order to take the cost of capital into account in establishing a hire rate, it is unrealistic to charge interest on the full capital cost of an item of equipment for each year of its life. Since an initial capital cost is recovered gradually through a hire rate, the interest should be calculated only on the outstanding capital balance, for the sake of simplicity, at each year-end. A simple example will illustrate the overall principle:

Capital cost of plant: £10 000.
Life: 3 years.
Cost of capital: 12 per cent.
Resale value: nil.

Table 14.4

Beginning of year	Capital outstanding, £	Interest for year at 12%, £	Total amount outstanding at year-end, £	Annual capital and interest recovery in hire rate, £	Amount outstanding at year end, £	Capital recovered during year, £
1	10 000	1 200	11 200	4 163	7 037	2 963
2	7 037	844	7 881	4 163	3 718	3 319
3	3 718	445	4 163	4 163	Nil	3 718
4	Nil					
	Totals	2 489		12 489		10 000

In this example a uniform annual capital recovery sum has been calculated, which, in itself, is convenient because it contributes to the uniformity and stability of a hire rate. From Table 14.4 it will be seen that the annual recovery of capital plus interest amounts to £4163, but that the actual capital recovered in each year varies, as does the interest. As is to be expected, the annual amount of interest paid diminishes in relation to the capital outstanding.

From an accounting point of view, it may be simpler to consider annual capital recovery and interest to be recovered at a uniform rate. Once the interest is calculated as in Table 14.4, the annual amounts can be arrived at as 2489/3 = £830 and 10 000/3 = £3333 respectively. The apparent rate of interest on the full capital is then 8.30 per cent, but the real rate of interest on capital outstanding is 12 per cent.

The effects of a change in interest rate in the above example are shown in Table 14.5.

The estimated life of an item of mechanical plant is another important factor in the calculation of a hire rate. Figure 14.17 shows the variation in annual capital recovery for a small loader costing £11 500 and having a range of estimated lives from 5 to 12 years. Three curves are shown—one each for a 20 per cent, a 10 per cent, and a zero residual value. The interest rate assumed is 12 per cent and capital recovery is as calculated earlier in this section.

It is seen in Fig. 14.17 that different residual values can have a considerable effect on the hire rate of equipment. In the case of a loader costing £11 500 and assuming a residual value of 20 per cent of its net capital cost, the residual value is £2300. With a life of 10 years, the capital recovery at an interest rate of 12 per cent is

$$(11\ 500 - 2300)\ 0.176\ 98 + 2300 \times 0.12 = £1904 \text{ p.a.}$$

Table 14.5

Interest (%)	6	8	10	12	14	16	18	20
Total annual charge (£)	3741	3880	4021	4163	4307	4453	4599	4747
Weekly charge (52 weeks) (£)	72	75	77	80	83	86	88	91

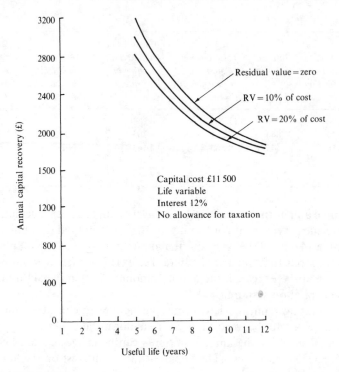

Figure 14.17

If the actual resale value at the end of the equipment's useful life is only half of that estimated above, that is £1150, the capital recovery becomes

$$(11\,500 - 1150)\,0.176\,98 + 1150 \times 0.12 = £1969 \text{ p.a.}$$

an increase of £65 p.a. or 3.4 per cent. If there is negligible resale value then the annual recovery becomes £2035 p.a.—a total increase of 6.88 per cent.

In order to be able to calculate the hire rate for an item of mechanical equipment, in addition to the above items, figures must be put to all the relevant items listed 1 to 14 above. Many, such as insurance and licence, are straightforward. Others will have to be the result of experience and the assiduous collection of the relevant data. A decision must be made, for example, about the number of weeks in a year over which the plant hire rate will be recovered. Forty is a useful guide, but individual policies and practices will clearly be influential in arriving at the ultimate figure to be used. A decision will need to be made concerning the charge to be made for standing plant. Capital charges clearly continue whether plant is working or not.

A typical calculation for the hire rate of a small loader having an initial cost of £11 500, a residual cost of £2300 and a life of 10 years with an interest rate of 12 per cent, is as shown in Table 14.6.

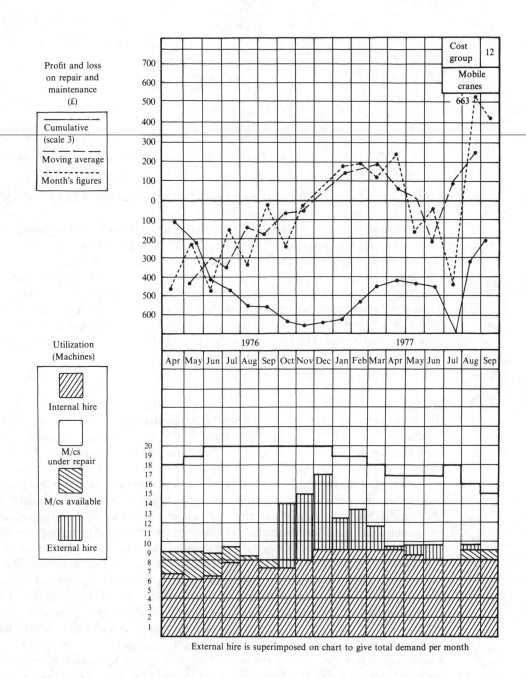

Profit and loss
on repair and
maintenance
(£)

Cumulative
(scale 3)

Moving average

Month's figures

Utilization
(Machines)

Internal hire

M/cs
under repair

M/cs available

External hire

Cost group   12

Mobile cranes

663

1976   1977

Apr May Jun Jul Aug Sep Oct Nov Dec Jan Feb Mar Apr May Jun Jul Aug Sep

External hire is superimposed on chart to give total demand per month

Figure 14.18

Table 14.6

	£/year
Capital recovery (see Fig. 14.17)	1904
Repair costs—historic percentage of capital cost, say, 9% of £11 500	1265
Overheads and profit, say, 6% of capital cost, £11 500	690
	£3859

On the basis of a 40 week per annum hire, hire rate becomes 3859/40 = £96.50 per week.

This calculation does not include the actual running costs for the equipment such as the operator's wages, fuel, oil and grease, and running maintenance. These would be charged separately, depending upon the circumstances of the hire.

Figure 14.18 shows the sort of information that is required in assessing plant requirements and costs. The upper figure takes as an example mobile cranes and gives a graph, with trends, of the profit and loss of the repair activities concerned with that equipment. The lower part of the figure gives statistical data concerned with plant utilization. The latter figure assists with the determination of the required size of plant fleet as well as its utilization.

14.13 Conclusions

To have a sound system of cost and time control for a project is becoming increasingly important. The problems faced in the *one-off project* are quite different from those that have been largely overcome in the production engineering industries. The variable nature of each contract presents great difficulties in the collection of standard data that will be of use for future occasions. When action is required in the control of a construction project, it is required immediately, since the opportunity for controlling that particular operation may have passed by in a comparatively short time and will not present itself again during the life of the contract. With the introduction of better and more rigorous methods of planning work, together with cost anaysis, the control of contract work has become more systematic. The needs of management are met more completely by providing a system whereby decisions can be readily taken whether to overspend in a particular area of operation, offsetting such overspending by a foreseeable saving elsewhere. Many managers are also faced continually with the situation in which they must take a decision whether to complete the project in the specified time, no matter how great the cost of the operation, or whether to treat the budgeted or estimated cost as the controlling feature and to optimize the project duration within this limitation.

Better methods of controlling cost must lead ultimately to more efficient and more profitable working. Unfortunately, in the contracting business, with the one-off project, it is impossible to obtain the 'before and after' figures. It is difficult to attribute direct benefits to the use of an improved cost control system, and, unfortunately, many less enlightened managers are interested only in looking at the additional cost of the implementation of such a system. The incorporation of improved systems in a company's activities can only be regarded in the long run as resulting in more successful operations, even though it is impossible to put a figure to the advantages which have accrued. The degree to which a cost control system should be implemented depends very much upon the characteristics of the company involved and the nature of the work they carry out. Another predominant factor in influencing such a decision must be the method of estimating that the firm adopts initially. If the detailed data cannot be obtained from the initial estimate then there is no measure but that of overall cost as to whether the contract is being carried out satisfactorily.

Because a cost control system is instituted, that should not draw attention away from the importance of efficient site supervision. Cost control is not a substitute for efficiency in the supervision and organization of activities on site. As has been said, its purpose is not only to control operations on site, but also to provide data for future work and feedback for the estimators. The importance of such a system lies in the fact that it is carried out systematically and at regular intervals. A system that makes purely spasmodic checks in the form of spot checks neither serves effectively as an adequate control system nor provides reliable feedback data on the types that will be of use in the future.

While it has been stressed that in cost control systems the most important resources to consider are labour and plant, the importance of collecting data with reference to the wastage of materials is not to be overlooked. This is an equally vital subject for feedback of data for use in future situations, since wastage rates can certainly make the difference between gaining or losing a contract, where several large items of expensive materials are to be used. In considering the wastage rate of material on a site, as calculated from data sent back from the site, care should be taken that allowance is made for variations in the quantity of the work that has been undertaken from that originally estimated. Not only in this context must the raw materials such as aggregate, sand, timber, bricks, etc., be investigated, but there is also a need to investigate, in certain situations, the use of composite materials such as concrete. Where concrete, for example, is used as a blinding material in the bottom of an excavation, a nominal thickness such as 75 mm will be stated in the billed rate and priced accordingly. The estimator will allow a margin over and above the 75 mm for what appears to be necessary to ensure a minimum of 75 mm on the site. Data concerned with the actual thickness or the actual quantities placed on the concrete should be fed back so that information about wastage or excesses of this nature can also be stored for future occasions. Another example of this type of information occurs in rock tunnelling, where there is a considerable overbreak that has to be made up between the internal face of the lining and the rock face. This is another important instance of excess materials being used in one form or another and control being required for future projects.

Summary

This chapter is concerned with the controlling of cost in construction. It starts with a general introduction and leads on to the question of unit costs. Unit costs are described and then a simple reporting system using unit costing is set out. Attention is drawn to trend analysis and the importance and relationship of weekly and monthly reporting. An alternative system to that of unit costing is described under standard costing and variance analysis. This has the distinct advantage that the cost analysis assigns reasons to over- or underspending and enables further investigation to be carried out. Material, labour, and overhead variance analysis is described with examples. Time–cost optimization or time–cost tradeoff is introduced. An activity that was almost impossible to carry out for other than very small networks when it was introduced can now be undertaken with ease because of the immense increase in the size and power of computers. Using a network to assist with the control of costs has proved to be a very satisfactory method of doing so and this follows on from time–cost optimization. Finally, before the conclusions, because of the ever increasing investment in mechanical equipment, some attention is paid to the cost control of using mechanical equipment in construction.

Problems

Problem 14.1   Table 14.7 lists the activities making up a network diagram for a small project, together with the costs of carrying out the work for both the normal

Table 14.7

| Activity | | Normal | | Crash | |
i	j	Duration, weeks	Cost, £	Duration, weeks	Cost, £
1	2	5	600	3	700
1	3	4	200	3	400
2	4	3	150	3	150
2	5	2	400	2	400
3	4	7	420	5	640
4	7	1	50	1	50
4	8	5	300	4	400
5	6	6	660	3	900
6	9	5	200	2	380
7	9	7	160	5	210
8	9	4	100	3	130

and crash durations. Establish the following:
(a) The total costs of both methods of carrying out the work.
(b) The optimized direct cost of carrying out the work in the crash overall duration.
(c) The optimum length of contract when overheads and other indirect costs of £120 per week are taken into account.
   In calculating the optimized direct cost under part (b), work from the all-crash diagram in the first instance and check the result by use of the normal diagram, crashing the optimum activities.

Problem 14.2   What is the direct cost of carrying out the project in Problem 14.1 in 16 weeks?

Problem 14.3   Plot the budgeted expenditure curve for the cost of work report for the normal cost condition of the simple project of Problem 14.1.

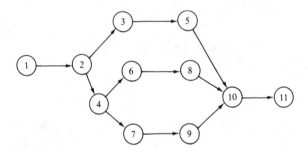

Figure 14.19

Table 14.8

i–j	Normal Duration, weeks	Normal Cost, £	Crash Duration, weeks	Crash Cost, £
1–2	5	300	3	600
2–3	10	—	10	—
2–4	14	500	13	610
3–5	12	600	8	1200
4–6	7	120	6	240
4–7	4	200	3	290
5–10	16	500	10	980
6–8	10	600	6	840
7–9	12	520	10	800
8–10	9	600	7	800
9–10	6	—	6	—
10–11	4	150	3	200

Problem 14.4   Figure 14.19 sets out the outline network programe for the construction of a small office block. The client, for whom the contractor is working, requires to know the variation in cost for the project, related to the overall contract duration (stated in the tender documents as being 49 weeks). The client has a particular wish for a contract duration of 40 weeks, but is apprehensive of the additional cost that shortening the work duration will incur.

Given the data of Table 14.8, draw a graph for the client relating direct cost to contract duration, the latter varying from 40 to 49 weeks.

Problem 14.5   Figure 14.20 shows a self-contained arrow diagram that describes the sequence of events for a small project. The durations and direct costs for each activity in the network under both normal and crash conditions are listed in Table 14.9.

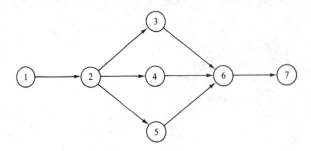

Figure 14.20

Table 14.9

		Normal		Crash	
		Duration,	Cost,	Duration,	Cost,
i	j	weeks	£	weeks	£
1	2	3	175	2	235
2	3	6	300	4	500
2	4	5	450	4	600
2	5	7	150	6	230
3	6	5	250	3	350
4	6	4	600	4	600
5	6	3	100	2	115
6	7	2	72	1	142

Calculate the minimum direct cost of the work if the overall duration for the project is to be 12 weeks. If the indirect costs for the project amount to £100 per week, what is the optimal duration of the contract when considering the total cost involved?

Problem 14.6 The durations and the direct costs for each activity in the network, Fig. 14.21, under both normal and crash conditions are listed in Table 14.10.

Establish the minimum direct cost for the project if it is desirable that the overall duration should be 24 weeks. Determine the duration of the contract at minimum total cost if the indirect costs amount to £150 per week.

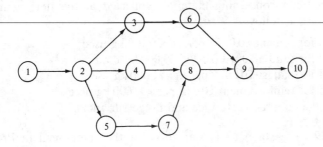

Figure 14.21

Table 14.10

| | Normal | | Crash | |
| | Duration, | Total direct, | Duration, | Total direct, |
Activity	weeks	cost £	weeks	cost £
1–2	4	500	3	750
2–3	4	100	2	300
2–4	2	200	2	200
2–5	5	600	4	760
3–6	6	700	5	830
4–8	4	200	3	300
5–7	7	140	5	200
6–9	4	200	2	300
7–8	2	80	2	80
8–9	1	100	1	100
9–10	7	600	6	670

Problem 14.7 The standard costs for the materials involved in mixing a compound are £0.60 per kg of material X, £0.75 per kg of material Y and £0.53 per kg of material Z. The prices paid for each material in practice are £0.65 per kg, £0.73 per kg, and £0.55 per kg respectively. The specification of the compound calls for a standard mix of 100 kg of X, 270 kg of Y, and 135 kg of Z. A check on the material stocks show that 105 kg, 267 kg, and 140 kg respectively are being used. Calculate the material variances for the actual operation.

Problem 14.8   The standard cost details for a single precast concrete component are as follows:

100 kg cement at £0.007 per kg
200 kg sand at £0.002 per kg
350 kg ballast at £0.0025 per kg
 30 kg reinforcement at £0.08 per kg

The actual data concerning materials collected at the time of making 200 components are as follows:

Price paid for cement £0.0075 per kg; 20 100 kg used
Price paid for sand £0.0019 per kg; 39 000 kg used
Price paid for ballast £0.0026 per kg; 71 000 kg used
Price paid for reinforcement £0.085 per kg; 600 kg used
Calculate the material cost, price and usage variances.

Problem 14.9   The standard labour costs for the component in Problem 14.8 above are:

Labour rate = £0.45 per hour
Hours   = 20 per component

Data collected at the time of production concerning actual costs are:

Components produced = 200
Labour rate   = £0.46 per hour
Hours worked   = 3950

Calculate the labour cost, rate, and efficiency variances.

Problem 14.10   The data below relate to the overhead costs concerned with the production of some standard formwork for a retaining wall in a joiners' shop:

	Budget	Actual
Overhead cost	£1000	£950
Production	300 m^2	280 m^2
Number of days	15	16
Number of hours	750	800

Calculate the following overhead variances:

(a) Budget. (b) Efficiency. (c) Volume. (d) Calendar

Problem 14.11   The data in Table 14.11 are provided for a small building modification. Produce a statement of actual cost against budgeted costs and a full variance analysis for the data collected.

In examining the cost variances for a series of overhead charges the statement

Table 14.11

	Budget	Actual
Material A, use	2 000 kg	2 200 kg
Material B, use	20 000 kg	19 800 kg
Material C, use	9 000 kg	9 100 kg
Material A, price	£0.06/kg	£0.065/kg
Material B, price	£0.13/kg	£0.125/kg
Material C, price	£0.09/kg	£0.092/kg
Labour, rate	£0.39/hour	£0.40/hour
Labour, output	600 m^3	620 m^3
Labour, hours	800	860
Overhead	£250	£240
Number of days	10	12

Table 14.12

	Expected budget	Likely range	Actual cost
Salaries	£12 000	£10 000–14 000	£13 000
Administration	10 000	9 000–11 000	10 700
Computer	15 000	12 500–17 500	13 750
Accounts	12 000	11 500–12 500	12 300

in Table 14.12 is revealed. Select the item that is most likely and that which is least likely have been influenced by random factors.

Further reading

American Society of Civil Engineers, *Construction Cost Control*, ASCE Manuals and Reports of Engineering Practice No. 65, revised edn, New York, 1985.

Antill, J. M. and R. W. Woodhead: *Critical Path Methods in Construction Practice*, 3rd edn, Wiley, New York, 1982.

Batty, J.: *Standard Costing*, 3rd edn, McDonald and Evans, London, 1970.

Gobourne, J.: *Site Cost Control in the Construction Industry*, Butterworths, London, 1982.

Kharbanda, O. P., E. A. Stallworthy and L. F. Williams: *Project Cost Control in Action*, Gower, Farnborough, England, 1980.

Moder, J. J., C. R. Phillips and E. W. Davis: *Project Management with CPM, PERT and Precedence Diagramming*, 3rd edn, Van Nostrand Reinhold, New York, 1983.

Owler, L. J.: *Wheldon's Cost Accounting and Costing Methods*, 15th edn, MacDonald and Evans, London, 1984.

Park, W. R., *The Strategy of Contracting for Profit*, Prentice-Hall, Englewood Cliffs, NJ, 1966.

Walker, T. M.: *Understanding Standard Costing*, Gee, London, 1980.

15. Decision theory and linear programming in construction

15.1 What is operational research?

The definition of operational research as originally prepared by the Council of the Operational Research Society is as follows:

Operational research is the attack of modern science on complex problems arising in the direction and management of large systems of men, machines, materials and money in industry, business, government and defence. The distinctive approach is to develop a scientific model of the system incorporating measurements of factors, such as chance and risk, with which to predict and compare the outcomes of alternative decisions, strategies or controls. The purpose is to help management determine its policy and actions scientifically.

The aim of operational research is the use of mathematics as an aid in arriving at a rational decision in situations where there are large numbers of likely alternative solutions. Making the decision may also be aided in part by an intuitive or human judgement approach. Nevertheless, in such circumstances, the area within which intuition is required is considerably narrowed. Operational research often makes use of probability theory. In these cases, the solution obtained will be the one that is most probable in the light of the data collected.

Like many of the other sciences, some of the basic methods of operational research have been in existence for a very long time, but have been recognized as a composite body of knowledge only since 1940, when the techniques first bore its name. Operational research was initiated in the United Kingdom, but was subsequently developed rapidly in the USA. As a management tool it is now widely used throughout the USA and Western Europe. Operational research had its origins during the Second World War as the result of the cooperation of scientists with senior officers of the Royal Air Force, to look into the problems arising from a belief that the first radar sets to be delivered were not being as effective in use as they might have been. The ineffectiveness did not appear to result from technical faults but rather from overall policy and tactics in use. It was found that the scientific-mathematical approach to this particular problem could also be applied in other military situations, and it was not long before operational research teams were set up to look at problems, not only in the Royal Air Force but in the Army and the Royal Navy as well.

On returning to industry in the postwar years, many of the personnel who had been practising operational research in the services found that the systematic appraisal methods of operational research had a sound application to industrial problems. While emphasis has been placed upon the mathematical approach to problems, it is not necessarily from this particular discipline that operational researchers are drawn. The original practitioners of operational research were biologists, and they have been followed by people from a large number of varied disciplines. The use of the operational researcher in the construction industry has been severely limited, though many of the techniques used more recently, particularly in planning, form part of operational research developments.

Operational research does not necessarily aim at the establishment of a solution that is better than the practice already in hand. Its stated object is to find the best possible solution to a problem. For one reason or another this may not always be practicable, but the aim is always evident in the work. The title of the science includes the word 'research' because this is, in fact, what the operational researcher is doing into the manager's task. The concern here is not so much with what has been done, or what is being done, but rather with how the particular job should be done in order to obtain the optimal result.

Operational research has now become divided into a number of fairly well-defined subject areas in which a considerable amount of research and development has been carried out. Much of this research and development occurred because certain problems had appeared before the operational researcher many times, and consequently had received far greater attention than other problems that make only very occasional and sporadic appearances. Standard techniques have therefore been developed for the solution of problems that fall into each of the standard categories.

15.2 OR method and models

Scientific method consists of a number of well-defined steps. Before the application of this method the need must exist to solve a problem or explain why a given situation occurs, and a carefully worded statement as to the precise nature of the problem and the objective of the research must be made. The steps which then follow are:

1. the observation of the problem situation and the collection of data, evidence, and information concerning it;
2. the analysis of the data, evidence, and information that have been collected;
3. the classification of these data;
4. the development of a hypothesis, which may be based on a theory, a guess, or an inference at this stage;

5. the testing of the hypothesis;
6. the formulation of a law;
7. the use of the law so formulated in order to make a prediction or forecast of what is likely to happen when the next similar event occurs.

Because the above series of steps postulates the use of experimentation in a narrow sense, modified methods have to be used in business. Operational research (OR) uses a model to represent a system and its operations, usually in the form of equations.

The seven steps of scientific method are therefore modified, when applied to an OR project, as follows:

1. formulation of the problem, establishing the objectives and any constraints that may apply;
2. building a model that represents the system under analysis;
3. using the model in order to obtain a solution to the problem;
4. comparing a solution obtained by means of the model with that in current use;
5. implementing the results of the study and monitoring the performance of the system through changing conditions.

Note that each of the above stages is not a watertight compartment and some parts of the study may well be carried out concurrently with others.

In OR studies, the continual recurrence of some problem types means that many standard techniques have been developed to model them and to assist in obtaining a solution. Examples are:

1. *Allocation* of resources to jobs to be carried out.
2. *Replacement* of plant and equipment.
3. *Queueing* and the selection of the next for service.
4. *Inventory* or stock control for holding and storing valuable resources.

New techniques are continually being developed, though many problems still exist that do not fall into one of these specified categories. It is very important that each problem is examined on its merits and is not immediately seen to be the opportunity to use a standard technique.

The second step in scientific method as described above and as applied to an OR project is that of building a model to represent the system under investigation. Two important aspects in such model building are as follows:

1. The model may not be obvious or immediately apparent. It needs to be created using the theory or hypothesis that has been developed.
2. The phases of scientific method are not always distinct in themselves, nor are they necessarily non-repetitive. The process of scientific method tends to be cyclic in nature.

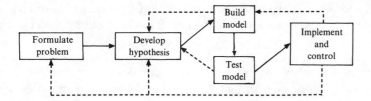

Figure 15.1

Figure 15.1 represents the flow model of scientific method. Recycling is shown by broken arrows.

The definition of a model is 'a representation of a real system, process, or device illustrating its operation and/or its properties'. Models must be capable of manipulation so as to predict the optimal values of the chosen variables, must be simpler than the reality they represent, must be easier to manipulate and control, and as a result must be cheaper because failure of the real thing can be expensive.

Models may be solved by experimentation or by mathematical analysis. By experimentation is meant the process of substituting a series of values for the variables in the model, viewing the results and deciding when an optimal solution has been obtained. Mathematical analysis means solving a mathematical equation or a series of equations to obtain a precise and optimal result.

Experience shows that simplicity in a model does not negate the value or usefulness of the model in providing an explanation of a real system. However, considerable skill and ingenuity is required in building a model and particularly in knowing which variables may be omitted without greatly influencing the representative functioning of the model.

There are three basic types of model:

1. *Iconic*—generally scale models of two- or three-dimensional form. The physical properties of the materials in the model may or may not represent the physical properties of those in the real system. Iconic models are frequently scaled down, as in the cases of ships, bridges, aeroplanes, framed structures, etc., or they can be scaled up, as in the cases of molecular structures, insects, or intricate mechanisms.

2. *Analogue*—these rely on representation by a more readily manipulated means than those properties of the real system that cannot readily be manipulated in practice. An example is the hydraulic model built at the London School of Economics and representing the UK economy: the flow of liquid represents the flow of money. Another example is the representation of a hydraulic system by an electrical network.

3. *Symbolic*—these use symbols, letters and/or numbers to represent the variables in the real system. Relationships are expressed by equations or inequations.

Either iconic or analogue models or both may be used as stepping stones to a symbolic model. A symbolic model is the most desirable because it is more precise than others and is less cumbersome in manipulation. On occasions symbolic models have become so complex that an analogue model has been built in order to facilitate their solution.

One important class of analogue models is that of simulation models. These differ from others in that they are dynamic and incorporate the element of time. A simulation model can be built so that a long period of time can be represented by a comparatively short duration of model manipulation using a computer. Simulation models are not required if the model can be expressed in symbolic form and can then be solved with relative ease. In this sense they are to be considered the last resort, although they have a ready application in situations where the variables tend to be stochastic in nature.

15.3 Decision analysis

Much of what has been written about in this book has been concerned with making *choices:* between investments, about schedules for projects, about whether to lease or buy equipment, and so on. In many cases, alternatives have been posed and the choice has been between them—a *decision* as to which choice has become necessary before more progress could be made. Decisions have only been necessary where there have been preferences for particular choices. If a decision maker is entirely indifferent as to which choice is selected, there is no *decision problem*. This leads to a statement that a decision situation is one in which there are:

1. several alternative outcomes from which a choice has to be made; and
2. preferences about each of the alternative outcomes.

Decision problems can be formalized by structuring them in a *payoff table* or *decision matrix*. Suppose a decision maker wishes to obtain the long-term use of some mechanical equipment for a construction business, and eventually the choice falls between purchasing the equipment outright or leasing it. If it is purchased outright then the owner will be responsible for its future repairs; if it is leased then the lessor will include the cost of the repairs within the lease. The choices are, therefore, two-fold and a decision must be made between them. The natural question to be asked is, How often will the owner need to repair the equipment and what will the repairs cost? Further research elicits the facts that if the machine proves to have *high reliability* then the outright purchase is likely to cost the owner (including repairs) the sum of £225 per week. With medium reliability it will cost £310 per week. If the machine has low reliability then this sum will increase to £350 per week. A lessor for equipment of the type required quotes £300 per week if the value of repairs falls below a certain level (this coincides with high and medium reliability), and £330 per week for repair levels above this (coinciding with low reliability). The *cost matrix* for this problem is as shown in Fig. 15.2. If the purchase/lease of the equipment will produce an income of £500 per week then the matrix can be transformed to that of Fig. 15.3, a *decision* or *payoff matrix*, by subtracting the cost of the machine from £500.

	High reliability	Medium reliability	Low reliability
Purchase	£225/week	£310/week	£350/week
Lease	£300/week	£300/week	£330/week

Figure 15.2

	(θ_1) High reliability	(θ_2) Medium reliability	(θ_3) Low reliability
(a_1) Purchase	275	190	150
(a_2) Lease	200	200	170

Figure 15.3

The choice in this simple example is between *purchase* (a_1) and *lease* (a_2). These are the two *feasible alternatives* known as *actions*. There can be as many actions in a matrix as there are feasible solutions between which a decision must be made, and which are required by the decision maker to be considered. The three conditions (there could be many more) that may arise once the decision is made, and need to be considered, are those of high, medium, and low reliability, θ_1, θ_2, and θ_3. These are the *uncertainties* of the decision and are alternatively referred to as the *states of nature* or *states of the world*. The decision maker has no control over the states of nature and they therefore introduce uncertainty into the decision making. The elements of the decision matrix are the *outcomes* or *pay-offs* of taking a particular decision and then a particular state of nature occurring. In the example, the outcomes are expressed in £/week. Outcomes are usually referred to as v_{ij}, where v_{ij} is the outcome of action, a_i under state of nature, θ_j. Note that choosing one or other of the alternatives does not necessarily determine the outcome of the decision. The outcome is determined by the decision taken *and* the state of nature that occurs.

A decision matrix displays the problem in an orderly and formal manner prior to an action being chosen. One first approach to a solution might be to evaluate each action in terms of its *best* outcome. In this case, to purchase gives a profit of £275/week and to lease gives £200/week, so that the choice would be for action, a_1, to purchase. Frequently, however, decision makers prefer to know what is the worst they are likely to meet and how can they minimize the effect of this. In the example, the worst outcome for purchase gives a profit of £150/week and for leasing it is £170/week. A decision maker would therefore opt for leasing.

The question as to which action to follow once the decision matrix has been constructed has received some attention in the past and a number of criteria have been offered. Some of these are set out below.

The *maximin criterion* is based upon identifying the minimum outcome for each action and then from these selecting the maximum. In the case of Fig. 15.3, the minimum outcome for a_1 is £150 per week and for a_2 it is £170 per week. The maximum of the two alternatives is £170 per week for *leasing*. This criterion is what is known as *risk-averse*—risk-avoiding. It is a very reserved approach to decision making and would rarely be applied in practice.

The *minimax criterion* uses the maximum selection from each action and selects the smallest of these. From Fig. 15.3, the maximums are £275 and £200, so that *leasing* would be chosen again. It happens that the same decision has been reached in both cases and this is not unlikely because of the nature of the problem—the acceptance of leasing releases the decision maker from much uncertainty. In many instances, the minimax criterion will result in quite different selections in comparison with the maximin.

The *Savage minimax criterion* minimizes the maximum *regret* or *opportunity loss* associated with each of the actions. *Regret*, as experienced by a decision maker, is, according to Savage, the feeling the individual has that results from hindsight. Knowing the state of nature that occurs, there is regret at not having selected one of the other actions. Savage believed that regret could be measured by the difference between the outcome that was actually received as a result of making the decision and that which could have been received if the state of nature that did in fact occur had been known. Therefore, from Fig. 15.3, if the decision maker had selected action *lease* and low reliability resulted, there would be satisfaction that the better result had been selected and no regret would be felt. If action *lease* had been selected and high reliability resulted then regret would be 75. Similarly, if the choice had been *purchase* and medium reliability occurred then regret would be $200 - 190 = 10$, whereas if *lease* had been selected there would have been no regret. The *regrets* are presented in Fig. 15.4 in what is called a *regret matrix*.

The maximin regret that can be experienced is 20 if the *purchase* action is taken and 75 if the *lease* action is accepted. The decision maker of Savage would accept the minimum of these so as to achieve protection against high regret. *Purchase* would be accepted.

The above approaches attempt to deal with the uncertainty in decision taking by providing a selection of possible outcomes under various of the uncertain conditions and then using a criterion by which the outcomes are measured relative to one another. The outcomes that meet the criteria are used to determine the action that should be taken. An alternative means of dealing with the uncertainty is by gathering more information about the states of nature, so as to be able to judge the likelihood of each occurring. For example, in the case of the purchase-versus-lease problem, it should be possible to gather information about the reliability of the equipment, i.e.

	High reliability	Medium reliability	Low reliability
Purchase	0	10	20
Lease	75	0	0

Figure 15.4

information as to the number of repairs, their cost, and their distribution through the life of the equipment. If it were known that 80 per cent of the items of equipment suffered from low reliability then this would influence strongly a preference for the leasing action. The use of probability theory in decision making helps an optimal decision to be made.

15.4 Probability theory in decision analysis

Probability theory and *expected values* were introduced in Sections 5.9 and 5.10. The probabilities in Section 5.9 were *unconditional probabilities*, where the events involved are independent of one another. However, more frequently, events are linked to each other, however strongly or weakly, and need to be dealt with in a different manner. A *conditional* probability occurs where events are not independent.

The rule for conditional probabilities may be stated:

For two events A and B, the probability of their joint occurrence,

$p(AB)$, equals the product $p(A)p(B|A) = p(B)p(A|B)$.

$p(B|A)$ signifies a conditional probability and may be read as 'the probability that B will occur given that A has occurred'. The above expression for $p(AB)$ may alternatively be written as follows:

$$p(A \cap B) = p(A)p(B|A) = p(B)p(A|B) \qquad (15.1)$$

where the symbol $\cap$ may be read simply as AND.

If event A is the failure by compression and shortening of a timber prop supporting formwork to a concrete beam and slab floor and event B is the use of a substandard prop drawn from a combination of props cut from both good and substandard timber, then A may well be *conditional* on B. If 30 per cent of the supporting props are fully loaded based on good timber and hence overloaded for substandard timber, and 25 per cent of the props in the store are substandard, then the probability that prop failure will occur by compression will be, from Eq. (15.1):

$$p(AB) = p(B)p(A|B) = (0.25)(0.30) = 0.075, \text{ or } 7.5 \text{ per cent}$$

The concept of *expected value* (*EV*) or *expected monetary value* (*EMV*) was introduced in Section 5.10. The principle of applying expected values to probabilities is due to Bayes, whose hypothesis is that if there is no reason for the probabilities of states of nature in a decision problem to be different then they should all be assumed to be equal. Thus, if that principle is applied to the decision matrix of Fig. 15.3 then

Expected value (purchase) $= 275/3 + 190/3 + 150/3 = 205$

Expected value (lease) $= 200/3 + 200/3 + 170/3 = 190$

and, on this basis, the equipment should be purchased because that action has the greater expected value.

	High reliability	Medium reliability	Low reliability	Expected value
Probability	0.45	0.25	0.30	
Purchase	123.75	47.5	45.0	216.25
Lease	90.0	50.0	51.0	191.00

Figure 15.5

The principle can be extended one stage further. If it is possible to predict approximate probabilities for each of the states of nature, say, gathering further information about the likelihood of them happening and collecting experts' opinions concerning them, then Bayes' hypothesis can be applied to them. Figure 15.5 is 15.3 extended to include the probabilities of the three states of nature and the expected values of each action are calculated as follows:

$$EV(purchase) = 0.45(275) + 0.25(190) + 0.30(150) = 216.25$$
$$EV(lease) = 0.45(200) + 0.25(200) + 0.30(170) = 191.00$$

Purchasing would be the decision to take on the basis of Bayes' hypothesis.

Bayes' hypothesis can also be applied to the regret matrix. Using the same probabilities gives an expected value of regret for purchasing as 8.5 and for leasing it will be 33.75, supporting a decision to purchase because regret should be minimized.

15.5 Decision trees

A decision process is often not a relatively simple one of making a single decision and then implementing it as the end of the problem. There may be several stages following the first decision, where states of nature will occur, requiring further consideration and subsequent decision making. Decision making is a dynamic process of deciding, reviewing results, redeciding and so on. The *decision tree* has been developed as a means of assisting with sequential decision making and the 'what-if?' approach that is part of it.

A decision tree consists of *nodes* and *branches*. Nodes are of two kinds and are used at junctions of the branches. Firstly, there are *decision* nodes, which represent points at which decisions must be made and are usually, though not exclusively, represented by a small numbered square. Paths emanating from decision nodes bear the alternative actions that can be taken. Secondly, there are *chance* or *uncontrollable event* nodes—state-of-nature nodes—from which spring branches bearing the alternative uncontrollable events and the probability of their occurrence. A simple example to illustrate the use of these components is shown in Fig. 15.6. The final branches of a tree, that is those at the extreme right in the examples, have at their ends a quantitative statement of the parameter that will be used to evaluate decisions, such as profit, loss, etc.

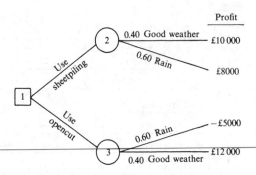

Figure 15.6

EXAMPLE 15.1

A decision needs to be made as to whether sheetpiling or timber strutting will be used in an excavation in soft clay. The states of nature that may occur are concerned with the weather—rain or fine weather are to be the two considered. The various outcomes are calculated in money terms. The tree commences with the decision point at node 1. The choices are represented by two branches, 'Use sheetpiling' and 'Use open cut'. If sheetpiling is used there is a chance that it will either rain or be fine weather; the same states of nature may occur if open cut is used. The chance nodes are indicated by a circle from which the outcomes of the chance event spring. Each outcome is shown with a sum of money that represents the profit made on the operation depending on the decision taken and the state of nature which occurs. Each chance node is evaluated for EMV. In Fig. 15.6, node 2 has an EMV of $(04.)(10\,000) + (0.60)(8000) = £8800$ and node 3 has an EMV $= (0.60)(-5000) + (0.40)(12\,000) = £1800$. The decision to use sheetpiling would be accepted. A more detailed example will illustrate the value of decision trees.

EXAMPLE 15.2

A civil engineering contractor, when considering the future of the business, has to decide whether to buy a large or a small fleet of mechanical equipment. A large fleet will cost £1 050 000 and a small fleet £200 000. Buying the small fleet entails no commitment to large expenditures except for the near future, and at some later date the needs can be reviewed in the light of market developments.

   In terms of probability, the contractor estimates that the chances of obtaining a large proportion of the work that will become available at 0.60 and those of not getting too much at 0.40 (these two probabilities must add up to 1). After reviewing all the possible decisions that will have to be considered and the states of nature that are likely to occur together with their probability of occurrence, the contractor arrives at the decision tree shown in Fig. 15.7. Altogether, three decisions must be made—those at nodes 1, 2, and 3. Decision 1 is made initially, decision 2 needs to be made after approximately 1 year, and decision 3, 2 years after decision 1. The decision tree sets out, in a clear and precise fashion, the overall view of the problem

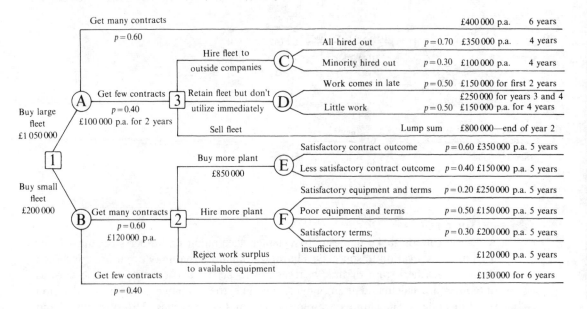

Figure 15.7

and what might be done over the next 6 years. Everything shown in the decision tree at this stage can be predetermined in the definition of the problem.

The analysis of the tree is carried out from right to left, using what is known as the *roll-back* technique. Decision points are eliminated as the roll-back takes place. Chance nodes are evaluated by calculating the expected monetary value of the branches that emerge from them towards the right.

It is preferable, though not essential, to take the time value of money into account, and 12 per cent has been assumed to be the required rate of return for this example. The appropriate values of the branches emanating from nodes C and D have been evaluated at the discount rate of 12 per cent in Table 15.1. The expected value at

Table 15.1

Chance event node	Annual outcome, £	Life, years	Discount factor, 12%	Present value, £
C	350 000	4	3.037	1 062 950
	100 000	4	3.037	303 700
D	$\left.\begin{array}{c} 150\,000 \\ 250\,000 \end{array}\right\}$ $\left.\begin{array}{c} 2 \\ 2 \end{array}\right\}4^{*}$		$\left.\begin{array}{c} 1.690 \\ 1.347 \end{array}\right\}$	590 250
	150 000	4	3.037	455 550

*In this case the cash flows change after the first two years. The discount factor for the second two years is obtained by subtracting (P|A, 12%, 2) from (P|A, 12%, 4).

Table 15.2

Decision	Chance event	Probability	Present value, £000	Present expected value, £000
1. Hire fleet to outside companies	All hired out	0.70	1062.95	744.07
	Not much hired out	0.30	303.70	91.11
			Total	835.18
2. Keep plant fleet but do not use immediately	Late work obtained	0.50	590.25	295.13
	Little worked obtained	0.50	455.55	227.78
			Total	522.91
3. Sell all plant	—	1.00	800.00	800.00

node C is, therefore, $0.70(1\,062\,950) + 0.30(303\,700) = £835\,180$, as calculated in the upper part of Table 15.2. The calculations for the other path into node C are also shown in Tables 15.1 and 15.2. The three paths (the third one has no chance event) entering decision point 3 from the right have values of £835 180, £522 910 and £800 000 respectively. The decision should be to accept the largest, that is, to 'Hire fleet to outside companies', bringing in £835 180.

Decision box 2 can be determined in similar fashion. Node E has an expected income of £973 332, from which must be subtracted £850 000 to purchase more plant; node F has an expected income of £666 870. The third path into decision point 2 results in an expected income of £432 564. The decision at node 2 is, therefore, to 'Hire more plant'.

Decisions 2 and 3 have now been determined and can be eliminated from the diagram by reducing it to that shown in Fig. 15.8. Table 15.3 shows the discounting and expected present value of the two remaining paths into decision point 1, resulting in a decision to 'Buy small fleet'.

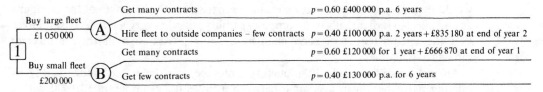

Buy large fleet £1 050 000 — (A) — Get many contracts — $p = 0.60$ £400 000 p.a. 6 years
Hire fleet to outside companies – few contracts — $p = 0.40$ £100 000 p.a. 2 years + £835 180 at end of year 2

Buy small fleet £200 000 — (B) — Get many contracts — $p = 0.60$ £120 000 for 1 year + £666 870 at end of year 1
Get few contracts — $p = 0.40$ £130 000 p.a. for 6 years

Figure 15.8

Table 15.3

Decision	Chance event	Probability	Present value, £000	Present expected value, £000
1. Buy large fleet	Get more contracts	0.60	1644.56	986.74
	Hire fleet to outside contractors	0.40	169.00 +665.80	
			834.80	333.92
		Total		1320.66
	less: Capital investment			1050.00
		Net total		270.66
2. Buy small fleet	Get more contracts	0.60	702.56	421.53
	Get few contracts	0.40	534.48	213.79
		Total		635.32
	less Capital investment			200.00
		Net total		435.32

15.6 Utility theory

Monetary outcomes have an intrinsic worth to decision makers known as *utility*. If a universal, accurate measure of utility was available then it would be found that each individual is likely to view the same monetary outcome with a different measure of its value to them as an individual—it has a different utility to them. The same can be said of business enterprises, though in the case of a large company organization it is not an easy concept to visualize because of the diverse decision making that most large organizations undertake. However, there are some areas of company activity that show clearly that the large, wealthy company places less utility on sums of money that would be highly valued by other, probably smaller companies. Insurance of company assets is one area where, if the company is large enough, it will carry the insurance for some assets rather than use outside insurers. This may be because the risk of loss is relatively low and infrequent, and, in any event, a loss would be a very small monetary sum in comparison with the assets of the company as a whole. An individual's or a company's utility will also vary as time passes. Given that the circumstances of each changes as time passes, so will their utility for monetary outcomes.

An individual's attitude to monetary outcomes can be tested quite readily by offering participation in a gamble between a prize, for example, of £10 000 with a probability of 0.50 or a loss of nothing (also with a probability of 0.50) *or* alternatively, instead of the gamble, an offer of £5000 for certain. The gamble could easily be settled by using the toss of a fair coin. It will be noted that the EMV of the gamble is $(0.50)(10\,000)+(0.50)(0)=£5000$, equal to the alternative certain offer. The monetary terms are of equal numerical value, but with the gamble there is an element of *uncertainty* or *risk*. Most people are naturally strongly *risk-averse*, and for the average person £5000 is a very attractive and useful sum of money, particularly if one has a low income. In fact, the lower the income the more attractive it becomes.

If the certain offer is then reduced to, say, £4000 and followed by further reductions, but at each reduction the option is again put to the individual, then a point will eventually be reached where they will change their mind and opt for the gamble, demonstrating that their utility for the gamble now exceeds that for the certain sum. The same will happen (unless the individual is quite impecunious) if the total amounts are reduced, say £100 to £0 for the gamble against £50 for certain. However, it is likely that many more people will accept the gamble at this lower scale, or their changeover point will come sooner on the reducing scale. Early analysts chose to explain this behaviour by whether the individuals were wealthy or not.

The point of this illustration is to propose the concept of utility in which different outcomes are not directly proportional to their expected monetary outcomes.

Utility theory provides an alternative way of establishing the payoffs in the decision matrices that were used earlier in the chapter. The attractiveness of the monetary outcome can be measured by using a *utility function* such as the function shown in Fig. 15.9. This function is typically risk-averse and it should be noted that it is an ever-increasing function; that is, the greater the sum of money, the greater is its attractiveness. The important point to note is that the slope of the curve decreases with increasing money values. This means that the *marginal utility* of money decreases, as is shown by the first £100 having an increase of 0.60 in utility value over zero money; £200 having 0.20 over £100; £300 having 0.10 over £200, and so on. The utility of £100 is referred to as $U(100)$.

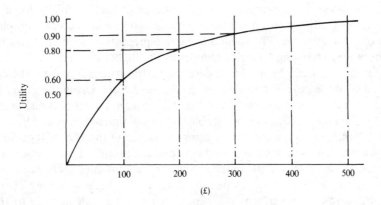

Figure 15.9

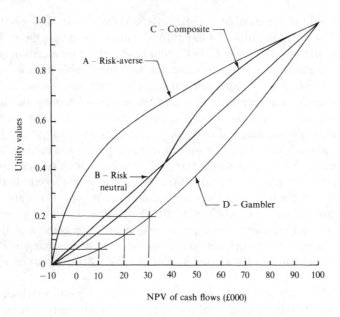

Figure 15.10

The concave downwards, or risk-averse, curve is not the only representative utility curve in existence. Figure 15.10 shows two other basic curves plus a composite curve. The opposite of being risk-averse is to be risk-seeking or a gambler. This curve is convex downwards, in which a gain of a specific sum of money increases the utility more than the loss of the same sum decreases the utility. As an example, in Fig. 15.10, if a risk seeker started with £20 000 the utility of which is 0.12 on the vertical scale, an increase in holding to £30 000 reflects an increase in utility of 0.09, whereas a decrease from £20 000 to 10 000 brings a reduction of utility of only 0.07.

The *risk-neutral* curve represents a 1:1 ratio of change of utility for change of money value. In other words, the risk-neutral decision maker accepts completely the concept of expected value. The *composite* curve represents a gambler at relatively low values of money changing to a risk avoider at the higher levels.

A utility function can be used in a decision matrix by reading from the curve the utility values of the payoffs and then proceeding as before. Using the utility function of Fig. 15.9 and the decision matrix of Fig. 15.3 gives the revised decision matrix of Fig. 15.11. Expected utility can be established in the same way as EMV.

A utility function can be created by using the device of the offer of a gamble versus a certain reward, as was described in the second paragraph of this section. Firstly, a utility scale must be set for the money values to be considered. The scale is usually from 0 for the lowest money value to 1.00 at the highest value. There is no compelling reason, however, why this should be so, and a scale from 23 to 162 could just as well be used, if that was perceived to be useful. Let it be assumed that the

	High reliability	Medium reliability	Low reliability
Purchase	0.89	0.78	0.72
Lease	0.90	0.90	0.75

Figure 15.11

maximum money involved is £100 and the minimum is −£20. The gamble can then be designed as a payment of £100, with a probability of p or a payment of −£20 with a probability of $(1-p)$.

If it is required to find the utility on the above scale of a sum of £50 then the alternative offers are either £50 for certain or a gamble in the terms described. The decision maker is questioned to arrive at a value of p. If $p=1$ then £100 will be received from the gamble and clearly the gamble will be more acceptable than the certain offer. In $p=0$ then the decision maker gets −£20 out of the gamble and this would be unsatisfactory. By adjusting the value of p up and down between 1 and 0, a point will be reached eventually where the decision maker is totally indifferent between the gamble and certain offer. At this point the value of p is the utility of £50 to the decision maker. Other points on the curve can be established accordingly.

15.7 Introduction to linear programming

There are many industrial situations—not least in the construction industry—in which scarce resources must be allocated among a number of demands for their employment. Resources in the context of construction are generally assumed to be people, machines, materials, and money. It is unusual if allocation can take place with complete freedom and without regard to the efficiency with which each demand can be satisfied. As a result, mathematicians have developed methods for making allocations with regard to optimizing some criterion that measures the efficiency of allocation. Such methods bear the generic title of mathematical programming. In arriving at an optimal solution to an allocation problem using mathematical programming, there is a need to express the constraints or restrictions within which the problem must be considered as a series of equations or inequalities, and at the same time to derive an objective function setting out the terms of the particular measure of performance to be used in the optimization. If all the relationships in both the constraints and the objective function are linear (or can be expressed as linear approximations), then, in particular, linear programming can be used in deriving a solution. Considerable success has been achieved in the application of linear programming methods to commercial practices, since the commencement of their development in the late forties.

In applying linear programming to allocation problems it is necessary to determine an objective by which each feasible solution can be ranked in an order of merit. Undoubtedly in the majority of problems there will be a great number of feasible solutions, and the problem then becomes one of selecting an optimum. Problems can be such that their solutions are finite in nature. For example, in having six jobs to allocate to six people, with no constraints of allocation, so that each person has one, there are 6! or 720 possible solutions. Each possible solution can be evaluated in turn, but clearly this method of solution quickly gets beyond practicable bounds. Problems that involve, for example, an allocation of money to various needs from a central fund or budget can have (in theory at least) an infinite number of solutions and are virtually impossible to solve by attempting to examine each feasible solution in this way. One possible answer to each allocation problem—the simplest—is to accept any feasible solution. It is a coarse criterion, but there are many extremely complex problems to which any feasible solution, if obtained, is a significant step forward.

Objectives by which an optimal solution may be selected commonly have a financial basis. One such objective is to make an allocation which minimizes the total cost of carrying out an operation; another is to allocate on the basis of maximal profit or return. Other criteria can be adopted and may include, for example, minimizing total man-hours, distance travelled or idle time, or maximizing production, product range, or storage capacity.

Small-scale problems in linear programming having two or three variables can either be solved graphically or mathematically, but solution by graphical means quickly becomes impracticable as the problem size increases. Graphical solutions are not possible for problems having more than three variables. Mathematical solutions are iterative and systematic. They lend themselves to solution by computer when the scale becomes unreasonable to manipulate quickly and efficiently by hand. There exist many standard programming packages for the solution of problems by linear programming.

15.8 Formulating linear programming problems

EXAMPLE 15.3

A contractor has one mechanical excavator and one bulldozer, which are available for work on either of two adjacent sites. On one site clay overburden is being excavated for a ballast pit owner, and on the other ballast is being removed under subcontract to another client. Experience tells the contractor that £50 profit can be made for every 1000 m^3 of clay overburden and £60 profit for every 1000 m^3 of ballast removed. A comprehensive work study assesses the resources required to remove 1000 m^3 of clay to be 8 hours' use of the excavator, 4 hours' use of the bulldozer and 50 man-hours of operatives' time. In the case of the excavation of 1000 m^3 of ballast, the resources are required for 4 hours, 5 hours, and 13 man-hours, respectively. The contractor's employees work a 40-hour week. The mechanical equipment is also available for a 40-hour week. In addition to the mechanical equipment, 5 labourers are available for up to 40 hours each in any one week in order to assist with the work. When they are not employed on the excavation, use can be made of the operatives' elsewhere. Question: How should the contractor's resources be used in order to maximize profit during one working week?

In answering the question the contractor has to decide how many cubic metres of clay and ballast are to be excavated in the one week. These are the unknowns.

Let $x_1 =$ units of $1000 \, m^3$ of clay excavated
and $x_2 =$ units of $1000 \, m^3$ of ballast excavated.

For every $1000 \, m^3$ of clay excavated the contractor requires the excavator for 8 hours and for every $1000 \, m^3$ of ballast it is required for 4 hours. Since the excavator is only available for 40 hours during a week the sum of these two requirements must be equal to or less than 40. Therefore, the first constraint on the use of one of the resources, the excavator, may be expressed as

$$8x_1 + 4x_2 \leqslant 40 \tag{15.2}$$

In like manner, the constraint on the use of the bulldozer is

$$4x_1 + 5x_2 \leqslant 40 \tag{15.3}$$

and that for labour is $\qquad 50x_1 + 13x_2 \leqslant 200 \tag{15.4}$

In addition to the above constraints (15.2)–(15.4) arising out of consideration of the limitations on the availability of the resources, x_1 and x_2 cannot be assigned negative values (the contractor cannot excavate negative quantities). Such constraints are usual in production and management problems and are known as *non-negativity constraints*. In this example they are

$$x_1 \geqslant 0 \tag{15.5}$$

and $\qquad\qquad\qquad\qquad x_2 \geqslant 0 \tag{15.6}$

The above constraints, (15.2)–(15.6), are the total constraints that apply to this problem in allocation. It is now necessary to arrive at the objective function, which expresses the objective by which each feasible allocation will be judged. In this case, P, profit, is to be maximized and must thus be expressed in terms of x_1 and x_2. The objective function in this example is

$$P = 50x_1 + 60x_2 \tag{15.7}$$

The problem has now been formulated as a linear programming problem and may be summarized as

Maximize $\qquad\qquad\qquad P = 50x_1 + 60x_2$

subject to the constraints:

$$8x_1 + 4x_2 \leqslant 40$$
$$4x_1 + 5x_2 \leqslant 40$$
$$50x_1 + 13x_2 \leqslant 200$$
$$x_1 \geqslant 0; \; x_2 \geqslant 0$$

15.9 Graphical solution of linear programming problems

Linear programming problems expressed in two variables can readily be solved using graphical methods. Simple problems in three variables can also be portrayed graphically for solution, though in such cases this method is of very restricted practical value. The graphical solution of simple linear programming problems is important because it provides a basis for the understanding of other methods of solution that extend into more complex problems having a larger number of variables. Example 15.3 will be used to demonstrate the graphical method of solution.

Any values of the variables x_1 and x_2 that satisfy all the constraints (15.2)–(15.3) constitute a feasible solution to the problem. That feasible solution that also optimizes the objective function is an optimal solution. In Fig. 15.12 are set out the x_1- and x_2-axes against which will be plotted the objective function and the various constraints. The non-negativity constraints confine the operations to the first quadrant because all points (x_1, x_2) lying on or to the right of the x_2 axis are such that $x_1 \geqslant 0$. Similarly, all points lying on or above the x_1 axis are such that $x_2 \geqslant 0$. If the equation $8x_1 + 4x_2 = 40$ is plotted in the first quadrant then a straight line is obtained. All points on this straight line satisfy the equation and the line divides the quadrant into two regions. One below and to the left of the line defines the region of the inequality $8x_1 + 4x_2 < 40$ and wherein any point (x_1, x_2) yields coordinates which are a feasible solution for this inequality. The other region above the line defines the inequality $8x_1 + 4x_2 > 40$. If the line $8x_1 + 4x_2 = 40$ is included with the lower region then we have the set of points that satisfies the inequality $8x_1 + 4x_2 \leqslant 40$ and the feasible region for this inequality with the non-negativity constraints is thus defined. It is bounded by and includes the x_1 and x_2 axes together with the graph of the equation $8x_1 + 4x_2 = 40$.

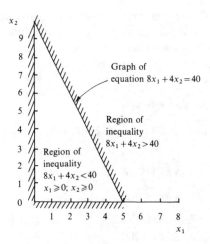

Figure 15.12

In similar fashion the remaining constraints can be plotted and the region of feasible solutions—the feasible regions—which satisfy all the constraints is thus defined. Any point (x_1, x_2) lying within the feasible region provides a feasible solution. The feasible region is cross-hatched in Fig. 15.13. The feasible region is a convex polygon, i.e. a polygon wherein a line which joins any two interior points is contained entirely within the polygon.

The solution to the problem posed above requires the objective function, $P = 50x_1 + 60x_2$, to be maximized. This postulates a selection of the largest feasible solution from the infinite number included in the feasible region. The objective function when plotted is a straight line and P may be assigned any positive value in the context of this problem. It should be noted that each value assigned to P results in a different graph, although each will be parallel to the others. (Reference to the general equation for a straight line, $y = mx + c$, will show that it is only the intercepts on the axes which are affected by altering P. The slope of the graph is not changed.)

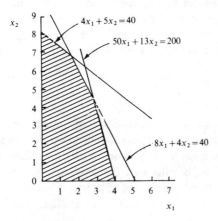

Figure 15.13

A series of lines representing the objective function for varying values of P is included in Fig. 15.14. By visualizing the movement of the contour for the objective function as P increases, it will be clear that wherever it continues to cut at least two of the boundary lines of the feasible region in distinct points it can still be moved further to the right. It can, therefore, be concluded that the maximum value for P will occur at one of the vertices of the convex polygon. The objective function in the case of the example attains a maximum value when it passes through vertex A in Fig. 15.14. The coordinates at this vertex can be established from the axes of x_1 and x_2. Alternatively, the exact values for the variables at the optimal value of P can be obtained by solving simultaneously the equations for the two lines that intersect at A. In this case the equations are $4x_1 + 5x_2 = 40$ and $8x_1 + 4x_2 = 40$ and their solution yields $x_1 = 1.67$ units and $x_2 = 6.67$ units.

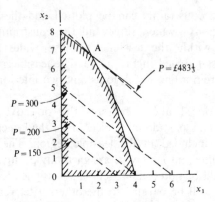

Figure 15.14

One solution to Example 15.3, therefore, is that the contractor should excavate 1667 m³ of clay and 6667 m³ of ballast during the week if he is to maximize his profit. The profit is £483⅓ under these circumstances, the excavator and bulldozer are fully utilized for the week of 40 hours (note that the use of these are represented by straight lines which intersect at vertex A) and 170 man-hours of the available labour resource are used.

15.10 The assignment problem

Allocation problems occur when there is a need to carry out a certain number of operations with a given quantity of resources within the limitation that an allocation of resources to operations cannot be completely without constraint. Each individual operation cannot necessarily, therefore, be allocated the resources to carry it out in the most effective manner, having in mind the overall objective. The overall objective will be to carry out an allocation in such a way that its total effectiveness over the whole situation is optimized.

Assignment problems belong to a special class of allocation problems. They are characterized by the requirement to allocate a number of 'units' to the same number of 'destinations' in such a way that a given measure of effectiveness is either maximized or minimized. Examples of such problems are:

1. The assignment of n gangs of operatives to n tasks such that they carry the work out in the minimum total time.
2. The assignment of n project managers to n projects so that the probable overall profit is maximized.
3. The assignment of n delivery trucks to n delivery routes so that the overall mileage covered is minimized.

EXAMPLE 15.4

A workshop has 4 tasks to be assigned to 4 skilled craftsmen. The tasks demand different blends of skill and experience and the manager's estimate of the units of time required for each craftsman to complete each task is shown in the matrix of Table 15.4. To which tasks should each craftsman be allocated if the total time taken for all the tasks is to be minimized?

Table 15.4

		Craftsmen			
		1	2	3	4
Tasks	A	16	20	18	17
	B	21	19	22	20
	C	18	20	22	21
	D	15	17	20	16

There are 4! possible sets of solutions and in this case it is not beyond the realm of practical possibility to write all 24 sets out, picking the optimum from amongst them. However, a procedure (or algorithm) has been developed to enable such problems and more complex problems to be solved without difficulty.

The method of solution depends on the fact that if a constant is subtracted from, or added to, each element in the matrix, then a solution that optimizes the objective function after the addition or subtraction process will also optimize it in its previous form. Constants are added or subtracted from the elements by row and then by column, until sufficient have zero value, thus making possible a zero-value solution in accordance with the objective function.

Step 1. Subtract the value of the minimum element in each row of the effectiveness matrix from all the elements in its row as Table 15.5.

Table 15.5

		Craftsmen			
		1	2	3	4
Tasks	A	0	4	2	1
	B	2	0	3	1
	C	0	2	4	3
	D	0	2	5	1

Step 2. Subtract the value of the minimum element in each column of the revised effectiveness matrix from all the elements in its column as Table 15.6.

Clearly any assignment cannot have a negative value. If a solution can be found whereby the total effectiveness of a complete assignment is zero it must necessarily be an optimal one. This leads to:

Table 15.6

		Craftsmen			
		1	2	3	4
Tasks	A	0	4	0	0
	B	2	0	1	0
	C	0	2	2	2
	D	0	2	3	0

Step 3. Examine each row in turn for a single zero and mark it by using a square, cancelling with a cross all other zeros in the column in which the assignment is made. Then examine each column in turn for a single unmarked zero and make an assignment where none has yet been made, at the same time deleting all unmarked zeros in the corresponding row. Repeat this step until there are no unmarked zeros left or those zeros which remain unmarked occur at least two in each row and column. Table 15.7 shows the result of this step on the example matrix.

Table 15.7

	Craftsmen			
	1	2	3	4
A	✗	4	0	✗
B	2	0	1	✗
C	0	2	2	2
D	✗	2	3	0

(Tasks)

The assignment is then A–3, B–2, C–1 and D–4. By referring to the original effectiveness matrix the total number of units of time to complete the four tasks is $18 + 19 + 18 + 16 = 71$. Note that an assignment of a craftsman has not always been made to what appears to be the obvious task. The overall time taken has nevertheless been minimized. In some instances more than one optimal solution may be present and, unlike the above example, an obvious solution may not always be present. However, having carried out the first three steps as above, a matrix will exist with at least one zero in each column and at least one zero in each row. All other elements in the matrix will be positive, though any further subtraction from these to produce more zeros and still maintain the relative effectiveness of each assignment will produce negative, and hence invalid, elements.

Assume that the matrix in Table 15.8 results from the application of the first three steps of the algorithm to an assignment problem. It will be noted that no complete assignment is yet possible (though with a simple, low dimensional matrix of this order inspection may well result in one). Further steps are now required.

Step 4. The above maximal assignment having been accepted (though it is not yet

Table 15.8

5	10	7	0	3
2	6	0	✗	1
0	1	3	2	✗
2	0	1	4	5
6	1	✗	4	9

an optimum, because the assignment is not yet complete), this step involves drawing a minimum number of lines, both horizontal and vertical, so as to cover all the zeros in the matrix at least once. This may be achieved either by inspection or by first marking, on the maximal assignment, all rows in which an assignment has not yet been made. All columns having zeros in rows previously marked are then themselves marked. The next marking is that of rows with assignments in marked columns, and these two latter markings are repeated until no further markings can be made. Lines can then be drawn through unmarked rows and marked columns which are the minimum number of lines to cover all the zeros in the matrix elements.

In an $n \times n$ matrix there should be the minimum number of n lines covering the zeros if there is a solution among them. If fewer than n lines can be used to cover all the zeros then there is no solution among them. In the example, four lines can be drawn covering all the zeros and therefore some other device needs to be used to obtain the fifth. The lines are shown in Table 15.9.

Table 15.9

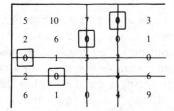

Step 5. Select the smallest element in the matrix that is not covered by a line (1 in this case). Add this smallest element to each of those elements under the intersection of two lines and subtract it from each element that is not covered by a line leaving other elements as they are. The modified matrix of Table 15.10 shows the optimal assignment.

Table 15.10

4	9	7	[0]	2
1	5	X	X	[2]
[0]	1	4	3	X
2	[0]	2	5	6
5	X	[0]	4	6

If the matrix, as modified by Steps 4 and 5, does not then yield an optimal solution to the problem then these steps should be repeated with the modified matrix until such a solution can be obtained.

The above problems have both sought to identify a minimal solution. If the matrix elements represent an objective that is to be maximized, for example profit, then such a matrix can be converted for treatment as a minimization problem by subtracting each of the elements from the largest element in the matrix. A modified matrix of differences will then be obtained.

There are variations to the regular $n \times n$ matrix problem in assignment. One is the case of a matrix that is not square. This condition may be overcome by adding sufficient dummy 'jobs' or 'resources', as appropriate, to make the matrix square and by then assigning zero value to each of the elements in the dummy. The solution can then be derived in the regular fashion.

Another situation is one in which particular assignments in the matrix must not be made. This may happen where it is a physical impossibility for certain combinations to be achieved. In a 'minimization' problem such assignments can be prevented by including a value of infinity for the appropriate element of the matrix.

15.11 The transportation problem

The transportation problem is another special case of a linear programming problem inasmuch as a less complicated method of solution is available than by the simplex method. The general characteristic of such problems is that goods, materials, etc., are required to be transported or distributed from a number of sources to a number of destinations in an optimal fashion. There are many variations within the general framework—sources and destinations may or may not be equal in number, despatches and receipts may or may not be balanced quantitatively, routes may be direct or through other destinations, etc. The technique may also be applied to linear programming problems that exhibit certain characteristics but have little or nothing to do with transportation or distribution. Within the general bounds of linear programing, the transportation method is a means of allocating scarce resources under certain given conditions.

A first essential stage is to find an initial feasible solution to a transportation problem. A simple, balanced problem will be used to illustrate the process. Five ballast pits are to provide four sites with filling material. The costs and quantities involved are set out in Table 15.11.

One means of finding a first feasible solution is by what is known as the *north-west corner rule*. A start is made in the top left-hand corner of a matrix allocating in the (1,1) position as large a quantity as possible without exceeding either of the constraints on quantity. In the example matrix, Table 15.12 shows this allocation. No more ballast is then available at Pit 1, but Site A still requires 8 units. The next move is to an adjacent square in order to satisfy the demand from A and the balance of 8 units required is supplied from Pit 2, leaving a supply of 7 units at that pit, which is then sent to Site B.

The process is carried on until all demands have been satisfied and all pits have been exhausted. At this stage there is no indication of how good the first feasible

Table 15.11

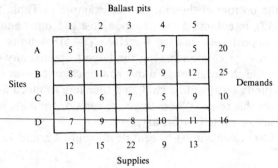

Ballast pits

	1	2	3	4	5	Demands
A	5	10	9	7	5	20
B	8	11	7	9	12	25
C	10	6	7	5	9	10
D	7	9	8	10	11	16
Supplies	12	15	22	9	13	

Sites

Table 15.12

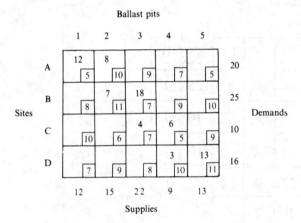

Ballast pits

	1	2	3	4	5	Demands
A	12 [5]	8 [10]	[9]	[7]	[5]	20
B	[8]	7 [11]	18 [7]	[9]	[10]	25
C	[10]	[6]	4 [7]	6 [5]	[9]	10
D	[7]	[9]	[8]	3 [10]	13 [11]	16
Supplies	12	15	22	9	13	

Sites

solution is in terms of minimizing the total cost, though the method will always generate a feasible solution.

$$
\begin{aligned}
\text{Total cost} &= (12 \times 5) + (8 \times 10) + (7 \times 11) + (18 \times 7) \\
&\quad + (4 \times 7) + (6 \times 5) + (3 \times 10) + (13 \times 11) \\
&= 60 + 80 + 77 + 126 + 28 + 30 + 30 + 143 \\
&= 574 \text{ units}
\end{aligned}
$$

In a transportation problem as described, there will be $(n+m-1)$ non-zero entries in a basic feasible solution for a matrix where there are n rows and m columns:

A second method of arriving at an initial feasible solution is by using *Vogel's approximation method*. Vogel's method uses *opportunity costs* to determine an initial feasible solution. The opportunity costs are established by subtracting the cheapest route in each column and row from the second cheapest in the same column or row.

The resulting costs are placed in the appropriate cell in the outermost columns to the right-hand side and the bottom of the matrix. For example, in Table 15.13 (the same matrix as Table 15.12), in column 5 the cheapest cost is 5 units and the next cheapest is 9 units. The opportunity cost is, therefore, $(9-5)=4$ units. The first allocation is made to the row or column having the largest opportunity cost. The reasoning for this is that, if the cheapest route in this row or column is *not* used, then the largest *penalty* (opportunity cost) will be paid, and therefore as much as possible should be transported down this route. Where there are rows or columns with equal opportunity costs the choice of route is arbitrary.

In Table 15.13 as much as possible must be sent down route A5, and 13 units are

Table 15.13

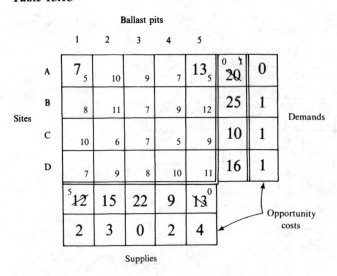

allocated, the totals of supplies and demands being adjusted. Because cell A1 also costs 5 units, the balance of 7 units may as well be sent down A1 at this stage to save one iteration through the process, though it is not essential to do this at this time. The balances of supply and demand in row A and column 5 are now zero and these two columns can be excluded from further operations on the matrix. This is accomplished in the matrix by shading the relevant columns, as illustrated in Table 15.14. The procedure now repeats and Tables 15.14–15.17 show the successive iterations. The final initial feasible solution is illustrated in Table 15.18. The procedure may be formalized as follows:

1. In the case of each column with an available supply and each row with an outstanding demand, the smallest cost is subtracted from the second smallest and recorded in the opportunity cost row and column.

Table 15.15

7			13		
8	11	7		25	1
10	1_6	7	9	$\cancel{1}^{\,0}$	1
7	9	8		16	1
5	$\cancel{15}^{\,14}$	22			
1	3	1			

Table 15.16

7			13		
8	11	7		25	1
	1		9		
7	14_9	8		$_2\cancel{16}$	1
5	$\cancel{14}^{\,0}$	22			
1	2	1			

Table 15.17

7				13	
3_8	22_7			$25^{0\ 3}$	1
	1	9			
2_7	14_8			2^0	1
5^0	22^0				
1	1				

Table 15.18

Ballast pits

Sites	1	2	3	4	5
A	7				13
B	3		22		
C		1		9	
D	2	14			

2. The row or column having the largest opportunity cost is identified.

3. If more than one row or column has an opportunity cost equal to the largest, select one of these rows or columns arbitrarily.

4. Allocate the maximum amount possible to the cheapest route in the row or column identified in 2 and 3 above.

5. Adjust the relevant supplies and demands by the amount allocated in 4 above.

6. Where a row or column has its required demand or available supply reduced to zero, remove it from any future analysis.

7. Return to 1 above and repeat the process until all demands have been satisfied or supplied exhausted.

Vogel's method yields an initial feasible solution of 469 units, which compares very favourably with that obtained with the north-west corner method. It must however be remembered that neither Vogel's method nor the noth-west corner method is guaranteed to yield an optimal solution, nor does either of them indicate when a solution is optimal, should that be the case. Their purpose is to provide an initial feasible solution only. (It will be seen that the solution provided by Vogel's method does in fact, in this case, yield an optimal solution, but this cannot be established without further testing.)

15.12 Establishing an optimal solution for the transportation problem

When an initial basic feasible solution is found it is not yet known whether it is optimal. It may or may not be so. The feasible solution must be tested to establish whether it is yet the best. The means for doing this is to test each zero cell or unused route in turn and find out whether by bringing it into use the total cost of transportation can be reduced. The initial basic feasible solution illustrated in Table 15.12 and established by the north-west corner rule will be used. It is reproduced in Table 15.19.

If a quantity θ is sent down route B1 then a similar quantity must be subtracted from B2 in order to maintain the row constraint. Likewise, θ must be added to route A2 and subtracted from A1 in order to maintain the balance. $8\theta - 11\theta + 10\theta - 5\theta = 2\theta$ will be added to the total cost. If θ consists of one unit, the total cost of the operation will be increased by 2 units, and therefore the new solution is not better than the initial basic feasible solution. This method can be used to test each unused route in turn. However, such an approach could prove to be tedious and instead the concept of *dummy costs* can be used to provide a short cut. (Elsewhere, these costs may be called *dual, variables, fictitious costs* or *row* and *column variables*.) The dummy costs are established by assuming that the cost of using each route in a basic feasible solution consists of a cost of despatch and a cost of receipt. There are, therefore, $n + m$ dummy costs—one to each row and one to each column. Because there are only $(n + m - 1)$ basic variables in a solution, one dummy cost (normally that at A) is assigned zero value enabling the others to be calculated using only used routes. Table 15.20 now has the dummy costs added to it.

Table 15.19

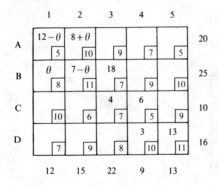

Table 15.20

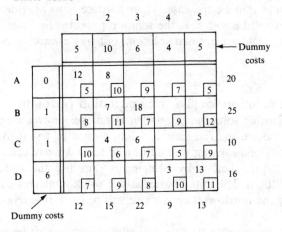

Each unused route cost is then examined and compared with the sum of the dummy costs for that route. For example, the dummy costs for route A3 amount to 6 units, 3 units less than the actual cost. For each unused route the sum of the dummy costs is subtracted from the actual cost and noted on the top left-hand corner of each cell. Table 15.21 summarizes the calculations.

On reverting to Table 15.19, it is seen that if a quantity θ was introduced down route B1, the chain of costs became $C_{B1} - C_{B2} + C_{A2} - C_{A1}$ for θ having unit value and C_{ij} being the cost of transport for any cell. In terms of dummy costs this becomes $C_{B1} - (C_B + C_2) + (C_A + C_2) - (C_A + C_1) = (C_{B1} - (C_B + C_1))$. This is equivalent to the actual cost of despatch down route B1 less the sum of the dummy costs for that route. In this case the difference is positive and the total cost increases by 2 units for each unit of material despatch down that route. In the case of a negative difference the total cost will decrease when the route in question is brought into use. It will be

Table 15.21

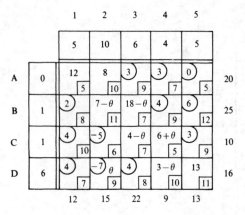

noted from Table 15.21 that for every unit sent down route D2 the total cost will be reduced by 7 units and therefore the route should be utilized as fully as possible though it still satisfies the constraints of the problem. If θ is sent down D2 the adjustments shown in Table 15.21 become necessary, θ cannot exceed 3 and when it is equal to 3 then the cost adjustment is 21, making a new total cost of 553 units. The new allocation plus revised dummy costs is shown in Table 15.22.

The process can now be repeated and θ units can be dispatched along route A5. The iteration can be repeated until no negative net cost results, when an optimal solution has been found. Such a solution is illustrated in Table 15.23. Note that where a zero difference exists between the sum of the dummy costs and the actual cost in the optimal solution, the optimal solution is not unique. The total cost of the optimal solution is 469 units.

Table 15.22

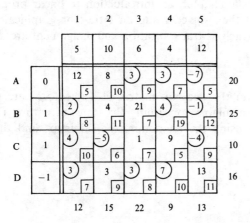

Table 15.23

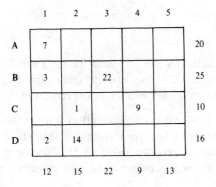

Summary

Chapter 15 is concerned with an introduction to operational research as applied to some construction situations. The introduction draws attention to definitions of operational research and the general methods used in solving analytical problems. These procedures are based on scientific method and therefore establish a base for almost all problem-solving activities. In spite of the general approach of operational research methods, problems recognized as having one of the commonly occurring structures may be solved by the application of standard techniques. Many of these techniques have been grouped under headings such as the first of those discussed in this chapter—decision analysis. Decision analysis provides a structure within which problems of all kinds may be set out for examination and solution. Probability is introduced and alternative philosophies in solution selection are exampled. Decision trees provide another structured response to problems and these are described using two examples. Utility theory, one of the best ways yet explored of dealing with the variability of the worth of money or returns by individuals, is dealt with. There then follows, for the remainder of the chapter, an introduction to linear programming. Problems are formulated and then solved using analytical or graphical methods. Standard iterative techniques such as transportation and assignment are described.

Problems

Problem 15.1 The entries in the decision matrix of Table 15.24 are profits in pounds. No information is available concerning the states of nature. Determine the optimal decision when using the maximum, the minimax, and the Savage regret minimax criteria.

Table 15.24

	States of Nature			
Actions	A	B	C	D
1	35	21	24	15
2	27	23	20	18
3	22	22	25	29
4	20	24	27	34

Problem 15.2 If the states of nature A, B, C, and D in Table 15.24 have probabilities of 0.25, 0.30, 0.20, and 0.25 respectively, determine the optimum decision that maximizes expected profit and minimizes expected regret.

Problem 15.3 A company head makes all the financial decisions in the company. Now decisions must be made between four new ventures for which potential profits are as set out in the decision matrix of Table 15.25. The states of nature are various states of the economy measured by a number of economic parameters,

Table 15.25

	States of Nature		
	A	B	C
1	0	20 000	20 000
2	− 5000	20 000	25 000
3	− 10 000	20 000	35 000
4	10 000	15 000	20 000

(Actions label on left side)

and each state is labelled A, B, or C. The company head has recently established a utility curve as shown in Fig. 15.15.

Firstly, if each state of nature is equally likely, determine which decision maximizes expected profit, which maximizes expected utility, and which decision minimizes expected regret.

Secondly, determine the same decisions if the probabilities of the states of nature A, B, and C are 0.40, 0.35, and 0.25 respectively.

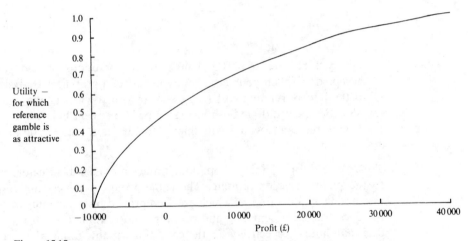

Figure 15.15

Problem 15.4 The fixed annual outgoings for an item of mechanical equipment, irrespective of whether it is put to use or not, are as shown in Table 15.26.

In addition to the fixed costs, for every hour that the equipment is working, the costs shown in Table 15.27 are included.

(a) Draw a graphical cost model from which can be obtained the total cost in any one year of working the mechanical equipment, up to a maximum of 2500 hours.

(b) Draw a graphical cost model showing how the actual cost per hour of the equipment will vary with the number of hours it works in any one year.

(c) Slick Scrapers Ltd offer an equivalent machine to the above at an inclusive

Table 15.26

Depreciation	£3500
Interest on finance	600
Insurance and tax	250
Total	£4350

Table 15.27

Fuel, oil and grease	£0.85
Air filters	0.04
Repairs and maintenance	1.75
Tyres	1.00
Tyre repairs	0.10
Operator	0.90
Total	£4.64

hourly hire rate of £9.00. Under what conditions should their offer be accepted? Explain your decision by means of a graphical analysis.

(d) If the hourly running cost (i.e., £4.64) of your own machine and hire rate of Slick Scrapers Ltd are both increased by 10 per cent but your own fixed costs remain the same, would you change the decision made in (c) above? Explain your answer.

Problem 15.5 In a joinery shop, door frames are assembled before passing them to the paint shop for priming. The same group of joiners are responsible for assembling window frames, which are then primed by the painter who also primes the door frames. Assembling and priming a door frame takes 0.5 man-hours and 0.25 man-hours respectively. In the case of a window frame, the respective times are 0.75 man-hours and 0.15 man-hours. Profit from the combined assembly and priming of door and window frames is in the proportion 3:5. The number of door and window frames to be produced in a given period of time so as to maximize profit is required. Formulate the problem as a linear programming problem on the basis that there are available 120 man-hours by joiners and that the one painter is available for 40 hours.

Problem 15.6 The sales director of a property development company considers that there are four main areas into which the company should put more working capital for its operations. The four areas are luxury housing estate development, high-rise flats, motels, and yachting marinas. The sales director shows that the average working capital required to build a luxury house is £10 000; to develop a block of high-rise flat is £200 000; to promote a motel is £10 000 and to initiate a yachting marina should be assumed to be £350 000. The profit from developing one of each of these four activities is £3000, £65 000, £42 000, and £780 000 respectively. The company has £6 000 000 to invest as working capital in its future programme, but its board of directors decides to limit investment in each area so as to spread the risks involved. The upper limit for investment in luxury housing is set at £800 000, for high-rise flats at £2 500 000, for motels at £17 500 000, and for yachting marinas at £2 500 000. Joint investment in housing and motels must not exceed half of the total working capital available. If profit is to be maximized, formulate the board's problem in linear programming form.

Problem 15.7 A firm producing joinery for builders has four principal products— doors, window frames, unit lengths of wooden fencing and roof trusses. The profit it makes on each product is £0.50, £1.50, £0.37, and £6.00 respectively. The labour required for each product in terms of man-hours is 4, 10, 2, and 36 respectively. If the firm buys all its timber prepared and is involved only in assembly, that is, its only production costs are for materials and labour with no machine costs, formulate the problem as a linear programming problem if the directors wish to maximize their profit when they employ 50 people on this work, each working a 40-hour week.

Problem 15.8 On carrying out a market survey for the producer in Problem 15.7, a forecast is made that the upper limit of the future sales of the firm is likely be 200 doors, 50 window frames, 100 units of fencing and 20 roof trusses. Similar sized timber is used for root trusses and fencing units and the manufacturer has a standing order for 2000 m per week. This order cannot be increased under 6 months' notice. A fencing unit requires 15 m of timber and a roof truss requires 80 m. A door requires £15.00 worth of timber and a window frame £30.00 worth. The total value of the manufacturer's order for timber to cover these two items is £4000 per week and the content of this order cannot be increased any more quickly than the other order. If the manufacturer now wishes to maximize the weekly profit, formulate his immediate problem as a linear programming problem.

Problem 15.9 Define, by drawing, the feasible region for the following constraints:

$$y+x \leqslant 7 \qquad 8y+5x \leqslant 40 \qquad 4y+6x \leqslant 24$$

where $x \geqslant 0$, $y \geqslant 0$.

Problem 15.10 Maximize, by using a graphical method, $P = 13x + 6y$, where $y \leqslant 19$, $x \leqslant 14$, $27y + 22x \leqslant 594$, $x \geqslant 0$, and $y \geqslant 0$. Indicate clearly the feasible region on your diagram and the values of x and y for which P is a maximum.

Problem 15.11 By using a graphical method maximize $Z = 12y + 7x$ subject to the following constraints:

$$7y+x \leqslant 112 \qquad 11y+8x \leqslant 220 \qquad 5y+22x \leqslant 330$$

where $x \geqslant 0$, $y \geqslant 0$.

Indicate clearly the feasible region on your diagram and the values of x and y for which Z is a maximum.

Problem 15.12 Solve the following assignment problem in order to minimize the following assignment values:

8	4	2	6	1
0	9	5	5	4
3	8	9	2	6
4	3	1	0	3
9	5	8	9	5

Problem 15.13 Four gangs of operatives must each be assigned to one of four jobs, A, B, C, and D on a building site. Each of the jobs has been carried out several times before and Table 15.28 shows the performance, in hours, of each gang at each task. How should the gangs be disposed such that the total time taken for completing all of the four jobs is at a minimum?

Table 15.28

		Gangs			
		1	2	3	4
	A	10	15	12	14
Jobs	B	16	12	11	9
	C	16	18	17	20
	D	16	12	15	14

If each gang starts at the same time, will the allocation of jobs to gangs which reduces the overall duration of the total work to a minimum be the same? If not what will be the revised allocation?

Problem 15.14 A civil engineering contractor has five similar excavators each at a different plant depot. Five of the company's sites each submit to the central office of the company a requisition for one such excavator. Table 15.29 gives the distances in miles between each site and each depot. What must be the allocation of excavators to sites in order to satisfy the demands and minimize the total distance over which the excavators must be transported? What is the total distance covered?

Table 15.29

		Excavators				
		1	2	3	4	5
	A	53	74	89	66	98
	B	67	10	32	87	52
Sites	C	11	81	43	24	93
	D	80	43	92	69	53
	E	18	76	63	19	72

Problem 15.15 Goodfellow and Co. Ltd wish to let five building contracts for

alterations within their offices and plant depot. The purchasing manager of the firm recommends to the board of Goodfellow's that firms A, B, C, D and E be invited to tender for the contracts since they are all good friends of the company. The board agrees subject to not more than one contract being placed with any one firm. Tenders for all five contracts were obtained from each firm and the purchasing manager summarizes them as shown in Table 15.30.

Table 15.30

| | Contracts | | | | |
	1	2	3	4	5
Firm A	20	13	25	36	24
Firm B	19	17	22	39	23
Firm C	21	16	27	41	24
Firm D	25	11	23	35	27
Firm E	23	12	19	38	22

(All figures are in £000)

How did the Goodfellow board place the orders so as to minimize their total spending, and what was the total sum of the orders placed?

Table 15.31

9	4	8	7	6	3	20
5	7	6	10	9	6	42
4	8	4	5	7	3	16
5	8	6	6	7	5	32
18	25	10	27	20	10	

Problem 15.16 Table 15.31 is a cost matrix for a transportation problem in which the cost is to be minimized. Find initial basic feasible solutions using four different methods, and compare the total costs of each.

Problem 15.17 Obtain an optimal solution to the transportation cost matrices shown in (a) Table 15.32 and (b) Table 15.33, where the total cost is to be minimized.

Table 15.32

	a	b	c	
A	6	7	8	20
B	7	9	5	15
C	6	10	7	10
	12	6	27	

Table 15.33

	1	2	3	4	5	
A	10	12	11	9	9	50
B	12	14	11	10	8	30
C	9	13	12	11	10	40
D	12	11	10	12	13	20
	25	35	45	18	17	

Further reading

Benjamin, I. R. and C. A. Cornell: *Probability, Statistics and Decision for Civil Engineers*, McGraw-Hill, New York, 1970.

Bunn, D.: *Analysis for Optimal Decisions*, Wiley, Chichester, 1982.

Hadley, G.: *Linear Programming*, Addison-Wesley, Reading, Mass., 1962.

Lindley, D. V.: *Making Decisions*, Wiley, London, 1971.

Raiffa, H.: *Decision Analysis*, Addison-Wesley, Reading, Mass., 1968.

Savage, L. J.: The theory of statistical decision, *Journal of the American Statistical Association*, vol. 46, 1951, pp. 55–67.

Schlaiffer, R.: *Analysis of Decisions Under Uncertainty*, New York, 1969.

Templeman, A. B.: *Civil Engineering Systems*, Macmillan, London, 1982.

Wagner, H.: *Principles of Operations Research*, Prentice-Hall, Englewood Cliffs, NJ, 1969.

16. Sequencing and simulation

16.1 Sequencing

Sequencing is the process of arranging the order in which customers in a queue receive a service so that some criterion of performance is optimized. The process may be differentiated from *scheduling*, which is concerned with the timing of the arrival of customers for a service such as one meets in construction planning. In scheduling problems the sequence of activities is fixed largely by other considerations. A typical example of a sequencing problem is that in which a number of jobs have to be processed through one or more machines in a machine shop.

While the concept of sequencing problems is simple and can be explained readily, solutions other than by enumeration are presently available for only three particular cases of the general problem.

16.2 A number of jobs, two stations

In this problem only two stations (or machines) are involved. Each job to be carried out must be processed through each machine in the same order, but there is no restriction on the order in which the jobs themselves must be processed, except that imposed by the need to optimize the objective function. The expected processing time for each job in each machine is known. The objective is to seek an order of processing jobs that will minimize the total time which elapses from when the first job enters the first machine until the last job emerges from the second machine.

S. M. Johnson (Optimal two- and three-stage production schedules with setup times included, *Naval Research Logistics Quarterly*, vol. 1, 1, 1954), proposed a method of computation which will be applied to the following problem.

EXAMPLE 16.1

The set-up times for the moulds of precast concrete building elements and their concreting times are set out in Table 16.1. If one element of each kind is required, in what order should the elements be cast so as to minimize the total time taken, on the assumption that only sufficient resources exist to produce one element at a time?

Procedure:

Step 1. Select the shortest time that appears in either list. If there are two or more similar times, the selection from them is arbitrary.

Table 16.1

Element	Mould set-up time, hours	Element concreting time, hours
A	15	9
B	10	4
C	8	2
D	10	11
E	10	8
F	4	4
G	9	3

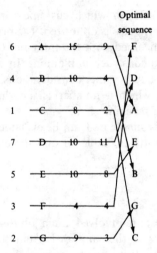

Figure 16.1

Table 16.2

Element	Mould set-up		Concreting	
	Start	Finish	Start	Finish
F	0	4	4	8
D	4	14	14	25
A	14	29	29	38
E	29	39	39	47
B	39	49	49	53
G	49	58	58	61
C	58	66	66	68

Step 2. If the minimum time selected appears in the first column, do the job first; if in the second column, place it last.

Step 3. Repeat Steps 1 and 2 on the remaining table after deleting the data referring to the job already sequenced.

Step 4. Repeat the process of assignment and deletion until all the jobs have been placed in sequence. This sequence will minimize the total elapsed time.

Example: The shortest time is 2 hours concreting for element C. Figure 16.1 illustrates the complete process.

The optimal elapsed time and any idle time may be calculated by use of a Gantt chart or by Table 16.2.

16.3   A number of jobs, three stations

Johnson also developed an algorithm for solving particular cases where n jobs have to be processed through three stations in such a way that the order of job processing is maintained through all three stations. The cases that can be solved by use of the algorithm must satisfy one or both of the following conditions:

1.   The minimum processing time at station 1 is greater than or equal to the maximum processing time at station 2.
2.   The minimum processing time at Station 3 is greater than or equal to the maximum processing time at station 2.

The method of solution is to form two columns of times, one out of the sum of those for stations 1 and 2 and the other out of the sum of those for stations 2 and 3. The solution can then follow that for n jobs, two stations.

EXAMPLE 16.2

It is decided to build a heavy supporting structure for a number of vessels in a petrochemical plant out of reinforced concrete. Owing to limitations on the area of site available, the structure is designed in such a way that it can be prefabricated in 12 pieces, each of which can be transported to the site. The structure is urgently required for erecting during a shutdown of the plant.

The 12 sections are basically similar, but each has a different number of cutouts, brackets, notches, and holes. Only one gang of carpenters is available to erect the formwork, and another different gang can be spared at the appropriate time to strip the formwork after pouring the concrete and curing is completed.

The estimated number of hours for the erection of the formwork, placing reinforcement after the formwork is erected, pouring concrete, finishing and for the curing and stripping of each member is as shown in Table 16.3. Determine the construction sequence that will minimize the time to complete the 12 members.

Combine the times for 'Formwork erection' and 'Positioning reinforcement, pouring concrete, and finishing' in one column and 'Curing and stripping' with the latter in another, as in Table 16.4.

Table 16.3

Member	Formwork erection	Positioning reinforcement, pouring concrete, and finishing	Curing and stripping
1	28	8	36
2	30	7	35
3	42	7	42
4	18	6	15
5	51	9	60
6	19	8	24
7	32	7	27
8	60	10	32
9	23	11	43
10	41	8	28
11	18	7	24
12	20	9	38

Table 16.4

Member	A	B	Optimal sequence
1	36	44	11
2	37	42	6
3	49	49	12
4	24	21	9
5	60	69	1
6	27	32	2
7	39	34	3
8	70	42	5
9	34	54	8
10	49	36	10
11	25	31	7
12	29	47	4

16.4 Two jobs, a number of stations

A simple method of obtaining a solution to this type of problem is by using a graphical approach. An optimal solution, however, is by no means assured. In essence, the detail of the problem to be solved is that there are two jobs which have to receive the attention of m different stations. The stations may be persons or machines, for example, each of which makes a contribution to the overall process. The order in which the jobs are processed through the stations is fixed for each job, but it is not necessarily the same for each one. The time to be devoted by each machine to each job is assumed to be known.

Table 16.5

Sequence	House 1	House 2
1	A–2	C–3
2	C–3	D–5
3	D–1	B–2
4	B–4	A–1
5	F–2	G–3
6	E–1	F–4
7	G–3	E–2

EXAMPLE 16.3

Two different houses remain to be completed on a housing estate. The jobs in each, labelled A–G, must be completed in the sequence shown in Table 16.5 and have the assigned durations in days. Each of the jobs is carried out by the same one group of operatives in either house, but different jobs are carried out by different groups.

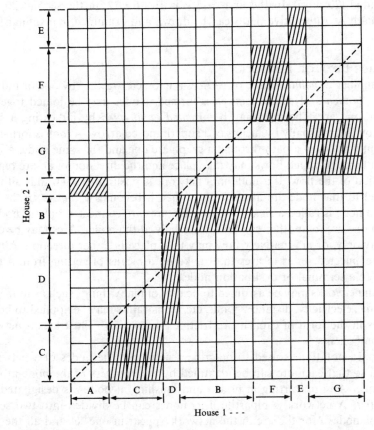

Figure 16.2

What is the minimum time in which both houses can be completed?

A horizontal and a vertical axis are drawn and graduated in such a way that the total duration of all the jobs in one house can be marked off on either of them. One is labelled 'House 1' and the other, 'House 2'. The times for each job are then set off in the appropriate sequence in each axis. Progress through the work will be represented, in Fig. 16.2, by horizontal movement in the case of House 1 and vertical movement in the case of House 2. Diagonal movement from the origin towards the top right-hand corner will represent progress on both houses simultaneously. However, since there is only one gang to do two jobs, similar jobs cannot be processed at the same time. Thus, to prevent this happening, areas bounded by these jobs are blocked out on the diagram and the progress line cannot encroach upon them. Clearly the optimal path is that which has as much diagonal travel as possible and can be drawn in by eye. Idle time on House 1 will be represented by the length of path at right angles to its axis and when added to the total time required (16 days) gives the overall duration of the work. Alternatively the idle time for House 2, which is represented by the length of path at right angles to it, can be added to the total time required (20 days). In the first case it is $16+6=22$; in the second $20+2=22$ days for Path A. Alternative paths can be drawn and examined in a similar fashion.

16.5 Shortest routes through networks

When a number of *points*, *nodes*, or *vertices* are joined together by one or more *lines*, *arcs*, *links*, *branches* or *edges*, a *graph* is formed. Nodes may be joined together by more than one arc. The arcs may be *oriented* or *directed* by indicating a sense of direction on them, usually by means of a superimposed arrow. A *loop* is formed when the extremity nodes of a path through a graph are one and the same node. A *network* is a graph through which flows may take place and the direction of an arc represents the direction of the flow. If a node in a network is positioned such that all flows in arcs joined to that node are away from it, then it becomes a *source*. In the reverse situation a node is called a *sink*. In a network, arcs are connected only at nodes, and flow from one arc to another can take place only at the nodes. Arcs may be directed in one way only, or alternatively flow may take place in either direction. A *route* or *path* is a connected series of arcs that make up a means of getting from a node at an origin to a terminal or destination node.

Flows in networks may be assumed to be that of many things such as money, time, fluids, traffic, electricity, distance, goods, etc. Constraints may be applied to both arcs and nodes in the form of capacity restrictions. Figure 16.3 illustrates a network of typical components.

A standard notation is used to describe a network. The nodes of a network are numbered so that each arc can be identified by the numbers of the node at its start and its finish. An arc starting at node i and finishing at node j is designated by the notation (i, j). A network is bipartite if its nodes can be divided into two sets such that all the nodes i for the arc in the network appear in one set and all the j nodes appear in the other. A transportation problem can be represented by a bipartite network as in Fig. 16.4.

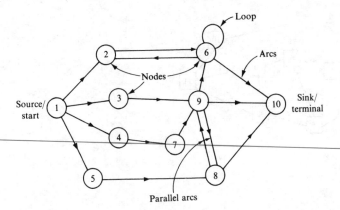

Figure 16.3

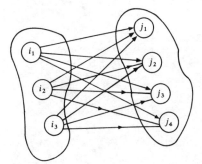

Figure 16.4

Establishing the shortest path through a network has many applications in traffic flow, in delivery and distribution problems, and in method study of complex materials movement. The objective in such problems is normally to minimize the distance travelled, to establish a route between two points at least cost or to select a route which minimizes the duration of travel between a starting point and a destination. A typical problem of this type involves movement from one point to another through other points on route, where there is a choice at the intermediate points. In a road network, the arcs of the network can be travelled over in either direction in the majority of cases. In the use of a conveyor system, there is usually a restriction on the direction of travel.

16.6 Graphical analysis of shortest-path problems

A road network is represented in Fig. 16.5, and it is required to establish the minimum-distance route between the source and the destination. An arrow superimposed on an arc indicates allowable movements in that direction only. Arcs without arrows may be traversed in either direction.

Step 1. Draw all arcs linking the source node, 1, with all other immediately

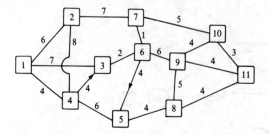

Figure 16.5

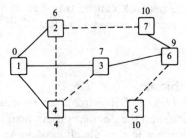

Figure 16.6

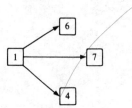

Figure 16.7

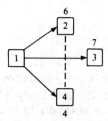

Figure 16.8

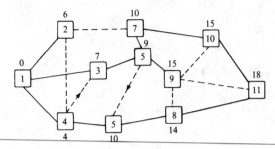

Figure 16.9

adjacent nodes, noting in the process the distance to be travelled at the *j* node as in Fig. 16.6.

Step 2. Examine the nodes drawn in Step 1 to see if there are arcs between them. Establish whether the direct or indirect routes are longest, and insert the shortest distance between node 1 and the node in question against each *j* node. Draw the shortest route as a full line, and alternative longer routes as a broken line as Fig. 16.7.

Step 3. Draw all arcs from the nodes established as a result of Step 1 and insert the minimum distances as before (Fig. 16.8). This process may be continued until analysis of the network is complete, as in Fig. 16.9.

The completed analysis, as depicted by Fig. 16.9, results in the establishment of the shortest path between node 1 and every other node in the network. It should be noted that there will be alternative routes of equal merit in some instances. For example, there are two routes from node 1 to node 11, both with a total of 18 units of distance – (1, 3, 6, 7, 10, 11) and (1, 4, 5, 8, 11).

16.7 Alternative method of solution

The Dijkstra algorithm can be used for solving a shortest path problem so long as the value against each arc is greater than or equal to zero. If this is not the case, the application of this algorithm can give incorrect results.

Each node in a network is assigned a two-part label, which may be either permanent or tentative. A label is applied to each node of a network and indicates two things. Firstly, it indicates the previous node or the path and secondly, its distance from the source node. A permanent label is one in which is established the best path to that point. A tentative label is one which records the best situation as far as the examination has so far gone. The method can be applied to the network diagram itself as a work sheet, or to a matrix layout. The network in Fig. 16.10 will be examined for the shortest path by using a distance matrix.

The Dijkstra algorithm may be stated in steps as follows. Figure 16.11 illustrates the flow diagram for the algorithm.

Step 1. A permanent label $\boxed{-, 0}$ is assigned to the source, indicating that the

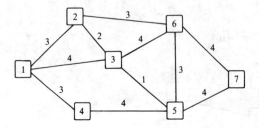

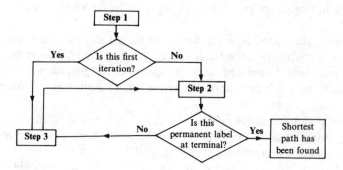

Figure 16.10

Figure 16.11

distance from 1 to 1 is zero and that there are no previous nodes on the path. Tentative labels are assigned to all other nodes in accordance with the convention $[-, \infty]$. The i node is therefore node 1.

Step 2. Consider the set of nodes having tentative labels and select i, the one where the length appearing in the second position of the label is least. Make this label permanent. If it is the label appearing at the terminal node then the shortest path is established; otherwise pass to Step 3.

Step 3. Each node j that is directly connected to the node that is acting as the i node for the time being and bears a tentative label is considered. If the length d in the permanent label of the i node, plus the arc length $l(i, j)$ is less than the distance appearing in the second place of the tentative label at any j node then a new tentative label is placed at j. In the new tentative label, the first place is changed to the i node in use and the second place assumes a value $[d + l(i, j)]$. If $[d + l(i, j)]$ is greater than or equal to the value appearing in the second position of the tentative label, no change is made. When each j node has been dealt with in his way, return to Step 2.

The final labelled matrix for the example network of Fig. 16.10 is shown in Fig. 16.12.

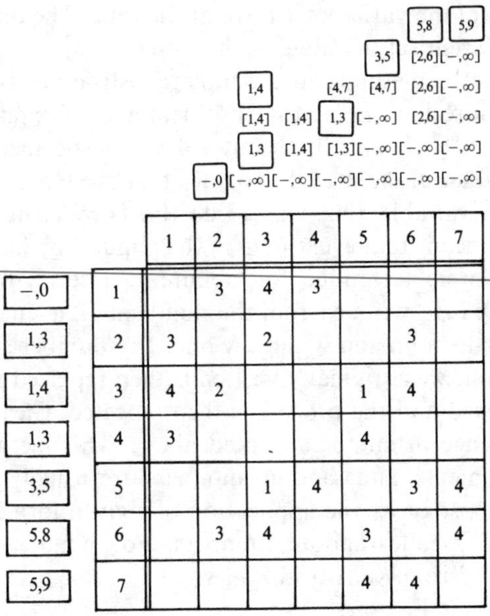

```
                              5,8   5,9
                         3,5  [2,6][-,∞]
            1,4    [4,7] [4,7] [2,6][-,∞]
           [1,4]  [1,4]  1,3 [-,∞]  [2,6][-,∞]
            1,3   [1,4]  [1,3][-,∞][-,∞][-,∞]
           -,0 [-,∞][-,∞][-,∞][-,∞][-,∞][-,∞]
```

		1	2	3	4	5	6	7
-,0	1			3	4	3		
1,3	2	3		2			3	
1,4	3	4	2			1	4	
1,3	4	3		.		4		
3,5	5			1	4		3	4
5,8	6		3	4		3		4
5,9	7					4	4	

Figure 16.12

16.8 Monte Carlo simulation

A *Monte Carlo simulation* is a dynamic model of a situation, which results in a picture of the likely operation of the subject being modelled over a period of time, In construction it can be applied to problems that involve queues of all types and that rarely lend themselves to ready solution by mathematical analysis. Queueing theory forms a large body of knowledge in operational research, and has resulted in a number of general solutions being provided to simple queueing situations. However, there are few situations in construction from which benefit can be derived as a result of the application of queueing theory. Simulation in general, and Monte Carlo simulation in particular, can be used to provide us with a better understanding of the situation being modelled, to assist in the selection of criteria for decision making, and to train personnel to use the real system in a proper fashion.

If the outcome of a situation is such that we can be certain of it, having selected or knowing the values of the variables involved, then the outcome is said to be *deterministic*. On the other hand, situations in which we do not know the values of some or all of the variables beforehand and in which we must allow for the fact that there is a varying probability that they will take certain values, are *probabilistic* or *stochastic situations*. A *stochastic simulation* is therefore one which takes into account the probability of variables taking certain values. These are commonly called Monte Carlo simulations. A *non-stochastic simulation* is one where the answer is arrived at on the basis of the single values of the variables known at the time. The output from the model changes only if a different set of values is then used.

The application of Monte Carlo methods to a simple construction situation is illustrated in Example 16.4. Before looking at that problem a brief explanation of the use of random numbers will be made. In the use of stochastic methods it is necessary to select a series of values of the model variables that are representative of the total data available for that variable. One way to do this is by using a process called random sampling, which means that every item in the population has an equal chance of being selected. A simple technique for obtaining a series of random numbers between, say, 0 and 99, is to write each of the numbers on a small piece of paper, put them in a hat, shake them up and withdraw one. The number is recorded and the piece of paper on which it was originally written is then replaced in the hat, the numbers are thoroughly mixed and the process is then repeated. Clearly such a process is not limited to the range of numbers quoted above. This method is not suitable for present-day requirements, and random numbers are usually generated by special machines for the purpose or by the application of a given formula. In the latter case they are known as pseudo-random numbers. For hand simulations, random numbers can be obtained from statistical tables.

EXAMPLE 16.4

A site concrete mixer produces 1 m^3 of concrete at each mixing. During a large pour of concrete, a wet hopper is used in front of the mixer into which the mixed concrete is discharged prior to being loaded into trucks. The wet hopper has a maximum capacity of 3 m^3. The cycle time from discharge to discharge in mixing 1 m^3 of concrete is studied and the frequency distribution shown in Table 16.6 results.

Table 16.6

Cycle time (min)	Frequency
3.0	10
3.5	27
4.0	47
4.5	32
5.0	4

A large pour of mass concrete in the foundations of a building is planned. The volume of the pour is 100 m^3 and it must be poured in a single continuous operation. A tower crane is used to lift a skip of concrete so that it can be discharged into the foundations. The frequency distribution of the crane's complete cycle time is, for one skip of concrete, as shown in Table 16.7. Either a 3-m^3 or a 2-m^3 skip can be used, since both are available. The frequency distribution for the placing of concrete using either skip is similar.

On the site are three truck mixers having a capacity of 4 m^3 of concrete each and

Table 16.7

Cycle time (min)	Frequency
2.00	3
2.25	10
2.50	24
2.75	37
3.00	18
3.25	9
3.50	2

five specially fitted tipper trucks having a capacity of 2 m³ each. Because the concrete mixer is a considerable distance by a devious route to the position of the pour, a large proportion of a truck's time is spent in travelling. The forward and the return journeys have similar frequency distributions for the time taken and are of the form shown in Table 16.8.

Select the number and size of trucks for concrete distribution that you believe to be the best combination and establish the total time it will take to complete the concreting.

The frequency distribution for the concrete mixer is as shown in Table 16.9. A

Table 16.8

Travelling time (min)	Frequency
20	5
22	15
24	30
26	17
28	3

Table 16.9

Cycle time	Frequency	Cumulative frequency	Cumulative relative frequency
3.0	10	10	0.08
3.5	27	37	0.31
4.0	47	84	0.70
4.5	32	116	0.97
5.0	4	120	1.00
	120		

Table 16.10

Cycle time (min)	Cumulative relative frequency	Randon numbers
3.0	0.08	00–07
3.5	0.31	08–30
4.0	0.70	31–69
4.5	0.97	70–96
5.0	1.00	97–99

series of two-digit random numbers ranging from 00 to 99 (i.e. 100 numbers) can be assigned to the cumulative relative frequency distribution of Table 16.10.

The cumulative relative frequency distribution and the allocation of random numbers to the various cycle times can be represented graphically as in Fig. 16.13.

Figure 16.13 highlights the main principle of using random numbers for selection from a frequency distribution. The number of individual random two-digit numbers is proportional to the frequency with which the cycle times have been observed to occur. Two-digit random numbers have been used in this example, but there is no reason why random numbers having one, three, four, or more digits should not be used if the degree of accuracy required demands it, e.g., if the cumulative relative frequency is calculated to three places beyond the decimal point, i.e., 0.083, it is possible to allocate random numbers 000 to 082 to this cycle time.

The problem, as stated, requires the production of $100 \, \text{m}^3$ of concrete, and therefore we need to generate 100 consecutive cycle times in a random fashion. Using a table of random numbers we can do this as shown in Table 16.11. This table gives

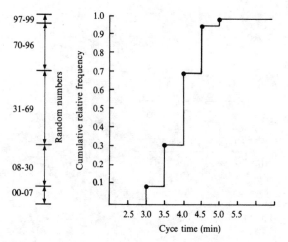

Figure 16.13

Table 16.11

Random number		Cycle time, min	Random number		Cycle time, min
1	78	4.5	11	68	4.0
2	39	4.0	12	17	3.5
3	44	4.0	13	42	4.0
4	02	3.0	14	07	3.0
5	71	4.5	15	97	5.0
6	14	3.5	16	85	4.5
7	27	3.5	17	60	4.0
8	21	3.5	18	33	4.0
9	34	4.0	19	42	4.0
10	43	4.0	20	32	4.0
					etc.

the cycle times for the production of the first $20\,m^3$ of concrete: the remainder can be generated in similar fashion. Likewise, it is possible to generate a series of cycle times for the crane and for the various travelling times of the trucks.

In this simple problem, there are four resources that interact one with the other, and unless we have an understanding of how they interact and the effect of the performance of each on the others then we cannot plan the work properly.

The first resource is that of the concrete mixer: if we list the states in which we might find it, in the context of the question, they are as follows:

1. Mixing concrete within mixer's cycle time.
2. Mixing, waiting to discharge.

These are the only two states that can affect the simulation model, and a flow diagram is not necessary in explanation.

The second resource is the wet hopper to store concrete and it can be found in one of three states:

1. Empty.
2. Part-full (i.e., with 1 or $2\,m^3$ of concrete in it).
3. Full (i.e., with $3\,m^3$ of concrete in it).

The flow diagram for the hopper is shown in Fig. 16.14.

The third resource is the trucks. The states in which we find them are:

1. Waiting to be filled with concrete.
2. Travelling loaded.
3. Waiting at crane.
4. Travelling unloaded.

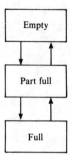

Figure 16.14

The flow diagram is as Fig. 16.15 (Note that this flow diagram is very simple, but there would be little purpose in having a more complex diagram, since no data are available for other states in which a truck might be found.)

The last resource to be considered is the crane. It can be found in only two different states: either actually within a lifting cycle or waiting to lift a skip.

The four resources can be linked together at the point of interaction in each flow diagram. Figure 16.16 shows the model for the whole process in a static form, and it now remains to introduce a time element to give the model a dynamic form.

The time base can be introduced by considering a clock, using intervals of, say, one minute. At the end of each interval and as necessary, each resource is 'inspected' in order to see what its state is at that time. Let us assume that we use four 2-m^3 tipper trucks (A, B, C, and D) and a 2-m^3 skip under the crane. The clock can be started at 00.00 when the mixer starts its first mix.

The schedule in Table 16.12 is written out as such in order to explain the principles of simulation. It will be seen that the process quickly becomes complicated and keeping track of each resource, even in a simple problem, is not easy.

From the above analysis it is possible to calculate the utilization of the resources. For example, truck 'D' is idle until $00.30\frac{1}{2}$, and thereafter its waiting time can be calculated and compared with its working time.

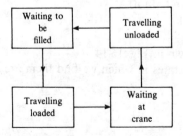

Figure 16.15

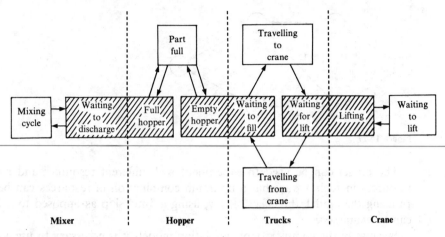

Figure 16.16

Table 16.12

Clock	Operation
00.00	Mixer starts.
00.04$\frac{1}{2}$	Mixer discharges 1 m³ to hopper.
00.08$\frac{1}{2}$	Mixer discharges 1 m³ to hopper—hopper now holds 2 m³ and fills truck 'A', which travels to crane to arrive at 00.08$\frac{1}{2}$ + 24 = 00.32$\frac{1}{2}$.
00.12$\frac{1}{2}$	Mixer discharges 1 m³ to hopper.
00.15$\frac{1}{2}$	Mixer discharges 1 m³ to hopper; truck 'B' filled and arrives at crane at 00.15$\frac{1}{2}$ + 24 = 00.39$\frac{1}{2}$.
00.20	Mixer discharges 1 m³ to hopper.
00.23$\frac{1}{2}$	Mixer discharges 1 m³ to hopper; truck 'C' filled and arrives at crane at 00.23$\frac{1}{2}$ + 22 = 00.45$\frac{1}{2}$.
00.27	Mixer discharges 1 m³ to hopper.
00.30$\frac{1}{2}$	Mixer discharges 1 m³ to hopper; truck 'D' filled and arrives at crane at 00.32$\frac{1}{2}$ + 26 = 00.58$\frac{1}{2}$.
00.32$\frac{1}{2}$	Truck 'A' arrives at crane, fills skip and departs getting back at mixer at 00.32$\frac{1}{2}$ + 24 = 00.56$\frac{1}{2}$.
00.32$\frac{1}{2}$	Crane deposits concrete—ready to lift again at 00.32$\frac{1}{2}$ + 3 = 00.35$\frac{1}{2}$.
00.34$\frac{1}{2}$	Mixer discharges 1 m³ to hopper.
00.38$\frac{1}{2}$	Mixer discharges 1 m³ to hopper.
00.39$\frac{1}{2}$	Truck 'B' arrives at crane; no waiting; fills skip and departs getting back at mixer at 00.39$\frac{1}{2}$ + 20 = 00.59$\frac{1}{2}$.
	Crane deposits concrete; ready to lift again at 00.39$\frac{1}{2}$ + 2$\frac{1}{2}$ = 00.42.
00.42$\frac{1}{2}$	Mixer discharges 1 m³ to hopper; hopper now full and mixer cannot discharge again until truck 'A' returns at 00.56$\frac{1}{2}$.
00.45$\frac{1}{2}$	Truck 'C' arrives at crane, fills skip and departs getting back at mixer at 00.45$\frac{1}{2}$ + 24 = 01.09$\frac{1}{2}$.
	Crane deposits concrete; ready to lift again at 00.45$\frac{1}{2}$ + 2$\frac{3}{4}$ = 00.48$\frac{1}{4}$.
00.56$\frac{1}{2}$	Truck 'A' arrives back at mixer; is loaded immediately and returns to crane arriving at 00.56$\frac{1}{2}$ + 24 = 01.20$\frac{1}{2}$.
00.58$\frac{1}{2}$	Truck 'D' arrives at crane, fills skip and departs getting back at mixer at 00.58$\frac{1}{2}$ + 28 = 01.26$\frac{1}{2}$.
	Crane deposits concrete; ready to lift again at 00.58$\frac{1}{2}$ + 2$\frac{3}{4}$ = 01.01$\frac{1}{4}$.
00.59$\frac{1}{2}$	Truck 'B' arrives back at mixer;
01.00$\frac{1}{2}$	Mixer discharges 1 m³ concrete; truck 'B' filled and arrives at crane at 01.00$\frac{1}{2}$ + 24 = 01.24$\frac{1}{2}$.
01.04	Mixer discharges 1 m³ concrete.

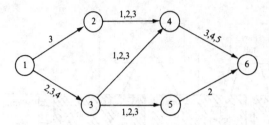

Figure 16.17

The model can be used to experiment with different resources and numbers of resources in such a way that a optimum combination of resources can be selected. In using the model, the effect of, say, using a 3-m^3 skip as opposed to a 2-m^3 skip can be examined.

Because of the complexity of simulation models it is necessary to use a computer in order to provide a practical solution. The printout from a computer can be designed to give, directly, statements of the resource utilization, etc.

16.9 The fallacy of averages

Figure 16.17 illustrates a network for a small construction project. Against each activity is shown its expected duration in time units. In some cases a single duration is shown, while in others there is a range of durations, each having an equal probability of being achieved.

If the minimum duration of each activity is used then the expected project duration is 7 units. Using the longest durations it is 12 units, and using the average it is 9 units. The use of the average duration of each activity does not therefore give the average project duration. Hence a justification for the use of stochastic simulation is such cases.

Summary

Chapter 16 deals with two further aspects of operational research as applied to construction. The first is sequencing, which is very effective in batch processing type activities. However the theory is limited and three limited cases are described and solution methods offered. This links to networks and the flows therein. Graphical solutions are discussed and also the use of one algorithm in particular is applied. Finally simulation methods are described. Stochastic simulation in particular has good applications in construction and enables realistic answers to be obtained where such problems are intractable using normal analysis methods.

Problems

Problem 16.1 On a building contract cutting and bending reinforcing steel is subcontracted. The work is planned in a number of well-defined sections, A–H inclusive. The order in which the steel is cut and bent and then fixed by the main contractor is immaterial, but it is required to minimize the overall duration of carrying out all the work from commencement to completion. If the estimated times for both processes are as set out in Table 16.14, answer the following questions:

(a) What is the sequence of work which will minimize the overall duration?
(b) What will be the idle time in both operations?
(c) What will be the overall duration?

Table 16.14

Work section	Cutting and bending, hours	Fixing, hours
A	15	22
B	27	16
C	18	17
D	12	22
E	10	23
F	24	13
G	19	14
H	16	26

Table 16.15

Work section	Cutting and bending, hours	Fixing, hours
J	15	20
K	12	12
L	18	16
M	16	14
N	15	14
O	14	15
P	18	16
Q	16	18
R	21	15
S	19	21

Given that, on another contract, the figures shown in Table 16.15 are expected, calculate (a), (b), and (c) as above. Are there alternative sequences?

Problem 16.2 Determine the shortest path through the two networks (a) and (b) shown in Fig. 16.18 by use of a graphical method, and in each case, check your solution by means of the Dijkstra algorithm.

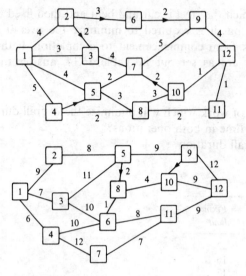

Figure 16.18

Problem 16.3   Seven machines, A–G inclusive, are available in a woodworking shop through which two projects must be processed in the following order:

Project 1	A, B, C, D, E, F, G
Project 2	A, B, D, C, E, G, F

The processing times for Project 1 are 3, 1, 1, 2, 4, 2, and 2 hours respectively. Those for Project 2 are 2, 2, 1, 1, 1, 2, and 1 hours respectively. Define an optimal programme for the work that minimizes the overall duration of it. What is that duration?

Problem 16.4   Table 16.16 lists a network of sites served by a head office wages department. The head office is located at node HO. A wages clerk picks up

Table 16.16

Path	Distance
HO–1	4
HO–4	7
HO–7	8
1–2	6
1–3	1
2–3	1
2–5	2
3–4	1
4–5	3
4–6	3
4–7	2
5–6	3
6–7	1

information from each site once a week but the weekday sometimes varies. The clerk needs to know the shortest route from the head office to each site and the other sites on that route. Using the Dijkstra algorithm, establish these routes and distances on the assumption that travel can take place in either direction along each route.

Problem 16.5 At the end of each working day, the number of bricks laid by one particular bricklayer is recorded. After some time, a frequency distribution of performance can be established. It is illustrated in Table 16.17.

Using the Monte Carlo method, demonstrate how to predict the number of days it should take the same bricklayer to lay 3000 bricks.

Table 16.17

Number of bricks laid	Number of days recorded
200	10
220	25
240	48
260	54
280	37
300	15
320	2

Describe an alternative method of sampling from a frequency distribution such as that given in the table above. Use sketches to amplify your description.

Problem 16.6 A mechanical shovel is loading topsoil into each of five trucks in turn, which as soon as they are loaded transport the topsoil to a tip and return empty for another load. The average capacity of each truck is $4\,m^3$. A work study officer observes the process and reports the data shown in Table 16.18.

Table 16.18

Shovel		Trucks	
Time to fill one truck, min	Number of observations	Time to run to tip and back after loading, min	Number of observations
8	12	16	5
9	20	17	15
10	38	18	30
11	20	19	25
12	10	20	15
		21	7
		22	3

Establish the probable working efficiency of the shovel and each truck over the first four-hour period in any day. In the light of your analysis suggest what steps might be taken to improve the efficiencies so established.

Problem 16.7 Four gangs of operatives are used to carry precast concrete beams from a storage compound to a tower crane. Each gang then loads the beams it has carried onto the crane and returns to the compound to get another load. The four gangs set up a cycle of operations in this way. Because of doubts about their efficiency, after a survey, a work study officer produces the data given in Table 16.19.

Table 16.19

Crane		Gangs	
Time to load, raise, place load and return, min	Number of observations	Cycle time to pick up load, travel to crane, load crane, and return to compound, min	Number of observations
6	18	16	20
8	30	18	33
10	45	20	42
12	30	22	39
14	12	24	15
		26	5

Simulate the first 3 hours' operations of the crane and the four gangs in a working day. From your result suggest possible action by a site manager to improve the efficiency of the work.

Problem 16.8 An excavator is used to load filling material from a borrow pit into dumptrucks, which then transport it to one of two sites. Equal numbers of dump trucks are used for supplying each site and any one truck travels only between one particular site and the borrow pit. The trucks leave each site in a random pattern.

The inter-arrival time distributions of trucks at the excavator from both sites are studied and found to have similar characteristics, which can be represented by the distribution shown in Table 16.20.

Table 16.20

Range, min	%
4–6	5
6–8	10
8–10	20
10–12	30
12–14	20
14–16	15

The trucks appear to arrive at the site in equal proportions from both sites, as well as at irregular intervals.

The times required for loading a truck with spoil are represented by the distribution shown in Table 16.21.

Table 16.21

Range, min	%
10–12	5
12–14	15
14–16	30
16–18	25
18–20	15
20–22	10

If the excavator does nothing else but load the trucks as and when they arrive at the borrow pit and deals with them in their order of arrival, calculate the total time it is likely that the trucks will be waiting to be loaded at the borrow pit during any one period of 4 hours selected at random.

Further reading

Ackoff, R. L. and M. W. Sasieni: *Fundamentals of Operations Research*, Wiley, New York, 1968.

Fishman, G. S.: *Concepts and Methods in Discrete Event Simulation*, Wiley–Interscience, New York, 1973.

Hammersley, J. M. and D. C. Handscomb: *Monte Carlo Methods*, Methuen, London, 1964.

Neelamkavil, F.: *Computer Simulation and Modelling*, Wiley, Chichester, 1987.

Nicholls, R. L.: Operations research in construction planning, *Journal of the Construction Division, ASCE*, vol. 89, no. CO2, September 1963, pp. 59–74.

Price, W. L.: *Graphs and Networks—an Introduction*, Butterworths, London, 1971.

Stark, R. M. and R. L. Nicholls: *Mathematical Foundations for Design, Civil Engineering Systems*, McGraw-Hill, New York, 1972.

Tocher, K. D.: *The Art of Simulation*, English Universities Press, London, 1963.

Wagner, H. M.: *Principles of Operational Research*, Prentice-Hall, Englewood Cliffs, NJ, 1969.

Appendix A: Case studies

A.1 'Project Power'

Purechem PLC, a chemical engineering company, decides to construct a small power generation plant to supply electricity and steam to a complex of chemical processing installations.

It is required to complete the construction of the civil engineering works for the power station in not more than 28 months. The approximate programme determined by Purechem for planning purposes is shown in Table A.1. Purechem's advisers, Powerbild & Co, estimate the costs and the durations of the civil works to be as shown in Table A.2.

Table A.1

		Year 1	Year 2	Year 3
A	Site works	*******		
B	Cooling water systems I	*******	**	
C	Boiler house foundations	**********	****	
D	Boiler house superstructure		**********	**
E	Turbine house foundations	***	**********	
F	Turbine house superstructure		***	***
G	Switchgear foundations		********	
H	Fuel storage		*********	*
J	Administration building	*****		
K	Welfare block		*****	
L	Cooling water systems II		**	
M	Landscaping			**

Purechem can finance the power plant by a loan costing 16 per cent per year but must pay a commitment fee of 0.5 per cent calculated on the amount of the total loan not yet taken up at the beginning of each 12-month period. Purechem decide to draw from the loan facility at the end of each quarter so as to meet the estimated expenditure in the previous quarter The managing director of Purechem requires to know the total cost of financing the loan before proceeding beyond this stage.

On giving the go-ahead to prepare the documents for seeking tenders for the work, Powerbild suggest to Purechem that the construction programme can be shortened by reducing the durations for the construction of the boiler and turbine houses together with any associated activities being brought forward as a result. They further suggest that this may result in a saving of interest payments on the loan. A

Table A.2

		Duration, months	Total estimated cost, £000
A	Site works	7	1 000
B	Cooling water systems I	9	3 000
C	Boiler house foundations	14	2 500
D	Boiler house superstructure	12	1 250
E	Turbine house foundations	14	3 200
F	Turbine house superstructure	6	2 100
G	Switchgear foundations	8	550
H	Fuel storage	10	600
J	Administration building	5	80
K	Welfare block	5	120
L	Cooling water systems II	2	700
M	Landscaping	2	300
			£15 400

reduction of 1 month in the overall programme is estimated to cost an additional £250 000 in acceleration charges, 2 months will be £350 000 and 3 months' reduction will cost approximately £520 000. The managing director wishes to resolve this question, since it affects the construction programme, before the documents are sent out to tender. It would be inconvenient to rearrange the programme at this stage in the work but this would be done if the saving in cost was estimated to be of the order of 5 per cent or more. Accordingly, a single graphical plot is requested on which are plotted the savings and costs for all three reductions so that the effect of any one of them can be gauged in comparison with the others.

After resolving this problem, Purechem's engineering consultant, Powerbild, sends out the tender documents. When tenders are received, the lowest (from Dooyor Wurst Construction) for the civil engineering works is at the estimated price of £15 400 000. However, some of the tendered prices for individual parts of the works do vary—the boiler house foundations will cost £2 190 000, the turbine house foundations will cost £3 400 000, the turbine house superstructure £2 000 000, the switchgear foundations £750 000, fuel storage £500 000, the administration block £120 000 and the welfare block £190 000. Negotiations proceed with Dooyor Wurst, in the course of which discussions concerning the methods of construction to be used and the technical constraints on the construction of the works are undertaken. The following constraints on the programme are established:

1. Landscaping can start only after the completion of all other works.
2. Boiler house foundations cannot start until the completion of the first two months of the Site works I programme.

3. Boiler house superstructure cannot start more than two months before the completion of the Boiler house foundations.
4. Turbine house superstructure cannot start before the completion of the turbine house foundations.
5. Cooling water system I must be complete before Switchgear foundations start.
6. Administration office is required before the end of the first year of construction but cannot start before Site works I is complete.
7. Welfare block cannot be started before the completion of the Administration block.
8. Cooling water system II cannot be started before the Switchgear foundations are completed.
9. Turbine house foundations cannot be started before the boiler house foundations are ready to start.
10. Fuel storage can be undertaken at any time between the start of the contract and the commencement of Landscaping—intermittently if that is desirable.
11. Switchgear foundations cannot start until the Turbine house foundations are complete.

The durations of the activities set out by Purechem are agreed without alteration by Dooyor Wurst Construction.

Purechem's project manager now requires you to set up the management control system for the construction of the works. It consists of the following items:

1. a precedence and an activity-on-the-arrow diagram, each with an analysis of the float, with a recommendation as to which should be used based on the advantages and disadvantages of each form;
2. a suitably coded work package structure to enable costs to be monitored;
3. an S-curve to monitor needs for finance and enable comparison with those predicted.

A.2 'Project Powerconstruct'

Dooyor Wurst Construction, a small construction company, received the contract from Purechem for the civil engineering works set out in Project Power above. This is a comparatively large contract for such a company and you are engaged to be the project manager in charge of the works. Dooyor Wurst require the company's annual financial budget, prepared by the financial director for the current year, to be updated. It is assumed that the Purechem contract will commence on 1 April of the current year. Table A.3 illustrates the budget as it was originally prepared. At that time the Purechem contract was anticipated in the budget under the contract reference number 1293. When the financial budget was prepared it was believed that the Purechem contract would start earlier than will be the case. It is now expected that contract 1293 will not contribute to the budget before 1 July and you need to check carefully the estimated cash flows in the light of the company's successful tender and the anticipated programme of work.

Table A.3

Contracts	Jan. £000	Feb. £000	Mar. £000	Apr. £000	May £000	June £000	July £000	Aug. £000	Sept. £000	Oct. £000	Nov. £000	Dec. £000	Totals £000
1246	102	140	150	190	180	180	200	140	100	40	0	0	1422
1271	234	240	200	150	100	100	0	0	200	0	0	0	1224
1274	130	130	130	180	180	200	210	200	210	180	180	180	2110
1280	120	80	80	80	90	90	60	60	70	70	100	100	1000
1293	0	0	0	400	400	400	800	800	800	700	700	700	5700
1301	0	0	0	0	0	0	60	100	150	150	150	150	760
1327	0	0	0	0	0	100	100	100	150	125	70	70	715
1334	50	50	50	70	80	80	80	70	60	20	20	20	650
Totals	636	640	610	1070	1030	1150	1510	1470	1740	1285	1220	1220	13 581

The company is concerned about the contribution being received from its current contracts and the late start of Contract 1293 has exacerbated the situation. ('Contribution' in this context, it will be remembered, is the amount of cash a contract or a sale contributes to the head office overheads and the profit of a company.) All of the costs so far used, including those in the financial budget, are inclusive of contribution. 7 per cent of the construction cost is added for head office overheads and 8 per cent of the net cost is then added for profit. The analysis of the annual budget total is therefore:

Direct costs + site oncosts	£11 752 336
Head office overheads (7 per cent)	822 664
Profit (8 per cent × 12 575 000)	1 006 000
Total	£13 581 000

So far this year, at the end of March, work to the value of £347 000 has been carried out on Contract 1246, but at a total cost of £363 000; work to the value of £692 000 has been completed on Contract 1271 at a total cost of £687 000; for Contract 1274, the respective figures are £385 000 and £350 000; for Contract 1280 the respective figures are £280 000 and £291 000. In addition Contract 1334, which was expected to start on 1 January, has not yet been awarded.

Having realized that Dooyor Wurst may be having some budget difficulties, you decide to analyse the current outcome of the budget in the light of the company's activities and make some recommendations to the board of directors as to the immediate action that should be taken.

It is clear from the contract documents that mixing and placing concrete is an important and major activity for the contract. From the contract documents it is established that the overall quantities of concrete in the works amount those shown in Table A.4.

Table A.4

	m^3
Site works I	400
Landscaping	200
Cooling water system I	6 000
Cooling water system II	2 000
Boiler house foundations	12 000
Turbine house foundations	14 500
Switchgear foundations	5 500
Fuel storage	1 000
Administration block	800
Welfare block	1 000
Total	43 400

For control purposes you draw up an overall S-curve for the quantities of concrete based on both the earliest and latest start programme for the work. Another diagram needs to be constructed to enable a concrete mixer with a suitable capacity/output to meet the demands of the programme for concrete production to be chosen. In both cases your staff need a brief written explanation of the ways in which the diagrams can be used. Select a suitable concrete mixer and investigate the possibility that it will fix the programme of work between the earliest and latest start programmes because of its capacity. Draw a histogram of planned concrete output over the life of the construction.

Dooyor Wurst is to be paid by the client monthly, receiving the money two months later than the end of the month in which work is carried out. Estimate the company's cost of working capital for the whole contract if finance costs 17 per cent per year. Assume a retention of 5 per cent of the value of the work carried out (up to a maximum of £500 000), half to be paid on completion of the works and the remainder 6 months later. The finance director requires a chart showing the variation in working capital requirements over the course of the contract with the maximum requirement and the point in time at which it occurs, clearly identified. Examine how this amount may change with a changing programme between the earliest and latest start limits.

Table A.5 lists the cost and value of work completed during the first 10 months of the contract. Costs are actual expenditures and charges; values are for sums as tendered, i.e., they include the direct costs, site oncosts, head office overheads, and markup. The finance needs a summary report on the current situation and the precise state of the time and cost situation at the date of the information in Table A.5. Recommendations as to remedial action, if required, should be included.

The cost and value figures for the first 10 months of the contract given above make an allowance for inflation at a rate of 0.5 per cent per month as required by the conditions of contract. At the end of month 10, an industrial dispute arises at the site and work ceases for 1 month. While Dooyor Wurst has to continue to pay

Table A.5

Month	Cumulative cost of work completed, £000	Cumulative value of work completed, £000
1	420	100
2	720	400
3	850	600
4	1500	1400
5	2100	2200
6	2900	2950
7	3500	3800
8	3800	4000
9	4400	4600
10	5800	5380

interest to the bank on working capital during that time, Purechem suspends all payments until work recommences. The additional costs to Dooyor Wurst will not be reimbursed on resumption of work. Examine this situation and its possible effect on the outcome of the contract.

After some 12 months of the programme, the contractor becomes concerned that the concrete production on the site has evidently become unprofitable. You decide to check the actual materials and labour usage and costs against those in the contractor's tender. The tendered costs are as follows:

Quantities and costs per cubic metre of concrete:

> Cement 350 kg at £75.00/tonne
> Sand 620 kg at £10.00/tonne
> Shingle 1200 kg at £12.00/tonne

In producing 300 m^3 of concrete, the following materials were actually used:

> Cement 109 tonnes at £72.00/tonne
> Sand 186 tonnes at £10.25/tonne
> Shingle 355 tonnes at £12.25/tonne

As to labour, the tender allowed for 8 operatives to work the concrete mixing plant at a uniform wage rate of £6.00 per hour.

In practice there are a total of 9 operatives employed for this purpose: 4 at £5.90 per hour and 5 at £5.75 per hour.

It was estimated at the time of the tender that the average output of the mixer would be 15 m^3 per hour. In checking the production over two working weeks, each of 40 hours, it was found to be at a rate of 1150 m^3 during continuous working. In order to establish a probable cause of any variance of cost from those assumed at the time of tender, you decide to undertake a variance analysis on the data. Comment on the results that you obtain and forecast the probable long-term effect of such variances on the financial outcome of the contract.

A.3 'Project Brickwork'

Dicey Contractors PLC were awarded a contract by Lethal Chemicals Ltd to construct the brickwork in the external cladding of a large chemical production plant. Because the plant produced a very highly corrosive chemical, a special type of chemically resistant cement was specified for use in the mortar for the brickwork. A stringent specification for the mortar was laid down by the client and it included 3-day compressive strength tests. Because of the special specification and their inexperience in using mortars having such stringent requirements, Dicey Contractors consulted the manufacturers of the cement about its proper and effective use.

Dicey Contractors commenced work on 1 August 1991 having a contract completion date of 31 October 1991. The contract duration was therefore established as 92 days. Laying brickwork commenced at a number of points on the cladding. Test cubes were made as the work proceeded, but on testing the cubes made on the first day of work, the mean crushing strength was found to be only 50 per cent of the specified requirement. A similar result was obtained with the mortar used on the second day; the test result for the third day's product was better but was still marginally under the specified requirements. Lethal Chemicals ordered Dicey Contractors to stop the work and to break out the brickwork so far laid, to remove it from the site, and to replace it with materials and workmanship that conformed to the specification. Dicey Contractors complied with the order and in addition engaged a firm of materials consultants to investigate and report on the defective mortar. At the same time, Dicey Contractors sought from Lethal Chemicals an extension of the contract duration to cover the period required to investigate the problem. Lethal Chemicals promptly refused this request.

Some time later, the consultants reported that the water:cement ratio being used for the mortar was inappropriate for the type of brick being laid, for attaining the compressive strength required by the specification and in the light of the climatic conditions under which the work was proceeding. Dicey Contractors quickly adjusted the conditions for using the mortar as recommended in the consultants' report and proceeded with the work, finally completing it on 10 December 1991.

Dicey Contractors had incurred considerable expense in correcting the work and decided to proceed against the cement manufacturer for the costs involved. Before preparing the claim, the following relevant information was collected:

Original total contract price	£987 000
Percentages used in tender:	
Head office overhead	8.82 per cent
Profit	8.00 per cent
Actual contract duration	132 days
All-in rate for bricklayers	£8.00/hour
for labourers	£6.30/hour
Overtime paid at 1.5 times normal rate	
Plant and equipment costs, including scaffolding,	
mixers, trucks, etc.	£1120/month

Site oncosts (including management
    and supervision)   £2150/month
Additional materials, including bricks, cement,
    sand, etc.   £7500

The following hours were spent on various aspects of the works:

    Bricklayers—300 hours in laying defective brickwork
    Bricklayers—450 hours replacing defective brickwork, including special pre-
             cautions, etc., 15 per cent of which were in overtime
    Labourers—150 hours in demolition and removing defective brickwork,
             10 per cent of which were in overtime
Cost of consultants' report   £2500

Prepare the claim to be presented to the cement manufacturers.

Appendix B: Interest tables

1% Discrete compound interest factors

	Single payment		Uniform series				Uniform
	Compound amount factor	Present worth factor	Sinking fund factor	Compound amount factor	Capital recovery factor	Present worth factor	Gradient conversion factor
n	F/P	P/F	A/F	F/A	A/P	P/A	A/G
1	1.0100	0.9901	1.0000	1.0000	1.0100	0.9901	0.0000
2	1.0201	0.9803	0.4975	2.0100	0.5075	1.9704	0.4975
3	1.0303	0.9706	0.3300	3.0301	0.3400	2.9410	0.9934
4	1.0406	0.9610	0.2463	4.0604	0.2563	3.9020	1.4876
5	1.0510	0.9515	0.1960	5.1010	0.2060	4.8534	1.9801
6	1.0615	0.9420	0.1625	6.1520	0.1725	5.7955	2.4710
7	1.0721	0.9327	0.1386	7.2135	0.1486	5.7282	2.9602
8	1.0829	0.9235	0.1207	8.2857	0.1307	7.6517	3.4478
9	1.0937	0.9143	0.1067	9.3685	0.1167	8.5660	3.9337
10	1.1046	0.9053	0.0956	10.4622	0.1056	9.4713	4.4179
11	1.1157	0.8963	0.0865	11.5668	0.0965	10.3676	4.9005
12	1.1268	0.8874	0.0788	12.6825	0.0888	11.2551	5.3815
13	1.1381	0.8787	0.0724	13.8093	0.0824	12.1337	5.8607
14	1.1495	0.8700	0.0669	14.9474	0.0769	13.0037	6.3384
15	1.1610	0.8613	0.0621	16.0969	0.0721	13.8651	6.8143
16	1.1726	0.8528	0.0579	17.2579	0.0679	14.7179	7.2886
17	1.1843	0.8444	0.0543	18.4304	0.0643	15.5623	7.7613
18	1.1961	0.8360	0.0510	19.6147	0.0610	16.3983	8.2323
19	1.2081	0.8277	0.0481	20.8109	0.0581	17.2260	8.7017
20	1.2202	0.8195	0.0454	22.0190	0.0554	18.0456	9.1694
21	1.2324	0.8114	0.0430	23.2392	0.0530	18.8570	9.6354
22	1.2447	0.8034	0.0409	24.4716	0.0509	19.6604	10.0998
23	1.2572	0.7954	0.0389	25.7163	0.0489	20.4558	10.5626
24	1.2697	0.7876	0.0371	26.9735	0.0471	21.2434	11.0237
25	1.2824	0.7798	0.0354	28.2432	0.0454	22.0232	11.4831
26	1.2953	0.7720	0.0339	29.5256	0.0439	22.7952	11.9409
27	1.3082	0.7644	0.0324	30.8209	0.0424	23.5596	12.3971
28	1.3213	0.7568	0.0311	32.1291	0.0411	24.3164	12.8516
29	1.3345	0.7493	0.0299	33.4504	0.0399	25.0658	13.3044
30	1.3478	0.7419	0.0287	34.7849	0.0387	25.8077	13.7557
31	1.3613	0.7346	0.0277	36.1327	0.0377	26.5423	14.2052
32	1.3749	0.7273	0.0267	37.4941	0.0367	27.2696	14.6532
33	1.3887	0.7201	0.0257	38.8690	0.0357	27.9897	15.0995
34	1.4026	0.7130	0.0248	40.2577	0.0348	28.7027	15.5441
35	1.4166	0.7059	0.0240	41.6603	0.0340	29.4086	15.9871
40	1.4889	0.6717	0.0205	48.8864	0.0305	32.8347	18.1776
45	1.5648	0.6391	0.0177	56.4811	0.0277	36.0945	20.3273
50	1.6446	0.6080	0.0155	64.4632	0.0255	39.1961	22.4363
55	1.7285	0.5785	0.0137	72.8525	0.0237	42.1472	24.5049
60	1.8167	0.5504	0.0122	81.6697	0.0222	44.9550	26.5333
70	2.0068	0.4983	0.0099	100.6763	0.0199	50.1685	30.4703
80	2.2167	0.4511	0.0082	121.6715	0.0182	54.8882	34.2492
90	2.4486	0.4084	0.0069	144.8633	0.0169	59.1609	37.8724
100	2.7048	0.3697	0.0059	170.4814	0.0159	63.0289	41.3426

2% Discrete compound interest factors

	Single payment		Uniform series				Uniform
	Compound amount factor	Present worth factor	Sinking fund factor	Compound amount factor	Capital recovery factor	Present worth factor	Gradient conversion factor
n	F/P	P/F	A/F	F/A	A/P	P/A	A/G
1	1.0200	0.9804	1.0000	1.0000	1.0200	0.9804	0.0000
2	1.0404	0.9612	0.4950	2.0200	0.5150	1.9416	0.4950
3	1.0612	0.9423	0.3268	3.0604	0.3468	2.8839	0.9868
4	1.0824	0.9238	0.2426	4.1216	0.2626	3.8077	1.4752
5	1.1041	0.9057	0.1922	5.2040	0.2122	4.7135	1.9604
6	1.1262	0.8880	0.1585	6.3081	0.1785	5.6014	2.4423
7	1.1487	0.8706	0.1345	7.4343	0.1545	6.4720	2.9208
8	1.1717	0.8535	0.1165	8.5830	0.1365	7.3255	3.3961
9	1.1951	0.8368	0.1025	9.7546	0.1225	8.1622	3.8681
10	1.2190	0.8203	0.0913	10.9497	0.1113	8.9826	4.3367
11	1.2434	0.8043	0.0822	12.1687	0.1022	9.7868	4.8021
12	1.2682	0.7885	0.0746	13.4121	0.0946	10.5753	5.2642
13	1.2936	0.7730	0.0681	14.6803	0.0881	11.3484	5.7231
14	1.3195	0.7579	0.0626	15.9739	0.0826	12.1062	6.1786
15	1.3459	0.7430	0.0578	17.2934	0.0778	12.8493	6.6309
16	1.3728	0.7284	0.0537	18.6393	0.0737	13.5777	7.0799
17	1.4002	0.7142	0.0500	20.0121	0.0700	14.2919	7.5256
18	1.4282	0.7002	0.0467	21.4123	0.0667	14.9920	7.9681
19	1.4568	0.6864	0.0438	22.8406	0.0638	15.6785	8.4073
20	1.4859	0.6730	0.0412	24.2974	0.0612	16.3514	8.8433
21	1.5157	0.6598	0.0388	25.7833	0.0588	17.0112	9.2760
22	1.5460	0.6468	0.0366	27.2990	0.0566	17.6580	9.7055
23	1.5769	0.6342	0.0347	28.8450	0.0547	18.2922	10.1317
24	1.6084	0.6217	0.0329	30.4219	0.0529	18.9139	10.5547
25	1.6406	0.6095	0.0312	32.0303	0.0512	19.5235	10.9745
26	1.6734	0.5976	0.0297	33.6709	0.0497	20.1210	11.3910
27	1.7069	0.5859	0.0283	35.3443	0.0483	20.7069	11.8043
28	1.7410	0.5744	0.0270	37.0512	0.0470	21.2813	12.2145
29	1.7758	0.5631	0.0258	38.7922	0.0458	21.8444	12.6214
30	1.8114	0.5521	0.0246	40.5681	0.0446	22.3965	13.0251
31	1.8476	0.5412	0.0236	42.3794	0.0436	22.9377	13.4257
32	1.8845	0.5306	0.0226	44.2270	0.0426	23.4683	13.8230
33	1.9222	0.5202	0.0217	46.1116	0.0417	23.9886	14.2172
34	1.9607	0.5100	0.0208	48.0338	0.0408	24.4986	14.6083
35	1.9999	0.5000	0.0200	49.9945	0.0400	24.9986	14.9961
40	2.2080	0.4529	0.0166	60.4020	0.0366	27.3555	16.8885
45	2.4379	0.4102	0.0139	71.8927	0.0339	29.4902	18.7034
50	2.6916	0.3715	0.0118	84.5794	0.0318	31.4236	20.4420
55	2.9717	0.3365	0.0101	98.5865	0.0301	33.1748	22.1057
60	3.2810	0.3048	0.0088	114.0515	0.0288	34.7609	23.6961
70	3.9996	0.2500	0.0067	149.9779	0.0267	37.4986	26.6632
80	4.8754	0.2051	0.0052	193.7720	0.0252	39.7445	29.3572
90	5.9431	0.1683	0.0040	247.1567	0.0240	41.5869	31.7929
100	7.2446	0.1380	0.0032	312.2323	0.0232	43.0984	33.9863

3% Discrete compound interest factors

	Single payment		Uniform series				Uniform
	Compound amount factor	Present worth factor	Sinking fund factor	Compound amount factor	Capital recovery factor	Present worth factor	Gradient conversion factor
n	F/P	P/F	A/F	F/A	A/P	P/A	A/G
1	1.0300	0.9709	1.0000	1.0000	1.0300	0.9709	0.0000
2	1.0609	0.9426	0.4926	2.0300	0.5226	1.9135	0.4926
3	1.0927	0.9151	0.3235	3.0909	0.3535	2.8286	0.9803
4	1.1255	0.8885	0.2390	4.1836	0.2690	3.7171	1.4631
5	1.1593	0.8626	0.1884	5.3091	0.2184	4.5797	1.9409
6	1.1941	0.8375	0.1546	6.4684	0.1846	5.4172	2.4138
7	1.2299	0.8131	0.1305	7.6625	0.1605	6.2303	2.8819
8	1.2668	0.7894	0.1125	8.8923	0.1425	7.0197	3.3450
9	1.3048	0.7664	0.0984	10.1591	0.1284	7.7861	3.8032
10	1.3439	0.7441	0.0872	11.4639	0.1172	8.5302	4.2565
11	1.3842	0.7224	0.0781	12.8078	0.1081	9.2526	4.7049
12	1.4258	0.7014	0.0705	14.1920	0.1005	9.9540	5.1485
13	1.4685	0.6810	0.0640	15.6178	0.0940	10.6350	5.5872
14	1.5126	0.6611	0.0585	17.0863	0.0885	11.2961	6.0210
15	1.5580	0.6419	0.0538	18.5989	0.0838	11.9379	6.4500
16	1.6047	0.6232	0.0496	20.1569	0.0796	12.5611	6.8742
17	1.6528	0.6050	0.0460	21.7616	0.0760	13.1661	7.2936
18	1.7024	0.5874	0.0427	23.4144	0.0727	13.7535	7.7081
19	1.7535	0.5703	0.0398	25.1169	0.0698	14.3238	8.1179
20	1.8061	0.5537	0.0372	26.8704	0.0672	14.8775	8.5229
21	1.8603	0.5375	0.0349	28.6765	0.0649	15.4150	8.9231
22	1.9161	0.5219	0.0327	30.5368	0.0627	15.9369	9.3186
23	1.9736	0.5067	0.0308	32.4529	0.0608	16.4436	9.7093
24	2.0328	0.4919	0.0290	34.4265	0.0590	16.9355	10.0954
25	2.0938	0.4776	0.0274	36.4593	0.0574	17.4131	10.4768
26	2.1566	0.4637	0.0259	38.5530	0.0559	17.8768	10.8535
27	2.2213	0.4502	0.0246	40.7096	0.0546	18.3270	11.2255
28	2.2879	0.4371	0.0233	42.9309	0.0533	18.7641	11.5930
29	2.3566	0.4243	0.0221	45.2189	0.0521	19.1885	11.9558
30	2.4273	0.4120	0.0210	47.5754	0.0510	19.6004	12.3141
31	2.5001	0.4000	0.0200	50.0027	0.0500	20.0004	12.6678
32	2.5751	0.3883	0.0190	52.5028	0.0490	20.3888	13.0169
33	2.6523	0.3770	0.0182	55.0778	0.0482	20.7658	13.3616
34	2.7319	0.3660	0.0173	57.7302	0.0473	21.1318	13.7018
35	2.8139	0.3554	0.0165	60.4621	0.0465	21.4872	14.0375
40	3.2620	0.3066	0.0133	75.4013	0.0433	23.1148	15.6502
45	3.7816	0.2644	0.0108	92.7199	0.0408	24.5187	17.1556
50	4.3839	0.2281	0.0089	112.7969	0.0389	25.7298	18.5575
55	5.0821	0.1968	0.0073	136.0716	0.0373	26.7744	19.8600
60	5.8916	0.1697	0.0061	163.0534	0.0361	27.6756	21.0674
70	7.9178	0.1263	0.0043	230.5941	0.0343	29.1234	23.2145
80	10.6409	0.0940	0.0031	321.3630	0.0331	30.2008	25.0353
90	14.3005	0.0699	0.0023	443.3489	0.0323	31.0024	26.5667
100	19.2186	0.0520	0.0016	607.2877	0.0316	31.5989	27.8444

5% Discrete compound interest factors

	Single payment		Uniform series				Uniform
	Compound amount factor	Present worth factor	Sinking fund factor	Compound amount factor	Capital recovery factor	Present worth factor	Gradient conversion factor
n	F/P	P/F	A/F	F/A	A/P	P/A	A/G
1	1.0500	0.9524	1.0000	1.0000	1.0500	0.9524	0.0000
2	1.1025	0.9070	0.4878	2.0500	0.5378	1.8594	0.4878
3	1.1576	0.8638	0.3172	3.1525	0.3672	2.7232	0.9675
4	1.2155	0.8227	0.2320	4.3101	0.2820	3.5460	1.4391
5	1.2763	0.7835	0.1810	5.5256	0.2310	4.3295	1.9025
6	1.3401	0.7462	0.1470	6.8019	0.1970	5.0757	2.3579
7	1.4071	0.7107	0.1228	8.1420	0.1728	5.7864	2.8052
8	1.4775	0.6768	0.1047	9.5491	0.1547	6.4632	3.2445
9	1.5513	0.6446	0.0907	11.0266	0.1407	7.1078	3.6758
10	1.6289	0.6139	0.0795	12.5779	0.1295	7.7217	4.0991
11	1.7103	0.5847	0.0704	14.2068	0.1204	8.3064	4.5144
12	1.7959	0.5568	0.0628	15.9171	0.1128	8.8633	4.9219
13	1.8856	0.5303	0.0565	17.7130	0.1065	9.3936	5.3215
14	1.9799	0.5051	0.0510	19.5986	0.1010	9.8986	5.7133
15	2.0789	0.4810	0.0463	21.5786	0.0963	10.3797	6.0973
16	2.1829	0.4581	0.0423	23.6575	0.0923	10.8378	6.4736
17	2.2920	0.4363	0.0387	25.8404	0.0887	11.2741	6.8423
18	2.4066	0.4155	0.0355	28.1324	0.0855	11.6896	7.2034
19	2.5270	0.3957	0.0327	30.5390	0.0827	12.0853	7.5569
20	2.6533	0.3769	0.0302	33.0660	0.0802	12.4622	7.9030
21	2.7860	0.3589	0.0280	35.7193	0.0780	12.8212	8.2416
22	2.9253	0.3418	0.0260	38.5052	0.0760	13.1630	8.5730
23	3.0715	0.3256	0.0241	41.4305	0.0741	13.4886	8.8971
24	3.2251	0.3101	0.0225	44.5020	0.0725	13.7986	9.2140
25	3.3864	0.2953	0.0210	47.7271	0.0710	14.0939	9.5238
26	3.5557	0.2812	0.0196	51.1135	0.0696	14.3752	9.8266
27	3.7335	0.2678	0.0183	54.6691	0.0683	14.6430	10.1224
28	3.9201	0.2551	0.0171	58.4026	0.0671	14.8981	10.4114
29	4.1161	0.2429	0.0160	62.3227	0.0660	15.1411	10.6936
30	4.3219	0.2314	0.0151	66.4388	0.0651	15.3725	10.9691
31	4.5380	0.2204	0.0141	70.7608	0.0641	15.5928	11.2381
32	4.7649	0.2099	0.0133	75.2988	0.0633	15.8027	11.5005
33	5.0032	0.1999	0.0125	80.0638	0.0625	16.0025	11.7566
34	5.2533	0.1904	0.0118	85.0670	0.0618	16.1929	12.0063
35	5.5160	0.1813	0.0111	90.3203	0.0611	16.3742	12.2498
40	7.0400	0.1420	0.0083	120.7998	0.0583	17.1591	13.3775
45	8.9850	0.1113	0.0063	159.7002	0.0563	17.7741	14.3644
50	11.4674	0.0872	0.0048	209.3480	0.0548	18.2559	15.2233
55	14.6356	0.0683	0.0037	272.7126	0.0537	18.6335	15.9664
60	18.6792	0.0535	0.0028	353.5837	0.0528	18.9293	16.6062
70	30.4264	0.0329	0.0017	588.5285	0.0517	19.3427	17.6212
80	49.5614	0.0202	0.0010	971.2288	0.0510	19.5965	18.3526
90	80.7304	0.0124	0.0006	1594.6073	0.0506	19.7523	18.8712
100	131.5013	0.0076	0.0004	2610.0252	0.0504	19.8479	19.2337

7% Discrete compound interest factors

	Single payment		Uniform series				Uniform
	Compound amount factor	Present worth factor	Sinking fund factor	Compound amount factor	Capital recovery factor	Present worth factor	Gradient conversion factor
n	F/P	P/F	A/F	F/A	A/P	P/A	A/G
1	1.0700	0.9346	1.0000	1.0000	1.0700	0.9346	0.0000
2	1.1449	0.8734	0.4831	2.0700	0.5531	1.8080	0.4831
3	1.2250	0.8163	0.3111	3.2149	0.3811	2.6243	0.9549
4	1.3108	0.7629	0.2252	4.4399	0.2952	3.3872	1.4155
5	1.4026	0.7130	0.1739	5.7507	0.2439	4.1002	1.8650
6	1.5007	0.6663	0.1398	7.1533	0.2098	4.7665	2.3032
7	1.6058	0.6227	0.1156	8.6540	0.1856	5.3893	2.7304
8	1.7182	0.5820	0.0975	10.2598	0.1675	5.9713	3.1465
9	1.8385	0.5439	0.0835	11.9780	0.1535	6.5152	3.5517
10	1.9672	0.5083	0.0724	13.8164	0.1424	7.0236	3.9461
11	2.1049	0.4751	0.0634	15.7836	0.1334	7.4987	4.3296
12	2.2522	0.4440	0.0559	17.8885	0.1259	7.9427	4.7025
13	2.4098	0.4150	0.0497	20.1406	0.1197	8.3577	5.0648
14	2.5785	0.3878	0.0443	22.5505	0.1143	8.7455	5.4167
15	2.7590	0.3624	0.0398	25.1290	0.1098	9.1079	5.7583
16	2.9522	0.3387	0.0359	27.8881	0.1059	9.4466	6.0897
17	3.1588	0.3166	0.0324	30.8402	0.1024	9.7632	6.4110
18	3.3799	0.2959	0.0294	33.9990	0.0994	10.0591	6.7225
19	3.6165	0.2765	0.0268	37.3790	0.0968	10.3356	7.0242
20	3.8697	0.2584	0.0244	40.9955	0.0944	10.5940	7.3163
21	4.1406	0.2415	0.0223	44.8652	0.0923	10.8355	7.5990
22	4.4304	0.2257	0.0204	49.0057	0.0904	11.0612	7.8725
23	4.7405	0.2109	0.0187	53.4361	0.0887	11.2722	8.1369
24	5.0724	0.1971	0.0172	58.1767	0.0872	11.4693	8.3923
25	5.4274	0.1842	0.0158	63.2490	0.0858	11.6536	8.6391
26	5.8074	0.1722	0.0146	68.6765	0.0846	11.8258	8.8773
27	6.2139	0.1609	0.0134	74.4838	0.0834	11.9867	9.1072
28	6.6488	0.1504	0.0124	80.6977	0.0824	12.1371	9.3289
29	7.1143	0.1406	0.0114	87.3465	0.0814	12.2777	9.5427
30	7.6123	0.1314	0.0106	94.4608	0.0806	12.4090	9.7487
31	8.1451	0.1228	0.0098	102.0730	0.0798	12.5318	9.9471
32	8.7153	0.1147	0.0091	110.2182	0.0791	12.6466	10.1381
33	9.3253	0.1072	0.0084	118.9334	0.0784	12.7538	10.3219
34	9.9781	0.1002	0.0078	128.2588	0.0778	12.8540	10.4987
35	10.6766	0.0937	0.0072	138.2369	0.0772	12.9477	10.6687
40	14.9745	0.0668	0.0050	199.6351	0.0750	13.3317	11.4233
45	21.0025	0.0476	0.0035	285.7493	0.0735	13.6055	12.0360
50	29.4570	0.0339	0.0025	406.5289	0.0725	13.8007	12.5287
55	41.3150	0.0242	0.0017	575.9286	0.0717	13.9399	12.9215
60	57.9464	0.0173	0.0012	813.5204	0.0712	14.0392	13.2321
70	113.9894	0.0088	0.0006	1 614.1342	0.0706	14.1604	13.6662
80	224.2344	0.0045	0.0003	3 189.0627	0.0703	14.2220	13.9273
90	441.1030	0.0023	0.0002	6 287.1854	0.0702	14.2533	14.0812
100	867.7163	0.0012	0.0001	12 381.6618	0.0701	14.2693	14.1703

9% Discrete compound interest factors

	Single payment		Uniform series				Uniform
	Compound amount factor	Present worth factor	Sinking fund factor	Compound amount factor	Capital recovery factor	Present worth factor	Gradient conversion factor
n	F/P	P/F	A/F	F/A	A/P	P/A	A/G
1	1.0900	0.9174	1.0000	1.0000	1.0900	0.9174	0.0000
2	1.1881	0.8417	0.4785	2.0900	0.5685	1.7591	0.4785
3	1.2950	0.7722	0.3051	3.2781	0.3951	2.5313	0.9426
4	1.4116	0.7084	0.2187	4.5731	0.3087	3.2397	1.3925
5	1.5386	0.6499	0.1671	5.9847	0.2571	3.8897	1.8282
6	1.6771	0.5963	0.1329	7.5233	0.2229	4.4859	2.2498
7	1.8280	0.5470	0.1087	9.2004	0.1987	5.0330	2.6574
8	1.9926	0.5019	0.0907	11.0285	0.1807	5.5348	3.0512
9	2.1719	0.4604	0.0768	13.0210	0.1668	5.9952	3.4312
10	2.3674	0.4224	0.0658	15.1929	0.1558	6.4177	3.7978
11	2.5804	0.3875	0.0569	17.5603	0.1469	6.8052	4.1510
12	2.8127	0.3555	0.0497	20.1407	0.1397	7.1607	4.4910
13	3.0658	0.3262	0.0436	22.9534	0.1336	7.4869	4.8182
14	3.3417	0.2992	0.0384	26.0192	0.1284	7.7862	5.1326
15	3.6425	0.2745	0.0341	29.3609	0.1241	8.0607	5.4346
16	3.9703	0.2519	0.0303	33.0034	0.1203	8.3126	5.7245
17	4.3276	0.2311	0.0270	36.9737	0.1170	8.5436	6.0024
18	4.7171	0.2120	0.0242	41.3013	0.1142	8.7556	6.2687
19	5.1417	0.1945	0.0217	46.0185	0.1117	8.9501	6.5236
20	5.6044	0.1784	0.0195	51.1601	0.1095	9.1285	6.7674
21	6.1088	0.1637	0.0176	56.7645	0.1076	9.2922	7.0006
22	6.6586	0.1502	0.0159	62.8733	0.1059	9.4424	7.2232
23	7.2579	0.1378	0.0144	69.5319	0.1044	9.5802	7.4357
24	7.9111	0.1264	0.0130	76.7898	0.1030	9.7066	7.6384
25	8.6231	0.1160	0.0118	84.7009	0.1018	9.8226	7.8316
26	9.3992	0.1064	0.0107	93.3240	0.1007	9.9290	8.0156
27	10.2451	0.0976	0.0097	102.7231	0.0997	10.0266	8.1906
28	11.1671	0.0895	0.0089	112.9682	0.0989	10.1161	8.3571
29	12.1722	0.0822	0.0081	124.1354	0.0981	10.1983	8.5154
30	13.2677	0.0754	0.0073	136.3075	0.0973	10.2737	8.6657
31	14.4618	0.0691	0.0067	149.5752	0.0967	10.3428	8.8083
32	15.7633	0.0634	0.0061	164.0370	0.0961	10.4062	8.9436
33	17.1820	0.0582	0.0056	179.8003	0.0956	10.4644	9.0718
34	18.7284	0.0534	0.0051	196.9823	0.0951	10.5178	9.1933
35	20.4140	0.0490	0.0046	215.7108	0.0946	10.5668	9.3083
40	31.4094	0.0318	0.0030	337.8824	0.0930	10.7574	9.7957
45	48.3273	0.0207	0.0019	525.8587	0.0919	10.8812	10.1603
50	74.3575	0.0134	0.0012	815.0836	0.0912	10.9617	10.4295
55	114.4083	0.0087	0.0008	1 260.0918	0.0908	11.0140	10.6261
60	176.0313	0.0057	0.0005	1 944.7921	0.0905	11.0480	10.7683
70	416.7301	0.0024	0.0002	4 619.2232	0.0902	11.0844	10.9427
80	986.5517	0.0010	0.0001	10 950.5741	0.0901	11.0998	11.0299
90	2 335.5266	0.0004	0.0000	25 939.1842	0.0900	11.1064	11.0726
100	5 529.0408	0.0002	0.0000	61 422.6755	0.0900	11.1091	11.0930

10% Discrete compound interest factors

	Single payment		Uniform series				Uniform
	Compound amount factor	Present worth factor	Sinking fund factor	Compound amount factor	Capital recovery factor	Present worth factor	Gradient conversion factor
n	F/P	P/F	A/F	F/A	A/P	P/A	A/G
1	1.1000	0.9091	1.0000	1.0000	1.1000	0.9091	0.0000
2	1.2100	0.8264	0.4762	2.1000	0.5762	1.7355	0.4762
3	1.3310	0.7513	0.3021	3.3100	0.4021	2.4869	0.9366
4	1.4641	0.6830	0.2155	4.6410	0.3155	3.1699	1.3812
5	1.6105	0.6209	0.1638	6.1051	0.2638	3.7908	1.8101
6	1.7716	0.5645	0.1296	7.7156	0.2296	4.3553	2.2236
7	1.9487	0.5132	0.1054	9.4872	0.2054	4.8684	2.6216
8	2.1436	0.4665	0.0874	11.4359	0.1874	5.3349	3.0045
9	2.3579	0.4241	0.0736	13.5795	0.1736	5.7590	3.3724
10	2.5937	0.3855	0.0627	15.9374	0.1627	6.1446	3.7255
11	2.8531	0.3505	0.0540	18.5312	0.1540	6.4951	4.0641
12	3.1384	0.3186	0.0468	21.3843	0.1468	6.8137	4.3884
13	3.4523	0.2897	0.0408	24.5227	0.1408	7.1034	4.6988
14	3.7975	0.2633	0.0357	27.9750	0.1357	7.3667	4.9955
15	4.1772	0.2394	0.0315	31.7725	0.1315	7.6061	5.2789
16	4.5950	0.2176	0.0278	35.9497	0.1278	7.8237	5.5493
17	5.0545	0.1978	0.0247	40.5447	0.1247	8.0216	5.8071
18	5.5599	0.1799	0.0219	45.5992	0.1219	8.2014	6.0526
19	6.1159	0.1635	0.0195	51.1591	0.1195	8.3649	6.2861
20	6.7275	0.1486	0.0175	57.2750	0.1175	8.5136	6.5081
21	7.4002	0.1351	0.0156	64.0025	0.1156	8.6487	6.7189
22	8.1403	0.1228	0.0140	71.4027	0.1140	8.7715	6.9189
23	8.9543	0.1117	0.0126	79.5430	0.1126	8.8832	7.1085
24	9.8497	0.1015	0.0113	88.4973	0.1113	8.9847	7.2881
25	10.8347	0.0923	0.0102	98.3471	0.1102	9.0770	7.4580
26	11.9182	0.0839	0.0092	109.1818	0.1092	9.1609	7.6186
27	13.1100	0.0763	0.0083	121.0999	0.1083	9.2372	7.7704
28	14.4210	0.0693	0.0075	134.2099	0.1075	9.3066	7.9137
29	15.8631	0.0630	0.0067	148.6309	0.1067	9.3696	8.0489
30	17.4494	0.0573	0.0061	164.4940	0.1061	9.4269	8.1762
31	19.1943	0.0521	0.0055	181.9434	0.1055	9.4790	8.2962
32	21.1138	0.0474	0.0050	201.1378	0.1050	9.5264	8.4091
33	23.2252	0.0431	0.0045	222.2515	0.1045	9.5694	8.5152
34	25.5477	0.0391	0.0041	245.4767	0.1041	9.6086	8.6149
35	28.1024	0.0356	0.0037	271.0244	0.1037	9.6442	8.7086
40	45.2593	0.0221	0.0023	442.5926	0.1023	9.7791	9.0962
45	72.8905	0.0137	0.0014	718.9048	0.1014	9.8628	9.3740
50	117.3909	0.0085	0.0009	1 163.9085	0.1009	9.9148	9.5704
55	189.0591	0.0053	0.0005	1 880.5914	0.1005	9.9471	9.7075
60	304.4816	0.0033	0.0003	3 034.8164	0.1003	9.9672	9.8023
70	789.7470	0.0013	0.0001	7 887.4696	0.1001	9.9873	9.9113
80	2 048.4002	0.0005	0.0000	20 474.0021	0.1000	9.9951	9.9609
90	5 313.0226	0.0002	0.0000	53 120.2261	0.1000	9.9981	9.9831
100	13 780.6123	0.0001	0.0000	137 796.1234	0.1000	9.9993	9.9927

12% Discrete compound interest factors

	Single payment		Uniform series				Uniform
	Compound amount factor	Present worth factor	Sinking fund factor	Compound amount factor	Capital recovery factor	Present worth factor	Gradient conversion factor
n	F/P	P/F	A/F	F/A	A/P	P/A	A/G
1	1.1200	0.8929	1.0000	1.0000	1.1200	0.8929	0.0000
2	1.2544	0.7972	0.4717	2.1200	0.5917	1.6901	0.4717
3	1.4049	0.7118	0.2963	3.3744	0.4163	2.4018	0.9246
4	1.5735	0.6355	0.2092	4.7793	0.3292	3.0373	1.3589
5	1.7623	0.5674	0.1574	6.3528	0.2774	3.6048	1.7746
6	1.9738	0.5066	0.1232	8.1152	0.2432	4.1114	2.1720
7	2.2107	0.4523	0.0991	10.0890	0.2191	4.5638	2.5515
8	2.4760	0.4039	0.0813	12.2997	0.2013	4.9676	2.9131
9	2.7731	0.3606	0.0677	14.7757	0.1877	5.3282	3.2574
10	3.1058	0.3220	0.0570	17.5487	0.1770	5.6502	3.5847
11	3.4785	0.2875	0.0484	20.6546	0.1684	5.9377	3.8953
12	3.8960	0.2567	0.0414	24.1331	0.1614	6.1944	4.1897
13	4.3635	0.2292	0.0357	28.0291	0.1557	6.4235	4.4683
14	4.8871	0.2046	0.0309	32.3926	0.1509	6.6282	4.7317
15	5.4736	0.1827	0.0268	37.2797	0.1468	6.8109	4.9803
16	6.1304	0.1631	0.0234	42.7533	0.1434	6.9740	5.2147
17	6.8660	0.1456	0.0205	48.8837	0.1405	7.1196	5.4353
18	7.6900	0.1300	0.0179	55.7497	0.1379	7.2497	5.6427
19	8.6128	0.1161	0.0158	63.4397	0.1358	7.3658	5.8375
20	9.6463	0.1037	0.0139	72.0524	0.1339	7.4694	6.0202
21	10.8038	0.0926	0.0122	81.6987	0.1322	7.5620	6.1913
22	12.1003	0.0826	0.0108	92.5026	0.1308	7.6446	6.3514
23	13.5523	0.0738	0.0096	104.6029	0.1296	7.7184	6.5010
24	15.1786	0.0659	0.0085	118.1552	0.1285	7.7843	6.6406
25	17.0001	0.0588	0.0075	133.3339	0.1275	7.8431	6.7708
26	19.0401	0.0525	0.0067	150.3339	0.1267	7.8957	6.8921
27	21.3249	0.0469	0.0059	169.3740	0.1259	7.9426	7.0049
28	23.8839	0.0419	0.0052	190.6989	0.1252	7.9844	7.1098
29	26.7499	0.0374	0.0047	214.5828	0.1247	8.0218	7.2071
30	29.9599	0.0334	0.0041	241.3327	0.1241	8.0552	7.2974
31	33.5551	0.0298	0.0037	271.2926	0.1237	8.0850	7.3811
32	37.5817	0.0266	0.0033	304.8477	0.1233	8.1116	7.4586
33	42.0915	0.0238	0.0029	342.4294	0.1229	8.1354	7.5302
34	47.1425	0.0212	0.0026	384.5210	0.1226	8.1566	7.5965
35	52.7996	0.0189	0.0023	431.6635	0.1223	8.1755	7.6577
40	93.0510	0.0107	0.0013	767.0914	0.1213	8.2438	7.8988
45	163.9876	0.0061	0.0007	1358.2300	0.1207	8.2825	8.0572
50	289.0022	0.0035	0.0004	2400.0182	0.1204	8.3045	8.1597
55	509.3206	0.0020	0.0002	4236.0050	0.1202	8.3170	8.2251
60	897.5969	0.0011	0.0001	7471.6411	0.1201	8.3240	8.2664

13% Discrete compound interest factors

	Single payment		Uniform series				Uniform
	Compound amount factor	Present worth factor	Sinking fund factor	Compound amount factor	Capital recovery factor	Present worth factor	Gradient conversion factor
n	F/P	P/F	A/F	F/A	A/P	P/A	A/G
1	1.1300	0.8850	1.0000	1.0000	1.1300	0.8850	0.0000
2	1.2769	0.7831	0.4695	2.1300	0.5995	1.6681	0.4695
3	1.4429	0.6931	0.2935	3.4069	0.4235	2.3612	0.9187
4	1.6305	0.6133	0.2062	4.8498	0.3362	2.9745	1.3479
5	1.8424	0.5428	0.1543	6.4803	0.2843	3.5172	1.7571
6	2.0820	0.4803	0.1202	8.3227	0.2502	3.9975	2.1468
7	2.3526	0.4251	0.0961	10.4047	0.2261	4.4226	2.5171
8	2.6584	0.3762	0.0784	12.7573	0.2084	4.7988	2.8685
9	3.0040	0.3329	0.0649	15.4157	0.1949	5.1317	3.2014
10	3.3946	0.2946	0.0543	18.4197	0.1843	5.4262	3.5162
11	3.8359	0.2607	0.0458	21.8143	0.1758	5.6869	3.8134
12	4.3345	0.2307	0.0390	25.6502	0.1690	5.9176	4.0936
13	4.8980	0.2042	0.0334	29.9847	0.1634	6.1218	4.3573
14	5.5348	0.1807	0.0287	34.8827	0.1587	6.3025	4.6050
15	6.2543	0.1599	0.0247	40.4175	0.1547	6.4624	4.8375
16	7.0673	0.1415	0.0214	46.6717	0.1514	6.6039	5.0552
17	7.9861	0.1252	0.0186	53.7391	0.1486	6.7291	5.2589
18	9.0243	0.1108	0.0162	61.7251	0.1462	6.8399	5.4491
19	10.1974	0.0981	0.0141	70.7494	0.1441	6.9380	5.6265
20	11.5231	0.0868	0.0124	80.9468	0.1424	7.0248	5.7917
21	13.0211	0.0768	0.0108	92.4699	0.1408	7.1016	5.9454
22	14.7138	0.0680	0.0095	105.4910	0.1395	7.1695	6.0881
23	16.6266	0.0601	0.0083	120.2048	0.1383	7.2297	6.2205
24	18.7881	0.0532	0.0073	136.8315	0.1373	7.2829	6.3431
25	21.2305	0.0471	0.0064	155.6196	0.1364	7.3300	6.4566
26	23.9905	0.0417	0.0057	176.8501	0.1357	7.3717	6.5614
27	27.1093	0.0369	0.0050	200.8406	0.1350	7.4086	6.6582
28	30.6335	0.0326	0.0044	227.9499	0.1344	7.4412	6.7474
29	34.6158	0.0289	0.0039	258.5834	0.1339	7.4701	6.8296
30	39.1159	0.0256	0.0034	293.1992	0.1334	7.4957	6.9052
31	44.2010	0.0226	0.0030	332.3151	0.1330	7.5183	6.9747
32	49.9471	0.0200	0.0027	376.5161	0.1327	7.5383	7.0385
33	56.4402	0.0177	0.0023	426.4632	0.1323	7.5560	7.0971
34	63.7774	0.0157	0.0021	482.9034	0.1321	7.5717	7.1507
35	72.0685	0.0139	0.0018	546.6808	0.1318	7.5856	7.1998
40	132.7816	0.0075	0.0010	1 013.7042	0.1310	7.6344	7.3888
45	244.6414	0.0041	0.0005	1 874.1646	0.1305	7.6609	7.5076
50	450.7359	0.0022	0.0003	3 459.5071	0.1303	7.6752	7.5811
55	830.4517	0.0012	0.0002	6 380.3979	0.1302	7.6830	7.6260
60	1 530.0535	0.0007	0.0001	11 761.9498	0.1301	7.6873	7.6531

14% Discrete compound interest factors

	Single Payment		Uniform series				Uniform
	Compound amount factor	Present worth factor	Sinking fund factor	Compound amount factor	Capital recovery factor	Present worth factor	Gradient conversion factor
n	F/P	P/F	A/F	F/A	A/P	P/A	A/G
1	1.1400	0.8772	1.0000	1.0000	1.1400	0.8772	0.0000
2	1.2996	0.7695	0.4673	2.1400	0.6073	1.6467	0.4673
3	1.4815	0.6750	0.2907	3.4396	0.4307	2.3216	0.9129
4	1.6890	0.5921	0.2032	4.9211	0.3432	2.9137	1.3370
5	1.9254	0.5194	0.1513	6.6101	0.2913	3.4331	1.7399
6	2.1950	0.4556	0.1172	8.5355	0.2572	3.8887	2.1218
7	2.5023	0.3996	0.0932	10.7305	0.2332	4.2883	2.4832
8	2.8526	0.3506	0.0756	13.2328	0.2156	4.6389	2.8246
9	3.2519	0.3075	0.0622	16.0853	0.2022	4.9464	3.1463
10	3.7072	0.2697	0.0517	19.3373	0.1917	5.2161	3.4490
11	4.2262	0.2366	0.0434	23.0445	0.1834	5.4527	3.7333
12	4.8179	0.2076	0.0367	27.2707	0.1767	5.6603	3.9998
13	5.4924	0.1821	0.0312	32.0887	0.1712	5.8424	4.2491
14	6.2613	0.1597	0.0266	37.5811	0.1666	6.0021	4.4819
15	7.1379	0.1401	0.0228	43.8424	0.1628	6.1422	4.6990
16	8.1372	0.1229	0.0196	50.9804	0.1596	6.2651	4.9011
17	9.2765	0.1078	0.0169	59.1176	0.1569	6.3729	5.0888
18	10.5752	0.0946	0.0146	68.3941	0.1546	6.4674	5.2630
19	12.0557	0.0829	0.0127	78.9692	0.1527	6.5504	5.4243
20	13.7435	0.0728	0.0110	91.0249	0.1510	6.6231	5.5734
21	15.6676	0.0638	0.0095	104.7684	0.1495	6.6870	5.7111
22	17.8610	0.0560	0.0083	120.4360	0.1483	6.7429	5.8381
23	20.3616	0.0491	0.0072	138.2970	0.1472	6.7921	5.9549
24	23.2122	0.0431	0.0063	158.6586	0.1463	6.8351	6.0624
25	26.4619	0.0378	0.0055	181.8708	0.1455	6.8729	6.1610
26	30.1666	0.0331	0.0048	208.3327	0.1448	6.9061	6.2514
27	34.3899	0.0291	0.0042	238.4993	0.1442	6.9352	6.3342
28	39.2045	0.0255	0.0037	272.8892	0.1437	6.9607	6.4100
29	44.6931	0.0224	0.0032	312.0937	0.1432	6.9830	6.4791
30	50.9502	0.0196	0.0028	356.7868	0.1428	7.0027	6.5423
31	58.0832	0.0172	0.0025	407.7370	0.1425	7.0199	6.5998
32	66.2148	0.0151	0.0021	465.8202	0.1421	7.0350	6.6522
33	75.4849	0.0132	0.0019	532.0350	0.1419	7.0482	6.6998
34	86.0528	0.0116	0.0016	607.5199	0.1416	7.0599	6.7431
35	98.1002	0.0102	0.0014	693.5727	0.1414	7.0700	6.7824
40	188.8835	0.0053	0.0007	1 342.0251	0.1407	7.1050	6.9300
45	363.6791	0.0027	0.0004	2 590.5648	0.1404	7.1232	7.0188
50	700.2330	0.0014	0.0002	4 994.5213	0.1402	7.1327	7.0714
55	1 348.2388	0.0007	0.0001	9 623.1343	0.1401	7.1376	7.1020
60	2 595.9187	0.0004	0.0001	18 535.1333	0.1401	7.1401	7.1197

16% Discrete compound interest factors

	Single payment		Uniform series				Uniform
	Compound amount factor	Present worth factor	Sinking fund factor	Compound amount factor	Capital recovery factor	Present worth factor	Gradient conversion factor
n	F/P	P/F	A/F	F/A	A/P	P/A	A/G
1	1.1600	0.8621	1.0000	1.0000	1.1600	0.8621	0.0000
2	1.3456	0.7432	0.4630	2.1600	0.6230	1.6052	0.4630
3	1.5609	0.6407	0.2853	3.5056	0.4453	2.2459	0.9014
4	1.8106	0.5523	0.1974	5.0665	0.3574	2.7982	1.3156
5	2.1003	0.4761	0.1454	6.8771	0.3054	3.2743	1.7060
6	2.4364	0.4104	0.1114	8.9775	0.2714	3.6847	2.0729
7	2.8262	0.3538	0.0876	11.4139	0.2476	4.0386	2.4169
8	3.2784	0.3050	0.0702	14.2401	0.2302	4.3436	2.7388
9	3.8030	0.2630	0.0571	17.5185	0.2171	4.6065	3.0391
10	4.4114	0.2267	0.0469	21.3215	0.2069	4.8332	3.3187
11	5.1173	0.1954	0.0389	25.7329	0.1989	5.0286	3.5783
12	5.9360	0.1685	0.0324	30.8502	0.1924	5.1971	3.8189
13	6.8858	0.1452	0.0272	36.7862	0.1872	5.3423	4.0413
14	7.9875	0.1252	0.0229	43.6720	0.1829	5.4675	4.2464
15	9.2655	0.1079	0.0194	51.6595	0.1794	5.5755	4.4352
16	10.7480	0.0930	0.0164	60.9250	0.1764	5.6685	4.6086
17	12.4677	0.0802	0.0140	71.6730	0.1740	5.7487	4.7676
18	14.4625	0.0691	0.0119	84.1407	0.1719	5.8178	4.9130
19	16.7765	0.0596	0.0101	98.6032	0.1701	5.8775	5.0457
20	19.4608	0.0514	0.0087	115.3797	0.1687	5.9288	5.1666
21	22.5745	0.0443	0.0074	134.8405	0.1674	5.9731	5.2766
22	26.1864	0.0382	0.0064	157.4150	0.1664	6.0113	5.3765
23	30.3762	0.0329	0.0054	183.6014	0.1654	6.0442	5.4671
24	35.2364	0.0284	0.0047	213.9776	0.1647	6.0726	5.5490
25	40.8742	0.0245	0.0040	249.2140	0.1640	6.0971	5.6230
26	47.4141	0.0211	0.0034	290.0883	0.1634	6.1182	5.6898
27	55.0004	0.0182	0.0030	337.5024	0.1630	6.1364	5.7500
28	63.8004	0.0157	0.0025	392.5028	0.1625	6.1520	5.8041
29	74.0085	0.0135	0.0022	456.3032	0.1622	6.1656	5.8528
30	85.8499	0.0116	0.0019	530.3117	0.1619	6.1772	5.8964
31	99.5859	0.0100	0.0016	616.1616	0.1616	6.1872	5.9356
32	115.5196	0.0087	0.0014	715.7475	0.1614	6.1959	5.9706
33	134.0027	0.0075	0.0012	831.2671	0.1612	6.2034	6.0019
34	155.4432	0.0064	0.0010	965.2698	0.1610	6.2098	6.0299
35	180.3141	0.0055	0.0009	1 120.7130	0.1609	6.2153	6.0548
40	378.7212	0.0026	0.0004	2 360.7572	0.1604	6.2335	6.1441
45	795.4438	0.0013	0.0002	4 965.2739	0.1602	6.2421	6.1934
50	1 670.7038	0.0006	0.0001	10 435.6488	0.1601	6.2463	6.2201
55	3 509.0488	0.0003	0.0000	21 925.3050	0.1600	6.2482	6.2343
60	7 370.2014	0.0001	0.0000	46 057.5085	0.1600	6.2492	6.2419

18% Discrete compound interest factors

	Single payment		Uniform series				Uniform
	Compound amount factor	Present worth factor	Sinking fund factor	Compound amount factor	Capital recovery factor	Present worth factor	Gradient conversion factor
n	F/P	P/F	A/F	F/A	A/P	P/A	A/G
1	1.1800	0.8475	1.0000	1.0000	1.1800	0.8475	0.0000
2	1.3924	0.7182	0.4587	2.1800	0.6387	1.5656	0.4587
3	1.6430	0.6086	0.2799	3.5724	0.4599	2.1743	0.8902
4	1.9388	0.5158	0.1917	5.2154	0.3717	2.6901	1.2947
5	2.2878	0.4371	0.1398	7.1542	0.3198	3.1272	1.6728
6	2.6996	0.3704	0.1059	9.4420	0.2859	3.4976	2.0252
7	3.1855	0.3139	0.0824	12.1415	0.2624	3.8115	2.3526
8	3.7589	0.2660	0.0652	15.3270	0.2452	4.0776	2.6558
9	4.4355	0.2255	0.0524	19.0859	0.2324	4.3030	2.9358
10	5.2338	0.1911	0.0425	23.5213	0.2225	4.4941	3.1936
11	6.1759	0.1619	0.0348	28.7551	0.2148	4.6560	3.4303
12	7.2876	0.1372	0.0286	34.9311	0.2086	4.7932	3.6470
13	8.5994	0.1163	0.0237	42.2187	0.2037	4.9095	3.8449
14	10.1472	0.0985	0.0197	50.8180	0.1997	5.0081	4.0250
15	11.9737	0.0835	0.0164	60.9653	0.1964	5.0916	4.1887
16	14.1290	0.0708	0.0137	72.9390	0.1937	5.1624	4.3369
17	16.6722	0.0600	0.0115	87.0680	0.1915	5.2223	4.4708
18	19.6733	0.0508	0.0096	103.7403	0.1896	5.2732	4.5916
19	23.2144	0.0431	0.0081	123.4135	0.1881	5.3162	4.7003
20	27.3930	0.0365	0.0068	146.6280	0.1868	5.3527	4.7978
21	32.3238	0.0309	0.0057	174.0210	0.1857	5.3837	4.8851
22	38.1421	0.0262	0.0048	206.3448	0.1848	5.4099	4.9632
23	45.0076	0.0222	0.0041	244.4868	0.1841	5.4321	5.0329
24	53.1090	0.0188	0.0035	289.4945	0.1835	5.4509	5.0950
25	62.6686	0.0160	0.0029	342.6035	0.1829	5.4669	5.1502
26	73.9490	0.0135	0.0025	405.2721	0.1825	5.4804	5.1991
27	87.2598	0.0115	0.0021	479.2211	0.1821	5.4919	5.2425
28	102.9666	0.0097	0.0018	566.4809	0.1818	5.5016	5.2810
29	121.5005	0.0082	0.0015	669.4475	0.1815	5.5098	5.3149
30	143.3706	0.0070	0.0013	790.9480	0.1813	5.5168	5.3448
31	169.1774	0.0059	0.0011	934.3186	0.1811	5.5227	5.3712
32	199.6293	0.0050	0.0009	1 103.4960	0.1809	5.5277	5.3945
33	235.5625	0.0042	0.0008	1 303.1253	0.1808	5.5320	5.4149
34	277.9638	0.0036	0.0006	1 538.6878	0.1806	5.5356	5.4328
35	327.9973	0.0030	0.0006	1 816.6516	0.1806	5.5386	5.4485
40	750.3783	0.0013	0.0002	4 163.2130	0.1802	5.5482	5.5022
45	1 716.6839	0.0006	0.0001	9 531.5771	0.1801	5.5523	5.5293
50	3 927.3569	0.0003	0.0000	21 813.0937	0.1800	5.5541	5.5428
55	8 984.8411	0.0001	0.0000	49 910.2284	0.1800	5.5549	5.5494
60	20 555.1400	0.0000	0.0000	114 189.6665	0.1800	5.5553	5.5526

20% Discrete compound interest factors

	Single payment		Uniform series				Uniform
	Compound amount factor	Present worth factor	Sinking fund factor	Compound amount factor	Capital recovery factor	Present worth factor	Gradient conversion factor
n	F/P	P/F	A/F	F/A	A/P	P/A	A/G
1	1.2000	0.8333	1.0000	1.0000	1.2000	0.8333	0.0000
2	1.4400	0.6944	0.4545	2.2000	0.6545	1.5278	0.4545
3	1.7280	0.5787	0.2747	3.6400	0.4747	2.1065	0.8791
4	2.0736	0.4823	0.1863	5.3680	0.3863	2.5887	1.2742
5	2.4883	0.4019	0.1344	7.4416	0.3344	2.9906	1.6405
6	2.9860	0.3349	0.1007	9.9299	0.3007	3.3255	1.9788
7	3.5832	0.2791	0.0774	12.9159	0.2774	3.6046	2.2902
8	4.2998	0.2326	0.0606	16.4991	0.2606	3.8372	2.5756
9	5.1598	0.1938	0.0481	20.7989	0.2481	4.0310	2.8364
10	6.1917	0.1615	0.0385	25.9587	0.2385	4.1925	3.0739
11	7.4301	0.1346	0.0311	32.1504	0.2311	4.3271	3.2893
12	8.9161	0.1122	0.0253	39.5805	0.2253	4.4392	3.4841
13	10.6993	0.0935	0.0206	48.4966	0.2206	4.5327	3.6597
14	12.8392	0.0779	0.0169	59.1959	0.2169	4.6106	3.8175
15	15.4070	0.0649	0.0139	72.0351	0.2139	4.6755	3.9588
16	18.4884	0.0541	0.0114	87.4421	0.2114	4.7296	4.0851
17	22.1861	0.0451	0.0094	105.9306	0.2094	4.7746	4.1976
18	26.6233	0.0376	0.0078	128.1167	0.2078	4.8122	4.2975
19	31.9480	0.0313	0.0065	154.7400	0.2065	4.8435	4.3861
20	38.3376	0.0261	0.0054	186.6880	0.2054	4.8696	4.4643
21	46.0051	0.0217	0.0044	225.0256	0.2044	4.8913	4.5334
22	55.2061	0.0181	0.0037	271.0307	0.2037	4.9094	4.5941
23	66.2474	0.0151	0.0031	326.2369	0.2031	4.9245	4.6475
24	79.4968	0.0126	0.0025	392.4842	0.3035	4.9371	4.6943
25	95.3962	0.0105	0.0021	471.9811	0.2021	4.9476	4.7352
26	114.4755	0.0087	0.0018	567.3773	0.2018	4.9563	4.7709
27	137.3706	0.0073	0.0015	681.8528	0.2015	4.9636	4.8020
28	164.8447	0.0061	0.0012	819.2233	0.2012	4.9697	4.8291
29	197.8136	0.0051	0.0010	984.0680	0.2010	4.9747	4.8527
30	237.3763	0.0042	0.0008	1 181.8816	0.2008	4.9789	4.8731
31	284.8516	0.0035	0.0007	1 419.2579	0.2007	4.9824	4.8908
32	341.8219	0.0029	0.0006	1 704.1095	0.2006	4.9854	4.9061
33	410.1863	0.0024	0.0005	2 045.9314	0.2005	4.9878	4.9194
34	492.2235	0.0020	0.0004	2 456.1176	0.2004	4.9898	4.9308
35	590.6682	0.0017	0.0003	2 948.3411	0.2003	4.9915	4.9406
40	1 469.7716	0.0007	0.0001	7 343.8578	0.2001	4.9966	4.9728
45	3 657.2620	0.0003	0.0001	18 281.3099	0.2001	4.9986	4.9877
50	9 100.4382	0.0001	0.0000	45 497.1908	0.2000	4.9995	4.9945
55	22 644.8023	0.0000	0.0000	113 219.0113	0.2000	4.9998	4.9976
60	56 347.5144	0.0000	0.0000	281 732.5718	0.2000	4.9999	4.9989

25% Discrete compound interest factors

	Single payment		Uniform series				Uniform
	Compound amount factor	Present worth factor	Sinking fund factor	Compound amount factor	Capital recovery factor	Present worth factor	Gradient conversion factor
n	F/P	P/F	A/F	F/A	A/P	P/A	A/G
1	1.2500	0.8000	1.0000	1.0000	1.2500	0.8000	0.0000
2	1.5625	0.6400	0.4444	2.2500	0.6944	1.4400	0.4444
3	1.9531	0.5120	0.2623	3.8125	0.5123	1.9520	0.8525
4	2.4414	0.4096	0.1734	5.7656	0.4234	2.3616	1.2249
5	3.0518	0.3277	0.1218	8.2070	0.3718	2.6893	1.5631
6	3.8147	0.2621	0.0888	11.2588	0.3388	2.9514	1.8683
7	4.7684	0.2097	0.0663	15.0735	0.3163	3.1611	2.1424
8	5.9605	0.1678	0.0504	19.8419	0.3004	3.3289	2.3872
9	7.4506	0.1342	0.0388	25.8023	0.2888	3.4631	2.6048
10	9.3132	0.1074	0.0301	33.2529	0.2801	3.5705	2.7971
11	11.6415	0.0859	0.0235	42.5661	0.2735	3.6564	2.9663
12	14.5519	0.0687	0.0184	54.2077	0.2684	3.7251	3.1145
13	18.1899	0.0550	0.0145	68.7596	0.2645	3.7801	3.2437
14	22.7374	0.0440	0.0115	86.9495	0.2615	3.8241	3.3559
15	28.4217	0.0352	0.0091	109.6868	0.2591	3.8593	3.4530
16	35.5271	0.0281	0.0072	138.1085	0.2572	3.8874	3.5366
17	44.4089	0.0225	0.0058	173.6357	0.2558	3.9099	3.6084
18	55.5112	0.0180	0.0046	218.0446	0.2546	3.9279	3.6698
19	69.3889	0.0144	0.0037	273.5558	0.2537	3.9424	3.7222
20	86.7362	0.0115	0.0029	342.9447	0.2529	3.9539	3.7667
21	108.4202	0.0092	0.0023	429.6809	0.2523	3.9631	3.8045
22	135.5253	0.0074	0.0019	538.1011	0.2519	3.9705	3.8365
23	169.4066	0.0059	0.0015	673.6264	0.2515	3.9764	3.8634
24	211.7582	0.0047	0.0012	843.0329	0.2512	3.9811	3.8861
25	264.6978	0.0038	0.0009	1 054.7912	0.2509	3.9849	3.9052
26	330.8722	0.0030	0.0008	1 319.4890	0.2508	3.9879	3.9212
27	413.5903	0.0024	0.0006	1 650.3612	0.2506	3.9903	3.9346
28	516.9879	0.0019	0.0005	2 063.9515	0.2505	3.9923	3.9457
29	646.2349	0.0015	0.0004	2 580.9394	0.2504	3.9938	3.9551
30	807.7936	0.0012	0.0003	3 227.1743	0.2503	3.9950	3.9628
31	1 009.7420	0.0010	0.0002	4 034.9678	0.2502	3.9960	3.9693
32	1 262.1774	0.0008	0.0002	5 044.7098	0.2502	3.9968	3.9746
33	1 577.7218	0.0006	0.0002	6 306.8872	0.2502	3.9975	3.9791
34	1 972.1523	0.0005	0.0001	7 884.6091	0.2501	3.9980	3.9828
35	2 465.1903	0.0004	0.0001	9 856.7613	0.2501	3.9984	3.9858
40	7 523.1638	0.0001	0.0000	30 088.6554	0.2500	3.9995	3.9947
45	22 958.8740	0.0000	0.0000	91 831.4962	0.2500	3.9998	3.9980

30% Discrete compound interest factors

	Single payment		Uniform series				Uniform
	Compound amount factor	Present worth factor	Sinking fund factor	Compound amount factor	Capital recovery factor	Present worth factor	Gradient conversion factor
n	F/P	P/F	A/F	F/A	A/P	P/A	A/G
1	1.3000	0.7692	1.0000	1.0000	1.3000	0.7692	0.0000
2	1.6900	0.5917	0.4348	2.3000	0.7348	1.3609	0.4348
3	2.1970	0.4552	0.2506	3.9900	0.5506	1.8161	0.8271
4	2.8561	0.3501	0.1616	6.1870	0.4616	2.1662	1.1783
5	3.7129	0.2693	0.1106	9.0431	0.4106	2.4356	1.4903
6	4.8268	0.2072	0.0784	12.7560	0.3784	2.6427	1.7654
7	6.2749	0.1594	0.0569	17.5828	0.3569	2.8021	2.0063
8	8.1573	0.1226	0.0419	23.8577	0.3419	2.9247	2.2156
9	10.6045	0.0943	0.0312	32.0150	0.3312	3.0190	2.3963
10	13.7858	0.0725	0.0235	42.6195	0.3235	3.0915	2.5512
11	17.9216	0.0558	0.0177	56.4053	0.3177	3.1473	2.6833
12	23.2981	0.0429	0.0135	74.3270	0.3135	3.1903	2.7952
13	30.2875	0.0330	0.0102	97.6250	0.3102	3.2233	2.8895
14	39.3738	0.0254	0.0078	127.9125	0.3078	3.2487	2.9685
15	51.1859	0.0195	0.0060	167.2863	0.3060	3.2682	3.0344
16	66.5417	0.0150	0.0046	218.4722	0.3046	3.2832	3.0892
17	86.5042	0.0116	0.0035	285.0139	0.3035	3.2948	3.1345
18	112.4554	0.0089	0.0027	371.5180	0.3027	3.3037	3.1718
19	146.1920	0.0068	0.0021	483.9734	0.3021	3.3105	3.2025
20	190.0496	0.0053	0.0016	630.1655	0.3016	3.3158	3.2275
21	247.0645	0.0040	0.0012	820.2151	0.3012	3.3198	3.2480
22	321.1839	0.0031	0.0009	1 067.2796	0.3009	3.3230	3.2646
23	417.5391	0.0024	0.0007	1 388.4635	0.3007	3.3254	3.2781
24	542.8008	0.0018	0.0006	1 806.0026	0.3006	3.3272	3.2890
25	705.6410	0.0014	0.0004	2 348.8033	0.3004	3.3286	3.2979
26	917.3333	0.0011	0.0003	3 054.4443	0.3003	3.3297	3.3050
27	1 192.5333	0.0008	0.0003	3 971.7776	0.3003	3.3305	3.3107
28	1 550.2933	0.0006	0.0002	5 164.3109	0.3002	3.3312	3.3153
29	2 015.3813	0.0005	0.0001	6 714.6042	0.3001	3.3317	3.3189
30	2 619.9956	0.0004	0.0001	8 729.9855	0.3001	3.3321	3.3219
31	3 405.9943	0.0003	0.0001	11 349.9811	0.3001	3.3324	3.3242
32	4 427.7926	0.0002	0.0001	14 755.9755	0.3001	3.3326	3.3261
33	5 756.1304	0.0002	0.0001	19 183.7681	0.3001	3.3328	3.3276
34	7 482.9696	0.0001	0.0000	24 939.8985	0.3000	3.3329	3.3288
35	9 727.8604	0.0001	0.0000	32 422.8681	0.3000	3.3330	3.3297

Activity	Independent	Float free	Total
3–5	—	—	1
4–7	—	—	—
5–6	—	—	1
6–7	—	1	1
6–9	10	11	11
7–8	—	—	—
8–9	—	—	—

Problem 9.4

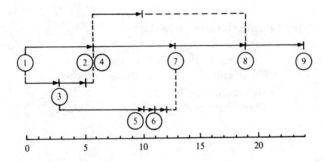

Problem 9.5   40 weeks

Problem 9.6   0.25

Chapter 10

Problem 10.1   Programme is now extended to 42 days

Problem 10.2   Yes

Problem 10.3

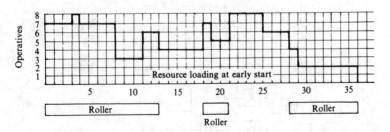

Problem 10.4   9 at early start, 16 at late start and 12 after smoothing

Problem 10.5 (a) Network diagram:

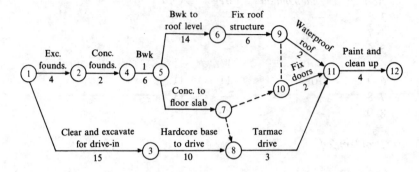

(b) Overall duration = 38 days
   Float:
        Concrete to floor slab 15 days
        Clear and excavate for drive in 6 days
        Hardcare base to drive   6 days
        Tarmacadam drive   6 days

Chapter 11

Problem 11.1

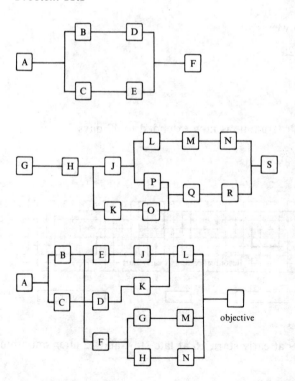

Problem 11.2

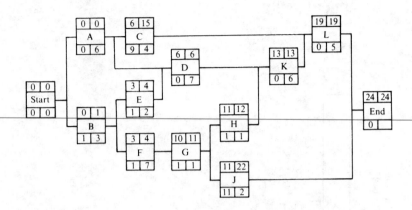

Problem 11.3

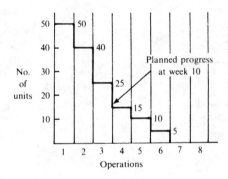

Problem 11.5 69 working days; increase m to 0.52

Problem 11.6 21; 39 weeks; 62 weeks; 27 weeks; $42\frac{1}{3}$ weeks

Problem 11.7

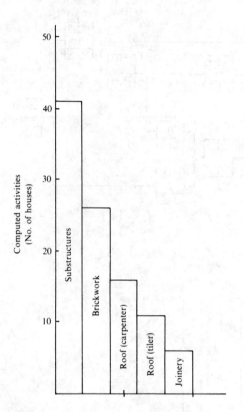

Problem 11.9 First and twelfth units completed after 2.8 and 8.3 weeks. At the end of week seven, 12, 11 and 9.4 units of A, B and C should have been completed.

Problem 11.10

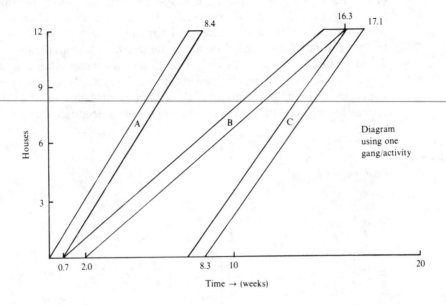

Diagram using one gang/activity

Problem 11.11 54 weeks; 92 gang weeks.

Chapter 12

Problem 12.4 Current ratio = 2.044.
Liquid ratio = 0.77.
Debtors ratio = 39.8 days.
Net asset turnover = 2.34.
Debt ratio = 0.25.
Return on equity = 12 per cent.
Return on net assets = 11 per cent.
Profit margin = 11.6 per cent.

The company is solvent but has poor liquidity; weak in liquid assets—inventory too high as a percentage of current assets. Debtors ratio is high and credit should be restricted; poor payers should be persuaded to speed up payments to boost liquidity. Reasonable net asset turnover but more sales should be generated. Debt ratio very good and more capital could be raised by a long-term loan. Profitability ratios not exceptional.

Problem 12.5 While the company is solvent it has a liquidity problem, with a low liquid capital ratio not improving significantly over the two years. Too much working capital is tied up in inventory and must be released. A funds flow statement shows that funds have been found predominantly by increasing the level of creditors (+£15 500), raising a long-term loan (+£5000), retaining earnings (+£6100) and reducing cash balances (+£3700). These funds have been applied to the purchase of plant and machinery, increasing inventory and funding increased trade debtors. Debt collection needs tightening up considerably because the debtors ratio has increased from 29 to 45 days between balance sheets. There is scope for raising loan finance and the profitability of the company is good. The management or administration of the company probably requires strengthening and this would certainly have to be a condition of lending money.

Chapter 13

Problem 13.1

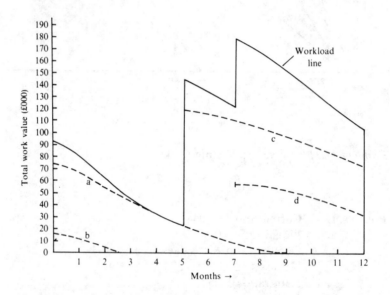

Problem 13.3   £608

Problem 13.4   (a) £6670
                  (b) 43.6%

Problem 13.5   Loss £200 000

Problem 13.6   (a) £725
                  (b) £119 shortfall

Chapter 14

Problem 14.1 (a) Normal cost, £3240
                        Crash cost, £4360
                   (b) Optimized cost, £3975
                   (c) 18 weeks

Problem 14.2 £3715

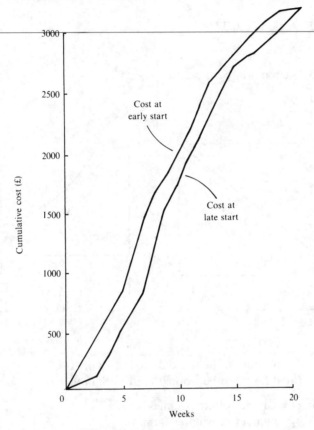

Problem 14.4

Weeks	Cost
40	5300
41	5030
42	4840
43	4690
44	4540
45	4400
46	4260
47	4200
48	4140
49	4090

Problem 14.5   (a) £2342
(b) 12 weeks

Problem 14.6   (a) £3520
(b) 23 weeks

Chapter 15

Problem 15.1   Action 3; 2; 3

Problem 15.2   Action 3; 3

Problem 15.3   Action 4 in each case

Problem 15.4

(a)   (b)

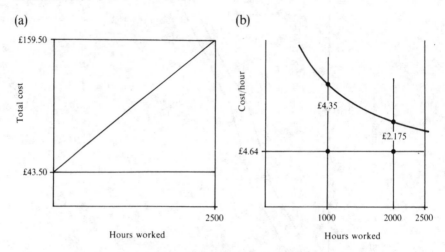

(c) If hours to be worked are less than 1000
(d) If hours to be worked are less than 960

Problem 15.5 x_1 = number of door frames
x_2 = number of window frames
Maximize $P = 3x_1 + 5x_2$
subject to: $0.5x_1 + 0.75x_2 \leqslant 120$
$0.25x_1 + 0.15x_2 \leqslant 40$
$x_1; x_2 \geqslant 0$

Problem 15.6 x_1 = number of housing developments
x_2 = number of high rise flats
x_3 = number of hotels
x_4 = number of yachting marinas
Maximize $P = 60\,000x_1 + 1\,300\,000x_2 + 860\,000x_3 + 1560\,000x_4$
subject to: $200\,000x_1 \leqslant 16\,000\,000$

$$4\,000\,000x_2 \leqslant 50\,000\,000$$
$$2\,000\,000x_3 \leqslant 35\,000\,000$$
$$7\,000\,000x_4 \leqslant 50\,000\,000$$
$$200\,000x_1 + 4\,000\,000x_2 + 2\,000\,000x_3 + 7\,000\,000x_4 \leqslant 120\,000\,000$$
$$200\,000x_1 + 2\,000\,000x_3 \leqslant 60\,000\,000$$
and
$$x_1;\, x_2;\, x_3;\, x_4 \geqslant 0.$$

Problem 15.7 $x_1 =$ number of doors
$x_2 =$ number of window frames
$x_3 =$ number of lengths of fencing
$x_4 =$ number of roof trusses
Maximize $P = 0.5x_1 + 1.50x_2 + 0.37x_3 + 6.00x_4$
subject to: $4x_1 + 10x_2 + 2x_3 + 36x_4 \leqslant 2000$
and $x_1;\, x_2;\, x_3$ and $x_4 \geqslant 0.$

Problem 15.8   In addition to restraints of 15.4 add
$x_1 \leqslant 4000,\ x_2 \leqslant 1000,\ x_3 \leqslant 2000,\ x_4 \leqslant 400$
$300x_3 + 1600x_4 \leqslant 40\,000$
$15x_1 + 30x_2 \leqslant 4000.$

Problem 15.9

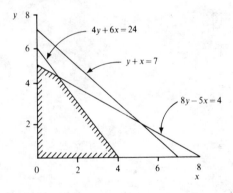

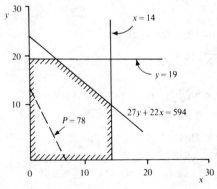

Problem 15.11

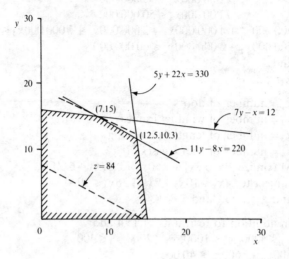

Problem 15.12

8	4	2	6	1
0	9	5	5	4
3	8	9	2	6
4	3	1	0	3
9	5	8	9	5

Problem 15.13   A-1, B-4, C-3, D-2
No
C-1, D-2, B-4, A-3 for example

Problem 15.14   A-1, B-2, C-3, D-5, E-4
178 miles

Problem 15.15   A-4, B-1, C-5, D-2, E-3
£109 000.

Problem 15.16   (a) Aa = 12   Bc = 15
                        Ab = 6   Cc = 10
                        Ac = 2

$$(b) \quad A2 = 15 \quad B5 = 17$$
$$A3 = 17 \quad C1 = 25$$
$$A4 = 18 \quad C3 = 15$$
$$B3 = 13 \quad D2 = 20.$$

Chapter 16

Problem 16.1 (a) EDAHCBGF

(b) Cut and bend: nil; fix: 10 hours

(c) 163 hours

(a) KOJQSPLRMN (P and L can be transposed, K can be first or last)

(b) Fix: 17 hours

(c) 178 hours

Problem 16.2 (a) 1—4—8—11—12 or 1—5—8—11—12 of 10 units length

(b) 1—4—6—8—10—12, 30 units

Problem 16.3 17 hours

Problem 16.4 Node 2 HO–1–3–2   6 miles

3 HO–1–3   5

4 HO–1–3–4   6

5 HO–1–3–2–5   8

6 HO–1–3–4–6   9

7 HO–7   8

Index